Year	Recipients	Nobel Prize	Research Topic
1970	N. Borlaug	Peace Prize	Genetic improvement of Mexican wheat
1969	M. Delbruck A. D. Hershey S. E. Luria	Medicine or Physiology	Replication mechanisms and genetic structure of bacteriophages
1968	H. G. Khorana M. W. Nirenberg	Medicine or Physiology	Deciphering the genetic code
	R. W. Holley	Medicine or Physiology	Structure and nucleotide sequence of transfer RNA
1966	P. F. Rous	Medicine or Physiology	Viral induction of cancer in chickens
1965	F. Jacob A. M. L'woff J. L. Monod	Medicine or Physiology	Genetic regulation of enzyme synthesis in bacteria
1962	F. H. C. Crick J. D. Watson M. H. F. Wilkins	Medicine or Physiology	Double helical model of DNA
	J. C. Kendrew M. F. Perutz	Chemistry	Three-dimensional structure of globular proteins
1959	A. Kornberg S. Ochoa	Medicine or Physiology	Biological synthesis of DNA and RNA
1958	G. W. Beadle E. L. Tatum	Medicine or Physiology	Genetic control of biochemical processes
	J. Lederberg	Medicine or Physiology	Genetic recombination in bacteria
	F. Sanger	Chemistry	Primary structure of proteins
1954	L. Pauling	Chemistry	Alpha helical structure of proteins
1946	H. J. Muller	Medicine or Physiology	X-ray induction of mutations in *Drosophila*
1933	T. H. Morgan	Medicine or Physiology	Chromosomal theory of genetics
1930	K. Landsteiner	Medicine or Physiology	Discovery of human blood groups

CONCEPTS OF

GENETICS

Fifth Edition

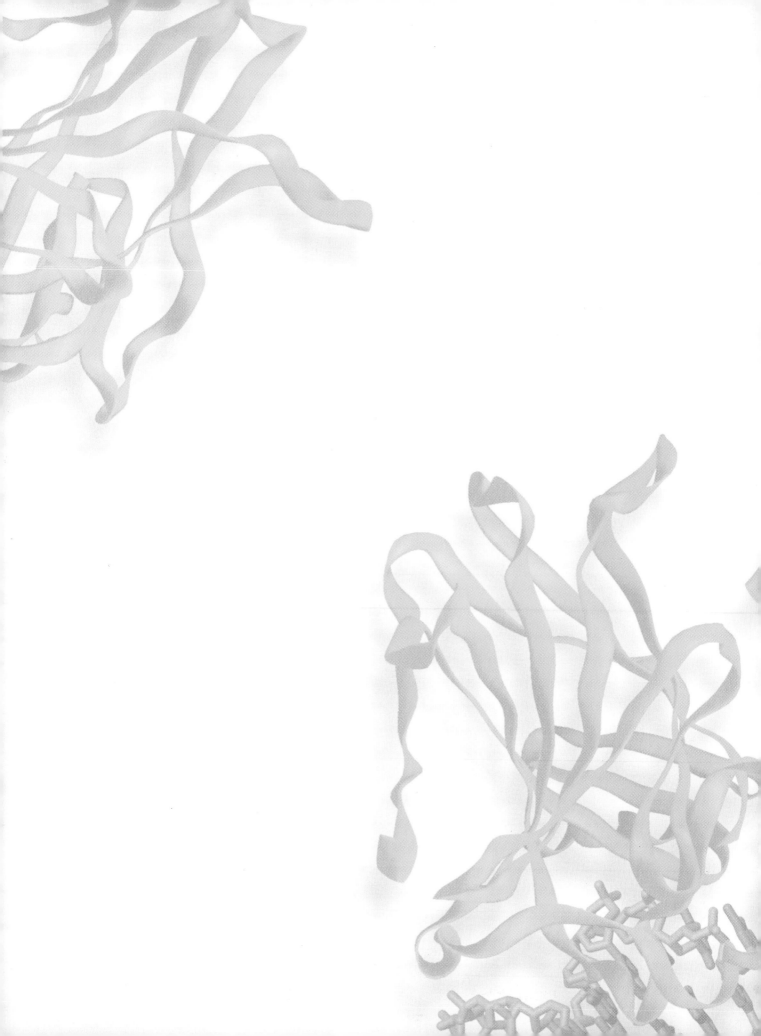

CONCEPTS OF
GENETICS
Fifth Edition

WILLIAM S. KLUG
The College of New Jersey

MICHAEL R. CUMMINGS
University of Illinois, Chicago

Essays contributed by

Mark Shotwell *Slippery Rock University*

Charlotte Spencer *University of Alberta, Edmonton*

 PRENTICE HALL Upper Saddle River, NJ 07458

Library of Congress Cataloging-in-Publication Data

Klug, William S.
   Concepts of genetics / William S. Klug, Michael R. Cummings. — 5th ed.
      p.   cm.
   Includes bibliographical references and index.
   ISBN 0-13-531062-8
   1. Genetics.   I. Cummings, Michael R.   II. Title.
QH430.K574   1997   96-41854
575.1—dc20   CIP

Executive Editor: Sheri L. Snavely
Production Editor/Page Layout: Donna Young
Art Director/Cover Design: Heather Scott
Interior Design: Lee Goldstein
Cover Art: Kenneth Eward at BIOGRAFX
Interior Art: Boston Graphics
Photo Researcher: Cindy-Lee Overton
Copy Editor: James Tully
Development Editors: Dick and Beth Morel
Creative Director: Paula Maylahn
Editorial Director: Tim Bozik
Editor-in-Chief: Paul Corey
Editor-in-Chief of Development: Ray Mullaney
Executive Managing Editor: Kathleen Schiaparelli
Manufacturing Manager: Trudy Pisciotti
Vice President of Production and Manufacturing: David W. Riccardi
Senior Marketing Manager: Jennifer Welchans
Editorial Assistants: Lisa Tarabokjia and Nancy Gross
Marketing Assistant: David Stack

© 1997 by Prentice-Hall, Inc.
Simon & Schuster/A Viacom Company
Upper Saddle River, New Jersey 07458

Previous editions copyright © 1983 by C. E. Merrill Publishing Company,
copyright © 1986 by Scott Foresman and Company, and copyright
© 1994, 1991 by Macmillan Publishing Company.

Printed in the United States of America

10 9 8 7 6 5 4 3 2

ISBN 0-13-531062-8

Prentice-Hall International (UK) Limited, *London*
Prentice-Hall of Australia Pty. Limited, *Sydney*
Prentice-Hall Canada Inc., *Toronto*
Prentice-Hall Hispanoamericana, S.A., *Mexico*
Prentice-Hall of India Private Limited, *New Delhi*
Prentice-Hall of Japan, Inc., *Tokyo*
Simon & Schuster Asia Pte. Ltd., *Singapore*
Editora Prentice-Hall do Brasil Ltda., *Rio de Janeiro*

D E D I C A T I O N

We wish to recognize the intended audience of *Concepts of Genetics* by dedicating its fifth edition to students of our text and of the courses that we have taught over the years:

"You have provided an important intellectual contribution to our efforts by being enthusiastic and inquisitive students in the classroom, and by collectively serving as a constant presence in our mind's eye as we write. This text could not have been successful without you, and for that, we are most grateful."

W. S. K.
M. R. C.

ABOUT THE AUTHORS

WILLIAM S. KLUG is currently Professor of Biology at The College of New Jersey (formerly Trenton State College) in Ewing, New Jersey. He served as Chairman of the Biology Department for 17 years, a position to which he was first elected in 1974. He received his B.A. degree in Biology from Wabash College in Crawfordsville, Indiana and his Ph.D. from Northwestern University in Evanston, Illinois. Prior to coming to Trenton State College, he returned to Wabash College as an Assistant Professor, where he first taught genetics as well as electron microscopy. His research interests have involved ultrastructural and molecular genetic studies of oogenesis in *Drosophila*. He has taught the undergraduate genetics course to Biology majors for each of the last 28 years.

MICHAEL R. CUMMINGS is currently Associate Professor in the Department of Biological Sciences and in the Department of Genetics at the University of Illinois at Chicago. He has also served on the faculty at Northwestern University and Florida State University. He received his B.A. from St. Mary's College in Winona, Minnesota, and his M.S. and Ph.D. from Northwestern University in Evanston, Illinois. He has also written textbooks in human genetics and general biology for non-majors. His research interests center on the molecular organization and physical mapping of human acrocentric chromosomes. At the undergraduate level, he teaches courses in Mendelian genetics, human genetics, and general biology for non-majors.

Concepts of Genetics is now well into its second decade of providing support to students as they study one of the most fascinating scientific disciplines. Certainly no subject area has had a more sustained impact on shaping our knowledge of the living condition. As a result of discoveries over the past 50 years, we now understand with reasonable clarity the underlying genetic mechanisms that explain how organisms develop into and then function as adults. We also better understand the basis of biological diversity and have greater insight into the evolutionary process.

Advances in genetic technology are now having a profound impact on our knowledge of human genetics. More than any other event, the launching of the Human Genome Project in 1990 symbolizes our commitment to the pursuit of such knowledge. As we near the twenty-first century, the application of genetics to the betterment of the human condition will be commonplace. While this era will be filled with the excitement of scientific discovery, many accompanying problems and controversies will also face us. Of these, how we utilize our knowledge of the nucleotide sequence of the human genome promises to be the most significant. This growing body of information has already generated many legal and ethical issues. Currently, the implications of genetic testing and gene therapy are becoming important societal concerns.

As geneticists and students of genetics, the thrill of being part of this era must be balanced by a strong sense of responsibility for providing careful attention to the many related issues that will undoubtedly arise. The formulation of proper laws and policies will depend on the comprehensive knowledge of genetics and measured responses to these issues. As a result, there has never been a higher premium on the need to provide a useful and up-to-date textbook that supports the study of genetics.

This edition of *Concepts of Genetics*, as with all past efforts, has been designed to achieve five major goals:

1. To establish a **conceptual framework** that represents a sound approach to learning, to facilitate the comprehension of the vast amount of information constituting the field of genetics.
2. To emphasize the **rich history of scientific discovery and analytical thought**, so prevalent in genetics, to provide a unique opportunity for students to hone their problem-solving abilities and to explore how we know what we know.
3. To cover material with a **clean, crisp organizational format**, both within each chapter and throughout the chapter sequence, to facilitate effective use of the text.
4. To present **figures that teach rather than simply illustrate** the topic at hand, and support explanations of complex, analytical experiments.
5. To provide students with **clearly written, straightforward explanations** which elucidate difficult, complex topics without having to oversimplify presentations.

Creating a text that achieves these goals and having the opportunity to refine it over five editions has been a labor of love for us, one that has been possible because of the feedback and encouragement provided over the past decade from students, adopters, and reviewers.

NEW TO THIS EDITION

New Organization

As the field of genetics has expanded, many topics have taken on greater importance and stature within the discipline. As with previous editions, this has necessitated substantial reorganization, which is apparent in the Table of Contents. This edition continues to reflect our belief that classical studies of heredity, which we refer to as Transmission Genetics, should initiate the text (Part One). At the urging of many reviewers, we have moved forward to this section our coverage of bacterial and phage genetics, as well as that of extranuclear inheritance. Then, in Part Two, we have followed this coverage with a group of chapters that center around DNA, its structure, expression, and variation, culminating with a discussion of genetic technology and applications of recombinant DNA research. With these topics as background, Part Three then presents numerous advanced topics that focus on genetic analysis, revealing how genetic expression is regulated, as well as the role of genetic information in development, cancer, immunity, and behavior. We conclude our coverage with a consideration of the genetics of populations and the role of genetics in the process of evolution.

While we realize that no Table of Contents in a diverse field such as genetics can satisfy everyone, the present organization can be used in all first courses in undergraduate genetics, whether offered in a semester, quarter, or two-quarter format. The Parts and Chapters within the text are written so as to be used interchangeably, providing flexibility for the instructor.

New Art Program

In the current edition, we have taken the opportunity to redesign the entire art program. Literally every figure has been scrutinized, revised, and refined. In conjunction with the many beautiful color photographs, the text provides an attractive and pedagogically superior visual program to enhance and support discussions of each major topic. As in past editions, our emphasis has been on the creation of figures that facilitate learning. In many cases, this has involved the development of "flowchart" figures, presenting experimental approaches that have led to the development of major findings and concepts.

Modernization

One of the major trends in genetics involves the utilization of recombinant DNA technology to study many areas of biology. In this edition, this trend is reflected in the revision and modernization of many aspects of the text, but particularly that of **Applications of Recombinant DNA Technology** (Chapter 16), which has been extensively updated. All areas of molecular genetics have been enhanced by the ability to clone and analyze DNA. However, the discipline that has benefited more than any other is human genetics. The expansion of knowledge involving our own species represents a second major trend in genetics. As a result, we have increased our coverage of human genetics throughout the text. This is particularly evident in two cases, **Genetics and Cancer** (Chapter 21) and **Genetic Basis of the Immune Response** (Chapter 22), which heavily emphasize human genetics. Other topics, where current findings have been added, and which reflect "cutting edge" information, include cell cycle regulation, euphenics, genomic imprinting, genetic anticipation, trinucleotide repeat mutations, gene therapy, human genetic disorders and their diagnosis, DNA fingerprinting and forensics, human gene mapping, and the many spin-offs of the vast amount of information provided by the Human Genome Project.

Major Revisions

With the advice and suggestions of several researchers whose studies directly involve the relevant topics, we have revised our coverage of **Population Genetics** (Chapter 24) and **Genetics and Evolution** (Chapter 25). These revisions reflect the progress in these fields over the past decade, as well as the influence of molecular genetics on these disciplines. This effort is in keeping with the important role genetics plays in the study of population and evolutionary biology.

New Essays

In keeping with the influence of genetics on our everyday lives, we have introduced a new series of 16 essays, each contained in a section referred to as **Genetics, Technology, and Society**. The essays appear near the end of most chapters, and provide accounts of how genetics interfaces with society. The presence of this new information is in keeping with the tremendous impact genetic technology has on our existence as we prepare to enter a new century.

Added Emphasis on Analysis

At the end of each chapter, we have expanded the section, **Insights and Solutions**. Providing detailed solutions to problems and insights into genetic analysis, this section enhances the development of analytical thinking skills by students.

The **Problems and Discussion Questions** section presents problems at various levels of difficulty. At the end of this section, we have added a new feature called **Extra-Spicy Problems**. There, the student is presented with several "high-end," multiple-step problems that call on the student to analyze actual experimental data, to design genetic experiments, and to engage in cooperative learning. As with the jalapeno pepper (our logo for this section), all of these experiences are designed to leave an aftertaste that is memorable and pleasing for those who indulge themselves.

Other Pedagogic Features

Several features have been continued that we believe enhance student learning. Sections entitled **Chapter Concepts** and **Chapter Summary** precede and conclude each chapter. The former underscores our belief that within each chapter there exists one or a few major concepts that provide the framework for the chapter. The latter section provides a concise summary that may serve as a quick review of the most important topics in each chapter. We also provide **Key Terms** at the end of each chapter, to aid the student in reviewing the important vocabulary present in each chapter. New to this edition, page references are provided for these terms.

Finally, mention should be made of the extent and nature of the **Selected Readings** section at the end of each chapter. These sections provide references to both historical experiments and modern discoveries. In addition, many review articles are cited. Our goal is to enhance the reference value of the text by providing students with an entree into the primary and secondary literature related to topics introduced in each chapter.

Appendices

The appendices include expanded coverage of **Experimental Methods, Solutions to Selected Problems and Discussion Questions**, and an extensive **Glossary**. All three have been updated.

For the Student

Student Handbook: A Guide to Concepts and Problem Solving

Harry Nickla, Creighton University

This valuable handbook provides a detailed step-by-step solution or lengthy discussion for every problem in the text in a chapter-by-chapter format. The handbook also contains extra study problems and a thorough review of concepts and vocabulary.

New York Times Themes of the Times Supplement

Coordinated by Harry Nickla, Creighton University

This exciting newspaper-format supplement brings together recent genetics and molecular biology articles from the pages of the world-renowned *New York Times*. This free supplement, available through your local representative, encourages students to make the connections between the genetic concepts and the latest research and breakthroughs in science.

Life on the Internet: Biology

Andrew Stull, California State University at Fullerton

The perfect guide to help your students take advantage of our *Concepts of Genetics* home page on the World Wide Web. This unique resource gives clear steps to access our regularly updated genetics resource area as well as an overview of general navigation and research strategies.

Concepts of Genetics World Wide Web Home Page

Available in January 1997, this unique tool is designed to launch student exploration of genetics resources on the Web. This page is regularly updated and linked specifically to text chapters.

For the Instructor

Instructor's Manual with Testbank

Harry Nickla, Creighton University

This manual/testbank contains over 800 questions and problems an instructor can use to prepare exams. The manual also provides optional course sequences, a guide to audiovisual supplements, and several "starter references" for term papers and special research projects. The testbank portion of the manual is also available in IBM Windows and Macintosh formats (see below).

Prentice Hall Custom Test–IBM
Prentice Hall Custom Test–Macintosh

Harry Nickla, Creighton University

Available for Windows and Macintosh, *Prentice Hall Custom Test* allows instructors to create and tailor exams to their own needs. With the Online Testing option, exams can also be administered online and data can then be automatically transferred for evaluation. A comprehensive desk reference guide is included, along with online assistance.

Transparencies

150 four-color large type transparencies from the text are available for adopters.

Prentice Hall CD-ROM Image Bank for Genetics

This unique image bank contains all illustrations from the fifth edition of *Concepts of Genetics* as well as animations and video in a digitized format for use in the classroom. The CD-ROM includes a navigational tool to allow instructors to customize lecture presentations. Additional features include keyword searches and the ability to incorporate lecture notes based on custom presentations.

ACKNOWLEDGMENTS

No text in its fifth edition can be the sole work of its authors. While we assume complete responsibility for any errors herein, we gratefully acknowledge the advice, contributions, and suggestions made by reviewers of all editions, and particularly those who were involved in this edition:

Gwen Acton	*Harvard University*
Marsha Altschuler	*Williams College*
Alan G. Atherly	*Iowa State University*
Brian Bradley	*University of Maryland–Baltimore County*
Kuo C. Chen	*Wayne State University*
Mary Clancy	*University of New Orleans*
Rebecca V. Ferrell	*Metropolitan State College of Denver*
William Firshein	*Wesleyan University*
David W. Foltz	*Louisiana State University*
Elliott S. Goldstein	*Arizona State University*
Mary F. Guest	*University of Central Florida*
Tamara Horton	*Princeton University*
Robert Ivarie	*University of Georgia*
Walter Kaczmarczyk	*West Virginia University*
Mark Kirkpatrick	*University of Texas–Austin*
Trip Lamb	*East Carolina University*
Eric Lambie	*Dartmouth College*
Rob McClung	*Dartmouth College*
Mark A. McPeek	*Dartmouth College*
Grant G. Mitman	*Montana Tech of the University of Montana*
Michelle A. Murphy	*University of Notre Dame*
John C. Osterman	*University of Nebraska–Lincoln*
Russell Ott	*Boise State University*
Patricia J. Pukkila	*The University of North Carolina at Chapel Hill*
James V. Robinson	*University of Texas at Arlington*
Douglas Ruden	*The University of Kansas*
Henry E. Schaffer	*North Carolina State University*
John M. Sedivy	*Brown University*
Ken Spitze	*University of Miami*
John L. Sternick	*Mansfield University*
Joann Tornow	*The University of Southern Mississippi*
Alan S. Waldman	*University of South Carolina*

In addition to those who reviewed various portions of the manuscript, we are very grateful to Harry Nickla at Creighton University. In his role of author of the *Student Handbook* and the *Instructor's Manual*, he has reviewed and edited the many problems that are presented at the end of each chapter. We are also especially indebted to several other colleagues. Mark Shotwell, at Slippery Rock University, and Charlotte Spencer, at the University of Alberta, wrote the **Genetics, Technology, and Society** essays at the end of most chapters. Carol Trent, at Western Washington University, contributed numerous problems that are included in the new section called **Extra-Spicy Problems**. Craig Almeida, at Stonehill College, and Anthony Maffia provided a thorough accuracy check of the new edition. We appreciate the contributions of all of these geneticists and, particularly, the pleasant manner and dedication to this text displayed during our many interactions.

We also offer special thanks to our faculty colleagues at our home institutions and our secretarial staff, who together have bolstered our efforts with encouragement, specific discussions, and endless technical support. In particular, at The College of New Jersey, Marcia O'Connell and Jim Bricker were always willing to read and discuss newly written material. Mrs. Monica Zrada was responsible for the day-to-day technical assistance so essential to a project of this magnitude. At the University of Illinois at Chicago, both Susan Liebman and Don Morrison contributed end-of-chapter problems, and Don made many suggestions to improve the accuracy of figures.

At Prentice Hall, we express appreciation and high praise for the senior editorial guidance of Sheri Snavely, whose ideas and efforts have helped to shape and refine the features of this and the previous edition of the text. She has worked tirelessly to provide us with reviews from leading specialists who are also dedicated teachers, and to ensure that the pedagogy and design of the book are at the cutting edge of a rapidly changing discipline. We were also blessed with the production efforts of Donna Young, whose high standards are apparent throughout the text. Without her work ethic and talent, the text would never have come to fruition. Marketing is being handled with talent and enthusiasm by Jennifer Welchans and Kelly McDonald.

Skillful developmental editing of both text and art was provided by Dick and Beth Morel at Strong House, whose efforts were directed by Ray Mullaney at Prentice Hall. Finally, the beauty and consistent presentation of the art work is the product of Paul Foti and his staff at Boston Graphics, who were also responsible for the art program in the first edition of *Concepts of Genetics* in 1983. We are most pleased to have had the opportunity to once again work with this talented group of individuals. We also thank several individuals whose efforts have contributed to the overall appearance of the text. Cindy-Lee Overton provided the photo research leading to the many striking photographs found throughout the text. We also wish to thank Heather Scott, art director, for creatively guiding the text and cover design. The molecular model that adorns the cover was created by Kenneth Eward at BIOGRAFX.

Despite the intensity and associated pressures of a project of this enormity, interactions with those involved with this edition have been pleasant and enjoyable. A text such as this is, most of all, a collective enterprise. All of the above individuals deserve to share in any success this text enjoys. We want them to know that our gratitude greatly exceeds the sentiments that we have expressed above. Many, many thanks to you all.

B R I E F   C O N T E N T S

CONTENTS

PART ONE
HEREDITY AND THE PHENOTYPE 17

8 Extranuclear Inheritance 208

9 Chromosome Variation and Sex Determination 221

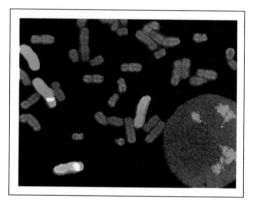

PART TWO
MOLECULAR BASIS OF HEREDITY 261

10   Structure and Analysis of DNA and RNA 262

11   DNA Replication and Recombination 298

12   Storage and Expression of Genetic Information 324

13 Proteins: The End Product of Gene Expression 364

14 Gene Mutation, DNA Repair, and Transposable Elements 389

PART THREE
ADVANCED TOPICS IN GENETIC ANALYSIS 521

18 Regulation of Gene Expression in Bacteria and Phages 522

GENETICS, TECHNOLOGY, AND SOCIETY
Antisense Oligonucleotides: Attacking the Messenger 538

19 Regulation of Gene Expression in Eukaryotes 544

GENETICS, TECHNOLOGY, AND SOCIETY
Entrapping Genes That Regulate Development 562

20 Developmental Genetics 567

21 Genetics and Cancer 591

22 Genetic Basis of the Immune Response 614

23 The Genetics of Behavior 637

1

An Introduction to Genetics

CHAPTER CONCEPTS

Genetics is the science of heredity. The discipline has a rich history and involves investigations of molecules, cells, organisms, and populations, using many different experimental approaches. Not only does genetic information play a significant role during evolution, but its expression influences the function of individuals at all levels. Thus, genetics unifies the study of biology and has a profound impact on human affairs.

Welcome to the study of genetics. You are about to explore a subject that many students before you have found to be the most interesting and fascinating in the field of biology. This is not surprising because an understanding of genetic processes is fundamental to the comprehension of life. Genetic information directs cellular function, determines an organism's external appearance, and serves as the link between generations in every species. Knowing how these processes occur is important in understanding the living world. The topics studied in genetics also overlap directly with molecular biology, cell biology, physiology, evolution, ecology, systematics, and behavior. The study of each of these disciplines is incomplete without the knowledge of the genetic components underlying each of them. Genetics thus unifies biology and serves as its "core."

Fascination with this discipline further stems from the fact that, in genetics, many initially vague and abstract concepts have been so thoroughly investigated that, subsequently, they have become clearly and definitively understood. As a result, genetics has a rich history that exemplifies the nature of scientific inquiry and the analytical approach used to acquire information. Scientific analysis, moving from the unknown to the known, is one of the major forces that attracts students to biology.

There is still another reason why the study of genetics is so appealing. Every year large numbers of new findings are made. Although it has been said that scientific knowledge doubles every ten years, one estimate holds that the doubling time in genetics is less than five years. Certainly, over the past five decades, no five-year period has passed without new discoveries in genetics having caused us to revise our thinking or to extend our knowledge beyond a major frontier. Each advance becomes part of an ever-expanding cornerstone upon which further progress is based. It is exciting to be in the midst of these developments, whether you are studying or teaching genetics.

1

The Historical Context of Genetics

In the chapters that follow, we will focus on the way in which genetic information is transmitted from generation to generation as well as the way it is stored, expressed, and regulated in the individual organism. The initial basis for such information was provided by Gregor Mendel in the middle of the nineteenth century. His findings, unrecognized for about half a century, were rediscovered at a time when other significant scientific information was becoming available. By the early twentieth century, several related ideas were gaining acceptance that would become cornerstones in the understanding of biology:

1. Matter is composed of atoms;
2. Cells are fundamental units of living organisms;
3. Nuclei somehow serve as the "life force" of cells; and
4. Chromosomes housed within nuclei somehow play an important role in heredity.

Together these beliefs provided the underlying basis for an important synthesis of ideas. When combined with the newly rediscovered findings of Gregor Mendel and integrated with Charles Darwin's theory of evolution and natural selection, the scene was clearly set for a major breakthrough in our comprehension of the living process in both individuals and populations. The era of modern-day biology was initiated on this foundation.

Before embarking further with our discussion of the transmission and expression of genetic information, we will backtrack well before the nineteenth century. We will briefly consider some of the ideas that preceded those of Mendel and Darwin and served as forerunners of nineteenth-century thought—several of which can be traced back well over 1000 years! As we will see, their influence was still apparent in the nineteenth century.

Prehistoric Domestication of Animals and Cultivation of Plants

We may never know when people first recognized the existence of heredity. However, a variety of archeological evidence (primitive art, preserved bones and skulls, dried seeds, etc.) has provided many insights. Such evidence documents the successful domestication of animals and cultivation of plants thousands of years ago. These efforts represent artificial selection of genetic variants within populations.

For example, between 8000 B.C. and 1000 B.C., horses, camels, oxen, and numerous breeds of dogs (derived from the wolf family) were domesticated and served

various roles. Plant groups, including maize, wheat, rice, and the date palm, are thought to have been cultivated about 5000 B.C. Assyrian art depicts artificial pollination of the date palm, which is thought to have originated in Babylonia (Figure 1.1). This deliberate selection of individual variants undoubtedly influenced the type of modern-day palms found in the region. Today, there are over 400 varieties of date palm in just four oases in the Sahara, differing from one another in traits such as fruit taste.

Prehistoric evidence of cultivated plants and domesticated animals documents our ancestors' successful attempts to manipulate the genetic composition of useful species. There is little doubt that people soon learned that desirable and undesirable traits were passed to successive generations and that they could select more desirable varieties of animals and plants. Human awareness of heredity seems to have existed even during prehistoric times.

The Greek Influence: Hippocrates and Aristotle

Although few, if any, ideas were put forward to explain heredity during prehistoric times, considerable attention was directed toward this subject during the Golden Age of Greek culture. This is particularly evident in the writings of the Hippocratic school of medicine (500–400 B.C.) and subsequently of the philosopher and naturalist Aristotle (384–322 B.C.) (Figure 1.2).

These ancient philosophers directed their attention toward an understanding of the source of the **physical substance**, the tangible material that gives rise to an individual, and the nature of the **generative force**, that operative energy that directs the physical substance as it ma-

FIGURE 1.1 Relief carving depicting artificial pollination of date palms during the reign of Assyrian King Assurnasirpal II (883–859 B.C.).

FIGURE 1.2 Illuminated manuscript page of the preface to the Latin translation of the *Book of Aristotle's Physics* by Johannes Argyropoulos (1416–1486), a Greek scholar.

terializes (develops) into a whole organism. For example, the Hippocratic school's treatise *On the Seed* argues that male semen is formed in numerous parts of the body and is transported through blood vessels to the testicles. Active "humors" act as the bearer of hereditary traits and are drawn from various parts of the body to the semen. These humors could be healthy or diseased. Diseased humors account for the appearance of newborns exhibiting congenital disorders or deformities. Furthermore, it was believed that these humors could be altered in individuals, and in their new form, be passed to their offspring. In this way, newborns could "inherit" traits that their parents had "acquired in their environment."

Aristotle, who had studied under Plato for some 20 years, was more critical and more expansive than the Hippocratic school in his analysis of human heredity. Aristotle proposed that male semen is formed from blood rather than from each organ, and that its generative power resides in a "vital heat" it contains. This vital heat has the capacity to produce offspring of the same "form" (i.e., basic structure and capacities) as the parent. He believed that it generated offspring by cooking and shaping the menstrual blood produced by the female, which was the "matter" for the offspring. The embryo would develop from the initial "setting" of the menstrual blood by the semen into a mature offspring, not because it already contained the parts in miniature (as some Hippocratics had thought), but because of the shaping power of the vital heat. These ideas constitute only one part of the Aristotelian philosophy of order in the living world.

Although to modern geneticists, the ideas of Hippocrates and Aristotle may sound naive, we should recall that prior to the early 1800s neither sperm nor eggs had yet been observed in mammals, let alone humans. Thus, in their own right, the explanations of the Greek philosophers were worthy ones in their time and for centuries to come. As we will see, their thinking was not so different from that of Charles Darwin in his formal proposal put forward in the nineteenth century on the theory of pangenesis.

The Dawn of Modern Biology: 1600–1850

During the ensuing 1900 years (300 B.C.–A.D. 1600), the theoretical understanding of genetics was not extended by significant, new ideas, but interest in applied genetics remained strong. By the Middle Ages, naturalists, well aware of the impact of heredity on organisms they studied, were faced with reconciling their findings with current religious beliefs. The theories of Hippocrates and Aristotle still prevailed and, when applied to humans, they no doubt conflicted with some of the religious doctrines of the time.

Between 1600 and 1900, major strides were made in experimental biology. The resultant knowledge provided much greater insights into the basis of life. As we shall see, although some historic ideas are now clearly recognized as incorrect, the discussions and debates surrounding them were the beginnings for the coalescence of today's explanations.

In the 1600s, the English anatomist William Harvey (1578–1657), better known for his experiments demonstrating that the blood is pumped by the heart through a circulatory system made up of arteries and veins, also wrote a treatise on reproduction and development. In it he is credited with the earliest statement of the **theory of epigenesis***—that an organism is derived from sub-

---

*Note that while scientists of this period described these as *theories* (often used to describe a body of fundamental principles, e.g., the theory of evolution), epigenesis and preformationism more accurately represented *hypotheses* (used to describe an unsupported set of assumptions that are only provisionally accepted) in the 1600s.

stances present in the egg, which are assembled and differentiate during embryonic development. Patterned after Aristotle's ideas, epigenesis holds that new structures, such as body organs, are not present initially, but instead, arise *de novo* during development.

The theory of epigenesis conflicts directly with the **theory of preformationism**, first advanced in the seventeenth century. Preformationists proposed that sex cells contain a complete miniature adult called the **homunculus** (Figure 1.3). These ideas were popular well into the eighteenth century. However, work by the embryologist Casper Wolff (1733–1794) and others clearly disproved this theory, strongly favoring epigenesis. Wolff was quite convinced that several structures, such as the alimentary canal, were not present in the earliest embryos he studied, but instead, were formed later during development.

During this period other significant findings in chemistry and biology impacted on future scientific thinking. In 1808, John Dalton expounded his **atomic theory**, which states that all matter is composed of small invisible units called atoms. Improved microscopes became available, and about 1830, Matthias Schleiden and Theodor Schwann proposed the **cell theory**. This theory states that all organisms are composed of basic units called cells, which are derived from preexisting cells. By this time, the idea of **spontaneous generation**, the idea that living organisms could spontaneously arise from nonliving components, had clearly been disproved by experiments of Francesco Redi (1621–1697), Lazzaro Spallanzani (1729–1799), and Louis Pasteur (1822–1895), among others. As a result of these various findings, all living organisms were considered to be derived from preexisting organisms and to consist of cells made up of atoms.

Another prevailing notion had a major influence on nineteenth-century thinking: the **fixity of species**. According to this doctrine, animal and plant groups remain unchanged from the moment of their initial appearance on earth. Embraced particularly by those also adhering to the religious belief of **special creation**, this doctrine was popularized by several people, including the Swedish physician and plant taxonomist Carolus Linnaeus (1707–1778), who is better known for devising the binomial system of nomenclature.

The influence of this tenet is illustrated by considering the work of the German botanist Joseph Gottlieb Kolreuter (1733–1806), who produced findings that were potentially quite far-reaching. In work with tobacco, he crossbred two groups and derived a new hybrid form, which he then converted back to one of the parental types by repeated backcrosses. In other breeding experiments, using carnations, he clearly observed segregation of traits, which was to become one of Mendel's principles of genetics. These results seemed to contradict the idea of species not changing with time. Because of Kolreuter's belief in both special creation and the fixity of species, he was puzzled about these outcomes, and he failed to recognize the real significance of his own findings.

Like Kolreuter, Karl Friedrich Gaertner (1772–1850), experimenting with peas, obtained results similar to those Mendel would later record in 1865. Whereas Mendel's data led him to propose the principles of dominance/recessiveness and segregation, Gaertner did not concentrate on the analysis of individual traits, and he too failed to grasp the significance of his own work.

Darwin: The Gap in His Theory of Evolution

With the above information as background, we conclude our coverage of the historical context of genetics with a brief discussion of the work of Charles Darwin, who in 1859 published the book-length statement of his evolutionary theory, *The Origin of Species*. Darwin's many geological, geographical, and biological observations convinced him that existing species arise by descent with modification from other ancestral species. Greatly influenced by his now famous voyage on the *Beagle* (1831–

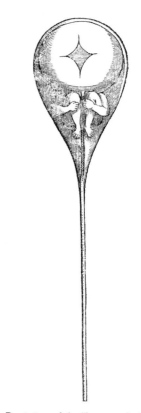

FIGURE 1.3   Depiction of the "homunculus," a sperm containing a miniature adult, perfect in proportion and fully formed.

1836), Darwin's thinking culminated in his formulation of the **theory of natural selection**, which attempted to explain the causes of evolutionary change. Formulated and proposed at the same time, but independently, by Alfred Russell Wallace, natural selection is based on the observation that populations tend to consist of more offspring than the environment can support, leading to a struggle for existence among organisms. In such a struggle, those organisms with heritable traits that better adapt them to their environment are better able to survive and reproduce than are those with less-adaptive traits. Over a long period of time, slight, but advantageous, variations will accumulate. If a population of organisms bearing these variations becomes reproductively isolated from other populations, a new species may be formed.

The primary gap in Darwin's theory was a lack of understanding of the genetic basis of variation and inheritance, leaving the theory open to reasonable criticism well into the twentieth century. Aware of this weakness in his theory of evolution, in 1868 Darwin published a second book, *Variation of Animals and Plants under Domestication*, in which he attempted to provide a more definitive explanation of how heritable variation arises gradually over time. Two of his major ideas, pangenesis and the inheritance of acquired characteristics, have their roots in the theories involving "humors," as put forward by Hippocrates and Aristotle.

In his provisional hypothesis of **pangenesis**, Darwin coined the term **gemmules** (rather than humors) to describe the physical units representing the various body parts that he thought were gathered by the blood into the semen. Darwin believed that these gemmules determine the nature or form of each body part. He further believed that gemmules could respond in an adaptive way to an individual's external environment. Once altered, such changes would be passed on to offspring, allowing for the **inheritance of acquired characteristics**. Lamarck had much earlier formalized this idea in his treatise *Philosophie Zoologique*. Lamarck's theory, which became known as the **doctrine of use and disuse**, proposed that when organisms acquire or lose characteristics, they become heritable.

The ideas expressed in Darwin's 1868 publication were not universally embraced by his colleagues. In 1863, August Weismann, a disciple of Darwin, was to take major issue with the concept of gemmules and the inheritance of acquired characteristics. In his treatise *The Germplasm: A Theory of Heredity*, Weismann proposed that living organisms consist of two kinds of materials, **somatoplasm** and **germplasm**. The former make up body tissues, constituting the major substance of an individual that undergoes development, growth, and ultimately, death. On the other hand, he envisioned the germplasm as constituting the immortal fragment of an organism that possesses the power of duplication of an individual. According to Weismann, the germplasm provides continuity among succeeding generations of individuals. Inconsistent with the theory of pangenesis, germplasm was not considered by Weismann to be derived from somatoplasm, nor was it formed anew with each individual; rather, it was considered to be a substance providing "a bridge of continuity" between generations.

Because offspring are not derived from somatoplasm, Weismann also rejected the idea of inheritance of acquired characteristics. Weismann's ideas were important ones that placed strong emphasis on germplasm (the hereditary material). His thinking represented a major advance leading to a more modern interpretation of inherited traits early in the twentieth century.

Even though Darwin never understood the basis for inherited variation, his ideas concerning evolution may be the most influential theory ever put forward in the history of biology. He was able to distill his extensive observations and synthesize his ideas into a cohesive description of the origin of the diversity of organisms populating the earth.

Mendel: An Experimental Biologist

It is against this backdrop that the work of Gregor Johann Mendel (Figure 1.4) performed his work between 1856 and 1863, forming the basis for his classic 1865 paper. In it, Mendel demonstrated for the first time clear quantitative patterns underlying inheritance, and he developed a theory involving hereditary factors in the germ cells that explained these patterns. The strength of Mendel's work is in his straightforward experimental design and in the quantitative analysis upon which he developed his postulates. His research was, however, decades ahead of its time. It was virtually ignored until it was partially duplicated and then cited by Carl Correns, Hugo de Vries, and Eric Von Tschermak in 1900, and championed by William Bateson.

In the interval between 1865 and 1900, it gradually became clear that Weismann's "germplasm" houses the genetic material and that both heredity and development are dependent on "information" contained in chromosomes, which are carried by gametes to individual offspring.

As we have seen, a rich history of scientific endeavor and thinking preceded and surrounded Mendel's work. In Chapter 3, we will return to a thorough analysis of his findings, which have served to this day as the foundation of genetics. We have, in this brief history of genetics, attempted to portray the ideas and inquiries that initiated the era of modern biological thought in the twentieth

FIGURE 1.4   Gregor Johann Mendel, who in 1865 put forward the major postulates of transmission genetics as a result of experiments with the garden pea.

century. The groundwork was clearly in place to appreciate Mendel's work and to explore meaningfully the subject of genetics.

Basic Concepts of Genetics

We turn now to a review of some of the simple but basic concepts in genetics that you have undoubtedly already studied. By reviewing them at the outset, we can establish an initial vocabulary and proceed through the text with a common foundation. We shall approach these basic concepts by asking and answering a series of questions. You may wish to write or think through an answer before reading the explanation of each question. Throughout the text, the answers to these questions will be expanded as more detailed information is presented.

What does "genetics" mean?

Genetics is the branch of biology concerned with heredity and variation. This discipline involves the study of cells, individuals, their offspring, and the populations

within which organisms live. Geneticists investigate all forms of inherited variation as well as the molecular basis underlying such characteristics.

What is the center of heredity in a cell?

Eukaryotic organisms are characterized by the presence of a **nucleus** that contains the genetic material (Figure 1.5). In prokaryotes, such as bacteria, the genetic material exists in an unenclosed but recognizable area of the cell called the **nucleoid region** (Figure 1.6). In viruses, which are not true cells, the genetic material is ensheathed in a protein coat referred to as the viral head or capsid.

What is the genetic material?

In eukaryotes and prokaryotes, **DNA** serves as the molecule storing genetic information. In viruses, either DNA or **RNA** serves this function.

What do DNA and RNA stand for?

DNA and RNA are abbreviations for **deoxyribonucleic acid** and **ribonucleic acid**, respectively. These are the two types of nucleic acids found in organisms. Nucleic acids, along with carbohydrates, lipids, and proteins, compose the four major classes of organic biomolecules that characterize life on earth.

How is DNA organized to serve as the genetic material?

DNA, although single stranded in a few viruses, is usually a double-stranded molecule organized as a **double helix**. Contained within each DNA molecule are

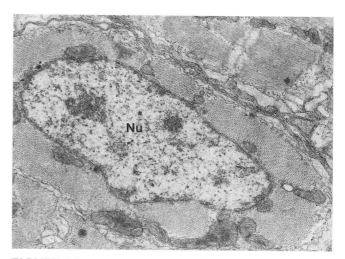

FIGURE 1.5   A transmission electron micrograph illustrating the nucleus (Nu). The micrograph was derived from muscle tissue of the mouse. Note the two prominent nucleolar areas present within the nucleus, as well as the more diffuse chromatin regions scattered throughout the nucleus.

hereditary units called **genes**, which are part of a larger element, the **chromosome**.

What is a gene?

In simplest terms, the gene is the functional unit of heredity. In chemical terms, it is a linear array of **nucleotides**—the chemical building blocks of DNA and RNA. A more conceptual approach is to consider it as an **informational storage unit** capable of undergoing **replication**, **mutation**, and **expression**. As investigations have progressed, the gene has been found to be a very complex genetic element.

What is a chromosome?

In viruses and bacteria, the chromosome is most simply thought of as a long, usually circular DNA molecule organized into genes. In eukaryotes, the chromosome is more complex. It is composed of a linear DNA molecule associated with proteins. In addition to an abundance of genes, the chromosome contains many nongenic regions. It is not yet clear what role, if any, is played by some of these regions. Our knowledge of the chromosome, like that of the gene, is continually expanding.

When and how can chromosomes be visualized?

If chromosomes are released from the viral head or the bacterial cell, they can be visualized under the electron microscope (Figure 1.7). In eukaryotes, chromosomes are most easily visualized under the light microscope when they are undergoing **mitosis** or **meiosis**. In these division processes, the material making up chromosomes is tightly coiled and condensed, giving rise to the characteristic image of chromosomes. Following division, this material, called **chromatin**, uncoils during interphase, where it is most easily studied under the electron microscope.

How many chromosomes does an organism have?

Although there are many exceptions, members of most species have a specific number of chromosomes present in each somatic cell called the **diploid number (2n)**. Upon close analysis, these chromosomes are found to occur in pairs, each member of which shares a nearly identical appearance when visible during cell division. Called **homologous chromosomes**, the members of each pair are identical in their length and in the location of the **centromere**, the point of spindle fiber attachment during division. They also contain the same sequence of gene sites, or **loci**, and pair with one another during meiosis. The number of different *types* of chromosomes in any diploid species is equal to half the diploid number and is called the **haploid (n)** number. Some organisms,

such as yeast, are haploid during most of their life cycle and contain only one "set" of chromosomes. Other organisms, especially many plant species, are sometimes characterized by more than two sets of chromosomes and are said to be **polyploid**.

What is accomplished during the processes of mitosis and meiosis?

Mitosis is the process by which the genetic material of eukaryotic cells is duplicated and distributed during cell division. **Meiosis** is the process whereby cell division produces gametes in animals and spores in most plants. While mitosis occurs in somatic tissue and yields two progeny cells with an amount of genetic material identical to that of the progenitor cell, meiosis creates cells with precisely one-half of the genetic material. Each gamete receives one member of each homologous pair of chromosomes and is haploid. This accomplishment is essential if offspring arising from two gametes are to maintain a constant number of chromosomes characteristic of their parents and other members of the species.

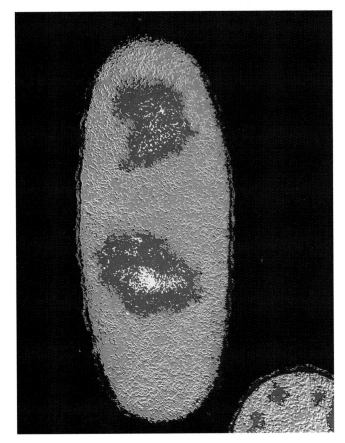

FIGURE 1.6   Color-enhanced electron micrograph of *E. coli*, demonstrating the nucleoid regions (shown in blue). The bacterium has replicated its DNA and is about to begin cell division.

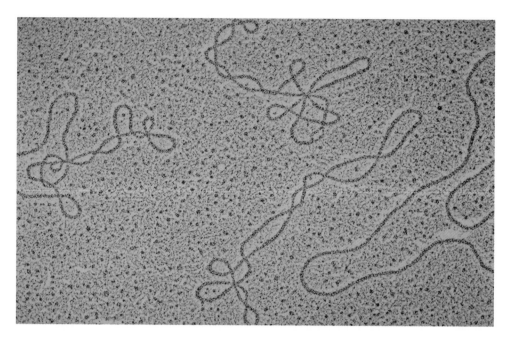

FIGURE 1.7   The DNA constituting the chromosomes of bacterial viruses (bacteriophages) viewed under the electron microscope.

What are the sources of genetic variation?

Classically, there are two sources of genetic variation: **chromosomal mutations** and **gene mutations**. The former, also called **chromosomal aberrations**, includes duplication, deletion, or rearrangement of chromosome segments. Gene mutations result from a change in the stored chemical information in DNA, collectively referred to as an organism's **genotype**. Such a change may include substitution, duplication, or deletion of nucleotides, which compose this chemical information. Alternative forms of the gene, which result from mutation, are called **alleles**. Genetic variation frequently, but not always, results in a change in some characteristic of an organism, referred to as its **phenotype**. Once part of an organism's genetic repertoire, such variation may be dispersed into the population through various reproductive mechanisms (Figure 1.8).

How does DNA store genetic information?

The sequence of nucleotides in a segment of DNA constituting a gene is present in the form of a **genetic code**. This code specifies the chemical nature (the amino acid composition) of proteins, which are the end product of genetic expression. Mutations are produced when the sequence of nucleotides is altered.

How is the genetic code organized?

There are four different nucleotides in DNA, each varying in one of its components, the **nitrogenous base**. The genetic code is a triplet; therefore, each combination of three nucleotides constitutes a code word. Almost all possible triplet codes specify one of 20 **amino acids**, the chemical building blocks of proteins.

How is the genetic code expressed?

The coded information in DNA is first transferred during a process called **transcription** into a **messenger RNA (mRNA)** molecule. The mRNA subsequently associates with the cellular organelle, the **ribosome**, where it is **translated** into a protein molecule.

Are there exceptions where proteins are not the end product of a gene?

Yes. For example, genes coding for **ribosomal RNA (rRNA)**, which is part of the ribosome, and for **transfer RNA (tRNA)**, which is involved in the translation process, are transcribed but not translated. Therefore, RNA is sometimes the end product of stored genetic information.

Why are proteins so important to living organisms that they serve as the end product of the vast majority of genes?

Many proteins serve as highly specific biological catalysts called **enzymes**. In this role, these proteins control cellular metabolism, determining which carbohydrates, lipids, nucleic acids, and other proteins are present in the cell. Many other proteins perform nonenzymatic roles. For example, hemoglobin transports oxygen, collagen provides structural support and flexibility in many tissues, immunoglobulins provide the basis for the immune response, and insulin is a hormone.

Why are enzymes necessary in living organisms?

As biological catalysts, enzymes lower the **activation energy** required for most biochemical reactions and speed the attainment of equilibrium. Otherwise, these reactions would proceed so slowly as to be ineffectual in organisms living under the conditions on earth. Thus, some genes control the variety of enzymes present in any cell type, which in turn dictates its overall biochemical composition.

Investigative Approaches in Genetics

The scope of topics encompassed in the field of genetics is enormous. Studies have involved viruses, bacteria, and a wide variety of plants and animals and have spanned all levels of biological organization, from molecules to populations. It is helpful, before we embark on a detailed study of genetics, to know the types of investigations that have been used most often in this field. Although some overlap exists, most have used one of four basic approaches.

The most classical investigative approach is the study of **transmission genetics**, in which the patterns of inheritance of traits are examined. Experiments are designed so that the transmission of traits (e.g., eye color, plant height, etc.) from parents to offspring can be analyzed through several generations. Patterns of inheritance are sought that will provide insights into more general genetic principles. The first significant experimentation of this kind to have a major impact on understanding heredity was performed by Gregor Mendel in the middle of the nineteenth century. Information derived from his work serves today as the foundation of transmission genetics. In human studies, where designed matings are neither possible nor desirable, **pedigree analysis** is used. In pedigree analysis, patterns of inheritance are traced through as many generations as possible, leading to inferences concerning the mode of inheritance of the trait under investigation.

The second approach involves **cytological studies** of the genetic material. The earliest such studies used the light microscope. The initial discovery in the early twentieth century of chromosome behavior during mitosis and meiosis was a critical event in the history of genetics. In addition to playing an important role in the rediscovery and acceptance of Mendelian principles, these observations served as the basis of the **chromosomal theory of inheritance**. This theory, which viewed the chromosome as the carrier of genes and the functional unit of transmission of genetic information, was the cornerstone for further studies in genetics throughout the first half of the twentieth century.

The light microscope continues to be an important research tool. It is useful in the investigation of chromosome structure and abnormalities and is instrumental in preparing **karyotypes**, which illustrate the chromosomes characteristic of any species arranged in a standard sequence.

With the advent of electron microscopy, the repertoire of investigative approaches in genetics has grown. In high-resolution microscopy, genetic molecules and their behavior during gene expression can be directly visualized.

The third general approach, **molecular and biochemical analysis**, has had the greatest impact on the recent growth of genetic knowledge. Molecular studies beginning in the early 1940s have consistently expanded our

FIGURE 1.8 A bumblebee pollinating a flower, achieving cross-fertilization and enhancing genetic variability within the species.

knowledge of the role of genetics in life processes. Although experimental sources were initially bacteria and the viruses that invade them, extensive information is now available concerning the nature, expression, replication, and regulation of the genetic information in eukaryotes as well. The precise nucleotide sequence has been determined for genes cloned in the laboratory. **Recombinant DNA studies** (Figure 1.9), where genes from any organism are literally spliced into bacterial or viral DNA and cloned en masse, are the most significant and far-reaching research technologies used in molecular genetic investigations. As a result, it is now possible to probe gene structure and function with a resolution heretofore impossible. Such molecular and biochemical analysis has had profound implications in medicine and agriculture.

The final approach involves the study of **population genetics**. In these investigations scientists attempt to define how and why certain genetic variation is preserved in populations, while other variation diminishes or is lost with time. Such information is critical to the understanding of the evolutionary process. Knowledge of the genetic structure of populations also allows us to predict gene frequencies in future generations.

Together these approaches have transformed a subject poorly understood in 1900 into one of the most advanced disciplines in biology today. As such, the impact of genetics on society has been immense. We shall discuss many examples of the applications of genetics in the following section and throughout the text.

Genetics and Society

Scientific information has the potential for wide-scale application in ways that improve society at large. The discipline of genetics has remained at the forefront of scientific

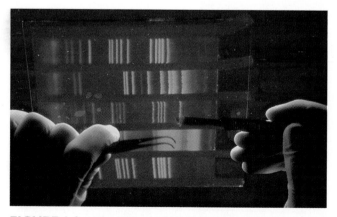

FIGURE 1.9 Visualization of DNA fragments under ultraviolet light. The bands were produced using recombinant DNA technology.

discovery from the time of the reexamination of Mendel's work early in the twentieth century. Since that time, the resultant findings have had a growing influence on society. Today, it is unusual to open a newspaper or magazine that doesn't make reference to some type of application of genetics for the improvement of human existence.

In this section, we provide a brief overview of the impact of genetics on society. As we shall see, the influence has not always been as positive as we perceive it today. We begin with two extremely negative cases.

Eugenics: The Misguided Application of Science

There is always the danger that scientific findings will be used to formulate policies and/or actions that are unjust or even tragic. This section reviews such a case, which began near the end of the nineteenth century. At that time, Darwin's theory of natural selection provided a major influence on some people's thinking concerning the human condition. Our story recounts the initial attempt to directly apply genetic knowledge for the improvement of human existence. Championed in England by Francis Galton, the general approach is called **eugenics**, a term Galton coined in 1883.

Francis Galton, a cousin of Charles Darwin, believed that many human characteristics were inherited and subject to artificial selection if human matings could be controlled. *Positive* eugenics encouraged parents displaying favorable characteristics to have large families. Superior intelligence, intellectual achievement, and artistic talent are examples. *Negative* eugenics attempted to restrict reproduction for parents displaying unfavorable characteristics. Low intelligence, mental retardation, and criminal behavior are examples.

In the United States, the eugenics movement was a significant social force and led to state and federal laws that required sterilization of those considered "genetically inferior." Over half of the states passed such laws, commencing in 1907 with Indiana. Sterilization was mandated for "imbeciles, idiots, convicted rapists, and habitual criminals." By 1931, involuntary sterilization also applied to "sexual perverts, drug fiends, drunkards, and epileptics." Immigration to the United States from certain areas of Europe and from Asia was also restricted, to prevent the influx of what were regarded as genetically inferior people.

In addition to the violation of individual human rights, such policies were seriously flawed by an inadequate understanding of the genetic basis of various characteristics. Formulation of eugenic policies was premised on the mistaken notion that "superior" and "inferior" traits are totally under genetic control, and that genes deemed unfavorable can be removed from a population by selecting against (sterilizing) individuals expressing those traits. The potential impact of the environment as well as genetic theory underlying population genetics were largely ignored as eugenic policies were developed.

In Nazi Germany in the 1930s, the concept of achieving a superior, racially pure group was an extension of the eugenics movement. Initially applied to those considered socially and physically defective, the underlying rationale of negative eugenics was soon to be applied to entire ethnic groups, including Jews and Gypsies. Fueled by various forms of racial prejudice, Adolf Hitler and the Nazi regime took eugenics to its extreme by instituting policies aimed at the extinction of these "impure" human populations. This deplorable disregard for human life was preceded by incremental policies involving forced sterilization and mercy killings. This movement, based on scientifically invalid premises, soon led to mass murder.

Even before the Nazi party came to power in 1933, English and American geneticists began separating themselves from the eugenics movement. They were concerned about the validity of the premises underlying the movement and the evidence in support of these premises. Thus, many geneticists chose not to study human genetics for fear of being grouped with those who supported eugenics.

However, since the end of World War II, tremendous strides have been made in human genetics research. Today, a new term, **euphenics**, has replaced eugenics. Euphenics refers to medical and/or genetic intervention designed to reduce the impact of defective genotypes on individuals. The use of insulin by diabetics and the dietary control of newborn phenylketonurics are long-standing examples. Today, "genetic surgery" to replace defective genes rests clearly on the horizon. Furthermore, social policies now have a solid genetic foundation upon which they may be based. Nevertheless, caution is still required to ensure that our expanded knowledge of human genetics does not obscure the role played by the environment in determining an individual's phenotype.

Soviet Science: The Lysenko Affair

One of the most famous examples of an interaction between science and politics occurred in the former Soviet Union. This instance happened to involve genetics. In particular, it had to do with the genetic and evolutionary theory of the inheritance of acquired characteristics. As we know from our earlier discussion, the theory originated with the ancient Greeks, was formalized by Lamarck in 1809, and considered important by Darwin. However, experimental evidence early in the twentieth century argued against it. Nevertheless, in the 1930s, Trofim Denisovich Lysenko, a young plant breeder from the Ukraine, espoused the idea that the efficiency of plant development (and thus agricultural productivity) could be improved by manipulating environmental conditions. Lysenko believed that such improvements would be incorporated into the genetic material and thus passed on to future generations of plants.

The political climate in the Soviet Union at that time provided a hospitable environment for the revival of Lamarck's ideas. First, the theory that environmental change could produce permanent genetic change was compatible with and paralleled the Marxist political thesis that the proper social conditions would induce permanent changes in human behavior. Second, Ivan Pavlov, the Soviet scientist known for his work on the learned or conditioned reflex, emphasized the importance of environmental stimuli. Pavlov claimed, although he later retreated from this statement, to have found an example of a conditioned reflex that was then inherited in mice.

Perhaps most significant in the acceptance of Lysenko's ideas was the poor condition of Soviet agriculture during this period. The Soviet government was desperate to improve agricultural production, since the traditional methods of selective breeding were not perceived as producing adequate harvests.

Facing a difficult situation, Josef Stalin was impressed with Lysenko's ideas. Lysenko proposed that germination of the winter wheat crop could be speeded up by **vernalization**, the practice of subjecting plants to an artificial cold period to shorten the dormant period of the seeds, which are usually shed in autumn. The early shedding of the seeds permitted the planting of an additional crop that could be harvested before autumn. Lysenko claimed that because the changes induced by vernalization were permanent ones, this technique need be applied only once.

In fact, this practice did not lead to a permanent increase in agricultural output, but by the time this became apparent, Lysenko had become director of the Institute of Genetics of the USSR Academy of Science. Appointed to this position in 1940, he managed to suppress throughout the Soviet Union all work in genetics that ran contrary to his own views. He was responsible for destroying research records, laboratory supplies, and experimental material of those who opposed him, and in some cases, he had his opponents arrested and sentenced to prison.

One of Russia's leading geneticists, Nikolai Ivanovich Vavilov was a favorite target of Lysenko. Director of the Lenin Academy of Agriculture, Vavilov was one of the world's experts on wheat. He was persecuted by Lysenko and eventually sentenced to prison for agricultural espionage in 1941. He died from malnutrition in 1943 (see the Popovsky reference in Selected Readings at the end of the chapter).

GENETICS, TECHNOLOGY, AND SOCIETY

A New Era in Plant Genetics: The Flavr Savr Tomato and Edible Vaccines

The genetic manipulation of plants by humans has been going on for thousands of years, and in fact it formed the foundation of Mendelian genetics. For the most part, this has involved crossing plants with desirable characteristics with other such plants to produce lines that breed true for these traits. Traditional plant breeding has been remarkably successful in creating improved varieties of crop plants, but it can be very complicated and time consuming, requiring many years of crossing and backcrossing to produce the desired combination of characteristics. In the last decade, new tools have become available that allow very specific alterations of the genetic constitution of a plant. But with this technology have come questions regarding the safety and propriety of genetically engineering plants, particularly those used for food.

Much of the initial excitement, as well as much of the controversy, focused on a tomato that has been genetically engineered to alter the ripening process. Tomatoes that ripen fully on the vine are too fragile to be transported long distances without being damaged. Up to now, the solution has

been to pick green, unripe tomatoes, ship them to market in chilled containers to prevent further ripening, and expose them to ethylene gas just before their marketing. The ethylene stimulates the completion of ripening, which involves synthesis of carotenoid pigments (color development), synthesis of sugars and other compounds (flavor development), and breakdown of cell walls (fruit softening). Unfortunately, although these tomatoes may look red and juicy, they lack the flavor and consistency of a tomato fully ripened on the vine.

Plant molecular geneticists realized that if fruit softening could be inhibited while the other phases of ripening proceeded, the tomato that would result could ripen almost completely on the vine but would still be firm enough to ship. The idea was to specifically inhibit the action of one enzyme, *polygalacturonase* (PG), which normally solubilizes the pectin fraction of the cell walls causing fruit softening. This would delay fruit softening while allowing the development of color and flavor to continue.

The procedure used to inhibit PG activity has been termed **antisense technology**; it involved inserting a segment of a cloned PG gene in the reverse (or complementary) orientation. When this reversed gene is expressed,

the RNA that is produced is complementary to the normal PG mRNA. In plant cells, this antisense RNA will therefore anneal to the PG mRNA, preventing the production of the PG enzyme. The beauty of antisense technology is its specificity: In theory, a single gene out of tens of thousands in the plant can be targeted for inhibition.

Using this strategy, scientists at Calgene, an agricultural biotechnology company based in Davis, California, created what has become known as the **Flavr Savr tomato**. Fruit softening was indeed delayed. Its look, its feel, and, most important, its taste were compared to tomatoes artificially ripened by exposure to ethylene. The Flavr Savr was found to be clearly superior, comparable to a vine-ripened tomato. After the completion of these tests, Calgene introduced the tomato to the market in May 1994. The arrival of the Flavr Savr

The Flavr Savr tomato.

During this time great advances were made in genetics, particularly in the United States. Scientifically based selective breeding programs had developed new varieties of hybrid grains whose yields were much greater than those of the old varieties. Yet none of this knowledge was made available to Soviet geneticists until 1964, when Lysenko finally fell from power and lost his stranglehold on Soviet genetics. One American botanist calculated that USSR corn production would have been increased by 6 million tons between 1947 and 1957 if only half the country's acreage had been planted with the new hybrid strains.

Unfortunately, Soviet genetics was held back for nearly a generation at a time when numerous significant advances were being made elsewhere in the world. Adherence to theories that were politically rather than scientifically acceptable was indeed costly to Soviet society. Even though the regions occupied by the former Soviet Union still experience agricultural shortfalls in the 1990s,

the application of modern plant breeding information has significantly improved agricultural productivity.

Genetic Advances in Agriculture and Medicine

The major benefits to society as a result of genetic study have been in the areas of agriculture and medicine. Although cultivation of plants and domestication of animals had begun long before, the rediscovery of Mendel's work in the early twentieth century spurred scientists to apply genetic principles to these human endeavors. The Lysenko era aside, selective breeding and hybridization techniques have had a most significant impact in agriculture.

Plants have been improved in four major ways: (1) enhanced potential for more vigorous growth and increased yields, including the unique genetic phenomenon of tomato was not universally hailed, however, and considerable opposition developed. One concern was that the engineered tomato plants contain a gene conferring resistance to the antibiotic kanamycin. It is introduced along with the PG antisense gene as a marker gene. Some critics feared that the *kan*R gene might somehow be picked up from plant debris by soil bacteria, making them kanamycin-resistant. Others were apprehensive that toxic or allergenic compounds resulting from the genetic manipulation, undetected by Calgene scientists, could be present in the tomatoes.

For some, however, the objection to the release of the tomato for public consumption was based not on scientific issues but rather on an emotional response. They felt that genetic engineering of plants is unnatural, potentially creating food that could have unpredictable effects on human health.

Ultimately, the success or failure of genetically engineered plants will be determined by the consumer. Unless rejected by consumers, other genetically engineered crop plants will shortly follow. Among these will be cotton plants that resist pests by incorporation of a bacterial gene encoding a protein toxic to insects, soybeans that produce a more healthful composition of fatty acids by alteration of key enzymes, and even flowers that stay fresh longer by inhibition of genes involved in senescence.

One of the most interesting possibilities involves the production of **edible vaccines**: Plants that when eaten trigger the production of antibodies against specific diseases. Leading the effort to develop edible vaccines in plants is the research team of Charles Arntzen at the Boyce Thompson Research Institute in Ithaca, New York. This group of researchers is focusing on intestinal diseases such as cholera and bacterial diarrhea, which remain the leading causes of death in infants and children in the developing countries of the world.

The bacteria that cause these diseases colonize the small intestine and produce proteins called enterotoxins, which bind to the surface of the mucosal cells lining the intestine. These stimulate the secretion of large amounts of fluid from the intestinal mucosa, causing acute diarrhea, tissue dehydration, and muscle cramps. When untreated, death may ensue. The enterotoxins are made up of two different polypeptides, the A and B subunits, which individually, are harmless to individuals. For the toxin to be active, A and B subunits must be complexed. Since it is the B subunit of this complex that is responsible for initial binding to cells of the intestinal mucosa, Arntzen's group decided use the B polypeptide as their potential antigen.

They began by inserting the gene encoding the B subunit of *E. coli* en-

terotoxin into the potato plant. They found that tubers of the engineered plants accumulated the enterotoxin B fragment. They next showed that mice fed a few grams of these tubers produced antibodies specific to the enterotoxin B subunit, including those that are secreted by the mucosa into the small intestine and are crucial for fighting bacterial infections. The final important step will be to find out whether this treatment protects the mice from subsequent infection with *E. coli*.

These results, though preliminary, have raised hopes that the production of vaccines in plants is feasible. Several technical hurdles will have to be overcome, however, before the promise of edible vaccines is realized. For example, since humans do not eat raw potatoes, and since cooking denatures proteins, potatoes may not be an especially good choice to deliver vaccines. Arntzen believes he has an excellent alternative, the banana, which is usually eaten raw. He and his colleagues have recently succeeded in introducing foreign genes into banana plants and are now attempting to introduce the *E. coli* enterotoxin gene. While much work remains before the feasibility of edible vaccines is established, immunizing the world population against devastating diseases and saving an untold number of lives by feeding them genetically engineered plants is an exciting prospect.

---

hybrid vigor (heterosis); (2) increased resistance to natural predators and pests, including insects and disease-causing microorganisms; (3) production of hybrids exhibiting a combination of superior traits derived from two different strains or even two different species (Figure 1.10); and (4) selection of genetic variants with desirable qualities such as increased protein value, increased content of limiting amino acids, which are essential in the human diet, or reduced plant size, lessening vulnerability to adverse weather conditions.

These improvements have resulted in a tremendous increase in yield and nutrient value in such crops as barley, beans, corn, oats, rice, rye, and wheat. It is estimated that in the United States the use of improved genetic strains has led to a threefold increase in crop yield per acre. In Mexico, where corn is the staple crop, a significant increase in protein content and yield has occurred. A substantial effort has also been made to improve the

growth of Mexican wheat. Led by Norman Borlaug, a team of researchers developed varieties of wheat that incorporated favorable genes from other strains found in various parts of the world. This resulted in the creation of superior varieties that revolutionized wheat production in many underdeveloped countries besides Mexico. Because of this effort, which led to the well-publicized "Green Revolution," Borlaug received the Nobel Peace Prize in 1970. There is little question that this application of genetics has contributed to the well-being of our own species by improving the quality of nutrition worldwide.

Applied research in genetics has also resulted in the development of superior breeds of animals. Enormous increases have occurred in usable meat supplies per unit of food intake. For example, selective breeding has produced chickens that grow faster, produce more high-quality meat per chicken, and lay greater numbers of larger eggs. In larger animals, including pigs and cows,

FIGURE 1.10   *Triticale,* a hybrid grain derived from wheat and rye, produced as a result of applied genetic research.

the use of artificial insemination has been particularly important (Figure 1.11). Sperm samples derived from a single male with superior genetic traits may now be used to fertilize thousands of females located in all parts of the world. Inherited animal disorders are also amenable to modern genetic analysis, creating the potential for their elimination through selective breeding.

Equivalent strides have been made in medicine as a result of advances in genetics, particularly since 1950. Numerous disorders in humans have been discovered to result from either a single mutation or a specific chromosomal abnormality. For example, the genetic basis of sickle-cell anemia, erythroblastosis fetalis, cystic fibrosis, hemophilia, Huntington disease, muscular dystrophy, Tay-Sachs disease, Down syndrome, and countless metabolic disorders is now well documented and understood at the molecular level. It is estimated that more than 10 million children or adults in the United States suffer from some form of genetic affliction. It is estimated that every childbearing couple stands an approximately 3 percent risk of having a child with some form of genetic anomaly.

Additionally, it has gradually become clear during the current decade that most, if not all, forms of cancer have a genetic basis. Although cancer may not necessarily be an inherited disorder, cancer is a genetic disorder at the

somatic cell level. That is, most cancers are derived from somatic cells that have undergone some type of genetic change; malignant tumors are then derived from the initial mutant cell.

Recognition of the genetic basis of these disorders has provided direction for the development of detection and treatment. In the case of inherited disorders, **genetic counseling** provides parents with objective information upon which they can base rational decisions about parenting. In the case of cancer, the discovery of the genetic basis of each type of malignancy will provide valuable insights into the cellular basis of this diverse group of diseases. This increased understanding has already led to more effective early detection. It is hoped that more efficient treatments and, ultimately, strategies for prevention will also result.

Applied research in genetics has provided other medical benefits. Increased knowledge in **immunogenetics** has made possible compatible blood transfusions as well as organ transplants. The discovery of genetically determined, tissue-bound antigens has led to the important concepts of **histocompatibility** and tissue typing. In conjunction with immunosuppressive drugs, transplant operations involving human organs, including the heart, liver, pancreas, and kidney, are increasing annually and are now considered routine surgery.

The most recent advances in human genetics have been dependent on the application of **recombinant DNA technology**. It is expected that the entire human genome will be sequenced by the year 2003. Cloned human genes that code for many medically important molecules such as insulin, interferon, and growth hormone already serve as the source for their mass production. Recombinant DNA techniques also play an essential role in **gene therapy**, which involves the direct manipulation of the genetic material. Currently in its infancy,

FIGURE 1.11   The effects of breeding and selection, as illustrated by the production of this Vietnamese pot-bellied pig.

this technology is now being used to alter the genetic constitution of individuals harboring genetic defects. Soon, the developing fetus will be targeted for gene therapy. Although such processes present ethical questions, they provide the potential for the correction of serious genetic errors in humans.

In later chapters, these and other examples in agriculture and medicine are discussed in great detail. Although other scientific disciplines are also expanding in knowledge, none have paralleled the growth of information that is occurring annually in genetics. By the end of this course, we are confident you will agree that the present truly represents the "Age of Genetics."

CHAPTER SUMMARY

1. Genetics, which emerged as a formal discipline of biology early in the twentieth century, has a rich history that dates back to prehistoric times.

2. Numerous concepts and a basic vocabulary play a fundamental role in understanding genetics.

3. Four investigative approaches are most often used in the study of genetics, including transmission genetic studies, cytogenetic analyses, molecular–biochemical experimentation, and inquiries into the genetic structure of populations.

4. Eugenics, the application of knowledge of genetics to the improvement of human existence, has a long and controversial history. Euphenics, genetic intervention designed to reduce or ameliorate the impact on individuals of defective genotypes, represents the modern approach.

5. Curtailment of the freedom of scientific inquiry had an extremely adverse effect in the USSR under the reign of Stalin, who supported the views and practices of Lysenko.

6. Applied research in genetics has had a profound effect on the human condition as a result of major advances in agriculture and medicine.

KEY TERMS

activation energy, 9
allele, 8
amino acid, 8
antisense technology, 12
atomic theory, 4
biochemical analysis, 9
cell theory, 4
centromere, 7
chromatin, 7
chromosomal mutation (aberration), 8
chromosomal theory of inheritance, 9
chromosome, 7
cytological study, 9
diploid number ($2n$), 7
DNA (deoxyribonucleic acid), 6
doctrine of use and disuse, 5
double helix, 6
edible vaccines, 13
enzyme, 8
epigenesis, 3
eugenics, 10
euphenics, 11
fixity of species, 4
Flavr Savr tomato, 13

gemmules, 5
gene, 7
gene mutation, 8
gene therapy, 14
generative force, 2
genetic code, 8
genetic counseling, 14
genetics, 6
genotype, 8
germplasm, 5
haploid number (n), 7
histocompatibility, 14
homologous chromosomes, 7
homunculus, 4
hybrid vigor (heterosis), 13
immunogenetics, 14
inheritance of acquired characteristics, 5
karyotype, 9
loci, 7
meiosis, 8
mitosis, 8
mRNA (messenger RNA), 8
natural selection, 5

nitrogenous base, 8
nucleoid region, 6
nucleotide, 7
nucleus, 6
pangenesis, 5
pedigree analysis, 9
phenotype, 8
physical substance, 2
polyploid, 7
population genetics, 10
preformationism, 4
recombinant DNA, 10
ribosome, 8
RNA (ribonucleic acid), 6
rRNA (ribosomal RNA), 9
somatoplasm, 5
special creation, 4
spontaneous generation, 4
transcription, 8
translation, 8
transmission genetics, 9
tRNA (transfer RNA), 8
vernalization, 11

PROBLEMS AND DISCUSSION QUESTIONS

1. Describe the ideas of Hippocrates and Aristotle related to the genetic basis of life.
2. Define and contrast the theories of epigenesis and preformationism. Into which theory did the concept of a homunculus fit?
3. Describe Darwin's and Wallace's evolutionary theory of natural selection. What information was lacking from it (i.e., what gap remained in it)?
4. Describe Darwin's proposal that attempted to bridge the "gap" in his theory of natural selection. How was this proposal related to ideas advanced much earlier within the historical context of genetics?
5. Contrast chromosomes and genes and describe their role in heredity.
6. Describe the four major investigative approaches used in studying genetics.
7. Contrast eugenics with euphenics. Which is the more modern term?
8. Norman Borlaug received the Nobel Peace Prize for his work in genetics. Why do you think he was awarded this prize?
9. How has genetic research been applied to the field of medicine?
10. Describe Lysenko's views of plant breeding. What was the major flaw in his approach to agricultural improvement?
11. What is the significance of Weismann's distinction between germplasm and somatoplasm?

SELECTED READINGS

ALLEN, G. E. 1996. Science misapplied: The eugenics age revisited. *Technology Review* 99:23–31.

BORLAUG, N. E. 1983. Contributions of conventional plant breeding to food production. *Science* 219:689–93.

BOWLER, P. J. 1989. *The Mendelian revolution: The emergence of hereditarian concepts in modern science and society.* London: Athlone.

COCKING, E. C., DAVEY, M. R., PENTAL, D., and POWER, J. B. 1981. Aspects of plant genetic manipulation. *Nature* 293: 265–70.

DAY, P. R. 1977. Plant genetics: Increasing crop yield. *Science* 197:1334–39.

DUNN, L. C. 1965. *A short history of genetics.* New York: McGraw-Hill.

GARDNER, E. J. 1972. *History of biology,* 3rd ed. New York: Macmillan.

GARVER, K. L., and GARVER, B. 1991. Eugenics: Past, present, and future. *Am. J. Hum. Genet.* 49:1109–18.

GASER, C. S. and FRALEY, R. T. 1989. Genetically engineering plants for crop improvement. *Science* 244:1293–99.

HAQ, A. H., MASON, H. S., CLEMENTS, J. D., and ARNTZEN, C. J. 1995. Oral immunization with a recombinant antigen produced in transgenic plants. *Science* 268:714–716.

HORGAN, J. 1993. Eugenics revisited. *Sci. Am.* (June) 268: 123–31.

JORAVSKY, D. 1970. *The Lysenko affair.* Cambridge: Harvard University Press.

KING, R. C., and STANSFIELD, W. D. 1990. *A dictionary of genetics,* 4th ed. New York: Oxford University Press.

MEDVEDEV, Z. A. 1969. *The rise and fall of T. D. Lysenko.* New York: Columbia University Press.

MOORE, J. A. 1993. *Science as a way of knowing.* Cambridge: Harvard University Press.

OELLER, P. W., et al. 1991. Reversible inhibition of tomato fruit senescence by antisense RNA. *Science* 254:437.

OLBY, R. C. 1966. *Origins of Mendelism.* London: Constable.

POPOVSKY, M. 1984. *The Vavilov affair.* Hamden, CT: Archon.

SOYFER, V. N. 1989. New light on the Lysenko era. *Nature* 339:415–20.

STUBBE, H. 1972. *History of genetics: From prehistoric times to the rediscovery of Mendel.* (Translated by T. R. W. Waters.) Cambridge, MA: MIT Press.

TORREY, J. G. 1985. The development of plant biotechnology. *Am. Sci.* 73:354–63.

VASIL, I. K. 1990. The realities and challenges of plant biotechnology. *Bio/Technology* 8:296–301.

WEINBERG, R. A. 1985. The molecules of life. *Sci. Am.* (Oct.) 253:48–57.

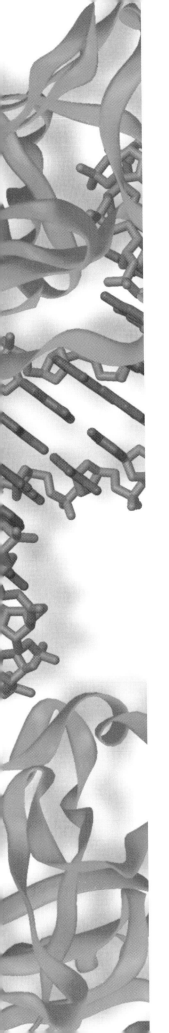

Part One

Heredity and the Phenotype

CHAPTER

2

Cell Division and Chromosomes

CHAPTER CONCEPTS

Genetic continuity between cells and organisms of any sexually reproducing species is maintained by the process of mitosis and meiosis. These processes are orderly and efficient, serving to produce diploid somatic cells and haploid gametes and spores, respectively. During these division stages, the genetic material is condensed into discrete, visible structures called chromosomes. These structures take on various forms and appearances, and their study has provided important insights into the nature of the genetic material.

In every living thing there exists a substance referred to as the **genetic material**. Except in certain viruses, this material is composed of the nucleic acid DNA. A molecule of DNA contains many units called **genes**, the products of which direct all metabolic activities of cells. DNA, with its array of genes, is organized into **chromosomes**, structures that serve as the vehicle for transmission of genetic information. The manner in which chromosomes are transmitted from one generation of cells to the next, and from organisms to their descendants, is exceedingly precise. In this chapter, we will consider just exactly how such transmission is accomplished as we pursue the topic of genetic continuity between cells and organisms.

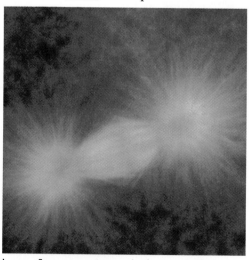
Immunofluorescent micrograph of microtubules.

Two major processes are involved in eukaryotes: **mitosis** and **meiosis**. Although the mechanisms of the two processes are similar in many ways, the outcomes are quite different. Mitosis leads to the production of two cells, each with the number of chromosomes identical to the parental cell. Meiosis, on the other hand, reduces the amount of genetic material and the number of chromosomes by precisely half. This reduction is essential if sexual reproduction is to occur without doubling the amount of genetic material at each generation. Strictly speaking, mitosis is that portion of the cell cycle during which the hereditary components are precisely and equally divided into daughter cells. Meiosis is part of a special type of cell division leading to the production of sex cells: gametes and spores. This process is an essential step in the transmission of genetic information from an organism to its offspring.

In most cases, chromosomes are visible only when cells are actually dividing, that is, during mitosis or meiosis. When cells are not undergoing division, the genetic material making up chromosomes unfolds and uncoils into a diffuse network within the nucleus, generally referred to as **chromatin**. In this chapter, we will examine this transition and also consider two specialized cases where chromosomes are extraordinarily large and amenable to investigation. The study of these examples, **polytene chromosomes** and **lampbrush chromosomes**, has greatly extended our knowledge of genetic organization and its relationship to genetic function.

Cell Structure

Before describing mitosis and meiosis, we will briefly review the structure of cells. As we shall see, many cellular components, such as the nucleolus, ribosome, and centriole, are involved directly or indirectly with genetic processes. Other components, the mitochondria and chloroplasts, contain their own unique genetic information. It is also useful for us to compare the structural differences of the bacterial prokaryotic cell with the eukaryotic cell. Variation in structure and function of cells is dependent on the specific genetic expression by each cell type.

Before 1940, knowledge of cell structure was based on information obtained with the light microscope. About 1940 the transmission electron microscope was in its early stages of development, and by 1950 many details of cell ultrastructure were unveiled. Under the electron microscope, cells were seen as highly organized, precise structures. A new world of whorling membranes, miniature organelles, microtubules, granules, and filaments was revealed. These discoveries revolutionized thinking in the entire field of biology. We will be concerned with those aspects of cell structure relating to genetic study. Many of the parts of the cell are described in Figure 2.1, which depicts a typical animal cell.

Cell Boundaries

The cell is surrounded by a **plasma membrane**, an outer covering that defines the cell boundary and delimits the cell from its immediate external environment. This membrane is not passive; rather, it controls the movement of material such as gases, nutrients, and waste products into and out of the cell. In addition to this membrane, plant cells have an outer covering called the **cell wall**. One of the major components of this rigid structure is a polysaccharide called **cellulose**. Bacterial cells also have a cell wall, but its chemical composition is quite different from that of the plant cell wall. The major component in bacteria is a complex macromolecule called a **peptidoglycan**. As its name suggests, the molecule consists of peptide and sugar units. Long polysaccharide chains are cross-linked with short peptides, which impart great strength and rigidity to the bacterial cell. Some bacterial cells have still another covering, a **capsule**. This mucus-like material protects these bacteria from phagocytic activity by the host during their pathogenic invasion of eukaryotic organisms. As we will see in Chapter 10, the presence of the capsule is under genetic control. Its loss due to mutation in the pneumonia-causing bacterium *Diplococcus pneumoniae* provided the underlying basis for a critical experiment proving that DNA is the genetic material.

Activities at cell boundaries are dynamic physiological processes. Both transport in and out of cells, and communication between cells, are critical to normal function. Because physiological processes are biochemical in nature, we would expect that many genes and their products are essential to these activities. This is indeed the case, and as such, mutations in these genes can alter or interrupt normal physiological functions, often with severe consequences. For example, the inherited disorder **Duchenne muscular dystrophy** is the result of complete loss of function of the product **dystrophin**, which is believed to function at the cell membrane of muscle cells. As shown in Figure 2.2, using immunofluorescent localization technology, dystrophin is completely absent from skeletal muscle of an afflicted individual compared to muscle from a control subject.

Many, if not most, animal cells have a covering over the plasma membrane called a **cell coat**. Consisting of glycoproteins (sometimes called the **glycocalyx**) and polysaccharides, the chemical composition of the cell coat differs from comparable structures in either plants or bacteria. One function served by the cell coat is to provide biochemical identity at the surface of cells. Among other forms of molecular recognition, various antigens are part of the cell coat. All forms of biochemical identity at the cell surface are under genetic control, and many have been thoroughly investigated. For example, the **AB** and **MN antigens**, which may elicit an immune response during blood transfusions, are found on the surface of red blood cells. In other cells, the **histocompatibility antigens**, which elicit an immune response during tissue and organ transplants, are part of the cell coat. Further, a variety of highly specific **receptor molecules** are integral components of the cell surface. These constitute recognition sites that receive chemical signals that are often

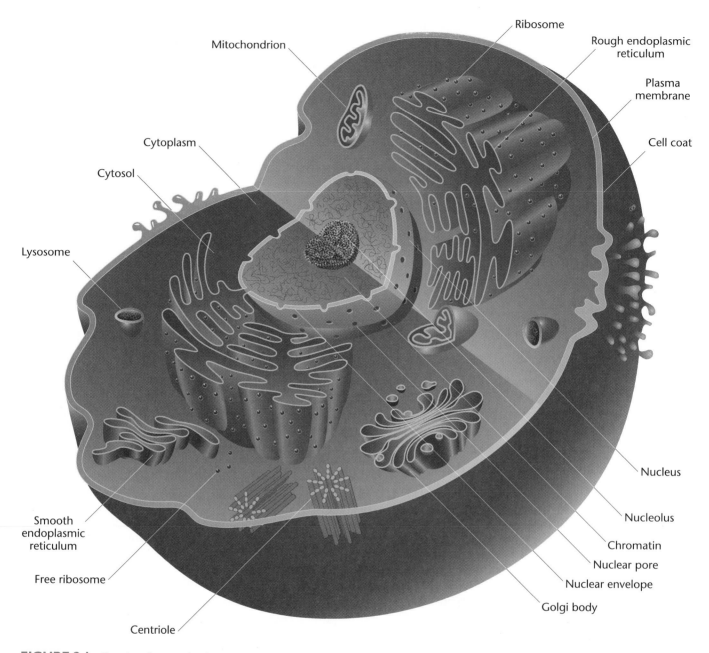

FIGURE 2.1   Drawing of a generalized animal cell. Emphasis has been placed on the cellular components discussed in the text.

transferred to the cell. Such signals may initiate a variety of chemical activities and may ultimately signal specific genes to turn on. We may conclude that the cell surface links its host cell to its outside world and is tied in numerous ways to genetic processes.

The Nucleus

The presence of the **nucleus** and other membranous organelles characterizes eukaryotic cells. The nucleus houses the genetic material, DNA, which is found in as-

sociation with large numbers of acidic and basic proteins. During nondivisional phases of the cell cycle this DNA/protein complex exists in an uncoiled, dispersed state called **chromatin**. As we will soon discuss, during mitosis and meiosis this material coils up and condenses into structures called **chromosomes**. Also present in the nucleus is the **nucleolus**, an amorphous component where ribosomal RNA (rRNA) is synthesized and where the initial stages of ribosomal assembly occur. The areas of DNA encoding rRNA are collectively referred to as the **nucleolar organizer region** or the **NOR**.

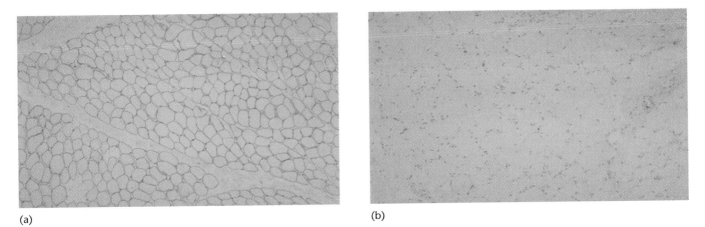

(a)   (b)

FIGURE 2.2   (a) Localization of dystrophin in the sarcolemmal areas of normal muscle using immunoperoxidase staining. (b) Complete lack of immunoreactive dystrophin in the skeletal muscle of a patient with Duchenne muscular dystrophy.

The lack of a nuclear envelope and membraneous organelles is characteristic of prokaryotes. In bacteria such as *E. coli* the genetic material is present as a long, circular DNA molecule that is compacted into an area referred to as the **nucleoid region**. Part of the DNA may be attached to the cell membrane, but in general the nucleoid region constitutes a large area throughout the cell. Although the DNA is compacted, it does not undergo the extensive coiling characteristic of the stages of mitosis where, in eukaryotes, chromosomes become visible. Nor is the DNA in these organisms associated as extensively with proteins as is eukaryotic DNA. Figure 2.3 shows the formation of two bacteria during cell division and illustrates the bacterial chromosomes in the nucleoid regions. Prokaryotic cells do not have a distinct nucleolus, but they do contain genes that specify rRNA molecules.

The Cytoplasm and Organelles

The remainder of the eukaryotic cell enclosed by the plasma membrane, and excluding the nucleus, is composed of **cytoplasm** and all associated **cellular organelles**. Cytoplasm consists of a nonparticulate, colloidal material referred to as the **cytosol**, which surrounds and encompasses the numerous types of cellular organelles. Beyond these components, an extensive system

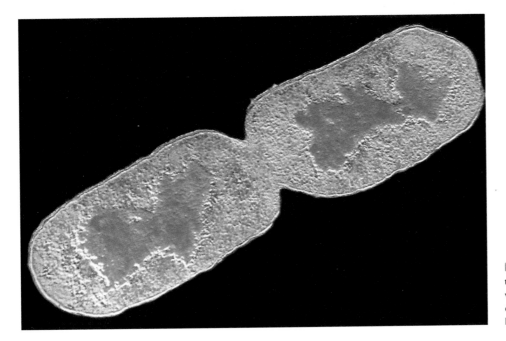

FIGURE 2.3   Color-enhanced electron micrograph of *E. coli* undergoing cell division. Particularly prominent are the two chromosomal areas (shown in red) that have been partitioned into the daughter cells.

of tubules and filaments comprising the **cytoskeleton** provides a lattice of support structures within the cytoplasm. Consisting primarily of tubulin-derived microtubules and actin-derived microfilaments, this structural framework maintains cell shape, facilitates cell mobility, and anchors the various organelles. Tubulin and actin are both proteins found abundantly in eukaryotic cells.

One organelle, the membranous **endoplasmic reticulum (ER)**, compartmentalizes the cytoplasm, greatly increasing the surface area available for biochemical synthesis. The ER may appear smooth, in which case it serves as the site for synthesis of fatty acids and phospholipids. Or, the ER may appear rough because it is studded with ribosomes. Ribosomes, which we will later discuss in detail (see Chapter 12), serve as sites for the translation of genetic information contained in messenger RNA (mRNA) into proteins.

Three other cytoplasmic structures are very important in the eukaryotic cell's activities: **mitochondria**, **chloroplasts**, and **centrioles**. Mitochondria are found in both animal and plant cells and are the sites of the oxidative phases of **cell respiration**. These chemical reactions generate large amounts of **adenosine triphosphate (ATP)**, an energy-rich molecule. The chloroplast is one type of plastid found in plants, algae, and some protozoans. This organelle is associated with **photosynthesis**, the major energy-trapping process on earth. Both mitochondria and chloroplasts contain a type of DNA distinct from that found in the nucleus. Furthermore, these organelles can duplicate themselves and transcribe and translate their genetic information. It is interesting to note that the genetic machinery of mitochondria and chloroplasts closely resembles that of prokaryotic cells. This and other observations have led to the proposal that these organelles were once primitive free-living organisms that established a symbiotic relationship with a primitive eukaryotic cell. This proposal, which describes the evolutionary origin of these organelles, is called the **endosymbiotic hypothesis**.

Animal and some plant cells also contain a pair of complex structures called the **centrioles**. These cytoplasmic bodies, found within a specialized region called the **centrosome**, are associated with the organization of those spindle fibers that function in mitosis and meiosis. In some organisms, the centriole is derived from another structure, the **basal body**, which is associated with the formation of cilia and flagella. Over the years, many reports have suggested that centrioles and basal bodies contain DNA, which is involved in the replication of these structures. However, it is the current consensus that this is not the case.

The organization of **spindle fibers** by the centrioles occurs during the early phases of mitosis and meiosis.

Composed of arrays of microtubules, these fibers play an important role in the movement of chromosomes as they separate during cell division. The microtubules consist of polymers of alpha and beta subunits of the protein tubulin. Interaction of the chromosomes and spindle fibers will be considered later in this chapter.

Homologous Chromosomes, Haploidy, and Diploidy

To discuss the processes of mitosis and meiosis, it is important to understand clearly the concept of **homologous chromosomes**. Such an understanding will also be critical to our future discussions of Mendelian genetics. Thus, before embarking further, we will address this topic and introduce other important terminology.

Chromosomes are most easily visualized during mitosis. When they are examined carefully, they are seen to take on distinctive lengths and shapes. Each contains a condensed or constricted region called the **centromere**, which establishes the general appearance of each chromosome. Figure 2.4 illustrates chromosomes with centromere placements at different points along their lengths. Extending from either side of the centromere are the arms of the chromosome. Depending on the position of the centromere, different arm ratios are produced. As Figure 2.4 illustrates, chromosomes are classified as **metacentric**, **submetacentric**, **acrocentric**, or **telocentric** on the basis of the centromere location. The shorter arm, by convention, is shown above the centromere and is called the **p arm** (*p* stands for "petite"). The longer arm is shown below the centromere and is called the **q arm** (*q* being the next letter in the alphabet).

When studying mitosis, we may make several other important observations. First, each somatic cell within members of the same species contains an identical number of chromosomes. This is called the **diploid number (2n)**. When the lengths and centromere placements of all such chromosomes are examined, a second general feature is apparent. Nearly all of the chromosomes exist in pairs with regard to these two criteria. The members of each pair are called **homologous chromosomes**. For each chromosome exhibiting a specific length and centromere placement, another exists with identical features. There are, of course, exceptions to the rule of chromosomes in pairs illustrated by organisms such as yeasts and molds that spend the predominant portion of their life cycle in the haploid stage.

Figure 2.5 illustrates the nearly identical physical appearance of members of homologous chromosome pairs.

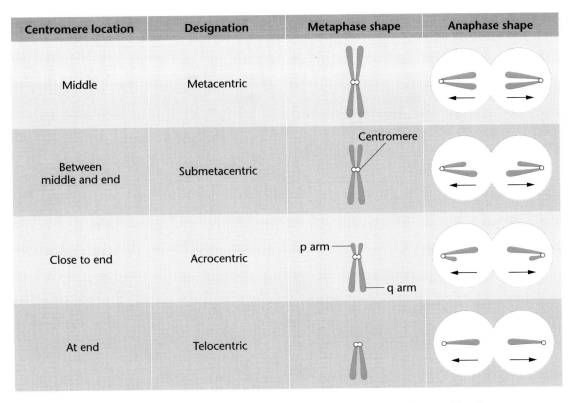

Centromere location	Designation	Metaphase shape	Anaphase shape
Middle	Metacentric		
Between middle and end	Submetacentric	Centromere	
Close to end	Acrocentric	p arm q arm	
At end	Telocentric		

FIGURE 2.4   Centromere locations and designations of chromosomes based on their location. Note that the shape of the chromosome during anaphase is determined by the position of the centromere.

There, the human mitotic chromosomes have been photographed (shown at the top of the figure), cut out of the print, and matched up, creating a **karyotype** (shown at the bottom of the figure). As you can see, humans have a *2n* number of 46 and exhibit a diversity of sizes and centromere placements. Note also that each one of the 46 chromosomes is clearly a double structure, consisting of two parallel **sister chromatids** connected by a common centromere. Had these chromosomes been allowed to continue dividing, each pair of sister chromatids, which are replicas of one another, would have separated into two new cells as division continued.

The **haploid number (*n*)** of chromosomes is one-half of the diploid number. Table 2.1 demonstrates the wide range of *n* values found in 32 diverse species of plants and animals. Collectively, the total set of genes contained on one member of each homologous pair of chromosomes constitutes the **haploid genome** of the species.

Homologous pairs of chromosomes have important genetic similarity. They contain identical gene sites along their lengths, each called a **locus** (pl. **loci**). Thus, they have identical genetic potential. In sexually reproducing organisms, one member of each pair is derived from the maternal parent (through the ovum) and one from the paternal parent (through the sperm). Therefore, each

diploid organism contains two copies of each gene as a consequence of **biparental inheritance**. As we will see in the following chapters on transmission genetics, the members of each pair of genes, while influencing the same characteristic or trait, need not be identical. Alternative forms of the same gene are called **alleles**. In a population of members of the same species, many different alleles of the same gene may exist.

The concepts of haploid number, diploid number, and homologous chromosomes may be related to the process of meiosis. During the formation of gametes or spores, meiosis converts the diploid number of chromosomes to the haploid number. As a result, haploid gametes or spores contain precisely one member of each homologous pair of chromosomes, that is, one complete haploid set. Following fusion of two gametes in fertilization, the diploid number is reestablished, that is, the zygote contains two complete haploid sets of chromosomes. The constancy of genetic material is thus maintained from generation to generation.

There is one important exception to the concept of homologous pairs of chromosomes. In many species, one pair, the **sex-determining chromosomes**, is often not homologous in size, centromere placement, arm ratio, or genetic potential. For example, in humans, males contain

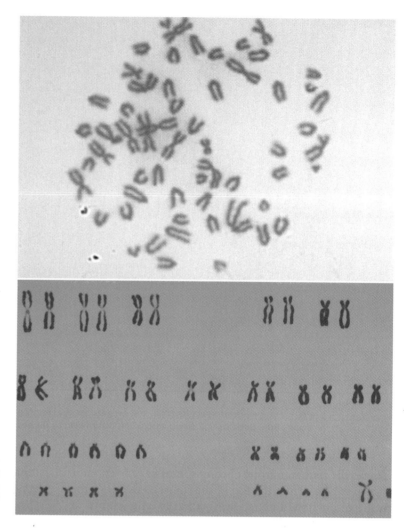

FIGURE 2.5 A metaphase preparation of chromosomes derived from a human male and the karyotype derived from it. All but the X and Y chromosomes are present in homologous pairs. Each chromosome is actually a double structure, constituting a pair of sister chromatids joined at a common centromere.

TABLE 2.1 The Haploid Number of Chromosomes for Many Diverse Organisms

Common Name	Scientific Name	Haploid No.	Common Name	Scientific Name	Haploid No.
Black bread mold	*Aspergillus nidulans*	8	House fly	*Musca domestica*	6
Broad bean	*Vicia faba*	6	House mouse	*Mus musculus*	20
Cat	*Felis domesticus*	19	Human	*Homo sapiens*	23
Cattle	*Bos taurus*	30	Jimson weed	*Datura stramonium*	12
Chicken	*Gallus domesticus*	39	Mosquito	*Culex pipiens*	3
Chimpanzee	*Pan troglodytes*	24	Pink bread mold	*Neurospora crassa*	7
Corn	*Zea mays*	10	Potato	*Solanum tuberosum*	24
Cotton	*Gossypium hirsutum*	26	Rhesus monkey	*Macaca mulatta*	21
Dog	*Canis familiaris*	39	Roundworm	*Caenorhabditis elegans*	6
Evening primrose	*Oenothera biennis*	7	Silkworm	*Bombyx mori*	28
Frog	*Rana pipiens*	13	Slime mold	*Dictyostelium discoidium*	7
Fruit fly	*Drosophila melanogaster*	4	Snapdragon	*Antirrhinum majus*	8
Garden onion	*Allium cepa*	8	Tobacco	*Nicotiana tabacum*	24
Garden pea	*Pisum sativum*	7	Tomato	*Lycopersicon esculentum*	12
Grasshopper	*Melanoplus differentialis*	12	Water fly	*Nymphaea alba*	80
Green alga	*Chlamydomonas reinhardi*	18	Wheat	*Triticum aestivum*	21
Horse	*Equus caballus*	32	Yeast	*Saccharomyces cerevisiae*	17

a Y chromosome in addition to one X chromosome (Figure 2.5), whereas females carry two homologous X chromosomes. The X and Y chromosomes are not strictly homologous. The Y is considerably smaller and lacks most of the loci contained on the X. Nevertheless, in meiosis they behave as homologs so that gametes produced by males receive either the X or Y chromosome.

Mitosis and Cell Division

The process of **mitosis** is critical to all eukaryotic organisms. In many single-celled organisms, such as protozoans, some fungi, and algae, mitosis, as a part of cell division, provides the mechanism underlying their asexual reproduction. Multicellular diploid organisms begin life as single-celled fertilized eggs or **zygotes**. The mitotic activity of the zygote and the subsequent daughter cells is the foundation for development and growth of the organism. In adult organisms, mitotic activity associated with cell division is prominent in wound healing and other forms of cell replacement in certain tissues. For example, the epidermal skin cells of humans are continuously being sloughed off and replaced. One estimate is that 100 billion (10^{11}) cells are sloughed off daily by each human! Cell division also results in a continuous production of reticulocytes, which eventually shed their nuclei and replenish the supply of red blood cells in vertebrates. In abnormal situations, somatic cells may exhibit uncontrolled cell divisions, resulting in cancer.

Usually following cell division, the initial size of new daughter cells is approximately one-half the size of their parent cell. However, the nucleus of each new cell is not appreciably smaller than the nucleus of the original cell. Quantitative measurements of DNA confirm that there are equivalent amounts of genetic material in the daughter nuclei as in the parent cell.

The process of cytoplasmic division is called **cytokinesis**. The division of cytoplasm requires a mechanism that results in a partitioning of the volume into two parts, followed by the enclosure of both new cells within a distinct plasma membrane. Cytoplasmic organelles either replicate themselves, arise from existing membrane structures, or are synthesized *de novo* (anew) in each cell. The subsequent proliferation of these structures is a reasonable and adequate mechanism for reconstituting the cytoplasm in daughter cells.

Nuclear division (**karyokinesis**), where the genetic material is partitioned into daughter cells, is more complex than cytokinesis and requires more precision. The chromosomes must first be exactly replicated and then accurately partitioned into daughter cells. The end result is the production of two daughter cells, each with a chromosome composition identical to the parent cell.

Interphase and the Cell Cycle

Many cells undergo a continuous alternation between division and nondivision. The interval between each mitotic division is called **interphase**. It was once thought that the biochemical activity during interphase was devoted solely to the cell's growth and its normal function. However, we now know that another biochemical step critical to the next mitosis occurs during interphase: **the replication of the DNA of each chromosome**. Occurring as the cell prepares to enter nuclear division (mitosis), this period during which DNA is synthesized is called the **S phase**. The initiation and completion of synthesis can be detected by monitoring the incorporation of radioactive DNA precursors such as ^{3}H-thymidine. Their incorporation can be monitored using the technique of autoradiography (Appendix A).

Investigations of this nature have demonstrated two periods during interphase, before and after S, when no DNA synthesis occurs. These are designated **G1 (gap1)** and **G2 (gap2)**, respectively. During both of these periods, as well as during S phase, intensive metabolic activity, cell growth, and cell differentiation occur. By the end of G2, the volume of the cell has roughly doubled, DNA has been replicated, and mitosis (M) is initiated. Following mitosis, continuously dividing cells then repeat this cycle (G1, S, G2, M) over and over. This concept of such a **cell cycle** is illustrated in Figure 2.6.

Much is known about the cell cycle based on *in vitro* (test tube) studies. When grown in culture, many cell types in different organisms traverse the complete cycle in about 16 hours. The actual process of mitosis occupies only a small part of the cycle, usually about an hour. The lengths of the S and G2 stages of interphase are fairly consistent among different cell types. Most variation is seen in the length of time spent in the G1 stage. Figure 2.7 illustrates the relative length of these periods in a typical cell.

The G1 period is of great interest in the study of cell proliferation and its control. At a point late in G1, all cells follow one of two paths. They either withdraw from the cycle and enter a resting phase in the **G0 stage**, or they become committed to initiate DNA synthesis and complete the cycle. The time when this decision is made is called the **G1 checkpoint**. This is one of three such restriction points, where under certain conditions, a cell may temporarily or permanently arrest. We will return to this topic momentarily when we discuss cell cycle regulation.

Cells that enter G0 remain viable and metabolically active but are nonproliferative. Cancer cells apparently avoid entering G0 or they pass through it very quickly. Other cells enter G0 and never reenter the cell cycle. Still others can remain quiescent in G0, but they may be stimulated to return to G1, reentering the cycle.

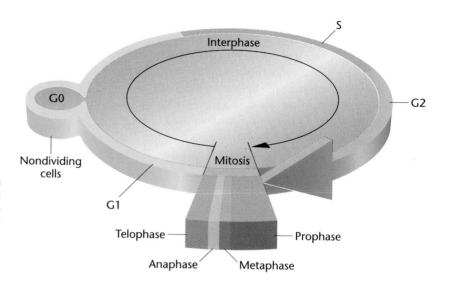

FIGURE 2.6   Diagrammatic representation of the stages comprising an arbitrary cell cycle. Following mitosis (M), cells initiate a new cycle (G1). Cells may become non-dividing (G0) or continue through G1, where they become committed to begin DNA synthesis (S) and complete the cycle (G2 and M). Following mitosis, two daughter cells are produced.

Cytologically, interphase is characterized by the absence of visible chromosomes. Instead, a distinct nucleus is evident, filled with chromatin that has formed as the chromosomes have unfolded and uncoiled following the previous mitosis. This is depicted diagrammatically in part (a) of Figure 2.8(A). When viewed under the light microscope [part (a) of Figure 2.8(B)], the nucleus appears to be filled with a speckled, granular material. This is the result of having sectioned through the uncoiled chromatin fibers.

Prophase

Once G1, S, and G2 are completed, mitosis is initiated. Mitosis is a dynamic period of vigorous and continual activity. For discussion purposes, the entire process is subdivided into discrete phases, and specific events are assigned

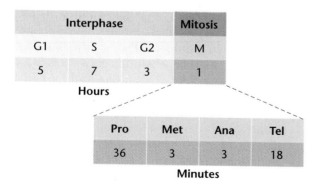

FIGURE 2.7   The time spent in each phase of one complete cell cycle of a human cell in culture. Times vary according to cell types and conditions.

to each stage. These stages, in order of occurrence, are **prophase**, **prometaphase**, **metaphase**, **anaphase**, and **telophase**. As in interphase, each of these stages is depicted in a drawing in Figure 2.8(A) and is shown as it actually occurs during plant mitosis in Figure 2.8(B).

A significant portion of mitosis is spent in prophase, a stage characterized by several essential activities. One of the early events in prophase of all animal cells involves the migration of two pairs of centrioles to opposite ends of the cell. These structures are found just outside the nuclear envelope in an area of differentiated cytoplasm called the **centrosome**. It is thought that each pair of centrioles consists of one mature unit and a smaller, newly formed centriole.

The direction of migration of the centrioles is such that two poles are established at opposite ends of the cell. Following their migration, the centrioles are responsible for the organization of cytoplasmic microtubules into a series of **spindle fibers** that are formed and run between these poles. This creates the axis along which chromosomal separation will occur. We will discuss the formation and role of these spindle fibers in greater detail below.

Interestingly, cells of most plants (there are a few exceptions), fungi, and certain algae seem to lack centrioles. Spindle fibers are nevertheless apparent during mitosis. Therefore, centrioles are not universally responsible for the organization of spindle fibers. If some other center that organizes microtubules into spindle fibers exists in the cells of the above organisms, it has yet to be discovered.

As the centrioles migrate, the nuclear envelope begins to break down and gradually disappears. In a similar fashion, the nucleolus disintegrates within the nucleus. While these events are taking place, the diffuse chromatin—the characteristic uncoiled form of the genetic material during interphase—begins to condense, a process that continues until distinct threadlike structures, or chromosomes, become visible. Further condensation occurs throughout prophase and by the completion of this stage, it first becomes apparent that each chromosome is a double structure split longitudinally except at a single point of constriction, the **centromere**. The two parts of each chromosome are called **chromatids**. Because the DNA contained in each pair of chromatids represents the duplication of a single chromosome during the S phase of the previous interphase, these chromatids are genetically identical and are called **sister chromatids**. In humans, with a diploid number of 46, a cytological preparation of late prophase will reveal 46 such chromosomal structures. At the completion of prophase, these are found randomly distributed in the area formerly occupied by the nucleus.

You will often see the term **kinetochore** used in conjunction with discussions of the centromere. The kinetochores of each chromosome consist of multilayered platelike structures that form on opposite sides of the centromere. Each kinetochore is intimately associated with one of the two sister chromatids. The external areas of the kinetochore structures ultimately attach to microtubules that make up the spindle fibers, and during the ensuing anaphase stage of mitosis the two kinetochores of each pair of sister chromatids are pulled to opposite poles of the cell. The internal areas of the kinetochore are closely aligned with the centromere. In this context, the centromere consists of specific DNA regions of each chromosome. The relationship among the kinetochore, the centromere, and attached microtubules is illustrated in the electron micrograph in Figure 2.9.

Prometaphase and Metaphase

The distinguishing event of the ensuing stage is the migration of the centromeric region of each chromosome to the equatorial plane. In some descriptions the term **metaphase** is applied strictly to the chromosome configuration following this movement. In such descriptions, **prometaphase** refers to the period of chromosome movement, as depicted in part (c) of Figures 2.8(A) and 2.8(B). The equatorial plane, also referred to as the **metaphase plate**, is the midline region of the cell, a plane that lies perpendicular to the axis established by the spindle fibers.

Migration is made possible by the binding of microtubules to the kinetochore regions associated with the centromere of the chromosomes. Spindle fibers actually consist of **microtubules**, which themselves consist of molecular subunits of the protein **tubulin**. Microtubules seem to originate and "grow" out of the two centrosome regions (containing the centrioles) at opposite poles of

the cell. They are dynamic structures that lengthen and shorten as a result of the addition or loss of polarized tubulin subunits. There are two major categories of microtubules. Those most directly responsible for chromosome migration during anaphase make contact to and adhere to kinetochores as they grow from the centrosome region. They are referred to as **kinetochore microtubules** and have one end near the centrosome region (at one of the poles of the cell) and the other anchored to the kinetochore. It is interesting to note that the number of microtubules that bind to the kinetochore varies greatly between organisms. Yeast (*Saccharomyces*) have only a single microtubule bound to each platelike structure of the kinetochore. Mitotic cells of mammals, at the other extreme, reveal 30 to 40 microtubules bound to each portion of the kinetochore.

Those microtubules that do not adhere to kinetochores make contact with growing microtubules from the opposite pole of the cell, often interdigitating with one another. The polarized nature of the tubulin subunits provides the force that joins them together. They are referred to as **polar microtubules** (and sometimes, spindle microtubules) and provide the cytoplasmic framework of spindle fibers that establish and maintain the separation of the two poles during chromosome separation. Kinetochore microtubules are apparent in Figure 2.9.

At the completion of metaphase, each centromere is aligned at the metaphase plate with the chromosome arms extending outward in a random array. This configuration is shown in part (d) of Figures 2.8(A) and 2.8(B).

Anaphase

Events critical to chromosome distribution during mitosis occur during its shortest stage, **anaphase**. It is during this phase that sister chromatids of each double chromosomal structure separate from each other and migrate to opposite ends of the cell. For complete separation to occur, each centromeric region must be divided into two. Once this has occurred, each chromatid is now referred to as a **daughter chromosome**.

As discussed above, movement of the chromosomes to the opposite poles of the cell is dependent upon the centromere-spindle fiber (kinetochore-microtubulin) attachment. Recent investigations have revealed that chromosome migration results from the activity of a series of specific proteins, generally called **motor proteins**. These proteins utilize the energy generated by the hydrolysis of ATP, and their activity is said to constitute **molecular motors** in the cell. These motors act at several positions within the dividing cell, but all are involved in the activity of microtubules and ultimately serve to propel the chromosomes to opposite ends of the cell. The centromeres

Mitosis

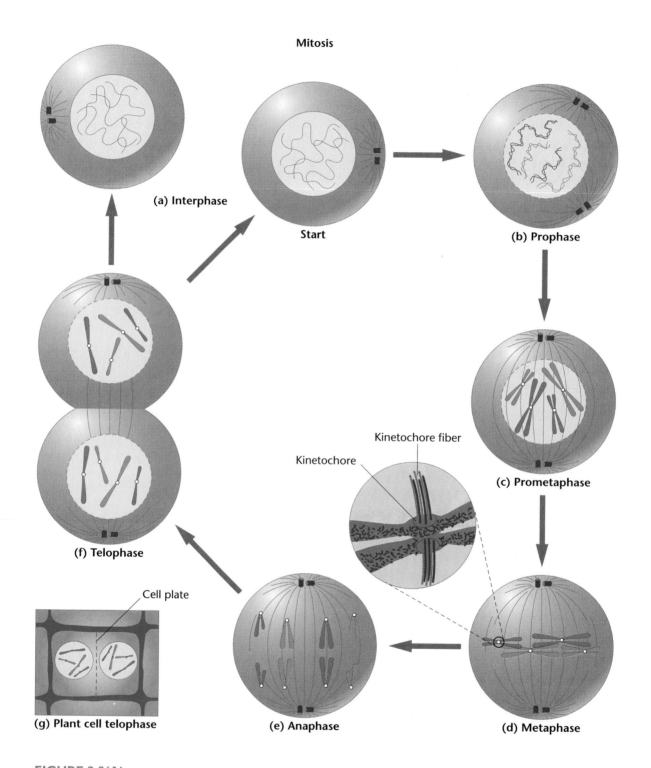

(a) Interphase

Start

(b) Prophase

(c) Prometaphase

Kinetochore fiber

Kinetochore

(f) Telophase

(d) Metaphase

Cell plate

(g) Plant cell telophase

(e) Anaphase

FIGURE 2.8(A) Mitosis in an animal cell with a diploid number of 4. The events occurring in each stage are described in the text. Of the two homologous pairs of chromosomes, one contains longer, metacentric members, and the other shorter, submetacentric members. The maternal chromosomes and the paternal chromosomes are shown in different colors. The inset (g), showing the telophase stage in a plant cell, illustrates the formation of the cell plate and lack of centrioles.

(a) Interphase

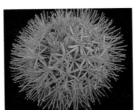

(g) Late telophase

(c) Prometaphase

(d) Metaphase

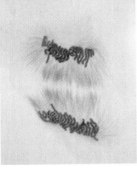

(f) Early telophase

(e) Anaphase

of each chromosome *appear* to lead the way during migration, with the chromosome arms trailing behind. Depending on the location of the centromere along the chromosome, different shapes are assumed during separation (review Figure 2.4).

The steps occurring during anaphase are critical in providing each subsequent daughter cell with an identical set of chromosomes. In human cells there would now be 46 chromosomes at each pole, one from each original sister pair. Part (e) of Figures 2.8(A) and 2.8(B) illustrates anaphase prior to its completion.

Telophase

Telophase is the final stage of mitosis. At its beginning, there are two complete sets of chromosomes, one at each pole. The most significant event is **cytokinesis**, the division or partitioning of the cytoplasm. Cytokinesis is essential if two new cells are to be produced from one. The

FIGURE 2.8(B) Light micrographs illustrating the stages of mitosis depicted in Figure 2.8(A). These stages are derived from the flower of *Haemanthus*, shown in the center of the figure.

mechanism differs greatly in plant and animal cells. In plant cells, a **cell plate** is synthesized and laid down across the dividing cell in the region of the metaphase plate. Animal cells, however, undergo a constriction of the cytoplasm in much the same way a loop might be tightened around the middle of a balloon. The end result is the same: Two distinct cells are formed.

It is not surprising that the process of cytokinesis varies among cells of different organisms. Plant cells, which are more regularly shaped and structurally rigid, require a mechanism for the deposition of new cell wall material around the plasma membrane. The cell plate, laid down during telophase, becomes the **middle lamella**. Subsequently, the primary and secondary layers of

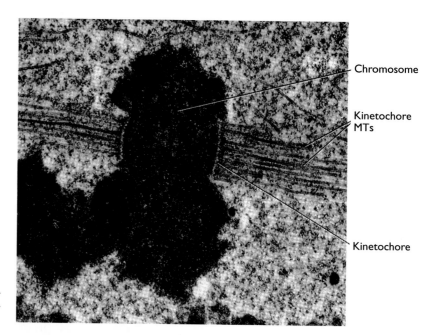

Chromosome

Kinetochore
MTs

Kinetochore

FIGURE 2.9 An electron micrograph showing the association between microtubules and the kinetochore (in association with a centromere) during mitosis.

the cell wall are deposited between the cell membrane and middle lamella on both sides of the boundary between the two daughter cells. In animals, complete constriction of the cell membrane produces the **cell furrow** characteristic of newly divided cells.

Other events necessary for the transition from mitosis to interphase are also initiated during late telophase. They represent a general reversal of those that occurred during prophase. In each new cell, the chromosomes begin to uncoil and become diffuse chromatin once again, while the nuclear envelope re-forms around them. The nucleolus gradually re-forms and is completely visible in the nucleus during early interphase. The spindle fibers also disappear. Telophase in animal and plant cells is illustrated in part (f) and part (g) of Figures 2.8(A) and 2.8(B).

Genetic Regulation of the Cell Cycle

The nature of the cell cycle, including mitosis, is fundamentally the same in all eukaryotic organisms. The similarity of the events leading to cell duplication in many evolutionarily diverse organisms suggests that the cell cycle is under tight genetic control, and that the genetic program regulating the cell cycle has been conserved throughout evolution. Elucidation of this genetic program not only provides information basic to our understanding of the nature of living organisms, but because disruption of this regulation may lead to uncontrolled cell division characterizing malignancy, interest in how genes regulate the cell cycle has been great.

A mammoth research effort over the past decade has paid high dividends, and we now have knowledge of many genes involved in the control of the cell cycle. As with

other studies of genetic input into essential biological processes, investigation has relied on the discovery of mutations that interrupt the cell cycle and the study of the subsequent effects of these mutations. As we shall return to this subject in even greater detail in Chapter 21 during our consideration of cancer, what follows is a brief overview of what has been learned.

First of all, gene products involved in regulation are quite numerous and exert their impact at every stage throughout the cell cycle. Genetic alterations in these genes are often designated as *cdc* **mutations** (**cell division cycle mutations**). The products of several of the most important genes represented by *cdc* mutations are enzymes called **protein kinases**, which serve as master control molecules. As kinases, these enzymes transfer phosphate groups from ATP to other proteins that themselves mediate various cell cycle and mitotic activities. As a result of phosphorylation, the chemical behavior of the involved proteins is changed, which causes cellular activities related to cell cycle regulation to heighten or diminish. The kinases often work in conjunction with other master molecules called **cyclins**. These are so named because their concentration in the cell "cycles" up and down as control of specific events is exerted throughout the cell cycle. When the catalytic activity of a kinase is dependent on cyclins, the kinase is called a **Cdk protein**, for "**cyclin-dependent kinase.**"

One major level of genetic control is the provision for checkpoints at critical junctures during the cell cycle. These checkpoints, illustrated in Figure 2.10, provide a way to monitor various conditions in the cell and to prevent premature activity that would damage the cell were it to proceed beyond the checkpoint. For example, as the cell passes through the G1 stage, perhaps the most critical point in the cycle is encountered—the transition into

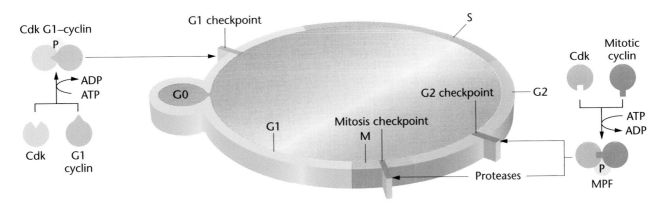

FIGURE 2.10 Illustration of the three major cell cycle checkpoints and the Cdk proteins and cyclins involved in their regulation.

the S phase, where DNA replication takes place. As mentioned earlier, this is called the **G1 checkpoint**. Under normal circumstances, once this transition occurs, the cell is committed to traverse the remainder of the cycle, including mitosis. As the cell moves toward S, the Cdk product of one of the *cdc* genes (*cdc2*) associates with a specific cyclin, forming a **Cdk G1–cyclin complex**. If certain conditions in the cell are not met, the concentration and activity of this complex fail to increase and entry to the S phase is restricted. For example, should the cell not have grown sufficiently, or should the DNA have incurred damage, the level and activity of Cdk-G1 cyclin fail to increase, and the cycle is arrested in G1 until the conditions are remedied by further growth or DNA repair, respectively. This regulatory step prevents the cell from progressing irreversibly down a path that it is ill prepared to traverse.

It is important to note another gene that plays an important role during the transition from G1 into S in the cell cycle of mammalian cells: the *p53* gene. This gene is of great significance in cell cycle control, because when the p53 gene product malfunctions as a result of mutations, a wide variety of human cancers are known to occur. As a result, the normal form of the gene has been referred to as a **tumor suppressor gene**. Experiments suggest that *p53* plays a role in "cell suicide (**apoptosis**)," a normal process activated by severe damage to the genetic material. The p53 gene product (a protein) functions as a transcriptional activator, whereby it regulates the expression of still other genes that control the cell cycle. Mutations in the *p53* gene appear to eliminate this control. As a result, damaged cells become immortalized (cancerous) rather than being eliminated.

A second important transition, the **G2 checkpoint** occurs as the cell completes DNA replication and prepares to move through G2 and enter mitosis. The same Cdk molecule (e.g., in yeast, the *cdc2* gene product discussed above) associates with a different cyclin (called mitotic cyclin or cyclin B) forming what is called **MPF**, standing for **maturation-promoting factor** (Figure

2.10). The cell progresses into mitosis only if the concentration of this complex and its concomitant kinase activity increases. If, on the other hand, DNA is insufficiently replicated or damaged, the activity of MPF does not increase and the cycle arrests until the replication or repair of DNA is accomplished. As with the G1 checkpoint, this regulation prevents the cell from proceeding with an activity (mitosis) that it is unprepared to complete. MPF has the ability to stimulate various activities during mitosis, including chromatin condensation (into chromosomes), the breakdown of the nuclear envelope, and spindle formation.

The final checkpoint (Figure 2.10) is during mitosis itself. Here, both the successful formation of the spindle fibers and their attachment to the kinetochores of chromatids are monitored prior to the initiation of chromosome migration during anaphase. If, for some reason, the spindle has not been formed completely or proper attachment does not occur, mitosis is arrested. This checkpoint appears to be controlled not by a Cdk-cyclin complex, but rather by specific proteolytic enzymes whose production have been stimulated by MPF.

The precise mechanisms by which the Cdk-cyclins and the molecules that are phosphorylated function to achieve the checkpoints discussed above have yet to be clearly defined. Nevertheless, a wealth of information has now become available that will provide the basis for a comprehensive understanding of the genetic regulation of the cell cycle. This body of knowledge will undoubtedly provide critical information in our understanding and treatment of cancer. We shall return to this topic in much greater detail in Chapter 21.

Meiosis and Sexual Reproduction

The process of meiosis, unlike mitosis, reduces the amount of genetic material. Whereas in diploids, mitosis produces daughter cells with a full diploid complement, meiosis produces **gametes** or **spores** with only one haploid set of chromosomes. During sexual reproduction, gametes then combine in fertilization to reconstitute the diploid complement found in parental cells. The process itself must be very specific, because it is insufficient to produce gametes with a random array of one-half of the total number of chromosomes. Instead, each gamete must receive one member of each homologous pair of chromosomes, ensuring genetic continuity from generation to generation.

The process of sexual reproduction also ensures genetic variety among members of a species. As you study meiosis, it will become apparent that this process results in gametes with many unique combinations of maternally and paternally derived chromosomes among the haploid complement. With a tremendous genetic variation among the gametes derived from two genetically unique individuals, a large number of chromosome combinations are possible at fertilization. Furthermore, we will see that the meiotic event referred to as **crossing over** results in genetic exchange between members of each homologous pair of chromosomes. This creates intact chromosomes that are mosaics of the maternal and paternal homologs from which they are derived. This has the effect of further enhancing the potential genetic variation in gametes and the offspring derived from them. Sexual reproduction, therefore, reshuffles the genetic material, producing offspring that often differ greatly from either parent. This process constitutes the major form of genetic recombination within species.

An Overview of Meiosis

We have already established what must be accomplished during meiosis. Before systematically considering the stages of this process (Figure 2.11) we will briefly describe how diploid cells are converted to haploid gametes or spores. Unlike mitosis, in which each paternally and maternally derived member of any given homologous pair of chromosomes behaves autonomously during division, in meiosis homologous chromosomes pair together; that is, they **synapse**. Each synapsed structure, called a **bivalent**, gives rise to a unit, the **tetrad**, consisting of four chromatids. The presence of four chromatids demonstrates that both chromosomes have duplicated. In order to achieve haploidy, two divisions are necessary. In the first, described as a **reductional division** (because the number of centromeres, each representing one chromosome, is "reduced" by one-half following this division), components of each tetrad representing the two homologs separate, yielding two **dyads**. Each dyad is composed of two sister chromatids joined at a common centromere. During the second division, described as **equational** (because the number of centromeres remains

"equal" following this division), each dyad splits into two **monads** of one chromosome each. Thus, the two divisions may potentially produce four haploid cells. As in mitosis, meiosis is a continuous process. We give names to the parts of each stage of division only for the convenience of discussion.

The First Meiotic Division: Prophase I

From a genetic standpoint, there are two critical events during prophase I. First, members of each homologous pair of chromosomes somehow find one another and undergo synapsis. Second, the exchange process referred to above as crossing over occurs between synapsed homologs. Because of the importance of these genetic events, this stage of meiosis has been subjected to continuous investigation. Most recently, the study of yeast has provided a model for our understanding of these critical events. As we discuss them, you should be aware that replication of DNA has preceded meiosis, even though chromosomes do not immediately reveal their duplication. Recall that this also occurs in mitosis. The first meiotic prophase is complex and has been further subdivided into five substages: leptonema,* zygonema,* pachynema,* diplonema,* and diakinesis [Figure 2.11(a)].

Leptonema

During the **leptotene stage**, the interphase chromatin material begins to condense, and the chromosomes, although still extended, become visible. Along each chromosome are **chromomeres**, localized condensations that resemble beads on a string. Recent evidence suggests that it is during leptonema that a process called **homology search** begins, one that precedes initial pairing of homologs.

Zygonema

The chromosomes continue to shorten and thicken during the **zygotene stage**. During the process of homology search, homologous chromosomes undergo initial alignment with one another. This alignment is considered to be a **rough pairing**, and by the end of zygonema it is complete. In yeast, homologs are separated by about 300 nm, and near the end of zygonema, structures referred to as *lateral elements* are visible between paired homologs. As meiosis proceeds, the overall length of the lateral elements increases and a more extensive ultrastructural component, the **synaptonemal complex**, begins to form between the homologs (see Figure 2.16).

---

*These are the noun forms of these stages. The adjective forms (leptotene, zygotene, pachytene, and diplotene) are also used in the text of this chapter.

At the completion of zygonema, the paired homologs represent structures referred to as **bivalents**. Although both members of each bivalent have already replicated their DNA, it is not yet visually apparent that each member is a double structure. The number of bivalents in each species is equal to the haploid (*n*) number.

Pachynema

In the transition from the zygotene to the **pachytene stage**, coiling and shortening of chromosomes continues, and further development of the synaptonemal complex occurs between the two members of each bivalent. This leads to a more intimate pairing referred to as **synapsis**. Compared to the rough-pairing characteristic of yeast pachynema, homologs are now separated by only 100 nm.

During pachynema, it first becomes evident that each homolog is a double structure, providing visual evidence of the earlier replication of the DNA of each chromosome. Thus, each bivalent contains four members, called **chromatids**. As in mitosis, replicates are called **sister chromatids**, while chromatids from maternal vs. paternal members of a homologous pair are called **nonsister chromatids**. The four-membered structure is also referred to as a **tetrad**, and each tetrad contains two pairs of sister chromatids.

Diplonema

During observation of the ensuing **diplotene stage** it is even more apparent that each tetrad consists of two pairs of sister chromatids. Within each tetrad, each pair of sister chromatids begins to separate. However, one or more areas remain in contact where chromatids are intertwined. Each such area, called a **chiasma** (pl. **chiasmata**), is thought to represent a point where nonsister chromatids have undergone genetic exchange through the process referred to above as **crossing over**. Although the physical exchange between chromosome areas occurred during the previous pachytene stage, the result of crossing over is visible only when the duplicated chromosomes begin to separate. Crossing over is an important source of genetic variability. New combinations of genetic material are formed during this process.

Diakinesis

The final stage of prophase I is **diakinesis**. The chromosomes pull farther apart, but nonsister chromatids remain loosely associated via the chiasmata. As separation proceeds, the chiasmata move toward the ends of the tetrad. This process, called **terminalization**, begins in late diplonema, and is completed during diakinesis. During this final period of prophase I, the nucleolus and nuclear envelope break down, and the two centromeres of each tetrad become attached to the recently formed spindle fibers.

Metaphase, Anaphase, and Telophase I

Following the first meiotic prophase stage, steps similar to those of mitosis occur. In the **metaphase stage of the first division**, the chromosomes have maximally shortened and thickened. The terminal chiasmata of each tetrad are visible and appear to be the only factor holding the nonsister chromatids together. Each tetrad interacts with spindle fibers, facilitating movement to the metaphase plate.

During the first division, a single centromere holds each pair of sister chromatids together. It does *not* divide. At the **first anaphase**, one-half of each tetrad (one pair of sister chromatids—called a **dyad**) is pulled toward each pole of the dividing cell. This separation process is the physical basis of what we refer to as "disjunction," meaning the separation of chromosomes from one another. Occasionally, errors in meiosis occur and separation is not achieved. The term **nondisjunction** describes such an error, the consequences of which will soon be discussed. At the completion of the normal anaphase I, there is a series of dyads equal to the haploid number present at each pole.

If no crossing over had occurred in the first meiotic prophase, each dyad at each pole would consist solely of either paternal or maternal chromatids. However, the exchanges produced by crossing over create mosaic chromatids of paternal and maternal origin. The alignment of each tetrad prior to this first anaphase stage is random. One-half of each tetrad will be pulled to one or the other pole at random, and the other half will move to the opposite pole. This random **segregation** of dyads is the basis for the Mendelian principle of **independent assortment**, which we will discuss in Chapter 3. You may wish to return to this discussion when you study this principle.

In many organisms, **telophase of the first meiotic division** reveals a nuclear membrane forming around the dyads. Next, the nucleus enters into a short interphase period. In other cases, the cells go directly from the first anaphase into the second meiotic division. If an interphase period occurs, the chromosomes do not replicate since they already consist of two chromatids. In general, meiotic telophase is much shorter than the corresponding stage in mitosis.

The Second Meiotic Division

A second division of the sister chromatids making up each dyad is essential if each gamete or spore is to receive only one chromatid from each original tetrad. During **prophase II**, each dyad is composed of one pair of sister

Meiosis I

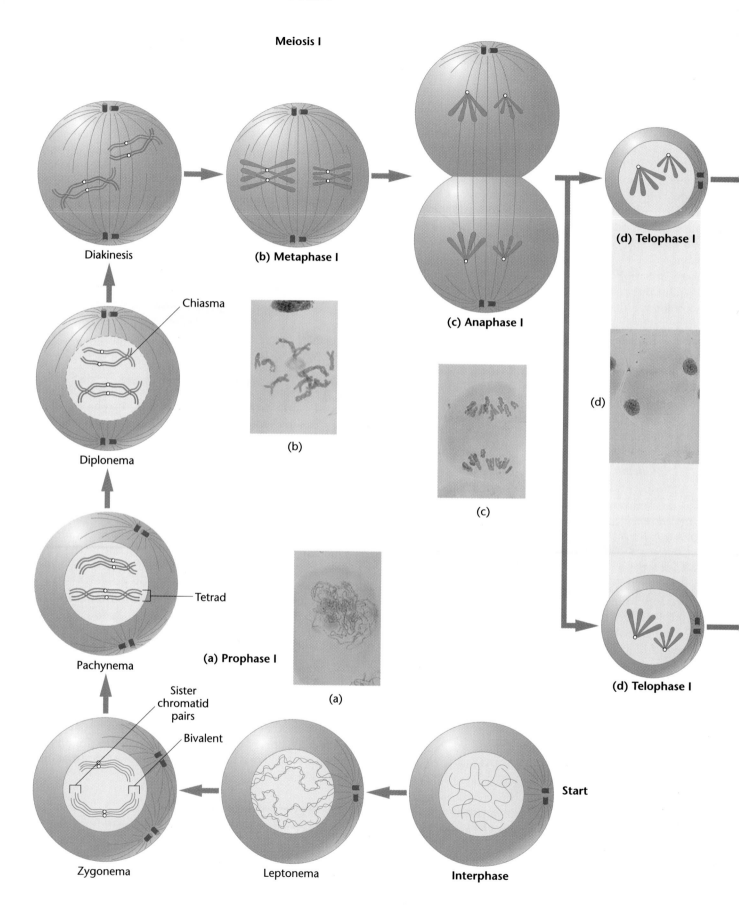

Diakinesis

(b) Metaphase I

Chiasma

(b)

(c) Anaphase I

(c)

(d) Telophase I

(d)

Diplonema

Tetrad

(a) Prophase I

(a)

Pachynema

(d) Telophase I

Sister
chromatid
pairs

Bivalent

Start

Zygonema   Leptonema   **Interphase**

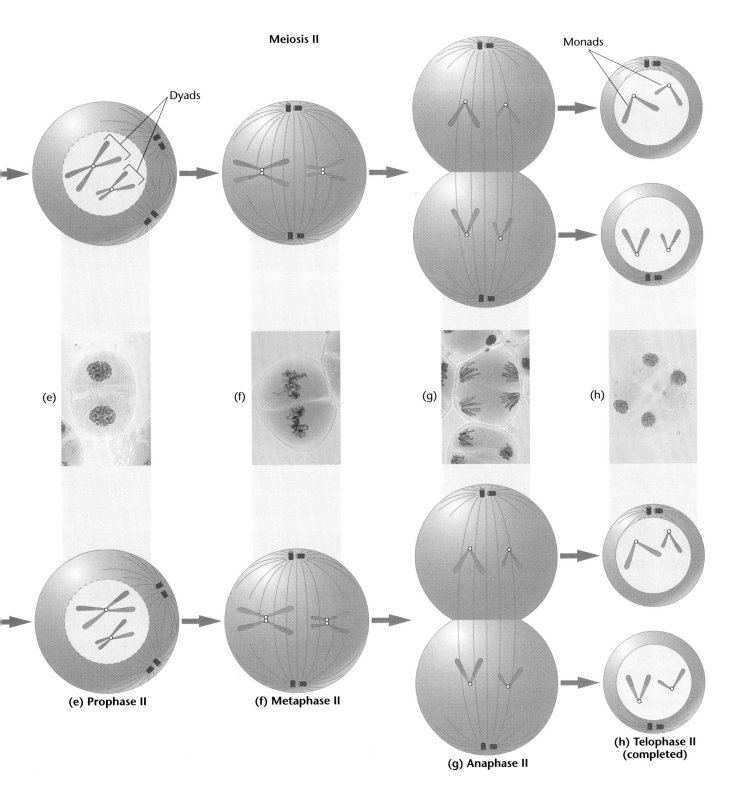

Meiosis II

Dyads

Monads

(e)

(f)

(g)

(h)

(e) Prophase II

(f) Metaphase II

(g) Anaphase II

(h) Telophase II (completed)

FIGURE 2.11   A diagrammatic representation of the major events occurring during meiosis in a male animal with a diploid number of 4. The same chromosomes described in Figure 2.8 are followed, as depicted in the legend there. Note that the combination of chromosomes contained in the cells produced following telophase II is dependent on the random alignment of each tetrad and dyad on the equatorial plate during metaphase I and metaphase II. Several other combinations (not shown) can be formed. The events depicted in this figure should be correlated with the description in the text. Light micrographs illustrate many of the stages of meiosis that are shown in the diagrams.

chromatids attached by a common centromere. During **metaphase II**, the centromeres are directed to the equatorial plate. Then, the centromeres divide, and during **anaphase II** the sister chromatids of each dyad are pulled to opposite poles. Because the number of dyads is equal to the haploid number, **telophase II** reveals one member of each pair of homologous chromosomes present at each pole. Each chromosome is referred to as a **monad**. At the conclusion of meiosis, not only has the haploid state been achieved, but if crossing over has occurred, each monad is a combination of maternal and paternal genetic information. As a result, the offspring produced by any gamete will receive from it a mixture of genetic information originally present in his or her grandparents. Potentially, following cytokinesis in telophase II, four haploid gametes may result from a single meiotic event.

Spermatogenesis and Oogenesis

Although events that occur during the meiotic divisions are similar in all cells that participate in gametogenesis, there are certain differences between the production of a male gamete (spermatogenesis) and a female gamete (oogenesis) in most animal species.

Spermatogenesis takes place in the testes, the male reproductive organs. The process begins with the expanded growth of an undifferentiated diploid germ cell called a **spermatogonium**. This cell enlarges to become a **primary spermatocyte**, which undergoes the first meiotic division. The products of this division are called **secondary spermatocytes**. Each secondary spermatocyte contains a haploid number of dyads. The secondary spermatocytes then undergo the second meiotic division, and each of these cells produces two haploid **spermatids**. Spermatids go through a series of developmental changes, **spermiogenesis**, and become highly specialized, motile **spermatozoa** or **sperm**. All sperm cells produced during spermatogenesis receive equal amounts of genetic material and cytoplasm. Figure 2.12 summarizes these steps.

Spermatogenesis may be continuous or occur periodically in mature male animals, with its onset determined by the nature of the species' reproductive cycle. Animals that reproduce year-round produce sperm continuously, whereas those whose breeding period is confined to a particular season produce sperm only during that time.

In animal **oogenesis**, the formation of **ova** (singular: **ovum**), or eggs, occurs in the ovaries, the female reproductive organs. The daughter cells resulting from the two meiotic divisions receive equal amounts of genetic material, but they do *not* receive equal amounts of cytoplasm. Instead, during each division, almost all the cytoplasm of the **primary oocyte**, itself derived from the **oogonium**, is concentrated in one of the two daughter cells. The concentration of cytoplasm is necessary because a major function of the mature ovum is to nourish the developing embryo following fertilization.

During the first meiotic anaphase in oogenesis, the tetrads of the primary oocyte separate, and the dyads move toward opposite poles. During the first telophase, the dyads present at one pole are pinched off with very little surrounding cytoplasm to form the **first polar body**. The other daughter cell produced by this first meiotic division contains most of the cytoplasm and is called the **secondary oocyte**. The first polar body may or may not divide again to produce two small haploid cells. The mature ovum will be produced from the secondary oocyte during the second meiotic division. During this division, the cytoplasm of the secondary oocyte again divides unequally, producing an **ootid** and a **second polar body**. The ootid then differentiates into the mature ovum. Figure 2.12 illustrates the steps leading to formation of the mature ovum and polar bodies.

Unlike the divisions of spermatogenesis, the two meiotic divisions of oogenesis may not be continuous. In some animal species the two divisions may directly follow each other. In others, including the human species, the first division of all oocytes begins in the embryonic ovary, but arrests in prophase I. Many years later, the first division is reinitiated in each oocyte upon its ovulation. The second division is completed only after fertilization.

The Significance of Meiosis

The process of meiosis is critical to the successful sexual reproduction of all diploid organisms. It is the mechanism by which the diploid amount of genetic information is reduced to the haploid amount. In animals, meiosis leads to the formation of gametes, whereas in plants haploid spores are produced, which in turn lead to the formation of haploid gametes.

The mechanism by which the above reduction to haploidy is accomplished serves as the basis for the production of extensive genetic variation within members of a population. As we have learned, each diploid organism contains its genetic information in the form of homologous pairs of chromosomes. Each pair consists of one member derived from the maternal parent and one from the paternal parent. Following the reduction to haploidy during meiosis, gametes or spores contain either the paternal or the maternal representative of every homologous pair of chromosomes. During sexual reproduction, this process has the potential for the production of huge

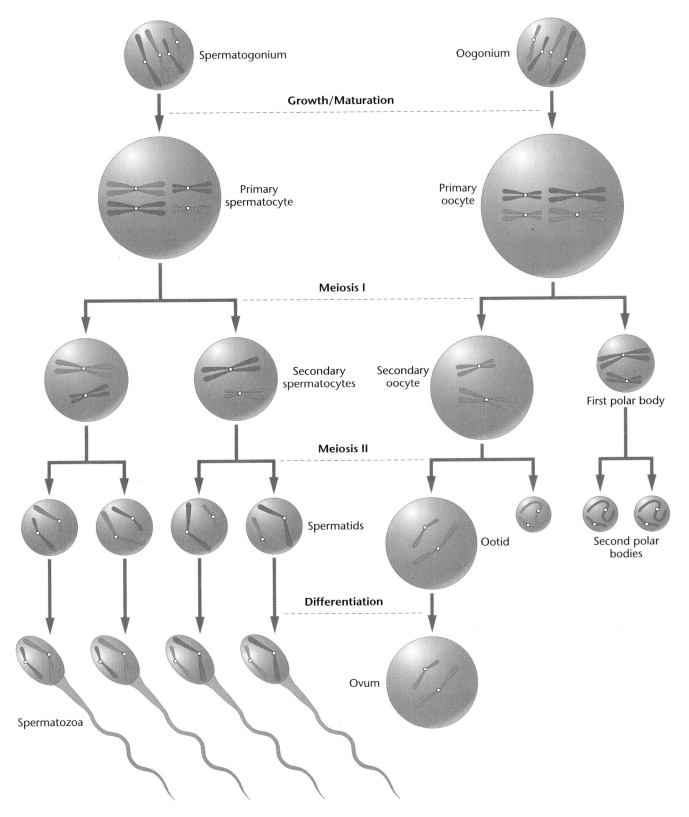

FIGURE 2.12   Spermatogenesis and oogenesis in animal cells.

quantities of genetically dissimilar gametes. As the number of homologous chromosomes (the haploid number) increases, the possibilities of different combinations of maternal and paternal chromosomes in any given gamete increase. For example, an organism with a haploid number of 10 can produce 2^n number of combinations, where n represents the haploid number. As you can calculate, 2^{10} is a substantial number (1024). If you continue this calculation to determine the number of different combinations of sperm or eggs in our own species (2^{23}), you cannot help but be impressed with the potential for genetic variation resulting from meiosis. To what does 2^{23} calculate?

Before the distribution of homologs occurs during meiosis, the process of crossing over occurs in the first meiotic prophase stage. This phenomenon further reshuffles the genetic information between the maternal and paternal members of each homologous pair. As a result, endless varieties of each homolog may occur in gametes, ranging from either intact maternal or paternal chromosomes, where no exchange occurred, to any mixture of maternal and paternal components, depending on where one or more exchanges occurred during crossing over.

In summary, the two most significant points about meiosis are that the process is responsible for:

1. The maintenance of a constancy of genetic information between generations; and

2. Extensive genetic variation within populations.

Two other topics involving meiosis are also important in the study of genetics. One involves the important role that meiosis plays in the life cycles of fungi and plants. In many fungi, the predominant stage of the life cycle consists of haploid cells. They arise through meiosis and proliferate by mitotic cell division. In multicellular plants, the life cycle alternates between the diploid **sporophyte stage** and the haploid **gametophyte stage**. While one or the other predominates in different plant groups during this "alternation of generations," the

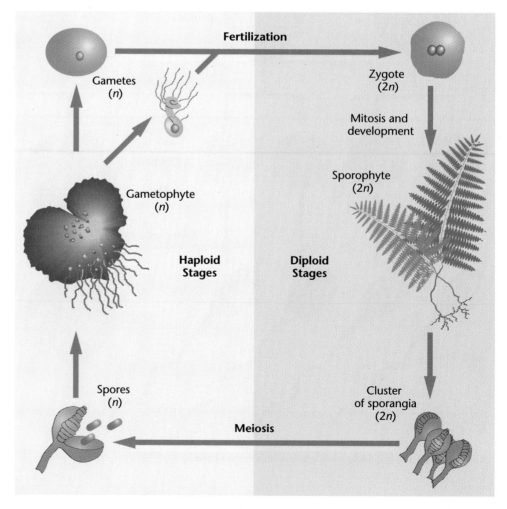

FIGURE 2.13   Alternation of generations between the diploid sporophyte (2n) and the haploid gametophyte (n) in a multicellular plant. The process of meiosis bridges the two phases of the life cycle.

processes of meiosis and fertilization constitute the "bridge" between the sporophyte and gametophyte generations (Figure 2.13). Therefore, meiosis is an essential component of the life cycle of plants.

Finally, it is important to know what happens when meiosis fails to achieve the expected outcome. As we pointed out earlier, rarely, at either the first or second division, separation or disjunction of the chromatids of a tetrad or dyad fails to occur. Instead, both members move to the same pole during anaphase. Such an event is called **nondisjunction**, because the two members fail to disjoin. If nondisjunction occurs at the first division stage, it is said to be a primary event; at the second division stage, it is called a secondary event.

The results of primary and secondary nondisjunction for just one chromosome of a diploid genome are shown in Figure 2.14. As can be seen, for the chromosome in question, some abnormal gametes may be formed containing either two members or none at all. Following fertilization of these with a normal gamete, a zygote is produced with either three members (**trisomy**) or only one member (**monosomy**) of this chromosome. While these conditions are more frequently tolerated in plants, they usually have severe or lethal effects in animals. Trisomy and monosomy will be described in greater detail in Chapter 9.

The Cytological Origin of the Mitotic and Meiotic Chromosome

Thus far in this chapter, we have focused on mitotic and meiotic chromosomes, emphasizing their behavior during cell division and gamete formation. Initially, biologists knew about these chromosomes only from routine observations made with the light microscope. Although chromosomes are invisible during interphase, they appear during the prophase stage of mitosis. Geneticists were curious as to how this could happen. Based on studies using electron microscopy, we are now quite clear as to why chromosomes are visible only during mitosis.

During interphase, only dispersed chromatin fibers are present in the nucleus [Figure 2.15(a)]. **Chromatin** consists of DNA and associated proteins, particularly proteins called **histones**. It is now believed that during interphase, starting in G1, mitotic chromosomes unwind to form these long chromatin fibers. It is in this physical arrangement that DNA can most efficiently function during transcription and can be replicated.

Once mitosis begins, however, the fibers coil and fold up, condensing into typical mitotic chromosomes [Figure 2.15(b)]. If the fibers making up the mitotic chromosome

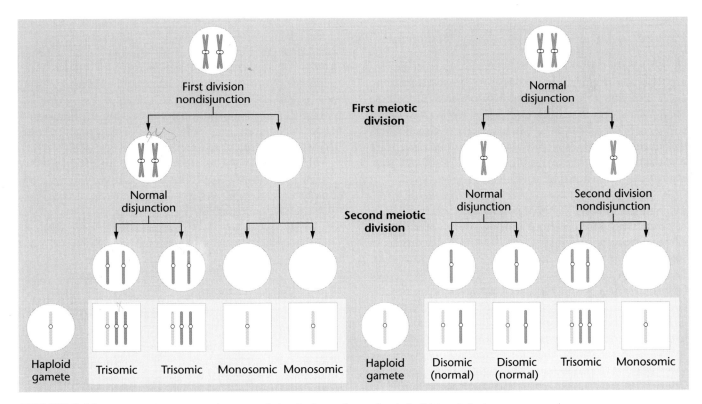

FIGURE 2.14 Diagram illustrating nondisjunction during the first and second meiotic divisions. In both cases, some gametes are formed either containing two members of a specific chromosome or lacking it altogether. Following fertilization by a normal haploid gamete, monosomic, disomic (normal), or trisomic zygotes result.

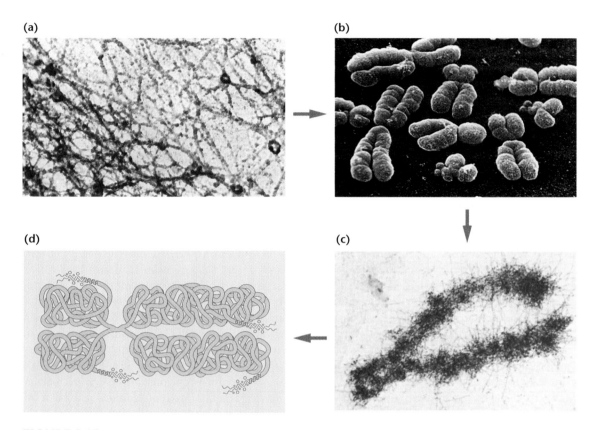

FIGURE 2.15   A comparison of (a) the chromatin fibers characteristic of the interphase nucleus with (b) and (c) metaphase chromosomes that are derived from chromatin during mitosis. Part (d) depicts the folded-fiber model showing how chromatin is condensed into a metaphase chromosome. Parts (a) and (c) are transmission electron micrographs, whereas part (b) is a scanning electron micrograph.

are loosened [Figure 2.15(c)], areas of greatest spreading reveal individual fibers similar to those seen in interphase chromatin. Very few fiber ends seem to be present. In some cases, none can be seen. Instead, individual fibers always seem to loop back into the interior. Such fibers are obviously twisted and coiled around one another, forming the regular pattern of the mitotic chromosome.

Electron microscopic observations of mitotic chromosomes in varying states of coiling led Ernest DuPraw to postulate the **folded-fiber model**, illustrated in Figure 2.15(d). During metaphase, each chromosome consists of two sister chromatids joined at the centromeric region. Each arm of the chromatid appears to consist of a single fiber wound up much like a skein of yarn. The fiber is composed of double-stranded DNA and protein tightly coiled together. An orderly coiling–twisting–condensing process appears to be involved in the transition of the interphase chromatin to the more condensed, genetically inert mitotic chromosomes. During the transition from interphase to prophase, it is estimated that a 5000-fold contraction occurs in the length of DNA within the chromatin fiber! This process must indeed be extremely precise, given the highly ordered nature and consistent ap-

pearance of mitotic chromosomes in all eukaryotes. Note particularly in the micrographs the clear distinction between the sister chromatids constituting each chromosome. They are joined only by the common centromere that they share prior to anaphase.

In a later chapter, after we have provided a more thorough description of DNA structure, we will return to this topic and explore the molecular basis of the chromatin fiber (see Chapter 17).

The Synaptonemal Complex

The electron microscope has also been used to visualize an additional ultrastructural component of the chromosome seen only in cells undergoing meiosis. This structure, first introduced during our earlier discussion of the first meiotic prophase stage, is found between synapsed homologs and is called the **synaptonemal complex.** * In

_____

*An alternative spelling of this term is *synaptinemal complex.*

1956, Montrose Moses observed this complex in spermatocytes of crayfish, and Don Fawcett saw it in pigeon and human spermatocytes. Because there was not yet any satisfactory explanation of the mechanism of synapsis or of crossing over and chiasmata formation, many researchers became interested in this structure. With few exceptions, the ensuing studies revealed the synaptonemal complex to be present in most plant and animal cells visualized during meiosis.

Figure 2.16(a) is an electron micrograph of the synaptonemal complex. It is composed primarily of a triplet set of parallel strands. The **central element** of this tripartite structure is usually less dense and thinner (100–150 Å) than the two identical outer elements (500 Å). The outer structures, called **lateral elements**, are intimately associated with the chromatin of the synapsed homologs on either side. Selective staining has revealed that these lateral elements consist primarily of DNA and protein. This is consistent with the interpretation that the lateral elements contain chromatin. Some DNA fibrils traverse these elements, making connections with the central element, which is composed primarily of protein. Figure 2.16(b) provides a diagrammatic interpretation of the electron micrograph consistent with the above description.

The formation of the complex is initiated prior to the pachytene stage. As early as leptonema of the first meiotic prophase, lateral elements are seen in association with sister chromatids. Homologs have yet to associate with one another and are randomly dispersed in the nucleus. By the next stage, zygonema, homologous chromosomes begin to align with one another in what is called **rough pairing**, but remain distinctly apart by some 300 nm. Then, during pachynema, the intimate association referred to as synapsis between homologs occurs as formation of the complex is completed. In some diploid organisms, this occurs in a zipperlike fashion beginning with the telomeres, which may be attached to the nuclear envelope.

It is now agreed that the synaptonemal complex is the vehicle for the pairing of homologs and their subsequent segregation during meiosis. However, some degree of synapsis can occur in certain cases where no synaptonemal complexes are formed. Thus, it is possible that the function of this structure may be more extensive than just its involvement in the formation of bivalents.

In certain instances where no synaptonemal complexes are formed during meiosis, synapsis is not as complete and crossing over is reduced or eliminated. For ex-

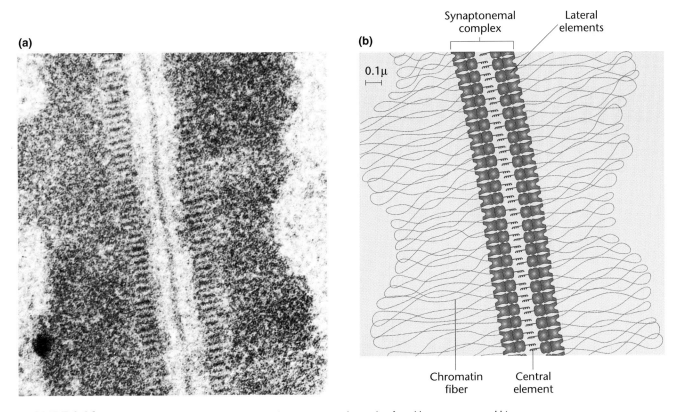

FIGURE 2.16   (a) Electron micrograph of a portion of a synaptonemal complex found between synapsed bivalents of *Neotiella rutilans*. (b) Schematic interpretation of the components making up the synaptonemal complex. The lateral elements, central element, and chromatin are labeled.

ample, in *Drosophila melanogaster* meiotic crossing over rarely, if ever, occurs in males. Synaptonemal complexes are not usually seen during male meiosis. This observation suggests that the synaptonemal complex may be important in order for chiasmata to form and crossing over to occur.

In the yeast *Saccharomyces cerevisiae*, the study of a mutation, *ZIP1*, has provided other insights concerning chromosome pairing. Cells bearing this mutation can undergo the initial alignment stage (rough pairing) and full-length axial element formation, but fail to achieve the intimate pairing characteristics of synapsis. It has been suggested that the gene product of the *ZIP1* locus is a protein that is a component of the central element of the synaptonemal complex. In its absence, no complex is formed. This observation suggests that the synaptonemal complex functions in the transition from the initial rough alignment stage to the intimate pairing characteristic of synapsis. While synapsis and subsequent chiasmata formation are essential to the normal disjunction of chromosomes during meiosis, the available evidence does not directly link the synaptonemal complex to these latter events.

The discovery of the synaptonemal complex is significant to the field of genetics. If the complex is critical to synapsis during meiosis, then all sexual reproduction is dependent on it.

Specialized Chromosomes

We conclude this chapter by introducing two unique types of chromosomes visible under the light microscope. It is useful to do so here in order to demonstrate several specialized forms that chromosomes may achieve besides

their mitoticlike structures. Both types were studied extensively long before we understood the rationale for how chromosomes "appear" during mitosis.

Polytene Chromosomes

Cells from a variety of organisms contain giant **polytene chromosomes**, which are found in various tissues in the larvae of some flies (salivary, midgut, rectal, and malpighian excretory tubules) and in several species of protozoans and plants. The large amount of information obtained from studies of these genetic structures has provided a model system for more recent investigations of chromosomes. Such structures were first observed by E. G. Balbiani in 1881 and are illustrated in Figure 2.17. What is particularly intriguing about these chromosomes is that they can be seen in the nuclei of interphase cells.

Each polytene chromosome observed under the light microscope reveals a linear series of alternating bands and interbands (see inset in Figure 2.17). The banding pattern is distinctive for each chromosome in any given species. Individual bands are sometimes called **chromomeres**, a more generalized term describing lateral condensations of material along the axis of a chromosome. Each polytene chromosome is 200 to 600 μm long.

Following extensive study using electron microscopy and radioactive tracers, the explanation for the unusual appearance of these chromosomes is now clear. First, polytene chromosomes represent paired homologs. This is highly unusual in most organisms since they are found in somatic cells. Second, their large size and distinctiveness result from the many DNA strands that compose them. The DNA of these paired homologs undergoes many rounds of replication, but without strand separation

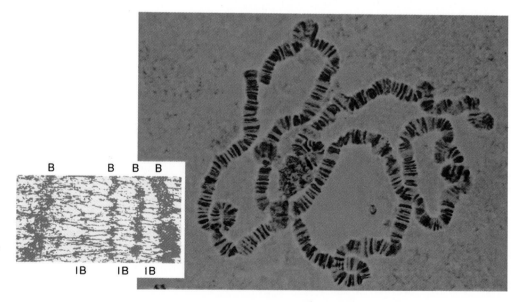

FIGURE 2.17 Polytene chromosomes derived from larval salivary gland cells of *Drosophila*. The inset depicts alternating band (B) and interband (IB) regions along the axis of these giant chromosomes.

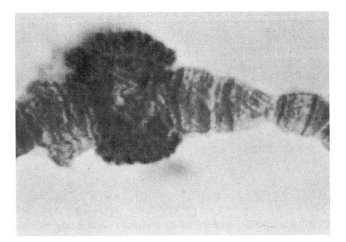

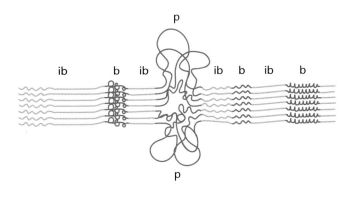

FIGURE 2.18   Photograph of a puff within a polytene chromosome. The schematic representation depicts the uncoiling of strands within a band (b) region to produce a puff (p) in polytene chromosomes. Interband regions (ib) are also labeled.

or cytoplasmic division. As replication proceeds, chromosomes are created having 1000 to 5000 DNA strands that remain in parallel register with one another. It is apparently the parallel register of so many DNA strands that gives rise to the distinctive band pattern along the axis of the chromosome.

The relationship between the structure of these chromosomes and the genes contained within them is intriguing. The presence of bands was initially interpreted as the visible manifestation of individual genes. When it became clear that the strands present in bands undergo localized uncoiling during genetic activity, this view was further strengthened. Each such uncoiling event results in what is called a **puff** because of its appearance (see Figure 2.18). That puffs are visible manifestations of gene activity (transcription that produces RNA) is evidenced by their high rate of incorporation of radioactivity-labeled RNA precursors, as assayed by **autoradiography**. Bands that are not extended into puffs incorporate much less radioactive precursors or none at all.

The study of bands during development in insects such as *Drosophila* and the midge fly *Chironomus* reveals differential gene activity. A characteristic pattern of band formation that is equated with gene activation is observed as development proceeds. This topic is pursued in more detail in Chapter 20. Despite many recent genetic investigations, it is still not clear how many genes are contained within each band. There is much more DNA present per band than is needed to encode a single gene.

Lampbrush Chromosomes

Another type of specialized chromosome that has provided insights into chromosomal structure is the **lampbrush chromosome**. It was given this name because its

appearance is similar to the brushes used to clean lamp chimneys in the nineteenth century. Lampbrush chromosomes were first discovered in 1892 in the oocytes of sharks and are now known to be characteristic of most vertebrate oocytes as well as spermatocytes of some insects. Therefore, they are meiotic chromosomes. Most experimental work has been done with material taken from amphibian oocytes.

These unique chromosomes are easily isolated from oocytes in the diplotene stage of the first prophase of meiosis, where they are active in directing the metabolic activities of the developing cell. The homologs are seen as synapsed pairs held together by chiasmata, but instead of condensing, as do most meiotic chromosomes, the lampbrush chromosomes are often extended to lengths of 500 to 800 μm. Later in meiosis they revert to their normal length of 15 to 20 μm. Based on these observations, lampbrush chromosomes are interpreted as uncoiled and unfolded versions of the normal meiotic chromosomes.

Figure 2.19 shows two views of these structures. Together, they provide significant insights into the morphology of these chromosomes. In part (a), the meiotic configuration, as described above, is seen under the light microscope. The linear axis of each structure contains large numbers of condensed areas that are repeated along the axis, referred to generally as **chromomeres**. Emanating from each chromomere is a pair of **lateral loops**, which give the chromosome its distinctive appearance. In part (b), the scanning electron microscope (SEM) reveals an extended axis and adjacent loops. As with bands in polytene chromosomes, there is much more DNA present in each loop than is needed to encode a single gene. This SEM provides a clear view of the chromomeres and the chromosomal fibers emanating from them.

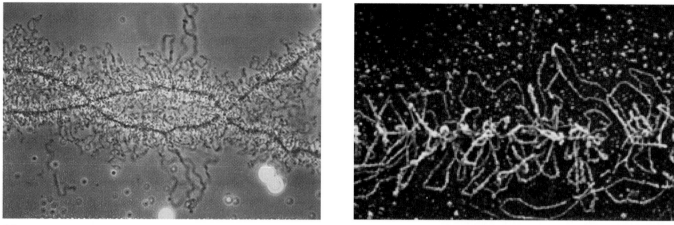

(a)   (b)

FIGURE 2.19 Lampbrush chromosomes derived from amphibian oocytes. Part (a) is a photomicrograph; parts (b) and (c) are scanning electron micrographs.

Each chromosomal loop is thought to be composed of one DNA double helix, while the central axis is made up of two DNA helices. This hypothesis is consistent with the belief that each chromosome is composed of a pair of sister chromatids. Studies using radioactive RNA precursors reveal that the loops are active in the synthesis of RNA. The lampbrush loops, in a way similar to puffs in polytene chromosomes, represent DNA that has been reeled out from the central chromomere axis during transcription. As with polytene chromosomes, the study of lampbrush chromosomes has provided many insights into the arrangement and function of the genetic information.

CHAPTER SUMMARY

1. The structure of cells is elaborate and complex. Many components of cells are involved directly or indirectly with genetic processes.

2. Chromosomes, in diploid organisms, exist in homologous pairs. Each pair shares the same size, centromere placement, and gene sites. One member of each pair is derived from the maternal parent and one is derived from the paternal parent.

3. Mitosis and meiosis are mechanisms by which cells distribute genetic information contained in their chromosomes to their descendants in a precise, orderly fashion.

4. Mitosis, or nuclear division, is part of the cell cycle and is the basis of cellular reproduction. Daughter cells are produced that are genetically identical to their progenitor cell.

5. Mitosis is subdivided into discrete phases: prophase, prometaphase, metaphase, anaphase, and telophase. Condensation of chromatin into chromosome structures occurs during prophase. During prometaphase, chromosomes take on the appearance of double structures, each represented by a pair of sister chromatids. In metaphase, chromosomes line up on the equatorial plane of the cell. During anaphase, sister chromatids of each chromosome are pulled apart and directed toward opposite poles. Telophase completes daughter cell formation and is characterized by cytokinesis, the division of the cytoplasm.

6. The cell cycle is characteristic of all eukaryotes and is under the control of an elaborate genetic program that regulates its activities. At least three checkpoints are present where cellular activities are monitored, whereby the cycle may be arrested if further progress jeopardizes the cell.

7. Meiosis, the underlying basis of sexual reproduction, results in the conversion of a diploid cell to a haploid gamete or spore. As a result of chromosome duplication and two subsequent divisions, each haploid cell receives one member of each homologous pair of chromosomes. Meiosis results in extensive genetic variation by virtue of the exchange between homologous chromosomes during crossing over and the random segregation of maternal and paternal chromatids.

8. A major difference exists between animal meiosis in males and females. Spermatogenesis partitions cytoplasmic volume equally and produces four haploid sperm cells. Oogenesis, on the other hand, accumulates the cytoplasm around one egg cell and reduces the other haploid sets of genetic material to polar bodies. The extra cytoplasm contributes to zygote development following fertilization.

9. Meiosis not only maintains the constancy of genetic material from generation to generation, but also results in extensive genetic variation. This is produced by virtue of the random distribution of either a maternal or a paternal member of every homologous pair of chromosomes into each gamete. Variation is further enhanced as a result of the exchange during crossing over between maternal and paternal homologs. In addition, meiosis plays an important role in the life cycle of fungi and plants, serving as the bridge between alternating generations.

10. Mitotic and meiotic chromosomes are produced as a result of the coiling and condensation of chromatin fibers characteristic of interphase. This transition is described by the "folded-fiber model."

11. The synaptonemal complex is an ultrastructural component of the cell present during the first meiotic prophase stage. It is important to the process of synapsis of homologs and may play a role in crossing over.

12. Polytene and lampbrush chromosomes are examples of specialized structures that have extended our knowledge of genetic organization and function.

KEY TERMS

AB antigens, 19
acrocentric chromosome, 22
allele, 23
anaphase, 26
apoptosis, 31
autoradiography, 43
basal body, 22
biparental inheritance, 23
bivalent, 32
capsule, 19
cdc mutation, 30
Cdk protein, 30
cell coat, 19
cell cycle, 25
cell furrow, 30
cell plate, 29
cell respiration, 22
cellular organelles, 21
cellulose, 19
cell wall, 19
centriole, 22
centromere, 27
centrosome, 22
chiasma (pl. chiasmata), 33
chloroplast, 22
chromatid, 27
chromatin, 19
chromomere, 32
chromosome, 18
crossing over, 32
cyclin, 30
cyclin-dependent kinase (Cdk), 30
cytokinesis, 25

cytoplasm, 21
cytoskeleton, 22
cytosol, 21
daughter chromosome, 27
diakinesis, 33
diploid number (2*n*), 22
diplotene stage, 33
disjunction, 33
Duchenne muscular dystrophy, 19
dyad, 32
dystrophin, 19
endoplasmic reticulum (ER), 22
endosymbiotic hypothesis, 22
equational division, 32
folded-fiber model, 40
G0 stage, 25
G1 checkpoint, 25
G1 stage, 25
G2 checkpoint, 31
G2 stage, 25
gamete, 31
gametophyte stage, 38
gene, 18
genetic analysis, 46
genetic material, 18
genome, haploid, 23
glycocalyx, 19
haploid number (*n*), 23
histocompatibility antigens, 19
histones, 39
homologous chromosomes, 22
homology search, 32
independent assortment, 33

interphase, 25
karyokinesis, 25
karyotype, 23
kinetochore, 27
lampbrush chromosome, 43
lateral loop, 43
leptotene stage, 32
locus (pl. loci), 23
meiosis, 18
metacentric chromosome, 22
metaphase, 27
metaphase plate, 27
microtubules, 27
middle lamella, 29
mitochondria, 22
mitosis, 18
MN antigens, 19
molecular motors, 27
monad, 32
monosomy, 39
motor proteins, 27
nondisjunction, 33
nonsister chromatids, 33
nucleoid region, 21
nucleolar organizer region (NOR), 20
nucleolus, 20
nucleus, 20
oocyte, 36
oogenesis, 36
oogonium, 36
ootid, 36
ovum (pl. ova), 36
p arm, 22

INSIGHTS AND SOLUTIONS

With this initial appearance of "Insights and Solutions," it is appropriate to provide a brief description of its value to you as a student. This section will precede the "Problems and Discussion Questions" in each chapter. One or more examples will be provided. Solutions to these problems and answers to these questions will illustrate approaches useful in **genetic analysis.** Our initial emphasis will be on insights that will help you arrive at correct solutions to ensuing problems.

1. In an organism with a haploid number of 3, how many individual chromosomal structures will align on the metaphase plate during (a) mitosis; (b) meiosis I; and (c) meiosis II? Describe each configuration.

   Solution:
   (a) In mitosis, where homologous chromosomes do not synapse, there will be 6 double structures, each con-

sisting of a pair of sister chromatids. The number of structures is equivalent to the diploid number.

(b) In meiosis I, the homologs have synapsed, reducing the number of structures to 3. Each is called a *tetrad* and consists of two pairs of sister chromatids.

(c) In meiosis II, the same number of structures exist (3), but in this case, they are called *dyads*. Each consists of a pair of sister chromatids. When crossing over has occurred, each chromatid may contain part of one of its nonsister chromatids obtained during exchange in prophase I.

2. For the chromosomes illustrated in Figure 2.11 (p. 34), draw all possible alignment configurations that may occur during metaphase of meiosis I.

   Solution: As shown in the illustration below, there are four configurations possible when $n = 2$.

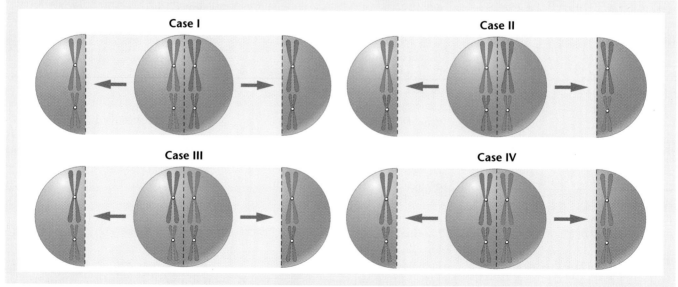

Case I   Case II

Case III   Case IV

3. As shown below, if we were to place one gene on both of the larger chromosomes and use two alleles (one on each large chromosome), *A* and *a*, and place a second gene with two alleles on the smaller chromosomes, using *B* and *b*, calculate the probability of generating each gene combination (*AB*, *Ab*, *aB*, *ab*) following meiosis I.

Solution:

Case I	*AB* and *ab*
Case II	*Ab* and *aB*
Case III	*aB* and *Ab*
Case IV	*ab* and *AB*

$$\text{Total:} \quad \begin{aligned} AB &= 2 \quad (p = 1/4) \\ Ab &= 2 \quad (p = 1/4) \\ aB &= 2 \quad (p = 1/4) \\ ab &= 2 \quad (p = 1/4) \end{aligned}$$

4. (a) How many different chromosome configurations can occur following meiosis I if three different pairs of chromosomes are present ($n = 3$)?

Solution: If $n = 3$, then eight different configurations would be possible. The formula 2^n, where n equals the haploid number, will allow you to calculate the number of potential alignment patterns. As we will see in the next chapter, these patterns are produced as a result of the Mendelian postulate called *segregation* and they serve as the physical basis of the Mendelian postulate of *independent assortment*.

(b) Assuming that comparable chromosomes in different individuals are genetically dissimilar because of different alleles, how many unique zygotic combinations are possible following fertilization in an organism where $n = 3$?

Solution: Assuming that no crossing over occurs (and that maternal and paternal chromatids remain intact), then each organism can produce 2^n or 2^3 different chromosome combinations in their gametes. Following random fertilization, $(2^3)(2^3) = 2^6 = 64$ unique combinations are possible in the offspring.

PROBLEMS AND DISCUSSION QUESTIONS

1. What role do the following cellular components play in the storage, expression, or transmission of genetic information: (a) chromatin, (b) nucleolus, (c) ribosome, (d) mitochondrion, (e) centriole, (f) centromere?

2. Discuss the concepts of homologous chromosomes, diploidy, and haploidy. What characteristics are shared between two chromosomes considered to be homologous?

3. If two chromosomes of a species are the same length and have similar centromere placements yet are *not* homologous, what *is* different about them?

4. Describe the events that characterize each stage of mitosis.

5. If an organism has a diploid number of 16, how many chromatids are visible at the end of mitotic prophase? How many chromosomes are moving to each pole during anaphase of mitosis?

6. How are chromosomes named on the basis of centromere placement?

7. Contrast telophase in plant and animal mitosis.

8. Outline and discuss the events, including regulatory checkpoints, of the cell cycle. What experimental technique was used to demonstrate the existence of the S phase? What is the role of Cdk proteins and cyclins in the regulation provided by checkpoints?

9. Examine Figure 2.12 showing oogenesis in animal cells. Will the genetic composition of the second polar bodies (derived from meiosis II) always be identical to that of the ootid? Why or why not?

10. Contrast the end results of meiosis with those of mitosis.

11. Define and discuss the following terms: (a) synapsis, (b) bivalents, (c) chiasmata, (d) crossing over, (e) chromomeres, (f) sister chromatids, (g) tetrads, (h) dyads, (i) monads, and (j) synaptonemal complex.

12. Contrast the genetic content and the origin of sister vs. nonsister chromatids during their earliest appearance in prophase I of meiosis. How might the genetic content of these change by the time tetrads have aligned at the equatorial plate during metaphase I?

13. Given the end results of the two types of division, why is it necessary for homologs to pair during meiosis and not desirable for them to pair during mitosis?

14. If an organism has a diploid number of 16 in an oocyte,
    (a) How many tetrads are present in the first meiotic prophase?
    (b) How many dyads are present in the second meiotic prophase?
    (c) How many monads migrate to each pole during the second meiotic anaphase?
    (d) What is the probability that a gamete will contain only paternal chromosomes?

15. Contrast spermatogenesis and oogenesis. What is the significance of the formation of polar bodies?

16. Explain why meiosis leads to significant genetic variation while mitosis does not.

17. During oogenesis in an animal species with a haploid number of 6, one dyad undergoes second division nondisjunction. Following the second meiotic division, the involved dyad ends up intact in the ovum. How many chromosomes are present in (a) the mature ovum, and (b) the second polar body? (c) Following fertilization by a normal sperm, what chromosome condition is created?

18. What is the probability that in an organism with a haploid number of 10 that a sperm will be formed that contains all 10 chromosomes whose centromeres were derived from maternal homologs?

19. During the first meiotic prophase, (a) when does crossing over occur? (b) When does synapsis occur? (c) During

which stage are the chromosomes least condensed? (d) When are chiasmata first visible?

20. What is the role of meiosis in the life cycle of a higher plant such as an angiosperm?

21. Describe the transition of a chromatin fiber into a mitotic chromosome. What model depicts this transition?

22. When during the cell cycle (Figure 2.6, page 26) does the transition state described in Problem 21 first occur?

23. How are giant polytene chromosomes formed?

24. What genetic process is occurring in a "puff" of a polytene chromosome? How do we know?

25. During what genetic process are lampbrush chromosomes present in vertebrates?

26. What ploidy values characterize polytene chromosomes and lampbrush chromosomes?

27. Discuss the role of the *p53* gene in the regulation of the cell cycle. What consequences result when this gene loses normal function as a result of mutation?

28. Contrast the roles of cyclins and cyclin-dependent kinases (CDKs) during the cell cycle. What are "checkpoints," and why are they important to multicellular organisms?

EXTRA-SPICY PROBLEMS

Following the "Problems and Discussion Questions" section in this and each ensuing chapter, we shall present you with one or more "extra-spicy" genetics problems. We have chosen to set these apart in order to identify problems that are particularly challenging. You may be asked to examine and assess actual data, to design genetics experiments, or to engage in cooperative learning. Like hot peppers, some of these experiences are just spicy and some are very hot. Hopefully, all of them will leave an aftertaste that is pleasing to those who indulge themselves.

29. A diploid cell contains three pairs of chromosomes designated *A*, *B*, and *C*. Each pair contains a maternal and paternal member, e.g., A^m and A^p, etc. Using these designations, demonstrate your understanding of mitosis and meiosis by drawing chromatid combinations in response to the following questions. Be sure to indicate when chromatids are paired as a result of replication and/or synapsis. You may wish to use a large piece of brown manila wrapping paper and work with a partner as you deal with this problem. Such "cooperative learning" may be a useful approach as you solve problems throughout the text.

(a) In mitosis, what chromatid combination(s) will be present during metaphase? What combination(s) will be present at each pole at the completion of anaphase?

(b) During meiosis I (assuming no crossing over), what chromatid combination(s) will be present at the completion of prophase? Draw all possible alignments of chromatids as migration begins during early anaphase.

(c) Are there any possible combinations present during prophase of meiosis II other than those that you drew in (b)? If so, draw them. If not, then proceed to (d).

(d) Draw all possible combinations of chromatids during the early phases of anaphase in meiosis II.

(e) Assume that during meiosis I none of the "C" chromosomes disjoin at metaphase, but they separate into dyads (instead of monads) during meiosis II. How would this change the alignments that you constructed during the anaphase stages in meiosis I and II? Draw them.

(f) Assume that each gamete resulting from (e) participated in fertilization with a normal haploid gamete. What combinations will result? What percentage of zygotes will be diploid, containing one paternal and one maternal member of each chromosome pair?

SELECTED READINGS

ALBERTS, B., et al. 1994. *Molecular biology of the cell*, 3rd ed. New York: Garland Publishing.

ANGELIER, N., et al. 1984. Scanning electron microscopy of amphibian lampbrush chromosomes. *Chromosoma* 89:243–53.

BAKER, B. A., et al. 1976. The genetic control of meiosis. *Annu. Rev. Genet.* 10:53–134.

BASERGA, R., and KISIELESKI, W. 1963. Autobiographics of cells. *Sci. Am.* (Aug.) 209:103–10.

BEERMAN, W., and CLEVER, U. 1964. Chromosome puffs. *Sci. Am.* (April) 210:50–58.

BRACHET, J., and MIRSKY, A. E. 1961. *The cell: Meiosis and mitosis*, Vol. 3. Orlando, FL: Academic Press.

CALLAN, H. G. 1986. *Lampbrush chromosomes*. New York: Springer-Verlag.

DUPRAW, E. J. 1970. *DNA and chromosomes*. New York: Holt, Rinehart & Winston.

GALL, J. G. 1963. Kinetics of deoxyribonuclease on chromosomes. *Nature* 198:36–38.

GLOVER, D. M., GONZALEZ, C. and RAFF, J. W. 1993. The centrosome. *Sci. Am.* (June) 268:62–68.

GOLOMB, H. M., and BAHR, G. F. 1971. Scanning electron microscopic observations of surface structures of isolated human chromosomes. *Science* 171:1024–26.

HALL, J. L., RAMANIS, Z., and LUCK, D. J. 1989. Basal body/centriolar DNA: Molecular genetic studies in *Chlamydomonas*. *Cell* 59:121–32.

HARTWELL, L. H., et al. 1974. Genetic control of the cell division cycle in yeast. *Science* 183:46–51.

HARTWELL, L. H. and KASTAN, M. B. 1994. Cell cycle control and cancer. *Science* 266:1821–28.

HARTWELL, L. H. and WEINERT, T. A. 1989. Checkpoint controls that ensure the order of cell cycle events. *Science* 246:629–34.

HAWLEY, R. S., and ARBEL, T. 1993. Yeast genetics and the fall of the classical view of meiosis. *Cell* 72:301–303.

HILL, R. J., and RUDKIN, G. T. 1987. Polytene chromosomes: The status of the band–interband question. *BioEssays* 7: 35–40.

MAZIA, D. 1961. How cells divide. *Sci. Am.* (Jan.) 205:101–20.

———. 1974. The cell cycle. *Sci. Am.* (Jan.) 235:54–64.

MCINTOSH, J. R., and MCDONALD, K. L. 1989. The mitotic spindle. *Sci. Am.* (Oct.) 261:48–56.

MOENS, P. B. 1973. Mechanisms of chromosome synapsis at meiotic prophase. *Int. Rev. Cytol.* 35:117–34.

MURRAY, A. W. and KIRSCHNER, M. 1993. *The cell cycle: An introduction.* New York: Oxford University Press.

PARDEE, A. B., et al. 1978. Animal cell cycle. *Annu. Rev. Biochem.* 47:715–50.

PRESCOTT, D. M. 1977. *Reproduction of eukaryotic cells.* Orlando, FL: Academic Press.

PRESCOTT, D. M., and FLEXER, A. S. 1986. *Cancer, the misguided cell*, 2nd ed. Sunderland, MA: Sinauer.

SAWIN, K. E., et. al. 1992. Mitotic spindle organization by a plus end-directed microtubular motor. *Nature* 359:540–43.

SWANSON, C. P., MERZ, T., and YOUNG, W. J. 1981. *Cytogenetics, the chromosome in division, inheritance, and evolution*, 2nd ed. Englewood Cliffs, NJ: Prentice-Hall.

WADSWORTH, P. 1993. Mitosis: Spindle assembly and chromosome movement. *Curr. Opin. Cell Biol.* 5:93–99.

WESTERGAARD, M., and vonWETTSTEIN, D. 1972. The synaptinemal complex. *Annu. Rev. Genet.* 6:71–110.

WHEATLEY, D. N. 1982. *The centriole: A central enigma of cell biology.* New York: Elsevier/North-Holland Biomedical.

YUNIS, J. J., and CHANDLER, M. E. 1979. Cytogenetics. In *Clinical diagnosis and management by laboratory methods*, ed. J. B. Henry, Vol. 1. Philadelphia: W. B. Saunders.

CHAPTER

3

Mendelian Genetics

CHAPTER CONCEPTS

Inherited characteristics are under the control of particulate factors called genes that are transmitted from generation to generation on vehicles called chromosomes, according to rules first described by Gregor Mendel. Genetic ratios, expressed as probabilities, are subject to chance deviation and may be evaluated using statistical analysis.

Although inheritance of biological traits has been recognized for thousands of years, the first significant insights into the mechanisms involved occurred less than a century and a half ago. In 1866, Gregor Mendel published the results of a series of experiments that would lay the foundation for the formal discipline of genetics. In the ensuing years the concept of the gene as a distinct hereditary unit was established, and the ways in which genes are transmitted to offspring and control traits were clarified. Research in these areas was accelerated in the first half of the twentieth century. The resultant findings served as the foundation for the tremendous interest and research effort in genetics since about 1940. It is safe to

Mendel's garden, as seen in the 1980s.

say that studies in genetics, most recently those at the molecular level, have remained continually at the forefront of biological research since the early 1900s.

In this chapter we focus on the development of the principles established by Mendel, now referred to as **Mendelian** or **transmission genetics**. These principles describe how genes are transmitted from parents to offspring and were derived directly from Mendel's experimentation.

When Mendel began his studies of inheritance using *Pisum sativum*, the garden pea, there was no knowledge of chromosomes nor of the role and mechanism of meiosis. Nevertheless, he was able to determine that distinct **units of inheritance** exist and to predict their behavior during the formation of gametes. Subsequent investigators, with access to cytological data, were able to relate their observations of chromosome behavior during meiosis to Mendel's principles of inheritance. Once this correlation was made, Mendel's postulates were accepted as the basis for the study of transmission genetics. Even today, they serve as the cornerstone of the study of inheritance.

Gregor Johann Mendel

In 1822, Johann Mendel was born to a peasant family in the European village of Heinzendorf, now part of the Czech Republic. An excellent student in high school, Mendel studied philosophy for several years afterward and was admitted to the Augustinian Monastery of St. Thomas in Brno in 1843. There, he took the name of Gregor and received support for his studies and research throughout the rest of his life. In 1849, he was relieved of pastoral duties and received a teaching appointment that lasted several years. From 1851 to 1853, he attended the University of Vienna, where he studied physics and botany. In 1854 he returned to Brno, where, for the next 16 years, he taught physics and natural science.

In 1856, Mendel performed his first set of hybridization experiments with the garden pea. The research phase of his career lasted until 1868, when he was elected abbot of the monastery. Although his interest in genetics remained, his new responsibilities demanded most of his time. In 1884, Mendel died of a kidney disorder. The local newspaper paid him the following tribute: "His death deprives the poor of a benefactor, and mankind at large of a man of the noblest character, one who was a warm friend, a promoter of the natural sciences, and an exemplary priest."

Mendel's Experimental Approach

In 1865, Mendel first reported the results of some simple genetic crosses between certain strains of the garden pea. Although, as we saw in Chapter 1, his was not the first attempt to provide experimental evidence pertaining to inheritance, Mendel's work is an elegant model of experimental design and analysis.

Mendel showed remarkable insight into the methodology necessary for good experimental biology. He chose an organism that was easy to grow and to hybridize artificially. The pea plant is self-fertilizing in nature, but is easy to cross-breed experimentally. It reproduces well and grows to maturity in a single season. Mendel worked with seven unit characters, visible features that were each represented by two contrasting forms or traits. For the char-

acter stem height, for example, he experimented with the traits *tall* and *dwarf*. He selected six other contrasting pairs of traits involving seed shape and color, pod shape and color, and pod and flower arrangement. True-breeding strains were available from local seed merchants. Each trait appeared unchanged generation after generation in self-fertilizing plants; that is, the strains exhibiting them "bred true."

Mendel's success in an area where others had failed can be attributed to several factors in addition to the choice of a suitable organism. He restricted his examination to one or very few pairs of contrasting traits in each experiment. He also kept accurate quantitative records, a necessity in genetic experiments. From the analysis of his data, Mendel derived certain postulates that have become the principles of transmission genetics.

The results of Mendel's experiments went unappreciated until the turn of the century, well after his death. Once Mendel's publications were rediscovered by geneticists investigating the function and behavior of chromosomes, the implications of his postulates were immediately apparent. He had discovered the basis for the transmission of hereditary traits!

The Monohybrid Cross

The simplest crosses performed by Mendel involved only one pair of contrasting traits. Each such breeding experiment involves a **monohybrid cross**. A monohybrid cross is constructed by mating individuals from two parent strains, each of which exhibits one of the two contrasting forms of the character under study. Initially we will examine the first generation of offspring of such a cross, and then we will consider the offspring of **selfing** or **self-fertilizing** individuals from this first generation. The original parents are called the P_1 or **parental generation**, their offspring are the F_1 or **first filial generation**, and the individuals resulting from the selfing of the F_1 generation are called the F_2 or **second filial generation**. We can, of course, continue to follow subsequent generations, if desirable.

The cross between true-breeding peas with tall stems and dwarf stems is representative of Mendel's monohybrid crosses. *Tall* and *dwarf* represent contrasting forms or traits of the character of stem height. Unless tall or dwarf plants are crossed together or with another strain, they will undergo self-fertilization and breed true, producing their respective trait generation after generation. However, when Mendel crossed tall plants with dwarf plants, the resulting F_1 generation consisted only of tall plants. When members of the F_1 generation were selfed, Mendel observed that 787 of 1064 F_2 plants were tall, while 277 of 1064 were dwarf. Note that in this cross

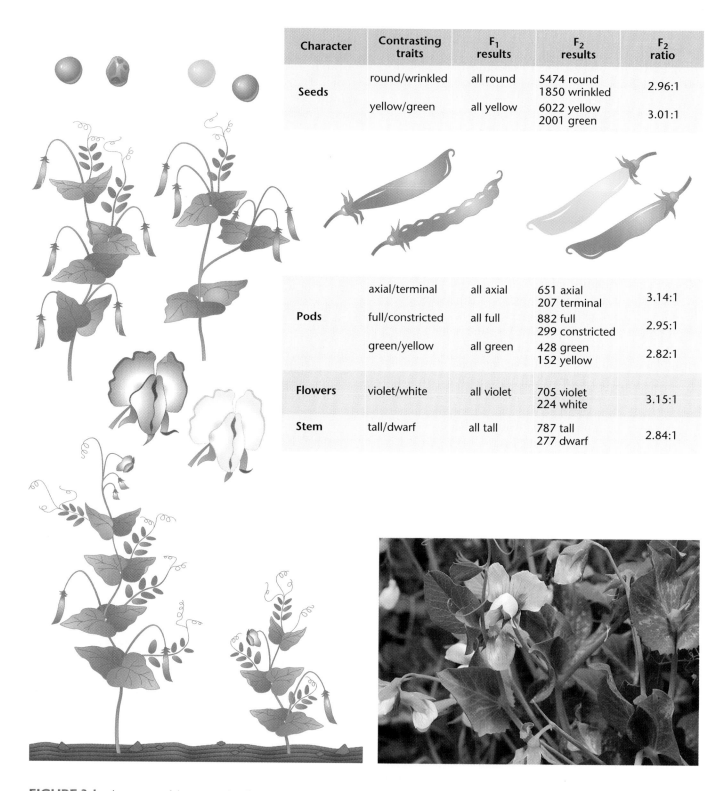

Character	Contrasting traits	F₁ results	F₂ results	F₂ ratio
Seeds	round/wrinkled	all round	5474 round 1850 wrinkled	2.96:1
	yellow/green	all yellow	6022 yellow 2001 green	3.01:1
Pods	axial/terminal	all axial	651 axial 207 terminal	3.14:1
	full/constricted	all full	882 full 299 constricted	2.95:1
	green/yellow	all green	428 green 152 yellow	2.82:1
Flowers	violet/white	all violet	705 violet 224 white	3.15:1
Stem	tall/dwarf	all tall	787 tall 277 dwarf	2.84:1

FIGURE 3.1   A summary of the seven pairs of contrasting traits and the results of Mendel's seven monohybrid crosses. In each case, pollen derived from plants exhibiting one contrasting trait was used to fertilize the ova of plants exhibiting the other contrasting trait. In the F₁ generation one of the two traits, referred to as dominant, was exhibited by all plants. The contrasting trait, referred to as recessive, then reappeared in approximately one-fourth of the F₂ plants. The garden pea (*Pisum sativum*) is shown in the photograph.

(Figure 3.1) the dwarf trait disappears in the F_1, only to reappear in the F_2 generation.

Genetic data are usually expressed and analyzed as ratios. In this particular example, many identical P_1 crosses were made and many F_1 plants—all tall—were produced. Of the 1064 F_2 offspring, 787 were tall and 277 were dwarf—a ratio of approximately 2.8:1.0, or about 3:1.

Mendel made similar crosses between pea plants exhibiting each of the other pairs of contrasting traits. Results of these crosses are also shown in Figure 3.1. In every case, the outcome was similar to the tall/dwarf cross just described. All F_1 offspring were identical to one of the parents. In the F_2, an approximate ratio of 3:1 was obtained. Three-fourths appeared like the F_1 plants, while one-fourth exhibited the contrasting trait, which had disappeared in the F_1 generation.

It is appropriate to point out one further aspect of the monohybrid crosses. In each, the F_1 and F_2 patterns of inheritance were similar regardless of which P_1 plant served as the source of pollen, or sperm, and which served as the source of the ovum, or egg. The crosses could be made either way—that is, pollen from the tall plant pollinating dwarf plants, or vice versa. These are called **reciprocal crosses**. Therefore, the results of Mendel's monohybrid crosses were not sex-dependent.

To explain these results, Mendel proposed the existence of particulate **unit factors** for each trait. He suggested that these factors serve as the basic units of heredity and are passed unchanged from generation to generation, determining various traits expressed by each individual plant. Using these general ideas, Mendel proceeded to hypothesize precisely how such factors could account for the results of the monohybrid crosses.

Mendel's First Three Postulates

Using the consistent pattern of results in the monohybrid crosses, Mendel derived the following three postulates or principles of inheritance.

1. UNIT FACTORS IN PAIRS
   Genetic characters are controlled by unit factors that exist in pairs in individual organisms.
   In the monohybrid cross involving tall and dwarf stems, a specific unit factor exists for each trait. Because the factors occur in pairs, three combinations are possible: two factors for tallness, two factors for dwarfness, or one of each factor. Every individual contains one of these three combinations, which determines stem height.

2. DOMINANCE/RECESSIVENESS
   When two unlike unit factors responsible for a single character are present in a single individual, one unit factor is dominant to the other, which is said to be recessive.
   In each monohybrid cross, the trait expressed in the F_1 generation results from the presence of the dominant unit factor. The trait that is not expressed in the F_1, but which reappears in the F_2, is under the genetic influence of the recessive unit factor. Note that this dominance/recessiveness relationship only pertains when unlike unit factors are present together in an individual. The terms **dominant** and **recessive** are also used to designate the traits. In the above case, the trait tall stem is said to be dominant to the recessive trait dwarf stem.

3. SEGREGATION
   During the formation of gametes, the paired unit factors separate or segregate randomly so that each gamete receives one or the other with equal likelihood.
   If an individual contains a pair of like unit factors (for example, both specify tall), then all gametes receive one tall unit factor. If an individual contains unlike unit factors (e.g., one for tall and one for dwarf), then each gamete has a 50 percent probability of receiving either the tall or dwarf unit factor.

These postulates provide a suitable explanation for the results of the monohybrid crosses. The tall/dwarf cross will be used to illustrate this explanation. Mendel reasoned that P_1 tall plants contained identical paired unit factors, as did the P_1 dwarf plants. The gametes of tall plants all received one tall unit factor as a result of segregation. Likewise, the gametes of dwarf plants all received one dwarf unit factor. Following fertilization, all F_1 plants received one unit factor from each parent, a tall factor from one and a dwarf factor from the other, reestablishing the paired relationship. Because tall is dominant to dwarf, all F_1 plants were tall.

When F_1 plants form gametes, the postulate of segregation demands that each gamete randomly receive $either$ the tall or dwarf unit factor. Following random fertilization events during F_1 selfing, four F_2 combinations will result in equal frequency:

(1) tall/tall
(2) tall/dwarf
(3) dwarf/tall
(4) dwarf/dwarf

Combinations (1) and (4) will clearly result in tall and dwarf plants, respectively. According to the postulate of dominance/recessiveness, combinations (2) and (3) will both yield tall plants. Therefore, the F$_2$ is predicted to consist of three-fourths tall and one-fourth dwarf, or a ratio of 3:1. This is approximately what Mendel observed in the cross between tall and dwarf plants. A similar pattern was observed in each of the other monohybrid crosses.

Modern Genetic Terminology

To illustrate the monohybrid cross and Mendel's first three postulates, we must introduce several new terms as well as a set of symbols for the unit factors. Traits such as tall or dwarf are visible expressions of the information contained in unit factors. The physical appearance of a trait is called the **phenotype** of the individual.

All unit factors represent units of inheritance called **genes** by modern geneticists. For any given character, such as plant height, the phenotype is determined by the presence of different combinations of alternative forms of

a single gene called **alleles**. For example, the unit factors representing tall and dwarf are alleles determining the height of the pea plant.

Depending on the organism being studied, there are many conventions used in assigning gene symbols. In Chapter 4 we will review some of these, but for now we will adopt an approach that we can use consistently through the next few chapters. By one convention, the first letter of the recessive trait may be chosen to symbolize the character in question. The lowercase letter designates the allele for the recessive trait, and the uppercase letter designates the allele for the dominant trait. Therefore, we let *d* stand for the dwarf allele and *D* represent the tall allele. When alleles are written in pairs to represent the two unit factors present in any individual (*DD*, *Dd*, or *dd*), these symbols are referred to as the **genotype**. This term reflects the genetic makeup of an individual whether it is haploid or diploid. By reading the genotype, it is possible to know the phenotype of the individual: *DD* and *Dd* are tall, and *dd* is dwarf. When identical alleles exist (*DD* or *dd*), the individual is said to be **homozygous**

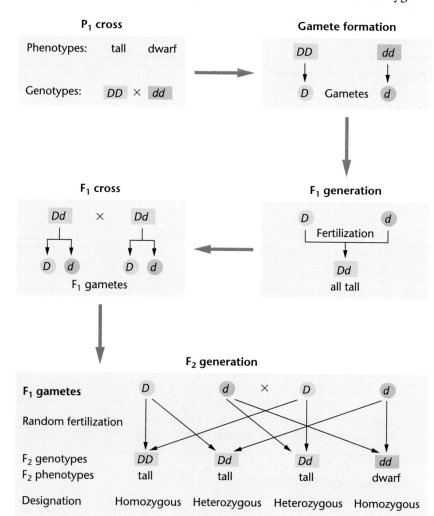

FIGURE 3.2 An explanation of the monohybrid cross between tall and dwarf pea plants. The symbols *D* and *d* are used to designate the tall and dwarf unit factors, respectively, in the genotypes of mature plants and gametes. All individuals are shown in rectangles. All gametes are shown in circles.

or a **homozygote**; when alleles are different (*Dd*), we use the term **heterozygous** or **heterozygote**. These symbols and terms are used in Figure 3.2 to illustrate the complete monohybrid cross.

Mendel's Analytical Approach

What led Mendel to deduce unit factors in pairs? Because there were two contrasting traits for each character, it seemed logical that two distinct factors must exist. However, why does one of the two traits or phenotypes disappear in the F_1 generation? Observation of the F_2 generation helps to answer this question. The recessive trait and its unit factor do not actually disappear in the F_1; they are merely hidden or masked, only to reappear in one-fourth of the F_2 offspring. Therefore, Mendel concluded that one unit factor for tall and one for dwarf were transmitted to each F_1 individual; but because the tall factor or allele is dominant to the dwarf factor or allele, all F_1 plants are tall. Finally, how is the 3:1 F_2 ratio explained? As shown in Figure 3.2, if the tall and dwarf alleles of the F_1 heterozygote segregate randomly into gametes, and if fertilization is random, this ratio is the natural outcome of the cross.

Because he operated without the hindsight that modern geneticists enjoy, Mendel's analytical reasoning must be considered a truly outstanding scientific achievement. On the basis of rather simple, but precisely executed breeding experiments, he proposed that discrete **particulate units of heredity** exist, and he explained how they are transmitted from one generation to the next.

Punnett Squares

The genotypes and phenotypes resulting from the recombination of gametes during fertilization can be easily visualized by constructing a **Punnett square**, so named after the person who first devised this approach, Reginald C. Punnett. Figure 3.3 illustrates this method of analysis for the $F_1 \times F_1$ monohybrid cross. Each of the possible gametes is assigned to an individual column or a row, with the vertical column representing those of the female parent and the horizontal row those of the male parent. After entering the gametes in rows and columns, the new generation is predicted by combining the male and female gametic information for each combination and entering the resulting genotypes in the boxes. This process represents all possible random fertilization events. The genotypes and phenotypes of all potential offspring are ascertained by reading the entries in the boxes.

The Punnett square method is particularly useful when one is first learning about genetics and how to solve problems. In Figure 3.3, note the ease with which the 3:1

phenotypic ratio and the 1:2:1 genotypic ratio may be derived in the F_2 generation.

The Test Cross: One Character

Tall plants produced in the F_2 generation are predicted to be of either the *DD* or *Dd* genotypes. We might ask if there is a way to distinguish the genotype. Mendel devised a rather simple method that is still used today in breeding procedures of plants and animals: the **test cross**. The organism of the dominant phenotype but unknown genotype is crossed to a **homozygous recessive individual**. For example, as shown in Figure 3.4(a), if a tall plant of genotype *DD* is test-crossed to a dwarf plant, which must have the *dd* genotype, all offspring will be tall phenotypically and *Dd* genotypically. However, as shown in Figure 3.4(b), if a tall plant is *Dd* and is crossed to a dwarf plant (*dd*), then one-half of the offspring will be tall (*Dd*) and the other half will be dwarf (*dd*). Therefore, a 1:1 tall/dwarf ratio demonstrates the heterozygous nature of

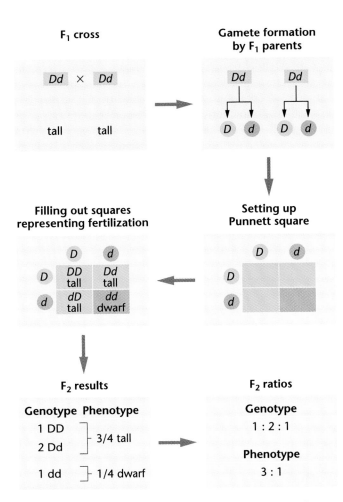

FIGURE 3.3 The use of a Punnett square in generating the F_2 ratio of the $F_1 \times F_1$ cross shown in Figure 3.2.

Test cross results

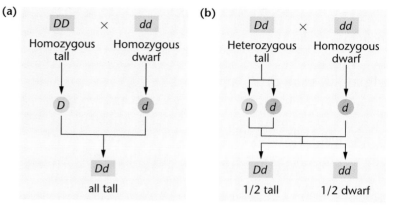

(a)

DD × dd

Homozygous tall | Homozygous dwarf

D   d

Dd

all tall

(b)

Dd × dd

Heterozygous tall | Homozygous dwarf

D d   d

Dd   dd

1/2 tall   1/2 dwarf

FIGURE 3.4 The test cross illustrated with a single character. In (a), the tall parent is homozygous. In (b), the tall parent is heterozygous. The genotypes of each tall parent may be determined by examining the offspring when each is crossed to the homozygous recessive dwarf plant.

the tall plant of unknown genotype. The results of test crosses reinforced Mendel's conclusion that separate unit factors control the tall and dwarf traits.

The Dihybrid Cross

A natural extension of performing monohybrid crosses was for Mendel to design experiments where two characters were examined simultaneously. Such a cross, involving two pairs of contrasting traits, is a **dihybrid cross**. It is also called a **two-factor cross**. For example, if pea plants having yellow seeds that are also round were bred with those having green seeds that are also wrinkled, the results shown in Figure 3.5 will occur. The F_1 offspring are all yellow and round. It is therefore apparent that yellow is dominant to green, and that round is dominant to wrinkled. In this dihybrid cross, the F_1 individuals are selfed,

and approximately 9/16 of the F_2 plants express yellow and round, 3/16 express yellow and wrinkled, 3/16 express green and round, and 1/16 express green and wrinkled.

A variation of this cross is also shown in Figure 3.5. Instead of crossing one P_1 parent with both dominant traits (yellow, round) to one with both recessive traits (green, wrinkled), plants with yellow, wrinkled seeds are crossed to those with green, round seeds. Despite the change in the P_1 phenotypes, both the F_1 and F_2 results remain unchanged. It will become clear in the next section why this is so.

Mendel's Fourth Postulate: Independent Assortment

We can most easily understand the results of a dihybrid cross if we consider it theoretically as consisting of two monohybrid crosses conducted separately. Think of the two sets of traits as being inherited independently of each

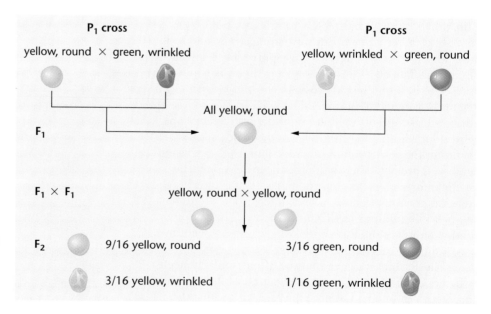

FIGURE 3.5 The F_1 and F_2 results of Mendel's dihybrid crosses between yellow, round and green, wrinkled pea plants, and between yellow, wrinkled and green, round pea plants.

P₁ cross
yellow, round × green, wrinkled

P₁ cross
yellow, wrinkled × green, round

F₁   All yellow, round

F₁ × F₁   yellow, round × yellow, round

F₂   9/16 yellow, round   3/16 green, round

3/16 yellow, wrinkled   1/16 green, wrinkled

other; that is, the chance of any plant becoming tall or dwarf is not at all influenced by the chance that this plant will also have round or wrinkled seeds. Thus, because yellow is dominant to green, all F_1 plants in the first theoretical cross would have yellow seeds. In the second theoretical cross, all F_1 plants would have round seeds because round is dominant to wrinkled. When Mendel examined the F_1 plants of the dihybrid cross, all were yellow and round, as predicted.

The predicted F_2 results of the first cross are 3/4 yellow and 1/4 green. Similarly, the second cross should yield 3/4 round and 1/4 wrinkled. Figure 3.5 shows that in the dihybrid cross, 12/16 of all F_2 plants are yellow while 4/16 are green, exhibiting the 3:1 ratio. Similarly, 12/16 of all F_2 plants have round seeds while 4/16 have wrinkled seeds, again revealing the 3:1 ratio.

Because the two pairs of contrasting traits are inherited independently, we can predict the frequencies of all possible F_2 phenotypes by applying the "product law" of probabilities: **When two independent events occur simultaneously, the combined probability of the two outcomes is equal to the product of their individual probabilities of occurrence.** For example, the probability of an F_2 plant having yellow *and* round seeds is (3/4)(3/4), or 9/16, because 3/4 of all F_2 plants should be yellow and 3/4 of all F_2 plants should be round.

In a like way, the probabilities of the other three F_2 phenotypes can be calculated: Yellow (3/4) *and* wrinkled (1/4) are predicted to be present together 3/16 of the time; green (1/4) *and* round (3/4) are predicted 3/16 of the time; and green (1/4) *and* wrinkled (1/4) are predicted 1/16 of the time. These calculations are illustrated in Figure 3.6. It is now apparent why the F_1 and F_2 results are identical whether the parents of the initial cross are yellow, round bred with green, wrinkled or if they are yellow, wrinkled bred with green, round. In both crosses, the F_1 genotype of all plants is identical. Each plant is heterozygous for both gene pairs. As a result, the F_2 generation is also identical in both crosses.

On the basis of similar results in numerous dihybrid crosses, Mendel proposed a fourth postulate:

4. INDEPENDENT ASSORTMENT
    During gamete formation, segregating pairs of unit factors assort independently of each other.

This postulate stipulates that segregation of any pair of unit factors occurs independently of all others. As a result of segregation, each gamete receives one member of every pair of unit factors. For one pair, whichever unit factor is received does not influence the outcome of segregation of any other pair. Thus, according to the postulate of **independent assortment**, all possible combinations of gametes will be formed in equal frequency.

Independent assortment is illustrated in the formation of the F_2 generation, shown in the Punnett square in Figure 3.7. Examine the formation of gametes by the F_1 plants. Segregation prescribes that every gamete receives either a G or g allele *and* a W or w allele. Independent assortment stipulates that all four combinations (GW, Gw, gW, and gw) will be formed with equal probabilities.

In every $F_1 \times F_1$ fertilization event, each zygote has an equal probability of receiving one of the four combinations from each parent. If a large number of offspring are produced, 9/16 are yellow and round, 3/16 are yellow and wrinkled, 3/16 are green and round, and 1/16 are green and wrinkled, yielding what is designated as **Mendel's 9:3:3:1 dihybrid ratio.** This ratio is based on probability events involving segregation, independent assortment, and random fertilization. Therefore, it is an ideal ratio. Because of deviation due strictly to chance, particularly if small numbers of offspring are produced, the ideal ratio will seldom be approached.

The Test Cross: Two Characters

The test cross may also be applied to individuals that express two dominant traits, but whose genotypes are unknown. For example, the expression of the yellow, round

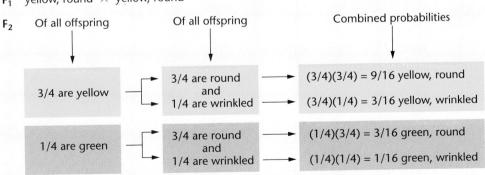

F_1 yellow, round $\times$ yellow, round

F_2 Of all offspring   Of all offspring   Combined probabilities

3/4 are yellow   →   3/4 are round and 1/4 are wrinkled   →   (3/4)(3/4) = 9/16 yellow, round
                                                                 (3/4)(1/4) = 3/16 yellow, wrinkled

1/4 are green   →   3/4 are round and 1/4 are wrinkled   →   (1/4)(3/4) = 3/16 green, round
                                                                 (1/4)(1/4) = 1/16 green, wrinkled

FIGURE 3.6 The determination of the combined probabilities of each F_2 phenotype for two independently inherited characters. The probability of each plant being yellow or green is independent of the probability of it being round or wrinkled.

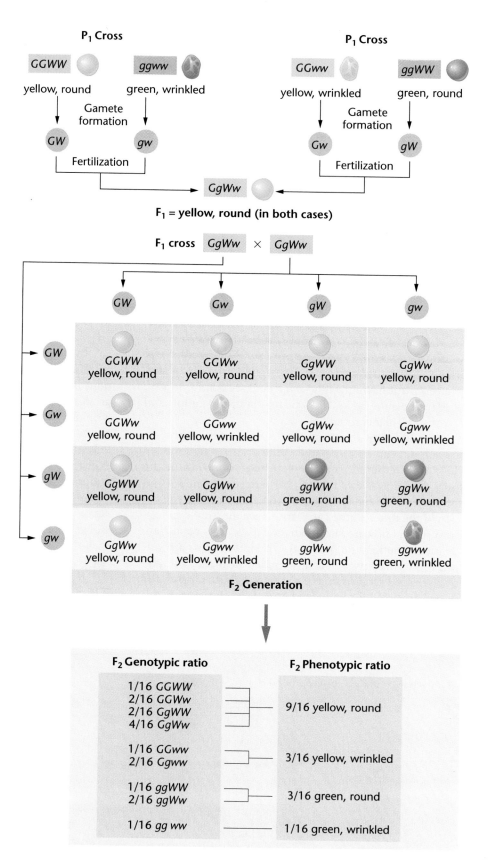

F_2 Generation

F_2 Genotypic ratio

1/16 *GGWW*
2/16 *GGWw*
2/16 *GgWW*
4/16 *GgWw*

1/16 *GGww*
2/16 *Ggww*

1/16 *ggWW*
2/16 *ggWw*

1/16 *gg ww*

F_2 Phenotypic ratio

9/16 yellow, round

3/16 yellow, wrinkled

3/16 green, round

1/16 green, wrinkled

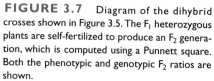

FIGURE 3.7 Diagram of the dihybrid crosses shown in Figure 3.5. The F_1 heterozygous plants are self-fertilized to produce an F_2 generation, which is computed using a Punnett square. Both the phenotypic and genotypic F_2 ratios are shown.

Test cross results of three yellow, round individuals

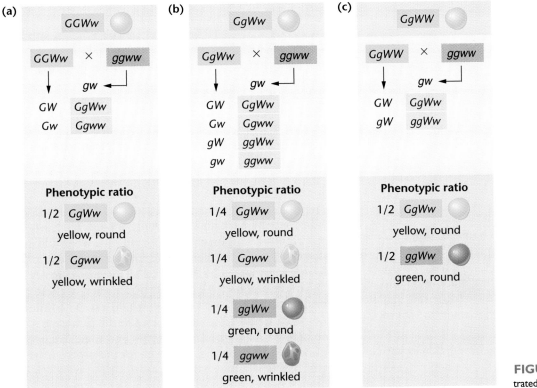

FIGURE 3.8 The test cross illustrated with two independent characters.

phenotype in the F_2 generation just described may result from the *GGWW*, *GGWw*, *GgWW*, and *GgWw* genotypes. If an F_2 yellow, round plant is crossed with the homozygous recessive green, wrinkled plant (*ggww*), analysis of the offspring will indicate the correct genotype of that yellow, round plant. Each of the above genotypes will result in a different set of gametes and, in a test cross, a different set of phenotypes in the resulting offspring. Three cases are illustrated in Figure 3.8.

The Trihybrid Cross

We have thus far considered inheritance by individuals of up to two pairs of contrasting traits. Mendel demonstrated that the identical processes of segregation and independent assortment apply to three pairs of contrasting traits in what is called a **trihybrid cross**, also referred to as a **three-factor cross**.

Although a trihybrid cross is somewhat more complex than a dihybrid cross, its results are easily calculated if the principles of segregation and independent assortment are followed. For example, consider the cross shown in Figure 3.9 where the gene pairs representing theoretical contrasting traits are symbolized *A/a*, *B/b*, and

C/c. In the cross between *AABBCC* and *aabbcc* individuals, all F_1 individuals are heterozygous for all three gene pairs. Their genotype, *AaBbCc*, results in the phenotypic expression of the dominant *A*, *B*, and *C* traits. When F_1 individuals are parents, each produces 8 different gametes in equal frequencies. At this point, we could construct a Punnett square with 64 separate boxes and read out the phenotypes. Because such a method is cumbersome in a cross involving so many factors, another method has been devised to calculate the predicted ratio.

The Forked-Line Method, or Branch Diagram

It is much less difficult to consider each contrasting pair of traits separately and then to combine these results using the **forked-line method**, which was first illustrated in Figure 3.6. This method, also called a **branch diagram**, relies on the simple application of the laws of probability established for the dihybrid cross. Each gene pair is assumed to behave independently during gamete formation.

When the monohybrid cross $AA \times aa$ is made, we know that:

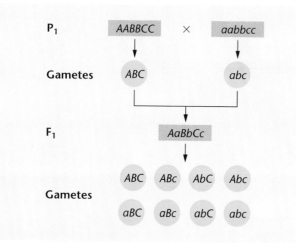

FIGURE 3.9 The formation of P₁ and F₁ gametes in a trihybrid cross.

1. All F₁ individuals have the genotype *Aa* and express the phenotype represented by the *A* allele, which is called the *A* phenotype in the following discussion.

2. The F₂ generation consists of individuals with either the *A* phenotype or the *a* phenotype in the ratio of 3:1, respectively.

The same generalizations may be made for the *BB* × *bb* and *CC* × *cc* crosses. Thus, in the F₂ generation, 3/4 of all organisms will express phenotype *A*, 3/4 will express *B*, and 3/4 will express *C*. Similarly, 1/4 of all organisms will express phenotype *a*, 1/4 will express *b*, and 1/4 will express *c*. The proportions of organisms expressing each phenotypic combination can be predicted by assuming that fertilization, following the independent assortment of these three gene pairs during gamete formation, is a random process. We must simply apply once again the product law of probabilities.

The phenotypic proportions of the F₂ generation calculated in this way using the forked-line method are illustrated in Figure 3.10. They fall into the trihybrid ratio of 27:9:9:9:3:3:3:1. The same method can be applied when solving crosses involving any number of gene pairs, *provided* that all gene pairs assort independently from each other. We will see later that this is not always the case. However, it appeared to be true for all of Mendel's characters.

Note in Figure 3.10 that only phenotypic ratios of the F₂ generation have been derived. It is possible to generate genotypic ratios as well. To do so, we again consider the *A/a*, *B/b*, and *C/c* gene pairs separately. For example, for the *A/a* pair the F₁ cross is *Aa* × *Aa*. Phenotypically, an F₂ ratio of 3/4 *A*:1/4 *a* is produced. Genotypically, however, the F₂ ratio is different; 1/4 *AA*:1/2 *Aa*:1/4 *aa* will result. Using Figure 3.10 as a model, we would enter these genotypic frequencies on the left side of the calculation. Each would be connected by three lines to 1/4 *BB*, 1/2 *Bb*, and 1/4 *bb*, respectively. From each of these nine designations, three more lines would extend to the 1/4 *CC*, 1/2 *Cc*, and 1/4 *cc* genotypes. On the right side of the completed diagram, 27 genotypes and their frequencies of occurrence would appear. One of the problems at the end of this chapter asks you to use the forked-line or branch diagram method to determine the genotypic ratios generated in a trihybrid cross (see Problem 17).

In crosses involving two or more gene pairs, the calculation of gametes and genotypic and phenotypic results is quite complex. Several simple mathematical rules will enable you to check the accuracy of various steps required in working genetic problems. First, you must determine the number of heterozygous gene pairs (*n*) involved in the cross. For example, where *AaBb* × *AaBb* represents the cross, $n = 2$; for *AaBbCc* × *AaBbCc*, $n = 3$; for *AaBBCcDd* × *AaBBCcDd*, $n = 3$ (because the *B* genes are not heterozygous). Once *n* is determined, 2^n is the number of different gametes that can be formed by each parent; 3^n is

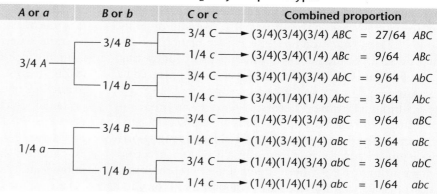

FIGURE 3.10 The generation of the F₂ trihybrid phenotypic ratio using the forked-line, or branch diagram, method.

Generation of F₂ trihybrid phenotypes

A or a	B or b	C or c	Combined proportion	
3/4 A	3/4 B	3/4 C	(3/4)(3/4)(3/4) *ABC* = 27/64	*ABC*
		1/4 c	(3/4)(3/4)(1/4) *ABc* = 9/64	*ABc*
	1/4 b	3/4 C	(3/4)(1/4)(3/4) *AbC* = 9/64	*AbC*
		1/4 c	(3/4)(1/4)(1/4) *Abc* = 3/64	*Abc*
1/4 a	3/4 B	3/4 C	(1/4)(3/4)(3/4) *aBC* = 9/64	*aBC*
		1/4 c	(1/4)(3/4)(1/4) *aBc* = 3/64	*aBc*
	1/4 b	3/4 C	(1/4)(1/4)(3/4) *abC* = 3/64	*abC*
		1/4 c	(1/4)(1/4)(1/4) *abc* = 1/64	*abc*

TABLE 3.1 Simple Mathematical Rules Useful in Working Genetics Problems

Crosses Between Organisms Heterozygous for Genes Exhibiting Independent Assortment			
Number of Heterozygous Gene Pairs	Number of Different Types of Gametes Formed	Number of Different Genotypes Produced	Number of Different Phenotypes Produced*
n	2^n	3^n	2^n
1	2	3	2
2	4	9	4
3	8	27	8
4	16	81	16

*The fourth column assumes that dominance and recessiveness are operational for all gene pairs.

the number of different genotypes that result following fertilization; and 2^n is the number of different phenotypes that are produced from these genotypes. Table 3.1 summarizes these rules, which may be applied to crosses involving any number of genes, provided that they assort independently from one another.

The Rediscovery of Mendel's Work

Mendel's work, initiated in 1856, was presented to the Brünn Society of Natural Science in 1865 and published the following year. However, his findings went largely unnoticed for about 35 years. Many reasons have been suggested to explain why the significance of his research was not immediately recognized.

First, Mendel's adherence to mathematical analysis of probability events was quite an unusual approach in biological studies. Perhaps his approach seemed foreign to his contemporaries. More important, his conclusions drawn from such analyses did not fit well with the existing hypotheses involving the cause of variation among organisms. The source of natural variation intrigued students of evolutionary theory. These individuals, stimulated by the proposal developed by Charles Darwin and Alfred Russell Wallace, believed in **continuous variation**, where offspring were a blend of their parents' phenotypes. As we have mentioned earlier, Mendel hypothesized that heredity was due to discrete or particulate units, therefore resulting in **discontinuous variation**. For example, Mendel proposed that the F_2 offspring of a dihybrid cross were merely expressing traits produced by new combinations of previously existing unit factors. As a result, Mendel's hypotheses did not fit well with the evolutionists' preconceptions about causes of variation.

Beyond this interpretation is still a further speculation as to why Mendel's contemporaries failed to grasp the significance of his findings. Perhaps they did not realize that Mendel's postulates explained *how* variation was transmitted to offspring. Instead, they may have at-

tempted to interpret his work in a way that addressed the issue of *why* certain phenotypes survive preferentially. It was this latter question that had been addressed in the theory of natural selection, but it was not addressed by Mendel. It may well be that the collective vision of Mendel's scientific colleagues was obscured by the impact of this extraordinary theory of organic evolution.

The Rebirth of Mendelian Genetics

Near the end of the nineteenth century, a remarkable observation set the scene for the rebirth of Mendel's work—Walter Flemming's discovery in 1879 of chromosomes in the nuclei of salamander cells. Flemming was able to describe the behavior of these threadlike structures during cell division. As a result of the findings of Flemming and many other cytologists, the presence of a nuclear component soon became an integral part of ideas surrounding inheritance. It was in this setting that scientists were able to reexamine Mendel's findings.

In the early twentieth century, research led to the rebirth of Mendel's work. Hybridization experiments similar to Mendel's were independently performed by three botanists, Hugo DeVries, Karl Correns, and Erich Tschermak. DeVries's work, for example, had focused on unit characters, and he demonstrated the principle of segregation in his experiments with several plant species. He had apparently searched the existing literature and found that Mendel's work had anticipated his own conclusions. Correns and Tschermak had independently reached findings similar to those of Mendel.

In 1902, two cytologists, Walter Sutton and Theodor Boveri, independently published papers linking their discoveries of the behavior of chromosomes during meiosis to the Mendelian principles of segregation and independent assortment. They pointed out that the separation of chromosomes during meiosis could serve as the cytological basis of these two postulates. Although they thought that Mendel's unit factors were probably chromosomes rather than genes on chromosomes, their findings re-

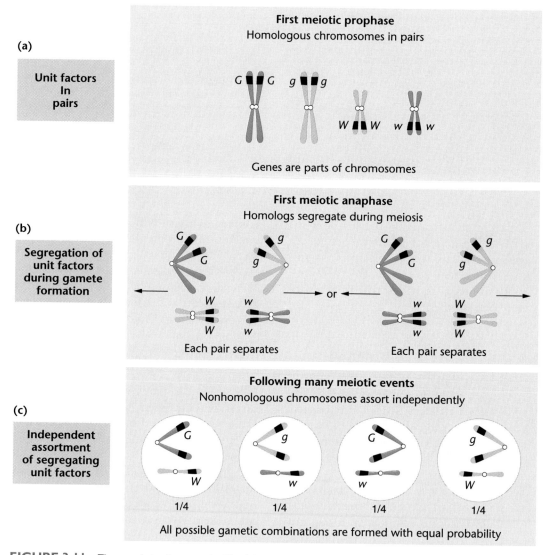

FIGURE 3.11   The correlation between the Mendelian postulates of (a) unit factors in pairs, (b) segregation, and (c) independent assortment, and the presence of genes located on homologous chromosomes and their behavior during meiosis.

established the importance of Mendel's work, which served as the foundation of ensuing genetic investigations.

Based on their studies, Sutton and Boveri are credited with initiating the **chromosomal theory of heredity**. As we will see in Chapters 4 and 5, subsequent work by Thomas H. Morgan, Alfred H. Sturtevant, Calvin Bridges, and others using fruit flies was to establish beyond a reasonable doubt that Sutton and Boveri's hypothesis was correct.

Unit Factors, Genes, and Homologous Chromosomes

Because the correlation between Sutton's and Boveri's observations and Mendelian principles is the foundation for the modern interpretation of transmission genetics, we will examine this correlation before moving to the topics of the next several chapters.

As pointed out in Chapter 2, each species possesses a specific number of chromosomes in each somatic (body) cell nucleus. For diploid organisms, this number is called the **diploid number (2n)** and is characteristic of that species. During the formation of gametes, this number is precisely halved (n), and when two gametes combine during fertilization, the diploid number is reestablished. During meiosis, the chromosome number is not reduced in a random manner, however. It was apparent to early cytologists that the diploid number of chromosomes is composed of homologous pairs identifiable by their morphological appearance and behavior. The gametes contain one member of each pair. The chromosome complement of a gamete is thus quite specific, and the number of chromosomes in each gamete is equal to the haploid number.

With this basic information, we can see the correlation between the behavior of unit factors and chromosomes and genes. Figure 3.11 shows three of Mendel's postulates and the accepted explanation of each. Unit factors are really genes located on homologous pairs of chromosomes [Figure 3.11(a)]. Members of each pair of homologs separate, or segregate, during gamete formation [Figure 3.11(b)]. Two different alignments are possible, both of which are shown.

To illustrate the principle of independent assortment, it is important to distinguish between members of any given homologous pair of chromosomes. One member of each pair is derived from the **maternal parent**, while the other comes from the **paternal parent**. We represent different parental origins by different colors. As shown in Figure 3.11(c) the two pairs of homologs segregate independently of one another during gamete formation. Each gamete receives one member from each pair. All possible combinations are formed. If we add the symbols used in Mendel's dihybrid cross (G, g and W, w) to the diagram, we see why equal numbers of the four types of gametes are formed. The independent behavior of Mendel's pairs of unit factors (G and W in this example) was due to the fact that they were on separate pairs of homologous chromosomes.

From observations of the phenotypic diversity of living organisms, we see that it is logical to assume that there are many more genes than chromosomes. Therefore, each homolog must carry genetic information for more than one trait. The currently accepted concept is that a chromosome is composed of a large number of linearly ordered, information-containing units called **genes**. Mendel's unit factors (which determine tall or dwarf stems, for example) actually constitute a pair of genes located on one pair of homologous chromosomes. The location on a given chromosome where any particular gene occurs is called its **locus** (pl. **loci**). The different forms taken by a given gene, called **alleles** (D or d), contain slightly different genetic information (tall or dwarf) that determines the same character (stem length). Alleles are alternative forms of the same gene. Although we have only discussed genes with two alternative alleles, most genes have *more* than two allelic forms. We discuss the concept of **multiple alleles** in Chapter 4.

We conclude this section by reviewing the criteria necessary to classify two chromosomes as a homologous pair:

1. During mitosis and meiosis, when chromosomes are visible as distinct figures, both members of a homologous pair are the same size and exhibit identical centromere locations.

2. During early stages of meiosis, homologous chromosomes pair together, or synapse.

3. Although not generally microscopically visible, homologs contain identical, linearly ordered, gene loci.

Independent Assortment and Genetic Variation

One of the major consequences of independent assortment is the production by an individual of genetically dissimilar gametes. Genetic variation results because the two members of any homologous pair of chromosomes are rarely, if ever, genetically identical. Because independent assortment leads to the production of all possible chromosome combinations, extensive genetic diversity results.

The number of possible gametes, each with different chromosome compositions, is 2^n, where n equals the haploid number. Thus, if a species has a haploid number of 4, then 2^4 or 16 different gamete combinations can be formed as a result of independent assortment. Although this number is not great, consider the human species, where $n = 23$. If 2^{23} is calculated, we find that in excess of 8×10^6, or over 8 million, different types of gametes are represented. Because fertilization represents an event involving only one of approximately 8×10^6 possible gametes from each of two parents, each offspring represents only one of $(8 \times 10^6)^2$, or 64×10^{12}, potential genetic combinations! It is no wonder that, except for identical twins, each member of the human species demonstrates a distinctive appearance and individuality. This number of combinations of chromosomes is far greater than the number of humans who have ever lived on earth! Genetic variation resulting from independent assortment has been extremely important to the process of organic evolution in all organisms.

Probability and Genetic Events

Genetic ratios are most properly expressed as probabilities (e.g., 3/4 tall:1/4 dwarf). These values predict the outcome of each fertilization event, such that the probability of each zygote having the genetic potential for becoming tall is 3/4, whereas the potential for becoming dwarf is 1/4. Probabilities range from 0, where an event is certain *not* to occur, to 1.0, where an event is certain *to* occur.

The Product Law and Sum Law

When two or more events occur independently of one another, but at the same time, we can calculate the probability that particular outcomes of the two events will both occur. This is accomplished by applying the **product law**. As mentioned in our earlier discussion of independent assortment, it states that the probability of two or more outcomes occurring simultaneously is equal to the *product* of their individual probabilities. Two or more events are independent of one another if the outcome of each one does not affect the outcome of any of the others under consideration.

To illustrate the use of the product law, consider the possible results of an event where you toss a penny (P) and a nickel (N) at the same time and examine all combinations of heads (H) and tails (T) that can occur. There are four possible outcomes:

$$(P_H{:}N_H) = (1/2)(1/2) = 1/4$$
$$(P_T{:}N_H) = (1/2)(1/2) = 1/4$$
$$(P_H{:}N_T) = (1/2)(1/2) = 1/4$$
$$(P_T{:}N_T) = (1/2)(1/2) = 1/4$$

The probability of obtaining a head or a tail with either coin is 1/2 and is unrelated to the outcome of the other coin. All four possible combinations are predicted to occur with equal probability.

If we were interested in calculating the probability of a generalized outcome that can be accomplished in more than one way, we would apply the **sum law** to the individual mutually exclusive outcomes that accomplish this general result. For example, we can ask: What is the probability of tossing our penny and nickel and obtaining one head and one tail? In such a case, we do not care whether it is the penny or nickel that comes up heads, provided that the other coin has the alternate outcome. As we can see above, there are two ways in which the desired outcome can be accomplished [($P_H{:}N_T$) and ($P_T{:}N_H$)], each with a probability of 1/4. Thus, according to the sum law, the overall probability is equal to

$$(1/4) + (1/4) = 1/2$$

One-half of all such tosses are predicted to yield the desired outcome.

These simple probability laws will be useful throughout our discussions of transmission genetics, and as you solve genetics problems. In fact, we have already applied the product law earlier when the forked-line method was used to calculate the phenotypic results of Mendel's dihybrid and trihybrid crosses. When we wish to know the results of a cross, we need only to calculate the probability of each possible outcome. The results of this calculation then allow us to predict the proportion of offspring expressing each phenotype or each genotype.

There is a very important point to remember when dealing with probability. Predictions of possible outcomes are usually realized only with large sample sizes. If we predict that 9/16 of the offspring of a dihybrid cross will express both dominant traits, it is unlikely that, in a small sample, that exactly 9 out of every 16 of them will do so. Instead, our prediction is that, of a large number of offspring, approximately 9/16 will express this phenotype. The deviation from the predicted ratio in small sample

sizes is attributed to deviation due to chance, a subject we deal with in our discussion of statistics in the next section. As we will see, the impact of deviation due strictly to chance is diminished as the sample size increases.

Conditional Probability

Sometimes we may wish to calculate the probability of an outcome that is dependent on a specific condition related to that outcome. For example, we might wonder what the probability is in the F_2 of Mendel's monohybrid cross involving tall and dwarf plants that a tall plant is heterozygous (and not homozygous). The "condition" we have set is to consider only tall F_2 offspring. Of any F_2 tall plant, what is the probability of it being heterozygous?

Because the outcome and specific condition are not independent, we cannot apply the product law of probability. The likelihood of such an outcome is referred to as a **conditional probability**. In its simplest terms, we are asking what is the probability that one outcome will occur, given the specific condition upon which this outcome is dependent. Let us call this probability p_c.

To solve for p_c, we must consider the probability of both the outcome of interest and that of the specific condition that includes the outcome. These are: (a) the probability of an F_2 plant being heterozygous as a result of receiving both a dominant and a recessive allele (p_a) and (b) the probability of the condition under which the event is being assessed, that is, being tall (p_b):

$$p_a = \text{probability of any } F_2 \text{ plant inheriting}$$
$$\text{one dominant and one recessive allele}$$
$$\text{(i.e., being a heterozygote)}$$
$$= 1/2$$

$$p_b = \text{probability of an } F_2 \text{ plant of a mono-}$$
$$\text{hybrid cross being tall}$$
$$= 3/4$$

To calculate the conditional probability (p_c), we divide p_a by p_b:

$$p_c = p_a/p_b$$
$$= (1/2)/(3/4)$$
$$= (1/2) \cdot (4/3)$$
$$= 4/6$$
$$p_c = 2/3$$

The conditional probability of any tall plant being heterozygous is two-thirds (2/3). Thus, on the average, two-thirds of the F_2 tall plants will be heterozygous. We can confirm this calculation by reexamining Figure 3.3.

Conditional probability has many applications in genetics. One application related to the case discussed above is utilized during genetic counseling. For example, the probability (p_c) may be calculated that an unaffected sibling of a brother or sister expressing a recessive disorder is a carrier of the disease-causing allele (i.e., a heterozygote). Assuming that both parents are unaffected (and are therefore carriers), the calculation of p_c is identical to the preceding example. The value of $p_c = 2/3$.

The Binomial Theorem

The final example of probability that we shall discuss involves cases where one of two alternative outcomes is possible during each of a number of trials. By applying the **binomial theorem**, we can rather quickly calculate the probability of any specific set of outcomes among a large number of potential events. For example, in families of any size, we can calculate the probability of any combination of male and female children: In a family of five, for instance, we can calculate the probability of having four children of one sex and one child of the other sex, and so on.

The expression of the binomial theorem is

$$(a + b)^n = 1$$

where a and b are the respective probabilities of the two alternative outcomes and n equals the number of trials. For each value of n, the binomial must be expanded:

n	Binomial	Expanded Binomial
1	$(a + b)^1$	$a + b$
2	$(a + b)^2$	$a^2 + 2ab + b^2$
3	$(a + b)^3$	$a^3 + 3a^2b + 3ab^2 + b^3$
4	$(a + b)^4$	$a^4 + 4a^3b + 6a^2b^2 + 4ab^3 + b^4$
5	$(a + b)^5$	$a^5 + 5a^4b + 10a^3b^2 + 10a^2b^3 + 5ab^4 + b^5$
		etc.

To expand any binomial, the various exponents (e.g., a^3b^2) are determined using the pattern

$$(a + b)^n = a^n, a^{n-1}b, a^{n-2}b^2, a^{n-3}b^3, \ldots, b^n$$

The numerical coefficients preceding each expression can be most easily determined using Pascal's triangle:

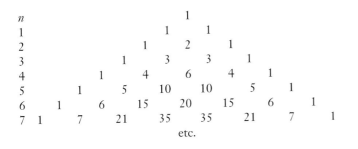

```
n
1                                 1
1                            1         1
2                        1       2       1
3                    1       3       3       1
4                1       4       6       4       1
5            1       5      10      10       5       1
6        1       6      15      20      15       6       1
7    1       7      21      35      35      21       7       1
                              etc.
```

Notice that all numbers other than the 1s are equal to the sum of the two numbers directly above them.

Using the above methods, the initial expansion of $(a + b)^7$ is

$$a^7 + 7a^6b + 21a^5b^2 + 35a^4b^3 + \cdots + b^7$$

If we apply the binomial theorem, we might ask: What is the probability that in a family of four children, two are male and two are female?

First, assign initial probabilities to each outcome:

$$a = \text{male} = 1/2$$
$$b = \text{female} = 1/2$$

Then locate the appropriate term in the expanded binomial, where $n = 4$:

$$(a + b)^4 = a^4 + 4a^3b + 6a^2b^2 + 4ab^3 + b^4$$

In each term, the exponent of a represents the number of males, and the exponent of b represents the number of females. Therefore, the correct expression of p is

$$
\begin{aligned}
p &= 6a^2b^2 \\
&= 6(1/2)^2(1/2)^2 \\
&= 6(1/2)^4 \\
&= 6(1/16) \\
&= 6/16 \\
p &= 3/8
\end{aligned}
$$

Thus, the probability of families of four children having two boys and two girls is 3/8. Of all families with four children, 3 out of 8 are predicted to have two boys and two girls.

Before examining one other example, we should note that a single formula can be applied in determining the numerical coefficient for any set of exponents:

$$\frac{n}{s!t!}$$

where n = the total number of events
  s = the number of times outcome a occurs
  t = the number of times outcome b occurs

Therefore, $n = s + t$.

The symbol ! means "factorial." For example,

$$5! = (5) \cdot (4) \cdot (3) \cdot (2) \cdot (1) = 120$$

Note that when using factorials that $0! = 1$.

Using the formula, let's determine the probability, in a family of seven, that five males and two females will occur. Thus, $s = 5$, $t = 2$, and $n = 7$. We first extend our equation to include five events of outcome a and two events of outcome b. The appropriate term is

$$p = \frac{n!}{s!t!} a^s b^t$$

$$= \frac{7!}{5!2!}(1/2)^5(1/2)^2$$

$$= \frac{(7) \cdot (6) \cdot (5) \cdot (4) \cdot (3) \cdot (2) \cdot (1)}{(5) \cdot (4) \cdot (3) \cdot (2) \cdot (1) \cdot (2) \cdot (1)} (1/2)^7$$

$$= \frac{(7) \cdot (6)}{(2) \cdot (1)} (1/2)^7$$

$$= \frac{42}{2} (1/2)^7$$

$$= 21(1/2)^7$$

$$= 21(1/128)$$

$$p = 21/128$$

Of families with seven children, on the average, 21/128 are predicted to have five males and two females.

Let's illustrate the use of the binomial theorem by examining one final example involving the autosomal recessive disorder albinism. Consider a family where both parents have normal pigmentation, but they have an albino child. This establishes that both of them are heterozygous carriers. If they have six more children, what is the probability that four will be normal (a) and two will have albinism (b)?

Using our formula, the appropriate expression is

$$p = \frac{6!}{4!2!} a^4 b^2$$

where, based on the cross of $Aa \times Aa$,

a = probability of a normal child = 3/4
b = probability of an albino child = 1/4

Therefore,

$$p = \frac{(6) \cdot (5) \cdot (4) \cdot (3) \cdot (2) \cdot (1)}{(4) \cdot (3) \cdot (2) \cdot (1) \cdot (2) \cdot (1)} (3/4)^4(1/4)^2$$

$$= \frac{(6) \cdot (5)}{(2) \cdot (1)} (81/256) \times (1/16)$$

$$= (15) \cdot (81/4096)$$

$$p = 1215/4096$$

The use of calculations involving the binomial theorem have various applications in genetics, including the analysis of polygenic traits (see Chapter 4) and in population equilibrium studies (see Chapter 24).

Evaluating Genetic Data: Chi-Square Analysis

Mendel's 3:1 monohybrid and 9:3:3:1 dihybrid ratios are hypothetical predictions based on the following assumptions: (1) each allele is dominant or recessive; (2) segregation is operative; (3) independent assortment occurs; and (4) fertilization is random. The last three assumptions are influenced by chance events and therefore are subject to random fluctuation. This concept, called **chance deviation**, is most easily illustrated by tossing a single coin numerous times and recording the number of heads and tails observed. In each toss, there is a probability of 1/2 that a head will occur and a probability of 1/2 that a tail will occur. Therefore, the expected ratio of many tosses is 1:1. If a coin were tossed 1000 times, usually *about* 500 heads and 500 tails would be observed. Any reasonable fluctuation from this hypothetical ratio (e.g., 486 heads and 514 tails) would be attributed to chance.

As the total number of tosses is reduced, the impact of chance deviation increases. For example, if a coin were tossed only 4 times, you wouldn't be too surprised if all 4 tosses resulted in only heads or only tails. But, for 1000 tosses, 1000 heads or 1000 tails would be most unexpected. In fact, you might believe that such a result would be impossible. Actually, all heads or all tails in 1000 tosses would be predicted to occur with a probability of only $(1/2)^{1000}$. Because $(1/2)^{20}$ is a probability of occurrence of less than 1 in 1 million times, an event occurring with a probability as small as $(1/2)^{1000}$ would be virtually impossible to occur.

Two major points are significant here:

1. The outcomes of segregation, independent assortment, and fertilization, like coin tossing, are subject to random fluctuations from their predicted occurrences as a result of chance deviation.

2. As the sample size increases, the average deviation from the expected fraction or ratio decreases on a proportional basis. Therefore, a larger sample size diminishes the impact of chance deviation on the final outcome.

It is important in genetics to be able to evaluate observed deviation. When we assume that data will fit a given ratio such as 1:1, 3:1, or 9:3:3:1, we establish what is called the **null hypothesis**. It is so named because the hypothesis assumes that there is no *real difference* between the **measured values** (or ratio) and the **predicted values** (or ratio). The *apparent* difference can be attributed purely to chance. Evaluation of the null hypothesis is accomplished by statistical analysis. On this basis, the null hypothesis may either: (1) be rejected, or (2) fail to be rejected. If it is rejected, the observed deviation from the expected is not attributed to chance alone. The null hypothesis and the underlying assumptions leading to it must be reexamined. If the null hypothesis fails to be rejected, any observed deviations can be attributed to chance.

Thus, statistical analysis provides a mathematical basis for examining how well observed data fit or differ from predicted or expected occurrences, testing what is called the **goodness of fit**. Assuming that the data do not "fit" exactly, just how much deviation can be allowed before the null hypothesis is rejected?

One of the simplest statistical tests devised to assess the goodness of fit of the null hypothesis is **chi-square analysis (χ^2)**. This test takes into account the observed deviation in each component of an expected ratio as well as the sample size and reduces them to a single numerical value. This value (χ^2) is then used to estimate how frequently the observed deviation, or even more deviation than was observed, can be expected to occur strictly as a result of chance. The formula used in chi-square analysis is

$$\chi^2 = \Sigma \frac{(o - e)^2}{e}$$

In this equation, o is the observed value for a given category and e is the expected value for that category. Σ (sigma) represents the "sum of" the calculated values for each category of the ratio. Because ($o - e$) is the deviation (d) in each case, the equation can be reduced to

$$\chi^2 = \Sigma \frac{d^2}{e}$$

Table 3.2(a) illustrates the step-by-step procedure necessary to make the χ^2 calculation for the F_2 results of a monohybrid cross. If you were analyzing these data, you would work from left to right, calculating and entering the appropriate numbers in each column. Regardless of whether the calculated deviation ($o - e$) is initially positive or negative, it becomes positive after the number is squared. Table 3.2(b) illustrates the analysis of the F_2 results of a hypothetical dihybrid cross. Based on your study of the calculations involved in the monohybrid cross, check to make certain that you understand how each number was calculated in the dihybrid example.

The final step in the chi-square analysis is to interpret the χ^2 value. To do so, you must initially determine the value of the **degrees of freedom (df)**, which in these analyses is equal to $n - 1$ where n is the number of different categories into which each datum point may fall. For the 3:1 ratio, plants can have one of *two* phenotypes. Thus, $n = 2$, so $df = 2 - 1 = 1$. For the 9:3:3:1 ratio, $df = 3$. Degrees of freedom must be taken into account because the greater the number of categories, the more deviation is expected as a result of chance.

TABLE 3.2   Chi-Square Analysis

(a) Monohybrid Cross					
Expected Ratio	Observed (o)	Expected (e)	Deviation ($o - e$)	Deviation^2 (d^2)	Deviation^2/ Expected (d^2/e)
3/4	740	3/4 (1000) = 750	740 − 750 = −10	$(-10)^2$ = 100	100/750 = 0.13
1/4	260	1/4 (1000) = 250	260 − 250 = +10	$(+10)^2$ = 100	100/250 = 0.40
	Total = 1000				χ^2 = 0.53
					p = 0.48

(b) Dihybrid Cross					
Expected Ratio	o	e	$o - e$	d^2	d^2/e
9/16	587	567	+20	400	0.71
3/16	197	189	+ 8	64	0.34
3/16	168	189	−21	441	2.33
1/16	56	63	− 7	49	0.78
	Total = 1008				χ^2 = 4.16
					p = 0.26

With this accomplished, the χ^2 value must now be interpreted in terms of a corresponding **probability value (p)**. Because this calculation is complex, the p value is usually located on a table or graph. Figure 3.12 shows the wide range of χ^2 and p values for numerous degrees of freedom in both forms. We will use the graph to explain how to determine the p value. The caption for Figure 3.12(b) explains how to use the table.

These simple steps must be followed to determine p:

1. Locate the χ^2 value on the abscissa (the horizontal axis).

2. Draw a vertical line from this point up to the line on the graph representing the appropriate df.

3. Extend a horizontal line from this point to the left until it intersects the ordinate (the vertical axis).

4. Estimate, by interpolation, the corresponding p value.

For our first example (the monohybrid cross) in Table 3.2, the p value of 0.48 may be estimated in this way and is illustrated in Figure 3.12(a). For the dihybrid cross, use this method to see if you can determine the p value. χ^2 is 4.16 and df equals 3. A p value of 0.26 is the approximate value. Use of the table rather than the graph confirms that both p values are between 0.20 and 0.50. Examine Table 3.2 to confirm this.

Thus far we have been concerned only with the determination of p. The most important aspect of χ^2 analysis is understanding what the p value actually means. We will use the example of the dihybrid cross ($p = 0.26$) to illustrate. In these discussions, it is simplest to think of the p value as a percentage (e.g., $0.26 = 26$ percent).

In our example, the p value indicates that, were the same experiment repeated many times, 26 percent of the trials would be expected to exhibit chance deviation as great as or greater than that seen in the initial trial. Conversely, 74 percent of the repeats would show less deviation as a result of chance than initially observed.

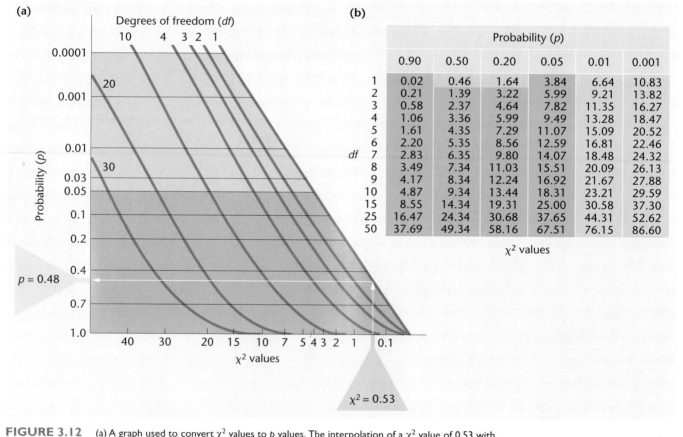

FIGURE 3.12 (a) A graph used to convert χ^2 values to p values. The interpolation of a χ^2 value of 0.53 with 1 degree of freedom (df) to an estimated probability value of 0.48 is illustrated. (b) A table showing χ^2 values for a variety of combinations of df and p. To use the table, locate the line with the appropriate df and along that line, determine the two values between which the calculated χ^2 value resides. These columns represent the correct range of the p value (shown along the top). In our example, a χ^2 value of 0.53 for a df of 1 is converted to a probability value between 0.20 and 0.50. From our graph in (a), the more precise value ($p = 0.48$) was estimated by interpolation. All values that serve to fail to reject the null hypothesis (<0.05) are shaded in both the graph and the chart.

These interpretations of the p value reveals that a hypothesis (a 9:3:3:1 ratio in this case) is never proved or disproved absolutely. Instead, a relative standard must be set to serve as the basis for either rejecting or failing to reject the hypothesis. This standard is often a probability value of 0.05. When applied to chi-square analysis, a p value less than 0.05 means the probability is only 5% or less that the observed deviation in the set of results could be obtained by chance alone. Such a p value indicates that the difference between the observed and predicted results is substantial and thus serves as the basis for rejecting the null hypothesis.

On the other hand, p values of 0.05 or greater $(1.0 - 0.05)$ indicate that the probability of the observed deviation being due to chance is 5% or more. The conclusion is not to reject the null hypothesis. In our example where $p = 0.26$, the hypothesis of independent assortment is not rejected by the experimental data. That is, the data do not provide any reason to reject the hypothesis. Therefore, the observed deviation can be reasonably attributed to chance.

A final note is relevant here concerning the case where the null hypothesis is rejected, i.e., $p = <0.05$. Suppose, for example, the null hypothesis being tested was that the data were representative of independent assortment, culminating in a 9:3:3:1 ratio, and that it was rejected. What is the alternative interpretation of the data? The researcher must first reassess the many assumptions that underlie the null hypothesis. In addition to the assumption that segregation and independent assortment are operable as Mendel described, several other assumptions have been made. First, we are assuming that fertilization is random, and that the viability of all gametes is equal irrespective of genotype. That is, we assume that all gametes are equally likely to participate in fertilization. Then, following fertilization, the assumption is that all preadult stages and adult offspring are equally viable, regardless of their genotype.

For example, if one of the mutations in a dihybrid cross resulted in a wingless fruit fly that is predicted to appear 3/16 of the time, that proportion of mutant zygotes may, in fact, occur at fertilization. However, these mutants may not survive as well during development or as young adults compared to flies whose genotype results in wings. As a result, when the data are gathered, fewer flies without wings are recovered than actually occurred at fertilization. Rejection of the null hypothesis would then not be cause for us to reject the notion that segregation and independent assortment are valid postulates.

The above discussion serves to point out that the assessment of statistical information must be made carefully on a case-by-case basis. When a null hypothesis is rejected, all underlying assumptions must be examined. If there is no concern about their validity, then other alternative hypotheses must be considered to explain the results.

Human Pedigrees

In all crosses discussed so far, one of the two traits for each character has been dominant to the other. Based on this observation, two significant questions can be asked:

1. Does the expression of all genes occur in this fashion?

2. Is it possible to ascertain the mode of inheritance of genes in organisms where designed crosses and the production of large numbers of offspring are not practical?

The answer to the first question is no. As we will see in Chapter 4, many modes of genetic expression exist that modify the monohybrid and dihybrid ratios observed by Mendel.

The answer to the second question is yes. Even in humans the pattern of inheritance of a specific phenotype can be studied.

The simplest way to study this pattern is to construct a family tree indicating the phenotype of the trait in question for each member. Such a family tree is called a **pedigree**. By analyzing the pedigree, we may be able to predict how the gene controlling the trait is inherited. If many similar pedigrees for the same trait are found, the prediction is strengthened.

Figure 3.13 shows the conventions used in constructing a pedigree. Circles represent females, and squares designate males. If the sex is unknown, a diamond may be used (II-2). If a pedigree traces only a single trait, as Figure 3.13 does, the circles, squares, and diamonds are shaded if the phenotype being considered is expressed. Those who fail to express a recessive trait, when known with certainty to be heterozygous, have only the left half of their square or circle shaded (see II-3 and II-4).

The parents are connected by a horizontal line, and vertical lines lead to their offspring. All such offspring are called **sibs** and are connected by a horizontal **sibship line**. Sibs are placed from left to right according to birth order and are labeled with Arabic numerals. Each generation is indicated by a Roman numeral.

Twins are indicated by diagonal lines stemming from a vertical line connected to the sibship line. For **monozygotic** or **identical twins**, the diagonal lines are linked by a horizontal line [see III-5,6 in Figure 3.13(a)]. **Dizygotic** or **fraternal twins** lack this connecting line [see III-8,9 in Figure 3.13(a)]. A number within one of the symbols [see II-10–13) in Figure 3.13(b)] represents numerous sibs of the same or unknown phenotypes. A male whose phenotype drew the attention of a physician or geneticist is called the **propositus**, while a female in the same circumstance is called a **proposita**. Such an individual is indicated by an arrow [see III-4) in Figure 3.13(a)].

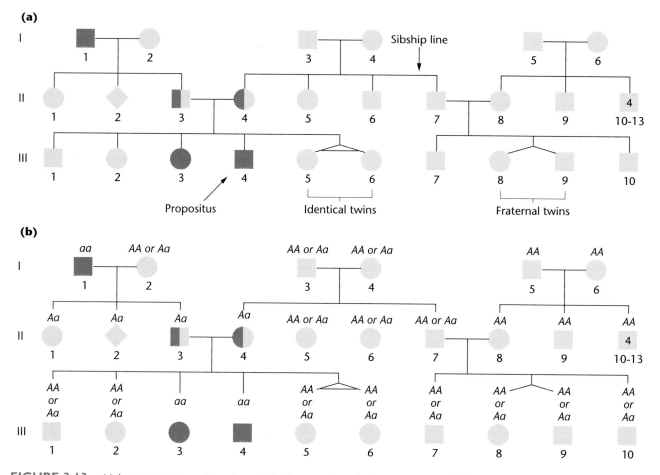

FIGURE 3.13   (a) A representative pedigree for a single character through three generations. (b) The most probable genotypes of each individual in the pedigree.

The pedigree shown in Figure 3.13 traces the pattern of inheritance of the human trait **albinism**. By analyzing the pedigree, we will see that albinism is inherited as a recessive trait.

One of the parents of the first generation, I-1, is affected. Because none of his offspring show the disorder, we might conclude that the unaffected female parent (I-2) was a homozygous normal individual. Had she been heterozygous, one-half of the offspring would be expected to exhibit albinism. However, such a small sample (three offspring) prevents any certainty in the matter.

An unaffected second generation is characteristic of a rare recessive trait. If albinism were inherited as a dominant trait, individual II-3 would have to express the disorder in order to pass it to his offspring (III-3 and III-4). He does not. Inspection of the offspring constituting the third generation [row III in Figure 3.13(a)] provides further support for the hypothesis that albinism is a recessive trait. If so, parents II-3 and II-4 are both heterozygous, and approximately one-fourth of their offspring should be affected. Two of the six offspring do show albinism. This deviation

from the expected ratio is not unexpected in crosses with few offspring.

Based on this pedigree analysis and the conclusion that albinism is a recessive trait, the most probable genotypes of all individuals can be predicted. For both the first and second generations, this can be done with some certainty in only a few cases. In the third generation, for most normal individuals, we cannot determine whether they are homozygous or heterozygous.

Pedigree analysis of many traits has been an extremely valuable research technique in human genetic studies. However, this approach does not usually provide the certainty in drawing conclusions that is afforded by designed crosses yielding large numbers of offspring. Nevertheless, when many independent pedigrees of the same trait or disorder are analyzed, consistent conclusions can often be drawn. Table 3.3 lists numerous human traits and classifies them according to their recessive or dominant expression. As we will see in Chapter 4, the genes controlling some of these traits are located on the sex-determining chromosomes.

TABLE 3.3   Representative Recessive and Dominant Human Traits

Recessive Traits	Dominant Traits
Albinism	Achondroplasia
Alkaptonuria	Brachydactyly
Ataxia telangiectasia	Congenital stationary night blindness
Color blindness	Ehler–Danlos syndrome
Cystic fibrosis	Fascio–scapulo–humeral muscular dystrophy
Duchenne muscular dystrophy	Huntington disease
Galactosemia	Hypercholesterolemia
Hemophilia	Marfan syndrome
Lesch–Nyhan syndrome	Neurofibromatosis
Phenylketonuria	Phenylthiocarbamide tasting (PTC)
Sickle–cell anemia	Porphyria
Tay–Sachs disease	Widow's peak

CHAPTER SUMMARY

1. Over a century ago, Gregor Mendel studied inheritance patterns in the garden pea, establishing the principles of transmission genetics.

2. Mendel's postulates help describe the basis for the inheritance of phenotypic expression. He showed that unit factors, later called alleles, exist in pairs and exhibit a dominant/recessive relationship in determining the expression of traits.

3. Mendel postulated that unit factors must segregate during gamete formation, such that each receives only one of the two factors with equal probability.

4. Mendel's final postulate of independent assortment states that each pair of unit factors segregates independently of other such pairs. As a result, all possible combinations of gametes will be formed with equal probability.

5. The discovery of chromosomes in the late 1800s and subsequent studies of their behavior during meiosis led to the rebirth of Mendel's work, linking the behavior of his unit factors with that of chromosomes during meiosis.

6. The Punnett square and the forked-line methods are used to predict the probabilities of phenotypes (and genotypes) from crosses involving two or more gene pairs.

7. Genetic ratios are expressed as probabilities. Thus, deriving outcomes of genetic crosses relies on an understanding of the laws of probability. The sum law, the product law, conditional probability, and the use of the binomial theorem have been described.

8. Statistical analysis is used to test the validity of experimental outcomes. In genetics, variations from the expected ratios are anticipated owing to chance deviation.

9. Chi-square analysis allows us to predict the probability of these variations being generated by chance alone. Calculating χ^2 provides the basis for assessing the null hypothesis, namely that there is no real difference between the expected and observed values.

10. Pedigree analysis provides a method for studying the inheritance pattern of human traits over several generations. This often provides the basis for determining the mode of inheritance of human characteristics and disorders.

KEY TERMS

albinism, 70
allele, 54
binomial theorem, 65
branch diagram, 59

carrier, 65
chance deviation, 66
chi-square (χ^2) analysis, 67
chromosomal theory of heredity, 62

conditional probability, 64
continuous variation, 61
degrees of freedom (*df*), 67
dihybrid cross, 56

INSIGHTS AND SOLUTIONS

Students demonstrate their knowledge of transmission genetics by solving genetics problems. Success at this task represents not only comprehension of theory but also its application to more practical genetic situations. Most students find problem solving in genetics to be challenging but rewarding. This section is designed to provide basic insights into the reasoning essential to this process.

Genetics problems are in many ways similar to algebraic word problems. The approach taken should be identical: (1) analyze the problem carefully; (2) translate words into symbols, defining each one first; and (3) choose and apply a specific technique to solve the problem. The first two steps are the most critical. The third step is largely mechanical.

The simplest problems are those that state all necessary information about the P_1 generation and ask you to find the expected ratios of the F_1 and F_2 genotypes and/or phenotypes. The following steps should always be followed when you encounter this type of problem:

1. Determine insofar as possible the genotypes of the individuals in the P_1 generation.

2. Determine what gametes may be formed by the P_1 parents.

3. Recombine gametes either by the Punnett square method, by the forked-line method, or, if the situation is very simple, by inspection. Read the F_1 phenotypes directly.

4. Repeat the process to obtain information about the F_2 generation.

Determining the genotypes from the given information requires an understanding of the basic theory of transmission genetics. For example, consider the following problem:

A recessive mutant allele, *black*, causes a very dark body in *Drosophila* when homozygous. The normal, wild-type color is described as gray. What F_1 phenotypic ratio is predicted when a black female is crossed to a gray male whose father was black?

To work this problem, you must understand dominance and recessiveness as well as the principle of segregation. Further, you must use the information about the male parent's father. You can work out the problem as follows:

1. Because the female parent is black, she must be homozygous for the mutant allele (*bb*).

2. The male parent is gray; therefore, he must have at least one dominant allele (*B*). Because his father was black (*bb*) and he received one of the chromosomes bearing these alleles, the male parent must be heterozygous (*Bb*).

From here, the problem is simple:

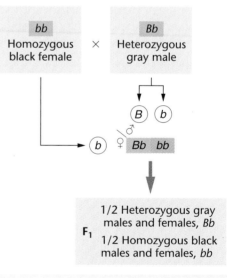

Apply this approach to the following problems.

1. In Mendel's work using peas, he found that full pods are dominant to constricted pods, whereas round seeds are dominant to wrinkled seeds. One of his crosses was between full, round plants and constricted, wrinkled

plants. From this cross, he obtained an F_1 that was all full and round. In the F_2, Mendel obtained his classic 9:3:3:1 ratio. Using the above information, determine the expected F_1 and F_2 results of a cross between homozygous constricted, round and full, wrinkled plants.

Solution: First, define gene symbols for each pair of contrasting traits. Select the lowercase forms of the first letter of the recessive traits to designate those traits, and use the uppercase forms to designate the dominant traits. For example, use C and c to indicate full and constricted, and use W and w to indicate the round and wrinkled phenotypes, respectively.

Now, determine the genotypes of the P_1 generation, form gametes, reconstitute the F_1 generation, and read off the phenotype(s):

P_1:   *ccWW*   ×   *CCww*
        constricted, round   full, wrinkled
              ↓   ↓
Gametes: *cW*   *Cw*
F_1:   *CcWw*
                    full, round

You can see immediately that the F_1 generation expresses both dominant phenotypes and is heterozygous for both gene pairs. We can expect that the F_2 generation will yield the classic Mendelian ratio of 9:3:3:1. Let's work it out anyway, just to confirm this, using the forked-line method. Because both gene pairs are heterozygous and can be expected to assort independently, we can predict the F_2 outcomes from each gene pair separately and then proceed with the forked-line method.

Every F_2 offspring is subject to the following probabilities:

$Cc \times Cc$ $Ww \times Ww$
      ↓   ↓
    CC WW
    Cc } 3/4 full Ww } 3/4 round
    cC wW
    cc 1/4 constricted ww 1/4 wrinkled

The forked-line method then allows us to confirm the 9:3:3:1 phenotypic ratio. Remember that this represents proportions of 9/16:3/16:3/16:1/16. Note that we are applying the product law as we compute the final probabilities:

3/4 full ⎰ 3/4 round $\xrightarrow{(3/4)(3/4)}$ 9/16 full, round
         ⎱ 1/4 wrinkled $\xrightarrow{(3/4)(1/4)}$ 3/16 full, wrinkled

1/4 constricted ⎰ 3/4 round $\xrightarrow{(1/4)(3/4)}$ 3/16 constricted, round
                ⎱ 1/4 wrinkled $\xrightarrow{(1/4)(1/4)}$ 1/16 constricted, wrinkled

2. Determine the probability that a plant of genotype *CcWw* will be produced from parental plants of the genotypes *CcWw* and *Ccww*.

Solution: Because the two gene pairs independently assort during gamete formation, we need only calculate the individual probabilities of the two separate events (*Cc* and *Ww*) and apply the product law to calculate the final probability:

$$Cc \times Cc \rightarrow 1/4\ CC : 1/2\ Cc : 1/4\ cc$$
$$Ww \times ww \rightarrow 1/2\ Ww : 1/2\ ww$$

$$p = (1/2\ Cc)(1/2\ Ww) = 1/4\ CcWw$$

3. In another cross, involving parent plants of unknown genotype and phenotype, the offspring shown below were obtained. Determine the genotypes and phenotypes of the parents.

Offspring: 3/8 full, round
              3/8 full, wrinkled
              1/8 constricted, round
              1/8 constricted, wrinkled

Solution: This problem is more difficult and requires keener insights since you must work backwards. The best approach is to consider the outcomes of pod shape separately from those of seed texture.

Of all plants, 6/8 (3/4) are full and 2/8 (1/4) are constricted. Of the various genotypic combinations that can serve as parents, which will give rise to a ratio of 3/4:1/4? Because this ratio is identical to Mendel's monohybrid F_2 results, we can propose that both unknown parents share the same genetic characteristic as the monohybrid F_1 parents: They must both be heterozygous for the genes controlling pod shape and thus are

Cc

Before accepting this hypothesis, let's consider the possible genotypic combinations that control seed texture. If we consider this characteristic alone, we can see that the traits are expressed in a ratio of 4/8 (1/2) round:4/8 (1/2) wrinkled. In order to generate such a ratio, the parents *cannot* both be heterozygous or their offspring would yield a 3/4:1/4 phenotypic ratio. They *cannot* both be homozygous or all offspring would express a single phenotype. Thus, we are left with testing the hypothesis that one parent is homozygous and one is heterozygous for the alleles controlling texture. The potential case of $WW \times Ww$ will not work, for it also would yield only a single phenotype. This leaves us with the potential case of $Ww \times ww$. Offspring in such a mating will yield 1/2 *Ww* (round):1/2 *ww* (wrinkled), exactly the outcome we are seeking.

Now, let's combine our hypotheses and predict the outcome of crossing. In our solution, we will use a "−"

to indicate that the second allele may be either dominant or recessive, since we are only predicting phenotypes.

$$CcWw \times Ccww$$

As we can see, this cross produces offspring in accordance with the information provided, solving the problem. Note that in this solution we have used *genotypes* in the forked-line method, in contrast to the use of *phenotypes* in the solution to the first problem.

4. In the laboratory, a genetics student crossed flies with normal long wings to flies with mutant *dumpy* wings, which she believed was a recessive trait. In the F_1, all flies had long wings. In the F_2, the following results were obtained:

792 long-winged flies
208 dumpy-winged flies

The student tested the hypothesis that the *dumpy* wing is inherited as a recessive trait by performing χ^2 analysis of the F_2 data.

(a) What ratio was hypothesized?

(b) Did the χ^2 analysis support the hypothesis?

(c) What do the data suggest about the *dumpy* mutation?

Solution:

(a) The student hypothesized that the F_2 data (792:208) fit Mendel's 3:1 monohybrid ratio for recessive genes.

(b) The initial step in χ^2 analysis is to calculate the expected results (e) if the ratio is 3:1 and the deviations (d) between the expected and observed data:

Ratio	o	e	d	d^2	d^2/e
3/4	792	750	42	1764	2.35
1/4	208	250	−42	1764	7.06

Total = 1000

$$\chi^2 = \sum \frac{d^2}{e}$$

$$= 2.35 + 7.06$$

$$= 9.41$$

Consulting Figure 3.12 allows us to determine the probability (p). This value will let us determine whether the deviations from the null hypothesis can be attributed to chance. There are two possible outcomes (n), so the degrees of freedom (df) = $n - 1$ or 1. The table in Figure 3.12 shows that $p = 0.01$ to 0.001. The graph gives an estimate of about 0.001. That p is less than 0.05 causes us to reject the null hypothesis. The data do not statistically fit a 3:1 ratio.

(c) When we accept Mendel's 3:1 ratio as a valid expression of the monohybrid cross, numerous assumptions are made. One of these may explain why the null hypothesis was rejected. We must assume *that all genotypes are equally viable*. That is, at the time the data are collected, genotypes yielding long wings are equally likely to survive from fertilization through adulthood as the genotype yielding *dumpy* wings. Further study would reveal that *dumpy* flies are somewhat less viable than normal flies. As a result, we would expect *less* than 1/4 of the total offspring to express dumpy. This observation is borne out in the data, although we have not proved this.

5. If two parents, both heterozygous carriers of the autosomal recessive gene causing cystic fibrosis, have five children, what is the probability that three will be normal?

Solution: First, the probability of having a normal child during each pregnancy is

$$p_a = \text{normal} = 3/4$$

while the probability of having an afflicted offspring is

$$p_b = \text{afflicted} = 1/4$$

Then apply the formula

$$\frac{n!}{s!t!} \, a^s b^t$$

where $n = 5$, $s = 3$, and $t = 2$:

$$p = \frac{(5) \cdot (4) \cdot (3) \cdot (2) \cdot (1)}{(3) \cdot (2) \cdot (1) \cdot (2) \cdot (1)} (3/4)^3 (1/4)^2$$

$$= \frac{(5) \cdot (4)}{(2) \cdot (1)} (3/4)^3 (1/4)^2$$

$$= 10(27/64) \cdot (1/16)$$

$$= 10(27/1024)$$

$$= 270/1024$$

$$p = {\sim}0.26$$

PROBLEMS AND DISCUSSION QUESTIONS

When working genetics problems in this and succeeding chapters, always assume that members of the P_1 generation are homozygous, unless the information given indicates or requires otherwise.

1. In a cross between a black and a white guinea pig, all members of the F_1 generation are black. The F_2 generation is made up of approximately 3/4 black and 1/4 white guinea pigs.
   (a) Diagram this cross, showing the genotypes and phenotypes.
   (b) What will the offspring be like if two F_2 white guinea pigs are mated?
   (c) Two different matings were made between black members of the F_2 generation with the results shown below. Diagram each of the crosses.

Cross	Offspring
Cross 1	All black
Cross 2	3/4 black, 1/4 white

2. Albinism in humans is inherited as a simple recessive trait. For the following families, determine the genotypes of the parents and offspring. When two alternative genotypes are possible, list both.
   (a) Two normal parents have five children, four normal and one albino.
   (b) A normal male and an albino female have six children, all normal.
   (c) A normal male and an albino female have six children, three normal and three albino.
   (d) Construct a pedigree of the families in (b) and (c). Assume that one of the normal children in (b) marries one of the albino children in (c) and that they have eight children.
3. Which of Mendel's postulates are illustrated by the pedigree in Problem 2? List and define these postulates.
4. Discuss the rationale relating Mendel's monohybrid results to his postulates.
5. What advantages were provided by Mendel's choice of the garden pea in his experiments?
6. Pigeons may exhibit a checkered or plain pattern. In a series of controlled matings, the following data were obtained:

P_1 Cross	F_1 Progeny	
	Checkered	Plain
(a) checkered × checkered	36	0
(b) checkered × plain	38	0
(c) plain × plain	0	35

Then, F_1 offspring were selectively mated with the following results. The P_1 cross giving rise to each F_1 pigeon is indicated in parentheses.

F_1 × F_1 Crosses	Progeny	
	Checkered	Plain
(d) checkered (a) × plain (c)	34	0
(e) checkered (b) × plain (c)	17	14
(f) checkered (b) × checkered (b)	28	9
(g) checkered (a) × checkered (b)	39	0

How are the checkered and plain patterns inherited? Select and define symbols for the genes involved and determine the genotypes of the parents and offspring in each cross.
7. Mendel crossed peas having round seeds and yellow cotyledons with peas having wrinkled seeds and green cotyledons. All the F_1 plants had round seeds with yellow cotyledons. Diagram this cross through the F_2 generation using both the Punnett square and forked-line, or branch diagram, methods.
8. Based on the above cross, in the F_2 generation, what is the probability that an organism will have round seeds and green cotyledons *and* be true-breeding?
9. Based on the same characters and traits as in Problem 7, determine the genotypes of the parental plants involved in the crosses shown below by analyzing the phenotypes of their offspring.

Parental Plants	Offspring
(a) round, yellow × round, yellow	3/4 round, yellow 1/4 wrinkled, yellow
(b) wrinkled, yellow × round, yellow	6/16 wrinkled, yellow 2/16 wrinkled, green 6/16 round, yellow 2/16 round, green
(c) round, yellow × round, yellow	9/16 round, yellow 3/16 round, green 3/16 wrinkled, yellow 1/16 wrinkled, green
(d) round, yellow × wrinkled, green	1/4 round, yellow 1/4 round, green 1/4 wrinkled, yellow 1/4 wrinkled, green

10. Which of the crosses in Problem 9 is a test cross?
11. Which of Mendel's postulates can only be demonstrated in crosses involving at least two pairs of traits? Define it.
12. Correlate Mendel's four postulates with what is now known about homologous chromosomes, genes, alleles, and the process of meiosis.
13. What is the basis for homology among chromosomes?
14. Distinguish between homozygosity and heterozygosity.
15. In *Drosophila*, *gray* body color is dominant to *ebony* body color, while *long* wings are dominant to *vestigial* wings. Work the following crosses through the F_2 generation and determine the genotypic and phenotypic ratios for each generation. Assume the P_1 individuals are homozygous.

(a) gray, long × ebony, vestigial
(b) gray, vestigial × ebony, long
(c) gray, long × gray, vestigial

16. How many different types of gametes can be formed by individuals of the following genotypes: (a) *AaBb*, (b) *AaBB*, (c) *AaBbCc*, (d) *AaBBcc*, (e) *AaBbcc*, and (f) *AaBbCcDdEe*? What are they in each case?

17. Using the forked-line, or branch diagram, method, determine the genotypic and phenotypic ratios of the trihybrid crosses (a) *AaBbCc* × *AaBBCC*, (b) *AaBBCc* × *aaBBCc*, and (c) *AaBbCc* × *AaBbCc*.

18. Mendel crossed peas with green seeds with those of yellow seeds. The F_1 generation produced only yellow seeds. In the F_2, the progeny consisted of 6022 plants with yellow seeds and 2001 plants with green seeds. Of the F_2 yellow-seeded plants, 519 were self-fertilized with the following results: 166 bred true for yellow and 353 produced a 3:1 ratio of yellow:green. Explain these results by diagramming the crosses.

19. In a study of black and white guinea pigs, 100 black animals were crossed individually to white animals and each cross was carried to an F_2 generation. In 94 of the cases, the F_1 individuals were all black, and an F_2 ratio of 3 black:1 white was obtained. In the other 6 cases, half of the F_1 animals were black and the other half were white. Why? Predict the results of crossing the black and white F_1 guinea pigs from the 6 exceptional cases.

20. Mendel crossed peas with round, green seeds to ones with wrinkled, yellow seeds. All F_1 plants had seeds that were round and yellow. Predict the results of test-crossing these F_1 plants.

21. Thalassemia is an inherited anemic disorder in humans. Individuals can be completely normal, they can exhibit a "minor" anemia, or they can exhibit a "major" anemia. Assuming that only a single gene pair and two alleles are involved in the inheritance of these conditions, which phenotype is recessive?

22. Below are shown F_2 results of two of Mendel's monohybrid crosses. State a null hypothesis to be tested using χ^2 analysis. Calculate the χ^2 value and determine the p value for both. Interpret the p values. Can the deviation in each case be attributed to chance or not? Which of the two crosses shows a greater amount of deviation?

(a)	Full pods	882
	Constricted pods	299
(b)	Violet flowers	705
	White flowers	224

23. In one of Mendel's dihybrid crosses, he observed 315 round yellow, 108 round green, 101 wrinkled yellow, and 32 wrinkled green F_2 plants. Analyze these data using the chi-square test to see if
(a) they fit a 9:3:3:1 ratio.
(b) the round:wrinkled data fit a 3:1 ratio.
(c) the yellow:green data fit a 3:1 ratio.

24. A geneticist, in assessing data that fell into two phenotypic classes, observed values of 250:150. She decided to perform chi-square analysis using two different null hypotheses: (a) the data fit a 3:1 ratio; and (b) the data fit a 1:1 ratio. Calculate the χ^2 values for each hypothesis. What can be concluded about each hypothesis?

25. The basis for rejection of any null hypothesis is arbitrary. The researcher can set more or less stringent standards by deciding to raise or lower the p value used to reject or fail to reject the hypothesis. In the case of chi-square analysis of genetic crosses, would the use of a standard of $p = 0.10$ be more or less stringent in failing to reject the null hypothesis? Explain.

26. For the following pedigree, predict the mode of inheritance and the most probable genotypes of each individual. Assume that the alleles *A* and *a* control the expression of the trait.

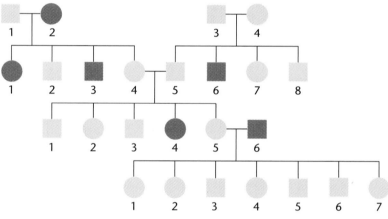

27. The following pedigree is for myopia (near-sightedness) in humans. Predict whether the disorder is inherited as the result of a dominant or recessive trait. Determine the most probable genotype for each individual based on your prediction.

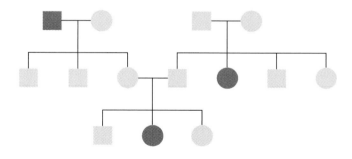

28. Draw all possible conclusions concerning the mode of inheritance of the trait denoted in each of the following limited pedigrees. (Each case is based on a different trait.)

(a)

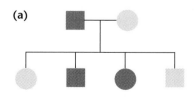

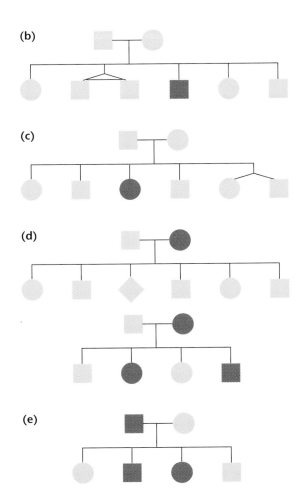

(b)

(c)

(d)

(e)

29. Consider three independently assorting gene pairs, *A/a*, *B/b*, and *C/c*, where each demonstrates typical dominance (*A–*, *B–*, *C–* and recessiveness (*aa*, *bb*, *cc*). What is the probability of obtaining an offspring that is *AABbCc* from parents that are *AaBbCC* and *AABbCc?*

30. What is the probability of obtaining a triply recessive individual from the parents shown in Problem 29?

31. Of all offspring of the parents in Problem 29, what proportion will express all three dominant traits?

32. When a die (one of a pair of dice) is rolled, it has an equal probability of landing on any of its six sides.
(a) What is the probability of rolling a 3 with a single throw?
(b) When a die is rolled twice, what is the probability that the first throw will be a 3 and the second will be a 6?
(c) If two dice are rolled together, what is the probability that one will be a 3 and the other will be a 6?
(d) If one die is rolled and it comes up as an odd number, what is the probability that it is a 5?

33. Consider the F_2 offspring of Mendel's dihybrid cross. Determine the conditional probability that F_2 plants expressing both dominant traits are heterozygous at both loci.

34. Cystic fibrosis is an autosomal recessive disorder. A male whose brother has the disease marries a female whose sister has the disease. It is not known if either the male or the female is a carrier. If the male and female have one child, what is the probability that the child will have cystic fibrosis?

35. In a family of five children, what is the probability that
(a) All are males?
(b) Three are males and two are females?
(c) Two are males and three are females?
(d) All are the same sex?
Assume that the probability of a male child is equal to the probability of a female child ($p = 1/2$).

36. In a family of eight children, where both parents are heterozygous for albinism, what mathematical expression predicts the probability that six are normal and two are albinos?

EXTRA-SPICY PROBLEMS

37. Two true-breeding pea plants were crossed. One parent is round, terminal, violet, constricted, while the other expresses the contrasting phenotypes of wrinkled, axial, white, full, respectively. The four pairs of contrasting traits are controlled by four genes, each located on a separate chromosome. In the F_1, only round, axial, violet, and full were expressed. In the F_2, all possible combinations of these traits were expressed in ratios consistent with Mendelian inheritance.
(a) What conclusion about the inheritance of the traits can be drawn based on the F_1 results?
(b) In the F_2 results, which phenotype appeared most frequently? Write a mathematical expression that predicts the probability of occurrence of this phenotype.
(c) Which F_2 phenotype is expected to occur least frequently? Write a mathematical expression that predicts this probability.
(d) In the F_2 generation, how often is either of the P_1 phenotypes likely to occur?
(e) If the F_1 plants were test-crossed, how many different phenotypes would be produced? How does this number compare to the number of different phenotypes in the F_2 generation discussed above?

38. Tay-Sachs disease (TSD) is an inborn error of metabolism that results in death by the age of two. You are a genetic counselor and one day you interview a phenotypically normal couple, where the male had a female first cousin (on his father's side) who died from TSD, and where the female had a maternal uncle with TSD. There are no other known cases in either of the families and none of the matings were/are between related individuals. Assume that this trait is very rare.
(a) Draw a pedigree of the families of this couple, showing the relevant individuals.
(b) Calculate the probability that both the male and female are carriers for TSD.
(c) What is the probability that neither of them are carriers?
(d) What is the probability that one of them is a carrier and the other is not? (*Hint:* The *p* values in (b), (c), and (d) should equal 1.)

39. The wild type (normal) fruit fly, *Drosophila melanogaster*, has straight wings and long bristles. Mutant strains have been isolated that have either curled wings or shaven bristles. The genes representing these two mutant traits are located on separate autosomes. Carefully examine the data from the five crosses below. (a) For each mutation, determine whether it is dominant or recessive. In each case, identify which crosses support your answer; and (b) define gene symbols, and for each cross, determine the genotypes of the parents.

40. What do you suppose the following mathematical expression applies to? Can you think of a genetic example where it might have application?

$$\frac{n!}{s!t!u!} (a^s b^t c^u)$$

Cross			Number of Progeny			
			straight wings, long bristles	straight wings, short bristles	curled wings, long bristles	curled wings, short bristles
#1	straight, short	straight, short	30	90	10	30
#2	straight, long	straight, long	120	0	40	0
#3	curled, long	straight, short	40	40	40	40
#4	straight, short	straight, short	40	120	0	0
#5	curled, short	straight, short	20	60	20	60

SELECTED READINGS

CUMMINGS, M. R. 1994. *Human heredity: Principles and issues,* 3rd ed. St. Paul, MN: West Publishing.

DUNN, L. C. 1965. *A short history of genetics.* New York: McGraw-Hill.

OLBY, R. C. 1985. *Origins of Mendelism,* 2nd ed. London: Constable.

PETERS, J., ed. 1959. *Classic papers in genetics.* Englewood Cliffs, NJ: Prentice-Hall.

SNEDECOR, G. W., and COCHRAN, W. G. 1980. *Statistical methods,* 7th ed. Ames: Iowa State University Press.

SOKAL, R. R., and ROHLF, F. J. 1987. *Introduction to biostatistics,* 2nd ed. New York: W. H. Freeman.

SOUDEK, D. 1984. Gregor Mendel and the people around him. *Am. J. Hum. Genet.* 36:495–98.

STERN, C., 1950. *The birth of genetics.* (Supplement to *Genetics* 35.)

STERN, C., and SHERWOOD, E. 1966. *The origins of genetics: A Mendel source book.* San Francisco: W. H. Freeman.

STUBBE, H. 1972. *History of genetics: From prehistoric times to the rediscovery of Mendel's laws.* Cambridge, MA: MIT Press.

STURTEVANT, A. H. 1965. *A history of genetics.* New York: Harper & Row.

VOELLER, B. R., ed. 1968. *The chromosome theory of inheritance: Classical papers in development and heredity.* New York: Appleton-Century-Crofts.

4

Modification of Mendelian Ratios

CHAPTER CONCEPTS

Specific phenotypes are often controlled by one or more gene pairs whose alleles exhibit modes of expression other than dominance and recessiveness. While classical Mendelian ratios are thus modified, the Mendelian principles of segregation and independent assortment are nevertheless operative during the distribution of the alleles into gametes.

In Chapter 3, we discussed the simplest principles of transmission genetics. We saw that genes are present on homologous chromosomes and that these chromosomes **segregate** from each other and **independently assort** with other segregating chromosomes during gamete formation. These two postulates are the fundamental principles of gene transmission from parent to offspring. However, when an offspring has received the total set of genes, it is the expression of genes that determines the organism's phenotype. When gene expression does not adhere to a simple dominant/recessive mode, or when more than one pair of genes influences the expression of a single character, the classic 3:1 and 9:3:3:1 F_2 ratios are usually modified. Although in this and the next several chapters we

A mutant *Drosophila* displaying white eyes.

consider more complex modes of inheritance, the fundamental principles set down by Mendel still hold true in these situations as well.

In this chapter, our discussion will initially be restricted to the inheritance of traits that are under the control of only one set of genes. In diploid organisms, where homologous pairs of chromosomes exist, two copies of each gene influence such traits. The copies need not be identical since alternative forms of genes, or **alleles**, occur within populations. How alleles act to influence a given phenotype will be our major consideration. Then we will proceed to consider how a single phenotype may be controlled by more than one gene. This general phenomenon is referred to as **gene interaction**, indicating that phenotypes are frequently under the influence of more than one gene pair. Numerous examples will be presented to illustrate a variety of heritable patterns observed in such situations.

We will conclude the chapter by examining cases where genes are present on the X chromosome illustrating **X-linkage**. Up to that point in the text, we will have restricted our discussion to chromosomes other than the X and Y pair. All such chromosomes are called **autosomes** to distinguish them from the X and Y chromosomes. As we will see, X-linkage provides yet another modification of Mendelian ratios.

Potential Function of an Allele

Following the rediscovery of Mendel's work in the early 1900s, research focused on the many ways in which genes can influence an individual's phenotype. This course of investigation, stemming from Mendel's findings, is called **neo-Mendelian genetics** (*neo* from the Greek word meaning *since* or *new*).

Each type of inheritance described in this chapter was investigated when observations of genetic data did not precisely conform to the expected Mendelian ratios. Hypotheses that modified and extended the Mendelian principles were proposed and tested with specifically designed crosses. Explanations for these observations were in accordance with the principle that a phenotype is under the control of one or more genes located at specific loci on one or more pairs of homologous chromosomes. If we adhere to the principles of segregation and independent assortment, we can predict accurately the transmission of any number of allele pairs.

To understand the various modes of inheritance, we must first examine the potential function of an allele. Alleles are alternative forms of the same gene. Therefore, they contain modified genetic information and often specify an altered gene product. For example, in human populations, there are many known alleles of the gene that encodes the β-chain of human hemoglobin. All such alleles store information necessary for the synthesis of the β-chain polypeptide, but each allele specifies a modification of the chemical structure of the β-chain. Once manufactured, however, the product of an allele may or may not have its function altered.

The allele that occurs most frequently in a population, or the one that is arbitrarily designated as normal, is often referred to as the **wild-type allele**. This common allele is usually dominant, such as the allele for tall plants in the garden pea, and its product is functional in the cell. Wild-type alleles are responsible, of course, for the corresponding wild-type phenotype and serve as standards for comparison against all mutations occurring at a particular locus.

The process of **mutation** is the source of new alleles. Each new allele *may* be recognized by a change in the phenotype. A new phenotype results from a change in functional activity of the cellular product controlled by that gene. Usually, the alteration or mutation is expressed as a loss of the specific wild-type function. For example, if a gene is responsible for the synthesis of a specific enzyme, a mutation in the gene may change the conformation of this enzyme, thus eliminating its affinity for the substrate. This mutation results in a total loss of function. Conversely, another organism may have a different mutation in this gene, which results in an enzyme with a reduced or increased affinity for binding the substrate. This mutation, representing a separate allele of this gene, may reduce or enhance rather than eliminate the functional capacity of the gene product. In either case, the phenotype may or may not be altered in a discernible way.

Although phenotypic traits may be affected by a single mutation, traits are often influenced by many gene products. In the case of enzymatic reactions, most are part of complex metabolic pathways: Therefore, phenotypic traits may be under the control of more than one gene and the allelic forms of each gene involved. Diploid organisms have two copies of each gene, which may be occupied by the same allele or two different alleles.

In the initial part of this chapter, examples will be restricted to only one gene and the alleles associated with it. Thus, in these cases, we see a modification of the 3:1 monohybrid ratio. Then we will consider traits controlled by two genes and the accompanying modification of the 9:3:3:1 dihybrid ratio.

Symbols for Alleles

In Chapter 3, we learned to symbolize alleles for very simple Mendelian traits. We used the lowercase form of the initial letter of the name of a recessive trait to denote the recessive allele and the same letter in uppercase form to refer to the dominant allele. Thus, for *tall* and *dwarf*, where *dwarf* is recessive, D and d represent the alleles responsible for these respective traits. Mendel also used upper and lowercase letters to symbolize alleles.

Another useful system was developed in *Drosophila* genetics to discriminate between wild-type and mutant traits. In this system, the initial letter, or a combination of two or three letters, of the name of the mutant trait is selected. If the trait is recessive, the lowercase form is used; if it is dominant, the uppercase form is used. The contrasting wild-type trait is denoted by the same letter, but with a + as a superscript. This system works nicely for other organisms as well, provided there is a distinct wild-type phenotype for the character under consideration.

For example, *ebony* is a recessive body color mutation in the fruit fly *Drosophila melanogaster*. The normal wild-type body color is gray. Using the above system, *ebony* is denoted by the symbol e while gray is denoted by e^+. If we focus on the *ebony* mutation, the responsible locus may be occupied by either the wild-type allele (e^+) or the mutant allele (e). A diploid fly may thus exhibit three possible genotypes:

e^+/e^+ : gray homozygote (wild type)
e^+/e : gray heterozygote (wild type)
e/e : ebony homozygote

The slash is used to indicate that the two allele designations represent the same locus on two homologous chromosomes. If we were instead considering a dominant mutation in *Drosophila* such as *Wrinkled* (*Wr*), the three possible designations would be Wr^+/Wr^+, Wr^+/Wr, and Wr/Wr. The latter two genotypes express the wrinkled-wing phenotype.

One advantage of this system is that further abbreviation may be used when convenient: The wild-type allele may simply be denoted by the + symbol. Using *ebony* as an example under consideration in a cross, the designations of the three possible genotypes become:

$+/+$: gray homozygote (wild type)
$+/e$: gray heterozygote (wild type)
e/e : ebony homozygote (mutant)

As we will see in Chapter 5, this abbreviation is particularly useful when two or three genes linked together on the same chromosome are considered simultaneously.

Still other allele designations are sometimes useful. The system just described works well with alleles that are either dominant or recessive to one another. However, if no dominance exists, we may simply use uppercase letters and superscripts to denote alleles (e.g., R^1 and R^2, L^M and L^N, I^A and I^B). Their use will become apparent in ensuing sections of this chapter.

Two other points are important to make before we proceed with the remainder of this chapter. First, although we have adopted a standard convention for assigning genetic symbols, as presented above, there are many diverse systems of genetic nomenclature used to identify genes in various organisms. Usually, the symbol selected reflects the function of the gene, or even a disorder caused by a mutant gene. For example, the *cdc* gene discussed in Chapter 2 refers to *cell division cycle* genes discovered in yeast. In bacteria, *leu*$^-$ refers to a mutation that interrupts the biosynthesis of the amino acid *leucine*, where the wild-type gene is designated *leu*$^+$. The symbol *dnaA* represents a bacterial gene involved in DNA replication (and DnaA designates the protein made by that

gene). In humans, capital letters are used to name genes: *BRCA1* represents a gene associated with susceptibility to *br*east *can*cer. Although these different systems may sometimes be confusing, they all represent different ways to symbolize genes.

Finally, note that in each of the many crosses discussed in the next few chapters, only one or a few gene pairs are involved. It may be useful for you to remember that in each cross, all other genes, which are not under consideration, are assumed to have no effect on the inheritance patterns described.

Incomplete, or Partial, Dominance

Incomplete, or **partial, dominance** in the offspring is based on the observation of intermediate phenotypes generated by a cross between parents with contrasting traits. For example, if plants such as four-o'clocks or snapdragons with red flowers are crossed with plants with white flowers, offspring may have pink flowers. It appears that neither red nor white flower color is dominant. Because some red pigment is produced in the F_1 intermediate-colored pink flowers, dominance appears to be incomplete or partial.

If this phenotype is under the control of a single gene and two alleles where neither is dominant, the results of the F_1 (pink) $\times$ F_1 (pink) cross can be predicted. The resulting F_2 generation is shown in Figure 4.1, confirming the hypothesis that only one pair of alleles determines these phenotypes. The *genotypic ratio* (1:2:1) of the F_2 generation is identical to that of Mendel's monohybrid cross. Because there is no dominance, however, the *phenotypic ratio* is identical to the genotypic ratio. Note here that because neither of the alleles is recessive, we have chosen not to use upper- and lowercase letters. Instead, we have chosen R^1 and R^2 to denote the red and white alleles. We could have chosen W^1 and W^2 or still other designations such as C^W and C^R, where C indicates "color."

Clear-cut cases of incomplete dominance, which result in intermediate expression of the overt phenotype, are relatively rare. However, even when complete dominance seems apparent, careful examination of the level of the gene product, rather than the phenotype, often reveals an intermediate level of gene expression. For example, in the human biochemical disorder **Tay-Sachs disease**, homozygous recessive individuals are severely affected with a fatal lipid storage disorder where neonates die during their first one to three years of life. Heterozygotes, with only a single copy of the mutant gene, are phenotypically normal. In afflicted individuals there is almost no activity of the responsible enzyme **hexosaminidase**, which is normally involved in lipid metabolism. Heterozygotes, on the other hand, express about 50 per-

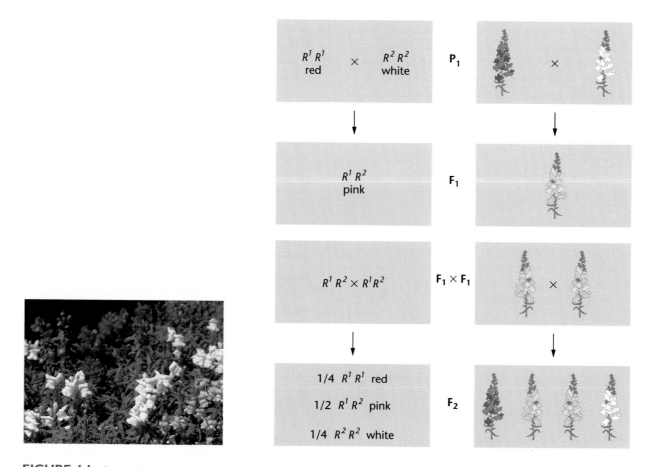

FIGURE 4.1   Incomplete dominance illustrated by flower color. The photograph illustrates red, white, and pink snapdragons.

cent of the enzyme activity found in homozygous normal individuals. Fortunately, this level of enzyme activity is adequate to achieve normal biochemical function. This situation is not uncommon in enzyme disorders. It illustrates the somewhat arbitrary nature of the terms *dominance* and *recessiveness*.

Codominance

If two alleles of a single gene are responsible for the production of two distinct and detectable gene products, a situation different from incomplete dominance or dominance/recessiveness arises. In such a case, the joint expression of both alleles in a heterozygote is called **codominance**. The **MN blood group** in humans illustrates this phenomenon and is characterized by a molecule called a **glycoprotein** found on the surface of red blood cells. Discovered by Karl Landsteiner and Philip Levine, these molecules are **native antigens** that provide biochemical and immunological identity to individuals.

In the human population, two forms of this glycoprotein exist, designated M and N. An individual may exhibit either one or both of them.

The MN system is under the control of an autosomal locus found on chromosome 4 and two alleles designated L^M and L^N. Because humans are diploid, three combinations are possible, each resulting in a distinct blood type:

Genotype	Phenotype
$L^M L^M$	M
$L^M L^N$	MN
$L^N L^N$	N

As predicted, a mating between two MN parents may produce children of all three blood types:

$$L^M L^N \times L^M L^N$$
$$\downarrow$$

1/4 $L^M L^M$
1/2 $L^M L^N$
1/4 $L^N L^N$

Codominant inheritance results in distinct evidence of the gene products of both alleles. Individual expression of each allele is apparent. This characteristic distinguishes it from other modes of inheritance, such as incomplete dominance, where heterozygotes express an intermediate, or blended, phenotype.

Multiple Alleles

Because the information stored in any gene is extensive, mutations can modify this information in many ways. Each change has the potential for producing a different allele. Therefore, for any gene, the number of alleles within a population of individuals need not be restricted to only two. When three or more alleles of the same gene are found, the mode of inheritance is called **multiple allelism**.

The concept of multiple alleles can only be studied in populations. Any individual diploid organism has, at most, two homologous gene loci that may be occupied by different alleles of the same gene. However, among members of a species, many alternative forms of the same gene may exist. The following examples illustrate the concept of multiple alleles. In several of the examples the relationship among genetics, immunology, and medicine is apparent.

The ABO Blood Groups

The simplest possible case of multiple alleles is that in which there are three alternative alleles of one gene. This situation is illustrated in the inheritance of the **ABO blood groups** in humans, discovered by Karl Landsteiner in the early 1900s. The A and B antigens, like the antigens responsible for the MN blood types, are characterized by the presence of antigens on the surface of red blood cells. However, the A and B antigens are distinct from the MN antigens and are under the control of a different gene, located on chromosome 9. As in the MN system, one combination of alleles in the ABO system exhibits a codominant mode of inheritance.

The ABO phenotype of any individual is ascertained by mixing a blood sample with antiserum containing type A or type B antibodies. If the antigen is present on the surface of the person's red blood cells, it will react with the corresponding antibody and cause clumping or agglutination of the red blood cells.

When individuals are tested in this way, four phenotypes are revealed. Each individual has either the A antigen (A phenotype), the B antigen (B phenotype), the A and B antigens (AB phenotype), or neither antigen (O phenotype). In 1924, it was hypothesized that these phenotypes were inherited as the result of three alleles of a single gene. This hypothesis was based on studies of the blood types of many different families.

Although different designations can be used, we will use the symbols I^A, I^B, and I^O for the three alleles. The I designation stands for **isoagglutinogen**, another term for antigen. If we assume that the I^A and I^B alleles are responsible for the production of their respective A and B antigens and that I^O is an allele that does not produce any detectable A or B antigens, the various genotypic possibilities can be listed and the corresponding phenotype assigned to each:

Genotype	Antigen	Phenotype
$I^A I^A$	A	A
$I^A I^O$	A	
$I^B I^B$	B	B
$I^B I^O$	B	
$I^A I^B$	A, B	AB
$I^O I^O$	Neither	O

Note that in these assignments the I^A and I^B alleles behave dominantly to the I^O allele, but codominantly to each other.

We can test the hypothesis that three alleles control ABO blood groups by examining potential offspring from many combinations of matings, as shown in Table 4.1. If we assume heterozygosity wherever possible, we can predict which phenotypes can occur. These theoretical predictions have been upheld in numerous studies examining the blood types of children of parents with all possible phenotypic combinations. The hypothesis that three alleles control ABO blood types in the human population is now universally accepted.

Our knowledge of human blood types has several practical applications. Compatibility of blood transfusions can be safely predicted and decisions about disputed parentage can be more accurately made. The latter cases can occur when newborns are inadvertently mixed up in hospitals, or when it is uncertain whether a specific male is the father of a child. In both cases, an examination of the ABO blood groups as well as other inherited antigens of the possible parents and the child may help to resolve the situation. Table 4.1 demonstrates a variety of parental genotypes and phenotypes as well as the potential offspring from each. In only one of those cases can offspring of all four blood types be realized. The only mating that can result in offspring of all four phenotypes is between two heterozygous individuals, one showing the A phenotype and the other showing the B phenotype. On genetic grounds alone, then, a male or female may be unequivocally ruled out as the parent of a certain child. On the other hand, this type of genetic evidence never proves parenthood.

The A and B Antigens

The biochemical basis of the ABO blood type system has now been carefully worked out. The A and B antigens are actually carbohydrate groups (sugars) that are bound to

TABLE 4.1 Potential Phenotypes in the Offspring of Parents with All Possible ABO Blood Group Combinations, Assuming Heterozygosity Whenever Possible

Parents		Potential Offspring			
Phenotypes	Genotypes	A	B	AB	O
A × A	$I^A I^O \times I^A I^O$	3/4	—	—	1/4
B × B	$I^B I^O \times I^B I^O$	—	3/4	—	1/4
O × O	$I^O I^O \times I^O I^O$	—	—	—	all
A × B	$I^A I^O \times I^B I^O$	1/4	1/4	1/4	1/4
A × AB	$I^A I^O \times I^A I^B$	1/2	1/4	1/4	—
A × O	$I^A I^O \times I^O I^O$	1/2	—	—	1/2
B × AB	$I^B I^O \times I^A I^B$	1/4	1/2	1/4	—
B × O	$I^B I^O \times I^O I^O$	—	1/2	—	1/2
AB × O	$I^A I^B \times I^O I^O$	1/2	1/2	—	—
AB × AB	$I^A I^B \times I^A I^B$	1/4	1/4	1/2	—

lipid molecules (fatty acids) protruding from the membrane of the red blood cell. The specificity of the A and B antigens is based on the terminal sugar of the carbohydrate group.

Almost all individuals possess what is called the **H substance**, to which one or two terminal sugars is added. As shown in Figure 4.2, the H substance itself consists of three sugar molecules, galactose, *N*-acetylglucosamine,

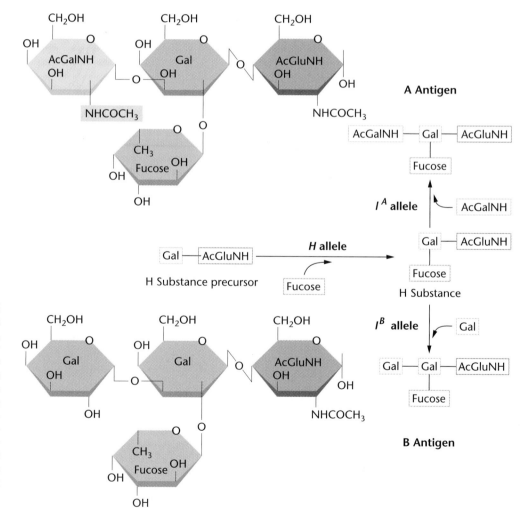

FIGURE 4.2 The biochemical basis of the ABO blood groups. The *H* allele, present in almost all humans, directs the conversion of a precursor molecule to the H substance by adding a molecule of fucose. Failure to do so results in the Bombay phenotype. The *I^A* and *I^B* alleles are then able to direct the addition of terminal sugar residues to the H substance. The *I^O* allele is unable to direct either of these terminal additions. Gal: galactose; AcGluNH: *N*-acetyl-D-glucosamine; AcGalNH: *N*-acetylgalactosamine.

and fucose, chemically linked together. The I^A allele is responsible for an enzyme that can add the terminal sugar N-acetylgalactosamine to the H substance. The I^B allele is responsible for a modified enzyme that cannot add N-acetylgalactosamine, but instead can add the terminal sugar galactose. Heterozygotes ($I^A I^B$) add either one or the other entity at the many sites available. This latter phenomenon illustrates the biochemical basis of codominance in individuals of the AB blood type. Persons of type O ($I^O I^O$) cannot add either terminal sugar, possessing only the H substance protruding from the surface of their red blood cells.

The molecular basis of the mutations leading to the I^A, I^B, and I^O alleles has recently been elucidated. We shall return to this topic in Chapter 14 when we discuss mutation and mutagenesis.

The Bombay Phenotype

In 1952, a most interesting situation was observed in a woman in Bombay that provided information concerning the genetic basis of the H substance. She displayed a unique genetic history that was inconsistent with her blood type. In need of a transfusion, she was found to lack both the A and B antigens and was thus typed as O. However, as shown in the partial pedigree in Figure 4.3, one of her parents was type AB and she was the obvious donor of an I^B allele to two of her offspring. Thus, she was genetically type B but functionally type O!

It was subsequently shown that this woman was homozygous for a rare recessive mutation, h, which prevented her from synthesizing the complete H substance. The terminal portion of the carbohydrate chain protruding from the red cell membrane was shown to lack fucose. In the absence of fucose, the enzymes specified by the I^A and I^B alleles are apparently unable to recognize the incomplete H substance as a proper substrate. Thus, nei-

ther the terminal galactose nor N-acetylgalactosamine can be added, even though the enzymes capable of doing so are present and functional. As a result, the ABO system genotype cannot be expressed in individuals of genotype hh, and they are functionally type O. To distinguish them from the rest of the population, they are said to demonstrate the **Bombay phenotype**. The frequency of the h allele is exceedingly low. Hence, the vast majority of the human population is of the HH or Hh genotype (almost all HH) and can synthesize the H substance.

The Secretor Locus

Still a third gene is known to affect the expression of the ABO blood type system, found at the **secretor locus**. In about 80 percent of the human population, the A and B antigens are present in various body secretions as well as on the membrane of red blood cells. The ability to secrete these antigens in body fluids such as saliva, gastric juice, semen, and vaginal fluids is under the influence of the dominant allele, Se (Se/Se or Se/se). The minority of the population who do not secrete the antigens (se/se) lack an enzyme that normally modifies the H substance, rendering it water soluble. Nonsecretors make the antigens as specified by the ABO loci but do not secrete them. Secretion of these antigens has great significance in forensic science (the application of scientific knowledge to legal proceedings).

The Rh Antigens

Another set of antigens thought by some geneticists to illustrate multiple allelism includes those designated Rh. Discovered by Landsteiner, Levine, and others about 1940, the Rh antigens have received a great deal of attention because of their direct involvement in the disorder **erythroblastosis fetalis**. Erythroblastosis fetalis, also referred to as **hemolytic disease of the newborn (HDN)**, is a form of anemia. It occurs in an Rh-positive fetus whose mother is Rh-negative and whose father is Rh-positive, contributing that allele to the fetus. Such a genetic combination results in a potential immunological incompatibility between the mother and fetus. If fetal blood passes through the ruptured placenta at birth and enters the maternal circulation, the mother's immune system recognizes the Rh antigen as foreign and builds antibodies against it. During a second pregnancy, the antibody concentration becomes high enough that when maternal antibodies, which can pass across the placenta, enter the fetus's circulation, they begin to destroy the fetus's red blood cells. This causes the hemolytic anemia.

About 10 percent of all human pregnancies demonstrate Rh incompatibility. However, for numerous reasons, less than 0.5 percent actually result in anemia. Cur-

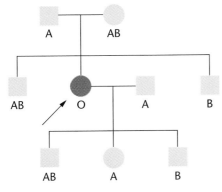

FIGURE 4.3 A partial pedigree of a woman displaying the Bombay phenotype. Functionally, her ABO blood group behaves as type O. Genetically, she is type B.

rently, incompatible mothers are given anti-Rh antisera immediately after giving birth to an Rh-positive baby. This destroys any Rh-positive cells that have entered the mother's circulation so that she does not produce her own anti-Rh antibodies. Before this treatment was developed, many fetuses failed to survive to term. For those that did, complete blood transfusions were often necessary. Even then, many newborns failed to survive.

The initial genetic investigations led to the belief that in the human population only two alleles controlled the presence or absence of the antigen. It was thought that the Rh^+ allele determined the presence of the antigen and behaved as a dominant gene. The Rh^- allele seemed to result in the absence of the antigen.

With the development of more refined antisera to test for the presence of the antigen, it became apparent that the genetic control of Rh antigens was much more complex than originally thought. For example, some presumed Rh-negative blood was found to contain the antigen, but in a different chemical form. Alexander Wiener has proposed the existence of at least eight alleles at a single Rh locus. Other workers, including Ronald A. Fisher, Robert R. Race, and Ruth Sanger, have proposed an alternative type of inheritance. These researchers believe there are three closely linked genes, each with two alleles involved in the inheritance of Rh factors. The term **linkage** is used to describe genes located on the same chromosome. We will discuss this concept in detail in Chapter 5.

The two systems of nomenclature designating the alleles are contrasted in Table 4.2. In the Wiener system, the presence of at least one of the four dominant alleles is sufficient to yield the Rh-positive blood type. Examination of the Fisher–Race system shows that the genetic locus bearing the set of D and d alleles is most critical. The presence of at least one dominant D allele results in the Rh-positive phenotype, whereas the dd genotype ensures the Rh-negative blood type. Although the C, c, E,

TABLE 4.3 Some of the Alleles Present at the *white* Locus of *Drosophila melanogaster* and Their Eye Color Phenotype

Allele	Name	Eye Color
w	*white*	pure white
w^a	*white-apricot*	yellowish orange
w^{bf}	*white-buff*	light buff
w^{bl}	*white-blood*	yellowish ruby
w^{cf}	*white-coffee*	deep ruby
w^e	*white-eosin*	yellowish pink
w^{mo}	*white-mottled orange*	light mottled orange
w^{sat}	*white-satsuma*	deep ruby
w^{sp}	*white-spotted*	fine grain, yellow mottling
w^t	*white-tinged*	light pink

and e alleles specify distinguishable antigens, they are not immunologically significant. Because of the complexity of the antigenic patterns, it is difficult to favor one system over the other.

The *white* Locus in *Drosophila*

Many other phenotypes in plants and animals are known to be controlled by multiple allelic inheritance. In *Drosophila*, for example, where the induction of mutations has been used extensively as a method of investigation, many alleles are known at practically every locus. The recessive eye mutation, *white*, discovered by Thomas H. Morgan and Calvin Bridges in 1912, represents only one of over 100 alleles that may occupy this locus. In this allelic series, eye colors range from complete absence of pigment in the *white* allele, to deep ruby in the *white-satsuma* allele, to orange in the *white-apricot* allele, to a buff color in the *white-buff* allele. These alleles are designated w, w^{sat}, w^a, and w^{bf}, respectively (Table 4.3). In each of these cases, the total amount of pigment in these mutant eyes is reduced to less than 20 percent of that found in the brick-red wild-type eye.

Lethal Alleles

Many gene products are essential to an organism's survival. Mutations resulting in the synthesis of a gene product that is nonfunctional can sometimes be tolerated in the heterozygous state; that is, one wild-type allele may be sufficient to produce enough of the essential product to allow survival. However, such a mutation behaves as a **recessive lethal allele**, and homozygous recessive individuals will not survive. The time of death will depend upon when the product is needed during development or even adulthood.

TABLE 4.2 A Comparison of the Alleles Involved in the Wiener System with Those of the Fisher–Race System to Explain the Genetic Basis of the Rh Blood Groups

Wiener Nomenclature	Fisher–Race Nomenclature	Phenotype
R^1	CDE	
R^2	CDe	
R^0	cDE	Rh^+
R^z	cDe	
r	CdE	
r'	Cde	
r''	cdE	Rh^-
r^y	cde	

In instances where one copy of the wild-type gene is not sufficient for normal development, even the heterozygote will not survive. In this case, the mutation is behaving as a **dominant lethal allele** because its presence somehow overrides the expression of the wild-type product, or the amount of wild-type product is simply insufficient to support its essential function.

In some cases, the allele responsible for a lethal effect when homozygous may result in a distinctive mutant phenotype when present heterozygously. Such an allele is behaving as a recessive lethal but is dominant with respect to the phenotype. For example, a mutation causing a yellow coat in mice was discovered in the early part of this century. The yellow coat varied from the normal agouti coat phenotype, as illustrated in Figure 4.4. Crosses between the various combinations of the two strains yielded unusual results:

Crosses		
A: agouti × agouti	⟶	all agouti
B: yellow × yellow	⟶	2/3 yellow : 1/3 agouti
C: agouti × yellow	⟶	1/2 yellow : 1/2 agouti

These results are explained on the basis of a single pair of alleles. The mutant *yellow* allele A^Y is dominant to the wild-type agouti allele A, so heterozygous mice will have yellow coats. However, the yellow allele also behaves as a homozygous lethal. Homozygous mice of the genotype $A^Y A^Y$ die before birth. As a result, no homozygous yellow mice are ever recovered. The genetic basis for these three crosses is provided in Figure 4.4.

Many genes are known to behave similarly in other organisms. In *Drosophila*, *Curly* wing (*Cy*), *Plum* eye (*Pm*), *Dichaete* wing (*D*), *Stubble* bristle (*Sb*), and *Lyra* wing (*Ly*) be-

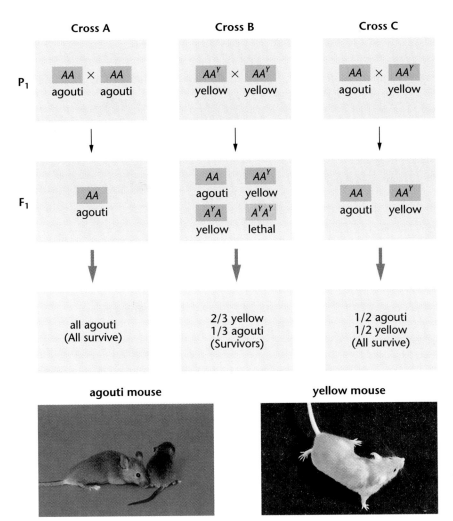

FIGURE 4.4 Inheritance patterns in three crosses involving the wild type *agouti* allele (*A*) and the mutant *yellow* allele (A^Y) in the mouse. Note that the mutant allele behaves dominantly to the normal allele (*A*) in controlling coat color, but it also behaves as a homozygous lethal allele. The genotype $A^Y A^Y$ does not survive.

have as homozygous lethals but are dominant with respect to the expression of the mutant phenotype when heterozygous.

Alleles in other organisms are known to behave as dominant lethals, where the presence of one copy will result in the death of the individual. For example, in humans, a disorder called **Huntington disease** (also referred to as Huntington's chorea) is due to a dominant allele, *H*, where the onset of the disease in heterozygotes (*Hb*) is delayed, usually well into adulthood. Affected individuals then undergo gradual neural degeneration until they die. This lethal disorder is particularly tragic because it has such a late onset, typically at about age 40. By that time, the affected individual may have produced a family. Each child has a 50 percent probability of inheriting the lethal allele, transmitting the allele to his or her offspring, and eventually developing the disorder. The American folk singer and composer Woody Guthrie died from this disease.

Dominant lethal alleles are rarely observed. For them to exist in a population, the affected individual must reproduce before the allele's lethality is expressed, as can occur in Huntington disease. If all affected individuals die before reaching the reproductive age, the mutant gene will not be passed to future generations and will disappear from the population unless it recurs again as a result of a new mutation.

Combinations of Two Gene Pairs

Each example discussed so far modifies Mendel's 3:1 F_2 monohybrid ratio. Therefore, combining any two of these modes of inheritance in a dihybrid cross will likewise modify the classical 9:3:3:1 ratio. Having established the foundation for the modes of inheritance of incomplete dominance, codominance, multiple alleles, and lethal alleles, we can now deal with the situation of two modes of inheritance occurring simultaneously. Mendel's principle of independent assortment applies to these situations when the genes controlling each character are not located on the same chromosome.

Suppose, for example, that a mating occurs between two humans who are both heterozygous for the auto-

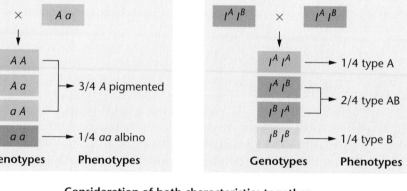

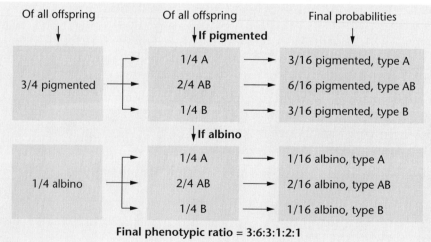

FIGURE 4.5   The calculation of the probabilities in a mating involving the ABO blood type and albinism in humans using the forked-line method.

somal recessive gene that causes albinism and who are both of blood type AB. What is the probability of any particular phenotypic combination occurring in each of their children? Albinism is inherited in the simple Mendelian fashion, and the blood types are determined by the series of three multiple alleles, I^A, I^B, and I^O. The solution to this problem is diagrammed in Figure 4.5, using the forked-line method.

Instead of this dihybrid cross yielding the classical four phenotypes in a 9:3:3:1 ratio, six phenotypes occur in a 3:6:3:1:2:1 ratio, establishing the expected probability for each phenotype. While Figure 4.5 solves the problem using the forked-line method first described in Chapter 3, we could solve the problem using the more conventional Punnett square. Recall that the forked-line method simply requires that the phenotypic ratios for each trait be computed individually (which can usually be done by inspection); all possible combinations can then be calculated.

The above example is just one of many variants of modified ratios possible when different modes of inheritance are combined. We can deal in a similar way with any combination of two modes of inheritance. You will be asked to determine the phenotypes and their expected probabilities for many of these combinations when you solve the problems at the end of the chapter. In each case, the final phenotypic ratio is a modification of the 9:3:3:1 dihybrid ratio.

Gene Interaction: Discontinuous Variation

Soon after the rediscovery of Mendel's work, experimentation revealed that individual characteristics displaying **discrete**, or **discontinuous**, **phenotypes** were often under the control of more than one gene. This was a significant discovery because it revealed that genetic influence on the phenotype is much more sophisticated than encountered by Mendel in his crosses with the garden pea. Instead of single genes controlling the development of individual parts of the plant and animal body, it soon became clear that phenotypic characteristics, such as eye color, hair color, or blood type, are influenced by many genes and their resultant products.

The concept of **gene interaction** does not mean that two or more genes, or their products, necessarily interact *directly* to influence a particular phenotype. Instead, this concept implies that the cellular function of numerous gene products is related to the development of a common phenotype. For example, the development of an organ like the eye of an insect is exceedingly complex and leads to a structure with multiple phenotypic manifestations.

The compound eye of an adult insect, most simply described, exhibits a specific size, shape, texture, and color. The development of the eye can best be envisioned as occurring as the result of a **complex cascade of developmental events** leading to its formation. This process illustrates the developmental concept of **epigenesis** (originally introduced in Chapter 1), whereby each ensuing step of development increases the complexity of this sensory organ and is under the control and influence of one or more genes. To clarify the above discussion, we will present several examples that directly illustrate gene interaction at the biochemical level.

Epistasis

Perhaps the best examples of gene interaction leading to discontinuous variation are those illustrating the phenomenon of **epistasis**. Derived from the Greek word meaning "stoppage," epistasis occurs when the expression of one gene pair *masks* or *modifies* the expression of another gene pair. This masking may occur under different conditions (Figure 4.6). The involved genes control the expression of the same general phenotypic character, sometimes in an antagonistic manner, as when masking occurs. In other cases, the genes may exert their influence on one another in a complementary, or cooperative, fashion.

For example, the homozygous presence of a recessive allele may prevent or override the expression of other alleles at a second locus (or several other loci). In this case, the alleles at the initial locus are said to be **epistatic** to those at the second locus. The alleles at the second locus, which are masked, are described as being **hypostatic** to those at the first locus. In another example, a single dominant allele at the first locus may influence the expression of the alleles at a second gene locus. In a third case, two gene pairs may complement one another such that at least one dominant allele at each locus is required to express a particular phenotype. Each of these three forms of epistasis will be examined in more detail.

An example of the homozygous recessive condition at one locus masking the expression of a second locus has been examined earlier in this chapter when we discussed the Bombay phenotype. There, the homozygous condition (*hh*) masked the expression of the I^A and I^B alleles. Only individuals with the *H–* genotype can form the A or B antigens. As a result, individuals with genotypes that include the I^A or I^B allele who are also *hh* express the type O phenotype.

An example of the outcome of matings between individuals heterozygous at both loci is illustrated in Figure 4.6. If many individuals of the genotype $I^A I^B H h$ have children, the phenotypic ratio of 3 A: 6 AB: 3 B: 4 O is expected in their offspring.

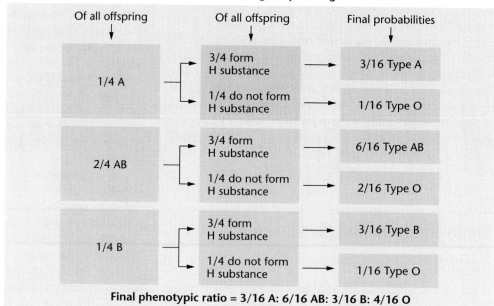

FIGURE 4.6 The outcome of a mating between individuals heterozygous at two genes determining their ABO blood type. Final phenotypes are calculated by considering both genes separately and then combining the results using the forked-line method.

Two important observations can be made when examining this cross and the predicted phenotypic ratio:

1. In illustrating gene interaction, an important distinction exists in this cross compared to the modified dihybrid cross illustrated in Figure 4.5: *only one characteristic—blood type—is being followed.* In the modified dihybrid cross in Figure 4.5, blood type *and* skin pigmentation are followed as separate phenotypic characteristics.

2. Even though only a single character was followed, the phenotypic ratio was expressed in 16s. If we knew nothing about the H substance and the genes controlling it, we could still be confident that a second gene pair, other than that controlling the A

and B antigens, was involved in the phenotypic expression. *When studying a single character, a ratio that is expressed in 16 parts (e.g., 3:6:3:4) suggests that two gene pairs are "interacting" during the expression of the phenotype under consideration.*

The study of gene interaction has revealed a number of inheritance patterns that modify the classical Mendelian dihybrid F_2 ratio (9:3:3:1) in other ways. In several of the subsequent examples that we shall consider, epistasis has the effect of combining one or more of the four phenotypic categories in various ways. The generation of these four groups is reviewed in Figure 4.7, along with several modified ratios.

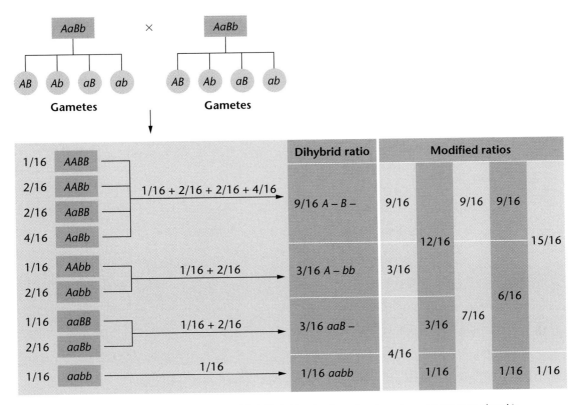

FIGURE 4.7   Generation of the various modified dihybrid ratios from the nine unique genotypes produced in a cross between individuals heterozygous at two genes.

As we discuss these and other examples, we will make several assumptions and adopt certain conventions:

1. In each case, distinct phenotypic classes are produced, each clearly discernible from all others. Such traits illustrate discontinuous variation, where phenotypic categories are discrete and qualitatively different from one another.

2. The genes considered in each cross are not linked and therefore assort independently of one another during gamete formation. So that you may easily compare the results of different crosses, we will designate alleles as *A, a* and *B,b* in each case.

3. When we assume that complete dominance exists between the alleles of any gene pair, such that *AA* and *Aa* or *BB* and *Bb* are equivalent in their genetic effects, the designations *A–* or *B–* will be used for both combinations. Therefore, the dash (–) indicates that either allele may be present, without consequence to the phenotype.

4. All P_1 crosses involve homozygous individuals (e.g., *AABB* × *aabb*, *AABb* × *aaBB*, or *aaBB* × *AAbb*). Therefore, each F_1 generation will consist of only heterozygotes of genotype *AaBb*.

5. In each example, the F_2 generation produced from these heterozygous parents will be the main focus of analysis. When two genes are involved (Figure 4.7), the F_2 genotypes fall into four categories: 9/16 *A–B–*, 3/16 *A–bb*, 3/16 *aaB–*, and 1/16 *aabb*. Because of dominance, all genotypes in each category are equivalent in their effect on the phenotype.

Our first example is seen in the inheritance of coat color in mice (Case 1 of Figure 4.8). As we saw in our discussion of lethal alleles, wild-type coat color is agouti, a grayish pattern formed by alternating bands of pigment on each hair. Agouti is dominant to black (non-agouti) hair, caused by the homozygous expression of a recessive allele, *a*. Thus, *A–* results in agouti, whereas *aa* yields black coat color. When homozygous, a recessive mutation, *b*, at a separate locus, eliminates pigmentation altogether, yielding albino mice, regardless of the genotype at the *a* locus. The presence of at least one *B* allele allows pigmentation to occur in much the same way that the *H* allele in humans allows the production of the A and B antigens. The black and agouti phenotypes are illustrated in Figure 4.9.

In a cross between agouti (*AABB*) and albino (*aabb*), members of the F_1 are all *AaBb* and have agouti coat color.

F₁	AaBb × AaBb ⟶		F₂ genotypes									Final phenotypic ratio
	Organism	Character	AABB 1/16	AABb 2/16	AaBB 2/16	AaBb 4/16	AAbb 1/16	Aabb 2/16	aaBB 1/16	aaBb 2/16	aabb 1/16	
Case	Pea	Mendel's dihybrid	9/16				3/16		3/16		1/16	9:3:3:1
1	Mouse	Coat color	agouti				albino		black		albino	9:3:4
2	Squash	Color	white						yellow		green	12:3:1
3	Pea	Flower color	purple				white					9:7
4	Squash	Fruit shape	disc				sphere				long	9:6:1
5	Chicken	Color	white						colored		white	13:3
6	Mouse	Color	white-spotted				white		colored		white-spotted	10:3:3
7	Shepherd's purse	Seed capsule	triangular								ovoid	15:1
8	Flour beetle	Color	red	sooty	red	sooty	black		jet		black	6:3:3:4

FIGURE 4.8 The basis of modified dihybrid F₂ phenotypic ratios, resulting from crosses between doubly heterozygous F₁ individuals. The four groupings of the F₂ genotypes shown in Figure 4.7 and across the top of this figure are combined in various ways to produce these ratios.

In the F₂ progeny of a cross between two F₁ double heterozygotes, the following genotypes and phenotypes are observed:

$$F_1: AaBb \times AaBb$$
$$\downarrow$$

F₂ Ratio	Genotype	Phenotype	Final Phenotypic Ratio
9/16	A– B–	agouti	9/16 agouti
3/16	A– bb	albino	4/16 albino
3/16	aa B–	black	
1/16	aa bb	albino	3/16 black

Gene interaction yielding the observed 9:3:4 F₂ ratio might be envisioned hypothetically as a two-step process:

	Gene B		**Gene A**	
Precursor Molecule (colorless)	↓ ⟶ B–	**Black Pigment**	↓ ⟶ A–	**Agouti Pattern**

In the presence of a B allele, black pigment can be made from a colorless substance. In the presence of an A allele, the black pigment is deposited during the development of hair in a pattern producing the agouti phenotype. If the aa genotype occurs, all of the hair remains black. If the bb

genotype occurs, no black pigment is produced, regardless of the presence of the A or a alleles, and the mouse is albino. Therefore, the bb genotype masks or suppresses the expression of the A allele, illustrating epistasis.

A second type of epistasis occurs when a dominant allele at one genetic locus masks the expression of the alleles of a second locus. For instance, Case 2 of Figure 4.8 deals with the inheritance of fruit color in summer squash. Here, the dominant allele A results in white fruit color regardless of the genotype at a second locus, B. In the absence of a dominant A allele (the aa genotype), BB or Bb results in yellow color while bb results in green color. Therefore, if two white-colored double heterozygotes (AaBb) are crossed together, an interesting genetic ratio occurs because of this type of epistasis:

$$F_1: AaBb \times AaBb$$
$$\downarrow$$

F₂ Ratio	Genotype	Phenotype	Final Phenotypic Ratio
9/16	A– B–	white	12/16 white
3/16	A– bb	white	
3/16	aa B–	yellow	3/16 yellow
1/16	aa bb	green	1/16 green

Of the offspring, 9/16 are A–B– and are thus white. The 3/16 bearing the genotypes A–bb are also white. Of the

FIGURE 4.9 Mice expressing the agouti, yellow, and black phenotypes.

remaining squash, 3/16 are yellow (*aaB–*), while 1/16 are green (*aabb*). Thus, the modified phenotypic ratio of 12:3:1 occurs.

Our third example (Case 3 of Figure 4.8), first discovered by William Bateson and Reginald Punnett (of Punnett square fame), is demonstrated in a cross between two strains of white-flowered sweet peas. Unexpectedly, the F_1 plants were all purple, and the F_2 occurred in a ratio of 9/16 purple to 7/16 white. The proposed explanation for these results suggests that the presence of at least one dominant allele of each of two gene pairs is essential in order for flowers to be purple. All other genotype combinations yield white flowers because the homozygous condition of either recessive allele masks the expression of the dominant allele at the other locus.

The cross is shown as follows:

$$P_1: \ AAbb \ \times \ aaBB$$
$$\text{white} \qquad \text{white}$$
$$\downarrow$$
$$F_1: \ \text{All } AaBb \ (\text{purple})$$
$$\downarrow$$

F_2 Ratio	Genotype	Phenotype	Final Phenotypic Ratio
9/16	*A– B–*	purple	9/16 purple
3/16	*A– bb*	white	
3/16	*aa B–*	white	7/16 white
1/16	*aa bb*	white	

We can envision the way in which two gene pairs might yield such results:

	Gene A		**Gene B**	
Precursor Substance (colorless)	↓ ───── *A–*	**Intermediate Product (colorless)**	↓ ───── *B–*	**Final Product (purple)**

At least one dominant allele from each pair of genes is necessary to ensure both biochemical conversions to the final product, yielding purple flowers. In the cross above, this will occur in 9/16 of the F_2 offspring. All other plants have flowers that remain white.

These three examples illustrate in a simple way how the products of two genes "interact" to influence the development of a common phenotype. In other instances, more than two genes and their products are involved in controlling phenotypic expression.

Novel Phenotypes

Other cases of gene interaction yield novel, or new, phenotypes in the F_2 generation, in addition to producing modified dihybrid ratios. Case 4 in Figure 4.8 depicts the inheritance of fruit shape in the summer squash *Cucurbita pepo*. When plants with disc-shaped fruit (*AABB*) are crossed to plants with long fruit (*aabb*), the F_1 generation all have disc fruit. However, in the F_2 progeny, fruit with a novel shape—sphere—appear as well as fruit exhibiting the parental phenotypes. These phenotypes are shown in Figure 4.10.

The F_2 generation, with a modified 9:6:1 ratio, is as follows:

$$F_1: \ AaBb \ \times \ AaBb$$
$$\text{disc} \qquad \text{disc}$$
$$\downarrow$$

F_2 Ratio	Genotype	Phenotype	Final Phenotypic Ratio
9/16	*A– B–*	disc	9/16 disc
3/16	*A– bb*	sphere	
3/16	*aa B–*	sphere	6/16 sphere
1/16	*aa bb*	long	1/16 long

In this example of gene interaction, both gene pairs influence fruit shape equivalently. A dominant allele at either locus ensures a sphere-shaped fruit. In the absence of dominant alleles, the fruit is long. However, if both dominant alleles (*A* and *B*) are present, the fruit is flattened into a disc shape.

Another interesting example of an unexpected phenotype arising in the F_2 generation is the inheritance of eye color in *Drosophila melanogaster*. The wild-type eye color is brick red. When two autosomal recessive mutants, *brown* and *scarlet*, are crossed, the F_1 generation consists of flies with wild-type eye color. In the F_2 generation wild, scarlet, brown, and white-eyed flies are found in a 9:3:3:1 ratio. While this ratio is numerically the same as Mendel's dihybrid ratio, the *Drosophila* cross involves only one character, eye color. The diagram of this cross

FIGURE 4.10 Summer squash exhibiting various fruit-shape phenotypes, where disc (white), long (orange gooseneck), and sphere (bottom left) are apparent.

uses the gene symbols *A* and *B* to maintain consistency in this section. (The actual symbols for the two genes are *st* and *bw*, respectively.)

$$P_1: \; AAbb \times aaBB$$
$$\text{brown} \quad \text{scarlet}$$
$$\downarrow$$
$$F_1: \; AaBb \text{ (wild type)}$$
$$\downarrow$$

F₂ Ratio	Genotype	Phenotype	Final Phenotypic Ratio
9/16	*A– B–*	wild type	9/16 wild
3/16	*A– bb*	brown	3/16 brown
3/16	*aa B–*	scarlet	3/16 scarlet
1/16	*aa bb*	white	1/16 white

This cross is an excellent example of gene interaction because the biochemical basis of eye color in this organism has been determined (Figure 4.11). *Drosophila*, as a typical arthropod, has compound eyes made up of individual visual units called **ommatidia**.

FIGURE 4.11 A theoretical explanation of the biochemical basis of the four eye-color phenotypes produced in a cross between flies with *brown* and *scarlet* eyes in *Drosophila melanogaster*. In the presence of at least one wild-type *bw⁺* allele, an enzyme is produced that converts substance *b* to *c*, and drosopterins are synthesized. In the presence of at least one wild-type *st⁺* allele, substance *e* is converted to *f*, and xanthommatins are synthesized. The homozygous presence of the recessive *bw* and *st* mutant alleles blocks the synthesis of these respective pigment molecules. Either none, one, or both of these pathways can be blocked, depending on the genotype.

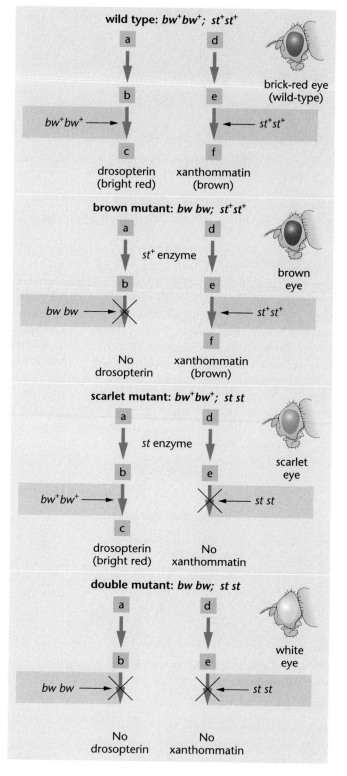

The wild-type eye color is due to the deposition and mixing of two separate pigments in each ommatidium. These include the bright red pigment **drosopterin** and the brown pigment **xanthommatin**. Each pigment is produced by separate biosynthetic pathways. Each step of each pathway is catalyzed by a separate enzyme and is thus under the control of a separate gene. As shown in Figure 4.11, the *brown* mutation, when homozygous, interrupts the pathway leading to the synthesis of the bright red pigment. Because only xanthommatin pigments are present, the eye is brown. The recessive mutation *scarlet*, affecting a gene located on a separate autosome, interrupts the pathway leading to the synthesis of the brown xanthommatins and renders the eye color bright red in homozygous mutant flies. Each mutation apparently causes the production of a nonfunctional enzyme. Flies that are double mutants and thus homozygous for both *brown* and *scarlet* lack both functional enzymes and can make neither of the pigments; they represent the novel white-eyed flies appearing in 1/16 of the F_2 generation.

Other Modified Dihybrid Ratios

The remaining cases (5–8) in Figure 4.8 illustrate additional modifications of the dihybrid ratio and provide still other examples of gene interactions. All cases (1–8) have two things in common. First, in arriving at a suitable explanation of the inheritance pattern of each one, we have not violated the principles of segregation and independent assortment. Therefore, the added complexity of inheritance in these examples does not detract from the validity of Mendel's conclusions. Second, the F_2 phenotypic ratio in each example has been expressed in sixteenths. When a similar observation is made in crosses where the inheritance pattern is unknown, it suggests to geneticists that two gene pairs are controlling the observed phenotypes. You should make the same inference in the analysis of genetics problems. Other insights into solving genetics problems are provided in the "Insights and Solutions" section at the conclusion of this chapter.

Gene Interaction: Continuous Variation

In the preceding section, examples in Figure 4.8 illustrated gene interaction leading to phenotypic variation that is easily classified into distinct categories or traits. Pea plants may be tall or dwarf; squash shape might be spherical, disc-shaped, or elongated; and fruit-fly eye color can be red, brown, scarlet, or white. These phenotypes are examples of discontinuous variation where dis-

crete phenotypic categories exist. There are many other traits in a population that demonstrate considerably more variation and are not easily categorized into distinct classes. Such phenotypes represent **continuous variation**. For example, in addition to his work with garden peas, Mendel experimented with beans. In a cross between a purple and a white-flowered strain the F_1 was purple. However, the F_2 contained not only purple-flowered and white-flowered bean plants, but also ones with numerous intermediate shades. Mendel was unable to explain these results satisfactorily, but he recognized that they were inconsistent with most of his data derived from peas. It was not until more than 50 years later that the inheritance of characters exhibiting continuous variation was explained.

It is now known that traits exhibiting continuous variation are often controlled by two or more genes that provide an additive component to the phenotype. Such traits are said to exhibit **continuous** or **quantitative variation** and are examples of **polygenic inheritance**. We will examine such patterns of inheritance in which a phenotypic trait is controlled by genes at two or more loci. Later in the text (Chapter 7), we will return to this general topic and outline the statistical tools used by geneticists to study traits that exhibit continuous variation.

Quantitative Inheritance: Polygenes

In the late eighteenth century, Josef Gottlieb Kölreuter showed that when tall and dwarf tobacco plants were crossed, the F_1 generation plants were not all tall, as Mendel found with garden peas. Rather, the individual plants were all intermediate in height. When the F_2 generation was examined, individuals showed continuous variation in height, ranging from tall to dwarf, including many heights between these extremes. The majority of the F_2 plants were intermediate like the F_1, whereas only a few were as tall or dwarf as the P_1 parents. These distributions are depicted in histograms in Figure 4.12. Note that the F_2 data demonstrate a normal distribution, as evidenced by the bell-shaped curve in the histogram.

At the beginning of the twentieth century, geneticists noted that many characters in different species had similar patterns of inheritance, such as height and physical stature in humans, seed size in the broad bean, grain color in wheat, and kernel number and ear length in corn. In each case, offspring in the succeeding generation seemed to be a blend of their parents' characteristics.

The issue of whether continuous variation could be accounted for in Mendelian terms caused considerable controversy in the early 1900s. William Bateson and Gudny Yule, who adhered to the Mendelian explanation of inheritance, suggested that a large number of factors or

FIGURE 4.12 Histograms showing the relative frequency of individuals expressing various height phenotypes derived from Kölreuter's cross between dwarf and tall tobacco plants carried to the F$_2$ generation. The photograph shows a field of tobacco plants.

genes were responsible for the observed patterns. This proposal, called the **multiple-factor** or **multiple-gene hypothesis**, implied that many factors or genes contribute to the phenotype in a *cumulative* or *quantitative* way. However, some geneticists argued that Mendel's unit factors could not account for the blending of parental phenotypes characteristic of these patterns of inheritance and were thus skeptical of these ideas.

By 1920, the conclusions of several critical experiments largely resolved the controversy and demonstrated that Mendelian factors could account for continuous variation. In one experiment, Edward M. East performed crosses between two strains of the tobacco plant *Nicotiana longiflora*. The fused inner petals of the flower, or corollas, of strain A were decidedly shorter than the corollas of strain B. With only minor variation, each strain was true breeding. Thus, the differences between them were clearly under genetic control.

When plants from the two strains were crossed, the F$_1$, F$_2$, and selected F$_3$ data (Figure 4.13) demonstrated a very distinct pattern. The F$_1$ generation displayed corollas that were intermediate in length, compared with the P$_1$ varieties, and showed only minor variability among individuals. While corolla lengths of the P$_1$ plants were about 40 mm or 94 mm, the F$_1$ generation contained plants with corollas that were all about 64 mm. In the F$_2$ generation, lengths varied much more, ranging from 52 mm to 82 mm. Most

individuals were similar to their F$_1$ parents, and as the deviation from this average increased, fewer and fewer plants were observed. When the data are plotted graphically (frequency vs. length), a bell-shaped curve results.

East further experimented with this population by selecting F$_2$ plants of various corolla lengths and allowing them to produce separate F$_3$ generations. Several are illustrated in Figure 4.13. In each case, a bell-shaped distribution was observed, with most individuals similar in height to the F$_2$ parents, but with considerable variation around this value.

East's experiments demonstrated that although the variation in corolla length seemed continuous, experimental crosses resulted in the segregation of distinct phenotypic classes as observed in the three independent F$_3$ categories. This finding strongly suggested that the multiple-factor hypothesis could account for traits that deviate considerably in their expression.

The multiple-factor hypothesis, suggested by the observations of East and others, embodies the following major points:

1. Characters under such control can usually be quantified by measuring, weighing, counting, etc.

2. Two or more pairs of genes, located throughout the genome, account for the hereditary influence

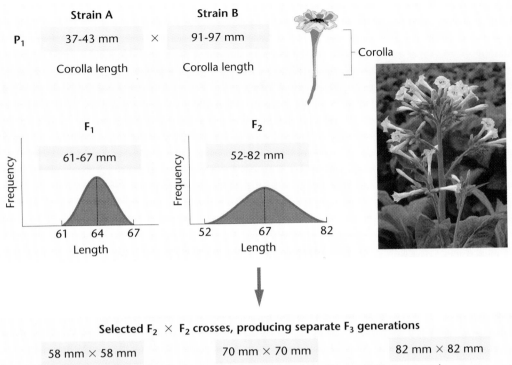

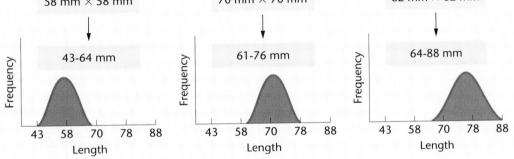

FIGURE 4.13   The F$_1$, F$_2$, and selected F$_3$ results of East's cross between two strains of *Nicotiana* with different corolla lengths. Plants of strain A vary from 37 to 43 mm, while plants of strain B vary from 91 to 97 mm. The photograph illustrates the flower and corolla of a tobacco plant.

on the phenotype in an *additive way*. Because many genes may be involved, inheritance of this type is often called *polygenic*.

3. Each gene locus may be occupied by either an **additive allele**, which contributes a set amount to the phenotype, or by a **nonadditive allele**, which does not contribute quantitatively to the phenotype.

4. The total effect of each additive allele at each locus, while small, is approximately equivalent to all other additive alleles at other gene sites.

5. Together, the genes controlling a single character produce substantial phenotypic variation.

6. Analysis of polygenic traits requires the study of large numbers of progeny from a population of organisms.

These points, as well as an explanation of the multiple-factor hypothesis, can be illustrated by examining Herman Nilsson-Ehle's experiments involving grain color in wheat performed in the early twentieth century. In one set of experiments, wheat with red grain was crossed to wheat with white grain (Figure 4.14). The F$_1$ generation demonstrated an intermediate color. In the F$_2$, approximately 15/16 of the plants showed some degree of red grain, while 1/16 of the plants showed white grain. Because the ratio occurred in sixteenths, we can hypothesize that two gene pairs control the phenotype and, if so, they segregate independently from one another in a Mendelian fashion.

Upon careful examination of the F$_2$, grain that exhibited various degrees of color could be classified into four different shades of red. If two gene pairs were operating, each with one potential additive allele and one potential

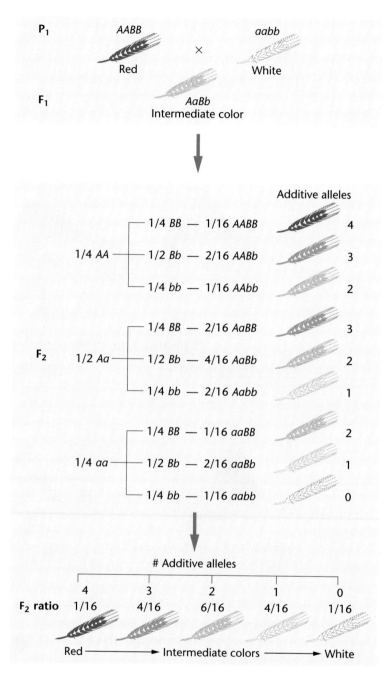

FIGURE 4.14 An illustration of how the multiple-factor hypothesis can account for the 1:4:6:4:1 phenotypic ratio of grain color when all alleles designated by an uppercase letter are additive and contribute an equal amount of pigment to the phenotype.

nonadditive allele, we can envision how the multiple-factor hypothesis could account for this variation. In the P_1, both parents were homozygous; the red parent contains only additive alleles (uppercase), while the white parent contains only nonadditive alleles (lowercase). The F_1, being heterozygous, contains only two additive alleles and expresses an intermediate phenotype. In the F_2, each offspring has either 4, 3, 2, 1, or 0 additive alleles (Figure 4.14). Wheat with no additive alleles (1/16) is white like one of the P_1 parents, whereas wheat with 4 additive alleles is red like the other P_1 parent. Plants with 3, 2, or 1 additive alleles constitute the other three categories of

red color observed in the F_2, with most (6/16) having 2 additive alleles like the F_1 plants.

Note that the above explanation of continuous variation describes the transmission of genes according to Mendelian principles. **Multiple-factor inheritance**, in which alleles contribute additively to a phenotype, is now an accepted mechanism to account for the inheritance of many phenotypes displaying continuous variation. Although the Nilsson-Ehle experiment involved two gene pairs, there is no reason why three, four, or more gene pairs cannot function in controlling various phenotypes. As we saw in Nilsson-Ehle's initial cross, if two gene pairs

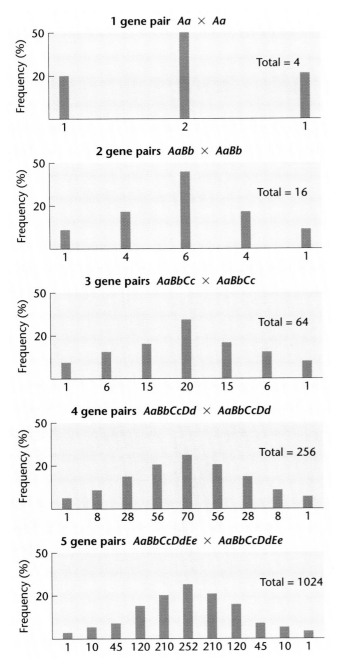

FIGURE 4.15   The results of crossing two heterozygotes where polygenic inheritance is in operation with one to five gene pairs. Each histogram bar indicates a distinct phenotypic class from one extreme (left end) to the other extreme (right end). Each phenotype results from a different number of additive alleles.

Calculating the Number of Polygenes

It is of interest to determine the number of genes that are involved when equally additive effects control polygenic traits. If the ratio of F_2 individuals resembling *either* of the two most extreme phenotypes can be determined, then the number of gene pairs involved (n) may be calculated using the following simple formula:

$$\frac{1}{4^n} = \text{ratio of } F_2 \text{ individuals expressing either extreme phenotype}$$

In our past example, the P_1 phenotypes represent these two extremes. In Figure 4.14, 1/16 of the F_2 are either red *or* white like the P_1 crosses; this ratio can be substituted on the right side of the equation prior to solving for n:

$$\frac{1}{4^n} = \frac{1}{16}$$

$$\frac{1}{4^2} = \frac{1}{16}$$

$$n = 2$$

Table 4.4 lists the ratio and the number of F_2 phenotypic classes produced in crosses involving up to five gene pairs.

For low numbers of gene pairs, it is sometimes easier to use the ($2n + 1$) rule. If n equals the number of gene pairs, $2n + 1$ will determine the total number of categories of possible phenotypic groups. Where $n = 2$, $2n + 1 = 5$, since each phenotypic category could have 4, 3, 2, 1, or 0 additive alleles. Where $n = 3$, $2n + 1 = 7$, since each phenotypic category could have 6, 5, 4, 3, 2, 1, or 0 additive alleles, and so on.

were involved, only five F_2 phenotypic categories, in a 1:4:6:4:1 ratio, would be expected. On the other hand, as three, four, five, or more gene pairs become involved, greater and greater numbers of classes would be expected to appear in more complex ratios. The number of phenotypes and the expected F_2 ratios of crosses involving up to five gene pairs are illustrated in Figure 4.15.

TABLE 4.4   Determination of the Number of Gene Pairs (*n*) Involved in Polygenic Crosses

n	Ratio of Individuals Expressing an Extreme Phenotype	Number of Distinct F_2 Phenotypic Classes
1	1/4	3
2	1/16	5
3	1/64	7
4	1/256	9
5	1/1024	11

The Significance of Polygenic Control

Polygenic control is a significant concept because it is believed to be the mode of inheritance for a vast number of traits. For example, height, weight, and physical stature in animals, size and grain yield in crops, beef and milk production in cattle, and egg output in chickens are thought to be under polygenic control. Thus, knowledge of this mode of inheritance is of prime importance in animal breeding and agriculture. In humans, the degree of skin pigmentation, intelligence, obesity, and predisposition to various diseases are thought to be under some form of polygenic control. In most cases, it is important to note that the genotype, which is fixed at fertilization, establishes the potential range in which a particular phenotype may fall. However, environmental factors determine how much of the potential will be realized. In the crosses described in this section we have assumed an optimal environment, which minimizes variation resulting from environmental sources.

Genes on the X Chromosome: X-Linkage

We near the conclusion of this chapter with a discussion of still one other mode of neo-Mendelian inheritance: **X-linkage**. This phenomenon results from the fact that one of the sexes in many animal and a few plant species contains a pair of *unlike* chromosomes, the X and Y, which are involved in sex determination. For example, in both fruit flies (*Drosophila*) and humans, males contain an X and a Y chromosome, whereas females contain two X chromosomes. As we will see shortly, the unique pattern of inheritance stems from the fact that the Y, while behaving as a homolog to the X during meiosis, contains only a few genes. X-linkage, therefore, involves the transmission and expression of the normal complement of genes located on the X chromosome. To distinguish this pair of sex chromosomes, all other pairs are referred to as **autosomal chromosomes** or just **autosomes**.

X-Linkage in *Drosophila*

One of the first cases of X-linkage was documented in 1910 by Thomas H. Morgan during his studies of the *white* eye mutation in *Drosophila* (Figure 4.16). We will use this case to illustrate X-linkage. The normal wild-type red eye color is dominant to white eye color.

Morgan's work established that the inheritance pattern of the white-eye trait was clearly related to the sex of the parent carrying the mutant allele. Unlike the outcome

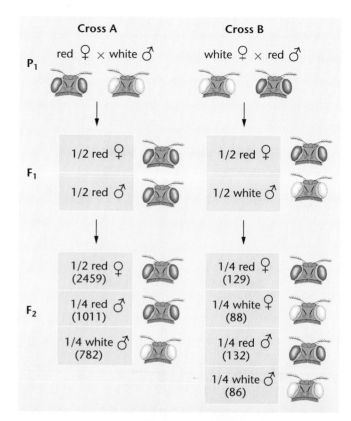

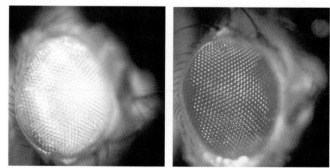

FIGURE 4.16   The F$_1$ and F$_2$ results of T. H. Morgan's reciprocal crosses involving the X-linked *white* mutation in *Drosophila melanogaster*. The actual F$_2$ results are shown in parentheses. The photograph contrasts white eyes with the brick red wild-type eye color.

of the typical monohybrid cross, reciprocal crosses between white-eyed and red-eyed flies did not yield identical results. In contrast, in all of Mendel's monohybrid crosses, F$_1$ and F$_2$ data were very similar regardless of which P$_1$ parent exhibited the recessive mutant trait. Morgan's analysis led to the conclusion that the *white* locus is present on the X chromosome rather than one of the autosomes and is said to be X-linked.

Results of reciprocal crosses between white-eyed and red-eyed flies are shown in Figure 4.16. The obvious differences in phenotypic ratios in both the F$_1$ and F$_2$ gener-

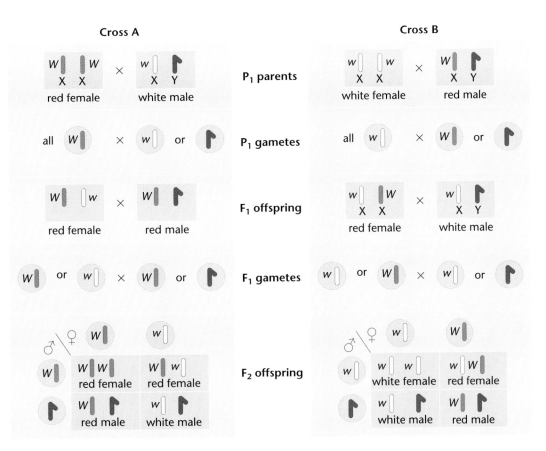

FIGURE 4.17 The chromosomal explanation of the results of the X-linked crosses shown in Figure 4.16.

ations are dependent on whether or not the P_1 white-eyed parent was male or female.

Morgan was able to correlate these observations with the difference found in the sex chromosome composition between male and female *Drosophila*. He hypothesized that in males with white eyes the recessive allele for white eye is found on the X chromosome, but its corresponding locus is absent from the Y chromosome. Females thus have two available gene sites, one on each X chromosome, while males have only one available gene site on their single X chromosome.

Morgan's interpretation of X-linked inheritance, shown in Figure 4.17, provides a suitable theoretical explanation for his results. Because the Y chromosome lacks homology with most genes on the X chromosome, whatever alleles are present on the X chromosome of the males will be directly expressed in the phenotype. As males cannot be either homozygous or heterozygous for X-linked genes, this condition is referred to as being **hemizygous**. In such cases, no alternative alleles are present, and the concept of dominance and recessiveness is irrelevant.

One result of X-linkage is the **crisscross pattern of inheritance**, whereby phenotypic traits controlled by re-

cessive X-linked genes are passed from homozygous mothers to all sons. This pattern occurs because females exhibiting a recessive trait must contain the mutant allele on both X chromosomes. Because male offspring receive one of their mother's two X chromosomes and are hemizygous for all alleles present on that X, all sons will express the same recessive X-linked traits as their mother.

In addition to documenting the phenomenon of X-linkage, Morgan's work has taken on great historical significance. By 1910, the correlation between Mendel's work and the behavior of chromosomes during meiosis had provided the basis for the **chromosome theory of inheritance**, as postulated by Sutton and Boveri (see Chapter 2). The work involving sex chromosomes by Morgan, and subsequently that of his student Calvin Bridges, is considered to be the first solid experimental evidence in support of this theory. The ability of Morgan and Bridges to provide direct evidence that genes are transmitted on specific chromosomes provided the basis for this support. In the ensuing two decades, these findings provided the impetus for further research, the findings of which provided indisputable evidence in support of this theory.

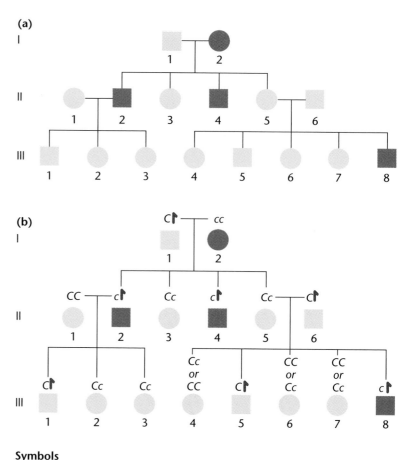

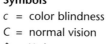

FIGURE 4.18   (a) A human pedigree of the X-linked color blindness trait. (b) The most probable genotypes of each individual in the pedigree.

Symbols

c = color blindness

C = normal vision

⚑ = Y chromosome

X-Linkage in Humans

In humans, many genes and the respective traits controlled by them are recognized as being linked to the X chromosome. These X-linked traits can be easily identified in pedigrees because of the crisscross pattern of inheritance. A pedigree for one form of human color blindness is shown in Figure 4.18. The mother in generation I passes the trait to all her sons but to none of her daughters. If the offspring in generation II marry normal individuals, the color-blind sons will produce all normal male and female offspring (III-1, 2, and 3); the normal-visioned daughters will produce normal-visioned female offspring (III-4, 6, and 7), as well as color-blind (III-8) and normal-visioned (III-5) male offspring.

Many X-linked human genes have now been identified, as shown in Table 4.5. For example, the genes controlling two forms of hemophilia and one form of muscular dystrophy are located on the X chromosome. Additionally, numerous genes whose expression yields

enzymes are X-linked. Glucose-6-phosphate dehydrogenase and hypoxanthine-guanine-phosphoribosyl transferase are two examples. In the latter case, the severe **Lesch–Nyhan syndrome** results from the mutant form of the X-linked gene product (see Table 4.5).

Because of the way in which X-linked genes are transmitted, unusual circumstances may be associated with recessive X-linked disorders in comparison to recessive autosomal disorders. For example, if an X-linked disorder debilitates or is lethal to the affected individual prior to reproductive maturation, the disorder occurs exclusively in males. This is the case because the only sources of the lethal allele in the population are heterozygous females who are "carriers" and who do not express the disorder. They pass the allele to one-half of their sons, who develop the disorder because they are hemizygous, but who rarely, if ever, reproduce. Heterozygous females also pass the allele to one-half of their daughters, who become carriers but do not develop the disorder. An example of such an X-linked disorder is the Duchenne form of

TABLE 4.5   Human X-Linked Traits

Condition	Characteristics
Color blindness, deutan type	Insensitivity to green light
Color blindness, protan type	Insensitivity to red light
Fabry's disease	Deficiency of galactosidase A; heart and kidney defects, early death
G-6-PD deficiency	Deficiency of glucose-6-phosphate dehydrogenase; severe anemic reaction following intake of primaquines in drugs and certain foods, including fava beans
Hemophilia A	Classical form of clotting deficiency; lack of clotting factor VIII
Hemophilia B	Christmas disease; deficiency of clotting factor IX
Hunter syndrome	Mucopolysaccharide storage disease resulting from iduronate sulfatase enzyme deficiency; short stature, clawlike fingers, coarse facial features, slow mental deterioration, and deafness
Ichthyosis	Deficiency of steroid sulfatase enzyme; scaly dry skin, particularly on extremities
Lesch–Nyhan syndrome	Deficiency of hypoxanthine-guanine phosphoribosyl transferase enzyme (HGPRT), leading to motor and mental retardation, self-mutilation, and early death
Muscular dystrophy (Duchenne type)	Progressive, life-shortening disorder characterized by muscle degeneration and weakness; sometimes associated with mental retardation; deficiency of the protein dystrophin

muscular dystrophy. The disease has an onset prior to age 6 and is often lethal prior to age 20. It normally occurs only in males.

Sex-Limited and Sex-Influenced Inheritance

Our final topics involve inheritance affected by the sex of the organism, but not necessarily by genes on the X chromosome. There are numerous examples in different organisms where the sex of the individual plays a determining role in the expression of certain phenotypes. In some cases, the expression of a specific phenotype is absolutely limited to one sex; in others, the sex of an individual influences the expression of a phenotype that is not limited to one sex or the other. This distinction differentiates **sex-limited inheritance** from **sex-influenced inheritance**.

In domestic fowl, tail and neck plumage is often distinctly different in males and females (Figure 4.19), demonstrating sex-limited inheritance. Cock-feathering is longer, more curved, and pointed, whereas hen-feathering is shorter and more rounded. The inheritance of feather type is due to a single pair of autosomal alleles whose expression is modified by the individual's sex hormones.

As shown in the following chart, hen-feathering is due to a dominant allele, *H;* but regardless of the homozygous presence of the recessive *h* allele, all females remain hen-feathered. Only in males does the *hh* genotype result in cock-feathering.

Genotype	Phenotype	
	♀	♂
HH	Hen-feathered	Hen-feathered
Hh	Hen-feathered	Hen-feathered
hh	Hen-feathered	Cock-feathered

In the development of certain breeds of fowl, one allele or the other has become fixed in the population. In the Leghorn breed, all individuals are of the *hh* genotype; as a result, males differ from females, who display distinctive plumage. Sebright bantams are all *HH*, showing no sexual distinction in feathering.

Cases of sex-influenced inheritance include pattern baldness in humans, horn formation in sheep, and certain coat patterns in cattle. In such cases, autosomal genes are responsible for the contrasting phenotypes displayed by both males and females, but the expression of these genes

FIGURE 4.19   Hen-feathering (left) versus cock-feathering (right) in domestic fowl. The feathers in the hen are shorter and less curved.

GENETICS, TECHNOLOGY, AND SOCIETY

The Uncertain Genetic Fate of Purebred Dogs

Dogs have been an important part of human society for thousands of years. It is generally agreed that dogs are domesticated versions of the gray wolf, which once lived throughout the world but is now restricted to small isolated areas due to hunting and habitat loss. Roughly 15,000 years ago, wolves became scavengers around human settlements, feeding on food scraps rather than hunting for their food. Gradually, the wolves lost their innate fear of humans, becoming tamer and tamer and eventually coming to live among them. Later, these domesticated wolves acquired roles in human society as hunters, herders, protectors, and ultimately companions. Genetic changes accompanied the domestication of the wolf, as the scavenger animals interbred and became reproductively isolated from their wild cousins. The domesticated dog and the gray wolf can still be interbred to produce fertile offspring,

however, and they are thus part of the same biological species, *Canis lupus*.

As dogs grew increasingly a part of human society, they were bred to maximize behaviors needed for performing jobs such as sheep-herding, hunting, racing, and serving as companions. Today, humans breed many dogs not so much for their behaviors but more for their appearance. In many cases, dogs are bred specifically for "show" competition, which has become a very big business. The goal of this breeding has been to attain a constancy of desired physical attributes, whether it be narrow snouts, long ears, or short legs, from generation to generation.

The simplest way to achieve this is to mate two dogs that each have the desired look, which often means mating close relatives. Thus, the creation of pure breeds with an idealized appearance has usually involved many generations of inbreeding. For example, matings between half- or even full siblings or between fathers and daughters or mothers and sons may have been carried out for many consecutive generations.

The establishment of purebred strains of dogs has come at a high price, because the many generations of inbreeding have led to the increased incidence of a variety of genetic diseases. By some estimates, as many as 25% of purebred dogs in the United States are afflicted with one or another genetic condition. For example, several breeds are plagued with hip dysplasia, a severe weakness in the joint that often leads to painful, crippling dislocations. This is especially prevalent in German shepherds, which have been bred to have sloping hips and short hind legs. Progressive retinal atrophy causes blindness in dachshunds, Cocker spaniels, and Labrador retrievers. Inherited deafness is common in Dalmatians. This list goes on and on: there are now over 400 genetic diseases known in dogs, and about 10 more are identified every year.

Not every genetic disease in dogs is the direct consequence of inbreeding, but clearly the high prevalence of congenital abnormalities in many breeds is due to such breeding practices. In

FIGURE 4.20 Pattern baldness, a sex-influenced autosomal trait in humans.

is dependent on the hormone constitution of the individual. Thus, the heterozygous genotype may exhibit one phenotype in one sex and the contrasting one in the other. For example, **pattern baldness** in humans, where the hair is very thin on the top of the head (Figure 4.20), is inherited in the following way:

Genotype	Phenotype	
	♀	♂
BB	Bald	Bald
Bb	Not bald	Bald
bb	Not bald	Not bald

Even though females may display pattern baldness, this phenotype is much more prevalent in males. When females do inherit the *BB* genotype, the phenotype is much less pronounced than in males and is expressed later in life.

many cases, the likely explanation for the increased incidence of a genetic disease is that deleterious recessive alleles were, by chance, closely linked to alleles contributing to some desired physical trait. Dogs bred to be homozygous for these desired alleles concomitantly became homozygous for the harmful recessive alleles. For example, such a linkage effect apparently accounts for the high incidence of the inherited retinal degeneration known as collie eye anomaly. The responsible gene is thought to be closely linked to a gene causing the long muzzle and closely set eyes that are selected for by collie breeders. The repeated mating of prizewinning dogs that carry harmful recessive alleles results in the rapid spread of such genetic diseases.

Geneticists, veterinarians, and dog breeders are now realizing that in creating purebred dogs, we have greatly reduced the quality of life of some members of these breeds. Some critics have questioned the morality of breeding purebred dogs at all if it must be at the expense of increased suffering of the animals. Is it ethically acceptable to breed dogs strictly for their appearance

if such programs increase the incidence of inherited disorders?

Just as unsound breeding practices have caused an increased incidence of inherited diseases in dogs, so can enlightened breeding practices help to undo the damage. As a first step, efforts are underway to identify heterozygous animals that "carry" recessive alleles but do not express the associated genetic disease, so that such animals may be eliminated from breeding programs. To that end, computerized registries of known carriers are being established. Dogs can then be certified free of specific inherited diseases before being bred further.

The mapping of the dog genome, while still in its infancy, may eventually permit the identification of genes responsible for many inherited disorders. This will make it possible to develop genetic tests to diagnose such disorders early in a dog's life and to identify carriers before they have produced affected offspring. In addition, veterinarians are now being trained to recognize genetic diseases in dogs so that their owners may be counseled on proper breeding practices.

Questionable breeding practices have created much suffering for "man's best friend," but through an organized effort to apply our knowledge of genetics and to develop new genetic tests, much of this misery may eventually be eliminated.

A purebred bulldog that illustrates generations of inbreeding.

CHAPTER SUMMARY

1. Ever since Mendel's work was rediscovered, the study of transmission genetics has been expanded to include many alternative modes of inheritance. In many cases, phenotypes may be influenced by two or more genes in a variety of ways.

2. Incomplete, or partial, dominance is exhibited when intermediate phenotypic expression of a trait occurs in a heterozygote.

3. Codominance is exhibited when distinctive expression of two alleles occurs in a heterozygous organism.

4. The concept of multiple allelism applies to populations, since a diploid organism may host only two alternative alleles at any given locus. Within a population, however, many alternative alleles of the same gene can occur.

5. Lethal mutations usually result in the inactivation or the lack of synthesis of gene products essential during development. Such mutations may be recessive or dominant. Some lethal genes, such as the gene causing Huntington disease, are not expressed until adulthood.

6. Mendel's classical F_2 ratio is often modified in instances where gene interaction results in either continuous or discontinuous variation.

7. Epistasis involves discontinuous variation, where two or more genes influence a single characteristic. Usually, the expression of one of the genes masks the expression of the other gene or genes.

8. Polygenic (or quantitative) inheritance results in continuous variation and is illustrated when products of several genes make additive contributions to phenotypes.

9. Genes located on the X chromosome display a unique mode of inheritance referred to as X-linkage.

10. Unique X-linked ratios result because hemizygous individuals (those with an X and a Y chromosome) express all alleles present on their X chromosome.

11. Both sex-limited and sex-influenced inheritance occur when the sex of the organism affects the phenotype under the control of an autosomal gene.

KEY TERMS

ABO blood groups, 83
additive alleles, 97
allele, 79
antigen, 82
autosomal chromosome (autosome), 100
Bombay phenotype, 85
chromosome theory of inheritance, 101
codominance, 82
continuous variation, 95
crisscross pattern of inheritance, 101
discontinuous variation, 89
drosopterin, 95
epistasis, 89
erythroblastosis fetalis (hemolytic disease of the newborn–HDN), 85

gene interaction, 79
glycoprotein, 82
H substance, 84
hemizygous, 101
hexosaminidase, 81
Huntington disease, 88
incomplete (partial) dominance, 81
isoagglutinogen, 83
Lesch–Nyhan syndrome, 102
lethal allele, 86
linkage, 86
MN blood groups, 82
multiple alleles, 83

multiple-factor (multiple gene) hypothesis, 96
mutation, 80
neo-Mendelian genetics, 80
pattern baldness, 104
polygenic inheritance, 95
quantitative variation, 95
sex-influenced inheritance, 103
sex-limited inheritance, 103
Tay-Sachs disease, 81
wild-type allele, 80
X-linkage, 80
xanthommatin, 95

INSIGHTS AND SOLUTIONS

Genetic problems take on an added complexity if they involve two independent characters and multiple alleles, incomplete dominance, or epistasis. The most difficult types of problems are those faced by pioneering geneticists in the laboratory. In these problems they had to determine the mode of inheritance by working backwards from the observations of offspring to parents of unknown genotype.

1. Consider the problem of comb shape inheritance in chickens, where walnut, rose, pea, and single are observed as distinct phenotypes. How is comb shape inherited, and what are the genotypes of the P_1 generation of each cross? Use the following data to answer these questions.

   Cross 1: single × single   ⟶ all single
   Cross 2: walnut × walnut   ⟶ all walnut
   Cross 3: rose × pea   ⟶ all walnut
   Cross 4: F_1 × F_1 of Cross 3
             walnut × walnut   ⟶ 93 walnut
                                28 rose
                                32 pea
                                10 single

   Solution: At first glance, this problem may appear quite difficult. The approach used in solving it must involve a multistep process. First, carefully analyze the data for any useful information. Once you determine something concrete, follow an empirical approach; that is, formulate an hypothesis and, in a sense, test it against the given data. Look for a pattern of inheritance that is consistent with all cases.

(a) In this problem there are two immediately useful facts. First, in Cross 1, P_1 singles breed true. Second, while P_1 walnut breeds true (Cross 2), a walnut phenotype is also produced in crosses between rose and pea (Cross 3). When these F_1 walnuts are mated (Cross 4), all four comb shapes are produced in a ratio that approximates 9:3:3:1. This observation should immediately suggest a cross involving two gene pairs because the resulting data closely resemble the ratio of Mendel's dihybrid crosses. Because only one character is involved and discontinuous phenotypes occur (comb shape), perhaps some form of epistasis is occurring. This may serve as a working hypothesis, and we must now propose how the two gene pairs interact to produce each phenotype.

(b) If we call the allele pairs, A, a and B, b, we might predict that, since walnut represents 9/16 in Cross 4, $A–B–$ will produce walnut. We might also hypothesize that in the case of Cross 2, the genotypes were $AABB × AABB$ where walnut was seen to breed true. (Recall that $A–$ and $B–$ mean AA or Aa and BB or Bb, respectively.)

(c) Since single is the phenotype representing 1/16 of the offspring of Cross 4, we could predict that this phenotype is the result of the $aabb$ genotype. This is consistent with Cross 1.

(d) Now we have only to determine the genotypes for rose and pea. A logical prediction would be that at least one dominant A or B allele combined with the

Walnut

Pea

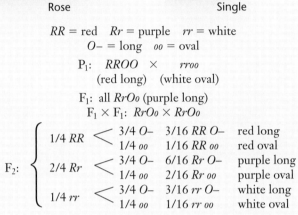

Rose

Single

double recessive condition of the other allele pair can account for these phenotypes. For example,

$$A\text{--}bb \longrightarrow \text{rose}$$
$$aa\ B\text{--} \longrightarrow \text{pea}$$

If, in Cross 3, *AAbb* (rose) were crossed with *aaBB* (pea), all offspring would be *AaBb* (walnut). This is consistent with the data, and we must now only look at Cross 4. We predict these walnut genotypes to be *AaBb* (as above), and from the cross

$$\begin{array}{ccc} AaBb & \times & AaBb \\ \text{(walnut)} & & \text{(walnut)} \end{array}$$

We expect

$$9/16\ A\text{--}B\text{--}\ \text{(walnut)}$$
$$3/16\ A\text{--}bb\ \text{(rose)}$$
$$3/16\ aa\ B\text{--}\ \text{(pea)}$$
$$1/16\ aa\ bb\ \text{(single)}$$

Our prediction is consistent with the data. The initial hypothesis of the epistatic interaction of two gene pairs proves consistent throughout, and the problem has been solved.

This example illustrates the need to have a basic theoretical knowledge of transmission genetics. Then, you must search for the appropriate clues so that you can proceed in a stepwise fashion toward a solution. Mastering problem solving requires practice, but it provides a great deal of satisfaction.

2. In radishes, flower color may be red, purple, or white. The edible portion of the radish may be long or oval. When only flower color is studied, red × white yields all purple. If these F_1 purples are interbred, no dominance is evident, and the F_2 generation consists of 1/4 red:1/2 purple:1/4 white. Regarding radish shape, long is dominant to oval in a normal Mendelian fashion.

(a) Determine the F_1 and F_2 phenotypes from a cross between a true-breeding red long radish and one that is white oval. Be sure to define all gene symbols initially.

Solution: This is a modified dihybrid cross where the gene pair controlling color exhibits incomplete dominance. Shape is controlled conventionally. First, establish gene symbols:

RR = red Rr = purple rr = white
$O\text{--}$ = long oo = oval

$$\text{P}_1: \quad \begin{array}{ccc} RROO & \times & rroo \\ \text{(red long)} & & \text{(white oval)} \end{array}$$

F_1: all *RrOo* (purple long)
$F_1 \times F_1$: *RrOo* × *RrOo*

$$F_2: \left\{ \begin{array}{l} 1/4\ RR \begin{cases} 3/4\ O\text{--} \\ 1/4\ oo \end{cases} \begin{array}{l} 3/16\ RR\ O\text{--} \\ 1/16\ RR\ oo \end{array} \begin{array}{l} \text{red long} \\ \text{red oval} \end{array} \\ 2/4\ Rr \begin{cases} 3/4\ O\text{--} \\ 1/4\ oo \end{cases} \begin{array}{l} 6/16\ Rr\ O\text{--} \\ 2/16\ Rr\ oo \end{array} \begin{array}{l} \text{purple long} \\ \text{purple oval} \end{array} \\ 1/4\ rr \begin{cases} 3/4\ O\text{--} \\ 1/4\ oo \end{cases} \begin{array}{l} 3/16\ rr\ O\text{--} \\ 1/16\ rr\ oo \end{array} \begin{array}{l} \text{white long} \\ \text{white oval} \end{array} \end{array} \right.$$

Note above that to generate the F_2 results, we have used the forked-line method. First, the outcome of crossing F_1 parents for the color genes is considered ($Rr \times Rr$). Then the outcome of shape is considered ($Oo \times Oo$).

(b) A red oval plant was crossed with a plant of unknown genotype and phenotype, yielding the data shown below. Determine the genotype and phenotype of the unknown plant.

Offspring: 103 red long:101 red oval:
              98 purple long:100 purple oval

Solution: Since the two characters are inherited independently, consider them separately. The data indicate a 1/4:1/4:1/4:1/4 proportion. First, consider color:

$$\text{P}_1: \quad \text{red} \times ???\ \text{(unknown)}$$
$$F_1: \quad 204\ \text{red}\ (1/2)$$
$$198\ \text{purple}\ (1/2)$$

Because the red parent *must* be *RR*, the unknown must have a genotype of *Rr* to produce these results. It is thus purple. Now, consider shape:

$$\text{P}_1: \quad \text{oval} \times ???\ \text{(unknown)}$$
$$F_1: \quad 201\ \text{long}\ (1/2)$$
$$201\ \text{oval}\ (1/2)$$

Because the oval plant *must* be *oo*, the unknown plant must have a genotype of *Oo* to produce these results. It is thus long. The unknown plant is thus

$$RrOo\ \text{purple long}$$

3. In a plant, height varies from 6 to 36 cm. When 6-cm and 36-cm plants are cross-fertilized, all F_1 plants are 21 cm. In the F_2 generation, about 3 of every 200 plants were as short as the 6-cm P_1 parent.

   (a) What mode of inheritance is illustrated and how many gene pairs are involved?

   Solution: This problem illustrates polygenic inheritance, where a continuous trait (height) is studied and where alleles contribute *additively* to the phenotype. When 3/200 F_2 plants are as extreme as either P_1 parent, we first must reduce the ratio:

   $$3/200 = 1/66.67$$

   If two gene pairs were involved, the ratio would be about 1/16 ($1/4^2$). If three gene pairs were involved, the ratio would be about 1/64 ($1/4^3$). If four gene pairs were involved, then the ratio would be about 1/256 ($1/4^4$), etc. On this basis, we conclude that three gene pairs are involved since 1/66.6 is very close to 1/64.

   (b) How much does each additive allele contribute to height?

   Solution: The variance between the two extremes equals

   $$36 - 6 = 30 \text{ cm}$$

   Since there are six potential additive alleles (*AABBCC*), then each contributes 30/6 = 5 cm to the base height of 6 cm, which occurs when no additive alleles are present (*aabbcc*).

   (c) List all genotypes that give rise to plants that are 31 cm.

   Solution: All combinations with five additive alleles result in a 31-cm phenotype:

   AABBCc
   AABbCC
   AaBBCC

   (d) In a cross separate from the F_1 above, a plant of unknown phenotype and genotype was test-crossed with the following results:

   1/4   11 cm
   2/4   16 cm
   1/4   21 cm

An astute genetics student realized that the unknown plant could be only one phenotype, but could be any of three genotypes. What were they?

Solution: When test-crossed (with *aabbcc*), the unknown plant must be able to contribute either one, two, or three additive alleles in its gametes in order to yield the three phenotypes in the offspring. Since no 6-cm offspring are observed, the unknown plant *never* contributes all nonadditive alleles (*abc*). Only plants that are homozygous at one locus and heterozygous at the other two loci will meet these criteria. Therefore, the unknown parent can be any of three genotypes, all of which have a phenotype of 21 cm:

   AABbCc
   AaBbCC
   AaBBCc

For example, in the first genotype (*AABbCc*):

   AABbCc × *aabbcc*

   1/4 *AaBbCc* $\longrightarrow$ 21 cm
   1/4 *AaBbcc* $\longrightarrow$ 16 cm
   1/4 *AabbCc* $\longrightarrow$ 16 cm
   1/4 *Aabbcc* $\longrightarrow$ 11 cm

which is the ratio of phenotypes observed.

4. In humans, red-green color blindness is inherited as an X-linked recessive trait. A woman with normal vision, but whose father is color-blind, marries a male who has normal vision. Predict the color vision of their male and female offspring.

   Solution: The female is heterozygous since she inherited an X chromosome with the mutant allele from her father. Her husband is normal. Therefore, the parental genotypes are

   $$Cc \times C\text{♂} \quad (\text{♂ is the Y chromosome.})$$

   All female offspring are normal (*CC* or *Cc*). One-half of the male children will be color-blind (*c*♂), and the other half will have normal vision (*C*♂).

PROBLEMS AND DISCUSSION QUESTIONS

1. In shorthorn cattle, coat color may be red, white, or roan. Roan is an intermediate phenotype expressed as a mixture of red and white hairs. The following data were obtained from various crosses:

red × red	$\longrightarrow$ all red
white × white	$\longrightarrow$ all white
red × white	$\longrightarrow$ all roan
roan × roan	$\longrightarrow$ 1/4 red: 1/2 roan: 1/4 white

How is coat color inherited? What are the genotypes of parents and offspring for each cross?

2. Contrast incomplete dominance and codominance.

3. In foxes, two alleles of a single gene, *P* and *p*, may result in lethality (*PP*), platinum coat (*Pp*), or silver coat (*pp*). What ratio is obtained when platinum foxes are interbred? Is the *P* allele behaving dominantly or recessively in causing lethality? In causing platinum coat color?

4. In mice, a short-tailed mutant was discovered. When it was crossed to a normal long-tailed mouse, 4 offspring were short-tailed and 3 were long-tailed. Two short-tailed mice from the F₁ generation were selected and crossed. They produced 6 short-tailed and 3 long-tailed mice. These genetic experiments were repeated three times with approximately the same results. What genetic ratios are illustrated? Hypothesize the mode of inheritance and diagram the crosses.

5. List all possible genotypes for the A, B, AB, and O phenotypes. Is the mode of inheritance of the ABO blood types representative of dominance? Of recessiveness? Of codominance?

6. With regard to the ABO blood types in humans, determine the genotype of the male parent and female parent below:

   Male Parent: Blood type B; mother type O
   Female Parent: Blood type A; father type B

   Predict the blood types of the offspring that this couple may have and the expected proportion of each.

7. In a disputed parentage case, the child is blood type O while the mother is blood type A. What blood type would exclude a male from being the father? Would the other blood types prove that a particular male was the father?

8. The A and B antigens in humans may be found in water-soluble form in secretions, including saliva, of some individuals (*Se/Se* and *Se/se*) but not in others (*se/se*). The population thus contains "secretors" and "nonsecretors."

   (a) Determine the proportion of various phenotypes (blood type and ability to secrete) in matings between individuals that are blood type AB and type O, both of whom are *Se/se*.

   (b) How will the results of such matings change if both parents are heterozygous for the gene controlling the synthesis of the H substance (*Hh*)?

9. Distinguish between epistasis and polygenic inheritance, and between discontinuous and continuous variation.

10. In rabbits, a series of multiple alleles controls coat color in the following way: *C* is dominant to all other alleles and causes full color. The chinchilla phenotype is due to the *c^cb* allele, which is dominant to all alleles other than *C*. The *c^b* allele, dominant only to *c^a* (albino), results in the Himalayan coat color. Thus, the order of dominance is *C > c^cb > c^b > c^a*. For each of the three cases below, the phenotypes of the P₁ generations of two crosses are shown, as well as the phenotype of one member of the F₁ generation. For each case, determine the genotypes of the P₁ generation and the F₁ offspring and predict the results of making each cross between F₁ individuals as shown.

	P₁ Phenotypes	F₁ Phenotypes
(a)	Himalayan × Himalayan ⟶ albino	
		× ⟶ ??
	full color × albino ⟶ chinchilla	
(b)	albino × chinchilla ⟶ albino	
		× ⟶ ??
	full color × albino ⟶ full color	
(c)	chinchilla × albino ⟶ Himalayan	
		× ⟶ ??
	full color × albino ⟶ Himalayan	

11. In the guinea pig, one locus involved in the control of coat color may be occupied by any of four alleles: *C* (black), *c^k* (sepia), *c^d* (cream), or *c^a* (albino). Like coat color in rabbits (Problem 10), an order of dominance exists: *C > c^k > c^d > c^a*. In the following crosses, write the parental genotypes and predict the phenotypic ratios that would result.

    (a) sepia × cream, where both guinea pigs had an albino parent

    (b) sepia × cream, where the sepia guinea pig had an albino parent and the cream guinea pig had two sepia parents

    (c) sepia × cream, where the sepia guinea pig had two full color parents and the cream guinea pig had two sepia parents

    (d) sepia × cream, where the sepia guinea pig had a full color parent and an albino parent and the cream guinea pig had two full color parents.

12. Three gene pairs located on separate autosomes determine flower color and shape as well as plant height. The first pair exhibits incomplete dominance where color can be red, pink (the heterozygote), or white. The second pair leads to personate (dominant) or peloric (recessive) flower shape, while the third gene pair produces either the dominant tall trait or the recessive dwarf trait. Homozygous plants that are red, personate, and tall are crossed to those that are white, peloric, and dwarf. Determine the F₁ genotype(s) and phenotype(s). If the F₁ plants are interbred, what proportion of the offspring will exhibit the same phenotype as the F₁ plants?

13. As in Problem 12, color may be *red*, *white*, or *pink*, and flower shape may be *personate* or *peloric*. For the following crosses, determine the P₁ and F₁ genotypes. What phenotypic ratios would result from crossing the F₁ of (a) to the F₁ of (b)?

(a) red peloric × white personate ⟶ F₁ = all pink personate
(b) red personate × white peloric ⟶ F₁ = all pink personate
(c) pink personate × red peloric ⟶ F₁ = 1/4 red personate
         1/4 red peloric
         1/4 pink personate
         1/4 pink peloric
(d) pink personate × white peloric ⟶ F₁ = 1/4 white personate
         1/4 white peloric
         1/4 pink personate
         1/4 pink peloric

14. Horses can be *cremello* (a light cream color), *chestnut* (a brownish color), or *palomino* (a golden color with white in the horse's tail and mane). Of these phenotypes, only palominos never breed true.
    (a) From the results shown below, determine the *mode of inheritance* by assigning gene symbols and indicating which genotypes yield which phenotypes.

$$
\begin{array}{l}
\text{cremello} \times \text{palomino} \longrightarrow
\begin{array}{l} 1/2 \ \text{cremello} \\ 1/2 \ \text{palomino} \end{array} \\[2ex]
\text{chestnut} \times \text{palomino} \longrightarrow
\begin{array}{l} 1/2 \ \text{chestnut} \\ 1/2 \ \text{palomino} \end{array} \\[2ex]
\text{palomino} \times \text{palomino} \longrightarrow
\begin{array}{l} 1/4 \ \text{chestnut} \\ 1/2 \ \text{palomino} \\ 1/4 \ \text{cremello} \end{array}
\end{array}
$$

    (b) Predict the F_1 and F_2 results of many initial matings between cremello and chestnut horses.

15. With reference to the eye color phenotypes produced by the recessive, autosomal, unlinked *brown* and *scarlet* loci in *Drosophila* (see Figure 4.11), predict the F_1 and F_2 results of the following P_1 crosses. Recall that when both the *brown* and *scarlet* alleles are homozygous, no pigment is produced, and the eyes are white.
    (a) wild type × white
    (b) wild type × scarlet
    (c) brown × white

16. Pigment in the mouse is only produced when the *C* allele is present. Individuals of the *cc* genotype have no color. If color is present, it may be determined by the *A*, *a* alleles. *AA* or *Aa* results in agouti color, while *aa* results in black coats.
    (a) What F_1 and F_2 genotypic and phenotypic ratios are obtained from a cross between *AACC* and *aacc* mice?
    (b) In three crosses between agouti females whose genotypes were unknown and males of the *aacc* genotype, the following phenotypic ratios were obtained:

    (1) 8 agouti   (2) 9 agouti   (3) 4 agouti
         8 colorless   10 black   5 black
                                            10 colorless

    What are the genotypes of these female parents?

17. In some plants a red pigment, cyanidin, is synthesized from a colorless precursor. The addition of a hydroxyl group (OH^-) to the cyanidin molecule causes it to become purple. In a cross between two randomly selected purple plants, the following results were obtained:

    94 purple
    31 red
    43 colorless

    How many genes are involved in the determination of these flower colors? Which genotypic combinations produce which phenotypes? Diagram the purple × purple cross.

18. In rats, the following genotypes of two independently assorting autosomal genes determine coat color:

A–B–	(gray)
A–Bb	(yellow)
aa B–	(black)
aa bb	(cream)

A third gene pair on a separate autosome determines whether or not any color will be produced. The *CC* and *Cc* genotypes allow color according to the expression of the *A* and *B* alleles. However, the *cc* genotype results in albino rats regardless of the *A* and *B* alleles present. Determine the F_1 phenotypic ratio of the following crosses:

    (a) *AAbbCC* × *aaBBcc*
    (b) *AaBBCC* × *AABbcc*
    (c) *AaBbCc* × *AaBbcc*
    (d) *AaBBCc* × *AaBBCc*
    (e) *AABbCc* × *AABbcc*

19. Given the inheritance pattern of coat color in rats, as described in Problem 18, predict the genotype and phenotype of the parents who produced the following F_1 offspring:
    (a) 9/16 gray:3/16 yellow:3/16 black:1/16 cream
    (b) 9/16 gray:3/16 yellow:4/16 albino
    (c) 27/64 gray:16/64 albino:9/64 yellow:9/64 black: 3/6 cream
    (d) 3/8 black:3/8 cream:2/8 albino
    (e) 3/8 black:4/8 albino:1/8 cream

20. In a species of the cat family, eye color can be gray, blue, green, or brown, and each trait is true-breeding. In separate crosses involving homozygous parents, the following data were obtained:

Cross	P_1	F_1	F_2
A	green × gray	all green	3/4 green: 1/4 gray
B	green × brown	all green	3/4 green: 1/4 brown
C	gray × brown	all green	9/16 green: 3/16 brown: 3/16 gray: 1/16 blue

    (a) Analyze the data: How many genes are involved? Define gene symbols and indicate which genotypes yield each phenotype.
    (b) In a cross between a gray-eyed cat and one of unknown genotype and phenotype, the F_1 generation was not observed. However, the F_2 resulted in the same F_2 ratio as in cross *C*. Determine the genotypes and phenotypes of the unknown P_1 and F_1 cats.

21. In a plant, a tall variety was crossed with a dwarf variety. All F_1 plants were tall. When $F_1 \times F_1$ plants were interbred, 9/16 of the F_2 were tall and 7/16 were dwarf.
    (a) Explain the inheritance of height by indicating the number of gene pairs involved and by designating which genotypes yield tall and which yield dwarf (use dashes where appropriate).
    (b) Of the F_2 plants, what proportion of them will be true-breeding if self-fertilized? List these genotypes.

22. In a unique species of plants, flowers may be yellow, blue, red, or mauve. All colors may be true-breeding. If plants

with blue flowers are crossed to red-flowered plants, all F_1 plants have yellow flowers. When carried to an F_2 generation, the following ratio was observed:

9/16 yellow:3/16 blue:3/16 red:1/16 mauve

In still another cross using true-breeding parents, yellow-flowered plants are crossed with mauve-flowered plants. Again, all F_1 plants had yellow flowers and the F_2 showed a 9:3:3:1 ratio, as shown above.
   (a) Describe the inheritance of flower color by defining gene symbols and designating which genotypes give rise to each of the four phenotypes.
   (b) Determine the F_1 and F_2 results of a cross between true-breeding red and true-breeding mauve-flowered plants.

23. Shown below are five maternal and paternal phenotypes (1–5), each designating the ABO, MN, and Rh blood-group antigens. Each combination resulted in one of the five offspring shown to the right (a–e). Arrange the offspring with the correct parents such that all five cases are consistent. Is there more than one set of correct answers?

Parental Phenotypes	Offspring
(1) A, M, Rh$^-$ × A, N, Rh$^-$	(a) A, N, Rh$^-$
(2) B, M, Rh$^-$ × B, M, Rh$^+$	(b) O, N, Rh$^+$
(3) O, N, Rh$^+$ × B, N, Rh$^+$	(c) O, MN, Rh$^-$
(4) AB, M, Rh$^-$ × O, N, Rh$^+$	(d) B, M, Rh$^+$
(5) AB, MN, Rh$^-$ × AB, MN, Rh$^-$	(e) B, MN, Rh$^+$

24. Assume that height in a plant is controlled by two gene pairs and that each additive allele contributes 5 cm to a base height of 20 cm (i.e., *aabb* is 20 cm).
   (a) What is the height of an *AABB* plant?
   (b) Predict the phenotypic ratios of F_1 and F_2 plants in a cross between *aabb* and *AABB*.
   (c) List all genotypes that give rise to plants that are 25 and 35 cm in height.
25. In a cross where three gene pairs determine weight in squash, what proportion of individuals from the cross *AaBbCC* × *AABbcc* will contain only two additive alleles? Which genotype or genotypes fall into this category?
26. An inbred strain of plants has a mean height of 24 cm. A second strain of the same species from a different geographical region also has a mean height of 24 cm. When plants from the two strains are crossed together, the F_1 plants are the same height as the parent plants. However, the F_2 generation shows a wide range of heights; the majority are like the P_1 and F_1 plants, but approximately 4 of 1000 are only 12 cm high, and about 4 of 1000 are 36 cm high.
   (a) What mode of inheritance is occurring here?
   (b) How many gene pairs are involved?
   (c) How much does each gene contribute to plant height?
   (d) Indicate one possible set of genotypes for the original P_1 parents and the F_1 plants that could account for these results.

   (e) Indicate three possible genotypes that could account for F_2 plants that are 18 cm high and F_2 plants that are 33 cm high.
27. In a series of crosses between plants of various heights to a 20-inch plant, the following results were obtained in the F_1 generations:

(a)	4″ × 20″	⟶	All 12″
(b)	8″ × 20″	⟶	All 14″
(c)	12″ × 20″	⟶	All 16″
(d)	16″ × 20″	⟶	All 18″

Propose an explanation for the inheritance of height in the above plant. Under the constraints of your explanation, predict the genotypes of plants of each height.
28. Erma and Harvey were a compatible barnyard pair, but a curious sight. Harvey's tail was only 6 cm while Erma's was 30 cm. Their F_1 piglet offspring all grew tails that were 18 cm. When inbred, an F_2 generation resulted in many piglets (Erma and Harvey's grandpigs) whose tails ranged in 4-cm intervals from 6 to 30 cm (6, 10, 14, 18, 22, 26, 30). Most had 18-cm tails, while 1/64 had 6-cm and 1/64 had 30-cm tails.
   (a) Explain how tail length is inherited by describing the mode of inheritance, indicating how many gene pairs are at work, and designating the genotypes of Harvey, Erma, and their 18-cm offspring.
   (b) If one of the 18-cm F_1 pigs were mated with the 6-cm F_2 pigs, what phenotypic ratio would be predicted if many offspring resulted? Diagram the cross.
29. Plants may be 10, 20, 30, 40, 50, 60, or 70 cm high where plant height is under polygenic control. A true-breeding plant that is 10 cm is crossed to another true-breeding plant that is 50 cm high. How many gene pairs are involved? What F_1 and F_2 results can be predicted?
30. A husband and wife have normal vision, although both of their fathers are red-green color-blind, which is inherited as an X-linked recessive condition. What is the probability that their first child will be (a) A normal son? (b) A normal daughter? (c) A color-blind son? (d) A color-blind daughter?
31. In humans, the ABO blood type is under the control of autosomal multiple alleles. Color blindness is a recessive X-linked trait. If two parents who are both type A and have normal vision produce a son who is color-blind and is type O, what is the probability that their next child will be a female who has normal vision and is type O?
32. In *Drosophila*, an X-linked recessive mutation, *scalloped* (*sd*), causes irregular wing margins. Diagram the F_1 and F_2 results if: (a) A *scalloped* female is crossed with a normal male; (b) a *scalloped* male is crossed with a normal female. Compare these results to those that would be obtained if *scalloped* were not X-linked.
33. Another recessive mutation in *Drosophila*, *ebony* (*e*), is on an autosome (chromosome 3) and causes darkening of the body compared with wild-type flies. What phenotypic F_1 and F_2 male and female ratios will result if a *scalloped*-winged female

with normal body color is crossed with a normal-winged *ebony* male? Work this problem by both the Punnett square method and the forked-line method.

34. In *Drosophila*, the X-linked recessive mutation *vermilion* (*v*) causes bright red eyes, which is in contrast to brick-red eyes of wild type. A separate autosomal recessive mutation, *suppressor of vermilion* (*su-v*), causes flies homozygous or hemizygous for *v* to have wild-type eyes. In the absence of vermilion alleles, *su-v* has no effect on eye color. Determine the F_1 and F_2 phenotypic ratios from a cross between a female with wild-type alleles at the *vermilion* locus, but who is homozygous for *su-v*, with a *vermilion* male who has wild-type alleles at the *su-v* locus.

35. While *vermilion* is X-linked and brightens the eye color, *brown* is an autosomal recessive mutation that darkens the eye. Flies carrying both mutations lose all pigmentation and are white eyed. Predict the F_1 and F_2 results of the following crosses:
(a) vermilion females × brown males
(b) brown females × vermilion males
(c) white females × wild-type males

36. In a cross in *Drosophila* involving the X-linked recessive eye mutation *white* and the autosomally linked recessive eye mutation *sepia* (resulting in a dark eye), predict the F_1 and F_2 results of crossing true-breeding parents of the following phenotypes:
(a) white females × sepia males
(b) sepia females × white males
Note that *white* is epistatic to the expression of *sepia*.

37. Consider the three pedigrees below, all involving a single human trait.

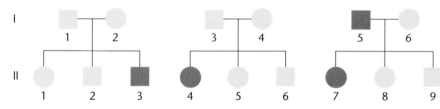

(a) Which conditions, if any, can be *excluded?*

Conditions: dominant *and* X-linked
dominant *and* autosomal
recessive *and* X-linked
recessive *and* autosomal

(b) For any condition that you excluded, indicate the *single individual* in generation II (e.g., II-1, II-2, etc.) that was most instrumental in your decision to exclude that condition. If none were excluded, answer "none apply."

(c) Given your conclusions above, indicate the *genotype* of the individuals listed below. If more than one possibility applies, list all possibilities. Use the symbols *A* and *a* for the genotypes.

II-1; II-6; II-9

38. In spotted cattle, the colored regions may be mahogany or red. If a red female and a mahogany male, both derived from separate true-breeding lines, are mated and the cross carried to an F_2 generation, the following results are obtained:

F_1: 1/2 mahogany males
1/2 red females
F_2: 3/8 mahogany males
1/8 red males
1/8 mahogany females
3/8 red females

When the reciprocal of the initial cross is performed (mahogany female and red male), identical results are obtained. Explain these results by postulating how the color is genetically determined. Diagram the crosses.

39. Predict the F_1 and F_2 results of crossing a male fowl that is cock-feathered with a true-breeding hen-feathered female fowl. Recall that these traits are sex-limited, as discussed in the text.

40. Two mothers give birth to sons at the same time at a busy urban hospital. The son of couple 1 is afflicted with hemophilia, a disease caused by an X-linked, recessive allele. Neither parent has the disease. Couple 2 have a normal son, despite the fact that the father has hemophilia. Several years later, couple 1 sues the hospital, claiming that these two newborns were swapped in the nursery following their birth. You are called, as a genetic counselor, to testify. What information can you provide the jury concerning the allegation?

41. In the parakeet, two autosomal genes (which are located on separate autosomes) control the production of feather pigment. Gene B controls the production of a blue pigment, and gene Y controls the production of a yellow pigment. Recessive mutations in each gene are known that interrupt the synthesis of the respective pigments.

A blue parakeet from a pet store in California is crossed to a yellow parakeet from a pet store in New Jersey. Both birds come from true-breeding strains. To the surprise of the amateur breeder, all the F_1 parakeets were green. Following many matings an F_2 generation yielded the following results:

green 92   blue 31   yellow 29   albino 11

(a) Explain the inheritance of feather color by indicating which genotypes yield which phenotypes as well as the genotypes of the P_1, F_1, and F_2 individuals.
(b) Predict the outcome of a test cross between the F_1 green bird and an albino bird.

EXTRA-SPICY PROBLEMS

42. Labrador retrievers may be black, brown, or golden in color. While each color may breed true, many different outcomes occur if many litters are examined from a variety of matings, where the parents are not necessarily true-breeding. Shown below are just some of the many possibilities. Propose a mode of inheritance that is consistent with these data, and indicate the corresponding genotypes of the parents in each mating. Indicate as well the genotypes of dogs that breed true for each color.

(a)	black × brown	⟶	all black
(b)	black × brown	⟶	1/2 black
			1/2 brown
(c)	black × brown	⟶	3/4 black
			1/4 golden
(d)	black × golden	⟶	all black
(e)	black × golden	⟶	4/8 golden
			3/8 black
			1/8 brown
(f)	black × golden	⟶	2/4 golden
			1/4 black
			1/4 brown
(g)	brown × brown	⟶	3/4 brown
			1/4 golden
(h)	black × black	⟶	9/16 black
			4/16 golden
			3/16 brown

43. Members of a strain of inbred skunks whose stripes are uniformly 20 cm long are crossed to those of another strain whose stripes are uniformly 24 cm. The F_1 hybrids all have stripes that are 22 cm. When the F_1 skunks were crossed, the F_2 generation produced animals with stripes of 16, 18, 20, 22, 24, 26, and 28 cm. The 22-cm variety was most frequent and the 16- and 28-cm variety least frequent. Of 1000 F_2 skunks, 15 were 16 cm and 16 were 28 cm.
    (a) What is the mode of inheritance illustrated above?
    (b) How many gene pairs are involved?
    (c) What is the contribution of each additive allele to stripe length?
    (d) What are the genotypes of the P_1 parents and F_1 offspring?
    (e) List four genotypes that give rise to skunks whose stripes are 20 cm long.

44. A true-breeding purple-leafed plant isolated from one side of the rain forest in Puerto Rico (El Yunca) was crossed to a true-breeding white variety found on the other side of the rain forest. The F_1 offspring were all purple. A large number of crosses resulted in the following results:

    purple: 4219   white: 5781   (Total = 10,000)

    Propose an explanation for the inheritance of leaf color. As a geneticist, how might you go about testing your hypothesis? Describe the genetic experiments that you would conduct.

45. In Dexter-Kerry cattle, animals may be polled or horned. The Dexter animals have short legs, whereas the Kerry animals have long legs. When many offspring were obtained from matings between polled Kerrys and horned Dexters, one-half were found to be polled Dexters and one-half polled Kerrys. When these F_1 cattle were interbred, the following F_2 data were obtained:

    3/8 polled Dexters
    3/8 polled Kerrys
    1/8 horned Dexters
    1/8 horned Kerrys

    A geneticist was puzzled by these data and interviewed farmers who had bred these cattle for decades. She learned that Kerrys were true-breeding. Dexters, on the other hand, were not true-breeding and never produced as many offspring as Kerrys. Provide a genetic explanation for these observations.

46. An invading alien geneticist from a planet where genetic research was prohibited brought with him to Earth two pure-breeding lines of pet frogs. One line croaked by uttering *rib-it rib-it*, and had purple eyes. The other line croaked more softly by muttering *knee-deep knee-deep*, and had green eyes. With a new-found sense of inquiry, he mated the two types of frogs. In the F_1, all frogs had blue eyes and uttered *rib-it rib-it*. He proceeded to make many $F_1 \times F_1$ crosses, and when he fully analyzed the F_2 data, he realized they could be reduced to the following ratio:

    27/64 blue-eyed, rib-it utterer
    12/64 green-eyed, rib-it utterer
     9/64 blue-eyed, knee-deep mutterer
     9/64 purple-eyed, rib-it utterer
     4/64 green-eyed, knee-deep mutterer
     3/64 purple-eyed, knee-deep mutterer

    (a) How many total gene pairs are involved in the inheritance of both traits? Support your answer.
    (b) Of these, how many are controlling eye color? How can you tell? How many are controlling croaking?
    (c) Assign gene symbols for all phenotypes and indicate the genotypes of the P_1 and F_1 frogs.
    (d) Indicate the genotypes of the six F_2 phenotypes.
    (e) After years of experiments, the geneticist isolated pure-breeding strains of all six F_2 phenotypes. Indicate the F_1 and F_2 phenotypic ratios of the following cross using these pure-breeding strains:

    blue-eyed, knee-deep mutterer
    ×
    purple-eyed, rib-it utterer

    (f) One set of crosses with his true-breeding lines initially caused the geneticist some confusion. When he crossed true-breeding purple-eyed, knee-deep mutterers with true-breeding green-eyed, knee-deep mutterers, he often got different results. In some matings, all offspring were blue-eyed, knee-deep mutterers, but in other matings all offspring were purple-eyed, knee-deep mutterers. In still a third mating, 1/2 blue-eyed, knee-deep mutterers and 1/2 purple-eyed, knee-deep mutterers were observed. Explain why the results differed.

(g) In another experiment, the geneticist crossed two purple-eyed, rib-it utterers together with the results shown below. What were the genotypes of the two parents?

> 9/16 purple-eyed, rib-it utterer
> 3/16 purple-eyed, knee-deep mutterer
> 3/16 green-eyed, rib-it utterer
> 1/16 green-eyed, knee-deep mutterer

47. The following pedigree is characteristic of an inherited condition known as **male precocious puberty**, where affected males show signs of puberty by age 4. Propose a genetic explanation of this phenotype. [*Reference:* A. Shenker, (1993).]

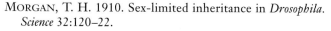

SELECTED READINGS

BRINK, R. A., ed. 1967. *Heritage from Mendel.* Madison: University of Wisconsin Press.

BULTMAN, S. J., MICHAUD, E. J., and WOYCHIK, R. P. 1992. Molecular characterization of the mouse *agouti* locus. *Cell* 71:1195–1204.

CHAPMAN, A. B. 1985. *General and quantitative genetics.* Amsterdam: Elsevier.

CLARKE, C. A. 1968. The prevention of "Rhesus" babies. *Sci. Am.* (Nov.) 219:46–52.

CORWIN, H. O., and JENKINS, J. B. 1976. *Conceptual foundations of genetics: Selected readings.* Boston: Houghton-Mifflin.

CROW, J. F. 1983. *Genetics notes*, 8th ed. New York: Macmillan.

DRAYNA, D., and WHITE, R. 1985. The genetic linkage map of the human X chromosome. *Science* 230:753–58.

DUNN, L. C. 1966. *A short history of genetics.* New York: McGraw-Hill.

EAST, E. M. 1910. A Mendelian interpretation of variation that is apparently continuous. *Am. Naturalist* 44:65–82.

———. 1916. Studies on size inheritance in *Nicotiana. Genetics* 1:164–76.

FALCONER, D. S. 1981. *Introduction to quantitative genetics*, 2nd ed. New York: Longman.

FOSTER, H. L., et al., eds. 1981. *The mouse in biomedical research: Vol. 1. History, genetics, and wild mice.* Orlando, FL: Academic Press.

FOSTER, M. 1965. Mammalian pigment genetics. *Adv. Genet.* 13:311–39.

GRANT, V. 1975. *Genetics of flowering plants.* New York: Columbia University Press.

LINDSLEY, D. C., and GRELL, E. H. 1967. *Genetic variations of Drosophila melanogaster.* Washington, DC: Carnegie Institute of Washington.

MCKUSICK, V. A. 1962. On the X chromosome of man. *Quart. Rev. Biol.* 37:69–175.

MORGAN, T. H. 1910. Sex-limited inheritance in *Drosophila. Science* 32:120–22.

NOLTE, D. J. 1959. The eye-pigmentary system of *Drosophila. Heredity* 13:233–41.

PAWELEK, J. M., and KÖRNER, A. M. 1982. The biosynthesis of mammalian melanin. *Am. Sci.* 70:136–45.

PETERS, J. A., ed. 1959. *Classic papers in genetics.* Englewood Cliffs, NJ: Prentice-Hall.

RACE, R. R., and SANGER, R. 1975. *Blood groups in man*, 6th ed. Oxford: Blackwell Scientific Publishers.

SHENKER, A. 1993. A constitutively activating mutation of the luteinizing hormone receptor in familial male precocious puberty. *Nature* 365:652–54.

SIRACUSA, L. D. 1994. The *agouti* gene: Turned on to yellow. *Cell* 10:423–28.

STERN, C. 1973. *Principles of human genetics.* 3rd ed. New York: W. H. Freeman.

SMITH, C. A. 1995. New hope for overcoming canine inherited disease. *J. Am. Veter. Med. Assoc.* 204:41–46.

TEARLE, R. G., et al. 1989. Cloning and characterization of the *scarlet* gene of *Drosophila melanogaster. Genetics* 122:595–606.

VOELLER, B. R., ed. 1968. *The chromosome theory of inheritance—Classic papers in development and heredity.* New York: Appleton-Century-Crofts.

VOGEL, F., and MOTULSKY, A. G. 1986. *Human genetics: Problems and approaches*, 2nd ed. New York: Springer-Verlag.

WATKINS, M. W. 1966. Blood group substances. *Science* 152:172–81.

WIENER, A. S., ed. 1970. *Advances in blood groupings, Vol. 3.* New York: Grune and Stratton.

YOSHIDA, A. 1982. Biochemical genetics of the human blood group ABO system. *Am. J. Hum. Genet.* 34:1–14.

ZIEGLER, I. 1961. Genetic aspects of ommochrome and pterin pigments. *Adv. Genet.* 10:349–403.

5

Linkage, Crossing Over, and Chromosome Mapping

CHAPTER CONCEPTS

Chromosomes contain many genes. Unless separated by crossing over, alleles present at the many loci on each homolog segregate as a unit during gamete formation. Recombinant gametes resulting from crossing over enhance genetic variability within a species and serve as the basis for constructing chromosomal maps.

As early as 1903, Walter Sutton, who along with Theodor Boveri united the fields of cytology and genetics, pointed out the likelihood that in organisms there are many more "unit factors" than chromosomes. Soon thereafter, genetic investigations with several organisms revealed that certain genes were not transmitted according to the law of independent assortment. When studied together in matings, these genes seemed to segregate as if they were somehow joined or linked together. Further investigations showed that such genes were part of the same chromosome, and they were indeed transmitted as a single unit.

We now know that most chromosomes consist of

Chiasmata present between synapsed homologs during the first meiotic prophase.

very large numbers of genes and, in fact, contain sufficient DNA to encode thousands of these units. Genes that are part of the same chromosome are said to be **linked** and to demonstrate **linkage** in genetic crosses.

Because the chromosome, not the gene, is the unit of transmission during meiosis, linked genes are not free to undergo independent assortment. Instead, the alleles at all loci of one chromosome should, in theory, be transmitted as a unit during gamete formation. However, in many instances this does not occur. During the first meiotic prophase, when homologs are paired (or synapsed), a reciprocal exchange of chromosome segments may take place. This event, called **crossing**

over, results in the reshuffling or **recombination** of the alleles between homologs.

The degree of crossing over between any two loci on a single chromosome is proportionate to the distance between them. Thus, the percentage of recombinant gametes varies, depending on which loci are being considered. This correlation serves as the basis for the construction of **chromosome maps**, which provide the relative locations of genes on the chromosomes.

Crossing over is currently viewed as an actual physical breaking and rejoining process that occurs during meiosis. This exchange of chromosome segments provides an enormous potential for genetic variation in the gametes formed by any individual. This type of variation, in combination with that resulting from independent assortment, ensures that all offspring will contain a diverse mixture of maternal and paternal alleles.

In this chapter, we will discuss linkage, crossing over, and chromosome mapping. We will also consider a variety of topics involving the exchange of genetic information. We will conclude the chapter by entertaining the rather intriguing question of why Mendel, who studied seven genes, did not encounter linkage. Or did he?

Linkage versus Independent Assortment

To provide a simplified overview of the major theme of this chapter, Figure 5.1 illustrates and contrasts the meiotic consequences of independent assortment, linkage *without* crossing over, and linkage *with* crossing over. Two homologous pairs of chromosomes are considered in the case of independent assortment and one homologous pair in the two cases of linkage.

Figure 5.1(a) illustrates the results of independent assortment of two pairs of nonhomologous chromosomes, each containing one heterozygous gene pair. No linkage is exhibited by the two genes. When a large number of meiotic events are observed, four genetically different gametes are formed in equal proportions.

We can compare these results with those that occur if the same genes are instead linked on the same chromosome. If no crossing over occurs between the two genes [Figure 5.1(b)], only two genetically different gametes are formed. Each gamete receives the alleles present on one homolog or the other, which has been transmitted intact as the result of segregation. This case illustrates **complete linkage**, which results in the production of only **parental** or **noncrossover gametes**. The two parental gametes are formed in equal proportions. While complete linkage between two genes seldomly occurs, consid-

eration of the theoretical consequences of this concept is useful when studying crossing over.

Figure 5.1(c) illustrates the results when crossing over occurs between two linked genes. As you will note, this crossover involves only two nonsister chromatids of the four chromatids present in the tetrad. This exchange generates two new allele combinations, called **recombinant** or **crossover gametes**. The two chromatids not involved in the exchange result in noncrossover gametes, as those in Figure 5.1(b).

The frequency with which crossing over occurs between any two linked genes is generally proportional to the distance separating the respective loci along the chromosome. In theory, two randomly selected genes can be so close to each other that crossover events are too infrequent to be easily detected. This circumstance, complete linkage, results in the production of only parental gametes, as shown in Figure 5.1(b). On the other hand, if a distinct but small distance separates two genes, few recombinant and many parental gametes will be formed. As the distance between the two genes increases, the proportion of recombinant gametes increases and that of the parental gametes decreases. As a result, as in Figure 5.1(c), the proportion of crossover gametes varies depending on the distance between the genes studied.

As will be explored again later in this chapter, when two linked genes whose loci are far apart are considered, the number of recombinant gametes approaches, but does not exceed, 50 percent. If 50 percent recombinants occurred, a 1:1:1:1 ratio of the four types (two parental and two recombinant gametes) would result. In such a case, transmission of two linked genes would be indistinguishable from that of two unlinked, independently assorting genes. That is, the proportion of the four possible genotypes would be identical, as shown in Figure 5.1(a,c).

The Linkage Ratio

If complete linkage exists between two genes because of their close proximity, and organisms heterozygous at both loci are mated, a unique F_2 phenotypic ratio results, which we shall designate the **linkage ratio**. To illustrate this ratio, we will consider a cross involving the closely linked recessive mutant genes *brown* (*bw*) eye and *heavy* (*hv*) wing vein in *Drosophila melanogaster* (Figure 5.2). The normal, wild-type alleles bw^+ and hv^+ are both dominant and result in red eyes and thin wing veins, respectively.

In this cross, flies with mutant brown eyes and normal thin veins are mated to flies with normal red eyes and mutant heavy veins. We can describe these flies in more concise terms by just referring to their mutant phenotypes and say that brown-eyed flies are crossed with heavy-veined flies. If we extend the system of using genetic symbols es-

(a)   **Independent assortment of two genes on two different homologous pairs of chromosomes**

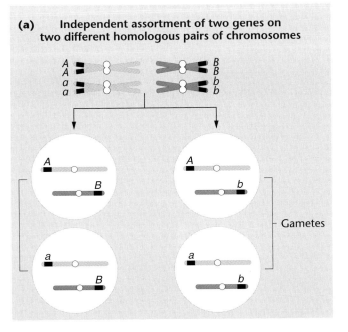

(b)   **Linkage between two genes on a single pair of homologs: no exchange occurs**

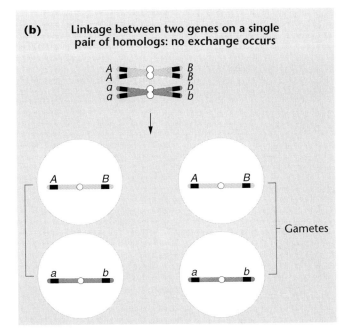

FIGURE 5.1   A comparison of the results of gamete formation where two heterozygous genes are (a) on two different pairs of chromosomes; (b) on the same pair of homologs, but with no exchange occurring between them; and (c) on the same pair of homologs, with an exchange between two nonsister chromatids.

tablished in Chapter 4, linked genes may be represented by placing their allele designations above and below a single or double horizontal line. Those placed above the line are located at loci on one homolog and those placed below at the homologous loci on the other homolog. Thus, we may represent the P_1 generation as follows:

$$P_1: \quad \frac{bw \quad hv^+}{bw \quad hv^+} \quad \times \quad \frac{bw^+ \quad hv}{bw^+ \quad hv}$$

brown, thin   red, heavy

Because the genes are located on an autosome, and not the X chromosome, no distinction for male and female is necessary.

In the F_1 generation each fly receives one chromosome of each pair from each parent; all flies are heterozygous for both gene pairs and exhibit the dominant traits of red eyes and thin veins:

$$F_1: \quad \frac{bw \quad hv^+}{bw^+ \quad hv}$$

red, thin

As shown in Figure 5.2, because of complete linkage, when the F_1 generation is interbred, each F_1 individual forms only parental gametes. Following fertilization, the F_2 generation will be produced in a 1:2:1 phenotypic and genotypic ratio. One-fourth of this generation will show brown eyes and thin veins; one-half will show both wild-

(c)   **Linkage between two genes on a single pair of homologs: exchange occurs between two nonsister chromatids**

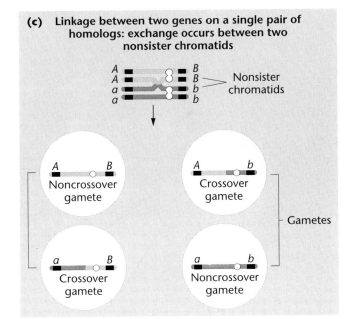

type traits, namely red eyes and thin veins; and one-fourth will show red eyes and heavy veins. In more concise terms, the ratio is 1 brown:2 wild:1 heavy. Such a ratio is characteristic of complete linkage, which is observed only when two genes are very close together and the number of progeny is relatively small.

Figure 5.2 also demonstrates the results of a test cross with the F_1 flies. Such a cross produces a 1:1 ratio of brown, thin and red, heavy flies. Had the genes controlling these traits been incompletely linked or located on separate autosomes, four phenotypes rather than two would have been produced.

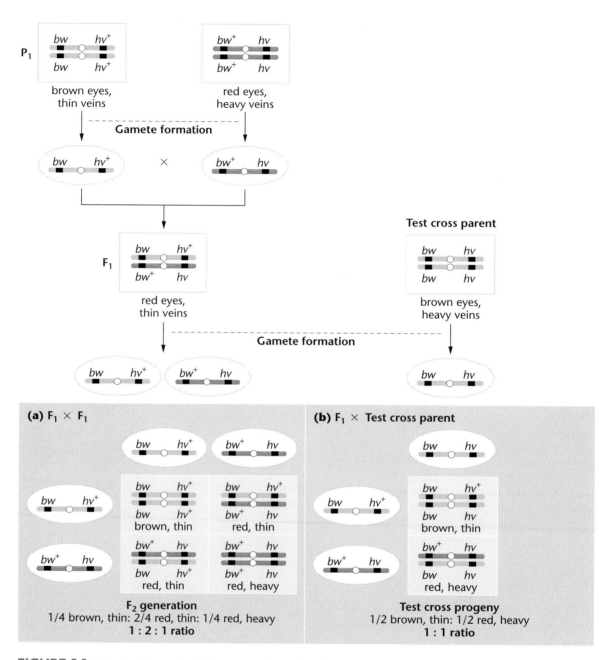

FIGURE 5.2   Results of a cross involving two genes located on the same chromosome, demonstrating complete linkage. Part (a) generates the F₂ results of the cross; Part (b) generates the results of a test cross involving the F₁ progeny.

When large numbers of mutant genes present in any given species are investigated, genes located on the same chromosome will show evidence of linkage to one another. As a result, **linkage groups** can be established, one for each chromosome. In theory, the number of linkage groups should correspond to the haploid number of chromosomes. In organisms where large numbers of mutant genes are available for genetic study, this correlation has been upheld.

Incomplete Linkage, Crossing Over, and Chromosome Mapping

If two genes linked on the same chromosome are selected randomly, it is highly improbable that they will be so close to one another along the chromosome that they demonstrate complete linkage. Instead, crosses involving two randomly selected linked genes will almost always produce a percentage of offspring resulting from recombinant gametes. This percentage is variable and depends upon the distance between the two genes along the chromosome. This phenomenon was first explained by two *Drosophila* geneticists, Thomas H. Morgan and his undergraduate student, Alfred H. Sturtevant.

Morgan and Crossing Over

As you may recall from our earlier discussion in Chapter 4, Morgan first discovered the phenomenon of X-linkage. In his studies, he investigated numerous *Drosophila* mutations located on the X chromosome. When he analyzed crosses involving only one trait, he was able to deduce the mode of X-linked inheritance. However, when he made crosses that simultaneously involved two X-linked genes, his results were at first puzzling. For example, as shown in Cross A of Figure 5.3, he crossed mutant *yellow* body (*y*) and *white* eyed (*w*) females with wild-type males (gray body and red eyes). The F$_1$ females were wild type, whereas the F$_1$ males expressed both mutant traits. In the F$_2$, 98.7 percent of the offspring showed the parental phenotypes—yellow-bodied, white-eyed flies and wild-type flies (gray-bodied, red-eyed). The remaining 1.3 percent of the flies were either yellow-bodied with red eyes or gray-bodied with white eyes. It was as if the genes had somehow separated from each other during gamete formation in the F$_1$ flies.

When Morgan made crosses involving other X-linked genes, the results were even more puzzling (Cross B of Figure 5.3). The same basic pattern was observed, but the proportion of F$_2$ phenotypes differed; for example, in a cross involving *white*-eye, *miniature*-wing mutants, only 62.8 percent of the F$_2$ showed the parental phenotypes, while 37.2 percent of the offspring appeared as if the mutant genes had been separated during gamete formation.

In 1911, Morgan was faced with two questions: (1) What was the source of gene separation? and (2) Why did the frequency of the apparent separation vary depending on the genes being studied? The proposed answer to the first question was based on his knowledge of earlier cytological observations made by F. Janssens and others. Janssens had observed that synapsed homologous chromosomes in meiosis wrapped around each other, creating **chiasmata** [sing., **chiasma**] where points of overlap are evident. Morgan proposed that these chiasmata could represent points of genetic exchange.

In the crosses shown in Figure 5.3, Morgan postulated that if an exchange occurred between the mutant genes on the two X chromosomes of the F$_1$ females, it would lead to the observed results. He suggested that such exchanges led to 1.3 percent recombinant gametes in the *yellow-white* cross and 37.2 percent in the *white-miniature* cross. On the basis of this and other experimen-

tation, Morgan concluded that linked genes exist in a linear order along the chromosome and that a variable amount of exchange occurs between any two genes.

As an answer to the second question, Morgan proposed that two genes located relatively close to each other along a chromosome are less likely to have a chiasma form between them than if the two genes are farther apart on the chromosome. Therefore, the closer two genes are, the less likely it is that a genetic exchange will occur between them. Morgan proposed the term **crossing over** to describe the physical exchange leading to recombination.

Sturtevant and Mapping

Morgan's student, Alfred H. Sturtevant, was the first to realize that his mentor's proposal could also be used to map the sequence of linked genes. According to Sturtevant, "In a conversation with Morgan . . . I suddenly realized that the variations in strength of linkage, already attributed by Morgan to differences in the spatial separation of the genes, offered the possibility of determining sequences in the linear dimension of a chromosome. I went home and spent most of the night (to the neglect of my undergraduate homework) in producing the first chromosomal map. . . ." For example, Sturtevant compiled data on recombination between the genes represented by the *yellow*, *white*, and *miniature* mutants initially studied by Morgan. Frequencies of crossing over between each pair of these three genes were observed to be:

(1)	*yellow, white*	0.5%
(2)	*white, miniature*	34.5%
(3)	*yellow, miniature*	35.4%

Because the sum of (1) and (2) is approximately equal to (3), Sturtevant argued that the recombination frequencies between linked genes are additive. On this basis, he predicted that the order of the genes on the X chromosome was *yellow-white-miniature*. In arriving at this conclusion, he reasoned as follows: The *yellow* and *white* genes are apparently close to each other because the recombination frequency is low. However, both of these genes are quite far apart from *miniature* because the *white, miniature* and *yellow, miniature* combinations show large recombination frequencies. Because *miniature* shows more recombination with *yellow* than with *white* (35.4 vs. 34.5), it follows that *white* is between the other two genes, not outside of them.

Sturtevant knew from Morgan's work that the frequency of exchange could be taken as an estimate of the relative distance between two genes or loci along the chromosome. He constructed a map of the three genes on the X chromosome, with one map unit being equated with 1 percent recombination between two

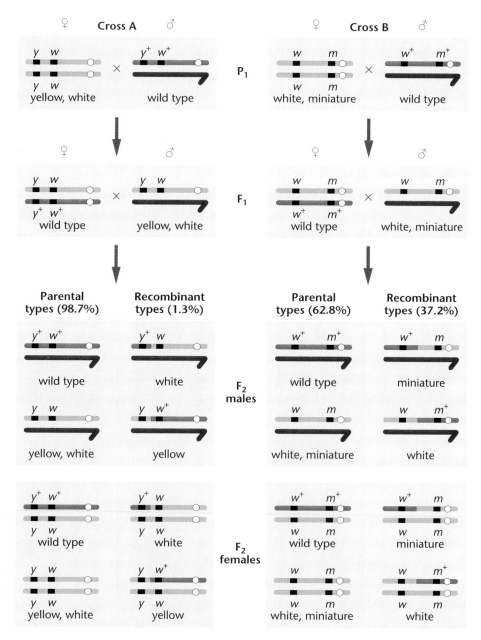

FIGURE 5.3 The F_1 and F_2 results of crosses involving the *yellow*-body, *white*-eye mutations and the *white*-eye, *miniature*-wing mutations. In the F_2 generation of Cross A, 1.3 percent of the flies demonstrate recombinant phenotypes, which express either *white* or *yellow*. In the F_2 generation of Cross B, 37.2 percent of the flies demonstrate recombinant phenotypes, which express either *miniature* or *white*.

genes.* In the preceding example, the distance between *yellow* and *white* would be 0.5 map unit, and between *yellow* and *miniature*, 35.4 map units. It follows that the distance between *white* and *miniature* should be (35.4 − 0.5) or 34.9. This estimate is close to the actual frequency of recombination between *white* and *miniature* (34.5). The simple map for these three genes is shown in Figure 5.4.

In addition to these three genes, Sturtevant considered two other genes on the X chromosome and produced a more extensive map including all five genes. He and a

colleague, Calvin Bridges, soon began a search for autosomal linkage in *Drosophila*. By 1923, they had clearly shown that linkage and crossing over were not restricted to X-linked genes. Rather, they had discovered linked genes on autosomes, between which crossing over occurred.

FIGURE 5.4 A simple map of the *yellow* (*y*), *white* (*w*), and *miniature* (*m*) genes on the X chromosome of *Drosophila melanogaster*. Each number represents the percentage of recombinant offspring produced in the three crosses, each involving two different genes.

*In honor of Morgan's work, one map unit is often referred to as a centimorgan (cM).

During this work, they made another interesting observation. In *Drosophila*, crossing over was shown to occur only in females. The fact that no crossing over occurs in males made genetic mapping much less complex to analyze in *Drosophila*. However, crossing over does occur in both sexes in most other organisms.

Although many refinements in chromosome mapping have developed since Sturtevant's initial work, his basic principles are accepted as correct. They have been used to produce detailed chromosome maps of organisms for which large numbers of linked mutant genes are known. In addition to providing the basis for chromosome mapping, Sturtevant's findings were historically significant to the field of genetics. In 1910, the **chromosomal theory of inheritance** was still being widely disputed. Even Morgan was skeptical of this theory prior to the time he conducted the bulk of his experimentation. Research has now firmly established that chromosomes contain genes in a linear order and that these genes are the equivalent of Mendel's unit factors.

Single Crossovers

Why should the relative distance between two loci influence the amount of recombination and crossing over observed between them? The basis for this variation is explained in the following analysis.

During meiosis, a limited number of crossover events occur in each tetrad. These recombinant events occur randomly along the length of the tetrad. Therefore, the closer that two loci reside along the axis of the chromosome, the less likely it is that any **single crossover** event will occur between them. The same reasoning suggests that the farther apart two linked loci are, the more likely it is that a random crossover event will occur between them.

In Figure 5.5(a), a single crossover occurs between two nonsister chromatids, but not between the two loci; therefore, the crossover goes undetected because no recombinant gametes are produced. In Figure 5.5(b), where two loci are quite far apart, the crossover occurs between them, yielding recombinant gametes.

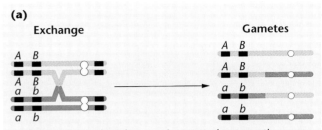

(a)

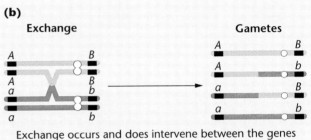

Exchange occurs but does not intervene between the genes

FIGURE 5.5 Two cases of exchange between two nonsister chromatids and the gametes subsequently produced. In Part (a) the exchange does not separate the alleles of the two genes, only parental gametes are formed, and the exchange goes undetected. In Part (b) the exchange separates the alleles, resulting in recombinant gametes, which are detectable.

When a single crossover occurs between two nonsister chromatids, the other two strands of the tetrad will not be involved in this exchange and may enter a gamete unchanged. Under these conditions, even if a single crossover occurs 100 percent of the time between two linked genes, recombination will subsequently be observed in only 50 percent of the potential gametes formed. This concept is diagrammed in Figure 5.6.

Multiple Crossovers

It is possible that in a single tetrad, two, three, or more exchanges will occur between nonsister chromatids as a result of several crossing over events. Double exchanges of genetic material result from **double crossovers**, as shown in Figure 5.7. For a double exchange to be studied,

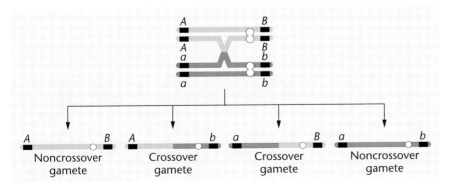

Noncrossover gamete   Crossover gamete   Crossover gamete   Noncrossover gamete

FIGURE 5.6 The consequences of a single exchange between two nonsister chromatids occurring in the tetrad stage. Two noncrossover (parental) and two crossover (recombinant) gametes are produced.

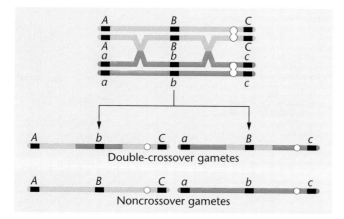

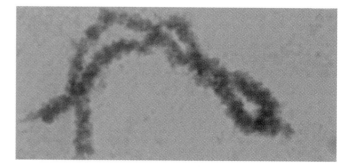

FIGURE 5.7 Results of a double exchange occurring between non-sister chromatids. Because the exchanges involve only two strands, two noncrossover gametes and two double-crossover gametes are produced. The photograph illustrates chiasmata found in a tetrad isolated during the first meiotic prophase stage.

three gene pairs must be investigated, each heterozygous for two alleles. Before we can determine the frequency of recombination among all three loci, we must review some simple probability calculations.

The probabilities of a single exchange occurring between the A and B or the B and C genes are directly related to the physical distance between each locus. The closer A is to B and B is to C, the less likely it is that a single exchange will occur between either of the two sets of loci. In the case of a double crossover, two separate and independent events or exchanges must occur simultaneously. The mathematical probability of two independent events occurring simultaneously is equal to the product of the individual probabilities. This is the product law previously introduced in Chapter 3.

Suppose that crossover gametes resulting from single exchanges between A and B are recovered 20 percent of the time ($p = 0.20$) and between B and C 30 percent of the time ($p = 0.30$). The probability of recovering a double-crossover gamete arising from two exchanges, between A and B and between B and C, is predicted to be $(0.20) \cdot (0.30) = 0.06$, or 6 percent. It is apparent from this calculation that the frequency of double-crossover gametes is always expected to be much lower than that of either single-crossover class of gametes.

If three genes are relatively close together along one chromosome, the expected frequency of double-crossover gametes is extremely low. For example, consider the A–B distance in Figure 5.7 to be 3 map units and the B–C distance in that figure to be 2 map units. The expected double-crossover frequency would be $(0.03) \cdot (0.02) = 0.0006$, or 0.06 percent. This translates to only 6 events in 10,000. In a mapping experiment such as just described, one involving closely linked genes, very large numbers of offspring are required to detect double-crossover events. In this example, it would be unlikely that a double crossover would be ob-

served even if 1000 offspring were examined. If these probability considerations are extended, it is evident that if four or five genes were being mapped, even fewer triple and quadruple crossovers could be expected to occur.

Three-Point Mapping in *Drosophila*

The information presented in the previous section serves as the basis for mapping three or more linked genes in a single cross. To illustrate this, we will examine a situation involving three linked genes.

Three criteria must be met for a successful mapping cross:

1. The genotype of the organism producing the crossover gametes must be heterozygous at all loci under consideration. If homozygosity occurred at any locus, all gametes produced would contain the same allele, precluding mapping analysis.

2. The cross must be constructed so that genotypes of all gametes can be accurately determined by observing the phenotypes of the resulting offspring. This is necessary because the gametes and their genotypes can never be observed directly. Thus, each phenotypic class must reflect the genotype of the gametes of the parents producing it.

3. A sufficient number of offspring must be produced in the mapping experiment to recover a representative sample of all crossover classes.

These criteria are met in the three-point mapping cross from *Drosophila melanogaster* shown in Figure 5.8. In this cross, three X-linked recessive mutant genes—*yellow* body color, *white* eye color, and *echinus* eye shape—are considered. To diagram the cross, we must assume some theoretical sequence, even though we do not yet know if it is correct. In Figure 5.8, we will initially assume the sequence of the three genes to be y–w–ec. If this is incorrect, our analysis will reveal the correct sequence.

In the P_1 generation, males hemizygous for all three wild-type alleles are crossed to females that are homozygous for all three recessive mutant alleles. Therefore, the P_1 males are wild type with respect to body color, eye color, and eye shape. They are said to have a wild-type

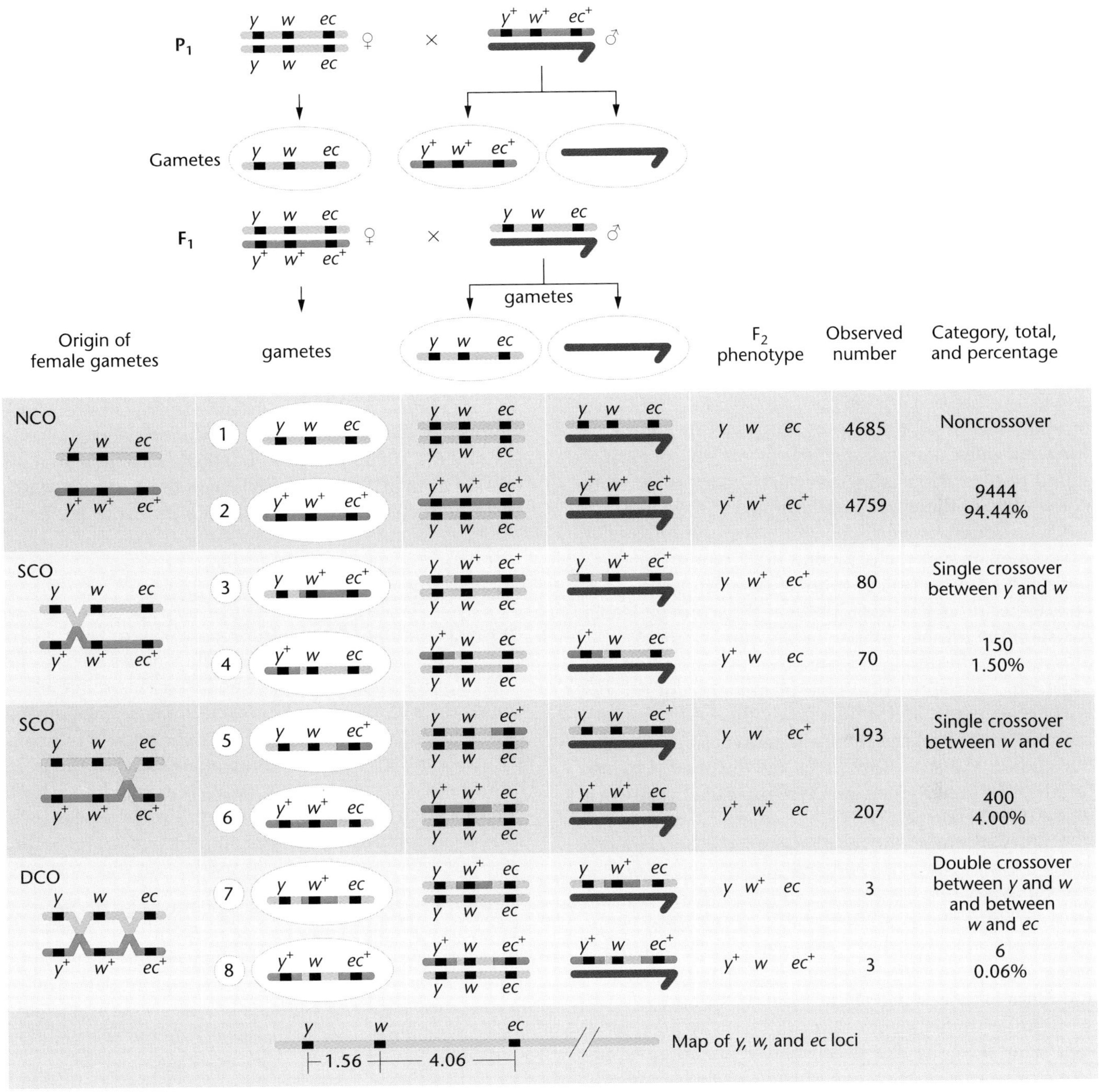

FIGURE 5.8   A three-point mapping cross involving the *yellow* (*y* or *y*$^+$), *white* (*w* or *w*$^+$), and *echinus* (*ec* or *ec*$^+$) genes in *Drosophila melanogaster*. NCO, SCO, and DCO refer to noncrossover, single-crossover, and double-crossover groups, respectively. Because of the complexity of this and several of the ensuing figures, centromeres have not been included on the chromosomes, and only two nonsister chromatids are initially shown. Crossing over always occurs in the four-strand tetrad stage.

phenotype. The females, on the other hand, exhibit the three mutant traits—*yellow* body color, *white* eyes, and *echinus* eye shape.

This cross produces an F$_1$ generation consisting of females heterozygous at all three loci and males that, because of the Y chromosome, are hemizygous for the three mutant alleles. Phenotypically, all F$_1$ females are wild type, while all

F$_1$ males are *yellow*, *white*, and *echinus*. The genotype of the F$_1$ females fulfills the first criterion for constructing a map of the three linked genes; that is, it is heterozygous at the three loci and may serve as the source of recombinant gametes generated by crossing over. Note that because of the genotypes of the P$_1$ parents, all three mutant alleles are on one homolog and all three wild-type alleles are on the other homolog. *Other arrangements are possible.* For example, the heterozygous F_1 female might have the *y* and *ec* mutant alleles on one homolog and the *w* allele on the other. This would occur if, in the P_1 cross, one parent was *yellow* and *echinus* and the other parent was *white*.

In our cross, the second criterion is met by virtue of the gametes formed by the F_1 males. Every gamete will contain either an X chromosome bearing the three mutant alleles or a Y chromosome, which is genetically inert for the three loci being considered. In both cases, following fertilization, the genotype of the gamete produced by the F_1 female will be expressed phenotypically in the F_2 female and male offspring. As a result, all noncrossover and crossover gametes produced by the F_1 female parent can be detected by observing the F_2 phenotypes.

With these two criteria met, we can construct a chromosome map from the crosses illustrated in Figure 5.8. First, we must determine which F_2 phenotypes correspond to the various noncrossovers and crossover categories. Two of these can be determined immediately.

The **noncrossover** F_2 phenotypes are determined by the combination of alleles present in the parental gametes formed by the F_1 female. In this case each gamete contains either three wild-type alleles *or* three mutant alleles, based on the presence of one or the other of the X chromosomes unaffected by crossing over. As a result of segregation, approximately equal proportions of the two types of gametes, and subsequently the F_2 phenotypes, are produced. Because the F_2 phenotypes complement one another (i.e., one is wild type and the other is mutant for all three genes), they are therefore called **reciprocal classes** of phenotypes.

The two noncrossover phenotypes are most easily recognized because *they exist in the greatest proportion.* Figure 5.8 shows that classes (1) and (2) are present in the greatest numbers. Therefore, flies that are *yellow, white,* and *echinus* and those that are normal or wild type for all three characters constitute the noncrossover category and represent 94.44 percent of the F_2 offspring.

The second category that can be easily detected is represented by the double-crossover phenotypes. Because of their probability of occurrence, *they must be present in the least numbers.* Remember that this group represents two independent but simultaneous single-crossover events. Two reciprocal phenotypes can be identified: class (7), which shows the mutant traits *yellow* and *echinus* but normal eye color; and class (8), which shows the mutant trait *white* but normal body color and eye shape. Together these double-crossover phenotypes constitute only 0.06 percent of the F_2 offspring.

The remaining four phenotypic classes represent two categories resulting from single crossovers. Classes (3) and (4), reciprocal phenotypes produced by single-crossover events occurring between the *yellow* and *white* loci, are equal to 1.50 percent of the F_2 offspring. Classes (5) and (6), constituting 4.00 percent of the F_2 offspring, represent the reciprocal phenotypes resulting from single-crossover events occurring between the *white* and *echinus* loci.

The map distances between the three loci can now be calculated. The distance between *y* and *w*, or between *w* and *ec*, is equal to the percentage of all detectable exchanges occurring between them. For any two genes under consideration, this will include all appropriate single crossovers as well as all double crossovers. The latter are included because they represent two simultaneous single crossovers. For the *y* and *w* genes this includes classes (3), (4), (7), and (8), totaling 1.50% + 0.06%, or 1.56 map units (mu). Similarly, the distance between *w* and *ec* is equal to the percentage of offspring resulting from an exchange between these two loci: classes (5), (6), (7), and (8), totaling 4.00% + 0.06%, or 4.06 map units (mu). The map of these three loci on the X chromosome, based on these data, is shown at the bottom of Figure 5.8.

Determining the Gene Sequence

In the preceding example, the order or sequence of the three genes along the chromosome was assumed to be *y–w–ec*. Our analysis established that this sequence is consistent with the data. However, in many mapping experiments, the gene sequence is not known, and this constitutes another variable in the analysis. Had the gene order been unknown in this example, it could have been determined in a straightforward way. Two methods can be used to accomplish this. You should select one or the other method and adhere to its use in your own analysis.

Method I   This method is based on the fact that there are only three possible orders, each containing one of the three genes in between the other two. While we assumed *y–w–ec*, *ec–w–y* is fully equivalent because the data provide no basis for knowing which is correct. Therefore, only three possibilities exist, and one of them must be correct:

(I)   *w–y–ec* (*y* is in the middle)
(II) *y–ec–w* (*ec* is in the middle)
(III) *y–w–ec* (*w* is in the middle)

If you use the following steps during your analysis, you will be able to determine gene order:

1. Assuming any of the three orders, first determine the **arrangement of alleles** along each homolog of the heterozygous parent giving rise to noncrossover and crossover gametes (the F_1 female in our example).

2. Determine whether a double-crossover event occurring within that arrangement will produce the **observed double-crossover phenotypes**. Remember that these phenotypes occur least frequently and can be easily identified.

3. If this order does not produce the correct phenotypes, try each of the other two orders. One of the three must work!

These steps are illustrated in Figure 5.9 for the cross previously shown in Figure 5.8. The three possible orders are labeled (I), (II), and (III). Either *y*, *ec*, or *w* must be in the middle.

1. Assuming *y* is between *w* and *ec*, the arrangement of alleles along the homologs of the F₁ heterozygote is:

$$(I) \quad \frac{w \quad y \quad ec}{w^+ \quad y^+ \quad ec^+}$$

We know this because of the way in which the P₁ generation was crossed. The P₁ female contributed an X chromosome bearing the *w*, *y*, and *ec* alleles, while the P₁ male contributed an X chromosome bearing the *w⁺*, *y⁺*, and *ec⁺* alleles.

2. A double crossover within the above arrangement would yield the following gametes:

$$\underline{w \quad y^+ \quad ec} \quad \text{and} \quad \underline{w^+ \quad y \quad ec^+}$$

Following fertilization, if *y* is in the middle, the F₂ double-crossover phenotypes will corre-

spond to the above gametic genotypes, yielding offspring that are *white, echinus* and offspring that are *yellow*. Instead, however, determination of the actual double crossovers reveals them to be *yellow, echinus* flies and *white* flies. Therefore, our assumed order is incorrect.

3. If we consider the other orders, one with *ec/ec⁺* alleles in the middle (II) or one with the *w/w⁺* alleles in the middle (III):

$$(II) \quad \frac{y \quad ec \quad w}{y^+ \quad ec^+ \quad w^+} \quad \text{or} \quad (III) \quad \frac{y \quad w \quad ec}{y^+ \quad w^+ \quad ec^+}$$

we see that arrangement II again provides *predicted* double-crossover phenotypes that do not correspond to the *actual* double-crossover phenotypes. The predicted phenotypes are *yellow, white* flies and *echinus* flies in the F₂ generation. Therefore, this order is also incorrect. However, arrangement III *will* produce the *observed* phenotypes, *yellow, echinus* flies and *white* flies. Therefore, this order, where the *w* gene is in the middle, is correct.

To summarize, this method is rather straightforward. Determine the arrangement of alleles on the homologs of the heterozygote yielding the crossover gametes. Then, test each of three possible orders to determine which one yields the observed (actual) double-crossover phenotypes. Whichever of the three does so represents the correct order. Testing the three possibilities in our example is summarized in Figure 5.9.

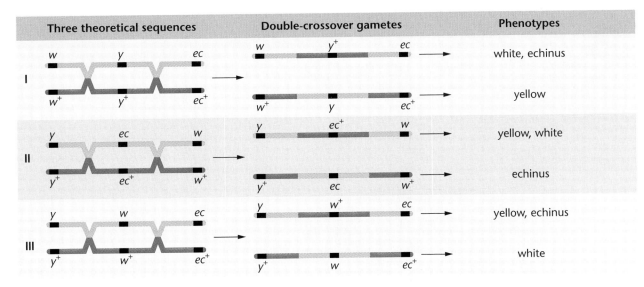

FIGURE 5.9   A summary of the three possible orders of the *white, yellow,* and *echinus* genes, the results of a double crossover in each case, and the resulting phenotypes produced in a test cross. The two noncrossover chromatids of each tetrad have been omitted to simplify the illustrations.

Method II   This method again requires that we determine the arrangement of alleles along each homolog of the heterozygous parent. It also assumes that one specific gene of the three is in the middle. It requires one further assumption:

Following a double-crossover event, the allele representing the middle gene will find itself present with the outside or flanking alleles that were present on the opposite parental homolog.

To illustrate, assume order (I), *w–y–ec*, in the arrangement:

$$(I) \quad \frac{w \quad y \quad ec}{w^+ \quad y^+ \quad ec^+}$$

Following a double-crossover event, the *y* and y^+ alleles would find themselves switched to the arrangement:

$$\frac{w \quad y^+ \quad ec}{w^+ \quad y \quad ec^+}$$

Following segregation, two gametes would be formed:

$$\underline{w \quad y^+ \quad ec} \quad \text{and} \quad \underline{w^+ \quad y \quad ec^+}$$

Because the genotype of the gamete will be expressed directly in the phenotype following fertilization, the double-crossover phenotypes will be:

white, echinus flies   and   *yellow* flies

Note that the *yellow* allele, assumed to be in the middle, is now associated with the two outside markers of the other homolog, w^+ and ec^+. However, these predicted phenotypes do not coincide with the observed double-crossover phenotypes. Therefore, the *yellow* gene is not in the middle.

This same reasoning can be applied to the assumption that the *echinus* gene or the *white* gene is in the middle. In the former case, a negative conclusion will be reached. If we assume that the *white* gene is in the middle, the *predicted* and *actual* double crossovers coincide. Therefore, we conclude that the *white* gene is located between the *yellow* and *echinus* genes.

To summarize, this method is also straightforward. Determine the arrangement of alleles on the homologs of the heterozygote yielding crossover gametes. Then determine the actual double-crossover phenotypes. Simply select the single allele that has been switched so that it is now no longer associated with its original neighboring alleles.

In our example above, *y*, *ec*, and *w* are together in the F$_1$ heterozygote, as are y^+, ec^+, and w^+. In the F$_2$ double-crossover classes, it is *w* and w^+ that have been switched.

The *w* allele is now associated with y^+ and ec^+, while the w^+ allele is now associated with the *y* and *ec* alleles. Therefore, the *white* gene is in the middle, and the *yellow* and *echinus* genes are the flanking markers.

A Mapping Problem in Maize

Having established the basic principles of chromosome mapping, we will now consider a related problem in maize (corn) where the gene sequence and interlocus distances are unknown.

This analysis differs from the preceding discussion in several ways and therefore will expand your knowledge of mapping procedures:

1. The previous mapping cross involved X-linked genes. Here, autosomal genes are considered.

2. In the previous discussion we initially knew the gene sequence, and we used this information to explain how to determine an unknown sequence. In this analysis of maize, the sequence is *initially* unknown.

3. In the discussion of this cross we will make a transition in the use of symbols, as first suggested in Chapter 4. Instead of using the symbols bm^+, v^+, and pr^+, we will simply use + to denote each wild-type allele. This symbol is less complex to manipulate, but requires a better understanding of mapping procedures.

When we consider three autosomally linked genes in maize, the experimental cross must still meet the same three criteria established for the X-linked genes in *Drosophila*: (1) One parent must be heterozygous for all traits under consideration; (2) the gametic genotypes produced by the heterozygote must be apparent from observing the phenotypes of the offspring; and (3) a sufficient sample size must be available.

In maize, the recessive mutant genes *bm* (brown midrib), *v* (virescent seedling), and *pr* (purple aleurone) are linked on chromosome 5. Assume that a female plant is known to be heterozygous for all three traits. Nothing is known about: (1) the arrangement of the mutant alleles on the maternal and paternal homologs of this heterozygote, (2) the sequence of genes, or (3) the map distances between the genes. What genotype must the male plant have to allow successful mapping? To meet the second criterion, the male must be homozygous for all three recessive mutant alleles. Otherwise, offspring of this cross showing a given phenotype might represent more than one genotype, making accurate mapping impossible. Note that this is equivalent to performing a test cross.

Figure 5.10 diagrams this cross. As shown, we do not know either the arrangement of alleles or the sequence of

(a) Possible allele arrangements and gene sequences in a heterozygous female

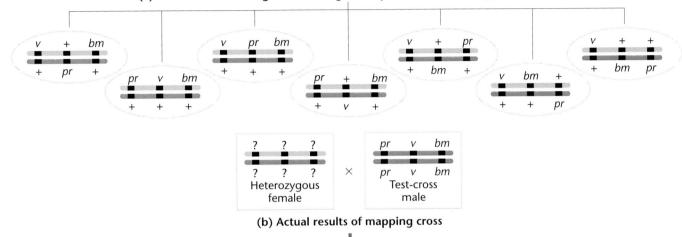

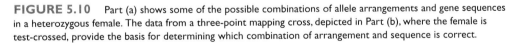

(b) Actual results of mapping cross

Phenotypes of offspring			Number	Total and percentage	Exchange classification
+	v	bm	230	467	Noncrossover
pr	+	+	237	42.1%	(NCO)
+	+	bm	82	161	Single crossover
pr	v	+	79	14.5%	(SCO)
+	v	+	200	395	Single crossover
pr	+	bm	195	35.6%	(SCO)
pr	v	bm	44	86	Double crossover
+	+	+	42	7.8%	(DCO)

FIGURE 5.10   Part (a) shows some of the possible combinations of allele arrangements and gene sequences in a heterozygous female. The data from a three-point mapping cross, depicted in Part (b), where the female is test-crossed, provide the basis for determining which combination of arrangement and sequence is correct.

loci in the heterozygous female. Some of the possibilities are shown, but we have yet to determine which is correct. In the test-cross male parent and the offspring, *we do not know the sequence*, but must write it down initially as a random selection. Note that we have chosen to place *v* in the middle. This may or may not be correct.

The offspring have been arranged in groups of two for each pair of reciprocal phenotypic classes. The two members of each reciprocal class are derived from either no crossing over (**NCO**), one of two possible single-crossover events (**SCO**), or a double crossover (**DCO**).

To solve this problem, consider the following questions. It will be helpful to refer to Figures 5.10 and 5.11 as you consider them.

1. *What is the correct heterozygous arrangement of alleles in the female parent?*

   Identify the two noncrossover classes, those that occur in the highest frequency. In this case, they are + *v bm* and *pr* + +. Therefore, the arrangement of alleles on the homologs of the female parent must be as shown in Figure 5.11(a). These

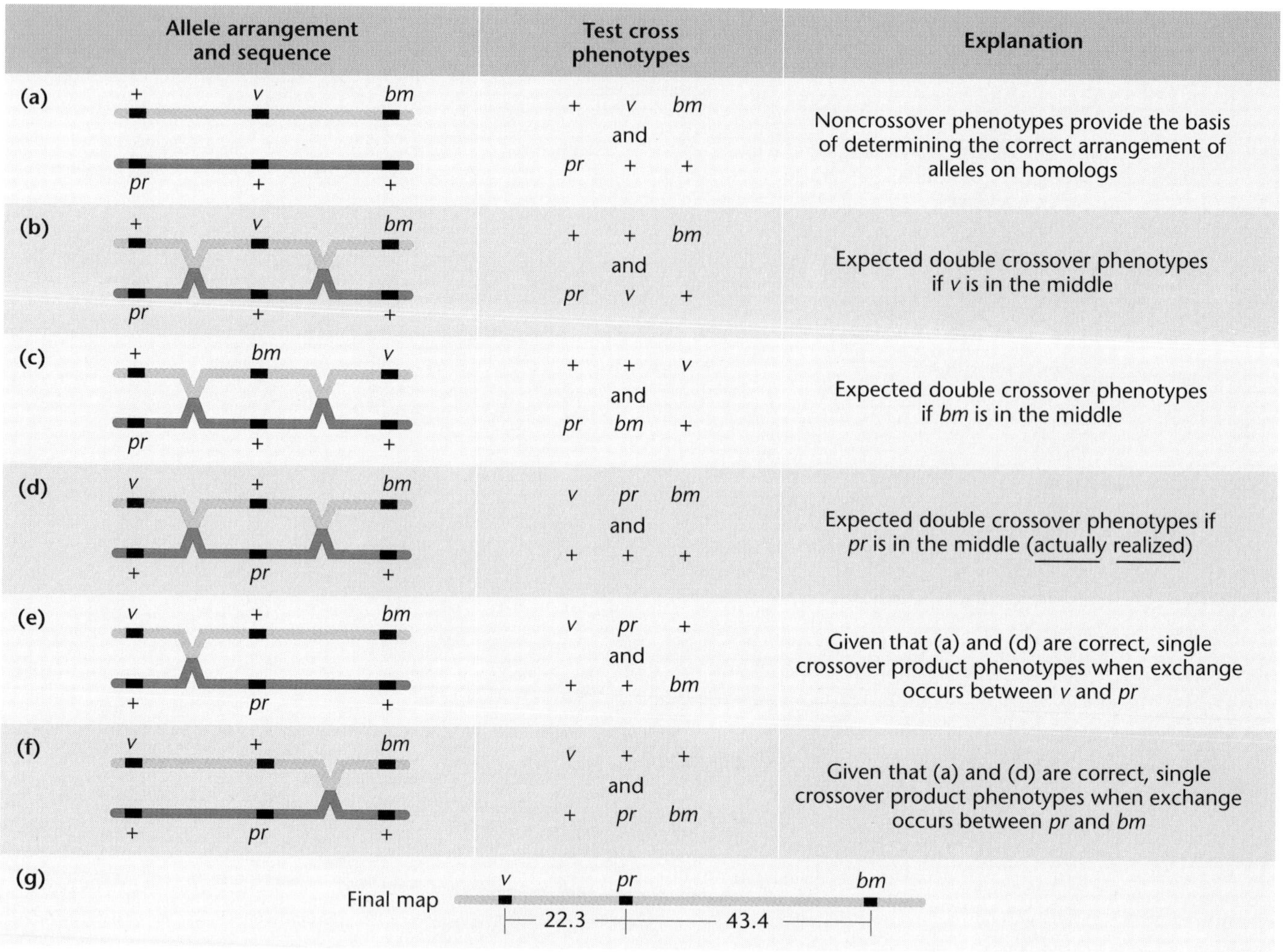

Allele arrangement and sequence	Test cross phenotypes	Explanation
(a) $+\ \ \ v\ \ \ bm$ / $pr\ \ \ +\ \ \ +$	$+\ \ \ v\ \ \ bm$ and $pr\ \ \ +\ \ \ +$	Noncrossover phenotypes provide the basis of determining the correct arrangement of alleles on homologs
(b) $+\ \ \ v\ \ \ bm$ / $pr\ \ \ +\ \ \ +$	$+\ \ \ +\ \ \ bm$ and $pr\ \ \ v\ \ \ +$	Expected double crossover phenotypes if v is in the middle
(c) $+\ \ \ bm\ \ \ v$ / $pr\ \ \ +\ \ \ +$	$+\ \ \ +\ \ \ v$ and $pr\ \ \ bm\ \ \ +$	Expected double crossover phenotypes if bm is in the middle
(d) $v\ \ \ +\ \ \ bm$ / $+\ \ \ pr\ \ \ +$	$v\ \ \ pr\ \ \ bm$ and $+\ \ \ +\ \ \ +$	Expected double crossover phenotypes if pr is in the middle (actually realized)
(e) $v\ \ \ +\ \ \ bm$ / $+\ \ \ pr\ \ \ +$	$v\ \ \ pr\ \ \ +$ and $+\ \ \ +\ \ \ bm$	Given that (a) and (d) are correct, single crossover product phenotypes when exchange occurs between v and pr
(f) $v\ \ \ +\ \ \ bm$ / $+\ \ \ pr\ \ \ +$	$v\ \ \ +\ \ \ +$ and $+\ \ \ pr\ \ \ bm$	Given that (a) and (d) are correct, single crossover product phenotypes when exchange occurs between pr and bm

(g) Final map

$$v \;\;|\!\!-\!\!- 22.3 -\!\!-\!| \;\; pr \;\;|\!\!-\!\!-\!\!- 43.4 -\!\!-\!\!-\!| \;\; bm$$

FIGURE 5.11 Steps used in producing a map of the three genes involved in the cross shown in Figure 5.10, where neither the arrangement of alleles nor the sequence of genes is initially known in the heterozygous female parent.

homologs segregate into gametes, unaffected by any recombination event. Any other arrangement of alleles could not yield the observed noncrossover classes. (Remember that $+\ v\ bm$ is equivalent to $pr^+\ v\ bm$, and that $pr\ +\ +$ is equivalent to $pr\ v^+\ bm^+$.)

2. *What is the correct sequence of genes?*

The approach described in Method I will first be used to answer this question. We know that the arrangement of alleles is

$$\frac{+\ \ v\ \ bm}{pr\ \ +\ \ +}$$

But, is the assumed sequence correct? That is, will a double-crossover event yield the observed double-crossover phenotypes following fertilization? Sim-

ple observation shows that it will not [Figure 5.11(b)]. Try the other two orders [Figure 5.11(c) and (d)], keeping the same arrangement:

$$\frac{+\ \ bm\ \ v}{pr\ \ +\ \ +} \quad \text{and} \quad \frac{v\ \ +\ \ bm}{+\ \ pr\ \ +}$$

Only the latter case will yield the observed double-crossover classes [Figure 5.11(d)]. Therefore, the pr gene is in the middle.

The same conclusion is reached if the problem is analyzed using Method II. In this case, no assumption of gene sequence is necessary. The arrangement of alleles in the heterozygous parent is

$$\frac{+\ \ v\ \ bm}{pr\ \ +\ \ +}$$

The double-crossover gametes are also known:

$$\underline{pr \quad v \quad bm} \quad \text{and} \quad \underline{+ \quad + \quad +}$$

We can see that the *pr* allele has shifted so as to be associated with *v* and *bm* following a double crossover. The latter two alleles were present together on one homolog, and they stayed together. Therefore, *pr* is the odd gene, so to speak, and it is in the middle.

3. *What is the distance between each pair of genes?*

Having established the correct sequence of loci as *v–pr–bm*, we can now determine the distance between *v* and *pr* and between *pr* and *bm*. Remember that the map distance between two genes is calculated on the basis of all detectable recombinational events occurring between them. This includes both the single- and double-crossover events involving the two genes being considered.

Figure 5.11(e) shows that the phenotypes $\underline{v\ pr\ +}$ and $\underline{+\ +\ bm}$ result from single crossovers between *v* and *pr*, accounting for 14.5 percent of the offspring. By adding the percentage of double crossovers (7.8%) to the number obtained for single crossovers, the total distance between *v* and *pr* is calculated to be 22.3 map units.

Figure 5.11(f) shows that the phenotypes $\underline{v\ +\ +}$ and $\underline{+\ pr\ bm}$ result from single crossovers between *pr* and *bm*, totaling 35.6 percent. With the addition of the double-crossover classes (7.8%) the distance between *pr* and *bm* is calculated to be 43.4 map units.

The final map for all three genes in this example is shown in Figure 5.11(g).

Interference and the Coefficient of Coincidence

An important phenomenon, called **interference**, affects the accuracy of mapping experiments by reducing the actual number of expected double crossovers when genes are quite close to one another along the chromosome. This reduction, due to factors that "interfere" with multiple crossovers, will be illustrated for three-point mapping.

We have already considered the probability relationships between single- and double-crossover events. In theory, the percentage of **expected double crossovers** is predicted by multiplying the percentage of the total crossovers between each pair of genes. Remember that a double crossover (DCO) represents two single-crossover events. For example, the expected double-crossover frequency of the cross illustrated in Figure 5.11(g) may be calculated in the following manner:

$$DCO_{exp} = (0.223) \times (0.434) = 0.097 = 9.7\%$$

Frequently, this predicted figure does not correspond precisely with the observed DCO frequency. Generally, there are fewer DCOs observed than predicted. In the maize cross, only 7.8 percent DCOs were observed. In some cases there are more DCOs than expected. These disparities, explained by the concept of interference, is quantified by calculating the **coefficient of coincidence (C)**:

$$C = \frac{\text{Observed DCO}}{\text{Expected DCO}}$$

In the maize cross, we have

$$C = \frac{0.078}{0.097} = 0.804$$

Once *C* is calculated, interference (*I*) may be quantified using the simple equation

$$I = 1 - C$$

In the maize cross, we have

$$I = 1.000 - 0.804 = 0.196$$

If interference is complete and no double crossovers occur, then $I = 1.0$. If fewer DCOs than expected occur, *I* is a positive number (as above) and **positive interference** has occurred. If more DCOs than expected occur, *I* is a negative number and **negative interference** has occurred. In the above example, *I* is a positive number (0.196), indicating that 19.6 percent fewer double crossovers occurred than expected.

In eukaryotic systems, positive interference is most often observed. It appears that a crossover event in one region of a chromosome inhibits a second crossover in neighboring regions of the chromosome. In general, the closer genes are to one another along the chromosome, the more positive interference is observed and the lower the *C* value. In *Drosophila*, when three genes are clustered within 10 map units, interference is often complete and no double-crossover classes are recovered. This observation suggests that interference may be explained by physical constraints that prohibit the formation of closely aligned chiasmata. Perhaps a mechanical stress is imposed on chromatids during crossing over such that one chiasma inhibits the formation of a second chiasma in the neighboring region. This interpretation is consistent with the finding that the impact of interference decreases as the genes in question are located farther apart. In the maize cross illustrated in Figures 5.10 and 5.11, the three genes are relatively far apart, and 80 percent of the expected double crossovers are observed.

The Inaccuracy of Mapping Experiments

Until now, we have considered that the proportion of crossover events that are detected in a mapping experiment are directly proportional to the actual distance between two genes. As we are about to see, this is not usually the case: *Experimentally derived mapping distances between any two randomly selected linked genes are almost always underestimates. The farther apart the two genes are, the greater is the inaccuracy.* The discrepancy is due primarily to potential multiple exchanges that are predicted to occur between the two genes, but which are not recovered during experimental mapping. As we will explain below, this inaccuracy is the result of probability events that can be described by the **Poisson distribution**. First, let us examine multiple crossover events during a mapping experiment that involve two exchanges between two genes that are far apart on a chromosome. As shown in Figure 5.12, there are three possible ways that a double crossover event can occur between nonsister chromatids within a tetrad. A **two-strand double exchange** yields no recombinant chromatids, a **three-strand double exchange** yields 50 percent recombinant chromatids, and a **four-strand double exchange** yields 100 percent recombinant chromatids. In the aggregate, therefore, these multiple events "even out" and two genes that are far apart on the chromosome theoretically yield the maximum of 50 percent recombination, essential for *accurate* gene mapping.

In such a mapping experiment, double exchanges (and, for that matter, all even numbered exchanges), are relatively infrequent in comparison to the total number of potential recombination events, most of which are single crossovers. As such, the *actual* occurrence of the infrequent events is subject to probability considerations based on the Poisson distribution. In our case, this distribution allows us mathematically to predict the frequency of samples that will *actually* undergo double exchanges. It is the failure of such exchanges to occur that leads to the underestimate of mapping distance.

To illustrate Poisson distribution, consider the analogy of an Easter egg hunt where 1000 children randomly search a large area for 1000 randomly hidden eggs. In one hour, all eggs are recovered. If all children are equally adept in the search, we can safely predict that many children will have one egg, but also that many will have either no eggs or more than one egg. The Poisson distribution allows us to mathematically predict the various frequencies of outcomes, i.e., the frequencies of children with 0, 1, 2, 3, 4, . . . eggs. Poisson distribution applies when the average number of events is small (one child finds one egg) while the total number of times the event that can occur within the sample is relatively large (1000 eggs can be found).

As Poisson distribution is applied to chromosome mapping, we are interested in the cases where double exchanges may potentially occur between two genes, but because of the predictions based on Poisson distribution, no such exchanges occur within the data sample. Such an analysis creates what is called a **mapping function** that

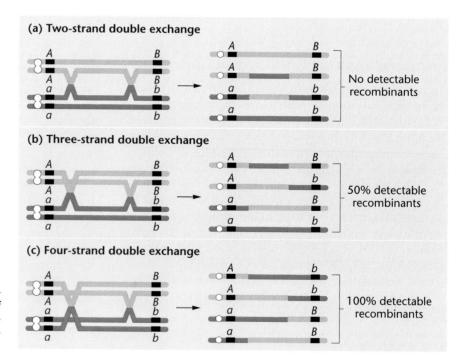

FIGURE 5.12 Three types of double exchanges that may occur between two genes. Two of them [parts (b) and (c)] involve more than two chromatids. In each case, the outcome regarding recombinant chromatids is illustrated.

relates recombination (crossover) frequency (RF) to map distance. If we assume that no interference occurs (see the previous section), then the Poisson terms used to calculate the predicted distributions of events are:

Distribution of Events	Probability
0	e^{-m}
1	me^{-m}
2	$(m^2/2)(e^{-m})$
3	$(m^3/6)(e^{-m})$
etc.	

where the mean number of independently occurring events is m, and e represents the base of natural logarithms (e = about 2.7).

Now, any class where m is one or more (one or more random crossovers) will yield, on average, 50 percent recombinant chromatids. Thus, we are interested in the zero term, which effectively reduces the number of recombinant chromatids. The proportion of meioses with one or more crossovers is equal to 1 − (fraction of zero crossovers), whereby 50 percent recombinant chromatids will occur. Therefore,

$$\% \text{ observed recombination (RF)} = 0.5(1 - e^{-m}) \times 100$$

If this equation is solved, the curve (mapping function) shown in Figure 5.13 is generated. This is compared to the hypothetical case where recombination is directly proportional to mapping distance, the case where interference is complete and no multiple exchanges occur.

Careful examination of this graph reveals two important observations. When the actual map distance is low (i.e, 0–7 mu), the two lines coincide. *When two genes are close together, the accuracy of a mapping experiment is very high! However, as the distance between two genes increases, the accuracy of the experiment diminishes.* The impact of the absence of multiple exchanges, as predicted by the application of the Poisson distribution, is indeed very significant. For example, when 25 percent recombinant chromatids are detected, actual map distance is almost 35 percent! When just over 30 percent recombinants are detected, the true distance, discounting any interference, is 50 mu! Such inaccuracy has been well documented in a number of studies involving various organisms, including maize, *Drosophila*, and *Neurospora*.

To return to and complete our Easter egg analogy, a similar calculation will reveal that over 300 children will fail to find an egg. Had we attempted to estimate the total number of youngsters in the hunt by tabulating the number who found at least one egg, we would have seriously underestimated the number of participants in the hunt, further emphasizing the significance of the consideration of probability on events subject to Poisson distribution.

The Genetic Map of *Drosophila*

In organisms such as *Drosophila*, maize, and the mouse, where large numbers of mutants have been discovered and experimental crosses are easy to perform, extensive maps of each chromosome have been made. Illustrated in Figure 5.14 are partial maps of the four chromosomes of *Drosophila*. Virtually every morphological feature of the fruit fly has been observed to be subject to mutation. The gene locus involved in determining an altered phenotype is first localized to one of the four chromosomes, or linkage groups, and then mapped in relation to other linked genes of that group. As can be seen, the genetic map of the X chromosome is somewhat less extensive than that of autosome 2 or 3. In comparison to these three, autosome 4 is minuscule. Based on cytological evidence, the relative lengths of the genetic maps have been found to correlate roughly with the relative physical lengths of these chromosomes.

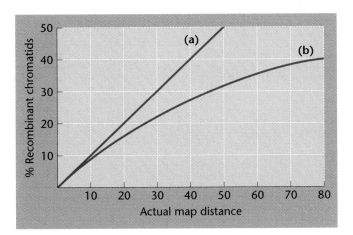

FIGURE 5.13   The average relationship between the frequency of detectable recombinant chromatids and actual map distance, as expected when (a) there is a direct relationship between recombination and map distance; and (b) Poisson distribution is used to predict the actual frequency of recombination in relation to map distance.

Other Aspects of Genetic Exchange

We have established that careful analysis of crossing over during gamete formation can serve as the basis for the construction of chromosome maps in both diploid and haploid organisms. We should not, however, lose sight of

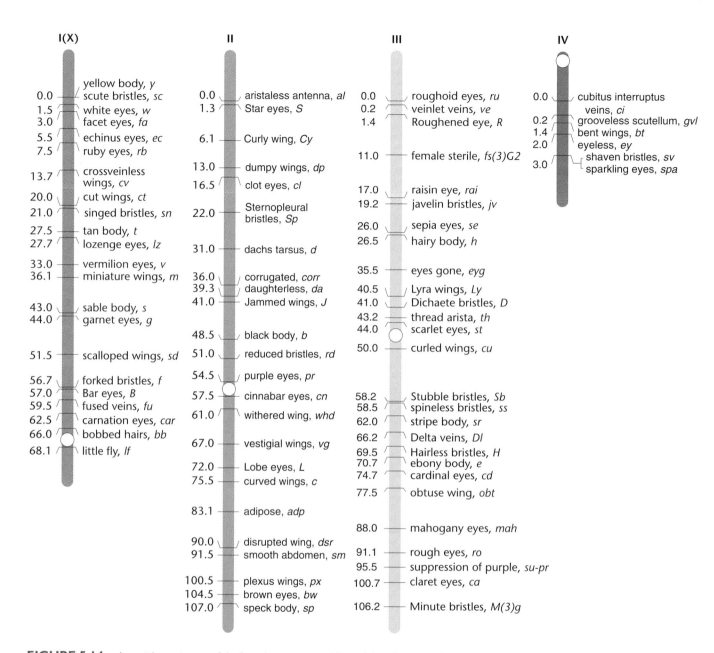

FIGURE 5.14 A partial genetic map of the four chromosomes of *Drosophila melanogaster*. The circle on each chromosome represents the position of the centromere.

the real biological significance of the process, which is to generate genetic variation in gametes and, subsequently, in the offspring derived from the resultant eggs and sperm. Because of the critical role of crossing over in generating variation, the study of genetic exchange has remained an important topic in genetics. Many questions need to be addressed. For example, does crossing over occur in the two- or four-strand stage of meiosis? Does crossing over involve an actual exchange of chromosome arms? Does exchange occur between paired sister chromatids during mitosis? We will briefly consider observations that attempt to answer these questions.

Crossing Over in the Four-Strand Stage

The question of when crossing over occurs during meiosis is important to understanding the process and consequence of genetic exchange in eukaryotes. Hypothetically there are two alternative times at which crossing over might occur. First, exchange could take place before the chromosomes have duplicated in what is referred to as the **two-strand stage**. Alternatively, exchange could occur following the pairing of homologs at the **four-strand stage**, after the chromosomes have duplicated. These alternative possibilities are contrasted in Figure 5.15.

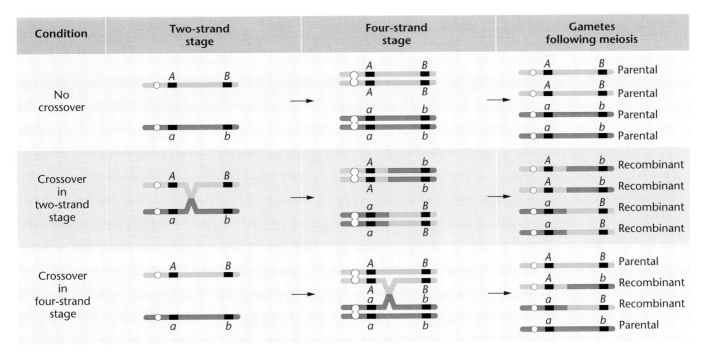

FIGURE 5.15 Comparison of the genotypes of gametes formed as a result of crossing over in the two-strand and four-strand stages with those formed when no crossing over occurs.

If crossing over occurs at the two-strand stage, all four products of a single meiotic event will be recombinant gametes because each pair of sister chromatids in the tetrad is derived from one of the members of the two-strand stage. If, on the other hand, crossing over occurs between two nonsister chromatids in the four-strand stage, two parental (noncrossover) chromatids and two recombinant chromatids will be formed.

While this topic is largely of historical interest, we can examine the results of an experiment performed by Carl L. Lindegren using an organism from which all four products of single meiotic events may be recovered and observed. The organism used in this experiment was the ascomycete *Neurospora*, a haploid bread mold. Fertilization occurs, and meiosis results in the formation of four haploid products, all retained in a sac called the **ascus** (pl., asci). Then, a mitotic division of each product occurs, producing eight haploid cells called **ascospores**. Most important, the entire process retains the haploid ascospore products in the order in which they are formed (see the photograph that is part of Figure 5.16). In addition to answering our original question about the time of crossing over, examination of the results of this experiment serves to introduce the topic of meiosis in a haploid organism such as *Neurospora*.

Lindegren's observations of meiotic segregation in *Neurospora* strongly support the theory that crossing over occurs in the four-strand stage. He examined the various possibilities of ascospore formation resulting from a cross

between an *albino* mutant strain (*a*) and one with normal pigmentation (+). His work focused on the results of a crossover that occurred in the region between the mutant *albino* locus and the centromere. Figure 5.16 illustrates the theoretical results for various alternatives of an exchange in this region for both the two- and four-strand stages. In (a), no exchange occurs. In (b), the results of crossing over in the two-strand stage are predicted. With or without a crossover event, the resulting ascus will always contain four pigmented ascospores and four unpigmented ascospores, in that sequence. The asci produced in Cases I and II cannot be distinguished from each other, i.e., the pattern of albino and pigmented ascospores is identical.

If, on the other hand, a crossover occurs in this region during the four-strand stage, an alternative arrangement of ascospores in the ascus is predicted, as seen in (c). The last case (d) shows still another arrangement that occurs as a result of a slightly different exchange during the four-strand stage.

Lindegren observed asci with arrangements shown in all four cases. His findings are consistent with the conclusion that crossing over indeed occurs in the four-strand stage and exclude crossing over in the two-strand stage, which cannot generate arrangements III and IV.

Similar findings have been drawn from studies of other organisms, including *Drosophila*. To date, no experiments have been reported that seriously dispute the conclusion that crossing over occurs in the four-strand stage in eukaryotic organisms.

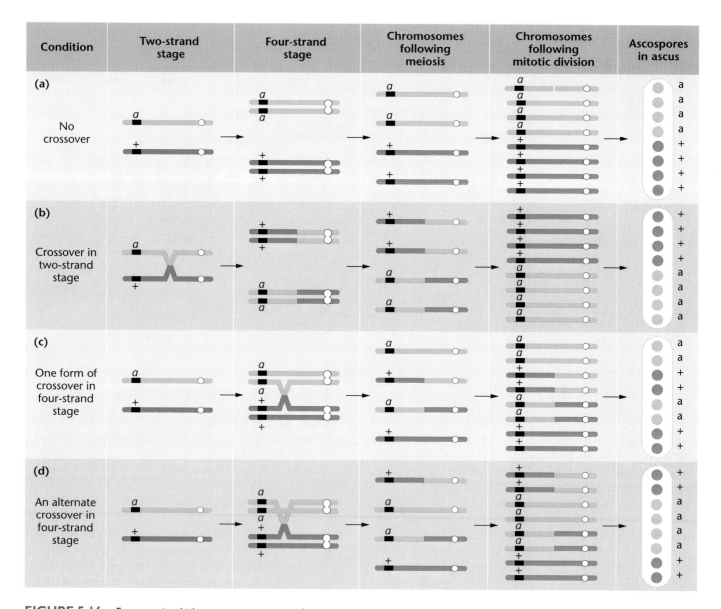

Condition	Two-strand stage	Four-strand stage	Chromosomes following meiosis	Chromosomes following mitotic division	Ascospores in ascus
(a) No crossover					a a a a + + + +
(b) Crossover in two-strand stage					+ + + + a a a a
(c) One form of crossover in four-strand stage					a a + + a a + +
(d) An alternate crossover in four-strand stage					+ + a a a a + +

FIGURE 5.16 Four ways in which ascospore patterns can be generated in the asci of *Neurospora*. Although the patterns produced in (a) and (b) cannot be distinguished from one another, these and the patterns produced in (c) and (d) were observed, leading to the conclusion that crossing over occurs in the four-strand stage. The photograph shows a variety of asci and ascospore patterns.

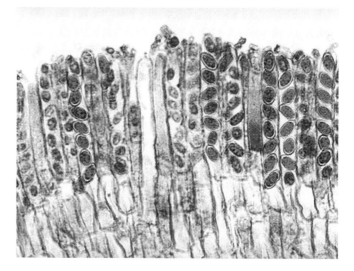

Cytological Evidence for Crossing Over

Visual proof that genetic crossing over in higher organisms is accompanied by an actual physical exchange between homologous chromosomes has been demonstrated independently by Curt Stern in *Drosophila* and by Harriet Creighton and Barbara McClintock in *Zea mays* (corn). Because the experiments are similar, we will consider only one of them, the work with corn. In Creighton and

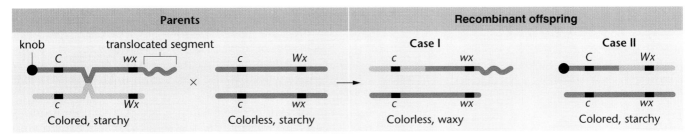

FIGURE 5.17   The phenotypes and chromosome compositions observed in Creighton and McClintock's experiment in maize demonstrating that crossing over involves a breakage and rejoining process. The knob and translocated segment served as cytological markers that established that crossing over involves an actual exchange of chromosome arms.

McClintock's work, two linked genes on chromosome 9 were studied. At one locus, the alleles *colorless* (*c*) and *colored* (*C*) control endosperm coloration. At the other locus, the alleles *starchy* (*Wx*) and *waxy* (*wx*) control the carbohydrate characteristics of the endosperm.

A corn plant was obtained that was heterozygous at both loci. The key to the experiment was that one of the homologs contained two unique cytological markers. The markers consisted of a densely stained knob at one end of the chromosome and a translocated piece of another chromosome (8) at the other end. The arrangements of alleles and cytological markers in this plant are shown in Figure 5.17. Creighton and McClintock crossed this plant to one homozygous for the color allele (*c*) and heterozygous for the endosperm alleles. They obtained a variety of different phenotypes in the offspring, but they were most interested in one that occurred as a result of crossing over involving the chromosome with the unique cytological markers. The chromosomes of this plant, with the colorless, waxy phenotype (Case I in Figure 5.17), were examined for the presence of the cytological markers. As expected, if genetic crossing over is accompanied by a physical exchange between homologs, the translocated chromosome will still be present, but the knob will not. This was the case! In a second plant (Case II), the phenotype colored, starchy can result from either nonrecombinant gametes or from crossing over. If so some of the cases ought to contain chromosomes with the dense knob but not the translocated chromosome. This was also the case, and again the findings supported the conclusion that a physical exchange had taken place. Similar findings by Stern using *Drosophila* leave no doubt about this conclusion involving the cytological basis of crossing over.

The Mechanism of Crossing Over

It has long been of interest to determine how and when during meiosis crossing over occurs. If we accept the evidence just discussed, then crossing over must be considered to be the result of an actual physical exchange between DNA molecules of two homologous chromosomes. Of particular interest is the relationship between the chiasmata observed cytologically during prophase I of meiosis and the breakage and reunions presumed to occur during genetic crossing over. Are chiasmata the cytological manifestations of crossover events?

Two main theories have been proposed, both of which involve chiasmata, but in quite different ways. The **classical theory** holds that crossing over events are the result of physical strains imposed by chiasmata, the presence and location of which occur randomly. While there is not necessarily a one-to-one relationship—a chiasma may or may not induce the breakage and rejoining that leads to crossing over—*this theory holds that chiasmata are responsible for crossover events and clearly precede them.* Since we know that chiasmata are first seen rather late in meiotic prophase I, during the diplotene stage, the classical theory predicts that crossing over occurs sometime after diplonema but before the chromosomes separate at meiotic anaphase I. The order of events presumed to occur and the resultant pairing relationships created according to the classical theory are illustrated in Figure 5.18(a).

The other proposal, called the **chiasmatype theory**, was first set forth by F. A. Janssens in the early twentieth century and later modified by John Belling and C. D. Darlington about 1930. It predicts that *crossing over precedes chiasma formation* and occurs early in meiotic prophase I, presumably in pachytene. *As a result, chiasmata are formed at points of genetic exchange.* Thus, a chiasma is the *consequence* of crossing over and, in diplotene, is a cytological manifestation of chromosome exchange [Figure 5.18(b)].

The chiasmatype theory has gained much support over the years even though it does not account for all related observations. Several findings derived from the use of modern research techniques are pertinent to this subject. The first involves the discovery of DNA synthesis during meiosis. Yasuo Hotta and Herbert Stern have performed careful analysis using *Lilium* (lily) anthers, where the stages of meiotic prophase can be easily identified and studied.

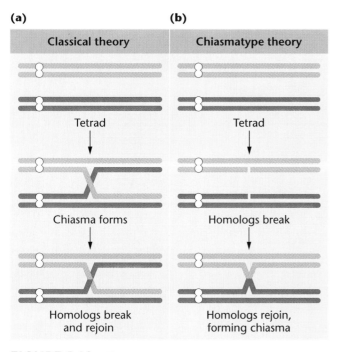

FIGURE 5.18 Two main theories proposed to account for the mechanism of crossing over.

They have shown that a small but measurable amount of DNA synthesis occurs during the zygotene stage of meiotic prophase I. Recall from our earlier discussion of mitosis and meiosis in Chapter 2 that DNA replication occurs during the S phase of the cell cycle, well before mitosis and meiosis begin. Thus, this finding has generated much interest. What is the role of this late-replicating DNA? It amounts to 0.3 percent of the total nuclear DNA and is distributed generally among all chromosomes. Furthermore, if this DNA synthesis is inhibited, then chromosomal synapsis is also inhibited and meiosis arrests. These latter findings suggest that the newly synthesized DNA plays an important role in chromosome alignment during meiosis.

Mitotic Recombination

In 1936, Curt Stern demonstrated that exchanges similar to crossing over, while very rare, do occur during mitosis in *Drosophila*. This finding, the first to demonstrate **mi-**totic recombination, was considered unusual because homologs do not normally pair up during mitosis in most organisms. However, such synapsis appears to be the rule in *Drosophila*. Since Stern's discovery, genetic exchange during mitosis has also been shown to be a general event in certain fungi as well.

Stern observed small patches of mutant tissue in females heterozygous for the X-linked recessive mutations *yellow* body and *singed* bristles. Under normal circumstances, a heterozygous female is completely wild type (gray-bodied with straight, long bristles). He explained the appearance of the mutant patches by postulating that, during mitosis in certain cells during development, homologous exchanges could occur between the loci for *yellow* and *singed* or between *singed* and the centromere, or that a double exchange could occur. These three possibilities are diagrammed in Figure 5.19. After these three types of exchanges occur, tissues derived from the progeny cells are produced with *yellow* patches, adjacent *yellow* and *singed* patches (**twin spots**), and *singed* patches, respectively. The last type of tissue, which represents the double exchange, was found in the lowest frequency, similar to the less frequent double-crossover frequency seen in meiosis. The frequency of twin spots argues strongly against the appearance of *yellow* or *singed* tissue being due to two spontaneous but independent mutational events. If the twin spots arose in such a way, their frequency would occur in only one in many million flies.

Table 5.1 compares the relative frequency of exchange leading to each of the three spotting types. These frequencies are correlated with the known genetic distances between the *yellow* and *singed* loci and the centromere, which effectively serves as an additional genetic marker. The data parallel the predicted frequencies if the exchanges occur according to the rules we established earlier for meiotic crossing over. This is additional evidence that the occurrence of mutant tissue spots is due to mitotic exchange in somatic cells.

In 1958 George Pontecorvo and others described a similar phenomenon in the fungus *Aspergillus*. Although the vegetative stage is normally haploid, some cells and their nuclei fuse, producing diploid cells that divide mitotically. Occasionally, crossing over occurs between linked genes during mitosis in this diploid stage so that

TABLE 5.1 Comparison of the Various Parameters Related to Mitotic Recombination Studies in *Drosophila*

Mutant Tissue Type	Region of Exchange	Relative Frequency of Occurrence	Type of Exchange	Relative Distance
Yellow	*y-sn*	High	Single	21 map units
Yellow-singed (twin spot)	*sn*-centromere	Highest	Single	49 map units
Singed	*y-sn* and *sn*-centromere	Lowest	Double	N/A

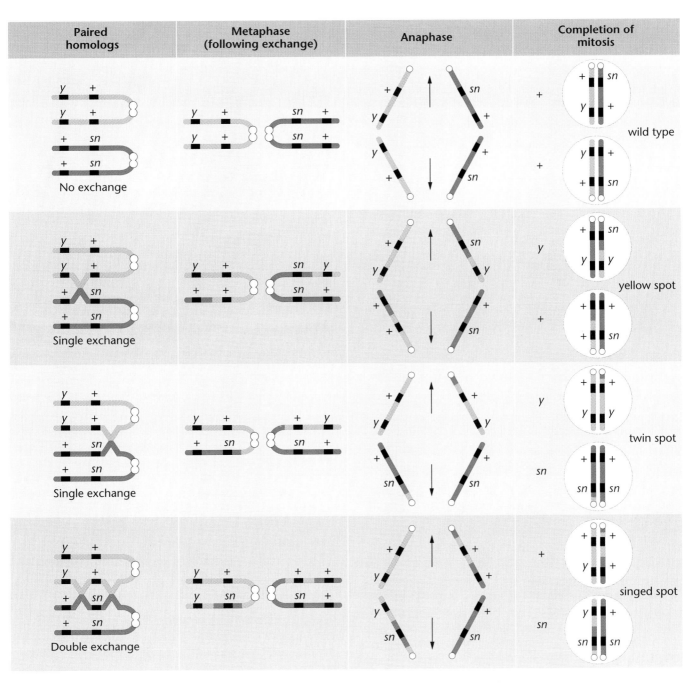

Paired homologs	Metaphase (following exchange)	Anaphase	Completion of mitosis

FIGURE 5.19   The production of mutant tissue in a female heterozygous for the recessive *yellow* and *singed* alleles as a result of mitotic recombination in *Drosophila*.

resulting cells are recombinant. Pontecorvo referred to these events that produce genetic variability as the **parasexual cycle**. On the basis of such exchanges, genes can be mapped by estimating the frequency of recombinant classes.

As a rule, if mitotic recombination occurs at all in an organism, it does so at a much lower frequency than meiotic crossing over. While it is assumed that there is always at least one exchange per meiotic tetrad, mitotic exchange occurs in 1 percent or less of mitotic divisions in organ-

isms that demonstrate it. Some researchers believe that the low frequency of exchange may be explained by a pairing of only some portions of homologous chromosomes.

Sister Chromatid Exchanges

Because homologous chromosomes do not usually pair up or synapse in somatic cells (*Drosophila* is an exception), each individual chromosome in prophase and metaphase of mitosis consists of two identical sister chromatids, joined at a common centromere. Surprisingly, several experimental approaches have demonstrated that reciprocal exchanges similar to crossing over occur between sister chromatids. While these **sister chromatid exchanges (SCEs)** do not produce new allelic combinations, evidence is accumulating that attaches significance to these events.

Identification and study of sister chromatid exchanges are facilitated by several modern staining techniques. In one approach, cells are allowed to replicate for two generations in the presence of the thymidine analogue bromodeoxyuridine (BUdR)*. Following two rounds of semiconservative replication, chromatids result with either one or with two strands containing chromatids with BUdR. Chromatids with both strands containing BUdR stain less brightly than do chromatids with only one strand of the double helix containing the analogue. Thus, for each pair of sister chromatids, one member contains both DNA strands "labeled" and one member has only one DNA strand "labeled." In Figure 5.20, numerous instances of SCE events can be detected. Additionally, this experiment demonstrates the validity of semiconservative replication.

Although the significance of sister chromatid exchanges is still uncertain, several observations have led to great interest in this phenomenon. It is known, for example, that agents that induce chromosome damage (such as viruses, X rays, ultraviolet light, and certain chemical mutagens) also increase the frequency of sister chromatid exchanges. Further, an elevated frequency of SCEs is characteristic of **Bloom syndrome**, a human disorder caused by a mutation in the chromosome 15 *BLM* gene. This rare, recessively inherited disease is characterized by prenatal and postnatal retardation of growth, a great sensitivity of the facial skin to the sun, immune deficiency, a predisposition to malignant and benign tumors, and abnormal behavior patterns. The chromosomes from cultured leukocytes, bone marrow cells, and fibroblasts derived from homozygotes are very fragile and unstable when compared to those derived from homozygous and heterozygous normal individuals. Increased breaks and rearrangements between nonhomologous chromosomes are observed in addition to excessive amounts of sister chromatid exchanges. Recent work by James German and colleagues suggest that the *BLM* gene encodes an enzyme called **DNA helicase**, which is best known for its role in DNA replication (see Chapter 11).

The mechanisms of exchange between nonhomologous chromosomes and between sister chromatids may prove to share common features because the frequency of both events increases substantially in individuals with genetic disorders. These findings suggest that further study

*The abbreviation BrdU is also used to denote bromodeoxyuridine.

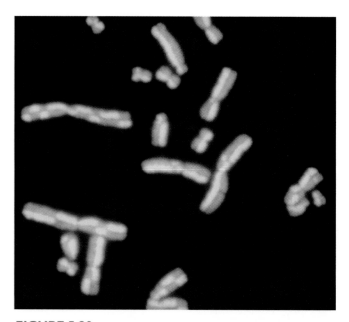

FIGURE 5.20 Demonstration of sister chromatid exchanges (SCEs) in mitotic chromosomes. Sometimes called **harlequin chromosomes** because of their patchlike appearance, chromatids containing the thymidine analogue BUdR in both DNA strands fluoresce *less* brightly than do those with the analogue in only one strand. These chromosomes were stained with 33258-Hoechst reagent and acridine orange and then viewed under fluorescence microscopy.

of sister chromatid exchange may contribute to an increased understanding of recombination mechanisms and the relative stability of normal and genetically abnormal chromosomes. We shall encounter still another demonstration of SCEs in Chapter 11 when we consider replication of DNA (see Figure 11.5).

Somatic Cell Hybridization and Human Chromosome Maps

For obvious reasons, our own species does not provide data necessary for the types of extensive linkage analysis performed with experimental organisms. Thus, in humans, the earliest linkage studies had to rely on pedigree analysis. Attempts were made to establish whether a trait was X-linked or autosomal. As we established in Chapter 4, traits determined by genes located on the X chromosome result in characteristic pedigrees, and some progress was made in identifying such genes. For autosomal traits, geneticists tried to distinguish clearly whether pairs of traits demonstrated linkage or independent assortment. When extensive pedigrees are available and the genes under consideration are closely linked (i.e., rarely separated by crossing over), linkage may be ascertained. In such cases, the two traits segregate together. For example, the genes encoding the **Rh antigens** and the phenotype referred to as **elliptocytosis** (where the shape of erythrocytes is oval) were determined to be linked in this way.

The difficulty arises, however, when two genes of interest are separated on a chromosome such that recombinant gametes are formed, obscuring linkage in a pedigree. In such cases, the demonstration of linkage is enhanced by the use of an approach relying on probability calculations called the *lod* **score method**. First devised by J. B. S. Haldane and C. A. Smith in 1947 and refined by Newton Morton in 1955, the *lod* score (standing for *log* of the *od*ds favoring linkage) assesses the probability that a particular pedigree involving two traits reflects linkage or not. First, the probability is calculated that family data (pedigrees) concerning two or more traits conform to the transmission of traits without linkage. Then the probability is calculated that the identical family data following these same traits result from linkage with a specified recombination frequency. The ratio of these probability values expresses the "odds" for and against linkage. Accuracy using the *lod* score method is limited by the extent of the family data, but nevertheless represented an important advance in assigning human genes to specific chromosomes and constructing preliminary human chromosome maps.

The initial results were discouraging because of the limitations of this approach and because of the relatively high haploid number of human chromosomes (23). By 1960, short of the assignment of some genes to the X chromosome, very little autosomal linkage or mapping information had become available.

However, in the 1960s, a new technique, **somatic cell hybridization**, was developed that aided immensely in the initial step in human gene mapping: assigning genes to their respective chromosomes. This technique, first discovered by Georges Barsky, relies on the fact that two cells in culture can be induced to fuse into a single hybrid cell. While Barsky used two mouse cell lines, it soon became evident that cells from different organisms will also fuse together. When this event occurs, an initial cell type called a **heterokaryon** is produced. The hybrid cell contains two nuclei in a common cytoplasm. Using the proper techniques, it is possible to fuse human and mouse cells, for example, and isolate the hybrids from the parental cells.

As the heterokaryons are cultured *in vitro*, two interesting changes occur. Eventually, the nuclei fuse together, creating what is termed a **synkaryon**. Then, as culturing is continued for many generations, chromosomes from one of the two parental species are gradually lost. In the case of the human–mouse hybrid, human chromosomes are lost randomly until, eventually, the synkaryon has a full complement of mouse chromosomes and only a few human chromosomes. As we will see, it is the preferential loss of human chromosomes rather than mouse chromosomes that makes possible the assignment of human genes to the chromosomes upon which they reside.

The experimental rationale is straightforward. For example, if a specific human gene product is synthesized in a synkaryon containing three human chromosomes, then the gene responsible for that product must reside on one of the three human chromosomes remaining in the hybrid cell. Or, if the human gene product is absent, the responsible gene cannot be present on any of the remaining three human chromosomes. Ideally, a panel of 23 hybrid cell lines, each with but one unique human chromosome, would allow the immediate assignment to a particular chromosome of any human gene for which the product could be characterized.

In practice, a panel of cell lines, each with several remaining human chromosomes, is most often utilized. The correlation of the presence or absence of each chromosome with the presence or absence of each gene product is called **synteny testing**. Consider, for example, the hypothetical data provided in Figure 5.21, where four gene products (*A, B, C,* and *D*) are tested in relationship to eight human chromosomes. Let us carefully analyze the gene that produces product *A*:

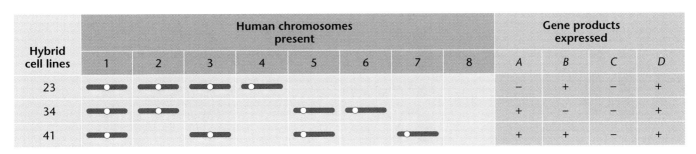

Hybrid cell lines	Human chromosomes present								Gene products expressed			
	1	2	3	4	5	6	7	8	*A*	*B*	*C*	*D*
23	●	●	●	●					−	+	−	+
34	●	●			●	●			+	−	−	+
41	●		●		●		●		+	+	−	+

FIGURE 5.21 A hypothetical grid of data used in synteny testing to assign genes to their appropriate human chromosomes. Three somatic hybrid cell lines, designated 23, 34, and 41, have each been scored for the presence or absence of human chromosomes 1 through 8, as well as for their ability to produce the hypothetical human gene products A through D.

1. Product *A* is not produced by cell line 23, but chromosomes 1, 2, 3, and 4 are present in cell line 23. Therefore, we can rule out the presence of gene *A* on those four chromosomes and conclude that it must be on chromosome 5, 6, 7, or 8.

2. Product *A* is produced by cell line 34, which contains chromosomes 5 and 6, but not 7 and 8. Therefore, gene *A* is on chromosome 5 or 6.

3. Product *A* is also produced by cell line 41, which contains chromosome 5 but not chromosome 6. Therefore, gene *A* is on chromosome 5, according to this analysis.

Using a similar approach, gene *B* can be assigned to chromosome 3. You should perform this analysis to demonstrate for yourself that this is correct. Gene *C* presents a unique situation. The data indicate that it is not present on any of the first seven chromosomes (1–7). While it might be on chromosome 8, no direct evidence supports this conclusion. Other panels are needed. We shall leave gene *D* for you to analyze. Upon what chromosome does it reside?

Using the approach described above, literally hundreds of human genes have been assigned to one chromosome or another. Some of the assignments shown in Figure 5.22 were derived in this way. For mapping still other genes, where the products have yet to be discovered, researchers have had to rely on other linkage techniques. For example, by combining a rather sophisticated approach using recombinant DNA technology with pedigree analysis, it has been possible to assign the genes responsible for **Huntington disease**, **cystic fibrosis**, and **neurofibromatosis** to their respective chromosomes, 4, 7, and 17. This approach will be discussed in Chapter 16.

We conclude this discussion by addressing the next step in human gene mapping: assigning genes to different regions of a given chromosome. Sometimes in hybrid cell lines, fragments of a particular chromosome become transferred to another chromosome, resulting in a **translocation**. It is possible using chromosome banding techniques to identify the exact origin of the translocation and correlate the presence of a chromosomal segment in hybrid cells with specific gene expression. In this way, gene maps of human chromosomes may be compiled. The partial maps of the X chromosome and chromosome 1 shown in Figure 5.22 illustrate this point. Although these maps are not as specific as the genetic map of *Drosophila*, we are beginning to learn a great deal about the chromosome locations of a multitude of human genes. As we will see in Chapters 15 and 16, modern technology involving recombinant DNA and the Human Genome Project have greatly extended our knowledge of gene locations within the human genome.

Did Mendel Encounter Linkage?

We conclude this chapter by examining a modern-day interpretation of the experiments that serve as the cornerstone of transmission genetics—the crosses with garden peas performed by Gregor Mendel.

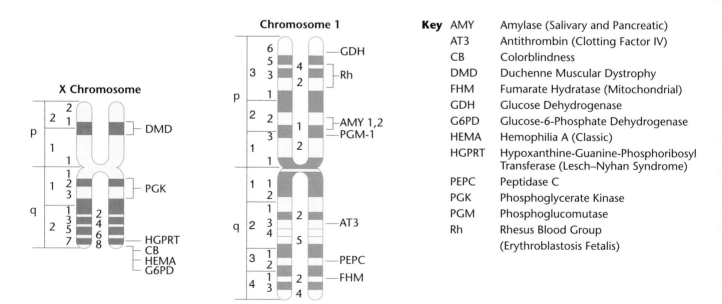

FIGURE 5.22 Representative regional gene assignments for human chromosome 1 and the X chromosome. Many assignments were initially derived using somatic cell hybridization techniques.

It has been said often that Mendel had extremely good fortune in his classical experiments with the garden pea. In none of his crosses did he encounter apparent linkage relationships between any of the seven mutant characters. Had Mendel obtained highly variable data characteristic of linkage and crossing over, these unorthodox observations might have hindered his successful analysis and interpretation.

The accompanying article by Stig Blixt, reprinted in its entirety, demonstrates the inadequacy of this hypothesis. As we shall see, some of Mendel's genes were indeed linked. We shall leave it to Stig Blixt to enlighten you as to why Mendel did not detect linkage.

*W*hy Didn't Gregor Mendel Find Linkage?

*I*t is quite often said that Mendel was very fortunate not to run into the complication of linkage during his experiments. He used seven genes, and the pea has only seven chromosomes. Some have said that had he taken just one more, he would have had problems. This, however, is a gross oversimplification. The actual situation, most probably, is shown in Table 1. This shows that Mendel worked with three genes in chromosome 4, two genes in chromosome 1, and one gene in each of chromosome(s) 5 and 7. It seems at first glance that, out of the 21 dihybrid combinations Mendel theoretically could have studied, no less than four (that is, *a-i*, *v-fa*, *v-le*, *fa-le*) ought to have resulted in linkages. As found, however, in hundreds of crosses and shown by the genetic map of the pea,* *a* and *i* in chromosome 1 are so distantly located on the chromosome that no linkage is normally detected. The same is true for *v* and *le* on the one hand, and *fa* on the other, in

chromosome 4. This leaves *v-le*, which ought to have shown linkage.

Mendel, however, seems not to have published this particular combination and thus, presumably, never made the appropriate cross to obtain both genes segregating simultaneously. It is therefore not so astonishing that Mendel did not run into the complication of linkage, although he did not avoid it by choosing one gene from each chromosome.

Weibullsholm Plant Breeding Institute, Landskrona, Sweden and Centro Energia Nucleare na Agricultura, Piraciaba, SP, Brazil STIG BLIXT

Received March 5; accepted June 4, 1975.
*Blixt, S. 1974. In *Handbook of genetics*, ed. R. C. King. New York: Plenum Press.

Table I Relationship between Modern Genetic Terminology and Character Pairs Used by Mendel

Character Pair Used by Mendel	Alleles in Modern Terminology	Located in Chromosome
Seed color, yellow-green	*l-i*	1
Seed coat and flowers, colored-white	*A-a*	1
Mature pods, smooth expanded-wrinkled indented	*V-v*	4
Inflorescences, from leaf axis-umbellate in top of plant	*Fa-fa*	4
Plant height, 1 m–around 0.5 m	*Le-le*	4
Unripe pods, green-yellow	*Gp-gp*	5
Mature seeds, smooth-wrinkled	*R-r*	7

CHAPTER SUMMARY

1. Genes located on the same chromosome are said to be linked. Alleles located on the same homolog, therefore, can be transmitted together during gamete formation. However, the mechanism of crossing over between homologs during meiosis results in the reshuffling of alleles, thereby contributing to genetic variability within gametes.

2. Early in this century, geneticists realized that crossing over could provide an experimental basis for mapping the location of linked genes relative to one another along the chromosome.

3. Experimental evidence using *Neurospora* has demonstrated that crossing over occurs in the four-strand tetrad stage of meiosis.

4. Cytological investigations of both corn and *Drosophila* have revealed that crossing over involves a physical exchange of segments between nonsister chromatids.

5. An exchange of genetic material between sister chromatids may occur during mitosis as well. These events are referred to as sister chromatid exchanges (SCEs). An elevated frequency of such events is seen in the human disorder called Bloom syndrome.

6. Somatic cell hybridization techniques have made possible linkage and mapping analysis of human genes.

7. Evidence now suggests that several of Mendel's seven genetic characters are linked. However, in each case studied, they are sufficiently far apart along the chromosome to prevent linkage from being detected.

KEY TERMS

INSIGHTS AND SOLUTIONS

1. In a series of two-point map crosses involving three genes linked on chromosome 3 in *Drosophila*, the following distances were calculated.

cd–sr	13 mu
cd–ro	16 mu

   (a) Determine the sequence and construct a map of these three genes.

   Solution: It is impossible to do so; there are two possibilities based on these limited data:

   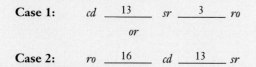

   Case 1: cd ——13—— sr ——3—— ro

   or

   Case 2: ro ——16—— cd ——13—— sr

   (b) What mapping data will resolve this?

   Solution: The map distance determined by crossing over between *ro* and *sr*. If Case 1 is correct, it should be 3 mu, and if Case 2 is correct, it should be 29 mu. In fact, this distance is 29 map units (mu), demonstrating that Case 2 is correct.

   (c) Can we tell which of the sequences shown below is correct?

   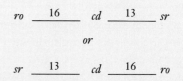

   ro ——16—— cd ——13—— sr

   or

   sr ——13—— cd ——16—— ro

   Solution: No; based on the mapping data, they are equivalent.

2. In rabbits, *black (B)* is dominant to *brown (b)*, while *full color (C)* is dominant to *chinchilla (C^{cb})*. The genes controlling these traits are linked. Rabbits that are heterozygous for both traits and express *black, full color* were crossed to rabbits that are *brown, chinchilla*, with the following results:

   31 *brown, chinchilla*
   35 *black, full*
   16 *brown, full*
   19 *black, chinchilla*

Determine the arrangement of alleles in the heterozygous parents and the map distance between the two genes.

Solution: This is a two-point map problem, where the two reciprocal noncrossover phenotypes are recognized as those present in the highest numbers (*brown, chinchilla* and *black, full*). The less frequent reciprocal phenotypes (*brown, full* and *black, chinchilla*) arise from a single crossover. The arrangement of alleles is derived from the noncrossover phenotypes because they enter gametes intact. The cross is shown as follows:

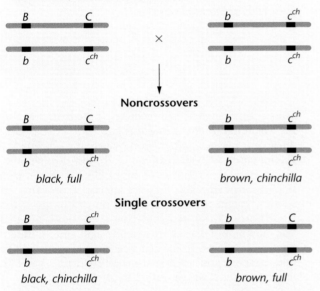

The single crossovers give rise to 35/100 offspring (35%). Therefore, the distance between the two genes is 35 map units (mu).

3. In *Drosophila*, *Lyra* (*Ly*) and *Stubble* (*Sb*) are dominant mutations located at locus 40 and 58, respectively, on chromosome 3. A recessive mutation with bright red eyes was discovered and shown also to be on chromosome 3. A map was obtained by crossing a female who was heterozygous for all three mutations to a male homozygous for the bright red mutation (which we will temporarily call *br*). The following data were obtained:

	Phenotype			Number
(1)	*Ly*	*Sb*	*br*	404
(2)	+	+	+	422
(3)	*Ly*	+	+	18
(4)	+	*Sb*	*br*	16
(5)	*Ly*	+	*br*	75
(6)	+	*Sb*	+	59
(7)	*Ly*	*Sb*	+	4
(8)	+	+	*br*	2
				1000

Determine the location of the bright red mutation on chromosome 3. By referring to Figure 5.14, predict what mutation has been discovered. How could you be sure?

Solution: First, determine the *arrangement* of the alleles on the homologs of the heterozygous crossover parent (the female in this case). This is done by locating the most frequent reciprocal phenotypes, which arise from the noncrossover gametes. These are phenotypes (1) and (2). Each one represents the arrangement of alleles on one of the homologs. Therefore, the arrangement is:

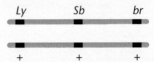

Second, determine the correct *sequence* of the three loci along the chromosome. This is done by determining which sequence will yield the *observed* double-crossover phenotypes, and which are the least frequent reciprocal phenotypes (7 and 8).

If the sequence is correct as written, then a double crossover depicted below:

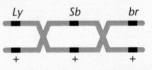

will yield $\underline{Ly + br}$ and $\underline{+ Sb +}$ as phenotypes. Inspection shows that these categories (5 and 6) are actually single crossovers, not double crossovers. Therefore, the sequence, as written, is incorrect. There are only two other possible sequences. The *br* gene is either to the left of *Ly*, or it is between *Ly* and *Sb*:

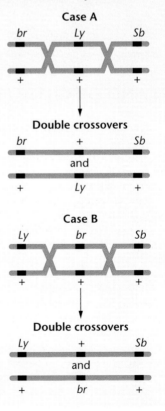

Comparison with the actual data shows that Case B is correct. The double-crossover gametes (7) and (8) yield flies that express *Ly* and *Sb*, but not *br*, *or* express *br*, but not *Ly* and *Sb*. Therefore, the correct arrangement *and* sequence is:

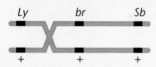

Once this has been determined, it is possible to determine the location of *br* relative to *Ly* and *Sb*. A single crossover between *Ly* and *br*, as shown below:

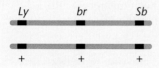

yields flies that are *Ly + +* and *+ br Sb* (categories 3 and 4). Therefore, the distance between the *Ly* and *br* loci is equal to

$$\frac{18 + 16 + 4 + 2}{1000} = \frac{40}{1000} = 0.04 = 4 \text{ map units}$$

Remember that we must add in the double crossovers, since they represent two single crossovers occurring simultaneously. Because we need to know the frequency of all crossovers between *Ly* and *br*, they must be included.

Similarly, the distance between the *br* and *Sb* loci is derived mainly from single crossovers between them:

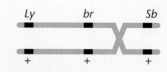

This event yields *Ly br +* and *+ + Sb* phenotypes (categories 5 and 6). Therefore, the distance equals

$$\frac{75 + 59 + 4 + 2}{1000} = \frac{140}{1000} = 0.14 = 14 \text{ map units}$$

The final map shows that *br* is located at locus 44, since *Lyra* and *Stubble* are known:

```
40 ——— (4) ——— 44 ————————— (14) ————————— 58
        |                    |                               |
       Ly                   br                              Sb
```

Inspection of Figure 5.14 reveals that the mutation *scarlet*, which has bright red eyes, is known to exist at locus 44, so it is reasonable to hypothesize that the bright red eye mutation is an allele of *scarlet*. To test this hypothesis, we could cross females of our bright red mutant with known *scarlet* males. If they are alleles, all progeny will reveal a bright red mutant eye. If not, all progeny will show normal brick-red wild-type eyes, indicating that our mutant and scarlet are actually at slightly different loci (very near 44.0). In such a case, all progeny will be heterozygous and will not show a bright red eye. This cross represents what is called an **allelism test**.

PROBLEMS AND DISCUSSION QUESTIONS

1. What is the significance of genetic recombination to the process of evolution?
2. Describe the cytological observation that suggests that crossing over occurs during the first meiotic prophase.
3. Why does more crossing over occur between two distantly linked genes than between two genes that are very close together on the same chromosome?
4. Why is a 50 percent recovery of single-crossover products the upper limit, even when crossing over *always* occurs between two linked genes?
5. Why are double-crossover events expected in lower frequency than single-crossover events?
6. What is the proposed basis for positive interference?
7. What two essential criteria must be met in order to execute a successful mapping cross?
8. The genes *dumpy wing* (*dp*), *clot eye* (*cl*), and *apterous wing* (*ap*) are linked on chromosome 2 of *Drosophila*. In a series of two-point mapping crosses, the following genetic distances were determined:

dp–ap	42
dp–cl	3
ap–cl	39

   What is the sequence of the three genes?
9. Consider two hypothetical recessive autosomal genes *a* and *b*. Where a heterozygote is test-crossed to a double-homozygous mutant, predict the phenotypic ratios under the following conditions:
   (a) *a* and *b* are located on separate autosomes.
   (b) *a* and *b* are linked on the same autosome but are so far apart that a crossover always occurs between them.
   (c) *a* and *b* are linked on the same autosome but are so close together that a crossover almost never occurs.
   (d) *a* and *b* are linked on the same autosome about 10 map units apart.
10. In corn, colored aleurone (in the kernels) is due to the dominant allele *R*. The recessive allele *r*, when homozygous, produces colorless aleurone. The plant color (not the kernel

color) is controlled by the gene pair *Y* and *y*. The dominant *Y* gene results in green color, whereas the homozygous presence of the recessive *y* gene causes the plant to appear yellow. In a test cross between a plant of unknown genotype and phenotype and a plant that is homozygous recessive for both traits, the following progeny were obtained:

Colored green	88
Colored yellow	12
Colorless green	8
Colorless yellow	92

Explain how these results were obtained by determining the exact genotype and phenotype of the unknown plant, including the precise association of the two genes on the homologs (i.e., the arrangement).

11. In the cross shown below involving two linked genes, *ebony* (*e*) and *claret* (*ca*), in *Drosophila*, where crossing over does not occur in males,

$$\frac{e \quad ca^+}{e^+ \quad ca} ♀ \times \frac{e \quad ca^+}{e^+ \quad ca} ♂$$

offspring were produced in a 2 +:1 *ca*:1 *e* phenotypic ratio. These genes are 30 units apart on chromosome 3. What contribution did crossing over in the female make to these phenotypes?

12. With two pairs of genes involved (*P/p* and *Z/z*), a test cross (*ppzz*) with an organism of unknown genotype indicated that the gametes produced were in the following proportions:

PZ, 42.4%; *Pz*, 6.9%; *pZ*, 7.1%; and *pz*, 43.6%

Draw all possible conclusions from these data.

13. In a series of two-point map crosses involving five genes located on chromosome II in *Drosophila*, the following recombinant (single crossover) frequencies were observed:

pr–adp	29
pr–vg	13
pr–c	21
pr–b	6
adp–b	35
adp–c	8
adp–vg	16
vg–b	19
vg–c	8
c–b	27

(a) If *adp* gene is present near the end of chromosome II (locus 83), construct a map of these genes.

(b) In another set of experiments, a sixth gene, *d*, was tested against *b* and *pr*:

d–b	17%
d–pr	23%

Predict the results of two-point maps between *d* and *c*, *d* and *vg*, and *d* and *adp*.

14. Two different female *Drosophila* were isolated, each heterozygous for the autosomally linked genes *b* (*black body*), *d* (*dachs tarsus*), and *c* (*curved wings*). These genes are in the order *d–b–c*, with *b* being closer to *d* than to *c*. Shown below is the genotypic arrangement for each female along with the various gametes formed by both. Identify which categories are noncrossovers (NCO), single crossovers (SCO), and double crossovers (DCO) in each case. Then, indicate the relative frequency in which each will be produced.

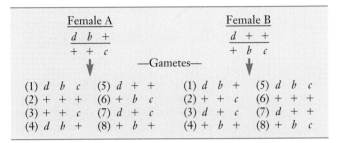

Female A			Female B		
d	*b*	+	*d*	+	+
+	+	*c*	+	*b*	*c*

—Gametes—

Female A		Female B	
(1) *d b c*	(5) *d + +*	(1) *d b +*	(5) *d b c*
(2) *+ + +*	(6) *+ b c*	(2) *+ + c*	(6) *+ + +*
(3) *+ + c*	(7) *d + c*	(3) *d + c*	(7) *d + +*
(4) *d b +*	(8) *+ b +*	(4) *+ b +*	(8) *+ b c*

15. In *Drosophila*, a cross was made between females expressing the three X-linked recessive traits, *scute* (*sc*) bristles, *sable body* (*s*), and *vermilion eyes* (*v*), and wild-type males. In the F₁, all females were wild type, while all males expressed all three mutant traits. The cross was carried to the F₂ generation and 1000 offspring were counted, with the results shown below. No determination of sex was made in the F₂ data.

Phenotype			Offspring
sc	*s*	*v*	314
+	+	+	280
+	*s*	*v*	150
sc	+	+	156
sc	+	*v*	46
+	*s*	+	30
sc	*s*	+	10
+	+	*v*	14

(a) Determine the genotypes of the P₁ and F₁ parents, using proper nomenclature.

(b) Determine the sequence of the three genes and the map distance between them.

(c) Are there more or fewer double crossovers than expected? Calculate the coefficient of coincidence. Does this represent positive or negative interference?

16. Another cross in *Drosophila* involved the recessive, X-linked genes *yellow* (*y*), *white* (*w*), and *cut* (*ct*). A female that was yellow-bodied and white-eyed with normal wings was crossed to a male whose eyes and body were normal but whose wings were cut. The F₁ females were wild type for all three traits, while the F₁ males expressed the yellow-body, white-eye traits. The cross was carried to an F₂, and only male offspring were tallied. On the basis of the data shown below, a genetic map was constructed.

Phenotype			Male Offspring
y	+	ct	9
+	w	+	6
y	w	ct	90
+	+	+	95
+	+	ct	424
y	w	+	376
y	+	+	0
+	w	ct	0

(a) Diagram the genotypes of the F$_1$ parents.
(b) Construct a map, assuming that *white* is at locus 1.5 on the X chromosome.
(c) Were any double-crossover offspring expected?
(d) Could the F$_2$ female offspring be used to construct the map? Why or why not?

17. In *Drosophila*, *Dichaete* (*D*) is a chromosome 3 mutation with a dominant effect on wing shape. It is lethal when homozygous. The genes *ebony* (*e*) and *pink* (*p*) are chromosome 3 recessive mutations affecting the body and eye color, respectively. Flies from a *Dichaete* stock were crossed to homozygous *ebony*, *pink* flies, and the F$_1$ progeny, with a *Dichaete* phenotype, were backcrossed to the *ebony*, *pink* homozygotes. The results of this backcross were as follows:

Phenotype	Number
Dichaete	401
ebony, pink	389
Dichaete, ebony	84
pink	96
Dichaete, pink	2
ebony	3
Dichaete, ebony, pink	12
wild type	13

(a) Diagram this cross, showing the genotypes of the parents and offspring of both crosses.
(b) What is the sequence and interlocus distance between these three genes?

18. *Drosophila* females homozygous for the third chromosomal genes *pink* and *ebony* (the same genes from the previous problem) were crossed with males homozygous for the second chromosomal gene *dumpy*. Because these genes are recessive, all offspring were wild type (normal). F$_1$ females were test-crossed to triply recessive males. If we assume that the two linked genes, *pink* and *ebony*, are 20 map units apart, predict the results of this cross. If the reciprocal cross were made (F$_1$ males—where no crossing over occurs—with triply recessive females), how would the results vary, if at all?

19. In *Drosophila*, two mutations, *Stubble* (*Sb*) and *curled* (*cu*), are linked on chromosome 3. *Stubble* is a dominant gene that is lethal in a homozygous state, and *curled* is a recessive gene. If a female of the genotype

$$\frac{Sb \quad cu}{+ \quad +}$$

is to be mated to detect recombinants among her offspring, what male genotype would you choose as a mate?

20. In *Drosophila*, a heterozygous female for the X-linked recessive traits *a*, *b*, and *c* was crossed to a male that was phenotypically *a b c*. The offspring occurred in the following phenotypic ratios:

+	b	c	460
a	+	+	450
a	b	c	32
+	+	+	38
a	+	c	11
+	b	+	9

No other phenotypes were observed.
(a) What is the genotypic arrangement of the alleles of these genes on the X chromosomes of the female?
(b) Determine the correct sequence and construct a map of these genes on the X chromosome.
(c) What progeny phenotypes are missing? Why?

21. Why were there more "twin spots" observed by Stern than *singed* spots in his study of somatic crossing over? If he had been studying *tan* body color (locus 27.5) and *forked* bristles (locus 56.7) on the X chromosome of heterozygous females, what relative frequencies of tan spots, forked spots, and "twin spots" would you predict might occur?

22. Are mitotic recombinations and sister chromatid exchanges effective in producing genetic variability in an individual? In the offspring of individuals?

23. What possible conclusions can be drawn from the observations that no synaptonemal complexes are observed in male *Drosophila* and female *Bombyx* and that no crossing over occurs in these respective organisms?

24. An organism of the genotype *AaBbCc* was test-crossed to a triply recessive organism (*aabbcc*). The genotypes of the progeny were as follows:

20 *AaBbCc*		20 *AaBbcc*	
20 *aabbCc*		20 *aabbcc*	
5 *AabbCc*		5 *Aabbcc*	
5 *aaBbCc*		5 *aaBbcc*	

(a) If these three genes were all assorting independently, how many genotypic and phenotypic classes would result in the offspring, and in what proportion, assuming simple dominance and recessiveness in each gene pair?
(b) Answer the same question assuming the three genes are so tightly linked on a single chromosome that no crossover gametes were recovered in the sample of offspring.
(c) What can you conclude from the *actual* data about the location of the three genes in relation to one another?

25. Based on our discussion of the potential inaccuracy of mapping (see Figure 5.13), would you revise your answer to Problem 24 above? If so, how?

26. In a plant, fruit was either red or yellow and was either oval or long, where red and oval are the dominant traits. Two separate plants, both heterozygous for these traits, were test-crossed, with the following results:

Phenotype	Progeny, Plant A	Progeny, Plant B
red, long	46	4
yellow, oval	44	6
red, oval	5	43
yellow, long	5	47
	100	100

Determine the location of the genes relative to one another and the genotypes of the two parental plants.

27. In a plant heterozygous for two gene pairs (Ab/aB), where the two loci are linked and 25 map units apart, two such individuals were crossed together. Assuming that crossing over occurs during the formation of both male and female gametes, and that the A and B alleles are dominant, determine the phenotypic ratio of the offspring.

EXTRA-SPICY PROBLEMS

28. A female of genotype

a	b	c
+	+	+

produces 100 meiotic tetrads. Of these, 68 show no crossover events. Of the remaining 32, 20 show a crossover between a and b, 10 show a crossover between b and c, and 2 show a double crossover between a and b and between b and c. Of the 400 gametes produced, how many of each of the 8 different genotypes will be produced? Assuming the order $a–b–c$ and the allele arrangement shown above, what is the map distance between these loci?

29. In *Drosophila*, a student was assigned an unknown mutation in lab that had a whitish eye. He crossed females from his true-breeding mutant stock to wild type (brick-red-eyed) males, recovering all wild-type F_1 flies. In the F_2 generation, the following offspring were recovered in the following proportions:

wild type	5/8
bright red	1/8
brown eye	1/8
white eye	1/8

The student was stumped until the instructor suggested that perhaps the whitish eye in the original stock was the result of homozygosity for a mutation causing brown eyes *and* a mutation causing bright red eyes, illustrating gene interaction (see Chapter 4). After much thought, the student was able to analyze the data, explain the results, and learn several things about the location of the two genes relative to one another. One key to his understanding was that crossing over occurs in *Drosophila* females but not in males. What did the student learn about the two genes based on his analysis?

30. *Drosophila melanogaster* has one pair of sex chromosomes (XX or XY) and three autosomes, referred to as chromosomes 2, 3, and 4. A male fly with very short legs was discovered by a genetics student. Using this male, the student was able to establish a pure breeding stock of this mutant and found that it was recessive. This mutant was then incorporated into a stock containing the recessive gene *black* (body color located on chromosome 2) and the recessive gene *pink* (eye color located on chromosome 3). A female from the homozygous *black, pink, short* stock was then mated to a wild-type male. The F_1 males of this cross were all wild type and were then backcrossed to the homozygous $b\ p\ sh$ females. The F_2 results appeared as shown in the following table. No other phenotypes were observed.

	Wild	Pink*	Black, Short	Black, Pink, Short
Females	63	58	55	69
Males	59	65	51	60

Pink indicates that the other two traits are wild type, and so on.

(a) Based on these results, the student was able to assign *short* to a linkage group (a chromosome). Which one was it? Include a step-by-step reasoning.

(b) The experiment was subsequently repeated making the reciprocal cross, F_1 females backcrossed to homozygous $b\ p\ sh$ males. It was observed that 85 percent of the offspring fell into the above classes, but that 15 percent of the offspring were equally divided among $b + p$, $b + +$, $+ sh\ p$, and $+ sh +$ phenotypic males and females. How can these results be explained and what information can be derived from the data?

SELECTED READINGS

Allen, G. E. 1978. *Thomas Hunt Morgan: The man and his science.* Princeton, NJ: Princeton University Press.

Blixt, S. 1975. Why didn't Gregor Mendel find linkage? *Nature* 256:206.

Catcheside, D. G. 1977. *The genetics of recombination.* Baltimore: University Park Press.

Chaganti, R., Schonberg, S., and German, J. 1974. A many-fold increase in sister chromatid exchange in Bloom's syndrome lymphocytes. *Proc. Natl. Acad. Sci. USA* 71:4508–12.

Creighton, H. S., and McClintock, B. 1931. A correlation of cytological and genetical crossing over in *Zea mays. Proc. Natl. Acad. Sci. USA* 17:492–97.

DOUGLAS, L., and NOVITSKI, E. 1977. What chance did Mendel's experiments give him of noticing linkage? *Heredity* 38:253–57.

ELLIS, N. A., et. al. 1995. The Bloom's syndrome gene product is homologous to RecQ helicases. *Cell* 83:655–66.

EPHRUSSI, B., and WEISS, M. C. 1969. Hybrid somatic cells. *Sci. Am.* (April) 220:26–35.

GARCIA-BELLIDO, A. 1972. Some parameters of mitotic recombination in *Drosophila melanogaster. Molec. Genet.* 115:54–72.

HOTTA, Y., TABATA, S., and STERN, H. 1984. Replication and nicking of zygotene DNA sequences: Control by a meiosis-specific protein. *Chromosoma* 90:243–53.

KING, R. C. 1970. The meiotic behavior of the *Drosophila* oocyte. *Int. Rev. Cytol.* 28:125–68.

LATT, S. A. 1981. Sister chromatid exchange formation. *Annu. Rev. Genet.* 15:11–56.

LINDSLEY, D. L., and GRELL, E. H. 1972. *Genetic variations of Drosophila melanogaster.* Washington, DC: Carnegie Institute of Washington.

MOENS, P. B. 1977. The onset of meiosis. In *Cell biology—A comprehensive treatise: Vol. 1, Genetic mechanisms of cells,* ed. L. Goldstein and D. M. Prescott, pp. 93–109. Orlando, FL: Academic Press.

MORGAN, T. H. 1911. An attempt to analyze the constitution of the chromosomes on the basis of sex-linked inheritance in *Drosophila. J. Exp. Zool.* 11:365–414.

MORTON, N. E. 1955. Sequential test for the detection of linkage. *Am. J. Hum. Genet.* 7:277–318.

———. 1995. *LODs*—past and present. *Genetics* 140:7–12.

NEUFFER, M. G., JONES, L., and ZOBER, M. 1968. *The mutants of maize.* Madison, WI: Crop Science Society of America.

PERKINS, D. 1962. Crossing-over and interference in a multiply marked chromosome arm of *Neurospora. Genetics* 47:1253–74.

RUDDLE, F. H., and KUCHERLAPATI, R. S. 1974. Hybrid cells and human genes. *Sci. Am.* (July) 231:36–49.

STERN, C. 1936. Somatic crossing over and segregation in *Drosophila melanogaster. Genetics* 21:625–31.

STERN, H., and HOTTA, Y. 1973. Biochemical controls in meiosis. *Annu. Rev. Genet.* 7:37–66.

———. 1974. DNA metabolism during pachytene in relation to crossing over. *Genetics* 78:227–35.

STURTEVANT, A. H. 1913. The linear arrangement of six sex-linked factors in *Drosophila,* as shown by their mode of association. *J. Exp. Zool.* 14:43–59.

———. 1965. *A history of genetics.* New York: Harper & Row.

TAYLOR, J. H., ed. 1965. *Selected papers on molecular genetics.* Orlando, FL: Academic Press.

VOELLER, B. R., ed. 1968. *The chromosome theory of inheritance: Classical papers in development and heredity.* New York: Appleton-Century-Crofts.

VON WETTSTEIN, D., RASMUSSEN, S. W., and HOLM, P. B. 1984. The synaptonemal complex in genetic segregation. *Annu. Rev. Genet.* 18:331–414.

WOLFF, S., ed. 1982. *Sister chromatid exchange.* New York: Wiley–Interscience.

CHAPTER

6

Recombination and Mapping in Bacteria and Bacteriophages

CHAPTER CONCEPTS

Bacteria and bacteriophages (viruses that have bacteria as their hosts) have been the subject of extensive genetic analysis. They demonstrate several mechanisms by which genetic recombination occurs, processes that may be utilized to perform genetic mapping. Bacteria often contain extrachromosomal DNA in the form of plasmids. Both plasmids and bacteriophage DNA may be involved in recombination and can be integrated into the bacterial chromosome. An extensive analysis of gene structure in bacteriophage T4 established the basis for genetic complementation and complementation testing, an assay that remains important in modern-day investigations.

In this chapter we shift from the consideration of transmission of genetic information in eukaryotes to a discussion of various genetic phenomena in **bacteria** (prokaryotes) and **bacteriophages**, viruses that have bacteria as their host. The study of bacteria and bacteriophages has been essential to the accumulation of knowledge in many areas of genetic study. For example, much of what is known about molecular genetics, recombinational phenomena, and gene structure has been initially derived from experimental work with these organisms. Furthermore, as we shall see in Chapters 15 and 16, our abundant knowledge of bacteria and their resident plasmids has served as the basis for their

widespread use in DNA cloning during recombinant DNA studies.

Their successful use in genetic investigations is due to numerous factors. Both bacteria and their viruses have extremely short reproductive cycles. Literally hundreds of generations, giving rise to billions of genetically identical bacteria or phages, can be produced in short periods of time. Furthermore, they can be studied in pure cultures. That is, a single species or mutant strain of bacteria or one type of virus can be isolated and investigated independently of other similar organisms.

Our major emphasis in this chapter will be on genetic recombination that

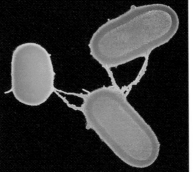

Transmission electron micrograph of conjugating *E. coli.*

occurs in both bacteria and bacteriophages. Detailed processes have evolved that facilitate genetic recombination within populations of these microorganisms and viruses. As we will see, these processes serve as the basis for performing chromosome mapping analysis. Such studies established the initial foundation of knowledge concerning the genetic identity of bacteria and bacteriophages that served as the cornerstone of subsequent molecular genetic investigations.

Bacterial Mutation and Growth

Genetic studies using bacteria depend upon our ability to isolate mutations in these organisms. Although it was known well before 1943 that pure cultures of bacteria could give rise to small numbers of cells exhibiting heritable variation, particularly with respect to survival under different environmental conditions, the source of the variation was hotly debated. The majority of bacteriologists believed that environmental factors induced changes in certain bacteria that led to their survival or adaptation to the new conditions. For example, strains of *E. coli* are known to be *sensitive* to infection by the bacteriophage T1. Infection by the bacteriophage leads to the reproduction of the virus at the expense of the bacterial cell, which is lysed or destroyed. If a plate of *E. coli* is homogeneously sprayed with T1, almost all cells are lysed. Rare *E. coli* cells, however, survive infection and are not lysed. If these cells are isolated and established in pure culture, all descendants are *resistant* to T1 infection. It can be argued that the mutations responsible for T1 resistance were "induced" by the presence of the T1 viruses, and that, in the absence of the T1 viruses, the mutations would not

have occurred. However, as we will explore in a subsequent chapter, such T1-resistant cells result from **spontaneous mutation**, which was elegantly proved by Salvador Luria and Max Delbruck in 1943 (see Chapter 14).

Bacterial cells that bear spontaneous mutations, such as T1 resistance, can be isolated and established independently from the parent strain by using various selection techniques. As a result, mutations for almost any desired characteristic can now be induced and isolated. Because bacteria and viruses that infect them are haploid, all mutations are expressed directly in the descendants of mutant cells, adding to the ease with which these microorganisms can be studied.

Bacteria are grown in either a liquid culture medium or in a Petri dish on a semisolid agar surface. If the nutrient components of the growth medium are very simple and consist only of an organic carbon source (such as a glucose or lactose) and a variety of ions, including Na^+, K^+, Mg^{++}, Ca^{++}, and NH_3^+ present as inorganic salts, it is called **minimal medium**. To grow on such a medium, a bacterium must be able to synthesize all essential organic compounds (e.g., amino acids, purines, pyrimidines, sugars, vitamins, fatty acids). A bacterium that can accomplish this remarkable biosynthetic feat—one that we ourselves cannot duplicate—is termed a **prototroph**. It is said to be wild type for all growth requirements. On the other hand, if a bacterium loses, through mutation, the ability to synthesize one or more organic components, it is said to be an **auxotroph**. For example, if it loses the ability to make histidine, then this amino acid must be added as a supplement to the minimal medium in order for growth to occur. The resulting bacterium is designated as a *his⁻* auxotroph, as opposed to its prototrophic *his⁺* counterpart. Note that medium which has been extensively supplemented is referred to as **complete medium**.

To study mutant bacteria in a quantitative fashion, an inoculum of bacteria is placed in liquid culture medium. A characteristic growth pattern is exhibited, as illustrated in Figure 6.1. Initially, during the **lag phase**, growth is slow. Then, a period of rapid growth ensues called the **log phase**. During this phase, cells divide many times with a fixed time interval between cell divisions, resulting in logarithmic growth. When a cell density of about 10^9 cells per milliliter is reached, nutrients and oxygen become limiting and cells enter the **stationary phase**. As the doubling time during the log phase may be as short as 20 minutes, an initial inoculum of a few thousand cells can easily achieve a maximum cell density in an overnight culture.

Cells grown in liquid medium can be quantitated by transferring them to semisolid medium in a Petri dish. Following incubation and many divisions, each cell gives rise to a visible colony on the surface of the medium. The number of colonies allows one to calculate the number of

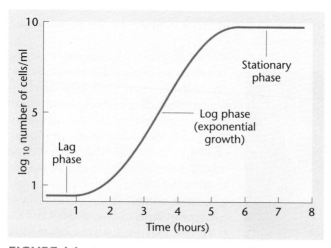

FIGURE 6.1   A typical bacterial population growth curve illustrating the initial lag phase, the subsequent log phase, where exponential growth occurs, and the stationary phase that occurs when nutrients are exhausted.

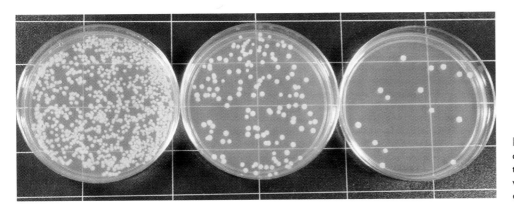

FIGURE 6.2 Results of the serial dilution technique and subsequent culture of bacteria. Each of the dilutions varies by a factor of 10. Each colony was derived from a single bacterial cell.

bacteria present in the original culture. If the number of colonies is too great to count, then serial dilutions of the original liquid culture can be made and plated, until the colony number is reduced to the point where it can be counted (Figure 6.2). Such calculations are useful in a variety of studies.

For example, in the three plates shown in the photograph in Figure 6.2, let us assume that a liquid culture of bacteria has been sampled. An initial milliliter (ml) is withdrawn, and it is our desire to determine how many cells are present per milliliter. If the initial milliliter is subjected to a series of serial dilutions, this can be accomplished easily.

Assume that the three Petri dishes in Figure 6.2 represent dilutions of 10^{-3}, 10^{-4}, and 10^{-5}, respectively (left to right). We need to select only the dish where the number of colonies can be accurately counted. Because each colony presumably arose from a single bacterium, that number times the dilution factor represents the number of bacteria in the initial milliliter. In our case, the dish farthest to the right contains 15 colonies. Since it represents a dilution of 10^{-5}, the initial number of bacterium is estimated to be 15×10^5 per milliliter.

Genetic Recombination in Bacteria: Conjugation

Development of techniques that allowed the identification and study of bacterial mutations led to detailed investigations of the arrangement of genes on the bacterial chromosome. In 1946, Joshua Lederberg and Edward Tatum initiated these studies. They showed that bacteria undergo **conjugation**, a parasexual process in which the genetic information from one bacterium is transferred to and recombined with that of another bacterium. Like meiotic crossing over in eukaryotes, genetic recombination in bacteria provided the basis for the development of

methodology for chromosome mapping. Note that the term **genetic recombination**, as applied to bacteria and bacteriophages, leads to the *replacement* of one or more genes present in one strain with those from a genetically distinct strain. While this is somewhat different from our use of genetic recombination in eukaryotes, where the term describes crossing over that results in *reciprocal exchange events*, the overall effect is the same: Genetic information is transferred from one organism to another, resulting in an altered genotype. Two other phenomena, **transformation** and **transduction**, also result in the transfer of genetic information from one bacterium to another and have also served as a basis for determining the arrangement of genes on the bacterial chromosome. We will return to a discussion of these processes in ensuing sections of this chapter.

Lederberg and Tatum's initial experiments were performed with two multiple auxotroph strains (nutritional mutants) of *E. coli* K12. Strain A required methionine (met) and biotin (bio) in order to grow, whereas strain B required threonine (thr), leucine (leu), and thiamine (thi), as shown in Figure 6.3. Neither strain would grow on minimal medium. The two strains were first grown separately in supplemented media, and then cells from both were mixed and grown together for several more generations. They were then plated on minimal medium. Any bacterial cells that grew on minimal medium would be prototrophs (wild-type bacteria). It was highly improbable that any of the cells that contained two or three mutant genes would undergo spontaneous mutation simultaneously at two or three locations. Therefore, any prototrophs recovered must have arisen as a result of some form of genetic exchange and recombination.

In this experiment, prototrophs were recovered at a rate of $1/10^7$ (10^{-7}) cells plated. The controls for this experiment involved separate plating of cells from strains A and B on minimal medium. No prototrophs were recovered. Based on these observations, Lederberg and Tatum proposed that genetic recombination had occurred!

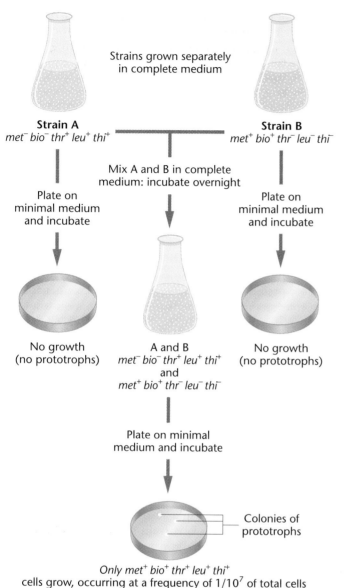

FIGURE 6.3 Genetic recombination involving two auxotrophic strains producing prototrophs. Neither auxotroph will grow on minimal medium, but prototrophs will, allowing their recovery.

F⁺ and F⁻ Bacteria

There soon followed numerous experiments designed to elucidate the genetic basis of conjugation. It became quickly evident that different strains of bacteria were involved in a unidirectional transfer of genetic material. Cells of one strain could serve as donors of parts of their chromosomes and are designated as **F⁺ cells** (F for "fertility"). Recipient bacteria, which undergo genetic recombination by receiving donor chromosome material, now known to be DNA, and recombining it with part of their own, are designated as **F⁻ cells**.

It was established subsequently that cell contact is essential to chromosome transfer. Support for this concept was provided by Bernard Davis, who designed a U-tube in which to grow F⁺ and F⁻ cells (Figure 6.4). At the base of the tube was a sintered glass filter with a pore size that allowed passage of the liquid medium, but was too small to allow the passage of bacteria. The F⁺ cells were placed on one side and F⁻ cells on the other side of the filter. The medium was moved back and forth across the filter so that the bacterial cells essentially shared a common medium during incubation. Samples from both sides of the tube were then plated independently on minimal medium, but no prototrophs were found. Davis concluded that *physical contact is essential to genetic recombination.* This physical interaction is the initial stage of the process of conjugation and is mediated through a structure called the **F** or **sex pilus**. Bacteria often have many pili, which are microscopic extensions of the cell. Different types of pili perform different cellular functions, but all pili are involved in some way with adhesion. After contact has been initiated between mating pairs via the F pili (Figure 6.5), transfer of DNA begins.

Evidence was provided subsequently that F⁺ cells contained a **fertility factor** (called the **F factor**), conferring their ability to donate part of their chromosome during conjugation. In experiments by Joshua and Esther Lederberg and by William Hayes and Luca Cavalli-Sforza, it

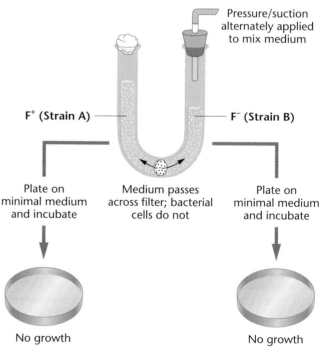

FIGURE 6.4 When strain A and B auxotrophs are grown in a common medium but are separated by a filter, no genetic recombination occurs and no prototrophs are produced. This apparatus is called a Davis U-tube.

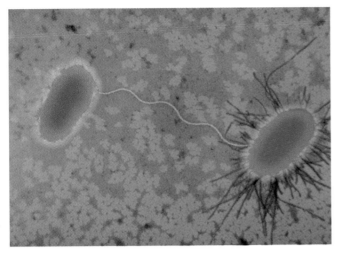

FIGURE 6.5 An electron micrograph of conjugation between an F⁺ *E. coli* cell and an F⁻ cell. The sex pilus is clearly visible.

one of the two strands into the recipient. The other remains in the donor cell. Both of these parental strands serve as templates for DNA replication, resulting in two intact F factors, one in each of the two F⁺ cells. This process is diagrammed in Figure 6.6.

To summarize, *E. coli* cells may or may not contain the F factor. When it is present, the cell is able to form a sex pilus and potentially serve as a donor of genetic information. During conjugation, a copy of the F factor is almost always transferred from the F⁺ cell to the F⁻ recipi-

was shown that certain conditions could eliminate the F factor in otherwise fertile cells. However, if these "infertile" cells were then grown with fertile donor cells, the F factor was regained.

This conclusion that the F factor is a mobile element was further supported by the observation that, following conjugation and genetic recombination, recipient cells always become F⁺. Thus, in addition to the *rare* case of transfer of genes from the bacterial chromosome, the F factor itself is passed to *all* recipient cells. On this basis, the initial crosses of Lederberg and Tatum (Figure 6.3) may be designated:

STRAIN A   STRAIN B

F⁺   ×   F⁻

DONOR   RECIPIENT

Confirmation of these conclusions is based on the isolation of the F factor. Like the bacterial chromosome, but distinct from it, the F factor has been shown to consist of a circular, double-stranded DNA molecule. It contains an amount of DNA equivalent to about 2 percent of that making up the bacterial chromosome (about 100,000 nucleotide pairs). Contained in the F factor, among others, are 19 genes, the products of which are involved in the transfer of genetic information (*tra* genes), including those essential to the formation of the sex pilus.

As we soon shall see, the F factor is in reality an autonomous genetic unit referred to as a **plasmid**. However, in our historical coverage of its discovery, we will continue to refer to it as a "factor."

It is believed that the transfer of the F factor during conjugation involves separation of two strands of the helical DNA making up the F factor and the movement of

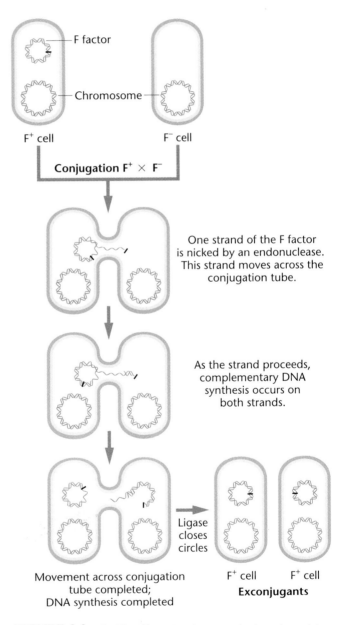

FIGURE 6.6 An F⁺ × F⁻ mating demonstrating how the recipient F⁻ cell is converted to F⁺. During conjugation, the DNA of the F factor is replicated with one new copy entering the recipient cell, converting it to F⁺. The black bar has been added to the F factors to follow their rotation.

ent, converting it to the F$^+$ state. The question remains as to exactly how a very low percentage of F$^-$ cells undergo genetic recombination. As we will discuss, the answer awaited further experimentation.

Hfr Bacteria and Chromosome Mapping

Subsequent discoveries not only clarified how genetic recombination occurs but also defined a mechanism by which the *E. coli* chromosome could be mapped. We shall first address chromosome mapping.

In 1950, Cavalli-Sforza treated an F$^+$ strain of *E. coli* K12 with nitrogen mustard, a potent chemical known to induce mutations. From these treated cells he recovered a strain of donor bacteria that underwent recombination at a rate of 1/10^4 (10^{-4}), 1000 times more frequently than the original F$^+$ strains. In 1953, William Hayes isolated another strain demonstrating a similar elevated frequency. Both strains were designated **Hfr**, or **high-frequency recombination**. Because Hfr cells behave as chromosome donors, they are a special class of F$^+$ cells.

Another important difference was noted between Hfr strains and the original F$^+$ strains. If the donor is an Hfr strain, recipient cells, while sometimes displaying genetic recombination, never become Hfr; that is, they remain F$^-$. In comparison, then,

$$F^+ \times F^- \longrightarrow F^+ \text{ (low rate of recombination)}$$

$$Hfr \times F^- \longrightarrow F^- \text{ (higher rate of recombination)}$$

Perhaps the most significant characteristic of Hfr strains is the nature of recombination. In any given strain, certain genes are more frequently recombined than others, and some not at all. This *nonrandom pattern of gene transfer* was shown to vary from Hfr strain to Hfr strain. While these results were puzzling, Hayes interpreted them to mean that some physiological alteration of the F factor had occurred, resulting in the production of Hfr strains of *E. coli*.

In the mid-1950s, experimentation by Ellie Wollman and François Jacob explained the differences between cells that are Hfr compared to those that are F$^+$ and showed how Hfr strains allow genetic mapping of the *E. coli* chromosome. In their experiments, Hfr and F$^-$ strains with suitable marker genes were mixed and recombination of specific genes assayed at different times. To accomplish this, a culture containing a mixture of an Hfr and an F$^-$ strain was first incubated and samples removed at various intervals and placed in a blender. The shear forces created in the blender separated conjugating bacteria so that the transfer of the chromosome was effectively terminated. The cells were then assayed for genetic recombination. To facilitate the recovery of only recombinants, the Hfr strain was sensitive to an antibiotic while the recipient strain was resistant. Following the blender treatment, the cells were grown on medium *containing* the antibiotic in order to initially ensure the recovery of only recipient cells.

This process, called the **interrupted mating technique**, demonstrated that specific genes of a given Hfr strain were transferred and recombined sooner than others. Figure 6.7 illustrates this point. During the first 8 minutes after the two strains were initially mixed, no genetic recombination could be detected. At about 10 minutes, recombination of the *azi^R* gene could be detected, but no transfer of the *ton^S*, *lac$^+$*, or *gal$^+$* genes was noted. By 15 minutes, 70 percent of the recombinants were *azi^R*; 30 percent were now also *ton^S*; but none was *lac$^+$* or *gal$^+$*. Within 20 minutes, the *lac$^+$* gene was found among the recombinants; and within 30 minutes, *gal$^+$* was also being transferred. Therefore, Wollman and Jacob had demonstrated an *oriented transfer of genes* that was correlated with the length of time conjugation was allowed to proceed.

It appeared that the chromosome of the Hfr bacterium was transferred linearly and that the gene order and distance between them, as measured in minutes, could be predicted from such experiments (Figure 6.8). This information served as the basis for the first genetic map of the *E. coli* chromosome. "Minutes" in bacterial mapping are equivalent to "map units" in eukaryotes.

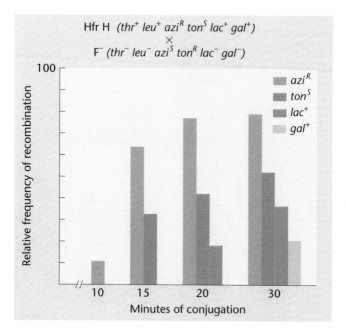

FIGURE 6.7 The progressive transfer during conjugation of various genes from a specific Hfr strain of *E. coli* to an F$^-$ strain. Certain genes (*azi* and *ton*) are transferred sooner than others and recombine more frequently. Others (*lac* and *gal*) take longer to be transferred and recombine with a lower frequency. Others (*thr* and *leu*) are always transferred and are used in the initial screen for recombinants.

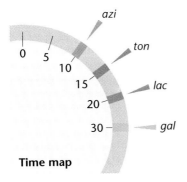

FIGURE 6.8   A time map of the genes studied in the experiment depicted in Figure 6.7.

Wollman and Jacob then repeated the same type of experimentation with other Hfr strains, obtaining similar results with one important difference. Although genes were always transferred linearly with time, as in their original experiment, which genes entered first and which followed later seemed to vary from Hfr strain to Hfr strain [Figure 6.9(a)]. When they reexamined the rate of entry of genes, and thus the different genetic maps for each strain, a definite pattern emerged. The major differences between all strains were simply the point of the origin and the direction in which entry proceeded from that point [Figure 6.9(b)].

To explain these results, Wollman and Jacob postulated that the *E. coli* chromosome is circular. If the point of origin (*O*) varied from strain to strain, a different sequence of genes would be transferred in each case. But what determines *O*? They proposed that in various Hfr strains, the F factor integrates into the chromosome at different points. Its position determines the *O* site. One such case of integration is shown in Figure 6.10. During subsequent conjugation between this Hfr and an F⁻ cell, the position of the F factor determines the initial point of transfer. Those genes adjacent to *O* are transferred first. *The F factor becomes the last part to be transferred.* Apparently, conjugation rarely, if ever, lasts long enough to allow the entire chromosome to pass across the conjugation tube. This proposal explains why recipient cells, when mated with Hfr cells, remain F⁻.

Figure 6.10 also depicts the way in which the two strands making up a DNA molecule unwind during transfer, allowing for the entry of one of the strands of DNA into the recipient. Following replication of the entering DNA, it now has the potential to recombine with its homologous region of the host chromosome. The DNA strand that remains in the donor also undergoes replication.

The use of the interrupted mating technique with different Hfr strains has provided the basis for mapping the entire *E. coli* chromosome. Mapped in time units,

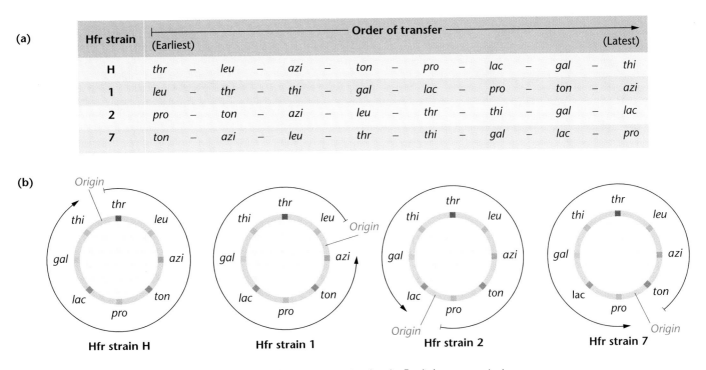

FIGURE 6.9   (a) The order of gene transfer in four Hfr strains, suggesting that the *E. coli* chromosome is circular. (b) The point where transfer originates (Origin) is identified in each strain. Note that transfer can proceed in either direction, depending on the strain. The origin is determined by the point of integration into the chromosome of the F factor, and the direction of transfer is determined by the orientation of the F factor as it integrates.

strain K12 (or *E. coli* K12) is 100 minutes long. Over 900 genes have now been placed on the map. In most instances, only a single copy of each gene exists.

Recombination in F⁺ × F⁻ Matings: A Reexamination

The above model has helped geneticists to understand better how genetic recombination occurs during the F⁺ × F⁻ matings. Recall that recombination occurs much less frequently in them than in Hfr × F⁻ matings, and that random gene transfer is involved. The current belief is that when F⁺ and F⁻ cells are mixed, conjugation occurs readily and that each F⁻ cell involved in conjugation with an F⁺ cell receives a copy of the F factor, *but that no genetic recombination occurs*. However, at an extremely low frequency in a population of F⁺ cells, the F factor integrates spontaneously from the cytoplasm to a random point in the bacterial chromosome. This integration converts the F⁺ cell to the Hfr state (Figure 6.10). Therefore, in F⁺ × F⁻ crosses, the extremely low frequency of genetic recombination (10^{-7}) is attributed to the rare, newly formed Hfr cells, which then undergo conjugation with F⁻ cells. Because the point of integration of the F factor is

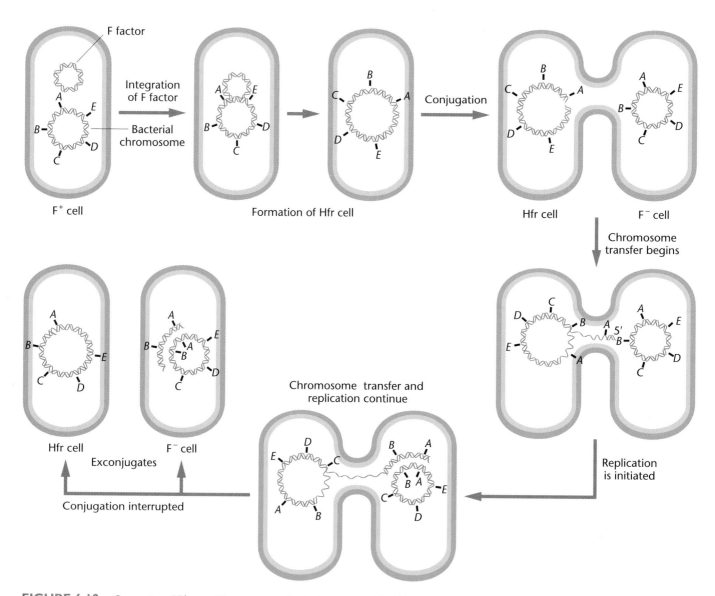

FIGURE 6.10 Conversion of F⁺ to an Hfr state occurs by the integration of the F factor into the bacterial chromosome. The point of integration determines the origin of transfer. During conjugation, the F factor is nicked by an enzyme, creating the origin of transfer of the chromosome. The major portion of the F factor is now on the end of the chromosome adjacent to the origin and is the last part to be transferred. Conjugation is usually interrupted prior to complete transfer. Above, only the A and B genes are transferred.

random, a nonspecific gene transfer ensues, leading to the low-frequency, random genetic recombination observed in the $F^+ \times F^-$ experiment. Unless the recipient cell simultaneously or subsequently undergoes conjugation with a separate F^+ cell, it will remain F^-. Most often, the recombinants become F^+.

The F′ State and Merozygotes

In 1959, during experiments with Hfr strains of *E. coli*, Edward Adelberg discovered that the F factor could lose its integrated status, causing reversion to the F^+ state.

When this occurs, the F factor frequently carries several adjacent bacterial genes along with it. He labeled this condition **F′** to distinguish it from F^+ and Hfr. F′ is thus a special case of F^+. Hence, this conversion is described as one from Hfr to F′, which is illustrated in Figure 6.11.

The presence of bacterial genes within a cytoplasmic F factor creates an interesting situation. An F′ bacterium behaves as an F^+ cell, initiating conjugation with F^- cells. When this occurs, the F factor, containing chromosomal genes, is transferred to the F^- cell. As a result, whatever chromosomal genes are part of the F factor are now present in duplicate in the recipient cell because the recipi-

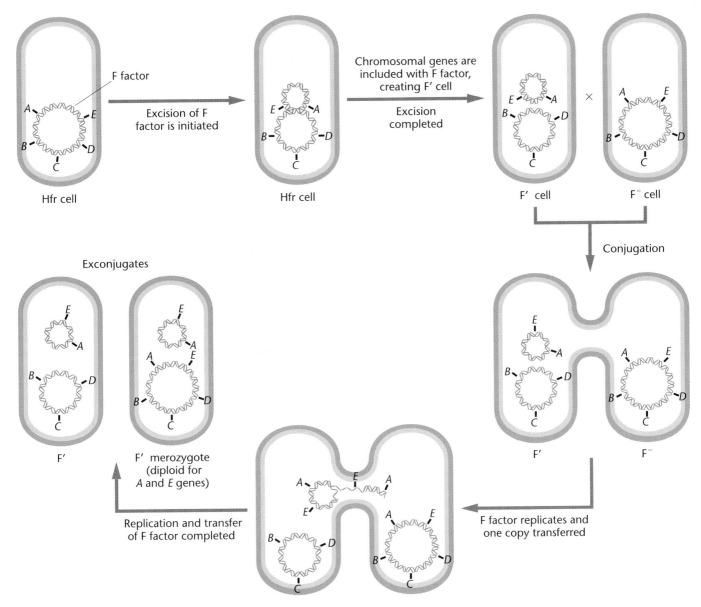

FIGURE 6.11   Conversion of an Hfr bacterium to F′ and its subsequent mating with an F^- cell. The conversion occurs when the F factor loses its integrated status. During excision from the chromosome, it carries with it one or more chromosomal genes (A and E). Following conjugation with such an F′ cell, an F^- recipient becomes partially diploid and is called a merozygote. It also becomes F′.

ent still has a complete chromosome. This creates a partially diploid cell called a **merozygote**. Pure cultures of F′ merozygotes can be established. They have been extremely useful in the study of bacterial genetics, particularly in genetic regulation.

The Rec Proteins and Bacterial Recombination

Having established that a unidirectional transfer of DNA occurs between bacteria, determining how the actual recombination event occurs in the recipient cell has been of interest. Just how does the donor DNA replace the comparable region in the recipient chromosome? As with many systems, the mechanism by which a biochemical event occurs is often deciphered as a result of genetic studies. Such an approach has been particularly fruitful in the case of the phenomenon of recombination between DNA molecules in bacteria. Major insights were gained as a result of the isolation of a group of mutations involving a group of genes called *rec*. This discovery illustrates quite well the value of isolating mutations, establishing their phenotypes, and determining the biological role of the normal, wild-type gene as a result of subsequent investigation.

The first relevant observation in this case involved a series of mutant genes labeled *recA*, *recB*, *recC*, and *recD*. The first mutant gene, *recA*, was found to diminish genetic recombination in bacteria 1000-fold, nearly eliminating it altogether. The other *rec* mutations reduced recombination by about 100 times! Clearly, the normal wild-type products of these genes play some essential role in the process.

By looking for a functional gene product present in normal cells, but missing in mutant cells, several gene products were subsequently isolated and were shown to play a role in genetic recombination. The first is called the **RecA protein.*** The second is a more complex protein called the **RecBCD product**, an enzyme consisting of polypeptide subunits encoded by three other *rec* **genes**.

The general scheme of **RecA-mediated recombination** is illustrated in Figure 6.12. The roles of these proteins have now been elucidated *in vitro*. As a result of this genetic research, our knowledge of the process of recombination has been extended considerably.

The RecA protein has a strong affinity *in vitro* for binding to DNA that is single-stranded (ssDNA). As we

saw in Figure 6.6, during conjugation only one of the two strands of donor DNA initially enters the recipient cell. The binding of RecA with this strand creates a DNA–protein complex that then invades and probes DNA of the recipient chromosome. The complex migrates along the chromosome until the region complementary to the ssDNA is located. When this homologous region, representing the same gene(s) as the invading DNA, is encountered, the invading DNA pairs with and eventually displaces its counterpart in the chromosome. The pairing and displacement, leading to genetic recombination, is facilitated by the RecA protein. The RecBCD enzyme is equally important to this process. It is thought to be essential to recombination in the role of "unwinding" and "cutting" DNA, which also facilitates the displacement of the recipient DNA with the invading molecule. This scheme will undoubtedly be clarified as further research is conducted. A recent finding, for example, has shown that the pairing and displacement step, performed *in vivo* by RecA can be accomplished *in vitro* by a 20-amino-acid peptide fragment derived from the parent molecule. This experimental observation pinpoints the region of the protein critical to its function.

Plasmids

The above sections have introduced and discussed extensively the extrachromosomal heredity unit called the F factor. When existing autonomously in the bacterial cytoplasm, this unit takes the form of a double-stranded closed circle of DNA. These characteristics place the F factor in the more general category of the genetic structures called **plasmids**. These structures contain one or more genes and, often, quite a few. Their replication depends on the same enzymes that replicate the chromosome of the host cell, and they are distributed to daughter cells along with the host chromosome during cell division.

Plasmids are generally classified according to the genetic information specified by their DNA. The F factor confers fertility and contains genes essential for sex pilus formation, upon which genetic recombination depends. Other examples of plasmids include the **R** and the **Col plasmids**.

Most R plasmids consist of two components: the **RTF (resistance transfer factor)** and one or more **r-determinants** (Figure 6.13). The RTF contains information essential to transfer of the plasmid between bacteria, and r-determinants are genes conferring antibiotic resistance. While the DNA of RTFs is quite similar in a variety of plasmids from different bacteria species, there is a wide variation in r-determinants. Each is specific for

---

*Note that the names of bacterial genes begin with lower-case letters and are italicized. The corresponding gene products (proteins) are designated beginning with an upper-case letter and are not italicized. For example, the *recA* gene encodes the RecA protein.

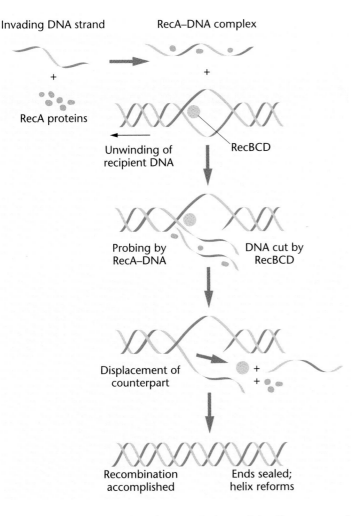

Invading DNA strand

RecA proteins

RecA–DNA complex

Unwinding of recipient DNA

RecBCD

Probing by RecA–DNA

DNA cut by RecBCD

Displacement of counterpart

Recombination accomplished

Ends sealed; helix reforms

FIGURE 6.12 A theoretical model illustrating the possible role of the RecA and RecBCD proteins in the process of genetic recombination between an invading single-stranded bacterial DNA molecule and the host DNA molecule. The RecA protein has an affinity to bind to single-stranded regions of DNA and facilitates the probing and displacement of the host DNA. The RecBCD enzyme unwinds and cuts the host DNA. One further round of DNA replication is needed to fully replace the host DNA.

resistance to one class of antibiotic. If a bacterial cell contains r-determinant plasmids but no RTF is present, the cell is resistant but cannot transfer genetic material to recipient cells. Therefore, resistance cannot be transferred. However, the most commonly studied plasmids consist of the RTF as well as one or more r-determinants. Resistances to tetracycline, streptomycin, ampicillin, sulfonamide, kanamycin, and chloramphenicol are most frequently encountered. Sometimes these occur in a single plasmid, conferring multiple resistance to several antibiotics (Figure 6.13). Bacteria bearing such plasmids are of great medical significance, not only because of their multiple resistance, but because of the ease with which the plasmids may be transferred to other bacteria.

Another plasmid, ColE1, derived from *E. coli*, encodes one or more proteins that are highly toxic to bacterial strains that do not harbor the same plasmid. These proteins, called **colicins**, may kill neighboring bacteria. Those that carry the plasmid are said to be colicinogenic. Present in 10 to 20 copies per cell, the plasmid also contains a gene encoding an immunity protein that protects the host cell from the toxin. Unlike the F and R plasmids,

the Col plasmid is not transmissible to other cells unless one of the former structures is present to confer the ability to engage in conjugation.

Interest in plasmids has increased dramatically because of their role in the genetic technology referred to as

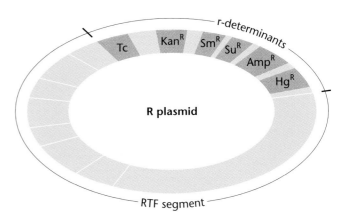

FIGURE 6.13 R plasmid containing resistance transfer factors (RTFs) and multiple r-determinants (Tc, tetracycline; Km, kanamycin; Sm, streptomycin; Su, sulfonamide; Ap, ampicillin; Hg, mercury).

recombinant DNA research (Chapters 15 and 16). Specific genes from any source can be inserted into a plasmid, which may then be inserted into a bacterial cell. As the cell replicates its DNA and undergoes division, the foreign gene is treated like one of the cell's own.

Bacterial Transformation

Still another process, **transformation**, provides a mechanism for the recombination of genetic information in some bacteria. This process involves the recombination of genetic material between bacteria as the result of trans-

fer of extracellular pieces of DNA that are taken up by a living bacterium, ultimately leading to a stable genetic change in the recipient cell. We are most interested in transformation in this chapter because, in those bacterial species where it occurs, the process can be used to map bacterial genes, although in a more limited way compared to conjugation. We will return to the topic of transformation in Chapter 10 because of the central role the process played in experiments providing evidence that DNA is the genetic material.

This process (Figure 6.14) consists of numerous steps that can be divided into two main categories: (1) entry of DNA into a recipient cell, and (2) recombination of the donor DNA with its homologous region in the recipient

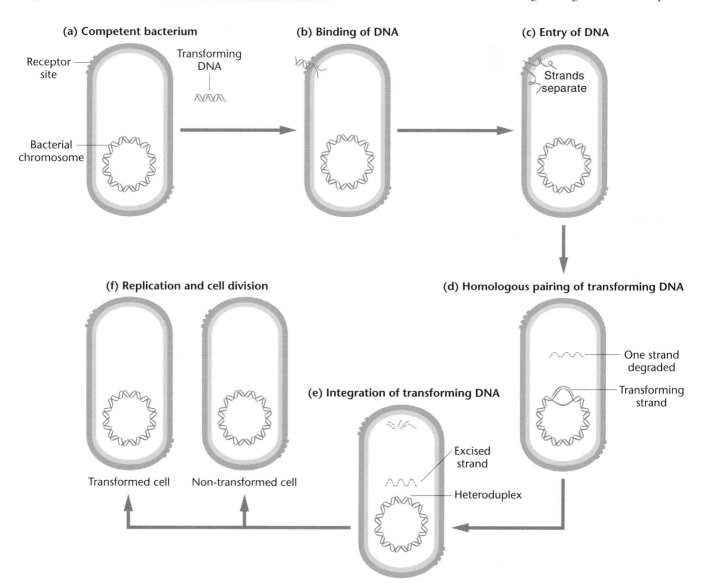

FIGURE 6.14   Proposed steps leading to transformation of a bacterial cell by exogenous DNA. Only one of the two strands of the entering DNA is involved in the transformation event, which is completed following cell division.

chromosome. In a population of cells, only those in a particular physiological state, referred to as **competence**, take up DNA. Entry is thought to occur at a limited number of receptor sites on the surface of the bacterial cell. The efficient length of transforming DNA is about 10,000 to 20,000 nucleotide pairs, an amount equal to about 1/200 of the *E. coli* chromosome. Passage across the cell wall and membrane is an active process requiring energy and specific transport molecules. This concept is supported by the fact that substances that inhibit energy production or protein synthesis in the recipient cell also inhibit the transformation process.

During the process of entry, one of the two strands of the invading DNA molecule is digested by nucleases, leaving only a single strand to participate in transformation. The surviving DNA strand then aligns with its complementary region of the bacterial chromosome. In a process involving several enzymes, this segment of DNA replaces its counterpart in the chromosome, which is excised and degraded (Figure 6.14).

For recombination to be detected, the transforming DNA must be derived from a different strain of bacteria, bearing some genetic variation. Once integrated into the chromosome, the recombinant region contains one DNA strand from the bacterial chromosome and one from the transforming DNA. Because these strands are not genetically identical, this helical region is referred to as a **heteroduplex**. Following one round of replication, one chromosome is restored to its original configuration, identical to that of the recipient cell, and the other contains the transformed gene. Following cell division, one host cell and one transformed cell are produced.

Transformation and Mapping

Because the size of DNA that is effective in transformation is between 10,000 to 20,000 nucleotide pairs, the DNA contains sufficient genetic information to encode several genes. Genes that are adjacent or very close to one another on the bacterial chromosome may be carried on a single piece of DNA of this size. Because of this fact, a single event may result in the **cotransformation** of several genes simultaneously. Genes that are close enough to each other to be cotransformed are said to be linked. Note that here *linkage* refers to the proximity of genes, in contrast to the use of this term in eukaryotes to indicate all genes on a single chromosome.

If two genes are not linked, simultaneous transformation can occur only as a result of two independent events involving two distinct segments of DNA. As in double crossing over in eukaryotes, the probability of two independent events occurring simultaneously is equal to the product of the individual probabilities. Thus, the frequency of two unlinked genes being transformed simultaneously is much lower than if they are linked.

Linked genes were first demonstrated in 1954 during studies of *Pneumococcus* by Rollin Hotchkiss and Julius Marmur. They were examining transformation at the *streptomycin* and *mannitol* loci. Recipient cells were *str^s* and *mtl^-*, meaning that they were sensitive to streptomycin and could not ferment mannitol. Cells that were *str^r mtl^+*, which are resistant to streptomycin and able to ferment mannitol, were used to derive the transforming DNA. If these genes are unlinked, simultaneous transformation for both genes will occur with such a minimal probability as to be nearly undetectable. However, as shown in Table 6.1, a low but detectable rate (0.17%) of double transformation did occur.

To confirm that the genes were indeed linked, a second experiment was performed. Instead of using *str^r mtl^+* DNA, a mixture of *str^r mtl^-* and *str^s mtl^+* DNA simultaneously served as the donor DNA. Thus, both the *str^r* and *mtl^+* genes were available to recipient cells, but on separate segments of DNA. If the observed rate of double transformants in the previous experiment had been due to two independent events and the genes unlinked, then a similar rate would also be observed under these conditions. Instead, the number of double transformants was

TABLE 6.1 The Results of Several Transformation Experiments, Which Establish Linkage Between the *str* and *mtl* Loci in *Pneumococcus*

Donor DNA	Recipient Cell Genotype	Percentage of Transformed Genotypes		
		str^r mtl^-	*str^s mtl^+*	*str^r mtl^+*
str^r mtl^+	*str^s mtl^-*	4.3	0.40	0.17
str^r mtl^- and *str^s mtl^+*		2.8	0.85	0.0066

Source: Data from Hotchkiss and Marmur, 1954, p. 55.

25 times fewer, thus confirming linkage (Table 6.1). Careful examination of the data in this table not only confirms the linkage of these loci, but raises other interesting questions. Why, for example, is the *str* locus always transformed more frequently than the *mtl* locus? This observation suggests that transformation of genes is not a totally random process.

Subsequent studies of transformation have shown that various bacteria readily undergo this process of recombination (e.g., *Bacillus subtilis* and *Shigella paradysenteriae*). In other cases, culture conditions can be adjusted to "induce" artificial transformation, an important component of recombinant DNA technology (Chapter 15). Under certain conditions, studies have shown that relative distances between linked genes can be determined from the recombination data provided by transformation experiments. While analysis is more complex, such data are interpreted in a manner analogous to chromosome mapping in eukaryotes. In the *Problems and Discussion Questions* section at the end of the chapter we consider data that provides mapping information.

The Genetic Study of Bacteriophages

There is still a third mode of genetic transfer between bacteria, called **transduction**. We shall temporarily delay our discussion of this topic until we have considered the genetics of bacteriophages, since transduction is a process mediated by these bacterial viruses. Aside from their role in transduction, genetic recombination also occurs between bacterial viruses, which can be observed when genetically distinguishable strains have simultaneously infected the same bacterial host cell. In this section, we will first review the life cycle of a typical **T-even bacteriophage**. We will then discuss how these phages are studied during their infection of bacteria. Finally, we will contrast two possible modes of behavior once initial phage infection occurs. This information will serve as background for our subsequent discussion of transduction and bacteriophage recombination. Furthermore, a great deal of genetic research has been done (and continues to be done) using bacteriophages as a model system, making them a worthy subject of discussion.

Phage T4 Infection and Reproduction

Infection and reproduction in phage T4 relies on the ability of the bacteriophage to inject its genetic material into the bacterial host cell. Its DNA is sufficient in quantity to encode more than 150 average-sized genes. The genetic material is enclosed by an icosahedral protein coat (a polyhedron with 20 faces) making up the head of the virus. This is connected to a tail that contains a collar and a contractile sheath surrounding a central core. Six tail fibers protrude from the tail; their tips contain binding sites that specifically recognize unique areas of the outer membrane forming the cell surface of the *E. coli* cell wall. Figure 6.15 illustrates the components described above, as well as the general scheme for their assembly into a mature bacteriophage.

Binding of tail fibers to these specific areas of the cell wall is the first step of infection. Then, an ATP-driven contraction of the tail sheath causes the central core to penetrate the cell wall. The DNA in the head is extruded where it then moves across the cell membrane into the bacterial cytoplasm. Within minutes, all bacterial DNA, RNA, and protein synthesis is inhibited, and synthesis of viral macromolecules begins. This initiates the **latent period**, in which the production of viral components occurs prior to the assembly and appearance of mature virus particles.

Bacteriophage gene activity is initiated using host cell enzymes. The viral genes are divided into three groups: (1) **immediate early genes**, (2) **delayed early genes**, and (3) **late genes**. One of the earliest gene products is a DNase enzyme that degrades the *E. coli* chromosome. The T4 DNA molecule is protected from digestion by this enzyme because its cytosine residues are modified to form **5-hydroxymethylcytosine** (Figure 6.16); as a result, this DNase is unable to use T4 DNA as a substrate. You may wish to jump ahead and examine Figure 9.9, which illustrates the chemical structure of the five most significant nitrogenous bases, including cytosine.

A plethora of gene products are subsequently synthesized using the host cell ribosomes. For example, about 20 different proteins are part of the protein capsid of the head. Many others are structural components of the tail and its fibers. One late gene product, **lysozyme**, digests the bacterial cell wall, leading to its rupture and release of mature phages once they are assembled. Prior to the synthesis of many viral proteins, DNA replication begins, leading to a pool of viral DNA molecules that are available to be packaged into phage heads. Prior to being packaged, genetic recombination may occur between DNA molecules.

The assembly of mature viruses is a complex process that has been well studied by William Wood, Robert Edgar, and others. Three major independent pathways occur, leading to: (1) DNA packaged into heads; (2) construction of tails; and (3) synthesis and assembly of tail fibers. As shown in Figure 6.15, once DNA is packaged into the head, this complex combines with the tail, and only then are the tail fibers added. Total construction is a combination of self-assembly and enzyme-directed processes. When approximately 200 viruses are assembled, the bacterial cell is ruptured by the action of lyso-

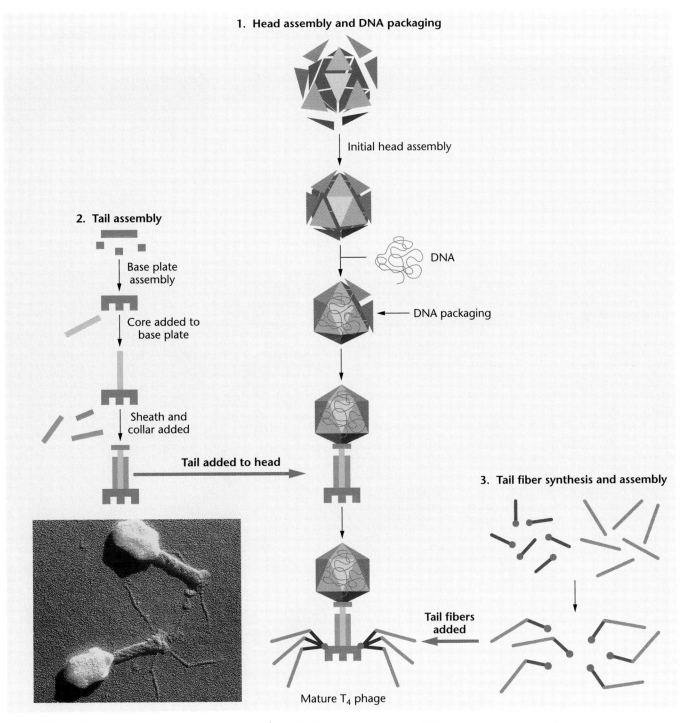

1. Head assembly and DNA packaging

Initial head assembly

DNA

DNA packaging

2. Tail assembly

Base plate
assembly

Core added to
base plate

Sheath and
collar added

Tail added to head

3. Tail fiber synthesis and assembly

**Tail fibers
added**

Mature T₄ phage

FIGURE 6.15 The structure and assembly of bacteriophage T4 following infection of *E. coli*. Three separate pathways lead to the formation of an icosahedral head filled with DNA, a tail containing a central core, and tail fibers. The tail is added to the head and then tail fibers are added. Each step of each pathway is influenced by one or more phage genes.

zyme and the mature phages are released. If this occurs on a lawn of bacteria, the 200 new phages will infect other bacterial cells and the process will be repeated over and over again. For each isolated phage on the lawn, this leads to a clear area where all bacteria have been lysed. This area is called a **plaque** and represents multiple clones of the single-infecting T4 bacteriophage.

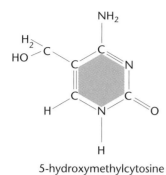

5-hydroxymethylcytosine

FIGURE 6.16 The chemical structure of 5-hydroxymethylcytosine, which is present in some bacteriophage DNA.

The Plaque Assay

The experimental study of bacteriophages and other viruses has played a critical role in our understanding of molecular genetics. During infection of bacteria, enormous quantities of bacteriophages may be obtained for investigation. Often, over 10^{10} viruses per milliliter of culture medium are produced. Many genetic studies have relied on the ability to quantitate the number of phages produced following infection under specific culture conditions. The technique utilized is called the **plaque assay**.

This assay is illustrated in Figure 6.17, where actual plaque morphologies are also shown. A serial dilution of the original virally infected bacterial culture is first performed. Then, a 0.1-ml aliquot from one or more dilutions is added to a small volume of melted nutrient agar (about 3 ml) to which a few drops of a healthy bacterial culture have been mixed. The solution is then poured evenly over a base of solid nutrient agar in a Petri dish and allowed to solidify prior to incubation. As described in the above section, viral plaques occur at each place where a single virus has initially infected one bacterium in the lawn that has grown up during incubation. If the dilution factor is too low, the plaques are plentiful and will fuse, lysing the entire lawn. This has occurred in the 10^{-3} dilution in Figure 6.17. On the other hand, if the dilution factor is increased, fewer phages are present in a given aliquot and plaques can be counted. From such data, the density of viruses in the initial culture can be estimated. The calculation is the same type as that used for determining bacterial density by counting colonies following serial dilution of an initial culture:

$$(\text{plaque number/ml}) \times (\text{dilution factor})$$

Using the results shown in Figure 6.17, it is observed that there are 23 phage plaques derived from the 0.1-ml aliquot of the 10^{-5} dilution. Therefore, we can estimate that there are 230 phages per milliliter at this dilution.

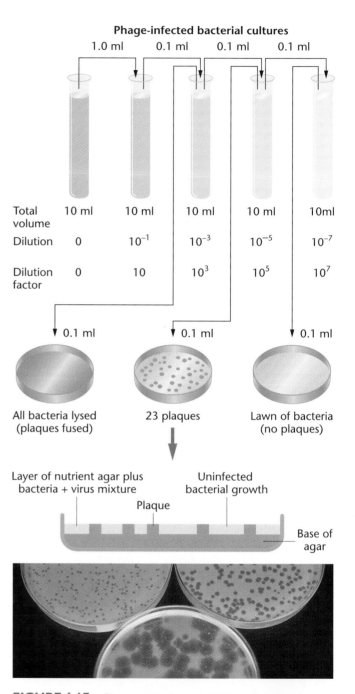

FIGURE 6.17 Diagrammatic illustration of the plaque assay for bacteriophage analysis. Serial dilutions of a bacterial culture infected with bacteriophages are first made. Then three of the dilutions (10^{-3}, 10^{-5}, and 10^{-7}) are analyzed using the plaque assay technique. In each case, 0.1 ml of the diluted culture is used. Each plaque represents the initial infection of one bacterial cell by one bacteriophage. In the 10^{-3} dilution, so many phages are present that all bacteria are lysed. In the 10^{-5} dilution, 23 plaques are produced. In the 10^{-7} dilution, the dilution factor is so great that no phages are present in the 0.1-ml sample, and thus no plaques form. From the 0.1-ml sample of the 10^{-5} dilution, the original bacteriophage density can be calculated as $23 \times 10 \times 10^5$ phages/ml (23×10^6 or 2.3×10^7), as described in the text. The photo illustrates a plaque assay involving phage T2 and *E. coli*.

This initial viral density in the undiluted sample, where 23 plaques are observed from 0.1 ml of the 10^{-5} dilution, is calculated as:

$$(230/\text{ml}) \times (10^5) = 230 \times 10^5/\text{ml} = 2.3 \times 10^7/\text{ml}$$

Because this figure is derived from the 10^{-5} dilution, we can estimate that there will be only 0.23 phage per 0.1 ml in the 10^{-7} dilution. As a result, when 0.1 ml from this tube is assayed, it is expected that no phage particles will be present. This possibility is borne out in Figure 6.17, where an intact lawn of bacteria exists. The dilution factor is simply too great.

The use of the plaque assay has been invaluable in mutational and recombinational studies of bacteriophages. We will apply this technique more directly later in this chapter, when Seymour Benzer's elegant genetic analysis of a single gene in phage T4 is discussed.

Lysis and Lysogeny

The relationship between virus and bacterium does not always result in viral reproduction and lysis. As early as the 1920s it was known that some bacteriophages could enter a bacterial cell and establish a symbiotic relationship with it. The precise molecular basis of this symbiosis is now well understood. Upon entry, the viral DNA, instead of replicating in the bacterial cytoplasm, is integrated into the bacterial chromosome, a step that characterizes the developmental stage referred to as **lysogeny**. Subsequently, each time the bacterial chromosome is replicated, the viral DNA is also replicated and passed to daughter bacterial cells following division. No new viruses are produced and no lysis of the bacterial cell occurs. However, in response to certain stimuli, such as chemical or ultraviolet light treatment, the viral DNA may lose its integrated status and initiate replication, phage reproduction, and lysis of the bacterium.

Several terms are used to describe this relationship. The viral DNA, integrated in the bacterial chromosome, is called a **prophage**. Viruses that can either lyse the cell or behave as a prophage are called **temperate**. Those that can only lyse the cell are referred to as **virulent**. A bacterium harboring a prophage has been **lysogenized** and is said to be **lysogenic**; that is, it is capable of being lysed as a result of induced viral reproduction. The viral DNA, which can replicate either in the bacterial cytoplasm or as part of the bacterial chromosome, is sometimes classified as an **episome**.

The relationship of a prophage and its host cell is symbiotic because lysogenic bacteria are immune to viral attacks by the same phage whose genetic information it harbors. This immunity is due to the synthesis of a prophage repressor molecule that regulates the expression of the viral DNA (Chapter 18).

Transduction: Virus-Mediated Bacterial DNA Transfer

In 1952, Joshua Lederberg and Norton Zinder were investigating possible recombination in the bacterium *Salmonella typhimurium*. Although they recovered prototrophs from mixed cultures of two different auxotrophic strains, subsequent investigations revealed that recombination was occurring in a manner different from that attributable to the presence of an F factor, as in *E. coli*. What they were to discover was a process of bacterial recombination mediated by bacteriophages and now called **transduction**.

The Lederberg–Zinder Experiment

Lederberg and Zinder mixed the *Salmonella* auxotrophic strains LA-22 and LA-2 together and recovered prototroph cells when the mixture was plated on minimal medium. LA-22 was unable to synthesize the amino acids phenylalanine and tryptophan (*phe⁻ trp⁻*), and LA-2 could not synthesize the amino acids methionine and histidine (*met⁻ his⁻*). Prototrophs (*phe⁺ trp⁺ met⁺ his⁺*) were recovered at a rate of about $1/10^5$ (10^{-5}) cells.

Although these observations at first appeared to suggest that the type of recombination involved was the kind observed earlier in conjugative strains of *E. coli*, experiments using the Davis U-tube soon showed otherwise (Figure 6.18). When the two auxotrophic strains were separated by a glass-sintered filter, thus preventing cell contact but allowing growth to occur in a common medium, a startling observation was made. When samples were removed from both sides of the filter and plated independently on minimal medium, prototrophs were recovered, but only from one side of the tube. Recall that if conjugation were responsible, the conditions in the Davis U-tube would be expected to prevent recombination.

Prototrophs were recovered only when cells from the side of the tube containing LA-22 bacteria were plated. Obviously, since LA-2 cells were the source of the new genetic information, the presence of LA-2 cells on the other side of the tube was essential for recombination. Because the genetic information responsible for recombination had somehow to pass across the filter, but in an unknown form, it was initially designated simply as a **filterable agent (FA)**.

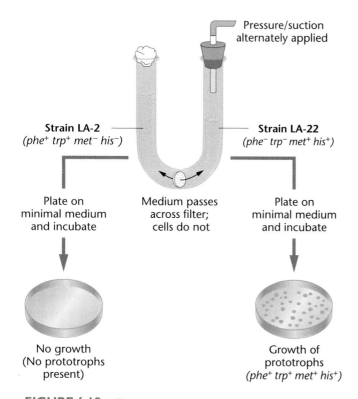

FIGURE 6.18 The Lederberg–Zinder experiment using *Salmonella*. After placing two auxotrophic strains on opposite sides of a Davis U-tube, Lederberg and Zinder recovered prototrophs from the side containing LA-22 but not from the side containing LA-2. These initial observations led to the discovery of the phenomenon called transduction.

Three subsequent observations made it clear that this recombination was quite distinct from any other form of recombination.

1. The FA would not pass across filters with a pore diameter of less than 100 nm, a size that normally allows passage of small DNA molecules.

2. Testing the FA in the presence of DNase, which will enzymatically digest DNA, showed that the FA was not destroyed. These first two observations demonstrate that FA is not naked DNA.

3. The third observation was particularly important. It was observed that FA was produced by the LA-2 cells only when they were grown in association with LA-22 cells. If LA-2 cells were grown independently and that culture medium was then added to LA-22 cells, recombination was not observed. Therefore, LA-22 cells play some role in the production of FA by LA-2 cells and do so only when sharing common growth medium. This was a key observation in determining the basis of genetic recombination.

These observations were explained by the presence of a prophage (P22) present in the LA-22 *Salmonella* cells.

Rarely, P22 prophages entered the vegetative or lytic phase, reproduced, and lysed some of the LA-22 cells. This phage, being much smaller than a bacterium, was then able to cross the filter and lyse some of the LA-2 cells, because this strain was not immune to attack. In the process of lysis, the P22 phages produced in LA-2 often acquired a region of the LA-2 chromosome along with their own genetic material. If this region contained the phe^+, and trp^+ genes present in LA-2, and if the phages subsequently passed back across the filter and reinfected LA-22 cells, prototrophs were produced. The exact nature of how this occurred was not immediately clear.

The Nature of Transduction

Further studies revealed the existence of transducing phages in other species of bacteria. For example, *E. coli*, *Bacillus subtilis*, and *Pseudomonas aeruginosa* can be transduced by the phages P1, SPO1, and F116, respectively. The precise mode of transfer of DNA during transduction has also been established. The process most often begins when a prophage enters the lytic cycle and progeny viruses subsequently infect and lyse other bacteria. During infection, the bacterial DNA is degraded into small fragments and the viral DNA is replicated. As packaging of the viral chromosomes in the protein head of the phage occurs, errors are sometimes made. Rarely, instead of viral DNA, segments of bacterial DNA are packaged.

Even though the initial discovery of transduction involved lysogenic bacteria, the same process can occur during the normal lytic cycle. Following infection, the bacterial chromosome is degraded into small pieces. If, during bacteriophage assembly, a small piece of bacterial DNA is packaged along with the viral chromosome, subsequent transduction is possible.

Sometimes, *only* bacterial DNA is packaged. Regions as large as 1 percent of the bacterial chromosome may become enclosed randomly in the viral head. Following lysis, these aberrant phages, lacking their own genetic material, are released in the culture medium. Because the ability to infect is a property of the protein coat, they can initiate infection of other unlysed bacteria. When this occurs, bacterial rather than viral DNA is injected into the bacterium and can either remain in the cytoplasm or recombine with its homologous region of the bacterial chromosome. If the bacterial DNA remains in the cytoplasm, it does not replicate but may remain in one of the progeny cells following each division. When this happens, only a single cell, partially diploid for the transduced genes, is produced—a phenomenon called **abortive transduction**. If the bacterial DNA recombines with its homologous region of the bacterial chromosome, the transduced genes are replicated as part of the chromosome and passed to all daughter cells. As in conjugation and transformation, an even number of crossovers is necessary for recombination to occur. This process is called **complete transduction**. Both abortive and complete transduction are subclasses of the broader category of **generalized transduction**. As described above, transduction is characterized by the random nature of DNA fragments and genes transduced. Each fragment has a finite but small chance of being packaged in the phage head. Most cases of generalized transduction are of the abortive type; some data suggest that complete transduction occurs 10 to 20 times less frequently. This finding may be related to the fact that double-stranded DNA is involved. In comparison with single-stranded DNA, which is integrated during transformation, it may be much more difficult for double-stranded DNA to become integrated.

Mapping and Specialized Transduction

As with transformation, transduction has been used in linkage studies and mapping of the bacterial chromosome. Cotransduced genes must be aligned closely to one another along the chromosome. By concentrating on two or three linked genes, transduction studies can also determine the precise order of genes. Such an analysis is predicated on the same rationale underlying other mapping techniques, where outcomes resulting from single events occur much more frequently than those relying on two or more independent events occurring simultaneously.

Another aspect of virally mediated bacterial recombination involves **specialized transduction**. Compared with generalized transduction, where all genes have an equal probability of being transduced, specialized transduction is restricted to certain genes. One of the best examples involves transduction of *E. coli* by the prophage λ. In this case, only the *gal* (galactose) or *bio* (biotin) genes are transduced.

The reason why specialized transduction occurs became clear when it was learned how the λ DNA integrates into the *E. coli* chromosome during lysogeny. A region of λ DNA, designated *att*, is some 15 nucleotides long and is responsible for integration. A precisely homologous region exists on the *E. coli* chromosome, which is flanked by the *gal* and *bio* loci on either end. Therefore, λ DNA always integrates at a location in the chromosome between these genes (Figure 6.19).

Phage λ DNA can be induced to excise (detach) from the chromosome and cause lysis. Sometimes the excision process occurs incorrectly and carries either the *gal* or *bio E. coli* genes in place of part of the viral DNA. This hap-

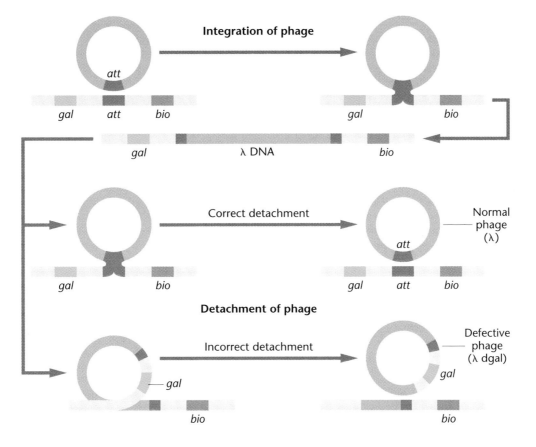

FIGURE 6.19   Production of a defective phage λ, leading to specialized transduction.

pens when the recombinational event leading to detachment occurs incorrectly, outside the *att* region of the λ chromosome. The resulting chromosome is defective because it has lost some of its genetic information, but it is nevertheless replicated and packaged during the formation of mature phage particles. Once the previously lysogenized cell is lysed, the virus can inject the defective chromosome into another bacterial cell.

In a process involving lysogeny by a nondefective chromosome, the defective viral chromosome is integrated into the bacterial chromosome and is replicated along with it. Such cells are diploid for the *gal* or *bio* genes. If the recipient cells are *gal⁻* and are unable to use galactose as a carbon source, the presence of the transducing *gal⁺* DNA will cause them to revert to a *gal⁺* phenotype, where they can use this carbohydrate. In a similar way, *bio⁻* cells can be transduced to a *bio⁺* phenotype.

As is evident from this discussion, specialized transduction occurs in quite a different way from generalized transduction. Because it is limited to certain genes, it is not as useful in linkage or mapping studies.

Mutation and Recombination in Viruses

Much of what is known concerning viral genetics has been derived from studies of bacteriophages. Phage mutations often affect the morphology of the plaques formed following lysis of the bacterial cells. For example, in 1946 Alfred Hershey observed unusual T2 plaques on plates of *E. coli* strain B. While the normal T2 plaques are small and have a clear center surrounded by a diffuse (nearly invisible) halo, the unusual plaques were larger and possessed a more distinctive outer perimeter (Figure 6.20). When the viruses were isolated from these plaques and replated on *E. coli* B cells, an identical plaque appearance was noted. Thus, the plaque phenotype was an inherited trait resulting from the reproduction of mutant phages. Hershey named the mutant *rapid lysis* (*r*) because the plaques were larger, apparently resulting from a more rapid or more efficient life cycle of the phage. It is now known that wild-type phages undergo an inhibition of reproduction once a particular-sized plaque has been formed. The *r* mutant T2 phages are able to overcome this inhibition, producing larger plaques.

Another bacteriophage mutation, *host range* (*h*), was discovered by Luria. This mutation extends the range of bacterial hosts that the phage can infect. Although wild-type T2 phages can infect *E. coli* B, they cannot normally attach or be adsorbed to the surface of *E. coli* B-2. The *h* mutation, however, provides the basis for adsorption and

subsequent infection of B-2 *E. coli*. When grown on a mixture of *E. coli* B and B-2, the center of the *h* plaque appears much darker than the *h⁺* plaque (Figure 6.20).

Table 6.2 lists other representative types of mutations that have been isolated and studied in the T-even series of bacteriophages (T2, T4, T6, etc.). These mutations are important to the study of genetic phenomena in bacteriophages.

Genetic Exchange between Bacterial Viruses

About 1947, several research teams demonstrated that recombination occurs between viruses. This discovery of genetic exchange was made during experiments in which two mutant strains of bacteriophages were allowed to simultaneously infect the same bacterial culture. These **mixed infection experiments** were designed so that the number of viral particles sufficiently exceeded the number of bacterial cells so as to ensure simultaneous infection of most cells by both viral strains.

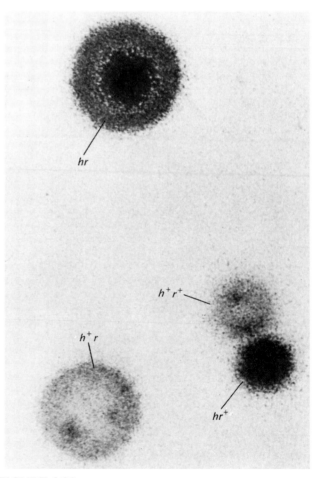

FIGURE 6.20 Actual plaque phenotypes observed following simultaneous infection of *E. coli* by two strains of phage T2, *h⁺r* and *hr⁺*. In addition to the parental genotypes, recombinant plaques *hr* and *h⁺r⁺* were also recovered.

TABLE 6.2 Some Mutant Types of T-Even Phages

Name	Description
minute	Small plaques
turbid	Turbid plaques on *E. coli* B
star	Irregular plaques
uv-sensitive	Alters UV sensitivity
acriflavin-resistant	Forms plaques on acriflavin agar
osmotic shock	Withstands rapid dilution into distilled water
lysozyme	Does not produce lysozyme
amber	Grows in *E. coli* K12, but not B
temperature-sensitive	Grows at 25°C, but not at 42°C

For example, in one study using the T2/*E. coli* system, the viruses were of either the h^+r or hr^+ genotype. If no recombination occurred, these two parental genotypes (wild-type host range restriction, rapid lysis; and extended host range, normal lysis) would be the only expected phage progeny. However, the recombinants h^+r^+ and hr were detected in addition to the parental genotypes (Figure 6.20). As with eukaryotes, the percentage of recombinant plaques divided by the total number of plaques reflects the relative distance between the genes:

$$\frac{(h^+r^+) + (hr)}{\text{total plaques}} \times 100 = \text{recombinational frequency}$$

Sample data for the *h* and *r* loci are shown in Table 6.3.

Similar recombinational studies have been performed with large numbers of mutant genes in a variety of bacteriophages. Data are analyzed in much the same way as they are in eukaryotic mapping experiments. Two- and three-point mapping crosses are possible, and the percentage of recombinants in the total number of phage progeny is calculated. This value is proportional to the relative distance between two genes along the DNA molecule constituting the chromosome.

An interesting observation in phage crosses is **negative interference**. Recall that in eukaryotic mapping crosses, positive interference is the rule, leading to the re-

covery of fewer than expected double-crossover events. In phage crosses, often just the opposite occurs. In three-point analysis, a greater than expected frequency of double exchanges is observed. Negative interference is explained on the basis of the dynamics of the conditions leading to recombination within the bacterial cell.

How phage recombination occurs has been investigated. Available evidence supports a model where recombination between phage chromosomes involves a breakage and reunion process similar to that of eukaryotic crossing over. The process is facilitated by nucleases that nick and reseal DNA strands. A fairly clear picture of the dynamics of viral recombination has emerged.

Following the early phase of infection, the chromosomes of the phages begin replication. As this stage progresses, a pool of chromosomes accumulates in the bacterial cytoplasm. If double infection by phages of two genotypes has occurred, then the pool of chromosomes initially consists of the two parental types. Genetic exchange between these two types will occur before, during, and after replication, producing recombinant chromosomes.

In the case of the h^+r and hr^+ example just discussed, h^+r^+ and hr chromosomes are produced. Each of these recombinant chromosomes may undergo replication and is also free to undergo new exchange events with the other and with the parental types. Furthermore, recombination is not restricted to exchange between two chromosomes—three or more may be involved. As phage development progresses, chromosomes are randomly removed from the pool and packed into the phage head, forming mature phage particles. Thus, parental and recombinant genotypes are produced.

As we will see in the following section, powerful selection systems have made it possible to detect *intragenic* recombination in viruses, where exchanges occur at points within a single gene. Such studies have led to the fine-structure analysis of the gene, as discussed in an ensuing section of this chapter.

TABLE 6.3 Results of a Cross Involving the *h* and *r* Genes in Phage T2 ($hr^+ \times h^+r$)

Genotype	Plaques	Designation
$h\ r^+$	42	Parental Progeny 76%
h^+r	34	
h^+r^+	12	Recombinants 24%
$h\ r$	12	

Source: Data derived from Hershey and Rotman, 1949.

Intragenic Recombination in Phage T4

We conclude this chapter with an account of an ingenious example of **genetic analysis**. In the early 1950s Seymour Benzer undertook a detailed examination of a single locus, *rII*, in phage T4. Benzer successfully designed experiments to recover the extremely rare genetic recombinants arising as a result of **intragenic exchange**. Such recombination is equivalent to eukaryotic crossing over, *but in this case, within the gene rather than between two genes.* Benzer demonstrated that such recombination occurs between the DNA of individual bacteriophages during simultaneous infection of the host bacterium *E. coli*.

The end result of Benzer's work was the production of a detailed map of the *rII* locus. Because of the extremely detailed information provided from his analysis, Benzer's work is often described as the **fine-structure analysis of the gene**. Because these experiments occurred decades before DNA sequencing techniques were developed, the insights concerning the internal structure of the gene took on added significance.

The *rII* Locus of Phage T4

The primary requirement in genetic analysis is the isolation of a large number of mutations in the gene being investigated. Mutants at the *rII* locus produce distinctive plaques when plated on *E. coli* strain B, allowing their easy identification (see Figure 6.20). Benzer's approach was to isolate many independent *rII* mutants—he eventually obtained about 20,000—and to perform recombinational studies so as to produce a genetic map of this locus.

The key to his analysis was that *rII* mutant phages, while capable of infecting and lysing *E. coli* B, could not successfully lyse a second related strain, *E. coli* K12(λ)*. However, wild-type phages could lyse both the B and the K12 strains. Benzer realized that these conditions provided the potential for a highly sensitive screening system. He reasoned that if phages from any two independent mutant strains were allowed to simultaneously infect *E. coli* B, any exchanges between the two mutant sites within the locus would produce rare wild-type recombinants (Figure 6.21). If the phage population, which contained over 99.9 percent *rII* phages and less than 0.1 percent wild-type phages, were then allowed to infect strain K12, only those wild-type recombinants would be successful in reproducing and producing wild-type plaques. *This is the critical step in recovering and quantifying rare recombinants.*

By using serial dilution techniques, it is possible to determine the total number of mutant *rII* phages produced on *E. coli* B and the total number of recombinant wild-type phages that would lyse *E. coli* K12. These data provide the basis for calculating the frequency of recombination. This value is considered to be proportional to the distance within the gene between the two mutations being studied. As we will quickly see, this experimental design was extraordinarily sensitive. It was possible for Benzer to detect as few as one recombinant wild-type phage among 100 million mutant phages! Such resolution is truly remarkable.

When information from many such experiments is combined, a detailed map of the locus is possible. We now know that what Benzer did was to determine the order of specific nucleotides in the *rII* locus, in other words, to genetically "sequence" the gene. This in itself was a truly remarkable accomplishment.

However, before we describe this mapping we need to describe an important discovery made by Benzer during the early development of his screen, for this discovery led to the development of a technique used widely in genetics labs today, the **complementation assay**.

Complementation by *rII* Mutations

Before Benzer was able to initiate intragenic recombinational studies, he encountered a problem in his experimentation that he had to resolve. He realized that, while doing a control for his experiment in which he infected K12 bacteria with pairs of *rII* mutant strains, in many cases the two mutant strains combined could infect and lyse the K12 strain! This was initially quite puzzling since only the wild-type *rII* was supposed to be capable of lysing K12 bacteria. How could two mutant strains of *rII*, each of which was thought to contain a defect in the same gene, show a wild-type function?

Benzer's initial proposal to explain this observation was that, although he believed the *rII* locus represented a

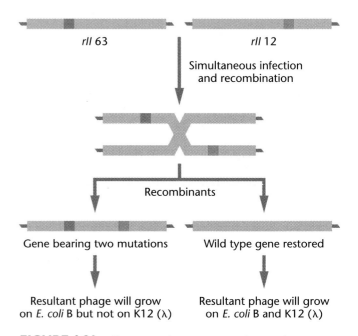

FIGURE 6.21 Illustration of intragenic recombination between two mutations in the *rII* locus of phage T4. The result is the production of a wild-type phage that will grow on *E. coli* B and K12 and a phage that has incorporated both mutations into the *rII* locus. It will grow on *E. coli* B, but not on *E. coli* K12.

---

*The inclusion of (λ) in the designation of K12 indicates that this bacterial strain is lysogenized by phage λ. This, in fact, is the reason that *rII* mutants cannot lyse such bacteria. Note that as we refer to this strain in future discussions, we will abbreviate it simply as *E. coli* K12.

single gene, somehow during simultaneous infection each mutant strain provided something that the other lacked. He called this phenomenon **complementation**, which is illustrated in Figure 6.22(a). We now know that in fact this is the case, each strain is providing something that the other lacks, because the *rII* locus actually contains *two separate genes*. Therefore, there is a simple explanation for what Benzer was observing; each mutant strain was producing the wild-type gene product that was mutant in the other strain, and so together the complete wild-type function was restored.

Benzer's experiments actually provided evidence for the existence of two genes in the *rII* locus because, upon testing pairs of mutations, they all fell into one of two possible **complementation groups**, A or B. Those that complemented each other were assigned to different complementation groups, while those that failed to complement one another were placed in the same complementation group. Benzer coined the term **cistron** to describe each complementation group, which he defined as the smallest functional genetic unit.

We now know that Benzer's A and B cistrons represent the two genes of what we refer to as the *rII* locus. Complementation results when K12 bacteria are infected with two *rII* mutants, one with a mutation in the A gene, and one with a mutation in the B gene. Therefore, there is a source of both wild-type gene products, since the A mutant provides wild-type B, and the B mutant provides

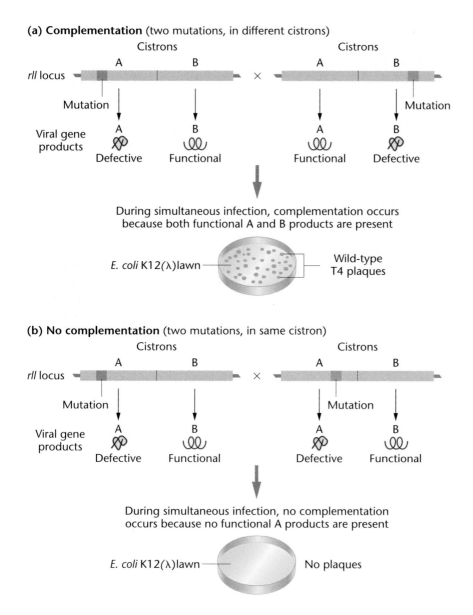

FIGURE 6.22 Comparison of two *rII* mutations that either complement one another (a), or do not (b). Complementation occurs when each mutation is in a separate cistron and results in lysis of the cell. Failure to complement occurs if the two mutations are in the same cistron and results in no lysis.

wild-type A. We can also explain why two strains that fail to complement, say two A cistron mutants, are actually mutations in the same gene. In this case, if two A cistron mutants are combined, there will be a source of the wild-type B product, but no source of wild-type A product.

Once Benzer was able to place all *rII* mutations in either the A or B cistron, he was set to return to his intragenic recombination studies, testing mutations in the A cistron against each other and testing mutations in the B cistron against each other.

Recombinational Analysis

Of the approximately 20,000 *rII* mutations, roughly half fell into each cistron. Benzer set about to map the mutations within each one. For example, focusing initially on the A cistron, if two *rIIA* mutants were first allowed to infect *E. coli* B in a liquid culture, and if a recombination event occurred between the mutational sites in the A cistron, then wild-type progeny viruses would be produced at low frequency. If samples of the progeny viruses from such an experiment were then plated on *E. coli* K12, only the recombinants would lyse the bacteria and produce plaques. The total number of nonrecombinant progeny viruses was also determined by plating samples on *E. coli* B.

This experimental protocol is illustrated in Figure 6.23. The percentage of recombinants can be determined by counting the plaques at the appropriate dilution in each case. As in eukaryotic mapping experiments, the frequency of recombination can be taken as an estimate of the distance between the two mutations within the cistron.

For example, if the number of recombinants is equal to 4×10^3/ml, and the total number of progeny is 8×10^9/ml, then the frequency of recombination between the two mutants is:

$$2\left(\frac{4 \times 10^3}{8 \times 10^9}\right) = 2(0.5 \times 10^{-6})$$

$$= 10^{-6}$$

$$= 0.000001$$

Multiplying by 2 is necessary because each recombinant event yields two reciprocal products. Only one of them—the wild type—is detected.

Deletion Testing of the *rII* Locus

Although the selective system of recombination just described was available to map mutations within each cistron, testing 1000 mutants two at a time in all combinations would have required millions of experiments.

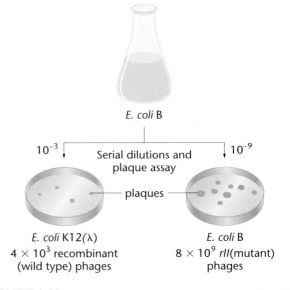

Simultaneous infection with two *rIIA* or two *rIIB* mutations

E. coli B

10^{-3} Serial dilutions and plaque assay 10^{-9}

plaques

E. coli K12(λ)
4×10^3 recombinant (wild type) phages

E. coli B
8×10^9 *rII*(mutant) phages

FIGURE 6.23 The experimental protocol for recombinational studies between pairs of *rIIA* or *rIIB* mutations.

Fortunately, Benzer was able to overcome this obstacle when he discovered that some of the *rII* mutations were in reality **deletions** of small parts of each cistron. That is, the genetic changes giving rise to the *rII* properties were not point mutations involving a single nucleotide change, but instead were due to the loss or deletion of a variable number of nucleotides within the cistron. These deletions could be identified by their failure to revert to wild type. Doing so is a characteristic of point mutations. Furthermore, *and this is the key point*, it was observed that a deletion, when tested during simultaneous infection with a point mutation located in the deleted part of the same cistron, *never* yielded any wild-type recombinants. The basis for the failure to do so is illustrated in Figure 6.24. Thus, a method was available that could roughly but quickly localize any mutation, provided it was contained within a region covered by a deletion.

As shown in Figure 6.25, **deletion testing** could serve as the basis for the initial localization of each mutation. For example, seven overlapping deletions spanning various regions of the A and B cistrons were used for initial screening of the point mutations. Based on the ability or the failure of the viral chromosome bearing a point mutation to undergo recombination with the chromosome bearing a deletion, each point mutation can be assigned to a specific area of the cistron. Then, further deletions within each of the seven areas can be used to localize or map each *rII* point mutation more specifically. Remember that, in each case, a point mutation is localized in the area of a deletion when it fails to give rise to any wild-type recombinants.

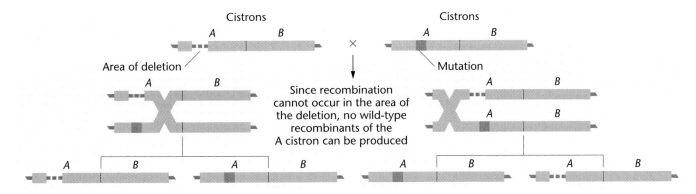

FIGURE 6.24   Demonstration that recombination between a phage chromosome with a deletion in the A cistron and another with a mutation overlapped by that deletion cannot yield a chromosome with wild-type A and B cistrons.

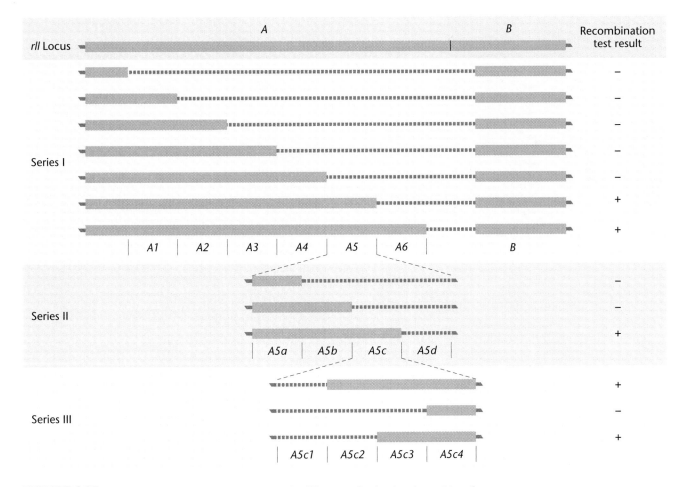

FIGURE 6.25   Three series of overlapping deletions in the *rII* locus used to localize the position of an unknown *r* mutation. For example, if a mutant strain tested against each deletion (dashed areas) in Series I for the production of recombinant wild-type progeny shows the results at the right (+ or −), the mutation must be in segment A5. In Series II, the mutation is further narrowed to segment A5c, and in Series III to segment A5c3.

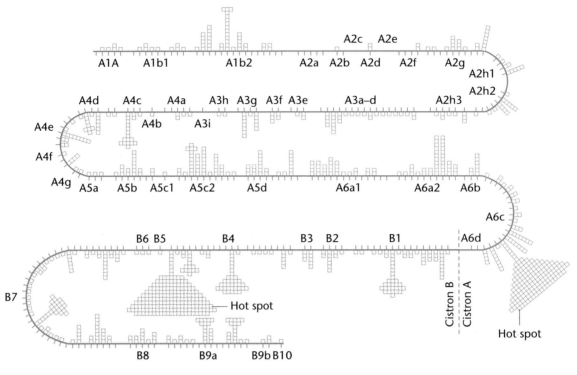

FIGURE 6.26

A partial map of mutations in the A and B cistrons of the *rII* locus of phage T4. Each square represents an independently isolated mutation. Note the two areas where the largest number of mutations are present, referred to as "hot spots" (A6cd and B5).

The *rII* Gene Map

After several years of work, Benzer produced a genetic map of the two cistrons composing the *rII* locus of phage T4 (Figure 6.26). Of the 20,000 mutations analyzed, 307 distinct sites within this locus had been mapped in relation to one another. Many mutations fell into the same site. Areas containing many mutations were designated as **hot spots**. It appeared that such areas were more susceptible to mutation than areas where only one or a few mutations are found. Additionally, Benzer found areas within the cistrons where no mutations were localized. He estimated that as many as 200 recombinational units had not been localized by his studies.

The significance of Benzer's work is his application of genetic analysis to what had previously been considered an abstract unit—the gene. Benzer demonstrated that a gene is not an indivisible particle but instead consists of mutational and recombinational units that are arranged in a specific order, that is, a sequence of nucleotides. His research was performed shortly after the publication of Watson and Crick's work on DNA and some time before the genetic code was unraveled in the early 1960s. Thus, his analysis is considered one of the classic examples of genetic experimentation.

CHAPTER SUMMARY

1. Genetic recombination in bacteria may be accomplished as a result of three different genetic modes: conjugation, transformation, and transduction.

2. Conjugation is initiated by a bacterium housing a plasmid called the F factor. If the F factor is in the cytoplasm (F^+) of a donor cell, one strand is transferred

to and replicated in the recipient cell, converting it to the F^+ status.

3. If the F factor is integrated into the donor cell chromosome (Hfr), recombination is initiated with the recipient cell, with genetic information flowing unidirectionally. Time mapping of the bacterial chro-

mosome is based on the orientation and position of the F factor in the donor chromosome.

4. The products of a group designated as the *rec* genes have been found to be directly involved in the process of recombination between the invading DNA and the recipient bacterial chromosome.

5. Plasmids, such as the F factor, are autonomously replicating DNA molecules found in the bacterial cytoplasm. Some plasmids may contain genes such as those conferring antibiotic resistance, as well as those necessary for their transfer during conjugation.

6. The phenomenon of transformation, which does not require cell contact, involves the entry of single-stranded exogenous DNA into the host chromosome of a recipient bacterial cell. Linkage mapping of closely aligned genes may be performed using this process.

7. Bacteriophages can be studied using the plaque assay. The discovery of mutations in plaques, such as those causing variation in plaque morphology and the alteration of the host range of infectivity, have allowed analysis of recombination between viruses. Such genetic exchange occurs during simultaneous (or mixed) infection of bacterial cells by two distinct mutant viral strains.

8. Transduction, or virus-mediated bacterial DNA recombination, requires the integration of the viral chromosome into the bacterial chromosome (lysogeny). When a lysogenized bacterium subsequently enters the lytic cycle, the new bacteriophages serve as the vehicle for the transfer of host (bacterial) DNA. In the process of generalized transduction, a random part of the bacterial chromosome is transferred. In specialized transduction, only specific genes adjacent to the point of insertion of the prophage into the bacterial chromosome are transferred. Transduction may also be used for bacterial linkage and mapping studies.

9. Various mutant phenotypes have been studied in bacteriophages. These have served as the basis for investigating genetic exchange between these viruses.

10. Genetic analysis of the *rII* locus in bacteriophage T4 allowed Seymour Benzer to explain the phenomenon of complementation and to study intragenic recombination. By isolating *rII* mutants, performing recombinational studies, and relying on deletion mapping, Benzer was able to locate over 300 distinct sites within the two cistrons of the *rII* locus.

KEY TERMS

INSIGHTS AND SOLUTIONS

1. Time mapping was performed in a cross involving the genes *his*, *leu*, *mal*, and *xyl*. The recipient cells were auxotrophic for all four genes. After 25 minutes, mating was interrupted with the following results in recipient cells.

$$90\% \text{ were } xyl^+$$
$$80\% \text{ were } mal^+$$
$$20\% \text{ were } his^+$$
$$\text{none were } leu^+$$

What are the positions of these genes relative to the origin (*O*) of the F factor and to one another?

Solution: Because the *xyl* gene was transferred most frequently, it is closest to *O* (very close). The *mal* gene is next and reasonably close to *xyl*, followed by the *his* gene. The *leu* gene is well beyond these three, since no recombinants are recovered that include it. The diagram below illustrates these relative locations along a piece of the circular chromosome.

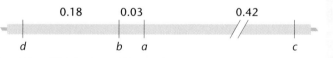

2. Three strains of bacteria, each bearing a separate mutation, a^-, b^-, or c^-, were used as the sources of donor DNA in a transformation experiment involving recipient cells that were wild type for those genes, but which expressed the mutant d^-.

(a) Based on the following data, and assuming that the location of the *d* gene precedes the *a*, *b*, and *c* genes, propose a linkage map for the four genes:

DNA Donor	Recipient	Transformants	Frequency of Transformants
$a^- d^+$	$a^+ d^-$	$a^+ d^+$	0.21
$b^- d^+$	$b^+ d^-$	$b^+ d^+$	0.18
$c^- d^+$	$c^+ d^-$	$c^+ d^+$	0.63

Solution: These data reflect the relative distances between each of the *a*, *b*, *c* genes and the *d* gene. The *a* and *b* genes are about the same distance away from the *d* gene and are thus tightly linked to one another. The *c* gene is more distant. Assuming that the *d* gene precedes the others, the map looks like this:

```
         0.18        0.03              0.42
    |_____|_____|___ //_____|
    d              b   a                    c
```

(b) If the donor DNA was wild type and recipient cells were either $a^- b^-$, $a^- c^-$, or $b^- c^-$, in which case would wild-type transformants be expected most frequently?

Solution: Because the *a* and *b* genes are closely linked, they are most likely to be cotransformed in a single event. Thus, recipient cells of $a^- b^-$ are most likely to be converted to wild type.

3. In four Hfr strains of bacteria, all derived from an original F^+ culture grown over several months, a group of hypothetical genes was studied and shown to be transferred in the following orders:

Hfr Strain	Order of Transfer
1	*E R I U M B*
2	*U M B A C T*
3	*C T E R I U*
4	*R E T C A B*

Assuming that *B* is the first gene along the chromosome, determine the sequence of all genes shown. One strain creates an apparent dilemma. Which one is it? Explain why the dilemma is only apparent and not real.

Solution: The solution is arrived at by overlapping the genes in each strain in the sequence in which they were transferred:

	2	*U M B A C T*
Strain:	3	*C T E R I U*
	1	*E R I U M B*

Starting with *B*, the sequence of genes is: *BACTERIUM*!

Strain 4 creates an apparent dilemma, which is resolved by realizing that the F factor is integrated in the opposite orientation; thus, the genes enter in the opposite sequence, starting with gene *R*.

$$MUIRETCAB$$
$$\longrightarrow$$

4. In Benzer's fine-structure analysis of the *rII* locus in phage T4, he was able to perform complementation testing of any pair of mutations once it was clear that the locus contained two cistrons. Complementation was assayed by simultaneously infecting *E. coli* K12 with two phage strains, each with an independent mutation, neither of which could alone lyse K12. From the following data, determine which mutations are in which cistron, assuming that mutation 1 (M-1) is in the A cistron and mutation 2 (M-2) is in the B cistron. Are there any cases where the mutation cannot be properly assigned?

Test Pair	Results
1, 2	+
1, 3	−
1, 4	−
1, 5	+
2, 3	−
2, 4	+
2, 5	−

Solution: M-1 and M-5 complement one another and are therefore not in the same cistron. Thus, M-5 must be in the B cistron. M-2 and M-4 complement one another. Using the same reasoning, M-4 is not with M-2 and therefore is in the A cistron.

M-3 fails to complement either M-1 or M-2 and so would seem to be in both cistrons. One explanation is that the physical basis of M-3 somehow overlaps both the A and B cistrons. It might be a double mutation with one in each cistron. It might also be a deletion that overlaps both cistrons, making it impossible for it to complement either M-1 or M-2.

5. Another mutation, M-6, was tested with the results shown below. Draw all possible conclusions about M-6.

Test Pair	Results
1, 6	+
2, 6	−
3, 6	−
4, 6	+
5, 6	−

Solution: These results are consistent with assigning M-6 to the B cistron.

6. Recombination testing was then performed for M-2, M-5, and M-6 so as to map the B cistron. Recombination analy-

sis using both *E. coli* B and K12 showed that recombination occurred between M-2 and M-5 and between M-5 and M-6, but not between M-2 and M-6. Why not?

Solution: Either M-2 and M-6 represent identical mutations, or one of them may be a deletion that overlaps the other, but not M-5. Furthermore, the data cannot rule out the possibility that both are deletions.

7. In recombination studies, what is the significance of the value determined by calculating growth on the K12 versus the B strains of *E. coli* following simultaneous infection in *E. coli*? Which value is always greater?

Solution: By performing plaque analysis on *E. coli* B, where wild-type and mutant phages are both lytic, the total number of phages per milliliter can be determined. Because almost all cells are *rII* mutants of one type or another, this value is much larger. To avoid total lysis of the plate, extensive dilutions are necessary. On K12, *rII* mutations will not grow, but wild-type phages will. Since wild-type phages are the rare recombinants, there are relatively few of them and extensive dilution is not required.

PROBLEMS AND DISCUSSION QUESTIONS

1. Distinguish among the three modes of recombination in bacteria.
2. With respect to F$^+$ and F$^-$ bacterial matings, answer the following questions:
   (a) How was it established that physical contact was necessary?
   (b) How was it established that chromosome transfer was unidirectional?
   (c) What is the physical–chemical basis of a bacterium being F$^+$?
3. List all major differences (a) between the F$^+$ × F$^-$ and the Hfr × F$^-$ bacterial crosses; and (b) between F$^+$, F$^-$, Hfr, and F′ bacteria.
4. Describe the basis for chromosome mapping in the Hfr × F$^-$ crosses.
5. Why are the recombinants produced from an Hfr × F$^-$ cross never F$^+$?
6. Describe the origin of F′ bacteria and merozygotes.
7. Describe what is known about the mechanisms of the transformation process.
8. In a transformation experiment involving a recipient bacterial strain of genotype a^-b^-, the following results were obtained:

Transforming DNA	% Transformants		
	a^+b^-	a^-b^+	a^+b^+
a^+b^+	3.1	1.2	0.04
a^+b^- and a^-b^+	2.4	1.4	0.03

What can you conclude about the location of the *a* and *b* genes relative to each other?

9. In a transformation experiment, donor DNA was obtained from a prototroph bacterial strain ($a^+b^+c^+$), and the recipi-

ent was a triple auxotroph ($a^-b^-c^-$). The following transformant classes were recovered:

$a^+b^-c^-$	180
$a^-b^+c^-$	150
$a^+b^+c^-$	210
$a^-b^-c^+$	179
$a^+b^-c^+$	2
$a^-b^+c^+$	1
$a^+b^+c^+$	3

What general conclusions can you draw about the linkage relationships among the three genes?

10. Explain the observations that led Zinder and Lederberg to conclude that the prototrophs recovered in their transduction experiments were not the result of F-mediated conjugation.
11. Define plaque, lysogeny, and prophage.
12. Differentiate between generalized and restricted transduction. Which can be used in mapping and why?
13. Two theoretical genetic strains of a virus ($a^-b^-c^-$ and $a^-b^+c^+$) are used to simultaneously infect a culture of host bacteria. Of 10,000 plaques scored, the following genotypes were observed:

$a^+b^+c^+$	4100	$a^-b^+c^-$	160
$a^-b^-c^-$	3990	$a^+b^-c^+$	140
$a^+b^-c^-$	740	$a^-b^-c^+$	90
$a^-b^+c^+$	670	$a^+b^+c^-$	110

Determine the genetic map of these three genes on the viral chromosome. Determine whether interference was positive or negative.

14. Describe the conditions under which genetic recombination may occur in bacteriophages.

15. If a single bacteriophage infects one *E. coli* cell present on a lawn of bacteria, and upon lysis yields 200 viable viruses, how many phages will exist in a single plaque if only three more lytic cycles occur?

16. A phage-infected bacterial culture was subjected to a series of dilutions, and a plaque assay was performed in each case, with the results shown below. What conclusion can be drawn in the case of each dilution?

	Dilution Factor	Assay Results
(a)	10^4	All bacteria lysed
(b)	10^5	14 plaques
(c)	10^6	0 plaques

17. In complementation studies of the *rII* locus of phage T4, three groups of three different mutations were tested. For each group, only two combinations were tested. On the basis of each set of data, predict the results of the third experiment.

Group A	Group B	Group C
$d \times e$—lysis	$g \times h$—no lysis	$j \times k$—lysis
$d \times f$—no lysis	$g \times i$—no lysis	$j \times l$—lysis
$e \times f$—?	$h \times i$—?	$k \times l$—?

18. In an analysis of other *rII* mutants, complementation testing yielded the following results:

Mutants	Results (lysis)
1, 2	+
1, 3	+
1, 4	−
1, 5	−

   (a) Predict the results of testing 2 and 3, 2 and 4, and 3 and 4 together.
   (b) If further testing yielded the following results, what would you conclude about mutant 5?

Mutants	Results
2, 5	−
3, 5	−
4, 5	−

   (c) Following mixed infection of mutants 2 and 3 on *E. coli* B, progeny viruses were plated in a series of dilutions on both *E. coli* B and K12 with the following results. What is the recombination frequency between the two mutants?

Strain Plated	Dilution	Colonies
E. coli B	10^{-5}	2
E. coli K12	10^{-1}	5

   (d) Another mutation, 6, was then tested in relation to mutations 1 through 5. In initial testing, mutant 6 complemented mutants 2 and 3. In recombination testing with 1, 4, and 5, mutant 6 yielded recombinants with 1 and 5, but not with 4. What can you conclude about mutation 6?

19. When the interrupted mating technique was used with five different strains of Hfr bacteria, the following order of gene entry and recombination was observed. On the basis of these data, draw a map of the bacterial chromosome. Do the data support the concept of circularity?

Hfr Strain	Order
1	*T C H R O*
2	*H R O M B*
3	*M O R H C*
4	*M B A K T*
5	*C T K A B*

EXTRA-SPICY PROBLEMS

20. During the analysis of seven *rII* mutations in phage T4, mutants 1, 2, and 6 were in cistron A, while mutants 3, 4, and 5 were in cistron B. Of these, mutant 4 was a deletion overlapping mutant 5. The remainder were point mutations. Nothing was known about mutant 7.
   (a) Predict the results of complementation (+ or −): 1 and 2; 1 and 3; 2 and 4; and 4 and 5.
   (b) In recombination studies between 1 and 2, the following results were obtained. Calculate the recombination frequency.

Strain	Dilution	Plaques	Phenotypes
E. coli B	10^{-7}	4	r
E. coli K12	10^{-2}	8	+

   (c) When mutant 6 was tested for recombination with mutant 1, the data were the same for strain B as shown above, but not for K12. The researcher lost the K12 data but remembered that recombination was 10 times more frequent than when mutants 1 and 2 were tested. What were the lost values (dilution and colony numbers)?
   (d) Mutant 7 failed to complement any of the other mutants (1–6). Define the nature of mutant 7.

21. In *Bacillus subtilis*, linkage analysis of two genes affecting the synthesis of the two amino acids tryptophan (trp_2^-) and tyrosine (tyr_1^-) was performed using transformation. Examine the data below and draw all possible conclusions regarding linkage. What is the role of Part B of the experiment? [*Reference*: E. Nester, M. Schafer, and J. Lederberg (1963).]

	Donor DNA	Recipient Cell	Transformants	No.
A.	$trp_2^+tyr_1^+$	$trp_2^-tyr_1^-$	trp^+tyr^-	196
			trp^-tyr^+	328
			trp^+tyr^+	367
B.	$trp_2^+tyr_1^-$ and $trp_2^-tyr_1^+$	$trp_2^-tyr_1^-$	trp^+tyr^-	190
			trp^-tyr^+	256
			trp^+tyr^+	2

22. An Hfr strain is used to map three genes in an interrupted mating experiment. The cross is $Hfr/a^+b^+c^+rif^s \times F^-/a^-b^-c^-rif^r$ (No map order is implied in the listing of the alleles; *rif* = the antibiotic rifampicin). The a^+ gene is required for the biosynthesis of nutrient A, the b^+ gene for nutrient B, and c^+ for nutrient C. The minus alleles are auxotrophs for these nutrients.

The cross is initiated at time = 0, and at various times the mating mixture is plated on three types of medium. Each plate contains minimal medium (MM) *plus* rifampicin *plus* specific supplements that are indicated below. The results for each time point are shown as number of colonies growing on each plate.

Supplements Added to MM	Time of Interruption			
	5 min	10 min	15 min	20 min
Nutrients A and B	0	0	4	21
Nutrients B and C	0	5	23	40
Nutrients A and C	4	25	60	82

(a) What is the purpose of rifampicin in the experiment?
(b) Based on these data, determine the approximate location on the chromosome of the *a*, *b*, and *c* genes relative to one another and relative to the F factor.
(c) Can the location of the *rif* gene be determined in this experiment? If not, design an experiment to determine the location of *rif* relative to the F factor and gene B.

SELECTED READINGS

ADELBERG, E. A. 1960. *Papers on bacterial genetics.* Boston: Little, Brown.

BENZER, S. 1955. Fine structure of a genetic region in bacteriophage. *Proc. Natl. Acad. Sci. USA* 42:344–54.

———. 1961. On the topography of the genetic fine structure. *Proc. Natl. Acad. Sci. USA* 47:403–15.

———. 1962. The fine structure of the gene. *Sci. Am.* (Jan.) 206:70–87.

BIRGE, E. A. 1988. *Bacterial and bacteriophage genetics—An introduction.* New York: Springer-Verlag.

BROCK, T. 1990. *The emergence of bacterial genetics.* Cold Spring Harbor, NY: Cold Spring Harbor Laboratory Press.

BRODA, P. 1979. *Plasmids.* New York: W. H. Freeman.

CAIRNS, J., STENT, G. S., and WATSON, J. D., eds. 1966. *Phage and the origins of molecular biology.* Cold Spring Harbor, NY: Cold Spring Harbor Laboratory Press.

CAMPBELL, A. M. 1976. How viruses insert their DNA into the DNA of the host cell. *Sci. Am.* (Dec.) 235:102–13.

FOX, M. S. 1966. On the mechanism of integration of transforming deoxyribonucleate. *J. Gen. Physiol.* 49:183–96.

HAYES, W. 1953. The mechanisms of genetic recombination in *Escherichia coli. Cold Spring Harbor Symp. Quant. Biol.* 18: 75–93.

———. 1968. *The genetics of bacteria and their viruses,* 2nd ed. New York: Wiley.

HERSHEY, A. D. 1946. Spontaneous mutations in a bacterial virus. *Cold Spring Harbor Symp. Quant. Biol.* 11:67–76.

HERSHEY, A. D., and CHASE, M. 1951. Genetic recombination and heterozygosis in bacteriophage. *Cold Spring Harbor Symp. Quant. Biol.* 16:471–79.

HERSHEY, A. D., and ROTMAN, R. 1949. Genetic recombination between host range and plaque-type mutants of bacteriophage in single cells. *Genetics* 34:44–71.

HOTCHKISS, R. D., and MARMUR, J. 1954. Double marker transformations as evidence of linked factors in deoxyribonucleate transforming agents. *Proc. Natl. Acad. Sci. USA* 40:55–60.

JACOB, F., and WOLLMAN, E. L. 1961a. *Sexuality and the genetics of bacteria.* Orlando, FL: Academic Press.

———. 1961b. Viruses and genes. *Sci. Am.* (June) 204:92–106.

LEDERBERG, J. 1986. Forty years of genetic recombination in bacteria: A fortieth anniversary reminiscence. *Nature* 324: 627–28.

LWOFF, A. 1953. Lysogeny. *Bacteriol. Rev.* 17:269–337.

MESELSON, M., and WEIGLE, J. J. 1961. Chromosome breakage accompanying genetic recombination in bacteriophage. *Proc. Natl. Acad. Sci. USA* 47:857–68.

MORSE, M. L., LEDERBERG, E. M., and LEDERBERG, J. 1956. Transduction in *Escherichia coli* K12. *Genetics* 41:141–56.

NESTER, E., SCHAFER, M., and LEDERBERG, J. 1963. Gene linkage in DNA transfer: A cluster of genes in *Bacillus subtilis. Genetics* 48:529–51.

NOVICK, R. P. 1980. Plasmids. *Sci. Am.* (Dec.) 243:102–27.

OZEKI, H., and IKEDA, H. 1968. Transduction mechanisms. *Annu. Rev. Genet.* 2:245–78.

SMITH, H. O., DANNER, D. B., and DEICH, R. A. 1981. Genetic transformation. *Annu. Rev. Biochem.* 50:41–68.

SMITH-KEARY, P. F. 1989. *Molecular genetics of Escherichia coli.* New York: Guilford Press.

STAHL, F. W. 1979. *Genetic recombination: Thinking about it in phage and fungi.* New York: W. H. Freeman.

———. 1987. Genetic recombination. *Sci. Am.* (Nov.) 256:91–101.

STENT, G. S. 1963. *Molecular biology of bacterial viruses.* New York: W. H. Freeman.

———. 1966. *Papers on bacterial viruses,* 2nd ed. Boston: Little, Brown.

TESSMAN, I. 1965. Genetic ultrafine structure in the T4 *rII* region. *Genetics* 51:63–75.

VISCONTI, N., and DELBRUCK, M. 1953. The mechanism of genetic recombination in phage. *Genetics* 38:5–33.

WOLLMAN, E. L., JACOB, F., and HAYES, W. 1956. Conjugation and genetic recombination in *Escherichia coli* K-12. *Cold Spring Harbor Symp. Quant. Biol.* 21:141–62.

ZINDER, N. D. 1953. Infective heredity in bacteria. *Cold Spring Harbor Symp. Quant. Biol.* 18:261–69.

———. 1958. Transduction in bacteria. *Sci. Am.* (Nov.) 199:38–47.

ZINDER, N. D., and LEDERBERG, J. 1952. Genetic exchange in *Salmonella. J. Bacteriol.* 64:679–99.

CHAPTER CONCEPTS

This chapter addresses a number of topics representing extensions of the general subject of trans-mission genetics introduced in earlier chapters. These topics are both analytical and somewhat more complex than those previously introduced. They include a consideration of variation in phenotypic expression, the quantitative analysis of polygenic traits, heritability, and mapping analysis per-formed using haploid eukaryotes. Expression of the phenotype is not always the direct reflection of the genotype and may be modified by many internal and external conditions. The quantitative analysis of polygenic traits requires the use of biometric and statistical methodology. Chromosome mapping in haploid organisms requires approaches different from those used in mapping diploid organisms, but adheres to the same general principles.

Thus far in the text, we have introduced a core of information that constitutes the general body of knowledge referred to as **trans-mission genetics**. In this chapter we turn to several advanced topics that are more accessible now that you have gained the foundation provided in the first six chapters. These constitute relevant exten-sions of our previous discussions and are highly analytical in nature.

We will begin by looking at the topic of **phenotypic expression** and reviewing factors that affect it. As we shall see, expression of genetic infor-mation is not always as clear-cut as most of our previous examples might suggest. An organism bearing a mu-tant genotype may exhibit consider-able variation in the expression of the

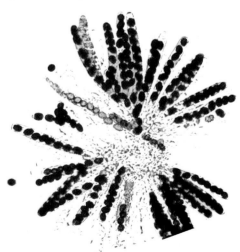

Ascospores contained in the asci of the bread mold *Neurospora crassa.*

related phenotype. In some cases, a percentage of the mu-tant organisms may fail to express the phenotype at all! As we shall see, many factors may mod-ify phenotypic expression.

Next we will extend our prior discussion of polygenic inheritance from Chapter 4 with a more thor-ough examination of **quantitative inheritance**. As we do so, we shall consider several statistical ap-proaches that are often used in ana-lyzing quantitative traits and ex-pressing variance in population data. Then we will introduce the concept of **heritability**—the assess-ment of the extent to which genetic factors contribute to phenotypic variation within populations. We will conclude this chapter by re-turning to the topics of linkage and crossing over and extending our discussion to include **chromosome mapping in haploid organisms**. We will examine this topic in eukaryotic fungus *Neurospora* and the green alga *Chlamydomonas*.

Phenotypic Expression

Gene expression is often discussed as though the genes operate in a closed, "black box" system in which the presence or absence of functional products directly determines the collective phenotype of an individual. The situation is actually much more complex. Most gene products function within the internal milieu of the cell; cells interact with one another in various ways, and the organism must survive under diverse environmental influences. Gene expression and the resultant phenotype are often modified through the interaction between an individual's particular genotype and the internal and external environment.

The degree of environmental influence can vary from inconsequential to subtle to very strong. Subtle interactions are the most difficult to detect and document, and they have led to unresolved "nature–nurture" conflicts in which scientists debate the relative importance of genes versus environment. In this section we will deal with some of the variables known to modify gene expression.

Penetrance and Expressivity

Some mutant genotypes are always expressed as a distinct phenotype, whereas others produce a proportion of individuals whose phenotypes cannot be distinguished from normal (wild type). The degree of expression of a particular trait may be quantitatively studied by determining the **penetrance** and **expressivity** of the genotype under investigation. The percentage of individuals that show at least some degree of expression of a mutant genotype defines the penetrance of the mutation. For example, the phenotypic expression of many mutant alleles in *Drosophila* may overlap with wild type. Flies homozygous for the recessive mutant gene *eyeless* yield phenotypes that range from the presence of normal eyes to a partial reduction in size to the complete absence of one or both eyes (Figure 7.1). If 15 percent of mutant flies show the wild-type appearance, the mutant gene is said to have a penetrance of 85 percent.

On the other hand, the *range of expression* defines the **expressivity** of the mutant genotype. In the case of *eyeless*, although the average reduction of eye size is one-fourth to one-half, the range of expressivity is from complete loss of both eyes to completely normal eyes (Figure 7.1).

Examples such as the expression of the *eyeless* phenotype have provided the basis for experiments designed to

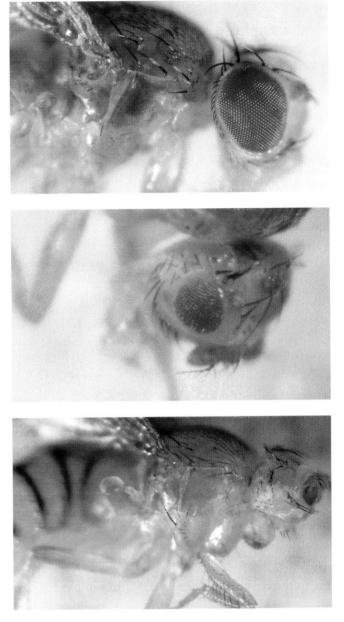

FIGURE 7.1 Gradations in phenotype, ranging from wild type to eyeless, associated with the *eyeless* mutation in *Drosophila*.

determine the causes of phenotypic variation. If the laboratory environment is held constant and extensive variation is still observed, the genetic background may be investigated. It is possible that other genes are influencing or modifying the *eyeless* phenotype. On the other hand, if the genetic background is not the cause of the phenotypic variation, environmental factors such as temperature, humidity, and nutrition may be tested. In the case of the *eyeless* phenotype, it has been determined experimentally that both genetic background and environmental factors influence its expression.

Genetic Background: Suppression and Position Effects

With only certain exceptions, it is difficult to assess the specific effect of the **genetic background** and the expression of a gene responsible for determining the potential phenotype. Two of the better characterized effects of genetic background are described below.

First, the expression of other genes throughout the genome may have an effect on the phenotype produced by the gene in question. The phenomenon of **genetic suppression** is an example. Mutant genes such as *suppressor of sable (su-s)*, *suppressor of forked (su-f)*, and *suppressor of Hairy-wing (su-Hw)* in *Drosophila* completely or partially restore the normal phenotype to an organism that is homozygous (or hemizygous) for *sable*, *forked*, and *Hairy-wing*, respectively. For example, flies hemizygous for both *forked* (a bristle mutation) and *su-f* have normal bristles.

The phenomenon of suppression occurs in a wide variety of organisms. In microorganisms, where molecular studies are easier to perform, some suppressor genes encode molecules that function during the genetic translation process. For example, there are mutations in transfer RNA (tRNA), genes that suppress the expression of mutations present in other genes. During translation, the altered tRNA misreads the mutant information present in mRNA, restoring wild-type function to the gene product. The impact is suppression of the mutant phenotype.

It has also been hypothesized that a suppressor gene product might provide or activate an alternative metabolic route that bypasses a block in a biosynthetic pathway caused by the primary mutation. Such situations have also been documented. Suppressor genes are excellent examples of the genetic background modifying primary gene effects.

Second, the physical location of a gene in relation to other genetic material may influence its expression. Such a situation is called a **position effect**. For example, if a gene is included in a **translocation** or **inversion** event (in which a region of a chromosome is relocated or rearranged), the expression of the gene may be affected. This is particularly true if the gene is relocated to or near certain areas of the chromosome that are genetically inert, referred to as **heterochromatin**.

An example of such a position effect involves female *Drosophila* heterozygous for the X-linked recessive eye-color mutant *white (w)*. Prior to translocation, the w^+/w genotype results in a wild-type brick-red eye color. However, if the region of the X chromosome containing the wild-type w^+ allele is translocated so that it is close to a heterochromatic region, expression of the w^+ allele is modified (Figure 7.2). Therefore, following translocation, the dominant effect of the normal w^+ allele is reduced. Instead of having a red color, the eyes are variegated, or mottled with red and white patches. A similar

(a)

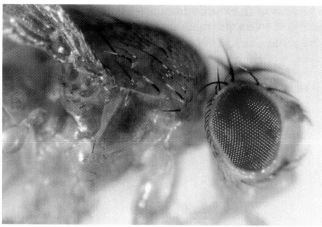

(b)

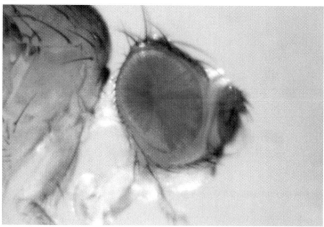

FIGURE 7.2   Eye phenotype in two female *Drosophila* heterozygous for the gene *white*. (a) Normal dominant phenotype showing brick-red eye color. (b) Variegated color of an eye caused by rearrangement of the *white* gene to another location in the genome.

position effect is produced if a heterochromatic region is relocated next to the *white* locus on the X chromosome. Apparently, heterochromatic regions inhibit the expression of adjacent genes. Loci in many other organisms also exhibit position effects, providing proof that alteration of the normal arrangement of genetic information can modify its expression.

Temperature Effects

Because chemical activity depends on the kinetic energy of the reacting substances, which in turn depends on the surrounding temperature, we can expect temperature to influence phenotypes. For example, the evening primrose produces red flowers when grown at 23°C and white flowers when grown at 18°C. Siamese cats and Himalayan rabbits exhibit dark fur in regions of the nose, ears, and paws because the body temperatures of these extremities are slightly cooler (Figure 7.3). In these cases, it appears that

(a)

(b)

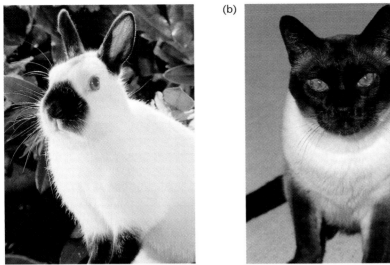

FIGURE 7.3 (a) A Himalayan rabbit. (b) A Siamese cat. Both show dark fur color at the extremities of the ears, nose, and paws. These patches are due to expression of a temperature-sensitive allele responsible for pigment production at the lower temperatures of the extremities.

the enzyme responsible for pigment production is functional at the lower temperatures present in the extremities, but it loses its catalytic function at the slightly higher temperatures found throughout the rest of the body.

Mutations that are affected by temperature are said to be **conditional** and are called **temperature sensitive**. Examples are known in a variety of organisms, including bacteria, viruses, fungi, and *Drosophila*. In extreme cases, an organism carrying a mutant allele may express a mutant phenotype when grown at one temperature, but express the wild-type phenotype when reared at another temperature. This type of temperature effect is useful in studying mutations that interrupt essential processes during reproduction and development and are thus normally detrimental to the organism. For example, bacterial viruses carrying certain **temperature-sensitive mutations** are allowed to infect bacteria cultured at 42°C; infection progresses until the essential gene product is needed and then arrests; this temperature is referred to as the **restrictive condition**. If cultured under **permissive conditions** of 25°C, the gene product is functional, infection proceeds normally, and new viruses are produced. The use of temperature-sensitive mutations, which may be induced and isolated, has added immensely to the study of viral genetics.

Similarly, many temperature-sensitive mutations have been discovered in *Drosophila* that affect development, morphology, and behavior. Most are recessive lethals and have proven to be invaluable in the genetic dissection of these processes.

Nutritional Effects

Another example of conditional mutations involves nutrition. As we saw in Chapter 6, in microorganisms, mutations that prevent synthesis of nutrient molecules are quite common. These **nutritional mutants** arise when

an enzyme essential to a biosynthetic pathway becomes inactive. In microorganisms, an organism bearing such a mutation is called an **auxotroph**. If the end product of a biochemical pathway can no longer be synthesized, and if that molecule is essential to normal growth and development, the mutation prevents growth and may be lethal. If, for example, the bread mold *Neurospora* can no longer synthesize the amino acid leucine, proteins cannot be synthesized unless leucine is added to the growth medium. If leucine is added, the detrimental effect is overcome. Nutritional mutants have been crucial to molecular genetic studies and also served as the basis for George Beadle and Edward Tatum's proposal, in the early 1940s, that one gene functions to produce one enzyme (see Chapter 13).

In humans, a slightly different set of circumstances is known. The presence or absence of certain dietary substances, which normal individuals may consume without harm, can adversely affect individuals with abnormal genetic constitutions. Often, a mutation may prevent an individual from metabolizing some substance commonly found in normal diets. For example, those afflicted with the genetic disorder **phenylketonuria** cannot metabolize the amino acid phenylalanine. Those with **galactosemia** cannot metabolize galactose. Other individuals are intolerant of the milk sugar lactose and are said to exhibit **lactose intolerance**. Those with **diabetes** cannot adequately metabolize glucose. In this case, excessive amounts of the molecule accumulate in the body and become toxic, and a characteristic phenotype results. However, if the dietary intake of the molecule is drastically reduced or eliminated, the associated phenotype may be reversed.

The case of lactose intolerance illustrates the general principles involved. Lactose is a disaccharide consisting of a molecule of glucose and a molecule of galactose. It is present as 7 percent of human milk and 4 percent of cow's milk. To metabolize lactose, humans require the enzyme **lactase**, which cleaves the disaccharide. Adequate amounts of lactase are produced during the first few years after birth. However, in many racial and ethnic groups, the levels of this enzyme soon drop drastically, and adults become intolerant of milk. The major phenotypic effect involves severe intestinal diarrhea, flatulence, and abdominal cramps. This condition, while not limited to, is particularly prevalent in Eskimos, Africans, Asiatics, and Americans with these heritages. In these cultures, milk is usually converted to other foods, including cheese, butter, and yogurt. In these forms, the amount of lactose is reduced significantly and adverse effects can be nearly eliminated. In the United States, milk low in lactose is commercially available, and to aid in the digestion of other lactose-containing foods, lactase is now a commercial product that can be ingested.

Onset of Genetic Expression

Not all genetic traits are expressed at the same time during an organism's life span. In most cases, the age at which a gene is expressed corresponds to the normal sequence of growth and development. In humans, the prenatal, infant, preadult, and adult phases require different kinds of genetic information. In a similar way, many genetic disorders can be expected to manifest themselves at different stages of life. Lethal genes account for many of the frequent spontaneous abortions and miscarriages occurring in the human population. These mutations are believed to alter genetic products essential to prenatal development.

In humans, many severe inherited disorders are often not manifested until after birth. For example, **Tay-Sachs disease**, inherited as an autosomal recessive, is a lethal lipid metabolism disease involving an abnormal enzyme, **hexosaminidase A**. However, newborns appear normal for the first five months. Then, progressive deterioration occurs, and the affected children die before age four.

The **Lesch–Nyhan syndrome**, inherited as an X-linked recessive, causes abnormal nucleic acid metabolism (biochemical salvage of nitrogenous purine bases), leading to the accumulation of uric acid in blood and tissues, mental retardation, palsy, and self-mutilation of the lips and fingers. The disorder is due to a mutation in the gene encoding **hypoxanthine-guanine phosphoribosyl transferase (HPRT)**. Newborns are normal for six to eight months prior to the onset of the first symptoms. Still another example involves **Duchenne muscular dystrophy (DMD)**, an X-linked recessive disorder associated with progressive muscular wasting. It is usually diagnosed between the ages of three and five years. Even with modern medical intervention, the disease is often fatal in the early twenties.

Perhaps the most variable of all inherited human disorders regarding age of onset is the tragic **Huntington disease**. Inherited as an autosomal dominant, Huntington disease affects the frontal lobes of the cerebral cortex where progressive cell death occurs over a period of more than a decade. Brain deterioration is accompanied by spastic uncontrolled movements, intellectual and emotional deterioration, and ultimately death. While onset has been reported at all ages, it most frequently occurs between ages 30 and 50, with a mean onset of 38 years. Most often, early onset is associated with an affected male parent, although such a response is not universal to offspring of males with the disorder.

Observations such as these support the concept that the critical expression of normal genes varies throughout the life cycle of organisms, including humans. Gene products may play more essential roles at certain times. Also, it would appear that the internal physiological environment of an organism changes with age.

Genetic Anticipation

Considerable insights have been gained concerning the onset of genetic expression by studying cases where clear-cut patterns of initial expression have been observed. The phenomenon of **genetic anticipation** refers to the occurrence of a genetic disorder with a progressively earlier age of onset in successive generations. As we will see, earlier onset is correlated with an increased severity of the disorder.

At least three human disorders clearly demonstrate genetic anticipation: **myotonic dystrophy (DM)**, **fragile-X mental retardation**, and **spinal and bulbar muscular atrophy (Kennedy disease)**. We will focus on the information derived from studies of DM to discuss genetic anticipation.

Myotonic dystrophy, the most common type of adult muscular dystrophy, is an autosomal dominant disorder of variable severity and age of onset. Mildly affected individuals develop cataracts as adults with little or no muscular weakness. Severely affected individuals demonstrate more severe myopathy and may be mentally retarded. In its most extreme form, the disease is fatal just after birth. Careful investigation during the past half-century has revealed that the increased severity is correlated with earlier onset, linking the disease with genetic anticipation.

A great deal of excitement has been generated recently among geneticists as a result of more recent studies involving myotonic dystrophy. In 1989, C. J. Howeler and colleagues reported on their study of 61 parent–child pairs. In 60 of 61 cases, age of onset was earlier in the child! By 1992, several research teams had discovered a possible molecular basis for this case of genetic anticipation. Located on chromosome 19 (19q1.3 contains the DM locus) is an unstable DNA sequence characterized by a specific trinucleotide (CTG) that is repeated a variable number of times. The remarkable finding is the correlation between the size of the **trinucleotide repeat** and both the severity and onset of the disorder. In successive generations the size of the repeated segment increases. Normal individuals average about five copies; minimally affected individuals reveal about 50 copies, while severely affected individuals demonstrate over 1000 copies!

Although it is not yet clear how an expansion of the size of this gene-associated region affects onset and phenotypic expression, the correlation is extremely strong. Of great interest is that both fragile-X syndrome and Kennedy disease also reveal an association between a gene-associated DNA amplification and disease severity. More recently, a similar finding has been made in the gene encoding Huntington disease. We will return to this general topic in Chapter 14. For now, we can only anxiously await further information linking the molecular defect to the phenomenon of genetic anticipation.

Genomic (Parental) Imprinting

One of the major exceptions to the assumptions underlying the laws of Mendelian inheritance involves the case where genetic expression varies, depending on the parental origin of the chromosome carrying a particular gene. The phenomenon is called **genomic** (or **parental**) **imprinting**. It appears that, in some species, certain regions of chromosomes and the genes contained within them somehow retain a memory, or an "imprint," of their parental origin that influences their genetic expression. As a result, specific genes either are expressed or remain genetically silent—i.e., are not expressed, based on their imprint.

For example, in Huntington disease in humans, as pointed out above, early onset of the disease most often occurs when the mutant gene is inherited from the father. In myotonic dystrophy, just the opposite is true. In this disorder, when early onset is observed, the affected offspring usually inherits the gene from the mother.

The imprinting step is thought to occur before or during gamete formation, leading to differentially marked genes (or chromosome regions) in sperm-forming versus egg-forming tissues. The process is clearly different from mutation because in the cases cited above, the imprint can be reversed in succeeding generations as these autosomal genes pass from mother to son to granddaughter, and so on.

Another example of imprinting involves the inactivation of one of the X chromosomes in mammalian females. As we will discuss in Chapter 9, a mechanism of **dosage compensation** exists whereby the random inactivation of either the paternal or maternal X chromosome occurs during embryonic development. However, in mice, prior to the development of the embryo proper, imprinting occurs in tissues such that the X chromosome of paternal origin is genetically inactivated in all cells while the genes on the maternal X chromosome remains genetically active. As embryonic development is subsequently initiated,

the imprint is "released" and random inactivation of either the paternal *or* the maternal X chromosome can occur.

In 1991, more specific information became available when it was established that three specific mouse genes undergo imprinting. One of them is the gene encoding insulinlike growth factor II (*Igf2*). A mouse that carries two nonmutant alleles of this gene is normal in size, whereas a mouse that carries two mutant alleles lacks a growth factor and is a dwarf. The size of a heterozygous mouse (one allele normal and one mutant—Figure 7.4) depends on the parental origin of the normal allele. The mouse is normal in size if the normal allele came from the father, but dwarf if the normal allele came from the mother. From this it can be deduced that the normal *Igf2* gene is imprinted to function poorly during the course of egg production in females but functions normally when it has passed through sperm-producing tissue in males.

Imprinting continues to depend on whether the gene passes through sperm-producing or egg-forming tissue leading to the next generation. For example, a heterozygous normal-sized male (above) will donate a "normal-functioning" wild-type allele to half his offspring that will counteract a mutant allele received from the mother.

In humans, two distinct genetic disorders are thought to be caused by differential imprinting of the same region of chromosome 15 (15q1). In both cases, the disorders *appear* to be due to an identical deletion of this region in one member of the chromosome 15 pair. The first disorder, **Prader-Willi syndrome (PWS)**, results when only an

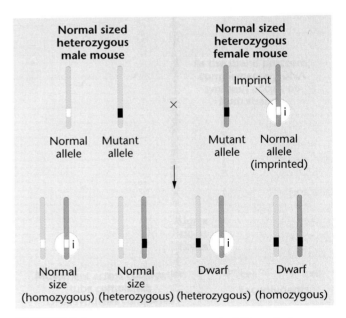

FIGURE 7.4 The effect of imprinting on the *Igf2* gene in the mouse, which produces dwarf mice in the homozygous condition. Heterozygous offspring that receive the normal allele from their father are normal in size. Heterozygotes that receive the normal allele from their mother, which has been imprinted (i), are dwarf.

undeleted maternal chromosome remains. If only an undeleted paternal chromosome remains, an entirely different disorder, **Angelman syndrome (AS)**, results.

Prader-Willi syndrome is clearly different from AS. In the former, mental retardation is noted as well as a severe eating disorder marked by an uncontrollable appetite, obesity, and diabetes. On the other hand, AS leads to distinct clinical manifestations involving behavior as well as mental retardation. One can conclude that region 15q1 is imprinted differently in male versus female gametes and that both a maternal and paternal region are required for normal development.

The area of genomic imprinting has received a great deal of recent research attention; many questions remain unanswered. It is not known how many genes are subject to imprinting, nor its developmental role. While it appears that regions of chromosomes rather than specific genes are imprinted, the molecular mechanism of imprinting is still a matter for conjecture. It has been hypothesized that **DNA methylation** may be involved. In vertebrates, methyl groups can be added to the carbon atom at position 5 in cytosine (see Chapter 10) as a result of the activity of the enzyme **DNA methyl transferase**. Methyl groups are added when the dinucleotide CpG or groups of CpG units (called CpG islands) are present along a DNA chain. DNA methylation is a reasonable mechanism for establishing a molecular imprint, since there is some evidence that a high level of methylation can inhibit gene activity and that active genes (or their regulatory sequences) are often undermethylated. Additionally, variation in methylation has been noted in the mouse genes that undergo imprinting. Whatever the cause of this phenomenon, it is a fascinating topic and one that will receive significant attention in future research studies.

Continuous Variation and Polygenes

In Chapter 4 we considered how the classical patterns of Mendelian inheritance (e.g., the 3:1 and 9:3:3:1 F_2 ratios) are modified because of gene interaction or the action of multiple genes on a single trait. At that time, we discussed examples of **polygenic traits** such as the inheritance of grain color in wheat, a trait that is controlled by three gene pairs. These traits were discussed to illustrate that even in complex modes of inheritance, the fundamental principles of segregation and independent assortment discovered by Mendel are operational.

Having considered the patterns of inheritance that are characteristic of polygenic traits in that earlier discussion, in this chapter we will describe the methods used by geneticists to study traits that are controlled by several genes. These methods are often statistical and involve analysis of traits using mathematical tools in addition to those of biochemistry or molecular biology.

Polygenic traits are at the heart of several disciplines in genetics, including plant breeding, livestock breeding, and wildlife management. Polygenic traits are also an important part of human genetics; traits such as intelligence, skin color, obesity, and predisposition to certain diseases are thought to be under polygenic control. In the following discussion, remember that in polygenic inheritance, as in monogenic inheritance, barring accidents, the genotype is fixed at the moment of fertilization, while the phenotype is more flexible and changes over the life span of the organism as the genotype interacts with the environment.

Continuous versus Discontinuous Variation

Traits studied by Mendel in the garden pea could be separated into distinct qualitative categories (e.g., tall vs. dwarf plants, green vs. yellow seeds) [Figure 7.5(a)]. Such variation is described as **discontinuous**. In contrast, polygenic traits are said to exhibit **continuous variation** between the most extreme phenotypes expressed in a population. For example, Sir Francis Galton studied the diameter of sweet peas. When he crossed plants with the largest diameter to those with the smallest diameter, the F_1 plants had peas with an intermediate diameter. In the F_2, continuous variation from the largest to the smallest was exhibited, but with most intermediate in size [Figure 7.5(b)].

Galton devised statistical methods to study these continuous traits, and the resulting field became known as **biometry**. Although these mathematical techniques provided an important tool for the analysis of experimental data, they offered no explanation for the underlying mechanisms of inheritance. Not surprisingly, this led to many erroneous conclusions about the characteristics of continuous variation and to a subsequent debate as to whether Mendelian principles could explain this form of inheritance. As outlined in Chapter 4, this debate was settled in experiments demonstrating the role of Mendelian factors in continuous variation.

Analysis of continuous traits always involves some quantitation step expressed in weight, stature, volume, height, color, or other form of measurement. As you may recall from our earlier discussion of this topic in Chapter 4, we make the assumption that alleles involved in polygenic inheritance are either *additive* or *nonadditive*. We further studied examples where the total effect of each additive allele was small and roughly equal to all others involved in the genetic control of a specific phenotype. In reality, this is seldom the case. Groups of additive genes affecting a single characteristic are believed to have variable effects, with some making major contributions and others making minor ones. While these alleles are subject

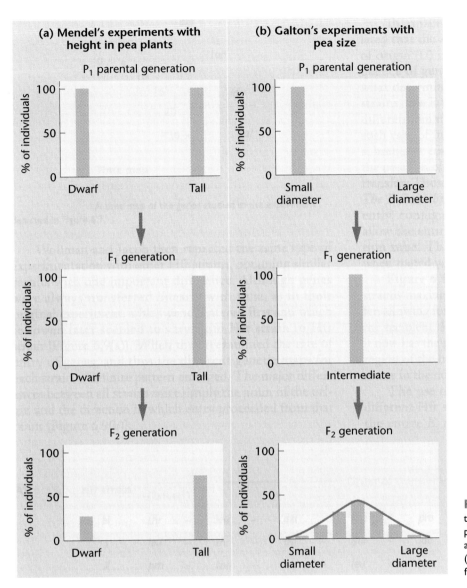

FIGURE 7.5 Histograms comparing phenotypic distributions in two crosses involving pea plants. (a) The distribution of phenotypes in the F_1 and F_2 for a trait exhibiting discontinuous variation. (b) The distributions of phenotypes in the F_1 and F_2 for a trait exhibiting continuous variation.

to the impact of genetic background and environmental influences, for simplicity we discounted these in our examples. The previous section, which considers variation in phenotypic variation, should make it clear that both genetic background and the environment will impact on the expression of additive alleles.

Mapping Quantitative Trait Loci

In discussing polygenic inheritance, it is useful to know how to approximate the number of genes that control a given trait. In a given polygenic cross, the phenotype of the F_2 generation will often show greater phenotypic variability than the P_1 or F_1 generation. In the most simple situations, the number of genes controlling a trait can be estimated by determining the frequency of either of the most extreme (parental) phenotypes in the F_2 genera-

tion (see Chapter 4). However, such calculations provide no information about the location of these genes, positions on chromosomes referred to as **quantitative trait loci (QTLs)**. Because alleles occupying QTLs act on a single trait, it is reasonable to ask whether such genes might be physically clustered on a portion of a single chromosome, or scattered throughout the genome.

In *Drosophila*, resistance to the insecticide DDT has been shown to be a polygenic trait. To find the loci responsible, strains selected for resistance to DDT and strains selected for sensitivity were crossed to a stock carrying dominant genes that serve as markers on each of *Drosophila*'s four chromosomes. Following a variety of crosses, offspring were produced that contained many different combinations of marker chromosomes and chromosomes from either sensitive or resistant strains (Figure 7.6). As shown in this figure, flies were then tested for resistance (survival) when exposed to DDT. Results indicate that each of the chromosomes in *Drosophila* contains genes that contribute to resistance. In other words, the loci bearing the genes that control DDT resistance are not clustered together on a single chromosome, but instead are scattered throughout the genome. This does not establish that in all cases polygenes controlling a trait are scattered throughout the genome, but it does show that polygenes need not be clustered together in order to affect a single trait.

It is possible to more specifically map the position of the QTLs because of the presence of identifiable molecular markers present on each chromosome. The locations of the QTLs are determined relative to known positions of these markers. The markers, called **restriction fragment-length polymorphisms (RFLPs)**, are molecular in nature and represent specific nuclease cleavage sites (see Chapter 16 for a more detailed description of RFLPs). They provide a new approach to the enumeration and mapping of QTLs. Interestingly, RFLPs were first used to map genes in humans. In many other organisms of agri-

cultural importance, RFLP markers are also now available, making possible systematic mapping of these loci.

For example, in the tomato, the location of hundreds of RFLP markers are known that are spaced along the 12 chromosomes present in the haploid genome of this plant. They are spaced about every 5 cM (one cM = one map unit) along each chromosome. Analysis is performed by crossing plants with extreme but opposite phenotypes (usually from two different varieties), and through several generations looking for the cosegregation of specific RFLPs along with the phenotypic expression of the trait controlled by the QTLs of interest. Whenever both an RFLP molecular marker and a QTL responsible for the trait in question are closely linked on a single chromosome, they are much more likely to demonstrate an association together throughout a pedigree than if they are not closely linked. When both the marker and the phenotypic trait are expressed together, they are said to cosegregate. When this occurs consistently, it establishes the presence of a QTL at or near the RFLP marker along the chromosome. When numerous QTLs are located, a genetic map is created for the involved genes.

Using RFLPs in conjunction with some newly derived analytical methods, QTLs for fruit weight, soluble solids, and acidity have been identified and mapped in the tomato. Six loci responsible for fruit weight were found on six different chromosomes, four loci for soluble solids were identified on five chromosomes, and five loci for acidity were found on four different chromosomes. Note that each locus may indeed house only a single gene, although the RFLP method simply allows the identification of chromosomal regions, not individual genes. Several chromosomes contained loci for all three traits. Further investigation has revealed that the genes present at these loci account for approximately 50 percent of the phenotypic variance expressed in these traits.

The determination of the locations of QTLs for agriculturally important characteristics has permitted increased efficiency in selection and will open the door to their future genetic manipulation and possible transfer between organisms. Establishing RFLP maps became fairly routine in the 1990s; this approach has since become an important tool in the repertoire of genetic engineering. Eventually, such methods can be used to isolate and characterize the genes controlling quantitative traits.

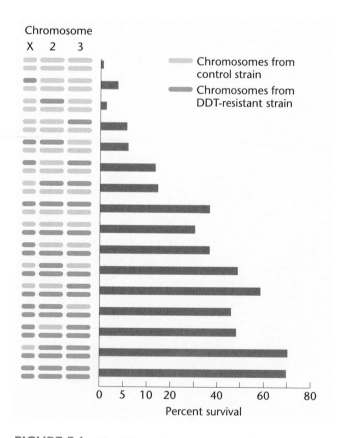

FIGURE 7.6 The differential survival of *Drosophila* carrying combinations of chromosomes from DDT-resistant and DDT-sensitive strains when exposed to DDT. Results indicate that DDT resistance is polygenic, with genes on each of the major chromosomes making a major contribution. Chromosome 4 carries few genes and was omitted from this analysis.

Analysis of Polygenic Traits

Analysis of any given polygenic trait involves quantitative measurements, usually from a series of crosses. The outcome can be expressed as a frequency diagram that most often takes on a normal (bell-shaped) distribution (Figure

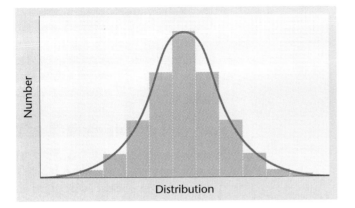

FIGURE 7.7 A normal frequency distribution characterized by a bell-shaped curve.

7.7). While it is hoped that each series of crosses is representative of the population at large, variation in small samples due strictly to chance may also influence the data that are gathered. To assess the experimental validity of the data, statistical techniques must be employed.

Statistical analysis serves three purposes:

1. Data can be mathematically reduced to provide a **descriptive summary** of the sample.
2. Data from a small but random sample can be utilized to infer information about groups larger than those from which the original data were obtained (**statistical inference**).
3. Two or more sets of experimental data can be compared to determine whether they represent significantly different populations of measurements.

Several terms used in **statistics** are helpful in the analysis of traits that exhibit a normal distribution, including the mean, variance, standard deviation, and standard error of the mean.

The Mean

The distribution of each set of phenotypic values displayed in Figure 7.8 tends to cluster around a central value. This clustering is called a **central tendency**, the most common measurement of which is the **mean** ($\overline{X}$). The mean is simply the arithmetic average of a set of measurements or data and is calculated as

$$\overline{X} = \frac{\sum \overline{X}_i}{n}$$

where $\overline{X}$ is the mean, $\sum \overline{X}_i$ represents the sum of all individual values in the sample, and n is the number of individual values.

Although the mean provides a descriptive summary of the sample, it is in itself of limited value. All values in the sample may be clustered near the mean, or they may be distributed widely around it. Figure 7.8(a) reflects two sets of measurements with identical means, but a widely different distribution of values. These contrasting conditions represent different types of variation within each sample, called the **frequency distributions**. Whether due to chance or to one or more experimental variables, such variation creates the need for methods to describe the sample measurements.

Variance

As seen in Figure 7.8(a), the range and distribution of values on either side of the mean will determine the shape of the distribution curve. The degree to which values within this distribution diverge from the mean is called the sample **variance** (s^2 or V) and is used as an estimate of the variation present in an infinitely large population. The variance for a sample is calculated as:

(a)

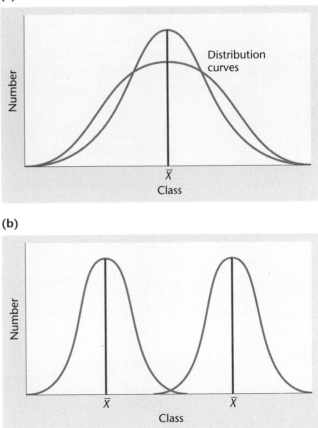

(b)

FIGURE 7.8 (a) Two normal frequency distributions with the same mean but different amounts of variation. (b) Two normal distributions with different means but the same amount of variation.

$$s^2 = V = \frac{\Sigma(X_i - \bar{X})^2}{n - 1}$$

where the sum (Σ) of the squared differences between each measured value (X_i) and the mean ($\bar{X}$) is divided by 1 less than the total sample size ($n - 1$). To avoid the numerous subtraction functions necessary in calculating s^2 for a large sample, we can convert the equation to its algebraic equivalent:

$$s^2 = V = \frac{\Sigma X_i^2 - n\bar{X}^2}{n - 1}$$

The variance is a valuable measure of sample variability. Two samples may have identical means ($\bar{X}$) yet vary considerably in their frequency distribution around the mean. The variance represents the average squared deviation of the measurements from the mean. The estimation of variance has been particularly valuable in determining the degree of genetic control of traits where the immediate environment also influences the phenotype.

Standard Deviation

Because the variance is a squared value, its unit of measurement is also squared (cm^2, mg^2, etc.). To express variation around the mean in the original units of measurement, it is necessary to calculate the square root of the variance, a term called the **standard deviation (s)**:

$$s = \sqrt{s^2}$$

Table 7.1 shows what percentage of the individual values within an ideal distribution is included with different multiples of the standard deviation. The mean plus or minus one standard deviation ($\bar{X} \pm 1s$) includes 68 percent of all values in the sample. Over 95 percent of all values are found within two standard deviations ($\bar{X} \pm 2s$). As such, the standard deviation provides an important descriptive summary of a set of data. Furthermore, s can be interpreted as a probability. The $\bar{X} \pm 1s$ indicates that there is a 68 percent probability that a measured value picked at random will fall within that range.

TABLE 7.1 Sample Inclusion for Various *s* Values

Multiples of *s*	Percent of Sample Included
$\bar{X} \pm 1s$	68.3%
$\bar{X} \pm 1.96s$	95.0
$\bar{X} \pm 2s$	95.5
$\bar{X} \pm 3s$	99.7

Standard Error of the Mean

To estimate how much the means of other similar samples drawn from the same population might vary, we can calculate the **standard error of the mean ($S_{\bar{X}}$)**:

$$S_{\bar{X}} = \frac{s}{\sqrt{n}}$$

where s is the standard deviation, and $\sqrt{n}$ is the square root of the sample size. The standard error of the mean is a measure of the accuracy of the sample mean, that is, the variation of sample means in replications of the experiment. It can be expected that the standard deviation of mean values will reflect less variance than the standard deviation of a set of individual measurements. Because the standard error of the mean is computed by dividing s by $\sqrt{n}$, it is always a smaller value than the standard deviation.

Analysis of a Quantitative Character

Many characteristics of importance in livestock and crop plants are controlled by polygenic systems. In Chapter 4 we discussed such phenotypes as ear length in corn and kernel color in wheat. To illustrate how biometric methods are used in the analysis of quantitative characters, we will consider a simplified example involving fruit weight in tomatoes. Let us assume that fruit weight is a quantitative character, and that one highly inbred strain produces tomatoes averaging 18 oz. in weight, and another highly inbred strain produces fruit averaging 6 oz. in weight. These two varieties are crossed and produce an F_1 generation with weights ranging from 10 oz. to 14 oz. The F_2 population contains individuals that produce fruit ranging from 6 oz. to 18 oz. The distribution of these quantitative phenotypes in both generations is shown in Table 7.2. Note that x represents the midpoint of a class interval. For example, all tomatoes weighing between 5.5 and 6.5 oz. are classified as 6 oz.

The mean value for the fruit weight in the F_1 generation can be calculated as:

$$\bar{X} = \frac{\Sigma \bar{X}_i}{n} = \frac{626}{52} = 12.04$$

Similarly, the mean value for fruit weight in the F_2 generation is calculated as:

$$\bar{X} = \frac{\Sigma \bar{X}_i}{n} = \frac{872}{72} = 12.11$$

Average fruit weight is 12.04 oz. in the F_1 generation and 12.11 oz. in the F_2 generation. Although these mean val-

TABLE 7.2 Distribution of F_1 and F_2 Progeny

								Weight in oz.						
		6	7	8	9	10	11	12	13	14	15	16	17	18
Number of Individuals	F_1:					4	14	16	12	6				
	F_2:	1	1	2	0	9	13	17	14	7	4	3	0	1

ues are similar, it is apparent from a comparison of the distribution of phenotypes (Table 7.2) that there is more variation present in the F_2 generation. Fruit weight ranges from 6 to 18 oz. in the F_2 generation, but only from 10 to 14 oz. in the F_1 generation.

To assist in quantifying the amount of variation present in each sample, we can calculate the variance (Table 7.3). As noted above, the sample variance can be calculated as the sum of the squared differences between each value and the mean, divided by 1 less than the total number of observations. However, in the case where a number of observations (f) have been grouped into representative classes (x), the variance can be calculated according to the formula:

$$s^2 = V = \frac{n\Sigma fx^2 - (\Sigma fx)^2}{n(n-1)}$$

As shown in Table 7.3, the value for the F_1 generation is 1.29, and for the F_2 generation, it is 4.28. When con-

verted to the standard deviation ($s = \sqrt{s^2}$), the values become 1.13 and 2.06, respectively. Thus, the distribution of tomato weight in the F_1 generation can be described as 12.04 ± 1.13, and in the F_2 as 12.11 ± 2.06. This analysis indicates that the mean fruit weight of the F_1 is nearly identical to that of the F_2, but that the F_2 generation shows greater variability in the distribution of weights than does the F_1.

Observations about the inheritance of fruit weight in crosses between these two strains of tomatoes meet the expectations for polygenic traits. For the sake of this example, if we assume that each parental strain is homozygous for additive or nonadditive alleles that control fruit weight, we can estimate the number of gene pairs involved in controlling this trait in these two strains of tomatoes. Since 1/72 of the F_2 offspring have a phenotype that overlaps one of the parental strains (72 total F_2 offspring, one weighs 6 oz., one weighs 18 oz.; see Table 7.2), the formula $1/4^n = 1/72$ indicates that most likely

TABLE 7.3 Calculation of Variance

	F_1				F_2		
x	f	fx	fx^2	x	f	fx	fx^2
6				6	1	6	36
7				7	1	7	49
8				8	2	16	128
9				9	0	0	0
10	4	40	400	10	9	90	900
11	14	154	1694	11	13	143	1573
12	16	192	2304	12	17	204	2448
13	12	156	2028	13	14	182	2366
14	6	84	1176	14	7	98	1372
15				15	4	60	900
16				16	3	48	768
17				17	0	0	0
18				18	1	18	324
	$n = 52$	$\Sigma fx = 626$	$\Sigma fx^2 = 7602$		$n = 72$	$\Sigma fx = 872$	$\Sigma fx^2 = 10{,}864$

For F_1:

$$s^2 = \frac{52 \times 7602 - (626)^2}{52(52-1)}$$

$$= \frac{395{,}304 - 391{,}876}{2652}$$

$$= 1.29$$

For F_2:

$$s^2 = \frac{72 \times 10{,}864 - (872)^2}{72(72-1)}$$

$$= \frac{782{,}208 - 760{,}384}{5112}$$

$$= 4.27$$

three genes (less likely four genes) control fruit weight in these tomato strains. If this experimental cross was repeated many times with similar results, our confidence in this conclusion would be bolstered considerably.

Heritability

Having just introduced several ways in which quantitative or continuous variation can be measured and characterized in populations, we now consider how we assess the extent to which genetic factors contribute to such phenotypic variation. Often, much of the variation can be attributed to genetic factors, with the total environment having less impact. In other cases, the environment may have a greater impact on phenotypic variation within a population. The following discussion considers how geneticists attempt to define the impact of heredity versus environment on phenotypic variation.

Broad-Sense Heritability

Provided that a trait can be quantitatively measured, experiments on many plants and animals can test the causes of variation. One approach is to use inbred strains containing individuals of a relatively homogeneous (highly homozygous) **genetic background**. Experiments are then designed to test the effects of the range of prevailing environmental conditions on phenotypic variability. Variation observed *between* different inbred strains reared in a constant environment is due predominantly to genetic factors. Variation observed *among* members of the same inbred strain reared under different environmental conditions is due to nongenetic factors, which are generally categorized as "environmental."

The relative importance of genetic versus environmental factors can be formally assessed by examining the **heritability (H^2)** of a trait, which can be calculated using an analysis of variance among individuals of a known genetic relationship. (In the ensuing discussion, we will assign the term V to designate variance, also designated as s^2.) This is an important approach when investigating organisms with long generation times. Also called **broad-sense heritability**, H^2 measures the degree to which **phenotypic variance** (V_P) is due to variation in genetic factors for a single population within the limits of environmental variation during the study. An important distinction can be made regarding H^2. Its calculation *does not* determine the *proportion of the total phenotype* attributed to genetic factors. It *does* estimate the *proportion of observed variation in the phenotype* attributed to genetic factors, in comparison to environmental factors.

Phenotypic variance is due to the sum of three components: **environmental variance (V_E), genetic vari-**ance **(V_G)**, and variance resulting from the interaction of genetics and environment (V_{GE}). Therefore, phenotypic variance (V_P), is theoretically expressed as

$$V_P = V_E + V_G + V_{GE}$$

Because V_{GE} is often negligible and difficult to measure, it is usually omitted. Therefore, the simpler equation is used:

$$V_P = V_E + V_G$$

Broad-sense heritability expresses that proportion of variance due to the genetic component:

$$H^2 = \frac{V_G}{V_P}$$

An H^2 value that approaches 1.0 indicates that environmental conditions have had little impact on phenotypic variance in the population studies. An H^2 value close to 0.0 indicates that variation in the environment has been almost solely responsible for the observed phenotypic variation within the population studied.

It is not possible to obtain an absolute H^2 value for any given character. If measured in a different population under a greater or lesser degree of environmental variability, H^2 might well change for that character. Additionally, broad-sense heritability is not very accurate in estimating the selection potential of a quantitative trait. Therefore, another type of calculation, narrow-sense heritability, has been devised that is of more practical use.

Narrow-Sense Heritability

Information regarding heritability is most useful in animal and plant breeding as a measure of potential response to selection. In this case, a different estimate of heritability must be used, based on a subcomponent of V_G, referred to as **additive variance (V_A)**:

$$V_G = V_A + V_D + V_I$$

where V_A represents additive variance that results from the average effect of additive components of genes, V_D represents **dominance variance**, that is, deviation from the additive components that results when phenotypic expression in heterozygotes at a locus is not precisely intermediate between the two homozygotes. V_I reflects **interaction variance**, deviation from the additive components occurring when two or more loci behave epistatically. V_I reflects the variance not associated with the average effect of V_A and is often negligible. Thus, it is often excluded from calculations.

When V_G is partitioned into V_A and V_D, a new assessment of heritability, h^2, or **narrow-sense heritability**, may be calculated. It is h^2 that is useful in assessing selection potential in randomly breeding animal and plant populations:

$$h^2 = \frac{V_A}{V_P}$$

Because $V_P = V_E + V_G$ and $V_G = V_A + V_D$, we obtain

$$h^2 = \frac{V_A}{V_E + V_A + V_D}$$

Artificial Selection

A relatively high h^2 value is a prediction of the impact selection will have in altering a population. As you can see from our discussion above, measuring the components necessary to calculate h^2 is a complex task. For the trait under study, a more simplified approach involves the measurement of the central tendencies (the means) from (1) a parental population exhibiting a bell-shaped distribution (M), (2) a "selected" segment of the parental population, usually the segment that expresses the desired phenotypes (M1), and (3) the offspring resulting from interbreeding the selected M1 group (M2). When this is accomplished, the following relationship of the three means and h^2 exists:

$$M2 = M + h^2(M1 - M)$$

Solving this equation for h^2 leads to the formula:

$$h^2 = \frac{M2 - M}{M1 - M}$$

For example, assume that we measured a population of corn kernels, whose mean diameter (M) was much larger than desirable (30 mm), and from that population we selected a group with the smallest diameters where the mean (M1) equaled 15 mm. If plants that yielded this selected population are interbred and the progeny kernels yield a mean (M2) of 20 mm, then we can calculate h^2 in order to estimate the potential for artificial selection on kernel size:

$$h^2 = \frac{20 - 30}{15 - 30}$$

$$h^2 = \frac{-10}{-15} = 0.67$$

On the basis of this calculation, we may conclude that selection will be effective in obtaining strains of corn with reduced kernel size as a result of selecting alleles that contribute to smaller kernel size.

Table 7.4 provides estimates of narrow heritability for a variety of traits in various organisms. These proportions are expressed as percentage values. As you can see, heritability varies considerably between traits. In general, heritability is lower for traits essential to an organism's survival, primarily because the genetic component has been largely optimized during evolution. Egg production, litter size, and conception rate are examples where such physiological limits on selection have already been established. Nevertheless, traits less critical to survival, such as body weight, tail length, and wing length, show higher heritability values. Narrow-sense heritability estimates are most valuable when they have been collected in many populations and environments and there is a clear trend established. Based on such estimates, selection techniques have led to vast improvements in the quality of animal and plant products.

Twin Studies in Humans

In humans, traditional heritability studies are not possible. However, twins are very useful subjects for studying the heredity versus environment question. **Monozygotic (MZ)** or **identical twins**, derived from the division and splitting of a single egg following fertilization, are identical in their genetic compositions. Although most identical twins are reared together and are exposed to very similar environments, some pairs are separated and raised in different settings. For any particular trait, average similarities or differences can be investigated. Such an analysis is particularly useful because characteristics that remain similar in different environments are considered to be inherited. These data can then be compared with a

TABLE 7.4 Estimates of Heritability for Traits in Different Organisms

Trait	Heritability (h^2)
Mice	
Tail length	60%
Body weight	37
Litter size	15
Drosophila	
Abdominal bristle number	52
Wing length	45
Egg production	18
Chickens	
Body weight	50
Egg production	20
Egg hatchability	15
Cattle	
Birth weight	51
Milk yield	44
Conception rate	3

TABLE 7.5 A Comparison of Concordance of Various Traits Between Monozygotic (MZ) and Dizygotic (DZ) Twins

| | Concordance | |
Trait	MZ	DZ
Blood types	100%	66%
Eye color	99	28
Mental retardation	97	37
Measles	95	87
Idiopathic epilepsy	72	15
Schizophrenia	69	10
Diabetes	65	18
Identical allergy	59	5
Tuberculosis	57	23
Cleft lip	42	5
Club foot	32	3
Mammary cancer	6	3

similar analysis of **dizygotic** (DZ) or **fraternal twins**, who originate from two separate fertilization events. Dizygotic twins are thus no more genetically similar than any two siblings.

A form of quantitative analysis of characteristics of twins reared together can also be pursued. Twins are said to be **concordant** for a given trait if both express it *or* neither expresses it, and **discordant** if one shows the trait *and* the other does not. Table 7.5 lists concordance values for various traits in both monozygotic and dizygotic twins reared together.

These data must be examined very carefully before any conclusions are drawn. If the concordance value approaches 90 to 100 percent with monozygotic twins, we might be inclined to interpret this value as indicating a large genetic contribution to the expression of the trait. In some cases—blood types and eye color, for example—we know this is indeed true. In the case of measles, however, a high concordance value merely indicates that the disease is not of genetic origin, but is induced by a factor in the environment—in this case, a virus.

It is more meaningful to compare the *difference* between the concordance values of monozygotic and dizygotic twins. If these values are significantly higher for monozygotic twins than for dizygotic twins, we suspect a genetic component is involved in the determination of the trait. We reach this conclusion because monozygotic twins, with identical genotypes, would be expected to show a greater concordance than genetically related, but not genetically identical, dizygotic twins. In the case of measles, where concordance is high in both types of twins, the environment is assumed to contribute significantly even though there may be some genetic contribution since the value for MZ is slightly higher than for DZ.

Even though a particular trait may demonstrate considerable genetically based variation, it is often difficult to formulate a precise mode of inheritance based on avail-

able data. In many cases the trait is considered to be controlled by multiple-factor inheritance. However, when the environment is also exerting a partial influence, such a conclusion is particularly difficult to prove.

The Use of Haploid Organisms in Linkage and Mapping Studies

We conclude this chapter by turning to still another topic that is an extension of our study of transmission genetics: **linkage analysis** and **chromosome mapping in haploid eukaryotes**. As we shall see, even though analysis of the location of genes relative to one another throughout the genome of haploid organisms may *seem* a bit more complex than in diploid organisms (Chapter 4), the basic underlying principles are the same. In fact, many basic principles of inheritance were established during the study of haploid fungi.

While many single-celled eukaryotes are haploid during the vegetative stages of their life cycle, they also form reproductive cells that fuse during fertilization, forming a diploid zygote. The zygote then undergoes meiosis and reestablishes haploidy. The haploid meiotic products are the progenitors of the subsequent members of the vegetative phase of the life cycle. Figure 7.9 illustrates this type of cycle in the green alga *Chlamydomonas*. Even though the haploid cells that fuse during fertilization *look* identical (and are thus called **isogametes**), a chemical identity that distinguishes two distinct types exists on their surface. As a result, all strains are either "+" or "−" and fertilization occurs only between unlike cells.

To perform genetic experiments with haploid organisms, genetic strains of different genotypes are isolated and crossed to one another. Following fertilization and meiosis, the meiotic products are retained together and can be analyzed. Such is the case in *Chlamydomonas* as well as in the fungus *Neurospora*, which we shall use as an example in the ensuing discussion. Following fertilization in *Neurospora* (Figure 7.10), meiosis occurs in a saclike structure called the **ascus** and the initial set of haploid products, called a **tetrad**, are retained within it. This term has a quite different meaning from when it is used to describe the four-stranded chromosome configuration characteristic of meiotic prophase I in diploids.

Following meiosis in *Neurospora*, each cell in the ascus divides mitotically, producing eight haploid **ascospores**. These can be dissected out and examined morphologically or tested to determine their genotypes and phenotypes. Because the eight cells reflect the *sequence* of their formation following meiosis, the tetrad is "ordered" and we can do **ordered tetrad analysis** or **tetrad analysis** for short. This process is critical to our subsequent discussion.

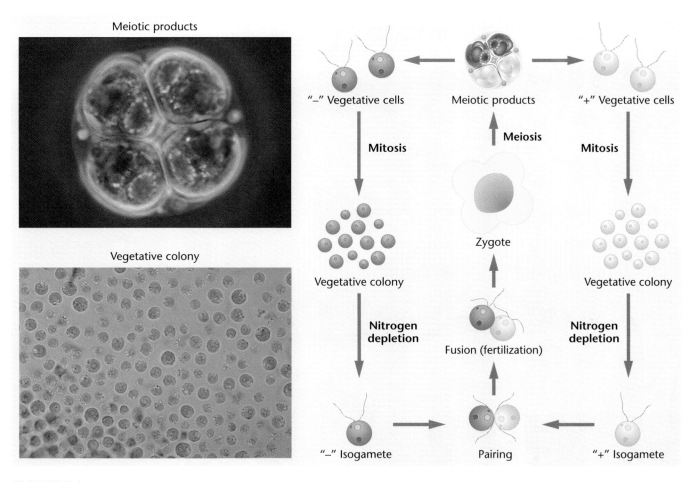

FIGURE 7.9 The life cycle of *Chlamydomonas*. The diploid zygote (in the center) undergoes meiosis, producing "+" or "−" haploid cells that undergo mitosis, yielding vegetative colonies. Unfavorable conditions stimulate them to form isogametes, which fuse in fertilization, producing a zygote, and repeating the cycle. Two of these stages are illustrated photographically.

Gene-to-Centromere Mapping

When a single gene (*a*) is analyzed in *Neurospora*, as diagrammed in Figure 7.11, the data can be used to calculate the map distance between that gene and the centromere. This process is sometimes referred to as **mapping the centromere**. It can be accomplished by experimentally determining the frequency of recombination using tetrad data.

When no crossover event occurs between the gene under study and the centromere, the pattern of ascospores in the ascus appears as shown in Figure 7.11(a) (*aaaa*++++). This pattern represents **first-division segregation** because the two alleles are separated during the first meiotic division. However, crossover events will alter this pattern, as shown in Figure 7.11(b) (*aa*++*aa*++) and 7.11(c) (++*aaaa*++). Two other recombinant patterns can also occur, depending on which chromatids are involved in the crossover event: ++*aa*++*aa* and *aa*++++*aa*. All four of these patterns resulting from a crossover event between the *a* gene and the centromere reflect **second-division**

segregation because the two alleles are not separated until the second meiotic division. Since the mitotic division simply replicates the patterns (from 4 to 8 ascospores) ordered tetrad data are usually condensed to reflect the genotypes reflected in ascospore pairs that may be distinguished from one another. Five unique combinations are possible:

First-Division Segregation

a a + +

Second-Division Segregation

a + a +
+ a + a
+ a a +
a + + a

To calculate the distance between the gene and the centromere, data must be tabulated from a large number of asci resulting from a controlled cross. Using these data, the distance is calculated:

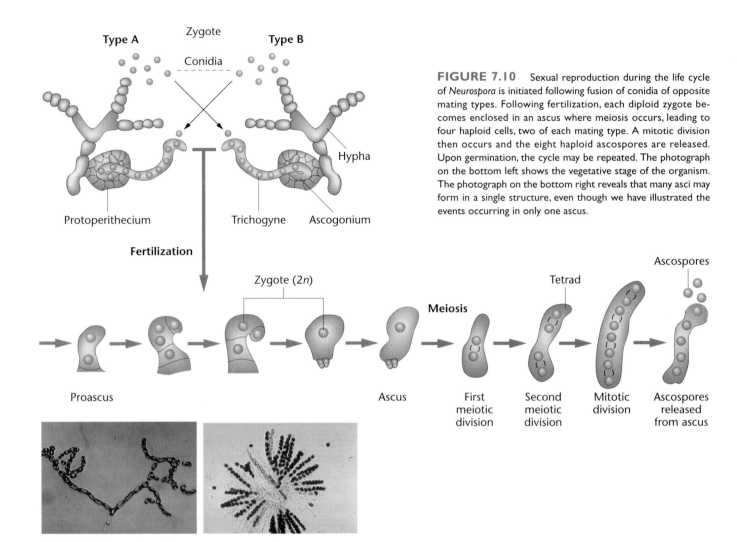

FIGURE 7.10 Sexual reproduction during the life cycle of *Neurospora* is initiated following fusion of conidia of opposite mating types. Following fertilization, each diploid zygote becomes enclosed in an ascus where meiosis occurs, leading to four haploid cells, two of each mating type. A mitotic division then occurs and the eight haploid ascospores are released. Upon germination, the cycle may be repeated. The photograph on the bottom left shows the vegetative stage of the organism. The photograph on the bottom right reveals that many asci may form in a single structure, even though we have illustrated the events occurring in only one ascus.

$$\frac{1/2(\text{second division segregant asci})}{\text{total asci scored}}$$

The recombination percentage is only one-half the number of second-division segregants because crossing over in each of them has occurred in only two of the four strands during meiosis.

To illustrate, assume that *a* represents albino and + represents wild type in *Neurospora*. In crosses between the two genetic types, suppose the following data were observed:

65 first-division segregants
70 second-division segregants

The distance between *a* and the centromere is thus:

$$\frac{(1/2)(70)}{135} = 0.259$$

or about 26 map units.

As the distance increases up to 50 units, in theory, all asci should reflect second-division segregation. However, numerous factors prevent this. As in diploid organisms, accuracy is greatest when the gene and the centromere are relatively close together.

Ordered versus Unordered Tetrad Analysis

In our previous discussion we assumed that the genotype of each ascospore and its position in the tetrad can be determined. To perform such an **ordered tetrad analysis**, individual asci must be dissected and each ascospore must be tracked as it germinates. This is a tedious process, but it is essential for two types of analysis:

1. To distinguish between first-division segregation and second-division segregation of alleles in meiosis.

2. To determine whether recombinational events are reciprocal or not.

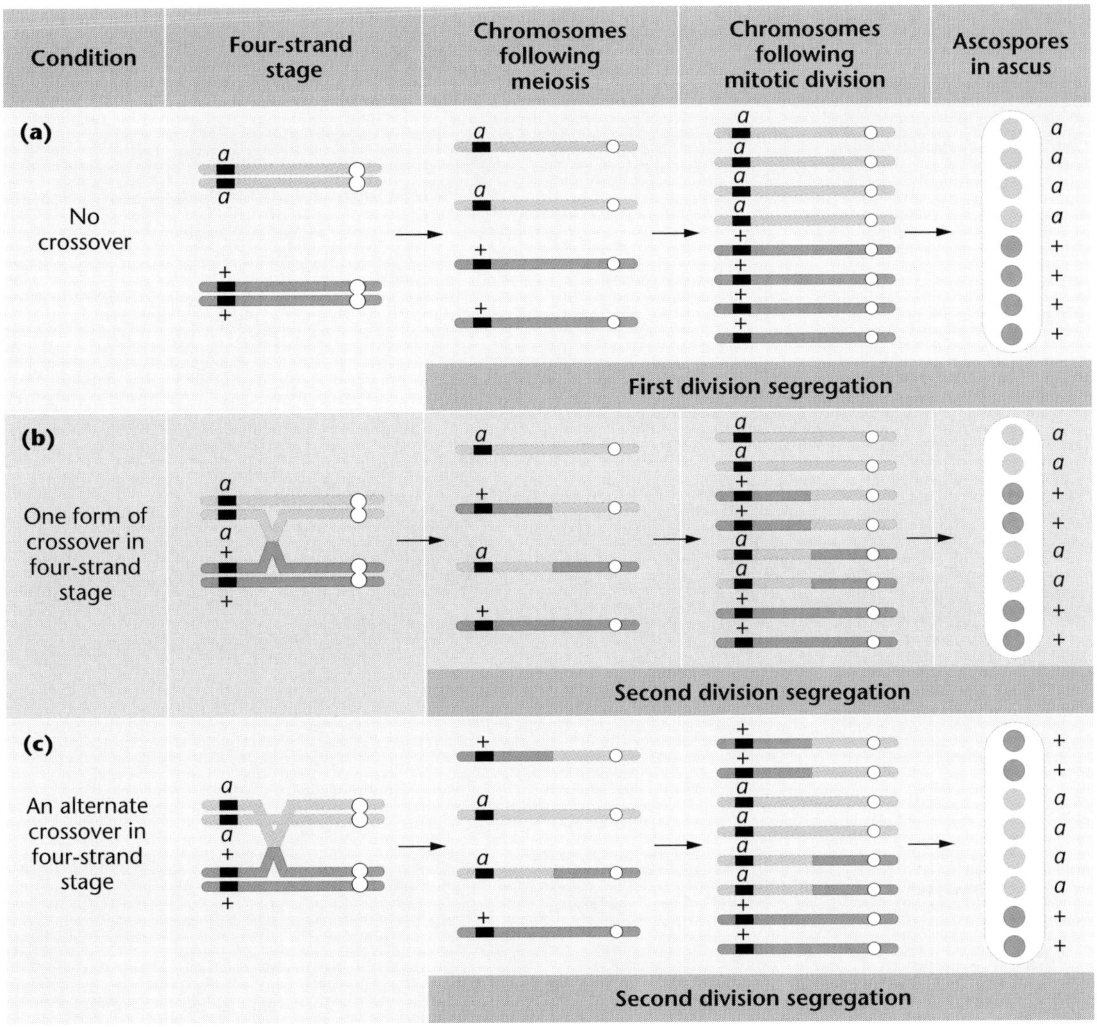

Condition	Four-strand stage	Chromosomes following meiosis	Chromosomes following mitotic division	Ascospores in ascus

FIGURE 7.11 Three ways in which different ascospore patterns can be generated in *Neurospora*. Analysis of these patterns can serve as the basis of "gene-to-centromere" mapping, as described in the text.

In the first case, such information is essential to "map the centromere" as we have just discussed. Thus, ordered tetrad analysis must be performed so as to map the distance between a gene and the centromere.

In the second case, ordered tetrad analysis has revealed that recombinational events are not always reciprocal, particularly when closely linked genes are studied in Ascomycetes. This observation has led to the investigation of the phenomenon called **gene conversion**. Because its discussion requires a background in DNA structure and analysis, we will return to this topic in Chapter 11.

It is much less tedious to isolate individual asci, allow them to mature, and then determine the genotypes of each ascospore, but not in any particular order. This approach is referred to as **unordered tetrad analysis**. As we shall see in the next section, this type of analysis can be used to determine whether two genes are linked on the same chromosome or not, and if so, to determine the map distance between them.

Linkage and Gene Mapping in Haploid Organisms

Analysis of genetic data derived from haploid organisms can be used to distinguish between linkage and independent assortment of two genes; it further allows mapping distances to be calculated between gene loci once linkage is established. In the following discussion we shall consider tetrad analysis in the alga *Chlamydomonas*. With the exception that the four meiotic products are not ordered and *do not* undergo a mitotic division following the completion of meiosis, the general principles discussed for *Neurospora* also apply to *Chlamydomonas*.

To compare independent assortment and linkage, we will consider two theoretical mutant alleles, *a* and *b*, rep-

TABLE 7.6   Tetrad Analysis in *Chlamydomonas*

Category	I	II	III
Tetrad type	Parental (P)	Nonparental (NP)	Tetratypes (T)
Genotypes present	+ + + + a b a b	a + a + + b + b	+ + a + + b a b
Number of tetrads	43	43	14

resenting two distinct loci in *Chlamydomonas*. Suppose that 100 tetrads derived from the cross $ab \times ++$ yield the tetrad data shown in Table 7.6. As you can see, all tetrads produce one of three patterns. For example, all tetrads in category I produce two $++$ cells and two ab cells and are designated as **parental ditypes (P)**. Category II tetrads produce two $a+$ cells and two $+b$ cells and are called **nonparental ditypes (NP)**. Category III tetrads produce one cell each of the four possible genotypes and are thus termed **tetratypes (T)**.

These data support the hypothesis that the genes represented by the a and b alleles are located on separate chromosomes. To understand why, you must refer to Figure 7.12. In parts (a) and (b) of Figure 7.12, the origin of parental (P) and nonparental (NP) ditypes is demonstrated for two unlinked genes. According to the Mendelian principle of independent assortment of unlinked genes, approximately equal proportions of these tetrad types are predicted. Thus, when the parental ditypes are equal to the nonparental ditypes, then the two genes are not linked. The data in Table 7.6 confirm this prediction. Because independent assortment has occurred, it can be concluded that the two genes are located on separate chromosomes.

The origin of category III, the **tetratypes**, is diagrammed in Figure 7.12(c) and (d). The genotypes of tetrads in this category can be generated in two possible ways. Both involve a crossover event between one of the genes and the centromere. In Figure 7.12(c) the exchange involves one of the two chromosomes and occurs between gene a and the centromere; in Figure 7.12(d), the other chromosome is involved and the exchange occurs between gene b and the centromere.

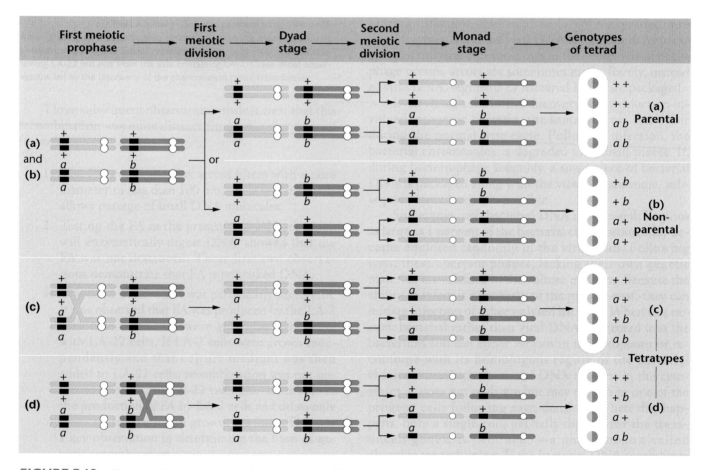

FIGURE 7.12   The origin of various genotypes found in tetrads in *Chlamydomonas* when two genes located on separate chromosomes are considered.

Production of tetratype tetrads does not alter the final ratio of the four genotypes present in all meiotic products. If the genotypes from 100 tetrads (which yield 400 cells) are computed, 100 of each genotype are found. This 1:1:1:1 ratio is predicted according to the expectation of independent assortment.

Now consider the case where the genes *a* and *b* are linked (Figure 7.13). The same categories of tetrads will be produced. However, parental and nonparental ditypes will not necessarily occur in equal proportions; nor will the four genotypic combinations be found in equal numbers if the genotypes of all meiotic products are computed. For example, the following data might be encountered:

Category I P	Category II NP	Category III T
64	6	30

Since the parental and nonparental categories are not produced in equal proportion, we can conclude that independent assortment is not in operation and that the two genes are linked. We can then proceed to determine the map distance between them.

In the analysis of these data, we are concerned with the determination of which tetrad types represent genetic exchanges between the two genes. The **parental ditype tetrads** (**P**) arise only when no crossing over occurs between the two genes. The **nonparental ditype tetrads** (**NP**) arise only when a double exchange involving all four chromatids occurs between two genes. The **tetratype tetrads** (**T**) arise when either a single crossover occurs or when an alternative type of double exchange occurs between the two genes. The various types of exchanges described here are diagrammed in Figure 7.13.

When the proportion of the three tetrad types has been determined, it is possible to calculate the map distance between the two linked genes. The following formula computes the exchange frequency, which is proportional to the map distance between the two genes:

$$\text{exchange frequency (\%)} = \frac{NP + 1/2(T)}{\text{total number of tetrads}} \times 100$$

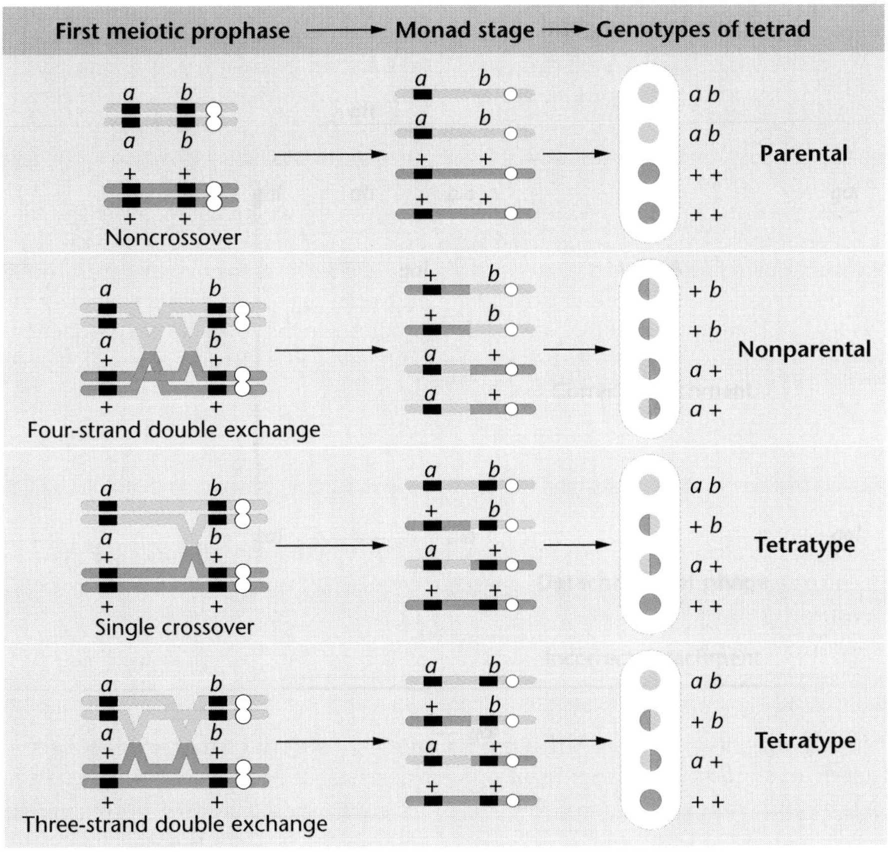

FIGURE 7.13 The various types of exchanges leading to the genotypes found in tetrads in *Chlamydomonas* when two genes located on the same chromosome are considered.

GENETICS, TECHNOLOGY, AND SOCIETY

Preserving Plant Germplasm: The Key to the Future of Agriculture

Of the quarter million species of flowering plants, perhaps 250 were domesticated at one time or another in some part of the world. Over the course of thousands of years of cultivation, each of these species evolved into a vast number of genetically distinct varieties, each adapted to specific conditions of local elevation, rainfall, temperature, soil quality, and resistance to pests and diseases. The native varieties that still exist are now referred to as **landraces**, which embody unique genetic collections of **plant germplasm**.

Intensive breeding of landraces, culminating in the Green Revolution of the 1960s, resulted in the development of many high-yielding varieties of major crop plants. High-yielding "elite" inbred lines that are genetically uniform were produced. These elite lines, originally derived from a small number of different varieties, thus lack the vast genetic diversity of the many existing landraces of each crop. This narrow genetic base was responsible for higher yields, but it has also resulted in increased susceptibilities to pests and diseases. Over time, pathogenic organisms mutate so that they may suddenly be able to infect a previously resistant plant host. The elite lines often lack the genetic variability within populations of plants required to respond to such repeated challenges.

As a result, thousands of acres planted in a single uniform line can be wiped out, resulting in massive crop failure.

Examples of such genetic vulnerability abound in every major crop. The most famous case is the Irish potato blight of 1845–1850, during which the genetically homogeneous potato crop was decimated by infestation with a fungal pathogen. More recently, brown spot disease wiped out the rice crop of India in 1943, triggering a famine. In 1946 Victoria blight attacked the U.S. oat crop, destroying 80 percent of the varieties in use, and in 1954 a wheat rust stem epidemic resulted in the loss of 65 percent of that year's crop. The Southern corn blight that struck in 1970 caused over $1 billion in losses in the United States.

To counter epidemics, breeders may turn to landraces throughout the world to find resistance to the pathogen, which can then be crossed back into the elite line. Many current wheat varieties, for example, contain genes for powdery mildew resistance extracted from a Japanese landrace. If that strategy fails, wild relatives of the crop plant, never under cultivation, are assayed for resistance. Resistance to potato blight was found in a wild potato species located in the Peruvian Andes. By now, wild relatives of almost every major crop have been used in breeding programs.

The landraces and wild relatives of crop plants can thus be viewed as repositories of genetic diversity that serve as the raw material for further development of crop varieties. Initially identified by the Russian plant geneticist N. I. Vavilov in the 1920s, such repositories are most often located along the Equator and in the Southern Hemisphere, primarily in Third World countries. Future crop improvement will thus depend upon the transfer of "exotic" plant germplasm from these landraces and their wild relatives that still grow in the Third World to "elite" varieties developed in industrialized countries.

Unfortunately, several factors have led to the loss of repositories of plant diversity at a time when their importance for world agriculture is being fully understood. For one, the replacement of landraces with high-yielding elite varieties in Third World countries has resulted in the extinction of many of these landraces, some that had been in continuous cultivation for thousands of years. In the past 40 years, many native varieties of rice, wheat, barley, sorghum, and potatoes have been permanently lost. And lost with them were potentially valuable genes, the products of those thousands of years of adaptation to specific environmental conditions. Moreover, accelerated habitat loss, exacerbated by population growth, is further eroding the genetic diversity of plants.

How are we to maintain the genetic diversity of plant germplasm that will be necessary to ensure the future success of agriculture? There is no way to predict which landraces may someday be needed to contribute genes for disease resistance, so the strategy must be

---

In this formula NP represents the nonparental tetrads; all meiotic products represent an exchange. The tetratype tetrads are represented by T; assuming only single exchanges, one-half of the meiotic products represents exchanges. The sum of the scored tetrads that fall into these categories is then divided by the total number of tetrads examined (P + NP + T). If this calculated number is multiplied by 100, it is converted to a percentage, which is directly equivalent to the map distance between the genes.

In our example, the calculation reveals that genes *a* and *b* are separated by 21 map units:

$$\frac{6 + 1/2(30)}{100} = \frac{6 + 15}{100} = \frac{21}{100} = 0.21 \times 100 = 21\%$$

Although we have considered linkage analysis and mapping of only two genes at a time, such studies often involve three or more genes. In these cases, both gene sequence and map distances can be determined.

to preserve the widest possible diversity of plant germplasm. To that end, crop breeding institutes began collecting established landraces in the 1970s, and in 1974 the International Board for Plant Genetic Resources was organized to guide conservation efforts. The strategy is twofold: to protect existing landraces in their natural habitats, primarily in Third World countries; and to preserve these plants in seed banks, mostly in the industrialized countries.

The transfer of plant genetic resources from the "gene-rich" Third World countries to the "gene-poor" industrialized countries has not been without conflict, however. Plant breeders and seed companies in the industrialized countries have traditionally viewed wild plant germplasm as a global resource enabling the development of improved crops that benefit all countries worldwide. Governments of Third World countries, on the other hand, increasingly believe that they should be able to maintain property rights on the germplasm removed from their borders and should realize an economic return from its use in crop development. This disagreement may intensify as molecular techniques permit the rapid introduction of "exotic" genes into cultivated varieties. Reconciling these competing claims may require the industrialized countries to compensate the developing countries for the collection of plant resources to provide an incentive for the preservation of the habitat of potentially valuable plants. Whatever the mechanism, we have a responsibility to preserve as much of the world's plant genetic resources as possible for future generations. Failing to do so could conceivably result in the catastrophic loss of major crop species and threaten the world's food supply.

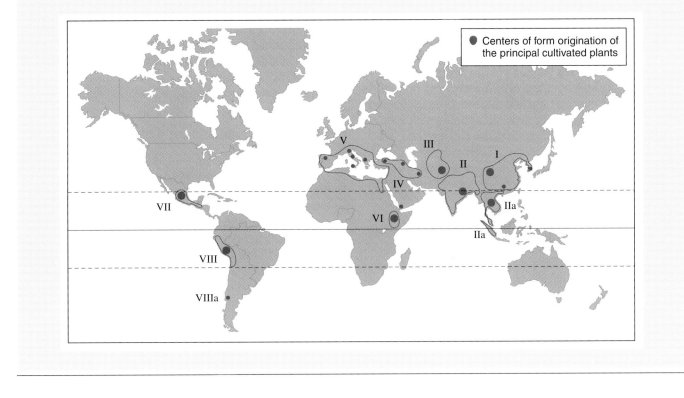

CHAPTER SUMMARY

1. The penetrance of a mutant allele is measured by the percentage of organisms in a population that exhibits evidence of the mutant phenotype. Expressivity, on the other hand, measures the range of mutant expression within a population.

2. Phenotypic expression can be modified by genetic background, temperature, nutrition, and other such environmental factors. The phenomena of genetic suppression and position effects have been used to illustrate the existence of genetic background factors. Together, all such factors constitute the total environment in which genetic information is expressed.

3. The time of onset of gene expression in all organisms varies as the need for certain gene products occurs at different periods during the processes of development, growth, and aging.

4. Variation due to genetic and environmental factors in the establishment of a given phenotype is difficult to ascertain. Heritability can be calculated for many characters and is especially useful in selective breeding of commercially valuable plants and animals.

5. Studies involving twins are aimed at resolving the question of heredity versus environment in human traits. The degree of concordance and discordance of a trait can be compared in monozygotic (identical) and dizygotic (fraternal) twins raised together or apart.

6. Continuous variation is exhibited in crosses involving traits under polygenic control. Such traits are quanti-

tative in nature and are inherited as a result of the cumulative impact of additive alleles.

7. Polygenic characteristics can be analyzed using statistical methods, which include the mean, the variance, the standard deviation, and the standard error of the mean. Such statistical analysis can be descriptive, can be used to make inferences about a population, or can be used to compare sets of data.

8. Linkage analysis and chromosome mapping are possible in haploid eukaryotes. Our discussion has included gene-to-centromere and gene-to-gene mapping as well as the consideration of how to distinguish between linkage and independent assortment.

KEY TERMS

additive variance, 192
Angelman syndrome (AS), 186
artificial selection, 193
ascospore, 194
ascus, 194
auxotroph, 183
biometry, 186
broad-sense heritability (H^2), 192
central tendency, 189
chromosome mapping in haploid organisms, 194
concordant, 194
conditional mutation, 183
continuous variation, 186
diabetes, 183
discontinuous variation, 186
discordant, 194
dizygotic (fraternal) twins, 194
DNA methylation, 186
DNA methyl transferase, 186
dominance variance, 192
dosage compensation, 185
Duchenne muscular dystrophy (DMD), 184
environmental variance, 192
expressivity, 181
first-division segregation, 195
fragile-X mental retardation, 184
frequency distribution, 189
galactosemia, 183

gene conversion, 197
genetic anticipation, 184
genetic background, 182
genetic suppression, 182
genetic variance, 192
genomic (parental) imprinting, 185
heritability, 192
heterochromatin, 182
hexosaminidase A, 184
Huntington disease, 184
hypoxanthine-guanine phosphoribosyl transferase (HPRT), 184
inversion, 182
interaction variance, 192
isogametes, 194
lactase, 184
lactose intolerance, 183
landraces, 200
Lesch–Nyhan syndrome, 184
linkage analysis in haploids, 194
mapping the centromere, 195
mean, 189
monozygotic (identical) twins, 193
myotonic dystrophy (DM), 184
narrow-sense heritability (h^2), 192
nonparental ditype (NP), 198
nutritional mutants, 183
ordered tetrad analysis, 194
parental ditype (P), 198
penetrance, 181

permissive condition (temperature), 183
phenotypic expression, 180
phenotypic variance, 192
phenylketonuria, 183
plant germplasm, 200
polygenic trait, 186
position effect, 182
Prader-Willi syndrome (PWS), 185
quantitative inheritance, 180
quantitative trait loci (QTL), 187
restriction fragment-length polymorphisms (RFLPs), 188
restrictive condition (temperature), 183
second-division segregation, 195
spinal and bulbar muscular atrophy (Kennedy disease), 184
standard deviation, 190
standard error of the mean, 190
statistical inference, 189
statistics, 189
Tay-Sachs disease, 184
temperature-sensitive mutation, 183
tetrad, 194
tetrad analysis, 194
tetratype (T), 198
translocation, 182
transmission genetics, 180
trinucleotide repeat, 184
unordered tetrad analysis, 197
variance, 189

INSIGHTS AND SOLUTIONS

1. The following results were recorded for ear length in corn:

							Length of ear in cm										
	5	6	7	8	9	10	11	12	13	14	15	16	17	18	19	20	21
Parent A	4	21	24	8													
Parent B									3	11	12	15	26	15	10	7	2
F_1					1	12	12	14	17	9	4						

(a) For each of the parental strains, and the F_1, calculate the mean values for ear length.

Solution: The mean values are calculated as follows:

$$\bar{X} = \frac{\sum \bar{X}_i}{n} \quad P_A: \quad \bar{X} = \frac{\sum \bar{X}_i}{n} = \frac{378}{57} = 6.63$$

$$P_B: \quad \bar{X} = \frac{\sum \bar{X}_i}{n} = \frac{1697}{101} = 16.80$$

$$F_1: \quad \bar{X} = \frac{\sum \bar{X}_i}{n} = \frac{836}{69} = 12.11$$

(b) Compare the mean of the F_1 with that of each parental strain. What does this tell you about the type of gene action involved?

Solution: The F_1 mean (12.11) is almost midway between the parental means of 6.63 and 16.80. This indicates that the genes in question are probably additive in effect.

2. In a cross in *Neurospora* where one parent expresses the mutant allele *a* and the other expresses a wild-type phenotype (+), the following data were obtained in the analysis of ascospores. Calculate the gene-to-centromere distance.

Asci Types

	1	2	3	4	5	6	
Sequence of ascospores in ascus	+	*a*	*a*	+	*a*	+	
	+	*a*	*a*	+	*a*	+	
	+	*a*	+	*a*	+	*a*	
	+	*a*	+	*a*	+	*a*	
	a	+	*a*	+	+	*a*	
	a	+	*a*	+	+	*a*	
	a	+	+	*a*	*a*	+	
	a	+	+	*a*	*a*	+	
	39	33	5	4	9	10	Total = 100

Solution: Ascus types 1 and 2 represent first-division segregants (fds) where no crossing over occurred between the *a* locus and the centromere. All others (3–6) represent second-division segregation (sds). By applying the formula

$$distance = \frac{1/2 \; sds}{total \; asci}$$

we obtain the following result:

$$d = \frac{1/2(5 + 4 + 9 + 10)}{100}$$

$$= \frac{1/2(28)}{100}$$

$$= 0.14$$

$$= 14 \; map \; units$$

3. The mean and variance of corolla length in two highly inbred strains of *Nicotiana* and their progeny are shown below. One parent (P_1) has a short corolla and the other (P_2) has a longer corolla.

Strain	Mean (mm)	Variance
P_1 short	40.47	3.12
P_2 long	93.75	3.87
F_1 ($P_1 \times P_2$)	63.90	4.74
F_2 ($F_1 \times F_1$)	68.72	47.70

Calculate the heritability (H^2) of corolla length in this plant.

Solution: The formula for estimating heritability is

$$H^2 = V_G/V_P$$

where V_G and V_P are the genetic and phenotypic components of variation, respectively. The main issue in this problem is obtaining some estimate of two components of phenotypic variation: genetic and environmental factors.

V_P is the combination of genetic and environmental variance. Because the two parental strains are true-breeding, they are assumed to be homozygous, and the variance of 3.12 and 3.87 is considered to be the result of environmental influences. The F_1 is also genetically homogeneous and gives us an additional estimation of the environmental factors. By averaging with the parents

$$\frac{3.12 + 3.87 + 4.74}{3} = 3.91$$

we obtain a relatively good idea of environmental impact on the phenotype. The phenotypic variance in the F_2 is the sum of the genetic (V_G) and environmental (V_E) components. We have estimated the environmental input as 3.91, so 47.71 minus 3.91 gives us an estimate of V_G, which is

43.80. Heritability then becomes 43.80/47.71 or 0.92. This value, when viewed in percentage form, indicates that about 92 percent of the variation in corolla length is due to genetic influences.

PROBLEMS AND DISCUSSION QUESTIONS

1. Distinguish between continuous and discontinuous variation.
2. List as many human traits as you can that are likely to be under the control of a polygenic mode of inheritance.
3. Describe the difference between penetrance and expressivity.
4. Define and discuss the significance of the following terms: (a) position effect, (b) suppressor genes, (c) monozygotic and dizygotic twins, (d) concordance and discordance, and (e) heritability.
5. In the following table, average differences of height and weight between monozygotic twins (reared together and apart), dizygotic twins, and siblings are compared. Draw as many conclusions as you can concerning the effects of genetics and the environment in influencing these human traits.

Trait	MZ Reared Together	MZ Reared Apart	DZ Reared Together	Sibs Reared Together
Height (cm)	1.7	1.8	4.4	4.5
Weight (kg)	1.9	4.5	4.5	4.7

Source: Newman, Freeman, and Holzinger, 1937.

6. Height in humans depends on the additive action of genes. Assume that this trait is controlled by the four loci *R, S, T, U* and that environmental effects are negligible. Dominant alleles contribute two units and recessive alleles contribute one unit to height.
   (a) Can two individuals of moderate height produce offspring that are much taller or shorter than either parent? If so, how?
   (b) If an individual with the minimum height specified by these genes marries an individual of intermediate or moderate height, could any of their children be taller than the tall parent? Why?
7. In a hypothetical situation, vitamin A content and cholesterol content of eggs from a large population of chickens were investigated thoroughly. The variances (V) were calculated as shown below:

Variance	Trait Vitamin A	Trait Cholesterol
V_P	123.5	862.0
V_E	96.2	484.6
V_A	12.0	192.1
V_D	15.3	185.3

(a) Calculate the narrow heritability (h^2) for both traits.
(b) Which, if either, is likely to respond to selection?

8. In the assessment of learning in *Drosophila*, flies may be trained to avoid certain olfactory cues. In one population, a mean of 8.5 trials was required. A subgroup of this population was trained more quickly (mean = 6.0). Members of this group were interbred and their progeny tested. These flies demonstrated a mean training value of 7.5. Calculate and interpret the narrow heritability (h^2) for olfactory learning in *Drosophila*.

9. A population of tomato plants was studied whose mean fruit weight is 60 g, where $h^2 = 0.3$ for this trait. Predict the results of artificial selection (the mean weight of the progeny) if tomato plants whose fruit weight averaged 80 g were selected from the original population and interbred.

10. In a cross in *Neurospora* involving two alleles *B* and *b*, the following tetrad patterns were observed. Calculate the distance between the gene and the centromere.

Tetrad Pattern	Number
BBbb	36
bbBB	44
BbBb	4
bBbB	6
BbbB	3
bBBb	7

11. In *Neurospora*, the cross *a+* × *+b* yielded only two types of ordered tetrads in approximately equal numbers:

	Spore Pair 1–2	Spore Pair 3–4	Spore Pair 5–6	Spore Pair 7–8
Tetrad Type 1	*a +*	*a +*	*+ b*	*+ b*
Tetrad Type 2	*+ +*	*+ +*	*a b*	*a b*

What can be concluded?

12. Below are two sets of data derived from crosses in *Chlamydomonas*, involving three genes represented by the mutant alleles *a, b,* and *c*. Determine as much as you can concerning genetic arrangement of these three genes relative to one another. Describe the expected results of Cross 3, assuming that *a* and *c* are linked and are 38 map units apart, and that 100 tetrads are produced.

Cross	Genes Involved	Tetrads		
		P	NP	T
1	*a* and *b*	36	36	28
2	*b* and *c*	79	3	18
3	*a* and *c*			

13. In *Chlamydomonas*, a cross *ab* × ++ yielded the following unordered tetrad data where *a* and *b* are linked.

(1)	+ + + + *a b* *a b*	38	(3)	*a* + *a* + + *b* + *b*	6	(5)	*a b* + + + *b* *a* +	2
(2)	+ + *a b* + + *a b*	5	(4)	*a b* *a* + + *b* + +	17	(6)	*a b* + *b* *a* + + +	3

(a) Identify the categories representing parental ditypes (P), nonparental ditypes (NP), and tetratypes (T).
(b) Explain the origin of category (2).
(c) Determine the map distance between *a* and *b*.

14. The following results are ordered tetrad pairs from a cross between strain *cd* and strain ++(c^+d^+). They are summarized by tetrad classes.

Tetrad Class						
1	2	3	4	5	6	7
c +	*c* +	*c d*	+ *d*	*c* +	*c d*	*c* +
c +	*c d*	*c d*	*c* +	+ +	+ +	+ *d*
+ *d*	+ +	+ +	*c* +	*c d*	*c d*	*c d*
+ *d*	+ *d*	+ +	+ *d*	+ *d*	+ +	+ +
1	17	41	1	5	3	1

(a) Name the ascus type of each class from 1 to 7 (P, NP, or T).
(b) The above data support the conclusion that the *c* and *d* loci are linked. State the evidence in support of this conclusion.
(c) Calculate the gene–centromere distance for each locus.
(d) Calculate the distance between the two linked loci.
(e) Draw a linkage map including the centromere and explain the discrepancy between the distances determined by the two different methods in parts (c) and (d).
(f) Describe the arrangement of crossovers needed to produce the ascus class 6 above.

15. In a cross in *Chlamydomonas*, *AB* × *ab*, 211 unordered asci were recovered:

10 *AB, Ab, aB, ab*
102 *Ab, aB, Ab, aB*
99 *AB, AB, ab, ab*

(a) Correlate each of the three tetrad types in the problem with their appropriate tetrad designations (names).
(b) Are genes *A* and *B* linked?
(c) If they are linked, determine the map distance between the two genes. If they are unlinked, provide the maximum information you can about why you drew this conclusion.

EXTRA-SPICY PROBLEMS

16. Corn plants from a test plot are measured, and the distribution of heights at 10-cm intervals is recorded below:

Height (cm)	Plants (no.)
100	20
110	40
120	90
130	130
140	180
150	120
160	70
170	50
180	40

Calculate (a) the mean height, (b) the variance, (c) the standard deviation, and (d) the standard error of the mean. Plot a rough graph of height vs. frequency. How would you describe the variation shown in the graph?

17. The mean and variance of plant height of two highly inbred parental strains (P_1 and P_2) and their F_1 and F_2 progeny are shown below. Calculate the broad heritability (H^2) of plant height in this species.

Strain	Mean (cm)	Variance
P_1	34.2	4.2
P_2	55.3	3.8
F_1	44.2	5.6
F_2	46.3	10.3

18. In 1988, Horst Wilkens investigated blind cavefish, comparing them to members of a sibling species with normal vision that are found in a lake (we will call them cavefish and lakefish). He found that cavefish eyes are about seven times smaller than lakefish eyes. F_1 hybrids have eyes of intermediate size. These data as well as the F_1 × F_1 cross and those from backcrosses (F_1 × cavefish and F_1 × lakefish) are depicted below. Examine Wilkens' results and respond to the following questions:

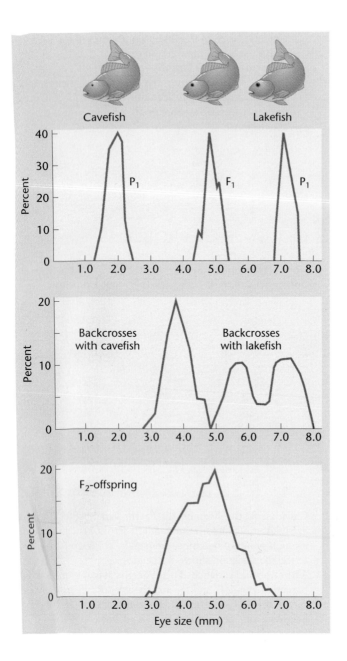

(a) Based strictly on the F_1 and F_2 results of Wilkens' initial crosses, what possible explanation concerning the inheritance of eye size seems most feasible?

(b) Based on the results of the F_1 backcross with cavefish, is your explanation supported? Explain.

(c) Based on the results of the F_1 backcross with lakefish, is your explanation supported? Explain.

(d) Wilkens examined about 1000 F_2 progeny and estimated that 6–7 genes are involved in determining eye size. Is this sample size adequate to justify this conclusion? (You may want to refer back to Chapter 4.) Propose an experimental prototcol to test this hypothesis.

(e) A comparison of the embryonic eye in cavefish and lakefish revealed that both reach approximately 4 mm in diameter. However, lakefish continue to grow, while cavefish eye size is greatly reduced. Speculate on the role of the genes involved in this problem.

[*Reference:* Wilkens, H. (1988). *Ecol. Biol.* 23:271–367.]

SELECTED READINGS

Brink, R. A., ed. 1967. *Heritage from Mendel.* Madison: University of Wisconsin Press.

Chapman, A. B. 1985. *General and quantitative genetics.* Amsterdam: Elsevier.

Corwin, H. O., and Jenkins, J. B. 1976. *Conceptual foundations of genetics: Selected readings.* Boston: Houghton Mifflin.

Crow, J. F. 1966. *Genetics notes*, 6th ed. Minneapolis: Burgess.

Dunn, L. C. 1966. *A short history of genetics.* New York: McGraw-Hill.

Falconer, D. 1989. *Introduction to quantitative genetics*, 3rd ed. Harlow, UK: Longman Scientific and Technical.

Farber, S. L. 1980. *Identical twins reared apart.* New York: Basic Books.

Feldman, M. W., and Lewontin, R. C. 1975. The heritability hang-up. *Science* 190:1163–68.

Foster, M. 1965. Mammalian pigment genetics. *Adv. Genet.* 13:311–39.

Haley, C. 1991. Use of DNA fingerprints for the detection of major genes for quantitative traits in domestic species. *Anim. Genet.* 22:259–77.

Harper, P. S., et al. 1992. Anticipation in myotonic dystrophy: New light on an old problem. *Am. J. Hum. Genet.* 51:10–16.

HOWELER, C. J., et al. 1989. Anticipation in myotonic dystrophy: Fact or fiction? *Brain* 112:779–97.

LANDER, E. S., and BOTSTEIN, D. 1989. Mapping Mendelian factors underlying quantitative traits using RFLP linkage maps. *Genetics* 121:185–99.

LEWONTIN, R. C. 1974. The analysis of variance and the analysis of causes. *Am. J. Hum. Genet.* 26:400–11.

LINDSLEY, D. C., and GRELL, E. H. 1967. *Genetic variations of Drosophila melanogaster.* Washington, DC: Carnegie Institute.

MAHEDEVAN, M., et al. 1992. Myotonic dystrophy mutation: An unstable CTG repeat in the 3′ untranslated region of the gene. *Science* 255:1253–58.

MATHER, K. 1965. *Statistical analysis in biology.* London: Methuen.

NEWMAN, H. H., FREEMAN, F. N., and HOLZINGER, K. J. 1937. *Twins: A study of heredity and environment.* Chicago: University of Chicago Press.

NOLTE, D. J. 1959. The eye-pigmentary system of *Drosophila. Heredity* 13:233–41.

PATERSON, A. H., DeVERNA, J. W., LANNI, B., and TANKSLEY, S. D. 1990. Fine mapping of quantitative trait loci using selected overlapping recombinant chromosomes in an interspecific cross of tomato. *Genetics* 124:735–42.

PATERSON, A. H., LANDER, E. S., HEWITT, J. D., PETERSON, S., LINCOLN, S. E., and TANKSLEY, S. D. 1988. Resolution of quantitative traits into Mendelian factors by using a complete linkage map of restriction fragment length polymorphisms. *Nature* 335:721–26.

PAWELEK, J. M., and KÖRNER, A. M. 1982. The biosynthesis of mammalian melanin. *Am. Sci.* 70:136–45.

PLOMIN, R., McCLEARN, G., GORA-MASLAK, G., and NEIDERHISER, J. 1991. Use of recombinant inbred strains to detect quantitative trait loci associated with behavior. *Behav. Genet.* 21:99–116.

PONTECORVO, G. 1958. *Trends in genetic analysis.* New York: Columbia University Press.

RAEBURN, P. 1995. *The last harvest: The genetic gamble that threatens to destroy American agriculture.* New York: Simon & Schuster.

RANSON, R., ed. 1982. *A handbook of Drosophila development.* New York: Elsevier Biomedical.

RAZIN, A., and CEDAR, H. 1994. DNA methylation and genomic imprinting. *Cell* 77:473–76.

SAPIENZA, C. 1990. Parental imprinting of genes. *Sci. Am.* (Oct.) 363:52–60.

STAHL, F. W. 1979. *Genetic recombination.* New York: W. H. Freeman.

SURANI, M. A. 1994. Genomic imprinting: Control of gene expression by epigenetic inheritance. *Curr. Opin. Cell Biol.* 6:390–95.

VOELLER, B. R., ed. 1968. *The chromosome theory of inheritance— Classic papers in development and heredity.* New York: Appleton-Century-Crofts.

WILHAM, R., and WILSON, D. 1991. Genetic predictions of racing performance in quarter horses. *J. Anim. Sci.* 69:3891–94.

Extranuclear Inheritance

CHAPTER CONCEPTS

Many traits are known in eukaryotes that do not yield genetic patterns normally associated with biparental inheritance. These traits illustrate what is called extranuclear inheritance. In some cases, such patterns are due to gene products that are stored in the egg prior to fertilization, creating a maternal effect. In other cases, inheritance patterns are based on the uniparental (usually female) transmission of genetic information contained in the cytoplasmic organelles, the mitochondria and chloroplasts. In still other cases, infectious particles are transmitted through gametes, also illustrating extranuclear inheritance.

Throughout the history of genetics, occasional reports have challenged the basic tenets of Mendelian transmission genetics—the production of the phenotype through the transmission of nuclear genes located on chromosomes of both parents. Instead, genetic results have been observed that do not reflect either Mendelian or neo-Mendelian principles. Some of these reports have indicated an apparent extranuclear or extrachromosomal influence on the phenotype. Such observations were often regarded with skepticism. However, with the increasing knowledge of molecular genetics and the discovery of DNA in mitochondria and chloroplasts, **extranuclear inheritance** is now recognized as an important aspect of genetics.

There are many diverse examples of these unusual modes of inheritance. In this chapter we will focus on three general types of genetic phenomena: (1) maternal influence resulting from the effect of stored products of nuclear genes of the female parent during early development; (2) organelle heredity resulting from the expression of DNA contained in mitochondria and chloroplasts; and (3) infectious heredity resulting from the symbiotic or parasitic association of microorganisms with eukaryotic cells. Each has the effect of producing inheritance patterns that vary from those predicted by the concepts of Mendelian and neo-Mendelian genetics.

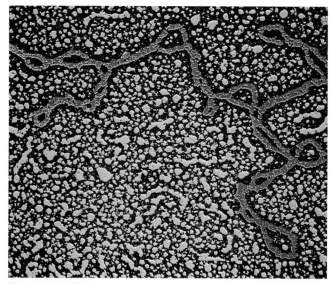

Mitochondrial DNA.

Maternal Effect

Maternal effect, also referred to as *maternal influence*, implies that an offspring's phenotype for a particular trait is strongly influenced by the nuclear genotype of the maternal parent. This is in contrast to most cases, where inheritance of traits is biparental. In cases of maternal effect, the genetic information of the female gamete is transcribed, and these genetic products (either proteins or yet untranslated mRNAs) are present in the egg cytoplasm. Following fertilization, these products influence patterns or traits established during early development. Two examples will illustrate the influence of the maternal genome on particular traits.

Ephestia Pigmentation

In the Mediterranean meal moth, *Ephestia kuehniella*, the wild-type larva has a pigmented skin and brown eyes as a result of the dominant gene *A*. The pigment is derived from a precursor molecule, kynurenine, which is in turn a derivative of the amino acid tryptophan. A mutation, *a*, interrupts the synthesis of kynurenine and, when homozygous, may result in red eyes and little pigmentation in larvae. As illustrated in Figure 8.1, however, different results are obtained in the cross *Aa* × *aa*, depending on which parent carries the dominant gene. When the male is the heterozygous parent, a 1:1 brown/red-eyed ratio is observed in larvae as predicted by Mendelian segregation. However, when the female is heterozygous for the *A* gene, all larvae are pigmented and have brown eyes, in spite of half of them being *aa*. As these larvae develop into adults, one-half of them gradually develop red eyes, reestablishing the 1:1 ratio.

One explanation of these results is that the *Aa* oocytes synthesize kynurenine or an enzyme necessary for its synthesis and accumulate it in the ooplasm prior to the completion of meiosis. Even in *aa* progeny (whose mothers were *Aa*), this pigment is distributed in the cytoplasm of the cells of the developing larvae—thus, they develop pigmentation and brown eyes. Eventually, the pigment is diluted among many cells and used up, resulting in the conversion to red eyes as adults. The *Ephestia* example demonstrates the maternal effect in which a cytoplasmically stored nuclear gene product influences the larval phenotype and, at least temporarily, overrides the genotype of the progeny.

Limnaea Coiling

Shell coiling in the snail *Limnaea peregra* is an excellent example of maternal effect and represents a permanent rather than a transitory phenotype. Some strains of this snail have left-handed or sinistrally coiled shells (*dd*), while others have right-handed or dextrally coiled shells (*DD* or *Dd*). These snails are hermaphroditic and may undergo either cross- or self-fertilization, providing a variety of types of matings.

Figure 8.2 illustrates the results of reciprocal crosses between true-breeding snails. As can be seen, these crosses yield different outcomes, even though both crosses are between sinistral and dextral organisms and produce offspring all of whom are heterozygous. Examination of the progeny reveals that their *phenotypes* depend on the *genotypes* of the female parents. If we adopt that conclusion as a working hypothesis, we can test it by examining the offspring in subsequent generations of self-fertilization events. In each case, the hypothesis is upheld. *The coiling pattern of the progeny snails is determined by the genotype of the parent producing the egg, regardless of the phenotype of that parent.* Maternal parents that are *DD* or *Dd* produce only dextrally coiled progeny. Maternal parents that are *dd* produce only sinistrally coiled progeny.

Investigation of the developmental events in these snails reveals that the orientation of the spindle in the first cleavage division after fertilization determines the direction of coiling. Spindle orientation appears to be controlled by maternal genes acting on the developing eggs in the ovary. The orientation of the spindle, in turn, influences cell divisions following fertilization and establishes the permanent adult coiling pattern.

The dextral allele (*D*) produces an active gene product that causes right-handed coiling. If ooplasm from dextral eggs is injected into uncleaved sinistral eggs, they cleave in a dextral pattern. However, in the converse experiment, sinistral ooplasm has no effect when injected into dextral eggs. Apparently the sinistral allele is the result of a classical recessive mutation that encodes an inactive gene product.

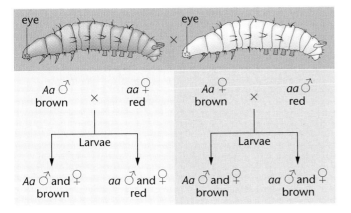

FIGURE 8.1   Illustration of maternal influence in the inheritance of eye pigment in the meal moth *Ephestia kuehniella*. Multiple light receptor structures (eyes) are present on each side of the anterior portion of larvae.

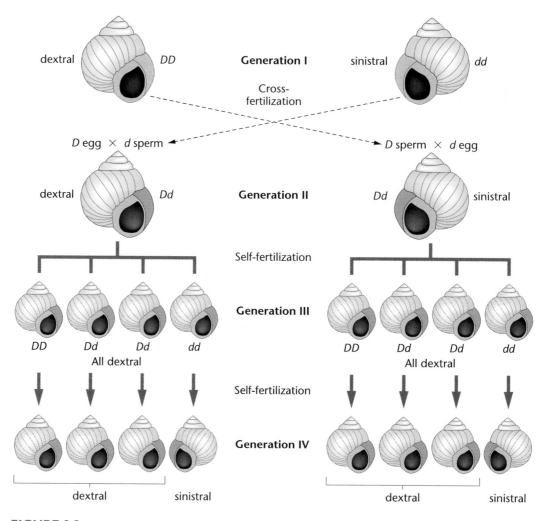

FIGURE 8.2 Inheritance of coiling in the snail *Limnaea peregra*. Coiling is either dextral (right-handed) or sinistral (left-handed). A maternal effect is evident in generations II and III, where the genotype of the maternal parent, rather than the offspring's own genotype, controls the phenotype of the offspring.

Based on these varied observations, we can conclude that females that are either *DD* or *Dd* produce oocytes that synthesize the *D* gene product, which is stored in the ooplasm. Even if the oocyte contains only the *d* allele following meiosis and is fertilized by a *d*-bearing sperm, the resulting *dd* snail will be dextrally coiled (right-handed).

Embryonic Development in *Drosophila*

A more recently documented example of maternal effect involves various genes that function to control embryonic development in *Drosophila melanogaster*. As a result of the extensive work by Edward B. Lewis, Christiane Nüsslein-Volhard, and Eric Wieschaus (who jointly shared the 1995 Nobel Prize for Physiology or Medicine for their findings), the role of these and other genes has become clear. Those that illustrate maternal effect are genes whose products are synthesized by the developing egg and stored in the oocyte prior to fertilization. Following fertilization, these products specify molecular gradients in the early embryo that determine spatial organization as development proceeds.

For example, the gene *bicoid* (*bcd*) plays an important role in specifying the development of the anterior portion of the fly. This conclusion is based on the observation that embryos derived from mothers who are homozygous for this mutation (*bcd⁻/bcd⁻*) fail to develop anterior areas that normally give rise to the head and thorax of the adult fly. Embryos whose mothers contain at least one wild-type allele (*bcd⁺*) develop normally, even if the genotype of the embryo is homozygous for the mutation. Consistent with the concept of *maternal effect*, the *genotype of the female parent*, not the *genotype of the embryo* determines the *phenotype* of the offspring.

The genetic control of embryonic development in *Drosophila* is a fascinating story. The protein products of the maternal-effect genes function to activate other genes, which may in turn activate still other genes. This cascade of gene activity leads to a normal embryo whose subsequent development yields a normal adult fly. We shall return to a more detailed discussion of this general topic in Chapter 20. There, we will see examples of other genes illustrating maternal effect, as well as many so-called "zygotic" genes whose expression occurs during early development and which behave genetically in a conventional Mendelian fashion.

Organelle Heredity

In this section we will examine examples of inheritance patterns of phenotypes related to chloroplast and mitochondrial function. Prior to the discovery of DNA in these organelles and extensive characterization of the genetic processes occurring within them, these patterns were grouped under the category of **cytoplasmic inheritance**. That is, certain mutant phenotypes seemed to be inherited through the cytoplasm rather than through the genetic information of chromosomes. Transmission was most often from the maternal parent through the ooplasm, causing the results of reciprocal crosses to vary. Such patterns are now more appropriately considered examples of **organelle heredity**, which is distinct from maternal effect.

Analysis of the hereditary transmission of mutant alleles of chloroplast and mitochondrial DNA has been difficult. First, the function of these organelles is dependent upon gene products of both nuclear and organelle DNA. Second, the number of organelles contributed to each progeny often exceeds one. If many chloroplasts and/or mitochondria are contributed and only one or a few of them contain a mutant gene, the corresponding mutant phenotype may not be revealed. Analysis is thus much more involved than for Mendelian characters.

In this section we shall discuss examples of inheritance patterns related to these organelles. We will return to this topic and provide a more detailed analysis of the molecular organization of the DNA in Chapter 17.

Chloroplasts: Variegation in Four O'Clock Plants

In 1908, Carl Correns (one of the rediscoverers of Mendel's work) provided the earliest example of inheritance linked to chloroplast transmission. Correns discovered a variety of the four o'clock plant, *Mirabilis jalapa*, that had branches with either white, green, or variegated leaves. As shown in Table 8.1, inheritance in all possible combinations of crosses is strictly determined by the *phenotype* of the ovule source. For example, if the seeds (representing the progeny) were derived from ovules on branches with green leaves, all progeny plants bore only green leaves regardless of the phenotype of the source of pollen.

Correns concluded that inheritance was through the cytoplasm of the maternal parent because the pollen, which contributes little or no cytoplasm to the zygote, had no apparent influence on the progeny phenotypes. Since leaf coloration involves the chloroplast, either genetic information in that organelle or in the cytoplasm, and influencing the chloroplast, could be responsible for this inheritance pattern.

Iojap in Maize

A phenotype similar to *Mirabilis* but with a different pattern of inheritance has been analyzed in maize by Marcus M. Rhoades. In this case, green, colorless, or green-and-colorless striped leaves are under the influence not only of the cytoplasm but also of a nuclear gene located on chromosome 7. This locus is called ***iojap***, after the recessive mutation *ij* located there. The wild-type allele is designated *Ij*. Plants homozygous for the mutation (*ij/ij*) may have green-and-white striped leaves. However, when reciprocal crosses are made between plants with striped leaves (*ij/ij*) and green leaves (*Ij/Ij*), the results are seen to vary, depending upon which parent is mutant (Figure 8.3). If the female is striped (*ij/ij*) and the male is green (*Ij/Ij*), plants with colorless, striped, and green leaves are observed as progeny. If the male parent is striped (*ij/ij*) and the female is green (*Ij/Ij*), only green plants are produced! In both types of crosses, all offspring have identical genotypes (*Ij/ij*). We may

TABLE 8.1 Crosses between Flowers from Various Branches of Variegated Four O'Clocks

	Location of Ovule		
Source of Pollen	White Branch	Green Branch	Variegated Branch
White branch	White	Green	White, green, or variegated
Green branch	White	Green	White, green, or variegated
Variegated branch	White	Green	White, green, or variegated

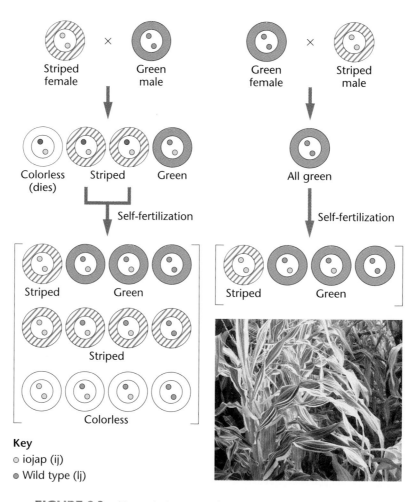

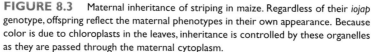

Key
○ iojap (ij)
● Wild type (lj)

FIGURE 8.3 Maternal inheritance of striping in maize. Regardless of their *iojap* genotype, offspring reflect the maternal phenotypes in their own appearance. Because color is due to chloroplasts in the leaves, inheritance is controlled by these organelles as they are passed through the maternal cytoplasm.

conclude that, although a nuclear gene seems to be involved, the inheritance pattern is strictly influenced maternally.

We can understand this pattern better by examining the offspring resulting from self-fertilization of heterozygous plants (Figure 8.3). The striped plant gives rise to progeny with colorless, striped, and green leaves, regardless of the genotype of the progeny. Green plants give rise to both green and striped progeny in a 3:1 ratio. Results of these self-fertilizations substantiate that the mutant chloroplasts are transmitted solely through the female cytoplasm, regardless of the plant's nuclear genotype.

Apparently, the nuclear genotype *ij/ij* somehow alters chloroplasts, which are then transmitted maternally. The colorless areas of the leaf are due to the lack of the green chlorophyll pigment in chloroplasts originally caused by the *ij* allele. Once acquired, chlorophyll-deficient (colorless) chloroplasts are transmitted through the maternal cytoplasm, establishing the phenotypes of leaves of progeny plants.

Chlamydomonas Mutations

The unicellular green alga *Chlamydomonas reinhardi* has provided an excellent system for the investigation of plastid and mitochondrial inheritance. The organism is eukaryotic and contains both a single large chloroplast and numerous mitochondria. Matings are followed by meiosis, and the various stages of the life cycle are easily studied in culture in the laboratory. The first cytoplasmic mutant, *streptomycin resistance* (*sr*), was reported in 1954 by Ruth Sager. Although *Chlamydomonas*' two mating types—*mt*⁺ and *mt*⁻—appear to make equal cytoplasmic contributions to the zygote, Sager determined that the *sr* phenotype is transmitted only through the *mt*⁺ parent.

Since Sager's discovery, a number of other *Chlamydomonas* mutations (including an acetate requirement as well as resistance to or dependence on a variety of bacterial antibiotics) have been discovered that show a similar uniparental inheritance pattern. These mutations have been linked to the transmission of the chloroplast, and their study has extended our knowledge of chloroplast inheritance.

Following fertilization, the single chloroplasts of the two mating types fuse. After the resulting zygote has undergone meiosis, it is apparent that the genetic information of the chloroplasts of progeny cells is derived only from the mt^+ parent. The genetic information present within the mt^- chloroplast has degenerated.

Furthermore, studies suggested that these chloroplast mutations, representing numerous gene sites, form a single circular linkage group—the first such nuclear unit to be established in a eukaryotic organism. All available evidence supports the hypothesis that this linkage group consists of DNA residing in the chloroplast.

Mitochondria: *poky* in *Neurospora*

Mitochondria, like chloroplasts, play a critical role in cellular bioenergetics, and they contain a distinctive genetic system. Mutants affecting mitochondrial function have been discovered and studied. As with chloroplast mutants, these are transmitted through the cytoplasm and result in extranuclear inheritance patterns. Furthermore, the mitochondrial genetic system has now been extensively characterized.

In 1952, Mary B. and Hershel K. Mitchell discovered a slow-growing mutant strain of the mold *Neurospora crassa* and called it *poky*. (It is also designated *mi-1* for *maternal inheritance*.) Studies have shown slow growth to be associated with impaired mitochondrial function specifically related to certain cytochromes essential to electron transport. Results of genetic crosses between wild-type and *poky* strains suggest that the trait is maternally inherited. If the female parent (the cytoplasmic-rich protoperithecium—see Figure 7.10) is *poky* and the male parent is wild type, all progeny colonies are *poky*. The reciprocal cross produces normal wild-type colonies.

Studies with *poky* mutants illustrate the use of **heterokaryon** formation during the investigation of maternal inheritance in fungi. Occasionally, hyphae from separate mycelia fuse with one another, giving rise to structures containing two or more nuclei in a common cytoplasm. If the hyphae contain nuclei of different genotypes, the structure is called a *heterokaryon*. The cytoplasm will contain mitochondria derived from both initial mycelia. A heterokaryon may give rise to haploid spores, or **conidia**, that produce new mycelia. The phenotypes of these structures may be determined.

Heterokaryons produced by the fusion of *poky* and wild-type hyphae initially show normal rates of growth and respiration. However, mycelia produced through conidia formation become progressively more abnormal until they show the *poky* phenotype. This occurs despite the presumed presence of both wild-type and *poky* mitochondria in the cytoplasm of the hyphae.

To explain the initial growth and respiration pattern, it is assumed that the wild-type mitochondria support the respiratory needs of the hyphae. The subsequent expression of the *poky* phenotype suggests that the presence of the *poky* mitochondria may somehow prevent or depress the function of these wild-type mitochondria. Perhaps the *poky* mitochondria replicate more rapidly and "wash out" or dilute wild-type mitochondria numerically. Another possibility is that *poky* mitochondria produce a substance that inactivates the wild-type organelle or interferes with the replication of its DNA (**mtDNA**). As a result of this type of interaction, *poky* is an example of a broader category referred to as **suppressive mutations**. This general phenomenon is characteristic of many other suspected mitochondrial mutations of *Neurospora* and yeast.

Petite in *Saccharomyces*

Another extensive study of mitochondrial mutations has been performed with the yeast *Saccharomyces cerevisiae*. The first such mutation, **petite**, was described by Boris Ephrussi and his co-workers in 1956. The mutant is so named because of the small size of the yeast colonies. Many independent *petite* mutations have since been discovered and studied, and they all have a common characteristic: deficiency in cellular respiration involving abnormal electron transport. Fortunately, this organism is a facultative anaerobe and can grow by fermenting glucose through glycolysis. Thus, although colonies are small, the organism may survive the loss of mitochondrial function by generating energy anaerobically.

The complex genetics of *petite* mutations is diagrammed in Figure 8.4. A small proportion of these mutants exhibit Mendelian inheritance and are called *segregational petites*, indicating that they are the result of nuclear mutations. The remainder demonstrates cytoplasmic transmission, producing one of two effects in matings. The *neutral petites*, when crossed to wild type, produce meiotic products (called ascospores) that give rise only to wild-type or normal colonies. The same pattern continues if progeny of this cross are backcrossed to *neutral petites*. This is because the majority of neutrals lack mtDNA completely or have lost a substantial portion of it. Thus, the wild-type cell is the effective source of normal mitochondria capable of reproduction.

A third type, the *suppressive petites*, behaves similarly to *poky* in *Neurospora*. Crosses between mutant and wild type give rise to mutant diploid zygotes, which, upon undergoing meiosis, immediately yield all mutant cells. Under these conditions, the *petite* mutation behaves "dominantly" and seems to suppress the function of the wild-type mitochondria. *Suppressive petites* also have dele-

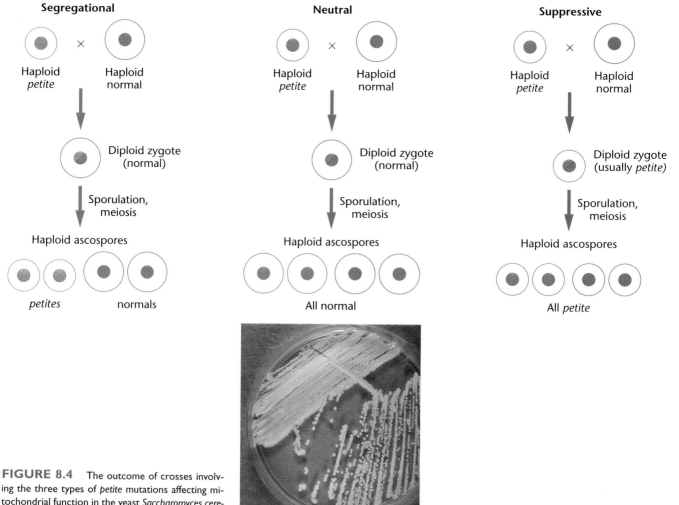

FIGURE 8.4 The outcome of crosses involving the three types of *petite* mutations affecting mitochondrial function in the yeast *Saccharomyces cerevisiae*. The photograph shows normal yeast colonies.

tions of mtDNA, but they are not nearly as extensive as deletions in the *neutral petites*.

Suppressiveness remains unexplained. Two major hypotheses have been advanced. One explanation suggests that the mutant (or deleted) mtDNA replicates more rapidly, and thus mutant mitochondria "take over" or dominate the phenotype by numbers alone. The second explanation suggests that recombination occurs between the mutant and wild-type mtDNA, introducing errors into or disrupting the normal mtDNA. It is not yet clear which one, if either, of these explanations is correct.

Mitochondrial DNA and Human Diseases

As we will see in Chapter 17, the DNA found in mitochondria (**mtDNA**) has been extensively characterized in a variety of organisms, including humans. Human mtDNA,

which is circular (see the photograph at the beginning of this chapter), contains 16,569 base pairs and is strictly inherited maternally. The mitochondrial gene products encoded include:

13 proteins, required for oxidative phosphorylation
22 transfer RNAs (tRNAs), required for translation
2 ribosomal RNAs (rRNAs), required for translation

Mitochondrial function is particularly vulnerable to mutations in mtDNA, since any such genetic alterations will potentially disrupt either the translation of all gene products *or* result in mutant proteins involved in oxidative phosphorylation. In the former case, diminution or complete loss of translation capability within the organelle would very likely have a lethal effect.

On the other hand, a zygote receives a large number of organelles through the egg, so if only one of them contains a mutation, its impact is severely diluted because there will be many more mitochondria that will function normally. During early development, cell division disperses the initial population of mitochondria present in the zygote, and in newly formed cells these organelles reproduce autonomously. Therefore, adults will exhibit cells with a variable mixture of both normal and abnormal organelles should a deleterious mutation arise or already be present in the initial population of organelles. This is a condition called **heteroplasmy**.

For a human disorder to be attributable to genetically altered mitochondria, several criteria must be met:

1. Inheritance must exhibit a maternal rather than a Mendelian pattern.
2. The disorder must reflect a deficiency in the bioenergetic function of the organelle.
3. A specific genetic mutation in one of the mitochondrial genes should be documented.

Thus far, several such cases are known that demonstrate these characteristics. For example, **myoclonic epilepsy and ragged red fiber disease (MERRF)** demonstrates a pattern of inheritance consistent with maternal inheritance. Only offspring of affected mothers inherit the disorder, whereas the offspring of affected fathers are all normal. Individuals with this rare disorder express deafness and dementia in addition to seizures. Both muscle fibers and mitochondria are abnormal in appearance. Such aberrant mitochondria, which appear empty and devoid of the internal cristae that characterize normal mitochondria (see inset photo), are evident in Figure 8.5. Upon analysis of mtDNA, a mutation has been found in one of the mitochondrial genes encoding a transfer RNA. This genetic alteration apparently interferes with translation within the organelle. Presumably, the deficiency in efficient mitochondrial function is related to the various manifestations of the disorder.

A second disorder, **Leber's hereditary optic neuropathy (LHON)**, also exhibits maternal inheritance as well as mtDNA lesions. The disorder is characterized by sudden bilateral blindness. The average age of vision loss is 27, but onset is quite variable. Four different mutations have been identified, all of which disrupt normal oxidative phosphorylation. Over 50 percent of cases are due to a mutation at a specific position in the mitochondrial gene encoding a subunit of NADH dehydrogenase. The amino acid arginine is converted to histidine. In large families, this mutation is transmitted to all maternal relatives. It is also of interest to note that in many instances of LHON, there is no family history. It appears that a significant number of cases may result from "new" mutations.

In a third disorder, **Kearns-Sayre syndrome (KSS)**, severely affected individuals lose their vision, undergo hearing loss, and display heart conditions. The genetic basis of KSS involves deletions at various positions within mtDNA. Many KSS patients are symptom-free as children but display progressive symptoms as adults. Analysis has revealed that the proportion of mtDNAs that reveal deletions increases as the severity of symptoms increases.

The study of hereditary, mitochondrial-based disorders provides insights into the importance and genetic basis of this organelle during normal development as well as the relationship between mitochondrial function and neuromuscular disorders. Furthermore, such study has suggested a hypothesis for aging based on the progressive accumulation of mtDNA mutations and the accompanying loss of mitochondrial function.

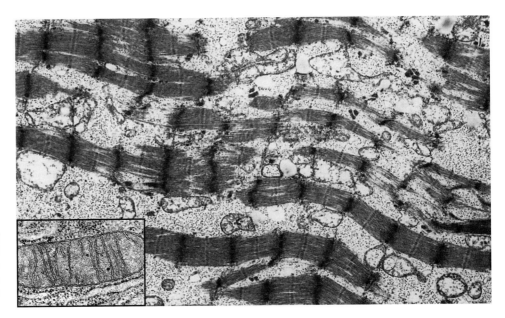

FIGURE 8.5 Electron micrograph of muscle tissue derived from an individual with MERFF (see text) that displays the abnormal mitochondria characterizing this human disorder. The photo inset shows a normal mitochondrion for comparison.

Infectious Heredity

Numerous examples abound of cytoplasmically transmitted phenotypes in eukaryotes that are due to an invading microorganism or particle. The foreign invader coexists in a symbiotic relationship, is usually passed through the maternal ooplasm to progeny cells or organisms, and confers a specific phenotype that may be studied. We shall consider several examples illustrating this phenomenon.

Kappa in *Paramecium*

First described by Tracy Sonneborn, certain strains of *Paramecium aurelia* are called **Killers** because they release a cytoplasmic substance called **paramecin** that is toxic and sometimes lethal to sensitive strains. This substance is produced by particles called **kappa** that replicate in the Killer cytoplasm. They contain DNA and protein, and they depend for their maintenance on a dominant nuclear gene *K*. One cell may contain 100 to 200 such particles. Paramecia are diploid protozoans that can undergo sexual exchange of genetic information through the process of **conjugation**. In some instances, cytoplasmic exchange also occurs. A variety of ways exist in which the *K* gene and kappa can be transmitted.

The genetic events occurring during conjugation in *Paramecium aurelia* are shown in Figure 8.6. Paramecia contain two diploid micronuclei. Early in conjugation,

both micronuclei in each mating pair undergo meiosis, resulting in eight haploid nuclei. However, seven of these degenerate, and the remaining one undergoes a single mitotic division. Each cell then donates one of the two haploid nuclei to the other, re-creating the diploid condition in both cells. As a result, exconjugates are of identical genotypes.

In a similar process involving only a single cell, **autogamy** occurs. Following meiosis of both micronuclei, seven products degenerate and one survives. This nucleus divides, and the resulting nuclei fuse to re-create the diploid condition. If the original cell was heterozygous, autogamy results in homozygosity because the newly formed diploid nucleus was derived solely from a single haploid meiotic product. In a population of cells originally heterozygous, half of the new cells express one allele and half express the other allele.

Figure 8.7 illustrates the results of crosses between *KK* and *kk* cells, without and with cytoplasmic exchange. When no cytoplasmic exchange occurs, even though the resultant cells may be *Kk* (or *KK* following autogamy), they remain sensitive if no kappa particles are transmitted. When exchange occurs, the cells become Killers provided the kappa particles are supported by at least one dominant *K* allele.

Kappa particles are bacterialike and may contain temperate bacteriophages. One theory holds that these viruses of kappa may become vegetative; during this mul-

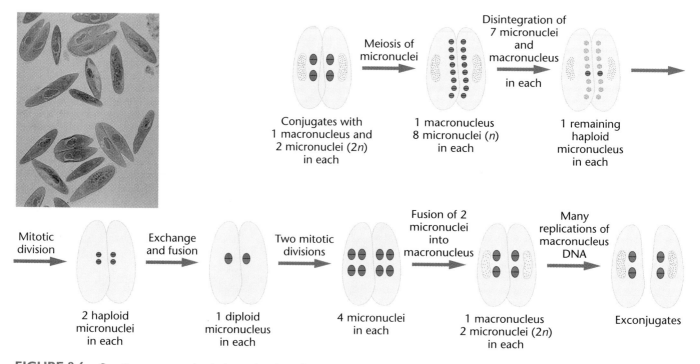

FIGURE 8.6 Genetic events occurring during conjugation in *Paramecium*. The photographic inset shows a pair of organisms undergoing conjugation.

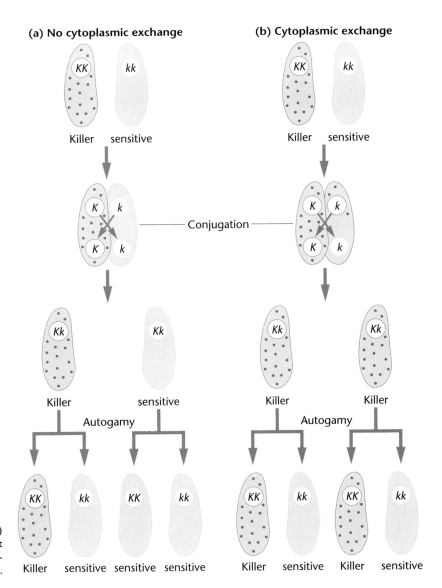

(a) No cytoplasmic exchange

(b) Cytoplasmic exchange

Killer sensitive

Killer sensitive

Conjugation

Killer sensitive Killer Killer

Autogamy Autogamy

Killer sensitive sensitive sensitive Killer sensitive Killer sensitive

FIGURE 8.7 Results of crosses between *Killer* (*KK*) and *sensitive* (*kk*) strains of *Paramecium*, with and without cytoplasmic exchange during conjugation. The kappa particles (dots) are maintained only when a *K* allele is present.

tiplication, they produce the toxic products that are released and that kill sensitive strains.

Infective Particles in *Drosophila*

Two examples of similar phenomena are known in *Drosophila*: **CO$_2$ sensitivity** and **sex-ratio**. In the former, flies that would normally recover from carbon dioxide anesthetization instead become permanently paralyzed and are killed by CO$_2$. Sensitive mothers pass this trait to all offspring. Furthermore, extracts of sensitive flies induce the trait when injected into resistant flies. Phillip L'Heritier has postulated that sensitivity is due to the presence of a virus, **sigma**. The particle has been visualized under the electron microscope and is smaller than kappa. Attempts to transfer the virus to other insects have been unsuccessful, demonstrating that specific genes support the presence of sigma in *Drosophila*.

A second example of infective particles comes from the study of *Drosophila bifasciata*. A small number of these flies were found to produce predominantly female offspring if reared at 21°C or lower. This condition, designated sex-ratio, was shown to be transmitted to daughters but not to the low percentage of males produced. This phenomenon was subsequently investigated in *Drosophila willistoni*. In these flies, the injection of ooplasm from sex-ratio females into normal females induced the condition. This observation suggests that an extrachromosomal element is responsible for the sex-ratio phenotype. The agent has now been isolated and shown to be a protozoan. While the protozoan has been found in both males and females, it is lethal primarily to developing male larvae. There is now some evidence that a virus harbored by the protozoan may be responsible for producing a male-lethal toxin.

CHAPTER SUMMARY

1. Patterns of inheritance sometimes vary from that expected of biparental transmission of nuclear genes. In such instances, phenotypes most often appear to result from genetic information transmitted through the egg.

2. Maternal effect patterns result when nuclear gene products controlled by the maternal genotype of the egg influence early development. *Ephestia* pigmentation and coiling in snails are examples.

3. Organelle heredity is based on the genotypes of chloroplast and mitochondrial DNA as these organelles are transmitted to offspring.

4. Chloroplast mutations affect the photosynthetic capabilities of plants, whereas mitochondrial mutations affect cells highly dependent on ATP generated through cellular respiration. The resulting mutants display phenotypes related to the loss of function of these organelles.

5. Another form of nuclear inheritance is attributable to the transmission of infectious microorganisms. Kappa particles and CO_2-sensitivity and sex-ratio determinants are examples.

KEY TERMS

autogamy, 216
bicoid (bcd) gene, 210
CO_2 sensitivity, 217
conidia, 213
conjugation, 216
cytoplasmic inheritance, 211
extranuclear inheritance, 208
heterokaryon, 213
heteroplasmy, 215
iojap locus (*ij*), 211

kappa, 216
Kearns-Sayre syndrome (KSS), 215
Killer strains, 216
Leber's hereditary optic neuropathy (LHON), 215
maternal effect, 209
mtDNA, 213
myoclonic epilepsy and ragged red fiber disease (MERRF). 215
neutral petites, 213

organelle heredity, 211
paramecin, 216
petite mutation, 213
poky mutation, 213
segregational petites, 213
sex-ratio, 217
sigma virus, 217
suppressive mutations, 213
suppressive petites, 213

INSIGHTS AND SOLUTIONS

1. Analyze the following hypothetical pedigree and determine the most consistent interpretation of how the trait is inherited and any inconsistencies.

Solution: The trait is passed from all male parents to all but one offspring but *never* passed maternally. Individual IV-7 (a female) is the only exception.

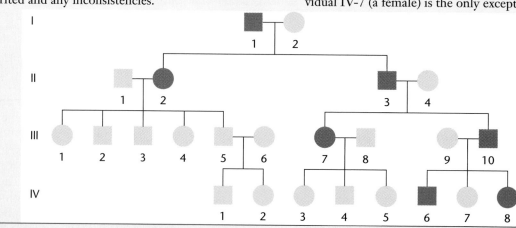

2. Can the explanation above be attributed to a gene on the Y chromosome? Defend your answer.

Solution: No, because male parents pass the trait to their daughters as well as to their sons.

3. Is the above case an example of a paternal effect or of paternal inheritance?

Solution: It has all the earmarks of paternal inheritance because males pass the trait to almost all of their offspring. To assess whether the trait is due to a paternal effect (resulting from a nuclear gene in the male gamete), analysis of further matings would be needed.

PROBLEMS AND DISCUSSION QUESTIONS

1. What genetic criteria distinguish a case of extrachromosomal inheritance from a case of Mendelian autosomal inheritance? From a case of X-linked inheritance?

2. In *Limnaea*, what results would be expected in a cross between a *Dd* dextrally coiled and a *Dd* sinistrally coiled snail, assuming cross-fertilization occurs as shown in Figure 8.2? What results would occur if the *Dd* dextral produced only eggs and the *Dd* sinistral produced only sperm?

3. Streptomycin resistance in *Chlamydomonas* may result from a mutation in a chloroplast gene or in a nuclear gene. What phenotypic results would occur in a cross between a member of an mt^+ strain resistant in both genes and a member of an mt^- strain sensitive to the antibiotic? What results would occur in the reciprocal cross?

4. A plant may have green, white, or green and white (variegated) leaves on its branches owing to a mutation in the chloroplast (which produces the white leaves). Predict the results of the following crosses.

	Ovule Source		Pollen Source
(a)	Green branch	×	White branch
(b)	White branch	×	Green branch
(c)	Variegated branch	×	Green branch
(d)	Green branch	×	Variegated branch

5. In an aerobically cultured yeast culture, a *petite* mutant is isolated. To determine the type of mutation causing this phenotype, a cross is made between the *petite* and wild-type strain. Shown below are three potential outcomes of such a cross. For each set of results, what conclusion about the type of *petite* mutation is justified?
   (a) all wild type
   (b) 1/2 *petite* : 1/2 wild type
   (c) all *petite*

6. In diploid yeast strains, sporulation and subsequent meiosis can produce haploid ascospores. These may fuse to reestablish diploid cells. When ascospores from a *segregational petite* strain fuse with those of a normal wild-type strain, the diploid zygotes are all normal. However, following meiosis, ascospores are 1/2 *petite* and 1/2 normal. Is the *segregational petite* phenotype inherited as a dominant or recessive trait?

7. Predict the results of a cross between ascospores from a *segregational petite* strain and a *neutral petite* strain. Indicate the phenotype of the zygote and the ascospores it may subsequently produce.

8. Described below are the results of three crosses between strains of *Paramecium*. Determine the genotypes of the parental strains.

(a)	Killer × sensitive ⟶ 1/2 Killer : 1/2 sensitive
(b)	Killer × sensitive ⟶ all Killer
(c)	Killer × sensitive ⟶ 3/4 Killer : 1/4 sensitive

9. *Chlamydomonas*, a eukaryotic green alga, is sensitive to the antibiotic erythromycin that inhibits protein synthesis in prokaryotes.

   (a) Explain why.
   (b) There are two mating types in this alga, mt^+ and mt^-. If an mt^+ cell sensitive to the antibiotic is crossed with an mt^- cell that is resistant, all progeny cells are sensitive. The reciprocal cross (mt^+ resistant and mt^- sensitive) yields all resistant progeny cells. Assuming that the mutation for resistance is in chloroplast DNA, what can be concluded?

10. In *Limnaea*, a cross where the snail contributing the eggs was dextral but of unknown genotype mated with another snail of unknown genotype and phenotype. All F_1 offspring exhibited dextral coiling. Ten of the F_1 snails were allowed to undergo self-fertilization. One-half produced only dextrally coiled offspring, whereas the other half produced only sinistrally coiled offspring. What were the genotypes of the original parents?

11. In *Drosophila subobscura*, the presence of a recessive gene called *grandchildless* (*gs*) causes the offspring of homozygous females, but not homozygous males, to be sterile. Can you offer an explanation as to why females but not males are affected by the mutant gene?

12. A male mouse from a true-breeding strain of hyperactive animals is crossed to a female mouse from a true-breeding strain of lethargic animals (these are hypothetical strains). All the progeny are lethargic. In the F_2 generation, all offspring are lethargic. What is the best genetic explanation for these observations? Propose a cross to test your explanation.

EXTRA-SPICY PROBLEMS

13. The specification of the anterior-posterior axis in *Drosophila* embryos is initially controlled by various gene products that are synthesized and loaded into the mature egg during oogenesis. Mutations in these genes result in abnormalities of the axis during embryogenesis. In the context of the current chapter, these mutations illustrate the concept of *maternal effect*. How do such mutations vary from those involved in organelle heredity that illustrate *cytoplasmic inheritance*? Devise a set of parallel crosses and expected outcomes involving mutant genes that contrast maternal effect and cytoplasmic inheritance.

14. A maternal-effect mutation (see Problem 13), *bicoid* (*bcd*), is recessive. In the absence of the bicoid protein product, embryogenesis is not completed. Consider a cross between a female heterozygous for bicoid (bcd^+/bcd^-) and a male homozygous for the mutation (bcd^-/bcd^-).
    (a) How is it possible for a male homozygous for the mutation to exist?
    (b) Predict the outcome (normal vs. failed embryogenesis) in the F_1 and F_2 generations of the cross described above.

15. Shown below are two pedigrees for a hypothetical human disorder. Analyze the pedigrees and propose a genetic explanation for the nature of the disorder. Consistent with your explanation, predict the outcome of a mating between individuals II-4 and II-5.

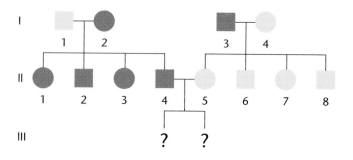

16. In the late 1950s, Yuichiro Hiraizumi, a postdoctoral fellow at the University of Wisconsin, crossed a laboratory strain of *Drosophila melanogaster* that was homozygous for the recessive second chromosomal mutations *cinnabar* (*cn*) and *brown* (*bw*) with a wild-type strain collected in Madison, WI. The lab strain had white eyes, while the wild strain had red eyes and was subsequently deisgnated *SD*. The resulting progeny, being heterozygous for *cn* and *bw*, had red eyes.

When Hiraizumi backcrossed F_1 females with *cn bw/cn bw* males, 50 percent of the offspring, as expected, had white eyes. However, when the reciprocal backcross was made (F_1 males with *cn bw/cn bw* females, less than 2 percent of the flies had white eyes.

(a) Propose an explanation that is consistent with these results.
(b) Design a genetic experiment to test your hypothesis.
(c) *SD* stands for Segregation Distortion. What is the significance of this description?
[*Reference: Genetics* 44:232–50 (1959).]

SELECTED READINGS

BOGORAD, L. 1981. Chloroplasts. *J. Cell Biol.* 91:256s–70s.

COHEN, S. 1973. Mitochondria and chloroplasts revisited. *Am. Sci.* 61:437–45.

FREEMAN, G., and LUNDELIUS, J. W. 1982. The developmental genetics of dextrality and sinistrality in the gastropod *Lymnaea peregra*. *Wilhelm Roux Arch.* 191:69–83.

GILLHAM, N. W. 1978. *Organelle heredity.* New York: Raven Press.

GOODENOUGH, U., and LEVINE, R. P. 1970. The genetic activity of mitochondria and chloroplasts. *Sci. Am.* (Nov.) 223:22–29.

GRIVELL, L. A. 1983. Mitochondrial DNA. *Sci. Am.* (March) 248:78–89.

LANDER, E. S., et al. 1990. Mitochondrial diseases: Gene mapping and gene therapy. *Cell* 61:925–26.

LEVINE, R. P., and GOODENOUGH, U. 1970. The genetics of photosynthesis and of the chloroplast in *Chlamydomonas reinhardi*. *Annu. Rev. Genet.* 4:397–408.

MARGULIS, L. 1970. *Origin of eukaryotic cells.* New Haven, CT: Yale University Press.

MITCHELL, M. B., and MITCHELL, H. K. 1952. A case of maternal inheritance in *Neurospora crassa*. *Proc. Natl. Acad. Sci. USA* 38:442–49.

NÜSSLEIN-VOLHARD, C. 1996. Gradients that organize embryo development. *Sci. Am.* (Aug.) 275:54–61.

PREER, J. R. 1971. Extrachromosomal inheritance: Hereditary symbionts, mitochondria, chloroplasts. *Annu. Rev. Genet.* 5:361–406.

ROSING, H. S., et al. 1985. Maternally inherited mitochondrial myopathy and myoclonic epilepsy. *Ann. Neurol.* 17:228–37.

SAGER, R. 1965. Genes outside the chromosomes. *Sci. Am.* (Jan.) 212:70–79.

———. 1985. Chloroplast genetics. *BioEssays* 3:180–84.

SCHWARTZ, R. M., and DAYHOFF, M. O. 1978. Origins of prokaryotes, eukaryotes, mitochondria and chloroplasts. *Science* 199:395–403.

SLONIMSKI, P. 1982. *Mitochondrial genes.* Cold Spring Harbor, NY: Cold Spring Harbor Laboratory Press.

SONNEBORN, T. M. 1959. Kappa and related particles in *Paramecium*. *Adv. Virus Res.* 6:229–56.

STRATHERN, J. N., et al., eds. 1982. *The molecular biology of the yeast* Saccharomyces: *Life cycle and inheritance.* Cold Spring Harbor, NY: Cold Spring Harbor Laboratory Press.

STURTEVANT, A. H. 1923. Inheritance of the direction of coiling in *Limnaea*. *Science* 58:269–70.

TZAGOLOFF, A. 1982. *Mitochondria.* New York: Plenum Press.

WALLACE, D. C. 1992. Mitochondrial genetics: A paradigm for aging and degenerative diseases. *Science* 256:628–32.

WALLACE, D. C., et al. 1988. Familial mitochondrial encephalomyopathy (MERRF): Genetic, pathophysiological and biochemical characterization of a mitochondrial DNA disease. *Cell* 55:601–10.

9

Chromosome Variation and Sex Determination

CHAPTER CONCEPTS

Genetic information of a diploid organism is delicately balanced in both content and location within the genome. As a result of studying variation in the sex chromosomes, valuable insights have been gained into the mode of sex determination in many organisms. Other studies have revealed that a change in chromosome number or in the arrangement of a chromosome region often results in phenotypic variation or disruption of development of an organism. Because the chromosome is the unit of transmission in meiosis, such variations are passed to offspring in a predictable manner, resulting in many informative genetic situations.

Up to this point in the text, we have emphasized how mutations and the resulting alleles affect an organism's phenotype, and how traits are passed from parents to offspring according to Mendelian principles. In this chapter, we shall look at phenotypic variation occurring as a result of changes in the genetic material that are more substantial than alterations of individual genes. These involve modifications at the level of the chromosome.

Although most members of diploid species normally contain precisely two haploid chromosome

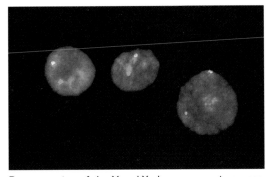

Demonstration of the X and Y chromosomes in a mammalian fetal cell using fluorescent *in situ* hybridization (FISH).

sets, many cases are known in which some variation from this pattern occurs. Modifications include variations in the number of individual chromosomes as well as rearrangements of the genetic material either within or among chromosomes. Taken together, such changes are called **chromosome mutations** or **chromosome aberrations**, to distinguish such genetic alterations from gene mutations. Because the chromosome is the unit transmitted according to Mendelian laws, chromosome aberrations are transmitted to offspring in a predictable manner, resulting in many interesting examples of heritable phenotypic variation.

In this chapter, we shall consider the many types of chromosomal aberrations, the consequences of them for organisms, and what we can learn from these situations. We will also discuss their role in the evolutionary process. We will see that the genetic component of a diploid organism is delicately balanced in content and location within the genome. Even minor alterations of content or location may result in some form of phenotypic variation, while more substantial changes may be lethal, particularly in animals.

We begin this chapter with the analysis of sex chromosome variation in humans and *Drosophila*. These studies have led to an understanding of how sex is determined in these organisms. We shall also pursue several other topics related to the genetic function of sex chromosomes.

Variation in Chromosome Number: An Overview

Before embarking on a discussion of variations involving the number of chromosomes in organisms and sex determination, it is useful to establish the terminology used that describes such changes. Variation in chromosome number ranges from the addition or loss of one or more chromosomes to the addition of one or more haploid sets of chromosomes. When an organism gains or loses one or more chromosomes, but not a complete set, the condition of **aneuploidy** is created. The loss of a single chromosome creates a condition called **monosomy**. The gain of one chromosome to an otherwise diploid genome results in **trisomy**. These are contrasted with the condition of **euploidy**, where complete haploid sets of chromosomes are present. If there are three or more sets, the more general term **polyploidy** is applicable. Those with three sets are specifically **triploid**; those with four sets are **tetraploid**, and so on. Table 9.1 provides a useful organizational framework for you to follow as we discuss each of these categories and the subsets within them.

The Diploid Chromosome Number in Humans

From the time when dividing cells of humans were first observed, geneticists tried to accurately determine the chromosome number of our own species. The first significant attempt was made in 1912, when H. von Winiwarter counted 47 chromosomes in a spermatogonial metaphase preparation. At that time, geneticists believed that the sex-determining mechanism in humans was

TABLE 9.1 Terminology for Variation in Chromosome Numbers

Term	Explanation
Aneuploidy	2n plus or minus chromosomes
Monosomy	$2n - 1$
Trisomy	$2n + 1$
Tetrasomy, pentasomy, etc.	$2n + 2, 2n + 3$, etc.
Euploidy	Multiples of n
Diploidy	2n
Polyploidy	3n, 4n, 5n, ...
Triploidy	3n
Tetraploidy, pentaploidy, etc.	4n, 5n, etc.
Autopolyploidy	Multiples of the same genome
Allopolyploidy (Amphidiploidy)	Multiples of different genomes

based on the presence of an extra chromosome in females; that is, females were thought to have 48 chromosomes. In the 1920s, Theophilus Painter observed between 45 and 48 chromosomes in cells of testicular tissue and also discovered the small Y chromosome, which is now known to occur only in males. In his original paper, Painter favored 46 as the diploid number in humans, but he later concluded incorrectly that 48 was the chromosome number in both males and females.

For 30 years, this number was accepted. Then, in 1956, Joe Hin Tjio and Albert Levan introduced improvements in chromosome preparation techniques. These improved techniques led to a strikingly clear demonstration of metaphase stages showing that 46 was indeed the human diploid number. Later that same year, C. E. Ford and John L. Hamerton, also working with testicular tissue, confirmed this finding.

Within the normal 23 pairs of human chromosomes, one pair was shown to vary in configuration in males and females. These two chromosomes were designated the X and Y sex chromosomes. The human female has two X chromosomes, and the human male has one X and one Y chromosome. As we shall see, however, these observations are insufficient to conclude that the Y chromosome determines maleness.

Chromosomes, Sex Differentiation, and Sex Determination in Humans

In our discussion of X-linkage in Chapter 4, we pointed out that, as part of the diploid chromosome composition of both humans and *Drosophila*, females possess two X chromosomes while males possess one X and one Y chromosome. This observation might lead us to conclude that the Y chromosome causes maleness in both species; however, this is not necessarily the case. Perhaps the lack of a second X chromosome somehow causes maleness while the Y plays no part whatsoever in sex determination. Perhaps the presence of two X chromosomes causes femaleness while the Y plays no role. The evidence that clarified which explanation was correct awaited the study of variations in the sex chromosome composition of both humans and flies. As we shall see, the first explanation, where the Y determines maleness, is valid in humans but invalid in *Drosophila*.

Klinefelter and Turner Syndromes

About 1940 it was observed that two human abnormalities, the **Klinefelter** and **Turner syndromes**,* are characterized by aberrant sexual development. Individuals with Klinefelter syndrome have genitalia and internal ducts that are usually male, but their testes are underdeveloped and fail to produce sperm. Although masculine development occurs, feminine sexual development is not entirely suppressed. Slight enlargement of the breasts is common, for example. This ambiguous sexual development, referred to as **intersexuality**, may lead to abnormal social development.

In Turner syndrome, the affected individual has female external genitalia and internal ducts, but the ovaries are rudimentary. Other characteristic abnormalities include short stature (usually under five feet); a webbed neck; and a broad, shieldlike chest.

In 1959, the karyotypes of individuals with these syndromes were independently determined to be abnormal with respect to the sex chromosomes. Individuals with Klinefelter syndrome most often are trisomic and have an XXY complement in addition to 44 autosomes [Figure 9.1(a)]. People with this karyotype are designated **47,XXY**. Individuals with Turner syndrome are most often monosomic and have only 45 chromosomes, including just a single X chromosome. They are designated **45,X** [Figure 9.1(b)]. Note the convention used in designating the above chromosome compositions. The number indicates how many chromosomes are present, and the information after the comma designates the deviation from the normal diploid content. Both conditions result from nondisjunction of the X chromosomes during meiosis (see Figure 2.14).

These karyotypes and their corresponding sexual phenotypes allow us to conclude that the Y chromosome determines maleness in humans. In its absence, the sex of the individual is female, even if only a single X chromo-

some is present. The presence of the Y chromosome in the individual with Klinefelter syndrome is sufficient to determine maleness, even though its expression is not complete. Similarly, in the absence of a Y chromosome, as in the case of individuals with Turner syndrome, no masculinization occurs.

Klinefelter syndrome occurs in about 2 of every 1000 male births. The karyotypes **48,XXXY**, **48,XXYY**, **49,XXXXY**, and **49,XXXYY** are similar phenotypically to **47,XXY**, but manifestations are often more severe in individuals with a greater number of X chromosomes.

Karyotypes other than 45,X also lead to Turner syndrome. These include individuals who display two different genetic cell lines in their tissues, each exhibiting a different karyotype. Such individuals are called **mosaics**. The two cell lines result from a mitotic error during early development, the most common chromosome combinations being **45,X/46,XY** and **45,X/46,XX**. Thus, an embryo that began life with a normal karyotype can give rise to an individual whose cells show a mixture of karyotypes and who expresses this syndrome.

Turner syndrome is observed in only about 1 in 3000 female births, a frequency much lower than that for Klinefelter syndrome. One explanation for this difference is the observation that a substantial majority of **45,X** fetuses die *in utero* and are aborted spontaneously. Thus, a similar frequency of the two syndromes may occur at conception.

47,XXX Syndrome

The presence of three X chromosomes along with a normal set of autosomes **(47,XXX)** results in female differentiation. This syndrome, which is estimated to occur in about 1 of 1200 female births, is highly variable in expression. Frequently, 47,XXX women are perfectly normal. In other cases, underdeveloped secondary sex characteristics, sterility, and mental retardation may occur. In rare instances, **48,XXXX** and **49,XXXXX** karyotypes have been reported. The syndromes associated with these karyotypes are similar to, but more pronounced than, the 47,XXX. Thus in many cases, the presence of additional X chromosomes appears to disrupt the delicate balance of genetic information essential to normal female development.

47,XYY Condition

Another human trisomy involving the sex chromosomes, **47,XYY**, has been discovered and intensively investigated. Studies of this condition, where the only deviation from diploidy is the presence of an additional Y chromosome in an otherwise normal male karyotype, have led to an interesting controversy.

---

*Although the possessive form of the names of most syndromes (eponyms) is often used (e.g., Klinefelter's), the current preference is to use the nonpossessive form. We have adopted this convention for all human disorders and syndromes.

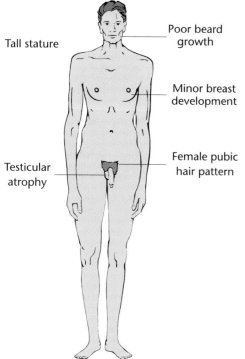

Tall stature

Poor beard growth

Minor breast development

Testicular atrophy

Female pubic hair pattern

(a) Klinefelter Syndrome (47,XXY)

Short stature

Webbed neck

Shield chest

Underdeveloped breasts and widely spaced nipples

Rudimentary ovaries

Brown nevi

(b) Turner Syndrome (45,X)

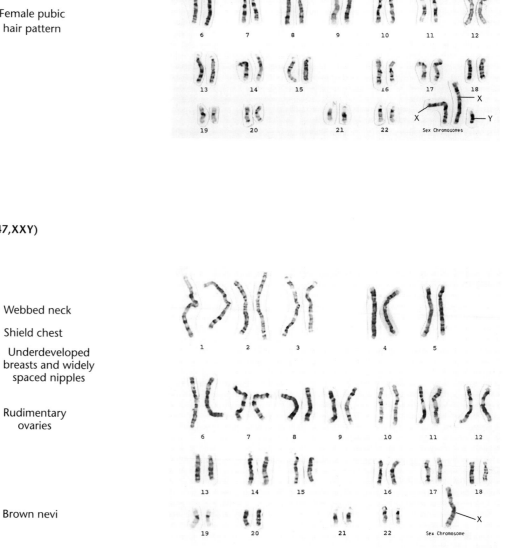

FIGURE 9.1   The karyotypes and phenotypic depictions of (a) Klinefelter syndrome (47,XXY) and (b) Turner syndrome (45,X). The sex chromosome compositions that vary from the normal human karyotype are labeled.

In 1965, Patricia Jacobs discovered 9 of 315 males in a Scottish maximum security prison to have the 47,XYY karyotype. These males were significantly above average in height and had been involved in criminal acts of serious social consequence. Of the nine males studied, seven were of subnormal intelligence, and all suffered personality disorders. In several other studies, similar findings were obtained.

Because of these investigations, the phenotype and frequency of the **47,XYY condition** in criminal and non-criminal populations have been examined more extensively (see Table 9.2). Above-average height (usually over six feet tall) and subnormal intelligence have been generally substantiated, and the frequency of males displaying this karyotype is indeed higher in penal and mental institutions compared with unincarcerated males.

The possible correlation between this chromosome composition and antisocial and criminal behavior has been of considerable interest. A particularly relevant question involves the characteristics displayed by XYY males who are not incarcerated. The only nearly constant association is that such individuals are over 6 feet tall!

A study that addressed this issue was initiated to identify 47,XYY individuals at birth and to follow their behavioral patterns during preadult and adult development. By 1974 the two primary investigators, Stanley Walzer and Park Gerald, had identified about 20 XYY newborns in 15,000 births at Boston Hospital for Women. However, they soon came under great pressure to abandon their research. Those opposed to the study argued that the investigation could not be justified and might cause great harm to those individuals who displayed this karyotype. They argued that (1) no association between the additional Y chromosome and abnormal behavior had been previously established in the population at large, and (2) by "labeling" these individuals, a self-fulfilling prophecy might be created. That is, as a result of participation in the study, parents, relatives, and friends might treat those identified as 47,XYY differently, perhaps leading to the type of antisocial behavior expected of them. Despite the support of a government funding agency and the faculty at Harvard Medical School, Walzer and Gerald abandoned the investigation in 1975.

Because it is now clear that many XYY males do not exhibit any form of antisocial behavior and lead normal lives, we must conclude that there is a high, but not constant correlation between the extra Y chromosome and behavior.

Sexual Differentiation in Humans

Once researchers had established that, in humans, it is the Y chromosome that houses genetic information necessary for maleness, efforts were made to pinpoint a specific gene or genes capable of providing the "signal" responsible for **sex determination**. Before pursuing a discussion of this topic, it is important to introduce the concept of **sexual differentiation** in order to better comprehend how adult male and female humans arise. During early development, every human embryo undergoes a period when it is potentially hermaphroditic or bisexual. Gonadal primordia arise as a pair of ridges associated with each embryonic kidney. Primordial germ cells migrate to these ridges, where an outer cortex and inner medulla form. The **cortex** is capable of developing into an ovary, while the inner **medulla** may develop into a testis. In addition, two sets of undifferentiated male (Wolffian) and female (Mullerian) ducts exist in each embryo.

The genital ridge is present after five weeks of gestation, and if these cells have the XY constitution, male development of the medullary region can be detected by the seventh week. However, in the absence of the Y chromosome, no male development is initiated, and the genital ridge is destined subsequently to form ovarian tissue. Parallel development of the appropriate male or female duct system then occurs, and the other duct system degenerates. A substantial amount of evidence indicates that in

TABLE 9.2 Frequency of XYY Individuals in Various Settings

Setting	Restriction	Number Studied	Number XYY	Frequency XYY
Control population	Newborns	28,366	29	0.10%
Mental–penal	No height restriction	4,239	82	1.93
Penal	No height restriction	5,805	26	0.44
Mental	No height restriction	2,562	8	0.31
Mental–penal	Height restriction	1,048	48	4.61
Penal	Height restriction	1,683	31	1.84
Mental	Height restriction	649	9	1.38

Source: Compiled from data presented in Hook, 1973, Tables 1–8. Copyright 1973 by the American Association for the Advancement of Science.

males, once testes differentiation is initiated, the embryonic testicular tissue secretes a hormone that is essential for continued male sexual differentiation.

In the absence of male development, the differentiation of the ovary and female duct system occurs. As the twelfth week of fetal development approaches, the oogonia within the ovaries begin meiosis and primary oocytes can be detected. By the twenty-fifth week of gestation, all oocytes become arrested in meiosis, where all will remain dormant until puberty is reached some 10 to 15 years later. Primary spermatocytes, on the other hand, are not formed in males until puberty is reached.

We can conclude that genetic information on the sex chromosomes is responsible for the primary sex determination event. Under normal conditions, once testicular or ovarian development has begun, subsequent early differentiation occurs under the influence of the male or female sex hormones. These hormones support, if not determine, expression of other secondary sexual characteristics during development.

The Y Chromosome and Male Development

Before turning to other types of chromosome variation and their effects in humans and other organisms, it is relevant to review information available concerning *how* the Y chromosome results in male development in humans. We have previously alluded to the fact that this chromosome, unlike the X, is nearly genetically blank. Although it shares only limited homology with loci on the X chromosome, it does carry genetic information that controls sexual development.

Therefore, some region of the Y chromosome exists, presumably a gene, that is responsible for encoding a product called the **testis-determining factor (TDF)**, a gene product that somehow triggers the undifferentiated gonadal tissue of the embryo to form testes. In its absence, female development occurs. Research has focused on just what constitutes the region and the product.

It is now clear that a small part of the human Y chromosome contains a gene called *SRY* **(sex-determining region Y)** that encodes TDF. Evidence proving that *SRY* is the responsible gene has relied on the molecular geneticist's ability to identify the presence or absence of DNA sequences in rare individuals whose expected sex chromosome composition does not correspond to their sexual phenotype. For example, there are human *males* who demonstrate two X and no Y chromosomes. They have attached to one of their Xs the region of the Y containing *SRY*. There are also *females* who have one X and one Y chromosome. Their Y is missing the *SRY* region. These observations argue strongly in favor of the role of *SRY* in male development.

The final proof that this gene is responsible for causing maleness involves an experiment using **transgenic mice**. Such animals are produced from fertilized eggs that have foreign DNA injected into them, which has been incorporated into the genetic composition of the developing embryo. In normal mice, a chromosome region designated *Sry* has been identified that is comparable to *SRY* in humans. When DNA containing only mouse *Sry* is injected into normal XX mouse eggs, most animals develop into males!

These studies suggest that the relevant gene has been identified. It is present in all mammals so far examined, having been conserved throughout evolution. How the product of this gene triggers the embryonic gonadal tissue to develop into testes rather than ovaries is now a reasonable question to ask and one that is amenable to investigation.

Sex Ratio in Humans

The presence of heteromorphic sex chromosomes in one sex of a species but not the other provides a potential mechanism for producing equal proportions of male and female offspring. The actual proportion of male to female offspring allows the calculation of what is called the **sex ratio**. It can be assessed in two ways. The **primary sex ratio** reflects the proportion of males and females *conceived* in a population. The **secondary sex ratio** reflects the proportion that are *born*. The secondary sex ratio is much easier to determine, but has the disadvantage of not accounting for disproportionate embryonic or fetal mortality, should it occur.

When the secondary sex ratio in the human population is determined using worldwide census data, it does not equal 1.0. For example, in the Caucasian population in the United States, the secondary ratio is 1.06, indicating that 106 males are born for each 100 females. In the black population in the United States, the ratio is 1.025. In other countries the excess of male births is even greater than reflected in these values. For example, in Korea, the secondary sex ratio is 1.15.

To account for the difference between the primary and secondary sex ratios, one possibility is that prenatal female mortality is greater than prenatal male mortality. If so, it is possible that the primary sex ratio is 1.0 and that it is altered before birth. However, this hypothesis has been shown to be false. In fact, just the opposite occurs. In a Carnegie Institute study, reported in 1948, the sex of approximately 6000 embryos and fetuses recovered from miscarriages and abortions was determined. On the basis of the data derived from this study, the primary sex ratio was estimated to be 1.079. More recent data have estimated that this figure is even higher—between 1.20 and 1.60. Therefore, many more males than females are conceived in the human population.

It is not clear why such a radical departure from the expected primary sex ratio of 1.0 occurs. A suitable explanation can be derived only from examining the assumptions upon which the theoretical ratio is based:

1. Because of segregation, males produce equal numbers of X- and Y-bearing sperm.
2. Each type of sperm has equivalent viability and motility in the female reproductive tract.
3. The egg surface is equally receptive to both X- and Y-bearing sperm.

There is no strong experimental evidence to suggest that any of these assumptions are invalid. However, it has been speculated that, because the human Y chromosome is smaller than the X chromosome, Y-bearing sperm are of less mass and therefore more motile. If this is true, then the probability of a fertilization event leading to a male zygote is increased, providing one possible explanation for the observed primary ratio.

Dosage Compensation in Humans

The presence of two X chromosomes in normal human females and only one X in normal human males is unique compared with the equal numbers of autosomes present in the cells of both sexes. On theoretical grounds alone, it is possible to speculate that this disparity should create a "genetic dosage" problem between males and females for all X-linked genes. Females have two copies and males only one. Therefore, there is the potential for producing twice as much of each gene product for all X-linked genes. The additional X chromosomes in both males and females exhibiting the various syndromes discussed earlier in this chapter should serve to compound this dosage problem. In this section, we will describe certain research findings regarding X-linked gene expression which demonstrate that a genetic mechanism underlying **dosage compensation** does indeed exist.

Barr Bodies

Murray L. Barr and Ewart G. Bertram's experiments with female cats, and Keith Moore and Barr's subsequent study with humans, demonstrate a genetic mechanism in mammals that compensates for X chromosome dosage disparities. Barr and Bertram observed a darkly staining body in interphase nerve cells of female cats. They found that this structure was absent in similar cells of males. In human females, this body can be easily demonstrated in cells derived from the buccal mucosa or in fibroblasts but not in

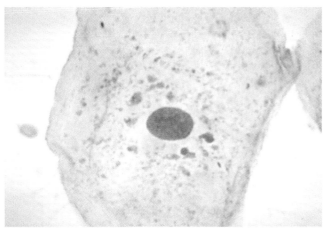
(a)

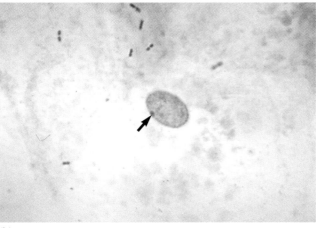
(b)

FIGURE 9.2 Photomicrographs comparing a cheek epithelial cell nucleus from a male that fails to reveal a Barr body (top) with a cheek cell from a female that demonstrates a Barr body (indicated by arrow in bottom image). This structure, also called a sex chromatin body, represents an inactivated X chromosome.

similar male cells (Figure 9.2). This highly condensed structure, about 1 μm in diameter, lies against the nuclear envelope of interphase cells. It stains positively in the Feulgen reaction for DNA.

Current experimental evidence demonstrates that this body, called a **sex chromatin body** or simply a **Barr body**, is an inactivated X chromosome. Susumo Ohno was the first to suggest that the Barr body arises from one of the two X chromosomes. This hypothesis is attractive because it provides a mechanism for dosage compensation. If one of the two X chromosomes is inactive in the cells of females, the dosage of genetic information that can be expressed in males and females is equivalent. Convincing but indirect evidence for this hypothesis comes from the study of the sex chromosome syndromes described earlier in this chapter. Regardless of how many X

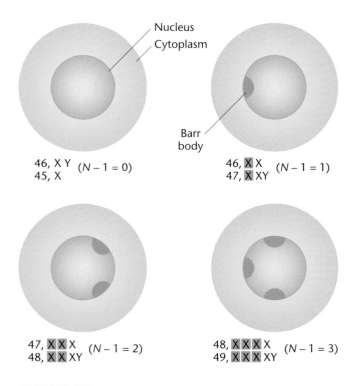

FIGURE 9.3 Diagrammatic representation of Barr body occurrence in various human karyotypes, where all X chromosomes except one ($N - 1$) are inactivated.

chromosomes exist, all but one of them appear to be inactivated and can be seen as Barr bodies. For example, none is seen in Turner 45,X females; one is seen in Klinefelter 47,XXY males; two in 47,XXX females; three in 48,XXXX females; and so on (Figure 9.3). Therefore, the number of Barr bodies follows an $N - 1$ rule, where N is the total number of X chromosomes present.

Although this mechanism of inactivation of all but one X chromosome increases our understanding of dosage compensation, it further complicates our perception of other matters. Because one of the two X chromosomes is inactivated in normal human females, why then is the Turner 45,X individual not entirely normal? Why aren't females with the triplo-X and tetra-X karyotypes (47,XXX and 48,XXXX) normal? Further, in Klinefelter syndrome (47,XXY), X chromosome inactivation effectively renders such individuals 46,XY. Why aren't they *unaffected* by the additional X chromosome in their nuclei?

One possible explanation is that chromosome inactivation does not normally occur in the very early stages of development of those cells destined to form gonadal tissues. Another possible explanation is that not all of each X chromosome forming a Barr body is inactivated. As a result, excessive expression of certain X-linked genes might still occur despite apparent inactivation of additional X chromosomes. Experimental support exists for both of these explanations.

The Lyon Hypothesis

In mammalian females one X chromosome is of maternal origin, and the other is of paternal origin. Which one is inactivated? Is the inactivation random? Is the same chromosome inactive in all somatic cells? In 1961, Mary Lyon and Liane Russell independently proposed a hypothesis that answers these questions. They postulated that the inactivation of X chromosomes occurs randomly in somatic cells at a point early in embryonic development, and that once inactivation has occurred, all progeny cells have the same X chromosome inactivated.

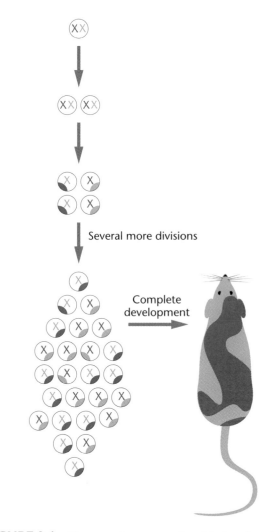

FIGURE 9.4 Diagrammatic representation of the Lyon hypothesis in a representative mammal. At an early stage in development, either the maternal or paternal X chromosome (shown in two different colors) is randomly inactivated in each cell, forming a Barr body. All progeny cells then inactivate the same X chromosomes as their progenitor cell, forming clones of like cells. In females heterozygous for the X-linked alleles, the adult is a mosaic for each such gene, with some groups of cells expressing one of the alleles and other groups expressing the other allele, as shown for coat color in the mouse. Although we have shown initial inactivation to occur at the four-cell stage, it actually occurs later during development.

FIGURE 9.5   A calico cat, where the random distribution of orange and black patches illustrates the Lyon hypothesis. The white patches are due to another gene, characterizing calicos from tortoiseshell cats.

This explanation, which has come to be called the **Lyon hypothesis**, was initially based on observations of female mice heterozygous for X-linked coat color genes (Figure 9.4). The pigmentation of these heterozygous females was mottled, with large patches of skin expressing the color allele on one X and other patches expressing the allele on the other X. Indeed, if different X chromosomes were inactive in adjacent patches of cells, such a phenotypic pattern would result. Similar mosaic patterns occur in the black and yellow-orange patches of female tortoiseshell and calico cats. Such X-linked coat color patterns do not occur in male cats because all cells contain the single maternal X chromosome, and are therefore hemizygous for only one X-linked coat color allele. A calico cat is shown in Figure 9.5.

The most direct evidence in support of the Lyon hypothesis comes from studies of gene expression in clones of human fibroblast cells. Individual cells may be isolated following biopsy and cultured *in vitro*. If each culture is derived from a single cell, it is referred to as a **clone**. The synthesis of the enzyme **glucose-6-phosphate dehydrogenase (*G6PD*)** is controlled by an X-linked gene. Numerous mutant alleles of this gene have been detected, and their gene products can be differentiated from the wild-type enzyme by their migration pattern in an electrophoretic field.

Fibroblasts have been taken from females heterozygous for different allelic forms of *G6PD* and studied. The Lyon hypothesis predicts the results of the examination of numerous clones derived from these female heterozygotes. If inactivation of an X chromosome occurs randomly early in development and is permanent in all progeny cells, such a female should show two types of clones, each showing only one electrophoretic form of *G6PD*, in approximately equal proportions.

In 1963, Ronald Davidson and colleagues performed an experiment involving 14 clones from a single heterozygous female. Seven showed only one form of the enzyme, and seven showed only the other form. What was most important was that none of the 14 showed both forms of the enzyme. Studies of *G6PD* mutants thus provide strong support for the random permanent inactivation of either the maternal or paternal X chromosome.

The Lyon hypothesis is generally accepted as valid; in fact, the inactivation of an X chromosome into a Barr body is sometimes referred to as *lyonization*. One extension of the hypothesis is that mammalian females are mosaics for *all heterozygous X-linked alleles*. Some areas of the body express only the maternally derived alleles, and others express only the paternally derived alleles. Two especially interesting examples involve **red-green color blindness** and **anhidrotic ectodermal dysplasia**, both X-linked recessive disorders. In the former case, hemizygous males are fully color-blind in all retinal cells. However, heterozygous females display mosaic retinas with patches of defective color perception and surrounding areas with normal color perception. Males hemizygous for anhidrotic ectodermal dysplasia show absence of teeth, sparse hair growth, and lack of sweat glands. The skin of females heterozygous for this disorder reveals patterns of tissue with and without sweat glands (Figure 9.6). In both examples, random inactivation of one or the other X chromosome early in the development of heterozygous females has led to these occurrences.

The Mechanism of Inactivation

The least understood aspect of the Lyon hypothesis is the mechanism of chromosome inactivation in mammals. How are most all genes of an entire chromosome inactivated? Recent investigations are beginning to clarify this issue. A single region of the human X chromosome, called the **X-inactivation center (*XIC*)**, is the major control unit. It is located on the proximal end of the p arm. Genetic expression of this region occurs *only* on the X chromosome that is inactivated. The constant association of expression and X chromosome inactivation sup-

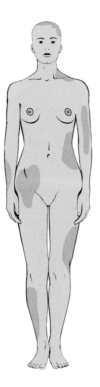

FIGURE 9.6 Depiction of the absence of sweat glands (shaded regions) in a female heterozygous for the X-linked condition ectodermal dysplasia. The locations vary randomly in such females based on the random pattern of X chromosome inactivation during early development, resulting in female mosaics.

port the conclusion that this region is an important genetic component in the process.

A gene, ***XIST*** (X-inactive specific transcript), is now believed to represent the critical locus within the *XIC*. A comparable region (*Xic*) and gene (*Xist*) exist in the mouse. Three interesting observations have been made regarding the RNA that is transcribed by the *Xist* gene. First, transcription of the genes is known to occur prior to chromosome inactivation. In the mouse, for example, the *Xist* gene becomes active about 24 hours before X-inactivation occurs in the embryo. Thus, one can argue that gene activation is more likely the cause rather than the consequence of inactivation. Second, the RNA that is made stays in the nucleus and associates only with the chromosome from which it was derived (i.e., it is *cis*-acting). This is in keeping with the possible role of initiating and/or maintaining inactivation.

The third observation is a most interesting one. The nucleotide sequence of the gene and its RNA product is quite large but lacks what is called an extended **open reading frame (ORF)**. An ORF includes the information necessary for translation of the RNA product into a protein. This observation suggests that the RNA is not translated, but rather the RNA itself serves a structural role in the nucleus, presumably in the mechanism of chromosome inactivation. This finding has led to speculation that the RNA products of *XIST* and *Xist* produce

some sort of molecular "cage" that entraps an X chromosome and inactivates it.

Recently, in 1996, a research group led by Graeme Penny provided the most convincing evidence available thus far that transcription of *Xist* is the critical event in chromosome inactivation. These researchers were able to introduce a targeted deletion (7 kb) into this gene that destroyed its activity and, as well, eliminated inactivation of the chromosome bearing this mutation. Several interesting questions remain unanswered. First, in cells with more than two chromosomes, what sort of "counting" mechanism exists that designates that all but one X chromosome are inactivated? Second, what "blocks" the *Xic* of the active chromosome, preventing transcription of *Xist*, thus allowing that X chromosome to remain active? Third, how is inactivation stably maintained in progeny cells? Whatever the answers to these questions, the discovery of *Xic* and *Xist* is exciting and pushes us closer to an understanding of how dosage compensation is accomplished in mammals.

As a final consideration of this topic, inactivation is sometimes said to be an **epigenetic event**, since the inactivation event is initially random, with either the maternal or paternal chromosome sharing an equal probability of being affected. Epigenetic events are those that involve gene expression during their occurrence, yet the final outcome does not depend on specific target sequences of DNA in order to be achieved. This description of epigenetic events applies to X chromosome inactivation because the chromosome that is inactivated does not contain unique DNA sequences specific to the *selection process* leading to inactivation. Nevertheless, the inactivation event involves gene expression (the *Xist* gene).

Chromosome Composition and Sex Determination in *Drosophila*

Because males and females in *Drosophila melanogaster* (and other *Drosophila* species) have the same general sex chromosome composition as humans, we might assume that the Y chromosome also causes maleness in these flies. However, the elegant work of Calvin Bridges in 1916 showed this not to be true. He studied flies with quite varied chromosome compositions, leading him to the conclusion that the Y chromosome is not involved in sex determination in this organism. Instead, Bridges proposed that both the X chromosomes and autosomes together play a critical role in sex determination.

His work can be divided into two phases: (1) a study of offspring resulting from nondisjunction of the X chromosomes during meiosis in females; and (2) subsequent work with progeny of triploid (3*n*) females. As we have seen previously, **nondisjunction** is the failure of paired chromosomes to segregate or separate during the anaphase stage of the first or second meiotic divisions. The result is the production of two abnormal gametes, one of which contains an extra chromosome ($n + 1$) and the other which lacks a chromosome ($n - 1$). Fertilization of such gametes with a haploid gamete produces ($2n + 1$) or ($2n - 1$) aneuploid zygotes. As in humans, if nondisjunction involves the X chromosome, in addition to the normal complement of autosomes, either an XXY or an X0 sex chromosome composition results. (The zero signifies that the second chromosome is absent.) Contrary to what was subsequently discovered in humans, Bridges found that the XXY flies were normal females, and the X0 flies were sterile males. The presence of the Y chromosome in the XXY flies did not cause maleness, and its absence in the X0 flies did not produce femaleness. From these data he concluded that the Y chromosome in *Drosophila* lacks male-determining factors, but apparently contains genetic information essential to male fertility as the X0 males were sterile.

Bridges was able to clarify the mode of sex determination in *Drosophila* by studying the progeny of triploid females ($3n$), which have *three* copies each of the haploid complement of chromosomes. These females apparently originate from rare diploid eggs fertilized by normal haploid sperm. Triploid females have heavy-set bodies, coarse bristles, and coarse eyes, and they may be fertile. Because of the odd number of each chromosome (three), during meiosis a wide range of chromosome complements is distributed into gametes that give rise to offspring with a variety of abnormal chromosome constitutions. A correlation among the sexual morphology, chromosome composition, and Bridges' interpretation is shown in Figure 9.7. *Drosophila* has a haploid number of four, thereby displaying three pairs of autosomes in addition to its pair of sex chromosomes.

Bridges realized that the critical factor in determining sex is the **ratio of X chromosomes to the number of haploid sets of autosomes present**. Normal (2X:2A) and triploid (3X:3A) females each have a ratio equal to 1.0, and both are fertile. As the ratio exceeds unity (3X:2A, or 1.5, for example), what was originally called a *superfemale* is produced. Because this female is rather weak and infertile and has lowered viability, this type is now more appropriately called a **metafemale**.

Normal (XY:2A) and sterile (X0:2A) males each have a ratio of 1:2, or 0.5. When the ratio decreases to 1:3, or 0.33, as in the case of an XY:3A male, infertile **metamales** result. Other flies recovered by Bridges in these studies contained an X:A ratio intermediate between 0.5 and 1.0. These flies were generally larger, and they exhibited a variety of morphological abnormalities and rudimentary bisexual gonads and genitalia. They were invariably sterile and were designated as **intersexes**.

These results indicate that in *Drosophila*, factors that cause a fly to develop into a male are not localized on the sex chromosomes, but are instead found on the autosomes. Some female-determining factors, however, are localized on the X chromosomes. Thus, with respect to primary sex determination, male gametes containing one of each autosome plus a Y chromosome result in male offspring *not* because of the presence of the Y chromosome but because of the lack of an X chromosome. This mode of sex determination is explained by the **genic balance theory**. Bridges proposed that a threshold for maleness is reached when the X:A ratio is 1:2 (X:2A), but that the presence of an additional X (XX:2A) alters this balance and results in female differentiation.

Bridges's conclusion received further support from the experiments of Jack Schultz and Theodosius Dobzhansky. They were able to obtain males, females, and intersexes that had gained or lost variously sized fragments of the X chromosome. Some of their results are shown in Table 9.3. The effects were proportional to the extent of the addition or subtraction of the X chromosome and were consistent with Bridges' genic balance interpretation of sex determination. This finding led to the conclusion that multiple factors critical to sex determination exist on the X chromosome.

Several mutant genes have been identified that are involved in sex determination in *Drosophila*. The recessive autosomal gene *transformer* (*tra*), discovered by Alfred H. Sturtevant, is especially interesting. Homozygous mutant females are transformed into sterile males, but males are unaffected when homozygous for *tra*. Experiments by Robert C. King and Dietrich Bodenstein showed that the transplantation of normal but immature female gonadal tissue into the abdomens of *tra/tra* sterile males results in normal ovarian development. Thus, it appears that the abdominal environment can support the production of eggs, but that the *transformer* gene has interfered with primary sex determination in these flies.

More recently, another gene, *Sex-lethal* (*Sxl*), has been shown to play a critical role in sex determination, serving as a "master switch" for the activation of a group of at least four other regulatory genes. Activation of the *Sxl* gene, located on the X chromosome, is essential to subsequent female development, and relies on a ratio of X chromosomes to sets of autosomes that equals 1.0. In the absence of activation, resulting from an X:A ratio of 0.5, male development occurs.

While it is not yet exactly clear how this ratio influences either the group of regulatory genes or the *Sxl* locus, some insights are available. The *Sxl* locus is part of a hierarchy of gene expression and exerts control over still other genes, including the *tra* gene (discussed above) and the *dsx* (*doublesex*) gene. The wild-type allele of *tra* is activated by the product of *Sxl* only in females, which in

Normal diploid male

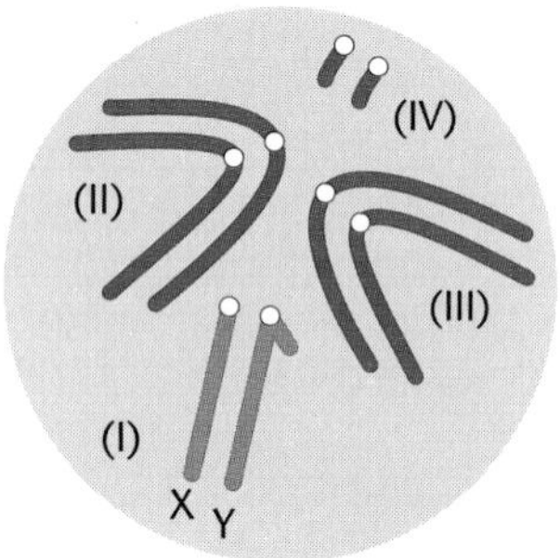

(IV)

(II)

(III)

(I)

X Y

2 sets of autosomes
+
X   Y

Chromosome composition	Chromosome formulation	Ratio of X chromosomes to autosome sets	Sexual morphology
	3X/2A	1.5	Metafemale
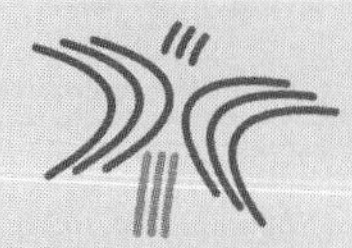	3X/3A	1.0	Female
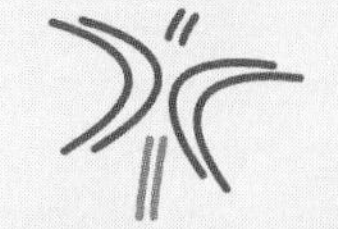	2X/2A	1.0	Female
	2X/3A	0.67	Intersex
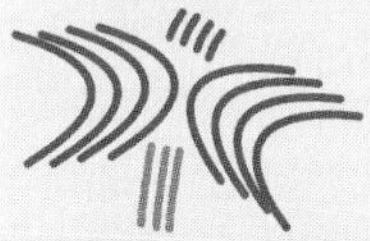	3X/4A	0.75	Intersex
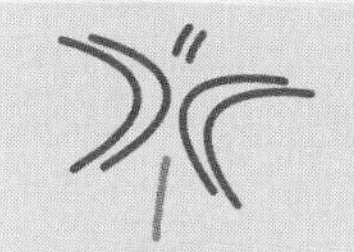	X/2A	0.50	Male
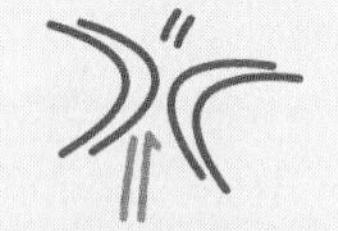	XY/2A	0.50	Male
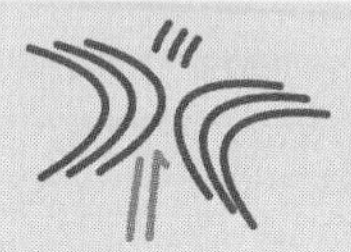	XY/3A	0.33	Metamale

FIGURE 9.7 Chromosome compositions, the ratios of X chromosomes to sets of autosomes, and the resultant sexual morphology in *Drosophila melanogaster*. The normal diploid male chromosome composition is shown as a reference on the left.

turn influences the expression of *dsx*. Depending on how the initial RNA transcript of *dsx* is processed, the resultant *dsx* protein activates either male- or female-specific genes required for sexual differentiation. Each step in this regulatory cascade requires a form of processing called **RNA splicing**, which will be discussed in more detail in Chapter 12. Portions of the RNA are removed and the remaining fragments "spliced" back together prior to translation. In the case of the *Sxl* gene, its transcript may be spliced in several different ways, a phenomenon called **alternative splicing**. Two different RNA transcripts are produced in females and males, respectively. In potential females, the transcript is active and initiates a cascade of regulatory gene expression, ultimately leading to female

differentiation. In potential males, the transcript is inactive, leading to different gene activity, whereby male differentiation occurs. Ultimately, different forms of the *dsx* protein are formed, accounting for sexual dimorphism.

Dosage Compensation in *Drosophila*

As is the case in mammals such as humans and mice, a dosage problem exists for X-linked genes in *Drosophila*. Females contain two copies whereas males contain only one copy of each gene. The mechanism of dosage compensation in *Drosophila* differs considerably from mammals, since no X chromosome inactivation is observed. Instead, male X-linked genes are transcribed at twice the

TABLE 9.3   The Effect of the Addition or Subtraction of Fragments of the *Drosophila* X Chromosome

Chromosome Composition	Sex	Addition or Subtraction	Effect
2X/3A	Intersex	+	Toward female
2X/2A	Female	+	Toward metafemale
2X/3A	Intersex	−	Toward male

level of the comparable genes in females. Interestingly, if groups of X-linked genes are moved (translocated) to autosomes, dosage compensation affects them even when they are no longer part of the X chromosome.

As in mammals, considerable gains have been made recently in understanding this phenomenon in *Drosophila*. At least four autosomal male-specific lethal genes are known to be involved in the process of dosage compensation. These are under the same master switch gene, *Sxl*, that controls sex determination. Mutations in any of the four autosomal genes reduce the increased expression of male X-linked genes and cause lethality in males, but not in females.

Evidence in support of a model for the mechanism of increased genetic activity is now available. The well-accepted model predicts that one of the autosomal genes, *mle* (*maleless*), encodes a protein that binds to numerous sites along the X chromosome, causing enhancement of genetic expression. The products of the other three autosomal genes also participate in and are required for *mle* binding.

The role of *Sxl* is also predicted in this model. In XX flies, *Sxl* turns on the *tra* gene, which leads ultimately to female differentiation, as we discussed above. When it is active, *Sxl* also functions to *inactivate* one or more of the male-specific autosomal genes, perhaps *mle*. In XY flies, *Sxl* is not expressed and the autosomal genes are *activated*, causing enhanced X chromosome activity. Although this model may yet be modified or refined, it is useful in guiding future research activity that will provide a more complete explanation.

The above discussion makes clear that an entirely different mechanism of dosage compensation exists in *Drosophila* (and probably many related organisms) compared to mammals. The development of elaborate mechanisms to equalize the expression of X-linked genes demonstrates the critical nature of gene expression. A delicate balance of gene products is necessary to maintain normal development of both males and females.

Drosophila Mosaics

The mode of sex determination and our knowledge of sex linkage in *Drosophila* (Chapter 4) both help to explain the appearance of the extremely unusual fruit fly shown in

Figure 9.8. This fly was recovered from a stock where all other females were heterozygous for the X-linked genes *white* eye (*w*) and *miniature* wing (*m*). It is a **bilateral gynandromorph**, which means that one-half of its body has developed as a male and the other half as a female. If a zygote heterozygous for *white* eye and *miniature* wing were to lose one of the X chromosomes during the first mitotic division, the two cells would be of the XX and X0 constitution, respectively. Thus, one cell would be female and the other would be male.

In the case of the bilateral gynandromorph, each of these cells is responsible for producing all progeny cells that make up either the right half *or* the left half of the body during embryogenesis. The original cell of X0 constitution apparently produced only identical progeny cells and gave rise to the left half of the fly, which was male. Because the male half demonstrated the *white*, *miniature* phenotype, the X chromosome bearing the w^+, m^+ alleles was lost, while the *w*, *m*-bearing homolog was retained. All female cells of the right side of the body remained heterozygous for both mutant genes and therefore contained normal eye–wing phenotypes. Depending on the orientation of the spindle during the first mitotic

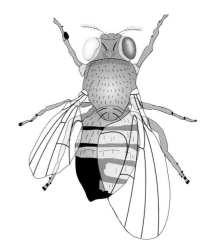

FIGURE 9.8   A bilateral gynandromorph of *Drosophila melanogaster* formed following the loss of one X chromosome in one of the two cells during the first mitotic division. The left side of the fly, composed of male cells containing a single X, expresses the mutant *white*-eye and *miniature*-wing alleles. The right side is composed of female cells containing two X chromosomes heterozygous for the two recessive alleles.

division, gynandromorphs can be produced where the "line" demarcating male versus female development occurs at almost any place along or across the fly's body.

Aneuploidy

We turn now to a consideration of variations in the number of autosomes and the genetic consequence of such changes. The most common examples of **aneuploidy**, where an organism has a chromosome number other than an exact multiple of the haploid set, are cases in which a single chromosome is either added to or lost from a normal diploid set. Such circumstances can arise as a result of primary or secondary nondisjunction (review Figure 2.14). The loss of one chromosome produces a $2n - 1$ complement and is called **monosomy**; the gain of one chromosome produces a $2n + 1$ complement and is described as **trisomy**. The $2n + 2$ and $2n + 3$ conditions are called **tetrasomy** and **pentasomy**, indicating that an individual has four or five copies of a particular chromosome in an otherwise diploid genome.

Monosomy

Although monosomy for the X chromosome occurs in humans, as we have seen in 45,X Turner syndrome, monosomy for any of the autosomes is not usually tolerated in humans or other animals. In *Drosophila*, flies monosomic for the very small chromosome 4—a condition referred to as **Haplo-IV**—survive, but they develop more slowly, exhibit a reduced body size, and have im-

paired viability. Monosomy for the larger chromosomes 2 and 3 is apparently lethal because such flies have never been recovered.

The failure of monosomic individuals to survive in many animal species is at first quite puzzling, since at least a single copy of every gene is present in the remaining homolog. However, if just one of those genes is represented by a lethal allele, the unpaired chromosome condition leads to the death of the organism. This occurs because monosomy creates hemizygosity for the involved chromosome and unmasks recessive lethals that are tolerated in heterozygotes carrying the corresponding wild-type alleles.

Another possible explanation is that the expression of genetic information during early development is carefully regulated such that a delicate equilibrium of gene products is required to ensure normal development. This requirement does not appear to be so stringent in the plant kingdom, where aneuploidy is tolerated. Monosomy for autosomal chromosomes has been observed in maize, tobacco, the evening primrose *Oenothera*, and the Jimson weed *Datura*, among many other plants. Nevertheless, such monosomic plants are usually less viable than their diploid derivatives. Haploid pollen grains, which undergo extensive development before participating in fertilization, are particularly sensitive to the lack of one chromosome and are seldom viable.

Partial Monosomy: Cri-du-Chat Syndrome

In humans, autosomal monosomy has not been reported beyond birth. Individuals with such chromosome complements are undoubtedly conceived, but none appar-

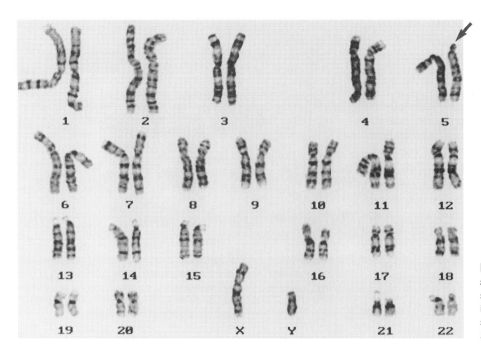
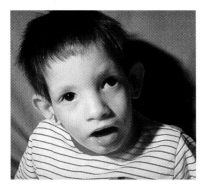

FIGURE 9.9 A representative karyotype and photograph of a child exhibiting cri-du-chat syndrome (46,5p−). In the karyotype, the arrow identifies the nearly complete absence of the short arm of one member of the chromosome 5 homologs.

ently survive embryonic and fetal development. There are, however, examples of survivors with **partial monosomy**, where only part of one chromosome is lost. These cases are also referred to as **segmental deletions**. One such case was first reported by Jérôme LeJeune in 1963 when he described the clinical symptoms of the **cri-du-chat syndrome** ("cry of the cat"). This syndrome is associated with the loss of about one-half of the short arm of chromosome 5 (Figure 9.9). Thus, the genetic constitution may be designated as **46,5p−**, meaning that such an individual has all 46 chromosomes but that some of the p arm (the petite or short arm) of chromosome 5 is missing.

Infants with this syndrome may exhibit anatomic malformations, including gastrointestinal and cardiac complications, and they are often mentally retarded. Abnormal development of the glottis and larynx is characteristic of individuals with this syndrome. As a result, the infant has a cry similar to that of the meowing of a cat, thus giving the syndrome its name.

Since 1963, hundreds of cases of cri-du-chat syndrome have been reported worldwide. An incidence of 1 in 50,000 live births has been estimated. The size of the deletion appears to influence the physical, psychomotor, and mental skill levels of these children. Although the effects of the syndrome are severe, many individuals achieve a level of social development in the trainable range. Those who receive home care and early special schooling are ambulatory, develop self-care skills, and learn to communicate verbally.

Trisomy

In general, the effects of trisomy parallel those of monosomy. However, the addition of an extra chromosome produces somewhat more viable individuals in both animal and plant species than does the loss of a chromosome.

As in monosomy, the sex chromosome variation of the trisomic type has a less dramatic effect on the phenotype than does autosomal variation. Recall from our previous discussion that *Drosophila* females with three X chromosomes and a normal complement of two sets of autosomes (3X:2A) survive and reproduce but are less viable than normal 2X:2A females. In humans, the addition of an extra X or Y chromosome to an otherwise normal male or female chromosome constitution (47,XXY, 47,XYY, and 47,XXX) leads to viable individuals exhibiting various syndromes. However, the addition of a large autosome to the diploid complement in both *Drosophila* and humans has severe effects and is usually lethal during development.

In plants, trisomic individuals are viable, but their phenotype may be altered. A classical example involves the Jimson weed *Datura*, whose diploid number is 24. Twelve different primary trisomic conditions are possible, and examples of each one have been recovered. Each

trisomy alters the phenotype of the capsule of the fruit sufficiently (Figure 9.10) to produce a unique phenotype.

In plants as well as animals, trisomy may be detected during cytological observations of meiotic divisions. Since three copies of one of the chromosomes are present, pairing configurations are irregular. At any particular region along the chromosome length, only two of the three homologs may synapse (Figure 9.11). At various re-

FIGURE 9.10   Drawings of capsule phenotypes of the fruits of *Datura stramonium*. In comparison with wild type, each phenotype is the result of trisomy of one of the 12 chromosomes characteristic of the haploid genome. The photograph illustrates the entire plant.

FIGURE 9.11   Diagrammatic representation of one possible pairing arrangement of three copies of a single chromosome, forming a trivalent configuration during meiosis I. Darker shaded areas are homologous with one another, as are lighter shaded areas.

gions, however, different members of the trio may be paired. When three copies of a chromosome are paired, the configuration is called a **trivalent**, which may be arranged on the spindle so that during anaphase one member moves to one pole, and two go to the opposite pole. In some cases, one bivalent and one univalent (an unpaired chromosome) may be present instead of a trivalent during the first metaphase stage. Meiosis thus produces gametes that can perpetuate the trisomic condition.

Down Syndrome

The only human autosomal trisomy in which a significant number of individuals survive longer than a year past birth was discovered in 1866 by Langdon Down. The condition is now known to result from trisomy of chromosome 21, one of the G group* (Figure 9.12), and is called **Down syndrome** or simply **trisomy 21** (designated **47,21+**). This trisomy is found in approximately 3 infants in every 2000 live births.

The external phenotype of these individuals is similar so that they bear a striking resemblance to one another. They display a prominent epicanthic fold in the corner of each eye and are characteristically short. They may have small, round heads; protruding, furrowed tongues, which cause the mouth to remain partially open; and short, broad hands with fingers showing characteristic palm and fingerprint patterns. Physical, psychomotor, and mental development is retarded, and the IQ is seldom above 70. Their life expectancy is shortened, and few individuals survive to be 50.

Down children are prone to respiratory disease and heart malformations and show an incidence of leukemia approximately 15 times higher than that of the normal population. However, careful medical scrutiny and treatment throughout their lives has extended their survival significantly. A striking observation is that death of older Down syndrome adults is frequently due to Alzheimer's disease.

One way in which this trisomic condition may originate is through nondisjunction of chromosome 21 during meiosis. Failure of paired homologs to disjoin during anaphase I or II can result in male or female gametes with the $n + 1$ chromosome composition. Following fertilization with a normal gamete, the trisomic condition is created. Chromosome analysis has shown that while the additional chromosome may be derived from either the mother or father, the ovum is most often the source.

Before the development of techniques that distinguish paternal from maternal homologs, this conclusion was supported by other indirect evidence derived from studies of the age of mothers giving birth to Down infants. Figure 9.13 shows the analysis of the distribution of maternal age and the incidence of Down syndrome newborns. The frequency of Down births increases dramatically as the age of the mother increases. Although the frequency is about 1 in 1000 at maternal age 30, a tenfold increase to a frequency of 1 in 100 is noted at age 40. The frequency increases still further to about 1 in 50 at age 45.

While the nondisjunctional event that produces Down syndrome seems more likely to occur during oogenesis in women between the ages of 35 and 45, we do not know with certainty why this is so. However, one observation may be relevant. In human females, all primary oocytes have been formed by birth. Therefore, once ovulation begins, each succeeding ovum has been arrested in meiosis for about a month longer than the one preceding it. As a result, women 30 or 40 years old produce ova that are significantly older and arrested longer than those they ovulated 10 or 20 years previously. However, it is not yet known whether ovum age is the cause of the increased incidence of nondisjunction leading to Down syndrome.

These statistics pose a serious problem for the woman who becomes pregnant late in her reproductive years. Genetic counseling early in such pregnancies serves two purposes. First, it informs the parents about the probability that their child will be affected and educates them about Down syndrome. Although some individuals with Down syndrome may be institutionalized, others benefit greatly from special-education programs and may be cared for at home. Further, they are noted as being affectionate, loving children. Second, a genetic counselor may recommend a prenatal diagnostic technique such as **amniocentesis** or **chorionic villus sampling (CVS)**. These techniques require the removal and culture of fetal cells. The karyotype of the fetus can then be determined by cytogenetic analysis. If the fetus is diagnosed as having Down syndrome, a therapeutic abortion is one option that parents may consider.

Because Down syndrome appears to be caused by a random error—nondisjunction of chromosome 21 during maternal or paternal meiosis—the occurrence of the

*On the basis of size and centromere placement, human autosomal chromosomes are divided into seven groups: A (1–3); B (4–5); C (6–12); D (13–15); E (16–18); F (19–20); and G (21–22).

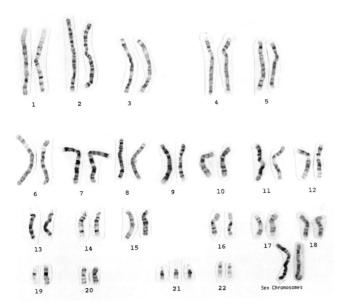

FIGURE 9.12 The karyotype and a photograph of a child with Down syndrome. In the karyotype, three members of the G-group chromosome 21 are present, creating the 47,21+ condition.

disorder is not expected to run in families. Nevertheless, in rare cases, it does so. These instances, referred to as **familial Down syndrome**, involve a **translocation** of chromosome 21, another type of chromosomal aberration, which we will discuss later in this chapter.

Patau Syndrome

In 1960, Klaus Patau and his associates observed an infant with severe developmental malformations with a karyotype of 47 chromosomes (Figure 9.14). The additional

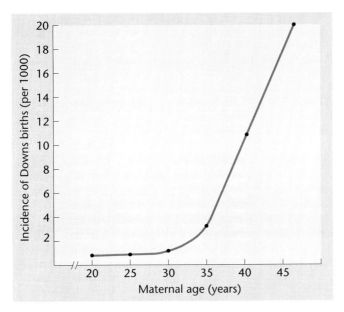

FIGURE 9.13 Incidence of Down syndrome births contrasted with maternal age.

chromosome was medium-sized, one of the acrocentric D group. It is now designated as chromosome 13. The trisomy 13 condition has since been described in many newborns and is called **Patau syndrome (47,13+)**. Affected infants are not mentally alert, are thought to be deaf, and characteristically have a harelip, cleft palate, and demonstrate polydactyly. Autopsy has revealed congenital malformation of most organ systems, a condition indicative of abnormal developmental events occurring as early as five to six weeks of gestation. The average survival of these infants is less than six months and a large majority are males.

The average maternal and paternal ages of parents of affected infants are higher than the ages of parents of normal children, but they are not as high as the average maternal age in cases of Down syndrome. Both male and female parents average about 32 years of age when the affected child is born. Because the condition is so rare, occurring as infrequently as 1 in 20,000 live births, it is not known whether the origin of the extra chromosome is more often maternal or paternal, or whether it arises equally from either parent.

Edwards Syndrome

In 1960, John H. Edwards and his colleagues reported on an infant trisomic for a chromosome in the E group, now known to be chromosome 18 (Figure 9.15). This aberration has been named **Edwards syndrome (47,18+)**. These infants are smaller than the average newborn. Their skulls are elongated in an anterior–posterior direction, and their ears are set low and malformed. A webbed neck, congenital dislocation of the hips, and a receding

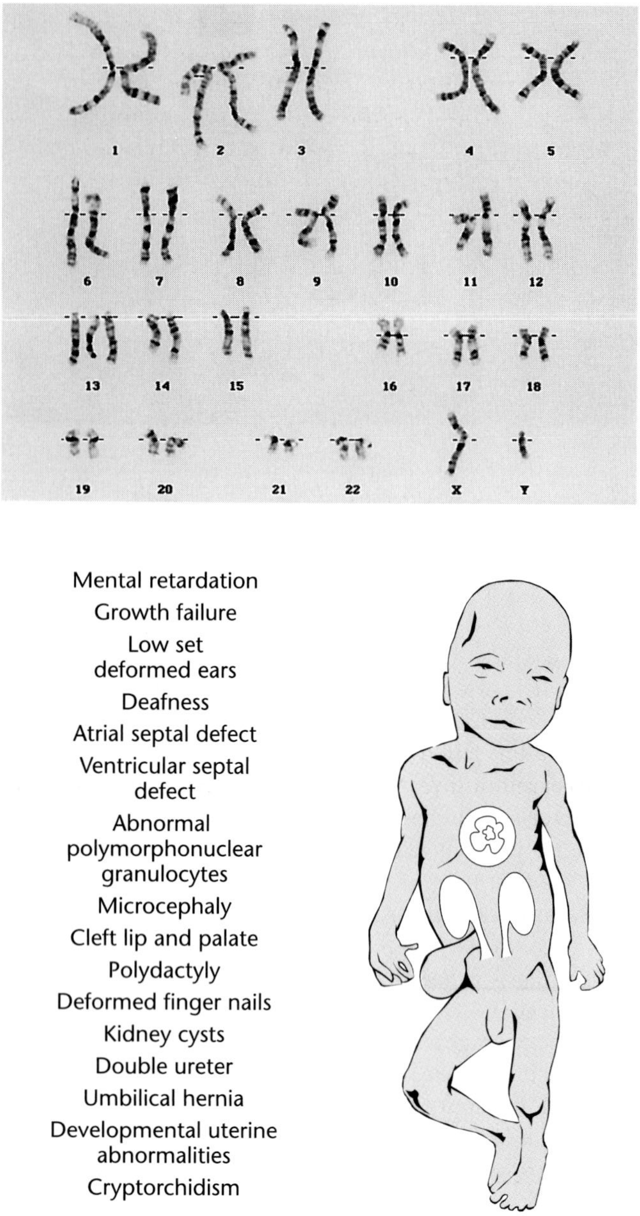

Mental retardation

Growth failure

Low set
deformed ears

Deafness

Atrial septal defect

Ventricular septal
defect

Abnormal
polymorphonuclear
granulocytes

Microcephaly

Cleft lip and palate

Polydactyly

Deformed finger nails

Kidney cysts

Double ureter

Umbilical hernia

Developmental uterine
abnormalities

Cryptorchidism

FIGURE 9.14   The karyotype and phenotypic depiction of an infant with Patau syndrome, where three members of the D-group chromosome 13 are present, creating the 47,13+ condition.

chin are often characteristic of such individuals. Although the frequency of trisomy 18 is somewhat greater than that of trisomy 13, the average survival time is about the same, less than four months. Death is usually caused by pneumonia or heart failure. The phenotype of this child, like that of individuals with Down and Patau syndromes, illustrates that the presence of an extra autosome produces congenital malformations and reduced life expectancy.

Again, the average maternal age is high—34.7 years by one calculation. In contrast to Patau syndrome, the preponderance of Edward syndrome infants are females. In one set of observations based on 143 cases, 80 percent were female. Overall, about 1 in 8000 live births exhibits this malady.

Viability in Human Aneuploidy

The reduced viability of individuals with recognized monosomic and trisomic conditions leads us to believe that many other aneuploid conditions may arise but that the affected fetuses do not survive to term. This observation has been confirmed by karyotypic analysis of spontaneously aborted fetuses.

These studies have revealed some striking statistics. At least 15 to 20 percent of all conceptions are terminated in spontaneous abortion (some estimates are considerably higher). About 30 percent of all spontaneous abortuses demonstrate some form of chromosomal anomaly, and approximately 90 percent of all chromosomal anomalies are terminated prior to birth as a result of spontaneous abortion.

A large percentage of spontaneous abortuses demonstrating chromosomal abnormalities are aneuploids. The aneuploid with highest incidence among abortuses is the 45,X condition, which produces an infant with Turner syndrome if the fetus survives to term.

An extensive review of this subject by David H. Carr has also revealed that a significant percentage of abortuses are trisomic for one of the autosomal chromosome groups. Trisomies for every human chromosome have been recovered. Monosomies were almost never found, however, even though nondisjunction should produce $n - 1$ gametes with a frequency equal to $n + 1$ gametes. This finding suggests that either gametes lacking a single chromosome are so functionally impaired that they never participate in fertilization, or that the embryo dies so early in its development that the pregnancy goes undetected. Various forms of polyploidy (see below) and other miscellaneous chromosomal anomalies were also found in Carr's study.

These observations support the hypothesis that normal embryonic development requires a precise diploid complement of chromosomes that maintains a delicate equilibrium of expression of genetic information. The prenatal mortality of most aneuploids provides a barrier against the introduction of a variety of chromosome-based genetic anomalies into the human population.

Polyploidy and Its Origins

The term **polyploidy** describes instances where more than two multiples of the haploid chromosome set are found. The naming of polyploids is based on the number

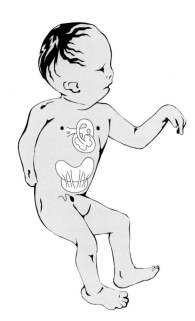

Growth failure
Mental retardation
Open skull sutures
    at birth
High arched eyebrows
Low set deformed ears
Short sternum
Ventricular septal defect
Flexion deformities
    of fingers
Abnormal kidneys
Persistent ductus
    arteriosus
Deformity of hips
Prominent external
    genitalia
Muscular hypertonus
Prominent heel
Dorsal flexion of big toes

FIGURE 9.15 The karyotype and phenotypic depiction of an infant with Edwards syndrome. Three members of the E-group chromosome 18 are present, creating the 47,18+ condition.

of sets of chromosomes found: a **triploid** has $3n$ chromosomes; a **tetraploid** has $4n$; a **pentaploid**, $5n$; and so forth. Several general statements may be made about polyploidy. This condition is relatively infrequent in most animal species, but is well known in lizards, amphibians, and fish. It is much more common in plant species. Odd numbers of chromosome sets are not usually maintained reliably from generation to generation because a polyploid organism with an uneven number of homologs usually does not produce genetically balanced gametes. For this reason, triploids, pentaploids, and so on are not usually found in species that depend solely upon sexual reproduction for propagation.

Polyploidy can originate in two ways: (1) The addition of one or more extra sets of chromosomes, identical to the normal haploid complement of the same species, results in **autopolyploidy**; and (2) the combination of chromosome sets from different species may occur as a consequence of interspecific matings resulting in **allopolyploidy** (from the Greek word *allo*, meaning other or different). The distinction between auto- and allopolyploidy is based on the genetic origin of the extra chromosome sets.

In our discussion of polyploidy, we will use the following symbols to clarify the origin of additional chromosome sets. For example, if A represents the haploid set of chromosomes of any organism, then

$$A = a_1 + a_2 + a_3 + a_4 + \cdots + a_n$$

where a_1, a_2, and so on represent individual chromosomes, and where n is the haploid number. Using this nomenclature, a normal diploid organism would be represented simply as AA.

Autopolyploidy

In **autopolyploidy**, each additional set of chromosomes is identical to the parent species. Therefore, **triploids** are represented as AAA, **tetraploids** are $AAAA$, and so forth.

Autotriploids may arise in several ways. A failure of all chromosomes to segregate during meiotic divisions (first-division or second-division nondisjunction) may lead to the production of a diploid gamete. If such a gamete survives and is fertilized by a haploid gamete, a zygote with three sets of chromosomes is produced. Or, occasionally two sperm may fertilize an ovum, resulting in a triploid zygote. Triploids can also be produced under experimental conditions by crossing diploids with tetraploids. Diploid organisms produce gametes with n chromosomes, whereas tetraploids produce $2n$ gametes. Upon fertilization, the desired triploid is produced.

Because they have an even number of chromosomes, **autotetraploids** ($4n$) are theoretically more likely to be found in nature than are autotriploids. Unlike triploids, which often produce genetically unbalanced gametes with odd numbers of chromosomes, tetraploids are more likely to produce balanced gametes when involved in sexual reproduction.

Tetraploid cells can be produced experimentally from diploid cells by applying cold or heat shock during meiosis or by applying colchicine to somatic cells undergoing mitosis. **Colchicine**, an alkyloid derived from the autumn crocus, interferes with spindle formation, and thus replicated chromosomes cannot be separated at anaphase and migrate to the poles. When colchicine is re-

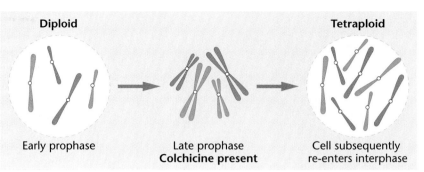

FIGURE 9.16 The potential involvement of colchicine in doubling the chromosome number, as occurs during the production of an autotetraploid. Two pairs of homologous chromosomes are followed. While each chromosome has replicated its DNA earlier during interphase, the chromosomes do not appear as double structures until late prophase. When anaphase fails to occur, the chromosome number doubles if the cell reenters interphase.

Diploid — Early prophase

Late prophase **Colchicine present**

Tetraploid — Cell subsequently re-enters interphase

moved, the cell may reenter interphase. When the paired sister chromatids separate and uncoil, the nucleus will contain twice the diploid number of chromosomes and is therefore 4*n*. This process is illustrated in Figure 9.16.

In general, autopolyploids are larger than their diploid relatives. Often, the flower and fruit of plants are increased in size. This increase seems to be due to larger cell size rather than greater cell number. Although autopolyploids do not contain new or unique information compared with the diploid relative, such varieties may be of greater commercial value. Economically important triploid plants include several potato species of the genus *Solanum*, Winesap apples, commercial bananas, seedless watermelons, and the cultivated tiger lily *Lilium tigrinum*. These plants are propagated asexually. Diploid bananas contain hard seeds, but the commercial, triploid, "seedless" variety has edible seeds. Tetraploid alfalfa, coffee, peanuts, and McIntosh apples are also of economic value because they are either larger or grow more vigorously than do their diploid or triploid counterparts. The commercial strawberry is an octoploid. These observations attest to the importance of autopolyploidy in domesticated plants.

Allopolyploidy

Polyploidy can also result from hybridization of two closely related species. If a haploid ovum from a species with chromosome sets *AA* is fertilized by a haploid sperm from a species with sets *BB*, the resulting hybrid is *AB*, where $A = a_1, a_2, a_3, \ldots, a_n$ and $B = b_1, b_2, b_3, \ldots, b_n$. The hybrid may be sterile because of its inability to produce viable gametes. Most often, this occurs because some or all of the *a* and *b* chromosomes cannot synapse in meiosis because *a* and *b* are not similar enough. As a result, unbalanced genetic conditions result. If, however, the new *AB* genetic combination undergoes a natural or induced chromosomal doubling two *a* chromosomes and two *b* chromosomes are available to pair during meiosis. As a result, a fertile *AABB* tetraploid is produced. These events are illustrated in Figure 9.17. Since this polyploid

contains the equivalent of four haploid genomes and since the hybrid contains unique genetic information compared with either parent, such an organism is called an **allotetraploid**. An equivalent term, **amphidiploid**, is also used to describe this situation in which a hybrid organism contains two complete diploid genomes. This latter term is preferred when the original species are known.

Allopolyploidy is the most common natural form of polyploidy in plants because the chance of forming bal-

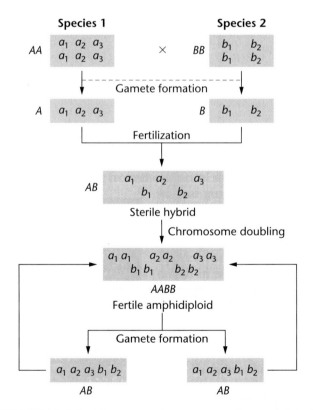

FIGURE 9.17 The origin and propagation of an amphidiploid. Species 1 contains genome A consisting of three distinct chromosomes, a_1, a_2, and a_3. Species 2 contains genome B consisting of two distinct chromosomes, b_1 and b_2. Following fertilization between members of the two species and chromosome doubling, a fertile amphidiploid containing two complete diploid genomes (*AABB*) is formed.

anced gametes is much greater than in other types of polyploidy. Since two homologs of each specific chromosome are present, meiosis can occur normally (Figure 9.17), and fertilization can successfully propagate the allotetraploid sexually. This discussion assumes the simplest situation, where none of the chromosomes in set *A* are homologous to those in set *B*. In hybrids of very closely related species, some homology between *a* and *b* chromosomes is likely. In this case, meiotic pairing is more complex. Multivalents, which are more complex than the trivalents shown in Figure 9.11, may be formed, resulting in the production of unbalanced gametes. In such cases, aneuploid varieties of amphidiploids may arise. Allopolyploids are rare in most animals because mating behavior is most often species-specific, and thus the initial step in hybridization is unlikely to occur.

A classical example of amphidiploidy in plants is the cultivated species of American cotton, *Gossypium* (Figure 9.18). This species has 26 pairs of chromosomes: 13 are large and 13 are much smaller. When it was discovered that Old World cotton had only 13 pairs of large chromosomes, allopolyploidy was suspected. After an examination of wild American cotton revealed 13 pairs of small chromosomes, this speculation was strengthened. J. O. Beasley was able to reconstruct the origin of cultivated cotton experimentally. He crossed the Old World strain with the wild American strain and then treated the hybrid with colchicine to double the chromosome number. The result of these treatments was a fertile amphidiploid variety of cotton. It contained 26 pairs of chromosomes and characteristics similar to the cultivated variety.

Amphidiploids often exhibit traits of both parental species. An interesting example, but one with no practical economic importance, is that of the hybrid formed between the radish *Raphanus sativus* and the cabbage *Brassica oleracea*. Both species have a haploid number of 9. The initial hybrid consists of 9 *Raphanus* and 9 *Brassica* chromosomes (9R + 9B). While hybrids are almost always sterile, some fertile amphidiploids (18R + 18B) have been produced. Unfortunately, the root of this plant is more like the cabbage and its shoot more like the radish. Had the converse occurred, the hybrid might have been of economic importance.

A much more successful commercial hybridization has been performed using the grasses wheat and rye. Wheat (genus *Triticum*) has a basic haploid genome of 7 chromosomes. In addition to normal diploids ($2n = 14$), cultivated allopolyploids exist, including tetraploid ($4n = 28$) and hexaploid ($6n = 42$) species. Rye (genus *Secale*) also has a genome consisting of 7 chromosomes. The only cultivated species is the diploid plant ($2n = 14$).

Using the technique outlined in Figure 9.17, geneticists have produced various allopolyploid hybrids. When tetraploid wheat is crossed with diploid rye, and the F_1 treated with colchicine, a hexaploid variety ($6n = 42$) is derived. The hybrid, designated *Triticale* (see Figure 1.10), represents a new genus. Fertile hybrid varieties derived from various wheat and rye species can be crossed together or backcrossed. These crosses have created many variations of the genus *Triticale*.

The hybrid plants demonstrate characteristics of both wheat and rye. For example, certain hybrids combine the high protein content of wheat with the high content of the amino acid lysine of rye. The lysine content is low in wheat and thus is a limiting nutritional factor. Wheat is considered a high-yielding grain, whereas rye is noted for its versatility of growth in unfavorable environments. *Triticale* species, combining both traits, have the potential of significantly increasing grain production. Programs designed to improve crops through hybridization have long been underway in several underdeveloped countries of the world, as discussed in Chapter 1.

Recall that in a previous chapter, we discussed the use of **somatic cell hybrids** to map human genes (see Chapter 5). This technique has also been applied to the production of allopolyploid plants (Figure 9.19). Cells from the developing leaves of plants can be treated to remove their cell wall, resulting in **protoplasts**. These altered cells can be maintained in culture and stimulated to fuse with other protoplasts, producing somatic cell hybrids.

If cells from different plant species are fused in this way, hybrid cells can be produced with unique chromosome combinations. Since protoplasts can be induced to divide and differentiate into stems that develop leaves, the potential for producing allopolyploids is available in the research laboratory. In some cases, entire plants can be derived from cultured protoplasts. If only stems and

FIGURE 9.18   The amphidiploid form of *Gossypium*, the cultivated cotton plant.

leaves are produced, these can be grafted onto the stem of another plant. If flowers are formed, fertilization may yield mature seeds, which upon germination yield an allopolyploid plant.

There are now many examples of allopolyploids that have been created commercially using the above approach. Although most of them are not true amphidiploids (one or more chromosomes are missing), precise allotetraploids are sometimes produced.

Endopolyploidy

Certain cells in an otherwise diploid organism have been observed to be polyploid. The term **endopolyploidy** describes this general condition. In such cells, replication and separation of chromosomes occur without nuclear division. The process leading to endopolyploidy is called **endomitosis**.

Vertebrate liver cell nuclei, including human ones, often contain 4*n*, 8*n*, or 16*n* chromosome sets. The stem and parenchymal tissue of apical regions of flowering plants are also often endopolyploid. Cells lining the gut of mosquito larvae attain a 16*n* ploidy, but during the pupal stages such cells undergo very quick reduction divisions, giving rise to smaller diploid cells. In the water strider *Gerris*, wide variations in chromosome numbers are found in different tissues, with as many as 1024 to 2048 copies of each chromosome in the salivary gland cells. Since the diploid number in this organism is 22, the nuclei of these cells may contain over 40,000 chromosomes!

Although the role of endopolyploidy is not clear, the proliferation of chromosome copies often occurs in cells where high levels of certain gene products are required. In fact, it is well established that certain genes, those whose product is in high demand in *every* cell, exist naturally in multiple copies in the genome. Ribosomal and transfer RNA genes are examples of multiple-copy genes. In certain cells of organisms, it may be necessary to replicate the entire genome, allowing an even greater rate of expression of various genes.

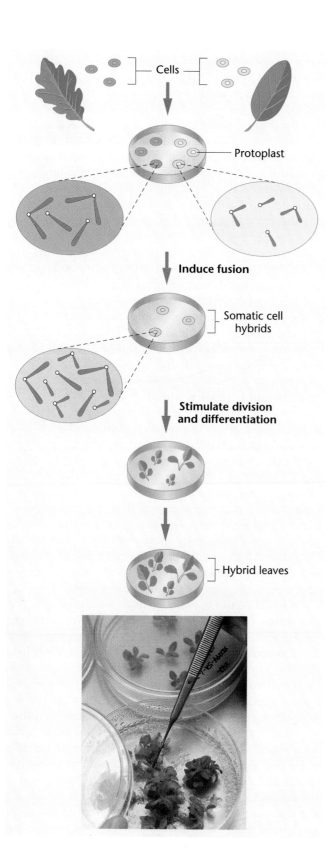

FIGURE 9.19 Application of the somatic cell hybridization technique in the production of an amphidiploid. Cells from the leaves of two species of plants are removed and cultured. The cell walls are digested away and the resultant protoplasts are induced to undergo cell fusion. The hybrid cell is selected and stimulated to divide and differentiate, as illustrated in the photograph. An amphidiploid has a complete set of chromosomes from each parental cell type and displays phenotypic characteristics of each. Two pairs of chromosomes from each species are depicted.

Variation in Chromosome Structure and Arrangement: An Overview

The second general class of chromosome aberrations includes structural changes that delete, add, or rearrange substantial portions of one or more chromosomes. Included in this category are **deletions** and **duplications** of genes or part of a chromosome and **rearrangements** of genetic material in which a chromosome segment is inverted, exchanged with a segment of nonhomologous chromosome, or merely transferred to one, leading to **translocations**. Before discussing these aberrations, we will present several general statements pertaining to them.

In most instances, these structural changes are due to one or more breaks along the axis of a chromosome, followed by either the loss or rearrangement of genetic material. Chromosomes can break spontaneously, but the rate of breakage may increase in cells exposed to chemicals or radiation. Although the actual ends of chromosomes, the **telomeres**, do not readily fuse with newly created ends of "broken" chromosomes or with other telomeres, the ends produced at points of breakage are "sticky" and can rejoin other broken ends. If breakage and rejoining does not merely reestablish the original relationship, and if the alteration occurs in germ plasm, the gametes will contain a structural rearrangement that will be heritable.

When the aberration is found in one homolog but not the other, the individual is said to be heterozygous for the aberration. In such cases, unusual but characteristic pairing configurations are formed during meiotic synapsis. These patterns are useful in identifying the type of change that has occurred. If no loss or gain of genetic material occurs, individuals bearing the aberration "heterozygously" will likely be unaffected phenotypically. However, the unusual pairing arrangements often lead to gametes that are duplicated or deficient for chromosomal regions. When this occurs, the offspring of "carriers" of certain aberrations often have an increased probability of demonstrating phenotypic manifestations.

Deletions

When a portion of a chromosome is lost, the missing piece is referred to as a **deletion** (or a **deficiency**). The deletion can occur either near one end or from the interior of the chromosome. These are called **terminal** or **intercalary deletions**, respectively. Both result from one or more breaks in the chromosome. The portion retaining the centromere region will usually be maintained when the cell divides, whereas the segment without the centromere will eventually be lost in progeny cells following mitosis or meiosis. For synapsis to occur between a chromosome with a large intercalary deficiency and a normal complete homolog, the unpaired region of the normal homolog must "buckle out." Such a configuration is called a **deficiency loop** or **compensation loop**. The origins of both types of deletion and the formation of such a loop are diagrammed in Figure 9.20.

If too much genetic information is lost as a result of a deletion, the aberration could be lethal and never become available for study. As seen in the cri-du-chat syndrome, where only part of the short arm of chromosome 5 is lost, a deficiency need not be very great before the effects become severe.

A final consequence of deletions can be noted in organisms heterozygous for a deficiency. Consider the mutant *Notch* phenotype in *Drosophila*. In these flies, the wings are notched on the posterior and lateral margins. Data from breeding studies seem to indicate that the phenotype is controlled by an X-linked dominant mutation since heterozygous females have notched wings and transmit this allele to one-half of their female progeny. The mutation also appears to behave as a homozygous

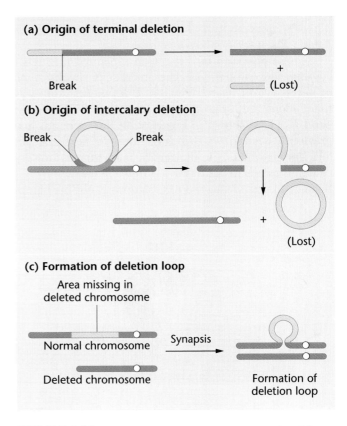

FIGURE 9.20 Origins of (a) a terminal and (b) intercalary deletion. In part (c), pairing occurs between a normal chromosome and one with an intercalary deletion by looping out the undeleted portion to form a deficiency (or a compensation) loop.

TABLE 9.4 *Notch* Genotypes and Phenotypes

Genotype	Phenotype
$\dfrac{N^+}{N}$	*Notch* female
$\dfrac{N}{N}$	Lethal
$\dfrac{N}{\Longrightarrow}$	Lethal
$\dfrac{N^+\,w}{N\ (w^+)^*}$	*Notch, white* female
$\dfrac{N^+fa\ \ spl}{N\ (fa^+spl^+)^*}$	*Notch, facet, split* female

*(deleted)

and hemizygous lethal because such females and males are never recovered. It has also been noted that if notched-winged females are also heterozygous for the closely linked recessive mutations *white*-eye, *facet*-eye, or *split*-bristle, they express these mutant phenotypes as well as *Notch*. Because these mutations are recessive, heterozygotes should express the normal, wild-type phenotypes. These genotypes and phenotypes are summarized in Table 9.4.

These observations have been explained through a cytological examination of the polytene X chromosome of heterozygous *Notch* females. A deficiency loop was found along the X chromosome from band 3C2 through band 3C11, as shown in Figure 9.21. These bands had previously been shown to include the *white*, *facet*, and *split* loci, among others. On the genetic map, this region corresponds to loci in the region 1.5 to 3.0. This region's deficiency in one of the two homologous X chromosomes

has two distinct effects. First, it results in the *Notch* phenotype. Second, by deleting the loci for genes whose mutant alleles are present on the other X chromosome, the deficiency creates a partially hemizygous condition so that the recessive *white*, *facet*, or *split* alleles are expressed. This type of phenotypic expression of recessive genes in association with a deletion is an example of the phenomenon called **pseudodominance**.

Many independently arising *Notch* phenotypes have been investigated. The common deficient band for all *Notch* phenotypes has now been designated as 3C7. In every case that *white* was also expressed pseudodominantly, the band 3C2 was also missing. The bands that cytologically distinguish the *Notch* locus from the *white* locus have been confirmed in this manner.

Duplications

When any part of the genetic material—a single locus or a large piece of a chromosome—is present more than once in the genome, it is called a **duplication**. As in deletions, pairing in heterozygotes may produce a compensation loop. Duplications can arise as the result of unequal crossing over between synapsed chromosomes during meiosis (Figure 9.22) or through a replication error prior to meiosis. In the former case, both a duplication and a deficiency are produced.

Three interesting aspects of duplications will be considered. First, they may result in gene redundancy. Second, as with deletions, duplications can produce phenotypic variation. Third, according to one convincing theory, duplications have also been an important source of genetic variability during evolution.

Gene Redundancy and Amplification: Ribosomal RNA Genes

Although many gene products are not needed in every cell of an organism, other gene products are known to be essential components of all cells. For example, ribosomal RNA must be present in abundance to support protein synthesis. The more metabolically active a cell is, the higher the demand for this molecule, which becomes a part of each ribosome. In theory, a single copy of the gene encoding rRNA may be inadequate in many cells. Studies using the technique of molecular hybridization, which allow the determination of the percentage of the genome coding for specific RNA sequences, show that in most organisms, there are multiple copies of the genes coding for rRNA. Such DNA is called **rDNA**, and the general phenomenon is called **gene redundancy**. For example, in the common intestinal bacterium *Escherichia coli* (*E. coli*)

Bands
3C2–3C11

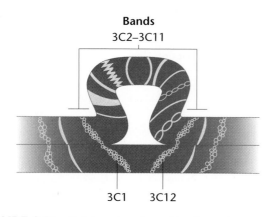

3C1   3C12

FIGURE 9.21 Deficiency loop formed in salivary chromosomes of *Drosophila melanogaster* where the fly is heterozygous for a deletion. The deletion encompasses bands 3C2 through 3C11, corresponding to the region associated with the *Notch* phenotype.

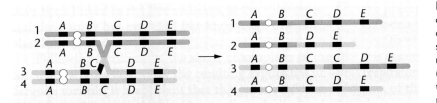

FIGURE 9.22 The origin of duplicated and deficient regions of chromosomes as a result of unequal crossing over. The tetrad at the left is mispaired during synapsis. A single crossover between chromatids 2 and 3 results in deficient and duplicated chromosomal regions (see chromosomes 2 and 3, respectively, on the right). The two chromosomes uninvolved in the crossover event remain normal in their gene sequence and content.

about 0.4 percent of the haploid genome consists of rDNA. This is equivalent to 5 to 10 copies of the gene. In *Drosophila melanogaster*, 0.3 percent of the haploid genome, equivalent to 130 copies, consists of rDNA. Although the presence of multiple copies of the same gene is not restricted to those coding for rRNA, we will focus on them in this section.

Studies of *Drosophila* have documented the need for the extensive amounts of rRNA and ribosomes made possible by multiple copies of these genes. In this organism, the X-linked mutation *bobbed* has, in fact, been shown to be due to a deletion of a variable number of the 130 genes coding for rRNA. Mutant flies have low viability, are underdeveloped, and have bristles reduced in size. Both general development and bristle formation, which occurs very rapidly during normal pupal development, apparently depend on a great number of ribosomes to support protein synthesis. Many *bobbed* alleles have been studied, and each has been shown to involve a deletion, often of a unique size. The extent to which both viability and bristle length are decreased correlates well with the relative number of rRNA genes deleted. We may conclude that the normal rDNA redundancy observed in wild-type *Drosophila* is near the minimum required for adequate ribosome production during normal development.

In some cells, particularly oocytes, even the normal redundancy of rDNA may be insufficient to provide adequate amounts of rRNA and ribosomes. Oocytes store abundant nutrients in the ooplasm for use by the embryo during early development. In fact, more ribosomes are included in the oocytes than in any other cell type. By considering how the amphibian *Xenopus laevis* (the South African clawed frog) acquires this abundance of ribosomes, we will see a second way in which the amount of rRNA is increased. This phenomenon is referred to as **gene amplification**.

The genes that code for rRNA are located in an area of the chromosome known as the **nucleolar organizer region (NOR)**. The NOR is intimately associated with the nucleolus, which is a processing center for ribosome production. Molecular hybridization analysis has shown that each NOR in *Xenopus* contains the equivalent of 400 redundant gene copies coding for rRNA. Even this number of genes is apparently inadequate to synthesize the vast

amount of ribosomes that must accumulate in the amphibian oocyte to support development following fertilization.

To amplify further the number of rRNA genes, the rDNA is selectively replicated, and each new set of genes is released from its template. Because each new copy is equivalent to the NOR, multiple small nucleoli are formed around each NOR in the oocyte. As many as 1500 of these "micronucleoli" have been observed in a single oocyte. If we multiply the number of micronucleoli (1500) by the number of gene copies in each NOR (400), we see that amplification in *Xenopus* oocytes can result in over half a million gene copies! If each copy is transcribed only 20 times during the maturation of a single oocyte, in theory, sufficient copies of rRNA will be produced to result in well over one million ribosomes.

The *Bar* Eye Mutation in *Drosophila*

Duplications can cause phenotypic variations that at first might appear to be caused by a simple gene mutation. The *Bar* eye phenotype (Figure 9.23) in *Drosophila* is a classical example. Instead of the normal oval eye shape, *Bar*-eyed flies have narrow, slitlike eyes. This phenotype appears to be inherited as a dominant X-linked mutation. However, because both heterozygous females and hemizygous males exhibit the trait, but homozygous females show a more pronounced phenotype than either of these two cases, the inheritance is more accurately described as **semidominant**.

In the early 1920s Alfred H. Sturtevant and Thomas H. Morgan discovered and investigated this "mutation." As illustrated in Figure 9.23(a), normal wild-type females (B^+/B^+) have almost 800 facets in each eye. Heterozygous females (B/B^+) have about 350 facets, whereas homozygous females (B/B) average fewer than 50 facets. Females are occasionally recovered with even fewer facets and are designated as *double Bar* (B^D/B^+).

Some 10 years later, Calvin Bridges and Herman J. Muller compared the polytene X chromosome banding pattern of the *Bar* fly with that of the wild-type fly. Such chromosomes contain specific banding patterns that have been well categorized into regions. Their studies revealed that one copy of region 16A of the X chromosome was present in wild-type flies and that this region was duplicated in *Bar* flies and triplicated in *double Bar* flies. These observations provided evidence that the *Bar* phenotype is

(a) Genotypes and phenotypes

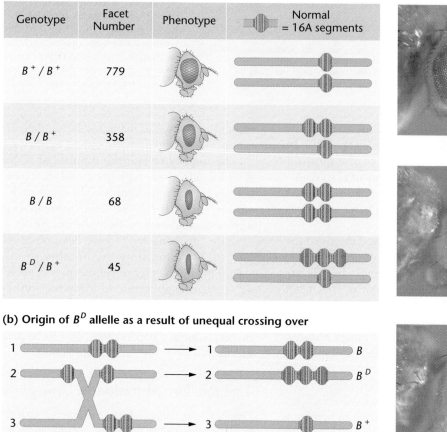

Genotype	Facet Number	Phenotype	Normal = 16A segments
B^+ / B^+	779		
B / B^+	358		
B / B	68		
B^D / B^+	45		

(b) Origin of B^D allele as a result of unequal crossing over

1	→	1	B
2	→	2	B^D
3	→	3	B^+
4	→	4	B

(c)

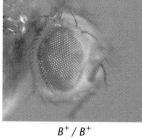

B^+ / B^+

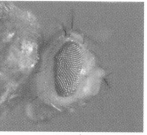

B / B^+

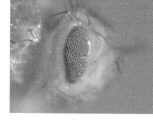

B / B

FIGURE 9.23 (a) Summary of the duplication genotypes and resultant *Bar* eye phenotypes. (b) The origin of the B^D (double *Bar*) allele. (c) Photographs illustrate the actual *Bar* eye phenotypes compared to wild type (B^+/B^+).

not the result of a simple chemical change in the gene, but is instead a duplication. The *double Bar* condition originates as a result of unequal crossing over, which produces the triplicated 16A region [Figure 9.23(b)].

Figure 9.23 also illustrates what is referred to as a **position effect**. You may recall from our introduction of this term in Chapter 7 that a position effect refers to altered gene expression resulting from new "positioning" of a gene within the genome. In this case, when the eye facet phenotypes of *B/B* and B^D/B^+ flies are compared, an average of 68 and 45 facets are found, respectively. In both cases, there are two extra 16A regions. However, when the repeated segments are distributed on the same homolog instead of being positioned on two homologs, the phenotype is more pronounced. Thus, the same amount of genetic information produces an altered *phenotype* depending on the *position* of the genes.

The Role of Gene Duplication in Evolution

One of the most intriguing aspects of the study of evolution is the consideration of the mechanisms for genetic variation. The origin of unique gene products present in phylogenetically advanced organisms but absent in less advanced, ancestral forms is a topic of particular interest. In other words, how do "new" genes arise?

In 1970, Susumo Ohno published the provocative monograph *Evolution by Gene Duplication* in which he elaborated on the importance of gene duplication to the origin of new genes during evolution. While he was not the first to suggest that duplications might provide a "reservoir" from which new genes might arise, Ohno provided a detailed account of this idea. Ohno's thesis was based on the supposition that the gene products of essential genes, present as only a single copy in the genome, are indispensable to the survival of members of any species during evolution. Therefore, these genes are not free to accumulate mutations that alter their primary function and potentially give rise to new genes.

However, if an essential gene were to become duplicated in a germ cell, major mutational changes in this extra copy would be tolerated because the original gene still provides the genetic information for its essential function. The duplicated copy would be free to acquire large numbers of mutational changes over extended periods of time. Over short intervals, the new genetic information may be of no practical advantage. However, over long evolutionary periods, the duplicated gene may change sufficiently so that its product assumes a divergent role in the cell. The new function may impart an "adaptive" advantage to organisms carrying such unique genetic information, enhancing their fitness. Ohno has outlined a mechanism through which sustained genetic variability may have originated.

Ohno's thesis is supported by the discovery of genes that have a substantial amount of their DNA sequence in common, but whose gene products are distinct. For example, trypsin and chymotrypsin fit this description, as do myoglobin and hemoglobin. The DNA sequence homology is great enough in these cases to conclude that members of each gene pair arose from a common ancestral gene through duplication. During evolution, the related genes diverged sufficiently so that their products became unique.

Other support includes the presence of **gene families**, regional groups of genes whose products perform the same general function. Again, members of a family show DNA sequence homology sufficient to conclude that they share a common origin. The various types of globin chains that are part of hemoglobin is one example.

As we will see in Chapter 13, various globin chains function at different times during development, but they are all a part of hemoglobin, functioning to transport oxygen. Other examples include the immunologically important **T-cell receptors** and antigens encoded by the **major histocompatibility complex (MHC)**.

Inversions

The **inversion**, another class of structural variation, is a type of chromosomal aberration in which a segment of a chromosome is turned around 180° within a chromosome. An inversion does not involve a loss of genetic information, but simply rearranges the linear gene sequence. An inversion requires two breaks along the length of the chromosome and subsequent reinsertion of the inverted segment. Figure 9.24 illustrates how one type of inversion might arise. By forming a chromosomal loop prior to breakage, the newly created "sticky ends" are brought close together and rejoined.

The inverted segment may be short or quite long and may or may not include the centromere. If the centromere is *not* part of the rearranged chromosome segment, the inversion is said to be **paracentric**. If the centromere *is* included in the inverted segment, the term **pericentric** describes the inversion, as in Figure 9.24.

Although the gene sequence has been reversed in the paracentric inversion, the ratio of arm lengths extending from the centromere is unchanged. In contrast (Figure 9.25), some pericentric inversions create chromosomes with arms of different lengths from those of the noninverted chromosome. As a result, the **arm ratio** is often changed when a pericentric inversion is produced. The

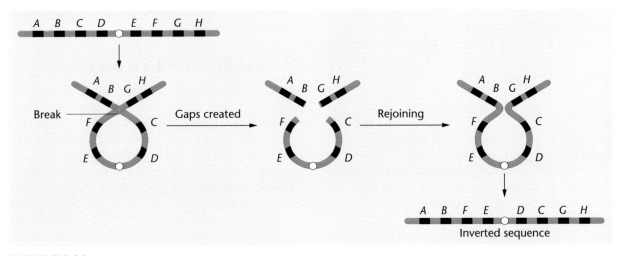

FIGURE 9.24   One possible origin of a pericentric inversion.

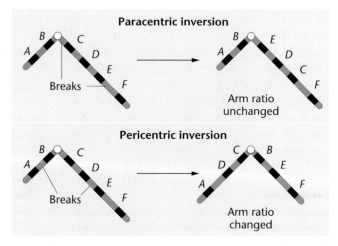

Paracentric inversion

Breaks

Arm ratio
unchanged

Pericentric inversion

Breaks

Arm ratio
changed

FIGURE 9.25 A comparison of the arm ratios of a submetacentric chromosome before and after the occurrence of a paracentric and pericentric inversion. Only the pericentric inversion can result in an alteration of the original ratio.

change in arm lengths may sometimes be detected during the metaphase stage of mitotic or meiotic divisions.

Although inversions may seem to have a minimal impact on the genetic material, their consequences are of interest to geneticists. Organisms heterozygous for inversions may produce aberrant gametes. Inversions may also result in position effects and might play an important role in the evolutionary process.

Consequences of Inversions during Gamete Formation

If only one member of a homologous pair of chromosomes has an inverted segment, normal linear synapsis during meiosis is not possible. Organisms with one inverted chromosome and one noninverted homolog are called **inversion heterozygotes**. Pairing between two such chromosomes in meiosis may be accomplished only if they form an **inversion loop**. In other cases, no loop can be formed and the homologs are seen to synapse everywhere but along the length of the inversion, where they appear separated.

If crossing over does not occur within the inverted segment of the inversion heterozygote, the homologs will segregate and result in two normal and two inverted chromatids that are distributed into gametes. However, if crossing over occurs within the inversion loop, abnormal chromatids are produced. The effect of single exchange events within a paracentric inversion is diagrammed in Figure 9.26(a).

As in any meiotic tetrad, a single crossover between nonsister chromatids produces two parental chromatids and two recombinant chromatids. One recombinant chromatid is **dicentric** (two centromeres), and one re-

combinant chromatid is **acentric** (lacking a centromere). Both contain duplications and deletions of chromosome segments as well. During anaphase, an acentric chromatid moves randomly to one pole or the other, or it may be lost, whereas a dicentric chromatid is pulled in two directions. This polarized movement produces a **dicentric bridge** that is cytologically recognizable. A dicentric chromatid will usually break at some point so that part of the chromatid goes into one gamete and part into another gamete during the reduction divisions. Therefore, gametes containing either broken chromatid are deficient in genetic material. When such a gamete participates in fertilization, the zygote most often develops abnormally, if at all.

A similar chromosomal imbalance is produced as a result of a crossover event between a chromatid bearing a pericentric inversion and its noninverted homolog [Figure 9.26(b)]. The recombinant chromatids that are directly involved in the exchange have duplications and deletions. However, no acentric or dicentric chromatids are produced. Gametes receiving these chromatids also produce inviable embryos following their participation in fertilization.

Because fertilization events involving these aberrant chromosomes do not produce viable offspring, it *appears* as if the inversion suppresses crossing over since offspring bearing crossover gametes are not recovered. *Actually*, in inversion heterozygotes, the inversion has the effect of suppressing the *recovery* of crossover products when chromosome exchange occurs within the inverted region. If crossing over always occurred within a paracentric or pericentric inversion, 50 percent of the gametes would be ineffective. The viability of the resulting zygotes is therefore greatly diminished. Furthermore, up to one-half of the viable gametes have the inverted chromosome, and the inversion will be perpetuated within the species. The cycle will be repeated continuously during meiosis in future generations.

Position Effects of Inversions

Another consequence of inversions involves the new positioning of genes relative to other genes and particularly to areas of the chromosome that do not contain genes, such as the centromere. If the expression of the gene is altered as a result of its relocation, a change in phenotype may result. Such a change is an example of what is called a **position effect**, as introduced in our earlier discussion of the *Bar* duplication.

In *Drosophila* females heterozygous for the X-linked recessive mutation *white* eye (w^+/w), the X chromosome bearing the wild-type allele (w^+) may be inverted such that the *white* locus is moved to a point adjacent to the centromere. If the inversion is not present, a heterozy-

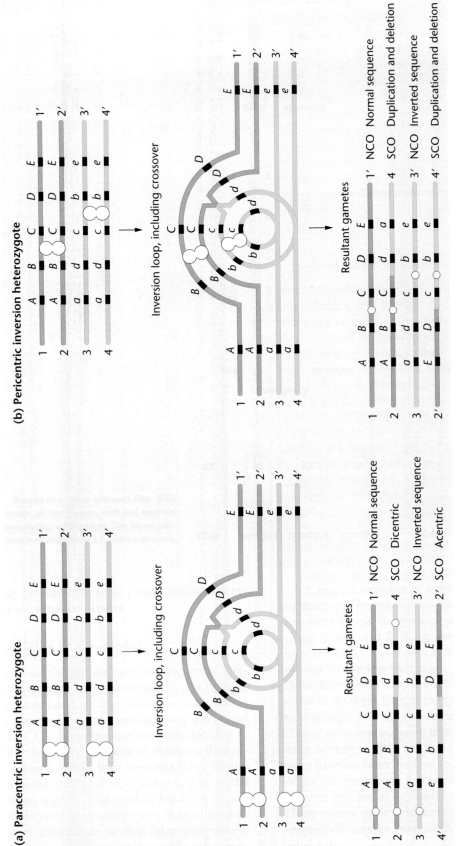

FIGURE 9.26 The effects of a single crossover with an inversion loop in cases involving (a) a paracentric inversion; and (b) a pericentric inversion. In (a), two altered chromosomes are produced, one that is acentric and one that is dicentric. Both chromosomes also contain duplicated and deficient regions. In (b), two altered chromosomes are produced, both with duplicated and deficient regions.

gous female has wild-type red eyes, since the *white* allele is recessive. Females with the X chromosome inversion described above have eyes that are mottled or variegated, i.e., with red and white patches (see Figure 7.2). Placement of the w^+ allele next to the centromere apparently inhibits wild-type gene expression, resulting in the loss of complete dominance over the *w* allele. Other genes, also located on the X chromosome, behave in the same manner when they are similarly relocated. Reversion to wild-type expression has sometimes been noted. When this has occurred, cytological examination has shown that the inversion has reestablished the normal sequence along the chromosome. We introduced the general topic of position effect in Chapter 7 during our discussion of phenotypic expression.

Evolutionary Consequences of Inversions

One major effect of an inversion is the maintenance of a set of specific alleles at a series of adjacent loci, provided that they are contained within the inversion. Because the recovery of crossover products is suppressed in inversion heterozygotes, a particular combination of alleles is preserved intact in the viable gametes. If these genes provide a survival advantage to organisms maintaining them, the inversion is beneficial to the evolutionary survival of the species.

For example, if the set of alleles *ABcDef* is more adaptive than the sets *AbCdeF* or *abcdEF*, the favorable set of genes will not be disrupted by crossing over if it is maintained within a heterozygous inversion. As we will see in Chapter 25, there are documented examples where inversions are adaptive in this way. Specifically, Theodosius Dobzhansky has shown that the maintenance of different inversions on chromosome 3 of *Drosophila pseudoobscura* through many generations has been highly adaptive to this species. Certain inversions seem to be characteristic of enhanced survival under specific environmental conditions.

Translocations

Translocation, as the name implies, involves the movement of a segment of a chromosome to a new place in the genome. For example, the exchange of segments between two nonhomologous chromosomes is a type of structural variation called a **reciprocal translocation**. The origin of a relatively simple reciprocal exchange is illustrated in Figure 9.27(a). The least complex way for this event to occur is for two nonhomologous chromosome arms to come close to each other so that an exchange is facilitated. For this type of translocation, only two breaks are required. If the exchange includes internal chromosome segments, four breaks are required, two on each chromosome.

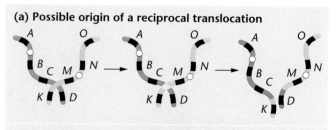

(a) Possible origin of a reciprocal translocation

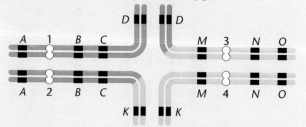

(b) Synapsis of translocation heterozygote

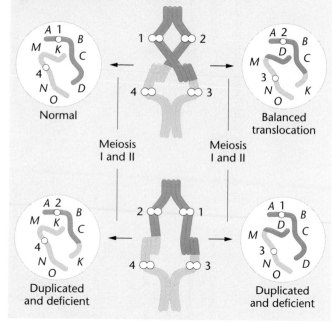

(c) Two possible segregation patterns leading to gamete formation

FIGURE 9.27 (a) The possible origin of a reciprocal translocation; (b) the synaptic configuration formed during meiosis in an individual that is heterozygous for the translocation; (c) two possible segregation patterns, one of which leads to a normal and a balanced gamete, and one that leads to gametes containing duplications and deletions (deficiencies).

The genetic consequences of reciprocal translocations are, in several instances, similar to those of inversions. For example, genetic information is not lost or gained. Rather, there is only a rearrangement of genetic material. The presence of a translocation does not, therefore, directly alter viability of individuals bearing it. Like an inversion, a translocation may also produce a position effect, because it may realign certain genes in relation to other genes. This exchange may create a new genetic linkage relationship that can be detected experimentally.

Homologs heterozygous for a reciprocal translocation undergo unorthodox synapsis during meiosis. As shown in Figure 9.27(b), pairing results in a crosslike configuration. As with inversions, genetically unbalanced gametes are also produced as a result of this unusual alignment during meiosis. In the case of translocations, however, aberrant gametes are not necessarily the result of crossing over. To see how unbalanced gametes are produced, focus on the homologous centromeres in Figure 9.27(b) and (c). According to the principle of independent assortment, the chromosome containing centromere 1 will migrate randomly toward one pole of the spindle during the first meiotic anaphase; it will travel with *either* the chromosome having centromere 3 *or* centromere 4. The chromosome with centromere 2 will move to the other pole along with *either* the chromosome containing centromere 3 *or* centromere 4. This results in four potential meiotic products. The 1,4 combination contains chromosomes uninvolved in the translocation. The 2,3 combination, however, contains translocated chromosomes. These contain a complete complement of genetic information and are balanced. The other two potential products, the 1,3 and 2,4 combinations, will contain chromosomes displaying duplicated and deleted segments.

When incorporated into gametes, the resultant meiotic products are genetically unbalanced. If they participate in fertilization, lethality often results. As few as 50 percent of the progeny of parents heterozygous for a reciprocal translocation may survive. This condition, called **semisterility**, has an impact on the reproductive fitness of organisms, thus playing a role in evolution. Furthermore, in humans, such an unbalanced condition results in partial monosomy or trisomy, leading to a variety of birth defects.

Translocations in Humans: Familial Down Syndrome

Research performed since 1959 has revealed numerous translocations in members of the human population. One common type of translocation involves breaks at the extreme ends of the short arms of two nonhomologous acrocentric chromosomes. These small acentric fragments are lost, and the larger chromosomal segments fuse at their centromeric region (Figure 9.28). This type of translocation produces a new, large submetacentric or metacentric chromosome and is called a **centric fusion translocation**. Such an occurrence in other organisms is called a **Robertsonian fusion**.

Such a translocation accounts for cases in which Down syndrome is inherited, or familial. Earlier in this chapter we pointed out that most instances of Down syn-

drome are due to trisomy 21. This chromosome composition results from nondisjunction during meiosis of one parent. Trisomy accounts for over 95 percent of all cases of Down syndrome. In such instances, the chance of the same parents producing a second afflicted child is extremely low. However, in the remaining families with a Down child, the syndrome occurs in a much higher frequency over several generations.

Cytogenetic studies of the parents and their offspring from these unusual cases explain the cause of **familial Down syndrome**. Analysis reveals that one of the parents contains a **14/21 D/G translocation** (Figure 9.29). That is, one parent has the majority of the G-group chromosome 21 translocated to one end of the D-group chromosome 14. This individual is normal even though he or she has only 45 chromosomes.

During meiosis, one-fourth of the individual's gametes will have two copies of chromosome 21: a normal chromosome and most of a second copy translocated to chromosome 14. When such a gamete is fertilized by a standard haploid gamete, the resulting zygote has 46 chromosomes but three copies of chromosome 21. These individuals exhibit Down syndrome. Other potential surviving offspring

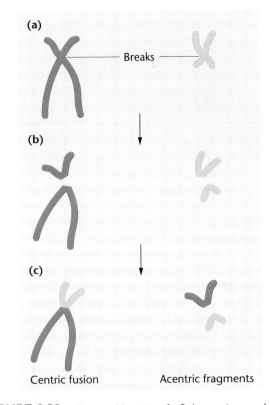

FIGURE 9.28 The possible origin of a Robertsonian translocation. Two independent breaks occur within the centromeric region on two nonhomologous chromosomes. Centric fusion of the long arms of the two acrocentric chromosomes creates the unique chromosome. Two acentric fragments remain.

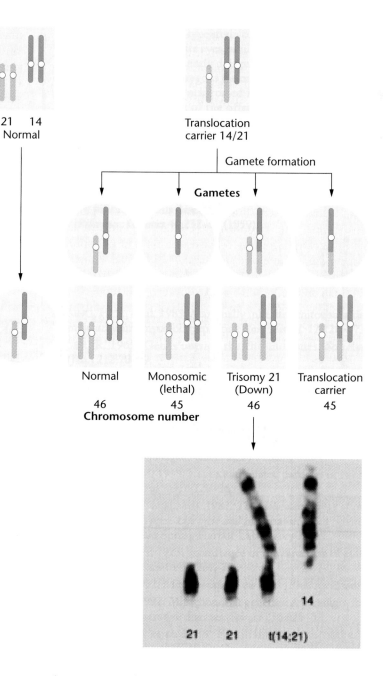

FIGURE 9.29 Illustration of the chromosomal involvement in familial Down syndrome, as described in the text. The photograph illustrates the relevant chromosomes from a trisomy 21 offspring produced by a translocation carrier parent.

contain either the standard diploid genome (without a translocation) or the balanced translocation like the parent. Both cases result in normal individuals. Knowledge of translocations has allowed geneticists to resolve the seeming paradox of an inherited trisomic phenotype in an individual with an apparent diploid number of chromosomes.

It is of interest to note that the "carrier," who has 45 chromosomes and exhibits a normal phenotype, does not contain the *complete* diploid amount of genetic material. A small region is lost from both chromosomes 14 and 21 during the translocation event. This occurs because the ends of both chromosomes have broken off prior to their fusion. These specific regions are known to be two of many chromosomal locations housing multiple copies of

the genes encoding rRNA, the major component of ribosomes. Despite the loss of up to 10 percent of these genes, the carrier individual is unaffected.

Fragile Sites in Humans

We conclude this chapter by briefly discussing the results of an interesting discovery made about 1970 during observations of metaphase chromosomes prepared from human cell cultures. In cells derived from certain individuals, a specific area along one of the chromosomes failed to stain, giving the appearance of a gap. In other individuals whose chromosomes displayed or expressed such morphology, the gap appeared in different positions within the set of chromosomes. Such areas eventually became known as **fragile sites** as they were considered susceptible to chromosome breakage when cultured in the absence of certain chemicals such as folic acid, which is normally present in the culture medium. Fragile sites were at first considered curiosities until a strong association was subsequently shown to exist between one of the sites and a form of mental retardation.

The cause of the fragility at these sites is of great interest. Because they appear to represent points along the chromosome susceptible to breakage, these sites may indicate regions where chromatin is not tightly coiled or compacted. It should be noted that almost all studies of fragile sites have been carried out *in vitro* using mitotically dividing cells, and it is unknown how this phenotype is expressed *in vivo*.

Fragile X Syndrome (Martin-Bell Syndrome)

Most fragile sites do not appear to be associated with any clinical syndrome. However, individuals bearing a *folate-sensitive site* on the X chromosome (Figure 9.30) exhibit the **fragile X syndrome** (or **Martin-Bell syndrome**), the most common form of inherited mental retardation. This syndrome affects about 1 in 1250 males and 1 in 2500 females. Because it is a dominant trait, females carrying only one fragile X chromosome can be mentally retarded. Fortunately, the trait is not fully expressed, as only about 30 percent of fragile X females are retarded, whereas about 80 percent of fragile X-bearing males are mentally retarded. In addition to mental retardation, affected males have characteristic long, narrow faces with protruding chins, enlarged ears, and increased testicular size.

A gene that spans the fragile site may be responsible for this syndrome. This gene, known as *FMR-1*, is one of a growing number of genes where a sequence of three nucleotides is repeated many times, expanding the size of one end of the gene. This phenomenon, called **trinucleotide repeats**, has been recognized in other human disorders, including Huntington disease. The number of repeats varies immensely within the human population, and this number correlates directly with expression of fragile X syndrome. Normal individuals have up to 50 repeats, whereas those with 50 to 200 repeats are considered to be "carriers" of the disorder. Above 200 repeats leads to expression of the syndrome.

The fragile X breakpoint is contained in this variable-length region, and full expression of the phenotype depends on both an increase in length and chemical modifications of the DNA nucleotides in this gene segment. The *FMR-1* gene is known to be expressed in the brain, but the exact nature and function of the encoded protein awaits further research. The gene may encode a DNA-binding protein based on the analysis of its nucleotide sequence. Although other genes may turn out to be involved in fragile X syndrome, both the structure and the function of the *FMR-1* gene make it the leading candidate to provide insights into the molecular basis of this common form of mental retardation.

You may recall that fragile X syndrome illustrates the phenomenon known as **genetic anticipation**, whereby expression of a malady becomes more severe in successive generations. In the case of this disorder, the number of trinucleotide repeats increases as the affected X chromosome passes through the gametes to offspring. We previously introduced this topic in Chapter 7.

FHIT Gene and Human Lung Cancer

A second significant association between a fragile site and a human condition was reported in 1996 by Carlo Croce, Kay Huebner, and their colleagues, who showed that a gene located on chromosome 3, *FHIT* (standing for *frag*ile *h*istidine *t*riad), is altered in cells taken from tumors recovered from individuals with lung cancer. In 80 percent of the tumors and cell lines derived from them, an abnormal *FHIT* gene was exhibited. This gene is part of the fragile area of the autosome designated *FRA3B*, which has also been linked to cancer of the esophagus, colon, and stomach.

FHIT may be more fragile in some people than in others, leading to a susceptibility to lung cancer following environmental insults to lung tissue by agents such as cigarette smoke. The research team found a variety of mutant sequences of the gene in cells derived from the tumors, where the gene had apparently been broken and incorrectly fused back together. These potentially lead to the synthesis of an altered gene product or to no product being synthesized.

FIGURE 9.30   A normal human X chromosome (left) contrasted with a fragile X chromosome (right). The "gap" region is associated with the fragile X syndrome, including mental retardation.

These are very recent findings, and ones that will undoubtedly yield many more significant details in the future. The impact of the discovery is indeed great. An explanation at the molecular genetic level may be provided that explains why some smokers are at risk for lung cancers, other smokers are not, and why some people who have never smoked nevertheless develop lung cancer. Furthermore, the potential for **genetic screening** is clearly on the scientific horizon.

CHAPTER SUMMARY

1. Investigations into the uniqueness of each organism's chromosomal constitution have further enhanced our understanding of genetic variation. Alterations of the precise diploid content of chromosomes are referred to as chromosomal aberrations or chromosomal mutations.

2. In humans, the study of individuals with altered sex chromosome compositions has established that the Y chromosome is responsible for male differentiation. The absence of the Y leads to female differentiation. Similar studies in *Drosophila* have excluded the Y in such a role, instead demonstrating that a balance between the number of X chromosomes and sets of autosomes is the critical factor.

3. The primary sex ratio in humans substantially favors males at conception. During embryonic and fetal development, male mortality is higher than that of females, whereas the secondary sex ratio at birth still favors males by a small margin.

4. Dosage compensation mechanisms limit the expression of X-linked genes in females who have two X chromosomes, as compared to males who have only one X. In mammals, compensation is achieved as a result of the inactivation of either the maternal or paternal X early in development. This process results in the formation of Barr bodies in female somatic cells.

5. The Lyon hypothesis states that, early in development, inactivation is random between the maternal and paternal X. All subsequent progeny cells inactivate the same X as their progenitor cell. Mammalian females thus develop as genetic mosaics with respect to their expression of heterozygous X-linked alleles.

6. Deviations from the expected chromosomal number, or mutations in the structure of the chromosome, are inherited in predictable Mendelian fashion; they often result in inviable organisms or substantial changes in the phenotype.

7. Aneuploidy is the gain or loss of one or more chromosomes from the diploid content, resulting in conditions of monosomy, trisomy, tetrasomy, etc. Studies of monosomic and trisomic disorders have increased our understanding of the delicate genetic balance that must exist for normal development to occur.

8. When complete sets of chromosomes are added to the diploid genome, polyploidy is created. These sets can have identical or diverse genetic origin, creating either autopolyploidy or allopolyploidy, respectively.

9. Large segments of the chromosome can be modified by deletions or duplications. Deletions can produce serious conditions such as the cri-du-chat syndrome in humans, whereas duplications can be particularly important as a source of redundant or new genes.

10. Inversions and translocations, while altering the gene order along chromosomes, initially cause little or no loss of genetic information or deleterious effects. However, heterozygous combinations may cause genetically abnormal gametes following meiosis, often causing lethality.

11. Fragile sites in human mitotic chromosomes have sparked research interest because one such site on the X chromosome is associated with the most common form of inherited mental retardation. Another fragile site, located on chromosome 3, has been linked to lung cancer.

KEY TERMS

acentric chromosome, 248
allopolyploidy, 239
allotetraploid, 240
alternative splicing, 232
amniocentesis, 236
amphidiploid, 240
aneuploidy, 222
anhidrotic ectodermal dysplasia, 229
arm ratio, 247

autopolyploidy, 239
autotetraploid, 239
autotriploid, 239
Barr body, 227
bilateral gynandromorph, 233
centric fusion translocation, 251
chorionic villus sampling (CVS), 236
chromosome aberration, 221
chromosome mutation, 221

clone, 229
colchicine, 239
compensation (deficiency) loop, 243
cri-du-chat syndrome, 235
deletion (deficiency), 243
dicentric bridge, 248
dicentric chromosome, 248
dosage compensation, 227
Down syndrome (trisomy 21), 236

INSIGHTS AND SOLUTIONS

1. In a cross using maize involving three linked genes, *a*, *b*, and *c*, a heterozygote (*abc*/+++) was test-crossed (to *abc*/*abc*). Even though the three genes were separated by at least 5 map units, only two phenotypes were recovered: *abc* and +++. Additionally, the cross produced significantly fewer viable plants than expected. Can you propose why no other phenotypes were recovered and why the viability was reduced?

   Solution: One of the two chromosomes contains an inversion that overlaps all three genes, effectively precluding the recovery of any "crossover" offspring. If this is a paracentric inversion and the genes are clearly separated (assuring that a significant number of crossovers will occur between them), then numerous acentric and dicentric chromosomes will be formed, resulting in the observed reduction in viability.

2. A male *Drosophila* from a wild-type stock was discovered with only 7 chromosomes, whereas the normal 2*n* number is 8. Close examination revealed that one member of chromosome IV (the smallest chromosome) was attached to (translocated to) the distal end of chromosome II and was missing its centromere, thus accounting for the reduction in chromosome number.

(a) Diagram all members of chromosomes II and IV during synapsis in Meiosis I.

Solution:

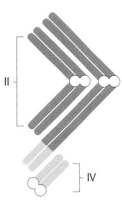

(b) If this male mates with a female with a normal chromosome composition who is homozygous for the recessive chromosome IV mutation *eyeless* (*ey*), what chromosome compositions will occur in the offspring regarding chromosomes II and IV?

Solution:

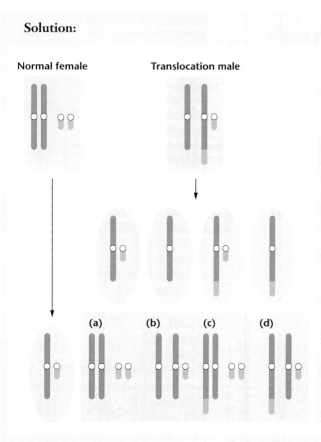

Normal female   Translocation male

(a)   (b)   (c)   (d)

(c) What phenotypic ratio will result regarding the presence of eyes, assuming all abnormal chromosome compositions survive?

Solution:

(a)—normal (heterozygous)
(b)—eyeless (monosomic, contains chromosome IV from mother)
(c)—normal (heterozygous)
(d)—normal (heterozygous)

The final ratio is 3/4 normal:1/4 eyeless.

3. If a Haplo-IV female *Drosophila* (containing only one chromosome 4, but an otherwise normal set of chromosomes) that has white eyes (an X-linked trait) and normal bristles is crossed with a male with a diploid set of chromosomes and normal red eye color, but who is homozygous for the recessive chromosome 4 bristle mutation, *shaven* (*sv*), what F_1 phenotypic ratio might be expected?

Solution: Let us first consider only the eye color phenotypes. This is a straightforward X-linked cross. Offspring will appear as 1/2 red females:1/2 white males as shown below:

$$P_1: \quad ww \quad \times \quad w^+/\text{♂}$$
white female   red male

$$F_1: \quad 1/2\ ww^+ \quad : \quad 1/2\ w/\text{♂}$$
red female   white male

The bristle phenotypes will be governed by the fact that the normal-bristled P_1 female produces gametes, one-half of which contain a chromosome 4 (sv^+) and one-half that have no chromosome 4 (− designates no chromosome). Following fertilization by sperm from the *shaven* male, one-half of the offspring will receive two members of chromosome 4 and be heterozygous for *sv*, expressing normal bristles. The other half will have only one copy of chromosome 4. Because its origin is from the male parent, where the chromosome bears the *sv* allele, these flies will express *shaven* since there is no wild-type allele present to mask this recessive allele.

$$P_1: \quad sv^+/- \quad \times \quad sv/sv$$
normal female   shaven male

$$F_1: \quad 1/2\ sv^+/sv \text{ males and females} = 1/2 \text{ wild type}$$
$$ \quad 1/2\ sv^+/- \text{ males and females} = 1/2 \text{ shaven}$$

Using the forked-line method, we can consider both eye color and bristle phenotypes together:

1/2 red-eyed females
— 1/2 normal bristles ⟶ 1/4 red-eyed, normal-bristled females
— 1/2 shaven bristles ⟶ 1/4 red-eyed, shaven-bristled females

1/2 white-eyed males
— 1/2 normal bristles ⟶ 1/4 white-eyed, normal-bristled males
— 1/2 shaven bristles ⟶ 1/4 white-eyed, shaven-bristled males

PROBLEMS AND DISCUSSION QUESTIONS

1. Contrast the evidence leading to the explanation of the different modes of sex determination in *Drosophila* and humans.

2. Devise a method of nondisjunction in human female gametes that would give rise to Klinefelter and Turner syndrome offspring following fertilization by a normal male gamete.

3. It has been suggested that any male-determining genes contained on the Y chromosome in humans should not be located in the limited region that synapses with the X chromosome during meiosis. What might be the outcome if such genes were located in this region?

4. What is a Barr body?
5. Indicate the expected number of Barr bodies in interphase cells of the following individuals: Klinefelter syndrome; Turner syndrome; and karyotypes 47,XYY, 47,XXX, and 48,XXXX.
6. Define the Lyon hypothesis.
7. Relate the potential effect of the Lyon hypothesis on the retina of a human female heterozygous for the X-linked red-green color-blindness trait.
8. Cat breeders are aware that kittens expressing the X-linked calico coat pattern are almost invariably females. Why?
9. What does the apparent need for dosage compensation mechanisms suggest about the expression of genetic information in normal diploid individuals?
10. The marine echiurid worm *Bonellia viridis* is an extreme example of the environment's influence on sex determination. Undifferentiated larvae either remain free-swimming and differentiate into females or they settle on the proboscis of an adult female and become males. If larvae that have been on a female proboscis for a short period are removed and placed in seawater, they develop as intersexes. If larvae are forced to develop in an aquarium where pieces of proboscises have been placed, they develop into males. Contrast this mode of sexual differentiation with that of mammals. Suggest further experimentation to elucidate the mechanism of sex determination in *Bonellia*.
11. Shown below are four graphs that plot the percentage of males occurring in various reptile groups versus the temperature that fertilized eggs encounter during early development. Interpret these data as they relate to reptilian sex determination.

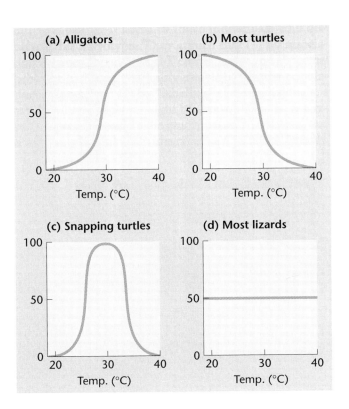

(a) Alligators

(b) Most turtles

(c) Snapping turtles

(d) Most lizards

12. Discuss the possible reasons why the primary sex ratio in humans is as high as 1.40 to 1.60.
13. Define and distinguish between the following pairs of terms:

> aneuploidy/euploidy
> monosomy/trisomy
> Patau syndrome/Edwards syndrome
> autopolyploidy/allopolyploidy
> autotetraploid/amphidiploid
> polyteny/endopolyploidy
> paracentric inversion/pericentric inversion

14. Contrast the relative survival times of individuals with Down syndrome, Patau syndrome, and Edwards syndrome. Why do you think such differences exist?
15. What evidence suggests that Down syndrome is more often the result of nondisjunction during oogenesis rather than during spermatogenesis?
16. Why do we believe that humans with aneuploid karyotypes occur, but are usually inviable?
17. What conclusions have been drawn about human aneuploidy as a result of karyotypic analyses of abortuses?
18. Contrast the fertility of an allotetraploid with an autotriploid and an autotetraploid.
19. When two plants belonging to the same genus but different species were crossed together, the F_1 hybrid was more viable and had more ornate flowers. Unfortunately, this hybrid was sterile and could only be propagated by vegetative cuttings. Explain the sterility of the hybrid. How might a horticulturist attempt to reverse its sterility?
20. For a species with a diploid number of 18, indicate how many chromosomes will be present in the somatic nuclei of individuals who are haploid, triploid, tetraploid, trisomic, and monosomic.
21. Discuss the origin of cultivated American cotton.
22. Predict how the synaptic configurations of homologous pairs of chromosomes might appear when one member is normal and the other member has sustained a deletion or duplication.
23. Inversions are said to "suppress crossing over." Is this terminology technically correct? If not, restate the description accurately.
24. Contrast the genetic composition of gametes derived from tetrads of inversion heterozygotes where crossing over occurs within a paracentric and pericentric inversion.
25. Contrast the *Notch* locus with the *Bar* locus in *Drosophila*. What phenotypic ratios would be produced in a cross between *Notch* females and *Bar* males? Under what circumstances has gene duplication been essential to an organism's survival?
26. Discuss Ohno's hypothesis on the role of gene duplication in the process of evolution.
27. What roles have inversions and translocations played in the evolutionary process?
28. A human female with Turner syndrome also expresses the X-linked trait hemophilia, as her father did. Which parent underwent nondisjunction during meiosis, giving rise to the gamete responsible for the syndrome?
29. The primrose, *Primula kewensis*, has 36 chromosomes that are similar in appearance to the chromosomes in two related species, *Primula floribunda* (2n = 18) and *Primula verticillata* (2n = 18). How could *P. kewensis* arise from these species? How would you describe *P. kewensis* in genetic terms?

30. Varieties of chrysanthemums are known that contain 18, 36, 54, 72, and 90 chromosomes; all are multiples of a basic set of 9 chromosomes. How would you describe these varieties genetically? What feature is shared by the karyotypes of each variety? A variety with 27 chromosomes was discovered, but it was sterile. Why?

31. What is the effect of a rare double crossover within a pericentric inversion present heterozygously? Within a paracentric inversion present heterozygously?

32. *Drosophila* may be monosomic or disomic for chromosome 4 and remain fertile. Contrast the F_1 and F_2 results of the following crosses involving the recessive chromosome 4 trait, *bent* bristles:
    (a) monosomic bent × disomic normal
    (b) monosomic normal × disomic bent

33. *Drosophila* may also be trisomic for chromosome 4 and remain fertile. Predict the F_1 and F_2 results of crossing:

    trisomic bent (*b/b/b*) × disomic normal (*B/B*)

34. Mendelian ratios are modified in crosses involving autotetraploids. Assume that one plant expresses the dominant trait green (seeds) and is homozygous (*WWWW*). This plant is crossed to one with white seeds that is also homozygous (*wwww*). If only one dominant allele is sufficient to produce green seeds, predict the F_1 and F_2 results of such a cross. Assume that synapsis between chromosome pairs is random during meiosis.

35. Having correctly established the F_2 ratio above, predict the F_2 ratio of a "dihybrid" cross involving two independently assorting characteristics (e.g., $P_1 = WWWWAAAA \times wwwwaaaa$).

36. A couple, in looking ahead to planning a family, were aware that through the past three generations on the male's side a substantial number of stillbirths had occurred and several malformed babies were born who died early in childhood. The female had studied genetics and urged her husband to visit a genetic counseling clinic, where a complete karyotype-banding analysis was subsequently performed. Although it was found that he had a normal complement of 46 chromosomes, banding analysis revealed that one member of the chromosome 1 pair (in Group A) contained an inversion covering 70 percent of its length. The homologue of chromosome 1 and all other chromosomes showed the normal banding sequence.
    (a) How would you explain the high incidence of past stillbirths?
    (b) What would you predict about the probability of abnormality/normality of their future children?
    (c) Would you advise the female that she would have to "wait out" each pregnancy to term so as to determine whether the fetus was normal? If not, what would you suggest she do?

37. The *Sry* gene (described in this chapter) is located on the Y chromosome very close to the region that pairs with the X chromosome during male meiosis. Given this information, propose a model to explain the generation of unusual males who have two X chromosomes (with an *Sry*-containing piece of the Y attached to one X chromosome).

38. The *Xg* cell-surface antigen is coded for by a gene that is located on the X chromosome. There is no equivalent gene on the Y chromosome. Two codominant alleles of this gene have been identified: *Xg1* and *Xg2*. A woman of genotype *Xg2/Xg2* marries a man of genotype *Xg1/♂*, and they produce a son with Klinefelter syndrome of genotype *Xg1/Xg2/♂*. Using proper genetic terminology, briefly explain how this individual was generated. In which parent and in which meiotic division did the mistake occur?

39. The genes encoding the red and green color-detecting proteins of the human eye are located next to one another on the X chromosome and probably arose during evolution from a common ancestral pigment gene. The two proteins demonstrate 96 percent homology in their amino acid sequences. A normal woman with one copy of each gene on each of her two X chromosomes has a red-color-blind son who was shown to contain one copy of the green gene and no copies of the red gene. Devise an explanation at the chromosomal level (during meiosis) that explains these observations.

EXTRA-SPICY PROBLEMS

40. An X-linked dominant mutation in the mouse, *Testicular feminization* (*Tfm*), eliminates the normal response to the testicular hormone testosterone during sexual differentiation. An XY animal bearing the *Tfm* allele on the X chromosome develops testes, but no further male differentiation occurs. The external genitalia of such an animal are female. From this information, what might you conclude about the role of the *Tfm* gene product and the X and Y chromosomes in sex determination and sexual differentiation in mammals? Can you devise an experiment, assuming you can "genetically engineer" the chromosomes of mice, to test and confirm your explanation?

41. In a cross in *Drosophila*, a female heterozygous for the autosomally linked genes *a*, *b*, *c*, *d*, and *e* (*abcde/+++++*) was test-crossed to a male homozygous for all recessive alleles. Even though the distance between each of the above loci was at least 3 map units, only four phenotypes were recovered, yielding the following data:

Phenotype	No. of Flies
+ + + + +	440
a b c d e	460
+ + + + *e*	48
a b c d +	52
Total	1000

Why are many expected crossover phenotypes missing? Can any of these loci be mapped from the data given here? If so, determine map distances.

42. A woman, seeking genetic counseling, was found to be heterozygous for a chromosomal rearrangement between the second and third chromosomes. Her chromosomes, compared to those in a normal karyotype, are diagrammed on the next page:

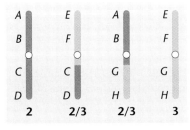

(a) What kind of chromosomal aberration is shown here?

(b) Using a drawing, demonstrate how these chromosomes would pair during meiosis. Be sure to label the different segments of the chromosomes.

(c) This woman is phenotypically normal. Does this surprise you? Why or why not? Under what circumstances might you expect a phenotypic effect of such a rearrangement?

(d) This woman has had two miscarriages due to the death of the developing fetus. She has come to you, an established genetic counselor, for advice. She raises the following questions. Provide an informed response to her concerns. "Is there a genetic explanation of her frequent miscarriages? Should she abandon her attempts to have a child of her own? If not, what is the chance that she could have a normal child?"

43. A son with Klinefelter syndrome is born to a mother who is phenotypically normal and a father who has the X-linked skin condition called anhidrotic ectodermal dysplasia. The mother's skin is completely normal with no signs of the skin abnormality. In contrast, her son has patches of normal skin and patches of abnormal skin.

(a) Which parent contributed the abnormal gamete?

(b) Using the appropriate genetic terminology, describe the meiotic mistake that occurred. Be sure to indicate which division the mistake occurred in.

(c) Using the appropriate genetic terminology, explain the son's skin phenotype.

SELECTED READINGS

BARR, M. L. 1966. The significance of sex chromatin. *Int. Rev. Cytol.* 19:35–39.

Beasley, J. O. 1942. Meiotic chromosome behavior in species, species hybrids, haploids, and induced polyploids of *Gossypium. Genetics* 27:25–54.

Blakeslee, A. F. 1934. New jimson weeds from old chromosomes. *J. Hered.* 25:80–108.

BORGAONKER, D. S. 1989. *Chromosome variation in man: A catalogue of chromosomal variants and anomalies,* 5th ed. New York: Alan R. Liss.

BOUE, A. 1985. Cytogenetics of pregnancy wastage. *Adv. Hum. Genet.* 14:1–58.

BURGIO, G. R., et al., eds. 1981. *Trisomy 21.* New York: Springer-Verlag.

CARR, D. H. 1971. Genetic basis of abortion. *Annu. Rev. Genet.* 5:65–80.

COURT-BROWN, W. M. 1968. Males with an XYY sex chromosome complement. *J. Med. Genet.* 5:341–59.

CROCE, C. M., et al. 1996. The *FHIT* gene at 3p14.2 is abnormal in lung cancer. *Cell* 85:17–26.

CUMMINGS, M. R. 1994. *Human heredity: Principles and issues.* 3rd ed. St. Paul, MN: West.

DAVIDSON, R., NITOWSKI, H., and CHILDS, B. 1963. Demonstration of two populations of cells in human females heterozygous for glucose-6-phosphate dehydrogenase variants. *Proc. Natl. Acad. Sci. USA* 50:481–85.

DEARCE, M. A., and KEARNS, A. 1984. The fragile X syndrome: The patients and their chromosomes. *J. Med. Genet.* 21:84–91.

ERICKSON, J. D. 1976. The secondary sex ratio of the United States, 1969–71: Association with race, parental ages, birth order, paternal education and legitimacy. *Ann. Hum. Genet. (London)* 40:205–12.

FELDMAN, M., and SEARS, E. R. 1981. The wild gene resources of wheat. *Sci. Am.* (Jan.) 244:102–12.

GORDON, J. W., and RUDDLE, F. H. 1981. Mammalian gonadal determination and gametogenesis. *Science* 211:1265–78.

GORMAN, M., KURODA, M., and BAKER, B. S. 1993. Regulation of sex-specific binding of maleness dosage compensation protein to the male X chromosome in *Drosophila. Cell* 72:39–49.

GUPTA, P. K., and PRIYADARSHAN, P. M. 1982. *Triticale:* Present status and future prospects. *Adv. Genet.* 21:256–346.

HASELTINE, F. P., and OHNO, S. 1981. Mechanisms of gonadal differentiation. *Science* 211:1272–78.

HASSOLD, T., et al. 1980. Effect of maternal age on autosomal trisomies. *Ann. Hum. Genet. (London)* 44:29–36.

HASSOLD, T. J., and JACOBS, P. A. 1984. Trisomy in man. *Annu. Rev. Genet.* 18:69–98.

HEAD, G., MAY, R., and PENDLETON, L. 1987. Environmental determination of sex in the reptiles. *Nature* 329:198–99.

HECHT, F. 1988. Enigmatic fragile sites on human chromosomes. *Trends Genet.* 4:121–22.

HODGKIN, J. 1990. Sex determination compared in *Drosophila* and *Caenorhabditis. Nature* 344:721–28.

HOOK, E. B. 1973. Behavioral implications of the humans XYY genotype. *Science* 179:139–50.

HSU, T. C. 1979. *Human and mammalian cytogenetics: A historical perspective.* New York: Springer-Verlag.

HULSE, J. H., and SPURGEON, D. 1974. Triticale. *Sci. Am.* (Aug.) 231:72–81.

JACOBS, P. A., et al. 1974. A cytogenetic survey of 11,680 newborn infants. *Ann. Hum. Genet.* 37:359–76.

KAISER, P. 1984. Pericentric inversions: Problems and significance for clinical genetics. *Hum. Genet.* 68:1–47.

KAY, G. F., et al. 1993. Expression of *Xist* during mouse development suggests a role in the initiation of X chromosome inactivation. *Cell* 72:171–82.

KHUSH, G. S. 1973. *Cytogenetics of aneuploids.* Orlando, FL: Academic Press.

KOOPMAN, P., et al. 1991. Male development of chromosomally female mice transgenic for *Sry. Nature* 351:117–21.

LEWIS, E. B. 1950. The phenomenon of position effect. *Adv. Genet.* 3:73–115.

LEWIS, W. H., ed. 1980. *Polyploidy: Biological relevance.* New York: Plenum Press.

LUCCHESI, J. 1983. The relationship between gene dosage, gene expression, and sex in *Drosophila. Dev. Genet.* 3:275–82.

LYON, M. F. 1961. Gene action in the X-chromosome of the mouse (*Mus musculus* L.). *Nature* 190:372–73.

———. 1962. Sex chromatin and gene action in the mammalian X chromosome. *Am. J. Hum. Genet.* 14:135–48.

———. 1972. X-chromosome inactivation and developmental patterns in mammals. *Biol. Rev.* 47:1–35.

———. 1988. X-chromosome inactivation and the location and expression of X-linked genes. *Am. J. Hum. Genet.* 42:8–16.

———. 1993. Epigenetic inheritance in mammals. *Trends Genet.* 9:123–28.

MANTELL, S. H., MATHEWS, J. A., and McKEE, R. A. 1985. *Principles of plant biotechnology: An introduction to genetic engineering in plants.* Oxford: Blackwell.

McLAREN, A. 1988. Sex determination in mammals. *Trends Genet.* 4:153–57.

McMILLEN, M. M. 1979. Differential mortality by sex in fetal and neonatal deaths. *Science* 204:89–91.

OBE, G., and BASLER, A. 1987. *Cytogenetics: Basic and applied aspects.* New York: Springer-Verlag.

OHNO, S. 1970. *Evolution by gene duplication.* New York: Springer-Verlag.

OOSTRA, B. A., and VERKERK, A. J. 1992. The fragile X syndrome: Isolation of the *FMR-1* gene and characterization of the fragile X mutation. *Chromosoma* 101:381–87.

PAGE, D. C., et al. 1987. The sex-determining region of the human Y chromosome encodes a finger protein. *Cell* 51:1091–1104.

PATTERSON, D. 1987. The causes of Down syndrome. *Sci. Am.* (Aug.) 257:52–61.

PENNY, G. D., et al. 1996. Requirement for *Xist* in X chromosome inactivation. *Nature* 379:131–37.

ROONEY, D. E., and CZEPULKOWSKI, B. H., eds. 1986. *Human cytogenetics: A practical approach.* Oxford: IRL Press.

SCHIMKE, R. T., ed. 1982. *Gene amplification.* Cold Spring Harbor, NY: Cold Spring Harbor Laboratory Press.

SHEPARD, J. F. 1982. The regeneration of potato plants from protoplasts. *Sci. Am.* (May) 246:154–66.

SHEPARD, J. F., et al. 1983. Genetic transfer in plants through interspecific protoplast fusion. *Science* 219:683–88.

SIMMONDS, N. W., ed. 1976. *Evolution of crop plants.* London: Longman.

SMITH, G. F., ed. 1984. *Molecular structure of the number 21 chromosome and Down syndrome.* New York: New York Academy of Sciences.

STEBBINS, G. L. 1966. Chromosome variation and evolution. *Science* 152:1463–69.

STRICKBERGER, M. W. 1996. *Evolution,* 2nd ed. Boston: Jones and Bartlett.

SUTHERLAND, G. 1984. The fragile X chromosome. *Int. Rev. Cytol.* 81:107–43.

———. 1985. The enigma of the fragile X chromosome. *Trends Genet.* 1:108–11.

SWANSON, C. P., MERZ, T., and YOUNG, W. J. 1981. *Cytogenetics: The chromosome in division, inheritance, and evolution,* 2nd ed. Englewood Cliffs, NJ: Prentice-Hall.

TAYLOR, A. I. 1968. Autosomal trisomy syndromes: A detailed study of 27 cases of Edwards syndrome and 27 cases of Patau syndrome. *J. Med. Genet.* 5:227–52.

THERMAN, E. 1980. *Human chromosomes.* New York: Springer-Verlag.

TJIO, J. H., and LEVAN, A. 1956. The chromosome number of man. *Hereditas* 42:1–6.

TURPIN, R., and LeJEUNE, J. 1969. *Human afflictions and chromosomal aberrations.* Oxford: Pergamon Press.

WESTERGAARD, M. 1958. The mechanism of sex determination in dioecious flowering plants. *Adv. Genet.* 9:217–81.

WHARTON, K. A., et al. 1985. *opa:* A novel family of transcribed repeats shared by the *Notch* locus and other developmentally regulated loci in *D. melanogaster. Cell* 40:55–62.

WILKINS, L. E., BROWN, J. A., and WOLF, B. 1980. Psychomotor development in 65 home-reared children with cri-du-chat syndrome. *J. Pediatr.* 97:401–5.

WITKIN, H. A., et al. 1976. Criminality in XYY and XXY men. *Science* 193:547–55.

YUNIS, J. J., ed. 1977. *New chromosomal syndromes.* Orlando, FL: Academic Press.

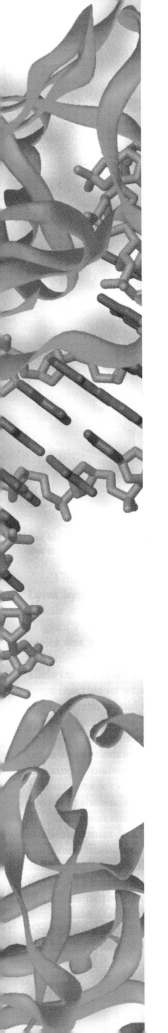

$Part$ Two

Molecular Basis of Heredity

C   H   A   P   T   E   R

10

Structure and Analysis of DNA and RNA

CHAPTER CONCEPTS

With few exceptions, the nucleic acid DNA serves as the genetic material in every living thing. The structure of DNA provides the chemical basis for storing and expressing genetic information within cells, as well as transmitting it to future generations. The molecule takes the form of a double-stranded helix united by hydrogen bonds formed between complementary nucleotides. RNA shares many similarities to DNA in its structure, but is single-stranded. RNA most often functions during the expression of genetic information, and in some viruses it serves as the genetic material.

In Part 1 of the text we discussed the presence of genes on chromosomes that control phenotypic traits and the way in which the chromosomes are transmitted through gametes to future offspring. Logically, some form of information must be contained in genes, which, when passed to a new generation, influences the form and characteristics of the offspring; this is called the **genetic information**. We might also conclude that this same information in some way directs the many complex processes leading to the adult form.

Until 1944 it was not clear what chemical component of the chromosome makes up genes and constitutes the genetic material. Because chromosomes were known to have both a nucleic acid and a protein component, both were considered candidates. In 1944, however, there emerged direct experimental evidence that the nucleic

James D. Watson and Francis H. C. Crick with one of their early DNA models.

acid, DNA, serves as the informational basis for the process of heredity.

Once the importance of DNA in genetic processes was realized, work was intensified with the hope of discerning not only the structural basis of this molecule but also the relationship of its structure to its function. Between 1944 and 1953, many scientists sought information that might answer one of the most significant and intriguing questions in the history of biology: How does DNA serve as the genetic basis for living processes? The answer was believed to depend strongly on the chemical structure of the DNA molecule, given the complex but orderly functions ascribed to it.

These efforts were rewarded in 1953 when James Watson and Francis Crick set forth their hypothesis for the double-helical nature of DNA. The assumption that the molecule's functions would be clarified more easily once its general structure was determined proved to be correct. This chapter initially reviews the evidence that DNA is the genetic material and then discusses the elucidation of its structure.

Characteristics of the Genetic Material

The genetic material has several characteristics: **replication**, **storage of information**, **expression of that information**, and **variation by mutation**. "Replication" of the genetic material is one facet of the cell cycle, a fundamental property of all living organisms. Once the genetic material of cells has been replicated, it must then be partitioned equally into daughter cells. During the formation of gametes, the genetic material is also replicated but is partitioned so that each cell gets only one-half of the original amount of genetic material. This process is called *meiosis*. Although the products of mitosis and meiosis are different, these processes are both part of the more general phenomenon of cellular reproduction.

The characteristic of "storage" may be viewed as genetic information that is present as a repository of all hereditary characteristics of an organism. However, that information may or may not be expressed. It is clear that, whereas most cells contain a complete complement of DNA, at any given point they express only a part of this genetic potential. For example, bacteria turn many genes on only in response to specific environmental conditions, only to turn them off when such conditions change. In vertebrates, skin cells may display active melanin genes but never activate their hemoglobin genes; digestive cells activate many genes specific to their function, but do not activate their melanin genes.

Inherent in the concept of storage is the need for the genetic material to be able to encode the nearly infinite variety of gene products found among the countless forms of life present on our planet. The chemical language of the genetic material must be capable of this potential task as it stores information and as it is transmitted to progeny cells and organisms.

"Expression" of the stored genetic information is a complex process and is the basis for the concept of **information flow** within the cell. Figure 10.1 shows a simplified illustration of this concept. The initial event is the **transcription** of DNA, resulting in the synthesis of three types of RNA molecules: **messenger RNA (mRNA)**, **transfer RNA (tRNA)**, and **ribosomal RNA (rRNA)**. Of these, mRNAs are translated into proteins. Each type of mRNA is the product of a specific gene and leads to the synthesis of a different protein. **Translation** occurs in conjunction with rRNA-containing ribosomes and involves tRNA, which acts as an adaptor to convert the chemical information in mRNA to the amino acids that make up proteins. Collectively, these processes serve as the foundation for the **central dogma of molecular genetics**: "DNA makes RNA, which makes proteins."

The genetic material is also the source of newly arising "variability" among organisms through the process of mutation. If a mutation—a change in the chemical composition of DNA—occurs, the alteration will be reflected during transcription and translation, often affecting the specified protein. If a mutation is present in gametes, it will be passed to future generations and, with time, may become distributed in the population. Genetic variation,

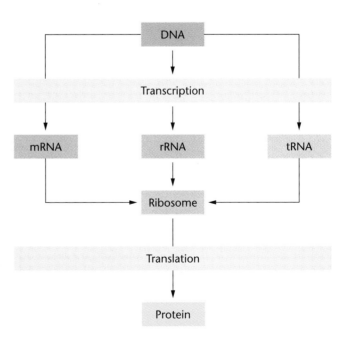

FIGURE 10.1   A simplified view of information flow involving DNA, RNA, and proteins within cells.

which also includes rearrangements within and between chromosomes (see Chapter 9), provides the raw material for the process of evolution.

The Genetic Material: 1900–1944

The idea that genetic material is physically transmitted from parent to offspring has been accepted for as long as the concept of inheritance has existed. Beginning in the late nineteenth century, research into the structure of biomolecules progressed considerably, setting the stage for the description of the genetic material in chemical terms. Although proteins and nucleic acid were both considered major candidates for the role of the genetic material, many geneticists, until the 1940s, favored proteins. Three factors contributed to this belief.

First, proteins are abundant in cells. Although the protein content may vary considerably, these molecules account for over 50 percent of the dry weight of cells. Because cells contain such a large amount and variety of proteins, it is not surprising that early geneticists believed that some of this protein could function as the genetic material.

The second factor was the accepted proposal for the chemical structure of nucleic acids during the early to mid-1900s. DNA was first studied in 1868 by Friedrick Miescher, a Swiss chemist. He was able to separate nuclei from the cytoplasm of cells and then isolate from these nuclei an acidic substance that he called **nuclein**. Miescher showed that nuclein contained large amounts of phosphorus and no sulfur, characteristics that differentiate it from proteins.

As analytical techniques were improved, nucleic acids, including DNA, were shown to be composed of four similar molecular building blocks called nucleotides. About 1910, Phoebus A. Levene proposed the **tetranucleotide hypothesis** to explain the chemical arrangement of these nucleotides in nucleic acids. He proposed that a very simple four-nucleotide unit, as shown in Figure 10.2, was repeated over and over in DNA. Levene based his proposal on studies of the composition of the four types of nucleotides. Although his actual data revealed proportions of the four nucleotides that varied considerably, he assumed a 1:1:1:1 ratio. The discrepancy was ascribed to inadequate analytical technique.

Because a single covalently bonded tetranucleotide structure was relatively simple, geneticists believed nucleic acids could not provide the large amount of chemical variation expected for the genetic material. Proteins, on the other hand, contain 20 different amino acids, thus providing the basis for substantial variation. As a result, attention was directed away from nucleic acids, strengthening the speculation that proteins served as the genetic material.

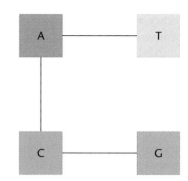

FIGURE 10.2 Diagrammatic depiction of Levene's proposed tetranucleotide containing one molecule each of the four nitrogenous bases: adenine (A), cytosine (C), guanine (G), and thymine (T). Each block represents a nucleotide. Other potential sequences of four nucleotides are possible.

The third contributing factor simply concerned the areas of most active research in genetics. Before 1940, most geneticists were engaged in the study of transmission genetics and mutation. The excitement generated in these areas undoubtedly diluted the concern for finding the precise molecule that serves as the genetic material. Thus, proteins were the most promising candidate and were accepted rather passively.

Between 1910 and 1930, other proposals for the structure of nucleic acids were advanced, but they were generally overturned in favor of the tetranucleotide hypothesis. It was not until the 1940s that the work of Erwin Chargaff led to the realization that Levene's hypothesis was incorrect. Chargaff showed that, for most organisms, the 1:1:1:1 ratio was inaccurate, disproving Levene's hypothesis.

Evidence Favoring DNA as the Genetic Material in Bacteria and Bacteriophages

The 1944 publication by Oswald Avery, Colin MacLeod, and Maclyn McCarty concerning the chemical nature of a "transforming principle" in bacteria marked the initial event leading to the acceptance of DNA as the genetic material. Along with the subsequent findings of other research teams, this work constituted direct experimental proof that, in the organisms studied, DNA, and not protein, is the biomolecule responsible for heredity. This period marked the beginning of an era of discovery in biology that has revolutionized our understanding of life on Earth. The impact of these findings parallels the work that followed the publication of Darwin's theory of evolution and that following the rediscovery of Mendel's postulates of transmission genetics. Together, these constituted three great revolutions in biology.

The initial evidence implicating DNA as the genetic material was derived from studies of prokaryotic bacteria and viruses that infect them. The reasons for the use of bacteria and bacterial viruses will become apparent as we discuss the experiments. Primarily, the reason is that bacteria and viruses are capable of rapid growth because they complete their life cycles in hours. They can also be experimentally manipulated, and mutations can be easily induced and selected, making them ideal for experimentation of this sort.

Transformation Studies

The research that provided the foundation for Avery, MacLeod, and McCarty's work was initiated in 1927 by Frederick Griffith, a medical officer in the British Ministry of Health. He performed experiments with several different strains of the bacterium *Diplococcus pneumoniae.** Some were **virulent** strains, which cause pneumonia in certain vertebrates (notably humans and mice), whereas others were **avirulent** strains, which do not cause illness.

The difference in virulence is related to the polysaccharide capsule of the bacterium. Virulent strains have this capsule, whereas avirulent strains do not. The nonencapsulated bacteria are readily engulfed and destroyed by phagocytic cells in the animal's circulatory system. Virulent bacteria, which possess the polysaccharide coat, are not easily engulfed; they are able to multiply and cause pneumonia.

The presence or absence of the capsule is the basis for another characteristic difference between virulent and avirulent strains. Encapsulated bacteria form a **smooth**, shiny-surfaced colony (*S*) when grown on an agar culture plate; nonencapsulated strains produce **rough** colonies (*R*) (Figure 10.3). This allows virulent and avirulent strains to be distinguished easily by standard microbiological culture techniques.

Each strain of *Diplococcus* may be one of dozens of different types called **serotypes**. The specificity of the serotype is due to the detailed chemical structure of the polysaccharide constituent of the thick, slimy capsule. Serotypes are identified by immunological techniques and are usually designated by Roman numerals. In the United States, types I and II are most common in causing pneumonia. Griffith used types II and III in the critical experiments that led to new concepts about the genetic material. Table 10.1 summarizes the characteristics of Griffith's two strains.

Griffith knew from the work of others that only living virulent cells would produce pneumonia in mice. If

*Note that this organism is now designated *Streptococcus pneumonia*.

TABLE 10.1 Strains of *Diplococcus pneumoniae* Used by Frederick Griffith in his Original Transformation Experiments

Serotype	Colony Morphology	Capsule	Virulence
II*R*	Rough	Absent	Avirulent
III*S*	Smooth	Present	Virulent

heat-killed virulent bacteria are injected into mice, no pneumonia results, just as living avirulent bacteria fail to produce the disease. Griffith's critical experiment (Figure 10.3) involved an injection into mice of living II*R* (avirulent) cells combined with heat-killed III*S* (virulent) cells. Since neither cell type caused death in mice when injected alone, Griffith expected that the double injection would not kill the mice. But, after five days, all mice receiving double injections were dead. Analysis of the blood of the dead mice revealed large numbers of living type III*S* (virulent) bacteria!

As far as could be determined regarding the capsular polysaccharide, these III*S* bacteria were identical to the III*S* strain from which the heat-killed cell preparation had been made. The control mice, injected only with living avirulent II*R* bacteria for this set of experiments, did not develop pneumonia and remained healthy. This finding ruled out the possibility that the avirulent II*R* cells had simply changed (or mutated) to virulent III*S* cells in the absence of the heat-killed III*S* fraction. Instead, some type of interaction was required between living II*R* and heat-killed III*S* cells.

Griffith thus concluded that the heat-killed III*S* bacteria were somehow responsible for converting live avirulent II*R* cells into virulent III*S* ones. Calling the phenomenon **transformation**, he suggested that the **transforming principle** might be some part of the polysaccharide capsule *or* some compound required for capsule synthesis, although the capsule alone did not cause pneumonia. To use Griffith's term, the transforming principle from the dead III*S* cells served as a "pabulum" for the II*R* cells.

Griffith's work led other physicians and bacteriologists to research the phenomenon of transformation. By 1931, Henry Dawson, at the Rockefeller Institute, had confirmed Griffith's observations and extended his work one step further. Dawson and his co-workers showed that transformation could occur *in vitro* (in a test tube). When heat-killed III*S* cells were incubated with living II*R* cells, living III*S* cells were recovered. Therefore, injection into mice was not necessary for transformation to occur. By 1933, Lionel J. Alloway had refined the *in vitro* system by using crude extracts of *S* cells and living *R*

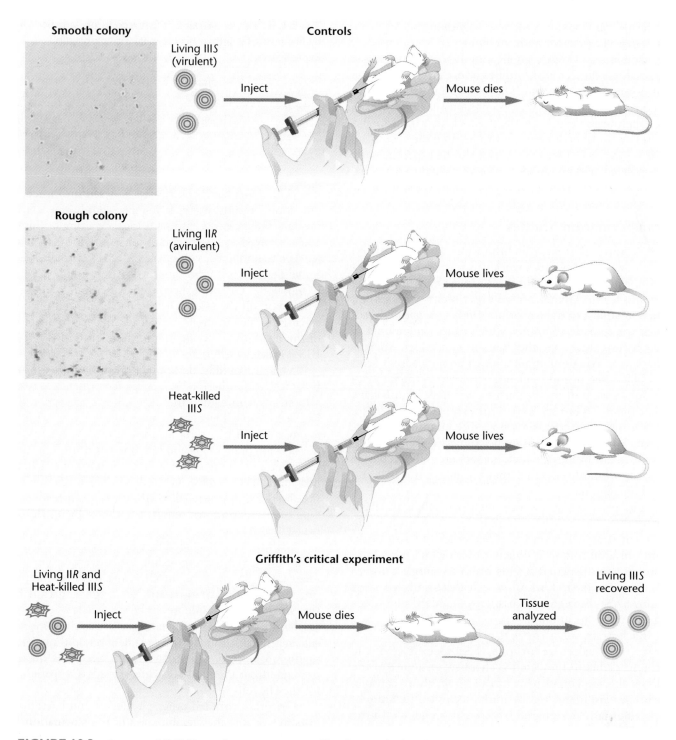

FIGURE 10.3   Summary of Griffith's transformation experiment. The photographs show bacterial colonies containing cells with capsules (type IIIS) and without capsules (type IIR).

cells. The soluble filtrate from the heat-killed *S* cells was as effective in inducing transformation as were the intact cells! Alloway and others did not view transformation as a genetic event, but rather as a physiological modification of some sort. Nevertheless, the experimental evidence that a chemical substance was responsible for transformation was quite convincing.

Then, in 1944, after 10 years of work, Avery, MacLeod, and McCarty published their results in what is now regarded as a classic paper in the field of molecular

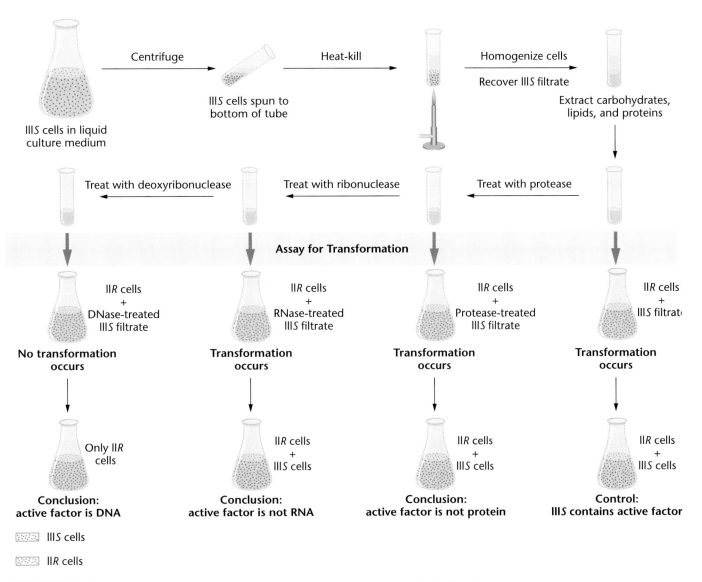

FIGURE 10.4   Summary of Avery, MacLeod, and McCarty's experiment demonstrating that DNA is the transforming principle.

genetics. They reported that they had obtained the transforming principle in a purified state, and that beyond reasonable doubt, the molecule responsible for transformation was DNA.

The details of the work, sometimes called the Avery, MacLeod, and McCarty experiment, are outlined in Figure 10.4. These researchers began their isolation procedure with large quantities (50 to 75 liters) of liquid cultures of type IIIS virulent cells. The cells were centrifuged, collected, and heat-killed. Following homogenization and several extractions with the detergent deoxycholate (DOC), they obtained a soluble filtrate that, when tested, still contained the transforming principle. Protein was removed from the active filtrate by several chloroform extractions, and polysaccharides were enzymatically digested

and removed. Finally, precipitation with ethanol yielded a fibrous mass that still retained the ability to induce transformation of type IIR avirulent cells. From the original 75-liter sample, the procedure yielded 10 to 25 mg of the "active factor."

Further testing established beyond a reasonable doubt that the transforming principle was DNA. The fibrous mass was first analyzed for its nitrogen/phosphorus ratio, which was shown to coincide with the ratio of "sodium desoxyribonucleate," the chemical name then used to describe DNA. To solidify their findings, they sought to eliminate, to the greatest extent possible, all probable contaminants from their final product. Thus, it was treated with the proteolytic enzymes trypsin and chymotrypsin and then with an RNA-digesting enzyme, called ribonuclease. Such treatments destroyed any remaining activity of proteins and RNA. Nevertheless, transforming activity still remained. Chemical testing of the final product gave strong positive reactions for DNA. The final confirmation came with experiments using crude samples of the DNA-digesting enzyme **deoxyribonuclease**, which was isolated from dog and rabbit sera. Digestion with this enzyme was shown to destroy transforming activity. There could be little doubt that the active transforming principle was DNA!

The great amount of work, the confirmation and reconfirmation of the conclusions drawn, and the unambiguous logic of the experimental design involved in the research of these three scientists are truly impressive. The conclusion to the 1944 publication was, however, very simply stated: "The evidence presented supports the belief that a nucleic acid of the desoxyribose* type is the fundamental unit of the transforming principle of *Pneumococcus* Type III."

Avery and his co-workers recognized the genetic and biochemical implications of their work when they observed that "nucleic acids of this type must be regarded not merely as structurally important but as functionally active in determining the biochemical activities and specific characteristics of pneumococcal cells." This suggested that the transforming principle interacts with the II*R* cell and gives rise to a coordinated series of enzymatic reactions culminating in the synthesis of the type III*S* capsular polysaccharide. They emphasized that, once transformation occurs, the capsular polysaccharide is produced in successive generations. Transformation is therefore heritable, and the process affects the genetic material.

Immediately after the publication of the report, several investigators turned to or intensified their studies of transformation in order to clarify the role of DNA in genetic mechanisms. In particular, the work of Rollin Hotchkiss was instrumental in confirming that the critical factor in transformation was DNA, and not protein. In 1949, in a separate study, Harriet Taylor isolated an **extremely rough (*ER*)** mutant strain from a rough (*R*) strain. This *ER* strain produced colonies that were more irregular than the *R* strain. The DNA from *R* accomplished the transformation of *ER* to *R*. Thus, the *R* strain, which served as the recipient in the Avery experiments, was shown also to be able to serve as the DNA donor in transformation.

Transformation has now been shown to occur in *Hemophilus influenzae*, *Bacillus subtilis*, *Shigella paradysenteriae*, and *Escherichia coli*, among many other microorganisms. Transformation of numerous genetic traits other than colony morphology has also been demonstrated, including ones involving resistance to antibi-

---

Desoxyribose is now spelled *deoxyribose*.

otics and the ability to metabolize various nutrients. These observations further strengthened the belief that transformation by DNA is primarily a genetic event, rather than simply a physiological change. This idea is pursued again in the "Insights and Solutions" section at the end of this chapter.

The Hershey–Chase Experiment

The second major piece of evidence supporting DNA as the genetic material was provided by the study of the infection of the bacterium *Escherichia coli* by one of the viruses to which it serves as a host, **bacteriophage T2**. Often referred to simply as a **phage**, the virus consists of a protein coat surrounding a core of DNA. Electron micrographs have revealed the phage's external structure to be composed of a hexagonal head plus a tail. The general aspects of the life cycle of a T-even bacteriophage such as T2, as known in 1952, is shown in Figure 10.5. Briefly, the phage adsorbs to the bacterial cell and some component of the phage enters the bacterial cell. Following this infection step, the viral information "commandeers" the cellular machinery of the host and proceeds to undergo viral reproduction. In a reasonably short time, many new phages are constructed and the bacterial cell is lysed, releasing the progeny viruses.

In 1952, Alfred Hershey and Martha Chase published the results of experiments designed to clarify the events leading to phage reproduction, as described above. Several of the experiments clearly established the independent functions of phage protein and nucleic acid in the reproduction process associated with the bacterial cell. Hershey and Chase knew from existing data that:

1. T2 phages consist of approximately 50 percent protein and 50 percent DNA.

2. Infection is initiated by adsorption of the phage by its tail fibers to the bacterial cell.

3. The production of new viruses occurs within the bacterial cell.

It appeared that some molecular component of the phage, DNA and/or protein, enters the bacterial cell and directs viral reproduction. Which was it?

Hershey and Chase used the radioisotopes ^{32}P and ^{35}S to follow the molecular components of phages during infection. Because DNA contains phosphorus (P) but not sulfur, ^{32}P effectively labels DNA, and because proteins contain sulfur (S) but not phosphorus, ^{35}S labels protein. *This was a key point in the experiment.* If *E. coli* cells are first grown in the presence of ^{32}P *or* ^{35}S and then infected with T2 viruses, the progeny phages will have either a radioactively-labeled DNA core *or* a radioactively-labeled protein coat,

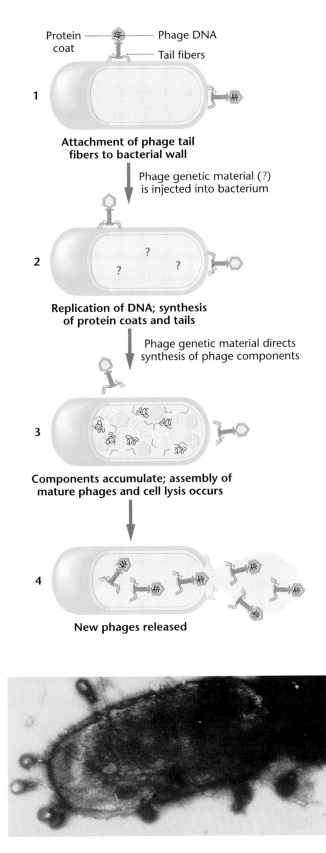

1

Protein coat — Phage DNA
Tail fibers

**Attachment of phage tail
fibers to bacterial wall**

Phage genetic material (?)
is injected into bacterium

2

? ? ?

**Replication of DNA; synthesis
of protein coats and tails**

Phage genetic material directs
synthesis of phage components

3

**Components accumulate; assembly of
mature phages and cell lysis occurs**

4

New phages released

FIGURE 10.5 The life cycle of a T-even bacteriophage. The electron micrograph shows *E. coli* during infection by phage T2.

respectively. These labeled phages can be isolated and used to infect unlabeled bacteria (Figure 10.6).

When labeled phages and unlabeled bacteria are mixed, an adsorption complex is formed as the phages attach their tail fibers to the bacterial wall. These complexes were isolated and subjected to a high shear force by placing them in a blender. This force stripped off the attached phages. When the mixture was centrifuged, the lighter phage particles separated from the heavier bacterial cells, allowing the isolation of both components (Figure 10.6). By tracing the radioisotopes, Hershey and Chase were able to demonstrate that most of the ^{32}P-labeled DNA had been transferred into the bacterial cell following adsorption; on the other hand, most of the ^{35}S-labeled protein remained outside the bacterial cell and was recovered in the phage "ghosts" (empty phage coats) after the blender treatment. Following this separation, the bacterial cells, which now contained viral DNA, were eventually lysed as new phages were produced. These progeny phages contained ^{32}P, but not ^{35}S.

Hershey and Chase interpreted these results to indicate that the protein of the phage coat remains outside the host cell and is not involved in directing the production of new phages. On the other hand, and most important, phage DNA enters the host cell and directs phage multiplication. They had demonstrated that in phage T2, DNA, not protein, is the genetic material.

This experimental work, along with that of Avery and his colleagues, provided convincing evidence to most geneticists that DNA was the molecule responsible for heredity. Since then, many significant findings have been based on this supposition. These many findings, constituting the field of molecular genetics, are discussed in detail in subsequent chapters.

Transfection Experiments

During the eight years following the publication of the Hershey–Chase experiment, additional research using bacterial viruses provided even more solid proof that DNA is the genetic material. In 1957, several reports demonstrated that if *E. coli* was treated with the enzyme **lysozyme**, the outer wall of the cell could be removed without destroying the bacterium. Enzymatically treated cells are naked, so to speak, and contain only the cell membrane as their outer boundary. Such structures are called **spheroplasts** (or **protoplasts**). John Spizizen and Dean Fraser independently reported that by using spheroplasts, they were able to initiate phage reproduction with disrupted T2 particles. That is, provided that the cell wall is absent, it is not necessary for a virus to be intact in order for infection to occur. Thus, the outer protein coat structure may be essential to the movement of

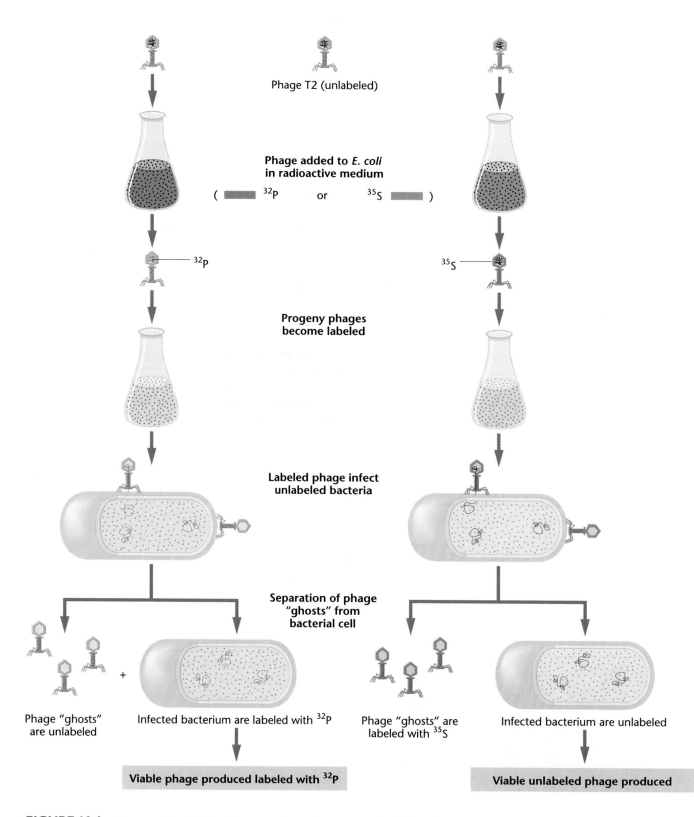

FIGURE 10.6 Summary of the Hershey–Chase experiment demonstrating that DNA, and not protein, is responsible for directing the reproduction of phage T2 during the infection of *E. coli*.

DNA through the intact cell wall, but it is not essential for infection when spheroplasts are used.

Similar but more refined experiments were reported in 1960 by George Guthrie and Robert Sinsheimer. DNA was purified from bacteriophage φX174, a small phage that contains a single-stranded, circular DNA molecule of some 5386 nucleotides. When added to *E. coli* protoplasts, the purified DNA resulted in the production of complete φX174 bacteriophages. This process of infection by only the viral nucleic acid, called **transfection**, proved conclusively that φX174 DNA alone contains all the necessary information for production of mature viruses. Thus, the evidence supporting the conclusion that DNA serves as the genetic material was further strengthened, even though all direct evidence thus far had been obtained from bacterial and viral studies.

Indirect Evidence Favoring DNA in Eukaryotes

In 1950, eukaryotic organisms were not amenable to the types of experiments performed to demonstrate that DNA is the genetic material in bacteria and viruses. Nevertheless, it was generally assumed that the genetic material would be a universal substance and also serve this role in eukaryotes. Support for this assumption initially relied on many diverse circumstantial (indirect) observations. Together, they provided support that DNA serves as the genetic material in eukaryotes. We shall look at several such observations.

Distribution of DNA

The genetic material should be found where it functions—in the nucleus as part of chromosomes. Both DNA and protein fit this criterion. However, protein is also abundant in the cytoplasm, whereas DNA is not. Both mitochondria and chloroplasts are known to perform genetic function, and DNA is also present in these organelles. Thus, DNA is found only where primary genetic function occurs. Protein, on the other hand, is found everywhere in the cell. These observations are consistent with the interpretation favoring DNA over proteins as the genetic material.

Because it had earlier been established that chromosomes within the nucleus contain the genetic material, a correlation was expected to exist between the ploidy of a cell (e.g., *n*, *2n*, etc.) and the quantitative amount of the molecule serving as the genetic material. Meaningful comparisons can be made between the amount of DNA and protein in gametes (sperm and eggs) and somatic (or

TABLE 10.2 DNA Content of Haploid Versus Diploid Cells of Various Species (in picograms)

Organism	*n*	*2n*
Human	3.25	7.30
Chicken	1.26	2.49
Trout	2.67	5.79
Carp	1.65	3.49
Shad	0.91	1.97

Note: Sperm (*n*) and nucleated precursors to red blood cells (*2n*) were used to contrast ploidy levels.

body) cells. The latter are recognized as being **diploid** (*2n*) and containing twice the number of chromosomes as gametes, which are **haploid** (*n*).

Table 10.2 compares the amount of DNA found in haploid sperm and diploid nucleated precursors of red blood cells from a variety of organisms. A close correlation exists between the amount of DNA and the number of sets of chromosomes. No such correlation can be observed between gametes and diploid cells for proteins. These data thus provide further circumstantial evidence favoring DNA over proteins as the genetic material of eukaryotes.

Mutagenesis

Ultraviolet (UV) light is one of a number of agents capable of inducing mutations in the genetic material. Bacteria and other organisms can be irradiated with various wavelengths of ultraviolet light and the effectiveness of each wavelength measured by the number of mutations it induces. When the data are plotted, an **action spectrum** of UV light as a mutagenic agent is obtained. This action spectrum can then be compared with the **absorption spectrum** of any molecule suspected to be the genetic material (Figure 10.7). The molecule serving as the genetic material is expected to absorb UV light at the wavelengths found to be mutagenic.

UV light is most mutagenic at the wavelength (λ) of 260 nanometers (nm). Both DNA and RNA absorb UV light most strongly at 260 nm. On the other hand, protein absorbs most strongly at 280 nm, yet no significant mutagenic effects are observed at this wavelength. This indirect evidence supports the idea that a nucleic acid is the genetic material and tends to exclude protein.

Direct Evidence for DNA: Eukaryotic Data

Although the circumstantial evidence just described does not constitute direct proof that DNA is the genetic material in eukaryotes, these observations spurred researchers

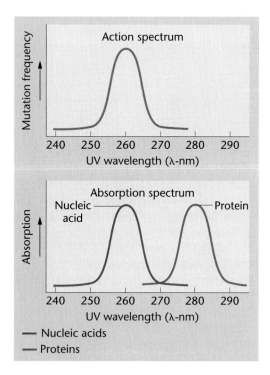

FIGURE 10.7 The "action spectrum" determining the most effective mutagenic wavelength contrasted with the "absorption spectrum" of nucleic acids vs. proteins in the UV range of the electromagnetic spectrum.

to forge ahead, basing their work on this hypothesis. Today, there is no doubt of the validity of this conclusion; DNA *is* the genetic material in eukaryotes.

The strongest direct evidence has been provided by the application of molecular analysis referred to as **recombinant DNA technology**. In such research, segments of eukaryotic DNA corresponding to specific genes are isolated and literally spliced into bacterial DNA. Such a complex can be inserted into a bacterial cell and its **genetic expression** monitored. If a eukaryotic gene is introduced, the presence of the corresponding eukaryotic protein product demonstrates directly that this DNA is not only present, but functional in the bacterial cell. This has been shown to be the case in countless instances. For example, the human gene products specifying insulin and interferon are produced by bacteria following the insertion of human genes that encode these proteins. As the bacterium divides, the eukaryotic DNA is replicated along with the host DNA and is distributed to the daughter cells, which also express the human genes and synthesize the corresponding proteins.

The availability of vast amounts of DNA coding for specific genes, available as a result of recombinant DNA research, has led to other direct evidence that DNA serves as the genetic material. Work in the laboratory of Beatrice Mintz has demonstrated that DNA encoding the human β-globin gene, when microinjected into a fertilized mouse egg, is later found to be present and expressed in adult mouse tissue and transmitted to and expressed in that mouse's progeny. These mice are examples of what are called **transgenic animals**.

More recent work has introduced *rat* DNA encoding a growth hormone into fertilized *mouse* eggs. About one-third of the resultant mice grew to twice their normal size. This indicated that foreign DNA was present *and* functional in the experimental mice. Subsequent generations of mice also grew to a large size.

We will pursue the topic of recombinant DNA again later (see Chapters 15 and 16). The point to be made here is that in eukaryotes, DNA has been shown directly to meet the requirement of expression of genetic information. Later we will see exactly how DNA is stored, replicated, expressed, and mutated.

RNA as the Genetic Material

Some viruses contain an RNA core rather than one composed of DNA. In these viruses, it would thus appear that RNA might serve as the genetic material—an exception to the general rule that DNA performs this function. In 1956, it was demonstrated that when purified RNA from **tobacco mosaic virus** (**TMV**) was spread on tobacco leaves, the characteristic lesions caused by TMV would appear later on the leaves. It was concluded that RNA is the genetic material of this virus.

Soon afterwards, another type of experiment with TMV was reported by Heinz Fraenkel-Conrat and B. Singer, as illustrated in Figure 10.8. These scientists discovered that the RNA core and the protein coat from wild-type TMV and other viral strains could be isolated separately. In their work, RNA and coat proteins were separated and isolated from TMV and a second viral strain, **Holmes ribgrass** (**HR**). Then, mixed viruses were reconstituted from the RNA of one strain and the protein of the other. When this "hybrid" virus was spread on tobacco leaves, the lesions that developed corresponded to the type of RNA in the reconstituted virus; that is, viruses with wild-type TMV RNA and HR protein coats produced TMV lesions and vice versa. Again, it was concluded that RNA serves as the genetic material in these viruses.

In 1965 and 1966, Norman R. Pace and Sol Spiegelman further demonstrated that RNA from the phage Qβ could be isolated and replicated *in vitro*. Replication was dependent on an enzyme, **RNA replicase**, which was isolated from host *E. coli* cells following normal infection. When the RNA replicated *in vitro* was added to *E. coli* protoplasts, infection and viral multiplication occurred. Thus,

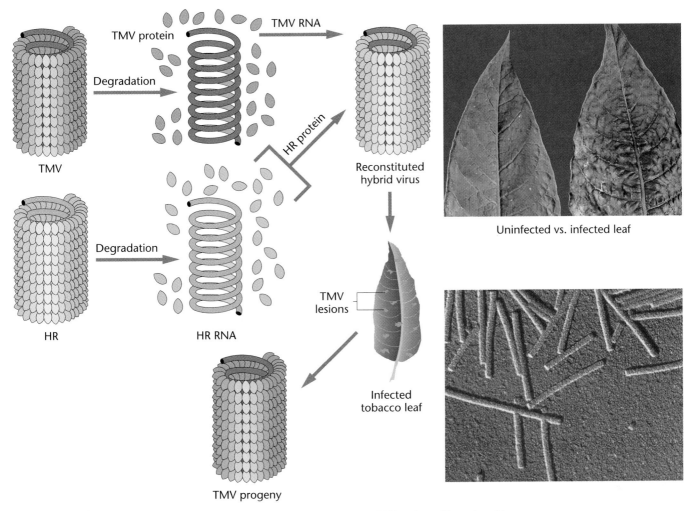

FIGURE 10.8 Reconstitution of hybrid tobacco mosaic viruses. In the hybrid, RNA is derived from the wild-type TMV virus, while the protein subunits are derived from the HR strain. Following infection, viruses are produced with protein subunits characteristic of the wild-type TMV strain and not those of the HR strain. The top photograph shows TMV lesions on a tobacco leaf compared to an uninfected leaf. At the bottom is an electron micrograph of mature viruses.

RNA synthesized in a test tube can amply serve as the genetic material in these phages by directing the production of all the components necessary for viral replication.

Finally, one other group of RNA-containing viruses bears mentioning. These are the **retroviruses**, which replicate in an unusual way. Their RNA serves as a template for the synthesis of the complementary DNA molecule. The process, designated as **reverse transcription**, occurs under the direction of an RNA-dependent DNA polymerase enzyme called reverse transcriptase. The viral genetic material is "represented" by this DNA intermediate, which may be incorporated into the genome of the host cell. Once present, if the DNA is expressed, transcription yields retroviral RNA genome. Retroviruses, including the human immunodeficiency virus (HIV) that causes AIDS, will be discussed at greater length in Chapter 22.

Nucleic Acid Chemistry

Having established thus far the critical importance of DNA and RNA in genetic processes, we shall now provide a brief introduction to the chemical basis of these molecules. As we shall see, the structural components of DNA and RNA are very similar. This chemical similarity is important in the coordinated functions played by these molecules during gene expression. Like the other major groups of organic biomolecules (proteins, carbohydrates, and lipids), nucleic acid chemistry is based on a variety of similar building blocks that are polymerized into chains of varying lengths.

Nucleotides: Building Blocks of Nucleic Acids

Nucleotides are the building blocks of all nucleic acid molecules. These structural units consist of three essential components: a **nitrogenous base**, a **pentose sugar** (a 5-carbon sugar), and a **phosphate group**. There are two kinds of nitrogenous bases: the nine-membered double-ringed **purines** and the six-membered single-ringed **pyrimidines**. Two types of purines and three types of pyrimidines are found commonly in nucleic acids. The two purines are **adenine** and **guanine**, abbreviated **A** and **G**. The three pyrimidines are **cytosine**, **thymine**, and **uracil**, abbreviated **C**, **T**, and **U**. The chemical structures of A, G, C, T, and U are shown in Figure 10.9(a). Both DNA and RNA contain A, C, and G; only DNA contains the base T, whereas only RNA contains the base U. Each nitrogen or carbon atom of the ring structures of purines and pyrimidines is designated by an unprimed number. Note that corresponding atoms in the two rings are numbered differently in most cases.

The pentose sugars found in nucleic acids give them their names. Ribonucleic acids (RNA) contain **ribose**, while deoxyribonucleic acids (DNA) contain **deoxyribose**. Figure 10.9(b) shows the ring structures for these two pentose sugars. Each carbon atom is distinguished by a number with a prime sign (e.g., C-1′, C-2′). As you can see, deoxyribose is missing one hydroxyl group at the C-2′ position compared with ribose. The presence of a hydroxyl group at the C-2′ position distinguishes RNA from DNA.

If a molecule is composed of a purine or pyrimidine base and a ribose or deoxyribose sugar, the chemical unit

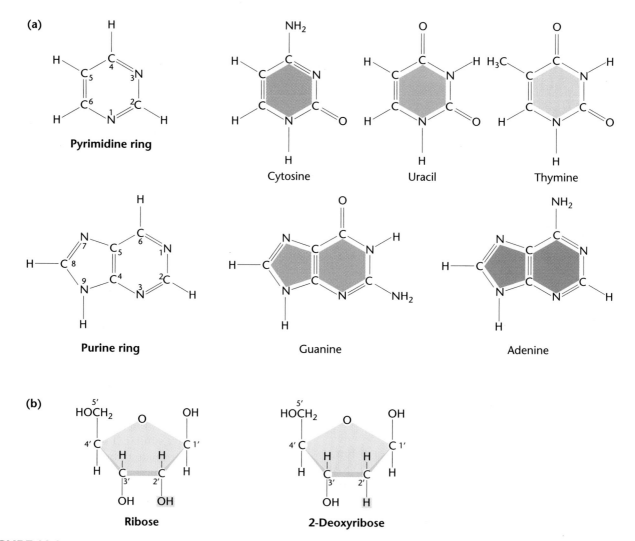

FIGURE 10.9 (a) Chemical structures of the pyrimidines and purines that serve as the nitrogenous bases in RNA and DNA. (b) Chemical ring structures of ribose and 2-deoxyribose, which serve as the pentose sugars in RNA and DNA, respectively.

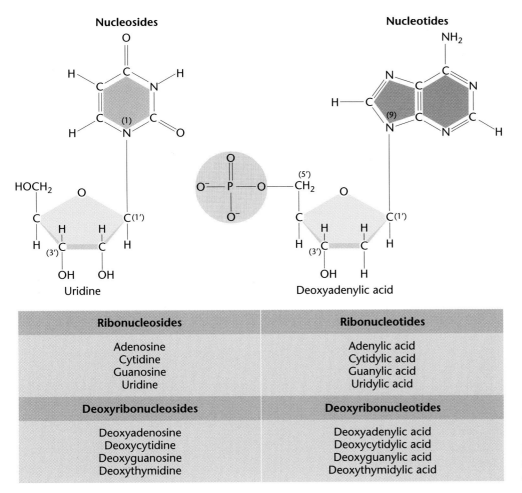

Nucleosides

Nucleotides

Uridine

Deoxyadenylic acid

Ribonucleosides	Ribonucleotides
Adenosine	Adenylic acid
Cytidine	Cytidylic acid
Guanosine	Guanylic acid
Uridine	Uridylic acid
Deoxyribonucleosides	**Deoxyribonucleotides**
Deoxyadenosine	Deoxyadenylic acid
Deoxycytidine	Deoxycytidylic acid
Deoxyguanosine	Deoxyguanylic acid
Deoxythymidine	Deoxythymidylic acid

FIGURE 10.10 The structures and names of the nucleosides and nucleotides of RNA and DNA.

is called a **nucleoside**. If a phosphate group is added to the nucleoside, the molecule is now called a **nucleotide**. Nucleosides and nucleotides are named according to the specific nitrogenous base (A, T, G, C, or U) that is part of the building block. The nomenclature and general structure of nucleosides and nucleotides are given in Figure 10.10.

The bonding among the three components of a nucleotide is highly specific. The C-1′ atom of the sugar is involved in the chemical linkage to the nitrogenous base. If the base is a purine, the N-9 atom is covalently bonded to the sugar. If the base is a pyrimidine, the bonding involves the N-1 atom. In a nucleotide, the phosphate group may be bonded to the C-2′, C-3′, or C-5′ atom of the sugar. The C-5′–phosphate configuration is shown in Figure 10.10. It is by far the most prevalent one in biological systems and that found in DNA and RNA.

Nucleoside Diphosphates and Triphosphates

Nucleotides are also described by the term **nucleoside monophosphate (NMP)**. The addition of one or two phosphate groups results in **nucleoside diphosphates**

(NDPs) and **triphosphates (NTPs)**, as illustrated in Figure 10.11. The triphosphate form is significant because it serves as the precursor molecule during nucleic acid synthesis within the cell (see Chapter 11). Additionally, the triphosphates **adenosine triphosphate (ATP)** and **guanosine triphosphate (GTP)** are important in the cell's bioenergetics because of the large amount of energy involved in the addition or removal of the terminal phosphate group. The hydrolysis of ATP or GTP to ADP or GDP and inorganic phosphate (P_i) is accompanied by the release of a large amount of energy in the cell. When the chemical conversion of ATP or GTP is coupled to other reactions, the energy produced may be used to drive the reactions. As a result, both ATP and GTP are involved in many cellular activities, including numerous genetic events.

Polynucleotides

The linkage between two mononucleotides consists of a phosphate group linked to two sugars. A **phosphodiester bond** is formed, because phosphoric acid has been joined

Nucleoside diphosphate (NDP)   **Nucleoside triphosphate (NTP)**

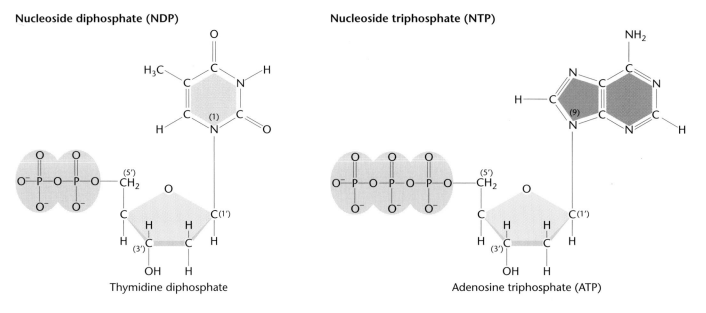

Thymidine diphosphate   Adenosine triphosphate (ATP)

FIGURE 10.11   The basic structures of nucleoside diphosphates and triphosphates, as illustrated by thymidine diphosphate and deoxyadenosine triphosphate.

to two alcohols (the hydroxyl groups on the two sugars) by an ester linkage on both sides. Figure 10.12(a) shows the resultant phosphodiester bond in DNA. The same bond is found in RNA. Each structure has a **C-5′ end** and a **C-3′ end**. The joining of two nucleotides forms a **dinucleotide**; of three nucleotides, a **trinucleotide**; and so forth. Short chains consisting of fewer than 20 nucleotides linked together are called **oligonucleotides**. Still longer chains are referred to as **polynucleotides**.

Because drawing the structures in Figure 10.12(a) is time-consuming and complex, a schematic shorthand method has been devised [Figure 10.12(b)]. The nearly vertical lines represent the pentose sugar; the nitrogenous base is attached at the top, or the C-1′ position. The diagonal line, with the (P) in the middle of it, is attached to the C-3′ position of one sugar and the C-5′ position of the neighboring sugar; it represents the phosphodiester bond. Several modifications of this shorthand method are in use, and they can be understood in terms of these guidelines.

Although Levene's tetranucleotide hypothesis (described earlier in this chapter) was generally accepted before 1940, research in subsequent decades revealed it to be incorrect. It was shown that DNA does not necessarily contain equimolar quantities of the four bases. Additionally, the molecular weight of DNA molecules was determined to be in the range of 10^6 to 10^9 daltons, far in excess of that of a tetranucleotide. The current view of DNA is that it consists of exceedingly long polynucleotide chains.

Long polynucleotide chains would account for the observed molecular weight and provide the basis for the most important property of DNA—storage of vast quan-

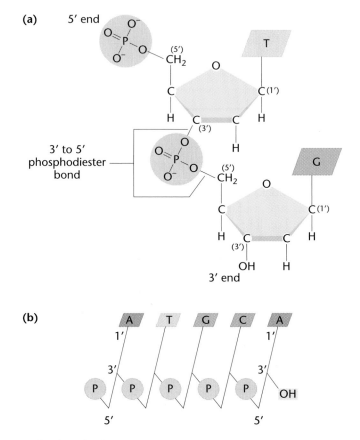

FIGURE 10.12   (a) The linkage of two nucleotides by the formation of a C-3′–C-5′ (3′–5′) phosphodiester bond, producing a dinucleotide. (b) A shorthand notation for a polynucleotide chain.

tities of genetic information. If each nucleotide position in this long chain is occupied by any one of four nucleotides, extraordinary variation is possible. For example, a polynucleotide that is only 1000 nucleotides in length can be arranged 4^{1000} different ways, each one different from all other possible sequences. This potential variation in molecular structure is essential if DNA is to serve the function of storing the vast amounts of chemical information necessary to direct cellular activities.

The Structure of DNA

The previous sections in this chapter have established that DNA is the genetic material in all organisms (with certain viruses being the exception) and have provided details as to the basic chemical components making up nucleic acids. What remained to be deciphered was the precise structure of DNA. That is, how are polynucleotide chains organized into DNA, which serves as the genetic material? Is DNA composed of a single chain, or more than one? If the latter is the case, how do the chains relate chemically to one another? Do the chains branch? And more important, how does the structure of this molecule relate to the various genetic functions served by DNA (i.e., storage, expression, replication, and mutation)?

From 1940 to 1953, many scientists were interested in solving the structure of DNA. Among others, Erwin Chargaff, Maurice Wilkins, Rosalind Franklin, Linus Pauling, Francis Crick, and James Watson sought information that might answer what many consider to be the most significant and intriguing question in the history of biology: "How does DNA serve as the genetic basis for the living process?" The answer was believed to depend strongly on the chemical structure and organization of the DNA molecule, given the complex but orderly functions ascribed to it.

In 1953, two young scientists, James Watson and Francis Crick, proposed that the structure of DNA is in the form of a double helix. Their proposal was published in a short paper in the journal *Nature*, which is reprinted in its entirety on pages 280–281. In a sense, this publication constituted the finish line in a highly competitive scientific race. This "race," as recounted in Watson's book *The Double Helix* (see Selected Readings), demonstrates the human interaction, genius, frailty, and intensity involved in the scientific effort that eventually led to the elucidation of DNA structure.

The data available to Watson and Crick, crucial to the development of their proposal, came primarily from two sources: base composition analysis of hydrolyzed samples of DNA, and X-ray diffraction studies of DNA. The analytical success of Watson and Crick can be attributed to model building that conformed to the existing data. If the structure of DNA can be analogized by a puzzle, Watson and Crick, working at the Cavendish Laboratory in Cambridge, England, were the first to put together successfully all of the pieces. Given the far-reaching significance of this discovery, you may be interested to know that Watson, who entered college (University of Chicago) at age 15, was a 24-year-old postdoctoral fellow in 1953. Crick, now considered one of the great theoretical biologists of our time, was still a graduate student.

Base Composition Studies

Between 1949 and 1953, Erwin Chargaff and his colleagues used chromatographic methods to separate the four bases in DNA samples from various organisms. Quantitative methods were then used to determine the amounts of the four bases from each source. Table 10.3(a) provides some of Chargaff's original data. Parts (b) and (c) of this table show more recently derived base-composition information from various organisms that reinforce Chargaff's findings.

On the basis of these data, which you should examine, the following conclusions may be drawn:

1. The amount of adenine residues is proportional to the amount of thymine residues in the DNA of any species (columns 1, 2, and 5). Also, the amount of guanine residues is proportional to the amount of cytosine residues (columns 3, 4, and 6).

2. Based on the above proportionality, the sum of the purines (A + G) equals the sum of the pyrimidines (C + T), as shown in column 7.

3. The percentage of C + G does not necessarily equal the percentage of A + T. As we can see, the ratio between the two values varies greatly among species, as shown in column 8, and as is apparent in Table 10.3(c).

These conclusions indicate definite patterns of base composition of DNA molecules. These data also served as the initial clue to "the puzzle." Additionally, they directly refute the tetranucleotide hypothesis, which stated that all four bases are present in equal amounts.

X-Ray Diffraction Analysis

When fibers of a DNA molecule are subjected to X-ray bombardment, these rays are scattered according to the molecule's atomic structure. The pattern of scatter can be captured as spots on photographic film and analyzed, particularly for the overall shape of and regularities within the molecule. This process, **X-ray diffraction** analysis,

TABLE 10.3 DNA Base Composition Data

(a) Chargaff's Data

| Source | Molar Proportions* | | | |
| | 1 | 2 | 3 | 4 |
	A	T	G	C
Ox thymus	26	25	21	16
Ox spleen	25	24	20	15
Yeast	24	25	14	13
Avian tubercle bacilli	12	11	28	26
Human sperm	29	31	18	18

Source: From Chargaff, 1950.

*Moles of nitrogenous constituent per mole of P (often, the recovery was less than 100 percent).

(c) G + C Content in Several Organisms

Organism	% G + C
Phage T2	36.0
Drosophila	45.0
Maize	49.1
Euglena	53.5
Neurospora	53.7

(b) Base Composition of DNAs from Various Sources

| Source | Base Composition | | | | Base Ratio | | | A + T/G + C Ratio |
| | 1 | 2 | 3 | 4 | 5 | 6 | 7 | 8 |
	A	T	G	C	A/T	G/C	(A + G)/(C + T)	(A + T)/(C + G)
Human	30.9	29.4	19.9	19.8	1.05	1.00	1.04	1.52
Sea urchin	32.8	32.1	17.7	17.3	1.02	1.02	1.02	1.58
E. coli	24.7	23.6	26.0	25.7	1.04	1.01	1.03	0.93
Sarcina lutea	13.4	12.4	37.1	37.1	1.08	1.00	1.04	0.35
T7 bacteriophage	26.0	26.0	24.0	24.0	1.00	1.00	1.00	1.08

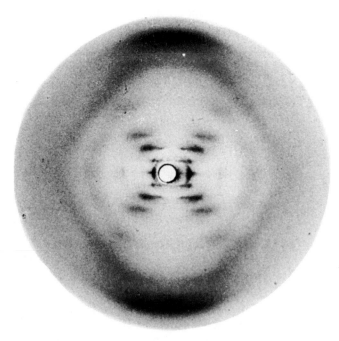

FIGURE 10.13 An X-ray diffraction photograph of the B form of crystallized DNA. The dark patterns at the top and bottom provide an estimate of the periodicity of nitrogenous bases, which are 3.4 Å apart. The central pattern is indicative of the molecule's helical structure.

was successfully applied to the study of protein structure by Linus Pauling and other chemists. The technique had been attempted on DNA as early as 1938 by William Astbury. By 1947, he had detected a periodicity of 3.4 Å* that suggested to him that the bases were stacked like pennies on top of one another.

Between 1950 and 1953, Rosalind Franklin, working in the laboratory of Maurice Wilkins, obtained improved X-ray data from more purified samples of DNA (Figure 10.13). Her work confirmed the 3.4 Å periodicity seen by Astbury and suggested that the structure of DNA was some sort of helix. However, she did not propose a definitive model. Pauling had analyzed the work of Astbury and others and incorrectly proposed that DNA was a triple helix.

The Watson–Crick Model

Watson and Crick published their analysis of DNA structure in 1953 (see pp. 280–281). By building models under the constraints of the information just discussed, they proposed the double-helical form of DNA, as shown in Figure 10.14(a). This model has the following major features:

---

*Today, measurement in nanometers (nm) is favored (1 nm = 10 Å). However, for historical purposes, we will use angstroms (Å), as reported in the original literature.

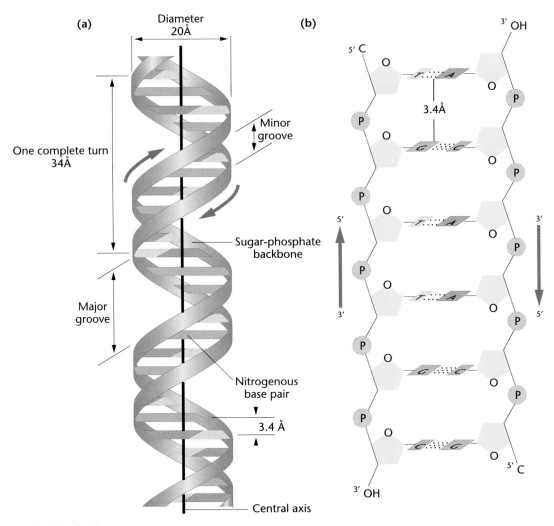

(a) Diameter 20Å

One complete turn 34Å

Minor groove

Sugar-phosphate backbone

Major groove

Nitrogenous base pair

3.4 Å

Central axis

(b) 3' OH

5' C

3.4Å

5'

3'

3'

5'

3' OH

5' C

FIGURE 10.14 (a) A schematic representation of the DNA double helix as proposed by Watson and Crick. The ribbonlike strands constitute the sugar-phosphate backbones, and the horizontal rungs constitute the nitrogenous base pairs, of which there are 10 per complete turn. The major and minor grooves are apparent. A solid vertical bar representing the central axis has been placed through the center of the helix. (b) A representation of the antiparallel nature of the two strands of the helix.

1. Two long polynucleotide chains are coiled around a central axis, forming a right-handed double helix.

2. The two chains are **antiparallel**; that is, their C-5'-to-C-3' orientations run in opposite directions.

3. The bases of both chains are flat structures, lying perpendicular to the axis; they are "stacked" on one another, 3.4 Å (0.34 nm) apart, and are located on the inside of the structure.

4. The nitrogenous bases of opposite chains are paired to one another as the result of the formation of **hydrogen bonds** (described in the following discussion); in DNA, only A-T and G-C pairs are allowed.

5. Each complete turn of the helix is 34 Å (3.4 nm) long; thus, 10 bases exist in each chain per turn.

6. In any segment of the molecule, alternating larger **major grooves** and smaller **minor grooves** are apparent along the axis.

7. The double helix measures 20Å (2.0 nm) in diameter.

The nature of base-pairing (Point 4 above) is the most genetically significant feature of the model. Before discussing it in detail, several other important features warrant emphasis. First, the antiparallel nature of the two chains is a key part of the double-helix model. While one chain runs in the 5'-to-3' orientation (what seems right-side-up to us), the other chain is in the 3'-to-5' orientation

*M*olecular Structure of Nucleic Acids: A Structure for Deoxyribose Nucleic Acid

*W*e wish to suggest a structure for the salt of deoxyribose nucleic acid (D. N. A.). This structure has novel features which are of considerable biological interest. A structure for nucleic acid has already been proposed by Pauling and Corey.[1] They kindly made their manuscript available to us in advance of publication. Their model consists of three intertwined chains, with the phosphates near the fibre axis, and the bases on the outside. In our opinion, this structure is unsatisfactory for two reasons: (1) We believe that the material which gives the X-ray diagrams is the salt, not the free acid. Without the acidic hydrogen atoms it is not clear what forces would hold the structure together, especially as the negatively charged phosphates near the axis will repel each other. (2) Some of the van der Waals distances appear to be too small.

Another three-chain structure has also been suggested by Fraser (in the press). In his model the phosphates are on the outside and the bases on the inside, linked together by hydrogen bonds. This structure as described is rather ill-defined, and for this reason we shall not comment on it.

We wish to put forward a radically different structure for the salt of deoxyribose nucleic acid. This structure has two helical chains each coiled round the same axis. We have made the usual chemical assumptions, namely, that each chain consists of phosphate diester groups joining β-D-deoxyribofuranose residues with 3′,5′ linkages. The two chains (but not their bases) are related by a dyad perpendicular to the fibre axis. Both chains follow right-handed helices, but owing to the dyad the sequences of the atoms in the two chains run in opposite directions. Each chain loosely resembles Furberg's[2] model No. 1; that is, the bases are on the inside of the helix and the phosphates on the outside. The configuration of the sugar and the atoms near it is close to Furberg's "standard configuration," the sugar being roughly perpendicular to the attached base. There is a residue on each chain every 3.4 Å in the z-direction. We have assumed an angle of 36° between adjacent residues in the same chain, so that the structure repeats after 10 residues on each chain, that is, after 34 Å. The distance of a phosphorus atom from the fibre axis is 10 Å. As the phosphates are on the outside, cations have easy access to them.

The structure is an open one, and its water content is rather high. At lower water contents we would expect the bases to tilt so that the structure could become more compact.

The novel feature of the structure is the manner in which the two chains are held together by the purine and pyrimidine bases. The planes of the bases are perpendicular to the fibre axis. They are joined together in pairs, a single base from one chain being hydrogen-bonded to a single base from the other chain, so that the two lie side by side with identical z-co-ordinates. One of the pair must be a purine and the other a pyrimidine for bonding to occur. The hydrogen bonds are made as follows: purine position 1 to pyrimidine position 1; purine position 6 to pyrimidine position 6.

If it is assumed that the bases only occur in the structure in the most plausible tautomeric forms (that is, with the keto rather than the enol configurations) it is found

(and thus appears upside-down). This is illustrated in Figure 10.14(b). Given the constraints of the bond angles of the various nucleotide components, the double helix could not be constructed easily if both chains ran parallel to one another.

Second, the right-handed nature of the helix is best appreciated by comparing such a structure to its left-handed counterpart, which is a mirror image, as shown in Figure 10.15. The conformation in space of the right-handed helix is most consistent with the data that were available to Watson and Crick. As we shall see momentarily, an alternative form of DNA (Z-DNA) does exist as a left-handed helix.

The key to the model proposed by Watson and Crick is the specificity of base pairing. Chargaff's data had suggested that the amounts of A equaled T and that G equaled C. Watson and Crick realized that if A pairs with T and C pairs with G, thus accounting for these proportions, the

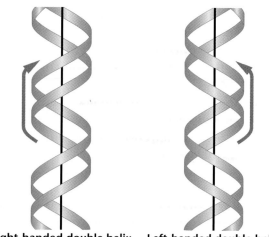

Right-handed double helix   **Left-handed double helix**

FIGURE 10.15   The right- and left-handed helical forms of DNA. Note that they are mirror images of one another.

that only specific pairs of bases can bond together. These pairs are: adenine (purine) with thymine (pyrimidine), and guanine (purine) with cytosine (pyrimidine).

In other words, if an adenine forms one member of a pair, on either chain, then on these assumptions the other member must be thymine; similarly for guanine and cytosine. The sequence of bases on a single chain does not appear to be restricted in any way. However, if only specific pairs of bases can be formed, it follows that if the sequence of bases on one chain is given, then the sequence on the other chain is automatically determined.

It has been found experimentally[3,4] that the ratio of the amounts of adenine to thymine, and the ratio of guanine to cytosine, are always very close to unity for deoxyribose nucleic acid.

It is probably impossible to build this structure with a ribose sugar in place of the deoxyribose, as the extra oxygen atom would make too close a van der Waals contact.

The previously published X-ray data[5,6] on deoxyribose nucleic acid are insufficient for a rigorous test of our structure. So far as we can tell, it is roughly compatible with the experimental data, but it must be regarded as unproved until it has been checked against more exact results. Some of these are given in the following communications. We were not aware of the details of the results presented there when we devised our structure, which rests mainly though not entirely on published experimental data and stereochemical arguments.

It has not escaped our notice that the specific pairing we have postulated immediately suggests a possible copying mechanism for the genetic material.

Full details of the structure, including the conditions assumed in building it, together with a set of co-ordinates for the atoms, will be published elsewhere.

We are much indebted to Dr. Jerry Donohue for constant advice and criticism, especially on interatomic distances. We have also been stimulated by a knowledge of the general nature of the unpublished experimental results and ideas of Dr. M. H. F. Wilkins, Dr. R. E. Franklin and their co-workers at King's College, London. One of us (J. D. W.) has been aided by a fellowship from the National Foundation for Infantile Paralysis.

J. D. Watson
F. H. C. Crick
Medical Research Council Unit for the Study of the
Molecular Structure of Biological Systems,
Cavendish Laboratory, Cambridge, England.

[1] Pauling, L., and Corey, R. B., *Nature*, 171, 346 (1953); *Proc. U.S. Nat. Acad. Sci.*, 39, 84 (1953).

[2] Furberg, S., *Acta Chem. Scand.*, 6, 634 (1952).

[3] Chargaff, E., for references see Zamenhof, S., Brawerman, G., and Chargaff, E., *Biochim. et Biophys. Acta*, 9, 402 (1952).

[4] Wyatt, G. R., *J. Gen. Physiol*, 36, 201 (1952).

[5] Astbury, W. T., *Symp. Soc. Exp. Biol. 1, Nucleic Acid*, 66 (Camb. Univ. Press, 1947).

[6] Wilkins, M. H. F., and Randall, J. T., *Biochim. et Biophys. Acta*, 10, 192 (1953).

members of each such base pair formed hydrogen bonds [Figure 10.14(b)], providing the chemical stability necessary to hold the two chains together. Arranged in this way, both **major** and **minor grooves** become apparent along the axis. Further, a purine (A or G) opposite a pyrimidine (T or C) on each "rung of the spiral staircase" of the proposed helix accounts for the 20-Å (2-nm) diameter suggested by X-ray diffraction studies.

The specific A-T and G-C base pairing is the basis for the concept of **complementarity**. This term describes the chemical affinity provided by the hydrogen bonds between the bases. As we will see, this concept is very important in the processes of DNA replication and gene expression.

Two questions are particularly worthy of discussion. First, why aren't other base pairs possible? Watson and Crick discounted the A-G and C-T pairs because these represent purine-purine and pyrimidine-pyrimidine pairings, respectively. These pairings would lead to alternating diameters of more than and less than 20 Å because of the respective sizes of the purine and pyrimidine rings; additionally, the three-dimensional configurations formed by such pairings do not produce the proper alignment leading to sufficient hydrogen-bond formations. It is for this reason that A-C and G-T pairings were also discounted, even though these pairs each consist of one purine and one pyrimidine.

The second question concerns hydrogen bonds. Just what is the nature of such a bond, and is it strong enough to stabilize the helix? A **hydrogen bond** is a very weak electrostatic attraction between a covalently bonded hydrogen atom and an atom with an unshared electron pair. The hydrogen atom assumes a partial positive charge,

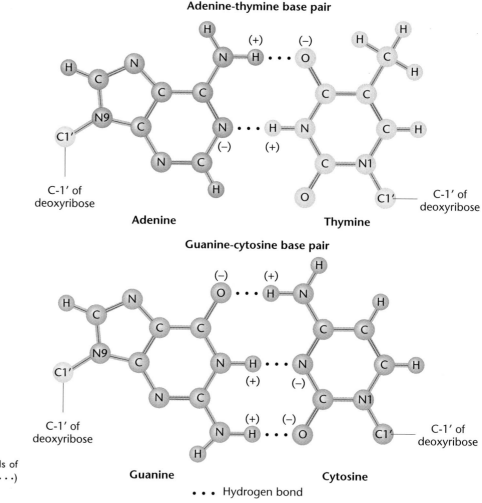

Adenine-thymine base pair

C-1′ of deoxyribose

Adenine

C-1′ of deoxyribose

Thymine

Guanine-cytosine base pair

C-1′ of deoxyribose

Guanine

C-1′ of deoxyribose

Cytosine

⋯ Hydrogen bond

FIGURE 10.16 Ball-and-stick models of A-T and G-C base pairs. The rows of dots (⋯) represent the hydrogen bonds that form.

while the unshared electron pair—characteristic of covalently bonded oxygen and nitrogen atoms—assumes a partial negative charge. These opposite charges are responsible for the weak chemical attraction. As oriented in the double helix (Figure 10.16), adenine forms two hydrogen bonds with thymine, and guanine forms three hydrogen bonds with cytosine. Although two or three hydrogen bonds taken alone are very weak, two or three thousand bonds in tandem (which would be found in two long polynucleotide chains) are capable of providing great stability to the helix.

Still another stabilizing factor is the arrangement of sugars and bases along the axis. In the Watson–Crick model, the hydrophobic or "water-fearing" nitrogenous bases are "stacked" almost horizontally on the interior of the axis, shielded from water. The hydrophilic sugar-phosphate backbone is on the outside of the axis, where both components may interact with water. These molecular arrangements provide significant chemical stabilization to the helix.

A more recent and accurate analysis of the form of DNA that served as the basis for the Watson–Crick model has revealed a minor structural difference. A precise measurement of the number of base pairs (bp) per turn has demonstrated a value of 10.4 rather than the 10.0 predicted by Watson and Crick. Where, in the classical model, each base pair is rotated around the helical axis 36°, relative to the adjacent base pair, the new finding requires a rotation of 34.6°. This results in slightly more than 10 base pairs per 360° turn.

The Watson–Crick model had an immediate effect on the emerging discipline of molecular biology. Even in their initial 1953 article, the authors observed, "It has not escaped our notice that the specific pairing we have postulated immediately suggests a possible copying mechanism for the genetic material." Two months later, in a second article in *Nature*, Watson and Crick pursued this idea, suggesting a specific mode of replication of DNA—the semiconservative model. The second article also alluded to two new concepts: the storage of genetic information in the sequence of the bases, and the mutation or genetic change that would result from alteration of the bases. These ideas have received vast amounts of experimental support since 1953 and are now universally accepted.

The "synthesis" of ideas by Watson and Crick was a remarkable feat and was highly significant to subsequent studies in genetics and biology. The nature of the gene and its role in genetic mechanisms could now be viewed and studied in biochemical terms. Recognition of their work, along with the contributions from Maurice Wilkins's laboratory, led to their receipt of the Nobel Prize in Physiology or Medicine in 1962. This was to be one of many such awards bestowed for work in the field of molecular genetics.

Other Forms of DNA

Under different conditions of isolation, several conformational forms of DNA have been recognized. At the time Watson and Crick performed their analysis, two forms—**A-DNA** and **B-DNA**—were known. Watson and Crick's analysis was based on X-ray studies of the B form by Rosalind Franklin, which is present under aqueous, low-salt conditions and is believed to be the biologically significant conformation.

Although DNA studies about 1950 relied on the use of X-ray diffraction, more recent investigations have been performed using **single-crystal X-ray analysis**. The earlier studies achieved resolution of about 5 Å, but single crystals diffract X rays at about 1 Å, near atomic resolution. As a result, every atom is "visible," and much greater structural detail is available during analysis.

Using these modern techniques, A-DNA has now been scrutinized. It is prevalent under high-salt or dehydration conditions. In comparison to B-DNA (Figure 10.17), A-DNA is slightly more compact, with 11 base pairs in each complete turn of the helix, which is 23 Å in diameter. While it is also a right-handed helix, the orientation of the bases is somewhat different. They are tilted and displaced laterally in relation to the axis of the helix. As a result of these differences, the appearance of the major and minor grooves is modified compared with those in B-DNA. It seems doubtful that A-DNA occurs under biological conditions.

Three other right-handed forms of DNA helices have been discovered when investigated under laboratory conditions. These have been designated C-, D-, and E-DNA. **C-DNA** is found to occur under even greater dehydration conditions than those observed during the isolation of A- and B-DNA. It has only 9.3 base pairs per turn and is, thus, less compact. Its helical diameter is 19 Å. Like A-DNA, C-DNA does not have its base pairs lying flat; rather, they are tilted

relative to the axis of the helix. Two other forms, **D-DNA** and **E-DNA**, occur in helices lacking guanine in their base composition. They have even fewer base pairs per turn: 8 and $7\frac{1}{2}$, respectively.

Still another form of DNA, called **Z-DNA**, was discovered by Andrew Wang, Alexander Rich, and their colleagues in 1979 when they examined a synthetic DNA oligonucleotide containing only C-G base pairs. Z-DNA takes on the rather remarkable configuration of a left-handed double helix (Figure 10.17). Like A- and B-DNA, Z-DNA consists of two antiparallel chains held together by Watson–Crick base pairs. Beyond these characteristics, Z-DNA is quite different. The left-handed helix is 18 Å (1.8 nm) in diameter, contains 12 base pairs per turn, and assumes a zigzag conformation (hence its name). The major groove present in B-DNA is nearly eliminated in Z-DNA.

Speculation has abounded over the possibility that regions of Z-DNA exist in the chromosomes of living organisms. The unique helical arrangement could provide an important recognition point for the interaction with other molecules. However, it is still not clear whether Z-DNA occurs *in vivo*.

The Structure of RNA

The second category of nucleic acids is the ribonucleic acids, or RNA. The structure of these molecules is similar to DNA, with several important exceptions. Although RNA also has as its building blocks nucleotides linked into polynucleotide chains, the sugar ribose replaces deoxyribose and the nitrogenous base uracil replaces thymine. Another important difference is that most RNA is usually considered to be single-stranded. However, RNA molecules sometimes fold back on themselves to form double-stranded regions following their synthesis; such a configuration results when regions of complementarity occur in positions that allow base pairs to form. Furthermore, some animal viruses that have RNA as their genetic material contain it in the form of double-stranded helices. Thus, there are several instances where RNA does not exist strictly as a linear, single-stranded molecule.

At least three major classes of cellular RNA molecules function during the expression of genetic information: **ribosomal RNA (rRNA)**, **messenger RNA (mRNA)**, and **transfer RNA (tRNA)**. These molecules all originate as complementary copies of one of the two strands of DNA segments during the process of transcription. That is, their nucleotide sequence is complementary to the deoxyribonucleotide sequence of DNA, which served as the template for their synthesis. Because uracil replaces thymine in RNA, uracil is complementary to adenine during transcription and during RNA base pairing.

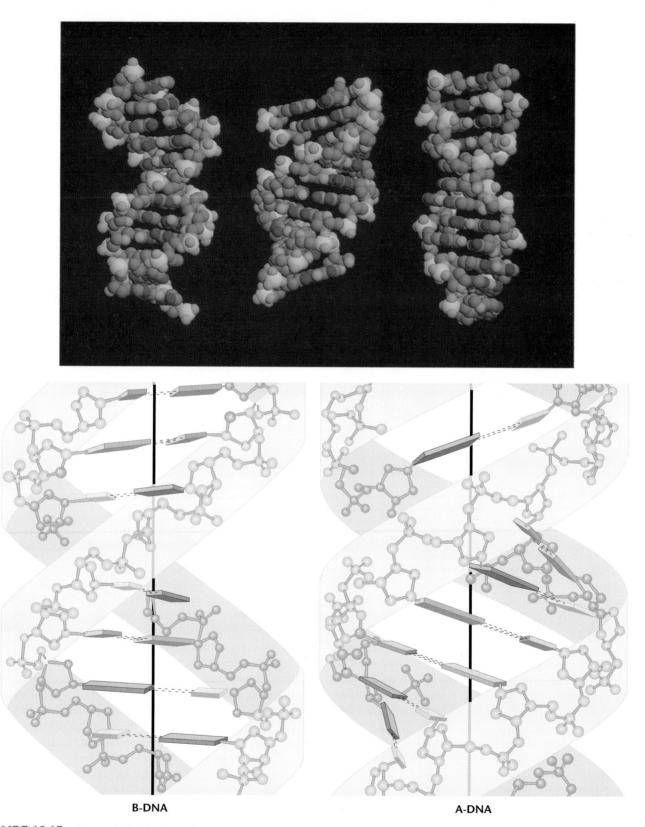

B-DNA

A-DNA

FIGURE 10.17   The top half of the figure shows computer-generated space-filling models of B-DNA (left), A-DNA (center), and Z-DNA (right). Below is an artist's depiction illustrating the orientation of the base pairs of B-DNA and A-DNA. (Note that in B-DNA the base pairs are perpendicular to the helix, while they are tilted and pulled away from the helix in A-DNA.)

TABLE 10.4 Sedimentation Coefficients, Molecular Weights, and Number of Nucleotides for Various RNAs

RNA Type	Abbreviation	Svedberg Coefficient (S)	Molecular Weight	Number of Nucleotides
Ribosomal RNA	rRNA	5S	35,000	120
		5.8S	47,000	160
		18S	700,000	1900
		28S	1,800,000	480
Transfer RNA	tRNA	5S	23,000–30,000	75–90
Messenger RNA	mRNA	5S	25,000–1,000,000	100–10,000

Each class of RNA can be characterized by its size, sedimentation behavior in a centrifugal field, and genetic function. Sedimentation behavior depends on a molecule's density, mass, and shape, and its measure is called the **Svedberg coefficient (S)**. Table 10.4 relates the S values, molecular weights, and approximate number of nucleotides of the major forms of RNA. While higher S values almost always designate molecules of greater molecular weight, the correlation is not direct; that is, a twofold increase in molecular weight does not lead to a twofold increase in S. This is because the size and the shape of the molecule impact on its rate of sedimentation (S). As you can see, a wide variation exists in the size of the three classes of RNA.

Ribosomal RNA is generally the largest of these molecules (as is generally reflected in its S values) and usually constitutes about 80 percent of all RNA in the cell. The various forms of rRNA found in prokaryotes and eukaryotes differ distinctly in size. The values in Table 10.4 are based on eukaryotic molecules. Ribosomal RNAs are important structural components of **ribosomes**, which function as a nonspecific workbench during the synthesis of proteins during the process of translation.

Messenger RNA molecules carry genetic information from the DNA of the gene to the ribosome, where translation occurs. They vary considerably in length, which is partially a reflection of the variation in the size of the gene serving as the template for transcription of mRNA species. Note that the values shown in Table 10.4 are only estimates, particularly at their upper limits. Precursors of many mRNAs, called **primary transcripts**, may demonstrate values considerably higher.

Transfer RNA, the smallest class of RNA molecules, carries amino acids to the ribosome during translation. Because more than one tRNA molecule interacts simultaneously with the ribosome, the molecule's smaller size facilitates these interactions.

We will discuss the functions of the three classes of RNA in much greater detail in Chapter 12. In addition, as we proceed through the text, we will encounter other unique RNAs that perform various roles. For example, **small nuclear RNA (snRNA)** participates in processing

mRNAs (Chapter 12). **Telomerase RNA** is involved in DNA replication at the ends of chromosomes (Chapter 11). And, **antisense RNA** is involved in gene regulation (Chapter 18). Our purpose in this section has been to contrast the structure of DNA, which stores genetic information, with that of RNA, which most often functions in the expression of that information.

Analysis of Nucleic Acids

Since 1953, the role of DNA as the genetic material and the role of RNA in transcription and translation have been clarified through detailed analysis of nucleic acids. We will consider several methods of analysis of these molecules in this chapter. Some of these, as well as other research procedures, are presented in greater detail in Appendix A.

Absorption of Ultraviolet Light (UV)

Nucleic acids absorb ultraviolet (UV) light most strongly at wavelengths of 254 to 260 nm due to the interaction between UV light and the ring systems of the purines and pyrimidines. Thus, any molecule containing nitrogenous bases (i.e., nucleosides, nucleotides, and polynucleotides) can be analyzed using UV light. This technique is especially important in the localization, isolation, and characterization of nucleic acids.

Ultraviolet analysis is used in conjunction with many standard procedures that separate molecules. As we shall see in the next section, the use of UV absorption is critical to the isolation of nucleic acids following their separation.

Sedimentation Behavior

Nucleic acid mixtures can be separated by subjecting them to one of several possible **gradient centrifugation** procedures (Figure 10.18). The mixture can be loaded on top of a solution prepared so that a concentration gradient has been formed from top to bottom. Then the entire

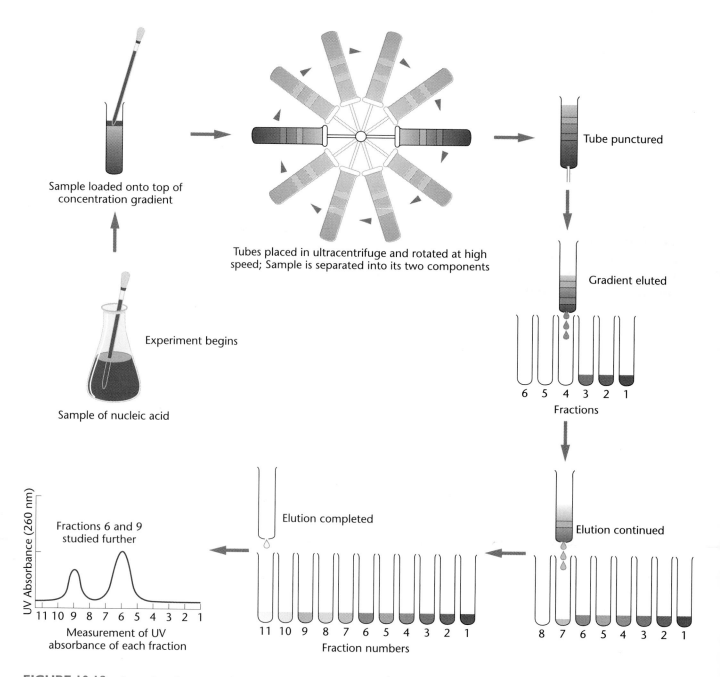

FIGURE 10.18   Separation of a mixture of two types of nucleic acid using gradient centrifugation. To fractionate the gradient, successive samples are eluted from the bottom of the tube. Each is measured for absorbance of ultraviolet light at 260 nm, producing a profile of the sample in graphic form.

mixture is centrifuged at high speeds in an ultracentrifuge. The mixture of molecules will migrate downward, with each component moving at a different rate. Centrifugation is stopped and the gradient eluted from the tube. Each fraction can then be measured spectrophotometrically for absorption at 260 nm. In this way, the previous position of a nucleic acid fraction along the gradient can be predicted and the fraction isolated and studied further.

The gradient centrifugations described rely on the sedimentation behavior of molecules in solution. Two major types of gradient centrifugation techniques are employed in the analysis of nucleic acids: sedimentation equilibrium and sedimentation velocity. Both require the use of high-speed centrifugation to create large centrifugal forces upon molecules in a gradient solution.

In **sedimentation equilibrium centrifugation** (sometimes called **density gradient centrifugation**), a

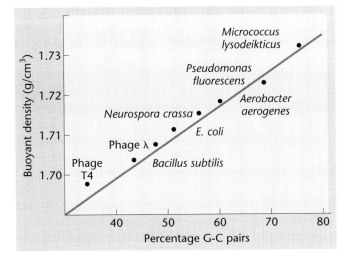

FIGURE 10.19   Percentage of guanine-cytosine (G-C) base pairs in DNA plotted against buoyant density for a variety of microorganisms.

density gradient is created that overlaps the densities of the individual components of a mixture of molecules. Usually, the gradient is made of a heavy metal salt such as cesium chloride (CsCl). During centrifugation the molecules migrate until they reach a point of neutral buoyant density. At this point, the centrifugal force on them is equal and opposite to the upward diffusion force, and no further migration occurs. If DNAs of different densities are used, they will separate as the molecules of each density reach equilibrium with the corresponding density of CsCl. The gradient may be fractionated and the components isolated (Figure 10.18). When properly executed, this technique provides high resolution in separating mixtures of molecules varying only slightly in density.

Sedimentation equilibrium centrifugation studies can also be used to generate data on the base composition of double-stranded DNA. The G-C base pairs, compared with A-T pairs, are more compact and dense. As shown in Figure 10.19, the percentage of G-C pairs in DNA is directly proportional to the molecule's buoyant density. By using this technique, we can make a useful molecular characterization of DNAs from different sources.

The second technique, **sedimentation velocity centrifugation**, employs an analytical centrifuge, which enables the migration of the molecules during centrifugation to be monitored with ultraviolet absorption optics. Thus, the "velocity of sedimentation" can be determined. This velocity has been standardized in units called Svedberg coefficients (S), as mentioned earlier.

In this technique, the molecules are loaded on top of the gradient, and the gravitational forces created by centrifugation drive them toward the bottom of the tube. Two forces work against this downward movement: (1) The vis-

cosity of the solution creates a frictional resistance, and (2) part of the force of diffusion is directed upward. Under these conditions, the key variables are the mass and shape of the molecules being examined. In general, the greater the mass, the greater the sedimentation velocity. However, the molecule's shape affects the frictional resistance. Therefore, two molecules of equal mass but different shape will sediment at different rates.

One use of the sedimentation velocity technique is the determination of **molecular weight (MW)**. If certain physical–chemical properties of a molecule under study are also known, the MW can be calculated based on the sedimentation velocity. The S values increase with molecular weight, but they are not directly proportional to it.

Denaturation and Renaturation of Nucleic Acids

When **denaturation** of double-stranded DNA occurs, the hydrogen bonds of the duplex structure break, the duplex unwinds, and the strands separate. However, no covalent bonds break. During strand separation, which can be induced by heat or chemical treatment, the viscosity of DNA decreases, and both the UV absorption and the buoyant density increase. Denaturation as a result of heating is sometimes referred to as **melting**. The increase in UV absorption of heated DNA in solution, called the **hyperchromic shift**, is easiest to measure. This effect is illustrated in Figure 10.20.

Because G-C base pairs have one more hydrogen bond than do A-T pairs, they are more stable to heat

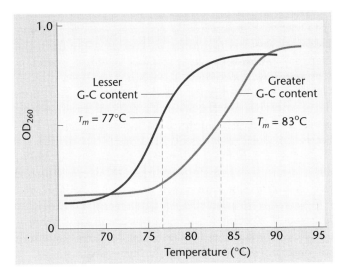

FIGURE 10.20   Comparison of the increase in UV absorbance with an increase in temperature (the hyperchromic effect) for two DNA molecules with different G-C contents. The molecule with a melting point (T_m) of 83°C has a greater G-C content than the molecule with a T_m of 77°C.

treatment. Thus, DNA with a greater proportion of G-C pairs than A-T pairs requires higher temperatures to denature completely. When absorption at 260 nm (OD_{260}) is monitored and plotted against temperature during heating, a **melting profile** of DNA is obtained. The midpoint of this profile, or curve, is called the **melting temperature** (T_m) and represents the point at which 50 percent of the strands are unwound or denatured (Figure 10.20). When the curve plateaus at its maximum optical density, denaturation is complete and only single strands exist. Analysis of melting profiles provides a characterization of DNA and an alternative method of estimating the base composition of DNA.

One might ask whether the denaturation process can be reversed; that is, can single strands of nucleic acids reform a double helix, provided that each strand's complement is present? Not only is the answer yes, but such reassociation provides the basis for several important analytical techniques that have provided much valuable information during genetic experimentation.

If DNA that has been denatured thermally is cooled slowly, random collisions between complementary strands will result in their reassociation. At the proper temperature, hydrogen bonds will re-form, securing pairs of strands into duplex structures. With time during cooling, more and more duplexes will form. Depending on the conditions, a complete match is not essential for duplex formation, provided there are at least stretches of base-pairing on two reassociating strands.

Molecular Hybridization

The property of renaturation of complementary single strands of nucleic acids is the basis for a powerful analytical technique in molecular genetics—**molecular hybridization**. This technique derives its name from the fact that renaturing single strands need not originate from the same nucleic acid source. For example, if DNA strands are isolated from different organisms and some degree of base complementarity exists between them, **molecular hybrids** will form during renaturation. Furthermore, when mixtures of DNA and RNA strands are cooled slowly, hybridization may also occur. A case in point would be when RNA and the DNA from which it has been transcribed are present together. The RNA will find its single-stranded DNA complement and renature. Since such experimental conditions are mixing *different* nucleic acid molecules, the resultant duplexes constitute still another example of molecular hybrids.

Figure 10.21 illustrates how the process of DNA-RNA hybridization occurs. In this example, the DNA strands are heated, causing strand separation, and then slowly cooled in the presence of single-stranded RNA. If the RNA has been transcribed on the DNA used in the experiment, and is therefore complementary to it, hybridization will occur. Several methods are available for monitoring the amount of double-stranded molecules produced following strand separation.

In the 1960s molecular hybridization techniques contributed to the increase in our understanding of transcriptional events occurring at the gene level. Refinements of this process have occurred continually and have been the forerunners of work in studies of molecular evolution as well as the organization of DNA in chromosomes. Hybridization can occur in solution or when DNA is bound either to a gel or to a special type of filter, facilitating recovery of the newly formed hybrids.

The technique can even be performed using cytological preparations, which is called ***in situ* molecular hybridization**. In this procedure mitotic or interphase cells are fixed to slides and subjected to hybridization conditions. Radioactive single-stranded DNA or RNA is added, and hybridization is monitored. The nucleic acid serves as a "probe," since it will hybridize only with the specific chromosomal areas for which it is complementary. It can be either radioactive or involve fluorescence to allow its detection on the slide. In the former case, the technique of **autoradiography** can be used (see Appendix A).

Fluorescent probes are prepared in a different way. When DNA is used, it is first coupled to the small organic molecule biotin (creating biotinylated DNA). Once *in situ* hybridization is completed, another molecule (avidin or streptavidin) that has a high binding affinity for biotin is utilized. A fluorescent molecule such as fluorescein is linked to avidin (or streptavidin) and the complex is reacted with the cytological preparation. The above procedure represents an extremely sensitive method for localizing the hybridized DNA. Because fluorescence is used, the technique is known by the acronym **FISH (fluorescent *in situ* hybridization)**.

Figure 10.22 illustrates the use of FISH in identifying the DNA specific to the centromeres of human chromosomes. The resolution of FISH is great enough to detect just a single gene within an entire set of chromosomes (see Appendix A). The use of this technique in identifying chromosomal locations housing specific genetic information has been a valuable addition to the repertoire of experimental geneticists.

Reassociation Kinetics and Repetitive DNA

One extension of molecular hybridization procedures is the technique that measures the *rate* of reassociation of complementary strands of DNA derived from a single source. This technique, called **reassociation kinetics**, was first refined and studied by Roy Britten and David Kohne.

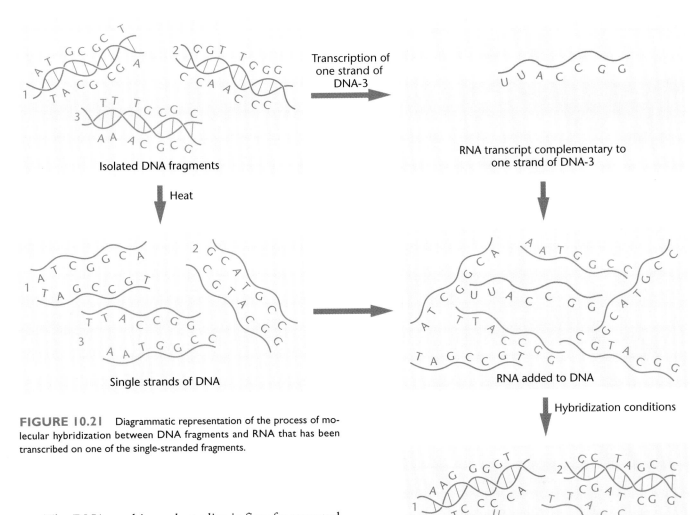

FIGURE 10.21 Diagrammatic representation of the process of molecular hybridization between DNA fragments and RNA that has been transcribed on one of the single-stranded fragments.

The DNA used in such studies is first fragmented into small pieces as a result of shearing forces introduced during isolation. The resultant DNA fragments cluster around a uniform average size of several hundred base pairs. These fragments of DNA are then dissociated into single strands by heating. Next, the temperature is lowered and reassociation is monitored. During reassociation, pieces of single-stranded DNA collide randomly. If they are complementary, a stable double strand is formed; if not, they separate and are free to encounter other DNA fragments. The process continues until all matches are made.

Results of such an experiment are presented in Figure 10.23. The percentage of reassociation of DNA fragments is plotted against a logarithmic scale of the product of C_0 (the initial concentration of DNA single strands in moles per liter of nucleotides), and t (the time, usually measured in minutes). The process of renaturation follows second-order rate kinetics according to the equation

$$\frac{C}{C_0} = \frac{1}{1 + kC_0t}$$

where C is the single-stranded DNA concentration remaining after time t has elapsed and k is the second-order rate constant. Initially, C equals C_0, and the fraction remaining single-stranded is 100 percent.

The initial shape of the curve reflects the fact that in a mixture of unique sequence fragments, each with one complement, initial matches take more time to make. Then, as many single strands are converted to duplexes, matches are made more quickly, reflecting an increase in the "slope" of the curve. Near the end of the reaction, the few remaining single strands require relatively greater time to make the final matches.

A great deal of information can be obtained from studies comparing the reassociation of DNA of different

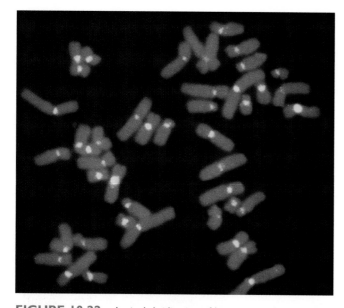

FIGURE 10.22 *In situ* hybridization of human metaphase chromosomes using a fluorescent technique (FISH). The probe, specific to centromeric DNA, produces a yellow fluorescence signal indicating hybridization. The red fluorescence is produced by propidium iodide counterstaining of chromosomal DNA.

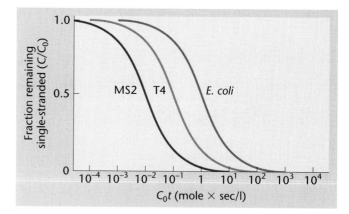

FIGURE 10.24 Comparison of the reassociation rate (C/C_0) of DNA derived from phage MS-2, phage T4, and *E. coli*. The genome of T4 is larger than MS-2, and that of *E. coli* is larger than T4.

organisms. For example, we may compare the point in the reaction when one-half of the DNA is present as double-stranded fragments. This point is called the $C_0t_{1/2}$, or **half reaction time**. Provided that all pairs of single-stranded DNA complements consist of unique nucleotide sequences and all are about the same size, $C_0t_{1/2}$ varies directly with the complexity of the DNA. Designated as X, complexity represents the length in nucleotide pairs of all unique DNA fragments laid end to end. If the DNA used in an experiment represents the

entire genome, and if all DNA sequences are different from one another, then X is equal to the size of the haploid genome.

Figure 10.24 illustrates what is found when DNA from various sources are compared. As can be seen, as genome size increases, the curves obtained are shifted farther and farther to the right, indicative of an increased reassociation time.

As shown in Figure 10.25, $C_0t_{1/2}$ is directly proportional to the size of the genome. If nearly the entire genome consists of unique DNA sequences, reassociation experiments can be used to determine the genome size of organisms. This method has been useful in assessing genome size in viruses and bacteria.

However, when reassociation kinetics of DNA from eukaryotic organisms were first studied, a surprising observation was made. The data showed that some of the DNA segments reassociated even more rapidly than did those derived from *E. coli*! The remainder, as expected because of its greater complexity, took longer to reassociate. For example, Britten and Kohne (cited earlier) examined DNA derived from calf thymus tissue (Figure 10.26). Based on these observations, they hypothesized that the rapidly reassociating fraction must represent **repetitive DNA sequences** present many times in the calf genome. This interpretation would explain why these DNA segments reassociate so rapidly. On the other hand, they hypothesized that the remaining DNA segments consist of unique nucleotide sequences present only once in the genome; because there are more of these unique sequences, increasing the DNA complexity in calf thymus (compared with *E. coli*), their reassociation takes longer. The *E. coli* curve has been added to Figure 10.26 for the sake of comparison.

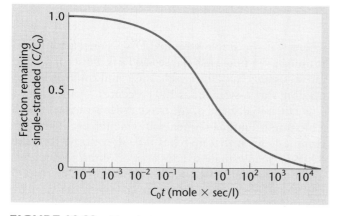

FIGURE 10.23 The ideal time course for reassociation of DNA (C/C_0) when, at time zero, all DNA consists of unique fragments of single-stranded complements. Note that the abscissa (C_0t) is plotted logarithmically.

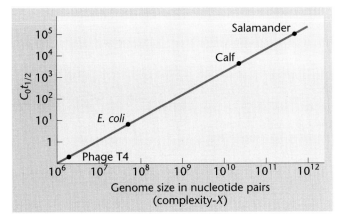

FIGURE 10.25   Comparison of $C_0t_{1/2}$ and genome size for phage T4, *E. coli*, calf, and salamander.

It is now clear that repetitive DNA sequences are prevalent in the genome of eukaryotes. When analyzed carefully, it is apparent that there are various levels of repetition. Cases are known where short DNA sequences are repeated over a million times, where longer DNA sequences are repeated only a few times, and where intermediate levels of sequence redundancy are present. We will return to this topic in Chapter 17, where we will discuss the organization of DNA in genes and chromosomes. For now, we conclude this chapter by pointing out that the discovery of repetitive DNA was one of the first clues that much of the DNA in eukaryotes is not contained in genes that encode proteins. This concept will be developed and elaborated on as we expand our coverage of the molecular basis of heredity.

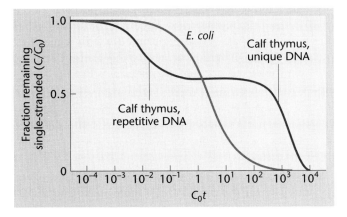

FIGURE 10.26   The C_0t curve of calf thymus DNA compared with *E. coli*. The repetitive fraction of calf DNA reassociates more quickly than that of *E. coli*, while the more complex unique calf DNA takes longer to reassociate than that of *E. coli*.

Electrophoresis of Nucleic Acids

We conclude our discussion by considering one of the most essential techniques involved in the analysis of nucleic acids: **electrophoresis**. This technique allows for the separation of different-sized fragments of DNA and RNA chains and is invaluable in current day-to-day research investigations in molecular genetics.

In general, electrophoresis separates, *or resolves*, molecules in a mixture by causing them to migrate under the influence of an electric field. A sample is placed on a porous substance (a piece of filter paper or a semisolid gel), which is placed in a solution that conducts electricity. If two molecules have approximately the same shape *and* mass, the one with the greatest net charge will migrate more rapidly toward the electrode of opposite polarity.

As electrophoretic technology was developed, as initially applied to the separation of proteins, it was discovered that the use of gels of varying pore sizes significantly improved the resolution of this research technique. This advance was particularly useful for mixtures of molecules with a similar charge:mass ratio, but of different sizes. For example, two polynucleotide chains of different *lengths* (e.g., 10 vs. 20 nucleotides) are both negatively charged based on the phosphate group of each nucleotide. While they both move to the positively charged pole (the anode), the charge:mass ratio is the same for both chains, and separation based strictly on the electric field is minimal. However, the use of a porous medium such as **polyacrylamide gels** or **agarose gels**, which can be prepared with various pore sizes, provides the basis for separation of these two molecules.

In such cases, *the smaller molecules migrate at a faster rate through the gel than the larger molecules* (Figure 10.27). The key to separation is based on the matrix (pores) of the gel, which *restricts migration of a larger molecule more than it restricts a smaller molecule*. The resolving power is so great that polynucleotides that vary by even one nucleotide in length are clearly separated. Once electrophoresis is complete, bands representing the variously-sized molecules are identified either by autoradiography (if a component of the molecule is radioactive), or by the use of a fluorescent dye that binds to nucleic acids.

Electrophoretic separation of nucleic acids is at the heart of a variety of commonly used research techniques discussed later in the text (Chapters 15 and 16). Of particular note are the various "blotting" techniques (e.g., Southern blots, Western blots, etc.), as well as DNA sequencing methods. We introduce electrophoresis here to complete our coverage of the analysis of nucleic acids.

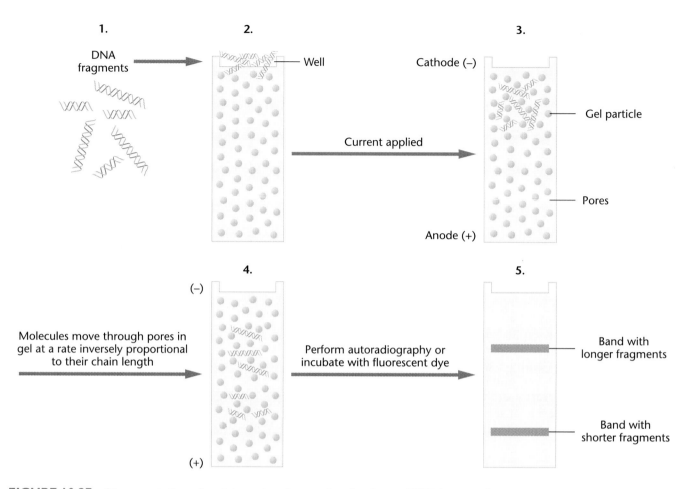

FIGURE 10.27 Diagrammatic illustration of electrophoretic separation of a mixture of DNA fragments that are 10 vs. 20 nucleotides in length.

CHAPTER SUMMARY

1. The existence of a genetic material capable of replication, storage, expression, and mutation is deducible from the observed patterns of inheritance in organisms.

2. Both proteins and nucleic acids were initially considered as the possible candidates for genetic material. Proteins are more diverse than nucleic acids, a requirement for the genetic material, and were favored owing to the advances being made in protein chemistry at the time. Additionally, Levene's tetranucleotide hypothesis had underestimated the magnitude of chemical diversity inherent in nucleic acids.

3. By 1952, transformation studies and experiments using bacteria infected with bacteriophages strongly suggested that DNA is the genetic material in bacteria and most viruses.

4. Initially, only circumstantial evidence supported the concept of DNA controlling inheritance in eukaryotes. This included the distribution of DNA in the cell, quantitative analysis of DNA, and studies of UV-induced mutagenesis. Recent recombinant DNA techniques, as well as experiments with transgenic mice, have provided direct experimental evidence that the eukaryotic genetic material is DNA.

5. Numerous viruses provide an important exception to this general rule, because many of them use RNA as their genetic material. These include some bacteriophages and some plant and animal viruses, as well as retroviruses.

6. Establishment of DNA as the genetic material paved the way for the expansion of molecular genetics research, and has served as the cornerstone for further important studies for nearly half a century.

7. During the late 1940s and early 1950s, a considerable effort was made to integrate accumulated information on the chemical structure of nucleic acids into a model of the molecular structure of DNA. The X-ray crystal-

lography data of Franklin and Wilkins suggested that DNA was some sort of helix. In 1953, Watson and Crick were able to assemble a model of the double-helical DNA structure based on these X-ray diffraction studies as well as on Chargaff's analysis of base composition of DNA.

8. The DNA molecule exhibits antiparallel orientation and adenine-thymine and guanine-cytosine base-pairing complementarity along the polynucleotide chains. This model of DNA presents an obvious straightforward mechanism for its replication. The structure assumed by the helix appears to be a function of the nucleotide sequence and its chemical environment. Several alternative forms of the DNA helical structure exist. Watson and Crick described a B configuration, one of several right-handed helices. Wang and Rich discovered the left-handed Z-DNA currently being investigated for its physiological and genetic significance.

9. The second category of nucleic acids important in genetic function is RNA, which is similar to DNA with the exceptions that it is usually single-stranded, the sugar ribose replaces the deoxyribose, and the pyrimidine uracil replaces thymine. Classes of RNA—ribosomal, transfer, and messenger—facilitate the flow of information from DNA to RNA to proteins, which are the end products of most genes.

10. The structure of DNA lends itself to various forms of analyses, which have in turn led to studies of the functional aspects of the genetic machinery. Absorption of UV light, sedimentation properties, denaturation–reassociation, and electrophoresis procedures are among the important tools for the study of nucleic acids. Reassociation kinetics analysis enabled geneticists to postulate the existence of repetitive DNA in eukaryotes, where certain nucleotide sequences are present many times in the genome.

KEY TERMS

absorption spectrum, 271
action spectrum, 271
adenine, 274
adenosine triphosphate (ATP), 275
A-DNA, 283
antiparallel chains, 279
antisense RNA, 285
autoradiography, 288
bacteriophage (phage), 268
B-DNA, 283
C-DNA, 283
central dogma of molecular genetics, 263
complementarity, 281
cytosine, 274
D-DNA, 283
denaturation, 287
deoxyribonuclease, 268
deoxyribose, 274
E-DNA, 283
electrophoresis, 291
FISH (fluorescent *in situ* hybridization), 288
genetic expression, 272
genetic information, 262
gradient centrifugation, 285
guanine, 274
guanosine triphosphate (GTP), 275
half reaction time. 290
haploid, 271
Hershey–Chase experiment, 268
Holmes ribgrass (HR) virus, 272

hydrogen bond, 279
hyperchromic shift, 287
information flow, 263
in situ molecular hybridization, 288
lysozyme, 269
major groove, 279
melting profile, 288
melting temperature (T_m), 288
messenger RNA (mRNA), 283
minor groove, 279
molecular hybridization, 288
molecular weight (MW), 287
nitrogenous base, 274
nuclein, 264
nucleoside, 275
nucleoside diphosphate (NDP), 275
nucleoside monophosphate (NMP), 275
nucleoside triphosphate (NTP), 275
nucleotide, 275
oligonucleotide, 276
pentose sugar, 274
phosphate group, 274
phosphodiester bond, 275
polynucleotide, 276
primary transcripts, 285
protoplast, 269
purine, 274
pyrimidine, 274
reassociation kinetics, 288
recombinant DNA technology, 272
repetitive DNA sequences, 290

replication, 263
retroviruses, 273
reverse transcription, 273
ribose, 274
ribosomal RNA (rRNA), 283
ribosome, 285
RNA replicase, 272
sedimentation equilibrium centrifugation (density gradient centrifugation), 286
sedimentation velocity centrifugation, 287
single-crystal X-ray analysis, 283
small nuclear RNA, 285
spheroplast, 269
Svedberg coefficient (*S*), 285
telomerase RNA, 285
tetranucleotide hypothesis, 264
thymine, 274
tobacco mosaic virus (TMV), 272
transcription, 263
transfection, 271
transfer RNA (tRNA), 283
transformation, 265
transgenic animals, 272
translation, 263
ultraviolet light (UV), 271
uracil, 274
variation by mutation, 263
X-ray diffraction, 277
Z-DNA, 283

INSIGHTS AND SOLUTIONS

In contrast to the preceding chapters, this one does not emphasize genetic problem solving. Instead, it recounts some of the initial experimental analyses that served as the cornerstone of modern genetics. Quite fittingly, then, our Insights and Solutions section shifts its emphasis to experimental rationale and analytical thinking, an approach that will continue through the remainder of the text whenever appropriate.

1. (a) Based strictly on the analysis of the transformation data of Avery, MacLeod, and McCarty, what objection might be made to the conclusion that DNA is the genetic material? What other conclusion might be considered?

    Solution: Based solely on their results, it may be concluded that DNA is essential for transformation. However, DNA might have been a substance that caused capsular formation by *directly* converting nonencapsulated cells to ones with a capsule. That is, DNA may simply have played a catalytic role in capsular synthesis, leading to cells displaying smooth type III colonies.

   (b) What observations argue against this objection?

    Solution: First, transformed cells pass the trait onto their progeny cells, thus supporting the conclusion that DNA is responsible for heredity, not for the direct production of polysaccharide coats. Second, subsequent transformation studies over a period of five years showed that other traits, such as antibiotic resistance, could be transformed. Therefore, the transforming factor has a broad general effect, not one specific to polysaccharide synthesis. This observation is more in keeping with the conclusion that DNA is the genetic material.

2. If RNA were the universal genetic material, how would this have affected the Avery experiment and the Hershey–Chase experiment?

    Solution: In the Avery experiment, digestion of the soluble filtrate RNase rather than DNase would have eliminated transformation. Had this occurred, Avery and his colleagues would have concluded that RNA was the transforming factor. The Hershey and Chase results would not have changed, since ^{32}P would also label RNA, but not protein. Had they been using a bacteriophage with RNA as its nucleic acid, and had they known this, they would have concluded that RNA was responsible for directing the reproduction of their bacteriophage.

3. Sea urchin DNA, which is double-stranded, was shown to contain 17.5 percent of its bases in the form of cytosine (C). What percentages of the other three bases are present in this DNA?

    Solution: The amount of C = G, so guanine is also present as 17.5 percent. The remaining bases, A and T, are

present in equal amounts and together they represent the rest of the bases (100 − 35). Therefore, A = T = 65/2 = 32.5 percent.

4. The quest to isolate an important disease-causing organism was successful and the molecular biologists were hard at work. The organism contained as its genetic material a remarkable nucleic acid with a base composition of A = 21 percent, C = 29 percent, G = 29 percent, U = 21 percent. When heated, it showed a major hyperchromic effect, and when kinetics were studied, the nucleic acid of this organism provided the C_0t curve shown below, in contrast to that of phage T4 and *E. coli*. T4 contains 10^5 nucleotide pairs and exhibits a $C_0t_{1/2}$ of 0.5. The unknown organism produced a $C_0t_{1/2}$ of 20. Analyze this information carefully and draw *all* possible conclusions about the genetic material of this organism, based strictly on the above observations. What important, straightforward information is missing and needed to confirm your hypothesis about the nature of this molecule?

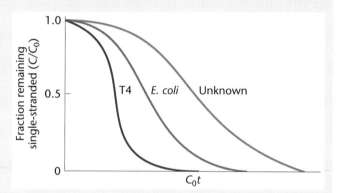

Solution: First of all, because of the presence of uracil (U), the molecule appears to be RNA. As A/U = G/C = 1, the molecule may be a double helix. The hyperchromic shift and reassociation kinetics support this hypothesis. The kinetic study demonstrates several things. First, the shape of the C_0t curve reveals that there is no repetitive sequence DNA. Further, the complexity (X), or total length of unique sequence DNA, is greater than that of either phage T4 or *E. coli*. X can, in fact, be calculated, since a direct proportionality between $C_0t_{1/2}$ and number of base pairs exists when there are only unique sequences present:

$$\frac{0.5}{10^5} = \frac{20}{X}$$

$$0.5X = 20(10^5)$$

$$X = 40(10^5)$$

$$= 4 \times 10^6 \text{ base pairs}$$

There *may* be a greater number of genes present compared to T4 or *E. coli*, but the excessive unique sequence DNA may serve some other role, or simply play no genetic role. The missing information concerns the sug-ars. Our model predicts that ribose rather than d-ribose should be present. If not, the organism contains a very unusual molecule as its genetic material.

PROBLEMS AND DISCUSSION QUESTIONS

1. The functions ascribed to the genetic material are replication, expression, storage, and mutation. What does each of these terms mean?
2. Discuss the reasons why proteins were generally favored over DNA as the genetic material before 1940. What was the role of the tetranucleotide hypothesis in this controversy?
3. Contrast the various contributions made to an understanding of transformation by Griffith, by Avery and his coworkers, and by Taylor.
4. When Avery and his colleagues had obtained what was concluded to be purified DNA from the IIIS virulent cells, they treated the fraction with proteases, RNase, and DNase, followed by the assay for retention or loss of transforming ability. What were the purpose and results of these experiments? What conclusions were drawn?
5. Why were ^{32}P and ^{35}S chosen for use in the Hershey–Chase experiment? Discuss the rationale and conclusions of this experiment.
6. Does the design of the Hershey–Chase experiment distinguish between DNA and RNA as the molecule serving as the genetic material? Why or why not?
7. Would an experiment similar to that performed by Hershey and Chase work if the basic design were applied to the phenomenon of transformation? Explain why or why not.
8. What observations are consistent with the conclusion that DNA serves as the genetic material in eukaryotes? List and discuss them.
9. What are the exceptions to the general rule that DNA is the genetic material in all organisms? What evidence supports these exceptions?
10. Draw the chemical structure of the three components of a nucleotide and then link the three together. What atoms are removed from the structures when the linkages are formed?
11. How are the carbon and nitrogen atoms of the sugars, purines, and pyrimidines numbered?
12. Adenine may also be named 6-amino purine. How would you name the other four nitrogenous bases using this alternative system? ($=$O is oxy, and $-CH_3$ is methyl.)
13. Draw the chemical structure of a dinucleotide composed of A and G. Opposite this structure, draw the dinucleotide TC in an antiparallel (or upside down) fashion. Form the possible hydrogen bonds.
14. Describe the various characteristics of the Watson–Crick double-helix model for DNA.
15. What evidence did Watson and Crick have at their disposal in 1953? What was their approach in arriving at the structure of DNA?
16. Had Chargaff's data from a single source indicated the following, what might Watson and Crick have concluded?

	A	T	G	C
%	29	19	21	31

Why would this conclusion be contradictory to Wilkins and Franklin's data?
17. How do covalent bonds differ from hydrogen bonds? Define base complementarity.
18. List three main differences between DNA and RNA.
19. What are the three types of RNA molecules? How is each related to the concept of information flow?
20. What component of the nucleotide is responsible for the absorption of ultraviolet light? How is this technique important in the analysis of nucleic acids?
21. Distinguish between sedimentation velocity and sedimentation equilibrium centrifugation.
22. What is the basis for determining base composition using density gradient centrifugation?
23. What is the physical state of DNA following denaturation?
24. Compare the following curves representing reassociation kinetics. What can be said about the DNAs represented by each set of data compared with *E. coli*?

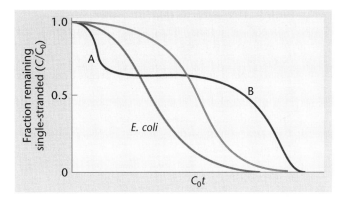

25. What is the hyperchromic effect? How is it measured? What does T_m imply?
26. Why is T_m related to base composition?
27. What is the chemical basis of molecular hybridization?
28. What did the Watson–Crick model suggest about the replication of DNA?
29. A genetics student was asked to draw the chemical structure of an adenine- and thymine-containing dinucleotide derived from DNA. His answer is shown below. The student made more than six major errors. One of them is circled, numbered ①, and explained. Find five others. Circle them,

number them ②–⑥, and briefly explain each, following the example below.

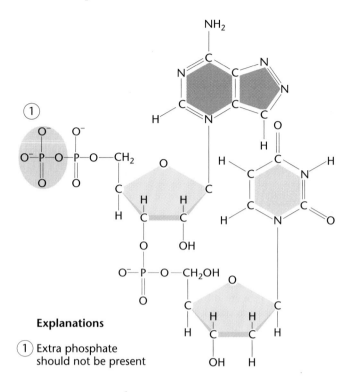

Explanations

① Extra phosphate should not be present

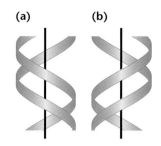

30. The DNA of the bacterial virus T4 produces a $C_0t_{1/2}$ of about 0.5 and contains 10^5 nucleotide pairs in its genome. How many nucleotide pairs are present in the genome of the virus MS2 and the bacterium *E. coli*, whose respective DNAs produce $C_0t_{1/2}$ values of 0.001 and 10.0?

31. A primitive eukaryote was discovered that displayed a unique nucleic acid as its genetic material. Analysis revealed the following observations:
    (i) X-ray diffraction studies display a general pattern similar to DNA, but with somewhat different dimensions and more irregularity.
    (ii) A major hyperchromic shift is evident upon heating and monitoring UV absorption at 260 nm.
    (iii) Base composition analysis reveals four bases in the following proportions.

    Adenine = 8%
    Guanine = 37%
    Xanthine = 37%
    Hypoxanthine = 18%

    (iv) About 75 percent of the sugars are d-ribose, while 25 percent are ribose.

    Attempt to solve the structure of this molecule by postulating a model that is consistent with the above observations.

32. Considering the information in this chapter on B- and Z-DNA and right- and left-handed helices, carefully analyze the structures below and draw conclusions about the helical nature of areas (a) and (b). Which is right-handed and which is left-handed?

33. One of the most common spontaneous lesions that occurs in DNA under physiological conditions is the hydrolysis of the amino group of cytosine, converting it to uracil. What would be the effect on DNA structure of a uracil group replacing cytosine?

34. In some organisms, cytosine is methylated at carbon 5 of the pyrimidine ring after it is incorporated into DNA. If a 5-methyl cytosine is then hydrolyzed as described in the previous problem (33), what base will be generated?

EXTRA-SPICY PROBLEMS

35. *Newsdate: March 1, 2015.* A unique creature has been discovered during exploration of outer space. Recently, its genetic material has been isolated and analyzed. This material is similar in some ways to DNA in its chemical makeup. It contains in abundance the 4-carbon sugar erythrose and a molar equivalent of phosphate groups. Additionally, it contains six nitrogenous bases: adenine (A), guanine (G), thymine (T), cytosine (C), hypoxanthine (H), and xanthine (X). These bases exist in the following relative proportions:

    $$A = T = H \quad and \quad C = G = X$$

    X-ray diffraction studies have established a regularity to the molecule and a constant diameter of about 30 Å.
    Together, these data have suggested a model for the structure of this molecule.
    (a) Propose a general model of this molecule. Describe it briefly.
    (b) What base-pairing properties must exist for H and for X in the model?
    (c) Given the constant diameter of 30 Å, do you think that H and X are *either* (i) both purines or both pyrimidines, *or* (ii) one is a purine and one is a pyrimidine?

36. You are provided with DNA samples from two newly discovered bacterial viruses. Based on the various analytical techniques discussed in this chapter, construct a research protocol that would be useful in characterizing and contrasting the DNA of both viruses. For each technique that you include in the protocol, indicate the type of information you hope to obtain.

SELECTED READINGS

ALLOWAY, J. L. 1933. Further observations on the use of pneumococcus extracts in effecting transformation of type *in vitro. J. Exp. Med.* 57:265–78.

AVERY, O. T., MACLEOD, C. M., and McCARTY, M. 1944. Studies on the chemical nature of the substance inducing transformation of pneumococcal types: Induction of transformation by a desoxyribonucleic acid fraction isolated from pneumococcus type III. *J. Exp. Med.* 79:137–58. (Reprinted in Taylor, J. H. 1965. *Selected papers in molecular genetics.* Orlando, FL: Academic Press.)

BRITTEN, R. J., and KOHNE, D. E. 1968. Repeated sequences in DNA. *Science* 161:529–40.

CAIRNS, J., STENT, G. S., and WATSON, J. D. 1966. *Phage and the origins of molecular biology.* Cold Spring Harbor, NY: Cold Spring Harbor Laboratory Press.

CHARGAFF, E. 1950. Chemical specificity of nucleic acids and mechanism for their enzymatic degradation. *Experientia* 6:201–9.

CRICK, F. H. C., WANG, J. C., and BAUER, W. R. 1979. Is DNA really a double helix? *J. Mol. Biol.* 129:449–61.

DAVIDSON, J. N. 1976. *The biochemistry of the nucleic acids*, 8th ed. Orlando, FL: Academic Press.

DAWSON, M. H. 1930. The transformation of pneumococcal types: I. The interconvertibility of type-specific *S. pneumococci. J. Exp. Med.* 51:123–47.

DEROBERTIS, E. M., and GURDON, J. B. 1979. Gene transplantation and the analysis of development. *Sci. Am.* (Dec.) 241:74–82.

DICKERSON, R. E. 1983. The DNA helix and how it is read. *Sci. Am.* (June) 249:94–111.

DICKERSON, R. E., et al. 1982. The anatomy of A-, B-, and Z-DNA. *Science* 216:475–85.

DUBOS, R. J. 1976. *The professor, the Institute and DNA: Oswald T. Avery, his life and scientific achievements.* New York: Rockefeller University Press.

FELSENFELD, G. 1985. DNA. *Sci. Am.* (Oct.) 253:58–78.

FRAENKEL-CONRAT, H., and SINGER, B. 1957. Virus reconstruction: II. Combination of protein and nucleic acid from different strains. *Biochem. Biophys. Acta* 24:530–48. (Reprinted in Taylor, J. H. 1965. *Selected papers in molecular genetics*, Orlando, FL: Academic Press.)

FRANKLIN, R. E., and GOSLING, R. G. 1953. Molecular configuration in sodium thymonucleate. *Nature* 171:740–41.

GEIS, I. 1983. Visualizing the anatomy of A, B and Z-DNAs. *J. Biomol. Struc. Dynam.* 1: 581–91.

GRIFFITH, F. 1928. The significance of pneumococcal types. *J. Hyg.* 27:113–59.

GUTHRIE, G. D., and SINSHEIMER, R. L. 1960. Infection of protoplasts of *Escherichia coli* by subviral particles. *J. Mol. Biol.* 2:297–305.

HERSHEY, A. D., and CHASE, M. 1952. Independent functions of viral protein and nucleic acid in growth of bacteriophage. *J. Gen. Phys.* 36:39–56. (Reprinted in Taylor, J. H. 1965. *Selected papers in molecular genetics.* Orlando, FL: Academic Press.)

HOTCHKISS, R. D. 1951. Transfer of penicillin resistance in pneumococci by the desoxyribonucleate derived from resistant cultures. *Cold Spring Harbor Symp. Quant. Biol.* 16:457–61. (Reprinted in Adelberg, E. A. 1960. *Papers on bacterial genetics.* Boston: Little, Brown.)

JUDSON, H. 1979. *The eighth day of creation: Makers of the revolution in biology.* New York: Simon & Schuster.

LEVENE, P. A., and SIMMS, H. S. 1926. Nucleic acid structure as determined by electrometric titration data. *J. Biol. Chem.* 70:327–41.

McCARTY, M. 1980. Reminiscences of the early days of transformation. *Annu. Rev. Genet.* 14:1–16.

———. 1985. *The transforming principle: Discovering that genes are made of DNA.* New York: W. W. Norton.

McCONKEY, E. H. 1993. *Human genetics—The molecular revolution.* Boston: Jones and Bartlett.

OLBY, R. 1974. *The path to the double helix.* Seattle: University of Washington Press.

PALMITER, R. D., and BRINSTER, R. L. 1985. Transgenic mice. *Cell* 41:343–45.

PAULING, L., and COREY, R. B. 1953. A proposed structure for the nucleic acids. *Proc. Natl. Acad. Sci. USA* 39:84–97.

RICH, A., NORDHEIM, A., and WANG, A. H.-J. 1984. The chemistry and biology of left-handed Z-DNA. *Annu. Rev. Biochem.* 53:791–846.

SCHILDKRAUT, C. L., MARMUR, J., and DOTY, P. 1962. Determination of the base composition of deoxyribonucleic acid from its buoyant density in CsCl. *J. Mol. Biol.* 4:430–43.

SPIZIZEN, J. 1957. Infection of protoplasts by disrupted T2 viruses. *Proc. Natl. Acad. Sci. USA* 43:694–701.

STENT, G. S., ed. 1981. *The double helix: Text, commentary, review, and original papers.* New York: W. W. Norton.

STEWART, T. A., WAGNER, E. F., and MINTZ, B. 1982. Human β-globin gene sequences injected into mouse eggs, retained in adults, and transmitted to progeny. *Science* 217:1046–48.

VARMUS, H. 1988. Retroviruses. *Science* 240:1427–35.

WATSON, J. D. 1968. *The double helix.* New York: Atheneum.

WATSON, J. D., and CRICK, F. C. 1953a. Molecular structure of nucleic acids: A structure for deoxyribose nucleic acids. *Nature* 171:737–38.

———. 1953b. Genetic implications of the structure of deoxyribose nucleic acid. *Nature* 171:964.

WEINBERG, R. A. 1985. The molecules of life. *Sci. Am.* (Oct.) 253:48–57.

WILKINS, M. H. F., STOKES, A. R., and WILSON, H. R. 1953. Molecular structure of desoxypentose nucleic acids. *Nature* 171:738–40.

ZIMMERMAN, B. 1982. The three-dimensional structure of DNA. *Annu. Rev. Biochem.* 51:395–428.

DNA Replication and Recombination

CHAPTER CONCEPTS

*Genetic continuity between parental and progeny cells is made possible by semiconservative repli-
cation of DNA, as predicted by the Watson–Crick model. Each strand of the parent helix serves as
a template for the production of its complement. Synthesis of DNA is a complex but orderly process,
orchestrated by a myriad of enzymes and other molecules. Together, they function with great fidelity to polymerize nucleotides into polynucleotide chains. Genetic recombination is also a process
that involves the interaction of DNA molecules with a group of enzymes.*

Following Watson and Crick's proposal for the structure of DNA, scientists focused their attention on
how this molecule is replicated. This process is an essential function of the genetic material and must be executed precisely if genetic continuity between cells is to
be maintained following cell division. This is an enormous, complex task. Con-
sider for a second that some
10^9 (three billion) base pairs
exist within the 23 chromosomes of the human ge-
nome. To duplicate faith-
fully the DNA of just one of
these chromosomes re-
quires a mechanism of ex-
treme precision. Even an
error rate of only 10^{-6} (one
in a million) will still create
3000 errors, obviously an
excessive number, during
each replication cycle of the

genome. While not error free, a much more accurate
system of DNA replication has evolved that functions
in all organisms.

As Watson and Crick suggested in their 1953
paper, the model of the double helix provided them
with the initial insight into how replication could
occur. This mode, called
**semiconservative replica-
tion**, has since received
strong experimental support
from studies of viruses,
prokaryotes, and eukaryotes.

Once the general mode
of replication was made clear,
research was intensified to
determine the precise details
of DNA synthesis. What has
since been discovered is that
numerous enzymes and other
proteins are needed to copy a
DNA helix. Because of the

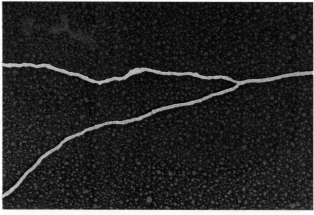

A replication fork formed following the initiation of DNA synthesis.

complexity of the chemical events during synthesis, this subject remains an extremely active area of research.

This chapter will examine the general mode of replication as well as the specific details of the synthesis of DNA. The research leading to this knowledge is still another link in our understanding of life processes at the molecular level.

The Mode of DNA Replication

It was apparent to Watson and Crick that because of the arrangement and nature of the nitrogenous bases, each strand of a DNA double helix could serve as a template for the synthesis of its complement (Figure 11.1). They proposed that if the helix were unwound, each nucleotide along the two parent strands would have an affinity for its complementary nucleotide. As we learned in Chapter 10, the complementarity is due to the potential hydrogen bonds that can be formed. If thymidylic acid (T) were present, it would "attract" adenylic acid (A); if guanidylic acid (G) were present, it would "attract" cytidylic acid (C); and so on. If these nucleotides were then covalently linked into polynucleotide chains along both templates,

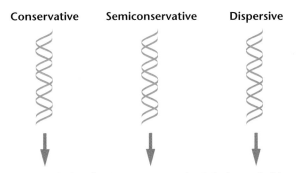

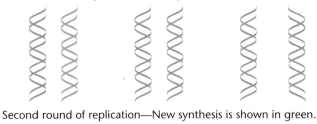

FIGURE 11.2   Results of two rounds of replication of DNA for each of the three possible modes of replication. Each round of synthesis is shown in a different color.

the result would be the production of two identical double strands of DNA. Each replicated DNA molecule would consist of one "old" and one "new" strand; hence, the reason for the name "semiconservative replication."

There are two other possible modes of replication (Figure 11.2) that also rely on the parental strands as a template. In **conservative replication**, synthesis of complementary polynucleotide chains would occur as described above. Following synthesis, however, the two newly created strands are brought back together, and the parental strands reassociate. The original helix is thus "conserved."

In the second alternative mode, called **dispersive replication**, the parental strands are seen to be dispersed into two new double helices following replication. Thus, each strand consists of both old and new DNA. This mode would involve cleavage of the parental strands during replication. Therefore, it is the most complex of the three possibilities and is therefore least likely to occur. It could not, however, be ruled out as an experimental model. The theoretical results of two generations of

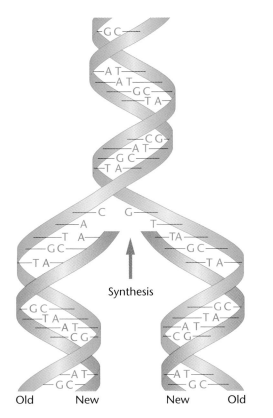

FIGURE 11.1   General model of semiconservative replication of DNA.

replication by the conservative and dispersive modes are compared with those of the semiconservative mode in Figure 11.2.

The Meselson–Stahl Experiment

In 1958, Matthew Meselson and Franklin Stahl published the results of an experiment providing strong evidence that semiconservative replication is the mode used by cells to produce new DNA molecules. *E. coli* cells were grown for many generations in a medium where $^{15}NH_4Cl$ (ammonium chloride) was the only nitrogen source. A "heavy" isotope of nitrogen, ^{15}N contains one more neutron than the naturally occurring ^{14}N isotope. Unlike "radioactive" isotopes, ^{15}N is stable and is thus not radioactive. After many generations, all nitrogen-containing molecules, including the nitrogenous bases of DNA, contained the isotope in the *E. coli* cells. DNA containing ^{15}N may be distinguished from ^{14}N-containing DNA by the use of **sedimentation equilibrium centrifugation**, where samples are "forced" by centrifugation through a density gradient of a heavy metal salt such as cesium chloride (see Chapter 10). The more dense ^{15}N-DNA will reach equilibrium in the gradient at a point closer to the bottom (where the density is greater) than will ^{14}N-DNA.

In this experiment, *E. coli* cells were grown for several generations in the presence of the heavy isotope, resulting in uniformly labeled ^{15}N cells. They were then transferred to a medium containing only $^{14}NH_4Cl$. Thus, all subsequent synthesis of DNA during replication contained the "lighter" isotope of nitrogen. The time of transfer was taken as time zero ($t = 0$). The *E. coli* cells were allowed to replicate during several generations with cell samples removed at various intervals. From each sample, DNA was isolated and subjected to sedimentation equilibrium centrifugation. The results are depicted in Figure 11.3.

After one generation, the isolated DNA was all present in a single band of intermediate density—the expected result for semiconservative replication. Each replicated molecule would be composed of one new ^{14}N-strand and one old ^{15}N-strand, as seen in Figure 11.4. This result was not consistent with the conservative replication mode, in which two distinct bands would have been predicted to occur.

After two cell divisions, DNA samples showed two density bands: One was intermediate and the other was lighter, corresponding to the ^{14}N position in the gradient. Similar results occurred after a third generation, except that the proportion of the ^{14}N-band increased. This was again consistent with the interpretation that replication is semiconservative.

Two observations served as the basis for ruling out the dispersive mode of replication. One involved analysis of DNA after the first generation of replication in an ^{14}N-containing medium. The observation of a molecule exhibiting intermediate density is also consistent with dispersive replication. However, Meselson and Stahl isolated this hybrid molecule and analyzed it after it had been heat-denatured. Recall from Chapter 10 that heating will separate a duplex into single strands. When the densities of the single strands of the hybrid were determined, they exhibited *either* an ^{15}N-profile *or* an ^{14}N-profile, but *not* an intermediate density. This observation is consistent with the semiconservative mode but inconsistent with the dispersive mode.

Furthermore, if replication were dispersive, *all* generations after $t = 0$ would demonstrate DNA of an intermediate density. In each subsequent generation after the first, the ratio of $^{15}N/^{14}N$ would decrease and the hybrid band would become lighter and lighter, eventually approaching the ^{14}N-band. This result was not observed. The Meselson–Stahl experiment provided very conclusive support in bacteria for semiconservative replication and tended to rule out the conservative and dispersive mode.

Semiconservative Replication in Eukaryotes

In 1957, the year before the work of Meselson and his colleagues was published, evidence was presented by J. Herbert Taylor, Philip Woods, and Walter Hughes that semiconservative replication also occurs in a eukaryotic organism. They experimented with root tips of the broad bean *Vicia faba*. Root tips are an excellent source of dividing cells. These researchers examined the chromosomes of these cells following replication of DNA. They were able to monitor the process of replication by labeling DNA with ^3H-thymidine, a radioactive precursor of DNA, and performing **autoradiography**. In this experiment, labeled thymidine is found only in association with chromosomes that contain newly synthesized DNA.

The technique of autoradiography, discussed in detail in Appendix A, is a cytological procedure that allows the isotope to be localized within the cell. In this procedure, a photographic emulsion is placed over a section of cellular material (root tips in this experiment), and the preparation is stored in the dark. The slide is then developed, much as photographic film is processed. Because the radioisotope emits energy, the emulsion turns black at the approximate point of emission following development. The end result is the presence of dark spots or "grains" on the surface of the section, identifying within the cell the location of newly synthesized DNA.

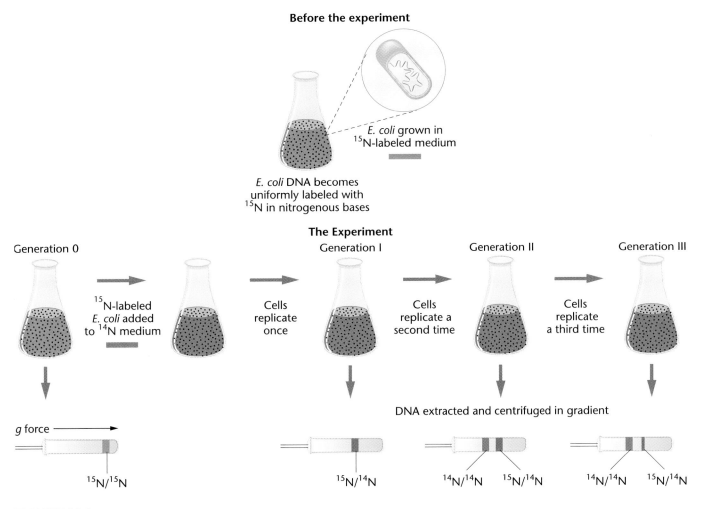

FIGURE 11.3   The Meselson–Stahl experiment.

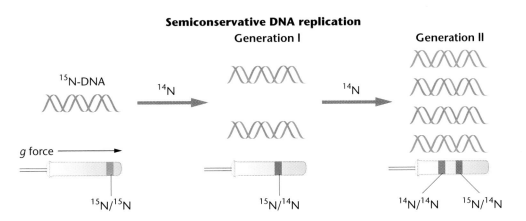

FIGURE 11.4   The expected results of two generations of semiconservative replication in the Meselson–Stahl experiment.

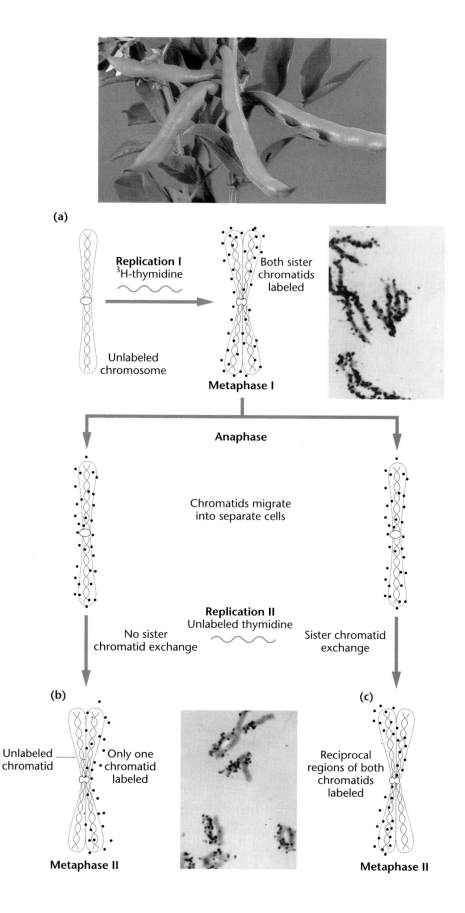

(a)

Replication I
³H-thymidine

Unlabeled chromosome

Both sister chromatids labeled

Metaphase I

Anaphase

Chromatids migrate into separate cells

Replication II
Unlabeled thymidine

No sister chromatid exchange

Sister chromatid exchange

(b)

Unlabeled chromatid

Only one chromatid labeled

Metaphase II

(c)

Reciprocal regions of both chromatids labeled

Metaphase II

FIGURE 11.5 Depiction of the experiment by Taylor, Woods, and Hughes demonstrating the semiconservative mode of replication of DNA in root tips of *Vicia faba*. The plant is shown in the top photograph. (a) An unlabeled chromatid proceeds through the cell cycle in the presence of ³H-thymidine. As it enters mitosis, both sister chromatids of the chomosome are labeled, as shown by autoradiography. After a second generation of replication, this time in the absence of ³H-thymidine, only one chromatid of each chromosome is expected to be surrounded by grains (b). Except where a reciprocal exchange has occurred between sister chromatids (c), the expectation was upheld. The micrographs are of the actual autoradiograms obtained in the experiment.

Root tips were grown for approximately one generation in the presence of the radioisotope and then placed in unlabeled medium, where cell division continued. At the conclusion of each generation, cultures were arrested at metaphase by the addition of colchicine (a chemical derived from the crocus plant that "poisons" the spindle fibers), and chromosomes were examined by autoradiography. Figure 11.5 illustrates replication of a single chromosome over two division cycles as well as the distribution of grains.

Results are compatible with the semiconservative mode of replication. After the first replication cycle, in the presence of the isotope, radioactivity is detected over both sister chromatids. This finding is expected because each chromatid will contain one "new" radioactive DNA strand and one "old" unlabeled strand. After the second replication cycle, which takes place in unlabeled medium, only one of the two new sister chromatids should be radioactive because half of the parent strands are unlabeled. With only minor exceptions of **sister chromatid exchanges** (see Chapter 5), this result was observed.

Together, the Meselson–Stahl experiment and the experiment by Taylor, Woods, and Hughes soon led to the general acceptance of the semiconservative mode of replication. The same conclusion has been reached in studies with other organisms. These experiments also strongly supported Watson and Crick's proposal for the double helix model of DNA.

Origins, Forks, and Units of Replication

The mode of replication just established represents the general pattern by which DNA is duplicated. Before turning to the details of how DNA is actually synthesized biochemically, we will briefly address several other issues relevant to the complete description of semiconservative replication. The first concerns the **origin of replication**. Where along the chromosome is DNA replication initiated? Is there but a single origin, or is there more than one point where DNA synthesis begins? Is any given point of origin random or is it located at a specific region along the chromosome?

A second issue involves the direction of replication. Once replication begins, does it proceed in a single direction or in both directions away from the origin? This consideration distinguishes between **unidirectional** and **bidirectional replication**, respectively.

To address these questions, we need to introduce two conceptual components integral to both the discussion and execution of DNA replication. The first concerns the actual point along the chromosome where replication is occurring. At such a point the strands of the helix must be unwound, creating what is called a **replication fork**. Such a "fork" will initially appear at the point of origin of synthesis. A replication fork will then move along the DNA duplex as replication proceeds. If replication is bidirectional, two such forks will be present that will migrate in opposite directions away from the origin.

The second conceptual component involves the length of DNA that is replicated following one initiation event at a single origin. This length of DNA is referred to as a unit called the **replicon**.

The evidence is reasonably clear regarding these issues raised initially above. In bacteria and most bacterial viruses, which have but one circular chromosome, there is only a single region where replication is initiated. In *E. coli* this specific region, called *oriC*, has been mapped along the chromosome. It consists of 245 base pairs, but only a small number are actually essential to the initiation of DNA synthesis. Since there is but a single point of origin of DNA synthesis in bacteriophages and bacteria, the entire chromosome constitutes one replicon.

Does replication proceed in a single direction, or in both directions? In *E. coli*, the evidence is clear. Replication is bidirectional from *oriC*, proceeding in both directions. Two replication forks are thus produced, as illustrated in Figure 11.6. These forks eventually merge as semiconservative replication of the entire chromosome is completed at a termination region, called *ter*.

Compared to bacteria, as described above, a major difference in replication exists in eukaryotes. While replication is bidirectional, creating two replication forks and a replication bubble during each replication cycle (Figure 11.7), there are multiple origins along each chromosome. As a result, during the S phase of interphase, numerous replicating events are occurring along each chromosome. Eventually, the numerous replication forks merge, completing replication of the entire chromosome. The presence of multiple replicons is undoubtedly related to the much greater length and complexity of a single eukaryotic chromosome compared to one from a bacterium. As a result, replication can be completed in a more reasonable period of time. We will return to the consideration of eukaryotic DNA synthesis later in this chapter.

Synthesis of DNA in Microorganisms

Determination that replication is semiconservative and bidirectional indicates only the pattern of DNA duplication and the association of finished strands with one another once synthesis is completed. A much more complex issue is how the *actual synthesis* of long complementary polynucleotide chains occurs on a DNA template. As in

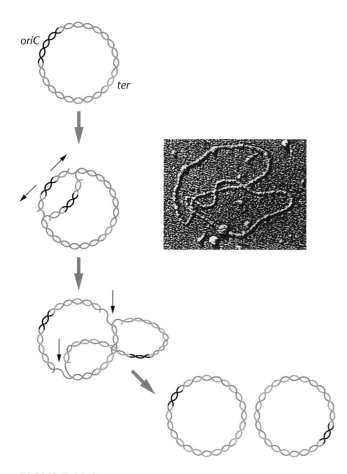

FIGURE 11.6 Bidirectional replication of the *E. coli* chromosome. The arrows identify the advancing replication forks. The micrograph is of a bacterial chromosome in the process of replication.

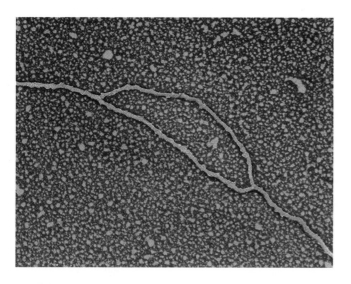

FIGURE 11.7 Transmission electron micrograph of human DNA from a HeLa cell, illustrating dual replication forks and the replication bubble that characterizes DNA replication within a single replicon.

most studies of molecular biology, this question was first approached by using microorganisms. Research began about the same time as the Meselson–Stahl work, and even today this topic is an active area of investigation. What is most apparent in this research is the tremendous chemical complexity of the biological synthesis of DNA.

DNA Polymerase I

Studies of the enzymology of DNA replication were first reported by Arthur Kornberg and colleagues in 1957. They isolated an enzyme from *E. coli* that was able to direct DNA synthesis in a cell-free (*in vitro*) system. The enzyme is now called **DNA polymerase I**, as it was the first of several to be isolated. Kornberg determined that there were two major requirements for *in vitro* DNA synthesis under the direction of the enzyme:

1. All four deoxyribonucleoside triphosphates (dATP, dCTP, dGTP, dTTP = dNTP)*
2. Template DNA

If any one of the four deoxyribonucleoside triphosphates was omitted from the reaction, no measurable synthesis occurred. If derivatives of these precursor molecules other than the nucleoside triphosphate were used (nucleotides or nucleoside diphosphates), synthesis did not occur. If no template DNA was added, synthesis of DNA occurred but was reduced greatly. Most synthesis directed by Kornberg's enzyme appeared to be exactly the type required for semiconservative replication. The reaction is summarized in Figure 11.8, where the addition of a single nucleotide is depicted. The enzyme has since been shown to consist of a single polypeptide containing 928 amino acids.

The way in which each nucleotide is added to the growing chain is a function of the specificity of DNA polymerase I. As shown in Figure 11.9, the precursor dNTP contains the three phosphate groups attached to the 5′-carbon of d-ribose. As the two terminal phosphates are cleaved during synthesis, the remaining phosphate attached to the 5′-carbon is covalently linked to the 3′-OH group of the d-ribose to which it is added. Each step *elongates* the growing chain by one nucleotide. Consistent with the above description and Figure 11.9, **chain elongation** occurs in the **5′-to-3′ direction** by the addition of nucleotides to the growing 3′ end. Each step provides a newly exposed 3′-OH group that can participate in the next addition of a nucleotide as DNA synthesis proceeds.

*dNTP designates the deoxyribose forms of the four nucleoside triphosphates; in a similar way, dNMP refers to the monophosphate forms.

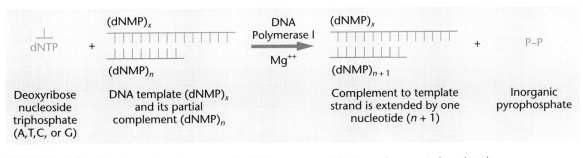

FIGURE 11.8   The chemical reaction catalyzed by DNA polymerase I. During each step, a single nucleotide is added to the growing complement of the DNA template using a nucleoside triphosphate as the substrate. The release of inorganic pyrophosphate drives the reaction energetically.

Fidelity of Synthesis

Having shown how DNA was synthesized, Kornberg sought to demonstrate the accuracy, or fidelity, with which the enzyme had replicated the DNA template. Because the nucleotide sequences of the template and the product could not be determined in 1957, he had to rely initially on several indirect methods.

One of Kornberg's approaches was to compare the nitrogenous base compositions of the DNA template with those of the recovered DNA product. Table 11.1 shows Kornberg's base composition analysis of three different DNA templates. These may be compared with the DNA product synthesized in each case. Within experimental error, the base composition of each product agreed with the template DNAs used. These data, along with other types of comparisons of template and product, suggested that the templates were replicated faithfully.

Kornberg also used the **nearest-neighbor frequency test** to assay the fidelity of copying (Figure 11.10). With this technique, Kornberg determined the frequency with which any two bases occur adjacent to each other along the polynucleotide chain. This test relies on the enzyme **spleen phosphodiesterase**, which cleaves the polynucleotide chain in a way that is different from the way in which it was assembled. As we have pointed out (see Figure 11.9), during synthesis of DNA, 5′-nucleotides are inserted; that is, each nucleotide is added with the phosphate on the C-5′ of deoxyribose. However, the phosphodiesterase enzyme cleaves between the phosphate and the C-5′ atom, thereby producing 3′-nucleotides. If the phosphates on only one of the four nucleotides (cytidylic acid, for example) are radioactive during DNA synthesis, then after enzymatic cleavage a radioactive phosphate will be transferred to the base that is the "nearest neighbor" on the 5′ side of all cytidylic acid nucleotides. Following four

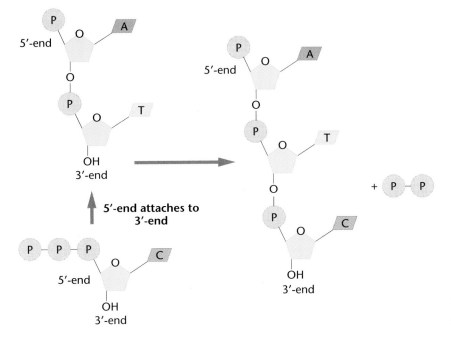

FIGURE 11.9   Demonstration of 5′–3′ synthesis of DNA.

TABLE 11.1 Base Composition of the DNA Template and the Product of Replication in Kornberg's Early Work

Organism	Template or Product	%A	%T	%G	%C
T2	Template	32.7	33.0	16.8	17.5
	Product	33.2	32.1	17.2	17.5
E. coli	Template	25.0	24.3	24.5	26.2
	Product	26.1	25.1	24.3	24.5
Calf	Template	28.9	26.7	22.8	21.6
	Product	28.7	27.7	21.8	21.8

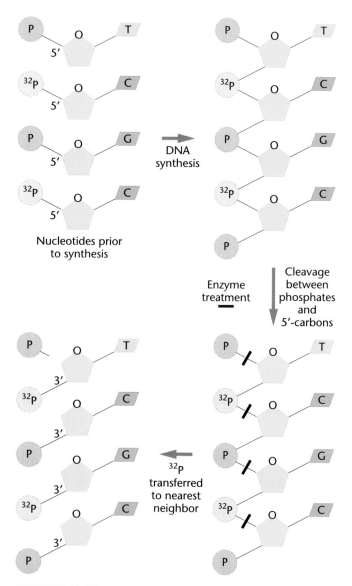

FIGURE 11.10 The theory of nearest-neighbor analysis. Initially, one of the four nucleotides (C) contains ^{32}P in the 5′-phosphate group. Following synthesis of a polynucleotide chain and enzymatic treatment with phosphodiesterase, the radioactive phosphate groups are transferred to the nearest neighbors (G and T). Subsequent analysis reveals the percentage of time each of the four nucleotides (A, T, C, and G) have become radioactive.

separate experiments, where in each case only one of the four nucleotide types is made radioactive, the frequency of all 16 possible nearest neighbors can be calculated. An example of such data is presented in one of the latter problems at the end of this chapter.

When this technique was applied to the DNA template and the resultant product from a variety of experiments, Kornberg found general agreement between the nearest-neighbor frequencies of the two. This type of analysis is a more stringent measure of the fidelity of copying than is the base composition analysis. Thus, DNA polymerase I seemed the likely candidate for the synthesis of DNA during replication within the cell, even though Kornberg's experiments were, by necessity, performed *in vitro*.

Synthesis of Biologically Active DNA

Despite Kornberg's extensive work, not all researchers were convinced that DNA polymerase I was the enzyme that replicates DNA within cells (*in vivo*). The primary reservations involved observations that the *in vitro* rate of synthesis was much slower than expected *in vivo*, that the enzyme was much more effective replicating single-stranded DNA than double-stranded DNA, and that the enzyme appeared to be able to *degrade* DNA as well as to *synthesize* it.

Faced with the uncertainty of the true cellular function of DNA polymerase I, Kornberg pursued another approach. He reasoned that if the enzyme could be used to synthesize **biologically active DNA** *in vitro*, then DNA polymerase I must be the major catalyzing force for DNA synthesis within the cell. The term *biological activity* means that the DNA synthesized is capable of supporting metabolic activities and directing reproduction of the organism from which it was originally duplicated.

In 1967, Mehran Goulian, Kornberg, and Robert Sinsheimer showed that the DNA of the small bacteriophage φX174 could be completely copied by DNA polymerase I *in vitro*, and that the new product could be isolated and used to successfully transfect *E. coli* protoplasts. Recall that in Chapter 10 we introduced the process of transfection, where viral DNA infects bacterial protoplasts. This process produced mature phages from the synthetic DNA, thus demonstrating biological activity! The ingenious experimental design is outlined in Figure 11.11.

The phage φX174 provided an ideal experimental system because it contains a very small (5386 nucleotides), circular, single-stranded DNA molecule as its genetic material. Because the molecule is a closed circle, the experiment depended on the isolation of a second enzyme, **DNA ligase** (also called the **polynucleotide joining enzyme**), which joins the two ends of the linear molecule following replication.

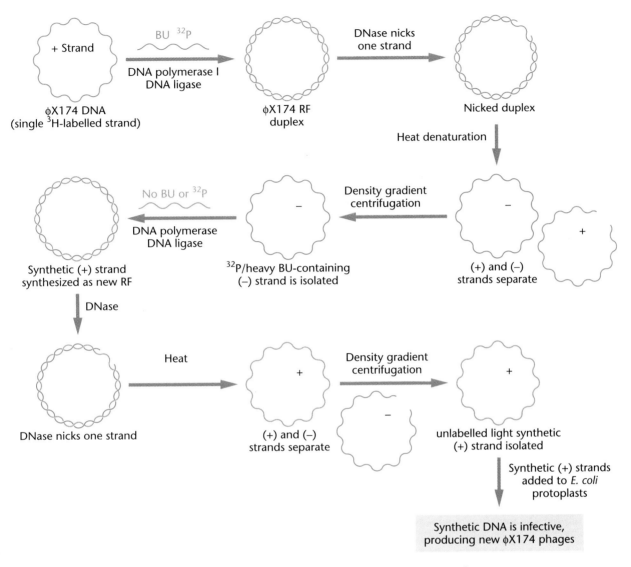

FIGURE 11.11   Schematic representation of Goulian, Kornberg, and Sinsheimer's experiment involving *in vitro* replication and isolation of synthetic (+) DNA strands of phage φX174. The DNA synthesized *in vitro* under the direction of DNA polymerase I successfully transfected *E. coli* protoplasts, thus demonstrating its biological activity.

In the normal course of φX174 infection, the circular, single-stranded DNA referred to as the **(+) strand** enters the *E. coli* cell and serves as a template for the synthesis of the complementary **(−) strand**. The two strands (+ and −) remain together in a circular double helix called the **replicative form (RF)**. The RF serves as the template for its own replication, and subsequently only (+) strands are produced. These strands are then packaged into viral coat proteins to form mature virus particles.

As shown in Figure 11.11, the experiment was carefully designed so that each newly synthesized strand could be distinguished and isolated from the template strand. Initially, the (+) strand was labeled with tritium (^{3}H)—a radioactive form of hydrogen. During synthesis, two "tags" were used to identify the new (−) strands.

First, radioactive ^{32}P was present in the precursor nucleotides. Second, the base analogue 5-bromouracil (BU) was used in place of thymine. In the chemical structure of this analogue, a bromine atom is substituted for a carbon atom in the methyl group at the C-5 position of the pyrimidine ring, increasing the mass and therefore the density. As a result, newly synthesized (−) strands were radioactive, and because their mass was greater, they could be isolated from the (+) strands using sedimentation equilibrium centrifugation.

Once the duplex was enzymatically "nicked" to open one strand, the BU-containing strands were isolated and represented newly synthesized DNA. The process was then repeated in the absence of any tags, using these strands as templates. The subsequent strands were then detectable based on being "lighter" and nonradioactive. Eventually, newly synthesized infectious (+) strands were isolated. The protocol of the experiment dictates that these (+) strands must have been synthesized *in vitro* under the direction of Kornberg's enzyme.

The critical test of biological activity was performed using the process of transfection. Newly synthesized (+) strands were added to bacterial protoplasts (bacterial cells minus their cell wall). Following infection by the synthetic DNA, mature phages were produced. Therefore, the synthetic DNA had successfully directed reproduction.

This demonstration of biological activity was viewed as a precise assessment of faithful copying. If even a single error had occurred to alter the base sequence of any of the 5386 nucleotides constituting the ϕX174 chromosome, the change might easily have caused a mutation that would prohibit the production of viable phages.

DNA Polymerase II and III

Although DNA synthesized under the direction of polymerase I demonstrated biological activity, a more serious reservation about the enzyme's true biological role was raised in 1969. Peter DeLucia and John Cairns reported the discovery of a mutant strain of *E. coli* that was deficient in polymerase I activity. The mutation was designated *polA1*. In the absence of the functional enzyme, this mutant strain of *E. coli* still duplicated its DNA and successfully reproduced. Other properties of the mutation led DeLucia and Cairns to conclude that in the absence of polymerase I, these cells were highly deficient in their ability to "repair" DNA. For example, the mutant strain is highly sensitive to ultraviolet light (UV) and radiation, both of which damage DNA and are mutagenic. Nonmutant bacteria are able to repair a great deal of UV-induced damage.

These observations led to two conclusions:

1. There must be at least one other enzyme present in *E. coli* cells that is capable of replicating DNA *in vivo*.

2. DNA polymerase I may serve only a secondary function *in vivo*. This function is now believed by Kornberg and others to be critical to the *fidelity* of DNA synthesis, but this enzyme is not the one that actually synthesizes the complementary strand.

To date, two unique DNA polymerases have been isolated from cells lacking polymerase I activity. These two enzymes have also been isolated from normal cells that also contain polymerase I.

The characteristics of the two enzymes, called **DNA polymerase II** and **III**, are contrasted with DNA polymerase I in Table 11.2. As is evident from that informa-

TABLE 11.2 Comparative Properties of the Three Bacterial DNA Polymerases

Properties	I	II	III
Initiation of chain synthesis	−	−	−
5′–3′ polymerization	+	+	+
3′–5′ exonuclease activity	+	+	+
5′–3′ exonuclease activity	+	−	−
Molecules of polymerase/cell	400	?	15

tion, all three polymerases share several characteristics. While none can *initiate* DNA synthesis on a template, all can *elongate* an existing DNA strand, called a **primer**. As we shall see, in addition to DNA, RNA is also an adequate primer and is, in fact, what is utilized initially.

The DNA polymerase enzymes are all large, complex proteins exhibiting a molecular weight in excess of 100,000 daltons. All three possess 3′–5′ **exonuclease activity**. This means that they can polymerize in one direction and then pause and excise nucleotides just added. As we will discuss later in this chapter, this activity provides a capacity to proofread newly synthesized DNA and to remove incorrect nucleotides, which may then be replaced.

DNA polymerase I demonstrates 5′–3′ exonuclease activity. This potentially allows the enzyme to excise nucleotides starting at the end where synthesis initiated, proceeding in the same direction of synthesis. Thus, DNA polymerase I has the potential ability to remove the RNA primer. Two final observations probably explain why Kornberg isolated polymerase I and not polymerase III: Polymerase I is present in much greater amounts than is polymerase III and it is also much more stable.

What then are the roles of the three polymerases *in vivo*? Polymerase III is considered to be the enzyme responsible for the polymerization essential to replication. Its 3′–5′ exonuclease activity provide its proofreading function, as described above. As we will soon see, gaps are a natural occurrence on one of the two strands during replication as RNA primers are removed. It is believed that polymerase I, originally studied by Kornberg, is responsible for removing the primer as well as for the synthesis that fills these gaps. Its exonuclease activity also allows proofreading to occur during this process. Alternatively, an RNase has been discovered that may remove the primer prior to polymerase I-directed synthesis.

Polymerase II appears to be involved in repair synthesis of DNA that has been damaged by external forces, such as ultraviolet light. It is encoded by a gene that may be activated by disruption of DNA synthesis at the replication fork.

We end this section by emphasizing the complexity of the DNA polymerase III molecule. Its active form, called a **holoenzyme**, consists of two sets (a dimer) of ten separate polypeptide chains (see Table 11.3) and exhibits

TABLE 11.3 Subunits Constituting DNA Polymerase III Holoenzyme

Subunit	Function	Groupings
α ε θ	5'–3' polymerization 3'–5' exonuclease ??	"Core" enzyme: Elongates polynucleotide chain and proofreads.
γ δ δ' χ ψ	Loads enzyme on template (Serves as clamp loader)	γ complex
β	Sliding clamp structure (Processivity Factor)	
τ	Dimerizes Core Complex	

a molecular weight in excess of 600,000 daltons. The largest subunit, α, has a molecular weight of 140,000 daltons and, along with two other subunits (ε and θ), constitutes the "core" enzyme responsible for the polymerization activity of the holoenzyme. The α subunit is responsible for nucleotide polymerization on the template strands, whereas the ε subunit of the core enzyme possesses the 3'–5' exonuclease activity.

A second group of five subunits (γ, δ, δ', χ, and ψ) forms what is called the γ complex, which is involved in "loading" the enzyme onto the template at the replication fork. This enzymatic function is energy-requiring and dependent on the hydrolysis of ATP. As we will discuss in more detail below, dimers of the β subunit serve as donut-shaped clamps that prevent the core enzyme from falling off the template during polymerization. Finally, the τ subunit functions to hold together the two core polymerases at the replication fork. Together with several other proteins at the replication fork, a complex nearly as large as a ribosome is present. This molecular entity is referred to as a **replisome**. Further details will be forthcoming below concerning the function of DNA polymerase III.

DNA Synthesis: A Model

We have thus far established that replication is semiconservative and bidirectional along a single replicon in bacteria and many viruses. Also, we know that synthesis is in the 5'-to-3' mode under the direction of DNA polymerase III, creating two replication forks. These move in opposite directions away from the origin of synthesis. We are now ready to pursue several other topics that describe how DNA synthesis is accomplished at a replication fork. In combination with the information provided above, we shall establish a coherent model of DNA replication that takes into account the following additional points:

1. A mechanism must exist by which the helix undergoes localized unwinding and is stabilized in this "open" configuration so that synthesis may proceed along both strands.

2. As unwinding and subsequent DNA synthesis proceeds, increased coiling creates tension farther down the helix, which must be reduced. *DNA gyrase*

3. A primer of some sort must be synthesized so that polymerization can commence under the direction of DNA polymerase III.

4. Once the RNA primers have been synthesized, DNA polymerase III commences synthesis of the complement of both strands of the parent molecule. Because the two strands are antiparallel to one another, continuous synthesis in the direction that the replication fork moves is possible along only one of the two strands. On the other strand, synthesis is discontinuous in the opposite direction.

5. The RNA primers must be removed prior to completion of replication. The gaps that are temporarily created must be filled with DNA complementary to the template at each location.

6. The newly synthesized DNA strand that fills each temporary gap must be ligated to the adjacent strand of DNA.

As we consider the above points, Figures 11.12, 11.13, and 11.14 will be used to illustrate how each issue is resolved. Figure 11.15 will summarize the model of DNA synthesis.

Unwinding the DNA Helix

As discussed earlier, there is but a single origin along the circular chromosome of most bacteria and viruses where DNA synthesis is initiated. This region of the *E. coli* chromosome has been particularly well studied. Called *oriC*, the origin of replication consists of 245 base pairs characterized by the presence of repeating sequences of 9 and 13 bases (called **9mers** and **13mers**). One particular protein, called **DnaA** (because it is encoded by the gene called *dnaA*), is responsible for the initial step in unwinding the helix. A number of subunits of the DnaA protein bind to each of several 9mers. This step is essential in facilitating the subsequent binding of **DnaB** and **DnaC** proteins that further open and destabilize the helix (Figure 11.12). Proteins such as these that require the energy normally supplied by the hydrolysis of ATP in order to break hydrogen bonds and denature the double helix are called **helicases**.

As unwinding proceeds, other proteins, called **single-stranded binding proteins (SSBPs)**, stabilize this conformation. As unwinding proceeds, a coiling tension is created ahead of the replication fork. Oftentimes, various forms of **supercoiling** occur. These may take the form of added twists and turns of the DNA in circular molecules, much like the coiling that would be created in a rubber band by holding one end and twisting the other. Such supercoiling can be relaxed by the action of an enzyme, **DNA gyrase**, a member of a larger group referred to as **DNA topoisomerases**. Depending on the form of the enzyme involved, either single- or double-stranded "cuts" are made by DNA gyrase. The enzyme also catalyzes localized manipulations of DNA strands that have the effect of "undoing" the twists and knots created during supercoiling. The strands are then resealed. To drive these various reactions, the energy released during ATP hydrolysis is required.

In combination with the polymerase complex, these proteins comprise an array of molecules that participate in DNA synthesis and are part of what we have previously called the replisome.

Initiation of Synthesis

Once a small portion of the helix is unwound, initiation of synthesis may occur. As previously mentioned, DNA polymerase III requires a free 3′ end as part of a primer in order to elongate a polynucleotide chain. This prompted researchers to investigate how the first nucleotide could be added, as no free 3′-hydroxyl group is initially present. There is now evidence that RNA is involved as the primer in initiating DNA synthesis.

It is thought that a short segment of RNA, complementary to DNA, is first synthesized on the DNA template. The RNA, about 5 to 15 nucleotides long, is made under the direction of a form of RNA polymerase called **primase**. The RNA polymerase does not require a free 3′ end to initiate synthesis. It is to this short segment of RNA that DNA polymerase III begins to add 5′-deoxyribonucleotides (Figure 11.13). After DNA synthesis occurs at an area adjacent to the RNA primer, the RNA segment is clipped off and replaced with DNA. Both steps are thought to be performed by DNA polymerase I. RNA priming has been recognized in viruses, bacteria, and several eukaryotic organisms and is thought to be a universal phenomenon.

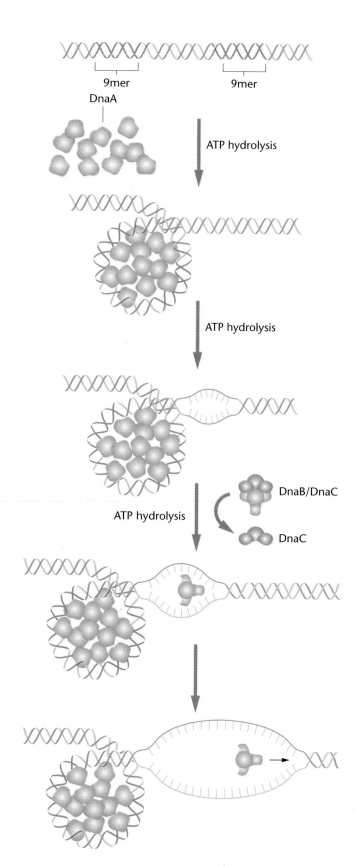

FIGURE 11.12 Helical unwinding of DNA during replication as accomplished by DnaA, DnaB, and DnaC proteins. Initial binding of many monomers of DnaA occurs at DNA sites containing repeating sequences of 9 nucleotides, called *9mers*. Not illustrated are 13mers that are also involved.

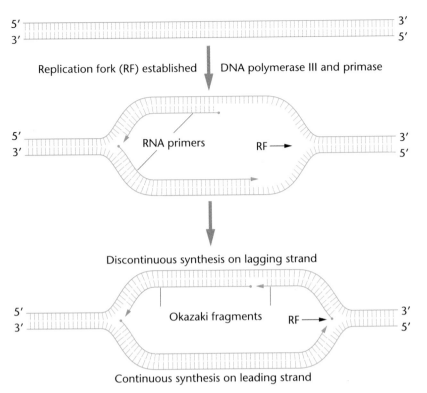

5′
3′

Replication fork (RF) established   DNA polymerase III and primase

5′
3′

RNA primers   RF →

3′
5′

Discontinuous synthesis on lagging strand

5′
3′

Okazaki fragments   RF →

3′
5′

Continuous synthesis on leading strand

FIGURE 11.13   Illustration of the opposite polarity of DNA synthesis along the two strands necessitated by the requirement of 5′–3′ synthesis of DNA polymerase III. On the lagging strand, synthesis must be discontinuous, resulting in the production of Okazaki fragments. On the leading strand, synthesis is continuous. RNA primers are utilized to initiate synthesis.

Continuous and Discontinuous DNA Synthesis

We must now reconsider the fact that the two strands of a double helix are antiparallel to each other. One runs in the 5′–3′ direction, while the other has the opposite 3′–5′ polarity. Because DNA polymerase III synthesizes DNA in only the 5′–3′ direction, simultaneous synthesis of antiparallel strands along an advancing replication fork occurs in one direction along one strand and in the opposite direction on the other.

As the strands unwind and the replication fork progresses down the helix, only one strand can serve as a template for **continuous DNA synthesis**. This strand is called the **leading DNA strand**. As the fork progresses, many points of initiation are necessary on the opposite, or **lagging DNA strand**, resulting in **discontinuous DNA synthesis** (Figure 11.13).

Evidence in support of discontinuous DNA synthesis was first provided by Reiji and Tuneko Okazaki and their colleagues. They discovered that when bacteriophage DNA is replicated in *E. coli*, some of the newly formed DNA that is hydrogen-bonded to the template strand is present as small fragments containing 1000 to 2000 nucleotides. RNA primers are part of each such fragment. These pieces, called **Okazaki fragments**, must then be enzymatically joined. As synthesis proceeds, the Okazaki fragments of low molecular weight are indeed converted into longer and longer DNA strands of higher molecular weight.

Discontinuous synthesis of DNA requires the removal of the RNA primer as well as an enzyme that can unite the smaller products into the longer continuous molecules that represent the lagging strand. While DNA polymerase I removes the primer and replaces the missing nucleotides, **DNA ligase** has been shown to be capable of catalyzing the formation of the phosphodiester bond, the last step in sealing the gap existing between discontinuously synthesized strands. The evidence that DNA ligase does perform this function during DNA synthesis is strengthened by the observation of a ligase-deficient mutant strain (*lig*) of *E. coli*. In this strain, Okazaki fragments accumulate in particularly large amounts. Apparently, they are not joined adequately.

Discontinuous synthesis on the lagging strand, as described above, is characteristic of both bacteria and eukaryotic cells.

Concurrent Synthesis on the Leading and Lagging Strands

Given the above model, the question has arisen as to how the holoenzyme of DNA polymerase III accomplishes synthesis on both the leading and lagging strands. Are the events totally separate, involving two copies of the enzyme? The available evidence suggests that this is not the case. Instead, it is believed that a mechanism exists to allow simultaneous synthesis at the replication fork on both strands, with nucleotide polymerization occurring on each template strand under the direction of one of the two monomers making up the dimeric holoenzyme (Figure 11.14). As Figure 11.14 illustrates, if the lagging strand forms a loop, it is possible for simultaneous synthesis to occur. In keeping with our knowledge of Okazaki fragments, after each 100 to 200 base pairs of synthesis, the lagging strand duplex is released by the enzyme. A new loop is then formed with the lagging template strand, and the process is repeated. Looping inverts the orientation of the template, but not the direction of actual synthesis on the lagging strand, which is always in the 5′ to 3′ direction.

Another important feature of the holoenzyme that facilitates synthesis at the replication fork is also illustrated in Figure 11.14. A dimer of one of the subunits called β forms a "clamp-like" structure around the newly formed DNA duplex. This β-subunit clamp prevents the **core enzyme** (the α, ζ, and θ subunits that are responsible for catalysis of nucleotide addition) from falling off the template as polymerization proceeds. Because the entire holoenzyme moves along the parent duplex, advancing the replication fork, the β-subunit dimer is often referred to as a "sliding-clamp." Note that we have illustrated this dimer structure on both the leading and lag-

ging strands in the enlargement on the right side of Figure 11.14.

Proofreading

The underpinning of semiconservative replication is the synthesis of a new strand of DNA that is precisely complementary to the template strand at each nucleotide position. Although the action of DNA polymerases is very accurate, synthesis is not perfect. Occasionally, a noncomplementary nucleotide is inserted erroneously. To compensate for such inaccuracies, both polymerases I and III possess **3′–5′ exonuclease activity**. They are capable of detecting, pausing, and excising a mismatched nucleotide (in the 3′–5′ direction). Once the mismatched nucleotide is removed, 5′–3′ synthesis can again proceed.

This process is called **exonuclease proofreading** and serves to increase the fidelity of synthesis. In the case of the holoenzyme form of DNA polymerase III, the epsilon (ε) subunit is responsible for enhancing the proofreading step. Strains of *E. coli* have been isolated where a mutation has occurred rendering the ε subunit nonfunctional. The error rate (the mutation rate) during DNA synthesis is increased in these strains as a result.

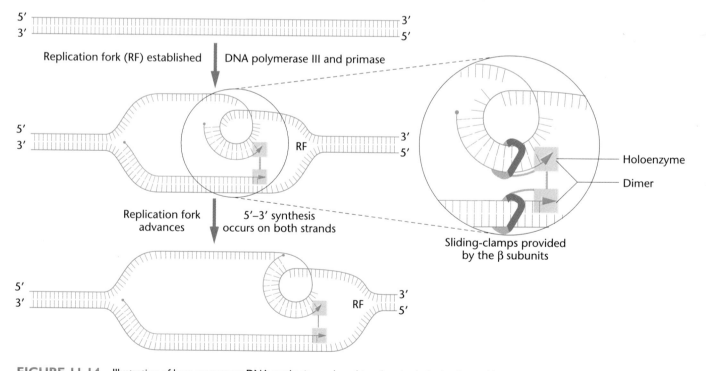

FIGURE 11.14 Illustration of how concurrent DNA synthesis may be achieved on both the leading and lagging strands at a single replication fork. The lagging template strand is "looped" in order to invert the physical direction of synthesis, but not the biochemical direction. Illustrated in the enlargement are the "sliding-clamp" structures provided by dimers of the β subunits of DNA polymerase III. The clamps function to prevent the core enzyme from slipping off the template.

Summary of DNA Synthesis

To summarize the process of DNA synthesis at the level of the replication fork in bacteria and bacteriophages, the following steps provide the biochemical basis for semiconservative replication. These steps are illustrated in Figure 11.15.

1. Unwinding proteins (called helicases) denature the helix at the origin of synthesis, creating the replication fork. Other proteins (SSBPs) stabilize the newly created single strands of denatured DNA.

2. As the replication fork moves away from the origin, increased coiling of the helix occurs ahead of the fork. The tension thus created is diminished by the enzymatic action of the topoisomerase enzyme DNA gyrase. This topoisomerase cuts DNA strands, allowing uncoiling to occur, and then reseals the DNA strands.

3. Initiation of DNA synthesis occurs concurrently on both template strands and involves an RNA primer synthesized under the direction of a unique RNA polymerase called *primase*. The resultant RNAs are complementary to their DNA templates.

4. DNA polymerase III polymerizes complementary DNA strands by simultaneously elongating the existing primer in the 5′–3′ direction.

5. As the replication fork moves away from the origin, synthesis is continuous on the leading strand, but is discontinuous on the lagging strand, producing short polynucleotides called *Okazaki fragments.*

6. The RNA primers are subsequently removed and the resulting gaps are filled with DNA under the direction of DNA polymerase I. The final gaps between the existing DNA and the replacement DNA are closed by DNA ligase.

7. Along the lagging strand, the Okazaki fragments are joined by DNA ligase.

8. As synthesis proceeds along both the leading and lagging strands, proofreading by DNA polymerase III occurs, and semiconservative replication is achieved.

Because the investigation of DNA synthesis is still an extremely active area of research, this model will no doubt be extended in the future. In the meantime, it provides a summary of DNA synthesis against which genetic phenomena can be interpreted.

Genetic Control of Replication

Much of what we know and have outlined in the previous section concerning the details of DNA replication in viruses and bacteria has been based on the genetic analysis of the process. For example, we have already discussed the *polA1* mutation, the study of which revealed that DNA polymerase I is not the major enzyme responsible for replication. Many other mutations have been isolated that interrupt or seriously impair some aspect of replication, such as the ligase-deficient and the proofreading-deficient mutations mentioned previously. These can be studied most easily if they are **temperature-sensitive mutations** (one type of **conditional mutation**). Such mutations are expressed under one condition (in this case, at a restrictive temperature), but are not expressed under a separate condition (in this case, the permissive temperature). As a result, a mutation that would otherwise be lethal may be maintained at the permissive temperature, while the genetic effects may be investigated by shifting mutant cells to the restrictive temperature. Investigation of such temperature-sensitive mutants has the potential to provide insights into the product and the associated function of the normal, nonmutant gene.

As shown in Table 11.4, the enzyme product or its general role in replication has been ascertained for a variety of genes in *E. coli*. For example, numerous mutations in genes specifying the subunits of polymerases I, II, and III have been isolated. Genes have also been identified that encode products involved in specification of the origin of synthesis, helix-unwinding and stabilization, initiation and priming, relaxation of supercoiling, repair, and ligation. The discovery of such a large group of genes attests to the complexity of the process of replication, even in the relatively simple prokaryote. This complexity is not unexpected, given the enormous quantity of DNA

TABLE 11.4 A List of Various *E. coli* Mutant Genes and Their Products or Role in Replication

Mutant Gene	Enzyme or Role
polA	DNA polymerase I
polB	DNA polymerase II
dnaE, N, Q, X, Z	DNA polymerase III subunits
dnaG	Primase
dnaA, I, P	Initiation
dnaB, C	Helicase at *oriC*
oriC	Origin of replication
gyrA, B	Gyrase subunits
lig	Ligase
rep	Helicase
ssb	Single-stranded binding proteins
rpoB	RNA polymerase subunit

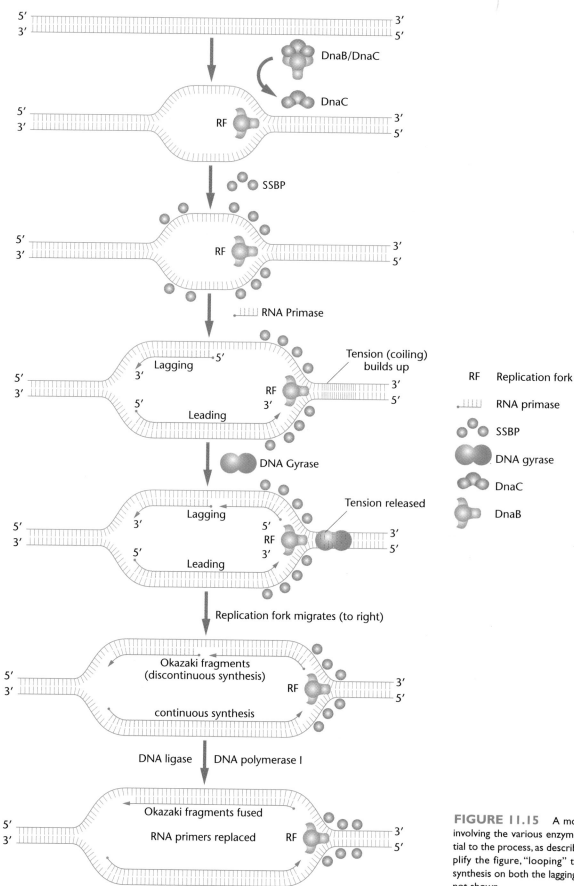

FIGURE 11.15   A model of DNA synthesis involving the various enzymes and proteins essential to the process, as described in the text. To simplify the figure, "looping" to achieve concurrent synthesis on both the lagging and leading strands is not shown.

that must be unerringly replicated in a very brief time. As we will see, the process is even more involved and therefore more difficult to investigate in eukaryotes.

Eukaryotic DNA Synthesis

The general scheme of DNA synthesis found in bacteria is now known to apply also to eukaryotic systems. However, because eukaryotic cells contain about 50 times as much DNA per cell and because this DNA is complexed with proteins, eukaryotes face many problems not encountered by bacteria. As we might expect, these complications make the process of DNA synthesis much more complex in eukaryotes. Unraveling the mysteries of replication in eukaryotes has thus been much more difficult.

Nevertheless, a great deal is now known about this process. As we will see in Chapter 17, eukaryotic DNA has complexed to it an array of histone proteins to form the chromatin fiber characteristic of interphase cells. The DNA–histone complex forms an orderly repeating structure called the **nucleosome**. The presence of these structures gives chromatin the appearance of beads on a string (Figure 11.16). Prior to the initiation of DNA synthesis at each replication fork, a process of dissociation must occur, where the histones are stripped away from the DNA. Then, immediately after DNA synthesis at the replication fork, histones reassociate with the newly formed duplexes, reestablishing the characteristic nucleosome pattern (Figure 11.16). Following replication of all chromosomes, because the total quantity of histone proteins has doubled, DNA synthesis is tightly coupled to histone synthesis in eukaryotic cells. Both occur during the S phase of the cell cycle.

Eukaryotic cells have been found to contain four different kinds of DNA polymerases, called α, β, γ, and δ. It appears that the α form is the major enzyme involved in the replication of nuclear DNA. The β form may be involved in DNA repair, while the γ form is the only DNA polymerase found in mitochondria. Presumably, it is unique in its replication function within that organelle. The δ form, like α, appears to function in nuclear DNA replication. DNA polymerases of eukaryotes have the same fundamental requirements for DNA synthesis as do bacterial and viral systems: four deoxyribonucleoside triphosphates, a template, and a primer.

Data and observations derived from autoradiographic and electron microscopic studies have provided many other insights. As mentioned earlier, both prokaryotic and eukaryotic DNA syntheses are bidirectional, creating two replication forks from each point of origin. As would be expected because of the increased amount of DNA in eukaryotes compared to prokaryotes, many more points of origin are present. In mammals, there are about 25,000 replicons present in the genome, each consisting on average of 100,000 to 200,000 base pairs (100–200 kb). In *Drosophila*, there are about 3500 replicons per genome of an average size of 40 kb.

To accommodate the increased number of replicons, many more DNA polymerase molecules are present in eukaryotic cells than are found in bacteria. While *E. coli* has about 15 copies of DNA polymerase III per cell, there may be up to 50,000 copies of the α form of DNA polymerase in animal cells. In theory, this increased number makes possible the simultaneous replication of all replicons. In practice, most, but not all, are replicated at approximately the same time.

The presence of smaller replicons in eukaryotes compensates for their slower rate of DNA synthesis compared to that in prokaryotes. In *E. coli*, 100 kb are added to a growing chain per minute, whereas eukaryotic synthesis ranges from only 0.5 to 5 kb per minute. Nevertheless, *E. coli* requires 20 to 40 minutes to replicate its chromosome, while *Drosophila*, with 40 times more DNA, accomplishes the same task in only 3 minutes during embryonic cell divisions.

Even though DNA synthesis is slower in eukaryotes, most general aspects of chain elongation are thought to be similar. The actions of a variety of proteins—DNA helicase and SSBPs—are believed to modulate strand separation that precedes RNA priming by a primase enzyme. On the lagging strand, synthesis is discontinuous, resulting in Okazaki fragments that are linked together by DNA ligase. These fragments are about 10 times smaller (100–150 nucleotides) than in prokaryotes. While synthesis may occur continuously on the leading strand, it is not clear whether or not it can always proceed uninterrupted for the full length of a replicon. In cases where it does not occur continuously, DNA synthesis on the leading strand is sometimes described as **semidiscontinuous**.

DNA Synthesis at the Ends of Linear Chromosomes

One final aspect of eukaryotic versus prokaryotic DNA synthesis involves the fact that eukaryotic chromosomes are linear compared to the circular forms displayed by bacteria, as well as most bacteriophages. A special prob-

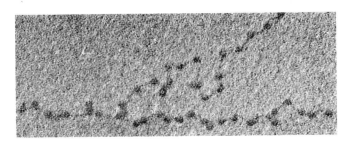

FIGURE 11.16 A replicating fork demonstrating the presence of histone protein-containing nucleosomes on both branches.

lem is encountered at the "ends" of linear molecules of DNA during replication. Such ends are part of each telomeric region of each chromosome.

While synthesis can proceed normally to the end of the leading strand, a problem is encountered on the **lagging strand**, as illustrated in Figure 11.17. The difficulty arises as the RNA primer is removed from the lagging strand. The newly created gap is normally filled by adding a nucleotide to the existing 3'-OH group provided during discontinuous synthesis (this would normally be present to the right of the gap in Figure 11.17). However, there is no strand present to provide the 3'-OH group because this is the end of the chromosome! As a result, each successive round of synthesis will theoretically shorten the chromosome by the length of the RNA primer. Because this is such a significant problem, we can predict that a molecular solution would have been forthcoming early in evolution, and be shared by all eukaryotes. Indeed, this appears to be the case.

In bacteria and viruses that contain circular DNA molecules, this is not a problem, as no free ends are encountered during replication. In eukaryotes, the discovery

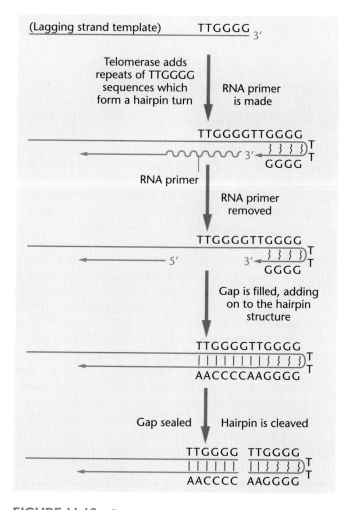

FIGURE 11.18 Diagram of the predicted solution to the hypothetical problem posed in Figure 11.17. The enzyme telomerase directs synthesis of the TTGGGG repeated sequence resulting in the formation of a hairpin structure. As described in the text, this process averts the creation of a gap during replication of the ends of linear chromosomes.

of a unique enzyme, **telomerase**, has helped us understand how more complex organisms solve this problem. In the ciliated protozoan *Tetrahymena*, the many telomeres all terminate in the sequence 5'-TTGGGG-3'. The enzyme is capable of adding repeats of TTGGGG to the ends of molecules already containing this sequence. This process, now known to occur in other eukaryotes under the direction of a similar enzyme, prevents the telomeric ends from shortening following each replication, as described below and illustrated in Figure 11.18.

Telomerase adds several copies of the 6-nucleotide repeat to the 3' end of the lagging strand (using 5'–3' synthesis). These repeats appear to be capable of forming a "hairpin loop," which is stabilized by unorthodox hydrogen bonding between opposite guanine residues (G ≈ G). This creates a free 3'-OH end that, following removal of the RNA primer, can serve as a substrate for DNA poly-

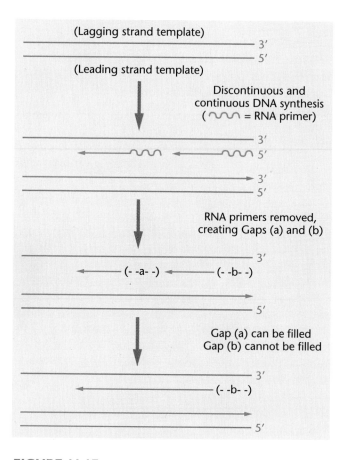

FIGURE 11.17 The hypothetical difficulty encountered during the replication of the ends of linear chromosomes. A gap is left following synthesis on the lagging strand.

merase I to fill the gap. If the hairpin is then cleaved off, the potential loss of DNA in each subsequent replication cycle is averted.

Further investigation of the *Tetrahymena* telomerase enzyme, isolated and studied extensively by Elizabeth Blackburn and Carol Greider, has yielded an extraordinary finding. This enzyme adds the same TTGGGG sequence to other DNA termini. Blackburn and Greider have now established how the enzyme works. They have discovered that the enzyme contains within its molecular structure a short piece of RNA that is essential to its catalytic activity! The functional enzyme is thus a ribonucleoprotein. The RNA component encodes the sequences used by the enzyme as a template. The RNA contains 159 bases, including the sequence 5′-AACCCC-3′, which is complementary to the sequence whose synthesis it directs. Analogous enzyme functions have now been found in still other single-celled organisms. The RNA-containing telomerase enzyme behaves like the retroviral enzyme, reverse transcriptase, in that it synthesize a DNA complement on an RNA template. In this case, the template is much shorter, and the enzyme supplies its own.

The analysis of telomeric DNA sequences, discussed in Chapter 17, has shown them to be highly conserved throughout evolution, reflecting the critical function of telomeres. We shall also see in Chapter 17 that the *absence* of telomerase in human somatic cells may actually "protect" them from the immortalization characterizing malignancy.

DNA Recombination

We conclude this chapter by returning to a topic discussed in Chapter 5—**genetic recombination**. It was pointed out there that the process of crossing over depends on breakage and rejoining of the DNA strands between homologs. Now that we have discussed the chemistry and replication of DNA, it is appropriate to consider how recombination occurs at the molecular level. In general, the following information pertains to genetic exchange between any two homologous double-stranded DNA molecules, whether they be viral or bacterial chromosomes or eukaryotic homologs during meiosis. Genetic exchange at equivalent positions along two chromosomes with substantial DNA sequence homology is referred to as **general** or **homologous recombination**.

Although several models are available that attempt to explain homologous recombination, they all share certain common features. First, all are based on the initial proposals put forth independently by Robin Holliday and Harold L. K. Whitehouse in 1964. They also depend on the complementarity between DNA strands for their precision of exchange. Finally, each model relies on a series of enzymatic processes in order to accomplish genetic recombination.

One such model is illustrated in Figure 11.19. It begins with two paired DNA duplexes or homologs (a), each of which has a single-stranded nick introduced (b) at an identical position by an endonuclease. The ends of the strands produced by these cuts are then displaced and subsequently pair with their complements on the opposite duplex (c). A ligase then seals the loose ends (d), creating hybrid duplexes called **heteroduplex DNA molecules**. The exchange creates a cross-bridged or **Holliday structure**. The position of this cross-bridge can then move down the chromosomes by a process referred to as branch migration (e). This occurs as a result of a zipper-like action as hydrogen bonds are broken and then reformed between complementary bases of the displaced strands of each duplex. This migration yields an increased length of heteroduplex DNA on both homologs.

If the duplexes now separate (f), and the bottom portions rotate 180° (g), an intermediate planar structure called a **chi form** is created. If the two strands on opposite homologs previously uninvolved in the exchange are now nicked by an endonuclease (h), and ligation occurs (i), recombinant duplexes are created. Note that in our illustration of the model that the arrangement of alleles is altered as a result of the recombination event.

Evidence supporting the above model includes the electron microscopic visualization of chi-form planar molecules from bacteria where four duplex arms are joined at a single point of exchange (Figure 11.19). Additionally, the discovery in *E. coli* of the **RecA protein** provides important evidence. This molecule promotes the exchange of reciprocal single-stranded DNA molecules as must occur in step (c) of the model. Further, RecA enhances the hydrogen bond formation during strand displacement, thus initiating heteroduplex formation. Finally, many other enzymes essential to the nicking and ligation process have also been discovered and investigated. The products of the *recB*, *recC*, and *recD* genes are thought to be involved in the nicking and unwinding of DNA. Numerous mutations that prevent genetic recombination have been found in viruses and bacteria. These mutations are thought to identify genes, whose products play an essential role in this process.

Gene Conversion

A modification of the above model has helped us to better understand a unique genetic phenomenon known as **gene conversion**. Initially found in yeast by Carl Lindegren and in *Neurospora* by Mary Mitchell, gene conversion is characterized by a *nonreciprocal* genetic exchange

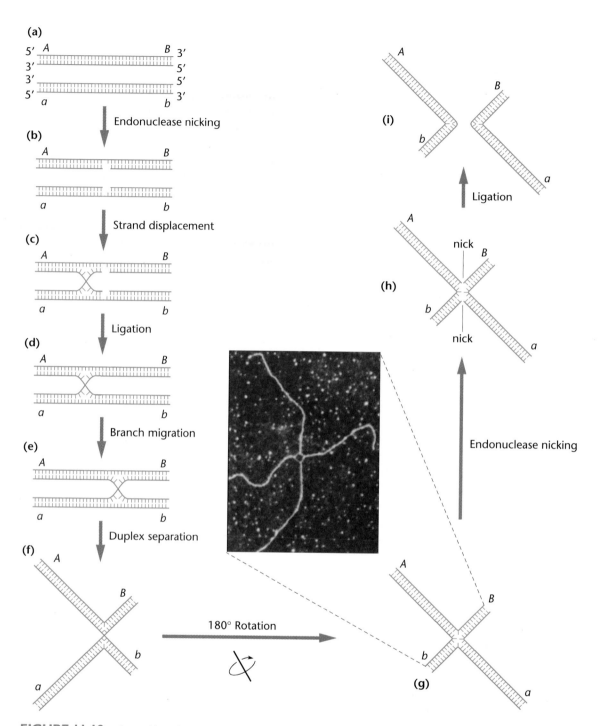

FIGURE 11.19 A possible molecular sequence depicting how genetic recombination occurs as a result of the breakage and rejoining of heterologous DNA strands. Each stage is described in the text. The electron micrograph shows DNA in a chi-form structure similar to the diagram in (g). The DNA, showing an extended Holliday structure, represents a recombination intermediate derived from the *Col*E1 plasmid of *E. coli.*

between two closely linked genes. For example, if one were to cross two *Neurospora* strains each bearing a separate mutation ($a+ \times +b$), a reciprocal recombination event between the genes would yield spore pairs of the

$++$ *and* the *ab* genotypes. However, a nonreciprocal exchange yields one pair without the other. Working with pyridoxine mutants, Mitchell observed several asci containing spore pairs with the $++$ genotype, but not the reciprocal product (*ab*). Because the frequency of these events was higher than the predicted mutation rate and could not be accounted for by that phenomenon, they were called *gene conversions*. They were so named because it appeared that one allele had somehow been "converted" to another during an event where genetic exchange also occurred. Similar findings are also apparent in the study of other fungi.

This genetic phenomenon is now considered to be a consequence of the process of DNA recombination, as discussed in the previous section. One possible explanation interprets the conversion event as a mismatch of base pairs during heteroduplex formation involved in genetic recombination as depicted in Figure 11.20. Mismatched regions of hybrid strands may be created during recombination, but they can be repaired by excision of one of the strands and synthesis of the complement using the remaining strand as a template. Excision of either one of the strands may occur in order to accomplish the repair, yielding two possible "corrections." One repairs the mismatched base pair and restores the original sequence; the other corrects the mismatch, but does so by copying the altered strand, creating a base-pair substitution. This conversion may have the effect of creating identical alleles on the two homologs that were different initially.

In our example in Figure 11.20, suppose the G ≡ C pair on one of the two homologs was responsible for the mutant allele while the A = T pair was part of the wild-type gene sequence on the other homolog. Conversion of the G ≡ C pair to A = T would have the effect of changing the mutant allele to wild type, just as Mitchell originally observed.

Gene-conversion events have helped to explain other puzzling genetic phenomena in fungi. For example, when mutant and wild-type alleles of a single gene are studied in a cross, asci should yield equal numbers of mutant and wild-type spores. However, exceptional asci with 3:1 or 1:3 ratios are sometimes observed. These ratios are more easily understood when interpreted in terms of gene conversion. The phenomenon has also been detected during mitotic events in fungi as well as during the study of unique compound chromosomes in *Drosophila*.

FIGURE 11.20 Illustration of a proposed mechanism that accounts for the phenomenon of gene conversion. A base-pair mismatch occurs in one of the two homologs (bearing the mutant allele) during heteroduplex formation, which accompanies recombination in meiosis. During excision repair, one of the two mismatches is removed and the complement is synthesized. In one case, the mutant base pair is preserved. When it is subsequently included in a recombinant spore, the mutant genotype will be maintained. In the other case, the mutant base pair is converted to the wild-type sequence. When included in a recombinant spore, the wild-type genotype will be expressed, leading to a nonreciprocal exchange ratio.

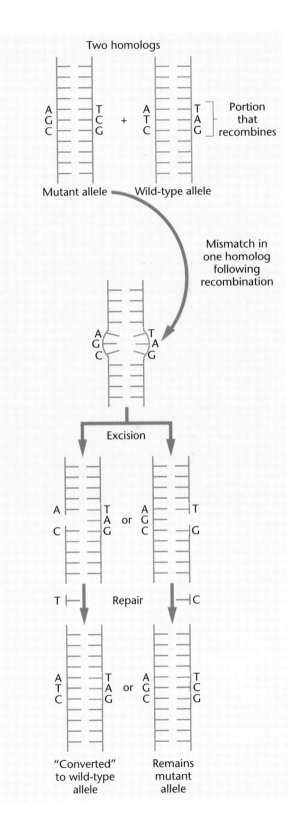

CHAPTER SUMMARY

1. In theory, three modes of DNA replication are possible: semiconservative, conservative, and dispersive. Although all three rely on base complementarity, semiconservative replication is the most straightforward and was predicted.

2. In 1958, Meselson and Stahl resolved this problem in favor of semiconservative replication in *E. coli*, showing that newly synthesized DNA consists of one old strand and one new strand. Taylor, Woods, and Hughes used root tips of the broad bean to demonstrate semiconservative replication in eukaryotes.

3. During the same period, Kornberg isolated DNA polymerase I from *E. coli* and demonstrated it to be an enzyme capable of *in vitro* DNA synthesis, provided that a template and precursor nucleoside triphosphates were supplied.

4. The subsequent discovery of the *polA1* mutant strain of *E. coli*, capable of DNA replication despite its lack of polymerase I activity, cast doubt on this enzyme's *in vivo* replicative function. DNA polymerases II and III were then isolated. Polymerase III has been identified as the enzyme responsible for DNA replication *in vivo*.

5. During the process of DNA synthesis, the double helix unwinds, forming a replication fork where synthesis begins. Proteins stabilize the unwound helix and assist in relaxing the coiling tension created ahead of the replication activity.

6. Synthesis is initiated at specific sites along each template strand by RNA primase, which results in a short segment of RNA that provides a suitable 3′ end, upon which DNA polymerase III can begin polymerization.

7. Because of the antiparallel nature of the double helix, polymerase III synthesizes DNA continuously on the leading strand in a 5′–3′ direction. On the opposite strand, called the lagging strand, synthesis results in short Okazaki fragments that are later joined by DNA ligase.

8. DNA polymerase I removes and replaces the RNA primer with DNA, which is joined to the adjacent polynucleotide by DNA ligase.

9. The isolation of numerous phage and bacterial mutant genes affecting many of the molecules involved in the replication of DNA has helped to define the complex genetic control of the entire process.

10. DNA replication in eukaryotes is similar to, but more complex than, replication in prokaryotes. For example, replication at the ends (telomeres) of linear molecules poses a special problem that can be solved by a unique RNA-containing enzyme called *telomerase*.

11. Homologous recombination between genetic molecules relies on a series of enzymes that can cut, realign, and reseal DNA strands. The phenomenon of gene conversion may be best explained in terms of mismatch repair synthesis during these exchanges.

KEY TERMS

autoradiography, 300
bidirectional replication, 303
biologically active DNA, 306
chain elongation, 304
chi form, 317
conditional mutation, 313
conservative replication, 299
continuous DNA synthesis, 311
core enzyme, 312
discontinuous DNA synthesis, 311
dispersive replication, 299
DnaA, 309
DnaB, 309
DnaC, 309
DNA gyrase, 310
DNA helicase, 315
DNA ligase (polynucleotide joining enzyme), 306
DNA polymerase (I, II, III, α, β, γ, δ), 304, 308

DNA topoisomerase, 310
exonuclease activity, 308
exonuclease proofreading (editing), 312
gene conversion, 317
genetic recombination, 317
helicase, 309
heteroduplex DNA molecules, 317
Holliday structure, 317
holoenzyme, 308
homologous (general) recombination, 317
lagging DNA strand, 311
leading DNA strand, 311
nearest-neighbor frequency test, 305
9mers, 309
nucleosome, 315
Okazaki fragments, 311
oriC, 309
origin of replication, 303
primase, 310

primer, 308
RecA protein, 317
replication fork, 303
replicative form (RF), 307
replicon, 303
replisome, 309
sedimentation equilibrium centrifugation, 300
semiconservative replication, 298
semidiscontinuous DNA synthesis, 315
single-stranded binding proteins (SSBPs), 309
sister chromatid exchanges, 303
spleen phosphodiesterase, 305
supercoiling, 310
telomerase, 316
temperature-sensitive mutation, 313
ter, 303
13mers, 309
unidirectional replication, 303

INSIGHTS AND SOLUTIONS

1. Predict the theoretical results of conservative and dispersive models of DNA synthesis using the conditions of the Meselson–Stahl experiment. Follow the results through two generations of replication after cells have been shifted to an ^{14}N-containing medium, using the following migration standards:

Solution:

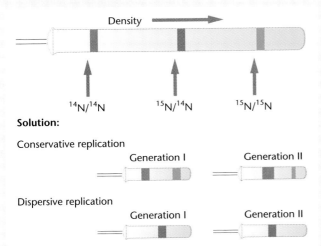

Solution:

2. Mutations in the *dnaA* gene of *E. coli* are lethal and can only be studied following the isolation of conditional, temperature-sensitive mutations. Such mutant strains grow nicely and replicate their DNA at the permissive temperature of 18°C, but they do not grow or replicate their DNA at the restrictive temperature of 37°C. Two observations were useful in determining the function of the DnaA protein product. First, *in vitro* studies using DNA templates that have been nicked (opened) do not require the DnaA protein. Second, if intact cells are grown at 18°C and then shifted to 37°C, DNA synthesis continues at this temperature until one round of replication is completed, and DNA synthesis stops. What do these observations suggest about the role of the product of the *dnaA* gene?

Solution: These observations suggest that *in vivo* the DnaA protein is essential to the initiation of DNA synthesis. At 18°C (the permissive temperature) the mutation is not expressed and DNA synthesis begins. Following the shift to the restrictive temperature, DNA synthesis already initiated continues, but no new synthesis can begin. Because the DnaA protein is not required for synthesis of "nicked" DNA, this observation suggests that the protein functions during initiation by interacting with the intact helix and somehow facilitating the localized denaturing necessary for synthesis to proceed. In fact, both conclusions are valid.

3. DNA is allowed to replicate in moderately radioactive ^{3}H-thymidine for several minutes and then switched to a highly radioactive medium for several more minutes. Synthesis is stopped and the DNA subjected to autoradiography and electron microscopy. Interpret as much as you can regarding DNA replication from the following drawing of the micrograph.

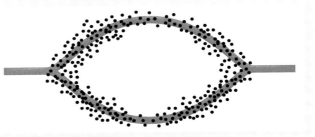

Solution: One interpretation is that there are two advancing replication forks proceeding in opposite directions and that replication is therefore bidirectional. The low density of grains represents synthesis that occurred during the first few minutes, beginning at the origin and proceeding outward in both directions. The higher density of grains represents the latter minutes of synthesis when the level of radioactivity was increased.

PROBLEMS AND DISCUSSION QUESTIONS

1. Compare conservative, semiconservative, and dispersive modes of DNA replication.
2. Describe the role of ^{15}N in the Meselson–Stahl experiment.
3. In the Meselson–Stahl experiment, which of the three modes of replication could be ruled out after one round of replication? After two rounds?
4. Predict the results of the experiment by Taylor, Woods, and Hughes if replication were (a) conservative and (b) dispersive.
5. Reconsider Problem 35 in Chapter 10. In the model you proposed, could the molecule be replicated semiconservatively? Why? Would other modes of replication work?
6. What are the requirements for the *in vitro* synthesis of DNA under the direction of DNA polymerase I?
7. In Kornberg's initial experiments, he actually grew *E. coli* in Anheuser-Busch beer vats (he was working at Washington University in St. Louis). Why do you think this was helpful?
8. How did Kornberg test the fidelity of copying DNA by polymerase I?
9. Which of Kornberg's tests is the more stringent assay? Why?
10. Which characteristics of DNA polymerase I led to doubts that its *in vivo* function is the synthesis of DNA leading to complete replication?

11. Explain the theory of nearest-neighbor frequency.

12. What is meant by "biologically active" DNA?

13. Why was the phage φX174 chosen for the experiment demonstrating biological activity?

14. Outline the experimental design of Kornberg's biological activity demonstration.

15. What was the significance of the *polA1* mutation?

16. Summarize and compare the properties of DNA polymerase I, II, and III.

17. List and describe the function of the 10 subunits constituting DNA polymerase III. Distinguish between the holoenzyme and the core enzyme.

18. Distinguish between (a) unidirectional and bidirectional synthesis and (b) continuous and discontinuous synthesis of DNA.

19. List the proteins that unwind DNA during *in vivo* DNA synthesis. How do they function?

20. Define and indicate the significance of (a) Okazaki fragments, (b) DNA ligase, and (c) primer RNA during DNA replication.

21. Outline the current model for DNA synthesis.

22. Why is DNA synthesis expected to be more complex in eukaryotes than in bacteria? How is DNA synthesis similar in the two types of organisms?

23. If the analysis of DNA from two different microorganisms demonstrated very similar nearest-neighbor frequencies, is the DNA of the two organisms identical in (a) amount, (b) base composition, and/or (c) nucleotide sequences?

24. Suppose that *E. coli* synthesizes DNA at a rate of 100,000 nucleotides per minute and takes 40 minutes to replicate its chromosome.
    (a) How many base pairs are present in the entire *E. coli* chromosomes?
    (b) What is the physical length of the chromosome in its helical configuration, that is, what is the circumference of the circular chromosome?

25. Several temperature-sensitive mutant strains of *E. coli* are isolated that display various characteristics. Predict what enzyme or function is being affected by each mutation:
    (a) Newly synthesized DNA contains many mismatched base pairs.
    (b) Okazaki fragments accumulate, and DNA synthesis is never completed.
    (c) No initiation occurs.
    (d) Synthesis is very slow.
    (e) Supercoiled strands are found to remain and replication is never completed.

26. Define gene conversion and describe how this phenomenon is related to genetic recombination.

27. Many of the gene products involved in DNA synthesis were initially defined by studying mutant *E. coli* strains that could not synthesize DNA.
    (a) The *dnaE* gene encodes the α subunit of DNA polymerase III. What effect is expected by a mutation in this gene? How could the mutant strain be maintained?
    (b) The *dnaQ* gene encodes the ε subunit of DNA polymerase. What effect is expected by a mutation in this gene?

EXTRA-SPICY PROBLEMS

28. The genome of the fruit fly *Drosophila melanogaster* consists of approximately 1.6×10^8 base pairs. DNA synthesis occurs at a rate of 30 base pairs per second. In the early embryo, the entire genome is replicated in five minutes. How many *bidirectional origins of synthesis* are required to accomplish this feat?

29. Analysis of nearest-neighbor data led Josse, Kaiser, and Kornberg in 1961 (*J. Biol. Chem.* 236:864–75) to conclude that the two strands of the double helix are in opposite polarity to one another. Demonstrate your understanding of the nearest-neighbor technique by determining the outcome of such an analysis if the strands of the following molecule are (a) antiparallel versus (b) parallel:

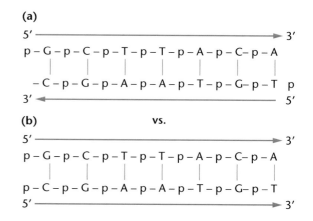

30. Assume a hypothetical organism where DNA replication is conservative. Design an experiment similar to Taylor, Woods, and Hughes that will unequivocally establish this. Using the format established in Figure 11.5, draw sister chromatids and illustrate the expected results establishing this mode of replication.

SELECTED READINGS

BLACKBURN, E. H. 1991. Structure and function of telomeres. *Nature* 350:569–72.

BRAMHILL, D., and KORNBERG, A. 1988. A model for initiation at origins of DNA replication. *Cell* 54:915–18.

CAMERINI-OTERO, R. D., and HSIEH, P. 1993. Parallel DNA triplexes, homologous recombination, and other homology-dependent DNA interactions. *Cell* 73:217–23.

DARNELL, J., LODISH, H., and BALTIMORE, D. 1990. *Molecular cell biology*, 2nd ed. New York: Scientific American Books.

DAVIDSON, J. N. 1976. *The biochemistry of nucleic acids*, 8th ed. Orlando, FL: Academic Press.

DELUCIA, P., and CAIRNS, J. 1969. Isolation of an *E. coli* strain with a mutation affecting DNA polymerase. *Nature* 224: 1164–66.

DENHARDT, D. T., and FAUST, E. A. 1985. Eukaryotic DNA replication. *Bioessays* 2:148–53.

DRESSLER, D., and POTTER, H. 1982. Molecular mechanisms in genetic recombination. *Annu. Rev. Biochem.* 51:727–61.

GELLERT, M. 1981. DNA topoisomerases. *Annu. Rev. Biochem.* 50:879–910.

GREIDER, C. W., and BLACKBURN, E. H. 1989. A telomeric sequence in the RNA of *Tetrahymena* telomerase required for telomere repeat synthesis. *Nature* 337:331–336.

———. 1996. Telomeres, telomerase, and cancer. *Sci. Am.* (Feb.) 274:92–97.

HERENDEEN, D. R., and KELLY, T. J. 1996. DNA polymerase III: Running rings around the fork. *Cell* 84:5–8.

HOLLIDAY, R. 1964. A mechanism for gene conversion in fungi. *Genet. Res.* 5:282–304.

HUBERMAN, J. C. 1987. Eukaryotic DNA replication: A complex picture partially clarified. *Cell* 48:7–8.

JOSSE, J., KAISER, A. D., and KORNBERG, A. 1961. Enzymatic synthesis of deoxyribonucleic acid: VIII. Frequencies of nearest-neighbor base sequences in deoxyribonucleic acid. *J. Biol. Chem.* 236:864–75.

KORNBERG, A. 1960. Biological synthesis of DNA. *Science* 131: 1503–8.

———. 1969. Active center of DNA polymerase. *Science* 163: 1410–18.

———. 1974. *DNA synthesis.* New York: W. H. Freeman.

———. 1979. Aspects of DNA replication. *Cold Spring Harbor Symp. Quant. Biol.* 43:1–10.

KORNBERG, A., and BAKER, T. 1992. *DNA replication*, 2nd ed. New York: W. H. Freeman.

LEHMAN, I. R. 1974. DNA ligase: Structure, mechanism, and function. *Science* 186:790–97.

LINDEGREN, C. C. 1953. Gene conversion in *Saccharomyces.* *J. Genet.* 51:625–37.

LODISH, H., et al. 1995. *Molecular cell biology*, 3rd ed. New York: Scientific American Books.

MESELSON, M., and STAHL, F. W. 1958. The replication of DNA in *Escherichia coli. Proc. Natl. Acad. Sci. USA* 44:671–82.

MITCHELL, M. B. 1955. Aberrant recombination of pyridoxine mutants of *Neurospora. Proc. Natl. Acad. Sci. USA* 41:215–20.

OGAWA, T., and OKAZAKI, T. 1980. Discontinuous DNA synthesis. *Annu. Rev. Biochem.* 49:421–57.

OKAZAKI, T., et al. 1979. Structure and metabolism of the RNA primer in the discontinuous replication of prokaryotic DNA. *Cold Spring Harbor Symp. Quant. Biol.* 43:203–22.

RADDING, C. M. 1978. Genetic recombination: Strand transfer and mismatch repair. *Annu. Rev. Biochem.* 47:847–80.

RADMAN, M., and WAGNER, R. 1988. The high fidelity of DNA duplication. *Sci. Am.* (Aug.) 259:40–46.

SCHEKMAN, R., WEINER, A., and KORNBERG, A. 1974. Multienzyme systems of DNA replication. *Science* 186:987–93.

STAHL, F. W. 1979. *Genetic recombination: Thinking about it in phage and fungi.* New York: W. H. Freeman.

———. 1987. Genetic recombination. *Sci. Am.* (Feb.) 256:90–101.

SZOSTAK, J. W., ORR-WEAVER, T. L., and ROTHSTEIN, R. J. 1983. The double-strand-break repair model for recombination. *Cell* 33:25–35.

TAYLOR, J. H., WOODS, P. S., and HUGHES, W. C. 1957. The organization and duplication of chromosomes revealed by autoradiographic studies using tritium-labeled thymidine. *Proc. Natl. Acad. Sci. USA* 48:122–28.

THÖMMES, P., and HÜBSCHER, U. 1992. Eukaryotic DNA helicases: Essential enzymes for DNA transactions. *Chromosoma* 101:467–73.

TOMIZAWA, J., and SELZER, G. 1979. Initiation of DNA synthesis in *E. coli. Annu. Rev. Biochem.* 48:999–1034.

WANG, J. C. 1982. DNA topoisomerases. *Sci. Am.* (July) 247:94–108.

———. 1987. Recent studies of DNA topoisomerases. *Biochim. Biophys. Acta* 909:1–9.

WATSON, J. D., et al. 1987. *Molecular biology of the gene:* Vol. 1. *General principles*, 4th ed. Menlo Park, CA: Benjamin/Cummings.

WHITEHOUSE, H. L. K. 1982. *Genetic recombination: Understanding the mechanisms.* New York: Wiley.

WOOD, W. B., WILSON, J. H., BENBOW, R. M., and HOOD, L. E. 1975. *Biochemistry—A problems approach.* Menlo Park, CA: Benjamin/Cummings.

ZUBAY, G. L., and MARMUR, J., eds. 1973. *Papers in biochemical genetics*, 2nd ed. New York: Holt, Rinehart & Winston.

ZYSKIND, J. W., and SMITH, D. W. 1986. The bacterial origin of replication, *oriC. Cell* 46:489–90.

12

Storage and Expression of Genetic Information

CHAPTER CONCEPTS

Genetic information, stored in DNA and transferred to RNA during the process of transcription, is present as three-letter codes. Using four different letters, corresponding to the four ribonucleotides in RNA, the 64 triplet codons not only specify the 20 different amino acids found in proteins but also provide signals that initiate and terminate protein synthesis. This unique language is the basis of life as we know it.

In previous chapters we established that DNA is the genetic material and we explored the structure and replication of DNA. In this chapter we explore two additional aspects of molecular genetics. First, we will discuss how DNA encodes genetic information. Second, we will examine how this information is decoded during expression of genetic information. As we shall see, the elaborate system that provides the physical and chemical basis for the storage and expression of genetic information has the potential to produce a nearly endless variety of protein molecules. To accom-

plish such synthesis, DNA of an active gene is first transcribed into an RNA complement that is then translated into a protein. The information that is stored in a sequence of deoxyribonucleotides in DNA is first transferred to the complementary sequence of ribonucleotides in RNA, which are then decoded as the RNA specifies the order of insertion of amino acids during protein synthesis. Proteins, as the end products of various gene sequences, are responsible for the normal and mutant phenotypes of organisms.

A fundamental question, then, is how an RNA molecule

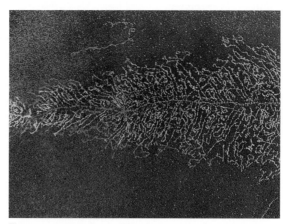
Electron micrograph visualizing the process of transcription.

consisting of only four different types of ribonucleotides (A, U, C, and G) can specify 20 different amino acids. This question poses an intriguing theoretical problem. When ingenious analytical research was applied to this problem, the genetic code was deciphered. It was established that the code is triplet in nature. Code words, or **codons**, consisting of three ribonucleotides direct the insertion of amino acids into a polypeptide chain during its synthesis.

The process by which RNA molecules are synthesized on a DNA template is called **transcription**. The ribonucleotide sequence of RNA is then capable of directing the process of **translation**. During translation, polypeptide chains—which mature into proteins—are synthesized. Protein synthesis is dependent on a series of **transfer RNA (tRNA)** molecules that serve as adaptors between the codons of mRNA (messenger RNA) and the amino acids specified by them. In addition, the process of translation occurs only in conjunction with an intricate cellular component, the **ribosome**.

The processes of transcription and translation are complex molecular events. Like the replication of DNA, both rely heavily on base-pairing affinities between complementary nucleotides. The initial transfer of stored information in DNA produces a molecule, usually mRNA, that is complementary to the nucleotide sequence of one of the two strands of the double helix. Then, each triplet codon of mRNA, which is complementary to the anticodon region of the tRNA that bears the corresponding amino acid, directs the sequential insertion of amino acids into a polypeptide chain.

In this chapter we will describe in detail how the genetic code was deciphered and how gene expression occurs. The work leading to these discoveries occurred most intensively in the late 1950s and early 1960s, one of the most exciting periods in the history of molecular genetics. This research revealed the intricacies of the specific chemical language that serves as the basis of all life on Earth.

An Overview of the Genetic Code

Before considering the various analytical approaches used in arriving at our current understanding of the genetic code, we shall provide a summary of the general features that characterize it:

1. The genetic code is written in linear form using the ribonucleotide bases that compose mRNA molecules as the letters. The ribonucleotide sequence is, of course, derived from its complement in DNA.

2. Each word within the mRNA contains three letters. Thus, the code is a **triplet**. Called a **codon**,

each group of *three* ribonucleotides specifies *one* amino acid.

3. The code is **unambiguous**, meaning that each triplet specifies only a single amino acid.

4. The code is **degenerate**, meaning that a given amino acid is specified by more than one triplet codon. This is the case for 18 of the 20 amino acids.

5. The code is **ordered**. Multiple codons specifying a given amino acid may be categorized together, most often varying by only the third base.

6. The code contains "start" and "stop" punctuation signals. Certain triplets are necessary to **initiate** and to **terminate** translation.

7. No commas (or internal punctuation) are used in the code. Thus, the code is said to be **commaless**. Once translation of mRNA begins, the three ribonucleotides are read in turn, one after the other.

8. The code is **nonoverlapping**. Once translation commences, any single ribonucleotide at a specific location within the mRNA is part of only one triplet.

9. The code is nearly **universal**. With only minor exceptions, a single coding dictionary is used by almost all viruses, prokaryotes, and eukaryotes.

Early Thinking about the Code

Before it became clear that mRNA serves as an intermediate in transferring genetic information from DNA to proteins, it was thought that DNA itself might directly encode proteins during their synthesis. Once mRNA was discovered, it was clear that even though genetic information is stored in DNA, the code that is translated into proteins resides in RNA. The central question was how only four letters—the four nucleotides—could specify 20 words—the amino acids.

As to the size of the code, in the early 1960s Sidney Brenner argued on theoretical grounds that it must be a triplet since three-letter words represent the minimal use of four letters to specify 20 amino acids. For example, four nucleotides, taken two at a time provide only 16 unique code words (4^2). Although a triplet code provides 64 words (4^3)—clearly more than the 20 needed—it is much simpler than a four-letter code, where 256 words (4^4) would be specified.

Brenner also argued that the code was nonoverlapping. Assuming a triplet code, Brenner considered restrictions that might be placed on it if it were overlapping. For example, he considered theoretical nucleotide sequences encoding a protein consisting of three amino acids. In the nucleotide sequence GTACA, parts of the central triplet,

TAC, are shared by the outer triplets, GTA and ACA, as shown below. Brenner reasoned that if this were the case, then only certain amino acids should be found adjacent to the one encoded by the central triplet:

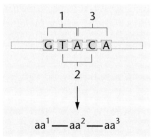

For any given central amino acid, only 16 combinations (2^4) of "three amino acid" sequences (a tripeptide) are theoretically possible.

Therefore, Brenner concluded that if the code were overlapping, tripeptide sequences within proteins should be somewhat limited. Looking at the available amino acid sequences of proteins that had been studied, he failed to find such restrictions in tripeptide sequences. For any central amino acid, many more than 16 different tripeptides were found.

A second major argument against an overlapping code involved the effect of a single nucleotide change characteristic of a point mutation. With an overlapping code, two adjacent amino acids would be affected. However, mutations in the genes coding for the protein coat of tobacco mosaic virus (TMV), human hemoglobin, and the bacterial enzyme tryptophan synthetase invariably revealed only single amino acid changes.

The third argument against an overlapping code was presented by Francis Crick in 1957, when he predicted that DNA does not serve as a direct template for the formation of proteins. Crick reasoned that any affinity between nucleotides and an amino acid would require hydrogen bonding. Chemically, however, such specific affinities seemed unlikely. Instead, Crick proposed that there must be an **adaptor molecule** that could covalently bind to the amino acid yet be capable of hydrogen bonding to a nucleotide sequence. Because various adaptors would somehow have to overlap one another at nucleotide sites during translation, Crick reasoned that physical constraints would make the process overly complex and, perhaps, even inefficient during translation. As we will see later in this chapter, Crick's prediction was correct; transfer RNA (tRNA) serves as the adaptor in protein synthesis. And, the ribosome accommodates two tRNA molecules at a time during translation.

Crick's and Brenner's arguments, taken together, strongly suggested that during translation, the genetic code is **nonoverlapping**. Without exception, this concept has been upheld.

The Code: Further Developments

Between 1958 and 1960, information related to the genetic code continued to accumulate. In addition to his adaptor proposal, Crick hypothesized, on the basis of genetic evidence, that the code is **commaless**; that is, he believed no internal punctuation occurs along the reading frame. Crick also speculated that only 20 of the 64 possible triplets specify an amino acid and that the remaining 44 carry no coding assignment.

At the time, however, there was as yet no experimental evidence that the code was indeed a triplet; nor had the concept of messenger RNA, as an intermediate between DNA and protein, been established. Because ribosomes had already been identified, the current thinking was that information in DNA is transferred in the nucleus to the RNA of the ribosome, which serves as the template for protein synthesis in the cytoplasm.

This concept soon became untenable as accumulating evidence demonstrated that the template intermediate was unstable. The RNA of ribosomes, on the other hand, was found to be extremely stable. As a result, in 1961 François Jacob and Jacques Monod postulated the existence of **messenger RNA (mRNA)**. The scene was thus set to demonstrate the triplet nature of the genetic code, as contained in the intermediate mRNA, and to decipher the specific codon assignments.

Many questions beyond the triplet assignments also remained. Are there start–stop **punctuation signals**? Is the code **ambiguous**, with one triplet specifying more than one amino acid? Was Crick wrong with respect to the 44 "blank" codes? That is, is the code **degenerate**, with more than one triplet assignment for many, if not all, amino acids? Is the code **universal**? As we shall see, these and other questions were answered in the next decade.

The Study of Frameshift Mutations

Before discussing the experimentation in which specific codon assignments were deciphered, we will consider the ingenious experimental work of Crick, Leslie Barnett, Brenner, and R. J. Watts-Tobin. Their work represented the first solid evidence for the triplet nature of the code.

In their work, these researchers induced insertion and deletion mutations in the B cistron of the *rII* locus of phage T4. The B cistron is one of two functional sites in this locus that, in mutant form, causes rapid lysis and distinctive plaques. Mutants in the *rII* locus will successfully infect strain B of *E. coli* but cannot reproduce on a separate strain of *E. coli*, designated K12. Crick and his colleagues used the acridine dye **proflavin** to induce mutations (see Figure 14.12, p. 405). This mutagenic agent intercalates within the double helix of DNA, often caus-

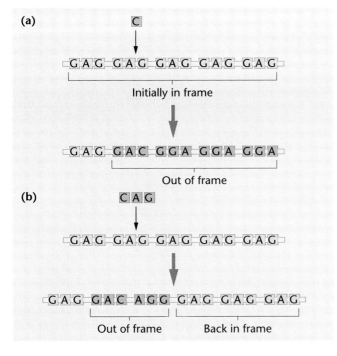

FIGURE 12.1   Schematic diagram of the effect of frameshift mutations on a DNA sequence repeating the triplet sequence GAG. In part (a) the insertion of a single nucleotide has shifted all subsequent reading frames. In part (b) the insertion of three nucleotides changes only two reading frames, but all remaining frames are retained in their original sequence.

ing the insertion or the deletion of one or more nucleotides during replication. As shown in Figure 12.1(a), an insertion of a single nucleotide causes the frame of reading to shift, changing the specific sequence of all subsequent triplets to the right of the insertion. Upon translation, the amino acid sequence of the encoded protein will be altered. Such mutations are called **frameshifts**. When they are present at the *rII* locus, T4 will not reproduce on *E. coli* K12.

Crick and colleagues reasoned that if phages with these induced mutations were treated again with proflavin, still other insertions or deletions would occur. A second change might result in a revertant phage, which would behave like wild type and successfully infect *E. coli* K12. For example, if the original mutant contained an insertion (+), a second event causing a deletion (−) close to the insertion would restore the original reading frame. In the same way, an event resulting in an insertion (+) might correct an original deletion (−).

In studying many mutations of this type, these researchers were able to compare various mutant combinations together within the same DNA sequence. They found that various combinations of one (+) and one (−) indeed caused reversion to wild-type behavior. Still other observations shed light on the number of nucleotides constituting the genetic code. When two (+)s were to-

gether or when two (−)s were together, the correct reading frame *was not* reestablished. This argued against a doublet (two-letter) code. However, when three (+)s [Figure 12.1(b)] or three (−)s were present together, the original frame *was* reestablished. These observations strongly supported the triplet nature of the code.

These data further suggested that the code is degenerate, which was contrary to Crick's earlier proposal. A degenerate code is one in which more than one codon specifies the same amino acid. The reasoning leading to this conclusion is as follows. In the cases where wild-type function is restored, i.e., (+) and (−), (+++) and (−−−), the original frame of reading is also restored. However, there may be numerous triplets between the various additions and deletions that would still be out of frame. If 44 of the 64 possible triplets were blank and did not specify an amino acid, one of these so-called **nonsense triplets** would very likely occur in the length of nucleotides still out of frame. If a nonsense code were encountered during protein synthesis, it was reasoned that the process would stop or be terminated at that point. If so, the product of the *rII*-B locus would not be made, and restoration would not occur. Because the various mutant combinations were able to reproduce on *E. coli* K12, Crick and his colleagues concluded that, in all likelihood, most if not all of the remaining 44 codes were not blank. It follows that the genetic code is **degenerate**. As we shall see, this reasoning proved to be correct.

Deciphering the Code: Initial Studies

In 1961 Marshall Nirenberg and J. Heinrich Matthaei published results characterizing the first specific coding sequences. These results served as a cornerstone for the complete analysis of the genetic code. Their success was dependent on the use of two experimental tools: an *in vitro* (cell-free) protein-synthesizing system, and an enzyme, **polynucleotide phosphorylase**, which allowed the production of synthetic mRNAs. These mRNAs served as templates for polypeptide synthesis in the cell-free system.

In the cell-free system, amino acids can be incorporated into polypeptide chains. This *in vitro* mixture, as might be expected, must contain the essential factors for protein synthesis in the cell: ribosomes, tRNAs, amino acids, and other molecules essential to translation. In order to follow (or trace) protein synthesis, one or more of the amino acids must be radioactive. Finally, an mRNA must be added, which serves as the template to be translated.

In 1961 mRNA had yet to be isolated. However, the use of the enzyme polynucleotide phosphorylase allowed artificial synthesis of RNA templates, which could be

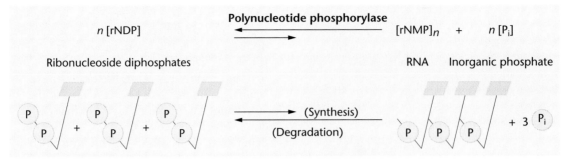

FIGURE 12.2 The reaction catalyzed by the enzyme polynucleotide phosphorylase. Note that the equilibrium of the reaction favors the degradation of RNA but can be "forced" in the direction favoring synthesis.

added to the cell-free system. This enzyme, isolated from bacteria, catalyzes the reaction shown in Figure 12.2. Discovered in 1955 by Marianne Grunberg-Manago and Severo Ochoa, the enzyme functions metabolically in bacterial cells to degrade RNA. However, *in vitro*, with high concentrations of ribonucleoside diphosphates, the reaction can be "forced" in the opposite direction to synthesize RNA, as illustrated.

In contrast to RNA polymerase, polynucleotide phosphorylase requires no DNA template. As a result, each addition of a ribonucleotide is random, based on the relative concentration of the four ribonucleoside diphosphates added to the reaction mixtures. The probability of insertion of a specific ribonucleotide is proportional to the availability of that molecule, relative to other available ribonucleotides. *This point is absolutely critical to understanding the work of Nirenberg and others in the ensuing discussion.*

Taken together, the cell-free system for protein synthesis and the availability of synthetic mRNAs provided a means of deciphering the ribonucleotide composition of various triplets encoding specific amino acids.

Nirenberg and Matthaei's Homopolymer Codes

In their initial experiments, Nirenberg and Matthaei synthesized **RNA homopolymers**, each consisting of only one type of ribonucleotide. Therefore, the mRNA added to the *in vitro* system was either UUUUUU..., AAAAAA..., CCCCCC..., or GGGGGG.... In testing each mRNA, they were able to determine which, if any, amino acids were incorporated into newly synthesized proteins. The researchers determined this by labeling one of the 20 amino acids added to the *in vitro* system and conducting a series of experiments, each with a different amino acid made radioactive.

For example, consider one of the initial experiments, where ^{14}C-phenylalanine was used (Table 12.1). From these and related experiments, Nirenberg and Matthaei

concluded that the message poly U (polyuridylic acid) directs the incorporation of only phenylalanine into the homopolymer polyphenylalanine. Assuming a triplet code, they had determined the first specific codon assignment. UUU codes for phenylalanine.

In the same way, they quickly found that AAA codes for lysine and CCC codes for proline. Poly G did not serve as an adequate template, probably because the molecule folds back on itself. Thus, the assignment for GGG had to await other approaches. Note that the specific triplet codon assignments were possible only because of the use of homopolymers. This method yields only the composition of triplets, not the sequence. However, three U's, C's, or A's can have only one possible sequence (i.e., UUU, CCC, and AAA).

The Use of Mixed Copolymers

With these techniques in hand, Nirenberg and Matthaei, and Ochoa and co-workers turned to the use of **RNA heteropolymers**. In this technique, two or more different ribonucleoside diphosphates are added in combination to form the message. These researchers reasoned that if the relative proportion of each type of ribonucleoside diphosphate is known, the frequency of any particular triplet codon occurring in the synthetic mRNA can be

TABLE 12.1 Incorporation of ^{14}C-Phenylalanine into Protein

Artificial mRNA	Radioactivity (counts/min)
None	44
Poly U	39,800
Poly A	50
Poly C	38

Source: After Nirenberg and Matthaei, 1961, p. 1595.

Possible compositions	Probability of occurrence of any triplet	Possible triplets	Final %
3A	$(1/6)^3 = 1/216 = 0.4\%$	AAA	0.4
1C:2A	$(1/6)^2(5/6) = 5/216 = 2.3\%$	AAC ACA CAA	3 x 2.3 = 6.9
2C:1A	$(1/6)(5/6)^2 = 25/216 = 11.6\%$	ACC CAC CCA	3 x 11.6 = 34.8
3C	$(5/6)^3 = 125/216 = 57.9\%$	CCC	57.9
			100.0%

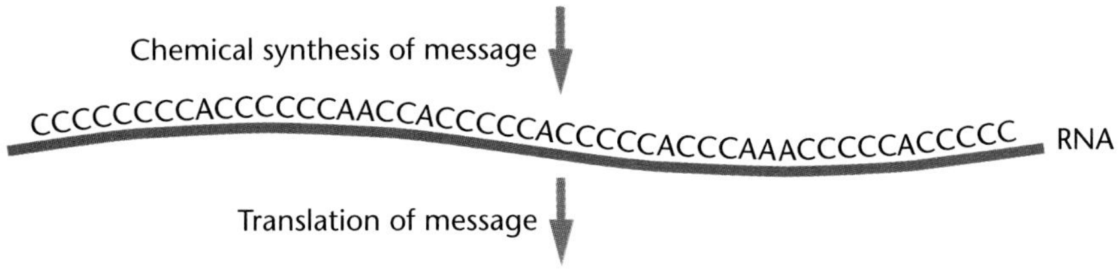

Chemical synthesis of message

CCCCCCCCACCCCCCAACCACCCCCACCCCCACCCAAACCCCCACCCCC RNA

Translation of message

Percentage of amino acids in protein		Probable base composition assignments
Lysine	1%	AAA
Glutamine	2%	1C:2A
Asparagine	2%	1C:2A
Threonine	12%	2C:1A
Histidine	14%	2C:1A, 1C:2A
Proline	69%	CCC, 2C:1A

FIGURE 12.3   Results and interpretation of a mixed copolymer experiment where a ratio of 1A:5C is used (1/6A:5/6C).

predicted. If the mRNA is then added to the cell-free system and the percentage of any particular amino acid present in the new protein is ascertained, correlations can be made and composition assignments predicted.

This approach is illustrated in Figure 12.3. Suppose that A and C are added in a ratio of 1A:5C. Now, the insertion of a ribonucleotide at any position along the RNA molecule during its synthesis is determined by the ratio of A:C. Therefore, there is a 1/6 possibility for an A and a 5/6 chance for a C to occupy each position. On this basis, we can calculate the frequency of any given triplet appearing in the message.

For AAA, the frequency is $(1/6)^3$, or about 0.4 percent. For AAC, ACA, and CAA, the frequencies are identical—that is, $(1/6)^2(5/6)$, or about 2.3 percent for each. Together, all three 2A:1C triplets account for 6.9 percent of the total three-letter sequences. In the same way, each of three 1A:2C triplets accounts for $(1/6)(5/6)^2$, or 11.6 percent (or a total of 34.8 percent). CCC is represented by $(5/6)^3$, or 57.9 percent of the triplets.

By examining the percentages of any given amino acid incorporated into the protein synthesized under the direction of this message, it is possible to propose probable base composition assignments (Figure 12.3). Because proline appears 69 percent of the time and because 69 percent is close to 57.9 percent + 11.6 percent, we can deduce that proline is coded by CCC and by one triplet of the 2C:1A variety. Histidine, at 14 percent, is probably coded by one 2C:1A (11.6 percent) and one 1C:2A (2.3 percent). Threonine, at 12 percent, is likely coded by only one 2C:1A. Asparagine and glutamine appear each to be coded by one of the 1C:2A triplets, and lysine appears to be coded by AAA.

Using as many as all four ribonucleotides to construct the mRNA, many similar experiments were conducted. Although determination of the *composition* of triplet code words corresponding to all 20 amino acids represented a very significant breakthrough, *specific sequences* of triplets were still unknown. Their determination awaited still other approaches.

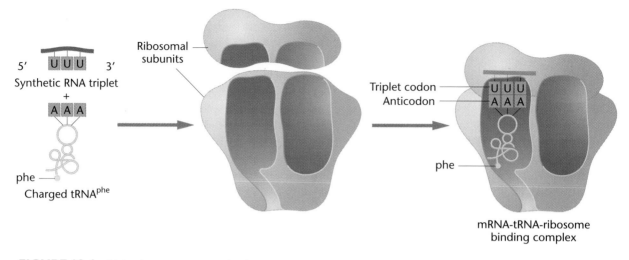

FIGURE 12.4 Molecular components used in the triplet binding assay.

The Triplet Binding Technique

It was not long before more advanced techniques were developed. In 1964 Nirenberg and Philip Leder developed the **triplet binding assay**, which led to specific assignments of triplets. The technique took advantage of the observation that ribosomes, when presented with an RNA sequence as short as three ribonucleotides, will bind to it and attract the charged tRNA (tRNA bonded to its amino acid) corresponding to the triplet codon. For example, if ribosomes are presented with an RNA triplet UUU and tRNA^phe, a complex will form that is similar to what actually occurs *in vivo*. The triplet acts like a codon in mRNA, and it attracts the complementary sequence called the **anticodon** found within tRNA (Figure 12.4). Anticodons are those triplet sequences in tRNAs that are complementary to the codons of mRNA. Although it was not yet feasible to chemically synthesize long stretches of RNA, triplets of known sequence could be synthesized in the laboratory to serve as templates.

All that was needed was a method to determine which tRNA-amino acid was bound to the triplet RNA-ribosome complex. The test system devised was quite simple. The amino acid to be tested was made radioactive, and a charged tRNA was produced. Because code compositions were known, it was possible to narrow the decision as to which amino acids should be tested for each specific triplet.

The radioactive charged tRNA, the RNA triplet, and ribosomes are incubated together on a nitrocellulose filter, which will retain the larger ribosomes but not the other smaller components, such as charged tRNA. If radioactivity is not retained on the filter, an incorrect amino acid has been tested. If radioactivity remains on the filter, it is retained because the charged tRNA has bound to the triplet associated with the ribosome. In such a case, a specific codon assignment can be made.

Work proceeded in several laboratories, and in many cases clear-cut, unambiguous results were obtained. Table 12.2, for example, shows 26 triplets assigned to 9 amino acids. However, in some cases the degree of binding was insufficient, and unambiguous assignments were not possible. Eventually, about 50 of the 64 triplets were assigned. The binding technique was a major innovation in deciphering the genetic code. Based on these specific assignments of triplets to amino acids, two major conclusions were drawn. The genetic code is **degenerate**; that is, one amino acid may be specified by more than one triplet. The code is also **unambiguous**; that is, a single triplet specifies only one amino acid. As we shall see later in this chapter, these conclusions have been upheld with only minor exceptions.

TABLE 12.2 Amino Acid Assignments to Specific Trinucleotides Derived from the Triplet Binding Assay

Trinucleotide	Amino Acid
UGU UGC	Cysteine
GAA GAG	Glutamic acid
AUU AUC AUA	Isoleucine
UUA UUG CUU	Leucine
CUC CUA CUG	Leucine
AAA AAG	Lysine
AUG	Methionine
UUU UUC	Phenylalanine
CCU CCC	Proline
CCG CCA	Proline
UCU UCC	Serine
UCA UCG	Serine

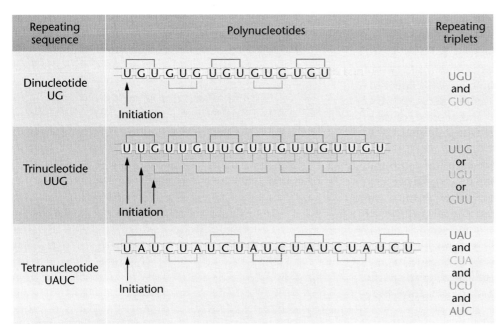

Repeating sequence	Polynucleotides	Repeating triplets
Dinucleotide UG	U G U G U G U G U G U G U Initiation	UGU and GUG
Trinucleotide UUG	U U G U U G U U G U U G U U G U U G U Initiation	UUG or UGU or GUU
Tetranucleotide UAUC	U A U C U A U C U A U C U A U C U A U C U Initiation	UAU and CUA and UCU and AUC

FIGURE 12.5   The conversion of di-, tri-, and tetranucleotides into repeating copolymers. The triplet codons produced in each case are shown.

The Use of Repeating Copolymers

Still another innovative technique used to decipher the genetic code was developed by Gobind Khorana, who was able to chemically synthesize long RNA molecules consisting of short sequences repeated many times, which could then be used in the cell-free protein-synthesizing system. First, he created shorter sequences (e.g., di-, tri-, and tetranucleotides), which were then replicated many times and finally joined enzymatically to form the long polynucleotides.

As illustrated in Figure 12.5, a dinucleotide made in this way is converted to a message with two repeating triplets. A trinucleotide is converted to one with three potential triplets, depending on the point at which initiation occurs. Similarly, a tetranucleotide creates four repeating triplets.

When these synthetic mRNAs were added to a cell-free system, the predicted number of different amino acids incorporated was upheld. Several examples are shown in Table 12.3. When such data were combined with conclusions drawn from other approaches (composition assignment, triplet binding), specific assignments were possible.

One example of specific assignments made from such data will illustrate the value of Khorana's approach. Consider the following three experiments in concert with one another. The repeating trinucleotide sequence UUCUUCUUC... produces three possible triplets: UUC, UCU, and CUU, depending on the first nucleotide to initiate reading. When placed in a cell-free translation system, the polypeptides containing phenylalanine (phe), serine (ser), and leucine (leu) are produced. On the other hand, the repeating dinucleotide sequence UCUCUCUC... produces the triplets UCU and CUC with the incorporation of leucine and serine into the polypeptide. Therefore, the triplets UCU and CUC specify leucine and serine, but it cannot be determined which is which. One can further conclude that *either* the CUU *or* the UUC triplet also encodes leucine *or* serine, while the other encodes phenylalanine.

TABLE 12.3   Amino Acids Incorporated Using Repeating Synthetic Copolymers of RNA

Repeating Copolymer	Codons Produced	Amino Acids in Polypeptide
UG	UGU GUG	Cysteine Valine
AC	ACA CAC	Threonine Histidine
UUC	UUC UCU CUU	Phenylalanine Serine Leucine
AUC	AUC UCA CAU	Isoleucine Serine Histidine
UAUC	UAU CUA UCU AUC	Tyrosine Leucine Serine Isoleucine
GAUA	GAU AGA UAG AUA	None

To derive more specific information, we can examine the results of using the repeating tetranucleotide sequence UUAC, which produces the triplets UUA, UAC, ACU, and CUU. The CUU triplet is one of the two in which we are interested. Three amino acids are incorporated: leucine, threonine, and tyrosine. Because CUU must specify only serine or leucine, and because, of these two, only leucine appears, we may conclude that CUU specifies leucine.

Once this is established, we can logically determine all other specific assignments. Of the two triplet pairs remaining (UUC and UCU from the first experiment *and* UCU and CUC from the second experiment), whichever triplet is common to both must encode serine. This is UCU. By elimination, UUC is determined to encode phenylalanine and CUC is determined to encode leucine. While the logic must be carefully followed, four specific triplets encoding three different amino acids have been assigned from these experiments.

From such interpretations, Khorana reaffirmed triplets already deciphered and filled in gaps left from other approaches. For example, the use of two tetranucleotide sequences, GAUA and GUAA, suggested that at least two triplets were termination signals. This conclusion was drawn because neither of these sequences directed the incorporation of any amino acids into a polypeptide. Because there are no triplets common to both messages, it was predicted that each repeating sequence contains at least one triplet that terminates protein synthesis. The possible triplets for the poly-(GAUA) sequence are shown in Table 12.3, one of which, UAG, is a termination codon.

Second position

First position (5′-end)	U	C	A	G	Third position (3′-end)
U	UUU *phe*	UCU	UAU *tyr*	UGU *cys*	U
	UUC	UCC *ser*	UAC	UGC	C
	UUA *leu*	UCA	UAA *Stop*	UGA *Stop*	A
	UUG	UCG	UAG *Stop*	UGG *trp*	G
C	CUU	CCU	CAU *his*	CGU	U
	CUC *leu*	CCC *pro*	CAC	CGC *arg*	C
	CUA	CCA	CAA *gln*	CGA	A
	CUG	CCG	CAG	CGG	G
A	AUU	ACU	AAU *asn*	AGU *ser*	U
	AUC *ile*	ACC *thr*	AAC	AGC	C
	AUA	ACA	AAA *lys*	AGA *arg*	A
	AUG *met*	ACG	AAG	AGG	G
G	GUU	GCU	GAU *asp*	GGU	U
	GUC *val*	GCC *ala*	GAC	GGC *gly*	C
	GUA	GCA	GAA *glu*	GGA	A
	GUG	GCG	GAG	GGG	G

Initiation Termination

FIGURE 12.6 The coding dictionary. AUG encodes methionine, which initiates most polypeptide chains. All other amino acids except tryptophan, which is encoded by UGG, are represented by two to six triplets. The triplets UAA, UAG, and UGA are termination signals and do not encode any amino acids.

The Coding Dictionary

The various techniques applied to decipher the genetic code have yielded a dictionary of 61 triplet codon–amino acid assignments. The remaining three triplets are termination signals, not specifying any amino acid. Figure 12.6 designates the assignments in a particularly illustrative form first suggested by Francis Crick.

Degeneracy, Wobble, and Order in the Code

A general pattern of triplet codon assignments becomes apparent when this type of presentation of the genetic code is inspected. Most evident is that the code is **degenerate**. That is, most all amino acids are specified by two, three, or four different codons. Three amino acids (serine, arginine, and leucine) are each encoded by six different codons. Only tryptophan and methionine are encoded by single codons.

What is also evident is the pattern of degeneracy. Most often in a set of codons specifying the same amino acid, the first two letters are the same, with only the third differing. This interesting pattern prompted Crick in 1966 to postulate the **wobble hypothesis**.

Crick concluded that the first two ribonucleotides of triplet codes are more critical than is the third member in attracting the correct tRNA. Thus, Crick proposed that hydrogen bonding at the third position of the codon–anticodon interactions need not adhere as specifically as the first two members to the established base-pairing rules. After examining the coding dictionary carefully, he proposed a new set of base-pairing rules at the third position of the codon (Table 12.4).

This relaxed base-pairing requirement, or "wobble," allows the anticodon of a single tRNA species to pair with more than one triplet in mRNA. As pointed out above, the degeneracy of the code often allows substitution at the position of the third base without changing the amino acid. Consistent with the wobble hypothesis and coding assignments, U at the third position (the 5′-end) of the anticodon of tRNA may pair with A or G at the third position (the 3′-end) of the triplet codon in mRNA, and that G may likewise pair with U or C. Inosine, one of the modified bases found in tRNA, may pair with C, U, or A.

TABLE 12.4   Codon-Anticodon Base-Pairing Rules

Base at Third Position (3′-end) of mRNA	Base at First Position (5′-end) of tRNA
A	U or I
C	G or I
G	C or U
U	A or G or I

Applying these wobble rules, a minimum of about 30 different tRNA species is necessary to accommodate the 61 triplets specifying an amino acid. If nothing more, wobble can be considered a potential economy measure, provided that fidelity of translation is not compromised. Current estimates are that 30–40 tRNA species are present in bacteria and up to 50 tRNA species in animal and plant cells.

More recently, another observation has become apparent in the pattern of triplet sequences and their corresponding amino acids, leading to the description referred to as an **ordered code**. Chemically similar amino acids often share one or two "middle" bases in the different triplets encoding them. For example, U or C are often present in the second position of triplets that specify hydrophobic amino acids, including valine and alanine, among others. Charged amino acids, such as lysine and aspartic acid or hydrophilic amino acids, such as glycine or threonine most often are specified by triplet codons with A or C in the second position. The chemical properties of amino acids will be discussed in more detail in Chapter 13. The end result of an "ordered" code is that it buffers the potential effect of mutation on protein function. While many mutations of the second base of triplet codons result in a change of one amino acid to another, the change is often to an amino acid with similar chemical properties. In such cases, protein function may not be noticeably altered.

Initiation, Termination, and Suppression

Initiation of protein synthesis is a highly specific process. In bacteria (in contrast to the *in vitro* experiments discussed earlier), the initial amino acid inserted into all polypeptide chains is a modified form of methionine—**N-formylmethionine (fmet)**. Only one codon, AUG, codes for methionine, and it is sometimes called the **initiator codon**. However, when AUG appears internally in mRNA, unformylated methionine is inserted into the polypeptide chain. Rarely, still another triplet, GUG, specifies methionine during initiation. It is not clear why this occurs, since GUG normally encodes valine.

In bacteria, either the formyl group is removed from the initial methionine upon completion of synthesis of a protein, or the entire formylmethionine residue is removed. In eukaryotes, methionine is also the initial amino acid during polypeptide synthesis. However, it is not formylated.

As mentioned in the preceding section, three other triplets (UAG, UAA, and UGA)* serve as **termination codons**, punctuation signals that do not code for any amino acid. They are not recognized by a tRNA molecule, and termination of translation occurs when they are encountered. Mutations that produce any of the three triplets internally in a gene also result in termination. Consequently, only a partial polypeptide is synthesized, since it is prematurely released from the ribosome. When such a change occurs, it is called a **nonsense mutation**.

Interestingly, a distinct mutation in a second gene may trigger **suppression** of premature termination. These mutations cause the chain-termination signal to be read as a "sense" codon. The "correction" usually inserts an amino acid other than that found in the wild-type protein. However, if the protein's structure is not altered drastically, it may function almost normally. Therefore, this second mutation has "suppressed" the mutant character resulting from the initial change to the termination codon.

Some suppressor mutations occur in genes specifying tRNAs. If the mutation results in a change in the anticodon such that it becomes complementary to a termination code, there is the potential for insertion of an amino acid and suppression.

Other types of suppression involve mutations in genes coding for aminoacyl synthetases, which are responsible for attaching the amino acid to tRNA (the process called *charging*), and in genes coding for ribosomal proteins. In both cases, suppression occurs as a result of misreading the mutant codon. That a ribosomal pro-

---

*Historically, the terms *amber* (UAG), *ochre* (UAA), and *opal* (UGA) have been used to distinguish the three mutation possibilities.

tein can cause ambiguity of translation demonstrates the intimate relationship among the ribosome, mRNA, and tRNA during translation.

Confirmation of Code Studies: Phage MS2

All aspects of the genetic code discussed so far yield a fairly complete picture. The code is triplet in nature, degenerate, unambiguous, and commaless, but it contains punctuation with respect to start and stop signals. These individual principles have been confirmed by the detailed analysis of the RNA-containing bacteriophage MS2 by Walter Fiers and his co-workers.

MS2 is a bacteriophage that infects *E. coli.* Its nucleic acid (RNA) contains only about 3500 ribonucleotides, making up only three genes. These genes specify a coat protein, an RNA-directed replicase, and a maturation protein (the A protein). This simple system of a small genome and few gene products allowed Fiers and his colleagues to sequence the genes and their products. The amino acid sequence of the coat protein was completed in 1970, and the nucleotide sequence of the gene and a number of nucleotides on each end of it were reported in 1972.

The coat protein contains 129 amino acids, and the gene contains 387 nucleotides, as expected for a triplet code. Each amino acid and triplet corresponds in linear sequence to the correct codon in the RNA code-word dictionary, confirming the earlier experimental work that served as the basis for deciphering the genetic code. These findings further provided direct proof of the **colinear relationship** between nucleotide sequence and amino acid sequence. The codon for the first amino acid is preceded by AUG, the common initiator codon; and the codon for the last amino acid is succeeded by two consecutive termination codons, UAA and UAG.

By 1976, the other two genes and their protein products were sequenced, providing similar confirmation. The analysis clearly shows that the genetic code, as established in bacterial systems, is identical in this virus. We shall now briefly consider other evidence suggesting that the code is also identical in eukaryotes.

Universality of the Code

Between 1960 and 1978, it was generally assumed that the genetic code would be found to be universal, applying equally to viruses, bacteria, and eukaryotes. Certainly, the nature of mRNA and the translation machinery seemed to be very similar in these organisms. For example, cell-free systems derived from bacteria could translate eukaryotic mRNAs. Poly U was shown to stimulate translation of polyphenylalanine in cell-free systems when the components were derived from eukaryotes. Many recent studies involving recombinant DNA technology (see Chapters 15 and 16) have revealed that eukaryotic genes can be inserted into bacterial cells and transcribed and translated. Within eukaryotes, mRNAs from mice and rabbits have been injected into amphibian eggs and efficiently translated. For the many eukaryotic genes that have been sequenced, notably those for hemoglobin molecules, the amino acid sequence of the encoded proteins adheres to the coding dictionary established from bacterial studies.

However, several 1979 reports on the coding properties of DNA derived from mitochondria of yeast and humans (mtDNA) altered the principle of universality of the genetic language. Since then, mtDNA has been examined in many other organisms.

Not only do mitochondria contain DNA, but transcription and translation also occur within these organelles. Cloned mtDNA fragments were sequenced and compared with the amino acid sequences of various mitochondrial proteins, revealing several exceptions to the coding dictionary (Table 12.5). Most surprising was that the codon UGA, normally causing termination, specifies the insertion of tryptophan during translation in yeast and human mitochondria. In human mitochondria, AUA, which normally specifies isoleucine, directs the internal insertion of methionine. In yeast mitochondria, threonine is inserted instead of leucine when CUA is encountered in mRNA.

More recently, in 1985, several other exceptions to the standard coding dictionary were discovered. These

TABLE 12.5 Exceptions to the Universal Code

Triplet	Normal Code Word	Altered Code Word	Source
UGA	termination	trp	Human and yeast mitochondria *Mycoplasma*
CUA	leu	thr	Yeast mitochondria
AUA	ile	met	Human mitochondria
AGA AGG	arg	termination	Human mitochondria
UAA	termination	gln	*Paramecium Tetrahymena Stylonychia*
UAG	termination	gln	*Paramecium*

and prior differences are also summarized in Table 12.5. Such altered code words have been observed in the bacterium *Mycoplasma capricolum*, and in the nuclear genes of the protozoan ciliates *Paramecium, Tetrahymena,* and *Stylonychia*. As shown in Table 12.5, each alteration converts one of the termination codons (UGA) to tryptophan. These changes are significant because both a prokaryote and several eukaryotes are involved, representing distinct species that have evolved over a long period of time.

Note the apparent pattern in several of the altered codon assignments. The change in coding capacity involves only a shift in recognition of the third, or wobble, position. For example, AUA specifies isoleucine during translation in the cytoplasm and methionine in the mitochondrion. In cytoplasmic translation, methionine is specified by AUG. In a similar way, UGA calls for termination in the cytoplasmic system but tryptophan in the mitochondrion. In the cytoplasm, tryptophan is specified by UGG. Although it has been suggested that such changes in codon recognition may represent an evolutionary trend toward reducing the number of tRNAs needed in mitochondria, the significance of these findings is not yet clear. It is known that only 22 tRNA species are encoded in human mitochondria. However, until still other examples are revealed, the differences must be considered as exceptions to the previously established general coding rules.

Reading the Code: The Case of Overlapping Genes

In this chapter we established that the genetic code is nonoverlapping. This means that each ribonucleotide in an mRNA is part of only one triplet. However, this characteristic of the code does not rule out the possibility that a single mRNA may have multiple initiation points for translation. If so, these points could theoretically create several different frames of reading within the same mRNA, thus specifying more than one polypeptide. This concept, which would create **overlapping genes**, is illustrated in Figure 12.7(a).

That this might actually occur in some viruses was suspected when phage φX174 was carefully investigated. The circular DNA chromosome consists of 5386 nucleotides, which should encode a maximum of 1795 amino acids, sufficient for five or six proteins. However, it was realized that this small virus in fact synthesizes 11 proteins consisting of more than 2300 amino acids. Comparison of the nucleotide sequence of the DNA and the amino acid sequences of the polypeptides synthesized has clarified this paradox. At least four cases of multiple initiation have been discovered, creating overlapping genes [Figure 12.7(b)].

The sequences specifying the K and B polypeptides are initiated with separate reading frames within the sequence specifying the A polypeptide. The K sequence overlaps into the adjacent sequence specifying the C polypeptide. The E sequence is out of frame with, but initiated in, that of the D polypeptide. Finally, the A′ sequence, while in frame, begins in the middle of the A sequence. They both terminate at the identical point. In all, seven different polypeptides are created from a DNA sequence that might otherwise have specified only three (A, C, and D).

A similar situation has been observed in other viruses, including phage G4 and the animal virus SV40. Like φX174, phage G4 contains a circular, single-stranded DNA molecule. The use of overlapping reading frames optimizes the use of a limited amount of DNA present in these small viruses. However, such an approach to storing information has the distinct disadvantage that a single mutation may affect more than one protein and thus increase the chances that the change will be deleterious or lethal. In the case discussed above, a single mutation at the junction of genes *A* and *C* could affect three proteins (the A, C, and K proteins). It may be for this reason that such an approach has not become common in other organisms.

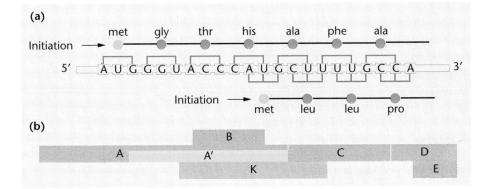

FIGURE 12.7 Illustration of the concept of overlapping genes. (a) An mRNA sequence initiated at two different AUG positions out of frame with one another will give rise to two distinct amino acid sequences. (b) The relative positions of the sequences encoding seven polypeptides of the phage φX174. Those encoding B, K, and E are out of frame.

Expression of Genetic Information: An Overview

Even while the genetic code was being studied, it was quite clear that proteins were the end products of many genes. Thus, while some geneticists were attempting to elucidate the code, other research efforts were directed toward the nature of genetic expression. The central question was how DNA, a nucleic acid, is able to specify a protein composed of amino acids. Put still another way: How is information transferred between DNA and protein? In the next chapter we shall examine the evidence supporting the conclusion that proteins are specified by genes; we shall also discuss protein structure and function. Here we will emphasize the concept of **information flow** as it occurs between DNA and protein. This topic was extremely interesting and exciting in the early 1960s, and it remains no less so in the 1990s.

Genetic information, stored in DNA, was shown to be transferred to RNA during the initial stage of gene expression. The process by which RNA molecules are synthesized on a DNA template is called **transcription**. The ribonucleotide sequence of RNA, written in a genetic code, is then capable of directing the process of **translation**. During translation, polypeptide chains—the precursors of proteins—are synthesized. Polypeptide synthesis is dependent on a series of **transfer RNA (tRNA)** molecules that serve as adaptors between the codons of mRNA and the amino acids specified by them. In addition, the process occurs only in conjunction with a large, intricate cellular organelle, the **ribosome**.

The processes of transcription and translation are complex molecular events. Like the replication of DNA, both rely heavily on base-pairing affinities between complementary nucleotides. The initial transfer from DNA to mRNA produces a molecule complementary to the gene sequence of one of the two strands of the double helix. Then, each triplet codon in mRNA is complementary to the anticodon region of its corresponding tRNA as the amino acid is correctly inserted into the polypeptide chain during translation. In the following sections we describe in detail how these processes were discovered and how they are executed.

Transcription: RNA Synthesis

The idea that RNA is involved as an intermediate molecule in the process of information flow between DNA and protein is suggested by the following observations:

1. DNA is, for the most part, associated with chromosomes in the nucleus of the eukaryotic cell. However, protein synthesis occurs in association with ribosomes located outside the nucleus in the cytoplasm. Therefore, DNA does not appear to participate directly in protein synthesis.
2. RNA is synthesized in the nucleus of eukaryotic cells, where DNA is found, and is chemically similar to DNA.
3. Following its synthesis, most RNA migrates to the cytoplasm, where protein synthesis occurs.
4. The amount of RNA is generally proportional to the amount of protein in a cell.

Collectively, these observations suggested that genetic information, stored in DNA, is transferred to an RNA intermediate, which directs the synthesis of proteins. As with most new ideas in molecular genetics, the initial supporting experimental evidence was based on studies of bacteria and their phages.

TABLE 12.6 Base Compositions (in mole percents) of RNA Produced Immediately Following Infection of *E. coli* by the Bacteriophages T2 and T7 in Contrast to the Composition of RNA of Uninfected *E. coli*

	Adenine	Thymine	Uracil	Cytosine	Guanine
Postinfection RNA in T2-infected cells	33	—	32	18	18
T2 DNA	32	32	—	17*	18
Postinfection RNA in T7-infected cells	27	—	28	24	22
T7 DNA	26	26	—	24	22
E. coli RNA	23	—	22	18	17

*5-hydroxymethyl cytosine.
Source: From Volkin and Astrachan, 1956; and Volkin, Astrachan, and Countryman, 1958.

Experimental Evidence for the Existence of mRNA

In two papers published in 1956 and 1958, Elliot Volkin and his colleagues reported their analysis of RNA produced immediately after bacteriophage infection of *E. coli*. Using the isotope ^{32}P to follow newly synthesized RNA, they found that its base composition closely resembled that of the phage DNA but was different from that of bacterial RNA (Table 12.6). Although this newly synthesized RNA was unstable, or short-lived, its production was shown to precede the synthesis of new phage proteins. Thus, Volkin and his co-workers considered the possibility that synthesis of RNA is a preliminary step in the process of protein synthesis.

Although ribosomes were known to participate in protein synthesis, their role in this process was not clear. One possibility was that each ribosome is specific for the protein synthesized in association with it. That is, perhaps genetic information in DNA is transferred to the RNA of a ribosome during its synthesis so that each class of ribosome specifies a particular protein. The alternative hypothesis was that ribosomes are nonspecific "workbenches" for protein synthesis and that specific genetic information rests with a "messenger" RNA.

In an elegant experiment using the *E. coli*–phage system, the results of which were reported in 1961, Sidney Brenner, François Jacob, and Matthew Meselson clarified this question. They labeled uninfected *E. coli* ribosomes with "heavy" isotopes (see Appendix A) and then allowed phage infection to occur in the presence of radioactive RNA precursors. By following these components during translation, the researchers demonstrated that synthesis of phage proteins (under the direction of newly synthesized RNA) occurred on bacterial ribosomes that were present prior to infection. Therefore, the ribosomes appeared to be nonspecific, strengthening the case that another type of RNA serves as an intermediary in the process of protein synthesis.

That same year, Sol Spiegelman and his colleagues reached the same conclusion when they isolated ^{32}P-labeled phage RNA following infection of bacteria and used it in molecular hybridization studies. They tried hybridizing this RNA to the DNA of both phages and bacteria in separate experiments. The RNA hybridized only with the phage DNA, showing that it was complementary in base sequence to the viral genetic information.

Results of these experiments agree with the concept of a **messenger RNA (mRNA)** being made on a DNA template and then directing the synthesis of specific proteins in association with ribosomes. This concept was formally proposed by François Jacob and Jacques Monod in 1961 as part of a model for gene regulation in bacteria.

Since then, mRNA has been isolated and thoroughly studied. There is no longer any question about its role in genetic processes.

RNA Polymerase

To prove that RNA can be synthesized on a DNA template, it was necessary to demonstrate that there is an enzyme capable of directing this synthesis. By 1959, several investigators, including Samuel Weiss, had independently discovered such a molecule from rat liver. Called **RNA polymerase**, it has the same general substrate requirements as does DNA polymerase, the major exception being that the nucleotides that are substrates contain the ribose rather than the deoxyribose form of the sugar. Unlike DNA polymerase, no primer is required to initiate synthesis. The initial base remains as a nucleoside triphosphate (NTP). The overall reaction summarizing the synthesis of RNA on a DNA template can be expressed as:

$$n(\text{NTP}) \xrightarrow[\text{enzyme}]{\text{DNA}} (\text{NMP})_n + n(\text{PP}_i)$$

As the equation reveals, nucleoside triphosphates (NTPs) serve as substrates for the enzyme, which catalyzes the polymerization of nucleoside monophosphates (NMPs), or nucleotides, into a polynucleotide chain $(\text{NMP})_n$. Nucleotides are linked during synthesis by 5′-to-3′ phosphodiester bonds (see Figure 11.9 on page 305). The energy created by cleaving the triphosphate precursor into the monophosphate form drives the reaction, and inorganic phosphates (PP_i) are produced.

A second equation summarizes the sequential addition of each ribonucleotide as the process of transcription progresses:

$$(\text{NMP})_n + \text{NTP} \xrightarrow[\text{enzyme}]{\text{DNA}} (\text{NMP})_{n+1} + \text{PP}_i$$

As this equation shows, each step of transcription involves the addition of one ribonucleotide (NMP) to the growing polyribonucleotide chain $(\text{NMP})_{n+1}$, using a nucleoside triphosphate (NTP) as the precursor.

RNA polymerase from *E. coli* has been extensively characterized and shown to consist of subunits designated α, β, β', and σ. The active form of the enzyme, the **holoenzyme**, contains the subunits $\alpha_2\beta\beta'\sigma$ and has a molecular weight of almost 500,000 daltons. Of these subunits, it is the β and β' polypeptides that provide the catalytic basis and active site for transcription. As we will see, the σ **(sigma) subunit** plays a regulatory function involving the initiation of RNA transcription.

TABLE 12.7 RNA Polymerases in Eukaryotes

Form	Product	Location
I	rRNA	Nucleolus
II	mRNA	Nucleoplasm
III	5S rRNA	Nucleoplasm
	tRNA	Nucleoplasm

While there is but a single form of the enzyme in *E. coli*, three forms of RNA polymerase are involved in the transcription of the three types of RNA in eukaryotes. The nomenclature used in describing these is summarized in Table 12.7. The three eukaryotic polymerases all consist of a greater number of polypeptide subunits than does the bacterial form of the enzyme.

Promoters, Template Binding, and the Sigma Subunit

Transcription results in the synthesis of a single-stranded RNA molecule complementary to a region along one of the two strands of the DNA double helix. For the purpose of future discussion, we will call the DNA strand that is transcribed the **template strand**. Its complement will be referred to as the **partner strand**. The initial step is referred to as **template binding**, where RNA polymerase interacts physically with DNA [Figure 12.8(b)]. The site of this initial binding in bacteria is established as a result of the recognition by the sigma subunit (σ) of the polymerase of specific DNA sequences called **promoters**.

These regions are located in the 5′-upstream region (to the left in the illustration) from the point of initial transcription of a gene. It is believed that the enzyme "explores" a length of DNA until the promoter region is recognized and a bound complex results. Once this occurs, the helix is denatured or unwound locally, making the DNA template accessible to the action of the enzyme. The enzyme is a large, complex molecule that binds to about 60 nucleotide pairs of the helix, 40 of which are upstream from the point of initial transcription.

The importance of promoter sequences cannot be overemphasized. They govern the efficiency of initiation of transcription. In bacteria, both strong promoters and weak promoters are recognized, leading to a variation of initiation from once every 1 to 2 seconds to only once every 10 to 20 minutes. Mutations in promoter sequences may have the effect of severely reducing the initiation of gene expression. Because the interaction of promoters with RNA polymerase governs transcription, the nature of the binding between them is at the heart of discussions concerning **genetic regulation**, the subject of Chapters 18 and 19. While we will pursue information involving promoter–enzyme interactions in greater detail in those chapters, several generalities are appropriate to introduce here.

First off is the concept of **consensus sequences** of DNA. These are sequences that are similar (homologous) in different genes of the same organism or in one or more genes of related organisms. Their conservation throughout evolution attests to the critical nature of their role in biological processes. In bacterial promoters, two such sequences have been found. One is located 10 nucleotides upstream from the site of initial transcription (the **−10 region**, or **Pribnow box**). The other is located 35 nucleotides upstream (the **−35 region**). Mutations in both regions diminish transcription, often severely. In most all eukaryotic genes studied, a consensus sequence comparable to the location of the −10 region has been recognized. Because it is rich in adenine and thymine residues, it is called the **TATA box**.

The second general point to be made involves the sigma subunit in bacteria. The major form is designated as σ^{70}, based on its molecular weight of 70 kilodaltons (kDa). While the promoters of most bacterial genes recognize this form, there are several alternative forms of RNA polymerase in *E. coli* that have unique σ subunits associated with them (e.g., σ^{28}, σ^{32}, σ^{38}, and σ^{54}. These recognize different promoter sequences and provide specificity to the initiation of transcription.

The Synthesis of RNA

Once the promoter has been recognized and bound by the enzyme complex, RNA polymerase catalyzes the insertion of the first 5′-ribonucleoside triphosphate, which is complementary to the first nucleotide at the start site of the DNA template strand. Unlike DNA synthesis, no primer is required. Subsequent ribonucleotide complements are inserted and linked together by phosphodiester bonds as RNA polymerization proceeds. This process, called **chain elongation** [Figure 12.8(c)], continues in the 5′-to-3′ direction, creating a temporary DNA/RNA duplex whose chains run antiparallel to one another.

After a few ribonucleotides have been added to the growing RNA chain, the σ subunit is dissociated from the holoenzyme, and elongation proceeds under the direction of the core enzyme [Figure 12.8(c)]. In *E. coli* this process proceeds at the rate of about 50 nucleotides/second at 37°C.

Eventually, the enzyme traverses the entire gene and encounters a termination signal, a specific nucleotide sequence. Such termination sequences are extremely important in prokaryotes because of the close proximity of the end of one gene and the upstream sequences of the adjacent gene. These sequences are about 40 base pairs in length. In some cases, termination of synthesis is dependent upon the **termination factor, rho (ρ)**. Rho is a large hexameric protein that physically interacts with the growing RNA transcript in achieving termination. At the point

(a) Transcription components

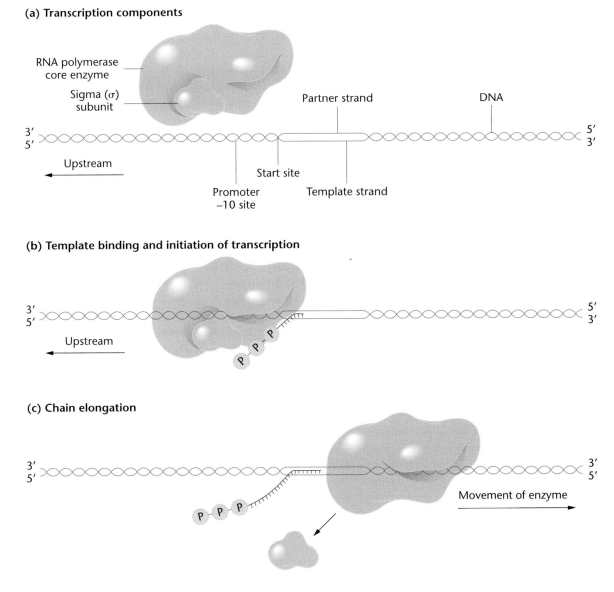

(b) Template binding and initiation of transcription

(c) Chain elongation

FIGURE 12.8 Schematic representation of the early stages of transcription in prokaryotes, including: (a) template binding and initiation involving the sigma subunit of RNA polymerase; and (b) chain elongation, after the sigma subunit has dissociated from the transcription complex.

of termination, the transcribed RNA molecule is released from the DNA template, and the core enzyme dissociates.

It is important to note that groups of genes whose products are related are often clustered together along the chromosome. In many such cases, they are contiguous and all but the last gene lack the encoded signals for termination of transcription. The result is that during transcription a large mRNA is produced, encoding more than one gene. Since genes in bacteria are sometimes called cistrons (see Chapter 6), the RNA is called a **polycistronic mRNA**. Since the products of genes transcribed in this fashion are usually all needed at the same time, this is an efficient way to transcribe and, subsequently, to translate the needed genetic information. In eukaryotes, **monocistronic** mRNAs are the rule.

The significance of the process of transcription is enormous, for it is the initial step in the process of information flow within the cell. Under the direction of RNA polymerase, an RNA molecule is synthesized that is precisely complementary to a DNA sequence representing the template strand of a gene. Wherever an A, T, C, or G residue existed, a corresponding U, A, G, or C residue has been incorporated into the RNA molecule, respectively. As we shall see shortly, this RNA molecule provides the information leading to the synthesis of all proteins present in the cell.

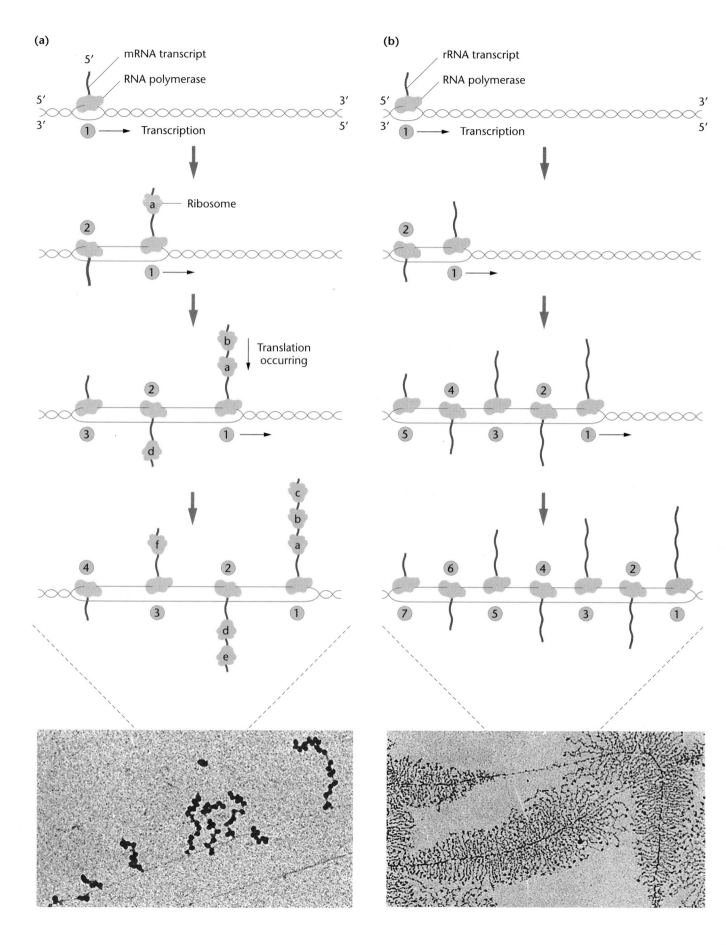

Visualization of Transcription

Electron microscope studies by Oscar Miller, Jr., Barbara Hamkalo, and Charles Thomas have provided striking visual demonstrations of the transcription process. Figure 12.9 shows micrographs and interpretive drawings from two organisms, the bacterium *E. coli* and the newt *Notophthalmus viridescens*. In both cases multiple strands of RNA are seen to emanate from different points along the more central DNA template. Many RNA strands result because numerous transcription events are occurring simultaneously along each gene. Progressively longer RNA strands are found farther downstream from the point of initiation of transcription, whereas the shortest strands are closest to the point of initiation.

An interesting picture emerges from the study of *E. coli* [Figure 12.9(a)]. Because prokaryotes lack nuclei, cytoplasmic ribosomes are not separated physically from the chromosome. As a result, ribosomes are free to attach to *partially* transcribed mRNA molecules and initiate translation. The longer RNA strands demonstrate the greatest number of ribosomes (see the later discussion of polyribosomes). In the case of the newt, the segment of DNA is derived from oocytes that are known to produce an enormous amount of rRNA. To accomplish this synthesis, the genes specific for RNA (**rDNA**) are replicated many times in the oocyte. This process is called **gene amplification**. The micrograph in Figure 12.9(b) shows several of these genes, each undergoing simultaneous transcription events. For each rRNA gene, longer and longer strands of incomplete rRNA molecules are produced as the enzymes move along the DNA strand. Visualization of transcription confirms our expectations based on the biochemical analysis of this process.

Transcription in Eukaryotes

Much of our knowledge of transcription has been derived from studies of prokaryotes. Many general aspects of the mechanics of these processes are similar in eukaryotes. There are, however, numerous notable differences, several of which are discussed below. We will first summarize some of the major differences and then expand on several of these in greater detail.

1. Transcription in eukaryotes occurs within the nucleus under the direction of three separate forms of RNA polymerase (see Table 12.7 on p. 338). Unlike prokaryotes, the RNA transcript is not free to associate with ribosomes prior to the completion of transcription. For the mRNA to be translated, it must move out of the nucleus into the cytoplasm.

2. Initiation and regulation of transcription involve a more extensive interaction between upstream DNA sequences and protein factors involved in stimulating and initiating transcription. There are, in addition to promoters, other control units called *enhancers* that may be located at positions other than in the 5'-regulatory region upstream from the point of initiation of transcription.

3. Maturation of eukaryotic mRNA from the primary RNA transcript involves many complex stages referred to generally as "processing." An initial step involves the addition of a 5'-cap and a 3'-tail to most transcripts destined to become mRNAs.

4. Most notably, extensive modifications occur to the internal sequence of nucleotides of eukaryotic RNA transcripts that eventually serve as mRNAs. The initial (or primary) transcripts are most often much larger than those that are eventually translated. Sometimes called **pre-mRNAs**, they are part of a group of molecules found only in the nucleus—a group referred to collectively as **heterogeneous nuclear RNA (hnRNA)**. Such RNA molecules are of variable size and are complexed with proteins, forming **heterogeneous nuclear ribonucleoprotein particles (hnRNPs)**. Only about 25 percent of hnRNA molecules are converted to mRNA. Those that are converted have substantial amounts of their ribonucleotide sequence excised, whereas the remaining segments are spliced back together prior to translation. This phenomenon has given rise to the concepts of **split genes** and **splicing** in eukaryotes.

As we proceed with our discussion in this chapter, we will elaborate on each of these differences. We will also return to other topics directly related to regulation of transcriptional activity in Chapters 18 and 19.

Eukaryotic Promoters, Enhancers, and Transcription Factors

Recognition of certain highly specific DNA regions by RNA polymerase is the basis of orderly genetic function in all cells. Both the polymerases and promoters leading

FIGURE 12.9 Electron micrographs and interpretive drawings of simultaneous transcription of genes in (a) *E. coli* and (b) *Notophthalmus* (*Triturus*) *viridescens*. In *E. coli*, a single gene has been isolated. During the process of transcription, each successive mRNA is engaged in translation by ribosomes. In *Notophthalmus*, several rRNA genes are present and many rRNA molecules are being transcribed along each gene. The diagrams illustrate the sequence of genetic events leading to what is visualized under the electron microscope.

to template binding have been found to be more complex in eukaryotes. Eukaryotic RNA polymerase, as mentioned earlier, exists in three forms (see Table 12.7), and each is larger and more complex than the prokaryotic counterpart of the enzyme. Each eukaryotic RNA polymerase consists of two large subunits and 10 to 15 smaller subunits. In regard to the initial template binding step and promoter regions, most is known about polymerase II, which transcribes all mRNAs in eukaryotes.

There are at least three *cis*-acting elements of a eukaryotic gene that function in the efficient initiation of transcription by polymerase II. The use of the term *cis* is drawn from organic chemistry nomenclature, meaning "next to, or on the same side as" in contrast to being "across from, or *trans* to" other functional groups. In molecular genetics, then, *cis*-elements are part of the same DNA molecule.

The first such element is called the **Goldberg–Hogness** or **TATA box**, which is found about 25 nucleotide pairs upstream (-25) from the start point of transcription. The consensus sequence is a heptanucleotide consisting solely of A and T residues. The sequence and function are analogous to that found in the -10 promoter region of prokaryotic genes. Because such a region is common to most eukaryotic genes, the TATA box is thought to be nonspecific and simply have the responsibility for fixing the site of initiation of transcription by facilitating the denaturation of the helix. Such a conclusion is supported by the fact that A = T base pairs are less stable than G ≡ C pairs.

The second *cis*-acting element of interest is found farther upstream from the TATA box. In different genes studied, positions anywhere from 5 to 500 nucleotides upstream appear to modulate transcription. These regions of DNA have been located on the basis of the effects of their deletion on transcription. Their loss appears to reduce *in vivo* transcription drastically. Some regions, such as those associated with globin and the viral SV40 genes, are about 50 to 100 nucleotides from the TATA sequence. Others, such as those associated with the sea urchin H2A gene and the *Drosophila* glue protein gene, are 200 to 500 nucleotides upstream. Because the sequence CCAAT is frequently part of these regions, they are sometimes called **CCAAT boxes**.

The third element is represented by DNA regions called **enhancers**. While their location may vary, enhancers often may be found even farther upstream than the regions discussed above, or even downstream or within the gene. They have the effect of modulating transcription from a distance. Although they may not participate directly in template binding, they are essential to highly efficient initiation of transcription. We will return to a discussion of these elements in Chapter 19.

Complementing the more extensive *cis*-acting regulatory sequences associated with eukaryotic genes are various **trans-acting factors** that serve in the role of facilitating template binding, and, therefore, the initiation of transcription. These are proteins referred to as **transcription factors**. They are essential because RNA polymerase II cannot bind directly to eukaryotic promoter sites and initiate transcription without their presence. The transcription factors involved with human RNA polymerase II-binding are well characterized and designated **TFIIA**, **TFIIB**, etc. Each may consist of protein subunits. Those that directly bind to the TATA-box sequence are sometimes called **TATA-binding proteins** (**TBPs**). For example, one such TBP is part of **TFIID**, which is responsible for initial binding to the TATA-box sequence. The TFIID consists of about 10 subunits. Once initial binding to DNA occurs, at least seven other transcription factors bind sequentially to TFIID, forming an extensive pre-initiation complex, which is then bound by RNA polymerase II.

Transcription factors with similar activity have been discovered in a variety of eukaryotes, including *Drosophila* and yeast. The nucleotide sequences in all organisms studied demonstrate a high degree of conservation. These factors appear to supplant the role of the sigma factor in the prokaryotic enzyme. We will return in Chapter 19 to a consideration of the role of other more specific transcription factors in eukaryotic gene regulation, as well as a discussion of the various DNA-binding domains that characterize some of these polypeptides.

Heterogeneous Nuclear RNA and Its Processing: Caps and Tails

Insights into other regions of DNA that do not directly encode proteins have come from the study of RNA. This research has provided detailed knowledge of eukaryotic gene structure. The genetic code is written in the ribonucleotide sequence of mRNA. This information originated, of course, in the template strand of DNA, where complementary sequences of deoxyribonucleotides exist. In bacteria, the relationship between DNA and RNA appears to be quite direct. The DNA base sequence is transcribed into an mRNA sequence, which is then translated into an amino acid sequence according to the genetic code.

However, in eukaryotes the situation is much more complex than in bacteria. It has been found that many internal base sequences of a gene may never appear in the mature mRNA that is translated. Other modifications occur at the beginning and the end of the mRNA prior to translation. These findings have made it clear that in eukaryotes, complex processing of mRNA occurs before it is transported to the cytoplasm to participate in translation.

By 1970, accumulating evidence showed that eukaryotic mRNA is transcribed initially as a much larger precursor molecule than that which is translated. This notion was based on the observation by James Darnell and co-workers of **heterogeneous nuclear RNA (hnRNA)** in mammalian cells. Heterogeneous RNA, complexed with an abundant variety of proteins (creating **hnRNPs—heterogeneous nuclear ribonucleoproteins**), is large but of variable size (up to 10^7 daltons). It is found only in the nucleus. Importantly, hnRNA was found to contain nucleotide sequences common to the smaller mRNA molecules present in the cytoplasm. Because of this observation, it was proposed that the initial transcript of a gene results in a large RNA molecule that must first be processed in the nucleus before it appears in the cytoplasm as a mature mRNA molecule. The various processing steps, to be discussed below, are summarized in Figure 12.10.

The initial **posttranscriptional modification** of eukaryotic RNA transcripts destined to become mRNAs involves the 5′-end of these molecules, where a **7-methylguanosine (7mG) cap** is added. The cap appears to be important to the subsequent processing within the nucleus, perhaps by protecting the 5′-end of the molecule from nuclease attack. Subsequently, this cap may be involved in transport of mature mRNAs across the nuclear membrane into the cytoplasm. The cap is fairly complex and distinguished by a unique 5′–5′ bonding between the cap and the initial ribonucleotide of the RNA. Some eukaryotes also contain a methyl group (—CH$_3$) on the 2′-carbon of the ribose sugars of the first two ribonucleotides of the RNA.

A subsequent discovery provided further insights into the processing of RNA transcripts during the maturation of mRNA. Both hnRNAs and mRNAs have been found to contain at their 3′-end a stretch of as many as 250 adenylic acid residues. Such **poly-A sequences** are added after the 5′-7mG cap has been added. The 3′-end of the initial transcript is first cleaved enzymatically at a point some 10 to 35 ribonucleotides from a highly conserved AAUAAA sequence. Then polyadenylation occurs by virtue of sequential addition of adenylic acid residues. Poly A has now been found at the 3′-end of almost all mRNAs studied in a variety of eukaryotic organisms. The exceptions seem to be the products of histone genes.

The importance of the AAUAAA sequence and the 3′-tail became apparent when mutations in the AAUAAA sequence were investigated. Cells bearing such mutations cannot add the poly-A sequence and, in the absence of this tail, the RNA transcripts are rapidly degraded. Therefore, the poly-A tail is critical if an RNA transcript is to be further processed and transported to the cytoplasm.

Intervening Sequences and Split Genes

One of the most exciting discoveries in the history of molecular genetics occurred in 1977 when Susan Berget, Philip Sharp, and Richard Roberts presented di-

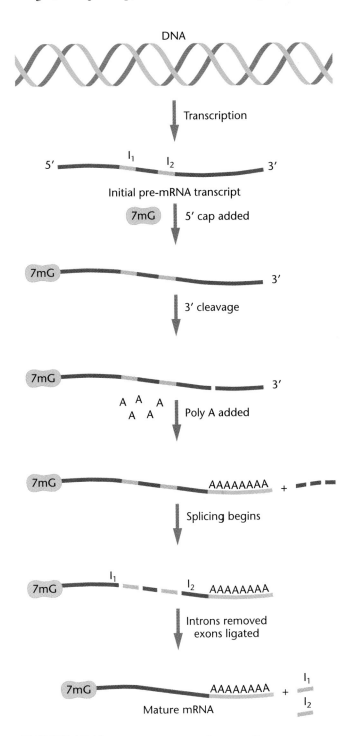

FIGURE 12.10 The conversion in eukaryotes of heterogeneous nuclear RNA (hnRNA) to messenger (mRNA), which contains a 5′-cap and a 3′-poly-A tail.

rect evidence that the genes of animal viruses contain *internal* nucleotide sequences that are not expressed in the amino acid sequence of the proteins they encode. This occurs because certain internal sequences present in DNA fail to appear in the mature mRNA that is translated into a protein.

Such nucleotide segments have been called **intervening sequences**, contained within **split genes**. Those DNA sequences that are not represented in the final mRNA product are also called **introns** ("int" for intervening), and those retained and expressed are called **exons** ("ex" for expressed). The removal of the sequences present in introns occurs as a result of an excision and rejoining process of RNA referred to as **splicing**. Intron sequences are present in initial RNA transcripts, but they are removed before the mature mRNA is translated.

Similar discoveries were soon made in a variety of eukaryotes. Two approaches have been most fruitful. The first involves molecular hybridization of purified, functionally mature mRNAs with DNA containing the genes specifying that message. When hybridization occurs be-

Mouse insulin

119
40   403

126   580
Rabbit β-globin
143   222   224

1550   246 576 398   860   370   1625
Chicken ovalbumin
188 53   132 118   142 155

■ Exons ■ Introns

FIGURE 12.12   Intervening sequences in various eukaryotic genes. The numbers indicate nucleotides present in various intron and exon regions.

tween nucleic acids that are not perfectly complementary, **heteroduplexes** are formed that can be visualized with the electron microscope. As illustrated in Figure 12.11, introns present in DNA but absent in mRNA must loop out and remain unpaired. Figure 12.11 shows an electron micrograph and an interpretive drawing derived from hybridization between the template strand of the gene encoding chicken ovalbumin and the mature RNA prior to its translation into that protein. There are seven introns (A–G) whose sequences are present in DNA but not in the final mRNA.

The second approach provides more specific information. It involves a comparison of nucleotide sequences of DNA with those of mRNA and the correlation with amino acid sequences. Such an approach allows the precise identification of all intervening sequences.

Thus far, a large number of genes from diverse eukaryotes have been shown to contain introns. One of the first so identified was the **beta-globin gene** in mice and rabbits, as studied independently by Philip Leder and Richard Flavell. The mouse gene contains an intron 550 nucleotides long, beginning immediately after the codon specifying the 104th amino acid. In the rabbit (Figure 12.12), there is an intron of 580 base pairs near the codon for the 110th amino acid. Additionally, a second intron of about 120 nucleotides exists earlier in both genes. Similar introns have been found in the beta-globin gene in all mammals examined.

As pointed out above, an extensive number of introns are present in the **ovalbumin gene** of chickens. The gene has been extensively characterized by Bert O'Malley in the United States and Pierre Chambon in France. As shown in Figure 12.12, the gene contains seven introns. Notice that the majority of the gene's DNA sequence is "silent," being composed of introns. The initial RNA

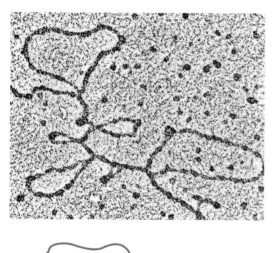

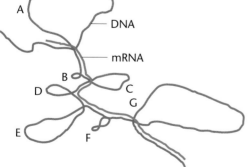

FIGURE 12.11   An electron micrograph and an interpretive drawing of the hybrid molecule formed between the template DNA strand of the ovalbumin gene and the mature ovalbumin mRNA. Seven DNA introns, A–G, produce unpaired loops.

transcript is four times the length of the mature mRNA. You should compare the information on the ovalbumin gene presented in Figures 12.11 and 12.12. Can you match the unpaired loops in Figure 12.11 with the sequence of introns specified in Figure 12.12?

The list of genes containing intervening sequences is growing rapidly. In fact, few eukaryote genes seem to be without introns. An extreme example of the number of introns in a single gene is that found in one of the chicken genes, *pro-α-2(1) collagen*. One of several genes coding for a subunit of this connective tissue protein, *pro-α-2(1) collagen* contains about 50 introns. The precision with which cutting and splicing occur must be extraordinary if errors are not to be introduced into the mature mRNA.

While the vast majority of eukaryotic genes thus far examined contain introns, there are several exceptions. Notably, those coding for histones and for interferon appear to contain no introns. It is not clear why or how the genes encoding these molecules have been maintained throughout evolution without acquiring the extraneous information characteristic of most all other genes.

Splicing Mechanisms: Autocatalytic RNAs

The discovery of split genes represents one of the most exciting genetic findings in molecular genetics. As a result, intensive investigation has been directed toward the elucidation of the mechanism by which introns of RNA are excised and exons are spliced back together. A great deal of progress has already been made. Interestingly, it appears that somewhat different mechanisms exist for different types of RNA as well as for RNAs produced in mitochondria and chloroplasts.

Based on splicing mechanisms, there are several groups of introns. One group (type I) includes those that are part of the primary transcript of rRNAs derived from the ciliate protozoan *Tetrahymena*. Contrary to what was expected, no additional components are required for intron excision; the intron itself is the source of the enzymatic activity necessary for its own removal. This **self-excision process** is illustrated in Figure 12.13. This amazing discovery, made in 1982 by Thomas Cech and his colleagues, revealed that RNA can demonstrate autonomous catalytic properties. As a result, the RNAs that are capable of splicing themselves are sometimes called **ribozymes**.

Chemically, two nucleophilic reactions (called transesterifation reactions) occur, the first involving an interaction between guanosine and the primary transcript [Figure 12.13(b)]. The 3'-OH group of guanosine is transferred to the nucleotide adjacent to the 5'-end of the intron. The second reaction involves the interaction

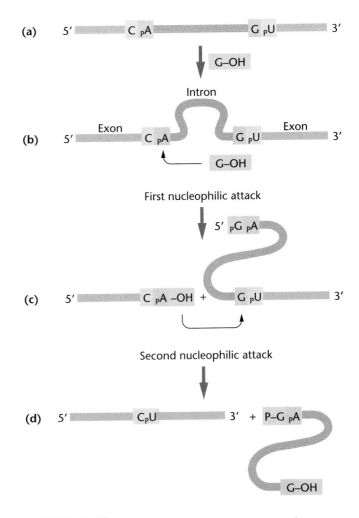

FIGURE 12.13 Splicing mechanism involved with group I introns removed from the primary transcript leading to rRNA. The process is one of self-excision involving two transesterification reactions.

of the newly acquired —OH group on the left hand exon and the nucleotide on the 3'-end of the intron [Figure 12.13(c)]. The intron is spliced out and the two exon regions are ligated, leading to the spliced RNA [Figure 12.13(d)].

Self-exision also seems to govern the removal of group II introns characteristic of the primary transcripts produced in mitochondria and chloroplasts. In these cases, the introns fold into secondary structures where regions of RNA sequences form base pairs with one another leading to the creation of stems and loops. Like group I molecules, two autocatalytic reactions then occur, leading to the excision of introns. There is some speculation that the RNA sequences contained in the stems and loops are the evolutionary forerunners of **small nuclear RNA molecules (snRNAs)** involved in splicing eukaryotic pre-mRNAs (see below).

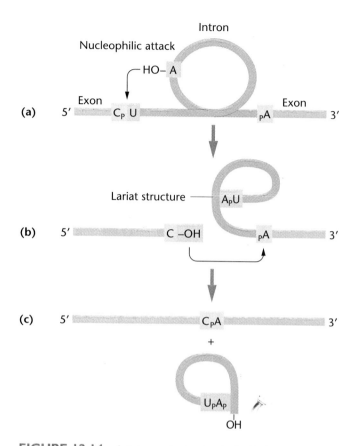

FIGURE 12.14 Splicing mechanism involved with introns removed from pre-mRNAs. Excision is dependent on the structure referred to as the *spliceosome*. The lariat structure in the intermediate stage is characteristic of this mechanism.

Splicing Mechanisms: The Spliceosome

Another major group of introns is found in nuclear-derived transcripts representing mRNAs. Introns in nuclear-derived mRNA, in comparison to other RNAs thus far discussed, can be much larger—up to 20,000 nucleotides—and they are more plentiful. Their removal appears to require a much more complex mechanism, which has been more difficult to define.

Many clues are now emerging. First, the nucleotide sequence around different introns is often similar. Most begin at the 5′-end with a GU dinucleotide sequence and terminate at the 3′-end with an AG dinucleotide sequence. These, as well as other consensus sequences shared by introns, may attract a molecular complex essential to ligation and splicing. Such a complex has been identified in extracts of yeast as well as mammalian cells. Called a **spliceosome**, it is very large, being $40\,S$ in yeast and $60\,S$ in mammals. One group of components of these complexes consists of a unique set

of **small nuclear RNAs (snRNAs)**, which we just mentioned above in conjunction with group II introns. These RNAs are usually 100 to 250 nucleotides or less and are often complexed with proteins. Found only in the nucleus, these complexes are called **small nuclear ribonucleoproteins (snRNPs—or snurps)**. Because they are rich in uridine residues, the snRNAs have been arbitrarily designated U1, U2,..., U6, and the list continues to grow.

The U1 snRNA bears a nucleotide sequence that is homologous to that of the 5′-end of the intron. Base pairing resulting from this homology promotes binding that represents the initial step in the formation of the spliceosome. Following the addition of the other snurps (U2, U4, U5, U6), splicing commences. Two transesterification reactions are involved. The first involves a 2′-OH group from an adenine (A) residue present within the intron [Figure 12.14(a)]. The position of the A residue represents what is called the **branch point** of the spliceosome structure. The second involves the —OH group, which has been added to the 3′-end of the exon [depicted on the left in Figure 12.14(b)]. The intermediate forms a characteristic lariat-like structure [Figure 12.14(b)] that is then excised [Figure 12.14(c)]. As ligation of the exon regions occurs, splicing is accomplished.

The finding of "genes in pieces," as split genes have been described, raises many interesting questions and has provided great insights into the organization of eukaryotic genes. In addition, the processing involved in splicing represents a potential regulatory step during gene expression. For example, several cases are known where introns present in pre-mRNAs *derived from the same gene* are spliced *in more than one way*, thereby yielding different collections of exons in the mature mRNA. This process, referred to as **alternative splicing**, yields a group of mRNAs, that upon translation, result in a series of related proteins called **isoforms**. A growing number of examples are now found in organisms ranging from viruses to *Drosophila* to humans.

Alternative splicing of pre-mRNAs provides the basis for producing related proteins from a single gene. We shall return to this topic in our discussion of the regulation of gene expression in eukaryotes (Chapter 19).

RNA Editing

In the late 1980s, still another quite unexpected form of posttranscriptional RNA processing was discovered. In this case, the nucleotide sequence of a pre-mRNA is actually changed prior to translation. This form of processing is referred to as **RNA editing**. As a result of the process, the ribonucleotide sequence of the mature RNA differs from the sequence encoded in the exons of the DNA from which the RNA was transcribed.

There are two main types of RNA editing: **substitution editing**, in which the identities of individual nucleotide bases are altered; and **insertion/deletion editing**, where nucleotides are added to or subtracted from the total number of bases. Substitution editing is used in some nuclear-derived eukaryotic RNAs and is highly prevalent in mitochondrial and chloroplast RNAs transcribed in plants. *Trypanosoma*, a parasite which causes African sleeping sickness, and its relatives use extensive insertion/deletion editing in mitochondrial RNAs. The number of uridines added to an individual transcript can make up more than 60 percent of the coding sequence, usually forming the initiation codon and placing the rest of the sequence into the proper reading frame. *Physarum polycephalum*, a slime mold, uses both substitution and insertion/deletion editing for its mitochondrial mRNAs.

Insertion/deletion editing in trypanosomes is directed by "**gRNA**" (**guide RNA**) templates which are also transcribed from the mitochondrial genome. These small RNAs are complementary to the edited region of the final, edited mRNAs. They base-pair with the pre-edited mRNAs to direct the editing machinery to make the correct changes.

Substitutional editing changes in RNAs are difficult to envision because they occur without any form of template information characteristic of all other forms of information flow in the cell. The mechanism by which such specific nucleotides are targetted and very precise modifications are achieved remains a mystery. Further, it is not yet clear just how extensive the phenomenon will prove to be. The finding that a single C to U RNA editing step is an important part of the developmental regulation of the apolipoprotein of mammals has undoubtedly spurred interest. A second system, the synthesis of subunits constituting the glutamate receptor channels (GluR) in mammalian brain tissue, is also affected by RNA editing. In this case, adenosine (A) to inosine (I) editing occurs in pre-mRNAs prior to their translation, where I is read as guanosine (G) during translation. A specific enzyme, **double-stranded RNA adenosine deaminase**, as been implicated in the editing. These changes alter the physiological parameters of the receptor containing "edited" subunits.

Findings such as these in mammals have established that RNA editing provides still another important mechanism of posttranscriptional modification, and that this process is not restricted to evolutionarily less advanced genetic systems such as mitochondria. It thus seems likely that RNA editing will be found to be more widespread as studies progress. Further, the process may have important implications in regulation of genetic expression.

Translation: Components Necessary for Protein Synthesis

Translation of mRNA is the biological polymerization of amino acids into polypeptide chains. The process, alluded to in our earlier discussion of the genetic code, occurs only in association with ribosomes. The central question in translation is how triplet ribonucleotides of mRNA direct specific amino acids into their correct position in the polypeptide. This question was answered once transfer RNA (tRNA) was discovered. This class of molecules adapts specific triplet codons in mRNA to their correct amino acids. This adaptor role of tRNA was postulated by Francis Crick in 1957.

In association with a ribosome, mRNA presents a triplet codon that calls for a specific amino acid. Because a specific tRNA molecule contains within its composition three consecutive ribonucleotides complementary to the **codon**, they are called the **anticodon** and can base-pair with the codon. Another region of this tRNA is covalently bonded to the amino acid called for. Inside the ribosome, hydrogen bonding of tRNAs to mRNA holds the amino acid in proximity so that a peptide bond can be formed. As this process occurs over and over as mRNA runs through the ribosome, amino acids are polymerized into a polypeptide.

In our discussion of translation, we will first consider the structure of the ribosome and transfer RNA, two of the major components essential for protein synthesis.

Ribosomal Structure

Because of its essential role in the expression of genetic information, the **ribosome** has been extensively analyzed. One bacterial cell contains about 10,000 of these structures, whereas a eukaryotic cell contains many times more. Electron microscopy has revealed that the bacterial ribosome is about 250 Å in its largest diameter and consists of two subunits, one large and one small. Both subunits consist of one or more molecules of rRNA and an array of **ribosomal proteins**.

The specific differences between prokaryotic and eukaryotic ribosomes are summarized in Figure 12.15. The subunit and rRNA components are most easily isolated and characterized on the basis of their sedimentation behavior (their rate of migration) in sucrose gradients (see Chapter 10). When the two subunits are associated with each other in a single ribosome, the structure is sometimes called a **monosome**. In prokaryotes the monosome is a 70S particle, and in eukaryotes it is approximately 80S. Sedimentation coefficients, which reflect the variable rate of migration of different-sized particles and molecules, are not additive. For example, the 70S monosome consists of a 50S and a 30S subunit, and the 80S monosome consists of a 60S and a 40S subunit.

The larger subunit in prokaryotes consists of a 23S RNA molecule, a 5S rRNA molecule, and 32 ribosomal proteins. In the eukaryotic equivalent, a 28S rRNA molecule is accompanied by a 5.8S and 5S rRNA molecule and about 50 proteins. The smaller prokaryotic subunits consist of a 16S rRNA component and 21 proteins. In the eukaryotic equivalent, an 18S rRNA component and about 33 proteins are found. The approximate molecular weights and number of nucleotides of these components are shown in Figure 12.15.

Molecular hybridization studies have established the degree of redundancy of the genes coding for the rRNA components. The *E. coli* genome contains seven copies of a single sequence that codes for all three components—23S, 16S, and 5S. The initial transcript of these genes produces a 30S RNA molecule that is enzymatically cleaved into these smaller components. Having the genetic information encoding these three rRNA components coupled together ensures that, following multiple transcription events, equal quantities of all three will be present as ribosomes are assembled.

In eukaryotes, many more copies of a sequence encoding the 28S and 18S components are present. In *Drosophila*, approximately 120 copies per haploid genome are each transcribed into a molecule of about 34S. This is processed to the 28S, 18S, and 5.8S rRNA species. These are homologous to the three rRNA components of *E. coli*. In *X. laevis*, over 500 copies per haploid genome are present. In mammalian cells, the initial transcript is 45S. The rRNA genes are part of the moderately repetitive DNA fraction and are present in clusters at various chromosomal sites. Each cluster consists of **tandem repeats**, with each unit separated by a noncoding **spacer DNA** sequence [see the micrograph in Figure 12.9(b)]. In humans, these gene clusters have been localized on the ends of chromosomes 13, 14, 15, 21, and 22.

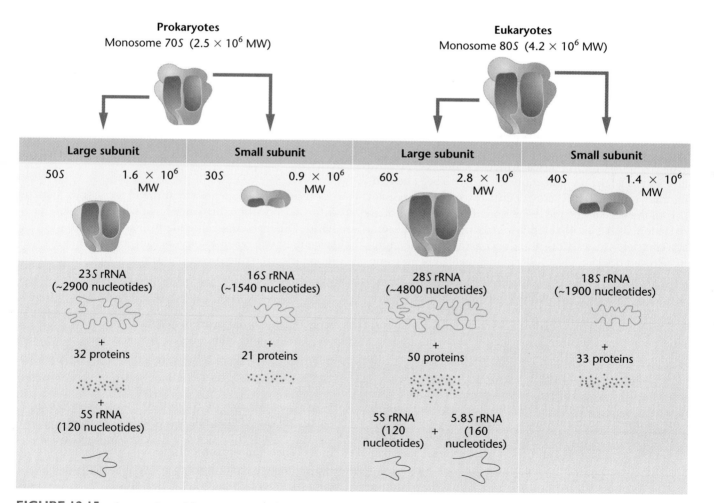

FIGURE 12.15   A comparison of the components in the prokaryotic and eukaryotic ribosome.

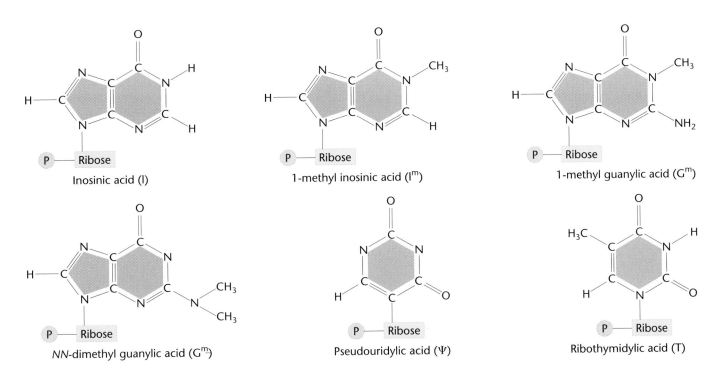

FIGURE 12.16 Unusual nitrogenous bases found in transfer RNA.

The unique 5S rRNA component of eukaryotes is not part of the larger transcript. Instead, genes coding for this ribosomal component are distinct and located separately. In humans, a gene cluster encoding them has been located on chromosome 1.

Despite the detailed knowledge available on the structure and genetic origin of the ribosomal components, a complete understanding of the function of these components has eluded geneticists. This is not surprising; the ribosome is perhaps the most intricate of all cellular organelles. In bacteria, the monosome has a combined molecular weight of 2.5 million daltons!

tRNA Structure

Because of their small size and stability in the cell, transfer RNAs (tRNAs) have been investigated extensively. In fact, tRNAs are the best characterized RNA molecules. They are composed of only 75 to 90 nucleotides, displaying a nearly identical structure in bacteria and eukaryotes. In both types of organisms, tRNAs are transcribed as larger precursors, which are cleaved into mature 4S tRNA molecules. In *E. coli*, for example, tRNA^{tyr} (the superscript identifies the specific tRNA and the amino acid that binds to it) is composed of 77 nucleotides, yet its precursor contains 126 nucleotides.

In 1965, Robert Holley and his colleagues reported the complete sequence of tRNA^{ala} isolated from yeast. Of great interest was the finding that a number of nucleotides are unique to tRNA. As shown in Figure 12.16, each is a modification of one of the four nitrogenous bases expected in RNA (G, C, A, and U). These include **inosinic acid**, which contains the purine **hypoxanthine**, **ribothymidylic acid**, and **pseudouridine** among others. These modified structures, sometimes referred to as *unusual*, *rare*, or *odd bases*, are created following transcription, illustrating the more general concept of **posttranscriptional modification**. In this case, the unmodified base is inserted during transcription, and subsequently, enzymatic reactions catalyze the chemical modifications to the base.

Holley's sequence analysis led him to propose the two-dimensional **cloverleaf model of tRNA**. It had been known that tRNA demonstrates secondary structure due to base pairing. Holley discovered that he could arrange the linear model in such a way that several stretches of base pairing would result. This arrangement created a series of paired stems and unpaired loops resembling the shape of a cloverleaf. Loops consistently contained modified bases that do not generally form base pairs. Holley's model is shown in Figure 12.17.

Because the triplets GCU, GCC, and GCA specify alanine, Holley looked for an anticodon sequence complementary to one of these codons in his tRNA^{ala} molecule. He found it in the form of CGI, in one of the loops of the cloverleaf. Recall from Crick's wobble hypothesis that I (inosinic acid) is predicted to pair with U, C, or A. Thus, the **anticodon loop** was established.

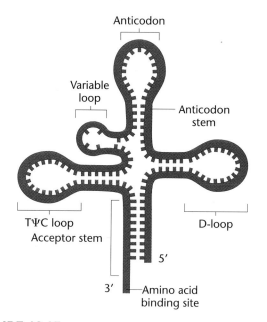

FIGURE 12.17 The two-dimensional cloverleaf model of transfer RNA.

Charging tRNA

Before translation can proceed, the tRNA molecules must be chemically linked to their respective amino acids. This activation process, called **charging**, occurs under the direction of enzymes called **aminoacyl tRNA synthetases**. Because there are 20 different amino acids, there must be at least 20 different tRNA molecules and as many different enzymes. In theory, because there are 61 triplet codes, there could be the same number of specific tRNAs and enzymes. However, because of the ability of the third member of a triplet code to "wobble," it is now thought that there are at least 32 different tRNAs; it is also believed that there are only 20 synthetases, one for each amino acid, regardless of a greater number of corresponding tRNAs.

The charging process is outlined in Figure 12.19. In the initial step, the amino acid is converted to an activated form, reacting with ATP to form an **aminoacyladenylic acid**. A covalent linkage is formed between the 5'-phosphate group of ATP and the carboxyl end of the amino acid. This molecule remains associated with the enzyme, forming a complex that then reacts with a specific tRNA molecule. In this second step, the amino acid

As other tRNA species were examined, numerous constant features were observed. First, at the 3'-end, all tRNAs contain the sequence ...**pCpCpA-3'**. It is to the terminal adenosine residue that the amino acid is joined covalently during charging. At the 5'-terminus, all tRNAs contain ...**pG-5'**.

Additionally, the lengths of various stems and loops are very similar. Each tRNA examined also contains an anticodon complementary to the known amino acid code for which it is specific. All anticodon loops are present in the same position of the cloverleaf.

Because the cloverleaf model was predicted strictly on the basis of nucleotide sequence, there was great interest in X-ray crystallographic examination of tRNA, which reveals three-dimensional structure. By 1974, Alexander Rich and his colleagues in the United States, and J. Roberts, B. Clark, Aaron Klug, and their colleagues in England, had been successful in crystallizing tRNA and performing X-ray crystallography at a resolution of 3Å. At such resolution, the pattern formed by individual nucleotides is discernible.

As a result of these studies, a complete three-dimensional model of tRNA is now available (Figure 12.18). At one end of the molecule is the anticodon loop and stem, and at the other end is the 3'-acceptor region where the amino acid is bound. It has been speculated that the shapes of the intervening loops may be recognized by specific enzymes responsible for adding the amino acid to tRNA, a subject to which we now turn our attention.

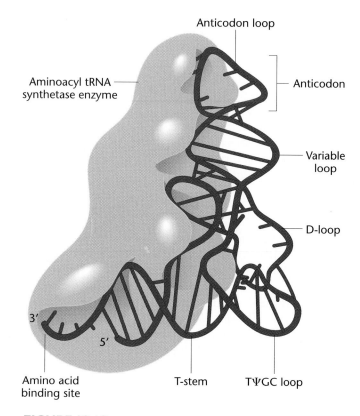

FIGURE 12.18 A three-dimensional model of transfer RNA as it associates with its corresponding aminoacyl tRNA synthetase, which catalyzes the addition of the appropriate amino acid.

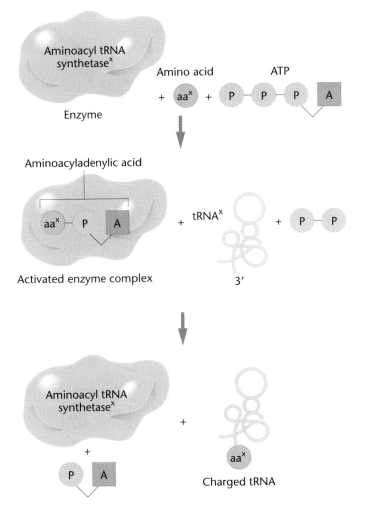

Enzyme

Amino acid   ATP

Aminoacyladenylic acid

Activated enzyme complex

Aminoacyl tRNA
synthetase^x

Charged tRNA

FIGURE 12.19   Steps involved in charging tRNA. The X denotes that for each amino acid, only the corresponding specific tRNA and specific aminoacyl tRNA synthetase enzyme are involved in the charging process.

is transferred to the appropriate tRNA and bonded covalently to the adenine residue at the 3′-end. The charged tRNA may participate directly in protein synthesis. Aminoacyl tRNA synthetases are highly specific enzymes because they recognize only one amino acid and only a subset of corresponding tRNAs called **isoaccepting tRNAs**. This is a crucial point if fidelity of translation is to be maintained. The basis for this recognition and subsequent binding has sometimes been referred to as the **second genetic code**, although this is not a particularly apt description.

Translation: The General Process

In a way similar to transcription, the process of translation can be best described by breaking it into discrete steps. Be aware, however, that translation is also a dynamic, ongoing process. Correlate the following discussion with the step-by-step characterization of the process

in Figure 12.20. Many of the protein factors and their roles in translation are summarized in Table 12.8.

Initiation (Steps 1–3)

Recall that the ribosome serves as a nonspecific workbench for the translation process. Most ribosomes, when they are not involved in translation, are dissociated into their large and small subunits. Initiation of translation in *E. coli* involves the small ribosome subunit, an mRNA molecule, a specific charged initiator tRNA, GTP, Mg^{++}, and at least three proteinaceous **initiation factors (IFs)**. The initiation factors are not part of the ribosome, but they are required to enhance the binding affinity of the various translational components, as described in Table 12.8. In prokaryotes, the initiation codon of mRNA—AUG—calls for the modified amino acid **formylmethionine**.

The small ribosomal subunit binds to several initiation factors, and this complex in turn binds to mRNA. In

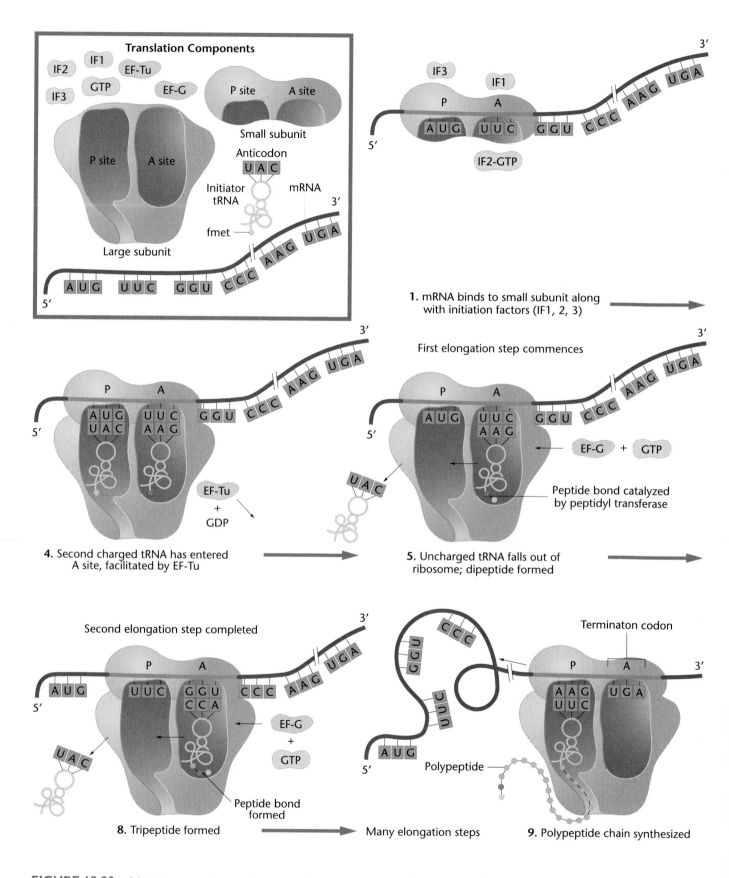

FIGURE 12.20 Schematic representation of the process of translation, depicting the steps involved in protein synthesis.

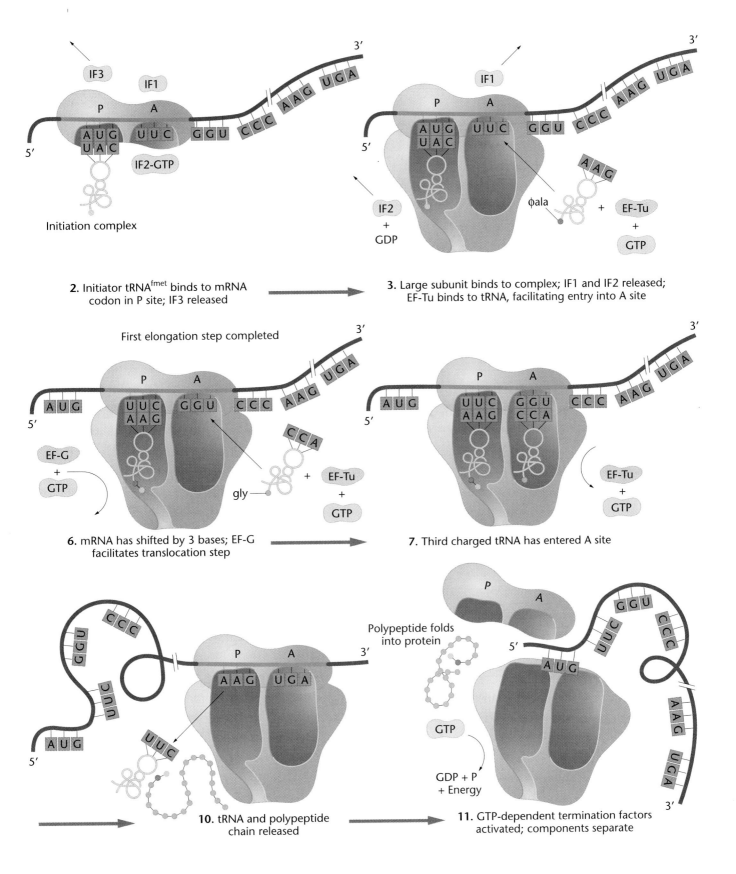

2. Initiator tRNA^fmet binds to mRNA codon in P site; IF3 released

3. Large subunit binds to complex; IF1 and IF2 released; EF-Tu binds to tRNA, facilitating entry into A site

First elongation step completed

6. mRNA has shifted by 3 bases; EF-G facilitates translocation step

7. Third charged tRNA has entered A site

Polypeptide folds into protein

10. tRNA and polypeptide chain released

11. GTP-dependent termination factors activated; components separate

TABLE 12.8 Various Protein Factors Involved During Translation in *E. coli*

Process	Factor	Role
Initiation of translation	IF1	Stabilizes 30*S* subunit
	IF2	Binds fmet-tRNA to 30*S*-mRNA complex; binds to GTP and stimulates hydrolysis
	IF3	Binds 30*S* subunit to mRNA; dissociates monosomes into subunits following termination
Elongation of polypeptide	EF-Tu	Binds GTP; brings aminoacyl-tRNA to the A site of ribosome
	EF-Ts	Generates active EF-Tu
	EF-G	Stimulates translocation; GTP-dependent
Termination of translation and release of polypeptide	RF1	Catalyzes release of the polypeptide chain from tRNA and dissociation of the translocation complex; specific for UAA and UAG termination codons
	RF2	Behaves like RF1; specific for UGA and UAA codons
	RF3	Stimulates RF1 and RF2

bacteria, this binding involves a sequence of up to six ribonucleotides (AGGAGG), which precedes the initial AUG codon of mRNA. This sequence (containing only purines and called the **Shine-Dalgarno sequence**) base-pairs with a region of the 16*S* rRNA of the small ribosomal subunit.

Another initiation protein then facilitates the binding of charged formylmethionyl-tRNA to the small subunit in response to the AUG triplet. This step "sets" the reading frame so that all subsequent groups of three ribonucleotides are translated accurately. This aggregate represents the **initiation complex**, which then combines with the large ribosomal subunit. In this process, a molecule of GTP is hydrolyzed, providing the required energy, and the initiation factors are released.

Elongation (Steps 4–9)

Once both subunits of the ribosome are assembled with the mRNA, binding sites for two charged tRNA molecules are formed. These are designated as the **P**, or **peptidyl**, and the **A**, or **aminoacyl**, **sites**. The initiation tRNA binds to the P site, provided that the AUG triplet of mRNA is in the corresponding position of the small subunit. The sequence of the second triplet in mRNA dictates which charged tRNA molecule will become positioned at the A site. Once it is present, **peptidyl transferase** catalyzes the formation of the peptide bond, which links the two amino acids together. This enzyme is part of the large subunit of the ribosome. At the same time, the covalent bond between the amino acid and the tRNA occupying the P site is hydrolyzed (broken). The product of this reaction is a dipeptide, which is attached to the 3′-end of tRNA at the A site. The step in which the growing polypeptide chain has increased in length by one amino acid is called **elongation**.

Before elongation can be repeated, the tRNA attached to the P site, which is now uncharged, must be released from the large subunit. The entire **mRNA–tRNA–aa₂–aa₁** complex then shifts in the direction of the P site by a distance of three nucleotides. This event requires several protein **elongation factors (EFs)** as well as the energy derived from hydrolysis of GTP (Table 12.8). The result is that the third triplet of mRNA is now in a position to accept another specific charged tRNA into the A site. One simple way to distinguish the two sites in your mind is to remember that, *following the shift,* the P site contains a tRNA attached to a peptide chain (*P* for peptide), whereas the A site contains a tRNA with an amino acid attached (*A* for amino acid).

The sequence of elongation is repeated over and over. An additional amino acid is added to the growing polypeptide chain each time the mRNA advances through the ribosome. The efficiency of the process is remarkably high; the observed error rate is only about 10^{-4}. An incorrect amino acid will occur only once in every 20 polypeptides of an average length of 500 amino acids! In *E. coli*, elongation occurs at a rate of about 15 amino acids per second at 37°C. The process can be likened to a tape moving through a cassette recorder. As the tape moves, sound is emitted sequentially from the recorder. Likewise, as mRNA moves, a growing polypeptide is produced by the ribosome.

Termination (Steps 10–11)

The termination of protein synthesis is signaled by one or more of three triplet codes: UAG, UAA, or UGA. These codons do not specify an amino acid, nor do they call for a tRNA in the A site. These codons are called **stop codons, termination codons**, or **nonsense codons**. The finished polypeptide is therefore still attached to the terminal tRNA at the P site, and the A site is empty. The termination codon signals the action of **GTP-dependent release factors** (Table 12.8), which cleave the polypeptide chain from the terminal tRNA, releasing it from the translation complex. Once this cleavage occurs, the tRNA is released from the ribosome, which then dissociates into its subunits. If a termination codon should appear in the middle of an mRNA molecule as a result of mutation, the same process occurs, and the polypeptide chain is prematurely terminated.

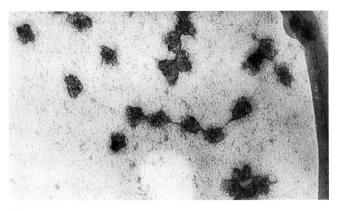

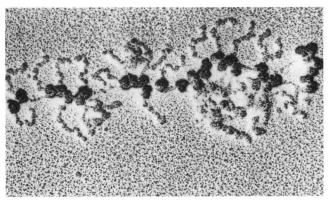

(a)   (b)

FIGURE 12.21   Polyribosomes visualized under the electron microscope. Those viewed by electron microscopy were derived in part (a) from rabbit reticulocytes engaged in the translation of hemoglobin mRNA and in part (b) from giant salivary gland cells of the midgefly, *Chironomus thummi*. In part (b), the nascent polypeptide chain is apparent as it emerges from each ribosome. Its length increases as translation proceeds from left (5′) to right (3′) along the mRNA.

Polyribosomes

As elongation proceeds and the initial portion of mRNA has passed through the ribosome, the message is free to associate with another small subunit to form another initiation complex. This process can be repeated several times with a single mRNA and results in what are called **polyribosomes** or just **polysomes**.

Polyribosomes can be isolated and analyzed following a gentle lysis of cells. Figure 12.21(a) and (b) illustrates these complexes as seen under the electron microscope; note the presence in part (a) of this figure of mRNA between the individual ribosomes. The micrograph in part (b) is even more remarkable, for it shows the polypeptide chains emerging from the ribosomes during translation. The formation of polysome complexes represents an efficient use of the components available for protein synthesis during a unit of time.

To complete the analogy with tapes (mRNA) and cassette recorders (ribosomes), in polysome complexes one tape would be played simultaneously by several recorders, but at any given moment, the recording (the polypeptide) would be at different stages of completion.

Translation in Eukaryotes

The general features of the model just presented were initially derived from investigations of the translation process in bacteria. During that discussion, we pointed out one of the main differences between translation in prokaryotes and eukaryotes. In the latter group, translation occurs on ribosomes that are larger and whose rRNA and protein components are more complex than prokaryotes (see Figure 12.15).

In addition, several other notable differences need to be mentioned. Eukaryotic mRNAs are much longer lived than are their prokaryotic counterparts. Most exist for hours rather than minutes prior to their degradation by nucleases in the cell, remaining available much longer to orchestrate protein synthesis.

Two aspects that involve the initiation of translation are different in eukaryotes. First, as we discussed during our consideration of mRNA maturation, the 5′-end is "capped" with a 7-methylguanosine residue. The presence of this "cap," absent in prokaryotes, is essential to efficient translation. RNAs lacking the cap are translated poorly. The cap seems to play a similar role to the Shine-Dalgarno sequence of prokaryotic mRNA in the initial binding of mRNA to the small subunit of the ribosome.

The second aspect related to initiation of translation involves the insertion of the first amino acid. Initiation of eukaryotic translation does not require the amino acid formylmethionine. However, as in prokaryotes, the AUG triplet, which encodes methionine, is essential to the formation of the translational complex, and a unique transfer RNA ($tRNA_i^{met}$) is used during initiation.

Finally, protein factors similar to those in prokaryotes guide initiation, elongation, and termination of translation in eukaryotes. Many of these eukaryotic factors are clearly homologous to their counterparts in prokaryotes. However, there may be a greater number of factors required during each of these steps, and they may be somewhat more complex in prokaryotes.

GENETICS, TECHNOLOGY, AND SOCIETY

Genetic Testing Dilemmas: Sickle-Cell Anemia and Breast Cancer

As the Human Genome Project nears its halfway point, one thing has become abundantly clear: the identification of genes responsible for human diseases will precede new therapies to alleviate the diseases by many years. What will become possible in the interim is the development of genetic tests, both to diagnose and to predict the future occurrence of an inherited disease. And the prediction of future genetic illness creates especially difficult issues that we may not be prepared to deal with.

Genetic testing is not new. In the early 1970s, a major effort was made to identify the roughly one in twelve African Americans in the U.S. who carried one recessive allele for **sickle-cell anemia** (and have the so-called *sickle-cell trait*). This involved examining the blood for the presence of sickled red blood cells, which are present in sickle-cell heterozygotes, although they generally do not show symptoms of sickle-cell disease, which develops only in homozygotes. The goal of this screening program was to advise all sickle-cell heterozygotes of their "carrier" status, information that presumably could be used in family planning. This, it was hoped, would lead to the gradual reduction in the incidence of sickle-cell disease.

The sickle-cell screening program had several unforeseen consequences, however. For one thing, those tested were not adequately counseled, leading to confusion between the usually harmless sickle-cell trait and the severe, often life-threatening sickle-cell disease. In addition, some people designated as sickle-cell carriers, although completely asymptomatic, suffered unwarranted job discrimination and difficulty in obtaining insurance, as well as social stigmatization. Finally, the implementation of this program by the predominantly white medical establishment raised the fear that the hidden goal of sickle-cell screening was to reduce the reproduction of African Americans. For these reasons, by the late 1970s most sickle-cell screening programs had been halted.

Several important lessons were learned from the abortive sickle-cell screening program. First, the test must have direct benefit to the person tested, for example by allowing a therapeutic treatment. This was not the case in sickle-cell screening. Second, the results of all genetic tests must be held in strictest confidence and not be divulged without permission to employers, insurance providers, governmental agencies, and so on. This will help prevent the kinds of discrimination that resulted from sickle-cell testing. Third, genetic testing must be accompanied by counseling about the nature of the test, the implications of a positive result, and the likelihood of the occurrence of the disease in the tested person's family members.

Will we have the foresight to apply these lessons to future genetic screening programs as more DNA-based tests become available as an outcome of the Human Genome Project? We won't have to wait long to find out. Genetic tests for mutations that result in a high risk of **breast cancer** are already available and are beginning to be used to screen selected populations.

Breast cancer is the most common form of cancer in women, striking one of every nine females. Roughly 200,000 new cases are diagnosed in this country every year. It had been known for several years that 5 to 10 percent of cases of breast cancer are hereditary, striking women generation after generation in some unfortunate families. This strongly suggested that mutations in specific genes are responsible for the development of at least some breast cancers. In 1990 this supposition was borne out by the work of Mary-Claire King at the University of California, who mapped a breast cancer susceptibility gene to the long arm chromosome 17. And in September of 1994, a group headed by Mark Skolnick at the University of Utah succeeded in isolating the first gene responsible for inherited breast cancer, which was named *BRCA1*.

It was soon demonstrated that mutations in the *BRCA1* gene are responsible for about half of all cases of inherited *b*reast *c*ancer, and later discovered that mutations in this gene are also associated with a significantly greater risk of ovarian cancer. It was hoped that by studying the protein product of this gene, new insights could be gained on the origins of these common cancers, and eventually treatments could be developed to prevent them. In the meantime, indentification of the gene has made it possible to test women in cancer-prone families to see whether they harbor mutations associated with a high risk of developing breast and ovarian cancer. Women with a positive test result could chose the option of bilateral mastectomy to vastly reduce their chance of developing cancer.

Contrary to expectations, however, about half the women in these families have declined to take the test. It is not simply a desire not to know their *BRCA1* status that is deterring these women, but justifiable fears that a positive test would lead to discrimination in employment, an increase in insurance costs, or denial of insurance coverage altogether. Some women carrying *BRCA1* mutations have gone to great lengths to conceal this fact from their insurance companies, while others have decided to have the test only using a false name.

Before high numbers of women in cancer-prone families choose to be tested, allowing them to take potentially life-saving measures if they have the mutation, they must be protected from the discrimination that may result from a positive test. As with all future screening programs, the results of genetic tests must be strictly confidential. Only with permission should they be divulged to employers, insurance providers, educational institutions, or governmental agencies.

Fortunately, steps are now being taken to ensure the confidentiality of genetic information and protect those tested from discrimination. Several states have already enacted genetic privacy laws, and the Genetic Privacy Act has been introduced in Congress to extend these protections nationwide, although its passage is not certain. This act is based on the recognition that genetic information is different from other types of personal information and requires special protection to prevent discrimination based on genotype.

Will these legal protections have the desired effect, or will fears of discrimination continue to keep those at risk from choosing to be tested? By the completion of the Human Genome Project in 2005, there will likely be many genetic tests available for a wide range of diseases and susceptibilities to diseases. These tests may have the potential to ease the suffering of countless people with inherited diseases, but only if they feel it in their best interests to take the tests.

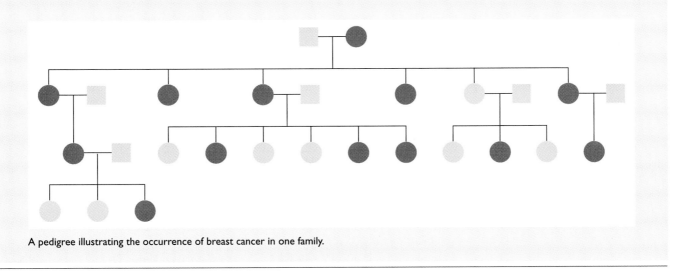

A pedigree illustrating the occurrence of breast cancer in one family.

CHAPTER SUMMARY

1. The genetic code, stored in DNA, is copied to RNA, where it is used to direct the synthesis of polypeptide chains. It is degenerate, unambiguous, nonoverlapping, and commaless.

2. The complete coding dictionary, determined using various experimental approaches, reveals that of the 64 possible codons, 61 encode the 20 amino acids found in proteins, while three triplets terminate translation. One of these 61 is the initiation codon and specifies methionine.

3. The observed pattern of degeneracy often involves only the third letter of a triplet series and led Francis Crick to propose the "wobble hypothesis."

4. Confirmation for the coding dictionary, including codons for initiation and termination, was obtained by comparing the complete nucleotide sequence of phage MS2 with the amino acid sequence of the corresponding proteins. Other findings support the be-

lief that, with only minor exceptions, the code is universal for all organisms.

5. In some bacteriophages, multiple initiation points may occur during the transcription of RNA, resulting in multiple reading frames and overlapping genes.

6. Transcription and translation—RNA and protein biosynthesis, respectively—are the fundamental processes essential to the expression of genetic information.

7. Transcription describes the synthesis, under the direction of RNA polymerase, of a strand of RNA complementary to a DNA template.

8. Translation is a complex energy-requiring process involving charged tRNA molecules, numerous protein factors, ribosomes, and mRNA. Transfer RNA (tRNA) serves as the adaptor molecule between an mRNA triplet and the appropriate amino acid. The ribosome serves as the workbench for translation.

9. The processes of transcription and translation, like DNA replication, can be subdivided into the stages of initiation, elongation, and termination. Also like DNA replication, both processes rely on base-pairing affinities between complementary nucleotides.

10. The processes of transcription and translation are more complex in eukaryotes than in prokaryotes.

The primary transcript representing mRNA in eukaryotes must be modified in various ways, including the addition of a 7mG cap and a poly-A tail, and the removal, through splicing, of intervening sequences, or introns. RNA editing of pre-mRNA prior to its translation also occurs in some systems.

KEY TERMS

adaptor molecule, 326
alternative splicing, 346
amber mutation, 333
ambiguous code, 326
aminoacyladenylic acid, 350
aminoacyl site (A site), 354
aminoacyl tRNA synthetase, 350
anticodon, 330
anticodon loop, 349
beta-globin gene, 344
breast cancer, 356
CCAAT box, 342
(cell-free) protein-synthesizing system, 327
chain elongation, 338
charging (tRNA), 350
cis-acting elements, 342
cloverleaf model of tRNA, 349
codon, 325
colinear relationship, 334
commaless code, 325
consensus sequence, 338
degenerate code, 325
double-stranded RNA adenosine deaminase, 347
elongation, 354
elongation factor (EF), 354
enhancer, 342
exon, 344
formylmethionine, 351
frameshift mutation, 327
gene amplification, 341
Goldberg–Hogness (TATA) box, 342
GTP-dependent release factor, 354
guide RNA (gRNA), 347
heteroduplex, 344
heterogeneous nuclear RNA (hnRNA), 341
heterogeneous nuclear ribonucleoprotein particles (hnRNPs), 341
holoenzyme, 337

hypoxanthine, 349
information flow, 336
initiation complex, 354
initiation factor (IF), 351
initiator codon (AUG), 333
insertion/deletion editing, 347
intervening sequence, 344
intron, 344
isoaccepting tRNAs, 351
isoforms, 346
messenger RNA (mRNA), 326
7-methylguanosine (7mG) cap, 343
monocistronic mRNA, 339
monosome, 347
MS2 phage, 334
N-formylmethionine (fmet), 333
nonoverlapping code, 325
nonsense codon, 333
nonsense mutation, 354
ochre mutation, 333
opal mutation, 333
ordered code, 325
ovalbumin gene, 344
overlapping genes, 335
partner strand, 338
peptidyl site (P site), 354
peptidyl transferase, 354
poly-A sequence, 343
polycistronic mRNA, 339
polynucleotide phosphorylase, 327
polyribosome (polysomes), 355
posttranscriptional modification, 343
pre-mRNA, 341
Pribnow box (−10 region), 338
proflavin, 326
promoter, 338
pseudouridine, 349
punctuation signal, 326
rDNA, 341
ribosomal proteins, 347

ribosome, 325
ribothymidylic acid, 349
ribozyme, 345
RNA editing, 346
RNA heteropolymer, 328
RNA homopolymer, 328
RNA polymerase, 337
second genetic code, 351
self-excision process, 345
Shine-Dalgarno sequence, 354
sickle-cell anemia, 356
σ (sigma) subunit, 337
small nuclear RNA molecules (snRNAs), 345
small nuclear ribonucleoprotein (snRNP), 346
spacer DNA, 348
spliceosome, 346
splicing, 341
split genes, 341
stop codons, 354
substitution editing, 347
suppression, 333
tandem repeats, 348
TATA-binding proteins, 342
TATA box, 338
template binding, 338
template strand, 338
termination factor (e.g., rho), 338
termination codons, 333
trans-acting factors, 342
transcription, 325
transcription factors, 342
transfer RNA (tRNA), 325
translation, 325
triplet code, 325
triplet binding assay, 330
unambiguous code, 325
universal code, 325
wobble hypothesis, 332

INSIGHTS AND SOLUTIONS

1. Had evolution seized on 6 bases (3 complementary base pairs) rather than 4 bases within the structure of DNA, calculate how many triplet codons would be possible. Would 6 bases accommodate a two-letter code, assuming 20 amino acids and start-and-stop codons?

   Solution: Six things taken three at a time will produce $(6)^3$ or 216 triplet codes. If the code was a doublet, there would be $(6)^2$ or 36 two-letter codes, more than enough to accommodate 20 amino acids and start–stop punctuation.

2. In a heteropolymer experiment using 1/2C:1/4A:1/4G, how many different triplets will occur in the synthetic RNA molecule? How frequently will the most frequent triplet occur?

   Solution: There will be $(3)^3$ or 27 triplets produced. The most frequent will be CCC, present $(1/2)^3$ or 1/8 of the time.

3. In a regular copolymer experiment, where UUAC is repeated over and over, how many different triplets will occur in the synthetic RNA, and how many amino acids will occur in the polypeptide when this RNA is translated? Be sure to consult Figure 12.6.

   Solution: The synthetic RNA will repeat four triplets—UUA, UAC, ACU, and CUU—over and over.

   Because both UUA and CUU encode leucine, while ACU and UAC encode threonine and tyrosine, respectively, polypeptides synthesized under the directions of such an RNA contain three amino acids in the repeating sequence leu-leu-thr-tyr.

4. Actinomycin D inhibits DNA-dependent RNA synthesis. This antibiotic is added to a bacterial culture where a specific protein is being monitored. Compared to a control culture, translation of the protein declines over a period of 20 minutes, until no further protein is made. Explain these results.

   Solution: The mRNA, which is the basis for the translation of the protein, has a lifetime of about 20 minutes. When actinomycin D is added, transcription is inhibited and no new mRNAs are made. Those already present support the translation of the protein for up to 20 minutes.

5. DNA and RNA base compositions were analyzed from a hypothetical bacterial species with the following results:

	(A + G)(T + C)	(A + T)(C + G)	(A + G)(U + C)	(A + U)(C + G)
DNA	1.0	1.2		
RNA			1.3	1.2

   On the basis of these data, what can you conclude about the DNA and RNA of the organism? Are the data consistent with the Watson–Crick model of DNA? Is the RNA single-stranded or double-stranded, or can't we tell? If we assume that the entire length of DNA has been transcribed, do the data suggest that RNA has been derived from the transcription of one or both DNA strands, or can't we tell from these data?

   Solution: This problem is a theoretical exercise designed to get you to look at the consequences of base complementarity as it affects base composition of DNA and RNA. The base composition of DNA is consistent with the Watson–Crick double helix. In a double helix, we expect A + G to equal T + C (the number of purines should equal the number of pyrimidines). In this case, there is a preponderance of A/T base pairs (120 A and T pairs to every 100 G and C pairs).

   Given what we know about RNA, there is no reason to expect the RNA to be double-stranded, but if it were double-stranded, then we would expect that A = U and C = G. If so, then (A + G)/(U + C) = 1. Since it doesn't equal unity, we can conclude that the RNA is not double-stranded.

   If all of the DNA is transcribed, from either one or both strands, the ratio of (A + U)/(C + G) in RNA should be 1.2, and as predicted, it is. Note that this ratio will not change, whether only one or both of the strands are transcribed. This is the case because for every A = T pair in DNA, for example, transcription of RNA will yield one A *and* one U if both strands are transcribed. If just one strand is transcribed, transcription will yield one A *or* one U. In either case, the (A + U)/(C + G) ratio in RNA will reflect the (A + T)/(C + G) ratio in the DNA from which it was transcribed. To prove this to yourself, draw out a DNA molecule with 12 A = T pairs and 10 C ≡ G pairs and transcribe *both* strands and then transcribe *either* strand. Count the bases in the RNAs produced in both cases and calculate the ratios. Thus, we cannot determine whether just one or both strands are transcribed from the (A + U)/(C + G) ratio.

   However, if both strands are transcribed, then the ratio of (A + G)/(U + C) should equal 1.0, and it doesn't. It equals 1.3. To verify this conclusion, examine the theoretical data you drew out on paper as you were asked to do above. One explanation for the observed ratio of 1.3 is that only one of the two strands is transcribed. If this is the case, then the (A + G)/(U + C) will reflect the proportion of A/T pairs that are A and the proportion of the G/C pairs that are G *on the DNA strand that is transcribed.* Another explanation is that transcription occurs only on one strand at any given point (e.g., for one gene), but on the other strand at other points (for other genes).

PROBLEMS AND DISCUSSION QUESTIONS

1. Early proposals regarding the genetic code considered the possibility that DNA served directly as the template for polypeptide synthesis (see the Gamow reference in Selected Readings). In eukaryotes, what difficulties would such a system pose? What observations and theoretical considerations argue against such a proposal?

2. Crick, Barnett, Brenner, and Watts-Tobin, in their studies of frameshift mutations, found that either 3 (+)s or 3 (−)s restored the correct reading frame. If the code were a sextuplet (consisting of six nucleotides), would the reading frame be restored by either of the above combinations?

3. In a mixed copolymer experiment using polynucleotide phosphorylase, 3/4G:1/4C was added to form the synthetic message. The resulting amino acid composition of the ensuing protein was determined:

Glycine	36/64	(56%)
Alanine	12/64	(19%)
Arginine	12/64	(19%)
Proline	4/64	(6%)

From this information:

(a) Indicate the percentage (or fraction) of the time each possible triplet will occur in the message.

(b) Determine one consistent base composition assignment for the amino acids present.

(c) Considering the wobble hypothesis, predict as many specific triplet assignments as possible.

4. When repeating copolymers are used to form synthetic mRNAs, dinucleotides produce a single type of polypeptide that contains only two different amino acids. On the other hand, using a trinucleotide sequence produces three different polypeptides, each consisting of only a single amino acid. Why? What will be produced when a repeating tetranucleotide is used?

5. The mRNA formed from the repeating tetranucleotide UUAC incorporates only three amino acids, but the use of UAUC incorporates four amino acids. Why?

6. In studies using repeating copolymers, ACA... incorporates threonine and histidine, and CAACAA... incorporates glutamine, asparagine, and threonine. What triplet code can definitely be assigned to threonine?

7. In a coding experiment using repeating copolymers (as shown in Table 12.3), the following data were obtained:

Copolymer	Codons Produced	Amino Acids in Polypeptide
AG	AGA, GAG	Arg, Glu
AAG	AGA, AAG, GAA	Lys, Arg, Glu

AGG is known to code for arginine. Taking into account the wobble hypothesis, assign each of the four remaining different triplet codes to its correct amino acid.

8. In the triplet binding technique, radioactivity remains on the filter when the amino acid corresponding to the triplet is labeled. Explain the basis of this technique.

9. When the amino acid sequences of insulin isolated from different organisms were determined, some differences were noted. For example, alanine was substituted for threonine, serine was substituted for glycine, and valine was substituted for isoleucine at corresponding positions in the protein. List the single-base changes that could occur in triplets of the genetic code to produce these amino acid changes.

10. In studies of the amino acid sequence of wild-type and mutant forms of tryptophan synthetase in *E. coli*, the following changes have been observed:

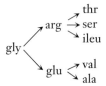

Determine a set of triplet codes in which only a single nucleotide change produces each amino acid change.

11. Why doesn't polynucleotide phosphorylase synthesize RNA *in vivo*?

12. Refer to Table 12.1. Can you hypothesize why a mixture of Poly U + Poly A would not stimulate incorporation of ^{14}C-phenylalanine into protein?

13. Predict the amino acid sequence produced during translation by the following short theoretical mRNA sequences. Note that the second sequence was formed from the first by a deletion of only one nucleotide.

   Sequence 1: AUGCCGGAUUAUAGUUGA
   Sequence 2: AUGCCGGAUUAAGUUGA

What type of mutation gave rise to Sequence 2?

14. A short RNA molecule was isolated that demonstrated a hyperchrome shift indicating secondary structure. Its sequence was determined to be

   AGGCGCCGACUCUACU

(a) Propose a two-dimensional model for this molecule.

(b) What DNA sequence would give rise to this RNA molecule through transcription?

(c) If the molecule were a tRNA fragment containing a CGA anticodon, what would the corresponding codon be?

(d) If the molecule were an internal part of a message, what amino acid sequence would result from it following translation? (Refer to the code chart in Figure 12.6.)

15. A glycine residue exists at position 210 of the tryptophan synthetase enzyme of wild-type *E. coli*. If the codon specifying glycine is GGA, how many single-base substitutions will result in an amino acid substitution at position 210? What are they? How many will result if the wild-type codon is GGU?

16. (a) Shown below is a theoretical viral mRNA sequence. Assuming that it could arise from overlapping genes, how many different polypeptide sequences can be produced? Using Figure 12.6, what are the sequences?

   5'-AUGCAUACCUAUGAGACCCUUGGGA-3'

(b) A base substitution mutation that altered the above sequence eliminated the synthesis of all but one polypeptide. The altered sequence is shown below. Using Figure 12.6, determine why.

5'-AUGCAUACCUAUGUGACCCUUGGA-3'

17. Most proteins have more leucine than histidine residues, but more histidine than tryptophan residues. Correlate the number of codons for these three amino acids with the above information.

18. Define and differentiate between transcription and translation. Where do these processes fit into the central dogma of molecular genetics?

19. What was the initial evidence for the existence of mRNA?

20. Describe the structure of RNA polymerase. What is the core enzyme? What is the role of the sigma subunit?

21. In a written paragraph, describe the abbreviated chemical reactions shown on page 337 that summarize RNA polymerase-directed transcription. What differences exist between the "visualization of transcription" studies of *E. coli* and *Notophthalmus?* Why?

22. List all of the molecular constituents present in a functional polyribosome.

23. Contrast the roles of tRNA and mRNA during translation and list all enzymes that participate in the transcription and translation process.

24. Francis Crick proposed the "adaptor hypothesis" for the function of tRNA. Why did he choose that description?

25. What molecule bears the codon? The anticodon?

26. The α chain of eukaryotic hemoglobin is composed of 141 amino acids. What is the minimum number of nucleotides in an mRNA coding for this polypeptide chain? Assuming that each nucleotide is 0.34 nm long in the mRNA, how many triplet codes can at one time occupy space in a ribosome that is 20 nm in diameter?

27. Summarize the steps involved in charging tRNAs with their appropriate amino acids.

28. For it to carry out its role, each transfer RNA requires at least four specific recognition sites that must be inherent in its tertiary structure. What are they?

29. Messenger RNA molecules are very difficult to isolate in prokaryotes because they are rather quickly degraded in the cell. Can you suggest a reason why this occurs? Eukaryotic mRNAs are more stable and exist longer in the cell than do prokaryotic mRNAs. Is this an advantage or disadvantage for a pancreatic cell making large quantities of insulin?

30. The following represent deoxyribonucleotide sequences derived from the template strand of DNA:

Sequence 1: CTTTTTTTGCCAT
Sequence 2: ACATCAATAACT
Sequence 3: TACAAGGGGTTCT

(a) For each strand, determine the mRNA sequence that would be derived from transcription.

(b) Using Figure 12.6, determine the amino acid sequence that would result from translation of these mRNAs.

(c) If we assume that each sequence has been derived from the same gene, which represents the initial part? The middle region? The terminal portion?

(d) For Sequence 1, what is the sequence of the partner strand?

(e) For Sequence 3, draw the structural formula of the polypeptide sequence resulting from translation (see Chapter 13).

EXTRA-SPICY PROBLEMS

31. In a mixed copolymer experiment, messengers were created with either 4/5C:1/5A or 4/5A:1/5C. These messages yielded proteins with the following amino acid compositions. Using these data, predict the most specific coding composition for each amino acid.

4/5C:1/5A		4/5A:1/5C	
Proline	63.0%	Proline	3.5%
Histidine	13.0%	Histidine	3.0%
Threonine	16.0%	Threonine	16.6%
Glutamine	3.0%	Glutamine	13.0%
Asparagine	3.0%	Asparagine	13.0%
Lysine	0.5%	Lysine	50.0%
	98.5%		98.5%

32. In 1962, F. Chapeville and others reported an experiment where they isolated radioactive ^{14}C-cysteinyl-tRNA^{cys} (charged tRNA^{cys} + cysteine). They then removed the sulfur group from the cysteine, creating alanyl-tRNA^{cys} (charged tRNA^{cys} + alanine). When alanyl-tRNA^{cys} was added to a synthetic mRNA calling for cysteine but not alanine, a polypeptide chain was synthesized containing alanine. What can you conclude from this experiment? [*Reference: Proc. Natl. Acad. Sci. USA* 48:1086–93 (1962).]

33. Shown below are the amino acid sequences of the wild-type and mutant forms of a short protein.

(a) Using Figure 12.6, determine the general type of mutation leading to each altered protein.

(b) For each mutant protein, determine the specific ribonucleotide change that led to its synthesis.

(c) The wild-type RNA consists of nine triplets. What is the role of the ninth triplet?

(d) For the first eight wild-type triplets, which, if any, can you determine specifically from an analysis of the mutant proteins? Explain why or why not in each case.

(e) Another mutation (Mutant #4) is isolated. Its amino acid sequence is unchanged, but mutant cells produce abnormally low amounts of the wild-type proteins. As specifically as you can, predict where in the gene this mutation exists.

Wild type:	met-trp-tyr-arg-gly-ser-pro-thr
Mutant #1:	met-trp
Mutant #2:	met-trp-his-arg-gly-ser-pro-thr
Mutant #3:	met-cys-ile-val-val-val-gln-his

SELECTED READINGS

ALBERTS, B., et al. 1994. *Molecular biology of the cell*, 3rd ed. New York: Garland.

BARALLE, F. E. 1983. The functional significance of leader and trailer sequences in eukaryotic mRNAs. *Int. Rev. Cytol.* 81:71–106.

BARRELL, B. G., AIR, G., and HUTCHINSON, C. 1976. Overlapping genes in bacteriophage φX174. *Nature* 264:34–40.

BARRELL, B. G., BANKER, A. T., and DROUIN, J. 1979. A different genetic code in human mitochondria. *Nature* 282: 189–94.

BEARDSLEY, T. 1996. Vital data. *Sci. Am.* (Mar.) 274:100–105.

BIRNSTIEL, M., BUSSLINGER, M., and STRUB, K. 1985. Transcription termination and 3′ processing: The end is in site. *Cell* 41:349–59.

BONITZ, S. G., et al. 1980. Codon recognition rules in yeast mitochondria. *Proc. Natl. Acad. Sci. USA* 77:3167–70.

BREITBART, R. E., and NADAL-GINARD, B. 1987. Developmentally induced, muscle-specific *trans* factors control the differential splicing of alternative and constitutive troponin T-exons. *Cell* 49:793–803.

BRENNER, S. 1989. *Molecular biology: A selection of papers.* Orlando, FL: Academic Press.

BRENNER, S., JACOB, F., and MESELSON, M. 1961. An unstable intermediate carrying information from genes to ribosomes for protein synthesis. *Nature* 190:575–80.

BRENNER, S., STRETTON, A. O. W., and KAPLAN, D. 1965. Genetic code: The nonsense triplets for chain termination and their suppression. *Nature* 206:994–98.

CATTANEO, R. 1991. Different types of messenger RNA editing. *Annu. Rev. Genet.* 25:71-88.

CECH, T. R. 1986. RNA as an enzyme. *Sci. Am.* (Nov.) 255(5):64–75.

———. 1987. The chemistry of self-splicing RNA and RNA enzymes. *Science* 236:1532–39.

CECH, T. R., ZAUG, A. J., and GRABOWSKI, P. J. 1981. *In vitro* splicing of the ribosomal RNA precursor of *Tetrahymena*. Involvement of a guanosine nucleotide in the excision of the intervening sequence. *Cell* 27:487–96.

CHAMBON, P. 1975. Eucaryotic nuclear RNA polymerases. *Annu. Rev. Biochem.* 44:613–38.

———. 1981. Split genes. *Sci. Am.* (May) 244:60–71.

CHAPEVILLE, F., et al. 1962. On the role of soluble ribonucleic acid in coding for amino acids. *Proc. Natl. Acad. Sci. USA* 48:1086–93.

CIGAN, A. M., FENG, L., and DONAHUE, T. F. 1988. tRNA^met functions in directing the scanning ribosome to the start site of translation. *Science* 242:93–98.

COLD SPRING HARBOR LABORATORY. 1966. The genetic code. *Cold Spring Harbor Symp. Quant. Biol.* Vol. 31.

CRICK, F. H. C. 1962. The genetic code. *Sci. Am.* (Oct.) 207:66–77.

———. 1966a. The genetic code: III. *Sci. Am.* (Oct.) 215:55–63.

———. 1966b. Codon-anticodon pairing: The wobble hypothesis. *J. Mol. Biol.* 19:548–55.

———. 1979. Split genes and RNA splicing. *Science* 204:264–71.

CRICK, F. H. C., BARNETT, L., BRENNER, S., and WATTS-TOBIN, R. J. 1961. General nature of the genetic code for proteins. *Nature* 192:1227–32.

DAHLBERG, A. E. 1989. The functional role of ribosomal RNA in protein synthesis. *Cell* 57:525–29.

DARNELL, J. E. 1983. The processing of RNA. *Sci. Am.* (Oct.) 249:90–100.

———. 1985. RNA. *Sci. Am.* (Oct.) 253:68–87.

DICKERSON, R. E. 1983. The DNA helix and how it is read. *Sci. Am.* (Dec.) 249:94–111.

DUGAICZK, A., et al. 1978. The natural ovalbumin gene contains seven intervening sequences. *Nature* 274:328–33.

FIERS, W., et al. 1976. Complete nucleotide sequence of bacteriophage MS2 RNA: Primary and secondary structure of the replicase gene. *Nature* 260:500–507.

GAMOW, G. 1954. Possible relation between DNA and protein structures. *Nature* 173:318.

GUTHRIE, C., and PATTERSON, B. 1988. Spliceosomal snRNAs. *Annu. Rev. Genet.* 22:387–419.

HALL, B. D., and SPIEGELMAN, S. 1961. Sequence complementarity of T2-DNA and T2-specific RNA. *Proc. Natl. Acad. Sci. USA* 47:137–46.

HALL, J. M., LEE, M. K., NEWMAN, B., MORROW, J. E., ANDERSON, L. A., HUEY, B., and KING, M.-C. 1990. Linkage of early-onset familial breast cancer to chromosome 17q21. *Science* 250:1684–89.

HAMKALO, B. 1985. Visualizing transcription in chromosomes. *Trends Genet.* 1:255–60.

HELMAN, J. D., and CHAMBERLIN, M. J. 1988. Structure and function of bacterial sigma factors. *Annu. Rev. Biochem.* 57:839–72.

HODGES, P., and SCOTT, J. 1992. Apolipoprotein B mRNA editing: A new tier for the control of gene expression. *Trends Biochem. Sci.* 17:77–81.

HOLLEY, R. W., et al. 1965. Structure of a ribonucleic acid. *Science* 147:1462–65.

HORTON, H. R., et al. 1996. *Principles of biochemistry*, 2nd ed. Upper Saddle River, NJ: Prentice-Hall.

HUMPHREY, T., and PROUDFOOT, N. J. 1988. A beginning to the biochemistry of polyadenylation. *Trends Genet.* 4:243–45.

JUDSON, H. F. 1979. *The eighth day of creation.* New York: Simon & Schuster.

JUKES, T. H. 1963. The genetic code. *Am. Sci.* 51:227–45.

KABLE, M. L., et al. 1996. RNA editing: A mechanism for gRNA-specified uridylate insertion into precursor mRNA. *Science* 273:1189–95.

KHORANA, H. G. 1967. Polynucleotide synthesis and the genetic code. *Harvey Lectures* 62:79–105.

LAKE, J. A. 1981. The ribosome. *Sci. Am.* (Aug.) 245:84–97.

LEHNINGER, A. H., NELSON, D. C., and COX, M. M. 1993. *Principles of biochemistry.* New York: Worth Publishers.

LODISH, H., et al. 1995. *Molecular cell biology,* 2nd ed. New York: Scientific American Books.

MANIATIS, T., and REED, R. 1987. The role of small nuclear ribonucleoprotein particles in pre-mRNA splicing. *Nature* 325:673–78.

MILLER, O. L., and BEATTY, B. R. 1969. Portrait of a gene. *J. Cell Physiol.* 74 (Suppl. 1): 225–32.

MILLER, O. L., HAMKALO, B., and THOMAS, C. 1970. Visualization of bacterial genes in action. *Science* 169:392–95.

MIN JOU, W., HAGEMAN, G., YSEBART, M., and FIERS, W. 1972. Nucleotide sequence of the gene coding for bacteriophage MS2 coat protein. *Nature* 237:82–88.

MOORE, P. B. 1988. The ribosome returns. *Nature* 331:223–27.

NIRENBERG, M. W. 1963. The genetic code: II. *Sci. Am.* (March) 190:80–94.

NIRENBERG, M. W., and LEDER, P. 1964. RNA codewords and protein synthesis. *Science* 145:1399–1407.

NIRENBERG, M. W., and MATTHAEI, H. 1961. The dependence of cell-free protein synthesis in *E. coli* upon naturally occurring or synthetic polyribosomes. *Proc. Natl. Acad. Sci. USA* 47:1588–1602.

NOLLER, H. F. 1973. Assembly of bacterial ribosomes. *Science* 179:864–73.

———. 1984. Structure of ribosomal RNA. *Annu. Rev. Biochem.* 53:119–62.

NOMURA, M. 1984. The control of ribosome synthesis. *Sci. Am.* (Jan.) 250:102–14.

O'MALLEY, B., et al. 1979. A comparison of the sequence organization of the chicken ovalbumin and ovomucoid genes. In *Eucaryotic gene regulation,* ed. R. Axel, et al., pp. 281–99. Orlando, FL: Academic Press.

PADGETT, R. A., GRABOWSKI, P. J., KONARSKA, M. M., SEILER, S., and SHARP, P. A. 1986. Splicing of messenger RNA precursors. *Annu. Rev. Biochem.* 55:1119–50.

REED, R., and MANIATIS, T. 1985. Intron sequences involved in lariat formation during pre-mRNA splicing. *Cell* 41:95–105.

RICH, A., and HOUKIM, S. 1978. The three-dimensional structure of transfer RNA. *Sci. Am.* (Jan.) 238:52–62.

RICH, A., WARNER, J. R., and GOODMAN, H. M. 1963. The structure and function of polyribosomes. *Cold Spring Harbor Symp. Quant. Biol.* 28:269–85.

ROSS, J. 1989. The turnover of mRNA. *Sci. Am.* (April) 260:48–55.

ROULD, M. A., et al. 1989. Structure of *E. coli* glutaminyl-tRNA synthetase complexed with tRNA^{gln} and ATP at 2.8 Å resolution. *Science* 246:1135–42.

SHARP, P. A. 1987. Splicing of messenger RNA precursors. *Science* 235:766–71.

——— 1994. Nobel Lecture: Split genes and RNA splicing. *Cell* 77:805–15.

SHARP, P. A., and EISENBERG, D. 1987. The evolution of catalytic function. *Science* 238:729–30.

SHATKIN, A. J. 1985. mRNA cap binding proteins: Essential factors for initiating translation. *Cell* 40:223–24.

SMITH, J. D. 1972. Genetics of tRNA. *Annu. Rev. Genet.* 6:235–56.

STEITZ, J. A. 1988. Snurps. *Sci. Am.* (June) 258(6):56–63.

VOLKIN, E., and ASTRACHAN, L. 1956. Phosphorus incorporation in *E. coli* ribonucleic acids after infection with bacteriophage T2. *Virology* 2:149–61.

VOLKIN, E., ASTRACHAN, L., and COUNTRYMAN, J. L. 1958. Metabolism of RNA phosphorus in *E. coli* infected with bacteriophage T7. *Virology* 6:545–55.

WARNER, J., and RICH, A. 1964. The number of soluble RNA molecules on reticulocyte polyribosomes. *Proc. Natl. Acad. Sci. USA* 51:1134–41.

WATSON, J. D. 1963. Involvement of RNA in the synthesis of proteins. *Science* 140:17–26.

WATSON, J. D., HOPKINS, N. H., ROBERTS, J. W., STEITZ, J. A., and WEINER, A. M. 1987. *Molecular biology of the gene,* 4th ed. Menlo Park, CA: Benjamin/Cummings.

WITTMAN, H. G. 1983. Architecture of prokaryotic ribosomes. *Annu. Rev. Biochem.* 52:35–65.

ZUBAY, G. L., and MARMUR, J. 1973. *Papers in biochemical genetics,* 2nd ed. New York: Holt, Rinehart & Winston.

13

Proteins: The End Product of Genetic Expression

CHAPTER CONCEPTS

The end products of most gene expression are polypeptide chains. They achieve a three-dimensional conformation that is based on their primary amino acid sequences, and they often interact with other such chains to create functional protein molecules. The function of any protein is closely tied to its structure, which can be disrupted by mutation, leading to a distinctive phenotypic effect.

In Chapter 12 we established that there is a genetic code that stores information in the form of triplet nucleotides in DNA, and that this information can be expressed through the orderly processes of transcription and translation. The final product of gene expression, in almost all instances, is a polypeptide chain consisting of a linear series of amino acids whose sequence has been prescribed by the genetic code. In this chapter we will review the evidence that confirmed that proteins are the end products of genes, and we will then discuss briefly the various levels of protein structure, diversity, and function. This information provides an important foundation in our understanding of how mutations, which arise in DNA, can result in the diverse phenotypic effects observed in organisms.

Collagen fibers, the most abundant protein in vertebrate organisms.

Garrod and Bateson: Inborn Errors of Metabolism

The first insight into the role of proteins in genetic processes was provided by observations made by Sir Archibald Garrod and William Bateson early in this century. Garrod was born into an English family of medical scientists. His father was a physician with a strong interest in the chemical basis of rheumatoid arthritis, and his eldest brother was a leading zoologist in London. It is not surprising, then, that as a practicing physician, Garrod became interested in several human disorders that seemed to be inherited. Although he also studied albinism and cystinuria, we will describe his investigation of the disorder **alkaptonuria**. Individuals afflicted with this disorder cannot metabolize the alkapton 2,5-dihydroxyphenylacetic acid, also known as **homogentisic acid.** As a result, an important metabolic pathway (Figure 13.1) is blocked. Homogentisic acid accumulates in cells and tissues and is excreted in the urine. The molecule's oxidation products are black and easily detectable in the diapers of newborns. The products tend to accumulate in cartilaginous areas, causing a darkening of the ears and nose. In joints, this deposition leads to a benign arthritic condition. This rare disease is not serious, but it persists throughout an individual's life.

Garrod studied alkaptonuria by increasing dietary protein or adding to the diet the amino acids phenylalanine or tyrosine, both of which are chemically related to homogentisic acid. Under such conditions, homogentisic acid levels increase in the urine of alkaptonurics but not in unaffected individuals. Garrod concluded that normal individuals can break down, or catabolize, this alkapton but that afflicted individuals cannot. By studying the patterns of inheritance of the disorder, Garrod further concluded that alkaptonuria was inherited as a simple recessive trait.

On the basis of these conclusions, Garrod hypothesized that the hereditary information controls chemical reactions in the body and that the inherited disorders he studied are the result of alternative modes of metabolism. While *genes* and *enzymes* were not familiar terms during Garrod's time, he used the corresponding concepts of *unit factors* and *ferments*. Garrod published his initial observations in 1902.

Only a few geneticists, including Bateson, were familiar with or referred to Garrod's work. Garrod's ideas

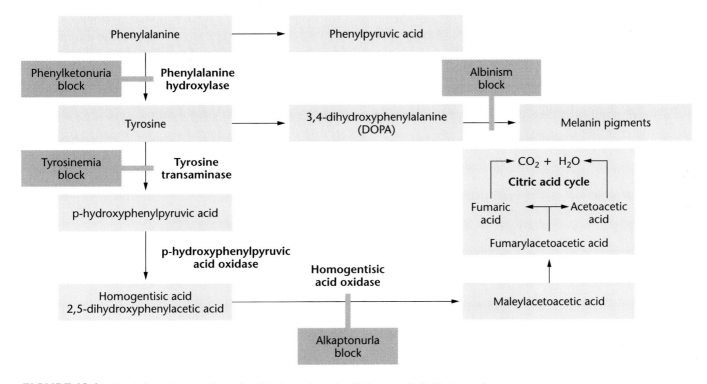

FIGURE 13.1   Metabolic pathway involving phenylalanine and tyrosine. Various metabolic blocks resulting from mutations lead to the disorders phenylketonuria, alkaptonuria, albinism, and tyrosinemia.

fit nicely with Bateson's belief that inherited conditions were caused by the lack of some critical substance. In 1909, Bateson published *Mendel's Principles of Heredity*, in which he linked ferments with heredity. However, for almost 30 years, most geneticists failed to see the relationship between genes and enzymes. Garrod and Bateson, like Mendel, were ahead of their time.

Phenylketonuria

Described first in 1934, **phenylketonuria (PKU)** can result in mental retardation and is inherited as an autosomal recessive disease. Afflicted individuals have a different step blocked in the metabolic pathway just discussed. They are unable to convert the amino acid phenylalanine to the amino acid tyrosine (see Figure 13.1). These molecules differ by only a single hydroxyl group (—OH), present in tyrosine, but absent in phenylalanine. The reaction is catalyzed by the enzyme **phenylalanine hydroxylase**, which is inactive in affected individuals and active at about a 30 percent level in heterozygotes. The enzyme functions in the liver. While the normal blood level of phenylalanine is about 1 mg/100 ml, phenylketonurics show a level as high as 50 mg/100 ml.

As phenylalanine accumulates, it may be converted to phenylpyruvic acid and, subsequently, other derivatives. These are less efficiently resorbed by the kidney and tend to spill into the urine more quickly than phenylalanine. Both phenylalanine and derivatives enter the cerebrospinal fluid, resulting in elevated levels in the brain. The presence of these substances during early development is thought to cause mental retardation.

As a result of early detection based on PKU screening of newborns, retardation can be prevented. When the condition is detected in the analysis of an infant's blood, a strict dietary regimen is instituted. A low phenylalanine diet can reduce such by-products as phenylpyruvic acid, and abnormalities characterizing the disease can be diminished. Screening of newborns occurs routinely in almost every state in this country. Phenylketonuria occurs in approximately 1 in 11,000 births.

Knowledge of inherited metabolic disorders such as alkaptonuria and phenylketonuria has caused a revolution in medical thinking and practice. No longer is human disease attributed solely to the action of invading microorganisms, viruses, or parasites. We know now that literally thousands of abnormal physiological conditions are caused by errors in metabolism that are the result of mutant genes. These human biochemical disorders are far-ranging and include all classes of organic biomolecules.

The One-Gene:One-Enzyme Hypothesis

In two separate investigations beginning in 1933, George Beadle was to provide the first convincing experimental evidence that genes are directly responsible for the synthesis of enzymes. The first investigation, conducted in collaboration with Boris Ephrussi, involved *Drosophila* eye pigments. Encouraged by these findings, Beadle then joined with Edward Tatum to investigate nutritional mutations in the pink bread mold *Neurospora crassa*. This investigation led to the **one-gene:one-enzyme hypothesis**.

Beadle and Ephrussi: *Drosophila* Eye Pigments

The studies of Beadle and Ephrussi involved **imaginal disks** in *Drosophila*. These are embryonic cells found in the larvae of insects. Upon metamorphosis, these cells differentiate into a variety of adult structures such as eyes, legs, antennae, and genital structures. Beadle and Ephrussi found that if an imaginal disk were removed from one larva and transplanted into the abdomen of another, that disk would differentiate during metamorphosis and its corresponding adult structure could be recovered from the abdomen of the adult fly. For example, if an eye disk were transplanted, it would develop into an eye-like structure that could be recovered and analyzed.

These investigators wondered whether a mutant eye disk, transplanted into a wild-type larva, would be altered by its new environment. Would it appear normally pigmented, or would it develop its characteristic mutant color? The first mutant used was *vermilion*, an X-linked, bright red eye color mutation. Mutant development was reversed! The *vermilion* (*v*) eye disk was altered to produce a normally pigmented wild-type eye. The normal wild-type color is brick red, produced as a combination of brown and bright red pigments. The *vermilion* mutant makes the bright red pigment, but lacks the brown pigment, indicating that its synthesis is inhibited. However, in the wild-type abdomen, synthesis of brown pigment also occurred normally in the mutant disk, producing wild-type color.

Beadle and Ephrussi then performed similar experimentation with 25 other mutants. Only one, *cinnabar* (*cn*), another bright red eye mutation missing the brown pigment, behaved like *vermilion* when transplanted; it developed a normal eye color. The remaining 24 behaved "autonomously" by exhibiting their respective mutant color when differentiated.

The two researchers concluded that the flies with either *vermilion* or *cinnabar* phenotypes have mutant eyes because of the *absence* of some *diffusible substance* in their disks during development. Because this substance is diffusible, it was present in the wild-type abdomen and thus accessible to the transplanted *v* or *cn* disks. Under these conditions, it would enable full pigment production in the mutant transplants. However, in the case of the 24 mutations that developed autonomously, the substance that might have permitted normal pigment production was not diffusible but confined to the host cells forming the eye. Thus, no "correction" was observed.

Beadle and Ephrussi then asked whether the same substance was lacking in both the *vermilion* and *cinnabar* mutants. To answer this question, they transplanted *vermilion* eye disks into *cinnabar* larvae, and vice versa. What

they found was intriguing. The *v* disks were "cured" or converted to wild-type color in *cn* abdomens, but *cinnabar* disks developed autonomously in *vermilion* hosts and retained their mutant phenotype. These experiments are diagrammed in Figure 13.2.

To explain such results, the researchers proposed that two sequential biochemical reactions are involved in normal synthesis of the brown pigment. One is inactive in flies with the *vermilion* mutation, and the other is inactive in flies with the *cinnabar* mutation. Each reaction produces a product that is diffusible in the tissues of wild type. As shown in Figure 13.2, the substance produced under the direction of the wild-type allele of the *vermilion* locus (substance Y) occurs first in the pathway. It is then

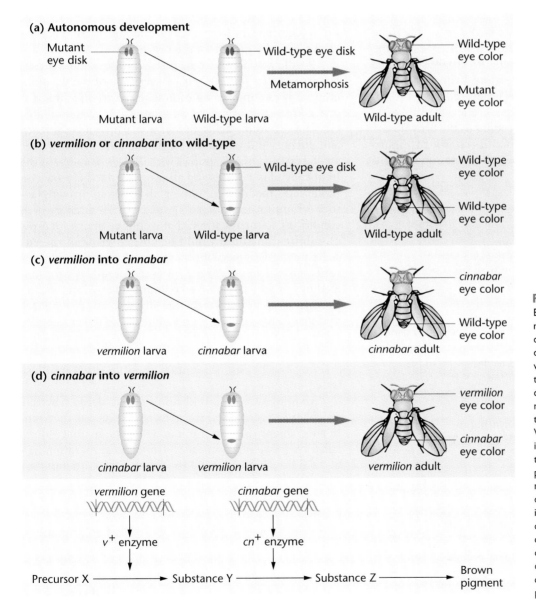

FIGURE 13.2 Beadle and Ephrussi's transplantation experiments involving *Drosophila* imaginal disks that develop into eyes. A mutant disk, transplanted into a wild-type larval abdomen, either develops autonomously, resulting in a mutant eye color, or the mutation is "cured" and normal eye color results, as illustrated in (a) and (b), respectively. When a *vermilion* disk is transplanted into a *cinnabar* host (c), normal wild-type eye color develops in the transplanted disk. The reciprocal experiment (d) results in autonomous development of the mutant eye color in the transplanted disk. On the basis of these results, the researchers concluded that the biosynthetic step controlled by the v^+ gene product occurs prior to that controlled by the cn^+ gene product within the same pathway.

converted to a second substance (Z) in the pathway. This conversion is controlled by the wild-type allele of the *cinnabar* locus.

Knowledge of the order of this pathway allows us to understand the results of the reciprocal transplantations. Consider first the *vermilion* (*v*) disk in the *cinnabar* (*cn*) host. The *vermilion* disk lacks the v^+ enzyme but has the cn^+ enzyme. As the host, the *cinnabar* tissue can make substance Y because it has the wild-type allele of the *vermilion* gene (v^+). Substance Y is diffusible and enters the *vermilion* disk where the cn^+ enzyme converts it to substance Z. Substance Z is then converted to the brown pigment, restoring the wild-type eye color.

Now consider the *cinnabar* disk in the *vermilion* host. The host tissue, lacking the v^+ gene, cannot make the v^+ enzyme, and so cannot make substance Y. Even though the cn^+ enzyme is present, because no Y is available to be converted to Z, no Z is produced either. As a result, the transplanted *cn* disk is still blocked. It can make substance Y but cannot complete the transition through Z to the brown pigment. Thus, it remains bright red!

Within a few years, other researchers had investigated the biochemical pathway leading to the brown xanthommatin pigment in *Drosophila*. As shown in Figure 13.3, the brown pigment is a derivative of the amino acid

tryptophan. The reactions controlled by the specific enzymes altered by the *vermilion* and *cinnabar* mutations have been identified. Other autosomal recessive mutations, including *scarlet* and *cardinal*, have also been studied and shown to affect enzymes in this pathway. Most relevant to our discussion was the confirmation in the 1940s that mutant genes controlling distinctive morphological phenotypes were linked directly to biochemical "errors," most likely due to a lack of enzyme function.

Beadle and Tatum: *Neurospora* Mutants

In the early 1940s, Beadle and Tatum chose to work with the fungus *Neurospora crassa* because much was known about its biochemistry, and mutations could be induced and isolated with relative ease. By inducing mutations, they produced strains that had genetic blocks of reactions essential to the growth of the organism.

Beadle and Tatum knew that *Neurospora* could manufacture nearly everything necessary for normal development. For example, using rudimentary carbon and nitrogen sources, this organism can synthesize 9 water-soluble vitamins, 20 amino acids, numerous carotenoid pigments, and purines and pyrimidines. Beadle and Tatum irradiated asexual conidia (spores) with X-rays to increase the frequency of mutations and allowed them to be grown on "complete" medium containing all necessary growth factors (vitamins, amino acids, etc.). Under such growth conditions, a mutant strain that would be unable to grow on minimal medium would be able to grow in the enriched complete medium by virtue of supplements present. All cultures were then transferred to minimal medium. If growth occurred on minimal medium, the organisms were able to synthesize all necessary growth factors themselves, and it was concluded that the culture did not contain a mutation. If no growth occurred, then it contained a nutritional mutation, and the only task remaining was to determine its type. Both cases are illustrated in Figure 13.4(a).

Many thousands of individual spores derived by this procedure were isolated and grown on complete medium. In subsequent tests on minimal medium, many cultures failed to grow, indicating that a nutritional mutation had been induced. To identify the mutant type, the mutant strains were tested on a series of different minimal media [Figure 13.4(b) and (c)], each containing groups of supplements, and subsequently on single vitamins, amino acids, purines, or pyrimidines until one specific supplement that permitted growth was found. Beadle and Tatum reasoned that the supplement that restores growth is the molecule that the mutant strain could not synthesize.

The first mutant strain isolated required vitamin B-6 (pyridoxine) in the medium, and the second one required

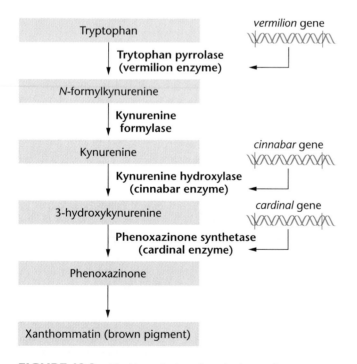

FIGURE 13.3 The biosynthetic pathway leading to the conversion of tryptophan to the brown eye pigment xanthommatin. Three different mutant blocks are shown, each interrupting synthesis at a different place in the pathway. In the absence of xanthommatin, all three mutations, when homozygous, result in the identical phenotype, bright red eyes.

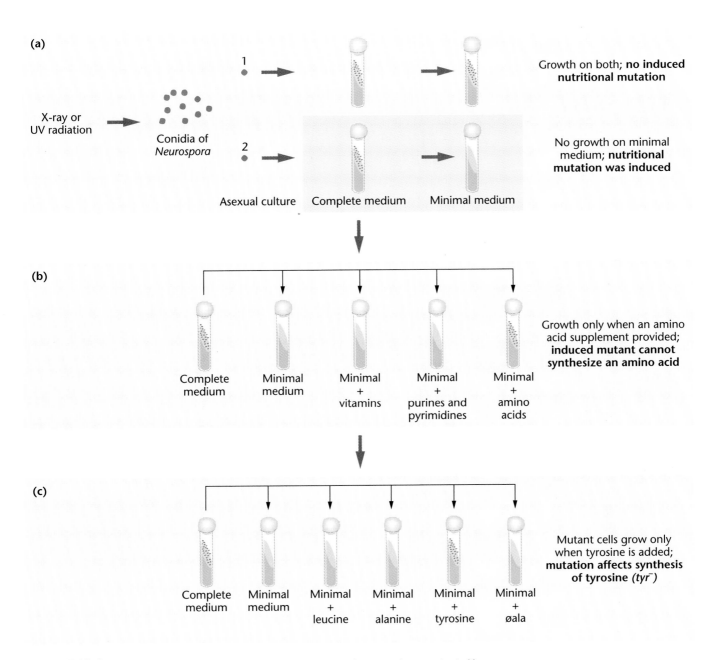

FIGURE 13.4   Induction, isolation, and characterization of a nutritional auxotrophic mutation in *Neurospora*. In (a), most conidia are not affected, but one conidium (shown in red) contains such a mutation. In (b) and (c), the precise nature of the mutation is determined to involve the biosynthesis of tyrosine.

vitamin B-1 (thiamine). Using the same procedure, Beadle and Tatum eventually isolated and studied hundreds of mutants deficient in the ability to synthesize other vitamins, amino acids, or other substances.

The findings derived from testing over 80,000 spores convinced Beadle and Tatum that genetics and biochemistry have much in common. It seemed likely that each nutritional mutation caused the loss of the enzymatic ac-

tivity that facilitates an essential reaction in wild-type organisms. It also appeared that a mutation could be found for nearly any enzymatically controlled reaction. Beadle and Tatum had thus provided sound experimental evidence for the hypothesis that **one gene specifies one enzyme**, an idea alluded to over 30 years earlier by Garrod and Bateson. With modifications, this concept was to become a major principle of genetics.

Genes and Enzymes: Analysis of Biochemical Pathways

The one-gene:one-enzyme concept and its attendant methods have been used over the years to work out many details of metabolism in *Neurospora*, *Escherichia coli*, and a number of other microorganisms. One of the first metabolic pathways to be investigated in detail was that leading to the synthesis of the amino acid arginine in *Neurospora*. By studying seven mutant strains, each requiring arginine for growth (*arg⁻*), Adrian Srb and Norman Horowitz were able to ascertain a partial biochemical pathway that leads to the synthesis of this molecule. The rationale followed in their work illustrates how genetic analysis can be used to establish biochemical information.

Srb and Horowitz tested each mutant strain's ability to reestablish growth if either **citrulline** or **ornithine**, two compounds with close chemical similarity to arginine, was used as a supplement to minimal medium. If either was able to substitute for arginine, they reasoned that it must be involved in the biosynthetic pathway of arginine. They found that both molecules could be substituted in one or more strains.

Of the seven mutant strains, four of them (*arg 4–7*) grew if supplied with either citrulline, ornithine, or arginine. Two of them (*arg 2* and *3*) grew if supplied with citrulline or arginine. One strain (*arg 1*) would grow only if arginine were supplied. Neither citrulline nor ornithine could substitute for it. From these experimental observations, the following pathway and metabolic blocks for each mutation were deduced:

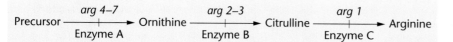

The reasoning supporting these conclusions is based on the following logic. If mutants *arg 4* through 7 can grow regardless of which of the three molecules is supplied as a supplement to minimal medium, the mutations preventing growth must cause a metabolic block that occurs *prior to* the involvement of ornithine, citrulline, or arginine in the pathway. When any one of these three molecules is added, its presence bypasses the block. As a result, it can be concluded that both citrulline and ornithine are involved in the biosynthesis of arginine. However, the sequence of their participation in the pathway cannot be determined on the basis of these data.

On the other hand, the *arg 2* and *3* mutations grow if supplied citrulline but not if they are supplied with only ornithine. Therefore, ornithine must occur in the pathway *prior to* the block. Its presence will not overcome the block. Citrulline, however, does overcome the block, so it

must be involved beyond the point of blockage. Therefore, the conversion of ornithine to citrulline represents the correct sequence in the pathway.

Finally, it can be concluded that *arg 1* represents a mutation preventing the conversion of citrulline to arginine. Neither ornithine nor citrulline can overcome the metabolic block because both participate earlier in the pathway.

Taken together, these reasons support the sequence of biosynthesis outlined here. Ever since Srb and Horowitz's work in 1944, the detailed pathway has been worked out and the enzymes controlling each step characterized. The metabolic pathway is shown in Figure 13.5.

The concept of one-gene:one-enzyme developed in the early 1940s was not accepted immediately by all geneticists. This is not surprising, because it was not yet clear how mutant enzymes could cause variation in many phenotypic traits. For example, *Drosophila* mutants demonstrated altered eye size, wing shape, wing vein pattern, and so on. Plants exhibited mutant varieties of seed texture, height, and fruit size. How an inactive, mutant enzyme could result in such phenotypes was puzzling to many geneticists. Another reason for their reluctance to accept this concept was the paucity of information then available in molecular genetics. It was not until 1944 that Avery, MacLeod, and McCarty showed DNA to be the transforming factor and not until the early 1950s that most geneticists believed that DNA serves as the genetic material (see Chapter 10). However, by that time, the evidence was overwhelming in support of the concept of enzymes as gene products, which was accepted as valid.

Nevertheless, the question of *how* DNA specifies the structure of enzymes remained unanswered at that time.

One-Gene:One-Protein/ One-Gene:One-Polypeptide

Two factors soon modified the one-gene:one-enzyme hypothesis. First, while nearly all enzymes are proteins, not all proteins are enzymes. As the study of biochemical genetics proceeded, it became clear that all proteins are specified by the information stored in genes, leading to the more accurate phraseology **one-gene:one protein**. Second, proteins were often shown to have a subunit structure consisting of two or more **polypeptide chains**. This is the basis of the **quaternary structure** of proteins, which we will discuss later in this chapter. Because each distinct polypeptide chain is encoded by a separate gene, a more modern statement of Beadle and Tatum's basic principle is **one-gene:one-polypeptide chain**. These

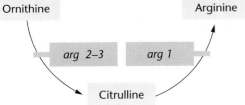

FIGURE 13.5   Abbreviated pathway resulting in the biosynthesis of arginine in *Neurospora*.

modifications of the original hypothesis became apparent during the analysis of hemoglobin structure in individuals afflicted with sickle-cell anemia.

Sickle-Cell Anemia

The first direct evidence that genes specify proteins other than enzymes came from the work on mutant hemoglobin molecules derived from humans afflicted with the disorder **sickle-cell anemia**. Affected individuals contain erythrocytes that, under low oxygen tension, become elongated and curved because of the polymerization of hemoglobin. The "sickle" shape of these erythrocytes is in contrast to the biconcave disc shape characteristic of normal individuals (Figure 13.6). Individuals with the disease suffer attacks when red blood cells aggregate in the venous side of capillary systems, where oxygen tension is very low. As a result, a variety of tissues may be deprived

of oxygen and suffer severe damage. When this occurs, an individual is said to experience a sickle-cell crisis. If untreated, a crisis may be fatal. The kidneys, muscles, joints, brain, gastrointestinal tract, and lungs may be affected.

In addition to suffering crises, these individuals are anemic because their erythrocytes are destroyed more rapidly than are normal red blood cells. Compensatory physiological mechanisms include increased red cell production by bone marrow and accentuated heart action. These mechanisms lead to abnormal bone size and shape as well as dilation of the heart.

In 1949, James Neel and E. A. Beet demonstrated that the disease is inherited as a Mendelian trait. Pedigree analysis revealed three genotypes and phenotypes controlled by a single pair of alleles, Hb^A and Hb^S. Normal and affected individuals result from the homozygous genotypes $Hb^A Hb^A$ and $Hb^S Hb^S$, respectively. The red blood cells of the heterozygote, which exhibits the **sickle-cell trait** but not the disease, undergo much less sickling because over half of their hemoglobin is normal. Although largely unaffected, such heterozygotes are "carriers" of the defective gene, which is transmitted, on average, to 50 percent of their offspring.

In the same year, Linus Pauling and his co-workers provided the first insight into the molecular basis of the disease. They showed that hemoglobins isolated from diseased and normal individuals differ in their rates of electrophoretic migration. In this technique (see Chapter 10 and Appendix A), charged molecules migrate in an electric field. If the net charge of two molecules is different, their rates of migration will be different. On this basis, Pauling and his colleagues concluded that a chemical difference exists between normal and sickle-cell hemoglobin. The two molecules are now designated **HbA** and **HbS**, respectively.

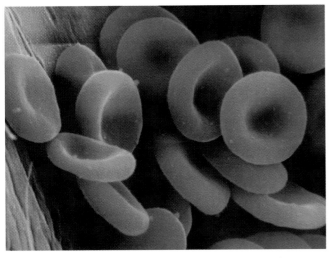

FIGURE 13.6   A comparison of erythrocytes from normal individuals (left) and from individuals with sickle-cell anemia (right).

Figure 13.7(a) illustrates the migration pattern of hemoglobin derived from individuals of all three possible genotypes when subjected to **starch gel electrophoresis**. The gel provides the supporting medium for the molecules during migration. In this experiment, samples are placed at a point of origin between the cathode ($-$) and the anode ($+$), and an electric field is applied. The migration pattern reveals that all molecules move toward the anode, indicating a net negative charge. However, HbA migrates farther than HbS, suggesting that its net negative charge is greater. The electrophoretic pattern of hemoglobin derived from carriers reveals the presence of both HbA and HbS, confirming their heterozygous genotype.

Pauling's findings suggested two possibilities. It was known that hemoglobin consists of four nonproteinaceous, iron-containing **heme groups** and a **globin portion** containing four polypeptide chains. The alteration in net charge in HbS could be due, theoretically, to a chemical change in either component.

Work carried out between 1954 and 1957 by Vernon Ingram resolved this question. He demonstrated that the chemical change occurs in the primary structure of the globin portion of the hemoglobin molecule. Using the **fingerprinting technique**, Ingram showed that HbS differs in amino acid composition compared to HbA. Human adult hemoglobin contains two identical alpha (α) chains of 141 amino acids and two identical beta (β) chains of 146 amino acids in its quaternary structure.

The fingerprinting technique involves enzymatic digestion of the protein into peptide fragments. The mixture is then placed on absorbent paper and exposed to an electric field, where migration occurs according to net charge. The paper is then turned at a right angle and placed in a solvent, where chromatographic action causes

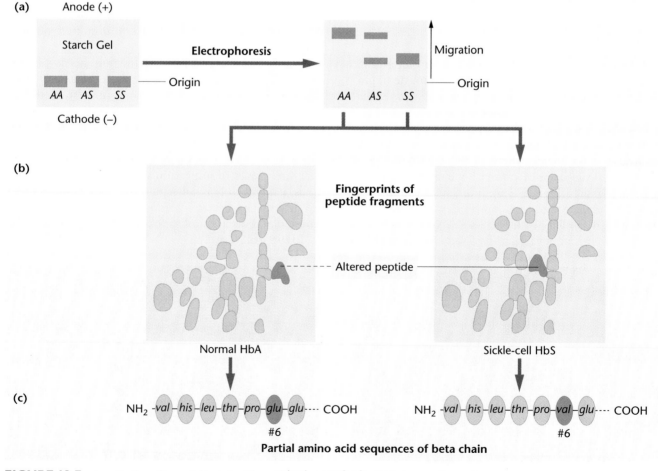

FIGURE 13.7 Investigation of hemoglobin derived from Hb^AHb^A and Hb^SHb^S individuals using electrophoresis, fingerprinting, and amino acid analysis. Hemoglobin from individuals with sickle-cell anemia (Hb^SHb^S): (a) migrates differently in an electrophoretic field, (b) shows an altered peptide in fingerprint analysis, and (c) shows an altered amino acid, valine, at the sixth position in the β chain. During electrophoresis, heterozygotes (Hb^AHb^S) reveal both forms of hemoglobin.

migration of the peptides in the second direction. The end result is a two-dimensional separation of the peptide fragments into a distinctive pattern of spots or "fingerprints." Ingram's work revealed that HbS and HbA differed by only a single peptide fragment [Figure 13.7(b)]. Further analysis then revealed a single amino acid change: Valine was substituted for glutamic acid at the sixth position of the β chain, accounting for the peptide difference [Figure 13.7(c)].

The significance of this discovery has been multifaceted. It clearly establishes that a single gene provides the genetic information for a single polypeptide chain. Studies of HbS also demonstrate that a mutation can affect the phenotype by directing a single amino acid substitution. Also, by providing the explanation for sickle-cell anemia, the concept of inherited **molecular disease** was firmly established. Finally, this work led to a thorough study of human hemoglobins, which has provided valuable genetic insights.

In the United States, sickle-cell anemia is found almost exclusively in the black population. It affects about one in every 625 black infants born in this country. Currently, about 50,000 to 75,000 individuals are afflicted. In about one of every 145 black married couples, both partners are heterozygous carriers. In these cases, each of their children has a 25 percent chance of having the disease. You may wish to review the essay, found on pages 356–357, extending the above discussion.

Human Hemoglobins

Having introduced human hemoglobins in an historical context, it may be useful to extend our discussion and provide an update of what is currently known about these molecules in our own species. Molecular analysis reveals that a variety of hemoglobin molecules are produced in humans. All are tetramers consisting of numerous combinations of seven distinct polypeptide chains, each encoded by a separate gene. Almost all of adult hemoglobin consists of **HbA**, which contains two **alpha (α)** and two **beta (β) chains**. Recall that the mutation in sickle cell anemia involves the β chain.

HbA represents about 98 percent of all hemoglobin found in an individual's erythrocytes after the age of six months. The remaining 2 percent consists of **HbA$_2$**, a minor adult component. This molecule contains two alpha and two **delta (δ) chains**. The delta chain is very similar to the beta chain, consisting of 146 amino acids.

During embryonic and fetal development, quite a different set of hemoglobins is found. The earliest set to develop is called **Gower 1** containing two **zeta (ζ) chains**, which are most similar to alpha chains, and two **epsilon (ε) chains**, which are most similar to beta

TABLE 13.1 Chain Composition of Human Hemoglobin from Conception to Adulthood

Hemoglobin Type	Chain Composition
Embryonic-Gower 1	$\zeta_2\varepsilon_2$
Fetal-HbF	$\alpha_2{}^G\gamma_2$ $\alpha_2{}^A\gamma_2$
Adult-HbA	$\alpha_2\beta_2$
Minor adult-HbA$_2$	$\alpha_2\delta_2$

chains. By eight weeks of gestation, this embryonic form is gradually replaced by still another hemoglobin molecule with still different chains. This molecule is called **HbF**, or **fetal hemoglobin**, and consists of two alpha chains and two **gamma (γ) chains**. There are two types of gamma chains designated $^G\gamma$ and $^A\gamma$. Both are most similar to beta chains and differ from each other by only a single amino acid.

The nomenclature and sequence of appearance of the five tetramers described so far are summarized in Table 13.1.

Colinearity

Once it was established that genes specify the synthesis of polypeptide chains, the next logical question was how genetic information contained in the nucleotide sequence of a gene can be transferred to the amino acid sequence of a polypeptide chain. It seemed most likely that a **colinear relationship** would exist between the two molecules. That is, the order of nucleotides in the DNA of a gene would correlate directly with the order of amino acids in the corresponding polypeptide.

The initial experimental evidence in support of this concept was derived from studies by Charles Yanofsky of the *trpA* gene that encodes the A subunit of the enzyme **tryptophan synthetase** in *E. coli*. Yanofsky isolated many independent mutants that had lost activity of the enzyme. He was able to map these mutations and establish their location with respect to one another within the gene. He then determined where the amino acid substitution had occurred in each mutant protein. When the two sets of data were compared, the colinear relationship was apparent. The location of each mutation in the *trpA* gene correlates with the position of the altered amino acid in the A polypeptide of tryptophan synthetase. This comparison is illustrated in Figure 13.8. Recall that we have already discussed other support for the concept of colinearity, as studied in the bacteriophage MS2 (see Chapter 12, page 334).

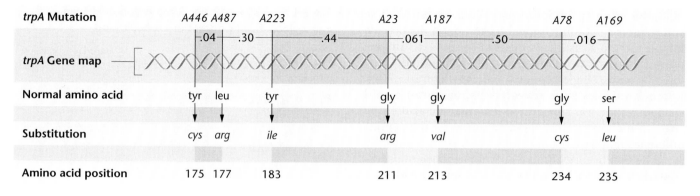

FIGURE 13.8 Demonstration of colinearity between the genetic map of various *trpA* mutations in *E. coli* and the affected amino acids in the protein product. The numbers shown between mutations represent linkage distances.

Protein Structure and Function

Having established that the genetic information is stored in DNA and influences cellular activities through the proteins it encodes, we turn now to a discussion of protein structure and function. How is it that these molecules play such a critical role in determining the complexity of cellular activities? As we will see, functional diversity of proteins is closely tied to the structure of these molecules.

Protein Structure

First, we should differentiate between the terms **polypeptides** and **proteins**. Both describe molecules composed of amino acids. The molecules differ, however, in their state of assembly and functional capacity.

Polypeptides are the precursors of proteins. As assembled on the ribosome during translation, the molecule is called a *polypeptide*. When released from the ribosome following translation, a polypeptide folds up and assumes a higher order of structure. When this occurs, a three-dimensional conformation in space emerges. In many cases, several polypeptides interact to produce this conformation. When the final conformation is achieved, the molecule, now fully functional, is appropriately called a *protein*. The three-dimensional conformation is essential to the function of the molecule.

The polypeptide chains of proteins, like nucleic acids, are linear nonbranched polymers. There are 20 amino acids that serve as the subunits (the building blocks) of proteins. Each amino acid has a **carboxyl group**, an **amino group**, and an **R (radical) group** (a side chain) bound covalently to a **central carbon atom**. The R group gives each amino acid its chemical identity. Figure 13.9 illustrates the 20 different R groups, which show a variety of configurations and can be divided into four main classes: (1) **nonpolar** or **hydrophobic**;

(2) **polar** or **hydrophilic**; (3) **negatively charged**; and (4) **positively charged**. Because polypeptides are often long polymers, and because each position may be occupied by any one of 20 amino acids with unique chemical properties, an enormous variation in chemical conformation and activity is possible. For example, if an average polypeptide is composed of 200 amino acids (molecular weight of about 20,000 daltons), 20^{200} different molecules, each with a unique sequence, can be created using 20 different building blocks.

About 1900, German chemist Emil Fischer determined the manner in which the amino acids are bonded together. He showed that the amino group of one amino acid can react with the carboxyl group of another amino acid during a dehydration reaction, releasing a molecule of H_2O. The resulting covalent bond is known as a **peptide bond** (Figure 13.10). Two amino acids linked together constitute a **dipeptide**, three a **tripeptide**, etc. Once 10 or more amino acids are linked by peptide bonds, the chain may be referred to as a polypeptide. Generally, no matter how long a polypeptide is, it will contain a free amino group at one end (the **N-terminus**) and a free carboxyl group at the other end (the **C-terminus**).

Four levels of protein structure are recognized: **primary (I°)**; **secondary (II°)**; **tertiary (III°)**; and **quaternary (IV°)**. The sequence of amino acids in the linear backbone of the polypeptide constitutes its **primary structure**. This sequence is specified by the sequence of deoxyribonucleotides in DNA via an mRNA intermediate. The primary structure of a polypeptide helps determine the specific characteristics of the higher orders of organization as a protein is formed.

The **secondary structure** refers to a regular or repeating configuration in space assumed by amino acids

FIGURE 13.9 Chemical structures and designations of the 20 amino acids found in living organisms, divided into four major categories.

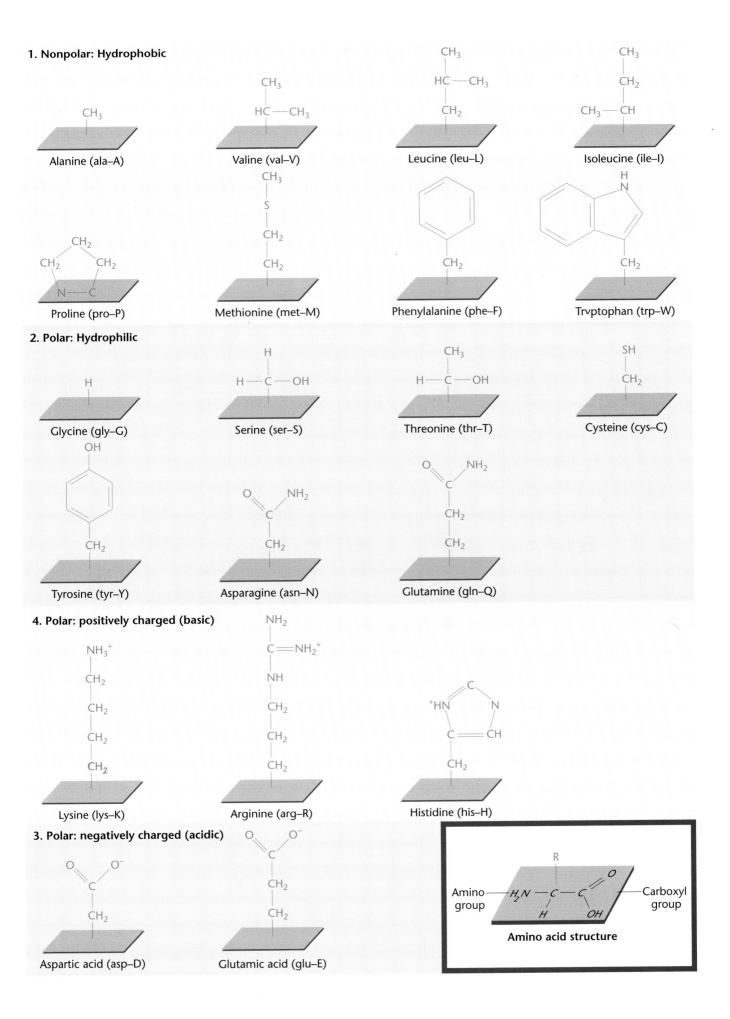

1. Nonpolar: Hydrophobic

Alanine (ala–A)

Valine (val–V)

Leucine (leu–L)

Isoleucine (ile–I)

Proline (pro–P)

Methionine (met–M)

Phenylalanine (phe–F)

Tryptophan (trp–W)

2. Polar: Hydrophilic

Glycine (gly–G)

Serine (ser–S)

Threonine (thr–T)

Cysteine (cys–C)

Tyrosine (tyr–Y)

Asparagine (asn–N)

Glutamine (gln–Q)

4. Polar: positively charged (basic)

Lysine (lys–K)

Arginine (arg–R)

Histidine (his–H)

3. Polar: negatively charged (acidic)

Aspartic acid (asp–D)

Glutamic acid (glu–E)

Amino group

Carboxyl group

Amino acid structure

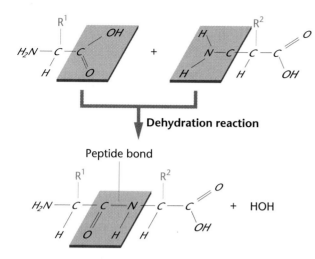

FIGURE 13.10 Peptide bond formation between two amino acids, resulting from a dehydration reaction.

aligned closely to one another in the polypeptide chain. In 1951, Linus Pauling and Robert Corey predicted, on theoretical grounds, an α (alpha) helix as one type of secondary structure. The α-helix model [Figure 13.11(a)] has since been confirmed by X-ray crystallographic studies. It is rodlike and has the greatest possible theoretical stability. The helix is composed of a spiral chain of amino acids stabilized by hydrogen bonds.

The side chains (the R-groups) of amino acids extend outward from the helix, and each amino acid residue occupies a vertical distance of 1.5 Å in the helix. There are 3.6 residues per turn. While left-handed helices are theoretically possible, all proteins demonstrating an α helix are right-handed.

Also in 1951, Pauling and Corey proposed a second structure, the β-pleated-sheet configuration. In this model, a single polypeptide chain folds back on itself, or several chains run in either parallel or antiparallel fashion next to one another. Each such structure is stabilized by hydrogen bonds formed between atoms present on adjacent chains [Figure 13.11(b)]. A single zigzagging plane is formed in space with adjacent amino acids 3.5 Å apart.

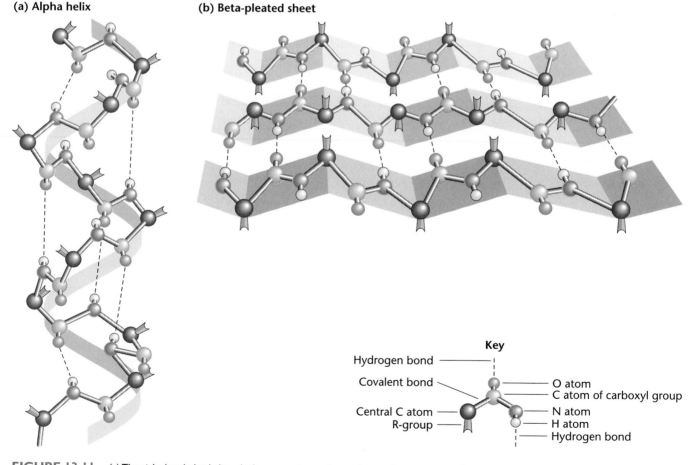

FIGURE 13.11 (a) The right-handed α helix, which represents one form of secondary structure of a polypeptide chain. (b) The β-pleated-sheet configuration, an alternative form of secondary structure of polypeptide chains. For the sake of clarity, some atoms are not shown.

As a general rule, most proteins demonstrate a mixture of α and β structure. Globular proteins, most of which are round in shape and water soluble, usually contain a core of β-pleated-sheet structure as well as many areas demonstrating a helical structure. The more rigid structural proteins, many of which are water insoluble, rely on more extensive β-pleated-sheet regions for their rigidity. For example, **fibroin**, the protein made by the silk moth, depends extensively on this form of secondary structure.

While the secondary structure describes the arrangement of amino acids within certain areas of a polypeptide chain, **tertiary (III°) protein structure** defines the three-dimensional conformation of the entire chain in space. Each protein twists and turns and loops around itself in a very specific fashion, characteristic of the specific protein. Three aspects of this III° structure are most important in determining this conformation and in stabilizing the molecule.

1. Covalent disulfide bonds form between closely aligned cysteine residues to form the unique amino acid cystine.

2. Nearly all of the polar, hydrophilic R groups are located on the surface, where they can interact with water.

3. The nonpolar, hydrophobic R groups are usually located on the inside of the molecule, where they interact with one another, avoiding interaction with water.

It is important to emphasize that the three-dimensional conformation achieved by any protein is a product of the primary (I°) structure of the polypeptide. Thus, the genetic code need only specify the sequence of amino acids in order to encode information that leads ultimately to the final assembly of proteins. The three stabilizing factors listed above depend on the location of each amino acid relative to all others in the chain. As folding occurs, the most thermodynamically stable conformation possible results.

A model of the three-dimensional tertiary structure of the respiratory pigment **myoglobin** is shown in Figure 13.12. This level of organization is extremely important because the specific function of any protein is directly related to its three-dimensional conformation.

The **quaternary level of organization** applies only to proteins composed of more than one polypeptide chain. The IV° structure indicates the conformation of the various chains in relation to one another. This type of protein is called **oligomeric**, and each chain is called a **protomer**, or less formally, a **subunit**. The individual protomers have conformations that fit together with other subunits in a specific complementary fashion. He-

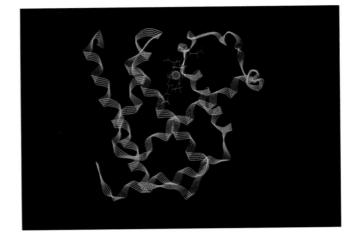

FIGURE 13.12   A ribbon drawing illustrating the tertiary (III°) level of protein structure in respiratory pigment myoglobin. The bound oxygen atom is shown in red.

moglobin, an oligomeric protein consisting of four polypeptide chains, has been studied in great detail. Its IV° structure is shown in Figure 13.13. Most enzymes, including DNA and RNA polymerase, demonstrate IV° structure.

Chaperones and Protein Folding

Because of the relationship of the three-dimensional structure to protein function, it has been of great interest to confirm how polypeptide chains fold into their final conformation. For many years, it was thought that **protein folding** was a spontaneous process, whereby the mol-

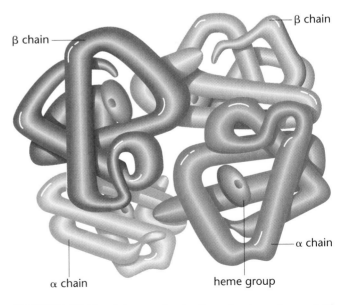

FIGURE 13.13   Schematic drawing illustrating the quaternary level (IV°) of protein structure as seen in hemoglobin. Four chains (2 alpha and 2 beta) interact with four heme groups to form the functional molecule.

ecule achieved maximum thermodynamic stability, based largely on the properties inherent in the amino acid sequence of the polypeptide chain(s) composing the protein as well as the interactions of the polypeptide subunits. However, numerous studies have shown that, for many proteins, folding is dependent upon members of a family of still other ubiquitous proteins called **chaperones**.

Chaperone proteins (sometimes called *molecular chaperones* or *chaperonins*) function to facilitate the folding of other proteins. Although the mechanism whereby they accomplish this role is still obscure, the interaction between chaperones and their target proteins is known not to be covalent. Furthermore, like enzymes, chaperones do not become part of the final product.

The initial discovery of chaperones was made in the 1970s, when they were first called **heat-shock proteins** because they were found to be induced in *Drosophila* following exposure of flies to elevated temperatures. Since then, this type of chaperone has been discovered in a variety of organisms, including bacteria, animals, and plants, where they are overproduced in response to heat-shock, presumably to minimize the denaturation process that proteins undergo when heated.

In cells that are not heat-shocked, other types of chaperones appear to be required for the normal assembly of many proteins. To facilitate assembly, they may bind to and stabilize polypeptide chains in a conformation essential for subsequent folding. This binding is sometimes envisioned as a process that prevents incorrect folding by restricting certain potential conformations. In other cases, chaperones are believed to bind to proteins that must partially unwind in order to be transported across intracellular membranes. The binding is thought to prevent potential aggregation that could occur during this unwinding. Still another type of chaperone is known to function to maintain histone proteins in a conformation accessible to interaction with DNA during the assembly of **nucleosomes** (see Chapter 17).

Even though the molecular mechanism of chaperone action remains obscure and a topic of intense research effort, it is clear that these molecules are of great importance to protein structure and function. Their discovery has expanded our view of protein folding.

Posttranslational Modification and Protein Targeting

Before turning to a discussion of protein function, it is important to point out that polypeptide chains, like RNA, are often modified once they have been synthesized. This additional processing is broadly described as **posttranslational modification**. Although many of these alterations are detailed biochemical transformations and be-

yond the scope of this discussion, you should be aware that they occur, and that they are critical to the functional capability of the final protein product. Furthermore, many such modifications are involved in the regulation of protein activity.

Several examples of posttranslational modification are presented below. The final example will be discussed in slightly more detail.

1. **The N-terminus and C-terminus amino acids are usually removed or modified.** For example, the initial N-terminal formylmethionine residue in bacterial polypeptides is usually removed enzymatically. Often, the amino group of the initial methionine residue is removed, and the amino group of the N-terminal residue is chemically modified (acetylated) in eukaryotic polypeptide chains.

2. **Individual amino acid residues are sometimes modified.** For example, phosphates may be added to the hydroxyl groups of certain amino acids such as tyrosine. Modifications such as these create negatively charged residues that may ionically bond with other molecules. The process of phosphorylation is extremely important in regulating many cellular activities as a result of the action of enzymes called **kinases**. In other proteins, methyl groups may be added enzymatically.

3. **Carbohydrate side chains are sometimes attached.** These are added covalently, producing **glycoproteins**, an important category of molecules that includes many antigenic determinants, such as those specifying the antigens in the ABO blood type system in humans.

4. **Polypeptide chains may be trimmed.** For example, insulin is first translated into a longer molecule that is enzymatically trimmed to its final 51-amino acid form.

5. **Signal sequences are removed.** At the N-terminal end of some proteins is found a sequence of up to 30 amino acids that plays an important role in directing the protein to the location in the cell where it functions. This is called a **signal sequence**, and it determines the final destination of a protein in the cell. This process is called **protein targeting**.

For example, proteins, whose fate involves secretion or which are to become part of the plasma membrane, are dependent on specific sequences for their initial transport into the lumen of the endoplasmic reticulum. While the signal sequence of various proteins with a common destination might differ in their primary amino acid sequence, they share many chemical properties. For example, those destined for secretion all contain a string of up to 15 hydrophobic amino acids preceded by a positively charged amino acid at the N-terminus of the signal sequence. Once transported, but prior to achieving their functional status as proteins, the signal sequence is enzymatically removed from these polypeptides.

6. **Polypeptide chains are often complexed with metals.** The tertiary and quaternary level of protein structure often includes and is dependent on metal atoms. The function of the protein is thus dependent on the molecular complex that includes both polypeptide chains and metal atoms. Hemoglobin, containing four iron atoms along with four polypeptide chains, is a good example.

These various types of posttranslational modifications are no doubt important in the conversion of newly translated polypeptide chains into their final three-dimensional conformation in space. Therefore, such modifications are critical in achieving the functional status specific to proteins.

Protein Function

The essence of life on earth rests at the level of diverse cellular function. One can argue that DNA and RNA simply serve as vehicles to store and express genetic information. However, proteins are at the heart of cellular function. And, it is the capability of cells to assume diverse structures and functions that distinguishes most eukaryotes from less evolutionarily advanced organisms such as bacteria. Therefore, an introductory understanding of protein function is critical to a complete view of genetic processes. Detailed coverage of this topic is an integral part of the study of **biochemistry**.

Proteins are the most abundant macromolecules found in cells. As the end products of genes, they play many diverse roles. For example, the respiratory pigments **hemoglobin** and **myoglobin** transport oxygen, which is essential for cellular metabolism. **Collagen** and **keratin** are structural proteins associated with the skin, connective tissue, and hair of organisms. **Actin** and **myosin** are contractile proteins, found in abundance in muscle tissue. Still other examples are the **immunoglobins**, which function in the immune system of vertebrates; **transport proteins**, involved in movement of molecules across membranes; some of the **hormones** and their **receptors**, which regulate various types of chemical activity; and **histones**, which bind to DNA in eukaryotic organisms.

The largest group of proteins with a related function are the **enzymes**. Since we have been referring to these molecules throughout this chapter, it may be useful to extend our discussion and include a more detailed description of their biological role. These molecules specialize in catalyzing chemical reactions within living cells. Enzymes increase the rate at which a chemical reaction reaches equilibrium, but they do not alter the end point of the chemical equilibrium. Their remarkable, highly specific catalytic properties largely determine the biomolecular nature of any cell type. The specific functions of many enzymes involved in the genetic and cellular processes of cells are described throughout the text.

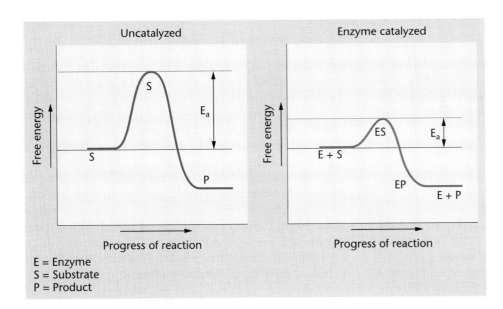

E = Enzyme
S = Substrate
P = Product

FIGURE 13.14 Energy requirements of an uncatalyzed versus an enzymatically catalyzed chemical reaction. The energy of activation (E_a) necessary to initiate the reaction is substantially lower as a result of catalysis.

Biological catalysis is a process whereby the **energy of activation** for a given reaction is lowered (Figure 13.14). The energy of activation is the increased kinetic energy state that molecules must usually reach before they react with one another. While this state can be attained as a result of elevated temperatures, enzymes allow biological reactions to occur at lower physiological temperatures. In this way, enzymes make possible life as we know it.

The catalytic properties and specificity of an enzyme are determined by the chemical configuration of the molecule's **active site**. This site is associated with a crevice, a cleft, or a pit on the surface of the enzyme, which binds the reactants, or substrates, facilitating their interaction. Enzymatically catalyzed reactions control metabolic activities in the cell. Each reaction is either **catabolic** or **anabolic**. Catabolism, as illustrated in the portion of the phenylalanine–tyrosine pathway interrupted in alkaptonuria (Figure 13.1), is the degradation of large molecules into smaller, simpler ones with the release of chemical energy. Anabolism, as illustrated in the production of eye pigments in *Drosophila* (Figure 13.3), is the synthetic phase of metabolism, yielding nucleic acids, proteins, lipids, and carbohydrates. Metabolic pathways that serve the dual function of anabolism and catabolism are **amphibolic**.

Protein Structure and Function: The Collagen Fiber

To provide a more in-depth example of the relationship between protein structure and function, we shall consider one of the most interesting proteins found in vertebrates: **collagen**. As the most abundant protein in mammals, it has been extensively studied and clearly illustrates this relationship.

Collagen is found in many places in the body, including tendons, ligaments, bone, connective tissue, skin, blood vessels, teeth, and the lens and cornea of the eye. In each case, the presence of collagen increases the tensile strength of, and provides support to, the specific tissue of which it is a part. A consideration of the above body parts and structures that contain or consist solely of collagen makes it evident that this protein must be a versatile protein, providing a variable degree of flexibility and support to these tissues.

Although several types of collagen are produced in vertebrates, we will restrict our discussion to just the major one, called *type I*. Its main subunit or building block is called **tropocollagen**, and it consists of three polypeptide chains wrapped around one another, forming a **triple helix**. This unit is extremely large, being 15 Å in diameter and 3000 Å long. Each polypeptide contains about 1000 amino acids, and the three interacting chains of the helix are stabilized by hydrogen bonds between them. There is no hydrogen bonding between amino acids within a single chain. Therefore, no α-helical or β-pleated-sheet secondary structure exists in collagen.

The maturation of the polypeptides into tropocollagen and the subsequent condensation of these triple helical structures into the more densely coiled collagen fibers is a fascinating tale. The polypeptides are first synthesized by fibroblasts (undifferentiated precursors of connective tissue cells) as even longer units called *procollagen*. These are secreted into extracellular spaces, where they are cleaved and shortened at both the N-terminus and C-terminus by specific enzymes called **procollagen peptidases** [Figure 13.15(a)].

Once tropocollagen has been formed, these triple-helical chains associate with one another spontaneously to form dense collagen fibers. The association follows an orderly pattern, where rows of end-to-end molecules line up in a staggered fashion next to one another [Figure 13.15(b)]. In each row a gap of approximately 400 Å exists between each tropocollagen unit. This complex structural arrangement creates protein fibers that strengthen and support a variety of tissues.

Although we have concentrated our discussion on type I collagen, there are about 20 different collagen genes, each encoding a slightly different primary chain. At least 10 different combinations of different chains have been discovered, each constituting a variant collagen molecule. It is likely that each imparts slightly different properties and characteristics to the different forms of collagen.

The Genetics of Collagen

We began this chapter with a discussion of an inherited biochemical disorder first studied early in this century by Sir Archibald Garrod. It is appropriate that we conclude with a short discussion of more modern discoveries of inherited biochemical disorders, this time related to collagen.

At least two inherited disorders known to involve defects in collagen synthesis and assembly are fairly well characterized genetically. For example, the inability to adequately transform procollagen to collagen leads to the human connective tissue disorder **Ehlers-Danlos syndrome**. Individuals exhibiting this disorder have fragile, stretchable skin and are loose-jointed, providing hypermobility. Some forms of the syndrome render individuals susceptible to arterial and colon ruptures as well

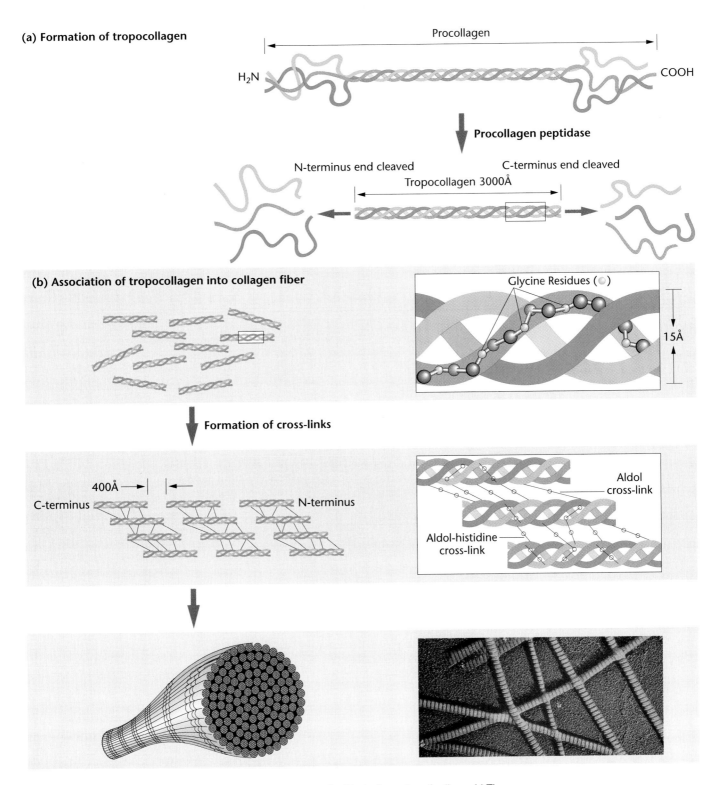

(a) Formation of tropocollagen

Procollagen

H_2N

COOH

Procollagen peptidase

N-terminus end cleaved

C-terminus end cleaved

Tropocollagen 3000Å

(b) Association of tropocollagen into collagen fiber

Glycine Residues ( )

15Å

Formation of cross-links

400Å

C-terminus

N-terminus

Aldol cross-link

Aldol-histidine cross-link

FIGURE 13.15 Various steps and intermediate structures involved in the formation of collagen. (a) The process of posttranslational modification of polypeptides during the formation of tropocollagen, the precursor of collagen. (b) The formation of cross-links leading to the assembly of the collagen fiber. An electron micrograph illustrates such fibers.

as periodontal problems. They exhibit elevated levels of procollagen and decreased activity of the enzyme procollagen peptidase.

Our second example involves the lethal human disorder **osteogenesis imperfecta**. There are many variations, but all affect long bone formation and result from some defect in collagen structure. In its most severe form (*type II*), widespread bone defects occur in the fetus and newborn; fractures occur spontaneously and death often occurs soon after birth.

Type I, in comparison, is fairly mild, with an onset sometimes as late as age 35 to 45. However, at that point, declining health is associated with the degeneration of collagen-rich tissues: Blood vessels weaken, bones become fragile, and they fracture frequently. A stroke or heart attack usually causes premature death.

In one instance, the specific genetic defect has been uncovered and found to involve but a single glycine residue in the primary procollagen chain. At position 988, a glycine residue has been altered to the amino acid cys-

GENETICS, TECHNOLOGY, AND SOCIETY

Prions, Mad Cows, and Heresies

One of the foundations of modern genetics is the "Central Dogma"—the concept that inherited information resides within the base sequence of nucleic acids—and that this information flows from nucleic acids toward proteins. Polypeptides are the ultimate products of the genetic code. Although the scientific world was willing to accept the exceptional idea that information in RNA could be transferred to DNA via reverse transcription, the notion that proteins could beget infectious proteins was harder to accept. Like many heretical ideas, the infectious proteins (or "prion") hypothesis arose from strange beginnings and remains highly controversial. It is a strange tale, with recent worldwide economic and medical implications.

Over the last three decades, scientists have identified a number of rare and fatal human neurodegenerative diseases. These diseases are known as spongiform encephalopathies and include Creutzfeldt-Jakob disease (CJD), Gerstmann-Sträussler-Scheinker disease, and Kuru. These diseases are characterized by long incubation times, progressive neurodegeneration, and death. Brain tissue resembles a sponge (hence spongiform) and often accumulates proteinaceous plaques. Over 80 percent

of CJD cases arise spontaneously and apparently randomly, at a rate of about 1 in a million per year worldwide. A less common form of CJD is inherited as an autosomal dominant condition. But the transmitted forms are the strangest. CJD has been transmitted through surgical grafts and from injection of cadaver-derived growth hormone. Kuru is a neurodegenerative illness of the Fore natives of New Guinea who transmitted the disease through ritualistic cannibalism, by eating the brains of infected relatives. Similar diseases afflict animals and include scrapie (sheep and goats) and bovine spongiform encephalopathy (BSE)—the infamous "mad cow disease."

Scrapie and BSE are thought to be transmitted between animals by ingestion of scrapie-infected and BSE-infected animal remains, particularly neural tissue. The recent mad cow disease scare in Britain arose following a decade-long epidemic of BSE among beef and dairy cattle. The epidemic was thought to arise as a result of adding insufficiently treated sheep (and later cow) entrails to cattle rations—as a protein additive. In 1988, the British government banned the use of cows and sheep as feed for other cows and sheep, and the epidemic is subsiding. However, the frightening question that triggered the worldwide ban on British beef, and the near collapse of the

British cattle industry, remains: Can BSE (or scrapie) be transmitted to humans, by eating contaminated meat? To address this question, we need to return to the prion hypothesis, and genetic heresies.

For many years, the spongiform encephalopathies defied analysis. The diseases are difficult to study, requiring injection of infected brain material into brains of experimental animals. In the early 1980s, Stanley Prusiner of the University of California, San Francisco, purified the infectious agent and discovered that it consisted only of protein. Scrapie infectious material was unaffected by treatments with radiation or nucleases that damage nucleic acids. However, it was destroyed by treatment with reagents that hydrolyze or modify proteins. Prusiner proposed that scrapie was spread by an infectious protein particle that he called a **prion**. His hypothesis was immediately dismissed by most scientists. However, over the last decade Prusiner and others have continued to present evidence supporting the prion hypothesis, and the notion that an infectious particle, devoid of nucleic acids, can cause disease has gained acceptance.

But the surprises were not over. Once the prion protein (PrP) was purified, scientists determined its amino acid sequence, designed a synthetic DNA probe based upon the protein's teine by a mutation of a single base change in the gene. This alteration of the primary structure inhibits the formation of highly ordered fibers that make up the bundles of collagen and leads to all of the aberrant phenotypic effects associated with the disease.

Collagen disorders are quite prevalent in the human population. The three disorders discussed here are inherited as autosomal dominant traits. It is estimated that over 100,000 individuals worldwide suffer from **osteogenesis imperfecta**. While **Ehlers-Danlos syndrome** is much rarer, there are about 20,000 people who suffer from still another collagen-based disorder, **Marfan syndrome**. Long, thin arms, legs, and digits of the hands and feet characterize the appearance of affected individuals. Ocular problems occur, and aortic heart disease most frequently is the cause of premature death. Abraham Lincoln is suspected to have suffered from this disorder. Given the complexity of the biochemistry of collagen and its wide distribution throughout the body, it is not surprising that these and other genetic disorders exist.

sequence, and used the probe to search for the gene that codes for PrP. To everyone's amazement, the gene that encodes PrP was found to reside in the genome of the host—in humans, on the short arm of chromosome 20. PrP is a normal protein, synthesized in neurons and found in the brains of all adult animals. How then could a normal and benign protein change to become a malignant, infectious prion? The key to answering the question came from studies of the structure of PrP. PrP from the brains of infected animals tends to aggregate, is insoluble in detergents, and is resistant to protease enzymes that degrade normal, noninfectious PrP. In addition, recent data on the secondary structure of PrP show that non-infectious PrP folds into α-helices whereas infectious PrP folds into β-sheets. It also appears that infectious PrP is created from normal cellular PrP that is synthesized by the host.

The prion diseases can now be understood as diseases of protein secondary structure. It is thought that non-infectious PrP can, under appropriate conditions, unfold its α-helices and refold them into β-sheets. The conversion is triggered by contact between infectious PrP and normal cellular PrP molecules. Once normal PrP is transformed into the infectious PrP form, it spreads its conformation to its neighboring PrP molecules, and the process proceeds like a chain reaction. The initial conversion can also occur spontaneously, particularly if the amino acid sequence of the normal cellular PrP is different from wild-type, making it more susceptible to refolding. This explains how the disease can be genetic and dominant, and how it can arise by rare spontaneous events triggered by somatic mutation in the PrP gene or by a random refolding of a PrP molecule. Once one PrP molecule makes the conversion in structure, a chain reaction begins and infectious PrP begins to accumulate and cause brain damage.

One of the most pressing current questions about prion diseases is whether they can spread from one species to another species. Scrapie has been transmitted to cattle but there is no evidence that it has ever spread to humans. BSE has been transmitted to mice and cats, but not to hamsters or dogs. There is as yet no conclusive evidence that BSE has spread to humans, although the possibility has not been excluded. It appears that the conversion of PrP from normal to infectious forms occurs more readily if two PrP molecules are very similar in amino acid sequence. PrPs of sheep and cows differ by only 7 amino acids whereas the PrP's of cows and humans differ by 30 amino acids. Transgenic mice that express both mouse and human PrP develop only mouse infectious PrP in response to injection of BSE prions. However, the question of whether the cow–human species barrier can be breached remains inconclusive.

Other still-unanswered and important questions surround the prion diseases. Can they be spread through milk, meat, or soil contamination, or is direct injection or ingestion of PrP required? Is mad cow disease now out of the food chain? Will a treatment be found? Will we develop tests to inform people of their genetic susceptibility? Are other neurodegenerative diseases such as Alzheimer's, Parkinson's, and amyotrophic lateral sclerosis (ALS) caused by prion-like agents? We can expect that research stimulated by Prusiner's heretical prion hypothesis will help to answer these questions.

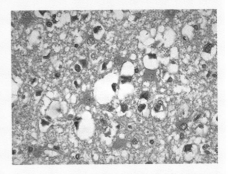

Holes (clear areas) in brain tissue, giving it a spongelike appearance, a frequent feature of prion diseases.

CHAPTER SUMMARY

1. Basic to Garrod's studies in the early 1900s of inborn, human metabolic disorders such as cystinuria, albinism, and alkaptonuria, was the concept that genes control the synthesis of specific metabolic products.

2. The investigation of eye pigments in *Drosophila* and nutritional requirements in *Neurospora* by Beadle and colleagues made it clear that mutations cause the loss of enzyme activity. Their work led to the concept of one gene:one enzyme.

3. The one gene:one enzyme hypothesis was later revised. Pauling and Ingram's investigations of hemoglobins from patients with sickle-cell anemia led to the discovery that one gene directs the synthesis of only one polypeptide chain.

4. Thorough investigations have revealed the existence of several major types of human hemoglobin molecules, found in the embryo, fetus, and the adult. Specific genes control each polypeptide chain constituting these various hemoglobin molecules.

5. The proposal suggesting that a gene's nucleotide sequence specifies in a colinear way the sequence of amino acids in a polypeptide chain was confirmed by Yanofsky's experiments involving mutations in the tryptophan synthetase gene in *E. coli*.

6. Proteins, the end products of genes, demonstrate four levels of structural organization that together provide the chemical basis for their three-dimensional conformation.

7. Of the myriad functions performed by proteins, the most influential role is assumed by enzymes. These highly specific, cellular catalysts play a central role in the production of all classes of molecules in living systems.

8. Collagen is an abundant, specialized protein in vertebrates, serving a structural role in a variety of tissues. Ehlers-Danlos syndrome, osteogenesis imperfecta, and Marfan syndrome are examples of inherited human disorders resulting from mutations that alter collagen structure and its function.

KEY TERMS

actin, 379
active site, 380
alkaptonuria, 365
alpha chain (of Hb), 373
α (alpha) helix, 376
amino group, 374
amphibolic reaction, 380
anabolic reaction, 380
arginine, biosynthesis, 370
beta chain (of Hb), 373
β-pleated-sheet, 376
biochemistry, 379
biological catalysis, 380
bovine spongiform encephalopathy (BSE), 382
carboxyl group, 374
catabolic reaction, 380
chaperones, 378
colinear relationship, 373
collagen, 379
Creutzfeldt-Jakob disease (CJD), 382
C-terminus, 374
dipeptide, 374
Ehlers-Danlos syndrome, 380
electrophoresis, starch gel, 372
energy of activation, 380
enzyme, 379

fetal hemoglobin, 373
fibroin, 377
fingerprinting technique, 372
Gower 1 hemoglobin, 373
heat-shock proteins, 378
heme groups, 372
hemoglobin, 373
histone, 379
homogentisic acid, 365
hormone, 379
hydrophilic group, 374
hydrophobic group, 374
imaginal disk, 366
immunoglobin, 379
keratin, 379
kinases, 378
Kuru, 382
Marfan syndrome, 383
molecular disease, 373
myoglobin, 377
myosin, 379
nucleosomes, 378
N-terminus, 374
oligomeric protein, 377
one-gene:one-enzyme hypothesis, 366
one gene:one polypeptide chain hypothesis, 370

one gene:one protein hypothesis, 370
osteogenesis imperfecta, 382
peptide bond, 374
phenylalanine hydroxylase, 366
phenylketonuria (PKU), 366
polypeptide chain, 370
polypeptides, 374
posttranslational modification, 378
primary (I°) protein structure, 374
prion, 382
procollagen peptidase, 380
protein, 374
protein folding, 377
protein targeting, 378
protomer, 377
quaternary (IV°) protein structure, 377
R (radical) group, 374
secondary (II°) protein structure, 374
sickle-cell anemia, 371
sickle-cell trait, 371
signal sequence, 378
tertiary (III°) protein structure, 374
transport proteins, 379
tropocollagen, 380
tryptophan synthetase, 373

INSIGHTS AND SOLUTIONS

1. In Beadle and Ephrussi's studies involving imaginal disk transplants, most eye color mutants behaved autonomously (i.e., they were not cured, or converted to wild-type color) when allowed to develop in a wild-type abdominal environment. One concludes that in such cases, the product influenced by the gene under study is *not* diffusible. The *vermilion (v)* and *cinnabar (cn)* eye disks were the exception, developing into wild-type eyes, because the products controlled by these genes are diffusible and supplied by the host. Recall that *cn* disks developing in *v* hosts remain as mutant *cn* eyes, but that *v* disks developing in *cn* hosts are cured and develop wild-type color. This is the basis of determining that the v^+ gene product acts at a point in the biosynthetic pathway prior to the cn^+ gene product.

   Suppose that a new bright red mutation is discovered and the product controlled by this gene is not diffusible. However, the gene product acts prior to the steps controlled by both the v^+ and cn^+ gene products in the biosynthesis of the brown xanthommatin pigment. Predict the outcome of a transplant experiment of the new mutant eye disk into a wild-type larva.

   Solution: Because the block in the new mutation precedes either the cn^+ or the v^+ steps, the wild-type host will make both of these diffusible products. As a result, the mutant disk will be cured and develop wild-type color. This happens because these diffusible products can enter the disk and allow the block in the new mutation to be bypassed.

2. The following growth responses were obtained using four mutant strains of *Neurospora* and the related compounds A, B, C, and D. None of the mutations grow on minimal medium. Draw all possible conclusions.

		Growth Product			
		C	D	B	A
Mutation	1	−	−	−	−
	2	−	+	+	+
	3	−	−	+	+
	4	−	−	+	−

   Solution: First, nothing can be concluded about mutation 1, except that it is lacking some essential growth factor, perhaps even unrelated to the biochemical pathway represented by mutations 2–4; nor can anything be concluded about compound C. If it is involved in the pathway, it is a product synthesized prior to the synthesis of A, B, and D.

   We must now analyze these three compounds and the control of their synthesis by the enzymes encoded by genes 2, 3, and 4. Because product B allows growth in all three cases, it may be considered the "end product." It bypasses the block in all three instances. Using similar reasoning, product A precedes B in the pathway since it allows a bypass in two or the three steps. Product D precedes B, yielding the more complete solution:

   $$C(?) \longrightarrow D \longrightarrow A \longrightarrow B$$

   Now determine which mutations control which steps. Since mutation 2 can be alleviated by products D, B, and A, it must control a step prior to all three products, perhaps the direct conversion to D, although we cannot be certain. Mutation 3 is alleviated by B and A, so its effect must precede them in the pathway. Thus, we will assign it as controlling the conversion of D to A. Likewise, we can assign mutation 4 to the conversion of A to B, leading to the more complete solution:

   $$C(?) \xrightarrow{2(?)} D \xrightarrow{3} A \xrightarrow{4} B$$

PROBLEMS AND DISCUSSION QUESTIONS

1. Discuss the potential difficulties involved in designing a diet to alleviate the symptoms of phenylketonuria.
2. Phenylketonurics cannot convert phenylalanine to tyrosine. Why don't these individuals exhibit a deficiency of tyrosine?
3. Phenylketonurics are often more lightly pigmented than are normal individuals. Can you suggest a reason why this is so?
4. Consider the following hypothetical circumstances involving reciprocal transplantations of three mutant imaginal eye disks. The three mutations, *a*, *b*, and *c*, are responsible for the enzymatic pathway shown below. However, it is not known which mutant gene controls which step in the reactions.

$$\begin{array}{ccc} (?) & (?) & (?) \\ \downarrow & \downarrow & \downarrow \\ w \longrightarrow x \longrightarrow y \longrightarrow z \end{array}$$

The results of the transplantations are shown below. Assuming that the *w*, *x*, *y*, and *z* substances are diffusible, determine which reactions are controlled by which genes.

Donor	Host	Transplant Eye Pigment	Development
a	*c*	Mutant	Autonomous
a	*b*	Mutant	Autonomous
c	*b*	Mutant	Autonomous
c	*a*	Wild type	Nonautonomous
b	*c*	Wild type	Nonautonomous
b	*a*	Wild type	Nonautonomous

5. What F_1 and F_2 ratios will occur in a cross between *vermilion* females and *cinnabar* males? Recall that *v* is X-linked and

cn is autosomal, and that these mutants produce identical bright red phenotypes. What ratios will occur in the reciprocal cross (*v* males and *cn* females)?

6. The synthesis of flower pigments is known to be dependent upon enzymatically controlled biosynthetic pathways. In the crosses shown below, postulate the role of mutant genes and their products in producing the observed phenotypes.

(a) P_1: white strain A × white strain B
F_1: all purple
F_2: 9/16 purple: 7/16 white

(b) P_1: white × pink
F_1: all purple
F_2: 9/16 purple: 3/16 pink: 4/16 white

7. A series of mutations in the bacterium *Salmonella typhimurium* result in the requirement of either tryptophan or some related molecule in order for growth to occur. From the data shown below, suggest a biosynthetic pathway for tryptophan.

	Growth Supplement				
Mutation	Minimal Medium	Anthranilic Acid	Indole Glycerol Phosphate	Indole	Tryptophan
trp-8	−	+	+	+	+
trp-2	−	−	+	+	+
trp-3	−	−	−	+	+
trp-1	−	−	−	−	+

8. The study of biochemical mutants in organisms such as *Neurospora* has demonstrated that some pathways are branched. The data shown below illustrate the branched nature of the pathway resulting in the synthesis of thiamine. Why don't the data support a linear pathway? Can you postulate a pathway for the synthesis of thiamine in *Neurospora*?

	Growth Supplement			
Mutation	Minimal Medium	Pyrimidine	Thiazole	Thiamine
thi-1	−	−	+	+
thi-2	−	+	−	+
thi-3	−	−	−	+

9. Explain why the one-gene:one-enzyme concept is not considered accurate today.

10. Why is an alteration of electrophoretic mobility interpreted as a change in the primary structure of the protein under study?

11. Contrast the polypeptide chain components of each of the hemoglobin molecules found in humans.

12. Using sickle-cell anemia as a basis, describe what is meant by a molecular or genetic disease. What are the similarities and dissimilarities between this type of a disorder and a disease caused by an invading microorganism?

13. Contrast the contributions of Pauling and Ingram to our understanding of the genetic basis for sickle-cell anemia.

14. Hemoglobins from two individuals are compared by electrophoresis and by fingerprinting. Electrophoresis reveals no difference in migration, but fingerprinting shows an amino acid difference. How is this possible?

15. Describe what colinearity means.

16. Certain mutations called *amber* in bacteria and viruses result in premature termination of polypeptide chains during translation. Many *amber* mutations found at different points along the gene coding for a head protein in phage T4 have been detected. How might this system be further investigated to demonstrate and support the concept of colinearity?

17. Define and compare the four levels of protein organization.

18. List as many different protein functions as you can, with an example of each.

19. How does an enzyme function? Why are enzymes essential for living organisms on earth?

20. Describe the assembly of the collagen fiber and the "collagen" diseases resulting from mutations that affect this assembly and/or the mature collagen fiber.

21. Why does Fiers' work with phage MS2, discussed in Chapter 12, constitute more direct evidence in support of colinearity than Yanofsky's work with the *trpA* locus in *E. coli* (discussed in this chapter)?

22. Shown below are several amino acid substitutions in the α and β chains of human hemoglobin. Using the code table (Figure 12.6), determine how many of them can occur as a result of a single nucleotide change.

Hb Type	Normal Amino Acid	Substituted Amino Acid
HbJ Toronto	ala	asp (α-5)
HbJ Oxford	gly	asp (α-15)
Hb Mexico	gln	glu (α-54)
Hb Bethesda	tyr	his (β-145)
Hb Sydney	val	ala (β-67)
HbM Saskatoon	his	tyr (β-63)

EXTRA-SPICY PROBLEMS

23. Three independently assorting genes are known to control the following biochemical pathway that provides the basis for flower color in a hypothetical plant. Homozygous recessive mutations, which interrupt each step, are known.

Colorless $\xrightarrow{\text{A}-}$ yellow $\xrightarrow{\text{B}-}$ green $\xrightarrow{\text{C}-}$ speckled

Determine the phenotypic results in the F_1 and F_2 generation resulting from the following P_1 crosses involving true breeding plants.

(a) speckled (*AABBCC*) × yellow (*AAbbCC*)
(b) yellow (*AAbbCC*) × green (*AABBcc*)
(c) colorless (*aaBBCC*) × green (*AABBcc*)

24. How would the results vary in cross (a) of Problem 23 if genes A and B were linked with no crossing over between them? How would cross (a) vary if genes A and B were linked and 20 map units apart?

25. Most wood lily plants have orange flowers and are true-breeding. A genetics student discovered both red and yellow true-breeding variants and proceeded to hybridize the three strains:

P$_1$	F$_1$	F$_2$
(1) orange × red	all orange	3/4 orange 1/4 red
(2) orange × yellow	all orange	3/4 orange 1/4 yellow
(3) red × yellow	all orange	9/16 orange 4/16 yellow 3/16 red

Being knowledgeable about transmission genetics, the student surmised that two gene pairs were involved: On this basis, she proposed how these genes account for the various colors.

(a) What outcome(s) of the crosses suggested two gene pairs were at work?

(b) Assuming that she was correct and using the mutant gene symbols y for yellow and r for red, propose which genotypes give rise to which phenotypes.

(c) The student then devised several possible biochemical pathways for production of these pigments as well as which steps are enzymatically controlled by which genes. These are listed below. Which of these (one or more) is/are consistent with your proposal in part (b); which was based on the original data? Assume that a mixture of red and yellow pigment results in orange flowers. Defend your answer.

I. white precursor I $\xrightarrow{y\,\text{gene}}$ red pigment

   white precursor II $\xrightarrow{r\,\text{gene}}$ yellow pigment

II. white precursor I $\xrightarrow{r\,\text{gene}}$ red pigment

   white precursor II $\xrightarrow{y\,\text{gene}}$ yellow pigment

III. white precursor $\longrightarrow$ red $\xrightarrow{y\,\text{gene}}$ yellow $\xrightarrow{r\,\text{gene}}$ orange

IV. white precursor $\longrightarrow$ red $\xrightarrow{r\,\text{gene}}$ yellow $\xrightarrow{y\,\text{gene}}$ orange

V. white precursor $\longrightarrow$ yellow $\xrightarrow{y\,\text{gene}}$ red $\xrightarrow{r\,\text{gene}}$ orange

VI. white precursor $\longrightarrow$ yellow $\xrightarrow{r\,\text{gene}}$ red $\xrightarrow{y\,\text{gene}}$ orange

26. Deep in a previously unexplored South American rain forest, a species of plants was discovered with true-breeding varieties whose flowers were either pink, rose, orange, or purple. A very astute plant geneticist made a single cross, carried to the F$_2$ generation, as shown below. Based solely on these data, he was able to propose both a mode of inheritance for flower pigmentation and a biochemical pathway for the synthesis of these pigments.

Carefully study the data. Create an hypothesis of your own to explain the mode of inheritance. Then propose a biochemical pathway consistent with your hypothesis. How could you test the hypotheses by making other crosses?

P$_1$: purple × pink

F$_1$: all purple

F$_2$: 27/64 purple
     16/64 pink
     12/64 rose
      9/64 orange

SELECTED READINGS

ANDERSON, R. M., et al. 1996. Transmission dynamics and epidemiology of BSE in British cattle. *Nature* 382:779–83.

ANFINSEN, C. B. 1973. Principles that govern the folding of protein chains. *Science* 181:223–30.

BARTHOLOME, K. 1979. Genetics and biochemistry of phenylketonuria—Present state. *Hum. Genet.* 51:241–45.

BATESON, W. 1909. *Mendel's principles of heredity.* Cambridge: Cambridge University Press.

BEADLE, G. W. 1945. Genetics and metabolism in *Neurospora*. *Physiol. Rev.* 25:643.

BEADLE, G. W., and EPHRUSSI, B. 1937. Development of eye colors in *Drosophila*: Diffusible substances and their interrelations. *Genetics* 22:76–86.

BEADLE, G. W., and TATUM, E. L. 1941. Genetic control of biochemical reactions in *Neurospora*. *Proc. Natl. Acad. Sci. USA* 27:499–506.

BEET, E. A. 1949. The genetics of the sickle-cell trait in a Bantu tribe. *Ann. Eugenics* 14:279–84.

BOYER, P. D., ed. 1974. *The enzymes*, 3rd ed. (Vol. 10). Orlando, FL: Academic Press.

BRAY, D. 1995. Protein molecules as computational elements in living cells. *Nature* 376:307–12.

BRENNER, S. 1955. Tryptophan biosynthesis in *Salmonella typhimurium*. *Proc. Natl. Acad. Sci. USA* 41:862–63.

BYERS, P. H. 1989. Inherited disorders of collagen gene structure and expression. *Am. J. Med. Genet.* 34:72–80.

COHN, D. H., BYERS, P. H., STEINMANN, B., and GELINAS, R. E. 1986. Lethal osteogenesis imperfecta resulting from a single nucleotide change in one human pro-α 1(I) collagen allele. *Proc. Natl. Acad. Sci. USA* 83:6045–47.

DICKERSON, R. E. 1964. X-ray analysis and protein structure. In *The proteins*, 2nd ed., ed. H. Neurath (Vol. 2). Orlando, FL: Academic Press.

DICKERSON, R. E., and GEIS, I. 1983. *Hemoglobin: Structure, function, evolution, and pathology.* Menlo Park, CA: Benjamin/Cummings.

DOOLITTLE, R. F. 1985. Proteins. *Sci. Am.* (Oct.) 253:88–99.

EZZELL, C. 1994. Evolutions: Molecular chaperones and protein folding. *J. NIH Res.* 6:103.

EPHRUSSI, B. 1942. Chemistry of eye color hormones of *Drosophila. Quart. Rev. Biol.* 17:327–38.

GARROD, A. E. 1902. The incidence of alkaptonuria: A study in chemical individuality. *Lancet* 2:1616–20.

———. 1909. *Inborn errors of metabolism.* London: Oxford University Press. (Reprinted 1963, Oxford University Press, London.)

GARROD, S. C. 1989. Family influences on A. E. Garrod's thinking. *J. Inher. Metab. Dis.* 12:2–8.

HOLLISTER, D. W., BYERS, P. H., and HOLBROOK, K. A. 1982. Genetic disorders of collagen metabolism. *Adv. Human Genet.* 12:1–88.

HORTON, H. R., et al. 1996. *Principles of biochemistry*, 2nd ed. Upper Saddle River, NJ: Prentice-Hall.

INGRAM, V. M. 1957. Gene mutations in human hemoglobin: The chemical difference between normal and sickle cell hemoglobin. *Nature* 180:326–28.

———. 1963. *The hemoglobins in genetics and evolution.* New York: Columbia University Press.

KOSHLAND, D. E. 1973. Protein shape and control. *Sci. Am.* (Oct.) 229:52–64.

LaDu, B. N., ZANNONI, V. G., LASTER, L., and SEEGMILLER, J. E. 1958. The nature of the defect in tyrosine metabolism in alkaptonuria. *J. Biol. Chem.* 230:251.

LEHNINGER, A. L., NELSON, D. L., and COX, M. M. 1993. *Principles of biochemistry*, 2nd ed. New York: Worth Publishers.

MANIATIS, T., et al. 1980. The molecular genetics of human hemoglobins. *Annu. Rev. Genet.* 14:145–78.

MECHANIC, G. 1972. Cross-linking of collagen in a heritable disorder of connective tissue: Ehlers-Danlos syndrome. *Biochem. Biophys. Res. Commun.* 47:267–72.

MURAYAMA, M. 1966. Molecular mechanism of red cell sickling. *Science* 153:145–49.

NEEL, J. V. 1949. The inheritance of sickle-cell anemia. *Science* 110:64–66.

PAULING, L., ITANO, H. A., SINGER, S. J., and WELLS, I. C. 1949. Sickle cell anemia: A molecular disease. *Science* 110:543–48.

PFEFFER, S. R., and ROTHMAN, J. E. 1987. Biosynthetic protein transport and sorting by the endoplasmic reticulum and golgi. *Annu. Rev. Biochem.* 56:829–52.

PROCKOP, D. J., and KIVIRIKKO, K. I. 1984. Heritable diseases of collagen. *New Engl. J. Med.* 311:376–86.

PROCKOP, D. J., and KIVIRIKKO, K. I. 1995. Collagens: Molecular biology, diseases, and potentials for therapy. *Annu. Rev. Biochem.* 64:403–34.

PRUSINER, S. B. 1995. The prion diseases. *Sci. Am.* (Jan.) 272:48–57.

RICHARDS, F. M. 1991. The protein folding problem. *Sci. Am.* (Jan.) 264:54–63.

SCOTT-MONCRIEFF, R. 1936. A biochemical survey of some Mendelian factors for flower colour. *J. Genet.* 32:117–70.

SCRIVER, C. R., and CLOW, C. L. 1980. Phenylketonuria and other phenylalanine hydroxylation mutants in man. *Annu. Rev. Genet.* 14:179–202.

SRB, A. M., and HOROWITZ, N. H. 1944. The ornithine cycle in *Neurospora* and its genetic control. *J. Biol. Chem.* 154:129–39.

SYKES, B. 1985. The molecular genetics of collagen. *BioEssays* 3:112–17.

WAGNER, R. P., and MITCHELL, H. K. 1964. *Genetics and metabolism*, 2nd ed. New York: Wiley.

YANOFSKY, C., DRAPEAU, G., GUEST, J., and CARLTON, B. 1967. The complete amino acid sequence of the tryptophan synthetase A protein and its colinear relationship with the genetic map of the *A* gene. *Proc. Natl. Acad. Sci. USA* 57:296–98.

ZIEGLER, I. 1961. Genetic aspects of ommochrome and pterin pigments. *Adv. Genet.* 10:349–403.

ZUBAY, G. L., and MARMUR, J., eds. 1973. *Papers in biochemical genetics*, 2nd ed. New York: Holt, Rinehart & Winston.

14

Gene Mutation, DNA Repair, and Transposable Elements

CHAPTER CONCEPTS

Aside from chromosomal mutations, the major basis of diversity among organisms, even those that are closely related, is variation at the level of the gene. The origins of such variation are gene mutation, whereby coding sequences are altered as a result of the substitution, addition, or deletion of one or more bases within those sequences. Relocation of mobile genetic elements within the genome also disrupts normal expression of genes, creating mutations. Fortunately, elaborate mechanisms have evolved to repair various errors and lesions within DNA. Gene mutations are the working tools of the geneticists and the raw material upon which evolution relies.

In our earlier discussion of DNA as the genetic material (Chapter 10), we defined the four characteristics or functions ascribed to the genetic information: replication, storage, expression, and variation by mutation. In a sense, *mutation* is a failure to store the genetic information faithfully. If a change occurs in the stored information, it may be reflected in the expression of that information and will be propagated by replication. Historically, the term mutation includes both chromosomal changes and changes within single genes. We have discussed the former

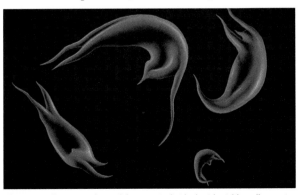

Mutant erythrocytes derived from an individual with sickle-cell anemia.

alterations in Chapter 9, referring to them collectively as chromosomal mutations or chromosomal aberrations. In this chapter we will be concerned with the so-called **gene mutations**. As we will see, a change may be a simple substitution of a single nucleotide or may involve the addition or deletion of one or more nucleotides within the normal sequence of DNA.

The term *mutation* was coined in 1901 by Hugo De-Vries to explain the variation he observed in crosses involving the evening primrose, *Oenothera lamarckiana.* Most of the variation was actually due to multiple translocations, but two cases were subsequently shown to be caused by gene mutations. As the studies of mutations progressed, it soon became clear that gene mutations serve as the source of most alleles and thus are the origin of much of the existing genetic variability within populations. As new alleles arise, they constitute the raw material to be tested by the evolutionary process of natural selection, which will determine whether they are detrimental, neutral, or beneficial.

Mutations provide the basis for genetic studies. The resulting phenotypic variability allows the geneticist to study the genes that control the traits that have been modified. In this regard, mutations serve as identifying "markers" of genes so that they may be followed during their transmission from parents to offspring. Without the phenotypic variability that mutations provide, genetic analysis would be impossible to conduct. For example, if all pea plants displayed a uniform phenotype, Mendel would have had no basis for his experimentation. Because of the importance of mutations, great attention has been given to their origin, induction, and classification.

Certain organisms lend themselves to induction of mutations that can be easily detected and studied throughout reasonably short life cycles. Viruses, bacteria, fungi, fruit flies, certain plants, and mice fit these criteria to various degrees. Thus, these organisms have often been used in studying mutation and mutagenesis, and through other studies they have also contributed to more general aspects of genetic knowledge.

Once we have completed our presentation of mutation, we will turn our attention to two related topics: DNA repair and transposable genetic elements. These topics are logical extensions of our consideration of gene mutation. Repair processes serve to counteract mutation. Transposable elements often disrupt the normal structure of the gene and therefore create mutations.

Random versus Adaptive Mutations

Although it was known well before 1943 that pure cultures of bacteria could give rise to small numbers of cells exhibiting heritable variation, particularly with respect to survival under different environmental conditions, the source of the variation was hotly debated. The majority of bacteriologists believed that environmental factors induced changes in certain bacteria that led to their survival or adaptation to the new conditions. For example, strains of *Escherichia coli* are known to be *sensitive* to infection by the bacteriophage T1. Infection by the bacteriophage leads to the reproduction of the virus at the expense of the bacterial cell, which is lysed or destroyed (see Figure 10.5). If a plate of *E. coli* is homogeneously sprayed with

T1, almost all cells are lysed. Rare *E. coli* cells, however, survive infection and are not lysed. If these cells are isolated and established in pure culture, all descendants are *resistant* to T1 infection. The **adaptation hypothesis**, put forth to explain this type of observation, implied that the interaction of the phage and bacterium was essential to the acquisition of immunity. In other words, the phage had somehow "induced" resistance in the bacteria.

The occurrence of **random (or spontaneous) mutations** provided an alternative model to explain the origin of T1 resistance in *E. coli*. In 1943, Salvador Luria and Max Delbruck presented the first direct evidence that mutation in bacteria occurs randomly within the chromosome of bacteria. That is, there is no way of predicting when or where in the chromosome mutation will occur. This experiment marked the initiation of modern bacterial genetic study.

The Luria–Delbruck Fluctuation Test

In a beautiful example of analytical and theoretical work, Luria and Delbruck carried out experiments to differentiate between the adaptation and random mutation hypotheses. In their work, they used the *E. coli*/T1 systems just described. Many small individual liquid cultures of phage-sensitive *E. coli* were grown up. Then, numerous aliquots from each culture were added to Petri dishes containing agar medium to which T1 bacteriophages had been previously added. Following incubation, each plate was scored for the number of phage-resistant bacterial colonies. This was easy to ascertain because only mutant cells were not lysed, and they survived to be counted. The precise number of total bacteria added to each plate prior to incubation was also determined so that quantitative data could be obtained.

The experimental rationale for distinguishing between the two hypotheses was as follows:

1. **Adaptation**: Every bacterium has a small but constant probability of acquiring resistance as a result of contact with the phages on the Petri dish. Therefore, the number of resistant cells will depend only on the number of bacteria and phages added to each plate. The final results should be independent of all other experimental conditions.

   The adaptation hypothesis predicts, therefore, that if a constant number of bacteria and phages is used for each culture and if incubation time is constant, then little fluctuation should be noted in the number of resistant cells on different plates, and from experiment to experiment.

2. **Random Mutation**: On the other hand, if resistance is acquired as a result of mutations that

occur randomly, resistance will occur at a low rate during the incubation in liquid medium *prior to plating*, before any contact with phage occurs. When mutations occur *early* during incubation, subsequent reproduction of the mutant bacteria will, relatively speaking, produce large numbers of resistant cells. When mutations occur relatively *late* during incubation, far fewer resistant cells will be present.

The random mutation hypothesis predicts, therefore, that significant fluctuation in the number of resistant cells will be observed from experiment to experiment, reflecting the various times when most spontaneous mutations occurred while in liquid culture.

Table 14.1 shows a representative set of data from the Luria–Delbruck experiments. The middle column shows the number of mutants recovered from a series of aliquots derived from one large individual liquid culture. In a large culture, experimental differences are evened out because the culture is constantly mixed. As a result, these data serve as a control because the number in each aliquot should be nearly identical. As predicted, little fluctuation is observed. The right-hand column, however, shows the number of resistant mutants recovered from a single aliquot from each of 10 independent liquid cultures. The amount of fluctuation in the data will support only one of the two alternative hypotheses. For this reason, the experiment has been designated the **fluctuation test**. A great fluctuation *is* observed between cultures, thus supporting the random nature of mutation. Fluctuation is measured by the amount of variance, a statistical calculation.

TABLE 14.1 The Luria–Delbruck Experiment Demonstrating That Spontaneous Mutations Are the Source of Phage-Resistant Bacteria

Sample No.	Number of T1-Resistant Bacteria	
	Same Culture (Control)	Different Cultures
1	14	6
2	15	5
3	13	10
4	21	8
5	15	24
6	14	13
7	26	165
8	16	15
9	20	6
10	13	10
Mean	16.7	26.2
Variance	15.0	2178.0

Source: After Luria and Delbruck, 1943.

The conclusion reached in the Luria–Delbruck experiment that random mutations account for inherited variation in bacteria received further support from other experimentation. Nevertheless, until the 1950s, staunch supporters of the adaptation theory were not convinced.

Adaptive Mutation in Bacteria

Although the concept of random mutation in viruses, bacteria, and even higher organisms is no longer disputed, the possibility that organisms such as bacteria might also be capable of "selecting" a specific set of mutations occurring as a result of environmental pressures has long intrigued geneticists. Two independent investigations, published in 1988 by John Cairns and Barry Hall and their colleagues, have provided preliminary evidence that this might be the case. While the findings presented below are still controversial, this general topic is the subject of a current investigation of potentially great significance.

Cairns devised an experimental protocol that improves on that used in the fluctuation test of Luria and Delbruck, discussed above. The procedures were designed to detect *nonrandom mutations* arising in response to factors in the environment in which bacteria are cultured. The results of this work suggest that some bacteria may select mutations that are "adaptive" to the environment.

Instead of using the characteristic of phage T1 resistance, where the only surviving cells are those that have mutated, the Cairns work involved a strain of bacteria that contained a lactose mutation (lac^-). Such bacteria cannot use lactose as a carbon source. (See the discussion of the *lac* operon in Chapter 18.) Cells first were grown in a rich liquid medium that provided an adequate carbon source other than lactose. Thus, the lac^- cells as well as any spontaneous lac^+ mutants were able to grow and reproduce quite well. This gene was subsequently followed by plating aliquots of cells on Petri plates containing minimal medium to which lactose had been added. The lac^- cells, present in the vast majority, will survive on the plates, but because they lack a carbon source that they can metabolize, they cannot proliferate and form colonies. On the other hand, cells that have mutated to lac^+, whether in the original liquid culture or as they sit on the plate, will form colonies and be detected. Those cells that have not mutated to lac^+ in liquid medium have the opportunity to do so *after* they are plated, and they may also be detected, since they will then form colonies. This is the major experimental difference between this approach and that of Luria and Delbruck some 45 years earlier.

Using a slightly more sophisticated mathematical analysis than Luria and Delbruck, Cairns attempted to identify the same sort of "fluctuation," or the lack of it, as an indication of either random (spontaneous) or adaptive mutations, respectively. Cairns found both types of data! The composite distribution was calculated for the case in which both spontaneous mutations occurred in the liquid cultures at random times (creating "fluctuation") and directed mutations arose as a response to the presence of lactose on the Petri plates. Indeed, such a composite distribution was observed.

But, how could it be proved that, in fact, the elevated frequency of mutations occurred on the plates in response to lactose? To answer this question, Cairns and colleagues asked whether another mutation, unrelated to the metabolism of lactose, had also occurred on the plates. They chose to assay for the production of Val^R mutations, which confer resistance to high concentrations of the amino acid valine. The assay was easily accomplished by overlaying the plates with agar containing valine and glucose. Only Val^R mutants will grow. Cairns and co-workers reasoned that if the lac^+ mutations were not necessarily in response to the lactose, then Val^R mutations should arise with a similar distribution as earlier observed for the lac mutations. The result was that plates accumulating lac^+ mutants were not, at the same time, accumulating Val^R mutations. The researchers concluded that the increased frequency of mutations affecting lactose metabolism was the result of the presence of lactose in the medium. This is, of course, an indirect inference.

Cairns' work suggests that the bacterial cell's genetic machinery can respond to its environment by producing adaptive mutations. This conclusion is contrary to the current thinking that mutations are completely spontaneous events, many of which occur as random errors during replication of DNA. As Cairns has pointed out, these findings, at first glance, seem to support the romantic notion that life is mysterious and that certain observations cannot be explained in simple, straightforward terms based on the laws of physics. Whatever the case may be, Cairns' observations are not isolated examples.

Barry Hall's work demonstrated a similar adaptive response by *E. coli* to growth on salicin. In this case, interestingly, a two-step genetic change was required. The first occurred in response to salicin even though that mutation offered no selective advantage without the second genetic alteration, the removal, or deletion, of a segment of nucleotides called an **insertion sequence (IS)**. Further, the changes must occur sequentially. They do so at a much higher frequency than predicted, provided salicin is present in the growth medium. The observed frequency is several orders of magnitude higher.

What possible explanation can account for observations such as these? As Barry Hall has pointed out, our failure to explain such phenomena adequately may be attributable to our ignorance of fundamental mechanisms and

rates of mutations *in nongrowing cells*. Such information and knowledge are extremely important because this physiological condition is much more akin to a natural environment as opposed to that found with cells grown in rich media under laboratory conditions. One suggestion is that under stressful nutritional conditions (starvation), bacteria may be capable of activating mechanisms that create a hypermutable state in genes that will enhance survival.

In the context of a textbook such as this, an important idea to be realized from this discussion is that knowledge may often be derived from observations that initially cannot be explained. This and the ensuing debates and surrounding controversy are what constantly maintain intrigue and interest in science.

Classification of Mutations

We turn now to a consideration of the way in which mutations are classified. There are various schemes by which mutations are organized. They are not mutually exclusive, but instead depend simply on which aspects of mutation are being investigated or discussed. In this section we describe three sets of distinctions concerning mutations.

Spontaneous versus Induced Mutations

Aside from our earlier discussion of the random nature of mutations, all mutations are described as either **spontaneous** or **induced**. Although these two categories overlap to some degree, **spontaneous mutations** are those that just happen in nature. No specific agents are associated with their occurrence, and they are generally assumed to be random changes in the nucleotide sequences of genes. Most such mutations are linked to normal chemical processes that alter the structure of nitrogenous bases that are part of existing genes. Most spontaneous mutations are thought to occur during the enzymatic process of DNA replication, an idea that we will discuss later in this chapter. Once an error is present in the genetic code, it may be reflected in the amino acid composition of the specified protein. If the changed amino acid is present in a part of the molecule critical to its structure or biochemical activity, a functional alteration can result.

In contrast to such spontaneous events, those that arise as a result of the influence of any artificial factor are considered to be **induced mutations**. It is generally agreed that any natural phenomenon that heightens chemical reactivity in cells will induce mutations. For example, radiation from cosmic and mineral sources and ultraviolet radiation from the Sun are energy sources to which most organisms are exposed. As such, these energy sources may be factors that cause spontaneous mutations. However, the earliest demonstration of the artificial induction of mutation occurred in 1927, when Herman J. Muller reported that X-rays could cause mutations in *Drosophila*. In 1928, Lewis J. Stadler reported the same finding in barley. In addition to various forms of radiation, a wide spectrum of chemical agents is also known to be mutagenic, the aspects of which we will examine later in this chapter.

Gametic versus Somatic Mutations

When considering the effects of mutation in eukaryotic organisms, it is important to distinguish whether the change occurs in somatic cells or in gametes. Mutations arising in somatic cells are not transmitted to future generations. Mutations occurring in somatic cells that create recessive autosomal alleles are rarely of any consequence to the organism. Expression of most such mutations is likely to be masked by the dominant allele. Somatic mutations will have a greater impact if they are dominant or if they are X-linked, since such mutations are most likely to be immediately expressed. Similarly, the impact will be more noticeable if such somatic mutations occur early in development, where undifferentiated cells will give rise to several differentiated tissues or organs. Mutations occurring in adult tissues are often masked by the thousands upon thousands of nonmutant cells performing the normal function.

Mutations in gametes or gamete-forming tissue are part of the germ line and are of greater significance because they are transmitted to offspring. They have the potential of being expressed in all cells of an offspring. **Dominant autosomal mutations** will be expressed phenotypically in the first generation. **X-linked recessive mutations** arising in the gametes of a heterogametic female may be expressed in hemizygous male offspring. This will occur provided that the male offspring receives the affected X chromosome. Because of heterozygosity, the occurrence of an **autosomal recessive mutation** in the gametes of either males or females (even one resulting in a lethal allele) may go unnoticed for many generations until the resultant allele has become widespread in the population. The new allele will become evident only when a chance mating brings two copies of it together in the homozygous condition.

Other Categories of Mutation

Aside from their point of origin (somatic versus gametic), various types of mutations may be classified on the basis of their phenotypic effects. A single mutation may well fall into more than one category. The most obvious mutations are those affecting a **morphological trait**. For example, all of Mendel's pea characters and many genetic variations encountered in the study of *Drosophila* fit this designation, since they cause obvious changes in the morphology of the organism. Such variations are recognized on the basis of their deviation from the normal or wild-type phenotype.

A second broad category of mutations include those that exhibit **nutritional** or **biochemical variations** in phenotype. In bacteria and fungi, the inability to synthesize an amino acid or vitamin is an example of a typical nutritional mutation. In humans, **hemophilia** is an example of a biochemical mutation. While such mutations in these organisms are not visible and do not always affect specific morphological characters, they can have a more general effect on the well-being and survival of the affected individual.

A third category consists of mutations that affect behavior patterns of an organism. For example, mating behavior or circadian rhythms of animals can be altered. The primary effect of **behavior mutations** is often difficult to discern. For example, the mating behavior of a fruit fly may be impaired if it cannot beat its wings. However, the defect may be in (1) the flight muscles; (2) the nerves leading to them; or (3) the brain, where the nerve impulses that initiate wing movements originate. The study of behavior and the genetic factors influencing it has benefited immensely from investigations of behavior mutations.

Still another type of mutation may affect the regulation of genes. A regulatory gene may produce a product that controls the transcription of another gene. In other instances, a region of DNA either close to or far away from a gene may modulate its activity. In either case, **regulatory mutations** can disrupt this process and permanently activate or inactivate a gene. Our knowledge of genetic regulation has been dependent on the study of mutations that disrupt this process.

Another group consists of **lethal mutations**. Nutritional and biochemical mutations can also fall into this category. A mutant bacterium that cannot synthesize a specific amino acid it needs will cease to grow if plated on a medium lacking that amino acid. Various human biochemical disorders, such as Tay-Sachs disease and Huntington disease, are lethal at different points in the life cycle of humans.

Finally, any of the above groups can exist as **conditional mutations**. Even though a mutation is present in the genome of an organism, it may not be evident except under certain conditions. Among the best examples are **temperature-sensitive mutations**, found in a variety of organisms. At certain "permissive" temperatures, a mutant gene product functions normally, only to lose its functional capability at a "restrictive" temperature. When shifted to this temperature, the impact of the mutation becomes apparent, even lethal, and is amenable to investigation. The study of conditional mutations has been extremely important in experimental genetics, particularly in understanding the function of genes essential to the viability of organisms.

Detection of Mutations

Before geneticists can study directly the mutational process or obtain mutant organisms for genetic investigations, they must be able to detect mutations. The ease and efficiency of detecting mutations in a particular organism has by and large determined the organism's usefulness in genetic studies. In this section we use several examples to illustrate how mutations are detected.

Detection in Bacteria and Fungi

Detection of mutations is most efficient in haploid microorganisms such as bacteria and fungi. Detection depends on a selection system where mutant cells are isolated easily from nonmutant cells. The general principles are similar in bacteria and fungi. To illustrate, we will describe how nutritional mutations in the fungus *Neurospora crassa* are detected. The life cycle of this organism was diagrammed in Figure 7.10.

Neurospora is a pink mold that normally grows on bread. It may also be cultured in the laboratory. This eukaryotic mold is haploid in the vegetative phase of its life cycle. Thus, recessive mutations may be detected without the complications generated by heterozygosity in diploid organisms.

Visible mutant traits such as *albino* have been well studied, but the full potential of *Neurospora* genetics was attained during the investigation of nutritional mutants. Wild-type *Neurospora* grows on a **minimal culture medium** of glucose, a few inorganic acids and salts, a nitrogen source such as ammonium nitrate, and the vitamin biotin. Induced nutritional mutants will not grow on minimal medium, but will grow on a supplemented or **complete medium** that also contains numerous amino acids, vitamins, nucleic acid derivatives, and so forth. Microorganisms that are nutritional wild types (requiring only minimal medium) are called **prototrophs**, whereas those mutants that require a specific supplement to the minimal medium are called **auxotrophs**.

The procedural details for the detection and characterization of nutritional mutations in *Neurospora* are illustrated in Figure 14.1. Haploid conidia bearing such a mutation can be detected and isolated by their failure to grow on minimal medium and their ability to grow on complete medium. The mutant cells can no longer synthesize some essential compound absent in minimal medium but present in complete medium [Figure 14.1(a)]. Once a nutritional mutant is detected and isolated, the missing compound can be determined by attempts to grow the mutant strain in a series of tubes, each containing minimal medium supplemented with a single compound [Figure 14.1(b) and (c)]. In this example, mutants can grow if tyrosine is supplied. Thus, the mutation is a tyrosine auxotroph (tyr^-). Recall that this was the procedure utilized by Beadle and Tatum in their pioneering research leading to the "one-gene: one-enzyme" hypothesis, as discussed in Chapter 13.

Detection in *Drosophila*

Herman J. Muller, in his studies demonstrating that X-rays are mutagenic, developed a number of detection systems in *Drosophila melanogaster*. These systems allow the estimation of the spontaneous and induced rates of X-linked and autosomal recessive lethal mutations. We will consider two techniques: the **ClB** and the **attached-X procedures**, which Muller devised. The ClB system (Figure 14.2) detects the rate of induction of X-linked recessive lethal mutations. The ClB stock involves C, an inversion that suppresses the recovery of crossover products; l, a recessive lethal allele; and B, the dominant gene duplication causing *Bar* eye, described in Chapter 9. All of these traits are located on the X chromosome.

In this technique, wild-type males are treated with a mutagenic agent—in this case, X-rays—and mated to untreated, heterozygous ClB females. In the F_1 generation, females expressing *Bar* are selected. They receive one X from their mother (the ClB chromosome) and one X from their father. Of the paternal X chromosomes, some will bear an induced X-linked lethal mutation. When a female that becomes heterozygous for such a lethal mutation is backcrossed to a wild-type male, no male progeny will result. One-half of them die because they are hemizygous for the ClB chromosome (remember, l is lethal) and the other half die because they are hemizygous for the newly induced lethal allele. Following many single-pair matings such as described, the percentage of vials bearing *only* females represents the frequency of induction of X-linked lethal mutations.

For example, if 500 such F_2 culture bottles were inspected, and 25 contained only females, the induced rate of such mutations would be 5 percent. This method also permits the detection of X-linked, morphological mutations. If a mutation of this type has been induced, all surviving F_2 males will show the trait.

The second technique, using females with attached-X chromosomes, is even simpler to use in detecting recessive morphological mutations because it requires only one generation. These females have two X chromosomes attached to a single centromere and one Y chromosome, in addition to the normal diploid complement of autosomes. When attached-X females are mated to males with normal sex chromosomes (XY), four types of progeny result: triplo-X females that die; viable attached-X females (XXY); YY males that also die; and viable XY males (who received their X from their father and their Y from their mother). Figure 14.3 illustrates how P₁ males that have been treated with a mutagenic agent produce F₁ male offspring that express any induced X-linked recessive mutation. In contrast to the *ClB* technique, which tests only a single X chromosome, the attached-X method tests numerous X chromosomes at one time in a single cross.

Detection techniques have also been devised for recessive autosomal lethals in *Drosophila*. In these tech-

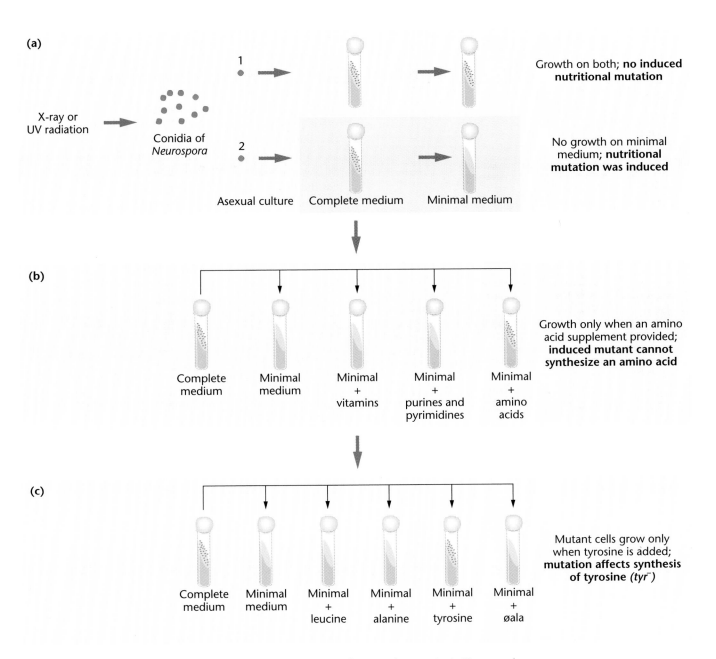

FIGURE 14.1   Induction, isolation, and characterization of a nutritional auxotrophic mutation in *Neurospora*. In (a), most conidia are not affected, but one conidium (shown in red) contains such a mutation. In (b) and (c), the precise nature of the mutation is determined to involve the biosynthesis of tyrosine.

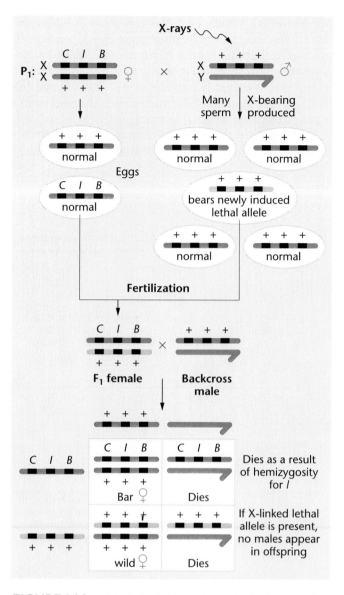

FIGURE 14.2 Muller's classical *ClB* technique for the detection of induced, X-linked recessive lethal mutations in *Drosophila*.

niques, dominant marker mutations are followed through a series of three generations. Although more cumbersome to perform, these techniques are also fairly efficient.

Detection in Plants

Genetic variation in plants is extensive. Mendel's peas, for example, were the basis for the fundamental postulates of transmission genetics. Studies of plants have also enhanced our understanding of gene interaction, polygenic inheritance, linkage, sex determination, chromosome rearrangements, and polyploidy. Many variations are de-

tected simply by visual observation. However, there are also techniques for the detection of biochemical mutations in plants.

One technique involves the analysis of the biochemical composition of plants. For example, the isolation of proteins from maize endosperm, hydrolysis of the proteins, and determination of the amino acid composition have revealed that the **opaque-2 mutant strain** contains significantly more lysine than do other, nonmutant lines. Because lysine content is usually low in maize protein, this mutation significantly improves the nutritional value of this plant.

As a result of this discovery, plant geneticists and other specialists have analyzed the amino acid compositions of various strains of other grain crops, including rice, wheat, barley, and millet. Results of such analyses are useful in combating malnutrition diseases resulting from inadequate protein or the lack of essential amino acids in the diet.

Another detection technique involves tissue culture of plant cell lines in defined medium. The plant cells are handled as microorganisms, and resistance to herbicides or disease toxins can be determined by adding these compounds to the culture medium. Other advantages accrue to the use of plant tissue culture. Techniques associated with conditional lethal mutants can be used on plant cells in tissue culture and then applied to the genetics of higher plants. Also, this method provides a detection system that is generally not useful with the intact plant.

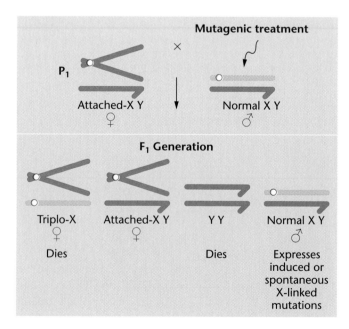

FIGURE 14.3 The attached-X method for detection of induced morphological mutations in *Drosophila*.

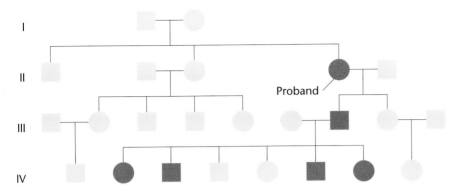

I

II

Proband

III

IV

FIGURE 14.4 A hypothetical pedigree of inherited cataract of the eye in humans.

These studies may add significantly to our understanding of plant growth, metabolism, and genetics.

Detection in Humans

Because humans are obviously not suitable experimental organisms, techniques developed for the detection of mutations in organisms such as *Drosophila* are not available to human geneticists. To determine the mutational basis for any human characteristic or disorder, geneticists must first analyze a pedigree that traces the family history as far back as possible. Once any trait has been shown to be inherited, it is possible to predict whether the mutant allele is behaving as a dominant or a recessive and whether it is X-linked or autosomal.

Dominant mutations are the simplest to detect. If they are present on the X chromosome, affected fathers pass the phenotypic trait to all their daughters. If dominant mutations are autosomal, approximately 50 percent of the offspring of an affected heterozygous individual are expected to show the trait. Figure 14.4 shows a pedigree illustrating the initial occurrence of an autosomal dominant allele for cataracts of the eye. The parents in Generation I were unaffected, but one of their three offspring (Generation II) developed cataracts. Presumably, the original mutation occurred in a gamete of one of the parents. The affected female, the proband, produced two children, of which the male child was affected (Generation III). Of his six offspring, four were also affected (Generation IV). These observations are consistent with, but do not prove, an autosomal dominant mode of inheritance. However, the high percentage of affected offspring in Generation IV favors this conclusion. Also, the presence of an unaffected daughter in this generation argues against X-linkage because she received her X chromosome from her affected father. Provided that the mutant allele is completely penetrant, such a conclusion is soundly based.

The X-linked recessive mutations can also be detected by pedigree analysis, as discussed in Chapter 5. The most famous case of an X-linked mutation in humans is that of **hemophilia,*** which was found in the descendants of Britain's Queen Victoria. The recessive mutation for hemophilia has occurred many times in human populations, but the political consequences of the mutation that occurred in the British royal family have been sweeping. Inspection of the pedigree in Figure 14.5 leaves little doubt that Victoria was heterozygous (*Hb*) for the trait. Her father was not affected, and there is no reason to believe that her mother was a carrier, as was Victoria.

In a similar manner, it is possible to detect autosomal recessive alleles. Because this type of mutation is "hidden" when heterozygous, it is not unusual for the trait to appear only intermittently through a number of generations. A mating between an affected individual and a homozygous normal individual will produce unaffected heterozygous carrier children. Matings between two carriers will produce, on the average, one-fourth affected offspring.

In addition to pedigree analysis, human cells may now be routinely cultured *in vitro*. This procedure has allowed the detection of many more mutations than any other form of analysis. Analysis of enzyme activity, protein migration in electrophoretic fields, and direct sequencing of DNA and proteins are among the techniques that have identified mutations and demonstrated wide genetic variation between individuals in human populations.

Spontaneous Mutation Rate

The types of detection systems just described allow geneticists to estimate mutation rates. It is of considerable interest to determine the rate of spontaneous mutation.

---

*Robert Massie's *Nicholas and Alexandra* and Robert and Suzanne Massie's *Journey* (see the list of Selected Readings at the end of this chapter) provide fascinating reading on the topic of hemophilia.

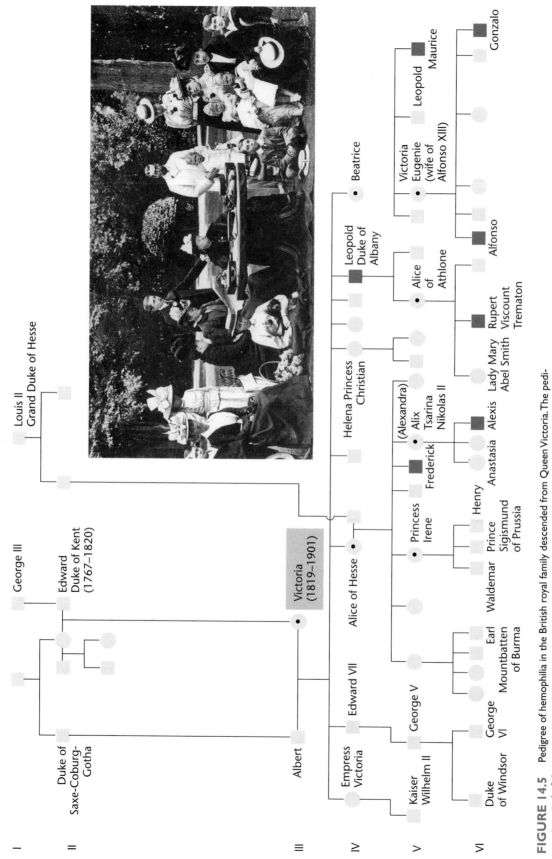

FIGURE 14.5 Pedigree of hemophilia in the British royal family descended from Queen Victoria. The pedigree is typical of the transmission of X-linked recessive traits. Circles with a dot in them indicate presumed female carriers heterozygous for the trait. The photograph shows Queen Victoria (seated front center) and some of her immediate family.

TABLE 14.2 Rates of Spontaneous Mutations at Various Loci in Different Organisms

Organism	Character	Gene	Rate	Units
Bacteriophage T2	Lysis inhibition	$r \rightarrow r^+$	1×10^{-8}	Per gene replication
	Host range	$h^+ \rightarrow h$	3×10^{-9}	
E. coli	Lactose fermentation	$lac^- \rightarrow lac^+$	2×10^{-7}	Per cell division
	Lactose fermentation	$lac^+ \rightarrow lac^-$	2×10^{-6}	
	Phage T1 resistance	$T1\text{-}s \rightarrow T1\text{-}r$	2×10^{-8}	
	Histidine requirement	$his^+ \rightarrow his^-$	2×10^{-6}	
	Histidine independence	$his^- \rightarrow his^+$	4×10^{-8}	
	Streptomycin dependence	$str\text{-}s \rightarrow str\text{-}d$	1×10^{-9}	
	Streptomycin sensitivity	$str\text{-}d \rightarrow str\text{-}s$	1×10^{-8}	
	Radiation resistance	$rad\text{-}s \rightarrow rad\text{-}r$	1×10^{-5}	
	Leucine independence	$leu^- \rightarrow leu^+$	7×10^{-10}	
	Arginine independence	$arg^- \rightarrow arg^+$	4×10^{-9}	
	Tryptophan independence	$try^- \rightarrow try^+$	6×10^{-8}	
Salmonella typhimurium	Tryptophan independence	$try^- \rightarrow try^+$	5×10^{-8}	Per cell division
Diplococcus pneumoniae	Penicillin resistance	$pen^s \rightarrow pen^r$	1×10^{-7}	Per cell division
Chlamydomonas reinhardi	Streptomycin sensitivity	$str^r \rightarrow str^s$	1×10^{-6}	Per cell division
Neurospora crassa	Inositol requirement	$inos^- \rightarrow inos^+$	8×10^{-8}	Mutant frequency among asexual spores
	Adenine independence	$ade^- \rightarrow ade^+$	4×10^{-8}	
Zea mays	Shrunken seeds	$sh^+ \rightarrow sh^-$	1×10^{-6}	Per gamete per generation
	Purple	$pr^+ \rightarrow pr^-$	1×10^{-5}	
	Colorless	$c^+ \rightarrow c$	2×10^{-6}	
	Sugary	$su^+ \rightarrow su$	2×10^{-6}	
Drosophila melanogaster	Yellow body	$y^+ \rightarrow y$	1.2×10^{-6}	Per gamete per generation
	White eye	$w^+ \rightarrow w$	4×10^{-5}	
	Brown eye	$bw^+ \rightarrow bw$	3×10^{-5}	
	Ebony body	$e^+ \rightarrow e$	2×10^{-5}	
	Eyeless	$ey^+ \rightarrow ey$	6×10^{-5}	
Mus musculus	Piebald coat	$s^+ \rightarrow s$	3×10^{-5}	Per gamete per generation
	Dilute coat color	$d^+ \rightarrow d$	3×10^{-5}	
	Brown coat	$b^+ \rightarrow b$	8.5×10^{-4}	
	Pink eye	$p^+ \rightarrow p$	8.5×10^{-4}	
Homo sapiens	Hemophilia	$h^+ \rightarrow h$	2×10^{-5}	Per gamete per generation
	Huntington disease	$Hu^+ \rightarrow Hu$	5×10^{-6}	
	Retinoblastoma	$R^+ \rightarrow R$	2×10^{-5}	
	Epiloia	$Ep^+ \rightarrow Ep$	1×10^{-5}	
	Aniridia	$An^+ \rightarrow An$	5×10^{-6}	
	Achondroplasia	$A^+ \rightarrow A$	5×10^{-5}	

Such information offers insights into evolution and provides the baseline for measuring the rate of experimentally induced mutation. Induction of mutation can only be ascertained when the induced rate clearly exceeds the spontaneous rate for the organism under study.

Examination of this rate in a variety of organisms reveals many interesting points (see Table 14.2). First, the rate is exceedingly low for all organisms studied. Second,

the rate is seen to vary considerably in different organisms. Third, even within the same species, the spontaneous mutation rate varies from gene to gene.

Viral and bacterial genes undergo spontaneous mutation on an average of about 1 in 100 million (10^{-8}) cell divisions. Although *Neurospora* exhibits a similar rate, maize, *Drosophila*, and humans demonstrate a rate several orders of magnitude higher. The genes studied in these

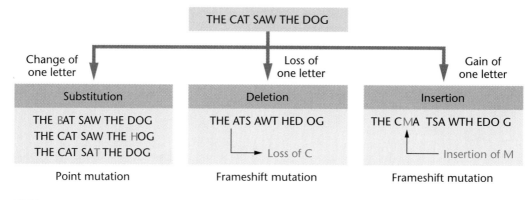

FIGURE 14.6 The impact of the substitution, deletion, and insertion of one letter in a sentence composed of three-letter words as analogies to point and frameshift mutations.

groups average between 1/1,000,000 and 1/100,000 (10^{-6} and 10^{-5}) mutations per gamete formed. Mouse genes are still another order of magnitude higher in their spontaneous mutation rate, 1/100,000 to 1/10,000 (10^{-5} to 10^{-4}). It is not clear why such a large variation occurs in the mutation rate. The variation might reflect the relative efficiency of enzyme systems whose function is to repair errors created during replication. Repair systems will be discussed later in this chapter.

Molecular Basis of Mutation

Even though we are aware that the gene is a reasonably complex genetic unit, particularly in eukaryotes (see Chapter 12), we will use a *simplified* definition of a gene in the following description of the molecular basis of mutation. In this context, it is easiest to consider a gene as a linear sequence of nucleotide pairs representing stored chemical information. Because the genetic code is a triplet, each sequence of three nucleotides specifies a single amino acid in the corresponding polypeptide. Any change that disrupts these sequences or the coded information provides sufficient basis for a mutation. The least complex change is the substitution of a single nucleotide. In Figure 14.6, such a change is compared with our own written language, using three-letter words to be consistent with the genetic code. A change of one letter can alter the meaning of the sentence "THE CAT SAW THE DOG," to "THE CAT SAW THE HOG" or the "THE BAT SAW THE DOG," creating what is called *missense*. These are analogies to what are most appropriately referred to as **base substitutions** or **point mutations**. The mutation has turned information that makes sense into various forms of missense.

Two other more formal terms are often used to describe nucleotide substitutions. If a pyrimidine replaces a pyrimidine or a purine replaces a purine, a **transition** has occurred. If a purine and a pyrimidine are interchanged, a **transversion** has occurred.

A second type of change that could occur is the insertion or deletion of one or more nucleotides at any point within the gene. As also illustrated in our analogy in Figure 14.6, the loss or addition of a single letter causes all subsequent three-letter words to be become changed. These examples are called **frameshift mutations** because the frame of reading has become altered. This will occur for any number of base deletions or additions other than ones that are a multiple of three, which would reestablish the initial frame of reading. We will return momentarily to a discussion of frameshift mutations as we introduce acridine dyes (see below).

The analogy in Figure 14.6 demonstrates that insertions and deletions have the potential to change all subsequent triplets in a gene. It is probable that, of the 64 possible triplets, one of the many altered triplets will be either UAA, UAG, or UGA. These are termination codons (see Chapter 12). When one of these triplets is encountered during translation, polypeptide synthesis is terminated. Obviously, the results of frameshift mutations can be very severe.

Tautomeric Shifts

In 1953, immediately after they had proposed a molecular structure of DNA, Watson and Crick published a paper in which they discussed the genetic implications of this structure. They recognized that the purines and pyrimidines found in DNA could exist in **tautomeric forms**; that is, a nitrogenous base can exist in alternate chemical forms called structural isomers, each differing by only a single proton shift in the molecule. Because such a shift changes the bonding structure of the molecule, Watson and Crick suggested that **tautomeric shifts**

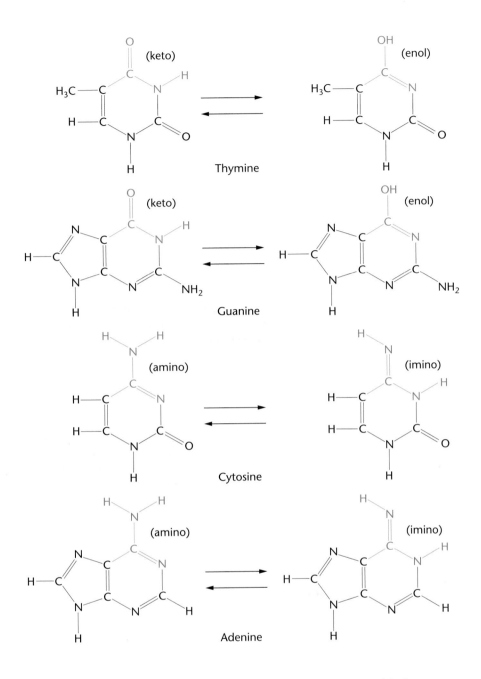

FIGURE 14.7   Rate tautomeric shifts that occur in the chemical structure of the four nitrogenous bases of DNA.

could result in base-pair changes or mutations. The biologically important tautomers involve keto-enol forms of thymine and guanine, and amino-imino forms of cytosine and adenine (Figure 14.7).

The most stable tautomers of the nitrogenous bases result in the standard hydrogen bonds that serve as the basis of the double helical model of DNA. The less frequently occurring, transient tautomers are capable of hydrogen bonding with noncomplementary bases. How-

ever, the pairing is always between a pyrimidine and a purine. Figure 14.8 compares the normal base-pairing relationships with the rare unorthodox pairings. As can be seen, anomalous $T \equiv G$ and $C = A$ pairs may be formed, among others.

The effect leading to mutation occurs during DNA replication when a rare tautomer in the template strand matches with a noncomplementary base. In the next round of replication, the "mismatched" members of the

(a) Standard base-pairing arrangements

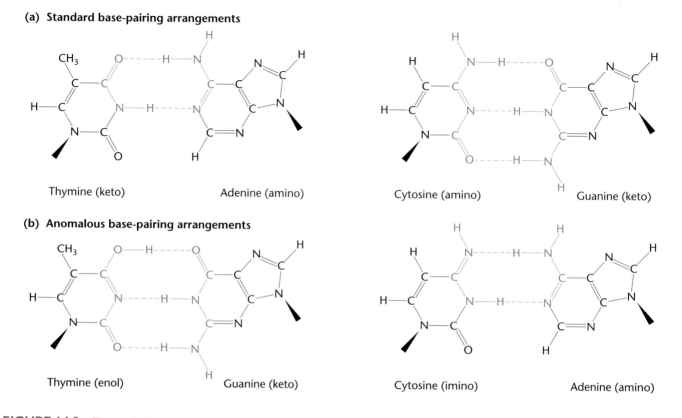

Thymine (keto)   Adenine (amino)   Cytosine (amino)   Guanine (keto)

(b) Anomalous base-pairing arrangements

Thymine (enol)   Guanine (keto)   Cytosine (imino)   Adenine (amino)

FIGURE 14.8   The standard base-pairing relationships compared with anomalous arrangements occurring as a result of tautomeric shifts. The dense arrowhead indicates the point of bonding to the pentose sugar.

base pair are separated, and each specifies its normal complementary base. The end result is a point mutation (see Figure 14.9).

Base Analogues

Base analogues, which are mutagenic chemicals, are molecules that may substitute for purines or pyrimidines during nucleic acid biosynthesis. The halogenated derivative of uracil in the number-5 position of the pyrimidine ring **5-bromouracil (5-BU)**[*] is a good example. Figure 14.10 compares the structure of this thymine analogue with the structure of thymine. The presence of the bromine atom in place of the methyl group increases the probability that a tautomeric shift will occur. If 5-BU is incorporated into DNA in place of thymine and a tautomeric shift to the enol form occurs, 5-BU base-pairs with guanine. After one round of replication, an A $=$ T to G $\equiv$ C transition results (see Figure 14.9). Furthermore, the presence of 5-BU within DNA increases the sensitivity of the molecule to ultraviolet (UV) light, which itself is mutagenic (see below).

There are other base analogues that are mutagenic. One, **2-amino purine (2-AP)**, can serve successfully as an analogue of adenine. In addition to its base-pairing affinity with thymine, 2-AP can also base pair with cytosine. As such, transitions from A $=$ T to G $\equiv$ C may result following replication.

Because of the specificity by which base analogues such as 2-AP induce transition mutations, base analogues may also be used to induce reversion to the wild-type nucleotide sequence. This alteration is called **reverse mutation.** The process can also occur spontaneously, but at a much lower rate.

Alkylating Agents

The sulfur-containing **mustard gases** were one of the first groups of chemical mutagens discovered. This discovery was made in studies involving chemical warfare during World War II. Mustard gases are **alkylating agents;** that is, they donate an alkyl group such as $CH_3—$ or $CH_3—CH_2—$ to amino or keto groups in nucleotides. **Ethylmethane sulfonate (EMS)**, for example, alkylates the keto groups in the number-6 position of guanine and in the number-4 position of thymine (Figure 14.11). As with base analogues, base-pairing affinities are altered

---

[*]If 5-BU is chemically linked to d-ribose, the nucleoside analogue bromodeoxyuridine (BUdR) is formed.

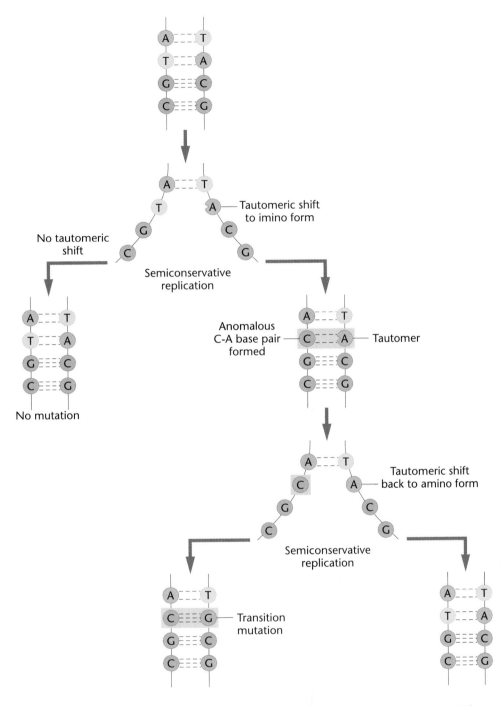

FIGURE 14.9   Formation of a T = A to a C ≡ G transition mutation as a result of a tautomeric shift in adenine.

and transition mutations result. In the case of **6-ethyl guanine**, this molecule acts like a base analogue of adenine, causing it to pair with thymine. Table 14.3 lists the chemical names and structures of several frequently used alkylating agents known to be mutagenic.

Acridine Dyes and Frameshift Mutations

Other chemical mutagens cause **frameshift mutations**. As illustrated in Figure 14.6 and introduced in the ensuing discussion, these result from the addition or removal of one or more base pairs in the polynucleotide sequence

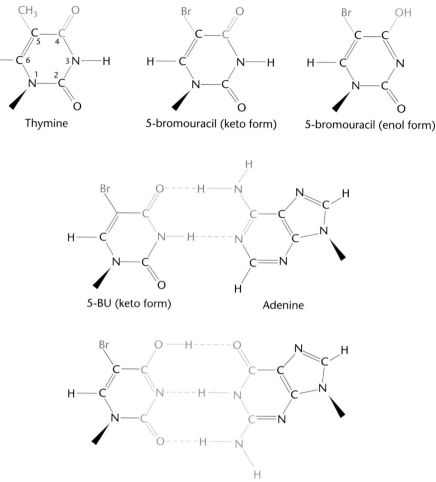

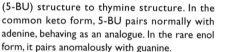

FIGURE 14.10 Similarity of 5-bromouracil (5-BU) structure to thymine structure. In the common keto form, 5-BU pairs normally with adenine, behaving as an analogue. In the rare enol form, it pairs anomalously with guanine.

of the gene. Inductions of frameshift mutations have been studied in detail with a group of aromatic molecules known as **acridine dyes**. The structures of **proflavin**, the most widely studied acridine mutagen, and **acridine orange** are shown in Figure 14.12. Acridine dyes are roughly the same dimension as a nitrogenous base pair and are known to intercalate or wedge between purines and pyrimidines of intact DNA. Intercalation of acridine

dyes induces contortions in the DNA helix, causing deletions and additions.

One model suggests that the resultant frameshift mutations are generated at gaps produced in DNA during replication, repair, or recombination. During these events, there is the possibility of slippage and improper base pairing of one strand with the other. The model suggests that intercalation of the acridine into an improperly

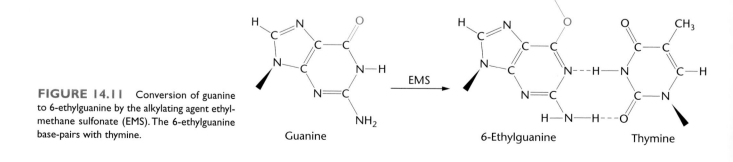

FIGURE 14.11 Conversion of guanine to 6-ethylguanine by the alkylating agent ethylmethane sulfonate (EMS). The 6-ethylguanine base-pairs with thymine.

TABLE 14.3 Alkylating Agents

Common Name or Symbol	Chemical Name	Chemical Structure
Mustard gas (sulfur)	Di-(2-chloroethyl) sulfide	$Cl-CH_2-CH_2-S-CH_2-CH_2-Cl$
EMS	Ethylmethane sulfonate	$CH_3-CH_2-O-\overset{\overset{O}{\parallel}}{\underset{\underset{O}{\vert}}{S}}-CH_3$
EES	Ethylethane sulfonate	$CH_3-CH_2-O-\overset{\overset{O}{\parallel}}{\underset{\underset{O}{\vert}}{S}}-CH_2-CH_3$

base-paired region can extend the existence of these slippage structures. If so, the probability increases that the mispaired configuration will exist when synthesis and rejoining occurs, thereby resulting in an addition or deletion of one or more bases in one of the strands.

The impact of a frameshift mutation is usually severe. The genetic code is read in three-letter groups, one after the other during translation. As pointed out earlier, the addition or deletion of any group of nucleotides other than a multiple of three shifts the frame of reading from that point on. Initially, this creates missense triplets (see Figure 14.6). Eventually, one of the new three-letter groups may be one of the three stop codons, resulting in premature termination of translation. This will result in an incomplete polypeptide chain. Loss of function usually results.

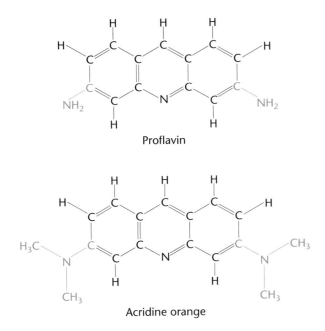

Proflavin

Acridine orange

FIGURE 14.12 Chemical structures of proflavin and acridine orange, which intercalate with DNA and cause frameshift mutations.

Apurinic Sites and Other Lesions

Still another type of mutation involves the spontaneous loss of one of the nitrogenous bases in an intact double-helical DNA molecule. Most frequently, such an event involves either guanine or adenine. These sites, created by the "breaking" of the glycosidic bond linking the 1′-C of d-ribose and the 9-N of the purine ring, are called **apurinic sites (AP sites)**. It has been estimated that thousands of such spontaneous lesions are formed daily in the DNA of mammalian cells in culture.

The absence of a nitrogenous base at an AP site will alter the genetic code if the involved strand is transcribed and translated. If replication occurs, the AP site is an inadequate template and may cause replication to stall. If a nucleotide is inserted, it is frequently incorrect, causing still another mutation! Fortunately, as we will soon see, cells contain repair systems that often counteract and correct this type of lesion.

Several other types of lesions are known to be the source of some mutations. In the process of **deamination**, an amino group is converted to a keto group in cytosine and adenine (Figure 14.13). In these two cases, cytosine is converted to uracil and adenine is changed to hypoxanthine.

The major effect of these changes is to alter the base-pairing specificities of these two molecules during DNA replication. For example, cytosine normally pairs with guanine. Following its conversion to uracil, which pairs with adenine, the original G ≡ C pair is converted to an A = U pair and, following an additional replication, to an A = T pair. When adenine is deaminated, an original A = T pair is converted to a G ≡ C pair because hypoxanthine pairs naturally with cytosine. Nitrous acid is a known mutagen capable of inducing deamination of bases in DNA.

The final type of mutational lesion that we shall mention is the group caused by oxidation reactions. Ac-

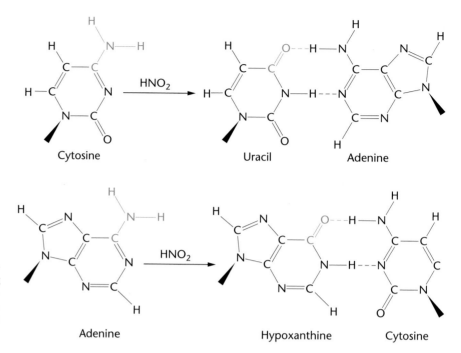

FIGURE 14.13 Deaminations caused by nitrous acid (HNO_2) leading to new base-pairing arrangements and mutations. Cytosine may be converted to uracil, which base pairs with adenine. Adenine may be converted to hypoxanthine, which base pairs with cytosine.

tive forms of oxygen radicals such as hydrogen peroxide (H_2O_2) and superoxides (O_2^-) can cause damage to DNA bases and result in mispairing during replication.

Ultraviolet Radiation, Thymine Dimers, and the SOS Response

In Chapter 10 we emphasized the fact that purines and pyrimidines absorb **ultraviolet (UV) radiation** most intensely at a wavelength of about 260 nm. This property has been used extensively in the detection and analysis of nucleic acids. In 1934, as a result of studies involving *Drosophila* eggs, it was discovered that UV radiation is muta-

genic. By 1960, several studies concerning the *in vitro* effect on the components of nucleic acids had been completed, with the following conclusions. The major effect of UV radiation is on pyrimidines, where dimers are formed, particularly between two thymine residues (Figure 14.14). While cytosine–cytosine and thymine–cytosine dimers may also be formed, they are less prevalent. It is believed that the dimers distort the DNA conformation and inhibit normal replication, which seems to be responsible, at least in part, for the killing effects of UV radiation on microorganisms.

For this UV-induced lesion to be mutagenic and not lethal, cells must somehow overcome the inhibition of

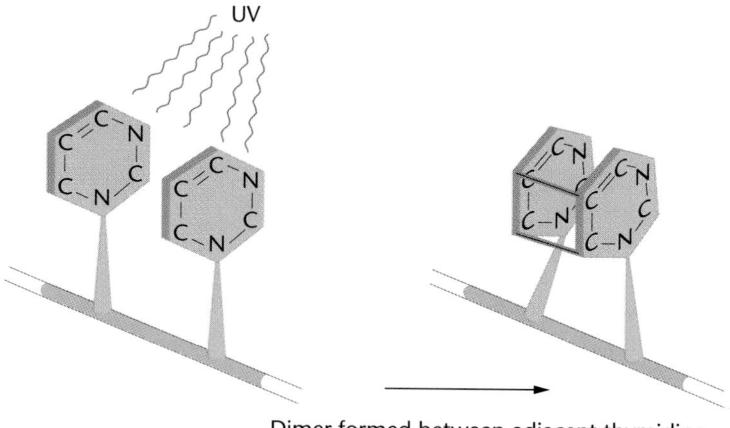

FIGURE 14.14 Depiction of the bonding and distortion of a thymine dimer induced by UV radiation. Only the atoms of the pyrimidine ring are shown.

Dimer formed between adjacent thymidine residues along a DNA strand replication, even if it means inserting an incorrect nucleotide during synthesis. A system in bacteria has been discovered that, when activated, appears to allow the "block" to be bypassed. The products of several genes in *E. coli*, including *lexA* and *recA*, somehow allow the strict adherence of the insertion of complementary bases to be relaxed. While this decreases the fidelity during replication, the process allows the survival of otherwise lethal effects of UV radiation. As a result, this error-prone system is called the **SOS response**.

This survival response is likely to be also activated by other types of lesions that block replication. For example, the AP sites discussed above undoubtedly block replication in a manner similar to pyrimidine dimers, and they also induce the SOS response, allowing error-prone replication to occur.

We will return to the subject of DNA repair and the SOS response, including the specific repair of UV-induced pyrimidine dimers, later in this chapter, where we will treat the topic more thoroughly.

Case Studies of Mutations in Humans

The preceding section summarizes the molecular basis of mutation. The information presented relies almost exclusively on the study of nucleic acid chemistry and has described numerous ways in which one nucleotide pair can be converted to another, either spontaneously or as a result of an induced change. Ways in which frameshift mutations occur have also been discussed. We know that these various types of mutations actually occur, primarily because analyses of amino acid sequences of many proteins within populations of a species show substantial diversity. This diversity has arisen during evolution and is a reflection of changes in the triplet codons following substitution, insertion, or deletion of one or more nucleotides in the DNA sequences constituting genes.

As our ability to analyze DNA more directly increases, we are better able to look specifically at the actual nucleotide sequence of genes and to gain greater insights into mutation. Several techniques capable of accurate, rapid sequencing of DNA are available (see Chapter 15 and Appendix A). These and other approaches involving the analysis of DNA have greatly extended our knowledge in molecular genetics.

ABO Blood Types versus Muscular Dystrophy

In this section we examine the results of two studies that have investigated the actual gene sequence of various mutations that seriously affect humans. The first provides an interesting insight into the molecular basis of the **ABO antigens**, originally presented in Chapter 4 as an example of multiple alleles. The second case involves the nature of the mutations that have led to the devastating X-linked disorder **muscular dystrophy**.

The ABO system is based on a series of antigenic determinants found on erythrocytes and other cells, particularly epithelial types. As previously discussed, three alleles of a single gene exist, the product of which is designed to modify the H substance. The modification involves glycosyltransferase activity, converting the H substance to either the A or B antigen, as a result of the product of the I^A or I^B allele, respectively, or failing to modify the H substance, as a result of the I^O allele (see Figure 4.2).

Using recombinant DNA technology, the responsible gene has been examined in 14 cases of varying ABO status. Four consistent nucleotide substitutions were found when the DNAs of the I^A and I^B alleles were compared. It is assumed that the changes in the amino acid sequence of the glycosyltransferase gene product resulting from these substitutions lead to the different modifications of the H substance.

The situation with the I^O allele is unique and interesting. Individuals homozygous for this allele are type O, lack glycosyltransferase activity, and fail to modify the H substance. Analysis of the DNA of this allele shows one consistent change compared to that of the other alleles: the deletion of a single nucleotide early in the coding sequence, causing a frameshift mutation. A complete messenger RNA is transcribed, but upon its translation, the frame of reading shifts at the point of deletion and continues out of frame for about 100 nucleotides before a stop codon is encountered. At this point, premature termination of the resulting polypeptide chain occurs, producing a nonfunctional product.

These results provide a direct molecular explanation of the ABO allele system and the basis for the biosynthesis of the corresponding antigens. The molecular basis for the antigenic phenotypes is clearly the result of structural alterations, or mutations, of the nucleotide sequence of the gene encoding the glycosyltransferase enzyme.

The second case of mutational analysis involves the severe disorder muscular dystrophy. It is characterized by progressive muscular degeneration, or myopathy, resulting in the death of affected individuals by early adulthood. Because the condition is recessive and X-linked, and because affected males do not reproduce, females are rarely affected by the disorder. The incidence of 1/3500 live male births makes muscular dystrophy one of the most common life-shortening hereditary disorders known. Two related forms exist: **Duchenne muscular** dystrophy (DMD) is more common and more severe than the allelic form, **Becker muscular dystrophy (BMD)**.

The region containing the gene has been analyzed extensively and consists of over 2 million base pairs. In unaffected individuals, transcription results in a messenger RNA containing about 14,000 bases (14 kb) that is translated into the protein **dystrophin**, consisting of 3685 amino acids. This protein can be detected in most cases of the less severe BMD, but is rarely found in DMD (see Figure 2.2). This has led to the hypothesis that most mutations causing BMD do not alter the reading frame, but that most DMD mutations change the reading frame early in the gene, resulting in premature termination of dystrophin translation. This hypothesis is consistent with the observed differences in severity of these two forms of the disorder.

In an extensive analysis of the DNA of 194 patients (160 DMD and 34 BMD), J. T. Den Dunnen and associates found that 128 of these mutations (65 percent) consisted of substantial deletions or insertions. Of 115 deletions, 17 occurred in BMD, and of 13 insertions, 1 was in BMD, with the remainder being found in DMD cases. In most cases, the above results were consistent with the "reading frame" hypothesis. With few exceptions, DMD mutations changed the frame of reading. BMD mutations usually did not alter the reading frame.

Perhaps the most noteworthy finding is the high percentage of deleterious mutations studied that represent the deletion or insertion of nucleotides within the gene. This observation reflects the fact that a mutation caused by a random single-nucleotide substitution within a gene is more likely to be tolerated without the devastating effect of muscular dystrophy than the addition or loss of numerous nucleotides that may alter the frame of reading. There are three reasons for this:

1. A nucleotide substitution may not change the encoded amino acid since the code is degenerate.

2. If an amino acid substitution does result, the change may not be present at a location within the protein that is critical to its function.

3. Even if the altered amino acid is present at a critical region, it may still have little or no effect on the function of the protein. For example, an amino acid might be changed to another with nearly identical chemical properties or to one with very similar recognition properties, such as shape.

As a result, single-base substitutions may have little or no effect on protein function, or they may simply reduce the efficiency but not eliminate the functional capacity of the gene product. It is clear that we cannot look at mutations with the oversimplified expectation that most of them are single-base substitutions. As more of them are analyzed directly, our picture of mutation will become increasingly clear.

Trinucleotide Repeats in Fragile-X Syndrome, Myotonic Dystrophy, and Huntington Disease

Beginning about 1990, molecular analysis of the DNA representing the genes responsible for a number of inherited human disorders provided a remarkable set of observations: Mutant genes were found to be characterized by an expansion of a simple **trinucleotide repeat sequence**, usually from fewer than 15 copies in normal individuals to large numbers in affected individuals. For example, in the cases of three separate genes responsible for the X-linked **fragile-X syndrome** and the autosomal disorders **myotonic dystrophy** and **Huntington disease**, each gene was found to contain a unique trinucleotide DNA sequence repeated many times. While each repeated sequence is also present in the nonmutant (normal) allele of each gene, what characterizes each mutation is a significant, variable increase in the number of times the trinucleotide is repeated.

You may recall that we introduced this topic earlier in Chapter 7 when we discussed the onset of expression of various phenotypes. In several cases, a correlation has been found between the number of repeats and the manifestation of mutant phenotypes. The greater the number of repeats, the earlier disease onset occurs.

Of great interest and significance is the fact that the number of repeats may increase in each subsequent generation. This general phenomenon, which we called **genetic anticipation** in Chapter 7, represents a unique form of mutation related to an instability of specific regions of the three different genes. Whereas other normal genes in humans, as well as many other species, are known to contain trinucleotide repeats, these repeats do not balloon in size, and the genes maintain normal function.

The gene responsible for fragile-X syndrome, *FMR-1*, may have several hundred to several thousand copies of the trinucleotide sequence CGG. Individuals with up to 50 copies are normal and do not display mental retardation associated with the syndrome. Individuals with 50 to 200 copies are considered "carriers." Although they are normal, their offspring may contain even more copies and express the syndrome.

Myotonic dystrophy, the most common form of adult muscular dystrophy, contains multiple copies of the sequence CTG. Fewer than 35 copies results in normal gene expression. Above this number, symptoms range from mild myopathy and cataracts to severe dystrophy, retardation, and even death in early childhood. Both severity and onset are directly correlated with the size of the repeated sequence.

Most recently, the gene responsible for Huntington disease has been found to demonstrate a similar pattern. The trinucleotide CAG repeat, present 10 to 35 times in normal individuals, is significantly increased in number in diseased individuals. Recall that the onset of this disease varies tremendously, most often being expressed in the mid- to late thirties. Much earlier onset occurs when very large numbers of copies are present. Interestingly, in still another disorder, **spinobulbar muscular atrophy** (**Kennedy disease**), the gene contains repeated copies of the same triplet. However, diseased individuals usually contain only 30 to 60 copies of the CAG sequence.

The role of such repeated sequences in normal and mutant genes remains a mystery. Their locations within the gene vary in each case. While Huntington and Kennedy diseases contain the repeat within the coding portion of the gene, this is not the case in the other two disorders. In the gene responsible for fragile-X syndrome, the repeat is upstream (5′) in an area that is most often involved in regulating gene expression. In the case of myotonic dystrophy, the repeat is downstream (the 3′ end).

In addition to the role of the repeat in normal and mutant genes, the mechanism by which the sequence expands from generation to generation is of great interest. How such an instability during DNA replication affects only specific areas of certain genes is currently an important research topic. This instability seems to be more prevalent in humans than in many other organisms. It is likely that other mutant genes will be found to parallel those discussed above.

Detection of Mutagenicity: The Ames Test

There is great concern about the possible mutagenic properties of any chemical that enters the human body, whether through the skin, digestive system, or respiratory tract. For example, great attention has been given to residual materials of air and water pollution, food preservatives and additives, artificial sweeteners, herbicides, pesticides, and pharmaceutical products. Although mutagenicity may be tested in various organisms, including *Drosophila*, mice, and cultured mammalian cells, the most common test involves bacteria and was devised by Bruce Ames.

The **Ames test** utilizes four tester strains of the bacterium *Salmonella typhimurium* that were selected for sen-

sitivity and specificity for mutagenesis. One strain is used to detect base-pair substitutions, and the other three detect various frameshift mutations. Each mutant strain is unable to synthesize histidine, and therefore requires histidine for growth (*his⁻*). The assay measures the frequency of reverse mutation, which yields wild-type (*his⁺*) bacteria. Greater sensitivity to mutagens occurs because these strains bear other mutations that eliminate both the DNA excision repair system (discussed later in this chapter) and the lipopolysaccharide barrier that coats and protects the surface of the bacteria.

It is very interesting to note that many substances entering the human body are relatively innocuous until activated metabolically to a more chemically reactive product. This usually occurs in the liver. Thus, the Ames test includes a step in which the test compound is incubated *in vitro* in the presence of a mammalian liver extract. Or, test compounds are modified by liver enzymes following their injection into a mouse, which is later sacrificed. Extracts are then tested.

In the initial use of Ames testing in the 1970s, a large number of known carcinogens were examined. Over 80 percent of these were shown to be strong mutagens! This is not surprising; transformation of cells to the malignant state undoubtedly occurs as a result of some alteration of DNA. Although a positive response as a mutagen does not prove the carcinogenic nature of a test compound, the Ames test is useful as a preliminary screening device. It is used extensively in conjunction with the industrial and pharmaceutical development of chemical compounds.

Repair of DNA

In previous sections of this chapter we established that replicating and nonreplicating DNA is vulnerable to various forms of errors and lesions that constitute or lead to gene mutations. Living systems have evolved a variety of very elaborate repair systems that are able to counteract many of the forms of DNA damage that lead to mutation. As we will see, such repair systems are essential to the survival of organisms on Earth. We will describe how, in humans, the loss of just one type of repair system leads to a devastating, life-shortening genetic disorder.

UV Radiation, Thymine Dimers, and Photoreactivation Repair

As established in Figure 14.14, UV light is mutagenic as a result of the creation of pyrimidine dimers. The study of mutagenicity of UV radiation paved the way for the discovery of many forms of natural repair of DNA damage.

The first relevant discovery concerning UV repair in bacteria was made in 1949 when Albert Kelner observed the phenomenon of **photoreactivation repair**. He showed that the UV-induced damage to *E. coli* DNA could be partially reversed if, following irradiation, the cells were exposed briefly to light in the blue range of the visible spectrum. The photoreactivation repair process has subsequently been shown to be temperature dependent, suggesting that the light-induced mechanisms involve an enzymatically controlled chemical reaction. Visible light appears to induce the repair process of the DNA damaged by UV radiation.

Further studies of photoreactivation have revealed that the process is due to a protein called the **photoreactivation enzyme** (**PRE**). This molecule can be isolated from extracts of *E. coli* cells. The enzyme's mode of action is to cleave the bonds between thymine dimers, thus reversing the effect of UV radiation on DNA [Figure 14.15(a)]. Although the enzyme will associate with a dimer in the dark, it must absorb a photon of light to cleave the dimer.

The gene(s) encoding PRE have been preserved throughout evolution. Activity of this repair system has been detected in both human cells in culture and in other eukaryotes. Conservation of the genetic components over millions of years suggests that this repair system is an important one to all organisms.

Excision Repair

Investigations in the early 1960s suggested that a light-independent repair system also exists in *E. coli*. Paul Howard-Flanders and co-workers isolated several independent mutants demonstrating increased sensitivity to UV radiation. One group of genes was designated *uvr* (ultraviolet repair) and included the *uvrA*, *uvrB*, and *uvrC* mutations. These genes and their protein products were subsequently shown to be involved in a process called **excision repair**. During this process, three steps have been shown to occur, as illustrated in Figure 14.15(b):

1. The distortion of the strand caused by the UV-induced dimer is recognized and enzymatically clipped out by a nuclease that cleaves the phosphodiester bonds. This "excision," which may include several nucleotides adjacent to the dimer as well, leaves a gap in the helix. The *uvr* gene products operate at this step.

2. **DNA polymerase I** fills this gap by inserting d-ribonucleotides complementary to those on the intact strand. The enzyme adds these bases to the 3′-OH end of the clipped DNA.

3. The joining enzyme **DNA ligase** seals the final "nick" that remains at the 3′-OH end of the last base inserted, closing the gap.

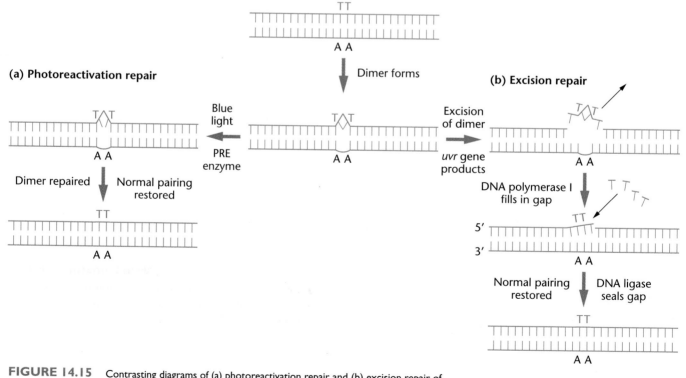

FIGURE 14.15   Contrasting diagrams of (a) photoreactivation repair and (b) excision repair of UV-induced thymine dimers. In actuality, 12 bases are excised in prokaryotes and 28 bases are excised in eukaryotes during repair.

DNA polymerase I, the enzyme discovered by Arthur Kornberg, was once assumed to be the universal DNA-replication enzyme (see Chapter 11). The discovery of the *polA1* mutation demonstrated that this is not the case. *E. coli* cells carrying the *polA1* mutation lack functional polymerase I. However, replication of DNA takes place normally. Cells with this mutation are unusually sensitive to UV radiation. Apparently, such cells are unable to fill the gap created by the excision of the thymine dimers. This finding demonstrates the importance of excision-repair mechanisms in counteracting the effects of UV radiation.

It has been shown subsequently that the process of excision repair can be activated in response to any damage to DNA that distorts the helix. For example, as we discussed earlier, the loss of a purine from d-ribose of one strand creates what is called an **apurinic site (AP site)**. The complementary pyrimidine on the opposite strand has nothing with which to form hydrogen bonds. Such a sugar with a missing base is recognized by an enzyme called **AP endonuclease**. The endonuclease makes a cut in the polynucleotide chain at the AP site. This creates the distortion that is recognized by the excision-repair system, which is then activated, leading ultimately to the correction of the error (the AP site).

Other enzymes recognize incorrect bases and stimulate repair. For example, one specific member of a group of enzymes called **DNA glycosylases** recognizes the presence of uracil when it is part of DNA. It cuts the glycosidic bond between the base and sugar, creating an AP site, which is then repaired as discussed above. Glycosylases are important repair components because, when created by deamination of cytosine, uracil will lead to a $C \equiv G$ to $T = A$ transition mutation after replication if it is not repaired.

In theory, excision repair may serve as the final step in a variety of repair processes, provided that the lesion or distortion in DNA may be recognized.

Proofreading and Mismatch Repair

As we pointed out in our discussion of DNA synthesis in Chapter 11, DNA polymerase III possesses a **proofreading** function. During polymerization, when an incorrect nucleotide is inserted, the enzyme complex has the potential to recognize the error and "reverse" it by cutting out the incorrect nucleotide and replacing it. In bacterial systems, proofreading is thought to increase fidelity during synthesis by two orders of magnitude. If initial mismatches occur in $1/10^5$ nucleotide pairs (a rate of 10^{-5}), proofreading decreases final mismatches to $1/10^7$ (a rate of 10^{-7}).

To cope with those errors that remain after proofreading, still another mechanism, called **mismatch repair**, may be activated. Proposed over 20 years ago by

Robin Holliday, the molecular basis of this process is now well established. Like other DNA lesions, (1) the alteration or mismatch must be detected, (2) the incorrect nucleotide must be removed, and (3) replacement with the correct base must occur. But a special problem exists with correction of a mismatch. How does the repair system recognize which strand is correct (the template strand) and which contains the mismatched base (the newly synthesized strand)? How the repair system discriminates and recognizes the "new" nucleotide puzzled geneticists for decades. If the mismatch is recognized, but no discrimination occurred and excision was random, half the time the strand bearing the correct base would be clipped out. The concept of strand discrimination by a repair enzyme is thus a critical step.

At least in some bacteria, including *E. coli*, this process has been elucidated and is based on the process of **DNA methylation**. These bacteria contain an enzyme, **adenine methylase**, that recognizes the DNA sequence

$$5' \ldots G\,A\,T\,C \ldots 3'$$
$$3' \ldots C\,T\,A\,G \ldots 5'$$

as a substrate. Upon recognition, a methyl group is added to each of the adenine residues. This modification is stable throughout the cell cycle.

Following a further round of replication, the newly synthesized strands remain temporarily unmethylated. It is at this point that the repair enzyme recognizes the mismatch and preferentially binds to the unmethylated strand. It is excised and the correct complement is inserted. Interestingly, the GATC sequence need only be within several thousand base pairs of the mismatch. A series of *E. coli* gene products, MutH, L, S, and U are involved in the discrimination step. Mutations in each result in strains deficient in mismatch repair.

Although it is agreed that mismatch repair undoubtedly also occurs in the DNA of higher organisms, the question of strand discrimination remains speculative in the absence of GATC methylation.

The SOS Response: Recombinational Repair

The final mode of repair to be discussed was discovered in an excision-defective strain of *E. coli* and first proposed by Miroslav Radman. Called **recombinational repair**, this system is thought to respond when damaged DNA has escaped repair and the damage disrupts the process of replication. Because this system "responds" to a signal of distress, so to speak (DNA damage), Radman initially referred to it as an **SOS response**. The cells that show this phenomenon are dependent on the product of a gene, *recA*, which is involved in several types of recombinational phenomena in *E. coli*.

When DNA bearing a lesion of some sort is being replicated, DNA polymerase at first stalls at and then skips over the distortion, creating a gap on one of the newly synthesized strands. To counteract this, the RecA protein directs a recombinational exchange process whereby this gap is filled as a result of the insertion of a segment initially present on the intact homologous strand. This creates a gap on the "donor" strand, which is filled by repair synthesis as replication proceeds.

Phil Hanawalt and Paul Howard-Flanders, among others, have established that as many as 20 different gene products are involved in this mode of repair. Of particular interest is the LexA protein product, which, when produced, serves to partially repress the transcription of the *recA* and *uvr* genes. However, when a RecA protein binds to single-stranded DNA in the area of a gap, this binding somehow activates a second function of the RecA protein—the ability to cleave the LexA repressor molecule, disrupting its regulatory capacity. The absence of a functional repressor molecule allows the activation of the *recA* and *uvr* genes, among others, leading to an increased production of the proteins for which they code. These products complete the repair process.

UV Radiation and Human Skin Cancer: Xeroderma Pigmentosum

The essential nature of any biological system can be assessed by examining the effects when the system fails. Regrettably, we can determine just how essential DNA repair is in humans by examining individuals who exhibit an inherited loss of function of one of the major repair systems. **Xeroderma pigmentosum (XP)**, a rare autosomal recessive disorder in humans, predisposes individuals to epidermal pigment abnormalities. Exposure to ultraviolet radiation present in sunlight results in malignant growth of the skin. Figure 14.16 contrasts two XP individuals, one of whom has been detected early and protected from sunlight.

The condition is very severe and may be lethal. Because sunlight contains UV radiation, a causal relationship has been predicted between thymine dimer production and XP. It has also been of great interest to determine which of the three forms of repair processes counteracting the effects of UV-induced damage to DNA (if any) are operating in humans. Furthermore, it was suspected that XP individuals might lack one or more repair systems, which may cause them to be susceptible to UV-induced skin damage.

The various modes of repair of UV-induced lesions have been investigated in human fibroblast cultures derived from XP and normal individuals. Fibroblasts are undifferentiated connective tissue cells. The results suggest that the XP phenotype may be caused by more than one mutant gene.

In 1968, James Cleaver showed that cells from XP patients were deficient in the **unscheduled DNA synthesis** (DNA synthesis other than occurring during chromosome replication) elicited in normal cells by UV radiation. This assay is thought to represent activity of the excision repair system, suggesting that XP cells are deficient in this form of repair. In 1974, the presence of a **photoreactivation enzyme (PRE)** was established in normal human cells. Betsy Sutherland identified the enzyme first in leukocytes and subsequently in fibroblast cells. Sutherland demonstrated further that some XP cultures contain a lower PRE activity than do control cultures. The activity in various XP cell strains ranges from 0 to 50 percent of normal in cultures established from different patients.

FIGURE 14.16 Two individuals with xeroderma pigmentosum. The 4-year-old boy on the left shows marked skin lesions induced by sunlight. Mottled redness (erythema) and irregular pigment changes that are a response to cellular injury are apparent. Two nodular cancers are present on his nose. The 18-year-old girl on the right has been carefully protected from sunlight since the diagnosis of xeroderma pigmentosum made in infancy. Several cancers have been removed and she has worked as a successful model.

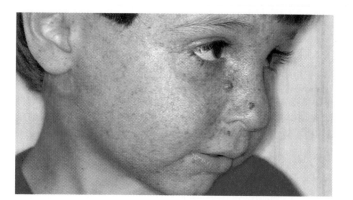
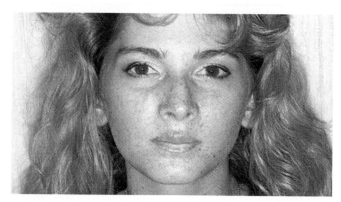

Somatic Cell Hybridization, XP, and Excision Repair

The link between xeroderma pigmentosum and inadequate excision repair has been strengthened by the use of **somatic cell hybridization** studies, a technique that we first introduced in Chapter 5. Cultured fibroblast cells from any two unrelated XP patients may be induced to fuse together, forming a **heterokaryon** where the two nuclei share a common cytoplasm. Once fusion is achieved, excision repair, as assayed by unscheduled DNA synthesis, is assessed. Sometimes repair is reestablished in the heterokaryon. When this occurs, the two variants are said to demonstrate **complementation**. Alone, neither cell type demonstrates excision repair, but together in a heterokaryon the process occurs. In genetic terms, this is strong evidence that the two patients from whom the cells were derived have different genes affected that led to the disease. Complementation occurs because the heterokaryon has at least one normal copy of each gene.

Based on many studies, most patients have been divided into seven complementation groups, which suggests that at least seven different genes may be involved in excision repair. These genes or their protein products have now been identified. They are found on disparate regions of the genome, and an homologous gene for each has been identified in yeast. Approximately 20 percent of XP patients do not fall into any of the seven groups. They manifest similar symptoms, but their fibroblasts do not demonstrate defective excision repair. There is some evidence that they are less efficient in normal DNA replication.

The study of xeroderma pigmentosum established that normal individuals are susceptible to UV-induced damage of DNA by exposure to sunlight. However, this damage activates the repair systems that counteract it. We can expect that future work will clarify the precise role and mechanism of repair of UV-induced damage in humans.

High-Energy Radiation

Within the electromagnetic spectrum, energy varies inversely with wavelength. Figure 14.17 compares the relative wavelengths of the various portions of the electromagnetic spectrum. **X-rays**, **gamma rays**, and **cosmic rays** have even shorter wavelengths than does UV radiation and are therefore more energetic. As a result, they are strong enough to penetrate deeply into tissues, causing ionization of the molecules encountered along the way. These sources of **ionizing radiation** are mutagenic, as established by Herman J. Muller and Lewis J. Stadler in the 1920s. Since that time, the effects of ionizing radiation, particularly X-rays, have been studied intensely.

As X-rays penetrate cells, electrons are ejected from the atoms of molecules encountered by the radiation. Thus, stable molecules and atoms are transformed into free radicals and reactive ions. Along the path of a high-energy ray, a trail of ions is left that can initiate a variety of chemical reactions. These reactions can affect directly or indirectly the genetic material, altering the purines and pyrimidines in DNA and resulting in point muta-

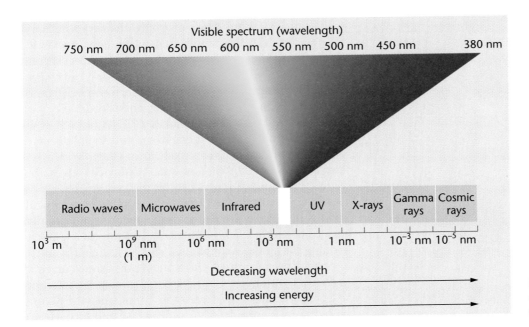

FIGURE 14.17 The components of the electromagnetic spectrum and their associated wavelengths.

tions. Such ionizing radiation is also capable of breaking phosphodiester bonds, disrupting the integrity of chromosomes. This results in a variety of aberrations.

Figure 14.18 shows a plot of induced X-linked recessive lethal mutations versus the dose of X rays administered. The graph shows a straight line that, if extrapolated, intersects near the zero axis. A linear relationship is evident between X-ray dose and the induction of mutation. For each doubling of dose, twice as many mutations are induced. Because the line intersects near the zero axis, this graph suggests that even very small doses of irradiation are mutagenic.

These observations can be interpreted in the form of the **target theory**, first proposed in 1924 by J. A. Crowther and F. Dessauer. The theory proposes that there are one or more sites, or targets, within cells and that a single event of irradiation at one site will bring about a damaging effect, or mutation. In a simple form, the target theory says that one "hit" of irradiation will cause one "event" or mutation, suggesting that the X rays interact directly with the genetic material.

Two other observations concerning irradiation effects are of particular interest. First, in some organisms studied, the *intensity of the dose* (dose-rate) administered seems to make little difference in mutagenic effect. That is, a total exposure of 100-roentgens (a **roentgen** is a measure of energy dose), whether administered in a single acute dose or spread out in time in several smaller chronic doses, seems to produce the same overall mutagenic effect. *Drosophila*, for example, shows this response. In mammals such as mice and humans, however, this appears not the case. This suggests that repair of the damage may occur during the intervals between irradiation.

The second observation is that certain portions of the cell cycle are more susceptible to radiation effects. As mentioned previously, in addition to a mutagenic effect, X rays can also break chromosomes, resulting in terminal or intercalary deletions, translocations, and general chromosome fragmentation. Damage occurs most readily when chromosomes are greatly condensed in mitosis. This property constitutes one of the reasons why radiation is used to treat human malignancy. Because tumorous cells are more often undergoing division than are their nonmalignant counterparts, they are more susceptible to the immobilizing effect of radiation.

Site-Directed Mutagenesis

This section introduces a useful experimental technique, **site-directed mutagenesis**, that allows researchers to introduce a designed mutation at a prescribed site within a gene of interest. The technique relies on the availability of a cloned gene and utilizes a number of manipulations involving recombinant DNA technology, which will be introduced in Chapter 15. However, the underlying principles are based on information previously presented.

The goal of the technique is to specifically alter a gene by changing, deleting, or inserting one or more nucleotides at a predetermined site. If, for example, a triplet code was altered, upon transcription and translation, the change will cause the insertion of a "mutant" amino acid in the protein encoded by the original gene. Such designed mutations are particularly useful in studying the effects of specific mutations on both gene expression and on protein function.

Various approaches can be used to accomplish the above goal; most are variations on the theme we are about to describe. The first step is to determine the nucleotide sequence of the gene being studied. This can be accomplished by DNA sequencing techniques, or it can be predicted if the amino acid sequence of the protein is known by utilizing our knowledge of the genetic code.

The next step is to isolate the DNA of known sequence and obtain from it one of the two complementary strands. A decision is then made as to which nucleotides are to be changed. Then, a small piece of DNA, a synthetic oligonucleotide, is chemically synthesized that is complementary to that region. It is complementary at all points, except in the triplet sequence or sequences that are to be altered. This sequence includes the triplet encoding the amino acid that will change in the protein.

As illustrated in Figure 14.19, this short piece of DNA is hybridized with the original parent strand, forming a partial duplex because of its complementarity along most of its length. If DNA polymerase and DNA ligase

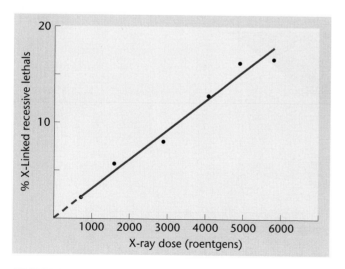

FIGURE 14.18 Plot of X-linked recessive mutations induced by increasing doses of X-rays. If extrapolated, the graph intersects the zero axis.

are then added to this hybrid complex, the short sequence is extended so that a duplex of the entire gene is formed. The two strands are perfectly complementary except at the point of alteration.

If this DNA is replicated, two types of duplexes are formed: One is like the original, unaltered gene and the other contains the newly designed sequence. Using recombinant DNA technology (see Chapter 15), it is possible not only to complete the above manipulations, but also to allow the altered gene to be expressed so that large amounts of the desired protein are available for study. Or, as we will see in the next section, the altered gene may be introduced into an organism and studied.

Knockout Genes and Transgenes

While a gene bearing a site-directed mutation may be investigated *in vitro*, the insertion of that gene into a living organism allows for the assessment of its *in vivo* function. If the insertion process involves the *replacement* of the comparable gene of the organism from which it originated, the technique is referred to as **gene knockout**. The genetically-altered organism is called a **knockout organism**, e.g., a "knockout mouse."

While specific gene replacement is difficult to achieve, it has been accomplished and is now fairly routine in research involving yeast and mice. In mice, the applica-

tion of gene knockout techniques has been particularly fruitful in studies involving the genetic control of early development and behavior. Most often, a "loss-of-function" mutation replaces the normal gene, and the effects are investigated. Furthermore, knockout mice now serve as models for studying human genetic disorders. In such cases, the comparable mouse gene that causes the human disorder is isolated, subjected to directed mutagenesis, and used to replace the nonmutant mouse gene. Cystic fibrosis and Duchenne muscular dystrophy have been investigated in this way. Applications of this general technology are discussed in more detail in Chapters 16 and 20.

If a gene is inserted into an organism *in addition* to its normal copies, it is called a **transgene**, and the organism is called a **transgenic organism** (e.g., a "transgenic plant"). In such cases, the gene may have undergone directed mutagenesis, or it may be a foreign gene isolated from another organism. This technology has been used more extensively than gene knockout, since it is easier to accomplish. This is so because specific replacement is not required.

The study of transgenic organisms has been routinely performed in plants, *Drosophila*, mice, and a variety of other organisms. Of particular note is the potential provided in agricultural studies as well as gene therapy techniques in our own species.

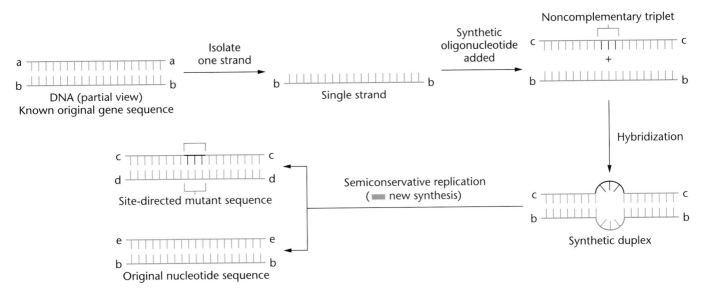

FIGURE 14.19 Flow diagram illustrating the principles of site-directed mutagenesis. A single strand of DNA from a gene of interest is initially isolated. This is hybridized with a synthetic oligonucleotide containing a triplet altered so as to encode an amino acid of choice. The partial duplex is completed under the direction of DNA polymerase and DNA ligase. Following semiconservative replication, a different complementary base pair is present in one of the new duplexes. Upon transcription and translation, a mutant protein, "designed" in the laboratory, will be produced.

Transposable Genetic Elements

We conclude this chapter by introducing the phenomenon of **transposable genetic elements**. Sometimes referred to as **transposons**, this topic encompasses a group of genetic units that are mobile. They can move or be "transposed" within the genome. Discussion of them in this chapter on mutation is appropriate because the movement of genetic units from one place in the genome to another often disrupts genetic function and results in phenotypic variation. As such, the impact of transposition of genetic units often fits into a broad definition of mutation.

As we shall see, transposable elements were first discovered almost 50 years ago in maize. However, even in the 1950s and 1960s the idea that genetic information was *not* fixed within the genome of an organism was slow to find acceptance. Such a notion was quite alien to the classical interpretation of genes on chromosomes. It was not until other transposable elements were discovered, and their molecular basis revealed, that the phenomenon was found to be nearly universal.

Insertion Sequences

Even though the presence of transposable elements in maize had been predicted earlier by Barbara McClintock, the first observation at the molecular level involved **insertion sequences (ISs)** in *E. coli*. Discovered in the early 1970s by a number of independent researchers, including Peter Starlinger and James Shapiro, insertion sequences were first visualized as a unique class of mutations affecting different genes in various bacterial strains. For example, the expression of a cluster of related genes involving galactose metabolism was repressed as a result of one such mutation.

This phenotypic effect was heritable, but was found not to be caused by a base pair change characteristic of conventional gene mutations. Instead, it was shown that a short, specific DNA segment had been inserted into the bacterial chromosome at the beginning of the galactose gene cluster. When this segment was spontaneously excised from the bacterial chromosome, wild-type function was restored.

It was subsequently revealed that several other distinct DNA segments could behave in a similar fashion, inserting into the chromosome and affecting gene function. These DNA segments are relatively short, not exceeding 2000 base pairs [2 kilobases (kb)]. For example, the first insertion sequence to be characterized in *E. coli* was IS1. It is about 800 base pairs long, whereas IS2, 3, 4, and 5 are about 1250 to 1400 base pairs in length.

Analysis of the DNA sequences of most IS units reveals a feature important to their mobility. At each end, the nucleotide sequences consist of **inverted terminal repeats (ITRs)** of one another. Though Figure 14.20 shows this terminal repeating unit to consist of only a few nucleotides, many more are actually involved. For example, in *E. coli*, the IS1 termini contain about 20 nucleotide pairs, IS2 and IS3 about 40 pairs, and IS4 about 18 pairs. It seems likely that these terminal sequences are an integral part of the mechanism of insertion of IS units into DNA. That insertion of IS units is more likely to occur at certain DNA regions than others suggests that IS termini can recognize certain target sequences in the DNA during the process of insertion.

Careful investigation has revealed that IS units are present in the wild-type *E. coli* chromosome as well as in other autonomous segments of bacterial DNA called plasmids. Thus, their presence does not always result in mutation. In the *E. coli* chromosome, five or more copies of IS1, IS2, and IS3 are known to be present. The exact number of each varies, depending on the strain examined.

Bacterial Transposons

In addition to their potential mutational effects, IS units play an even more significant role in the formation and movement of the larger **transposon (Tn) elements**. Transposons in bacteria consist of IS units that contain within their internal DNA sequence genes whose functions are unrelated to the insertion process. Like IS units, Tn elements are mobile in both bacterial and viral chromosomes and in plasmids. The Tn elements provide a mechanism for movement of genetic information from place to place both within and between organisms. Transposons were first discovered to move between DNA molecules as a result of observations of antibiotic-resistant bacteria. In the mid-1960s, Susumu Mitsuhashi first suggested that genes responsible for resistance to several antibiotics were mobile and could move between bacterial plasmids and chromosomes.

Electron microscopic studies may be used to confirm the presence of the terminal-inverted repeat sequences

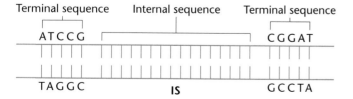

FIGURE 14.20 Diagrammatic representation of an insertion sequence (IS). The terminal sequences are perfect inverted repeats of one another.

within plasmids harboring transposons. When double-stranded DNA from such a plasmid is separated into single strands and each is allowed to reanneal separately, the inverted repeat units are complementary, forming what is called a **heteroduplex**. As might be predicted, all areas other than the terminal repeat units remain single stranded and form loops on either end of the double-stranded stem (Figure 14.21).

Transposons have become the focus of increased interest, particularly because they have been found in organisms other than bacteria. Bacteriophages that demonstrate the ability to insert their genetic material into the host chromosome behave in a similar fashion. The **bacteriophage mu**, consisting of over 35,000 nucleotides, can insert its DNA at various places in the *E. coli* chromosome. Like IS units, if insertion occurs within a gene, mutant behavior at that locus results. Transposons have also been discovered in higher organisms, including yeast, corn, *Drosophila*, and humans.

The *Ac–Ds* System in Maize

With our knowledge of insertion sequences and transposons in bacteria, it comes as no surprise that mobile genetic units also exist in eukaryotes. They are, in fact, more widespread, and they often have the effect of altering the expression of genetic information.

About 20 years before the discovery of transposons in prokaryotic organisms, Barbara McClintock analyzed the genetic behavior of two mutations, **Dissociation (*Ds*)** and **Activator (*Ac*)**, in corn plants (maize). Analysis involved an examination of the phenotypes of the kernels of maize resulting from genes expressed in either the endosperm or aleurone layers (Figure 14.22). These observations were correlated with cytological examination of the maize chromosomes. Initially, McClintock determined that *Ds* is located on chromosome 9. Provided that *Ac* is also present in the genome, *Ds* has the effect of inducing breakage at a point on the chromosome adjacent to its location. If breakage occurs in somatic cells during their development, progeny cells often lose part of chromosome 9, causing a variety of phenotypic effects.

Subsequent analysis suggested to McClintock that both the *Ds* and *Ac* genes are sometimes transposed to different chromosomal locations. While *Ds* moves only if *Ac* is also present, *Ac* is capable of autonomous movement. Where *Ds* comes to reside determines its genetic effect. Although it might cause chromosome breakage, it might instead inhibit gene expression. In cells where expression is inhibited, *Ds* might move again, releasing this inhibition. In these cases, the *Ds* element is believed to insert into a gene and subsequently to depart from it, causing changes in gene expression.

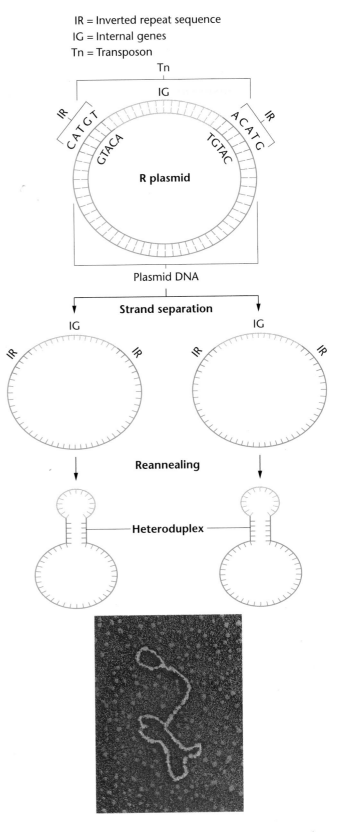

FIGURE 14.21   Heteroduplex formation as a result of inverted repeat sequences within a transposon inserted into a bacterial plasmid. The micrograph illustrates the final heteroduplex.

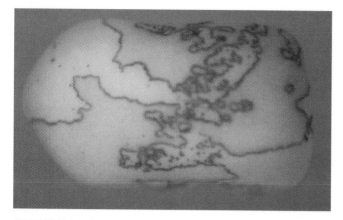

FIGURE 14.22 Corn kernel showing spots of colored aleurone produced by genetic transposition involving the *Ac–Ds* system.

Figure 14.23 illustrates the sort of movements and effects of the *Ds* and *Ac* elements described above. In McClintock's original observation, when the *Ds* element jumped out of chromosome 9, this excision event restored normal gene function. In such cells, pigment synthesis was restored. McClintock concluded that the *Ds* and *Ac* genes are **transposable controlling elements**.

It was not until many years later that anything comparable to the controlling elements in maize was recognized in other organisms. When bacterial insertion sequences and transposons were discovered, many parallels were evident. Transposons and insertion sequences were seen to move into and out of chromosomes, to insert at different positions, and to affect gene expression at the point of insertion.

Several *Ac* and *Ds* elements have now been isolated and carefully analyzed (Figure 14.24). As a result of this information, the relationship between the two elements has been clarified. The first *Ac* element sequence is 4563 bases long and strikingly similar to one of the known bacterial transposons. This sequence contains two 11-base-pair imperfect terminal-inverted repeats, two **open reading frames (ORFs)**, and three noncoding regions. Open reading frames contain initiation and termination sequences and are considered to encode genetic products. The first *Ds* element studied (*Ds-a*) is nearly identical in structure to *Ac* except for a 194-base segment that has been deleted from the largest open reading frame (Figure 14.24). There is some evidence that this gene encodes a **transposase enzyme**, essential to transposition of both *Ac* and *Ds* elements. The deletion of part of this gene in the *Ds-a* element explains its dependence on the *Ac* element for transposition. Several other *Ds* elements have also been sequenced, and each reveals an even larger deletion in the same region. In each case, however, the terminal repeats are retained and seem to be essential for transposition, provided that a functional transposase enzyme is supplied by the gene in the *Ac* element.

Although the validity of Barbara McClintock's proposed mobile elements was questioned following her initial observations, molecular analysis has since verified her conclusions. For her work, Barbara McClintock was awarded the Nobel Prize in 1983.

Other Mobile Genetic Elements in Plants: Mendel Revisited

More recent work on transposable elements in plants has led us full circle to the union of Mendel's own observations with molecular genetics. Recall that one of the first phenotypes investigated by Mendel involved the inheri-

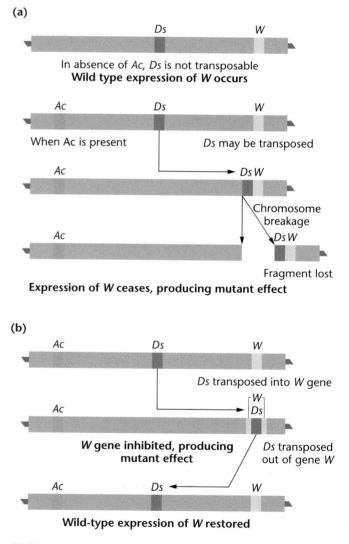

FIGURE 14.23 Two consequences of the influence of the *Activator* (*Ac*) element on the *Dissociation* (*Ds*) element. In (a), *Ds* is transposed to a region adjacent to a theoretical gene *W*. Subsequent chromosomal breakage is induced, the *W*-bearing segment is lost, and mutant gene expression occurs. In (b), *Ds* is transposed to a region within the *W* gene, causing immediate mutant expression. *Ds* may also "jump" out of the *W* gene, with the accompanying restoration of *W* gene activity and its wild-type expression.

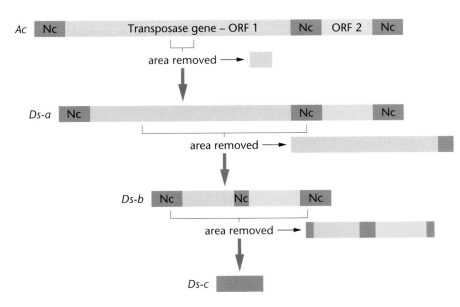

FIGURE 14.24 A comparison of the structure of an *Ac* element with three *Ds* elements, all of which have been isolated and sequenced. The imperfect and inverted repeats are at the ends of the *Ac* element. The transposase gene is in an open reading frame (ORF 1). No function has yet been assigned to ORF 2. Noncoding regions are designated by Nc. As this scheme shows, *Ds-a* appears to be simply an *Ac* element containing a small deletion in the gene encoding the transposase enzyme.

tance of round and wrinkled peas. The two phenotypes are produced by alleles of a single gene, *rugosus*. It is now known that the wrinkled phenotype is associated with the absence of an enzyme, **starch-branching enzyme (SBEI)**, that controls the formation of branch points in starch molecules. The lack of starch synthesis leads to the accumulation of sucrose and a higher water content and osmotic pressure in the developing seeds. As the seeds mature, those that are wrinkled (genotype *rr*) lose more water than do the smooth seeds (*RR* or *Rr*), producing the wrinkled phenotype (Figure 14.25).

The structural gene for SBEI has now been cloned and characterized in both wild-type and mutant genotypes. In the *rr* genotype, the SBEI protein is nonfunctional, presumably because the SBEI gene is interrupted by a 0.8-kb insertion, resulting in the production of an abnormal RNA transcript. The inserted DNA has 12-bp inverted repeats at each end that are highly homologous to the terminal sequences in the transposable element *Ac* from maize, and other *Ac*-like elements from snapdragons and parsley. Terminal repeated sequences and the genetic information encoding a transposase enzyme appear to be universal components of transposons in all organisms studied.

Copia and P Elements in *Drosophila*

Transposable elements have been discovered in other eukaryotic organisms, notably in yeast, in *Drosophila*, and in primates, including humans. For example, in 1975 David Hogness and his colleagues David Finnigan, Gerald Rubin, and Michael Young identified a class of genes in *Drosophila melanogaster*, which they designated as *copia*. These genes transcribe "copious" amounts of RNA (thus, their name), which can be isolated in the mRNA fraction. Present up to 30 times in the genome of cells, *copia* genes

are nearly identical in nucleotide sequence. Mapping studies show that they are transposable to different chromosomal locations and are dispersed throughout the genome.

Copia appears to be only one of the approximately 30 families of transposable elements in *Drosophila*, each of which is present from a few copies up to 20 to 50 times in the genome. Ever since the discovery of *copia*, many other families of transposable elements have been recognized. Some are referred to as *copia*-like. Together, the many families constitute about 5 percent of the *Drosophila* genome and over half of the middle repetitive DNA of this organism.

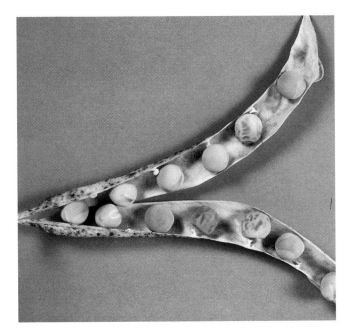

FIGURE 14.25 The *wrinkled* trait in garden peas, studied by Gregor Mendel, is caused by the insertion of a transposable element into the structural gene for starch-branching enzyme.

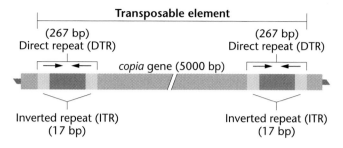

FIGURE 14.26   Structural organization of a *copia* transposable element in *Drosophila melanogaster,* demonstrating the terminal repeats.

Despite the variability in DNA sequence among the members of different families, they share a common structural organization thought to be related to the insertion and excision processes of transposition. Each *copia* gene consists of approximately 5000 to 8000 base pairs of DNA, including a long family-specific **direct terminal repeat (DTR)** sequence of 276 base pairs at each end. Within each repeat is a short **inverted terminal repeat (ITR)** of 17 base pairs. These features are illustrated in Figure 14.26. The DTR sequences are found in other transposons in other organisms, but they are not universal. However, the shorter ITR sequences are considered universal in *copia* elements.

Insertion of *copia*, as with other transposable elements, appears to be dependent on ITR sequences and seems to occur at specific target sites in the genome. The *copia*-like elements demonstrate regulatory effects at the point of their insertion in the chromosome. Certain mutations, including ones affecting eye color and segment formation, have been found to be due to insertions within genes. For example, the eye color mutation *white-apricot* (*w^a*), an allele of the *white* (*w*) gene, contains a *copia* insertion element within the gene. Removing the transposable element sometimes restores the wild-type allele.

Still another interesting category of transposable elements in *Drosophila* is the family called **P elements**. These were discovered while studying the phenomenon of **hybrid dysgenesis**, a condition causing sterility, elevated mutation rate, and chromosome rearrangement in the offspring of crosses between certain strains of fruit flies. P elements and their transposition are discussed in the essay found at the end of this chapter.

Transposable Elements in Humans

The final class of transposable elements that we shall discuss is represented by the ***Alu* family** of short interspersed elements (SINEs), which are characteristic of moderately repetitive DNA in mammals. Originally detected using reassociation kinetic analysis (see Chapter 10), human DNA contains some 300,000 copies of this 200 to 300 base-pair sequence in the genome. They have received their name because they contain specific nucleotide sequences that are cleaved by a restriction endonuclease called *Alu*I.

They are considered to be significant elements in the genome based on the fact that they have now been demonstrated in the DNA of all primates and rodents studied. Additionally, there is conserved within the *Alu* elements in mammals a specific 40 base-pair DNA sequence. Finally, *Alu* sequences are represented in nuclear transcripts.

These sequences are considered transposable based on several lines of evidence. Most important is the fact that they contain a 300-bp sequence flanked on either side by direct repeat sequences consisting of 7 to 20 bp. This observation parallels that made in bacterial insertion sequences. These flanking regions are related to the insertion process during transposition. The second line of evidence involves the observation that clustered regions of *Alu* sequences vary in the DNA of both normal and diseased individuals and in different tissues of the same individual. Additionally, the sequences have been found extrachromosomally.

The potential mobility and mutagenic effects of these and other elements have far-reaching implications. A recent example involves a situation where a transposon has been "caught in the act." The case involves a male child with **hemophilia**, as investigated by Haig Kazazian and his colleagues. One cause of hemophilia is a defect in blood-clotting factor VIII, the product of an X-linked gene. Kazazian found two cases where, inserted within this gene, there was a transposable element much longer than *Alu* sequences. Also present elsewhere in the genome, this sequence was shown to be an example of a type of repetitive DNA referred to as a **long interspersed element (LINE)**. Both LINEs and SINEs are discussed in more detail in Chapter 17.

There has been great interest in determining if one of the mother's X chromosomes also contains this specific LINE. If so, the unaffected mother would be heterozygous and pass the LINE-containing chromosome to her son. The startling finding is that the LINE sequence is *not* present on either of her *X chromosomes*, but *was* detected on *chromosome 22* of both parents. This suggests that this mobile element may have "moved" from one chromosome to another in the gamete-forming cells of the mother, prior to being transmitted to the son.

Many questions remain concerning this and other transposable elements. What is their origin? Were they once some sort of retrovirus (see Chapters 21 and 22)? Exactly how do they move, and what has been their role during evolution? These and other questions will intrigue researchers for many years to come.

GENETICS, TECHNOLOGY, AND SOCIETY

P Element Transposons: New Age Mutagens

Molecular geneticists have made great progress in understanding the function and regulation of many genes. This understanding has emerged through the use of classic genetic techniques such as analysis of mutant phenotypes in genetic crosses, as well as through direct manipulation of genes in the test tube. In *Drosophila* molecular genetics, one of the most revolutionary tools for deciphering gene function has been the **P element transposon**. Geneticists have harnessed P elements for a remarkable array of uses, including gene transfer, insertion mutagenesis, gene cloning by tagging, and enhancer trapping. The domestication of P elements has led to significant advances in our understanding of *Drosophila* development, gene regulation, DNA repair and recombination.

P elements were originally discovered because of their ability to cause **hybrid dysgenesis**—a syndrome characterized by high mutation rates, chromosomal rearrangements and sterility. Hybrid dysgenesis is caused by high rates of P element transposition in the germ line, where the mobile DNA elements insert themselves into or near genes, thereby causing mutations. P elements are 2.9 kb long, with 31 bp terminal inverted repeats. The elements encode at least two proteins, one of which is the transposase enzyme that is required for transposition. The transposase is expressed only in the germ line, accounting for the tissue specificity of P element transposition. Strains of flies that contain P elements inserted into their genomes are resistant to further transpositions, due to the presence of a repressor protein, also encoded by the P elements.

P elements are used as insertional mutagens. A common method of generating P element-induced mutations is to inject *Drosophila* embryos with a mixture of two cloned and purified P elements. One P element lacks the transposase gene and cannot move by itself.

This P element contains a gene, such as *rosy*, which allows researchers to select flies that have incorporated the P element into their DNA (their eyes will be rosy red). The other P element contains the transposase gene, but cannot insert into the genome. The two injected P elements will be taken up by the embryo's cells, transposase will be expressed in the germ line of the fly, and the *rosy*/P element will insert itself into the fly's DNA. The next generation will contain stable copies of P element DNA, inserted at random locations.

Mutations may arise from several kinds of insertional events. If a P element inserts into the coding region of a gene, it can destroy the normal gene product. If it inserts into the promoter region of a gene, it can affect the level of expression of the gene. Insertions into introns can affect splicing or cause premature termination of transcription. Researchers screen the flies for mutant phenotypes of interest, then isolate the fly's DNA and clone the gene that was mutated. The cloning is simplified, because the gene is tagged by the presence of a P element in or near it. This technology has been used to identify genes involved in *Drosophila* development, behavior, and regulation of gene expression.

A variation of the insertional mutagenesis technique involves the use of *Drosophila* strains that stimulate transposon jumping when crossed with another P element strain. The initiating strain, called a "jumpstarter," carries in its genome a P element that produces transposase but cannot transpose. The other strain carries a defective P element that can transpose but cannot produce its own transposase. Crosses between these two strains produce some flies that carry both types of P element. These flies initiate transposition in their germ line cells and the defective P element moves to a new site, perhaps creating a new mutation. Further crosses eliminate the chromosome that carries the transposase P element, and the P element remains stable in its new location.

Another equally important P element technique is germ line transforma-

tion. One of the most powerful ways to analyze a gene's activity and expression pattern is to clone the gene, manipulate its DNA sequences in a test tube and reintroduce the mutated or modified gene into an organism. P elements provide a vehicle (or "vector") for inserting a cloned gene into a fly's genome. The method is similar to that described for insertional mutagenesis. Embryos are injected with a mixture of two cloned and purified P elements. One contains the transposase gene (but cannot insert). The other contains the marker gene (such as *rosy*) and the cloned gene of interest, but lacks a transposase gene. Flies that have been transformed are selected on the basis of their rosy eye color, and the behavior of the cloned and transposed gene is observed.

Various other elegant techniques, based on P element transposition, have been developed. These have allowed researchers to study the mechanics of DNA repair and DNA recombination. In addition, researchers are presently developing methods to target P element insertions to precise single chromosomal sites. This should increase the precision of germ line transformation in the analysis of gene activity.

The sophistication of P element technology provides an example of how molecular genetics has harnessed naturally occurring genetic phenomena to tackle fundamental problems of biology.

The fruit fly, *Drosophila melanogaster.*

CHAPTER SUMMARY

1. The phenomenon of mutation not only provides the basis for most of the inherent variation present in living organisms but it also serves as the working tool of the geneticist in studying and understanding the nature of genes and the mechanisms governing genetic processes.

2. Mutations are distinguished by the tissues affected. Somatic mutations are those that may affect the individual, but are not heritable. Mutations arising in the gametic tissues may produce new alleles that can be passed on to offspring and enter the gene pool.

3. Another classification of mutations relies on their effect. Morphological mutations, for example, can be detected visibly. Other types include biochemical, lethal, conditional, and regulatory mutations, groups that are not mutually exclusive.

4. Organisms in which mutations can be easily induced and detected are most often used in genetic studies. Viruses, bacteria, fungi, and *Drosophila* are frequently used because of these properties, as well as the fact that they have short life cycles.

5. Spontaneous mutations may arise naturally as a result of rare chemical rearrangements of atoms, or tautomeric shifts, and as the result of errors occurring during DNA replication. While spontaneous mutations are very rare, the rate of mutagenesis can be increased experimentally by a variety of mutagenic agents.

6. Mutagenic agents such as nitrous acid, hydroxylamine, alkylating agents, and base analogues cause chemical changes in nucleotides that alter their base-pairing affinities. As a result, base substitution mutations arise following DNA replication.

7. Frameshift mutations arise when the addition or deletion of one or more nucleotides (but not multiples of three) occurs. Acridine dyes, which intercalate with DNA, are potent inducers of frameshift mutations.

8. Direct analysis of DNA from individuals carrying specific alleles that have arisen through mutations, including the ABO blood types and muscular dystrophy, has been informative. When complete loss of function occurs, as is the case in blood type O and the Duchenne form of muscular dystrophy, deletions or duplications of nucleotides have resulted in a shift in reading frame, eventually causing premature termination of translation of the resulting messenger RNA.

9. Another form of mutation, discovered in several human disorders, includes unstable trinucleotide units that balloon in size during DNA replication and that are inherited in this form through successive generations. Increasing numbers of repeats often correlate with early onset and severity of disease.

10. Ultraviolet radiation and high-energy radiation from gamma, cosmic, or X-ray sources are other forms of potent mutagenic agents. UV radiation induces the formation of pyrimidine dimers in DNA, whereas high-energy radiation is able to penetrate deeper into tissues, causing the ionization of molecules in its path, including DNA.

11. Various forms of repair of DNA lesions, such as pyrimidine dimers, have been discovered, including photoreactivation, excision, mismatch, and recombinational repair. Loss of repair function by mutation in humans results in the severe disorder xeroderma pigmentosum.

12. Site-directed mutagenesis is a technique allowing researchers to create specific alterations in the nucleotide sequence of the DNA of genes.

13. Insertion sequences in bacteria and other transposable elements in eukaryotes have a profound effect on genetic expression, thus serving as a distinct category of mutagenic agents. A recent example involves a mutation causing hemophilia in humans.

KEY TERMS

ABO antigens, 407
acridine dyes, 404
acridine orange, 404
Activator (*Ac*) element, 417
adaptation hypothesis, 390
adenine methylase, 411
alkylating agents, 402
Alu family, 420
Ames test, 409
AP endonuclease, 411
apurinic sites (AP sites), 405
attached-X procedure, 394

autosomal recessive mutation, 393
auxotroph, 394
bacteriophage mu, 417
base analogue, 402
base substitution, 400
Becker muscular dystrophy (BMD), 408
behavior mutation, 393
biochemical mutation, 393
ClB procedure, 394
complementation, 413
complete medium, 394
conditional mutation, 393

cosmic rays, 413
deamination, 405
direct terminal repeat (DTR), 420
Dissociation (*Ds*) element, 417
DNA glycosylase, 411
DNA ligase, 410
DNA methylation, 411
DNA polymerase I, 410
dominant autosomal mutation, 393
Duchenne muscular dystrophy (DMD), 407
dystrophin, 408
ethylmethane sulfonate (EMS), 402

INSIGHTS AND SOLUTIONS

1. How could you isolate a mutant strain of bacteria that is resistant to penicillin, an antibiotic that inhibits cell wall synthesis?

   Solution: Grow a culture of bacterial cells in liquid medium and plate the cells on agar medium to which penicillin has been added. Only penicillin-resistant cells will reproduce and form colonies. Each colony will, in all likelihood, represent a cloned group of cells with the identical mutation. Isolate members of each colony. To enhance the chance of such a mutation arising, you might want to add a mutagen to the liquid culture.

2. The base analogue 2-amino purine (2-AP) substitutes for adenine during DNA replication, but it may base-pair with cytosine. The base analogue 5-bromouracil (5-BU) substitutes for thymidine, but it may base pair with guanine. Follow the double-stranded trinucleotide sequence shown below through three rounds of replication, assuming that in the first round, both analogues are present and become incorporated wherever possible. In the second and third round of replication, they are removed. What final sequences occur?

   Solution: (see next column)

3. A rare dominant mutation was studied in humans that was expressed at birth. Records showed that six cases were discovered in 40,000 live births. Family histories revealed that in two cases, the mutation was already present in one of the parents. Calculate the spontaneous

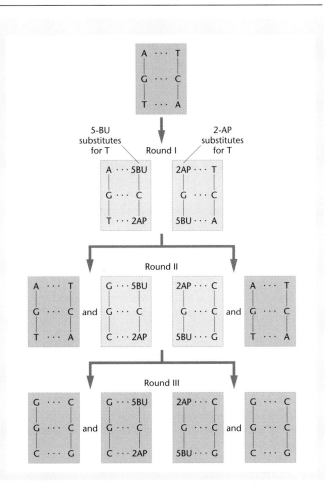

mutation rate for this mutation. What are some underlying assumptions that may affect our conclusions?

Solution: Only four cases represent a new mutation. Because each live birth represents two gametes, the sample size is from 80,000 meiotic events. The rate is equal to

$$\frac{4}{80,000} = \frac{1}{20,000} = 0.5 \times 10^{-4}$$

There are various reasons why this mutation rate may be inaccurate. We have assumed that the mutant gene is fully penetrant and is expressed in each individual bearing it. If it is not fully penetrant, our calculation may be an underestimate because one or more mutations may have gone undetected. We have also assumed that the screening was 100 percent accurate. One or more mutant individuals may have been "missed," again leading to an underestimate. Finally, we assume that the viability of mutant and nonmutant individuals is equivalent and that they survive equally *in utero*. Therefore, our assumption is that the number of mutant individuals at birth is equal to the number at conception. If this were not true, our calculation would again be inaccurate.

4. Consider the following estimates:
   (i)   There are 4×10^9 humans living on this planet.
   (ii) Each individual has 10^5 genes.
   (iii) The average mutation rate at the 10^5 loci is 10^{-5}.

How many spontaneous mutations are currently present in the human population? Assuming that these mutations are equally distributed among all genes, how many new alleles have arisen at each locus in the human population?

Solution: First, since each individual is diploid, there are two copies of each gene per person, each arising from a separate gamete. Therefore, the total number of spontaneous mutations is

$$\left(\frac{2 \times 10^5 \text{ genes}}{\text{individual}} \right) \left(\frac{4 \times 10^9 \text{ individuals}}{} \right) \left(\frac{10^{-5} \text{ mutations}}{\text{gene}} \right)$$
$$= (2 \times 10^5) \cdot (4 \times 10^9) \cdot (10^{-5}) \text{ mutations}$$
$$= 8 \times 10^9 \text{ mutations}$$

On the average, in the entire human population there are

$$\frac{8 \times 10^9 \text{ mutations}}{10^5 \text{ genes}} = \frac{8 \times 10^4 \text{ mutations}}{\text{gene}}$$

PROBLEMS AND DISCUSSION QUESTIONS

1. What is the difference between a chromosomal aberration and a gene mutation?
2. Discuss the importance of mutations to the successful study of genetics.
3. Describe the technique for detection of nutritional mutants in *Neurospora*.
4. Most mutations are thought to be deleterious. Why, then, is it reasonable to state that mutations are essential to the evolutionary process?
5. Why do you suppose that a random mutation is more likely to be deleterious than beneficial?
6. Most mutations in a diploid organism are recessive. Why?
7. What is meant by a conditional lethal mutation?
8. Contrast the concerns about mutation in somatic and gametic tissue.
9. In the *ClB* technique for detection of mutation in *Drosophila*:
   (a) What type of mutation may be detected?
   (b) What is the importance of the *l* gene?
   (c) Why is it necessary to have the crossover suppressor (*C*) in the genome?
   (d) What is *C*?
10. In an experiment with the *ClB* system, a student irradiated wild-type males and subsequently scored 500 F_2 cultures, finding all of them to have both males and females present. What conclusions can you draw about the induced mutation rate?
11. In *Drosophila*, induced mutations on chromosome 2 that are recessive lethals may be detected using a second chromosomal stock, *Curly, Lobe/Plum* (*Cy L/Pm*). These alleles are all dominant and lethal in the homozygous condition. Detection is performed by crossing *Cy L/Pm* females to wild-type males that have been subjected to a mutagen. Three generations are required. In the F_1, *Cy L* males are selected and individually backcrossed to *Cy L/Pm* females. In the F_2 of each cross, flies expressing *Cy* and *L* are mated to produce a series of F_3 generations. Diagram these crosses and predict how an F_3 culture will vary if a recessive lethal was induced in the original test male compared to the case where no lethal mutation resulted. In which F_3 flies would a recessive morphological mutation be expressed?
12. Describe tautomerism and the way in which this chemical event may lead to mutation.
13. Contrast and compare the mutagenic effects of deaminating agents, alkylating agents, and base analogues.
14. Acridine dyes induce frameshift mutations. Is such a mutation likely to be more detrimental than a point mutation where a single pyrimidine or purine has been substituted?
15. Why is X-radiation a more potent mutagen than UV radiation?
16. Contrast the induction of mutations by UV radiation and X-radiation.

17. Contrast the various types of DNA repair mechanisms known to counteract the effects of UV radiation and other DNA damage. What is the role of visible light in photo-reactivation?

18. Mammography is an accurate screening technique for the early detection of breast cancer in humans. Because this technique uses X rays diagnostically, it has been highly controversial. Can you explain why? What reasons justify the use of X rays for this type of medical diagnosis?

19. Compose a short essay that relates the molecular basis of fragile-X syndrome, myotonic dystrophy, and Huntington disease to the severity of the disorders, as well as to the phenomenon called genetic anticipation.

20. Describe the Ames assay for screening potential environmental mutagens. Why is it thought that a compound that tests positively in the Ames assay may also be carcinogenic?

21. Describe the general approach utilized in site-directed mutagenesis.

22. Describe the disorder xeroderma pigmentosum (XP) in humans. What genetic defect results in XP, and how does this defect relate to the phenotype associated with the disorder?

23. Differentiate between point mutations that are transitions versus transversions. Using the DNA bases (A, T, C, and G), list the four types of transitions and the eight types of transversions.

24. In a bacterial culture where all cells were unable to synthesize leucine (*leu⁻*) a potent mutagen was added, and the cells were allowed to undergo one round of replication. At that point, samples were taken and a series of dilutions was made prior to plating the cells on either minimal medium or minimal medium to which leucine was added. The first culture condition (minimal medium) allows the detection of mutations from *leu⁻* to *leu⁺*, while the second culture condition allows the determination of total cells, since all bacteria can grow. From the following results, determine the frequency of mutant cells. What is the rate of mutation at the locus involved with leucine biosynthesis?

Culture Condition	Dilution	Colonies
minimal medium	10^{-1}	18
minimal + leucine	10^{-7}	6

25. Contrast the various transposable genetic elements in bacteria, maize, *Drosophila*, and humans. What properties do they share?

26. The initial discovery of IS units in bacteria involved their presence upstream (5′) to three genes controlling galactose metabolism. All three genes were affected despite no direct physical interaction with the IS unit. Speculate as to why this might occur.

27. Ty, a transposable element in yeast, has been found to contain an open reading frame (ORF) that encodes the enzyme reverse transcriptase. This enzyme synthesizes DNA from an RNA template. Speculate as to the role of this gene product in the transposition of Ty within the yeast genome.

EXTRA-SPICY PROBLEMS

28. Presented below are hypothetical findings from studies of heterokaryons formed from seven human xeroderma pigmentosum cell strains:

	XP1	XP2	XP3	XP4	XP5	XP6	XP7
XP1	0						
XP2	0	0					
XP3	0	0	0				
XP4	+	+	+	0			
XP5	+	+	+	+	0		
XP6	+	+	+	+	0	0	
XP7	+	+	+	+	0	0	0

Note: + = complementing; 0 = noncomplementing.

These data represent the occurrence of unscheduled DNA synthesis in the fused heterokaryon when neither of the strains alone showed synthesis. What does unscheduled DNA synthesis represent? Which strains fall into the same complementation groups? How many different groups are revealed based on these limited data? How does one interpret the presence of these complementation groups?

29. Demonstrate your insights into both chromosomal and gene mutation by projecting yourself as one of the team of geneticists who immediately launched the study of the genetic effects of high energy radiation on the surviving Japanese population following the atomic bomb attacks at Hiroshima and Nagasaki in 1945. Outline a comprehensive short-term and long-term study that would address this topic. Be sure to include stategies for considering the effects on both somatic and germ line tissues.

30. Cystic fibrosis (CF) is a severe autosomal recessive disorder in humans that results from a chloride ion channel defect in epithelial cells. Over 300 sequence alterations have been identified in the 24 exons of the responsible gene (CFTR for cystic fibrosis conductance regulator), including dozens of different missense mutations and frameshift mutations, as well as numererous splice-site defects. Although all affected CF individuals demonstrate chronic obstuctive lung disease, there is variation in pancreatic enzyme insufficiency (PI). Speculate on which types of observed mutations are likely to give rise to less severe symptoms of CF, including only minor PI.

Of the 300 sequence alterations, a number of them found within the exon regions of the CFTR gene do not give rise to cystic fibrosis. Taking into account your accumulated knowledge of the genetic code, gene expression, protein function, and mutation, explain to a freshman biology major how this might be.

SELECTED READINGS

AMES, B. N., MCCANN, J., and YAMASAKI, E. 1975. Method for detecting carcinogens and mutagens with the *Salmonella/* mammalian microsome mutagenicity test. *Mut. Res.* 31:347–64.

AUERBACH, C. 1978. Forty years of mutation research: A pilgrim's progress. *Heredity* 40:177–87.

AUERBACH, C., and KILBEY, B. J. 1971. Mutations in eukaryotes. *Annu. Rev. Genet.* 5:163–218.

BALTIMORE, D. 1981. Somatic mutation gains its place among the generators of diversity. *Cell* 26:295–96.

BEADLE, G. W., and TATUM, E. L. 1945. *Neurospora* II. Methods of producing and detecting mutations concerned with nutritional requirements. *Am. J. Bot.* 32:678–86.

BECKER, M. M., and WANG, Z. 1989. Origin of ultraviolet damage in DNA. *J. Mol. Biol.* 210:429–38.

BERG, D., and HOWE, M., eds. 1989. *Mobile DNA.* Washington, DC: American Society of Microbiology.

BRENNER, S., BARNETT, L., CRICK, F. H. C., and ORGEL, L. 1961. The theory of mutagenesis. *J. Mol. Biol.* 3:121–24.

CAIRNS, J., OVERBAUGH, J., and MILLER, S. 1988. The origin of mutants. *Nature* 335: 142–45.

CAPECCHI, M. R. 1989. Altering the genome by homologous recombination. *Science* 244:1288–92.

CARTER, P. 1986. Site-directed mutagenesis. *Biochem. J.* 237:1–7.

CLEAVER, J. E. 1968. Defective repair of replication of DNA in xeroderma pigmentosum. *Nature* 218:652–56.

———. 1990. Do we know the cause of xeroderma pigmentosum? *Carcinogenesis* 11:875–82.

CLEAVER, J. E., and KARENTZ, D. 1986. DNA repair in man: Regulation by a multiple gene family and its association with human disease. *Bioessays* 6:122–27.

COHEN, S. N., and SHAPIRO, J. A. 1980. Transposable genetic elements. *Sci. Am.* (Feb.) 242:40–49.

CROW, J. F., and DENNISTON, C. 1985. Mutation in human populations. *Adv. Hum. Genet.* 14:59–216.

CULOTTA, E. 1994. A boost for "adaptive" mutation. *Science* 265:318–19.

DEERING, R. A. 1962. Ultraviolet radiation and nucleic acids. *Sci. Am.* (Dec.) 207:135–44.

DEN DUNNEN, J. T., et al. 1989. Topography of the Duchenne muscular dystrophy (DMD) gene. *Am. J. Hum. Genet.* 45:835–47.

DEVORET, R. 1979. Bacterial tests for potential carcinogens. *Sci. Am.* (Aug.) 241:40–49.

DORING, H. P., and STARLINGER, P. 1984. Barbara McClintock's controlling elements: Now at the DNA level. *Cell* 39:253–59.

DRAKE, J. W. 1970. *Molecular basis of mutation.* San Francisco: Holden-Day.

———. 1991. A constant rate of spontaneous mutation in DNA-based microbes. *Proc. Natl. Acad. Sci. USA* 88:7160–64.

DRAKE, J. W., et al. 1975. Environmental mutagenic hazards. *Science* 187:505–14.

DRAKE, J. W., GLICKMAN, B. W., and RIPLEY, L. S. 1983. Updating the theory of mutation. *Am. Sci.* 71:621–30.

FEDOROFF, N. V. 1984. Transposable genetic elements in maize. *Sci. Am.* (June) 250:85–98.

FINNEGAN, D. J. 1985. Transposable elements in eukaryotes. *Int. Rev. Cytol.* 93:281–326.

FRIEDBERG, E. C., WALKER, G. C., and SIEDE, W. 1995. *DNA repair and mutagenesis.* Washington, DC: ASM Press.

HALL, B. G. 1988. Adaptive evolution that requires multiple spontaneous mutations: I. Mutations involving an insertion sequence. *Genetics* 120:887–97.

———. 1990. Spontaneous point mutations that occur more often when advantageous than when neutral. *Genetics* 126: 5–16.

HANAWALT, P. C., and HAYNES, R. H. 1967. The repair of DNA. *Sci. Am.* (Feb.) 216:36–43.

HASELTINE, W. A. 1983. Ultraviolet light repair and mutagenesis revisited. *Cell* 33:13–17.

HOWARD-FLANDERS, P. 1981. Inducible repair of DNA. *Sci. Am.* (Nov.) 245:72–80.

KELNER, A. 1951. Revival by light. *Sci. Am.* (May) 184:22–25.

KNUDSON, A. G. 1979. Our load of mutations and its burden of disease. *Am. J. Hum. Genet.* 31:401–13.

KRAEMER, F. H., et al. 1975. Genetic heterogeneity in xeroderma pigmentosum: Complementation groups and their relationship to DNA repair rates. *Proc. Natl. Acad. Sci. USA* 72:59–63.

LITTLE, J. W., and MOUNT, D. W. 1982. The SOS regulatory system of *E. coli. Cell* 29:11–22.

LURIA, S. E., and DELBRUCK, M. 1943. Mutations of bacteria from virus sensitivity to virus resistance. *Genetics* 28: 491–511.

MASSIE, R. 1967. *Nicholas and Alexandra.* New York: Atheneum.

MASSIE, R., and MASSIE, S. 1975. *Journey.* New York: Knopf.

MCCANN, J., CHOI, E., YAMASAKI, E., and AMES, B. 1975. Detection of carcinogens as mutagens in the *Salmonella/* microsome test: Assay of 300 chemicals. *Proc. Natl. Sci. USA* 72:5135–39.

MCCLINTOCK, B. 1956. Controlling elements and the gene. *Cold Spring Harbor Symp. Quant. Biol.* 21:197–216.

MACDONALD, M. E., et al. 1993. A novel gene containing a trinucleotide repeat that is expanded and unstable in Huntington's disease chromosome. *Cell* 72:971–80.

MCKUSICK, V. A. 1965. The royal hemophilia. *Sci. Am.* (Aug.) 213:88–95.

MULLER, H. J. 1927. Artificial transmutation of the gene. *Science* 66:84–87.

——— 1955. Radiation and human mutation. *Sci. Am.* (Nov.) 193:58–68.

NEWCOMBE, H. B. 1971. The genetic effects of ionizing radiation. *Adv. Genet.* 16:239–303.

O'HARE, K. 1985. The mechanism and control of P element transposition in *Drosophila. Trends Genet.* 1:250–54.

OSSANA, N., PETERSON, K. R., and MOUNT, D. W. 1986. Genetics of DNA repair in bacteria. *Trends Genet.* 2:55–58.

RADMAN, M., and WAGNER, R. 1988. The high fidelity of DNA duplication. *Sci. Am.* (Aug.) 259(2):40–46.

SHERRATT, D. J. (ed.)1995. *Mobile genetic elements.* New York: Oxford University Press.

SHORTLE, D., DIMARIO, D., and NATHANS, D. 1981. Directed mutagenesis. *Annu. Rev. Genet.* 15:265–94.

SIGURBJORHSSON, B. 1971. Induced mutations in plants. *Sci. Am.* (Jan.) 224:86–95.

SPRADLING, A. C., STERN, D. M., KISS, I., ROOTE, J., LAVERTY, T., and RUBIN, G. M. 1995. Gene disruptions using P transposable elements: An integral component of the *Drosophila* Genome Project. *Proc. Natl. Acad. Sci. USA* 92:10824–10830.

STADLER, L. J. 1928. Mutations in barley induced by X-rays and radium. *Science* 66:84–87.

SUTHERLAND, B. M. 1981. Photoreactivation. *Bioscience* 31:439–44.

TOMLIN, N. V., and APRELIKOVA, O. N. 1989. Uracil DNA glycosylases and DNA uracil repair. *Int. Rev. Cytol.* 114:81–124.

TOPAL, M. D., and FRESCO, J. R. 1976. Complementary base pairing and the origin of substitution mutations. *Nature* 263:285–89.

VOGEL, F. 1992. Risk calculations for hereditary effects of ionizing radiation in humans. *Hum. Genet.* 89:127–46.

WELLS, R. D. 1994. Molecular basis of genetic instability of triplet repeats. *J. Biol. Chem.* 271: 2875–78.

WILLS, C. 1970. Genetic load. *Sci. Am.* (March) 222:98–107.

WOLFF, S. 1967. Radiation genetics. *Annu. Rev. Genet.* 1: 221–44.

YAMAMOTO, F., et al. 1990. Molecular genetic basis of the histo-blood group ABO system. *Nature* 345:229–33.

15

Recombinant DNA Technology

CHAPTER CONCEPTS

Recombinant DNA technology depends in part on the ability to cleave and rejoin DNA segments at specific base sequences. Using this methodology, individual DNA segments can be transferred to viruses or bacteria and amplified, isolated, and identified. The use of this technology has brought about significant advances in gene mapping, disease diagnosis, the commercial production of human gene products, and the transfer of genes between species in plants and animals.

The independent rediscovery of Mendel's work by de Vries, Correns, and von Tschermak in 1900 marked the beginning of genetics as an organized discipline. During the course of the growth and development of this science, several key discoveries have served as turning points, each accelerating the rate at which our knowledge of genetics has grown, and, in turn, opening new fields of investigation.

One of the first of these landmarks was the chromosome theory of inheritance. This concept, proposed by Sutton and Boveri in 1902, was developed by Morgan and his colleagues using the fruit fly, *Drosophila*. From these studies came our un-

derstanding of transmission genetics, sex determination, linkage, and the use of polytene chromosomes to map genes to specific cytological loci.

A second landmark was the discovery by Avery, MacLeod, and McCarty that DNA is the macromolecular carrier of genetic information. This work stimulated the use of viruses and bacteria as organisms for genetic research and led to the Watson–Crick model for the structure of DNA. From this model has come knowledge of the molecular basis for genetic coding, transcription, translation, and gene regulation.

We are now undergoing another and perhaps the

A sample tube containing a pellet of DNA.

most profound transition in the history of genetics—the development and application of **recombinant DNA technology**. This technology is used in basic research and in the development and production of vaccines, therapeutic proteins, and genetically modified plants and animals. It has also raised fears about epidemics or widespread ecological changes that might result from the release of genetically engineered organisms into the environment. This chapter reviews the basic methods of recombinant DNA technology that are used to isolate, replicate, and analyze genes. In the chapter that follows we will discuss some of the applications of this technology to agriculture, medicine, and industry.

Recombinant DNA Technology: An Overview

The term **recombinant DNA** refers to the creation of a new combination of DNA segments or DNA molecules that are not found together naturally. Although genetic processes such as crossing over technically produce recombinant DNA, the term is generally reserved for DNA molecules produced by joining segments derived from different biological sources.

Recombinant DNA technology uses techniques derived from the biochemistry of nucleic acids coupled with genetic methodology originally developed for the study of bacteria and viruses. As described below, the use of recombinant DNA is a powerful tool for the isolation of pure populations of specific DNA sequences from a mixed population of sequences. The basic procedures involve a series of steps:

1. DNA fragments are generated by using enzymes called **restriction endonucleases**, enzymes that recognize and cut DNA molecules at specific nucleotide sequences.

2. The fragments produced by digestion with restriction enzymes are joined to other DNA molecules that serve as **vectors**. Vectors can replicate autonomously in host cells and facilitate the manipulation of the newly created recombinant DNA molecule.

3. The recombinant DNA molecule, consisting of a vector carrying an inserted DNA segment, is transferred to a host cell. Within this cell, the recombinant DNA molecule replicates, producing dozens of identical copies known as **clones**.

4. As host cells replicate, the recombinant DNA is passed on to all progeny cells, creating a population of identical cells, all carrying the cloned sequence.

5. The cloned DNA segments can be recovered from the host cell, purified, and analyzed.

6. Potentially, the cloned DNA can be transcribed, its mRNA translated, and the gene product isolated and studied.

Making Recombinant DNA

The development of recombinant DNA techniques affords new opportunities for research, making it easier to obtain large amounts of DNA encoding specific genes and facilitating studies of gene organization, structure, and expression. This methodology has also given impetus to the development of a burgeoning biotechnology industry that is delivering a growing number of products to the marketplace. Recombinant DNA works by making large numbers of copies of specific DNA segments, including genes. The process is outlined in the following sections.

Restriction Enzymes

The cornerstone of recombinant DNA technology is a class of enzymes called **restriction endonucleases**. These enzymes, isolated from bacteria, received their name because they restrict or prevent viral infection by degrading the invading nucleic acid. Restriction enzymes recognize a specific nucleotide sequence (called a restriction site) on a double-stranded DNA molecule, and cut the DNA at that sequence. The 1978 Nobel Prize was given to Werner Arber, Hamilton Smith and Daniel Nathans for their work on restriction enzymes. To date, almost 200 different types of restriction enzymes have been isolated and characterized.

Restriction enzymes are named for the organism in which they were discovered, using a system of letters and numbers. The enzyme *Eco*RI is from *Escherichia coli*, and is pronounced "echo-r-one." *Hin*dIII was discovered in *Hemophilus influenzae* and is pronounced "hindee-three." There are two classes of restriction enzymes. Type I enzymes cut both strands of the DNA at a random location at some distance from the recognition site. Because the cutting site in type I enzymes is not precise, they are not widely used in recombinant DNA research. Type II enzymes recognize a specific sequence and precisely cut both strands of a DNA molecule within the recognition sequence. Because they cut at specific sites, type II enzymes are widely used in recombinant DNA research.

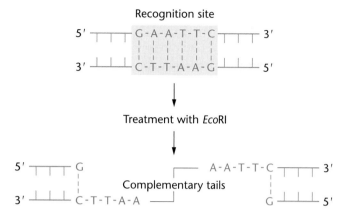

FIGURE 15.1 The restriction enzyme *Eco*RI recognizes and binds to the palindromic nucleotide sequence GAATTC. Cleavage of the DNA at this site produces complementary single-stranded tails. The resulting single-stranded tails can anneal with complementary tails from other DNA fragments to form recombinant DNA molecules.

The recognition sequence for type II enzymes are symmetrical. The sequence read in the 5′-to-3′ direction on one strand is the same as the sequence read in the 5′-to-3′ direction on the complementary strand. Sequences that read the same in both directions are known as **palindromes**. The palindromic recognition site and points of cleavage for *Eco*RI are shown in Figure 15.1.

The enzyme *Eco*RI cuts in a staggered fashion within the recognition site, leaving single-stranded ends. These tails, with identical nucleotide sequences, are "sticky" because they can hydrogen-bond to complementary tails of other DNA fragments. If DNAs from different sources share the same palindromic recognition sites, both will contain complementary single-stranded tails when treated with a restriction endonuclease (Figure 15.2). If the cut fragments are placed together under proper conditions, the DNA fragments from these two sources can form recombinant molecules by hydrogen bonding of the sticky ends. The enzyme DNA ligase can be used to covalently link the phosphate-sugar backbones of the two fragments, producing a recombinant DNA molecule (Figure 15.2).

Other Type II restriction enzymes such as *Sma*I cleave DNA to produce blunt-end fragments (Figure 15.3). DNA fragments with blunt ends can also be joined together to create recombinant molecules, after the DNA has been modified (Figure 15.3). The enzyme **terminal deoxynucleotidyl transferase** is used to created single-stranded ends by the addition of nucleotide "tails." If a poly dA tail is added to DNA fragments from one source, and a poly dT tail is added to DNA from another source, complementary tails are created, and the fragments can hydrogen-bond. Recombinant molecules can then be created by ligation (Figure 15.3). Figure 15.4 depicts some

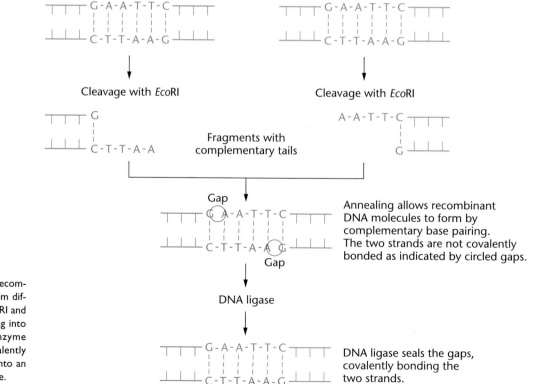

FIGURE 15.2 In forming recombinant DNA molecules, DNA from different sources is cleaved with *Eco*RI and mixed together to allow annealing into recombinant molecules. The enzyme DNA ligase is then used to covalently bond these annealed fragments into an intact recombinant DNA molecule.

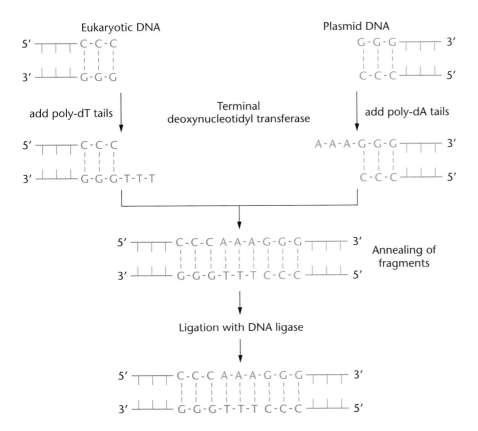

Eukaryotic DNA   Plasmid DNA

5′ ┬┬┬┬ C - C - C   G - G - G ┬┬┬┬ 3′
         ¦   ¦   ¦   ¦   ¦   ¦
3′ ┴┴┴┴ G - G - G   C - C - C ┴┴┴┴ 5′

add poly-dT tails ↓   Terminal   ↓ add poly-dA tails
            deoxynucleotidyl transferase

5′ ┬┬┬ C - C - C   A - A - A - G - G - G ┬┬┬┬ 3′
        ¦   ¦   ¦   ¦   ¦   ¦
3′ ┴┴┴┴ G - G - G - T - T - T   C - C - C ┴┴┴ 5′

            ↓   ↓

5′ ┬┬┬┬ C - C - C A - A - A - G - G - G ┬┬┬┬ 3′   Annealing of
         ¦   ¦   ¦   ¦   ¦   ¦   ¦   ¦   ¦   fragments
3′ ┴┴┴┴ G - G - G - T - T - T C - C - C ┴┴┴ 5′

            ↓

            Ligation with DNA ligase

            ↓

5′ ┬┬┬┬ C - C - C A - A - A - G - G - G ┬┬┬┬ 3′
         ¦   ¦   ¦   ¦   ¦   ¦   ¦   ¦   ¦
3′ ┴┴┴┴ G - G - G - T - T - T C - C - C ┴┴┴ 5′

FIGURE 15.3   Recombinant DNA molecules can be formed from DNA cut with enzymes that leave blunt ends. In this method, the enzyme terminal deoxynucleotidyl transferase is used to create complementary tails by the addition of poly dA and poly dT to the cut fragments. These tails serve to anneal DNA from different sources and to create recombinant DNA molecules that are covalently linked by treatment with DNA ligase.

common restriction enzymes and their recognition sequences, many of which yield cohesive or "sticky" ends, while others generate blunt ends.

Vectors

After being joined with a vector or cloning vehicle, a DNA segment can gain entry into a host cell and be replicated or cloned. Vectors are, in essence, carrier DNA molecules. To serve as a vector, a DNA molecule must have several properties:

1. It must be able to independently replicate itself and the DNA segments it carries.

2. It should contain a number of restriction enzyme cleavage sites that are present only once in the vector. This site is cleaved with a restriction enzyme and used to insert DNA segments cut with the same enzyme.

3. It should carry a selectable marker (usually in the form of antibiotic resistance genes or genes for enzymes missing in the host cell) to distinguish host cells that carry vectors from host cells that do not contain a vector.

4. It should be easy to recover from the host cell.

There are three main types of vectors currently in use: **plasmids**, **bacteriophages**, and **cosmids**.

Plasmid Vectors

Plasmids are naturally occurring, extrachromosomal double-stranded DNA molecules that carry an origin of replication site (ori^+) and replicate autonomously within bacterial cells (Figure 15.5). The genetics of plasmids and their host bacterial cells will be discussed in Chapter 18. In this section, emphasis will be on the role of plasmids as vectors. For use in genetic engineering, many plasmids have been modified or engineered to contain a limited number of restriction sites and specific antibiotic resistance genes.

The vector pBR322 was one of the first genetically engineered plasmids to be used in recombinant DNA (Figure 15.6). This plasmid carries an origin of replication (ori^+), two selectable genes (resistance to the antibiotics ampicillin and tetracycline), and a number of unique sites for restriction cleavage. Sites for the enzymes *Bam*HI, *Sph*I, *Sal*I, *Xma*III, and *Nru*I are within the tetracycline gene, and unique sites for *Rru*I, *Pvu*I, and *Pst*I are within the ampicillin resistance gene. If pBR322 is introduced into a plasmid-free, antibiotic-sensitive bacterial cell, the cell will become resistant to

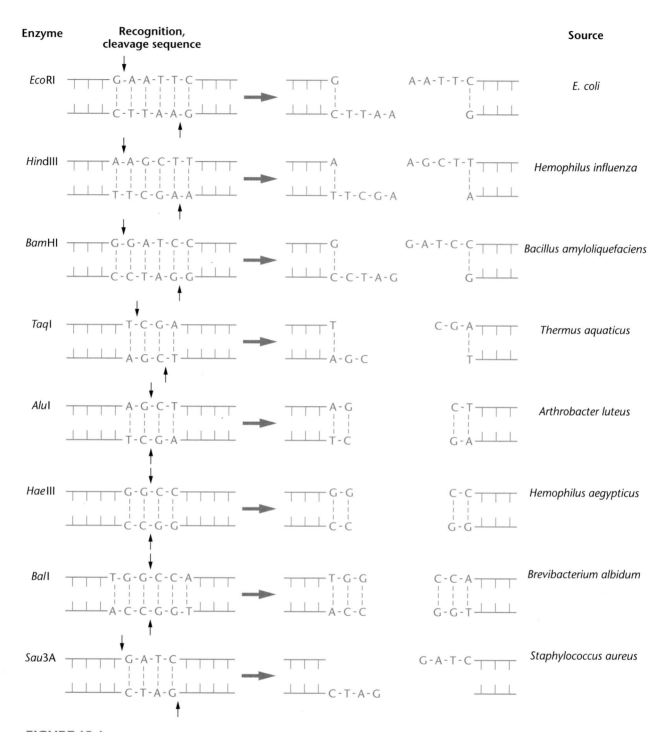

FIGURE 15.4   Some common restriction enzymes, with their recognition and cutting sites, cleavage patterns, and sources.

both tetracycline and ampicillin. If a DNA fragment is inserted into the *Rru*I, *Pvu*I, or *Pst*I sites, the ampicillin gene will become inactivated, but the gene for tetracycline resistance will remain active. If this recombinant plasmid is transferred into a plasmid-free bacterial cell, cells carrying the recombinant plasmid can be identified

because they will be resistant to tetracycline and sensitive to ampicillin.

Over the years, more sophisticated plasmid vectors have been developed, offering a number of useful features. One such plasmid, derived from pBR322, is pUC18 (Figure 15.7), which has the following properties:

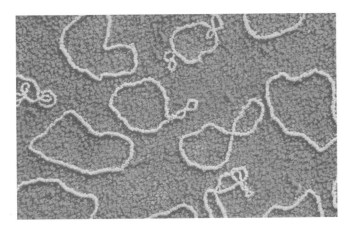

FIGURE 15.5 A color-enhanced electron micrograph of circular plasmid molecules isolated from the bacterium *E. coli*. Genetically engineered plasmids are used as vectors for cloning DNA segments.

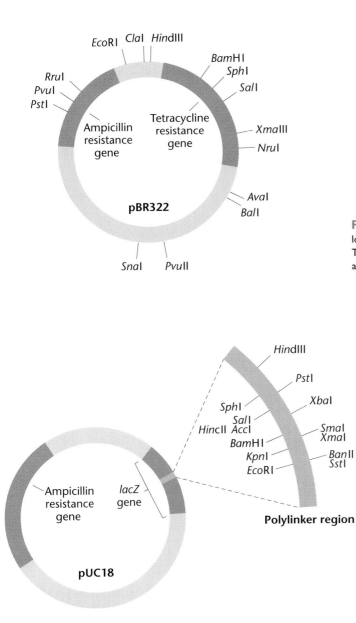

FIGURE 15.6 Restriction map of the plasmid pBR322 showing the locations of the restriction enzyme sites that cleave the plasmid only once. These sites can be used to insert DNA fragments for cloning. Also shown are the locations of the genes for antibiotic resistance.

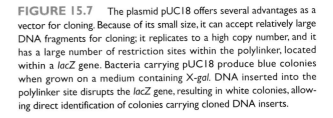

FIGURE 15.7 The plasmid pUC18 offers several advantages as a vector for cloning. Because of its small size, it can accept relatively large DNA fragments for cloning; it replicates to a high copy number, and it has a large number of restriction sites within the polylinker, located within a *lacZ* gene. Bacteria carrying pUC18 produce blue colonies when grown on a medium containing X-gal. DNA inserted into the polylinker site disrupts the *lacZ* gene, resulting in white colonies, allowing direct identification of colonies carrying cloned DNA inserts.

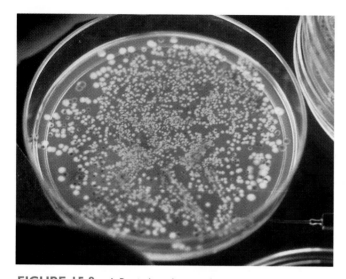

FIGURE 15.8 A Petri plate showing the growth of host cells after uptake of recombinant plasmids. The medium on the plate contains a compound called *X-gal.* DNA inserts in the pUC vector disrupt the gene responsible for the formation of blue colonies. Cells in blue colonies do not carry any cloned DNA inserts, whereas white colonies contain vectors carrying DNA inserts.

1. The plasmid is about half the size of pBR322, allowing larger DNA fragments to be inserted.

2. The pUC18 replicates to produce about 500 copies per cell, 5 to 10 times the number of copies produced by pBR322.

3. A large number of unique restriction sites are present in pUC18, and these are clustered in one region, known as a **polylinker.**

4. The plasmid carries a fragment of a bacterial gene called *lacZ,* and the polylinker is inserted into this gene fragment. The normal *lacZ* gene encodes the enzyme beta-galactosidase, which cleaves sugar molecules. When pUC18 is inserted into a bacterial host cell that carries a mutant *lacZ* gene, functional beta-galactosidase is produced. The presence of functional beta galactosidase in a bacterial colony can be detected by a color test. Bacterial cells carrying pUC18 form blue colonies when grown on medium containing a compound called *X-gal.* If DNA is inserted into the polylinker, the *lacZ* gene is disrupted, and white colonies form. White colonies that are resistant to ampicillin contain plasmids carrying DNA inserts (Figure 15.8).

Lambda and M13 Bacteriophage Vectors

Lambda is a bacteriophage widely used in recombinant DNA work (Figure 15.9). The genes of lambda have all been identified and mapped, and the nucleotide sequence of the entire genome is known. The central third of the lambda chromosome is dispensible, and this region can be replaced with foreign DNA without affecting the ability of the phage to infect cells and form plaques (Figure 15.10). Over 100 vectors based on lambda phages have been developed by removing various portions of the central gene cluster. To clone using the lambda vector, the phage DNA is cut with a restriction enzyme—for example, *Eco*RI—resulting in a left arm, a right arm, and the central region. The arms are isolated and ligated (using DNA ligase) to DNA segments generated by cutting genomic DNA with *Eco*RI.

The resulting recombinant molecules can be introduced into bacterial host cells by **transfection.** First, the host cells are made permeable by a chemical treatment, and then mixed with the ligated molecules. Vector molecules carrying inserts are taken into the host cells, where they direct the synthesis of infective phage, each of which carries a DNA insert. Alternatively, the lambda DNA carrying inserts can be mixed with phage protein components; from this mixture, infective phage particles are formed. Each phage can be amplified by growth on plated bacterial cells to form plaques or by infecting cells in liquid medium and harvesting the lysed cells.

Other bacteriophages are used as vectors, including the single-stranded phage known as M13 (Figure 15.11). When M13 infects a bacterial cell, the single strand (+ strand) replicates to produce a double-stranded molecule known as the **replicative form (RF).** RF molecules can be regarded as similar to plasmids, and foreign DNA can be inserted into single restriction enzyme cleavage sites present in the phage DNA. When reinserted into bacterial cells, the RF molecules replicate to produce single

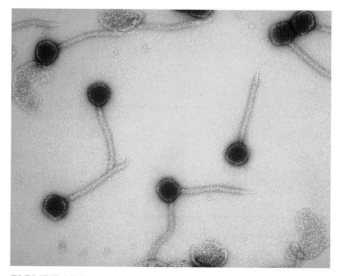

FIGURE 15.9 A colorized electron micrograph showing a cluster of the bacteriophage lambda widely used as a vector in recombinant DNA work.

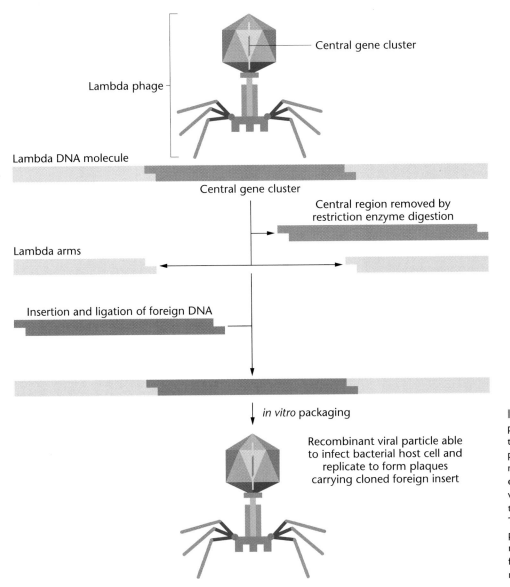

Lambda phage

Central gene cluster

Lambda DNA molecule

Central gene cluster

Central region removed by restriction enzyme digestion

Lambda arms

Insertion and ligation of foreign DNA

in vitro packaging

Recombinant viral particle able to infect bacterial host cell and replicate to form plaques carrying cloned foreign insert

FIGURE 15.10 Steps in cloning phage lambda as a vector. DNA is extracted from a preparation of lambda phage, and the central gene cluster is removed by treatment with a restriction enzyme. The DNA to be cloned is cut with the same enzyme and ligated into the arms of the lambda chromosome. The recombinant chromosome is then packaged into phage proteins to form a recombinant virus. This virus is able to infect bacterial cells and replicate its chromosome, including the DNA insert.

(+) strands, including one strand of the inserted DNA. The cloned single-stranded DNA produced by M13 are extruded from the cell and can be recovered and used directly for DNA sequencing or as a template for mutational alteration of the cloned sequence.

Cosmid Vectors and Shuttle Vectors

Cosmids are hybrid vectors constructed using parts of the lambda chromosome and plasmid DNA. Cosmids contain the *cos* sequence of phage lambda, necessary for packing phage DNA into the phage protein coats, and plasmid sequences for replication and an antibiotic resistance gene to identify host cells carrying cosmids (Figure 15.12). Cosmid DNA containing DNA inserts is packaged into

lambda protein heads to form infective phage particles. After the cosmid enters the bacterial host cell, it replicates as a plasmid. Because almost the entire lambda genome has been deleted, cosmids can carry DNA inserts that are much larger than those carried by lambda vectors. Cosmids can carry almost 50 kb of insert DNA; phage vectors can accommodate DNA inserts of about 15 kb; plasmids are usually limited to inserts of 5 to 10 kb in length.

Other hybrid vectors constructed with origins of replication derived from different sources (e.g., plasmids and animal viruses such as SV40) can replicate in more than one type of host cell. These **shuttle vectors** usually contain genetic markers that are selectable in both host systems, and can be used to shuttle DNA inserts back and forth between *E. coli* and another host cell, such as yeast.

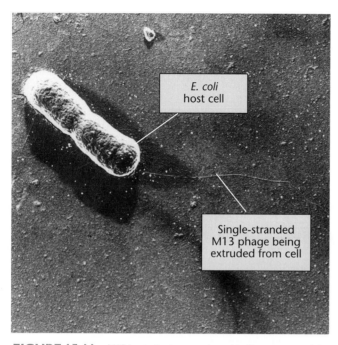

FIGURE 15.11   M13 bacteriophage are long thin filaments containing a single strand of DNA. They enter the *E. coli* cells that carry long thin filaments (thicker than M13) called *F pili* and do so by attaching to the end of the F pilus.

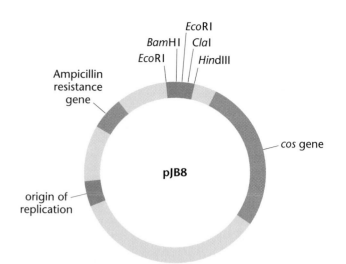

FIGURE 15.12   The cosmid pJB8 contains a bacterial origin of replication (*ori*), a single *cos* site (*cos*), an ampicillin resistance gene (*amp*, for selection of colonies that have taken up the cosmid), and a region containing four restriction sites for cloning (*Bam*HI, *Eco*RI, *Cla*I, and *Hind*III). Because the vector is small (5.4 kb long), it can accept foreign DNA segments between 33 and 46 kb in length. The *cos* site allows cosmids carrying large inserts to be packaged into lambda viral coat proteins as though they were viral chromosomes. The viral coats carrying the cosmid can be used to infect a suitable bacterial host, and the vector, carrying a DNA insert, will be transferred into the host cell. Once inside, the *ori* sequence allows the cosmid to replicate as a bacterial plasmid.

Often, such vectors are employed in studying gene expression. Other vectors, those used in specific applications, will be described in following sections.

Bacterial Artificial Chromosomes

A multipurpose vector that has been developed for the mapping and analysis of complex eukaryotic genomes is based on the F factor of bacteria and is called a **bacterial artificial chromosome (BAC).** Recall from Chapter 6 that F factors are independently replicating plasmids that are involved in the transfer of genetic information during bacterial conjugation. Because F factors can carry fragments of the bacterial chromosome up to 1 Mb in length, they have been engineered to act as vectors for eukaryotic DNA. BAC vectors carry the F factor genes for replication and copy number, and incorporate an antibiotic resistance marker, and restriction enzyme sites for inserting

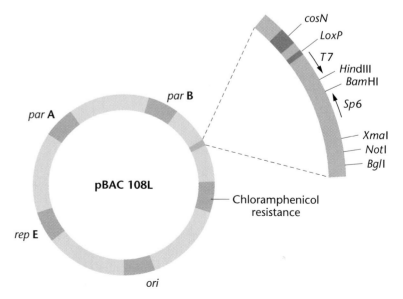

FIGURE 15.13   A bacterial artificial chromosome (BAC). The polylinker carries a number of unique restriction sites for the insertion of foreign DNA. The arrows labeled *T7* and *Sp6* are promoter regions that allow expression of genes cloned between these regions.

foreign DNA to be cloned. In addition, the cloning site is flanked by promoter sites that can be used to generate RNA molecules for expression of the cloned gene, or for use as probes in chromosome walking, and for DNA sequencing of the cloned insert (Figure 15.13).

Cloning DNA in *E. coli*

A variety of cell types can be used as hosts for replication of cloned DNA. One of the most commonly used hosts is the laboratory *E. coli* strain K12. This and other strains of *E. coli* are used as hosts because they are genetically well characterized and can accept a wide range of vectors, including plasmids, phages, and cosmids.

Several steps are required to create recombinant DNA molecules (Figure 15.14). Using a plasmid as a vector, the procedure is as follows:

1. The DNA to be cloned is isolated and treated with a restriction enzyme to create fragments ending in specific sequences.

2. The fragments are ligated to plasmid molecules that have been cut with the same restriction enzyme, creating a recombinant vector.

3. The recombinant vector is transferred into bacterial host cells, usually by transformation, a process in which DNA molecules are transferred across the cell membrane and into the host cell.

4. The host cells are grown on a nutrient plate, where they form colonies. Because the cells in each colony are derived from a single ancestral cell, all cells in the colony, and the plasmids they contain, are genetically identical, or **clones**. The colonies are screened to identify those that have taken up recombinant plasmids (Figure 15.15).

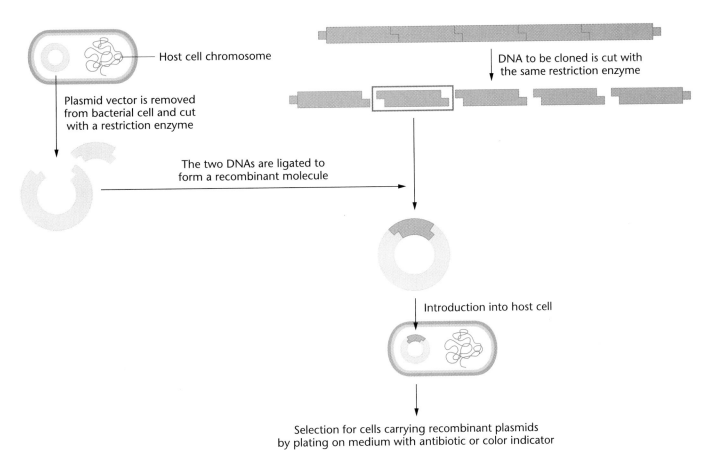

FIGURE 15.14 A summary of steps involved in cloning with a plasmid vector. Plasmid vectors are isolated and cut with a restriction enzyme. The DNA to be cloned is cut with the same restriction enzyme, producing a collection of fragments. These fragments are spliced into the vector and transferred to a bacterial host for replication. Bacterial cells carrying plasmids with DNA inserts can be identified by growth on selective medium and isolated. The cloned DNA can be recovered from the bacterial host for further analysis.

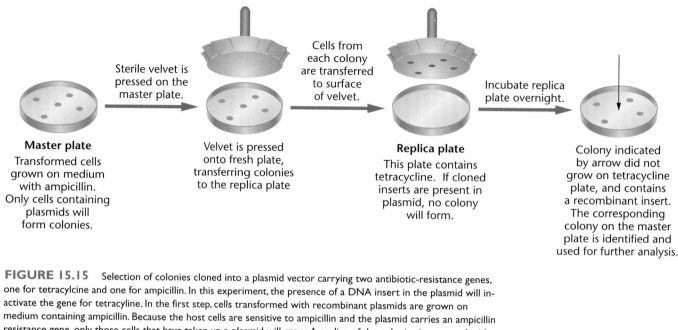

FIGURE 15.15 Selection of colonies cloned into a plasmid vector carrying two antibiotic-resistance genes, one for tetracylcine and one for ampicillin. In this experiment, the presence of a DNA insert in the plasmid will inactivate the gene for tetracyline. In the first step, cells transformed with recombinant plasmids are grown on medium containing ampicillin. Because the host cells are sensitive to ampicillin and the plasmid carries an ampicillin resistance gene, only those cells that have taken up a plasmid will grow. A replica of the colonies is prepared with a piece of sterile velvet. As the velvet is pressed onto the plate, some cells from each colony stick to the velvet. The velvet is then pressed onto the surface of a plate containing tetracycline. Cells carrying plasmids with inserts will not grow because the tetracycline gene has been inactivated. By comparing the pattern of colonies on the tetracycline plate with the master plate, colonies containing inserts can be identified and recovered from the master plate for growth and further analysis.

Various methods are used to select colonies that contain plasmids carrying DNA inserts. This process is called *screening*. For vectors such as pBR322, screening is a two-step process. For example, if the DNA to be cloned is inserted into the gene for tetracycline resistance, this gene will be inactivated. After a culture of host cells has been transformed with the recombinant plasmid, the cells are grown on plates containing the antibiotic ampicillin (Figure 15.15). All cells that took up a plasmid (with or without an insert) will grow and form colonies, but cells that did not take up a plasmid are sensitive to ampicillin and will die.

In the second step, colonies carrying plasmids with DNA inserts are identified. In this step, colonies are transferred from the plate containing ampicillin to a plate containing tetracycline via a procedure called *replica plating*. Colonies carrying plasmids with inserts will not grow on the tetracycline plate, because the insert has inactivated the tetracycline gene. The pattern of colony growth on the tetracycline plate is compared to the pattern of colony growth on the ampicillin plate, and the colonies on the ampicillin plate that did *not* grow on the tetracycline plate are identified. Cells from these colonies contain vectors with DNA inserts, and they are transferred to medium for growth and further analysis.

Other vectors such as pUC18 have been engineered to produce blue colonies when in bacterial cells plated on medium containing a compound called **X-gal**. The pUC18 polylinker restriction sites are in the *lacZ* gene responsible for the ability to form blue colonies, and insertion of DNA into the polylinker destroys this ability. (For a discussion of the *lac* gene, see Chapter 18.) As discussed earlier, plasmids carrying inserted segments of DNA produce white colonies, whereas those without inserts produce blue colonies.

Similarly, phages containing foreign DNA can be used to infect *E. coli* host cells, and when plated on solid medium, the resulting plaques each represent a cloned descendant of a single ancestral phage. Cosmids infect bacterial cells like phage, but then replicate as plasmids inside the host cell. Cells carrying cosmids with cloned DNA can be identified and recovered in the same way as can plasmids.

Cloning in Eukaryotic Hosts

We have described the use of *E. coli* as a host for cloning. Other species of bacteria can also be used as hosts for cloning, including *B. subtilis* and *Streptomyces*. These bacterial host systems and vectors are similar to those described for *E. coli*. However, to study expression and regulation of eukaryotic genes, it is often convenient and even necessary to use eukaryotic hosts. Several cloning systems using eukaryotic hosts have been developed, and in this section we describe cloning systems that use eukaryotic cells as hosts.

Yeast Vectors

Although yeast is a eukaryotic organism, it can be manipulated and grown in much the same way as bacterial cells. Further, the genetics of yeast has been intensively studied, providing a large catalog of mutations and highly developed genetic maps of the yeast chromosomes. A naturally occurring plasmid termed the **2-micron plasmid** is present in yeast, and it has been used to construct a number of yeast cloning vectors. By combining bacterial plasmid sequences with those of the yeast 2-micron plasmid, vectors with many useful properties can be produced.

Yeast Artificial Chromosomes

A second type of yeast vector is the **yeast artificial chromosome (YAC)**. In linear form, a YAC contains yeast telomeres at each end, an origin of replication (called an autonomously replicating sequence or ARS), and a yeast centromere that allows the distribution of replicated YACs to daughter cells at cell division. The YAC also contains a selectable marker on each arm (TRP1 and URA3), and a cluster of unique restriction sites for DNA inserts (Figure 15.16).

Segments of DNA more than 1 megabase long (1 Mb = 10^6 base pairs) can be inserted into YACs. The ability to clone large pieces of DNA into these vectors has made them an important tool for constructing physical maps of eukaryotic genomes, including the human genome. Much of the work on the Human Genome Project (discussed in Chapter 16) depends on the use of YAC vectors.

Although cloning in yeast is currently the most advanced eukaryotic host system, other fungal hosts are being developed, and other eukaryotes including plant and mammalian tissue culture cells show great promise as host systems.

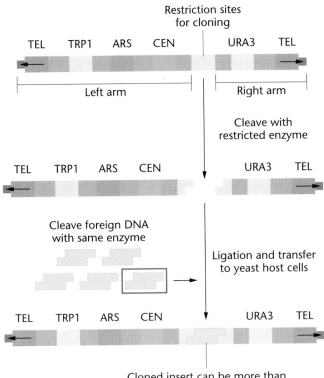

FIGURE 15.16 Cloning into a yeast artificial chromosome. The chromosome contains telomere sequences (*TEL*) derived from a ciliated protozoan, *Tetrahymena*, a centromere (*CEN*), and an origin of replication (*ARS*). These elements give the cloning vector the properties of a chromosome. *TRP1* and *URA3* are yeast genes that are selectable markers for the left and right arms of the chromosome, respectively. Near the centromere is a region containing several restriction sites. Cleavage with an enzyme in this region breaks the artificial chromosome into two arms. The DNA to be cloned is treated with the same enzyme, producing a collection of fragments. The arms and fragments can be ligated together, and the artificial chromosome can be inserted into yeast host cells. Because yeast chromosomes are large, the artificial chromosome can accept inserts up to several million base pairs (Mb = megabases).

Constructing DNA Libraries

Because each cloned DNA segment is relatively small, many separate clones must be constructed to include even a small portion of the genome of an organism. A cloned set of all the genomic sequences from a single individual is called a **library**. Cloned libraries can be derived from the entire genome of an individual, the DNA from a single chromosome, or the set of genes actively transcribed in a single cell type.

Genomic Libraries

Genomic libraries are usually constructed using phage vectors, which can carry large chromosomal fragments. To prepare a library in a phage lambda vector, the lambda DNA is cut with a restriction enzyme to remove the central gene cluster (review Figure 15.10). The genomic DNA to be cloned is cut with the same enzyme, and DNA fragments of an optimum size for packaging (15 to 17 kb) are purified by gel electrophoresis or centrifugation. These DNA fragments are ligated with the arms of the lambda chromosome to form the library.

All the genes in a single organism are represented in a genomic library. The library serves as a resource from which each gene in the organism's genome can be recovered and studied in detail, along with the adjacent regulatory sequences. Ideally, a genomic library contains at least one copy of all sequences represented in the genome. Each vector molecule can contain only a relatively few kilobases of insert DNA, however, so one of the primary tasks in preparing a genomic library is selection of the proper vector to carry the entire genome in the smallest number of clones. The number of clones required to contain all the sequences in a genome is dependent on the average size of the cloned inserts and the size of the genome to be cloned. This number can be represented by the following formula:

$$N = \frac{\ln(1 - P)}{\ln(1 - f)}$$

where N is the number of required clones, P is the probability of recovering a given sequence, and f represents the fraction of the genome present in each clone.

Suppose that we wish to prepare a library of the human genome using a lambda phage vector. The human genome is 3.0×10^6 kb, and if the average size of cloned inserts in the vector is 17 kb, a library of about 8.1×10^5 phages would be required (where $f = 1.7 \times 10^4/3.0 \times 10^9$) to have a 99 percent ($P = 0.99$) probability that any human gene is present in at least one copy. If we had selected pBR322 as the vector, with an average insert of about 5 kb, several million clones would be needed to contain the library.

Chromosome-Specific Libraries

A library made from a subgenomic fraction such as a single chromosome can be of great value in the selection of specific genes and the study of chromosome organization. For example, DNA from a small segment of the X chromosome of *Drosophila* corresponding to a region of about 50 polytene bands was isolated by microdissection. The DNA in this chromosomal fragment was extracted, cut with a restriction endonuclease, and cloned into a lambda vector. This region of the chromosome contains the genes *white*, *zeste*, and *Notch*, as well as the original site of a transposing element that can translocate a chromosomal segment to more than a hundred other loci. Although technically difficult, this procedure produces a library that contains only the genes of interest and their adjacent sequences.

Libraries derived from individual human chromosomes have been prepared using a technique known as **flow cytometry**. In this procedure, chromosomes from mitotic cells are stained with two fluorescent dyes, one that binds to A-T pairs, the other to G-C pairs. The stained chromosomes flow past a laser beam that stimulates them to fluoresce, and a photometer sorts and frac-

tionates the chromosomes by differences in dye binding and light scattering (Figure 15.17). Using this technique, the National Laboratories at Los Alamos and Lawrence-Berkeley have prepared cloned libraries for each of the human chromosomes. Having cloned libraries for each human chromosome has greatly facilitated the mapping and analysis of individual chromosomes as part of the Human Genome Project.

A modification of electrophoresis known as **pulse field gel electrophoresis** has been used to isolate yeast chromosomes for construction of chromosome-specific libraries (Figure 15.18). The isolation and construction of a cloned library of yeast chromosome III (315 kb) was the starting point for the formation of a consortium of 35 laboratories to determine the nucleotide sequence of this chromosome. One of the unexpected results of this project was the finding that about half of all protein-coding sequences on this chromosome were previously unknown. It was difficult for many geneticists to accept that the conventional methods of systematic mutagenesis and mapping of genes was so inefficient. However, this finding was confirmed and extended in 1994 with the publication of the sequence of yeast chromosome XI (664 kb), and the sequencing of the rest of the yeast genome in 1996. These results indicate that single chromosome libraries can be used to gain access to genetic loci where conventional methods such as mutagenesis and genetic analysis have proven unsuccessful, and where other probes such as mRNA or gene products are unavailable

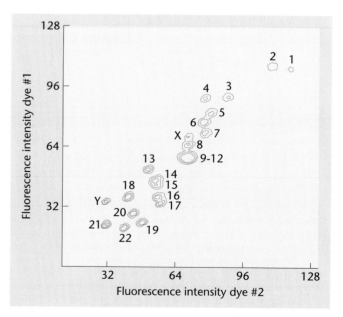

FIGURE 15.17 Each human chromosome has a unique profile resulting from the absorption of two fluorescent dyes. Based on this differential fluorescence, chromosomes can be separated from each other by flow sorting.

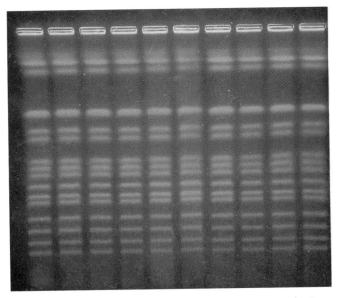

FIGURE 15.18 Intact yeast chromosomes separated using a method of electrophoresis employing contour-clamped homogeneous electric field (CHEF). In each lane, 15 of the 16 yeast chromosomes are visible, separated by size, with the largest chromosomes at the top.

or unknown. In addition, chromosome-specific libraries provide a means for studying the molecular organization and even the nucleotide sequence in a defined region of the genome.

cDNA Libraries

A library representing the genes that are being transcribed in a eukaryotic cell type at a specific time can be constructed using mRNA isolated from that cell type. Almost all eukaryotic mRNA molecules contain a poly-A tail at their 3'-ends. Beginning with a population of mRNA molecules containing 3' poly-A tails, oligo-dT (short, single-stranded DNA containing only deoxythymidine) is hybridized to the poly-A sequences (Figure 15.19). The oligo-dT sequences serve as the primer for the synthesis of a DNA strand using the enzyme **reverse transcriptase**. This enzyme is an RNA-dependent DNA polymerase that copies a single-stranded RNA template into a single-stranded DNA. The result is an RNA–DNA double-stranded duplex. The RNA strand is removed by treatment with the enzyme **ribonuclease H**. The single strand of DNA is then used as a template to synthesize a complementary strand of DNA using the enzyme **DNA polymerase I**. Because the 3'-end of the single strand of DNA often folds back upon itself to form a hairpin loop, it can serve as the primer for the synthesis of the second strand. The result is a DNA duplex with the strands joined together at one end. The hairpin loop can be opened using the enzyme **S₁ nuclease**, producing a double-stranded DNA molecule (called **complementary DNA or**

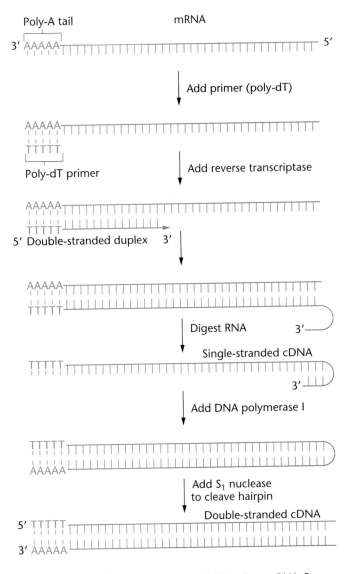

FIGURE 15.19 The production of cDNA from mRNA. Because many eukaryotic mRNAs have a polyadenylated tail of variable length (A_n) at their 3'-end, a short poly-dT oligonucleotide can be annealed to this tail. The poly-dT acts as a primer for the enzyme reverse transcriptase, which uses the mRNA as a template to synthesize a complementary DNA strand. A characteristic hairpin loop formed as synthesis is terminated on the template. The mRNA is removed by alkaline treatment of the complex, and DNA polymerase is used to synthesize the second DNA strand. The S₁ nuclease is used to open the hairpin loop; the result is a double-stranded cDNA molecule that can be cloned into a suitable vector, or used as a probe for library screening.

cDNA), with a nucleotide sequence derived from an mRNA molecule.

The cDNA can be cloned by attaching a short double-stranded piece of DNA containing a restriction site to each end. The linkers are cleaved with the appropriate restriction enzyme, producing sticky ends, thus allowing the cDNA to be inserted at a restriction site in a plasmid or phage vector.

A cDNA library is different from a genomic library in that it represents only a small a subset of all genes in a genome. cDNA libraries use mRNA as a starting point, and thus represent only the expressed sequences in a given cell type, tissue or stage of embryonic development. The decision whether to construct a genomic or a cDNA library depends on the question at hand. If you are interested in a particular gene, it may be easier to prepare a cDNA library from a tissue that expresses that gene. For example, red blood cells make large amounts of hemoglobin, and most of the mRNA in these cells is β-globin mRNA. A cDNA library prepared using mRNA isolated from red blood cells is a direct way to isolate a globin gene. If the regulatory sequences adjacent to the globin gene are of interest, then a genomic library would need to be constructed, since these regulatory sequences are not present in globin mRNA, and would not be represented in the cDNA library.

Identifying Specific Cloned Sequences

A genomic library can contain several hundred thousand clones. The problem is to identify and select only the clone or clones that contain a gene of interest and to determine whether a given clone contains all or only a part of this gene. There are several ways to do this, and the choice often depends on circumstances and available information. The methods described below use various approaches to finding a specific cloned DNA sequence in a library.

Probes to Screen for Specific Clones

Many procedures employ a **probe** to screen the library and identify the clone containing the gene of interest. Often, probes are radioactive polynucleotides containing a base sequence complementary to all or part of the gene of interest. Other methods use probes that depend on chemical reactions or color reactions to indicate the location of a specific clone. Probes can be derived from a variety of sources; even related genes isolated from other species can serve as probes if enough of the DNA sequence has been conserved. For example, extrachromosomal copies of the ribosomal genes of the clawed frog *Xenopus laevis* were isolated by centrifugation (Figure 15.20), cut with restriction enzymes, and cloned into plasmid vectors. Because ribosomal gene sequences have been highly conserved during eukaryotic evolution, the cloned *Xenopus* ribosomal genes were labeled with radioisotopes and used as probes to isolate the human ribosomal genes from a genomic library.

If the gene to be selected from the library is transcriptionally active in certain cell types, a cDNA probe

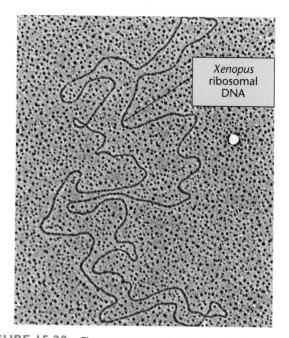

FIGURE 15.20 Electron micrograph showing extrachromosomal circles of *Xenopus* ribosomal DNA isolated by centrifugation. These circles can be treated with restriction enzymes, and the resulting fragments cloned into plasmid vectors.

can be used to find this gene. This technique is particularly useful when purified or enriched mRNA can be obtained. As mentioned earlier, β-globin mRNA is the predominant messenger RNA produced at a certain stage of red blood cell development. To make a probe, mRNA from these cells is isolated, purified, and copied by reverse transcriptase into a cDNA molecule. This cDNA can then be used as a probe to recover the β-globin gene, including its introns and adjacent control regions, from a cloned genomic library.

A different approach to generating probes uses a technique termed **reverse genetics**. This method goes from the protein to the gene. If the protein product of the gene is available, and part or all of its amino acid sequence is known, a polynucleotide probe can be constructed. Beginning with the amino acid sequence, the nucleotide sequence encoding these amino acids can be deduced (Figure 15.21). Because the genetic code is degenerate (up to six codons can specify an amino acid), several combinations of nucleotide sequences are possible. In the example given, eight different oligonucleotides could encode the amino acid sequence. Each of these oligonucleotides is chemically synthesized using radioactive nucleotides. A mixture of these probes is used to identify the clone encoding the gene of interest. Note that it is not necessary to know the entire amino acid sequence of the protein or to synthesize the nucleotide sequence of the entire gene. An oligonucleotide about 18 to 21 nucleotides long, corresponding to 6 or 7 amino acids, is usually sufficient to produce a probe.

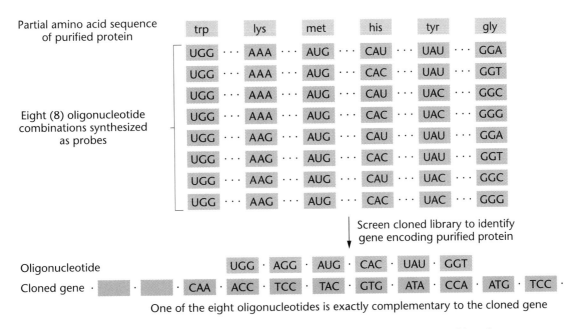

Partial amino acid sequence of purified protein

Eight (8) oligonucleotide combinations synthesized as probes

Screen cloned library to identify gene encoding purified protein

Oligonucleotide

Cloned gene

One of the eight oligonucleotides is exactly complementary to the cloned gene

FIGURE 15.21   The process of reverse translation. The coding DNA sequence can be deduced from the amino acid sequence of a small portion of a protein. In this example, two amino acids (*trp* and *met*) have only one codon each. Three others (*lys, his,* and *tyr*) are encoded by two codons. The sixth amino acid (*gly*) is encoded by codons that include all base combinations at the third position. For this protein fragment, eight coding sequences encompass all the possible combinations. The exact coding sequence of the gene must be one of the eight. Using a radioactive mixture of these sequences as probes, a cloned library can be screened to isolate the structural gene for the complete protein.

Screening a Library

To screen a plasmid library, clones to be screened are grown on nutrient plates, forming hundreds or thousands of colonies (Figure 15.22). A replica of the colonies on the plate is made by gently pressing a nitrocellulose or nylon filter onto the surface of the plate to transfer bacterial cells from the colonies to the filter. The filter is transferred through solutions to lyse the bacteria, denature the double stranded DNA and convert it to single strands, and bind these strands to the filter.

The colonies on the filter are screened by incubation with a nucleic acid probe. This method requires some knowledge about the gene or DNA sequence being screened for. If a radioactive probe is to be used for screening the library, the probe DNA is denatured to form single-stranded molecules and is added to a solution containing the filter. If the sequence of the probe is complementary to any of the cloned sequences in the colonies, a DNA-DNA hybrid molecule will form between the probe and the insert DNA (review the discussion of hybridization in Chapter 10). After unbound and excess probe is washed away, the filter is overlaid with a piece of photographic film. If the probe is complementary to the DNA of one of the lysed colonies, the radioactivity in the probe will expose part of the film, and the hybridized colony will be identified as a dark spot on the developed film (Figure

15.22). That colony can be recovered from the initial plate, and the cloned DNA it carries can be used in further experiments. Nonradioactive probes use chemiluminescence or colorimetric methods to detect the clone(s) of interest.

To screen a phage library, a slightly different method, called **plaque hybridization**, is used. A solution of phage carrying DNA inserts is plated onto a lawn of bacteria growing on a plate. The phage then replicate, forming plaques, which appear as clear spots on the plate. Each plaque represents the progeny of a single phage, and are clones. The plaques are transferred by pressing a filter to the surface of the plate and then lysed. The DNA is denatured into single strands and screened with a labeled probe as for a plasmid library. Phage plaques are much smaller than bacterial colonies, and many more plaques can be grown on a plate. Consequently, more plaques can be screened on a single filter, making this method more efficient for screening large genomic libraries.

Chromosome Walking

In some cases, when the approximate location of a gene is known, it is possible to clone the gene by first cloning nearby sequences. Often these nearby sequences are identified by linkage analysis. This sequence serves as the beginning point for **chromosome walking,** the isolation

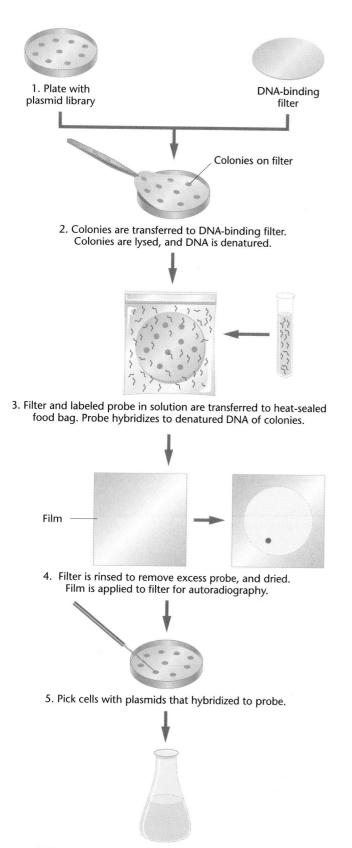

1. Plate with plasmid library

DNA-binding filter

Colonies on filter

2. Colonies are transferred to DNA-binding filter. Colonies are lysed, and DNA is denatured.

3. Filter and labeled probe in solution are transferred to heat-sealed food bag. Probe hybridizes to denatured DNA of colonies.

Film

4. Filter is rinsed to remove excess probe, and dried. Film is applied to filter for autoradiography.

5. Pick cells with plasmids that hybridized to probe.

6. Transfer cells to medium for growth, further analysis.

of adjacent clones from a genomic library. In a chromosome walk, the end piece of a cloned DNA fragment is subcloned, and used as a probe to recover overlapping clones from the genomic library (Figure 15.23). The overlapping clones are analyzed by restriction mapping to determine the amount of overlap. A subfragment from one end of the overlapping clone is used as a probe to recover another set of overlapping clones, and the analysis is repeated. In this way, it is possible to "walk" along the chromosome, clone by clone. The gene in question can be identified by nucleotide sequencing of the recovered clones and searching for an open reading frame (ORF), a stretch of nucleotides that begins with a start codon and has a stretch of amino acid-encoding codons followed by one or more stop codons. Although laborious and time-consuming, chromosome walking has been used to recover genes associated with human genetic disorders; the genes for cystic fibrosis and muscular dystrophy were isolated by chromosome walking.

There are several limitations to chromosome walking in complex eukaryotic genomes. If a probe contains a repetitive sequence, it will hybridize to many other clones in the genomic library, most of which are not adjacent segments, and the walk will be terminated. In these cases, a related technique called chromosome jumping can be used to skip over the region containing repetitive sequences and continue the walk.

FIGURE 15.22 Screening a plasmid library to recover a cloned gene. The library, present on nutrient plates, is overlaid with a DNA binding filter, and colonies are transferred to the filter. Colonies on the filter are lysed, and the DNA is denatured to single strands. The filter is placed in a hybridization bag along with buffer and a labeled single-stranded DNA probe. During incubation, the probe forms a double-stranded hybrid with complementary sequences on the filter. The filter is removed from the bag and washed to remove excess probe. Hybrids are detected by placing a piece of X-ray film over the filter and exposing for a short time. The film is developed, and hybridization events are visualized as spots on the film. From the orientation of spots on the film, colonies containing the insert that hybridized to the probe can be identified. Cells are picked from this colony for growth and further analysis.

Initial adjacent sequence

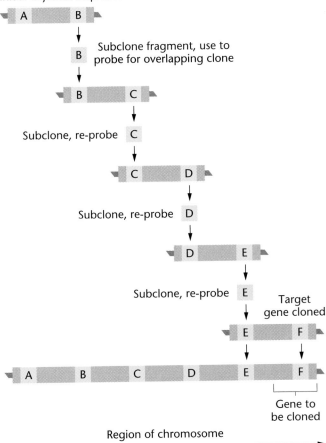

Subclone fragment, use to probe for overlapping clone

Subclone, re-probe

Subclone, re-probe

Subclone, re-probe

Subclone, re-probe

Target gene cloned

Gene to be cloned

Region of chromosome

FIGURE 15.23 In chromosome walking, the approximate location of a gene to be cloned is known. A subcloned fragment of a linked adjacent sequence is used to recover overlapping clones from a genomic library. This process of subcloning and probing the genomic library is repeated to recover overlapping clones until the gene in question has been reached.

Methods for the Analysis of Cloned Sequences

The identification and recovery of specific cloned DNA sequences is a powerful tool for the analysis of gene structure and function. Techniques described in the following sections are used to answer experimental questions about the organization and expression of cloned sequences.

Restriction Mapping

A **restriction map** is a compilation of the number, order, and distance between restriction enzyme cutting sites along a cloned segment of DNA. Map units are expressed in **base pairs (bp)** or for longer distances, **kilobase pairs (kb)**. Restriction maps provide information that can be used for subcloning fragments of a gene, or for compar-

ing the organization of a gene and its cDNA so as to identify exons and introns in the genomic copy of the gene.

Fragments generated by cutting DNA with restriction enzymes can be separated by **gel electrophoresis**, a method that separates fragments by size, with the smallest pieces moving farthest (see Appendix A for a detailed description of electrophoresis). The fragments appear as a series of bands that can be visualized by staining the DNA with ethidium bromide and viewing under ultraviolet illumination (Figure 15.24). The size of individual fragments can be determined by running a set of marker fragments of known size on the same gel in another lane.

Figure 15.25 shows the steps in constructing a restriction map from a cloned DNA segment that contains cutting sites for two restriction enzymes. For the construction of this map, let us begin with a cloned DNA segment 7.0 kb in length (Figure 15.25). Three samples of the cloned DNA are digested with restriction enzymes: one with HindIII, one with SalI, and one with both HindIII and SalI. The fragments generated by digestion with the restriction enzymes are separated by electrophoresis (Figure 15.25). The sizes of the separated fragments can be estimated by comparison to a set of standards run in adjacent lanes. To construct the map, the fragments generated by the restriction enzymes are analyzed.

1. When the DNA is cut with HindIII, two fragments are produced, one 0.8 kb, and one 6.2 kb, indicating that there is only one cutting site for this enzyme (located 0.8 kb from one end).

2. When the DNA is cut with SalI, two fragments result, one of 1.2 kb, and one of 5.8 kb, meaning there is one cutting site, located 1.2 kb from one end.

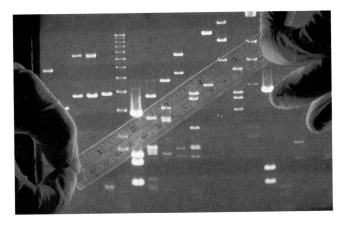

FIGURE 15.24 An agarose gel containing separated DNA fragments, stained with a dye (ethidium bromide) and visualized under ultraviolet light. Smaller fragments migrate faster and farther than do larger fragments, resulting in the distribution shown.

1. A population of cloned DNA fragments is prepared

2. DNA fragments are cut with restricted enzymes

3. The restricted fragments are separated by gel electrophoresis

4. Results

5. Results are used to construct models, and compared with results of double enzyme digests

6. Conclusion: model 1 is correct

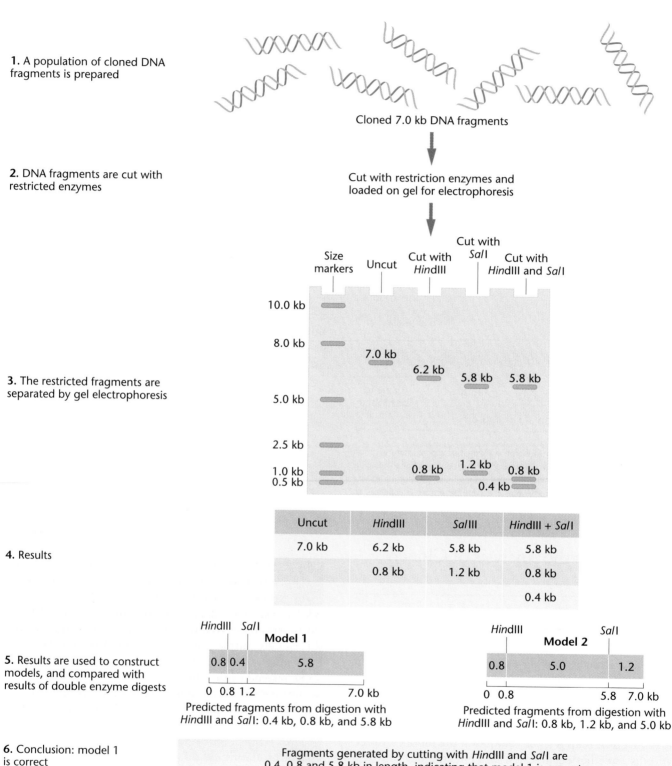

Cloned 7.0 kb DNA fragments

Cut with restriction enzymes and loaded on gel for electrophoresis

| Size markers | Uncut | Cut with HindIII | Cut with SalI | Cut with HindIII and SalI |

10.0 kb
8.0 kb   7.0 kb
                        6.2 kb   5.8 kb   5.8 kb
5.0 kb

2.5 kb

1.0 kb   0.8 kb   1.2 kb   0.8 kb
0.5 kb   0.4 kb

Uncut	HindIII	SalII	HindIII + SalI
7.0 kb	6.2 kb	5.8 kb	5.8 kb
	0.8 kb	1.2 kb	0.8 kb
			0.4 kb

HindIII SalI   **Model 1**
0.8 | 0.4 | 5.8
0 0.8 1.2   7.0 kb
Predicted fragments from digestion with HindIII and SalI: 0.4 kb, 0.8 kb, and 5.8 kb

HindIII   **Model 2**   SalI
0.8   5.0   1.2
0 0.8   5.8   7.0 kb
Predicted fragments from digestion with HindIII and SalI: 0.8 kb, 1.2 kb, and 5.0 kb

Fragments generated by cutting with HindIII and SalI are 0.4, 0.8 and 5.8 kb in length, indicating that model 1 is correct.

Taken together, the results show that there is one restriction site for each enzyme, but the relationship between the two sites is unknown. From the information available at this point, two different maps are possible. In one model, the *Hin*dIII site is located 0.8 kb from one end (Model 1), and the *Sal*I site is located 1.2 kb from the same end. In the alternative model (Model 2), the *Hin*dIII site is located 0.8 kb from one end, and the *Sal*I site is located 1.2 kb from the other end.

The correct model can be selected by considering the results from the digestion with both *Hin*dIII and *Sal*I. Model 1 predicts that digestion with both enzymes will generate three fragments of 0.4, 0.8, and 6.2 kb; the second model predicts that digestion with both enzymes will generate three fragments of 0.8, 1.2, and 5.0 kb. The actual fragment pattern observed after cutting with both enzymes indicates that Model 1 is correct (Figure 15.25).

In most cases, restriction mapping is more complex and involves a larger number of sites for each enzyme, and a greater number of enzymes. Restriction maps provide an important way of characterizing a DNA segment, and it can be constructed in the absence of any information about the coding capacity or function of the mapped DNA. In conjunction with other techniques, restriction mapping can be used to define the boundaries of a gene, and it provides a way of dissecting the molecular organization within a gene and its flanking regions. Mapping can also serve as a starting point for the isolation of an intact gene from cloned segments of DNA, and it provides a means for locating mutational sites within genes.

Restriction maps can also be used to refine genetic maps. To a large extent, the accuracy of maps constructed by genetic analysis is based both on the frequency of recombination between genetic markers and on the number of genetic markers used in construction of the map. If there is a large distance between markers and/or variation in recombination frequency, the genetic map may not correspond to the physical map of the chromosome region. In humans, for example, the genome size is large (3.2×10^9 bp), and the number of mapped genes is small (a few thousand), meaning that each map unit is composed of millions of base pairs of DNA. The result is a poor correlation between the genetic and physical maps of chromosomes. Restriction enzyme cutting sites can be used as genetic markers, thus reducing the distance between sites on a map, increasing the accuracy of maps, and providing reference points for the correlation of genetic and physical maps.

Restriction sites have played an important role in mapping genes to specific human chromosomes and to defined regions of individual chromosomes. In addition, if a restriction site maps close to a mutant gene, it can be used as a marker in a diagnostic test. These sites, known as restriction fragment length polymorphisms, or RFLPs, will be described in Chapter 16. The use of RFLPs has proven especially useful because the mutant genes underlying many human genetic diseases are poorly characterized at the molecular level, and closely linked restriction sites have been successfully used in the detection of afflicted individuals and heterozygotes at risk for having affected children.

Southern and Northern Blots

The DNA inserts cloned into vectors can be used in hybridization experiments to characterize the identity of specific genes, to locate coding regions or flanking regulatory regions within cloned sequences, and to study the molecular organization of genomic sequences.

The use of cloned DNA segments separated by electrophoresis, transferred to filters and screened by probes was developed by Edward Southern. Known as a **Southern blot**, this procedure has many applications. To make a Southern blot, cloned DNA is cut into fragments with one or more restriction enzymes, and the fragments are separated by gel electrophoresis (Figure 15.26). The DNA in the gel is denatured into single-stranded fragments and transferred to a sheet of DNA-binding material, usually nitrocellulose or a nylon derivative. The transfer is made by placing the sheet of membrane on top of the gel and causing buffer to flow through the gel and the nitrocellulose or nylon sheet by capillary action. The buffer flows through both, pulling the DNA out of the gel and immobilizing it in the membrane.

In practice, this is done by placing the gel with the denatured DNA onto a thick sponge that acts as a wick. The sponge is partially immersed in a tray of buffer. A sheet of membrane filter is placed on top of the gel and covered by sheets of blotting paper or paper towels and a weight. The buffer in the tray is drawn by capillary action

FIGURE 15.25 Construction of a restriction map. Samples of the 7.0-kb DNA fragments are digested with restriction enzymes: One sample is digested with *Hin*dIII, one with *Sal*I, and one with both *Hin*dIII and *Sal*I. The resulting fragments are separated by gel electrophoresis. The sizes of the separated fragments can be measured by comparison with molecular-weight standards in an adjacent lane. *Hin*dIII generates two fragments: 0.8 kb and 6.2 kb. *Sal*I produces two fragments: 1.2 kb and 5.8 kb. Models are constructed to explain the digestion patterns with *Hin*dIII and with *Sal*I. Two models are constructed, based on the data from the cuts with *Hin*dIII and with *Sal*I. These models can be used to predict the fragment sizes generated by digestion with both *Hin*dIII and *Sal*I. Model 1 predicts that the following fragments would result from cutting with both enzymes: 0.4 kb, 0.8 kb, 5.8 kb; model 2 predicts that the fragments would be: 0.8 kb, 1.2 kb, 5.0 kb. When the DNA is cut with both *Hin*dIII and *Sal*I, three fragments are produced: 0.4 kb, 0.8 kb, and 5.8 kb. Comparison of the predicted fragments with those observed on the gel indicate that Model 1 is the correct restriction map.

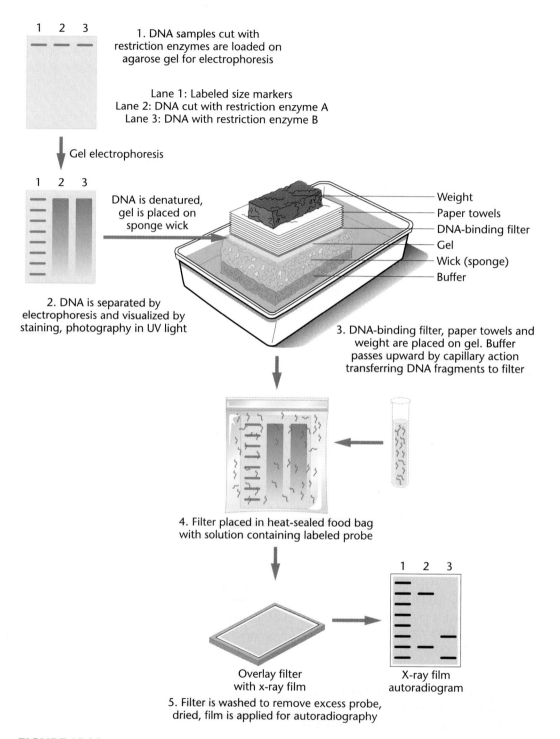

1. DNA samples cut with restriction enzymes are loaded on agarose gel for electrophoresis

Lane 1: Labeled size markers
Lane 2: DNA cut with restriction enzyme A
Lane 3: DNA with restriction enzyme B

Gel electrophoresis

DNA is denatured, gel is placed on sponge wick

Weight
Paper towels
DNA-binding filter
Gel
Wick (sponge)
Buffer

2. DNA is separated by electrophoresis and visualized by staining, photography in UV light

3. DNA-binding filter, paper towels and weight are placed on gel. Buffer passes upward by capillary action transferring DNA fragments to filter

4. Filter placed in heat-sealed food bag with solution containing labeled probe

Overlay filter with x-ray film

X-ray film autoradiogram

5. Filter is washed to remove excess probe, dried, film is applied for autoradiography

FIGURE 15.26   The Southern blotting technique. Samples of the DNA to be probed are cut with restriction enzymes and the fragments separated by gel electrophoresis. The pattern of fragments is visualized and photographed under ultraviolet illumination by staining the gel with ethidium bromide. The gel is then placed on a sponge wick in contact with a buffer solution and covered with a DNA-binding filter. Layers of paper towels or blotting paper are placed on top of the filter and held in place with a weight. Capillary action draws the buffer through the gel, transferring the pattern of DNA fragments from the gel to the filter. The DNA fragments on the binding filter are then denatured into single strands and hybridized with a labeled probe, washed, and overlaid with a piece of X-ray film for autoradiography. The hybridized fragments show up as bands on the X-ray film.

through the sponge, gel, membrane, and the stack of blotting paper. As the buffer passes through the gel, DNA fragments are transferred to the membrane, where they stick. After transfer, the single-stranded DNA is fixed to the membrane by baking at 80°C or by exposing to ultraviolet light to crosslink the fragments to the membrane.

The DNA fragments on the filter are then hybridized with a probe. Only those single-stranded DNA fragments embedded in the membrane that are complementary to the nucleotide sequence of the probe will form hybrids. The unbound probe is washed away, and the hybridized fragments are visualized. If a radioactive probe is used, the position of the probe is done by autoradiography, using a piece of film (see Appendix A for a detailed discussion of autoradiography).

In addition to characterizing cloned DNAs, Southern blots are used for many other purposes, including the mapping of restriction sites within and near a gene, the identification of DNA fragments carrying a single gene from a mixture of many fragments, and the identification of related genes in different species. Southern blots are also used to detect rearrangements and duplications in genes associated with human genetic disorders and cancers.

A related blotting technique can be used to determine whether a cloned gene is transcriptionally active in a given cell or tissue type by probing for the presence of RNA that is complementary to a cloned DNA segment. This is accomplished by extracting RNA from one or several cell and/or tissue types. The RNA is fractionated by gel electrophoresis and the pattern of RNA bands is transferred to a sheet of RNA-binding membrane as in the Southern blot. The sheet is then hybridized to a single-stranded DNA probe, derived from cloned genomic DNA or cDNA. If RNA complementary to the DNA probe is present, it will be detected by autoradiography as a band on the film. Because the original procedure using DNA bound to a filter became known as a Southern blot, this inverse procedure using RNA bound to a filter was called a **northern blot**. (Following this somewhat perverse logic, another procedure involving proteins bound to a filter is known as a **western blot**.)

Northern blots provide information about whether RNA transcripts complementary to cloned genes are present in a given cell or tissue, and are used to study patterns of gene expression in embryonic and adult tissues. Northern blots can also be used to detect alternatively spliced mRNAs and multiple types of transcripts derived from a single gene (discussed in Chapter 19). Northern blots also provide other information about transcribed mRNAs. If marker RNAs of known size are run in an adjacent lane, the size of the mRNA of a gene of interest can be calculated. In addition, the amount of transcribed RNA present in the cell or tissue being studied is related to the density of the RNA band on the x-ray film. This can be quantified by measuring the density of the band, providing a relative measurement of transcriptional activity. Thus, northern blots can be used to characterize and quantify the transcriptional activity of a given gene in different cells, tissues, or organisms.

DNA Sequencing

In a sense, the ultimate characterization of a cloned DNA segment is the determination of its nucleotide sequence. The ability to sequence cloned DNA has added immensely to our understanding of gene structure and the mechanisms of gene regulation. Although techniques were available in the 1940s to determine the base composition of DNA (see Chapter 10), it was not until the 1960s that methods for the determination of nucleotide sequences were developed. In 1965 Robert Holley determined the sequence of a tRNA molecule consisting of 74 nucleotides. This work required about one year of concentrated effort. In the 1970s, more efficient methods of sequencing were developed, and it is now possible for a molecular biology laboratory to sequence more than a thousand bases in a week. In the near future, with widespread automation of this process, it will be commonplace to sequence thousands of nucleotides in a single day.

One method of DNA sequencing devised by Alan Maxam and Walter Gilbert uses chemicals to cleave DNA, and this method was used to determine the base sequence of the 4362 nucleotides in the plasmid pBR322. The second method, which is more commonly used, was developed by Fredrick Sanger and his colleagues. The Sanger method is based on the 5′-to-3′ elongation of DNA molecules by DNA polymerase. Both sequencing strategies generate a series of single-stranded DNA molecules of differing lengths. Each method uses a series of four reactions, each in a separate tube, to determine the sequence of the nucleotides in a DNA segment. These sequences, differing in length by as little as one nucleotide, are separated from each other by gel electrophoresis in four adjacent lanes. The result is a series of bands forming a ladderlike pattern. The sequence can be read directly from the band pattern in the four lanes (Figure 15.27). Automated DNA sequencers employ fluorescent colored dyes instead of radioactive probes. Four different colored dyes are used, producing a colored pattern of peaks that can be read to provide the sequence (Figure 15.28). Details of each of the methods of DNA sequencing are presented in Appendix A.

The DNA sequencing method provides information about the organization of genes and the nature of mutational events that alter both genes and gene products, con-

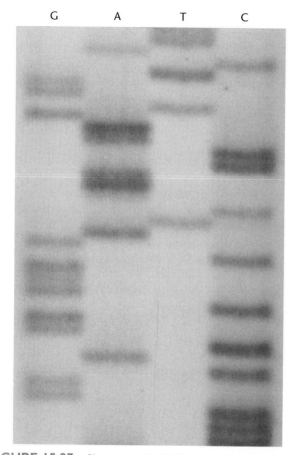

G	A	T	C

FIGURE 15.27   Photograph of a DNA sequencing gel showing the separation of bases in the four sequencing reactions (one per lane). The gel is analyzed to reveal the base sequence of the DNA fragment. To do this, the gel is read from the bottom, beginning with the lowest band in any lane, then the next lowest, then the next, etc. For example, the sequence of the DNA on this gel begins with TTAACCCGG, etc.

firming the conclusion that genes and proteins are colinear molecules. Sequencing has also been used to study the organization of regulatory regions that flank prokaryotic and eukaryotic genes, and to derive the amino acid sequence of proteins. The gene for cystic fibrosis (CF), an autosomal recessive human genetic disorder, was identified by cloning and sequencing DNA from a region on the long arm of chromosome 7. Since the protein product of the gene was unknown, DNA sequencing was used to identify a protein coding region. The DNA sequence from this region was used to generate a putative amino acid sequence of 1480 amino acids. This sequence was used to search data bases containing amino acid sequences of known proteins, to infer that the CF protein has properties similar to membrane proteins that play a role in ion transport. Further work confirmed that the product of the cystic fibrosis locus is a membrane protein that regulates the transport of chloride ions across the plasma membrane. In most cases of CF, the mutant gene has an altered DNA sequence that causes the production of a defective protein.

In addition to identifying DNA defects that cause mutant phenotypes, DNA sequencing is used to study the organization of a gene (the number of introns and exons and their boundaries), to provide information about the nature and function of proteins encoded by genes, including the size, number and type of domains (membrane-spanning, DNA-binding) and relationship to similar proteins and proteins from other organisms.

PCR Analysis

Recombinant DNA techniques were developed in the early 1970s, and in subsequent years they revolutionized the way geneticists and molecular biologists conduct research. In 1986, a new technique called the **polymerase chain reaction (PCR)** was developed, which has greatly extended the power of recombinant DNA research. It has even replaced some of the earlier methods. PCR analysis has found applications in a wide range of disciplines, including molecular biology, human genetics, evolution, and even forensics.

One of the prerequisites for many recombinant DNA techniques is the availability of large amounts of a specific DNA segment. Often these are obtained through tedious and labor-intensive efforts in cloning and recloning. PCR allows a direct amplification of *specific* DNA segments without the use of cloning, and it can be used on fragments of DNA that are initially present in infinitesimally small quantities. The PCR method is based on the amplification of a DNA segment using DNA polymerase and oligonucleotide primers that hybridize to opposite strands of the sequence to be amplified (Figure 15.29). There are three basic steps in the PCR reaction, and the amount of amplified DNA produced is limited theoretically only by the number of times these steps are repeated.

1. In the first step, the DNA to be amplified is denatured into single strands. This DNA does not have to be purified or cloned, and it can come from any number of sources, including genomic DNA, forensic samples such as dried blood or semen, samples stored as part of medical records, single hairs, mummified remains, and fossils. The double-stranded DNA is denatured by heating (at about 90°C) until it dissociates into single strands (usually about 5 minutes).

2. Primers are annealed to the single-stranded DNA. These primers are synthetic oligonucleotides that anneal to sequences flanking the segment to be amplified. Two different primers are most often used in PCR. Each primer has a sequence complementary to only one of the two strands of DNA. The

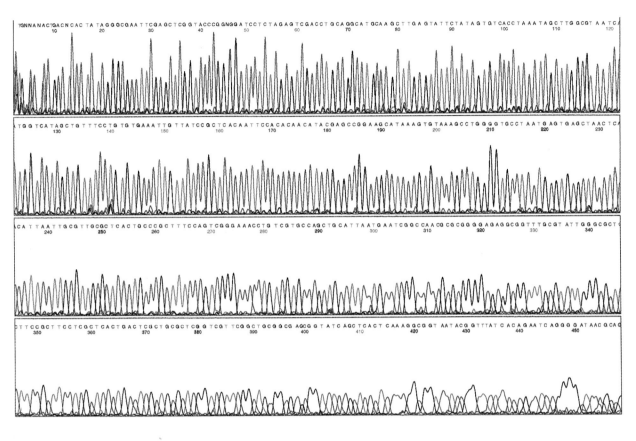

FIGURE 15.28 DNA sequencing has been automated with the use of fluorescent dyes, one for each base; the separated bases are read in order along the axis from left to right.

primers align themselves with their 3′-ends facing each other because they anneal to opposite strands (Figure 15.29). The use of synthetic primers means that some DNA sequence information must be known about the DNA to be amplified.

3. A heat-resistant DNA polymerase (Taq polymerase) is added to the reaction mixture (Figure 15.29). The polymerase extends the oligonucleotide primers in the 5′-to-3′ direction, using the single-stranded DNA bound to the primer as a template. The product is a double-stranded DNA molecule with the primers incorporated into the final product (Figure 15.29).

Each set of three steps—denaturation of the double-stranded product, annealing of primers, and extension of polymerase—is referred to as a *cycle*. Generally, each step of the cycle is performed at a different temperature. The cycle can be repeated by carrying out each of the steps again. Beginning with one DNA molecule, the first cycle produces two molecules of DNA, two cycles produce four, three cycles produce eight, and so forth. Twenty-five cycles of amplification result in a several-million-fold

amplification of the target DNA. The process is automated and uses a machine called a *thermocycler*, which can be programmed to carry out a predetermined number of cycles, yielding large amounts of amplified DNA segments in a few hours.

PCR is widely used in research laboratories and has far-ranging applications in fields that include infectious disease, archeology, food safety, human genetics, cancer therapy, molecular evolution, and others. In clinical applications, PCR is used to identify infectious agents such as the tuberculous bacillus and other bacteria, and HIV and other viruses. PCR has the advantage of being faster and less labor intensive than conventional cloning techniques, and has largely replaced the use of cloned probes in areas such as prenatal diagnosis. Although the development of PCR represents a major advance, the technique has limitations. Because PCR amplifies DNA sequences, even minor contamination of the starting material with DNA from other sources can cause difficulties. Cells shed from the skin of a laboratory worker while performing PCR can contaminate samples gathered from a crime scene, making it difficult to obtain accurate results. PCR must always be run with appropriate controls.

Transferring DNA in Eukaryotes

Both animal cells and plant cells can take up DNA from their environment, a process called *transfection*. Moreover, vectors, including YACs, can be used to transfer DNA into eukaryotic cells.

Plant Cells

As strange as it might seem, cloning using higher plants as hosts has been achieved using a bacterial plasmid. The infectious soil bacterium *Agrobacterium tumifaciens* produces tumors (called plant galls) in many species of plants. Tumor formation is associated with the presence of a tumor-inducing (Ti) plasmid in the bacteria (Figure 15.30). When these bacteria infect plant cells, a segment of the Ti plasmid, known as T-DNA, is transferred into the chromosomal DNA of the host plant cell. The T-DNA controls tumor formation and the synthesis of small molecules known as *opines* that are required for growth of the infecting bacteria. Foreign genes can be inserted into the T-DNA segment of Ti, and the altered Ti can be transferred into a plant cell by the infecting bacteria. The foreign DNA is inserted into the plant genome when the T-DNA is integrated into a host cell chromosome. Such genetically altered single cells can be induced to grow in tissue culture to form a cell mass known as a **callus**. By manipulating the culture medium, the callus can be induced to form roots and shoots, and eventually to develop into mature plants carrying a foreign gene. Plants (or animals) carrying a foreign gene are said to be **transgenic**. In the next chapter we will see how gene transfer in plants has been used genetically to alter agriculturally important crop plants.

Mammalian Cells

DNA can be taken up into mammalian cells by several methods, including coprecipitation with calcium phosphate and endocytosis, direct microinjection, exposure of cells to short pulses of high voltage (electroporation), and

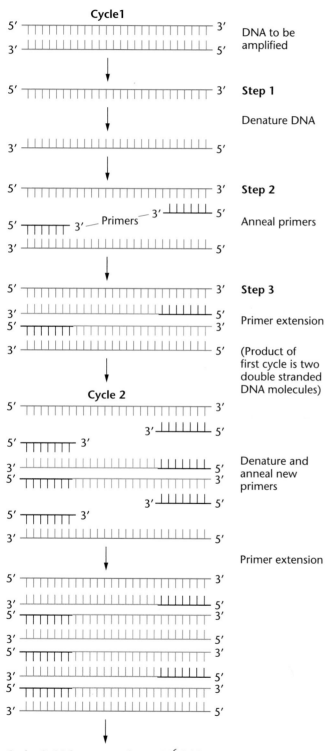

Cycles 3–25 for greater than a 10^6-fold increase in DNA

FIGURE 15.29 PCR amplification. In the polymerase chain reaction (PCR), the DNA to be amplified is denatured into single strands; each strand is then annealed to a short, complementary primer. The primers are synthetic oligonucleotides complementary to sequences flanking the region to be amplified. DNA polymerase and nucleotides are added to extend the primers in the 3′ direction, using the single-stranded DNA as a template. The result is a double-stranded DNA molecule with the primers incorporated into the newly synthesized strand. In a second PCR cycle, the products of the first cycle are denatured into single strands, primers are added, and they are then extended by DNA polymerase. Repeated cycles can amplify the original DNA sequence by more than a million-fold.

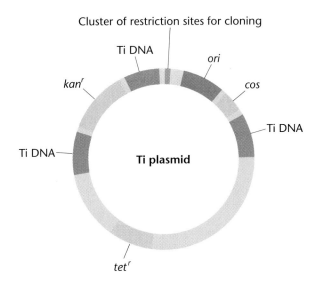

Cluster of restriction sites for cloning

Ti DNA

ori

*kan*ʳ

cos

Ti DNA

Ti DNA

Ti plasmid

*tet*ʳ

FIGURE 15.30 A Ti plasmid designed for cloning in plants. Segments of Ti DNA, including those necessary for opine synthesis and integration, are combined with bacterial segments that incorporate cloning sites and antibiotic resistance genes (*kan*ʳ and *tet*ʳ). The vector also contains an origin of replication (*ori*) and a lambda *cos* site that permits recovery of cloned inserts from the host plant cell.

by encapsulation of DNA into artificial membranes (liposomes) followed by fusion with cell membranes. Also, DNA can be transferred using YACs and vectors based on retroviruses. The DNA introduced into a mammalian cell by any of these methods is usually integrated into the host genome. DNA transfer methods have been used to transfer genes to fertilized eggs, thus producing transgenic animals; to study molecular aspects of gene expression; and to replicate cloned genes using mammalian cells as hosts.

YACs can be used to increase greatly the efficiency of transferring genes into the germ line of mice. Gene transfer is dependent on the introduction of DNA sequences into the nucleus of an appropriate mouse cell such as a fertilized egg or an embryonic stem cell, followed by the integration of the DNA into a chromosomal site.

Earlier experiments with transgenes in mice were hampered by the relatively short length of DNA that could be introduced using plasmid cloning vectors. In many cases, mammalian genes to be transferred were longer than the cloning capacity of the vectors, and in other cases, where the gene could be transferred, its expression was inhibited or inappropriate because adjacent or distant regulatory sequences were not transferred.

The YACs are transferred to mice in several ways. The first involves the microinjection of purified YAC DNA into the pronucleus of a mouse zygote (Figure 15.31). The zygotes are then transferred to the uteri of foster mothers for normal development. Other techniques use mouse embryonic stem (ES) cells as hosts. One of these methods fuses a yeast cell carrying a YAC with a stem cell, transferring the YAC and all or a portion of the yeast genome to the stem cell. The transgenic ES cells are then transferred into blastocyst stage mouse embryos, where they participate in the formation of adult tissues, including those that form germ cells.

The ability to transfer large DNA segments into mice has applications in many areas of research. Some of these will be described in the next chapter.

Vectors for the delivery of functional genes into mammalian cells have been developed using avian and mouse retroviruses. The genome of such retroviruses is a single-stranded RNA that is transcribed by reverse tran-

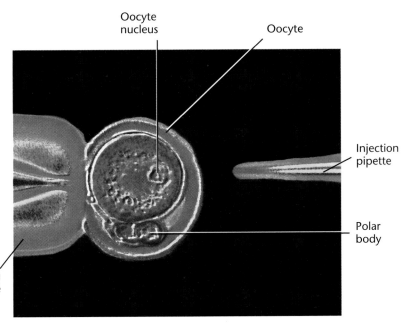

Oocyte nucleus

Oocyte

Injection pipette

Polar body

Holding pipette

FIGURE 15.31 Cloned DNA can be transferred to mammalian embryos by direct injection.

scriptase into a double-stranded DNA. This DNA is integrated into the host genome and passed on to daughter cells as part of the host chromosome. The retroviral genome can be engineered to remove some viral genes, creating vectors that accept foreign DNA, including human structural genes (Figure 15.32). These vectors are being used in the treatment of genetic disorders by gene therapy, a topic that will also be discussed in the following chapter.

The development of techniques for producing millions of copies of DNA segments by cloning opened the way for structural and functional studies of DNA from plants and animals. In turn, it led to the isolation of specific genes from organisms including humans, the generation of the array of research and commercial applications that has revolutionized the life sciences, medicine, and biotechnology, and the international effort to analyze the human genome at the nucleotide level known as the Human Genome Project.

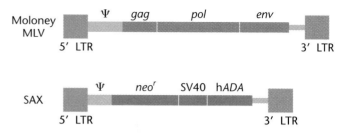

FIGURE 15.32 Retroviral vectors constructed from the Moloney murine leukemia virus (Moloney MLV). The native MLV genome contains a *psi* sequence required for encapsulation and genes that encode viral coat proteins (*gag*), an RNA-dependent DNA polymerase (*pol*), and surface glycoproteins (*env*). At each end, the genome is flanked by long terminal repeat (LTR) sequences that control transcription and integration into the host genome. The SAX vector retains the LTR and *psi* sequences, and it includes a bacterial neomycin resistance (*neo^r*) gene that can be used as a selective marker. As shown, the vector carries a cloned human adenosine deaminase (*hADA*) gene, fused to an SV40 early region promoter/enhancer. The SAX construct is typical of retroviral vectors that will be used in human gene therapy.

GENETICS, TECHNOLOGY, AND SOCIETY

Reporter Gene Technology: Tracking Gene Expression

Genes influence personality, behavior, and physical traits. Genes direct the processes of development and differentiation, and define each cell's response to its environment. They also cause sickle-cell anemia, cancer, hemophilia, and cystic fibrosis. One of society's goals is to control these negative incarnations of normal genes, and conquer genetic diseases. To achieve this goal, we must first identify those genes that are involved in normal and abnormal processes and, once identified, determine how those genes are regulated.

Genes exert their effects through gene expression—regulating the amount of gene product that is synthesized, in which tissue it is synthesized and at what times. The primary, and often most important, mechanism for controlling gene expression is to regulate that gene's transcription. Expression is regulated by *trans*-acting factors that bind to DNA sequences near the gene's start site of transcription. These sequences include promoter elements (usually located near the start site) and enhancer elements (often located far-

ther from the start site). One of the most intensely pursued questions in modern molecular biology is how promoters and enhancers, along with the proteins that bind to them, act to regulate transcription.

The first step towards understanding a gene's transcriptional regulation is to identify which sequences comprise its promoter and enhancers. Once identified, it is important to test how these elements respond to stimuli such as the presence of hormones or growth factors, whether the elements act in some cell types but not in others, and whether they act at certain times in development. Although it is often straightforward to clone a gene and its putative promoter/enhancer region, it is much less straightforward to define the essential sequences within that region. Deletions and point mutations can be inserted into the cloned promoter DNA in a test tube, and the *in vitro* mutated gene transferred back into cells. However, measuring the amount of gene product is complicated, because assays for RNA levels are laborious and difficult to interpret against a background of RNA that is already present in the cell.

An innovative method is now available that provides a sensitive and rapid way to measure changes in transcription. This method involves linking the cloned DNA of the putative promoter/enhancer region to cloned DNA of the coding region of an unrelated gene— the **reporter gene**. This chimeric reporter construct is then introduced into cultured cells or into transgenic animals. The cells or the tissues of the transgenic animal are then assayed for the amount or location of the reporter gene product (usually a protein or enzymatic activity) rather than the RNA. The amount of reporter gene product provides an estimate of the amount of gene transcription that was stimulated by the promoter/enhancer region that was cloned in front of the reporter gene.

Reporter genes must display several characteristics. The reporter gene product must be absent from the host cells, the assay for reporter gene product must be simple, sensitive, and exhibit a wide linear range, and the reporter gene product must not have adverse effects on the host cells. Several reporter genes have these characteristics and are used extensively for promoter analysis in a wide range of eu-

CHAPTER SUMMARY

1. Recombinant DNA refers to the creation of a new association between DNA molecules or segments not usually found together. The cornerstone of recombinant DNA technology is a class of enzymes called *restriction endonucleases* that cut DNA at specific recognition sites. The fragments produced are joined with DNA vector segments using the enzyme DNA ligase to form recombinant DNA molecules. Vectors have been constructed from many sources, including bacterial plasmids, viruses, and artificial vectors such as cosmids and synthetic yeast chromosomes.

2. Recombinant DNA molecules are transferred into a host, and cloned copies are produced during host cell replication. A variety of host cells may be used for replication, including yeast, bacteria, cells of higher plants, and mammalian cells in tissue culture. However, the most common host is *E. coli*. Cloned copies of foreign DNA sequences can be recovered, purified, and analyzed.

3. Cloned libraries have been constructed containing DNA sequences from entire genomes, single chromosomes, or chromosome segments. Researchers use libraries to select probes complementary to all or part of a gene being surveyed.

4. Once cloned, DNA sequences can be analyzed using a variety of methods, including restriction mapping and DNA sequencing. Other methods such as blotting and hybridization can be used to identify genes and flanking regulatory regions within the cloned sequences.

5. For studies of gene regulation in cloned eukaryotic genes, eukaryotic host systems have been developed, including yeast and the cells of higher plants and animals.

karyotic cells. These genes include the bacterial chloramphenicol acetyltransferase (*CAT*) gene, the firefly luciferase gene and the bacterial *lacZ* gene. *CAT* is the most widely used reporter gene and its activity is measured by a simple biochemical assay, in lysates prepared from *CAT*-transfected cells. Luciferase is measured by its ability to act on the substrate luciferin and produce a flash of light. Because some luciferase substrates can enter intact cells, luciferase activity is detectable in both live cells and cell extracts. The product of the *lacZ* gene (β-galactosidase) reacts with the substrate X-gal, to produce a dark blue color. This characteristic makes *lacZ* the reporter gene of choice in studies of tissue-specific gene expression during early development. The location of β-galactosidase-expressing cells can be detected in tissue sections and whole fixed embryos. Another β-galactosidase substrate (FDG) is able to penetrate whole cells, allowing detection of reporter gene activity in live cells. Another recent development in reporter gene technology involves the use of green fluorescent proteins (GFPs), which are encoded by genes found in bioluminescent jellyfish. Living cells which express GFP glow with a bright green fluorescence when they are illuminated with blue or UV light. GFP genes, under control of neuron-specific promoters have allowed researchers to observe the formation of neurons as they develop in individual organisms.

Many different vectors exist that allow the insertion of reporter genes into cells. A commonly used vector consists of a closed circular piece of DNA (a plasmid) that will replicate in bacteria. It contains the bacterial origin of replication (*ori*), which allows the plasmid to propagate to high numbers in bacterial cells. An antibiotic resistance gene is also part of the vector. It allows bacteria that contain the plasmid to grow on antibiotic-containing growth medium, thus providing for their selection. The reporter gene coding region and its promoter/enhancer DNA is the final part of the vector. The use of bacterial plasmids provides a mechanism for producing large quantities of the reporter gene construct, which may then be inserted into eukaryotic cells.

Recombinant DNA technology has become such a pervasive methodology that biotechnology companies develop and sell many types of reporter gene vectors, each vector designed for a different type of study. The use of reporter genes has been pivotal to many of the advances of molecular biology. Their use has contributed vital information on the behavior of genes involved in human diseases such as thalassemia and cancer, as well as those involved in the normal processes of cell growth and mammalian development.

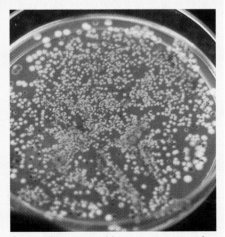

Genetically engineered bacteria carrying a plasmid with the *lac* operon are used to select cells carrying recombinant DNA inserts.

KEY TERMS

INSIGHTS AND SOLUTIONS

1. The recognition site for the restriction enzyme *Sau*3A is GATC (Figure 15.4). The recognition site for the enzyme *Bam*HI is GGATCC, where the four internal bases are identical to the *Sau*3A site. This means that the single-stranded ends produced by the two enzymes are identical. Suppose that you have a cloning vector containing a *Bam*HI site, and foreign DNA that you have cut with *Sau*3A. Can this DNA be ligated into the *Bam*HI site of the vector? Why?

   Solution: DNA cut with *Sau*3A can be ligated into the *Bam*HI site of the vector because the single-stranded ends generated by the two enzymes are compatible.

2. Can the DNA segment cloned into this site be cut from the vector with *Sau*3A? With *Bam*HI? What potential problems do you see with the use of *Bam*HI?

   Solution: The DNA can be cut from the vector with *Sau*3A because the recognition site for this enzyme is maintained. Recovering the cloned insert with *Bam*HI is more problematic. In the ligated vector, the conserved bases are GGATC (left) and GATCC (right). Only about 25 percent of the time will the correct base follow (GGATCC on the left), and only about 25 percent of the time will the correct base lead the conserved sequence (GATCC on the right). Thus, only about 6 percent of the time (0.25 × 0.25), will *Bam*HI be able to cut the insert from the vector.

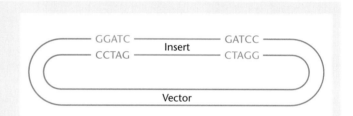

3. Calculate the average size of the fragments generated when cutting DNA with *Sau*3A and with *Bam*HI.

   Solution: According to the Watson–Crick model of DNA, the sequence of bases in a DNA molecule can be at random. Therefore, if an *A* is present in a strand of DNA, the next base in that strand can be any base, even another *A*. Thus, the chance that any given base in DNA can be an A is 1/4, and the chance that the next base is a *T* is also 1/4. As a result, the chance that the four-base sequence GATC will occur is 4^4, or 1 in every 256 bases. The chance of encountering the six-base sequence GGATCC is 4^6, or 1 in every 4096 bases. Accordingly, cuts with *Sau*3A should generate fragments that average 256 bp in length, whereas cuts with *Bam*HI should generate fragments that average 4096 base pairs in length. This, of course, is only an approximation, and it assumes that each base is present about 25 percent of the time. In reality, however, many genomes depart from this ideal and may have only 40 or 42 percent G + C content, instead of the 50 percent ideal.

PROBLEMS AND DISCUSSION QUESTIONS

1. In recombinant DNA studies, what is the role of each of the following: restriction endonucleases, terminal transferases, vectors, calcium chloride, and host cells?
2. Why is poly-dT an effective primer for reverse transcriptase?
3. An ampicillin-resistant, tetracycline-resistant plasmid is cleaved with *Eco*RI, which cuts within the ampicillin gene. The cut plasmid is ligated with *Eco*RI-digested *Drosophila* DNA to prepare a genomic library. The mixture is used to transform *E. coli* K12.
    (a) Which antibiotic should be added to the medium to select cells that have incorporated a plasmid?
    (b) What antibiotic-resistance pattern should be selected to obtain plasmids containing *Drosophila* inserts?
    (c) How can you explain the presence of colonies that are resistant to both antibiotics?
4. Clones from the preceding question are found to have an average length of 5 kb. Given that the *Drosophila* genome is 1.5×10^5 kb long, how many clones would be necessary to give a 99 percent probability that this library contains all genomic sequences?
5. Type II restriction enzymes recognize palindromic sequences in intact DNA molecules and cleave the double-stranded helix at these sites. Inasmuch as the bases are internal in a DNA double helix, how is this recognition accomplished?
6. In a control experiment, a plasmid containing a *Hin*dIII site within a kanamycin-resistant gene is cut with *Hin*dIII, religated, and used to transform *E. coli* K12 cells. Kanamycin-resistant colonies are selected, and plasmid DNA from these colonies is subjected to electrophoresis. Most of the colonies contain plasmids that produce single bands that migrate at the same rate as the original, intact plasmid. A few colonies, however, produce two bands, one of original size, and one that migrates much higher in the gel. Diagram the origin of this slow band during the religation process.
7. When making cDNA, the single-stranded DNA produced by reverse transcriptase can be made double stranded by treatment with DNA polymerase I. However, no primer is required with the DNA polymerase. Why is this?
8. What facts should you consider in deciding which vector to use in constructing a genomic library of eukaryotic DNA?
9. Using DNA sequencing on a cloned DNA segment, you recover the following nucleotide sequence:

    CAGTATCCTAGGCAT

    Does this segment contain a palindromic recognition site for a restriction enzyme? What is the double-stranded sequence of the palindrome? What enzyme would cut at this site? (Consult Figure 15.4 for a list of restriction enzyme recognition sites.)
10. Figure 15.4 lists restriction enzymes that recognize sequences of four bases and six bases. How frequently should four- and six-base recognition sequences occur in a genome? If the recognition sequence consists of eight bases, how fre-

quently would such sequences be encountered? Under what circumstances would you select such an enzyme for use?
11. List the steps involved in cloning a DNA insert into the *Sal*I site of the plasmid pBR322, up to and including the identification of host cells carrying recombinant plasmids.
12. You are given a cDNA library of human genes prepared in a bacterial plasmid vector. You are also given the cloned yeast gene that encodes EF-1alpha, a protein which is highly conserved in protein sequence among eukaryotes. Outline how you would use these resources to identify the human cDNA clone encoding EF-1alpha.
13. You have recovered a cloned DNA segment of interest, and determined that the insert is 1300 bp in length. To characterize this cloned segment, you have isolated the insert and decide to construct a restriction map. Using enzyme I and enzyme II, followed by gel electrophoresis, you determine the number and size of the fragments produced by enzymes I and II alone and in combination as follows:

Enzymes	Restriction Fragment Sizes (bp)
I	350 bp, 950 bp
II	200 bp, 1100 bp
I and II	150 bp, 200 bp, 950 bp

Construct a restriction map from these data, showing the positions of the restriction sites relative to one another, and the distance between them in base pairs.
14. Although the potential benefits of cloning in higher plants are obvious, the development of this field has lagged behind cloning in bacteria, yeast, and mammalian cells. Can you think of any reason for this?
15. cDNA can be cloned into vectors to create a cDNA library. In analyzing cDNA clones, it is often difficult to find clones that are full length—that is, extend to the 5′-end of the mRNA. Why is this so?
16. List the steps involved in screening a genomic library by reverse translation. What needs to be known before starting such a procedure? What are the potential problems with such a procedure? How can they be overcome or minimized?
17. Although the capture and trading of great apes has been banned in 112 countries since 1973, it is estimated that about 1000 chimpanzees are removed annually from Africa and smuggled into Europe, the United States, and Japan. This illegal trade is often disguised by private (such as zoo or circus) owners by simulating births in captivity. Until recently, genetic identity tests to uncover these illegal activities have not been used because of the lack of availability of highly polymorphic markers and the difficulties of obtaining chimp blood samples. Recently a study was reported in which DNA samples were extracted from freshly plucked chimp hair roots and used as templates for the polymerase

chain reaction. The primers used in these studies flank highly polymorphic sites in human DNA resulting from variable numbers of tandem nucleotide repeats. Several suspect chimp offspring and their supposed parents were tested to determine if the offspring were "legitimate" or were the "product of the illegal trading" and not the offspring of the putative parents. A sample of the data is shown below. Examine this data carefully and choose the best conclusion:

(a) None of the offspring are legitimate.

(b) Offspring B and C are not the products of these parents and were probably purchased on the illegal market. The data are consistent with offspring A being legitimate.

(c) Offspring A and B are products of the parents shown, but C is not and was therefore probably purchased on the illegal market.

(d) There is not enough data to draw any conclusions. Additional polymorphic sites should be examined.

(e) No conclusion can be drawn because "human" primers were used.

Lane 1: father chimp
Lane 2: mother chimp
Lanes 3–5: putative offspring A, B, C

18. Briefly describe the problem that a stretch of repeated sequences would cause in a chromosome walk and name the procedure that is used to overcome this problem.

19. You have obtained a clone of a human gene *A* that is linked (within 200 kb) to a gene that causes breast cancer. Choose and correctly order six of the items from the following list to outline how you could use the technique of chromosome walking to obtain clones of all genes within 200 kb of gene *A*.

(1) Partially digest DNA with restriction enzyme *Bam*HI to obtain overlapping fragments of about 20 kb.

(2) Completely digest DNA with restriction enzyme *Eco*RI.

(3) Isolate human genomic DNA.

(4) Isolate *E. coli* genomic DNA.

(5) Probe northern blot with *A*.

(6) Repeat above step until the desired number of clones is obtained.

(7) Insert fragment mixture into lambda to make a phage bank.

(8) Screen bank (or library) with gene probe *A* and isolate a hybridizing clone.

(9) Subclone a small fragment from the end of this clone and use it to rescreen the library and isolate a hybridizing clone.

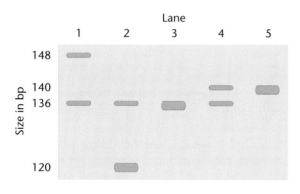

EXTRA-SPICY PROBLEMS

20. The partial restriction map at the right shows a recombinant plasmid, pBIO220, formed by cloning a piece of *Drosophila* DNA (striped box) including the gene *rosy* into the vector pBR322, which also contains the penicillin-resistance gene, *pen*. The vector part of the plasmid contains only the two E sites shown, and no A or B sites. The gel at the top of the next page shows several restriction digests of pBIO220.

The enzymes used were:
E—*Eco*RI
A—*Apo*I
B—*Bst*II

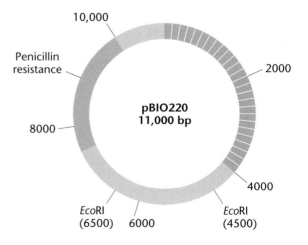

(a) Use the gel pattern to deduce where restriction sites are located in the cloned fragment.

(b) A PCR-amplified copy of the entire 2000-bp *rosy* gene was used to probe a Southern blot of the same gel. Use the results, shown at the bottom right, to deduce the locations of *rosy* in the cloned fragment. Redraw the map, showing the location of *rosy*.

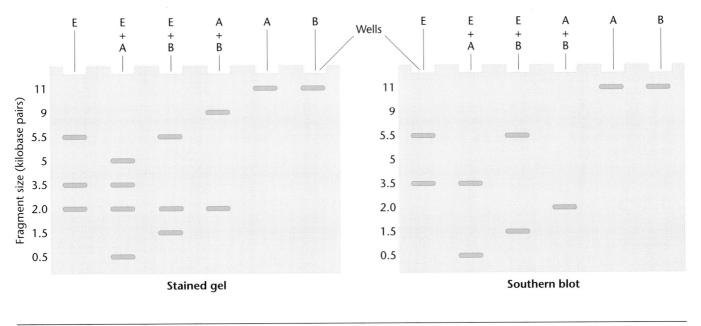

Stained gel   Southern blot

---

21. One of the restriction maps shown at the right of the figure below is consistent with the pattern of bands shown in the gel at the left after digestion with several restriction endonucleases.

The enzymes used were:
E—*Eco*RI
N—*Nco*I
A—*Aat*II

(a) From your analysis of the pattern of bands on the gel, select the correct map, and explain your reasoning.

(b) In a Southern blot prepared from this gel, the highlighted bands (purple) hybridized with the gene *pep*. Where is the *pep* gene located?

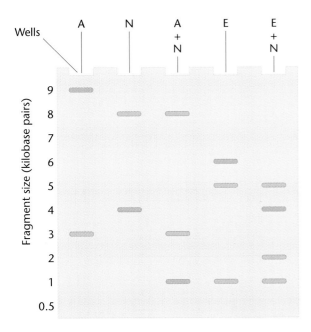

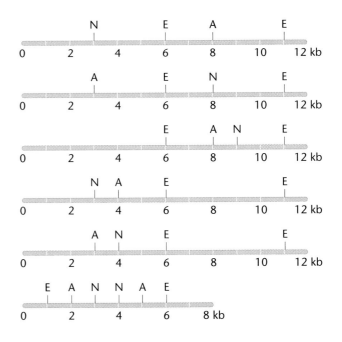

SELECTED READINGS

ALWINE, J. C., KEMP, D. J., and STARK, G. R. 1977. Method for detection of specific RNAs in agarose gels by transfer to diazobenzyloxymethyl paper and hybridization with DNA probes. *Proc. Natl. Acad. Sci. USA* 74:5350–54.

ANDERSON, W. F., and DIACUMAKOS, E. G. 1981. Genetic engineering in mammalian cells. *Sci. Am.* (July) 245:106–21.

BOYER, H. W. 1971. DNA restriction and modification mechanisms in bacteria. *Annu. Rev. Microbiol.* 25:153–76.

CAI, L., TAYLOR, J. F., WING, R. A., GALLAGHER, D. S., WOO, S. S., and DAVIS, S. K. 1995. Construction and characterization of a bovine bacterial artificial chromosome library. *Genomics* 29:413–425.

COHEN, S., et al. 1973. Construction of biologically functional bacterial plasmids *in vitro*. *Proc. Natl. Acad. Sci. USA* 70:3240–44.

COLLINS, J., and HOHN, B. 1978. Cosmids: A type of plasmid gene cloning vector that is packageable *in vitro* in bacteriophage heads. *Proc. Natl. Acad. Sci. USA* 75:4242–46.

DAVIES, K. E., et al. 1981. Cloning of a representative genomic library of the human X chromosome after sorting by flow cytometry. *Nature* 293:374–76.

DeSALLE, R., and BIRSTEIN, V. J. 1996. PCR identification of black caviar. *Nature* 381:197–98.

GLOVER, D. M. 1984. *Gene cloning: The mechanism of DNA manipulation.* London: Chapman and Hall.

GUBLER, U., and HOFFMAN, B. J. 1983. A simple and very effective method for generating cDNA libraries. *Gene* 25:263–69.

HOCHMEISTER, M. N. 1995. DNA technology in forensic applications. *Mol. Aspects Med.* 16:315–437.

KRUMLAUFF, R., JEANPIERRE, M., and YOUNG, B. D. 1982. Construction and characterization of genomic libraries from specific human chromosomes. *Proc. Natl. Acad. Sci. USA* 79:2971–75.

LeBAR, W. D. 1996. Keeping up with new technology: new approaches to diagnosis of *Chlamydia* infection. *Clin. Chem.* 42:809–812.

MESSING, J., and VIERIA, J. 1982. The pUC plasmids, an M13mp7-derived system for insertion mutagenesis and sequencing with synthetic universal primers. *Gene* 19:259–68.

MOLLIS, K. B. 1990. The unusual origin of the polymerase chain reaction. *Sci. Am.* (April) 262:56–65.

MONACO, A. P. 1994. YACs, BACs, and MACs, artificial chromosomes as research tools. *Trends Biotech.* 12:280–286.

OLD, R. W., and PRIMROSE, S. B. 1985. *Principles of genetic manipulation: An introduction to genetic engineering.* Palo Alto, CA: Blackwell Scientific.

SAMBROOK, J., FRITSCH, E., and MANIATIS, T. 1992. *Molecular cloning: A laboratory manual.* 2nd ed. Cold Spring Harbor, NY: Cold Spring Harbor Laboratory Press.

SHIZUA, H., BIRREN, B., KIM, U-J., MANCINO, V., SLEPAK, T., TACHIIRI, Y., and SIMON, M. 1992. Cloning and stable maintenance of 300 kilobase-pair fragments of human DNA in *Escherichia coli* using an F-factor-based vector. *Proc. Natl. Acad. Sci. USA* 89:8794–8797.

SOUTHERN, E. M. 1975. Detection of specific sequences among DNA fragments separated by gel electrophoresis. *J. Mol. Biol.* 98:503–17.

WATSON, J., GILMAN, M., WITKOWSKI, J., and ZOLLER, M. 1992. *Recombinant DNA*, 2nd ed. New York: Scientific American Books.

16

Applications of Recombinant DNA Technology

CHAPTER CONCEPTS

The use of recombinant DNA technology has brought about significant advances in experimental genetics, gene mapping, and the diagnosis and treatment of disease. These techniques have also formed the basis for the biotechnology industry and the commercial production of human gene products for therapeutic uses, and the transfer of genes across species in agricultural plants and animals.

In 1971, a paper published by Hamilton Smith, Daniel Nathans, and Walter Arber marked the beginning of the recombinant DNA era. The paper described the isolation of an enzyme from a strain of bacteria and the use of this enzyme to cleave viral DNA. It contained the first published photograph of DNA cut with a restriction enzyme. From this modest beginning, recombinant DNA technology has revolutionized all fields of experimental biology and spread beyond the research laboratory to many other fields. In the intervening years, this technology has expanded to become part of everyday life; recombinant DNA techniques are now being used to help produce medicines, milk, and soon, even the food at the supermarket.

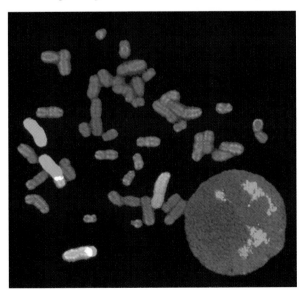

Fluorescence *in situ* hybridization (FISH) showing a translocation between chromosome 2 and chromosome 3 in a cancer cell.

In the last chapter we discussed the methods used to create and analyze recombinant DNA molecules. This chapter describes how these tools are being used in research and commercial applications. We will consider a cross section of applications illustrating the power of recombinant DNA technology to map and identify human genes, diagnose and treat disease, and to generate new plants and animals. We will begin by considering how recombinant DNA has changed some of the basic methods of genetic mapping. Next, we will examine the impact of recombinant DNA on human genetics and medicine, from the prenatal analysis of genotypes to the treatment of genetic disorders by repairing or replacing defective genes, to the mapping of the human genome. Finally, we will discuss some of the products and methods being used in the multibillion dollar biotechnology industry.

Mapping Human Genes

RFLPs as Genetic Markers

Although an increasing number of the 50,000 to 100,000 human genes have been localized to chromosomal sites, in the majority of human genetic disorders the primary defect is unknown, and without a marker, the gene cannot be mapped. Defective gene products have been identified in only a few hundred single-gene disorders, and the nature of the mutation is known in fewer than 100 of these. As a result, classical genetic methods cannot be used to identify the chromosomal locus of many human genetic disorders.

The development of recombinant DNA technology has led to new methods for mapping genes without the need to first have knowledge of the normal or mutant gene products. This approach is based on the observation that naturally occurring variations in nucleotide sequence occur throughout the human genome, mostly in noncoding regions, with a frequency of about 1 in every 200 bases. These random variations in nucleotide sequence can create or destroy restriction enzyme recognition sites. When this happens, it alters the pattern of cuts made in DNA by restriction enzymes. If one or more nucleotides at a restriction recognition site are altered, the enzyme may fail to cut at that site. Conversely, a nucleotide change can also create a new restriction site. If a restriction site is present on the DNA from one chromosome, but absent on its homolog, the two chromosomes can be distinguished from one another by their pattern of restriction fragments on a DNA blot (Figure 16.1). The region of chromosome *A* shown in the figure contains three *Bam*HI sites; its homolog, chromosome *B*, contains two such sites. When chromosome *A* is cut with *Bam*HI, 3-kb and 7-kb fragments are generated, whereas only a single 10-kb fragment is generated when chromosome *B* is cut. These specific fragments can be detected on a Southern blot using a probe from this chromosomal region.

These detectable variations in DNA fragment length generated by cutting with a restriction enzyme are called **restriction fragment length polymorphisms (RFLPs)**. Analysis indicates that RFLPs are quite common, and represent a natural variation brought about by changes in a single nucleotide pair, or by deletions or insertions of one or more nucleotide pairs. Several thousand RFLPs

have been detected in the human genome, and many of these have been assigned to individual chromosomes. These variations are inherited in a Mendelian fashion as codominant alleles. They can be mapped to individual chromosomes and used to follow the inheritance of specific chromosomes from generation to generation.

RFLPs are distributed at random throughout the human genome. They occur in regions between genes,

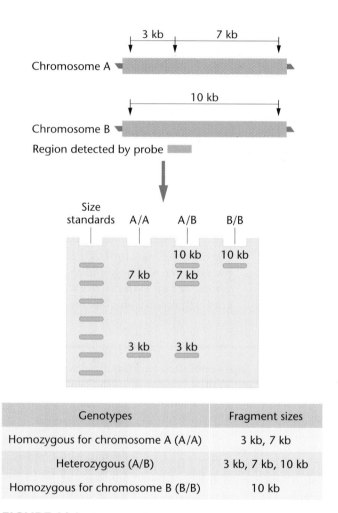

Genotypes	Fragment sizes
Homozygous for chromosome A (A/A)	3 kb, 7 kb
Heterozygous (A/B)	3 kb, 7 kb, 10 kb
Homozygous for chromosome B (B/B)	10 kb

FIGURE 16.1 Restriction fragment-length polymorphisms (RFLPs). The alleles on chromosome *A* and chromosome *B* represent DNA segments from homologous chromosomes. The region that hybridizes to a probe is shown below. Arrows indicate the location of restriction enzyme cutting sites that define the alleles. On chromosome *A*, three cutting sites generate fragments of 7 kb and 3 kb. On chromosome *B*, only two cutting sites are present, generating a fragment 10 kb in length. The absence of the cutting site in *B* could be the result of a single base mutation within the enzyme recognition/cutting site. Because these differences in restriction cutting sites are inherited in a codominant fashion, there are three possible genotypes: *AA*, *AB*, and *BB*. The allele combination carried by any individual can be detected by restriction digestion of genomic DNA (obtained from a blood sample or skin fibroblasts), followed by gel electrophoresis, transfer to a DNA binding filter, and hybridization to the appropriate probe. The fragment patterns for the three possible genotypes are shown as they would appear on a Southern blot.

and within genes. They are found in the regulatory regions adjacent to transcriptional units adjacent to genes, and within introns and exons. In 1980 it was proposed that these markers, in combination with known genes, could be used to construct a linkage map of the human genome. This proposal, made by David Botstein, Mark Skolnick, and Ray White, accelerated the mapping of human genes, and a partial linkage map of the human genome was produced in the next few years. This may not seem like rapid progress, but consider that from the 1930s to about 1970, only five cases of autosomal linkage were known in humans. From 1970 to 1980, additional examples of linkage were discovered through the use of somatic cell hybrids, but by 1980 only a relative handful of linked genes had been identified. Presently, RFLP analysis has made it possible to obtain high-resolution linkage maps for each human chromosome, and maps of RFLP markers are now available for each of the 24 different human chromosomes.

Using RFLPs to Make Linkage Maps

Linkage maps using chromosomally-assigned RFLP markers are constructed using multigenerational families (three generations or more) in which an RFLP marker and a genetic disorder are segregating. Selecting an RFLP to use in mapping a disease gene in a family is done by trial and error. The most useful RFLPs are those for which most members of the family are heterozygous, so that each member of a chromosome pair can be identified. Members of the family are tested using a collection of RFLPs, to generate a set that will be used as codominantly inherited genetic markers of known chromosomal locations (Figure 16.2). For mapping, the inheritance of a given RFLP and the genetic disorder are traced through the family. If the RFLP and the genetic disorder are on the same chromosome, they will show linkage. Most RFLPs will not show linkage: these will indicate the chromosome that does *not* carry the disease locus. In an-

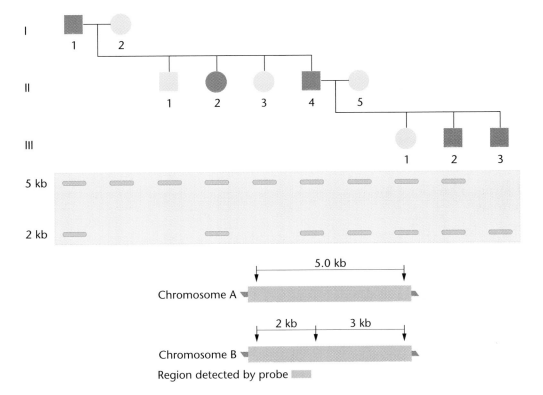

FIGURE 16.2 RFLP analysis and the inheritance of a dominant trait. The family shown in the pedigree has members affected by a dominant trait, as indicated by the filled symbols. Members of this family also carry two alleles for a restriction site: the A allele is a 5.0-kb fragment, and the B allele is two fragments, one of 2 kb and the other of 3 kb. The Southern blot pattern of RFLP alleles for each family member is shown below their pedigree symbol. In analyzing this pedigree, note that individual II-5 is from outside the family and is homozygous for the normal alleles, even though she carries the RFLP B allele. Her daughter, III-1, who is unaffected, probably received the A allele from her father and the B allele from her mother. Her oldest son (III-2) is affected, and probably received an A allele from his mother, and the B allele from his father. The youngest son (III-3) who is affected, received a B allele from each parent. Examination of the pedigree and the results of the Southern blot indicates that the mutant allele responsible for the disorder is carried on a chromosome with the B (2.0-kb) allele. Establishing linkage to a chromosome is the first step in mapping a gene using RFLP analysis.

alyzing linkage between an RFLP marker and a genetic disorder, the probability that the observed pattern of inheritance (of the gene and the RFLP) could occur by chance alone is calculated. This analysis begins with the assumption that the gene and the RFLP marker are unlinked. Then the analysis is repeated, calculating the probability of the observed pattern of inheritance if the gene and the marker have a certain degree of linkage. The ratio of the two probabilities (no linkage/certain degree of linkage) is computed and expressed as the odds for that degree of linkage. This calculation is known as the **logarithm of the odds** or **lod score**. A lod score of 3 or greater can be taken as evidence of linkage between the gene and the RFLP marker. A lod score of 3 means that the odds are 1000:1 that the gene and the marker are linked.

In mapping human chromosomes, the unit of linkage is the **centimorgan (cM)**, named after the geneticist T. H. Morgan. One centimorgan is equal to a recombination frequency of 1 percent between two markers. Genetic maps indicate the distance between markers in centimorgans. The genetic distance between two genes expressed in centimorgans is not directly correlated with the physical distance expressed in nucleotides, but in human chromosomes, there are about 1×10^6 bp of DNA for every centimorgan. In mapping human genes, a distance of 2 cM corresponds to roughly 2 million nucleotides of DNA. By compiling many family studies using RFLP markers and genetic traits, a genetic map for a human chromosome can be constructed (Figure 16.3).

Positional Cloning: The Gene for Neurofibromatosis

The value of RFLP analysis in gene mapping can be illustrated by considering the search for the chromosomal locus of the gene for type 1 neurofibromatosis (NF1). This disorder is inherited as an autosomal dominant condition, with an incidence of about 1 in 3000. NF1 is associated with a range of nervous system defects, including benign tumors and an increased incidence of learning disorders. The spontaneous mutation rate at this locus is very high, and it is estimated that 30–50 percent of all cases represent new mutations. To map this autosomal gene by conventional methods in a systematic fashion would be an almost impossible task requiring the identification of large families carrying genes mapped to each of the 22 autosomes, and also carrying neurofibromatosis.

Mapping of the NF1 gene was accomplished in several steps. First, a number of laboratories compared the inheritance of NF1 and dozens of RFLP markers in multigenerational families. Each RFLP marker represented a specific human chromosome or chromosome region. This produced an **exclusion map**, indicating the chromosomes and chro-

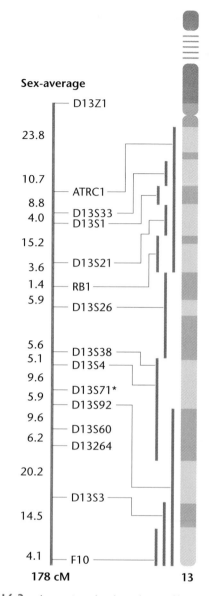

Sex-average

	D13Z1
23.8	
10.7	
8.8	ATRC1
4.0	D13S33
15.2	D13S1
3.6	D13S21
1.4	RB1
5.9	D13S26
5.6	
5.1	D13S38
9.6	D13S4
5.9	D13S71*
9.6	D13S92
6.2	D13S60
20.2	D13264
14.5	D13S3
4.1	F10

178 cM   **13**

FIGURE 16.3 A genetic and a physical map of human chromosome 13. The genetic map for females is 203 cM, and that for males is 158 cM, reflecting the difference in recombination frequencies between females and males. When the two maps are averaged together, the result is the sex-averaged map of 178 cM shown at left. The location of markers on the physical map is indicated by the brackets adjacent to the chromosome.

mosome regions where NF1 was not located. This work also produced some evidence of linkage, and pointed to chromosomes 5, 10, and 17 as candidates for carrying the NF1 gene. Subsequent work determined that the disorder segregated with an RFLP near the centromere of chromosome 17 (Figure 16.4). Then, more than 30 RFLP markers from this region of chromosome 17 were used to analyze 13,000 individuals from families in which NF1 was segregating, and the gene was assigned to region 17q11.2, on the long arm near the centromere. Finally, using clones that spanned a subregion of 17q11.2, the locus for NF1 was identified in 1990 by

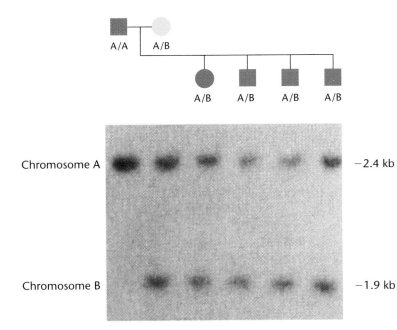

FIGURE 16.4   The segregation of a 2.4-kb RFLP allele with type 1 neurofibromatosis (NF1) in each of four affected offspring and their father. This RFLP is detected by probe pA10-41, which is known to be a DNA segment near the centromere of human chromosome 17. On the basis of this and results from the use of other probes, the locus for NF1 was assigned to chromosome 17.

chromosome walking and DNA sequencing of the recovered clones to identify coding regions. The identification of the gene was confirmed by finding mutations of the gene in individuals with NF1. Analysis of the DNA sequencing information indicates that the NF1 gene is 130 kb in length, has 11 exons and encodes a protein of 2485 amino acids.

Once the gene was identified, the amino acid sequence of the gene product was reconstructed from the DNA sequence. By searching protein sequence information in data bases, the NF1 gene product was found to be similar to proteins in signal transduction pathways. Further analysis confirms that the NF1 protein, neurofibromin, is involved in the transduction of intracellular signals and the down-regulation of a gene that controls cell growth. Mutation in the NF1 gene leads to a loss of control of cell growth, causing the production of small tumors that are characteristic of this disorder.

The mapping, cloning, and sequencing of this gene took a little over three years, beginning with no direct knowledge of the nature of the gene product or the mutational events that result in the production of the NF1 phenotype. The mapping and cloning of the gene for NF1 described above is an example of **positional cloning**. This recombinant DNA-based method is a departure from the previous methods, which worked from an identified gene product to the gene locus. In positional cloning, the gene can be mapped, isolated, and cloned with no knowledge of the gene product. Using this strategy, an ever-increasing number of human genes are being mapped and isolated. Where RFLP markers do not show close linkage to a gene, other methods, or a combination of methods, have been used to map and isolate genes associated with genetic disorders.

Candidate Genes: The Gene for Marfan Syndrome

Another approach to identifying and mapping human genes is the selection and investigation of candidate genes. This method can be used only when some information about the molecular basis of a genetic disorder is already available. This information provides clues about the function of the defective gene product: Is it likely to be an enzyme, a structural protein, a transport protein, or a hormone? Once the type of protein has been identified, genes that encode similar proteins are identified and examined to see whether mutant alleles are involved in the disorder.

The gene for Marfan syndrome was identified using the candidate gene approach. Marfan syndrome is an autosomal dominant disorder of connective tissue that affects about 1 in 10,000 individuals. It has been speculated that Abraham Lincoln, the 16th president of the United States, had this disorder. Given the nature of the defect, attention was focused on genes encoding structural proteins of connective tissue as the primary candidates. Although this approach seems straightforward, many connective tissue proteins are chemically modified after translation, making it difficult to identify the primary gene product.

Fibrillin is a connective tissue protein first identified in 1986 as a component of the eye, aorta, and other elastic tissues, all of which are affected in Marfan syndrome. Fluorescence *in situ* hybridization (FISH) techniques (Figure 16.5) had mapped the gene for fibrillin to the long arm of chromosome 15. When RFLP studies using large, multigenerational families with Marfan syndrome established linkage between markers on chromosome 15 and Marfan

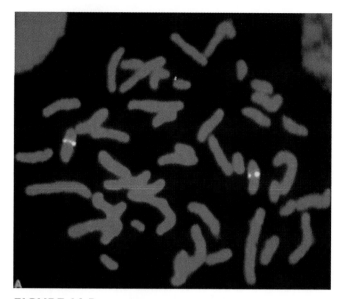

FIGURE 16.5 Localization of a gene by fluorescence *in situ* hybridization (FISH).

syndrome, the fibrillin gene became a candidate for Marfan syndrome. In the next step, the fibrillin gene was identified, cloned, and sequenced. Additionally, a mutation in the fibrillin gene was identified in three unrelated families with Marfan syndrome. In all three families, the mutation was a missense mutation leading to the substitution of a proline for arginine at position 239 in fibrillin. Unaffected family members carried fibrillin genes with arginine at position 239. Finding a mutant fibrillin gene in affected family members was the final link in establishing the relationship between Marfan syndrome and fibrillin.

Diagnosing and Screening Genetic Disorders

The most widely used methods for prenatal detection of genetic disorders are **amniocentesis** and **chorionic villus sampling (CVS)**. In amniocentesis, a needle is used to withdraw amniotic fluid (Figure 16.6). The fluid and the cells it contains can be analyzed for chromosomal or single-gene disorders. In CVS, a catheter is inserted into the uterus and used to retrieve a small tissue sample of the fetal chorion. This tissue is used for cytogenetic and biochemical prenatal diagnosis.

Coupled with these sampling methods, recombinant DNA technology has proven to be a highly sensitive and accurate tool for the detection of genetic disorders. The use of cloned DNA sequences has the advantage of allowing direct examination of the genotype rather than a po-

tentially unknown or unexpressed gene product. The prenatal detection of two different defects in the β-globin gene serves to illustrate the application of recombinant DNA technology to disease detection. Disorders of beta globin cannot be detected prenatally by other means because the β-globin gene is not transcribed until a few days after birth.

Deletions in Thalassemia

Beta thalassemia is an autosomal recessive disorder associated with decreased or absent β-globin production. At the molecular level, this condition can be caused by a variety of deletions or point mutations. One such deletion covers about 600 nucleotides, and includes the third exon of the gene (Figure 16.7). Restriction enzyme digestion and Southern blot hybridization to a cloned β-globin probe can be used to diagnose the genetic status of a fetus and other members of an affected family (Figure 16.7). Because the Southern blot detects fragments separated on the basis of size, this method is useful in studying mutations involving deletions, insertions, and rearrangements.

Deletions, however, are relatively rare mutational events. More commonly, mutations are associated with changes in one or a small number of nucleotides (point mutations). However, if changes in a single nucleotide take place in an exon, they can have devastating clinical and phenotypic effects, as discussed below.

Sickle-Cell Anemia and Prenatal Genotyping

Sickle-cell anemia is an autosomal recessive condition common in people with family origins in areas of West Africa and the Mediterranean basin, and also parts of India. Sickle-cell anemia is caused by a single nucleotide substitution in the first exon of the β-globin gene. The nucleotide change eliminates a cutting site for several restriction enzymes, including *Mst*II and *Cvn*I, altering the pattern of restriction fragments seen in Southern blots. These characteristic RFLP patterns can be used for prenatal diagnosis of sickle-cell anemia, and to determine the genotypes of parents and other family members who may be heterozygous carriers of this condition. Thus, RFLPs can be used for prenatal diagnosis as well as genetic mapping.

For prenatal diagnosis, fetal cells are obtained by amniocentesis or CVS. DNA is extracted from these cells and digested with a restriction enzyme such as *Mst*II. This enzyme cuts twice within the normal β-globin gene, producing two small DNA fragments. In the mutant allele, the second *Mst*II site has been destroyed by the mutation, producing one large restriction fragment (Figure 16.8). The restriction-digested DNA fragments are separated by gel electrophoresis and transferred to a nitrocellulose or nylon

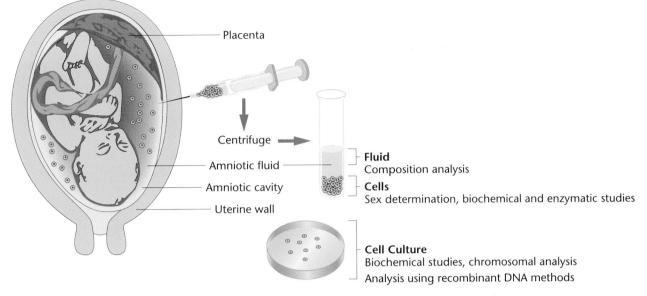

FIGURE 16.6   The technique of amniocentesis. The position of the fetus is first determined by ultrasound, and then a needle is inserted through the abdominal and uterine wall to recover fluid and fetal cells for cytogenetic and/or biochemical analysis.

membrane. The fragments are visualized by hybridization using a probe that spans part of the β-globin gene.

In the example shown in Figure 16.8, the parents (I-1 and I-2) are both heterozygous for the mutation. Digestion of DNA from each parent produces a large band (the mutant allele) and two smaller bands (the normal allele). Their first child (II-1) is homozygous normal because she has only the two smaller bands. The second child (II-2) has sickle-cell anemia; he has only one large

band, and is homozygous for the mutant allele. The fetus (II-3) has a large band and two small bands and is, therefore, heterozygous for sickle-cell anemia. He or she will be unaffected, but will be a carrier (Figure 16.8).

Not every point mutation involving a single-base change alters a restriction site. It is estimated that only about 5 to 10 percent of all point mutations can be detected by restriction analysis. However, if a mutant gene has been well characterized and the mutated region has

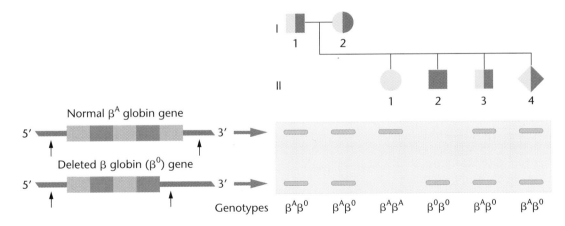

FIGURE 16.7   Diagnosis of beta thalassemia caused by a partial deletion of the β-globin gene. The family pedigree (□ = male, ○ = female, ◇ = fetus) is shown positioned above the individuals' respective genotypes on a Southern blot. The normal β-globin gene (β^A) contains three exons and two introns. The deleted β-globin gene (β^0) has the third exon deleted. Arrows indicate the cutting sites for restriction enzymes used in this analysis. The normal gene produces a larger fragment (shown as the top row of fragments on the Southern blot); the smaller fragments produced by the deleted gene are represented at the bottom of the gel. The genotypes of each individual in the pedigree can be determined from the pattern of bands on the blot, and they are shown below the blot.

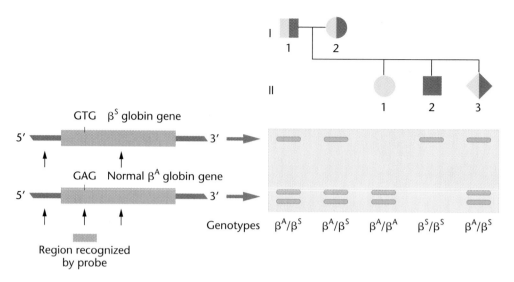

FIGURE 16.8 Southern blot diagnosis of sickle-cell anemia. Arrows represent the location of restriction enzyme cutting sites. In the mutant (β^S) globin gene, a point mutation (GAG → GTG) has destroyed a restriction enzyme cutting site, resulting in a single large fragment on a Southern blot. In the pedigree, the family has one unaffected homozygous normal daughter (II-1), an affected son (II-2), and an unaffected fetus (II-3). The genotypes of each family member can be read directly from the blot, and they are shown below the blot.

been sequenced, it is possible to use synthetic oligonucleotides as probes to detect mutant alleles.

Allele-Specific Nucleotides and Genetic Screening

Under the proper conditions, a probe known as an **allele-specific oligonucleotide (ASO)** will hybridize only with its complementary sequence, and not with other sequences that might vary by as little as a single nucleotide. A newer method using ASOs and PCR analysis is used to screen for sickle-cell anemia. In this method, white blood cells from a blood sample are lysed and heated to denature the DNA. A region of the β-globin gene from these cells is amplified by PCR. An aliquot of the amplified DNA is spotted onto filters, and each filter is hybridized to an ASO (Figure 16.9). After visualization, the genotype can be read directly from the filters. Using an ASO for the normal sequence (Figure 16.9A), homozygous normal (AA) genotypes produce a dark spot (they carry two copies of the normal allele), and heterozygous genotypes (AS) produce a light spot (they carry one copy of the normal allele). Homozygous sickle-cell genotypes will not bind the probe, and no spot will be visualized. Using a probe for the mutant allele (Figure 16.9B), the pattern is reversed. This rapid, inexpensive, and highly accurate technique will probably be the method of choice for diagnosis of a wide range of genetic disorders caused by point mutations.

In cases where the nucleotide sequence of the normal gene is known, and where the molecular nature of the mutant gene is known, ASOs can be used in screening for heterozygous carriers of genetic disorders. About 70 percent of the mutant alleles in cystic fibrosis (CF) are caused by a three-nucleotide deletion (called delta 508), causing the deletion of phenylalanine at position 508 in

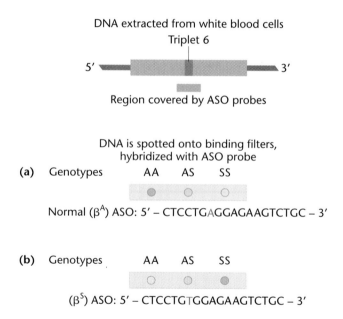

FIGURE 16.9 Genotype determinations using allele-specific oligonucleotides (ASOs). In this technique, the β-globin gene is amplified by PCR using DNA extracted from blood cells. The amplified DNA is denatured and spotted onto strips of DNA binding filters. Each strip is hybridized to a specific ASO and visualized on X-ray film after hybridization and exposure. If all three genotypes are hybridized to an ASO from the normal β-globin gene, the pattern in (a) would be observed: *AA*—homozygous normal hemoglobin has two copies of the normal β-globin gene and would show heavy hybridization; *AS*—heterozygous individuals carry one normal β-globin gene and one mutant gene, and would show weaker hybridization; *SS*—sickle-cell homozygotes carry no normal copy of the β-globin gene, and would show no hybridization to the ASO probe for the normal β-globin gene. (b) The same genotypes hybridized to the probe for the sickle-cell β-globin gene would show the reverse pattern: no hybridization by the *AA* genotype, weak hybridization by the heterozygote (*AS*), and strong hybridization by the homozygous sickle-cell genotype (*SS*).

the gene product. Cystic fibrosis is an autosomal recessive disorder associated with a defect in a protein called the **cystic fibrosis transmembrane conductance regulator (CFTR)**, which regulates chloride ion transport across the plasma membrane. To detect heterozygote carriers for the Δ508 mutation, nucleotide primers are used to make allele-specific oligonucleotides by PCR from the normal allele and the mutant allele. DNA samples prepared from white blood cells of individuals to be tested are applied to a nylon or nitrocellulose filter and hybridized to each of the ASOs (Figure 16.10). In affected individuals, only the ASO made from the mutant allele will hybridize. In heterozygotes, both ASOs will hybridize, and in normal homozygotes, only the ASO from the normal allele will hybridize.

Using ASOs from the Δ508 deletion and two other mutations, heterozygotes for 85 to 95 percent of all CF

mutations can be detected. Because CF affects approximately 1 in 2000 individuals of northern European descent, this technique may be used in population screening to detect and advise heterozygote carriers of their status for CF. However, the cost-effectiveness of undertaking such screening at the present time is in question. A negative result does not eliminate possible carrier status, for not all mutations can be assayed, and it is possible that not all mutations of the CF gene are known. Such screening will probably be commonplace once screening tests can cover 98 to 99 percent of all possible CF mutations.

Animal Models of Human Genetic Disease: Knockout Mice

Knockout mice are genetically altered so that a particular gene (a target gene) has been disrupted and made nonfunctional. The technique is complex and uses a combination of recombinant DNA technology, cell culture, and embryo manipulations. It takes about a year to construct a knockout mouse, and analyzing the effects of the knockout can take even longer. The method was developed by several groups in 1989, including Mario Cappechi at the University of Utah, Oliver Smithies at the University of North Carolina, and Elizabeth Robertson at Harvard Medical School; to date, over 100 different genes have been knocked out in mice.

The technique begins with a cloned target gene from a normal mouse. A marker gene for antibiotic resistance is inserted into this gene, disrupting its function and providing a marker to follow the gene through subsequent procedures. The altered gene is transfected into cultured embryonic stem cells (ES) (Figure 16.11). In some of these cells, the altered gene will replace the normal gene via a recombination event. Cells carrying one copy of the recombined gene and one normal allele are selected for and grown in culture. These genetically altered ES cells are injected into a mouse embryo where they will participate in the formation of adult tissues and organs.

The chimeric embryo is then transferred to a foster mother to complete development. By using coat-color genes as markers, it is possible to select chimeric mice that carry germ cells derived from the genetically altered ES cells. The heterozygous mice can be used to breed mice homozygous for the disrupted gene; these are called knockout mice because a single gene has been mutated or knocked out.

This method makes it possible to create mouse models for human genetic disorders. For example, knockout mice have been created that lack RAG-1 and RAG-2, two genes shown to be important in recombination in anti-

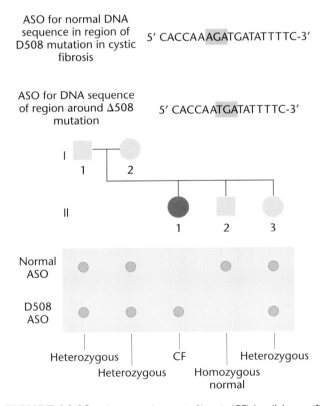

FIGURE 16.10 Screening for cystic fibrosis (CF) by allele-specific oligonucleotides (ASOs). ASOs for the region spanning the most common mutation in CF, a three-nucleotide deletion (Δ508), are prepared from normal CF genes and Δ508 CF genes. In screening, the CF gene is amplified by PCR using DNA extracted from blood samples and spotted on a DNA-binding membrane. The membrane is hybridized to a mixture of the two ASOs. The genotypes of each family member can be read directly from the filter. DNA from I-1 and I-2 hybridizes to both ASOs, indicating that they carry a normal allele and a mutant allele and are therefore, heterozygous. The DNA from II-1 hybridizes only to the Δ508 ASO, indicating that she is homozygous for the mutation and has cystic fibrosis. The DNA from II-2 hybridizes only to the normal ASO, indicating that he carries two normal alleles. II-3 has two hybridization spots, and is heterozygous.

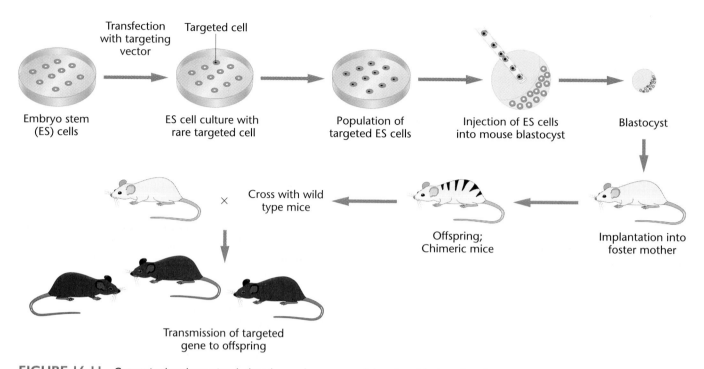

FIGURE 16.11 Generating knockout mice. A cloned normal mouse gene is inactivated by insertion of an antibiotic-resistance gene, which serves as a marker. The altered gene is transferred to cultured embryonic stem cells (ES). In some cells, the altered gene will replace the gene in the ES genome by a recombination event. Cells carrying the altered gene are selected by means of their antibiotic resistance and are grown in culture. These cells are injected into the blastocyst stage of a normal mouse embryo, where they participate in the formation of adult tissues and organs. The chimeric embryo is transferred to a host female to complete development. Coat color genes are used to select chimeric mice that carry germ cells derived from the altered ES cells. These heterozygous mice are bred to normal mice to eventually yield mice that are homozygous for the knocked-out gene. The developmental and functional effects of being homozygous for a mutation in this gene can then be studied in the homozygous progeny.

body genes in the immune system, and the basis for generating antibody diversity. These mice are being studied in both germ-free and normal environments to provide a clearer understanding of how these genes work during the maturation and operation of the immune system.

This technique has also been important in research on the understanding and treatment of human genetic disorders. For example, a mouse model has been created for Gaucher's disease (a progressive and fatal autosomal recessive disorder of lysosomal enzymes). Somewhat surprisingly, the homozygous recessive phenotype is more severe in the mice and leads to death hours after birth. This finding led to the realization that some affected humans also die as newborns, providing new insight into this disorder. A mouse model of cystic fibrosis has been created by the knockout technique, and this will prove valuable for testing gene therapy methods before they are applied to humans.

However, some surprises are also resulting from knockout studies. A knockout mouse lacking the HGPRT enzyme responsible for Lesch–Nyhan syndrome (an X-linked fatal disorder in humans) is perfectly normal and has no detectable behavioral defects. Further work indicates that these mice are more reliant on another purine salvage pathway, one that uses the enzyme adenine phosphoribosyltransferase (APRT), and are therefore more tolerant of HPRT deficiency than humans. If the HPRT-deficient mice are treated with an inhibitor of APRT, they exhibit a pattern of self-injurious behavior. Other work with knockout mice reflect similar findings; it is possible to knock out single genes with little or no phenotypic effect, even though two copies of the mutant gene have serious phenotypic consequences in humans. These finding suggest that species-specific alternative biochemical and developmental pathways can compensate for the loss of single genes.

Now under development are a second generation of mice carrying multiple gene knockouts, involving several genes in a biochemical pathway, and mice that will have genes knocked out in selective tissues, or will carry disabled instead of inactivated genes.

Recombinant DNA techniques have also been used to accomplish the reverse of gene knockouts, a technique called *gene therapy*.

Gene Therapy

Methods for the isolation and cloning of specific genes originally developed as a research tool are now being used to treat genetic disorders by the transfer of normal human alleles in a process known as *gene therapy*. Gene products, such as insulin, have been used for decades in therapeutic treatment. Gene therapy takes this process one step further, and seeks to modify the genome of somatic cells by transferring normal copies of genes that will produce adequate amounts of a normal gene product, the action of which will correct a genetic disorder. The delivery of these structural genes and their regulatory sequences is accomplished using a vector or gene transfer system.

As described in Chapter 15, several methods can be used to transfer genes and their regulatory sequences into human cells. These include the use of viruses as vectors, chemically assisted transfer that promotes the transfer of genes across cell membranes, and the fusion of cells with artificially constructed vesicles containing cloned DNA sequences.

At present, the most common method of gene transfer uses the Maloney murine leukemia virus as a vector to carry out gene transfer (review Figure 15.32). The *gag, pol* and *env* genes are removed from the virus, and the cloned human gene is inserted. The resulting construct is packaged into a viral protein coat; the retroviral vector can infect cells, but is replication-deficient because of the missing viral genes. The vector is mixed with a suspension of target cells, and enters the cell by means of a cell-surface receptor. Once in the cytoplasm, the viral genome carrying the cloned human gene integrates into one of the chromosomes in the target cell.

This experimental form of treatment is not undertaken without extensive review. At the local level, a review board at the medical center or hospital where the gene therapy will take place must review and approve the proposal and monitor the trial to protect the interests of the patient. At the national level, there are several levels of review at the National Institutes of Health (NIH). Proposals are reviewed by panels composed of scientists, lawyers, ethicists, and others who review and approve the trials. Finally, the director of NIH must approve the procedure.

The guidelines for undertaking gene therapy are well established and include several requirements. First, the gene must be isolated and available for transfer, usually through cloning. Second, there must be an effective means of transferring the gene. At present, most trials use retroviral vectors, although other methods, including adenovirus vectors and physical and chemical techniques, are also employed. Third, the target tissue must be accessible for gene transfer. The first generation of gene therapy trials used white blood cells or their precursors as an available target tissue. Fourth, there must be no other form of effective therapy available, and the gene therapy must not harm the patient.

Several heritable disorders are currently being treated with gene therapy, including severe combined immunodeficiency (SCID), familial hypercholesterolemia, cystic fibrosis, and muscular dystrophy. Trials to treat other disorders are currently being developed.

Severe Combined Immunodeficiency (SCID)

Severe combined immunodeficiency (SCID) is a genetic disorder in which affected individuals have no functioning immune system and usually die from otherwise minor infections (see the "Boy in the bubble" described in Chapter 22). An autosomal form of SCID is caused by a defect in a gene that encodes the enzyme **adenosine deaminase (ADA)**. Affected individuals lack functional T and B cells, which are important components of the immune system. Beginning in 1990, a young girl with this form of SCID was the first patient ever to receive gene therapy. A second patient began treatment a short time later. Treatment starts with the isolation of a subpopulation of T cells from the patient (Figure 16.12). These cells are mixed in solution with a genetically modified retrovirus that carries a normal copy of the human ADA gene. The virus infects the T cells, inserting a functional copy of the ADA gene into the T-cell's genome (Figure 16.12). The modified T cells are grown in the laboratory to ensure that the gene is active, and the patient is treated by injecting a billion or so genetically altered T cells into the bloodstream.

Familial hypercholesterolemia is another genetic disorder being treated by gene therapy. This autosomal dominant condition affects about 1 in 500 individuals, and it is characterized by the inability to metabolize dietary fats properly. It results in elevated levels of blood cholesterol, increased deposition of cholesterol in arterial plaques, and premature death from coronary heart disease. Affected individuals have defective or absent cell-surface receptors that remove cholesterol from the bloodstream.

For gene therapy, a lobe of the liver is removed from an affected individual and dissociated into single cells. A genetically modified Maloney retrovirus, carrying a human gene for cell-surface cholesterol receptors, is used to infect the liver cells. The treated cells are then injected into the patient through a vein that supplies the liver. The injected cells are carried by the circulating blood into the liver, where they take up residence, and express the gene for the cell-surface receptor. This is followed by a reduction in blood cholesterol levels, often to the normal range.

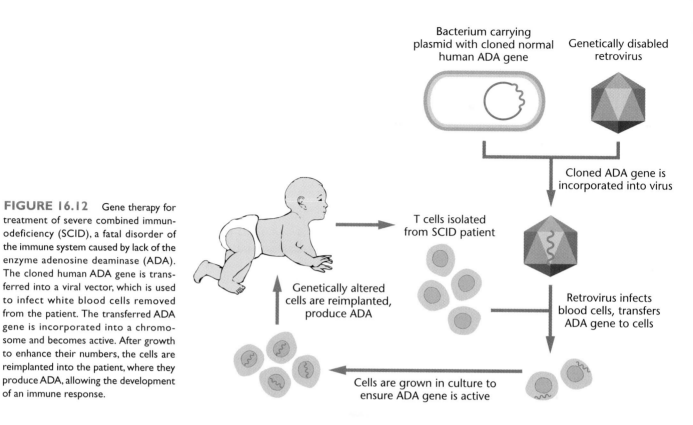

FIGURE 16.12 Gene therapy for treatment of severe combined immunodeficiency (SCID), a fatal disorder of the immune system caused by lack of the enzyme adenosine deaminase (ADA). The cloned human ADA gene is transferred into a viral vector, which is used to infect white blood cells removed from the patient. The transferred ADA gene is incorporated into a chromosome and becomes active. After growth to enhance their numbers, the cells are reimplanted into the patient, where they produce ADA, allowing the development of an immune response.

New Vectors and Target-Cell Strategies

The first round of gene therapy trials used genetically altered retroviruses as vectors for gene transfer. These vectors were constructed to allow the virus to retain its ability to infect cells, but not to replicate after introduction into the body. However, gene therapy using these vectors has been only marginally successful. In addition, these vectors have several drawbacks that limit their widespread use in gene therapy. First, integration of the retroviral genome (including the cloned human gene) into the host cell genome occurs only if the host cells are replicating their DNA. Under *in vivo* conditions, there is little DNA synthesis in many highly differentiated cell types that are suitable target tissues. Second, insertion of viral genomes into the host chromosome can inactivate or mutate an indispensable gene. Third, retroviruses have a low cloning capacity and cannot carry inserted sequences much larger than 8 kb. Many human genes, even without

introns, exceed this size. Finally, there is the possibility of producing an infectious virus if a recombination event takes place between the vector and retroviral genomes already present in the host cell. For these reasons, other viral vectors and strategies for targeting cells are being developed. Some of the second-generation vectors are listed in Table 16.1.

Vectors derived from adenoviruses have the advantage of being able to transform cells in the absence of DNA replication. In addition, adenovirus vectors selectively target cells in the lung, and they are being used in trial therapies for cystic fibrosis and emphysema. At the moment, the central issue concerning the use of adenovirus vectors is the lack of information about safety. Among the dangers is that co-infection of a genetically engineered adenovirus along with an intact adenovirus might spread the transferred gene to cells in nontarget tissues.

TABLE 16.1 New Viral Vectors for Gene Therapy

	Adenovirus	Adeno-Associated Virus	Herpes Virus
CELL TARGETS	Lung cells, cells of respiratory tract	Fibroblasts, T cells	Nerve cells, glial cells
CLONING CAPACITY	7–35 kb	3 kb	Up to 150 kb
INTEGRATION INTO CHROMOSOMAL DNA	No	Yes	No, but retained in nucleus

Genetically engineered herpes viruses are being developed as vectors to target selective cells of the nervous system. Because these cells do not divide, it is not necessary for this vector to integrate into the host genome to be retained in cells. This will prevent the danger of insertional mutagenesis associated with the use of retroviruses.

Gene therapy is being used or contemplated as a treatment for skin cancer, breast cancer, brain cancer, and AIDS. In the last few years, at least a dozen biotech companies have been founded specifically to develop products for gene therapy, and a number of such products should reach the marketplace soon. In the twenty-first century, gene therapy will be a commonplace method of treatment for a large number of disorders.

DNA Fingerprints

Restriction fragment-length polymorphisms (RFLPs) have been used to distinguish normal from mutant copies of single genes and can be used as genetic markers, since they are inherited in a codominant fashion.

Minisatellites and VNTRs

A second type of restriction fragment-length polymorphism arises from variations in the number of tandemly repeated DNA sequences present between two restriction

enzyme sites. These sequences are derived from **minisatellites**, which are clusters of sequences from 2 to 100 nucleotides in length. For example, the base sequence

GGAAGGGAAGGGAAGGGAAG

consists of four tandem repeats of the 5-nucleotide sequence GGAAG. Clusters of such sequences are widely dispersed in the human genome. Typically, each repeat contains between 14 to 100 nucleotides, and the number of repeats at each locus ranges from 2 to more than 100. These loci are known as **variable-number tandem repeats (VNTRs)**. The number of repeats at a given locus is variable, and each variation is a VNTR allele. Many loci have dozens of alleles each; as a result, heterozygosity is common.

The pattern of bands produced when VNTR sequences are cut with restriction enzymes and visualized by Southern blotting is known as a **DNA fingerprint** (Figure 16.13). These patterns are the equivalent of fingerprints because the pattern of bands is always the same for a given individual, no matter what tissue is used as the source of DNA, but the pattern varies from individual to individual. There is so much individual variation in the band pattern that, theoretically, each person's pattern is unique, like the individualized pattern in a fingerprint (Figure 16.13). DNA fingerprint analysis can be performed on very small samples of material (less than 60 microliters [µl] of blood) and on samples that are quite old (VNTR analysis has

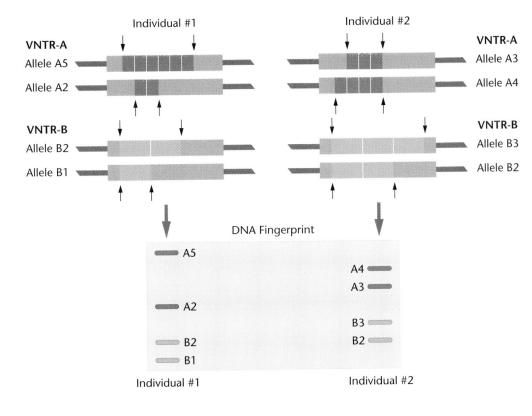

FIGURE 16.13 VNTR loci and DNA fingerprints. VNTR alleles at two loci (A and B) are shown for each individual. Arrows mark restriction cutting sites flanking the VNTRs. Restriction digestion produces a series of fragments that can be detected as bands on a Southern blot (below). Because of differences in the number of repeats at each locus, the overall pattern of bands is distinct for each individual, even though one band is shared (the band representing the B2 allele). Such a pattern is known as a DNA fingerprint.

been performed on Egyptian mummies over 2400 years old), thus increasing its usefulness in legal cases.

Forensic Applications

In the United States, DNA fingerprints have been used as evidence in criminal trials since 1988 and have also been used to settle immigration cases, disputes involving pure-bred dogs, paternity cases, and in animal conservation studies. At present, just over a dozen different probes are widely used in forensic testing (Figure 16.14). Most of these are on different chromosomes, and each probe can be used to visualize the VNTR pattern at a particular locus. Standard forensic tests employing four to six different VNTR probes are used in developing a detailed DNA fingerprint profile.

Results of DNA fingerprints are analyzed and interpreted using statistics, probability, and population genetics. The frequency for each VNTR allele is calculated, and these are multiplied together to give a combined frequency, which is a calculation of how often the observed DNA fingerprint might occur in the general population. The use of a number of VNTR probes increases the accuracy of the combined frequency. For example, if the frequency of locus 1 is determined to be 1 in 333, and the frequency of locus 2 is 1 in 83, their combined frequency is 1 in 28,000. This overall frequency is not very low, and may not serve as convincing evidence. If however, the frequency of a third locus (1 in 100) and a fourth locus (1 in 25) are included in the calculations, the combined frequency becomes 1 in 70 million.

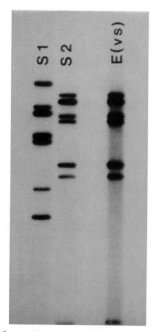

FIGURE 16.14 DNA fingerprinting in a forensic case. The DNA profile of suspect 2 (S2) matches that of the blood sample obtained as evidence E(VS).

For a time, the use of DNA fingerprints in criminal trials was the subject of some dispute. The controversy centered on two aspects of DNA testing, one technical and the other scientific. On the technical side, there was, at one time, little standardization of the methods used to perform DNA fingerprinting, and the quality control in some commercial laboratories was called into question. Scientifically, the chance that a given VNTR pattern is unique enough to be the sole source of such a pattern among individuals in a population depends on the frequency of specific VNTR alleles in various populations. There is now an extensive collection of population data on the frequency of VNTR alleles in unrelated individuals belonging to a large number of racial and ethnic groups. By the end of 1994, these issues were largely resolved, and today there is little dispute in the scientific community over the validity of DNA fingerprints as forensic evidence.

Genome Analysis

Over the past 90 years, geneticists have developed efficient methods for generating and mapping mutants. One drawback to these methods is that genes can be characterized only if mutants are isolated. To exhaustively characterize a genome, at least one mutation for each of the genes in the genome is required, and this is often a difficult, labor-intensive task.

Geneticists are now using recombinant DNA techniques in a direct approach to genetic analysis. Instead of screening for mutants and constructing genetic maps, a genomic library is established, and overlapping clones are assembled to establish genetic and physical maps encompassing the entire genome. The final step involves sequencing the entire genome. The result is a physical map, with all genes in the genome identified by location and their nucleotide sequence. Analysis of derived amino acid sequence and the generation of gene-specific mutants can be used to study the function of the gene product.

Genome projects are underway for several species (Table 16.2). These programs are now incorporated as part of the Human Genome Project (discussed below). These organisms were selected for several reasons, including their genome size, the availability of detailed genetic maps, and the value of making comparisons about gene number, location and function across species ranging from bacteria to humans.

Two basic approaches are used in genome projects: (1) the *bottom-up* approach, which starts with small, overlapping clones and assembles them into larger segments that eventually cover the entire genome; and (2) the *top-down* approach, which isolates very large, randomly cloned DNA segments that cover the entire genome; these are broken down into smaller units for mapping and

TABLE 16.2 Organisms Studied by Genome Projects

Organism	Genome Size (bp)	Estimated Number of Genes
E. coli (bacterium)	4×10^6	4,000
Saccharomyces cerevisciae (yeast)	1.5×10^7	6,000
Arabidopsis thaliana (plant)	1×10^8	25,000
Caenorhabditis elegans (nematode)	1×10^8	13,000
Drosophila melanogaster (insect)	1.2×10^8	10,000
Mus musculus (mouse)	3×10^9	80,000
Homo sapiens (human)	3.2×10^9	80,000

sequence analysis. A brief discussion of each of these approaches will serve as an introduction to the Human Genome Project and its goal of isolating and identifying the 50,000 to 100,000 genes in the human genome.

Model Organisms: The *E. coli* Project

The bottom-up method is being used to analyze the *E. coli* genome. A cloned genomic library of the *E. coli* K12 strain was prepared in lambda vectors by Yuji Kohara and his colleagues. Clones from this library were extensively characterized by restriction mapping, and computer analysis of the maps was used to identify overlapping clones by the presence of shared restriction sites (Figure 16.15). The resulting contiguous segment of DNA covered by the overlapping clones is called a **contig**. In the initial analysis, some 70 contigs were identified, covering 94 percent of the genome. Finding clones that spanned the small gaps between contigs eventually closed the gaps. The contigs were arranged in a physical map that covers the entire 4700 kilobases (kb) of DNA in the *E. coli* genome.

The second stage of the project involves sequencing the clones in the contig library. To date, about 30 percent of the genome has been sequenced, covering about 1.5 million base pairs of DNA. At this stage, it is fair to ask what has been found in this analysis. In a recent study, Fred Blattner and colleagues reported the nucleotide sequence of 91.4 kb of *E coli* DNA. They found that genes (identified by the presence of open reading frames within their nucleotide sequence) accounted for 84 percent of the nucleotides in this region. About half of the genes identified had already been identified by conventional mutant screening and mapping, but the rest were previously unknown. This result indicates that years of intensive efforts to map bacterial genes by conventional genetic analysis have identified only about half the genes in the *E. coli* genome, and that many more genes remain to be identified. In addition, this work demonstrates that a recombinant DNA-based approach using physical maps at the nucleotide level is effective in finding and mapping all the genes in a genome.

The *Drosophila* Genome

For the top-down approach being used to analyze the *Drosophila* genome, a cloned genomic library has been constructed using yeast artificial chromosome (YAC) vectors. The library contains the equivalent of 1.7 genomes carried in 965 YACs. The average clone in each YAC is about 200 kb, a span that encompasses many genes. The location of the *Drosophila* DNA in each YAC has been determined by *in situ* hybridization to the banded polytene chromosomes of larval salivary glands (Figure 16.16). The average YAC covers six to eight chromosome bands; contigs are constructed by identifying YACs that have

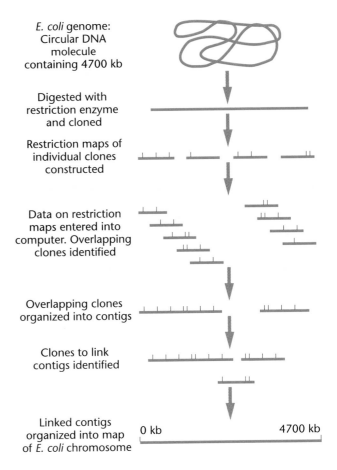

FIGURE 16.15 Bottom-up strategy for the *E. coli* genome project. The 4700-kb *E. coli* chromosome is digested with restriction enzymes into small fragments that are cloned into a vector. Individual clones are characterized by restriction mapping. Information about these restriction maps is analyzed by computer to identify overlapping clones. These overlapping clones are organized into contiguous segments of DNA called *contigs*. Next, more clones are screened to identify those spanning contigs. These are used to reduce gaps, bringing the number of contigs from 70 to 7 and finally to 1, which corresponds to the entire *E. coli* chromosome. Individual clones are now being sequenced, and eventually the nucleotide sequence of the entire chromosome will be known.

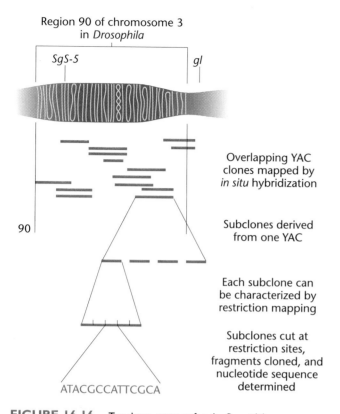

Region 90 of chromosome 3
in *Drosophila*

SgS-5   gl

90

Overlapping YAC
clones mapped by
in situ hybridization

Subclones derived
from one YAC

Each subclone can
be characterized by
restriction mapping

Subclones cut at
restriction sites,
fragments cloned, and
nucleotide sequence
determined

ATACGCCATTCGCA

FIGURE 16.16   Top-down strategy for the *Drosophila* genome project. A genomic library is constructed using large (~200 kb) fragments in YACs. The physical location of YACs are mapped to polytene chromosomes by *in situ* hybridization. Shown is region 90, located on chromosome 3. This region contains about 890 kb of DNA and approximately 34 genes. Approximately 14 percent of the region is transcribed as RNA. Each YAC covers several polytene bands and presumably includes several genes. Individual YACs are digested with restriction enzymes and the individual fragments are subcloned. The subclones are characterized by restriction mapping, and restriction fragments from these subclones are used for nucleotide sequence analysis.

two or more chromosome bands in common. The YACs examined to date cover about 80 percent of the euchromatic region of the genome in 161 contigs, with just over 150 gaps to be closed.

To study individual genes or regions, YACs are broken down by restriction digestion and subcloned into other vectors. Identifying genes by DNA sequencing in *Drosophila* is more difficult than in *E. coli*, in part because the relatively large size of the genome (see Table 16.2 to compare the genome sizes) makes sequencing a less efficient method of mapping. To map genes at the nucleotide level, each *Drosophila* gene is defined as a stretch of unique DNA sequence (each gene should occur only once in a genome). Therefore, any unique short DNA segment within a gene can be used as a marker or **sequence tagged site (STS)** to place a gene on the physical map. Note that this can be accomplished without knowing the nature of the gene, the gene product, or its phenotype.

Three sources of STS markers are used in this approach. First, a large number (over 3800) of mutant genes have been identified and mapped in *Drosophila* by conventional analysis, and many of these have been completely or partially sequenced. STS markers have been selected from the sequenced genes already available, and these markers can be placed on the physical map by *in situ* hybridization to serve as important reference points. A second set of STS markers has been derived from cDNA clones made from genes expressed at high levels in certain tissues. These markers, called **expressed sequence tags (ESTs)**, can also be placed on the physical map by *in situ* hybridization. This generates map positions for some genes that have not been recovered as mutants. Finally, *Drosophila* strains from the wild contain transposable DNA segments called **P elements**. Introduced into laboratory strains, the sites where these elements insert into the genome can be used as STS markers.

A computerized data base of information about the *Drosophila* mapping project that includes genes, gene products, assigned functions, map positions, and chromosome aberrations is available on the World Wide Web and can be accessed by investigators and students. At its completion, the *Drosophila* genome project will provide a molecular map of the *Drosophila* genome at the nucleotide level for research and teaching purposes.

Human Genome Project

The **Human Genome Project** is an international, coordinated effort to construct a physical map of the 3 billion base pairs in the haploid human genome. In the United States, consideration of a genome project began in 1986, and in 1988 the National Institutes of Health and the U.S. Department of Energy created a joint committee to develop a five-year plan. Called the Human Genome Project, the effort got under way in 1990. Other countries, notably France, Britain, and Japan, began similar projects, which are now coordinated by an international organization, the Human Genome Organization (HUGO).

This complex task is proceeding in a series of well-defined stages. The first stage involves the construction of high-resolution genetic maps for each of the 22 autosomes and the sex chromosomes (Figure 16.17). These maps are constructed using identified genes, RFLP markers, STS sites, and ESTs. A genetic map for all the chromosomes, incorporating about 15,000 markers, was completed in 1995. This genetic map, with an average distance of 2 million base pairs of DNA between markers, is being used to sort out and organize the contigs being generated by physical mapping.

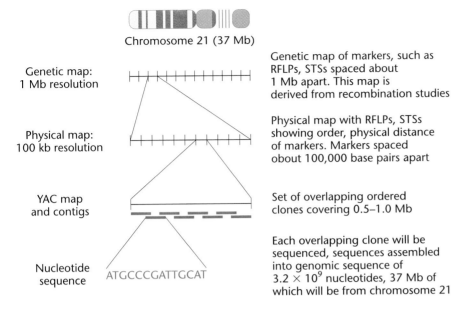

Chromosome 21 (37 Mb)

Genetic map:
1 Mb resolution

Genetic map of markers, such as RFLPs, STSs spaced about 1 Mb apart. This map is derived from recombination studies

Physical map:
100 kb resolution

Physical map with RFLPs, STSs showing order, physical distance of markers. Markers spaced obout 100,000 base pairs apart

YAC map
and contigs

Set of overlapping ordered clones covering 0.5–1.0 Mb

Each overlapping clone will be sequenced, sequences assembled into genomic sequence of 3.2×10^9 nucleotides, 37 Mb of which will be from chromosome 21

Nucleotide
sequence

ATGCCCGATTGCAT

FIGURE 16.17 An overview of the strategy used in the Human Genome Project. The first goal, achieved in 1995, was to have a genetic map of each chromosome, with markers spaced at distances of about 1 Mb (1 million base pairs of DNA). This work was accomplished by finding markers such as RFLPs and STSs and assigning them to chromosomes. Once assigned to chromosomes, the markers' inheritance was observed in heterozygous families to establish the order and distance between them (a genetic map). In the second stage, the goal is to prepare a physical map of each chromosome (our example uses chromosome 21, the smallest chromosome) containing the location of markers spaced about 100,000 base pairs apart. This goal will probably be achieved early in 1997. The third stage involves the construction of a set of overlapping clones, in yeast artificial chromosomes (YACs) or other vectors that cover the length of the chromosome. The last stage will be the sequencing of the entire genome. Sequencing on selected parts of the genome has started, with about 3 percent of the genome expected to be sequenced in the next few years.

The second level of the Human Genome Project is the construction of physical maps, using either the top-down approach beginning with chromosomes isolated in somatic cell hybrids, or the bottom-up approach of assembling contigs from DNA segments randomly cloned in cosmids or YACs. The goal is the creation of a physical map of the entire genome consisting of 30,000 STSs spaced at intervals of about 100 kb.

An STS-based physical map covering 95 percent of the genome, and encompassing over 23,000 STS sites, has been recently assembled. The STSs in this map contain 200 to 500 nucleotides, and once identified, they can be stored as a nucleotide sequence rather than cloned fragments of DNA. The sequences are available on the World Wide Web and, for use as a probe, any STS can be generated from the sequence information by polymerase chain reaction (PCR). Sequence tagged sites can serve as landmarks for researchers using YACs, phage or plasmid clones, allowing information from many sources to be integrated into maps with higher and higher resolution. The current map is more than halfway to the goal of 30,000 STSs, and that goal should be reached by 1997 or 1998.

The ultimate goal of the project is sequencing the 3.2 billion nucleotides in the human genome, a task that in itself may take a decade or more and require the development of new technology for sequencing DNA, as well as new technology for information storage, analysis, and retrieval. This last stage of the project has started, with the establishment of six centers in the United States to improve technology related to sequencing, and to begin large-scale sequencing of the genome. Regions on eight chromosomes have been selected for sequencing, covering about 3 percent of the genome. It is expected that these regions will be completely sequenced by 1999. A similar project, covering other regions of the genome, is underway in Europe.

After the Genome Projects

Several genome projects have already reached their goals. The genomes of *Hemophilus influenzae* and *Mycoplasma genitalium* were both completely sequenced in 1995. The *H. influenzae* genome is 1830 kb long, and that of *M. genitalium* is 580 kb long. Early in 1996, sequencing was completed on the 12,500-kb genome of the yeast, *Saccharomyces cerevisiae*. The nucleotide sequence of the *E. coli* genome (4720 kb) is not complete at this writing, but it should be finished in the near future. Work on sequencing the human genome is proceeding, and by 1998, regions on several chromosomes, covering some 3 percent of the genome, should be completely sequenced.

The work on *M. genitalium* is of interest for several reasons, some of them technical, some of them biological. First, the project employed a new method of assembling the genome sequence. In conventional genome sequencing schemes, a genomic library is first organized into a series of ordered, overlapping clones (contigs). Each of these larger clones is then broken down into smaller clones, which are sequenced and put into the correct order. For sequencing the *M. genitalium* genome, a genomic library was prepared from restriction digests of DNA. From this library, randomly selected fragments were sequenced, and the nucleotide sequence of 8472 such fragments was analyzed by an assembler software program, and organized into a complete genome sequence.

This "shotgun" sequencing method, also used on the *H. influenzae* genome, greatly shortens the time required for sequencing a genome, and by-passes the need to prepare physical maps. The genome of *M. genitalium* was sequenced in about eight months, from DNA extraction to completion of the manuscript for publication. In fact, all the sequencing reactions were carried out in eight weeks. Investigators are now working to determine whether the shotgun method can be adapted to sequencing the human genome, which would reduce the estimates for the time and cost of generating this information.

From a biological perspective, this sequencing project is of interest because *M. genitalium* has one of the smallest genomes of any free-living organism, with 482 genes. It uses 90 different proteins for translation, and about 30 gene products for DNA replication. Because of the small size of the genome, these may represent close to the minimum number of genes necessary to complete such tasks. Somewhat surprisingly, 140 of the 482 genes (about 30% of the genome) encode membrane-inserted proteins, emphasizing the importance of interaction between the cell and its environment for survival. Further work on this organism and those with similar-sized genomes may help define the minimum number of genes necessary for life.

Although the flow of information on the human genome is still in its early stages, several insights into human biology and genetic disorders have already been gained. Previously unknown mechanisms of mutation, including the expansion of trinucleotide repeats and an increase in gene dosage by duplication of small regions of chromosomes have been identified. In addition, it is now known that different mutations in a single gene can give rise to different genetic disorders, and that a single phenotype can be produced by mutations in different genes. For example, mutations in a gene called RET can give rise to any of four different genetic disorders, and mutation in any of several, related keratin genes can cause diseases with skin blistering as a phenotype.

In addition to new insights into human biology, the genome project has focused attention on the ethical, legal and social consequences of genomic research. These issues include the application of genetic testing, privacy and the fair use of genetic information in insurance, employment and health care. The ELSI (Ethical, Legal and Social Implications) Program of the Human Genome Project has been set up to address these issues and to involve scientists, health professionals, policy makers and the public in formulating policy recommendations and legislation.

Biotechnology

Although recombinant DNA techniques were originally developed to facilitate basic research into gene organization and regulation of expression, scientists have not been blind to the commercial possibilities of this technology. As a result, researchers have participated in the formation of biotechnology companies using recombinant DNA technology to develop products including hormones, clotting factors, herbicide-resistant plants, enzymes for food production, and vaccines. In the last decade, the biotechnology industry has grown into a multibillion dollar segment of the economy.

Insulin Production

The first human gene product manufactured using recombinant DNA and licensed for therapeutic use was human insulin, which became available in 1982. Insulin is a protein hormone that regulates sugar metabolism, and an inability to produce insulin results in diabetes, a disease that in its more severe form affects more than 2 million individuals in the United States.

Clusters of cells embedded in the pancreas synthesize a precursor peptide known as **preproinsulin.** As this polypeptide is secreted from the cell, amino acids are cleaved from the end and the middle of the chain to produce the mature insulin molecule, which consists of two polypeptide chains (the A and B chains), joined by disulfide bonds.

The initial method for producing insulin by recombinant DNA methods is instructive, as it shows both the promises and difficulties of this technology. Oligonucleotides were used to construct synthetic genes for the A and B subunits. The A subunit has 21 amino acids, and the B subunit has 30; the oligonucleotides encoding these peptides are 63 and 90 nucleotides, respectively. Each synthetic oligonucleotide was inserted into a vector at a position adjacent to a gene encoding the bacterial form of the enzyme, β-galactosidase. When transferred to a bacterial host, the modified β-galactosidase gene was transcribed and translated. The product is a fusion polypeptide, consisting of the amino acid sequence for β-galactosidase attached to the amino acid sequence for one of the insulin subunits (Figure 16.18). The fusion proteins were purified from bacterial extracts and treated with cyanogen bromide, which cleaves the fusion protein to produce β-galactosidase and one of the insulin subunits.

The fusion gene was genetically engineered to insert a methionine at the junction between β-galactosidase and the insulin subunit. Treatment with cyanogen bromide cleaves proteins at methionine sites, releasing an intact insulin subunit. Each insulin subunit was produced sepa-

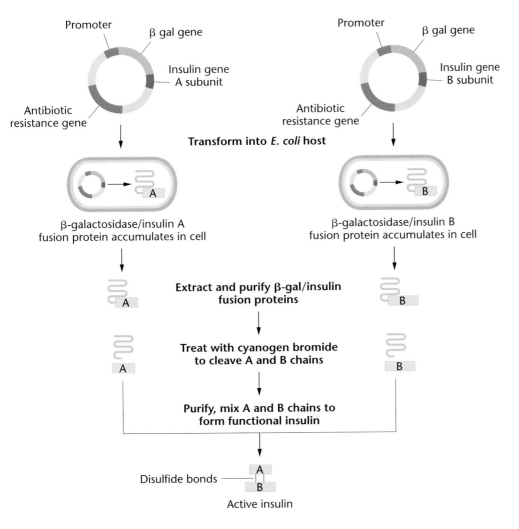

Promoter
β gal gene
Insulin gene
A subunit
Antibiotic
resistance gene

Promoter
β gal gene
Insulin gene
B subunit
Antibiotic
resistance gene

Transform into *E. coli* host

A

B

β-galactosidase/insulin A
fusion protein accumulates in cell

β-galactosidase/insulin B
fusion protein accumulates in cell

A

B

Extract and purify β-gal/insulin
fusion proteins

A

B

Treat with cyanogen bromide
to cleave A and B chains

A

B

Purify, mix A and B chains to
form functional insulin

A
B
Disulfide bonds

Active insulin

FIGURE 16.18 Method used to synthesize recombinant human insulin. Synthetic oligonucleotides encoding the insulin A and B chains were inserted at the tail end of a cloned *E. coli* β-galactosidase (β-gal) gene. These recombinant plasmids were transferred to *E. coli* hosts where the β-gal/insulin fusion protein was synthesized and accumulated in the host cells. Fusion proteins were extracted from the host cells and purified. Insulin chains were released from the β-galactosidase by treatment with cyanogen bromide. The insulin subunits were purified and mixed to produce a functional insulin molecule.

rately by this method. When mixed together, the two subunits spontaneously unite, forming an intact, active insulin molecule.

A number of genetically engineered proteins for therapeutic use have been produced by similar methods or are in clinical trials (Table 16.3). In most cases, these human proteins are produced by cloning a human gene into a plasmid vector and inserting the construct into a bacterial host. After ensuring that the transferred gene is active, large quantities of the transformed bacteria are produced, and the human protein is recovered and purified.

Pharmaceutical Products in Animal Hosts

Bacterial hosts were used to produce the first generation of recombinant proteins, even though there are some disadvantages in using prokaryotic hosts to synthesize eukaryotic proteins. Bacterial cells are unable to process and modify eukaryotic proteins, and they cannot add sugars and phosphate groups that are often needed for full biological activity. In addition, eukaryotic proteins produced

in prokaryotic cells often do not fold into the proper three-dimensional configuration, and as a result, they are inactive. To overcome these difficulties and to increase yields, second-generation products are being made in eu-

TABLE 16.3 Genetically Engineered Pharmaceutical Products Available or in Clinical Testing

Gene Product	Condition Being Treated
Atrial natriuretic factor	Heart failure, hypertension
Epidermal growth factor	Burns, skin transplants
Erythropoietin	Anemia
Factor VIII	Hemophilia
Gamma interferon	Cancer
Granulocyte colony-stimulating factor	Cancer
Hepatitis B vaccine	Hepatitis
Human growth hormone	Dwarfism
Insulin	Diabetes
Interleukin-2	Cancer
Superoxide dismutase	Transplants
Tissue plasminogen activator	Heart attacks

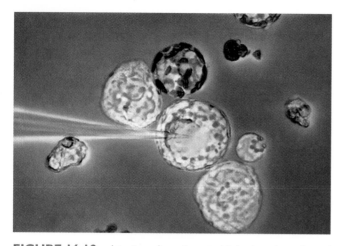

FIGURE 16.19 Injection of genetic material. A micropipette is used to transfer cloned genes into the nucleus of a mammalian zygote. The injected zygote will then be transferred to the uterus of a surrogate mother for development.

karyotic hosts. As an alternative to bacterial hosts or even mammalian cell cultures, human proteins such as alpha-1-antitrypsin are being produced in the milk of livestock, as described below.

A deficiency of the enzyme alpha-1-antitrypsin is associated with the heritable form of emphysema, a progressive and fatal respiratory disorder common among those of European ancestry. To produce alpha-1-antitrypsin by genetic engineering, the human gene was cloned into a vector at a site adjacent to a sheep DNA sequence that regulates expression of milk-associated proteins. This sequence, called a *promoter*, limits expression of its adjacent gene to mammary tissue. This fusion gene was then microinjected into *in vitro* fertilized sheep oocytes (Figure 16.19), which in turn were implanted into foster mothers. The resulting transgenic sheep developed normally and, after mating, produced milk that contained high concentrations of functional alpha-1-antitrypsin. This human protein is present in concentrations up to 35 grams per liter of milk. It is easy to envision that a small herd of lactating sheep could easily provide an adequate supply of this protein and that herds of other transgenic animals, acting as biofactories, might become part of the pharmaceutical industry.

Herbicide-Resistant Crop Plants

In agriculture, gene transfer vectors have been used to transfer traits for herbicide resistance to crop plants. The herbicide glyphosate is widely used for weed control, but it cannot be used on fields containing crop plants because it kills them along with the weeds. Weed growth is a serious agricultural problem, and damage from weeds reduces the yield on many crops by more than 10 percent.

As a herbicide, glyphosate is effective at very low concentrations, is not toxic to humans, and is rapidly degraded by soil microorganisms. At the molecular level, glyphosate works by inhibiting the action of a chloroplast enzyme called EPSP synthase, active in amino acid synthesis. Deprived of these vital amino acids, plants wither and die.

Glyphosate resistance can be generated by increasing the synthesis of EPSP synthase. To do this, a fusion gene was created by cloning the EPSP synthase gene into a vector under control of a promoter sequence from a plant virus. This fusion gene was put into a Ti vector, which was in turn transferred into the bacterium *A. tumifaciens* (Figure 16.20). Plasmid-carrying bacteria were used to infect cells in discs cut from plant leaves. Calluses formed from these discs were then selected for their ability to grow on glyphosate. Transgenic plants generated from glyphosate-resistant calluses were grown and sprayed with glyphosate at concentrations four times higher than needed to kill wild-type plants. The transgenic plants overproducing the EPSP synthase grew and developed, while the control plants withered and died. Glyphosate-resistant corn is expected to be available for planting in 1997.

Similar methods have been used to transfer resistance to virus infection, insects, and drought to crop plants. Other work has been directed at improving the nutritional value of crops such as soybeans and corn. Many of these projects are in the developmental stage, but one of the first genetically engineered plants that has reached the marketplace is a transgenic tomato that has improved flavor and ripening characteristics. Other transgenic products will reach the marketplace over the next few years.

Transgenic Plants and Vaccines

One of the potentially most valuable applications of recombinant DNA technology is the production of vaccines. Vaccines stimulate the immune system to produce antibodies against a disease-causing organism and thereby confer immunity against the disease (Chapter 22). Two types of vaccines are commonly used: **inactivated vaccines**, prepared from killed samples of the infectious virus or bacteria, and **attenuated vaccines**, which are live viruses or bacteria that can no longer reproduce and cause disease when present in the body.

Using recombinant DNA technology, a new type of vaccine called a **subunit vaccine** is being produced. These vaccines consist of one or more surface proteins of the virus or bacterium. The protein acts as an antigen to stimulate the immune system to make antibodies against the virus or bacterium. One of the first licensed subunit vaccines is for a surface protein of hepatitis B, a virus that

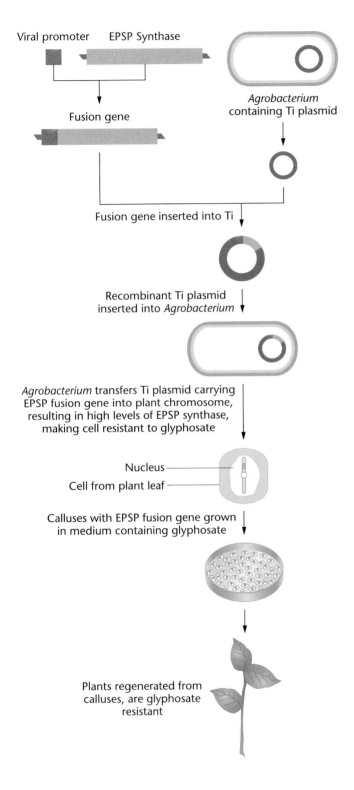

Viral promoter   EPSP Synthase

Fusion gene

Agrobacterium containing Ti plasmid

Fusion gene inserted into Ti

Recombinant Ti plasmid inserted into *Agrobacterium*

Agrobacterium transfers Ti plasmid carrying EPSP fusion gene into plant chromosome, resulting in high levels of EPSP synthase, making cell resistant to glyphosate

Nucleus

Cell from plant leaf

Calluses with EPSP fusion gene grown in medium containing glyphosate

Plants regenerated from calluses, are glyphosate resistant

FIGURE 16.20 Gene transfer of glyphosate resistance. The EPSP gene is fused to a promoter from cauliflower mosaic virus. This chimeric or fusion gene is then transferred to a Ti plasmid vector, and the recombinant vector is inserted into an *Agrobacterium* host. *Agrobacterium* infection of cultured plant cells transfers the EPSP fusion gene into a plant cell chromosome. Cells that acquire the gene are able to synthesize large quantities of EPSP synthase, making it resistant to the herbicide glyphosate. Resistant cells are selected by growth in herbicide-containing medium. Plants regenerated from these cells are herbicide-resistant.

In research currently under way, recombinant DNA technology is being used to produce the hepatitis B surface protein in plants. The idea is to have plant leaves or fruits serve as the source of an oral vaccine that can be eaten instead of injected. Plant-produced vaccines offer several advantages. They would be inexpensive, as industrial investment for production and extensive purification would be unnecessary. In addition, such vaccines would not require injection and would be useful in developing countries where medical services and delivery of health care are not highly developed.

The antigenic subunit of hepatitis B vaccine has been transferred to tobacco plants and expressed in the leaves (Figure 16.21). The tobacco plant is being used as a host system for the *Agrobacterium*-based vectors described in

FIGURE 16.21 Tobacco leaves expressing the hepatitis B antigen. Transgenic tobacco plants carrying an antigenic subunit of hepatitis B virus were generated. Leaves from transgenic plants were treated with antibodies against hepatitis B antigen, showing that the plants produced the antigen. The central leaf in the photograph is a leaf from a normal tobacco plant, and is unstained.

causes liver damage and cancer (see Chapter 21 for a discussion of hepatitis B). The gene for this hepatitis B protein was cloned into a yeast-expression vector and is grown in commercial quantities using yeast as a host. It is extracted and purified from the host cells and then packaged for use.

Chapter 15. For use as a source of vaccine, the gene would be inserted into food plants such as grain or vegetable plants.

The hepatitis B surface protein produced in the plant is similar if not identical to the antigens found in the serum of people infected with hepatitis B. If confirmed, this finding indicates that the plant-produced antigen may stimulate the immune system and confer immunity in the same way as does the viral antigen. Other experiments are directed at producing an edible vaccine against cholera toxin using alfalfa plants as hosts. In these experiments, the antigenic B subunit (Figure 16.22) of the bacillus *Vibrio* has been cloned and inserted into alfalfa plants.

Experiments are now under way to determine whether the viral protein antigens produced by genetically engineered plants will act as a vaccine and induce antibody production when ingested. Even if these tests are successful, other obstacles need to be overcome before we can be vaccinated against disease by eating our vegetables, but there is widespread enthusiasm for this approach.

As discussed in Chapter 1, plants and animals were domesticated some 8000 to 10,000 years ago, and by selective breeding we have been modifying these organisms ever since, producing the diversity of domesticated plants and animals present today. The development of recombinant DNA technology has changed the rate at which new

GENETICS, TECHNOLOGY, AND SOCIETY

PCR and DNA Fingerprinting in Forensics: The Case of the Telltale Palo Verde

The **polymerase chain reaction (PCR)** technique revolutionized many fields of biology, but nowhere has its impact been greater than in forensics where evidence is collected and used in court proceedings. One application of PCR that has proven very useful is called **DNA fingerprinting**, since under the right conditions it generates patterns of DNA fragments unique to each individual.

DNA fingerprinting using PCR involves the amplification of a set of DNA fragments of unknown sequence. To begin the procedure, an amplification primer is mixed with a sample of genomic DNA. As little as 1 nanogram of DNA is enough to serve as a template for the reaction. This amount can easily be isolated from a single spot of dried blood, from cells clinging to the base of a hair shaft, or from a variety of other tissue samples collected at a crime scene. The primer will anneal to all sites in the DNA sample that have a matching base sequence. For primers 10 nucleotides in length, binding will occur at a few

thousand sites scattered randomly throughout the human genome. When primers bind to DNA sites that are not too far apart (within about 2000 nucleotides), amplification of the DNA between these sites occurs, eventually producing hundreds of millions of copies of the DNA in each such case. These copies of each DNA segment will appear as a distinct band after electrophoresis on an agarose gel.

The locations of the primer binding sites in the genome are likely to differ between any two individuals in a population, due to minor DNA sequence differences. Because of this, the number and locations of the DNA bands (the electrophoretic pattern) will vary from one individual to another. This pattern of bands is referred to as a DNA fingerprint since, in theory, it represents a characteristic "snapshot" of the genome of an individual, allowing it to be distinguished from other individuals in population.

When a single DNA fingerprint generated from a tissue sample recovered at a crime scene matches that of a suspect, it *does not prove* that the tissue belongs to the suspect. However, it *does exclude* all those having a different pattern. The strategy, therefore, is to

generate five or more DNA fingerprints from the same sample using different amplification primers. The more patterns that match between the sample and the suspect, the more unlikely it is that the sample at the crime scene came from someone other than the suspect. Using statistical formulae, the likelihood that a complete match of all the banding patterns occurred simply by chance is estimated, culminating, for example, in a 1-in-100,000 or a 1-in-1,000,000 chance of a "random match."

One of the most interesting forensic uses of DNA fingerprinting in a criminal case did not involve the suspect's own DNA, but rather the DNA of some plants found at the crime scene. On the night of May 2, 1992, a Phoenix woman was strangled and her body dumped near an abandoned factory in the surrounding desert. Police discovered a pager near the body, making its owner the prime suspect in the murder. When questioned, this man admitted being with the woman the day of the murder, but claimed he had never been near the factory and suggested that the woman must have stolen his pager from his pick-up truck. A search of the pick-up provided the crucial clue placing the suspect at the factory: In the plants and animals can be developed, and to some degree this has altered the types of changes that can be made. Unlike selective breeding programs, biotechnology has generated concerns about the release of genetically modified organisms into the environment, and about the safety of eating such products. If biotechnology is to achieve a new green revolution, these concerns need to be addressed through prudent research and education.

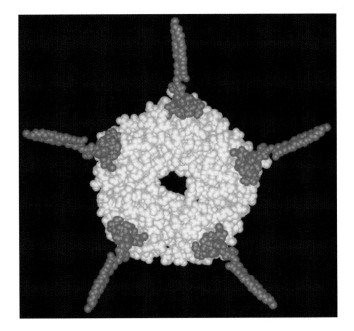

FIGURE 16.22 A molecular model of the cholera toxin B subunit. The gene encoding this subunit has been transferred to alfalfa plants to see whether the cholera B protein will induce antibody formation when eaten, making possible an edible vaccine against cholera.

truck bed were two seed pods from a palo verde tree.

The palo verde tree (*Cercidium floridum*) is an unusual plant native to the desert of the Southwest. A member of the bean family, it develops into a small shrubby tree. Well adapted to the hot, dry climate, it puts out very small leaves. To compensate for this, its stems become green and carry out photosynthesis. This adaptation gives the plant its name: *palo verde* is Spanish for "green stem." The plant may not resemble beans in some respects, but, like a typical bean, it produces seeds in pods that mature and drop to the ground.

The homicide investigators assigned to the case asked themselves the following question: Could it be proved that the seed pods found in the bed of the suspect's truck had fallen from one of the palo verde trees close to where the body was found? If so, it would be a key piece of evidence, placing the suspect at the scene and destroying his alibi. But how could this be demonstrated? The investigators turned to Dr. Timothy Helentjaris, then at the University of Arizona, in nearby Tucson. Helentjaris proceeded to determine whether the DNA fingerprint of the seed pods from

The palo verde tree.

the pick-up truck matched that of the palo verde trees at the scene. Helentjaris knew that for a match between a seed pod and a tree to have significance, he would first have to demonstrate that palo verdes show genetic variation from one tree to the next. Otherwise, it would not be possible to unambiguously assign a given pod to one particular tree. Fortunately, he found considerable differences in DNA fingerprints between individual plants.

Helentjaris was then given the two seed pods from the suspect's truck along with pods collected by investigators from 12 palo verde trees present in the general vicinity of the factory. The

investigators knew which of the 12 trees was the key one at the crime scene, but did not tell Helentjaris. He extracted DNA from seeds taken from each pod and performed PCR on each sample using one amplification primer he had earlier found to reproducibly yield fingerprints revealing 10–15 distinct bands. The results of this "blind" experiment were unmistakable: The pattern from one of the pods found in the truck exactly matched the pattern from only one of the 12 trees, the one nearest the body. In an important additional test, Helentjaris found that this pattern was different from that of pods collected from 18 trees located at random sites around Phoenix. From these results, he was able to estimate the likelihood of a "random match" at a little less than 1-in-1,000,000.

Helentjaris's analysis was admitted as evidence in the trial, the first time a DNA fingerprint of a plant had been used in a court case. This information was critical in placing the suspect at the crime scene. At the completion of the five-week trial, the suspect was found guilty of first-degree murder. His conviction was upheld upon appeal, and he is currently serving a life sentence without parole.

CHAPTER SUMMARY

1. Recombinant DNA technology offers a new approach to genetic analysis. Instead of relying on isolation and mapping of mutant genes, it is now possible to begin by manipulating large cloned segments of the genome to establish genetic and physical maps using molecular markers rather than phenotypes visible at the level of the organism.

2. Genome projects are under way for several organisms, including *E. coli*, yeast, *Drosophila*, the mouse, and humans. The goals of these genome projects are to identify and map all genes in the genome and to determine the complete nucleotide sequence of the genome.

3. Cloned DNA is being used in a wide variety of applications, including gene mapping and the identification and isolation of genes responsible for genetic disorders. The method known as *positional cloning*, based on recombinant DNA methodology, allows the mapping and identification of a gene without any knowledge of the nature or function of the gene product.

4. Recombinant DNA techniques are being used in the prenatal diagnosis of human genetic disorders. This method allows direct examination of the genotype, whereas previous methods relied on gene expression and the identification of the gene product. These methods can also be used to identify carriers of genetic disorders, and they are the basis of proposals to screen the population for a number of genetic disorders, including sickle-cell anemia and cystic fibrosis.

5. The availability of cloned human genes has led to their use in replacement of mutant genes in somatic tissues. This somatic gene therapy is carried out by transferring a cloned normal copy of a gene into a vector and using the vector as a means of transferring the gene to a target tissue that takes up and expresses the cloned copy of the gene, altering the mutant phenotype. The development of new and more effective vector systems probably means that gene therapy will become a standard method for the treatment of genetic disorders in the near future.

6. The use of recombinant DNA techniques in detecting allelic variants of variable tandem nucleotide repeats (DNA fingerprints) has found applications in forensics, although there are issues yet to be resolved before this method is universally accepted as evidence in criminal cases.

7. The biotechnology industry is using recombinant DNA methods to produce human gene products in a variety of hosts, ranging from bacteria to farm animals. In addition, gene-transfer techniques are being used to improve crop plants by transfer of herbicide resistance and to improve produce such as tomatoes. In the near future, it may be possible to offer vaccination against infectious disease through food plants, making resistance to infectious agents almost universal.

KEY TERMS

adenosine deaminase (ADA), 471
allele-specific oligonucleotide (ASO), 468
amniocentesis, 466
attenuated vaccine, 480
centimorgan (cM), 464
chorionic villus sampling (CVS), 466
contig, 475
cystic fibrosis transmembrane
    conductance regulator (CFTR), 469
DNA fingerprint, 473

exclusion map, 464
expressed sequence tag (EST), 476
Human Genome Project, 476
inactivated vaccine, 480
knockout mice, 469
logarithm of the odds (lod) score, 464
minisatellite, 473
P element, 476
polymerase chain reaction (PCR), 482
positional cloning, 465

preproinsulin, 478
restriction fragment-length
    polymorphism (RFLP), 462
sequence tagged site (STS), 476
severe combined immunodeficiency
    (SCID), 471
subunit vaccine, 480
T cell, 471
variable-number tandem repeat
    (VNTR), 473

INSIGHTS AND SOLUTIONS

1. DNA fingerprints have been used to test forensic specimens, to identify criminal suspects, and to settle immigration cases and paternity disputes. Probes for DNA fingerprinting can be derived from a single locus or multiple loci. Two multiple-loci probes have been widely employed in both criminal and civil cases and derive from minisatellite loci on chromosome 1 (1cen-q24) and chromosome 7 (7q31.3). These probes, which are widely used because they produce a highly individual fingerprint, have determined paternity in thousands of cases over the last few years.

   (a) Shown on the next page are the results of DNA fingerprinting of a mother (M), putative father (F), and child (C) using these probes.

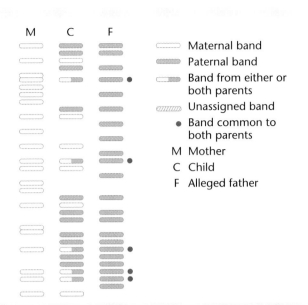

Maternal band
Paternal band
Band from either or both parents
Unassigned band
● Band common to both parents
M Mother
C Child
F Alleged father

As shown, the child has 6 maternal bands, 11 paternal bands, and 5 bands shared between the mother and the alleged father. Based on this fingerprint, can you conclude that the man tested is the father of the child?

Solution: All bands present in the child can be assigned as coming from either the mother or the father. In other words, all the bands in the child's DNA fingerprint that are not maternal are present in the father. Because the father and the child share 11 bands in common and the child has no unassigned bands, paternity can be assigned with confidence. In fact, the chance that the man tested is not the father is on the order of 10^{-13}.

(b) In a second case, the fingerprints of the mother, the alleged father, and the child are shown below.

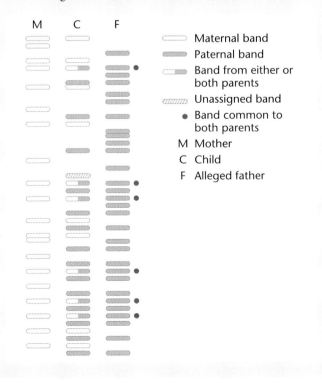

Maternal band
Paternal band
Band from either or both parents
Unassigned band
● Band common to both parents
M Mother
C Child
F Alleged father

In this case, the child has 8 maternal bands, 15 paternal bands, 6 bands that are common to both the mother and the alleged father, and 1 band that is not present in either the mother or the alleged father. What are the possible explanations for the presence of this band? Based on your analysis of the band pattern, which explanation is most likely?

Solution: In this case, one band in the child cannot be assigned to either parent. Two possible explanations for this finding are that the child is mutant for one band, or that the man tested is not the father. In estimating probability of paternity, the mean number of resolved bands (n) is determined, and the mean probability (x) that a band in individual A matches that in a second unrelated individual, B, is calculated. In this case because the child and the father share 15 bands in common, the probability that the man tested is not the father is very low (probably 10^{-7} or lower). As a result, the most likely explanation is that the child is a mutant for a single band. In fact, in 1419 cases of genuine paternity resolved by the minisatellite probes on chromosomes 1 and 7, single-mutant bands in the children were recorded in 399 cases, accounting for 28 percent of all cases.

2. Infection by HIV-1 (human immunodeficiency virus) is responsible for destruction of cells in the immune system and results in the symptoms of AIDS (acquired immunodeficiency syndrome). HIV infects and kills cells of the immune system that carry a cell-surface receptor known as CD4. An HIV surface protein known as gp120 binds to the CD4 receptor and allows the virus entry into the cell. The gene encoding the CD4 protein has been cloned. How can this clone be used along with recombinant DNA techniques to combat HIV infection?

Solution: Several methods utilizing the CD4 gene are being explored to combat HIV infection. First, because infection depends on an interaction between the viral gp120 protein and the CD4 protein, the cloned CD4 gene has been modified to produce a soluble form of the protein (sCD4). The idea is that HIV can be prevented from infecting cells if the gp120 protein of the virus is bound up with the soluble form of the CD4 protein, and thus unable to bind to CD4 proteins on the surface of immune system cells. Studies in cell-culture systems indicate that the presence of sCD4 effectively prevents HIV infection of tissue culture cells. However, studies in HIV-positive humans has been somewhat disappointing, mainly because the strains of HIV used in the laboratory are different from those found in infected individuals.

HIV-infected cells carry the viral gp120 protein on their surface. To kill such cells, the CD4 gene has been fused with those encoding bacterial toxins. The resulting fusion protein contains the CD4 regions that bind to gp120 and the toxin regions that kill the infected cell. In tissue-culture experiments, cells infected with HIV are killed by the fusion protein, whereas noninfected cells survive. It is hoped that the targeted delivery of drugs and toxins can be used in therapeutic applications to treat HIV infection.

PROBLEMS AND DISCUSSION QUESTIONS

1. In attempting to vaccinate against diseases by eating antigens, the antigen (such as the cholera toxin) must be presented to the cells of the small intestine. What are some potential problems in this method? Why don't absorbed food molecules stimulate the immune system and make you allergic to the food you eat?

2. Outline the steps involved in transferring glyphosate resistance to a crop plant. Do you envision that this trait can escape from the crop plant and make weeds glyphosate resistant? Why or why not?

3. Although not yet completed, what lessons from the *E. coli* genome project can be applied to the Human Genome Project?

4. Gene therapy for human genetic disorders involves transferring a copy of the normal human gene into a vector and using the vector to transfer the cloned human gene into target tissues. Presumably, the gene enters the target tissue, becomes active, and the gene product relieves the symptoms. Although now being used to treat several disorders, there are some unresolved problems with this method.
   (a) Why are disorders such as muscular dystrophy difficult to treat by gene therapy?
   (b) What are the potential problems in using retroviruses as vectors?
   (c) In the long run, should gene therapy involve germ tissue instead of somatic tissue? What are some of the potential ethical problems associated with this approach?

5. In producing physical maps of markers and cloned sequences, what advantage does *Drosophila* offer that other organisms including humans do not?

6. Outline the steps in identifying a gene by positional cloning. What steps may cause difficulty in this process? Once a region on a chromosome has been identified as containing a given gene, what kind of mutations would speed the process of identifying the locus?

7. The phenotype of many behavior traits such as manic depression or schizophrenia may be controlled by several genes, each at a different locus. Can positional cloning be used to map and isolate such genes? What if a trait is controlled by six genes, each contributing equally in an additive way to the phenotype? Can positional cloning be used in this case? Why or why not?

8. Suppose that you develop a screening method for cystic fibrosis that allows you to identify the predominant mutation (delta 508) and the next six most prevalent mutations. What do you need to consider before using this method in screening the population at large for this disorder?

9. The DNA sequence surrounding the site of the sickle-cell mutation in the β-globin is shown for normal and mutant genes:

5′ G A C T C C T G A G G A G A A G T 3′

3′ C T G A G G A C T C C T C T T C A 5′

Normal DNA

5′ G A C T C C T G T G G A G A A G T - 3′

3′ C T G A G G A C A C C T C T T C A - 5′

Sickle-cell DNA

Each type of DNA is denatured into single strands and applied to a filter. The paper containing the two spots is hybridized to an ASO of the following sequence: 5′–GACTC-CTGAGGAGAAGT–3′. Which (if either) spot will hybridize to this probe? Why?

Normal   Sickle-cell
DNA   DNA

10. Dominant mutations can be categorized according to whether they increase or decrease overall activity of a gene or gene product. Although a "loss-of-function" mutation (that is, a mutation that inactivates the gene product) is usually recessive, for some genes one dose of the gene product is not sufficient to produce a normal phenotype. In this case, a loss-of-function mutation in the gene will be dominant and the gene is said to be haploinsufficient. A second category of dominant mutations are "gain-of-function" mutations that result in an increased activity or expression of the gene or gene product. The phenotype of such a mutation results from too much gene product.

   The gene therapy technique currently used in clinical trials involves the "addition" to somatic cells of a normal copy of a gene. In other words, a normal copy of the gene is inserted into the genome of the mutant somatic cell, but the mutated copy of the gene is not removed or replaced. Will this strategy work for either of these two types of dominant mutations?

11. One form of hemophilia, an X-linked disorder of blood clotting, is caused by mutation in clotting factor VIII. Many single-nucleotide mutations of this gene have been described, making detection of mutant genes by Southern blots inefficient. There is, however, an RFLP for the enzyme *Hind*III contained in an intron of the factor VIII gene that can often be used in screening, as shown below.

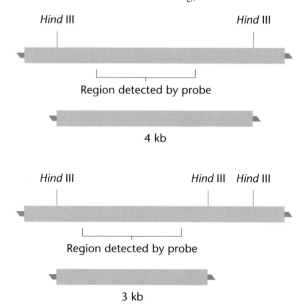

Hind III   *Hind* III

Region detected by probe

4 kb

Hind III   *Hind* III *Hind* III

Region detected by probe

3 kb

A female whose brother has hemophilia is at 50 percent risk of being a carrier of this disorder. To test her status, DNA is obtained from white blood cells from family members, cut with *Hin*dIII, and the fragments probed and visualized by Southern blotting. The results are shown below.

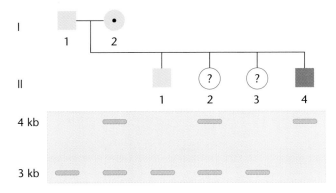

Determine whether any of the females in generation II are carriers for hemophilia.

12. The human insulin gene contains a number of introns. Despite the fact that bacterial cells will not excise introns from mRNA, how can a gene like this be cloned into a bacterial cell and produce insulin?

13. In mice transfected with the rabbit beta-globin gene, the rabbit gene is active in a number of tissues including spleen, brain, and kidney. In addition, some mice suffer from thalassemia, caused by an imbalance in the coordinate production of α- and β-globins. What problems associated with gene therapy are illustrated by these findings?

EXTRA-SPICY PROBLEMS

14. The following is a pedigree that shows the inheritance of a rare disease state.
(a) Which mode or modes of inheritance are excluded by or consistent with this pedigree?

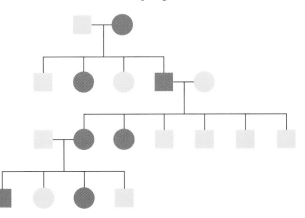

DNA samples from generations II and III (in the above pedigree) are obtained and subjected to RFLP linkage analysis. One RFLP examined is found on chromosome 10 (identified by probe A) and the other is found on chromosome 21 (identified by probe B). The results of the analysis are shown below. In answering the questions, assume that additional data was gathered on this family and that it was consistent with the data shown and statistically significant.

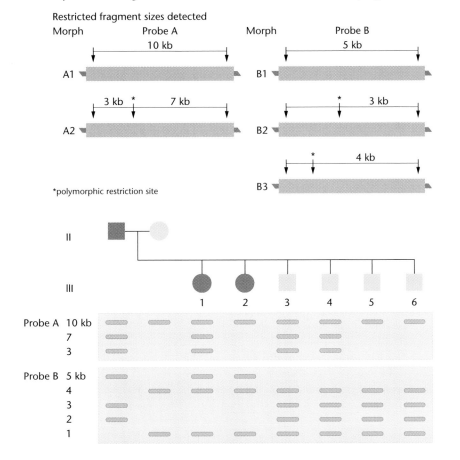

(b) On which chromosome is the disease gene located?

(c) Individual III-1 is married to a normal man whose RFLP genotype is A1A2 and B1B1. What kind of prenatal diagnostic test can be done to determine if the child they are expecting will be normal? Be sure to describe what you can conclude given this result or that result. Indicate the accuracy of such a test.

(d) Individual III-4 marries a woman of genotype B1B1. They have a child who is B1B1. The father assumes the child is illegitimate, but the mother, who has taken a genetics course, argues that the child could be the result of an event that occurred in the father's germ line, and she cites two possibilities. What are they?

15. You are asked to help in prenatal genetic testing of a couple found to be carriers for a deletion in the β-globin gene that produces β-thalassemia when homozygous. The couple already has one child who is unaffected and is not a carrier. The woman is pregnant, and they wish to know the status of the fetus. You receive DNA samples obtained from the fetus by amniocentesis and from the rest of the family by extraction from white blood cells. Using a probe that detects the deletion, the following blot is obtained:

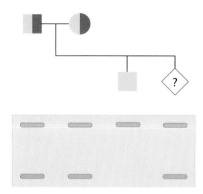

Is the fetus affected? What is its genotype for the β-globin gene?

SELECTED READINGS

AMOS, B., and PEMBERTON, J. 1993. DNA fingerprinting in non-human populations. *Curr. Opin. Genet. Dev.* 2:857–860.

BARKER, D., et al. 1987. Gene for von Recklinghausen neurofibromatosis is in the pericentromeric region of chromosome 17. *Science* 236:1001–1102.

CAWTHON, R. M., et al. 1990. A major segment of the neurofibromatosis type 1 gene: cDNA sequence, genomic structure, and point mutations. *Cell* 62:193–201.

CHANG, J. C., and KAN, Y. W. 1981. Antenatal diagnosis of sickle-cell anemia by direct analysis of the sickle mutation. *Lancet* 2:1127–29.

COLLINS, F. S. 1995. Ahead of schedule and under budget: The Genome Project passes its fifth birthday. *Proc. Natl. Acad. Sci. USA* 92:10821–10823.

CRYSTAL, R. G. 1995. Transfer of genes to humans: early lessons and obstacles to success. *Science* 270:404–410.

DANNA, K., and NATHANS, D. 1971. Specific cleavage of simian virus 40 DNA by restriction endonuclease of *Hemophilus influenzae. Proc. Natl. Acad. Sci. USA* 68:2913–2917.

DAVIDSON, B., ALLEN, E., KOZARSKY, K., WILSON, J., and ROESSLER, B. 1993. A model system for in vivo gene transfer into the central nervous system using an adenoviral vector. *Nat. Genet.* 3:219–23.

DIETRICH, W. F., COPELAND, N. G., GILBERT, D. J., MILLER, J. C., JENKINS, N. A., and LANDER, E. S. 1995. Mapping the mouse genome: current status and future prospects. *Proc. Natl. Acad. Sci. USA* 92: 10849–10853.

DUJON, B. 1996. The yeast genome project: What did we learn? *Trends Genet.* 12:263–270.

FLOTTE, T. 1993. Prospects for virus-based gene therapy for cystic fibrosis. *J. Bioenerg. Biomembr.* 25:37–42.

FRASER, C. M., GOCAYNE, J. D., WHITE, O., ADAMS, M. D., CLAYTON, R. A., FLEISCHMANN, R. D., BULT, C. J., KERLAVAGE, A. R., et al. 1995. The minimal gene complement of *Mycoplasma genitalium. Science* 270:397–403.

FLEISCHMANN, R. D., ADAMS, M. D., WHITE, O., CLAYTON, R. A., KIRKNESS, E. F., KERLAVAGE, A. R., BULT, C. J., TOMB, J-F., et al. 1995. Whole genome random sequencing and assembly of *Haemophilus influenzae Rd. Science* 269:496–512.

GOODMAN, H. M., ECKER, J. R., and DEAN, C. 1995. The genome of *Arabadopsis thaliana. Proc. Natl. Acad. Sci. USA* 92:10831–10835.

GUYER, M. and COLLINS, F. S. 1995. How is the Human Genome Project doing, and what have we learned so far? *Proc. Natl. Acad. Sci. USA* 92:10841–10848.

HARTL, D., and LOZOVSKAYA, E. 1992. The *Drosophila* genome project: Current status of the physical map. *Comp. Biochem. Physiol.* [B] 103:1–8.

HUDSON, T. J., STEIN, L., GERETY, S. S., MA, J., CASTLE, A. B., SILVA, J., SLONIM, D., BAPTISTA, R., et al. 1995. An STS-based map of the human genome. *Science* 270: 1945–1954.

INBAL, A., ENGLANDER, T., KORNBROT, N., RANDI, A., CASTAMAN, G., MANNUCCI, P., and Sadler, J. 1993. Identification of three candidate mutations causing type IIA von Willebrand disease using a rapid, nonradioactive, allele-specific hybridization method. *Blood* 82:830–836.

KRUMLAUFF, R., JEANPIERRE, M., and YOUNG, B. D. 1982. Construction and characterization of genomic libraries from specific human chromosomes. *Proc. Natl. Acad. Sci. USA* 79:2971–75.

LEWONTIN, R., and HARTL, D. 1991. Population genetics in forensic DNA typing. *Science* 254:1745–1750.

MANIATIS, T., FRITSCH, E. F., and SAMBROOK, J. 1988. *Molecular cloning: A laboratory manual*, 2nd ed. Cold Spring Harbor, NY: Cold Spring Harbor Laboratory Press.

MARSHALL, E., and PENNISI, E. 1996. NIH launches the final push to sequence the genome. *Science* 272:188–189.

MARSHALL, E. 1995. Gene therapy's growing pains. *Science* 269:1050–1055.

MASON, H., LAM, D., and ARNTZEN, C. 1992. Expression of hepatitis B surface antigen in transgenic plants. *Proc. Natl. Acad. Sci. USA* 89:11745–11749.

MOLDOVEANU, Z., NOVAK, M., HUANG, W., GILLEY, R., STAAS, J., SCHAFER, D., et al. 1993. Oral immunization with influenza virus in biodegradable microspheres. *J. Infect. Dis.* 167:84–90.

MOLLIS, K. B. 1990. The unusual origin of the polymerase chain reaction. *Sci. Am.* (April) 262:56–65.

NAHREINI, P., WOODY, M., ZHOU, S., and SRIVASTAVA, A. 1993. Versatile adeno-associated virus 2-based vectors for constructing recombinant virions. *Gene* 124:257–62.

REISS, J., and COOPER, D. N. 1990. Application of the polymerase chain reaction to the diagnosis of human genetic disease. *Hum. Genet.* 85:1–8.

SAIKI, R. K., et al. 1986. Analysis of enzymatically amplified beta-globin and HLA-DQ alpha DNA with allele-specific oligonucleotide probes. *Nature* 324:163–66.

SOUTHERN, E. M. 1975. Detection of specific sequences among DNA fragments separated by gel electrophoresis. *J. Mol. Biol.* 98:503–17.

TAYLOR, R. 1993. Food for thought: Seropositive plants may yield cheap oral vaccines. *J. N.I.H. Res.* 5:49–53.

VILLA-KOMAROFF, L., et al. 1978. A bacterial clone synthesizing proinsulin. *Proc. Natl. Acad. Sci. USA* 75:3727–31.

WATERSON, R., and SULSTON, J. 1995. The genome of *Caenorhabditis elegans*. *Proc. Natl. Acad. Sci.* 92:10836–10840.

WILLIAMS, J. G. K., KUBELIK, A. R., LIVAK, K. J. RAFALSKI, J. A., and TINGEY, S. V. 1990. DNA polymorphisms amplified by arbitrary primers are useful as genetic markers. *Nucl. Acids Res.* 18:6531–35.

WU, C.-L., and MELTON, D. W. 1993. Production of a model for Lesch-Nyhan syndrome in hypoxanthine phosphoribosyltransferase-deficient mice. *Nat. Genet.* 3:235–39.

ZUO, J., ROBBINS, C., BAHARLOO, S., COX, D., and MYERS, R. 1993. Construction of cosmid contigs and high-resolution restriction mapping of the Huntington disease region of human chromosome 4. *Hum. Mol. Genet.* 2:889–99.

17

Genomic Organization of DNA

CHAPTER CONCEPTS

The DNA composing the genome of organisms is organized in a manner consistent with the complexity of the host structure with which it is associated. Viruses, bacteria, mitochondria, and chloroplasts contain a shorter, often circular, DNA molecule relatively free of proteins. Eukaryotic cells contain greater amounts of DNA organized into nucleosomes and present as chromatin fibers. This increase in complexity is related to the larger amount of genetic information present as well as the greater complexity associated with their genetic function. In eukaryotes, genes are organized in various ways, ranging from single copies to families of related genes repeated in tandem arrays. The eukaryotic genome contains large amounts of noncoding DNA, some of which interrupts the coding portions of genes.

Once it was understood that DNA houses genetic information, it was very important to determine how DNA is organized into genes and how these basic units of genetic function are organized into chromosomes. In short, the major question was how the genetic material is organized as it makes up the genome of organisms. There has been much interest in this question because knowledge of the organization of the genetic material and associated molecules is important to the understanding of many other areas of genetics. For example, how the genetic information is stored, expressed, and regulated must be related to the molecular organization of

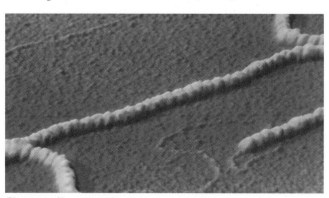

Chromatin fibers viewed using a scanning transmission electron microscope (STEM).

the genetic molecule, DNA. How genomic organization varies in different organisms—from viruses to bacteria to eukaryotes—will undoubtedly provide a better understanding of the evolution of organisms on earth. In eukaryotic genetics, how DNA is complexed with proteins to form chromatin, and how the chromatin fibers characteristic of interphase are condensed into chromosome structures visible during mitosis and meiosis, are both important questions.

In this chapter, we first provide a survey of the various ways DNA is organized into chromosomes, including examples from viruses, bacteria, and eukaryotes. The genetic material has been studied using numerous approaches, including molecular analysis and direct visualization by light and electron microscopy.

We will then turn to the consideration of how genes are organized as part of chromosomes. As we shall see, such organization is least complex in bacteriophages and bacteria. In their genomes, a chromosome consists largely of an array of tightly packed contiguous genes. The vast majority of DNA encodes genetic information or consists of control elements involved in the regulation of gene expression.

In comparison, the vast majority of DNA in most eukaryotes fails to encode genetic information or be involved in genetic regulation. Nor are genes tightly packed together. As we saw earlier in the text (see Chapter 12) a large portion of many eukaryotic genes consists of noncoding DNA sequences called *introns* and flanking sequences that do not become part of the final RNA transcript. Further, we shall see that related genes are often arranged and regulated in groups or families and that a large portion of the total DNA comprising the genome of eukaryotes consists of nongenic regions that often are occupied by repetitive DNA sequences.

Many of the findings related to genomic organization over the past several decades, which we shall recount in this chapter, represent some of the most exciting and most unexpected discoveries made in the field of genetics. For all geneticists, but particularly those trained prior to the 1970s, these discoveries have served as a constant source of wonderment. Perhaps we would have all been better prepared to learn of the complexity of genes and genomes, particularly those of eukaryotes, had we paid more attention to the quotation from Lewis Carroll's *Through the Looking Glass*, which was insightfully included by Richard B. Goldschmidt in his look into the future during his Presidential Address to the IX International Congress of Genetics in 1954:

"I can't believe that," said Alice. "Can't you?" the Queen said in a pitying tone. "Try again: draw a long breath, and shut your eyes."

Alice laughed. "There's no use trying," she said. "One can't believe impossible things."

"I dare say you haven't had much practice," said the Queen. "When I was your age I did it for half-an-hour a day. Why, sometimes I've believed as many as six impossible things before breakfast."

For all of us interested in genetics, it is certain that the next "impossible thing" regarding the organization of the genetic material is just around the corner!

Viral and Bacterial Chromosomes

In comparison with eukaryotes, the chromosomes of viruses and bacteria are much less complicated. They usually consist of a single nucleic acid molecule, largely devoid of associated proteins. Contained within the single chromosome of viruses and bacteria is much less genetic information than in the multiple chromosomes comprising the genome of higher forms. These characteristics have greatly simplified analysis, providing a fairly comprehensive view of the structure of viral and bacterial chromosomes.

The chromosomes of viruses consist of a nucleic acid molecule—either DNA or RNA—that can be either single or double stranded. They may exist as circular structures (closed loops), or they may take the form of linear molecules. The single-stranded DNA of the **φX174 bacteriophage** and the double-stranded DNA of the **polyoma virus** are ring-shaped molecules housed within the protein coat of the mature virus. The **bacteriophage lambda (λ)**, on the other hand, possesses a linear double-stranded DNA molecule prior to infection, which closes to form a ring upon infection of the host cell. Still other viruses, such as the **T-even series of bacteriophages**, have linear, double-stranded chromosomes of DNA, which do not form circles inside the bacterial host. Thus, circularity is not an absolute requirement for replication in some viruses.

Viral nucleic acid molecules have been visualized with the electron microscope. Figure 17.1 shows a mature bacteriophage lambda with its double-stranded DNA molecule in the circular configuration. One constant feature shared by viruses, bacteria, and eukaryotic cells is the ability to package an exceedingly long DNA molecule into a relatively small volume. In λ, the DNA is 17 μm long and must fit into the phage head, which is less than 0.1 μm on any side.

Table 17.1 compares the length of the chromosomes of several viruses to the size of their head structure. In each case, a similar packaging feat must be accomplished. The dimensions given for phage T2 may be compared with the micrograph of both the DNA and the viral particle shown in Figure 17.2. Seldom does the space available in the head of a virus exceed the chromosome volume by more than a factor of two. In many cases, almost all space is filled, indicating nearly perfect packing. Once packed within the head, the genetic material is functionally inert until released into a host cell.

Bacterial chromosomes are also relatively simple in form. They always consist of a double-stranded DNA molecule, compacted into a structure sometimes referred to as the **nucleoid**. *Escherichia coli*, the most extensively

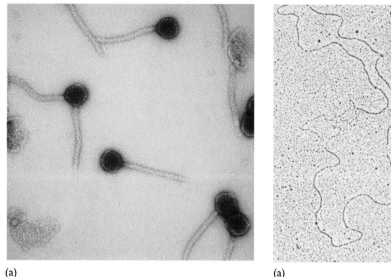

(a)

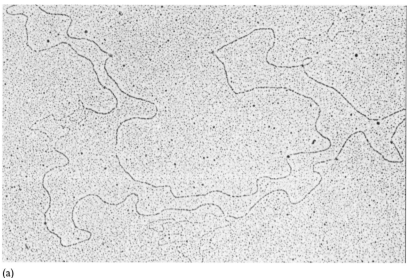

(a)

FIGURE 17.1 Electron micrographs of (a) phage {lambda} and (b) the DNA isolated from it. The chromosome is 17 µm long. Note that the phages in part (a) are magnified about five times more than the DNA in part (b).

studied bacterium, has a large, circular chromosome, measuring approximately 1200 µm (1.2 mm) in length. When the cell is gently lysed and the chromosome released, it can be visualized under the electron microscope (Figure 17.3).

This DNA is found to be associated with several types of **DNA-binding proteins**, including those called **HU** and **H**. They are small but abundant in the cell and contain a high percentage of positively charged amino acids that can bond ionically to the negative charges of the phosphate groups in DNA. As we will soon see, these proteins resemble structurally similar molecules called **histones** that are found associated with eukaryotic DNA. Unlike the tightly packed chromosome of a virus, the bacterial chromosome is not functionally inert. Despite the compacted condition of the bacterial chromosome, replication and transcription readily occur.

Supercoiling and Circular DNA

One major insight into the way in which DNA is organized and packaged has come from the discovery of **supercoiled DNA**, characteristic of covalently closed circular molecules and chromosomal loops. Supercoiled DNA was first proposed as a result of a study of double-stranded DNA molecules derived from the polyoma virus, which causes tumors in mice. In 1963, it was observed that when such DNA was subjected to high-speed centrifugation, it was resolved into three distinct components, each of different density and compactness. That which was least compact, and thus least dense, demon-

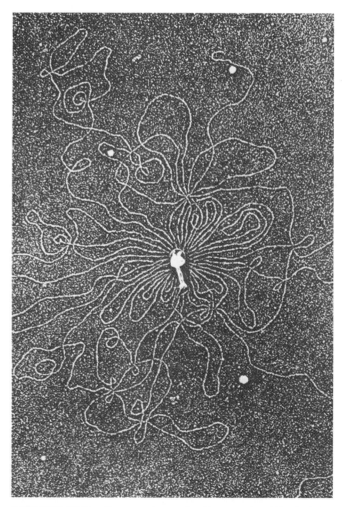

FIGURE 17.2 Electron micrograph of bacteriophage T2, which has had its DNA released by osmotic shock. The chromosome is 52 µm long.

TABLE 17.1   The Genetic Material of Representative Viruses and Bacteria

	Organism	Nucleic Acid			Overall Size of Viral Head or Bacteria (μm)
		Type	SS or DS	Length (μm)	
Viruses	φX174	DNA	SS	2.0	0.025 × 0.025
	Tobacco mosaic virus	RNA	SS	3.3	0.30 × 0.02
	Lambda phage	DNA	DS	17.0	0.07 × 0.07
	T2 phage	DNA	DS	52.0	0.07 × 0.10
Bacteria	*Hemophilus influenzae*	DNA	DS	832.0	1.00 × 0.30
	Escherichia coli	DNA	DS	1200.0	2.00 × 0.50

SS = single-stranded; DS = double-stranded.

strated a decreased sedimentation velocity; the other two fractions each showed greater velocities owing to their greater compaction and density. All three were of identical molecular weight.

In 1965 Jerome Vinograd proposed an explanation for the above observations. He postulated that the two fractions of greatest sedimentation velocity both consisted of polyoma DNA molecules that are circular, whereas the fraction of lower sedimentation contained polyoma DNA molecules that are linear. Closed circular

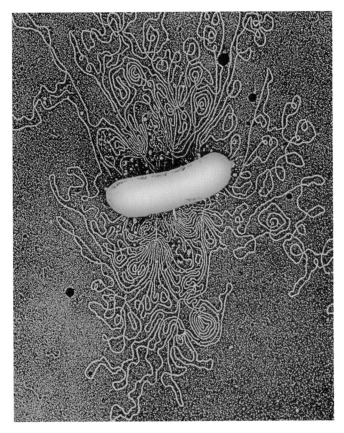

FIGURE 17.3   Electron micrograph of the bacterium *Escherichia coli*, which has had its DNA released by osmotic shock. The chromosome is 1200 μm long.

molecules are more compact, and they sediment more rapidly than do linear molecules of the same length and molecular weight.

Vinograd proposed further that the more dense of the two fractions of circular molecules consisted of covalently closed DNA helices that are slightly *underwound* in comparison to the less dense circular molecules. Energetic forces stabilizing the double helix resist this underwinding, causing it to **supercoil** in order to retain normal base pairing. Vinograd proposed that it is the supercoiled shape that causes tighter packing and thus the increase in sedimentation velocity.

The transitions described above can be visualized in Figure 17.4. Consider a double-stranded linear molecule that exists in the normal Watson–Crick right-handed helix [Figure 17.4(a)]. This helix contains 20 complete turns, which defines the **linking number** ($L = 20$) of this molecule. If the ends of the molecule are sealed, a closed circle is formed [Figure 17.4(b)], which is energetically *relaxed*. Suppose, however, that the circle is cut open, underwound by several full turns, and resealed [Figure 17.4(c)]. Such a structure, where L is equal to 18, is energetically "strained," and as a result it will exist only temporarily in this form.

In order to assume a more energetically favorable conformation, the molecule can form supercoils in the direction opposite of the underwound helix. In our case [Figure 17.4(d)], two negative supercoils are introduced spontaneously, reestablishing the total number of original "turns" in the helix. The use of the term "negative" refers to the fact that, by definition, the supercoils are left-handed. The end result is the formation of a more compact structure with enhanced physical stability.

In most closed circular DNA molecules in bacteria and their phages, DNA is slightly underwound [as in Figure 17.4(c)]. For example, the virus **SV40** contains 5200 base pairs, where 10.4 base pairs occupy each complete turn of the helix. The linking number can be calculated as

$$L = 5200/10.4 = 500$$

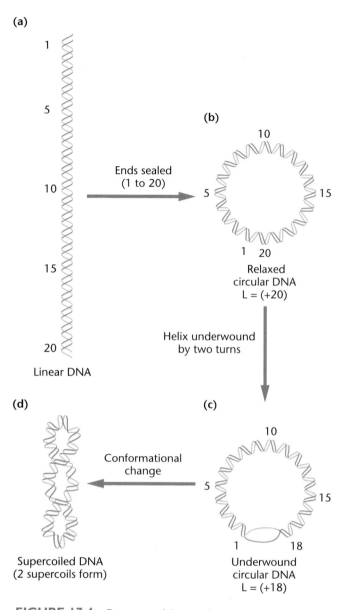

(a)

1

5

10

15

20

Linear DNA

**Ends sealed
(1 to 20)**

(b)

10

5 · 15

1 · 20

Relaxed
circular DNA
$L = (+20)$

**Helix underwound
by two turns**

(c)

10

5 · 15

1 · 18

Underwound
circular DNA
$L = (+18)$

**Conformational
change**

(d)

Supercoiled DNA
(2 supercoils form)

FIGURE 17.4 Depictions of the transformations leading to the supercoiling of circular DNA, as described in the text. L equals the linking number.

When circular SV40 DNA is analyzed, it is underwound by 25 turns and L is equal to only 475. Predictably, the presence of 25 negative supercoils is observed. In *E. coli*, an even larger number of supercoils is observed, greatly facilitating chromosome condensation in the nucleoid region.

When two otherwise identical molecules demonstrate a difference in their linking number, they are said to be **topoisomers** of one another. The only way in which these two topoisomers whose ends are constrained (not free to rotate) can be interconverted is by cutting one or both of the strands and winding or unwinding the helix before resealing the ends. Biologically, this may be accomplished by any one of a group of enzymes, appropri-

ately called **topoisomerases**. First discovered by Martin Gellert and James Wang, these catalytic molecules are either type I or II, depending on whether they cleave one or both strands in the helix, respectively. In *E. coli*, topoisomerase I serves to reduce the number of negative supercoils in a closed-circular DNA molecule. Topoisomerase II introduces negative supercoils into DNA. This enzyme is thought to bind to DNA, twist it, cleave both strands, and then pass them through the "loop" that it has created. Once the phosphodiester bonds are reformed, the linking number is decreased and one or more supercoils form spontaneously.

Supercoiled DNA and topoisomerases are also found in eukaryotes. While the chromosomes in these organisms are not usually circular, supercoils are made possible when areas of DNA are embedded in a lattice of proteins associated with the chromatin fibers. This association creates "anchored" ends that provide the stability for the maintenance of supercoils once they are introduced by topoisomerases.

Transcription of eukaryotic DNA into RNA faces several obstacles not encountered in prokaryotes. Most notable is the way in which DNA is packaged into structures called *nucleosomes*. In most studies to date, the formation of supercoils induced by topoisomerases has been implicated as a major factor in facilitating transcription. These enzymes may well play still other genetic roles involving DNA conformational changes.

Mitochondrial and Chloroplast DNA

Numerous observations have demonstrated that both **mitochondria** and **chloroplasts** contain their own genetic information. This was suggested by the discovery of mutations in yeast, other fungi, and plants that alter the function of these organelles (see Chapter 8). Furthermore, transmission of these mutations did not always demonstrate biparental inheritance patterns characteristic of nuclear genes. Instead, a **uniparental mode of inheritance** was observed, with traits being passed to offspring only from their mother. Because the inheritance of both mitochondria and chloroplasts is also uniparental, these observations suggested that these organelles might house their own DNA that influences their function.

Thus, geneticists set out to look for more direct evidence of DNA in these organelles. Electron microscopists not only documented the presence of DNA in both organelles, but they also saw DNA in a form quite unlike that seen in the nucleus of the eukaryotic cells that house these organelles. This DNA looked remarkably similar to that seen in viruses and bacteria! As we shall see, this similarity, among other observations, led to the

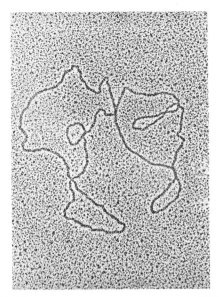

FIGURE 17.5   Electron micrograph of mitochondrial DNA (mtDNA) derived from *Xenopus laevis*.

TABLE 17.3   Sedimentation Coefficients of Mitochondrial Ribosomes

Kingdom	Examples	Sedimentation Coefficient (S)
Animals	Vertebrates	55–60
	Insects	60–71
Protists	*Euglena*	71
	Tetrahymena	80
Fungi	*Neurospora*	73–80
	Saccharomyces	72–80
Plants	Maize	77

idea that both mitochondria and chloroplasts are derived from primitive organisms that were free-living and much like bacterial organisms.

Molecular Organization and Function of Mitochondrial DNA

Extensive information is now available about the molecular aspects of **mitochondrial DNA (mtDNA)** and related gene function. In most eukaryotes, mtDNA (Figure 17.5) is a circular duplex that replicates semiconservatively and is free of the chromosomal proteins characteristic of eukaryotic DNA. In size, mtDNA differs among organisms. This can be seen by examining the information presented in Table 17.2. In a variety of animals, mtDNA consists of about 16,000 to 18,000 base pairs (16–18 kb). In vertebrates, there are 5 to 10 DNA molecules per organelle. A considerably greater amount of mtDNA is present in plant mitochondria, where 100 kb is not unusual.

Several general statements can now be made concerning mtDNA. There appear to be few or no gene repetitions, and replication is dependent upon enzymes encoded by nuclear DNA. Mitochondrial genes have been identified that code for the ribosomal RNAs, over 20 tRNAs, and numerous products essential to the cellular respiratory functions of the organelles.

The protein-synthesizing apparatus and the molecular components for cellular respiration are jointly derived from nuclear and mitochondrial DNA. Ribosomes found in the organelle are different from those present in the cytoplasm. The sedimentation coefficient of these particles varies among different species. The data in Table 17.3 show that mitochondrial ribosomes vary considerably in their coefficients (55S to 60S in vertebrates, to 70S in some algae and fungi, 80S in certain protozoans and fungi).

Many nuclear-coded gene products are essential to biological activity within mitochondria: DNA and RNA polymerases, initiation and elongation factors essential for translation, ribosomal proteins, aminoacyl tRNA synthetases, and some tRNA species. As in chloroplasts, these imported components are distinct from their cytoplasmic counterparts, even though both sets are coded by nuclear genes. For example, the synthetase enzymes essential for charging mitochondrial tRNA molecules (a process essential to translation) show a distinct affinity for the mitochondrial tRNA species as compared to the cytoplasmic tRNAs. Similar affinity has been shown for the initiation and elongation factors. Furthermore, while bacterial and nuclear RNA polymerases are known to be composed of numerous subunits, the mitochondrial variety consists of only one polypeptide chain. This polymerase is generally susceptible to antibiotics that inhibit bacterial RNA synthesis but not to eukaryotic inhibitors. The contributions of nuclear and mitochondrial gene products are contrasted in Figure 17.6.

The above information, along with that about to be presented concerning chloroplast DNA function, supplements our earlier discussion of modes of inheritance that are "extranuclear," which was earlier presented in Chapter 8.

TABLE 17.2   The Size of mtDNA in Different Organisms

Organism	Size in Kilobases
Human	16.6
Mouse	16.2
Xenopus (frog)	18.4
Drosophila (fruit fly)	18.4
Saccharomyces (yeast)	84.0
Pisum sativum (pea)	110.0

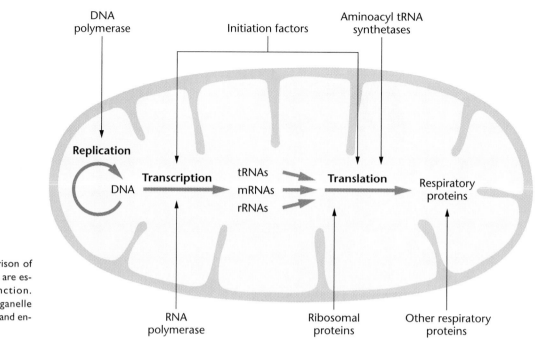

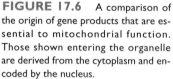

FIGURE 17.6 A comparison of the origin of gene products that are essential to mitochondrial function. Those shown entering the organelle are derived from the cytoplasm and encoded by the nucleus.

Molecular Organization and Function of Chloroplast DNA

A substantial amount of molecular information is now available about **chloroplast DNA (cpDNA)** and about the genetic function of this organelle. Chloroplasts, like mitochondria, contain an autonomous genetic system distinct from that found in the nucleus and cytoplasm. This system includes DNA as a source of genetic information and a complete protein-synthesizing apparatus. However, the molecular components of the translation apparatus are jointly derived from both nuclear and chloroplast genetic information. Chloroplast DNA, shown in Figure 17.7, is much larger than mitochondrial DNA, but it is nevertheless very similar to that found in prokaryotic cells.

The DNA isolated from chloroplasts is found to be circular, double stranded, replicated semiconservatively, and free of the associated proteins characteristic of eukaryotic DNA. Compared with nuclear DNA of the same organism, it invariably shows a different buoyant density and base composition.

In *Chlamydomonas*, there are about 75 copies of the chloroplast DNA molecule per organelle. Each copy consists of a length of DNA that contains 195,000 bases (195 kb). In higher plants such as the sweet pea, multiple copies of the DNA molecule are present in each organelle, but the molecule is considerably smaller than that in *Chlamydomonas*, consisting of 134 kb.

Some chloroplast gene products function to synthesize proteins. In a variety of higher plants (beans, lettuce, spinach, maize, and oats), two sets of the genes coding for the ribosomal RNAs—5S, 16S, and 23S rRNA—are

present. Additionally, chloroplast DNA codes for at least 25 tRNA species and a number of ribosomal proteins specific to the chloroplast ribosomes. These ribosomes have a sedimentation coefficient slightly less than 70S, similar to that of bacteria.

Chloroplast ribosomes are sensitive to the same protein-synthesis-inhibiting antibiotics as bacterial ribosomes: chloramphenicol, erythromycin, streptomycin, and spectinomycin. Even though chloroplast ribosomal proteins are encoded by both nuclear and chloroplast DNA, most, if not all, such proteins are distinct from their counterparts in cytoplasmic ribosomes.

Still other chloroplast genes have been identified that are specific to photosynthetic function. Mutations in these genes may have the effect of inactivating photosyn-

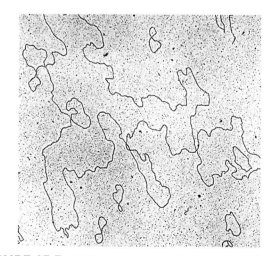

FIGURE 17.7 Electron micrograph of chloroplast DNA derived from lettuce.

thesis in those chloroplasts bearing such a mutation. One of the major photosynthetic enzymes is **ribulose-1-5-bisphosphate carboxylase (RuBP)**. Interestingly, the small subunit of this enzyme is encoded by a nuclear gene, whereas the large subunit is encoded by cpDNA.

Based on the observation that mitochondrial and chloroplast DNA and their genetic apparatus are similar to their counterparts in bacteria, a theory of the origin of these organelles has been proposed. Championed by Lynn Margulis and others, this hypothesis, called the **endosymbiotic hypothesis**, states that mitochondria and chloroplasts originated as distinct bacteria-like particles that became incorporated into primitive eukaryotic cells. In the evolution of this symbiotic relationship, the particles lost their ability to function independently and underwent an extensive evolution, while the eukaryotic host cell became dependent on them. Although there are many unanswered questions, the basic tenets of this theory are difficult to dispute.

Organization of DNA in Chromatin

The structure and organization of the genetic material in **eukaryotic cells** is much more intricate than in viruses or bacteria. This complexity is due to the greater amount of DNA per chromosome and the presence of large numbers of proteins associated with DNA in eukaryotes. For example, while DNA in the *E. coli* chromosome is 1200 μm long, the DNA in human chromosomes ranges from 14,000 to 73,000 μm in length. In a single human nucleus, all 46 chromosomes contain sufficient DNA to extend almost 2 meters. This genetic material, along with its associated proteins, is contained within a nucleus that usually measures about 5 μm in diameter.

For many reasons this intricacy is to be expected. It parallels the structural and biochemical diversity of the many types of eukaryotic cells present in a multicellular organism. Different cells assume specific functions based upon highly specific biochemical activity. While all cells carry a full genetic complement, different cells activate different sets of genes. A highly ordered regulatory system governing the readout of this information must exist if dissimilar cells performing different functions are present. Such a system must in some way be imposed on or related to the molecular structure of the genetic material.

Although bacteria can reproduce themselves, they never exhibit a complex process similar to mitosis. As described in Chapter 2, eukaryotic cells exhibit a highly organized cell cycle. During interphase, the genetic material and associated proteins are uncoiled and dispersed throughout the nucleus as **chromatin**. When mitosis begins, the chromatin condenses greatly, and during prophase it is compressed into recognizable chromosomes.

This condensation represents a contraction in length of some 10,000 times for each chromatin fiber and is the basis of the **folded-fiber model** of chromosome structure (Figure 2.14). This highly regular condensation-uncoiling cycle poses special organizational problems in eukaryotic genetic material.

Early studies of the structure of eukaryotic genetic material concentrated on intact chromosomes, preferably large ones, because of the limitation of light microscopy. Discussion of two such structures—polytene and lampbrush chromosomes—was presented in Chapter 2. Subsequently, new techniques for biochemical analysis, as well as the examination of relatively intact eukaryotic chromatin and mitotic figures under the electron microscope, have greatly enhanced our understanding of chromosome structure.

As established earlier, the genetic material of viruses and bacteria consists of strands of DNA or RNA nearly devoid of proteins. In eukaryotic chromatin, a substantial amount of protein is associated with the chromosomal DNA in all phases of the eukaryotic cell cycle. The associated proteins are divided into basic, positively charged **histones** and less positively charged **nonhistones**.

Nucleosome Structure

The general model for chromatin structure is based on the assumption that chromatin fibers, composed of DNA and protein, undergo extensive coiling and folding as they are condensed within the cell nucleus. Of the proteins associated with DNA, the histones clearly play the most essential structural role. Histones contain large amounts of the positively charged amino acids lysine and arginine, making it possible for them to bond electrostatically to the negatively charged phosphate groups of nucleotides. Recall that a similar interaction has been proposed for several bacterial proteins. There are five main types of histones (Table 17.4).

X-ray diffraction studies confirm that histones play an important role in chromatin structure. Chromatin produces regularly spaced diffraction rings, suggesting that repeating structural units occur along the chromatin axis. If the histone molecules are chemically removed

TABLE 17.4   Categories and Properties of Histone Proteins

Histone Type	Lysine-Arginine Content	Molecular Weight (daltons)
H1	Lysine-rich	23,000
H2A	Slightly lysine-rich	14,000
H2B	Slightly lysine-rich	13,800
H3	Arginine-rich	15,300
H4	Arginine-rich	11,300

from chromatin, the regularity of this diffraction pattern is disrupted.

The basic model for chromatin structure was worked out in the mid-1970s. Several observations are relevant to the development of this model:

1. Digestion of chromatin by certain endonucleases, such as micrococcal nuclease, yields DNA fragments that are approximately 200 base pairs in length, or multiples thereof. This demonstrates that enzymatic digestion is not random, for if it were, we would expect a wide range of fragment sizes. Thus, chromatin consists of some type of repeating unit, each of which is protected from enzymatic cleavage, except where any two units are joined. It is the area between all units that is attacked and cleaved by the nuclease.

2. Electron microscopic observations of chromatin have revealed that chromatin fibers are composed of linear arrays of spherical particles (Figure 17.8). Discovered by Ada and Donald Olins, the particles occur regularly along the axis of a chromatin strand and resemble beads on a string. These particles are now referred to as **ν-bodies** or **nucleosomes** (ν is the Greek letter *nu*). These findings conform nicely to the earlier proposal, which suggests the existence of repeating units.

3. Results of the study of precise interactions of histone molecules and DNA in the nucleosomes constituting chromatin show that histones H2A, H2B, H3, and H4 occur as two types of tetramers: $(H2A)_2 \cdot (H2B)_2$, and $(H3)_2 \cdot (H4)_2$. First demonstrated by Roger Kornberg, he predicted that each repeating nucleosome unit consists of one of each tetramer in association with about 200 base pairs of DNA. Such a structure is consistent with

previous observations and provides the basis for a model that explains the interaction of histone and DNA in chromatin.

4. When nuclease digestion time is extended, DNA is removed from both the entering and exiting strands, creating a **nucleosome core particle** consisting of 146 base pairs. This number is consistent in all organisms studied. The DNA lost in this prolonged digestion is responsible for linking nucleosomes together. This **linker DNA** is associated with histone H1.

5. On the basis of the above information as well as X-ray and neutron scattering analysis of crystallized core particles by John T. Finch, Aaron Klug, and others, a detailed model of the nucleosome was put forward (Figure 17.9). In this model, the 146 base-pair DNA core is coiled around an octamer of histones. The coiled DNA does not quite complete two full turns. The entire nucleosome is ellipsoidal in shape, measuring about 100 Å (10 nm) at its greatest dimension.

The extensive investigation of nucleosomes now provides the basis for predicting how the packaging of chromatin within the nucleus occurs (Figure 17.9). In its most extended state under the electron microscope, the chromatin fiber is about 100 Å in diameter, a size consistent with the longer dimension of the ellipsoidal nucleosome. It is believed that chromatin fibers consist of long strings of repeating nucleosomes [Figure 17.9(b)].

The 100-Å chromatin fiber is apparently further folded into a thicker fiber, sometimes called a **solenoid**, that is 300 Å (30 nm) in diameter. This larger fiber appears to consist of numerous nucleosomes coiled closely together.

In the overall transition from a fully extended chromatin fiber to the extremely condensed status of the mi-

(a)

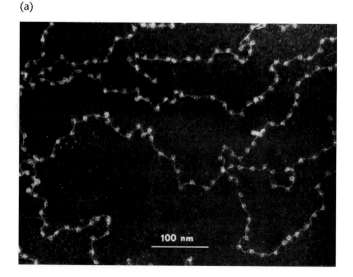

(b)

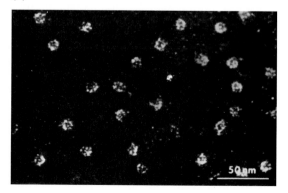

FIGURE 17.8 (a) Dark-field electron micrograph of nucleosomes present in chromatin derived from a chicken erythrocyte nucleus. (b) Dark-field electron micrograph of nucleosomes produced by micrococcal nuclease digestion.

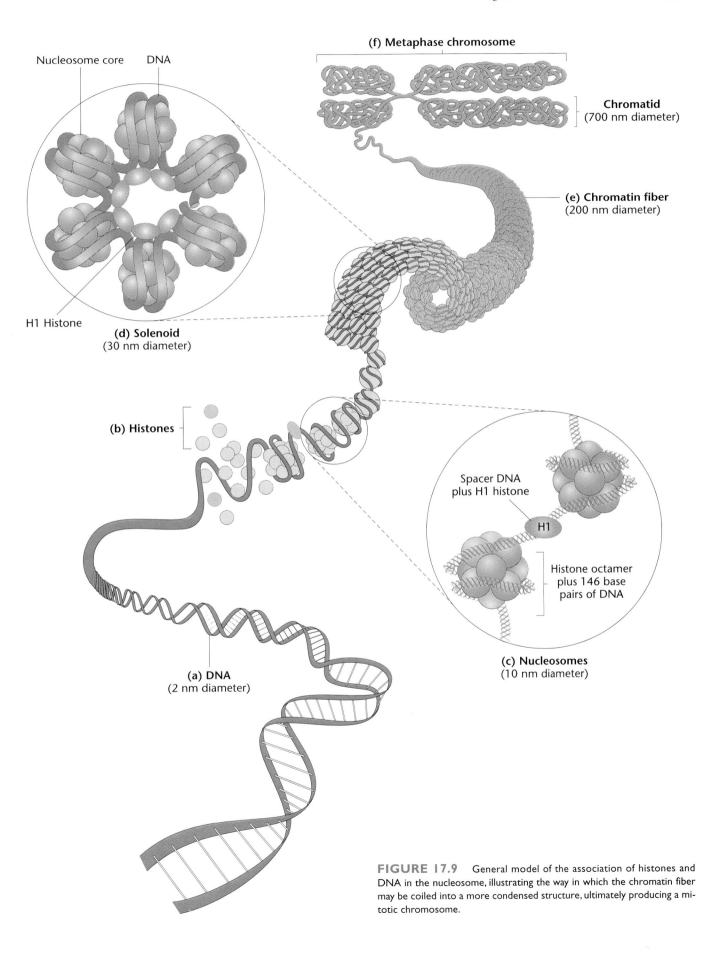

Nucleosome core

DNA

H1 Histone

(d) Solenoid
(30 nm diameter)

(b) Histones

(a) DNA
(2 nm diameter)

(f) Metaphase chromosome

Chromatid
(700 nm diameter)

(e) Chromatin fiber
(200 nm diameter)

Spacer DNA
plus H1 histone

H1

Histone octamer
plus 146 base
pairs of DNA

(c) Nucleosomes
(10 nm diameter)

FIGURE 17.9 General model of the association of histones and DNA in the nucleosome, illustrating the way in which the chromatin fiber may be coiled into a more condensed structure, ultimately producing a mitotic chromosome.

totic chromosome, a packing ratio (the ratio of DNA length to the length of the structure containing it) of about 500 must be achieved. Our model accounts for a ratio of about 50. Obviously, the larger fiber can be further bent, coiled, and packed as even greater condensation occurs during the formation of a mitotic chromosome.

Nuclear Scaffolds

There is now strong evidence that proteins other than histones play a role in the organization of chromatin in the cell. If histones are chemically removed from mitotic chromosomes, the folded loops of chromatin are seen under the electron microscope to emanate from a darkly staining central core of **nonhistone proteins** constituting a **nuclear scaffold**. One of the proteins identified in the scaffold of metaphase preparations is the enzyme **topoisomerase II**. Because this enzyme can manipulate DNA topologically by making double-stranded cuts, its role in the scaffold may be in the maintenance of organized folding characteristic of mitotic chromosomes.

Topoisomerase is seen to remain associated with the chromatin fiber after the cell passes through mitosis and enters metaphase. Its presence is thought to provide organization to the higher order of nucleosome folding associated with solenoid formation. The presence of such a nuclear scaffold during interphase may be critical to the replication of DNA during the S phase of the cell cycle. Binding sites along the DNA called **scaffold attachment regions** (SARs) have been identified in several organisms, including *Drosophila*. These sites exist between but not within transcriptional units. Perhaps SARs and associated scaffold proteins, including topoisomerase, are responsible for orienting chromatin in a way that facilitates replication and transcription. Given the size of the holoenzymes of DNA and RNA polymerases, some mechanism seems important in allowing these enzymes to "make their way" through the complex path of the chromatin fiber. As such, protein scaffolds are essential components in the organization and functional capabilities of the genetic material in eukaryotic cells.

Heterochromatin

All recent evidence supports the concept that the DNA of each eukaryotic chromosome consists of one continuous double-helical fiber along its entire length. A continuous fiber is the basis of the **unineme model of DNA** within a chromosome. Along with our knowledge of nucleosomes, we might be led to suspect that the chromosome would demonstrate structural uniformity along its entire length. However, in the early part of this century, it was observed that some parts of the chromosome remain condensed and stain deeply during interphase, but most are uncoiled and do not stain. In 1928, the terms **hetero-**

chromatin and **euchromatin** were coined to describe the parts of chromosomes that remain condensed and those that are uncoiled, respectively.

Subsequent investigation has revealed a number of characteristics of heterochromatin that distinguish it from euchromatin. Heterochromatic areas are genetically inactive, because they either lack genes or they contain genes that are repressed. Also, heterochromatin replicates later during the S phase of the cell cycle than does euchromatin. The discovery of heterochromatin provided the first clues that parts of eukaryotic chromosomes do not always encode proteins. Instead, some chromosome regions are thought to be involved in the maintenance of the chromosome's structural integrity and in other functions such as chromosome movement during cell division.

Heterochromatin is characteristic of the genetic material of eukaryotes. Early cytological studies showed that areas of the centromeres are composed of heterochromatin. The ends of chromosomes, called telomeres, are also heterochromatic. In some cases, whole chromosomes are heterochromatic. Such is the case with the mammalian Y chromosome, which for the most part is genetically inert. And, as we discussed in Chapter 9, the inactivated X chromosome in mammalian females is condensed into an inert heterochromatic Barr body. In some species, such as mealy bugs, all chromosomes of one entire haploid set are heterochromatic.

When certain heterochromatic areas from one chromosome are translocated to a new site on the same or another nonhomologous chromosome, genetically active areas sometimes become genetically inert if they now lie adjacent to the translocated heterochromatin. This influence on existing euchromatin is one example of what is more generally referred to as a **position effect**. That is, the position of a gene or groups of genes relative to all other genetic material may affect their expression. We first introduced this term in Chapter 7.

Satellite DNA and Repetitive DNA

Chapter 10 and Appendix A describe techniques useful in the analysis of DNA. Two of these—sedimentation equilibrium centrifugation and reassociation kinetics—have provided important information about the nature of DNA in chromatin. Specifically, the DNA of heterochromatin and euchromatin exhibits different molecular characteristics. The nucleotide composition of the DNA (e.g., the percentage of $G \equiv C$ versus $A = T$ pairs) of a particular species is reflected in its density, which can be measured with sedimentation equilibrium centrifugation. When sheared DNA of any eukaryotic species is analyzed in this way, the majority of it shows up as a single main peak or band of uniform density. However, one or more satellite peaks represent DNA that differs slightly in nucleotide composition and density from main-band DNA.

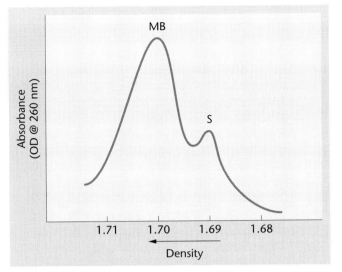

FIGURE 17.10 Separation of main-band (MB) and satellite (S) DNA from the mouse using ultracentrifugation in a CsCl gradient.

This component, called **satellite DNA**, represents a variable proportion of the total DNA, depending on the species. A profile of main-band and satellite DNA from the mouse is shown in Figure 17.10. It should be noted that bacteria, as representative prokaryotes, contain only main-band DNA.

The significance of satellite DNA remained an enigma until the mid 1960s, when the techniques for measuring the reassociation kinetics of DNA became available. These techniques, developed by Roy Britten and David Kohne, allowed geneticists to determine the rate of reassociation of fragmented DNA. Britten and Kohne showed that when complementary strands of DNA fragments were separated by heating and then reassociated by cooling, certain portions were capable of reannealing more rapidly than others. They concluded that rapid reannealing was characteristic of multiple DNA fragments composed of identical or nearly identical nucleotide sequences—the basis for the descriptive term **repetitive DNA**.

As discussed in Chapter 10, repeating nucleotide sequences are classified as either highly or moderately repetitive DNA. Evidence has now accumulated linking satellite DNA to repetitive DNA. Further, this specific DNA fraction has been shown to be present in heterochromatic regions of chromosomes. The following paragraphs review information supporting and clarifying these conclusions.

When satellite DNA is subjected to analysis by reassociation kinetics, it falls into the category of highly repetitive DNA. It therefore consists of short sequences repeated a large number of times. The available evidence further suggests that these sequences are present as tandem repeats clustered at various positions in the genome.

Satellite DNA is found in very specific chromosome areas known to be heterochromatic—the **centromeric**

regions. This was discovered in 1969 when several researchers, including Mary Lou Pardue and Joe Gall, applied the technique of *in situ* **hybridization** to the study of satellite DNA. This technique (see Figure 10.23 and Appendix A) involves the molecular hybridization between an isolated fraction of radioactively or fluorescently labeled DNA or RNA and the DNA contained in the chromosomes of a cytological preparation. Following the hybridization procedure, autoradiography is performed to locate the chromosome areas complementary to the fraction of DNA or RNA.

In their work, Pardue and Gall demonstrated that RNA probes made from mouse satellite DNA hybridizes with DNA of centromeric regions of mouse mitotic chromosomes (Figure 17.11). Based on this conclusion, several conclusions can be drawn. Satellite DNA differs from main-band DNA in its molecular composition, as established by buoyant density studies. It is also composed of short repetitive sequences. Finally, satellite DNA is found in the heterochromatic centromeric regions of chromosomes.

These observations establish one basis of the structural and functional differences between heterochromatin and euchromatin: The nucleotide composition is quite different. They further reflect the complexity of the genetic material in eukaryotes. Instead of viewing the chromosome as a linear series of genes, we must now see it as quite variable in its molecular organization and functional capacity. For example, there are nucleotide sequences that do not specify gene products. Are these genetically inert DNA sequences of heterochromatin related to the maintenance of the structural integrity of chromatin? Or might they be involved in a regulatory role of transcription? Such sequences may play more than one role. There is no doubt that more extensive investi-

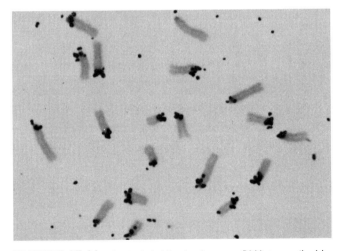

FIGURE 17.11 *In situ* hybridization between RNA transcribed by mouse satellite DNA and mitotic chromosomes. The grains in the autoradiograph localize the chromosome regions (the centromeres) containing satellite DNA sequences.

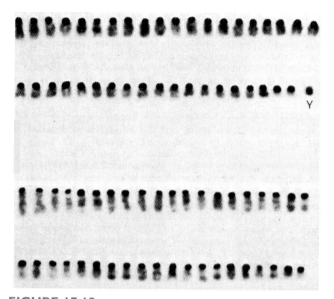

FIGURE 17.12 Karyotypes of a male (top) and female (bottom) mouse where chromosome preparations were processed to demonstrate C-banding, where only the centromeres stain.

gation will continually expand and modify our concept of eukaryotic genetic organization.

Chromosome Banding

Until about 1970, mitotic chromosomes viewed under the light microscope could be distinguished only by their relative sizes and the positions of their centromeres. Even in organisms with a low haploid number, two or more chromosomes are often indistinguishable from one another. However, about 1970, differential staining along the longitudinal axis of mitotic chromosomes was made possible by new cytological procedures. These methods are now called **chromosome banding techniques** because the staining patterns resemble the bands of polytene chromosomes.

One of the first chromosome banding techniques was devised by Mary Lou Pardue and Joe Gall. They found that if chromosome preparations were heat denatured and then treated with Giemsa stain, a unique staining pattern emerged. The centromeric regions of mitotic chromosomes preferentially took up the stain! Thus, this cytological technique stained a specific area of the chromosome composed of heterochromatin. A diagram of the mouse karyotype treated in this way is shown in Figure 17.12. The staining pattern is referred to as **C-banding**.

Still other chromosome banding techniques were developed about the same time. A group of Swedish researchers, led by Tobjorn Caspersson, used a technique that provided even greater staining differentiation of metaphase chromosomes. They employed fluorescent dyes that bind to nucleoprotein complexes and produce unique banding patterns. When the chromosomes are treated with the fluorochrome quinacrine mustard and viewed under a fluorescent microscope, precise patterns of differential brightness are seen. Each of the 23 human chromosome pairs can be distinguished by this technique. The bands produced by this method are called **Q-bands**.

Another banding technique produces a staining pattern nearly identical to the Q-bands. This method, producing **G-bands** (Figure 17.13), involves the digestion of the mitotic chromosomes with the proteolytic enzyme trypsin followed by Giemsa staining. Another technique results in the reverse G-band staining pattern, called an **R-band** pattern. In 1971, a meeting was held in Paris to establish the nomenclature for these various patterns. Figure 17.14 on pages 504–505 diagrams G-bands at their highest resolution (the 1000-band stage).

Intense efforts are currently under way to elucidate the molecular mechanisms involved in producing these banding patterns. The variety of staining reactions under different conditions reflects the heterogeneity and complexity of chromosome composition.

Organization of the Eukaryotic Genome

Having established the general view of the chromatin fiber, we now turn to the more specific consideration of the organization of the **genome** of eukaryotic organisms. The genome includes the entire haploid set of chromosomes constituting the genetic material of an organism.

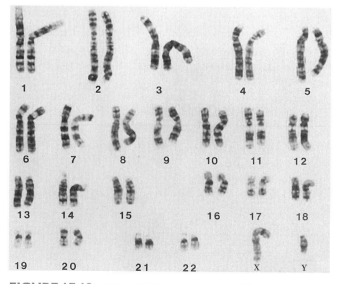

FIGURE 17.13 G-banded karyotype of a normal human male showing approximately 400 bands per haploid set of chromosomes. Chromosomes were derived from cells in metaphase.

Prior to 1970, little was known about eukaryotic gene structure and organization at the molecular level. Nevertheless, many geneticists believed that when the structure of genes and the organization of the genome of eukaryotes were revealed, fundamental differences from viruses and bacteria would be evident. Multicellular eukaryotes, it was reasoned, have several unique requirements distinguishing them from phages and bacteria. For example, cell differentiation, tissue organization, and coordinated development and function depend on genetic expression and its regulation. Do these requirements also depend on different modes of gene structure and organization?

When the technologies of DNA analysis were developed (Chapter 15), the answer to this question began to emerge. The eukaryotic gene is far more complex than we could ever have imagined. In this section, we will review many findings related to the organization of the eukaryotic genome and the structure of genes.

Eukaryotic Genomes and the *C* Value Paradox

The amount of DNA contained in the haploid genome of a species is called the *C* value. When such values were determined for a large variety of eukaryotic organisms (Figure 17.15) several trends were apparent. The most notable trends are that:

1. Eukaryotes contain substantially more DNA in their genomes than do viruses or prokaryotes and exhibit a wide variation among different groups.

2. Evolutionary progression has been accompanied by increased amounts of DNA. In many major phylogenetic groups, more evolutionarily complex forms generally contain more DNA than do most less advanced forms.

It might be argued that increasing *C* values are simply the result of a greater need for increased amounts and varieties of gene products in more complex organisms. However, several observations make this explanation unacceptable. First, the increases are very dramatic. Whereas viruses and bacteria contain 10^4 to 10^7 nucleotide pairs in their single chromosomes, eukaryotes contain 10^7 to 10^{11} nucleotide pairs in each set of chromosomes. Does a eukaryote require 10^4 (10,000) times as many genes than the most complex bacterium, as these figures might suggest? Second, closely related organisms with the same degree of complexity in body form and tissue and organ types often vary tenfold or more in DNA content. Third, the DNA contents of amphibians and flowering plants, which vary as much as 100-fold within their taxo-

nomic classes, are often greater than those of other more recently evolved eukaryotes. Despite this observation, there is no correlation between increased genome size and the morphological complexity of these groups.

Therefore, it is doubtful that the development of greater complexity during evolution can account for the amount of DNA found in eukaryotic genomes. This conclusion is the basis of what has been called the *C* value paradox: *Excess DNA is present that does not seem to be essential to the development or evolutionary divergence of eukaryotes.* Is such DNA vital to the organisms carrying it in their genomes, or is it simply DNA that has somehow accumulated during evolution and has no function?

Repetitive DNA: Centromeres and Telomeres

Careful analysis of the eukaryotic genome is beginning to partially unravel the mystery surrounding the *C* value paradox. For example, in Chapter 10 we established that noncoding sequences are prevalent in the form of various types of repetitive DNA. Earlier in this chapter, we demonstrated that satellite DNA of mammals such as the mouse is highly repetitive and is found to be associated with the centromere. As we will see below, an understanding of the molecular organization of repetitive DNA provides important insights into the organization of the genome.

We shall begin our discussion with a consideration of the **centromere**, the region of the chromosome that attaches to one or more spindle fibers and moves to one of the poles during the anaphase stages of mitosis and meiosis (see Chapter 2).

Separation of chromatids is essential to the fidelity of chromosome distribution during mitosis and meiosis. Most estimates of infidelity during mitosis are exceedingly low: 1×10^{-5} to 1×10^{-6}, or 1 error per 100,000 to 1 million cell divisions. As a result, it has been generally assumed that analysis of the DNA sequence of centromeric regions will provide insights into the rather remarkable features of this chromosomal region. This DNA region is designated the **CEN**.

Analysis of the CEN regions of yeast *Saccharomyces cerevisiae* chromosomes provided the basis for a model system first described by John Carbon and Louis Clarke. Because each centromere serves an identical function, it is not surprising that all CENs were found to be remarkably similar in their organization. The CEN region of yeast chromosomes consists of about 225 base pairs, which can be divided into three regions (Figure 17.16). The first and third regions (I and III) are relatively short and highly conserved, consisting of only 8 and 26 base pairs, respectively. Region II, which is larger (80–85 base pairs)

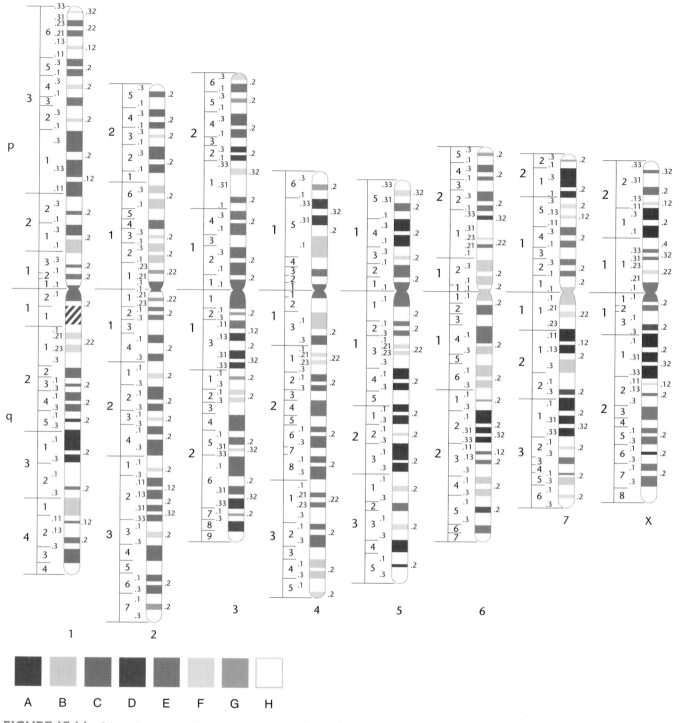

FIGURE 17.14 Schematic representation and nomenclature of human chromosomes at the 1000-band stage demonstrating G bands. Different shades represent varying intensities of bands, as shown in the key. The cross-hatched areas (chromosomes 1, 9, 16, and Y) represent heterochromatic regions.

and extremely A-T rich (up to 95%), varies in sequence between different chromosomes.

Mutational analysis suggests that regions I and II are less critical to centromere function than is region III. Mutations in the former regions are often tolerated, but mutations in region III usually disrupt centromere func-

tion. The DNA sequence of this region may be essential to binding to spindle fibers.

The amount of DNA associated with centromeres of multicellular eukaryotes is much more extensive than in yeast. Recall from an earlier discussion in this chapter that highly repetitive "satellite" DNA is localized in the

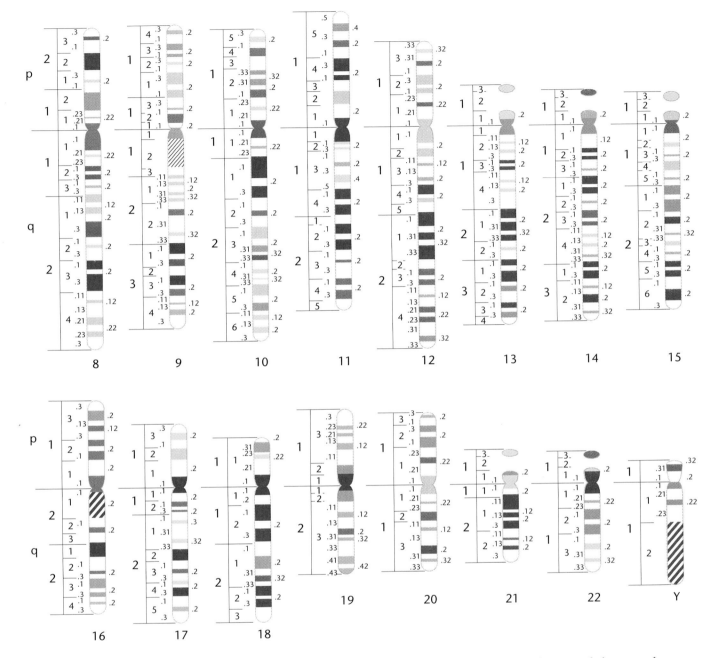

centromere regions of mice. Such sequences, absent from yeast, but characteristic of most multicellular organisms, vary considerably in size. For example, the 10 base-pair sequence AATAACATAG is tandemly repeated many times in the centromeres of all four chromosomes of *Drosophila*. In humans, one of the most recognized satellite DNA sequences is the **alphoid family**. Found mainly in the centromere regions, alphoid sequences, each about 170 base pairs in length, are present in tandem arrays of up to 1 million base pairs.

The role of this highly repetitive DNA in centromere function remains unclear; it is known that the sequences are not transcribed. There is some sequence variation among members of the alphoid family; the number of repeats is specific to each human chromosome. Perhaps there is a shorter sequence common to the alphoid family that, like yeast CEN DNA, is absolutely critical to centromere function.

The second important structure of chromosomes is the **telomere,** found at the ends of linear chromosomes. The function of telomeres is to provide stability to the chromosome by rendering chromosome ends generally inert in interactions with other chromosome ends. In contrast to broken chromosomes, whose ends may rejoin other such ends, telomere regions do not fuse with one another or with broken ends. It is thought that some aspect of the molecular structure of telomeres must be unique compared with most other chromosome regions.

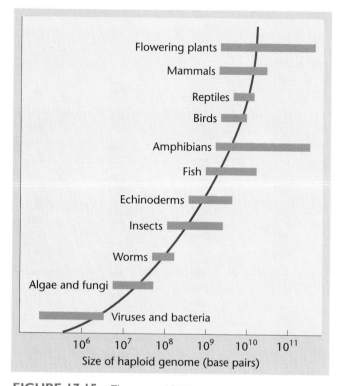

FIGURE 17.15 The range of DNA content of the haploid genome of many representative groups of organisms.

mosome. In the ciliate *Tetrahymena*, over 50 tandem repeats of the hexanucleotide sequence GGGGTT occur. In humans, the sequence GGGATT is repeated numerous times. The analysis of telomeric DNA sequences has shown them to be highly conserved throughout evolution, reflecting the critical role they play in maintaining the integrity of chromosomes.

The second type, **telomere-associated sequences**, are also repetitive and are found both adjacent to and within the telomere. These sequences vary between organisms, and their significance remains unknown.

As discussed in Chapter 11, replication of the telomere requires a unique RNA-containing enzyme **telomerase**. In its absence, the DNA at the ends of chromosomes becomes shorter during each replication. Because single-celled eukaryotes are immortalized cells, telomerase is critical for the survival of such species. In multicellular organisms, such as humans, telomerase is essential in germ line cells, but is inactive in somatic cells. Chromosome shortening is considered part of the natural process of cell aging, serving as an internal clock. In human cancer cells, which have become immortalized, the transition to malignancy appears to require the activation of telomerase in order to overcome the normal senescence associated with chromosome shortening.

Repetitive DNA: SINEs, LINEs, and VNTRs

A brief review of other prominent categories of repetitive DNA sheds more light on our understanding of the eukaryotic genome. In this discussion, we will also account for more of the genomic DNA that fails to encode proteins.

In addition to highly repetitive DNA, which constitutes about 5 percent of the human genome (and 10% of

As with centromeres, the problem was first approached by investigating the smaller chromosomes of simple eukaryotes, such as protozoans and yeast. The idea that all telomeres in a given species might share a common nucleotide sequence has now been borne out.

Two types of telomere sequences have been discovered. The first type, called simply **telomeric DNA sequences**, consists of short tandem repeats. It is this group that contributes to the stability and integrity of the chro-

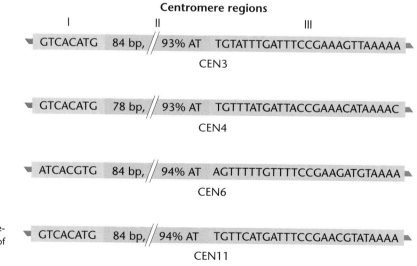

FIGURE 17.16 Nucleotide sequence information derived from DNA of the three major centromere regions of chromosomes 3, 4, 6, and 11 of yeast.

the mouse genome), a second category, **moderately** (or **middle**) **repetitive DNA**, is fairly well characterized. Because a great deal is being learned about the human genome, we will use our own species to illustrate the part played by this category of DNA in genome organization.

Moderately repetitive DNA consists of either interspersed or tandemly repeated sequences. That which is interspersed is either short or long. **Short interspersed elements**, called **SINEs**, are less than 500 base pairs long and are present as much as 500,000 times.

The best characterized human SINE is a set of closely related sequences called the *Alu* **family** (the name is based on the presence of DNA recognition sequences by the restriction endonuclease *Alu*I). Members of this DNA family, also found in other mammals, are 200 to 300 base pairs long and are present 700,000 to 900,000 times throughout the genome, both between and within genes. In humans, this family encompasses almost 10 percent of the entire genome.

Alu sequences are particularly interesting, although their function, if any, is yet undefined. Members of the *Alu* family are sometimes transcribed. The role of this RNA is unclear, but in some cases it may be related to the observation that *Alu* elements are potentially transposable within the genome. *Alu* sequences are thought to have arisen from an RNA element whose DNA complement was dispersed throughout the genome as a result of the activity of reverse transcriptase (an enzyme that synthesizes DNA on an RNA template).

Long interspersed elements (LINEs) represent still another category of moderately repetitive DNA. In humans, the most prominent example is a family of LINEs designated **L1**. Members of this sequence family are about 6400 base pairs long and are present up to 40,000 times, according to one estimate. Their 5′ end is highly variable, and their role within the genome has yet to be defined.

Moderately repetitive DNA also may be clustered as tandem repeats rather than interspersed throughout the genome. In some cases, this category includes functional genes present in multiple copies. For example, many of the copies of genes encoding 5.8*S*, 18*S*, and 28*S* rRNA in

humans are clustered on the p arm of the acrocentric chromosomes 13, 14, 15, 21, and 22. The genes encoding 5.0*S* rRNA are clustered together on the terminal portion of the p arm of chromosome 1. We will return momentarily to discuss this arrangement of tandemly repeated genes within the genome.

A second type of tandem repeats includes those called **variable number tandem repeats (VNTRs)**. The repeating DNA sequence or VNTRs may be 15 to 100 base pairs long. The number of tandem copies of any specific sequence varies in individuals, creating localized regions of 1000 to 5000 base pairs in length. As we saw in Chapter 16, such regions in humans are the basis for the forensic technique referred to as **DNA fingerprinting**. VNTRs may be found within and between genes.

Together, the various forms of moderately repetitive DNA comprise up to 30 percent of the human genome. Even including the portion of DNA that is highly repetitive, over 60 percent of the human genome is still unaccounted for. Such observations are not uncommon in eukaryotes. While the proportion of the genome consisting of repetitive DNA varies between organisms, one feature seems to be shared: Only a very small part of the genome codes for proteins. For example, the 20,000 to 30,000 genes encoding proteins in sea urchin occupy less than 10 percent of the genome. In *Drosophila*, only 5 to 10 percent of the genome is occupied by genes coding for proteins. In humans, it appears that only 1 to 2 percent of the genome encodes proteins! We must conclude that, although there is sufficient DNA to code for over 1 million genes in many eukaryotic organisms, *the vast majority of DNA does not represent genes encoding proteins.*

Eukaryotic Gene Structure

We turn now to a brief review of the structure of the eukaryotic gene, which was first introduced in Chapter 12. As our discussion ensues, you should refer to Figure 17.17, which presents a generalized model of the eukaryotic gene.

An important insight has been derived from a comparison of mRNA molecules and the DNA of the genes from which they are derived. It has been found that many

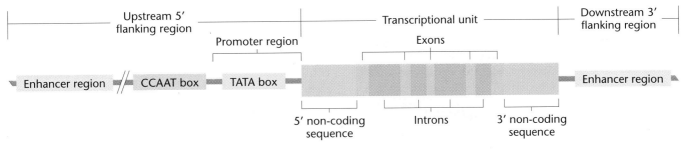

FIGURE 17.17   Modern concept of the eukaryotic gene.

TABLE 17.5 A Comparison of mRNAs and the Initial Transcripts from Which They Are Derived

Gene	mRNA Size	Minimum Transcript Length	Ratio
Rabbit β-globin	589	1295	2.2
Mouse β^{maj}-globin	620	1382	2.2
Mouse β^{min}-globin	575	1275	2.2
Mouse α-globin	585	850	1.5
Rat insulin I	443	562	1.3
Rat insulin II	443	1061	2.4
Chick ovalbumin	1859	7500	4.0
Chick ovomucoid	883	5600	6.3
Chick lysozyme	620	3700	6.0

internal base sequences of genes, referred to as *intervening sequences*, or **introns**, are not represented in the mature mRNAs, the final version of mRNA that is translated into protein. The remaining areas of each gene that are ultimately translated into the amino acid sequence of the encoded protein are called **exons**. While the entire gene is transcribed into a large precursor mRNA, which is part of a pool of other such molecules called **heterogeneous nuclear RNA (hnRNA)**, the intron regions of the initial transcript are excised and the exon regions of the transcript are spliced back together prior to translation.

Given this information about internal gene sequences, we can ask whether the nucleotide sequences found in introns solve the *C* value paradox. That is, do introns account for the remainder of the noncoding sequences beyond those of repetitive DNA? The answer is emphatically no! As shown in Table 17.5, the presence of introns may substantially increase the size of the average eukaryotic gene, However, introns represent no more than 10 percent of the genome, and usually much less. Therefore, we have yet to account for a great deal of noncoding DNA.

To complete our review of the eukaryotic gene, we shall discuss briefly both the 5′-upstream and 3′-downstream noncoding flanking regions. It is now clear that the 5′-region prior to the coding sequence is critical to efficient RNA transcription. Closest to the coding sequence is the **promoter** with its **TATA box**, where RNA polymerase ultimately binds as part of the initiation complex of transcription. The TATA box is so named because a sequence of that nature has been conserved in most eukaryotic genes. Farther upstream is the **CCAAT box**, which is somehow involved in the regulation of transcription. In numerous cases, an **enhancer** region is also present. Enhancers may also be found within the coding sequence or as part of the 3′-downstream region, beyond the coding sequence. These regions play a role in modulating transcription. Note also that in Figure 17.17 there are other 3′ and 5′ noncoding sequences that are part of the transcriptional unit that are trimmed off as the mRNA matures.

Since flanking regions are critical to specific gene function, their discovery has extended our knowledge of gene structure considerably.

Exon Shuffling and Protein Domains

At first glance, the discovery of introns is most puzzling. Why, during the evolution of eukaryotes, would their genes acquire noncoding sequences between the coding portions of genes whose presence requires highly precise splicing following transcription and whose replication requires a substantial expenditure of energy? While we do not yet have an exact answer, important clues are emerging primarily as a result of our ability to determine the nucleotide sequence of many genes. With the knowledge of the sequence of the exons of genes, the amino acid sequences of the corresponding proteins can be predicted accurately. When composite data bases are established and compared by computer analysis, the similarities of genes can be examined.

As a result of this approach, some very interesting findings have been forthcoming, as first recognized and proposed by Walter Gilbert in 1977. Gilbert suggested that genes in higher organisms may consist of collections of exons originally present in ancestral genes that are brought together through recombination during the course of evolution. Referring to this process as **exon shuffling**, Gilbert proposed that exons are modular in the sense that each might encode a **protein domain** specific to the function of the molecule (also called a **functional domain**). For example, a particular amino acid sequence encoded by a single exon might always create a specific type of fold in a protein. This domain may impart a characteristic type of molecular interaction to all molecules containing such a fold. Serving as the basis of useful parts of a protein, many similar domains might function in a variety of proteins. In Gilbert's proposal, during evolution, many exons could be mixed and matched to form unique genes in eukaryotes.

Several observations lend support to this proposal. Most exons are fairly small, averaging about 150 nucleotides and encoding about 50 amino acids. This size is consistent with the production of functional domains in proteins. Second, recombinational events that lead to exon shuffling would be expected to occur within areas of genes represented by introns. Because introns are free to accumulate mutations without harm to the organism, recombinational events would tend to further randomize their nucleotide sequences. Over extended evolutionary periods, diverse sequences would tend to accumulate. This is, in fact, what is observed. Introns range from 50 to 20,000 bases and exhibit fairly random base sequences.

Since 1977, a vast research effort has been directed toward the analysis of gene structure. In 1985, more di-

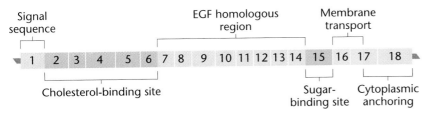

rect evidence in favor of Gilbert's proposal of exon modules was presented. For example, the human gene encoding the membrane receptor for low density lipoproteins (LDL) was isolated and sequenced. The **LDL receptor protein** is essential to the transport of plasma cholesterol into the cell. It mediates **endocytosis** and is expected to have numerous functional domains. These include the capability of this protein to bind specifically to the LDL substrates and to interact with other proteins at different levels of the membrane during transport across it. In addition, this receptor molecule is modified posttranslationally by the addition of a carbohydrate; a domain must exist that links to this carbohydrate.

Detailed analysis of the gene encoding this protein supports the concept of exon modules and their shuffling during evolution. The gene is quite large—45,000 nucleotides—and contains 18 exons. These represent only slightly less than 2600 nucleotides. These exons are related to the functional domains of the protein *and* appear to have been recruited from other genes during evolution.

Figure 17.18 illustrates these relationships. The first exon encodes a signal sequence that is removed from the protein before the LDL receptor becomes part of the membrane. The next five exons represent the domain specifying the binding site for cholesterol. This domain is made up of a 40-amino-acid sequence repeated seven times. The next domain consists of a sequence of 400 amino acids bearing a striking homology to the mouse peptide hormone **epidermal growth factor (EGF)**. This region is encoded by eight exons and contains three repetitive sequences of 40 amino acids. A similar sequence is also found in three blood-clotting proteins. The fifteenth exon specifies the domain for the posttranslational addition of the carbohydrate, while the remaining two exons specify regions of the protein that are part of the membrane, anchoring the receptor to specific sites called *coated pits* on the cell surface.

The above observations concerning the LDL exons are fairly compelling in favor of the theory of exon shuffling during evolution. Certainly, there is no disagreement concerning the concept of protein domains being responsible for specific molecular interactions.

What remains controversial and evocative in the exon shuffling theory is the question of when introns first appeared on the evolutionary scene. In 1978, W. Ford Doolittle proposed that these intervening sequences were part of the genome of the most primitive ancestors of

modern-day eukaryotes. In support of this idea, Gilbert has argued that if intron similarities in DNA sequence are found in identical positions within genes shared by totally unrelated eukaryotes (such as humans, chickens, and corn), they must have also been present in primitive ancestral genomes.

If Doolittle's proposal is correct, why are introns absent in most prokaryotes? Gilbert argues that they were present at one point during evolution, but as the genome of these primitive organisms evolved, they were lost. This occurred as a result of strong selection pressure to streamline their chromosomes to minimize energy expenditure supporting replication and gene expression. Further, streamlining leads to more error-free mRNA production. However, supporters of the opposing "intron-late" school, including Jeffrey Palmer, argue that introns first appeared much later during evolution, when they became a part of a single group of eukaryotes that are ancestral to modern-day members but not prokaryotes. Although it may be difficult to resolve, it seems certain that this controversy will persist for many years to come.

Multigene Families: The alpha- and beta-Globin Genes

Still another aspect of the organization of the eukaryotic genome includes the distribution of related genes. Called **multigene families**, members share DNA sequence homology, and their gene products are functionally related. Often, but not always, they are found together in a single location along a chromosome. The examination of multigene families provides further insights into the *C* value paradox.

We will first examine cases where groups of genes encode very similar, but not identical, polypeptide chains that become part of proteins that are very closely related in function. The globin gene families, responsible for encoding the various polypeptides that are part of hemoglobin molecules, are examples that provide many insights into the organization of the genome. Next, we will consider the case of identical, tandemly repeated genes, as represented by histone and ribosomal RNA genes.

The human **alpha-** and **beta-globin gene families** are two of the most extensive and best characterized groups. The alpha family resides on the short arm of chromosome 16 and contains five genes, whereas the beta

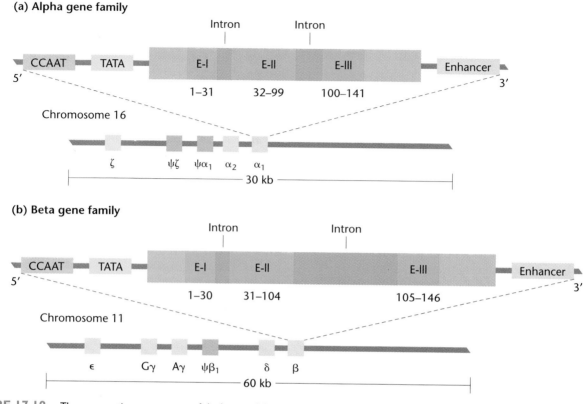

FIGURE 17.19 The comparative arrangement of the human alpha- and beta-gene families, depicted in parts (a) and (b), respectively. Also shown is internal organization of the functional genes from each family. All genes contain three exons (E—I, II, and III) and two introns, as well as 5′ and 3′ noncoding regions. The arabic numbers designate the amino acid positions encoded by each exon.

family resides on the short arm of chromosome 11 and contains six genes. Members of both families show homology to one another, but not nearly to the extent that members within the same family exhibit homology to one another. Members of both families encode globin polypeptides that combine into a single tetrameric molecule, which interacts with heme groups that can reversibly bind to oxygen (see Figure 13.13). And, within each group of genes, members are coordinately turned on and off during the embryonic, fetal, and adult stages of development in the precise order in which they occur along the chromosome within each family.

The alpha family [Figure 17.19(a)] spans more than 30,000 nucleotide pairs (30 kb) and contains five genes: the zeta gene (ζ), expressed only in the embryonic stage, two nonfunctional **pseudogenes**, and two copies of the alpha (α) gene, expressed during the fetal ($\alpha2$) and adult ($\alpha1$) stages. Pseudogenes are found in both families, are similar in sequence to the other genes in their family, but contain significant nucleotide substitutions, deletions, and duplications that prevent their transcription. The first pseudogene in the alpha family shows greater homology to the zeta gene, whereas the second shows greater similarity to the alpha gene. Pseudogenes are designated by

the prefix ψ (the Greek letter psi), followed by the symbol of the gene they most resemble. Thus, the designation $\psi\alpha1$ indicates a pseudogene of the adult alpha gene.

Examination of the organization of the members of the alpha family as well as their intragenic distribution of introns and exons [Figure 17.19(a)] reveals several interesting features. First, only a small portion of the chromosomal region housing the entire family is occupied by the three functional genes. The region consists almost entirely of intergenic regions. Second, the functional genes in this family contain two introns at precisely the same positions within the genes. The nucleotide sequences within corresponding exons are nearly identical when the zeta and alpha genes are compared. Both encode polypeptide chains that are 141 amino acids long. However, the intron sequences are extremely divergent, even though they are about the same size. Significantly, a high percentage of the nucleotide sequence within each gene is contained in these noncoding introns. The above information bears on both (a) the structural organization of a gene family and (b) our earlier discussion of the *C* value paradox.

The beta-globin family in humans is even more extensive than the alpha-globin family, containing six genes and occupying a 60-kb sequence of DNA [Figure 17.19(b)].

TABLE 17.6 The Number of Tandemly Repeated Histone Gene Clusters Found in a Variety of Eukaryotic Organisms

Organism	Repeats
Chickens	10
Mammals	22
Xenopus	40
Drosophila	100
Sea urchins	300–600

As with the alpha family, the sequence of genes along the chromosome parallels the order of expression during development.

Of the six genes, three are expressed prior to birth: the epsilon (ε) gene is embryonic, while two nearly identical gamma genes ($^G\gamma$ and $^A\gamma$) are fetal. The two gamma gene products vary by only a single amino acid. The two remaining functional genes, delta (∂) and beta (β), are expressed following birth. Finally, a single pseudogene ψβ1 is present within the family.

All five functional genes encode products that are 146 amino acids long and contain two similarly sized introns at exactly the same positions. The second intron is significantly larger than its counterpart in the functional alpha genes. These introns and their locations within the gene are compared in Figure 17.19(b).

The flanking regions of each gene have also been studied. Both a TATA box (-30) and a CCAAT region (-75 to -78) have been observed upstream from the transcriptional unit. Downstream, globin enhancers have been identified.

Several observations of the β-globin genes pertain to the *C* value paradox. Only about 5 percent of the 60-kb region consists of coding sequences. The remaining 95 percent includes the introns, flanking regions, and spacer DNA found between genes. Of this percentage, only about 11 percent consists of introns. The remainder of the DNA serves no known function and consists of sequences not found elsewhere in the genome.

The Histone Gene Family

A variation on the theme established in the globin gene families is illustrated by the **histone genes**. Here, a cluster of five related but nonidentical genes, separated from each other by highly divergent nontranscribed spacer sequences, is tandemly repeated many times. As you may recall, histones are positively charged (basic) proteins that interact with the negatively charged phosphate groups of DNA to form the nucleosome characteristic of the chromatin fiber. The need for histone proteins is extremely great because of the large amount of DNA in eukaryotic chromosomes. In rapidly dividing cells of some organisms, not only is the need great, but the necessary quantity must appear rapidly as one cell cycle runs into the next, time and time again, without pause to catch up with the cell's biochemical needs. To accommodate this need, evolution appears to have seized upon the duplication of the entire cluster in order to increase the number of copies of each histone gene.

Many interesting correlations exist that support this contention. Yeast cells, with much less DNA than other eukaryotes, have only two copies of each of four histone genes (they lack histone H1). Contrast this with the number of times the entire cluster is tandemly repeated in many evolutionarily advanced organisms, as displayed in Table 17.6. From 10 to 600 repeats occur. Sea urchins, which display the largest number of duplications of the cluster, are noted for their extremely rapid rate of cell division during development.

Beyond the repeat nature of the histone gene family, there are several other differences in comparison to the globin families. First, almost all histone genes lack introns. The entire cluster of genes is considerably smaller than either globin cluster. While some variation exists between organisms, due primarily to the different sizes of the noncoding spacer sequences, most clusters are less than 10 kb in size. The other interesting difference is the observation that the individual genes within the cluster of a given species are often oriented in opposite directions.

As shown in Figure 17.20, the arrangement of genes and their polarity of transcription vary in the sea urchin and *Drosophila*, two organisms where this cluster has been well studied. Polarity differences also exist in other organisms studied. Despite polarity differences, the amino acid sequences of the various histones (reflecting the DNA encoding each one) are remarkably similar in highly divergent organisms. Homology of histone genes is one of the best examples known of sequence conservation throughout evolution.

In mice and humans, members of this gene family appear to be clustered but not present as a tandem array of repeat units. For example, a region of the distal tip of the long arm of the human chromosome 7 contains all members of the histone gene family, but they may be interspersed with other nonhistone genes. Therefore, no single pattern of organization applies to all organisms.

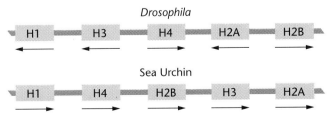

FIGURE 17.20 The histone gene clusters in *Drosophila* and the sea urchin. Arrows indicate the polarity of transcription.

Tandem Repeat Families: rRNA Genes

We conclude our discussion of the organization of multi-gene families by considering the case where numerous identical copies of a gene exist in a contiguous arrangement within the genome of an organism. Such an arrangement is referred to as a **tandem repeat gene family**. A good example is the gene family encoding rRNA molecules present in ribosomes. Each member of this gene family encodes three separate molecules in the following order along the chromosome: 18*S*, 5.8*S*, and 28*S* rRNAs (Figure 17.21). These RNAs are transcribed as a single unit and are not translated. Each repeated sequence of DNA contains substantial regions between each gene that are part of the initial RNA transcript, but that are not found in the final rRNA gene products. The initial transcript is processed in a manner *analogous* to the removal of introns from the primary transcript of most mRNAs. In addition, there is a nontranscribed spacer region (clearly visible in the electron micrographic inset in Figure 17.21) between each gene family unit. Note that one other RNA component—5.0*S* rRNA—is encoded elsewhere in the genome.

The basic repeating unit that is transcribed varies in length in different organisms from almost 7 kb in yeast to 13 kb in humans (Figure 17.21). Together, each gene family unit and its accompanying spacer regions occupy a substantial portion of DNA. In the human rRNA gene family, a length of about 43 kb of DNA is utilized in each tandem repetition. This amount of DNA is as large as the smaller globin gene cluster.

Multiple copies of a single transcriptional unit encoding rRNAs also exist in bacteria. In *E. coli* seven dispersed copies are present, presumably to accommodate the need for the rapid production of the approximately 15,000 ribosomes found in each cell. It is not surprising, therefore, to find many more than seven copies in the larger, more complex eukaryotic cells. Yeast contains over 100, while *Xenopus, Drosophila*, and humans have up to 400 copies per genome. A single eukaryotic cell may have a million or more ribosomes if it is actively producing proteins.

If all of these genes were present in a single cluster in an organism such as humans, a substantial part of one chromosome would be occupied by them (400 × 45 kb = 18,000 kb or 18 million base pairs). This amount of DNA is almost five times greater than the entire *E. coli* chromosome (4 × 10^6 base pairs)! As it turns out, they are present in clusters dispersed throughout the genome. In humans, as we mentioned in our earlier discussion of LINEs, clusters are found near the ends of chromosomes 13, 14, 15, 21, and 22. In *Drosophila*, they are present on both the X and Y chromosomes. Within the nucleus, these various gene regions are collectively referred to as the **nucleolus organizer regions (NORs)**, because of their association with the production of ribosomes in the nucleolus.

This information extends our knowledge and overall picture of how some genes are organized within the genome. It also yields insights into the vast amount of DNA that may be utilized in order to provide cells with an adequate quantity of specific gene products. Examination of the same gene family in different organisms also yields information valuable to the study of evolution at the molecular level.

Genomic Analysis

We conclude this chapter with a brief overview of what is unquestionably one of the most exciting areas of current investigation in the field of genetics: *characterization of the DNA sequences constituting the comparative genomes of a variety of model species, including humans.* The organisms that are being studied have been selected based on the rich ge-

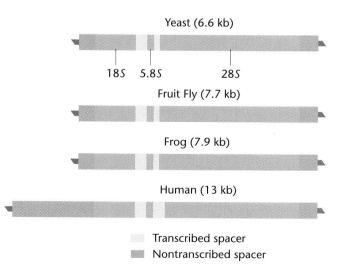

FIGURE 17.21 Comparison of the repeating rRNA gene unit in various eukaryotic organisms. The electron micrograph shows several rRNA genes undergoing transcription. The arrow identifies an intergenic region of DNA.

netic background available for each of them. As a result, knowledge of DNA sequences from these genomes will provide the basis for confirming, fine-tuning, and integrating previous genetic findings as well as opening new lines of investigation based on previously undiscovered information. Additionally, the organisms represent many stages of evolution, thus promising to provide valuable insights into this important biological process.

Up to this point in the chapter we have discussed information involving various aspects of the organization of DNA in many different organisms. The complete characterization of representative genomes will allow us to extend our knowledge by pursuing two fundamental questions concerning our understanding of life: (1) What is the minimal genetic requirement to exist as a free-living organism?; and (2) What is the genetic difference between less biologically complex organisms and more biologically complex organisms? Additionally, at the completion of these research efforts, we shall have available the specific genetic blueprints that prescribe the organisms under investigation, providing the basis to assess the genetic (and thus the evolutionary) relationships among them.

The Genome Project

When the technology to rapidly clone and sequence DNA was developed, as previously described in Chapter 15, it was only logical to consider sequencing the entire genome of organisms, including humans. With regard to our own species, discussion about the feasibility began in the mid-1980s. Such discussions soon turned into a debate concerning the political and ethical implications of acquiring such information. By 1990, the goals were established, an international effort was agreed upon, and the project was launched. As Table 17.7 reveals, six previously well-studied organisms, in addition to humans, have been included in the formal study. In the United States, approximately $200 million in government funding has been provided during each of the first five years primarily by the National Institutes of Health and the Department of Energy in support of the project, which is scheduled for completion about the year 2003.

In addition to the seven organisms studied under the auspices of the Genome Project, Table 17.7 includes information for two additional organisms whose genomes have now been completely sequenced as a result of separate research initiatives—*Mycoplasma genitalium* and *Hemophilus influenzae*. Examination of the information involving the nine organisms reveals an interesting comparison of the genome sizes and the estimated number of genes in each organism. The number of nucleotide pairs ranges from about one-half million (0.5 Mb) in *Mycoplasma* to 3 billion (3000 Mb) in humans. The number of genes ranges from a low of 470 in the above bacterium to 80,000 in humans. It is fascinating to dissect this information. The complexity of the three different bacterial forms is clearly related to an increase in genome size and gene number. Yeast, as the prototypic single-celled eukaryote, has further expanded its genome size and gene number. The plant and the two invertebrates have almost 10 times more DNA than yeast, but only two to four times as many genes! The two mammals (the mouse and humans) reveal about a 30-fold jump in DNA content compared to the invertebrates, but less that a tenfold increase in gene number.

When these figures for all nine organisms are compared, the most apparent finding involves the density of coding regions in each genome. For the three bacteria (*Mycoplasma*, *Hemophilus*, and *E. coli*) there are about 1000 genes per Mb unit of DNA. In the unicellular yeast, the density decreases to less than one-half this value. There is a further progressive decline in gene density as one com-

TABLE 17.7 A Comparison of the Genome Size and Estimated Number of Genes in Organisms Whose Genome Has Been or Is Being Sequenced

Organism/Type	Genome Size Base Pairs (Megabases [Mb])	Estimated Number of Genes
Mycoplasma genitalium (bacterium)*	0.58×10^6 (0.58)	470
Hemophilus influenzae (bacterium)*	1.8×10^6 (1.80)	1,727
Escherichia coli (bacterium)	4.2×10^6 (4.20)	4,000
Saccharomyces cerevisiae (yeast)	1.5×10^7 (15)	6,000
Arabidopsis thaliana (plant)	1.0×10^8 (100)	25,000
Caenorhabditis elegans (roundworm)	1.0×10^8 (100)	13,000
Drosophila melanogaster (fruit fly)	1.2×10^8 (120)	10,000
Mus musculus (mouse)	3.0×10^9 (3000)	80,000
Homo sapiens (human)	3.0×10^9 (3000)	80,000

*Organisms not part of the formal Genome Project.

TABLE 17.8   A Partial List of Gene Content Classified According to Function in *Mycoplasma genitalium* vs. *Hemophilus influenzae*

Biological Function	Number of Genes	
	Mycoplasma	*Hemophilus*
DNA Replication	32	87
Transcription	12	27
Translation	101	141
Cell Processes	21	53
Transport	34	123
Energy Metabolism	31	112
Nucleic Acid Metabolism	19	53
Fatty Acid and Phospholipid Metabolism	6	25
Amino Acid Biosynthesis	1	68

pares the model plant species (*Arabidopsis*), to the roundworm (*Caenorhabditis*), to the fruit fly (*Drosophila*), and then to mice and humans, which are roughly equivalent, averaging only about 25 genes per Mb of DNA. These findings represent a definite evolutionary trend, where more advanced genomes have acquired extensive regions of DNA that fail to encode gene products. In *Mycoplasma* and *Hemophilus*, 85–90 percent of all DNA constitutes functional gene areas, whereas in humans it is doubtful if as much as 10 percent will be found to represent functional coding regions.

The Minimal Coding Set

We now turn to the question of what minimal set of genes is essential to survival by a free-living organism. To answer this question we shall draw from the complete genomic sequence information now available for *Mycoplasma genitalium* and then make some comparisons with *Hemophilus influenzae*. *Mycoplasma* is one of the simplest self-replicating prokaryotes known. It is a member of a group of bacteria that lack a cell wall, have a characteristically low G + C content in their DNA, and invade and often cause disease states in a wide range of hosts, including humans (genital and respiratory tracts), insects, and plants.

The complete nucleotide sequence consists of 580,070 base pairs containing 470 predicted coding regions (genes). This is in comparison to 1,800,000 base pairs and 1727 coding regions in the more complex genome of the bacterium *Hemophilus*. Table 17.8 summarizes the biological role ascribed to the genes of both of these bacteria. As one might suspect, genes must exist for DNA replication and repair; for transcription and translation, including rRNA, ribosomal proteins, many tRNAs, and aminoacyl synthetases; for transport proteins; for general cellular processes, including cell division and secretion; for countless biochemical pathways involving energy metabolism; and for biosynthetic pathways ranging from the synthesis of all nucleic acid components to the synthesis of fatty acids and phospholipids. Genes for all such functions have been identified in *Mycoplasma* and establish the minimal complement necessary to exist as a self-reproducing organism that can thrive independently in nature.

Despite having nearly four times as many genes, the physiological capabilities of *Hemophilus* are very similar to *Mycoplasma*, although there appears to be many more genes involved in each general category of function. The greatest advance in *Hemophilus* involves its marked increase in biosynthetic capability. For example, this more complex bacterium demonstrates 68 genes involved in amino acid biosynthesis, whereas *Mycoplasma* has only one such gene. Without this capability *Mycoplasma* must rely on numerous metabolic products from its host.

Within several years we can expect similar comparisons to be forthcoming regarding eukaryotic genomes. At this writing, the complete genome of yeast has now been sequenced. Within a year or two, the genome of *E. coli* will become available. Without question, our understanding of the living process will be extended, and we will be able to define in great detail the genomic characterization of many species, including our own. We will then be capable of approaching other relevant questions concerning the organization of the components that make up the genome of more complex eukaryotes.

Until then, we are left to wonder why evolution has led to such an increase in DNA content in higher forms, and whether some functional role will be ascribed to what currently appears to be excess genetic material. We can also continue to contemplate and appreciate the ingenuity and brilliance that has led to the technology that is about to provide the nucleotide sequence of 3 billion base pairs constituting our own genome.

CHAPTER SUMMARY

1. Knowledge of the organization of the molecular components forming chromosomes is essential to the understanding of the function of the genetic material. Largely devoid of associated proteins, bacteriophage and bacterial chromosomes contain DNA molecules in a form equivalent to the Watson–Crick model.

2. Mitochondria and chloroplasts contain DNA that encodes products essential to their biological function. This DNA is remarkably similar in form and appearance to some bacterial and phage DNA, lending support to the endosymbiotic hypothesis, which suggests that these organelles were once free-living organisms.

3. The eukaryotic chromatin fiber is a nucleoprotein organized into repeating units called nucleosomes. Composed of about 200 base pairs of DNA and an octamer of four types of histones, the nucleosome is important in facilitating the conversion of the extensive chromatin fiber characteristic of interphase into the highly condensed chromosome seen in mitosis.

4. The structural heterogeneity of the chromosome axis has been established as a result of both biochemical and cytological investigation. Heterochromatin, prematurely condensed in interphase, is genetically inert. The centromeric and telomeric regions, the Y chromosome, and the Barr body are examples.

5. DNA analysis has unique nucleotide sequences in both the heterochromatic centromere and telomere regions. These sequences distinguish these structures from other parts of the chromosome and provide the basis for the critical features imparted to the chromosome by these regions.

6. The *C* value paradox, made apparent during the study of the organization of eukaryotic DNA, suggests that eukaryotic organisms contain much more DNA than necessary to encode only gene products essential to the development and normal functions of an organism.

7. Detailed analysis of eukaryotic gene structure has revealed numerous categories of noncoding DNA sequences, much of which include various forms of repetitive DNA. Introns, which are internal noncoding sequences, are represented in the initial mRNA molecules or heterogeneous RNA, but they never appear in the mature mRNAs from which proteins are synthesized.

8. The flanking regions of genes, particularly those giving rise to the 5′-initiation point of mRNA, have also been analyzed. Three upstream regions that appear to be essential to efficient transcription of mRNA are the TATA box, the CCAAT box, and the enhancer element.

9. Multigene families are composed of structurally related genes that are often clustered together and encode functionally similar products. The human alpha- and beta-globin families are two examples.

10. Histone and ribosomal RNA genes provide an example of tandem repeat families, where the same gene is duplicated many times.

11. A massive research initiative is now under way to provide comparative genome sequencing information for a variety of organisms, from bacteria to humans. When complete, the data promise to provide comprehensive insights into the organization of DNA as it serves as the genetic material.

KEY TERMS

alpha- and beta-globin gene families, 509
alphoid family, 505
Alu family, 507
bacteriophage lambda, 491
bacteriophage φX174, 491
CCAAT box, 508
C-banding, 502
CEN region, 503
central scaffold, 500
centromere, 503
chloroplast, 494
chloroplast DNA (cpDNA), 496
chromatin, 497

chromosome banding techniques, 502
C value paradox, 503
DNA-binding proteins, 492
DNA fingerprinting, 507
E. coli, 513
endocytosis, 509
endosymbiotic hypothesis, 497
enhancer, 508
epidermal growth factor (EGF), 509
euchromatin, 500
eukaryotic cells, 497
exon, 508
exon shuffling, 508
folded-fiber chromosome model, 497

functional domain, 508
G-bands, 502
genome, 502
Hemophilus influenzae, 513
heterochromatin, 500
heterogeneous nuclear RNA (hnRNA), 508
histone genes, 511
H and HU proteins, 492
in situ hybridization, 501
intron, 508
LDL receptor protein, 509
linker DNA, 498
linking number (*L*), 493

INSIGHTS AND SOLUTIONS

A previously undiscovered single-celled organism was found living at a great depth on the ocean floor. Its nucleus contained only a single linear chromosome containing 7×10^6 nucleotide pairs of DNA coalesced with three types of histonelike proteins.

1. A short micrococcal nuclease digestion yielded DNA fractions consisting of 700, 1400, and 2100 base pairs. What do these fractions represent? What conclusions can be drawn?

Solution: The chromatin fiber may consist of a variation of nucleosomes containing 700 base pairs of DNA. The 1400- and 2100-bp fractions represent two and three nucleosomes, respectively, linked together. Enzymatic digestion may have been incomplete, leading to the latter two fractions.

2. Analysis of individual nucleosomes revealed that each unit contained one copy of each protein, and that the short linker DNA contained no protein bound to it. If the entire chromosome consists of nucleosomes (discounting any linker DNA), how many are there, and how many total proteins are needed to form them?

Solution: Since the chromosome contains 7×10^6 base pairs of DNA, the number of nucleosomes, each containing 7×10^2 base pairs, is equal to

$$\frac{7 \times 10^6}{7 \times 10^2} = 10^4 \text{ nucleosomes}$$

The chromosome thus contains 10^4 copies of each of the three proteins, for a total of 3×10^4 molecules.

3. Analysis then revealed the above organism's DNA to be a double helix similar to the Watson–Crick model, but containing 20 base pairs per complete turn of the right-handed helix. The physical size of the nucleosome was exactly double the volume occupied by that found in all other known eukaryotes, by virtue of increasing the distance along the fiber axis by a factor of 2. Compare the degree of compaction of this organism's nucleosome to that found in other eukaryotes.

Solution: The unique organism compacts a length of DNA consisting of 35 complete turns of the helix (700 base pairs per nucleosome/20 base pairs per turn) into each nucleosome. The normal eukaryote compacts a length of DNA consisting of 20 complete turns of the helix (200 base pairs per nucleosome/10 base pairs per turn) into a nucleosome one-half the volume of that in the unique organism. The degree of compaction is therefore less in the unique organism.

4. No further coiling or compaction of this unique chromosome occurs in the unique organism. Compare this to a eukaryotic chromosome. Do you think an interphase human chromosome 7×10^6 base pairs in length would be a shorter or longer chromatin fiber?

Solution: The eukaryotic chromosome contains still another level of condensation in the form of "solenoids," which are dependent on the H1 histone molecule associated with linker DNA. Solenoids condense the eukaryotic fiber by still another factor of 5.

The length of the unique chromosome is compacted into 10^4 nucleosomes, each containing an axis length twice that of the eukaryotic fiber. The eukaryotic fiber consists of $7 \times 10^6/2 \times 10^2 = 3.5 \times 10^4$ nucleosomes, 3.5 more than the unique organism. However, they are compacted by the factor of 5 in each solenoid. Therefore, the chromosome of the unique organism is a longer chromatin fiber.

PROBLEMS AND DISCUSSION QUESTIONS

1. Compare and contrast the chemical nature, size, and form assumed by the genetic material of viruses and bacteria.
2. Contrast the DNA associated with mitochondria and chloroplasts.
3. Why might it be predicted that the organization of eukaryotic genetic material would be more complex than that of viruses or bacteria?
4. Describe the sequence of research findings leading to the model of chromatin structure. What is the molecular composition and arrangement of the nucleosome? What is a solenoid?
5. When chloroplasts and mitochondria are isolated, they are found to contain ribosomes that are similar to prokaryotic cells. How does this observation relate to the endosymbiotic hypothesis?
6. Provide a comprehensive definition of heterochromatin and list as many examples as you can.
7. Mammals contain a diploid genome consisting of at least 10^9 base pairs. If this amount of DNA is present as chromatin fibers where each group of 200 base pairs of DNA is combined with 9 histones into a nucleosome, and each group of 5 nucleosomes is combined into a solenoid, achieving a final packing ratio of 50, determine:
   (a) The total number of nucleosomes in all fibers.
   (b) The total number of solenoids in all fibers.
   (c) The total number of histone molecules combined with DNA in the diploid genome.
   (d) The combined length of all fibers.
8. Assume that a viral DNA molecule is in the form of a 50-μm-long circular strand of a uniform 20-Å diameter. If this molecule is contained within a viral head that is a sphere with a diameter of 0.08 μm, will the DNA molecule fit into the viral head, assuming complete flexibility of the molecule? Justify your answer mathematically.
9. How many base pairs are in a molecule of phage T2 DNA, which is 52 μm long?
10. Describe what is meant by exon shuffling.
11. The β-globin gene family consists of 60 kb of DNA yet only 5 percent of the DNA encodes β-globin gene products. Account for as much of the remaining 95 percent of the DNA as you can.
12. Compare the histone gene family and the rRNA gene family with the globin gene families.
13. Describe the rRNA gene family in eukaryotes. What accounts for the major difference in the size of tandem repeats in different organisms?
14. "In 1997, the C value paradox is not so paradoxical." Agree or disagree with this statement, and support your position.

EXTRA-SPICY PROBLEMS

15. An undergraduate genetics student in the year 2025 was provided DNA samples from three unknown organisms and asked to sequence each genome, which she did. The preliminary information concerning genome size and the number of genes is provided below. The student was able to make a reasonable prediction as to what type of organism was involved in two of the three cases. Which two were these and what type of organisms do you think were being examined? Discuss the third case and indicate why the student had a difficult time predicting what organism was involved.

Organism	Genome Size (Mb)	Number of Genes
A	120	142,000
B	0.72	925
C	300	50,000

16. The existence of introns was first revealed in the 1970s by so-called R-looping experiments. In such experiments, purified, mature mRNA is allowed to hybridize with cloned DNA of the gene from which the mRNA originated under conditions where an RNA–DNA double strand is more stable than a DNA–DNA double strand. The results are then analyzed using electron microscopy.

    Study the gene shown below. Then draw the RNA–DNA hybrid that will result from an R-looping experiment. Be sure to identify the DNA and RNA strands and to label the 5'- and 3'-ends of each strand. Make your drawing accurate and approximately to scale.

5'   3'

- Promoter
- Non-coding (untranslated) regions
- Introns
- Exons

17. A graphic method for comparing the chemical structure of two genes is shown in the dot matrix plots on the next page. In both cases, the nucleotide sequences for the two genes are compared, and for each interval of 10 bases, similarity is assessed. If 8 of the 10 bases are identical in a given interval, a dot is placed on the plot at the matching coordinate of the two genes. If fewer than 8 bases are identical, no dot is placed on the plot.

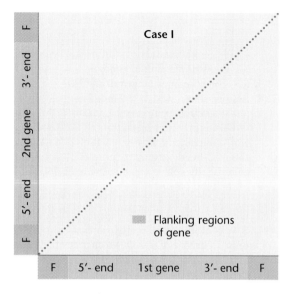

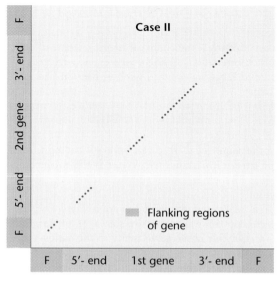

Shown above are two hypothetical cases. One compares the sequence of a fungal gene (encoding the same protein), but isolated from two closely related species. The other compares the sequence of a mammalian gene (encoding the same protein), but isolated from the mouse and from humans.

(a) Given what you have learned in this chapter about the organization of eukaryotic genes, how do you interpret these two plots? How do they compare, and how do they differ? How do you interpret the differences?

(b) Which plot appears to be derived from the comparison of the fungal gene, and which is derived from the comparison of the mammalian gene? Explain.

SELECTED READINGS

BAUER, W. R., CRICK, F. H. C., and WHITE, J. H. 1980. Supercoiled DNA. *Sci. Am.* (July) 243:118–33.

BLACKBURN, E. H. 1991. Structure and function of telomeres. *Nature* 350:569–73.

BLACKBURN, E. H., and SZOSTAK, J. W. 1984. The molecular structure of centromeres and telomeres. *Annu. Rev. Biochem.* 53:163–94.

BLOOM, K., HILL, A., and YEH, E. 1986. Structural analysis of a yeast centromere. *BioEssays* 4:100–05.

CARBON, J. 1984. Yeast centromeres: Structure and function. *Cell* 37:352–53.

CHEN, T. R., and RUDDLE, F. H. 1971. Karyotype analysis utilizing differential stained constitutive heterochromatin of human and murine chromosomes. *Chromosoma* 34:51–72.

COLLINS, F. S. 1995. Ahead of schedule and under budget: The Genome Project passes its fifth birthday. *Proc. Natl. Acad. Sci. USA* 92: 10821–23.

CORNEO, G., et al. 1968. Isolation and characterization of mouse and guinea pig satellite DNA. *Biochemistry* 7:4373–79.

DORIT, R. L., and GILBERT, W. 1991. The limited universe of exons. *Curr. Opin. Genet. Dev.* 1:464–69.

FRASER, C. M., et al. 1995. The minimal gene complement of *Mycoplasma genitalium. Science* 270:397–403.

FRITSCHE, E. F., LAWN, R. M., and MANIATIS, T. 1980. Molecular cloning and characterization of the human β-like globin gene cluster. *Cell* 19:959–72.

GALL, J. G. 1981. Chromosome structure and the *C*-value paradox. *J. Cell Biol.* 91:3s–14s.

GILBERT, W., MARCHIONNA, M., AND McKNIGHT, G. 1986. On the antiquity of introns. *Cell* 46:151–53.

GREEN, B. R., and BURTON, H. 1970. *Acetabularia* chloroplast DNA: Electron microscopic visualization. *Science* 168: 981–82.

GREIDER, C. N., and BLACKBURN, E. H. 1996. Telomeres, telomerases, and cancer. *Sci. Am.* (Feb.) 274:92–97.

HENTSCHEL, C. C., and BIRNSTIEL, M. L. 1981. The organization and expression of histone gene families. *Cell* 25: 301–13.

HEWISH, D. R., and BURGOYNE, L. 1973. Chromatin substructure. The digestion of chromatin DNA at regularly spaced sites by a nuclear deoxyribonuclease. *Biochem. Biophys. Res. Comm.* 52:504–10.

HSU, T. C. 1973. Longitudinal differentiation of chromosomes. *Annu. Rev. Genet.* 7:153–77.

KORENBERG, J. R., and RYKOWSKI, M. C. 1988. Human genome organization: *Alu*, LINEs and the molecular structure of metaphase chromosome bands. *Cell* 53:391–400.

KORNBERG, R. D. 1975. Chromatin structure: A repeating unit of histones and DNA. *Science* 184:868–71.

KORNBERG, R. D., and KLUG, A. 1981. The nucleosome. *Sci. Am.* (Feb.) 244:52–64.

MANIATIS, T., et al. 1980. The molecular genetics of human hemoglobins. *Annu. Rev. Genet.* 14:145–78.

MOYZIS, R. K. 1991. The human telomere. *Sci. Am.* (Aug.) 265:48–55.

OLINS, A. L., and OLINS, D. E. 1974. Spheroid chromatin units (v bodies). *Science* 183:330–32.

———. 1978. Nucleosomes: The structural quantum in chromosomes. *Am. Sci.* 66: 704–11.

PALMER, J. D., and LOGSDON, J. 1991. The recent origins of introns. *Curr. Opin. Genet. Dev.* 1:470.

SCHMID, C. W., and JELINEK, W. R. 1982. The *Alu* family of dispersed repetitive sequences. *Science* 216:1065–70.

SINGER, M. F. 1982. SINEs and LINEs: Highly repeated short and long interspersed sequences in mammalian genomes. *Cell* 28:433–34.

STOLTZFUS, A., et al. 1994. Testing the exon theory of genes: The evidence from protein structure. *Science* 265:202–207.

SUDHOF, T. C., et al. 1985. The LDL receptor gene: A mosaic of exons shared with different proteins. *Science* 228:815–22.

VAN HOLDE, K. E. 1989. *Chromatin.* New York: Springer-Verlag.

VERMA, R. S., ed. 1988. *Heterochromatin: Molecular and structural aspects.* Cambridge: Cambridge University Press.

WILLARD, H. F. 1990. Centromeres of mammalian chromosomes. *Trends Genet.* 6:410–416.

WILLARD, H. F., and WAYE, J. S. 1987. Hierarchical order in chromosome-specific human alpha satellite DNA. *Trends Genet.* 3:192–198.

YUNIS, J. J. 1976. High resolution of human chromosomes. *Science* 191:1268–70.

———. 1981. Chromosomes and cancer: New nomenclature and future directions. *Hum. Pathol.* 12:494–503.

YUNIS, J. J., and PRAKASH, O. 1982. The origin of man: a chromosomal pictorial legacy. *Science* 215:1525–30.

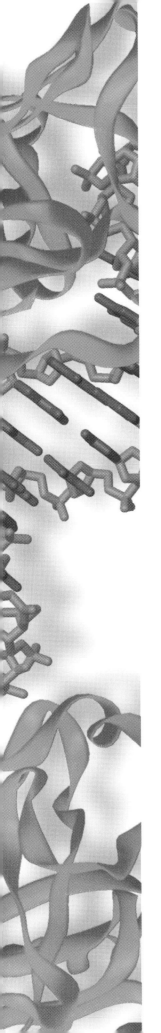

Advanced Topics in Genetic Analysis

18

Regulation of Gene Expression in Bacteria and Phages

CHAPTER CONCEPTS

Efficient expression of genetic information is dependent on regulatory mechanisms that either activate or repress gene activity. In bacteria, mechanisms exist that regulate genes in ways that meet metabolic needs of the cell. Gene regulation in bacteriophages determines the mode of existence of the virus within the bacterial host. In bacteria and their phages, these mechanisms provide responses to the prevailing cellular and extracellular conditions.

In earlier chapters we established how DNA is organized into genes, how genes store genetic information, and how this information is expressed. We now consider one of the most fundamental issues in molecular genetics: *How is genetic expression regulated?* The evidence in support of the idea that genes can be turned on and off is very convincing. Detailed analysis of proteins in *Escherichia coli* has shown that concentrations of the 4000 or so polypeptide chains encoded by the genome vary widely. Some proteins may be present in as few as 5 to 10 molecules per cell, whereas others, such as ribosomal proteins and the many proteins involved in the glycolytic pathway, are present in as many as 100,000 copies per cell. While a basal level (a few copies) of most gene products exists continuously,this level can be dramatically increased and subsequently decreased in prokaryotes. This makes it clear that fundamental regulatory mechanisms must exist to control the expression of the genetic information.

In this chapter, we will explore what is known about the regulation of ge-

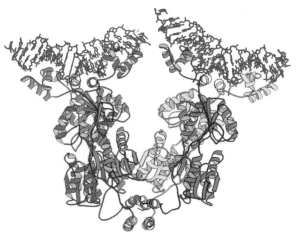

Model of the *lac* repressor bound to the DNA double helix representing the operator region of the *lac* operon.

netic expression in bacteria and bacteriophages. In numerous instances, highly detailed information has now been obtained. In the following chapter, we will focus on the regulation of gene expression in eukaryotes.

Genetic Regulation in Prokaryotes: An Overview

The regulation of gene expression has been extensively studied in prokaryotes, particularly in *E. coli*. What has been found is that highly efficient genetic mechanisms have evolved that turn genes on and off, depending on the cell's metabolic need for the respective gene products. The activity of enzymes may also be regulated once they have been synthesized in the cell, but we will focus primarily on what is known about regulation at the level of gene transcription. It is important to remember that it is the resulting levels of proteins ultimately present or absent that are critical to efficient cell function under varying environmental conditions.

It is not a particularly new concept that microorganisms regulate the synthesis of gene products. As early as 1900 it was shown that when **lactose** (a galactose-glucose-containing disaccharide) is present in the growth medium of yeast, enzymes specific to lactose metabolism are produced. When lactose is absent, the enzymes are not manufactured. It was soon shown that bacteria "adapt" to their chemical environment, producing certain enzymes only when specific substrates are present. Such enzymes were thus referred to as **adaptive**. In contrast, enzymes that are produced continuously regardless of the chemical makeup of the environment were called **constitutive**. Since then, the term *adaptive* in descriptive enzymology has been replaced with a more accurate term. We now call them **inducible** enzymes, reflecting the role of the substrate, which serves as the **inducer** in their production.

More recent investigation has revealed other cases where the presence of a specific molecule causes inhibition of genetic expression. This is usually true for molecules that are end products of biosynthetic pathways. For example, an amino acid such as tryptophan can be synthesized by bacterial cells. If an exogenous supply of this amino acid is present in the environment or culture medium, it is energetically inefficient to synthesize the enzymes necessary for tryptophan production. A mechanism has evolved whereby tryptophan plays a role in repressing transcription of RNA essential to the production of the appropriate biosynthetic enzymes. In contrast to the inducible system controlling lactose metabolism, that governing tryptophan is said to be **repressible**.

As we will soon see, instances of regulation, whether inducible or repressible, may be under either **negative** or **positive control**. Under negative control, genetic expression occurs unless it is shut off by some form of regulator. This is in contrast to positive control, where transcription does not occur unless a regulator molecule directly stimulates RNA production. In theory, each type of control can govern inducible or repressible systems. Examples discussed in the ensuing sections of this chapter will help distinguish between these possible mechanisms.

Lactose Metabolism in *E. coli*: An Inducible Gene System

In prokaryotes, structural genes tend to be organized in clusters controlled from a single regulatory site. Together, this group of genes constitutes what is called an **operon**. The regulatory site is linked to the gene cluster it controls and is known as a *cis*-acting element. The *cis*-acting sites are usually located upstream (at the 5′-end) from the gene cluster they regulate. Usually, the genes in the cluster produce products with related functions such as the reactions involved in a single metabolic pathway. Interactions at the regulatory site involve the binding of molecules that control transcription of the gene cluster. Such molecules are called *trans*-acting elements, because they are not physically linked to the genes they regulate. Interactions at the regulatory site determine whether the genes are expressed or not. In this section, we will discuss how such a bacterial gene cluster is coordinately regulated.

The gene cluster involved in the metabolism of lactose in *E. coli* has been extensively investigated and today serves as the paradigm for the study of genetic regulation. Beginning in 1946 with the studies of Jacques Monod and continuing through the next decade with significant contributions by Joshua Lederberg, François Jacob, and Andre L'woff, genetic and biochemical evidence was amassed. This research provided insights into the way in which the gene activity responsible for lactose metabolism is repressed when lactose is absent but induced when it is available. In the presence of lactose, the concentration of the enzymes responsible for its metabolism increases rapidly from 5 to 10 molecules to thousands per cell. The enzymes are described as **inducible**, and lactose serves as the **inducer**.

Paramount to the understanding of how gene expression is controlled in this system was the discovery of regulatory units that are part of the gene cluster. They consist of a regulatory gene and a regulatory site, neither of

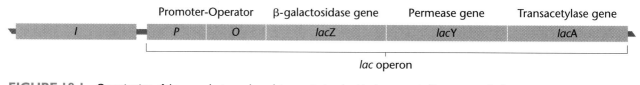

Promoter-Operator   β-galactosidase gene   Permease gene   Transacetylase gene

| *I* | *P* | *O* | *lacZ* | *lacY* | *lacA* |

lac operon

FIGURE 18.1   Organization of the gene cluster and regulatory units involved in the control of lactose metabolism.

which encodes enzymes necessary for lactose metabolism. Three other genes are responsible for the production of enzymes involved in lactose metabolism. The entire cluster is illustrated in Figure 18.1. As shown, the three structural genes and the adjacent regulatory site constitute the operon. Together, the entire cluster functions in an integrated fashion to provide a rapid response to the presence or absence of lactose.

Structural Genes

Genes coding for the primary structure of the enzymes are called **structural genes**. The so-called *lacZ* gene specifies the amino acid sequence of the enzyme **β-galactosidase**, which converts lactose to glucose and galactose (Figure 18.2). This conversion is essential if lactose is to serve as the primary energy source in glycolysis. The second gene, *lacY*, specifies the primary structure of **β-galactoside permease**, an enzyme that facilitates the entry of lactose into the bacterial cell. The third gene, *lacA*, codes for the enzyme

transacetylase whose physiological role is still not completely clear.

Studies of the genes coding for these three enzymes relied on the isolation of numerous mutations that eliminated the function of one or the other enzyme. Such *lac⁻* mutants were first isolated and studied by Joshua Lederberg. Mutant cells that fail to produce active β-galactosidase (*lacZ⁻*) or permease (*lacY⁻*) are unable to utilize lactose as an energy source. Mutations also were found in the transacetylase gene. Mapping studies by Lederberg established that all three genes are closely linked or contiguous to one another in the order Z-Y-A (see Figure 18.1).

Two other observations are relevant to what became known about the structural genes. First, knowledge of their close linkage led to the discovery that all three genes are transcribed as a single unit, resulting in what is called a **polycistronic mRNA** (Figure 18.3). As a result, all three genes are coordinately regulated since a single message serves as the basis for translation of all three gene products. Additionally, it has been shown that upon induction by lactose, the rapid appearance of the enzymes results from the *de novo* synthesis of this mRNA. Although this finding might seem obvious or expected, it is in contrast to the proposal that induced enzyme activity may result from the activation of existing but inactive forms of the enzymes.

The Discovery of Regulatory Mutations

How does lactose activate structural genes and induce the synthesis of the related enzymes? The discovery and study of **gratuitous inducers** ruled out one possibility. These molecules, which are chemical analogues of lactose, behave as inducers but they do not serve as substrates for the enzymes that are subsequently synthesized. One such gratuitous inducer is the sulfur analogue **isopropylthiogalactoside (IPTG)**, shown in Figure 18.4. The discovery that such molecules exist is strong evidence that the primary induction event does *not* depend on the interaction between the inducer and the enzyme. What, then, is the role of lactose in induction?

The answer to this question required the study of a class of mutations called **constitutive mutants**. In this type of mutant, the enzymes are produced regardless of the

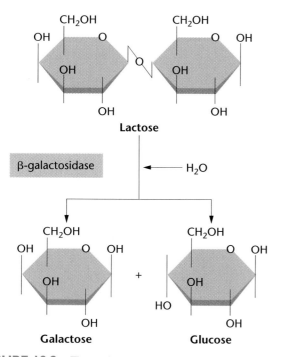

FIGURE 18.2   The catabolic conversion of the disaccharide lactose into its monosaccharide units, galactose and glucose.

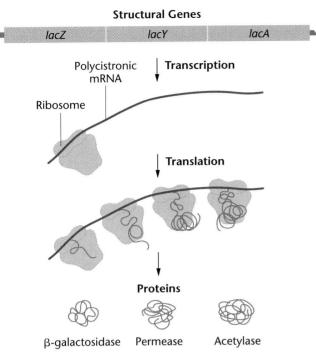

Structural Genes

| lacZ | lacY | lacA |

Transcription

Polycistronic mRNA

Ribosome

Translation

Proteins

β-galactosidase   Permease   Acetylase

FIGURE 18.3 The structural genes of the *lac* operon in *E. coli*. The three genes are transcribed into a single polycistronic mRNA, which is sequentially translated into the three enzymes encoded by the operon.

presence or absence of lactose. These mutations were used to study the regulatory scheme for lactose metabolism.

Maps of the first type of constitutive mutation, *lacI⁻*, showed that it is located at a site on the DNA close to, but distinct from, the structural genes. As we will soon see, the *lacI* gene is appropriately called a **repressor gene**. A second set of mutations produced identical effects but was found in a region immediately adjacent to the structural genes. This class of mutations is designated *lacO^C* and identifies the **operator region** of the operon. Because inducibility has been eliminated (the enzymes are continually produced) in both types of constitutive mutation, they clearly represent genetic changes that have disrupted regulation.

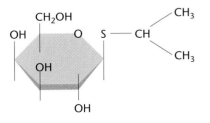

FIGURE 18.4 The gratuitous inducer isopropylthiogalactoside (IPTG).

The Operon Model: Negative Control

In 1961, Jacob and Monod proposed a scheme of negative control of regulation called the **operon model**, whereby a group of genes is regulated and expressed together as a unit. In their specific model (Figure 18.5), the **operon** consists of the *Z*, *Y*, and *A* structural genes as well as the adjacent sequences of DNA referred to as the operator region. They argued that the *lacI* gene regulates the transcription of the structural genes by producing a **repressor molecule**. The repressor was hypothesized to be **allosteric**, meaning that the molecule reversibly interacts with another molecule, causing both a conformational change in three-dimensional shape and a change in chemical activity.

Jacob and Monod suggested that the repressor normally interacts with the DNA sequence of the operator region. When it does so, it inhibits the action of RNA polymerase, effectively repressing the transcription of the structural genes [Figure 18.5(b)]. However, when lactose is present, this disaccharide binds to the repressor, causing the allosteric conformational change. This change alters the binding site of the repressor, rendering it incapable of interacting with operator DNA [Figure 18.5(c)]. In the absence of the repressor–operator interaction, RNA polymerase transcribes the structural genes, and the enzymes necessary for lactose metabolism are produced. Since transcription occurs only when the repressor fails to bind to the operator region, **negative control** is exerted.

The operon model uses these potential molecular interactions to explain the efficient regulation of the structural genes. In the absence of lactose, the enzymes encoded by the genes are not needed and are repressed. When lactose is present, it indirectly induces the activation of the genes by binding with the repressor.* If all lactose is metabolized, none is available to bind to the repressor, which is again free to bind to operator DNA and repress transcription.

Both the I^- and O^C constitutive mutations interfere with these molecular interactions, allowing continuous transcription of the structural genes. In the case of the I^- mutant, the repressor product is altered and cannot bind to the operator region, so the structural genes are always turned on. In the case of the O^C mutant, the nucleotide sequence of the operator DNA is altered and will not bind with a normal repressor molecule. The result is the same: Structural genes are always transcribed. Both types of constitutive mutations are illustrated diagrammatically in Figure 18.5(d) and (e).

---

*Technically, allolactose, an isomer of lactose, is the inducer. Allolactose is produced during the initial step in the metabolism of lactose by β-galactosidase.

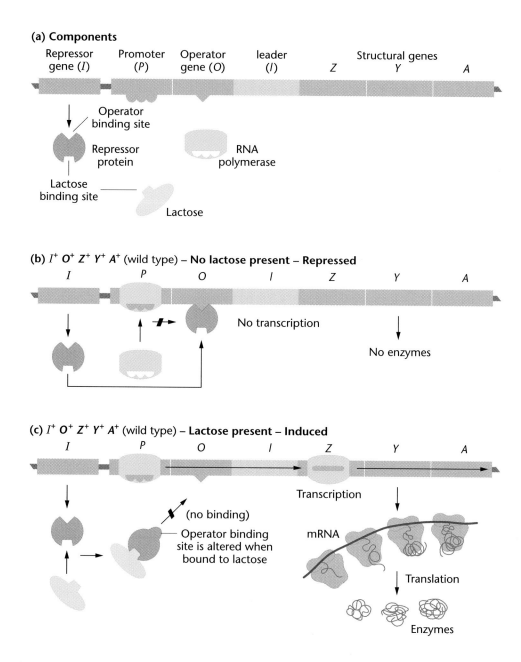

(a) Components

Repressor gene (I) Promoter (P) Operator gene (O) leader (l) Structural genes Z Y A

Operator binding site

Repressor protein

RNA polymerase

Lactose binding site

Lactose

(b) $I^+ O^+ Z^+ Y^+ A^+$ (wild type) – **No lactose present – Repressed**

I P O l Z Y A

No transcription

No enzymes

(c) $I^+ O^+ Z^+ Y^+ A^+$ (wild type) – **Lactose present – Induced**

I P O l Z Y A

Transcription

(no binding)

Operator binding site is altered when bound to lactose

mRNA

Translation

Enzymes

Genetic Proof of the Operon Model

The operon model is a good one because there are predictions derived from it that can be tested to determine its validity. The major predictions to be tested are that (1) the I gene produces a diffusible cellular product; (2) the O region is involved in regulation but does not produce a product; and (3) the O region must be adjacent to the structural genes in order to regulate transcription.

The construction of partial diploid bacteria (see Chapter 6) allows an assessment of these assumptions, particularly those that predict *trans*-acting regulatory elements. For example, certain mating schemes make it possible to construct genotypes where an I^+ gene has been introduced into an I^- host, or where an O^+ region is added to an O^C host. In such cases, the entire host chromosome is present plus one or a few genes of choice. They are inserted as part of a plasmid called the **F factor**, designated F' (see Chapter 6). The Jacob-Monod operon model predicts that adding an I^+ gene to an I^- cell should restore inducibility, because a normal repressor would again be produced. Adding an O^+ region to an O^C cell should have no effect on constitutive enzyme production, since regulation depends on an O^+ region immediately adjacent to the structural genes.

Results of these experiments are shown in Table 18.1, where Z represents the structural genes. The inserted genes are listed after the designation F'. In both cases de-

(d) *I⁻ O⁺ Z⁺ Y⁺ A⁺* (mutant repressor gene) – **No lactose present – Constitutive**

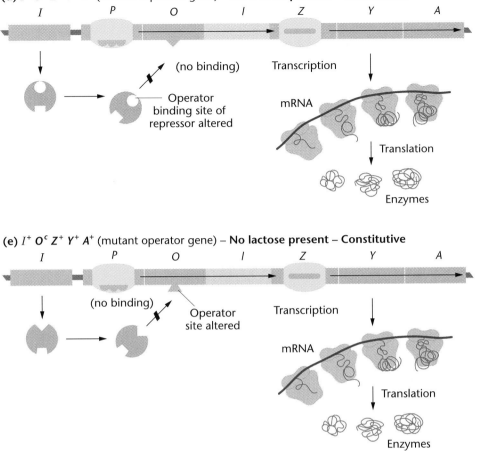

(e) *I⁺ Oᶜ Z⁺ Y⁺ A⁺* (mutant operator gene) – **No lactose present – Constitutive**

FIGURE 18.5   The components involved in the regulation of the *lac* operon and their interaction under various genotypic conditions, as described in the text.

TABLE 18.1   A Comparison of Gene Activity (+ or −) in the Presence or Absence of Lactose for Various *E. coli* Genotypes

		Presence of β-Galactosidase Activity	
	Genotype	Lactose Present	Lactose Absent
A.	$I^+O^+Z^+$	+	−
	$I^+O^+Z^-$	−	−
	$I^-O^+Z^+$	+	+
	$I^+O^CZ^-$	+	+
B.	$I^-O^+Z^+/F'I^+$	+	−
	$I^+O^CZ^+/F'O^+$	+	+
C.	$I^+O^+Z^+/F'I^-$	+	−
	$I^+O^+Z^+/F'O^C$	+	−
D.	$I^SO^+Z^+$	−	−
	$I^SO^+Z^+/F'I^+$	−	−

Note: In parts B to D, most genotypes are partially diploid, containing an F factor plus attached genes (F′).

scribed above, the Jacob-Monod model is upheld (part B of Table 18.1). Part C shows the reverse experiments, where either an I^- gene or an O^C region is added to cells of normal inducible genotypes. The model predicts that inducibility will be maintained in these partial diploids, and it is.

Another prediction of the operon model is that certain mutations in the I gene should have the opposite effect of I^-. That is, instead of being constitutive by failing to interact with the operator, mutant repressor molecules should be produced that cannot interact with the inducer, lactose. As a result, the repressor would always bind to the operator sequence, and the structural genes would be permanently repressed. If this were the case, the presence of an additional I^+ gene would have little or no effect on repression.

In fact, as shown in part D of Table 18.1, such a mutation, I^S, was discovered where the operon is "super-repressed." An additional I^+ gene does not effectively relieve repression of gene activity. These observations again support the operon model for gene regulation.

Isolation of the Repressor

Although the operon theory of Jacob and Monod succeeded in explaining many aspects of genetic regulation in prokaryotes, the nature of the repressor molecule was not known when their landmark paper was published in 1961. Although they had assumed that the allosteric repressor was a protein, RNA was also a candidate because activity of the molecule required the ability to bind to DNA. Despite many attempts to isolate and characterize the hypothetical repressor molecule, no direct chemical evidence was immediately forthcoming. A single *E. coli* cell contains no more than 10 or so copies of the *lac* repressor. Direct chemical identification of 10 molecules in a population of millions of proteins and RNAs in a single cell presented a tremendous challenge.

In 1966, Walter Gilbert and Benno Müller-Hill reported the isolation of the *lac* repressor in partially purified form. To achieve the isolation, they used a *regulator quantity* (I^q) mutant strain that contains about 10 times as much repressor as do wild-type *E. coli* cells. Also instrumental in their success were the use of the gratuitous inducer, IPTG (Figure 18.4), which binds to the repressor, and the technique of **equilibrium dialysis**. Using this technique, extracts of I^q cells were placed in a dialysis bag and allowed to attain equilibrium with an external solution of radioactive IPTG, which is small enough to diffuse freely in and out of the bag. At equilibrium, the concentration of IPTG was higher inside the bag than in the external solution, indicating that an IPTG-binding material was present in the cell extract and that this material was too large to diffuse across the wall of the bag.

Ultimately the IPTG-binding material was purified; it was shown to be heat labile and to have other characteristics of a protein as well. Extracts of I^- constitutive cells having no *lac* repressor activity did not exhibit IPTG-binding activity, strongly suggesting that the isolated protein was the repressor molecule.

To demonstrate this further, Gilbert and Müller-Hill grew *E. coli* cells in a medium containing radioactive sulfur and isolated the IPTG-binding protein, which was labeled in its sulfur-containing amino acids. This protein was mixed with DNA from a strain of phage lambda (λ) which carries the *lacO⁺* gene. The DNA sediments at $40S$, while the IPTG-binding protein sediments at $7S$. The DNA and protein were mixed and sedimented in a gradient using ultracentrifugation. If the radioactive protein binds to the DNA, then it should sediment at a much faster rate, moving in a band with the DNA. This was found to be the case. Further experiments showed that the IPTG-binding or repressor protein binds only to DNA containing the *lac* region. It was also shown not to bind to *lac* DNA containing an operator-constitutive (O^C) mutation.

Crystallographic Analysis of the Repressor

Very recently, in 1996, Mitchell Lewis, Ponzy Lu, and their colleagues succeeded in determining the crystal structure of the *lac* repressor as well as the structure of the repressor bound to IPTG and to operator DNA. The repressor monomer consists of 360 amino acids and functions as a homotetramer. The tetramer can be cleaved with a protease under controlled conditions and yields five fragments. Four are derived from the N-terminal ends of the tetramer and bind to operator DNA. The fifth fragment is the remaining core of the tetramer, derived from the COOH-terminus-ends; it binds to lactose and gratuitous inducers such as IPTG.

The operator DNA that was previously defined by mutational studies (*lacO^C*) and DNA sequencing analysis consists of 27 base pairs and is located 11 base pairs upstream from the start of the *lacZ* gene. In the crystallographic studies, each tetramer binds to two symmetrical operator DNA helices. Binding involves 21 of the 27 base pairs originally defined. Except for the central base pair, the two antiparallel strands of the helix are palindromic, a characteristic common to other regulatory sites. Binding by the repressor distorts the conformation of DNA, causing it to bend away from the repressor. A generalized image of this binding is illustrated in the opening photograph at the beginning of this chapter (page 522). In this depiction, an auxiliary operator region located 82 base pairs upstream from the *lacZ* gene is also bound, creating a "repression loop" of DNA.

These studies have also defined the three-dimensional conformational changes that accompany the allosteric transitions occurring during the interactions with the inducer molecules. Many of the related findings demonstrate that the molecular mechanisms giving rise to the transitions are prototypes for similar transitions known to occur in eukaryotic systems.

The crystallographic studies bring us to a new level of understanding in the regulatory process occurring within the *lac* operon. Remarkably, this current information has confirmed at the level of molecular visualization so many of the things that Jacob and Monod were able to predict in their model some 40 years ago based strictly on genetic grounds.

The Catabolite Activating Protein (CAP): Positive Control of the *lac* Operon

When the *lac* repressor is bound to the inducer, the *lac* operon is activated and RNA polymerase transcribes the structural genes. As we have learned in earlier chapters, this process is initiated as a result of the binding that occurs between the polymerase and the nucleotide sequence of the **promoter region**, found upstream (5′) from the initial coding sequences. Within the *lac* operon, the promoter is found between the *I* gene and the operator region (O^C)—(see Figure 18.1). Careful examination has revealed that, in this case, polymerase binding is not very efficient unless another protein is present to facilitate the process. This protein is called the **catabolite activating protein (CAP)**.

The discovery of CAP and the region within the promoter where it binds (the **CAP binding site**) occurred as a result of investigation of a most interesting observation. Even when cellular conditions are inducible (in the presence of lactose), *transcription of the operon is inhibited if glucose (a catabolic by-product of the metabolism of lactose) is present*. This inhibition, called **catabolite repression**, is a reflection of the greater simplicity with which glucose may be metabolized in comparison to lactose. The cells "prefer" glucose, and if present, they do not activate the *lac* operon, even when lactose is present.

Regulation of the *lac* operon by CAP and the role of glucose in catabolite repression is a fascinating story, summarized in Figure 18.6. In the absence of glucose, and under inducible conditions, CAP exerts **positive control** by binding to the CAP site, facilitating RNA polymerase binding at the promoter, and thus transcription. Therefore, for maximal transcription, repressor must be bound by lactose, *and* CAP must be bound to the CAP-binding site.

Still another complexity must be added to this story. CAP binding has been shown to be dependent on still another molecule, **cyclic adenosine monophosphate (cAMP)**. For binding to occur, CAP must be linked to cAMP. The level of cAMP is itself dependent on an enzyme **adenyl cyclase**, which catalyzes the conversion of ATP to cAMP.

The role of glucose in catabolite repression is now clear. It inhibits the activity of adenyl cyclase, causing a decline in the level of cAMP in the cell. Under this condition, CAP cannot form the CAP–cAMP complex essential to the positive control of transcription.

Like the *lac* repressor, CAP and cAMP–CAP have been examined using X-ray crystallography. CAP is a dimer that inserts into adjacent regions of a specific nucleotide sequence of DNA. When bound to DNA, the cAMP–CAP complex bends DNA, causing it to arc around itself at an angle of almost 90°.

Binding studies in solution further clarify the mechanism of gene activation. Alone, neither cAMP–CAP nor RNA polymerase has a strong affinity to bind to *lac* promoter DNA. Nor does either one have a strong affinity to bind to each other. However, when both are placed together along with the *lac* promoter DNA, a tightly bound complex is formed. This regulatory phenomenon illustrates what, in biochemical terms, is called **cooperative binding**. In the case of cAMP–CAP and the *lac* operon, the phenomenon illustrates the high degree of specificity that is involved in the genetic regulation of just one small group of genes.

Regulation of the *lac* operon by catabolite repression results in efficient energy utilization, because the presence of glucose will override the need for the metabolism of lactose, should it also be available to the cell. Catabolite repression involving CAP has also been observed for other inducible operons, including those controlling the metabolism of galactose and arabinose.

The *ara* Regulator Protein—Positive and Negative Control

Before turning to a consideration of repressible systems, we want to discuss briefly an interesting inducible operon in *E. coli*, the **arabinose operon**, in which the same regulatory protein exerts both positive and negative control (Figure 18.7).

In *E. coli* the metabolism of the sugar arabinose is under the direction of the enzymatic products of the *ara B*, *A*, and *D* genes. These genes are regulated by the protein encoded by the *ara C* gene. In the absence of arabinose, the protein behaves as a repressor, binding to regulatory sites within the *ara O* and *ara I* regions of the operon. Control is negative, and transcription is inhibited.

If arabinose *is* present, the AraC protein binds to it, and this complex becomes an activator. Attaching to only the *ara I* site on the operon stimulates the transcription of the B, A, and D genes. Under this condition, control is positive. A product must bind to induce gene expression.

A further complication is of interest in this system. Like the *lac* operon, the *ara* gene system is sensitive to glucose and under the positive control of the CAP protein. When glucose is present and the level of cyclic AMP is low, the CAP binding site is empty. This prevents the AraC–arabinose complex from activating the BAD gene promoter. Thus, both cyclic-AMP and arabinose must be present for the expression of the *BAD* genes in this operon.

Tryptophan Operon in *E. coli*: A Repressible Gene System

Although induction had been known for some time, it was not until 1953 that Monod and colleagues discovered **enzyme repression**. Wild-type *E. coli* are capable

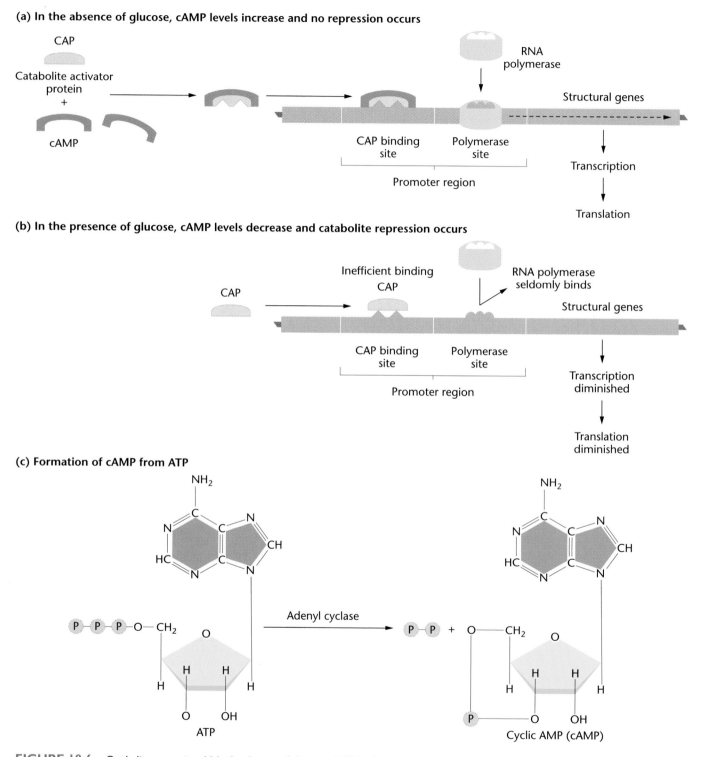

(a) In the absence of glucose, cAMP levels increase and no repression occurs

(b) In the presence of glucose, cAMP levels decrease and catabolite repression occurs

(c) Formation of cAMP from ATP

FIGURE 18.6 Catabolite repression. (a) In the absence of glucose, cAMP levels increase, resulting in cAMP–CAP binding and the ensuing stimulation of transcription. (b) In the presence of glucose, cAMP levels decrease, cAMP–CAP binding diminishes, and CAP fails to stimulate transcription. In (c) the formation of cAMP from ATP, catalyzed by adenyl cyclase, is diagrammed.

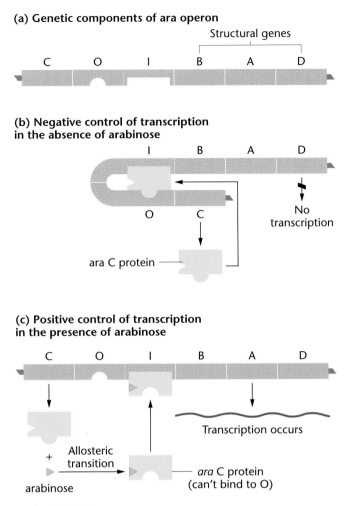

(a) Genetic components of ara operon

(b) Negative control of transcription in the absence of arabinose

(c) Positive control of transcription in the presence of arabinose

FIGURE 18.7 Gene regulation of the *ara* operon.

of producing the necessary enzymes essential to the biosynthesis of amino acids as well as other essential macromolecules. Monod focused his studies on the amino acid tryptophan and the enzyme **tryptophan synthetase**. He discovered that if tryptophan is present in sufficient quantity in the growth medium, the enzymes necessary for its synthesis are **repressed**. Energetically, such enzyme repression is highly economical to the cell, because synthesis is unnecessary in the presence of exogenous tryptophan.

Further investigation showed that a series of enzymes encoded by five contiguous genes on the *E. coli* chromosome is involved in tryptophan synthesis. These genes are part of an operon, and in the presence of tryptophan, all of these genes are coordinately repressed. As a result, none of the enzymes are produced. Because of the great similarity between this repression and the induction of enzymes for lactose metabolism, Jacob and Monod proposed a model of gene regulation analogous to the *lac* system (Figure 18.8).

To account for repression, they suggested that an *inactive repressor is normally made* that alone cannot interact with the operator region of the operon. However, it is an allosteric molecule and can interact with tryptophan, if present. The complex of repressor and tryptophan attains a new conformation that binds to the operator, repressing transcription. Thus, when tryptophan, the end product of this anabolic pathway, is present, the system is repressed and enzymes are not made. Because the regulatory complex inhibits transcription of the operon, this **repressible system** is under **negative control**. And, as tryptophan participates in repression, it is referred to as a **co-repressor** in this regulatory scheme.

Evidence for and Concerning the *trp* Operon

Support for the concept of a repressible operon was soon forthcoming, based primarily on the isolation of two distinct categories of constitutive mutations. The first class, *trpR⁻*, maps at a considerable distance from the structural genes. This locus represents the gene coding for the repressor. Presumably, the mutation either inhibits the interaction of the repressor with tryptophan or inhibits repressor formation entirely. Whichever the case, no repression ever occurs in cells with the *trpR⁻* mutation. As expected if the *trpR* gene encodes a repressor molecule, the presence of an additional *trpR⁺* gene restores repressibility.

The second constitutive mutant is analogous to that of the operator of the lactose operon because it maps immediately adjacent to the structural genes. Furthermore, the addition of a wild-type operator gene into mutant cells (as a *trans*-acting element) does not restore enzyme repression. This is predictable if the mutant operator can no longer interact with the repressor–tryptophan complex.

The entire *trp* operon has now been well defined, as shown in Figure 18.8. Five contiguous structural genes (*trp E, D, C, B,* and *A*) are transcribed as a polycistronic message that directs translation of the enzymes that catalyze the biosynthesis of tryptophan. As in the *lac* operon, a promoter region (*trpP*), representing the binding site for RNA polymerase, and an operator region (*trpO*), which binds the repressor, have been demonstrated. In the absence of binding, transcription is initiated within the overlapping *trpP–trpO* region and proceeds along a **leader sequence**. Transcription is initiated 162 nucleotides prior to the first structural gene (*trpE*). Within this leader sequence, still another regulatory site has been demonstrated. This component, called an attenuator, has been investigated extensively by Charles Yanofsky and his colleagues. As we shall see, this regulatory unit is an integral part of the control mechanism of this operon.

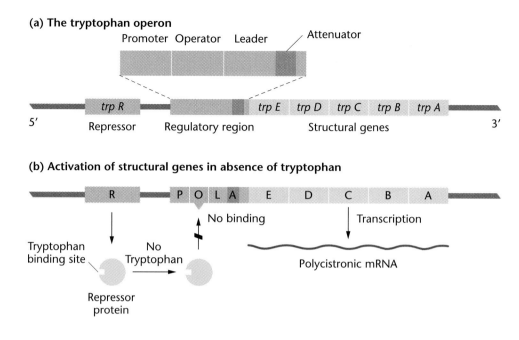

(a) The tryptophan operon

(b) Activation of structural genes in absence of tryptophan

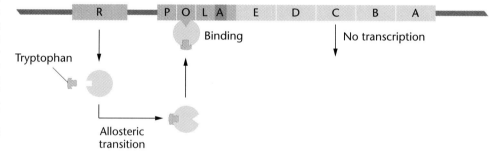

(c) Repression of structural genes in presence of tryptophan

FIGURE 18.8 (a) The components involved in the regulation of the tryptophan operon. (b) Regulatory conditions involving either activation or (c) repression of the structural genes. In the absence of tryptophan, an inactive repressor is made that cannot bind to the operator (*O*), thus allowing transcription to proceed. In the presence of tryptophan, it binds to the repressor, causing an allosteric transition to occur. This complex binds to the operator region, resulting in repression of the operon.

The Attenuator

Yanofsky observed that, even in the presence of tryptophan, initiation of transcription of this leader sequence usually occurs. The activated repressor, even when bound to the operator region, does not strongly inhibit expression of the operon. This suggests that there must be additional mechanisms by which tryptophan inhibits synthesis of the enzymes. In fact, as transcription of the leader sequence proceeds, if tryptophan is present in high concentration, mRNA synthesis is terminated at a point about 140 nucleotides along the transcript. It is this region that is called the **attenuator**. If, however, tryptophan is absent or present in low concentrations, not only is repressor unbound, but the process of **attenuation** is overcome. Transcription of the entire operon occurs, with the subsequent production of the enzymes essential to the biosynthesis of tryptophan.

Identification of the site of the attenuator was made possible by the isolation of deletion mutations in the region 123 to 150 nucleotides into the leader sequence.

Such mutations abolish attenuation. The phenomenon of attenuation appears to be common to other bacterial operons, including those regulating the biosynthesis of histidine and leucine.

An explanation of how attenuation occurs and how it is overcome has been put forward by Yanofsky and is summarized in Figure 18.9. Present in the region of the attenuator is a DNA sequence that, when transcribed, gives rise to an RNA molecule that immediately folds back on itself and forms a **hairpin loop**. The loop region is then followed by eight uridine residues. Such a structure is typical of that found at the 3′-end of many prokaryotic RNA transcripts, leading to the termination phase of their transcription. Thus, regardless of whether repressor is bound, transcription begins, but in the presence of tryptophan, the process is immediately terminated.

The question then remains as to how the absence (or a low concentration) of tryptophan allows attenuation somehow to be bypassed. Yanofsky's proposal is based on the discovery that the leader transcript contains a small open reading frame that includes two triplets (UGG) en-

(a) mRNA regions/selected base sequences

UGGUGG

5′

mRNA leader   trp codons   1   2   3   4   3′

(b) Potential base pairing regions

5′

1 2 3 4

(c) High concentrations of tryptophan: No stalling occurs during translation. Hairpin forms at regions 3 and 4, signalling termination of transcription.

Translation

5′

1   2   3   4

Hairpin signals termination of transcription

(d) Low concentrations of tryptophan: Ribosome stalls during translation. Transcription proceeds through regions 3 and 4.

Ribosome stalls

2   3

Transcription continues

5′

1   4

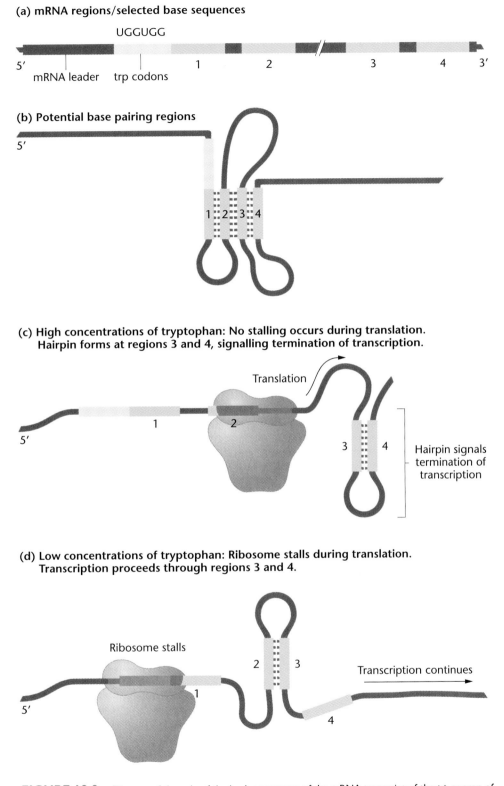

FIGURE 18.9   Diagram of the role of the leader sequence of the mRNA transcript of the *trp* operon of *E. coli* during attenuation. In (a) and (b), the critical nucleotide sequences and two potential base pairing regions are shown. In (c) where tryptophan is plentiful, no ribosome stalling occurs during translation of the trp codons. As a result, the ribosome proceeds through regions 1 and 2 and a "termination hairpin" forms in regions 3 and 4, leading to attenuation of transcription. In (d) where tryptophan is scarce, ribosome stalling occurs, the termination hairpin does not form, and transcription continues unabated. The model assumes simultaneous transcription and translation, as is known to occur in bacteria.

coding tryptophan. Even though this leader sequence is not part of the structural genes in the operon, following its transcription, the initiating AUG sequence prompts the initiation of its translation by ribosomes. Recall that in bacteria, translation of an mRNA is initiated well before transcription is completed.

The key to Yanofsky's model is that the leader transcript must be translated in order for the hairpin to form. Therefore, when adequate tryptophan is present, charged tRNA^{trp} is also present. As a result, translation proceeds, and the termination hairpin is formed [Figure 18.9(c)]. If cells are starved of tryptophan, charged tRNA^{trp} is unavailable, and the ribosome "stalls" during the translation of triplets calling for charged tRNA^{trp}. This event influences the secondary structure of the transcript such that the termination hairpin does not form [Figure 18.9(d)], attenuation is bypassed, and transcription and translation of the entire set of structural genes then proceed.

Details of the secondary structures of the transcript have now been worked out and described. They satisfy conditions leading either to continued transcription or termination (attenuation). Additionally, other mutations in the leader sequence have been isolated that are predicted to alter the secondary structure of the transcript and its impact on attenuation. In each case, the predicted result has been upheld. The model, while complex, has significantly extended our knowledge of genetic regulation. Furthermore, the supporting evidence represents an integrated approach involving genetic and biochemical analysis, which is becoming commonplace in molecular genetic research.

Genetic Regulation in Phage Lambda: Lysogeny or Lysis?

Our understanding of genetic regulation at the transcriptional level has benefited from studies of bacteriophage lambda as well as from studies of operons in bacteria.

Lambda DNA contains about 45,000 base pairs, enough bioinformation for about 35 to 40 average-sized genes. After injection of lambda DNA into *E. coli*, either the **lysogenic** or **lytic** pathway may be followed (see Chapter 6). In the lysogenic pathway, the phage DNA becomes integrated into the host genome and is almost totally repressed; in the lytic pathway, the phage DNA is expressed and viral reproduction ensues.

If the lysogenic pathway is followed, the genes responsible for phage reproduction and host cell lysis are turned off by the **λ repressor protein**. Since it is the product of one of the virus's own genes, *cI*, we will refer to this protein as the ***cI* repressor** in our subsequent discussions. If the lytic pathway is followed, the expression of the *cI* gene is repressed by a second protein, **Cro**. This protein is produced by still another viral gene of the same name. Both the *cI* repressor and the Cro protein have been isolated and characterized. The *cI* repressor, isolated by Mark Ptashne in 1967, is a protein consisting of 236 amino acids. The Cro protein, isolated more recently, consists of 66 amino acids.

The interaction of these two proteins with λ DNA has also been determined. The various sites of the regulatory system are diagrammed in Figure 18.10. Development of this information, which enhances our understanding of gene regulation, is a fascinating, but complex, story. When the *cI* repressor is produced, it recognizes two different operator regions in the λ DNA, O_L and O_R, found on either side (left or right) of the *cI* gene. The O_R region consists of 80 DNA base pairs and is located between the *cI* gene and the *cro* gene. There are three binding sites in O_R, each with a distinctive nucleotide sequence. The three regions are labeled O_R1, O_R2, and O_R3, and each consists of 17 nucleotides. Overlapping this region are the promoters for the *cI* and *cro* genes, called P_{RM} and P_R, respectively. Transcription of the *cI* and *cro* genes occurs in opposite directions, using opposite strands of the helix to achieve 5′-to-3′ RNA synthesis.

When the repressor is bound to both the O_R and O_L regions, two sets of so-called early genes are repressed, causing the remainder of the λ genes to be turned off as well. In this case, the repressor behaves as a negative control element. No lytic infection can occur and the lysogenic pathway is followed.

It is now clear that during this same binding, the *cI* repressor also stimulates transcription of its own *cI* gene. As shown in Figure 18.11, *cI* is self-regulating. There are three O_R binding sites. At low levels of the *cI* product, O_R1 and O_R2 are bound, stimulating *cI* transcription and inhibiting *cro*. However, as concentrations of the *cI* repressor increase, O_R3 is also bound and transcription of *cI* is inhibited. As the levels of the *cI* repressor decline, this autoregulatory cycle is repeated.

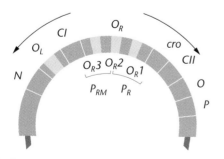

FIGURE 18.10 The regulatory sites controlling lysogeny and lysis in phage λ.

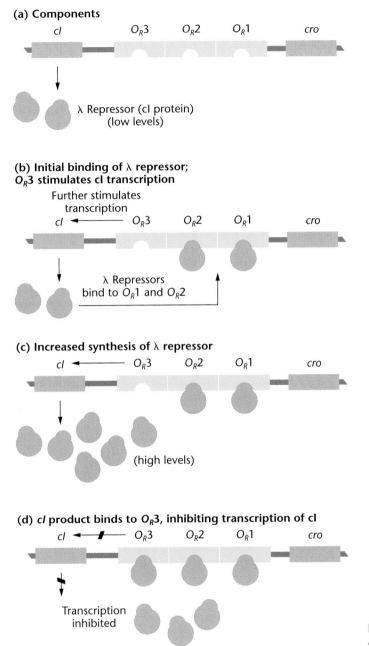

(a) Components

(b) Initial binding of λ repressor;
O_R3 stimulates cl transcription

Further stimulates
transcription

λ Repressors
bind to O_R1 and O_R2

(c) Increased synthesis of λ repressor

(high levels)

(d) cl product binds to O_R3, inhibiting transcription of cl

Transcription
inhibited

FIGURE 18.11 Self-regulation of the *cl* repressor gene, based on the concentration of the *cl* repressor.

When the repressor binding to O_L and O_R is absent, initiation at the respective promoters (P_{RM} and P_R) proceeds, resulting in the production of two proteins, N and Cro. The N protein functions as an antiterminator, allowing completion of transcription of genes essential to reproduction and lysis. The Cro protein, as stated earlier, serves as a repressor of *cl* gene transcription. Mutational analysis supports this model. Mutations in either O_L or O_R prevent repressor binding and abolish the potential for lysogeny. Mutations in the *cro* gene abolish the potential for lysis.

The greatest surprise was that the Cro protein, as well as the *cl* repressor, *also* demonstrates a binding affinity to the three sites of O_R. However, the binding of the two proteins to the three sites varies, depending on the concentration of the proteins. In addition, when either protein is bound to these sites, the other cannot bind. The *cl* repressor binds initially to O_R1 and O_R2 coordinately and then to O_R3, but only when present at very high concentrations. The Cro protein behaves in the opposite fashion. It initially binds O_R3 and subsequently binds O_R1 and O_R2.

Following initial infection, the determination of lysogeny versus lysis appears to depend on which operator region is bound first. If O_R1 and O_R2 are bound, P_R is inhibited and the *cro* gene is not transcribed. On the other hand, P_{RM} is available to RNA polymerase, and the *cI* gene is transcribed, resulting in increased amounts of the repressor. Lysogeny is then established. However, if O_R3 is bound by the Cro protein, P_{RM} is inhibited and the *cI* gene is repressed. This allows initiation at the P_R site, and transcription essential to lysis proceeds.

Though this description explains how the determination between lysis and lysogeny may be made, it does not address why one pathway is favored over the other. Still another gene, *cII*, and its protein product are critical to this decision. Both the cII protein and the Cro protein accumulate during early infection. The product of the *cII* gene affects P_{RM}, which activates the *cI* gene, causing production of the repressor. This favors the lysogenic pathway. Apparently, the activity of the cII protein changes under certain cellular conditions. For example, if glucose is plentiful, its activity is reduced, thus favoring the lytic pathway. This suggests that the cII protein may serve as a monitor of prevailing growth conditions. When its activity is reduced, the Cro protein prevails and successfully represses the *cI* gene, activating the lytic pathway.

To pursue this question further, Ptashne and others have found it useful to examine the molecular events that occur when an induction event transforms the lysogenic pathway to the lytic sequence. Discovered in 1946 by L'woff, treatment of lysogenized *E. coli* by various agents (including UV light and mutagens) induces a lytic response. While such an event occurs spontaneously about once in a million bacterial cell divisions, virtually all lysogenic bacteria may potentially be induced to enter the lytic cycle.

Induction events have one common property: They all involve interaction with DNA. In so doing, they stimulate what has been called the SOS response (see Chapter 14). During this response, the RecA protein, normally essential to recombination, is synthesized. Under inductive conditions, it behaves as a proteolytic enzyme and plays an essential role in the induction of lysis.

The cI repressor protein contains two structural domains and may exist in monomer or dimer form. It is the dimer form that binds strongly to O_R. The RecA protein cleaves the monomers at a sensitive region between the two domains, preventing their dimerization; as a result, the repressor protein can no longer bind to O_R. This allows expression of the *cro* gene. The Cro protein, also active in dimer form, then binds to O_R, inhibiting further transcription of the *cI* gene. Genetic events essential to viral reproduction then proceed, new viruses are constructed, and lysis of the bacterial cell occurs.

Research leading to the information just described is remarkable. It illustrates the depth of analysis attainable with regard to what at first may appear as a simple question: How is lysogeny regulated? Important insights have also been gained from this research regarding protein/nucleic acid interactions during genetic regulation. Molecular interactions such as these undoubtedly play major roles in many other genetic phenomena.

Phage Transcription during Lysis

When a bacteriophage invades its host bacterium and initiates the lytic pathway, it faces a novel problem. How can its genes be transcribed without the presence of virus-specific RNA polymerase, itself a product of transcription? The strategy that has evolved is an interesting example of parasite–host interaction.

The study of phage T7, which invades *E. coli*, has provided some answers to this question. This phage has evolved DNA sequences that are recognized as promoter sites by the *E. coli* RNA polymerase. As a result, upon the phage's entry to the host cell, a set of so-called **early genes** is transcribed into mRNAs by the bacterial polymerase, which are translated on bacterial ribosomes. One of these gene products is a viral-specific RNA polymerase that recognizes the promoters for the remaining T7 genes. A second early gene product, a protein kinase, inhibits bacterial transcription. The site of inhibitory interaction is presumably the bacterial RNA polymerase.

With the host cell's genetic apparatus secured, the remaining viral genes are expressed. The major products are DNA-replicating enzymes, head and tail components, enzymes involved in DNA packaging, and, finally, a lytic enzyme to break open the host cell.

Studies of the phage **SPO1**, which invades *Bacillus subtilis*, have revealed an even finer control or regulation of transcription during lysis (Figure 18.12). As in T7, **early genes** are read by the host RNA polymerase. One of the early gene products is a protein subunit (a type of sigma subunit) that associates with the host RNA polymerase. This complex can now interact with the promoter regions of the **middle genes** of the phage. This association between the early gene subunit and the bacterial polymerase curbs transcription of the early genes, and the middle genes associated with DNA replication are expressed. Two of the middle genes produce additional polypeptides that can interact with the host polymerase, again altering its specificity. As a result, the polymerase now directs the expression of the **late genes** responsible for producing components that package the new phage. Thus, sequential gene action results, providing a finer mechanism for the regulation of gene expression during the lytic pathway.

SPO1 phage gene control

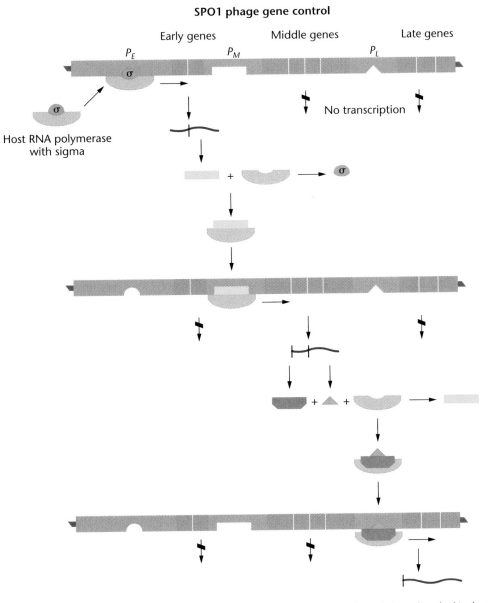

FIGURE 18.12   Initiation and regulation of phage SPO1 gene expression during lysis, as described in the text.

CHAPTER SUMMARY

1. Genetic analysis of bacteria and viruses has been pursued successfully since the 1940s. The ease of obtaining large quantities of pure cultures of bacteria and mutant strains of viruses as experimental material, and the efficient selection and isolation of naturally occurring and induced mutations, have made these the organisms of choice in numerous types of genetic studies.

2. A system of genetic regulation must exist if the complete genome is not to be continuously active in tran-
scription throughout the life of every cell of all species. Highly refined mechanisms have evolved that regulate transcription, maintaining genetic efficiency.

3. Both inducible and repressible operons, illustrated by the *lac* and *trp* gene complexes, respectively, have been documented and studied in bacteria. They involve genes of a regulatory nature in addition to structural genes that code for the enzymes of the system. Both operons are under the negative control of a repressor molecule.

GENETICS, TECHNOLOGY, AND SOCIETY

Antisense Oligonucleotides: Attacking the Messenger

Standard chemotherapies for diseases such as cancer and HIV are often accompanied by toxic side effects. Therapeutic drugs target both normal and diseased cells, with diseased cells or viruses being only slightly more susceptible than the patient's normal cells. Scientists have long wished for a magic bullet that could seek out and destroy the virus or the cancer cell, leaving normal cells alive and healthy. Over the last decade, one particularly promising candidate for magic bullet status has emerged—the **antisense oligonucleotide**.

Antisense therapies have arisen from an understanding of the molecular biology of gene expression. Gene expression is a two-step process. First, a single-stranded messenger RNA (mRNA) is copied from the template strand of the duplex DNA molecule. Second, the mRNA is transported to the cytoplasm, complexed with ribosomes, and its genetic information is translated into the amino acid sequence of a polypeptide.

Although genetic information is transcribed into RNA from one strand of the DNA duplex, it is sometimes possible for the other DNA strand (the partner strand) of a gene to be copied into RNA. The RNA produced by transcription of the partner strand is called antisense RNA. The base sequence of the antisense RNA is complementary to that of the gene's normal mRNA. Just as with complementary strands of DNA molecules, complementary strands of RNA can form double-stranded molecules. Naturally-occurring antisense RNAs have been detected in both bacteria and eukaryotes, where it is believed that they play a role in regulating gene expression.

The formation of duplex structures between a sense RNA and an antisense RNA has several potential effects on the sense RNA. First, if the antisense RNA hybridizes to the 5'-end of the sense RNA, it may physically block ribosome binding and hence inhibit translation. Second, the presence of antisense RNA may trigger the degradation of the sense RNA, as double-stranded RNA molecules are susceptible to attack by intracellular ribonucleases. Third, the antisense RNA may cleave the sense RNA into pieces—if the antisense RNA is a member of a special class of catalytic RNAs known as ribozymes. Ribozymes are RNA molecules that fold into precise structures and act like enzymes, hydrolyzing the phosphodiester bonds of RNA (their own, or others') at specific base sequences. Scientists can design ribozymes to recognize and cleave their target RNAs and can sequence these ribozymes in the test tube. They can also synthesize single-stranded complementary strands of DNA (oligodeoxynucleotides or ODNs), which also bind to mRNA and inhibit its expression.

It was this realization that sense and antisense RNAs form duplexes and affect gene expression that led to the idea of antisense therapeutics. What makes the antisense approach so exciting is its potential specificity. Because each gene (and hence its mRNA) has a unique nucleotide sequence, and since scientists now have the technology to manufacture oligonucleotides of precise base sequence in a test tube, it should be possible to treat cells with synthetic oligonucleotides of known sequence, to have the oligonucleotides enter the cell and bind to a precise target mRNA, and to turn off the synthesis of one specific gene product. If that gene is necessary for viral reproduction or cancer cell growth, the anti-

---

4. The catabolite activating protein (CAP) facilitates the binding of RNA polymerase to the promoter in the *lac* and other operons. Catabolite repression, a mechanism that represses the operon when glucose is present, has evolved presumably because glucose can be utilized more efficiently than lactose.

5. An additional regulatory step, referred to as *attenuation*, has been studied in the *trp* operon, whereby transcription begins, but is interrupted under appropriate cellular conditions.

6. The regulator protein of the *ara* gene exerts both positive and negative control over expression of the genes specifying the enzymes that metabolize the sugar arabinose.

7. Repression of transcription in phage lambda has been shown to be due to the product of one of its own genes, *cI*. Another gene product, Cro, acts to inhibit the transcription of *cI* and promotes the lytic pathway when the activity of cII protein is reduced during the presence of glucose.

8. During the lytic pathway, bacteriophages must rely on the host RNA polymerase to initiate transcription of their genome. Certain viral promoter sequences are recognized by the host polymerase. One of the early gene products in phage T7 is the RNA polymerase that transcribes the remainder of the genome.

sense oligonucleotide should inhibit these processes.

In the last few years, researchers have rushed to test antisense therapies and there have been some promising initial results. A particularly ingenious application of antisense technology is presently being applied in the fight against **dengue hemorrhagic fever**—a severe and often fatal disease that is spreading throughout tropical regions of the world. Dengue viruses contain a sense RNA genome which encodes 10 viral proteins. The virus is transmitted to humans by mosquitoes, which sequester the replicating virus in their salivary glands. Researchers at Colorado State University injected mosquitoes with a vector that expresses antisense dengue RNA. The presence of the antisense RNA prevented the growth of the virus in the salivary glands, and eliminated transmission of the virus to humans. If mosquitoes can be engineered to carry the dengue antisense gene in their genomes and to express the antisense RNA throughout their lives, it may be possible to release engineered mosquitoes to breed with natural populations, and thereby decrease the spread of dengue hemorrhagic fever. Antisense oligonucleotides are also being tested as

anti-viral drugs against herpes simplex virus and HIV.

Antisense oligonucleotides also have potential as cancer drugs. For example, mice transformed to carry human leukemia cells were injected with synthetic ODNs complementary to the RNA transcribed from the *c-myb* gene. The protein product of the *c-myb* gene stimulates growth of white blood cells and causes cancer when the gene is mutated. Treatment with *c-myb* antisense ODNs prolonged survival of treated mice up to 3.5 times longer than control animals. Antisense oligonucleotides have successfully suppressed other human tumors, such as melanomas, lymphomas, and glioblastomas, when tested in laboratory models. Laboratory tests of the efficacy of antisense treatments have been so promising that several clinical trials are now in progress. The trials involve antisense therapies against leukemia, AIDS, human papillomavirus, and cytomegalovirus.

The prospect of treating humans with antisense oligonucleotides is so new that many questions regarding their use have not yet been answered. There are obvious questions about side effects, safety, and cost. In addition, some researchers question whether the promising initial results were due to

true antisense effects, or resulted from secondary effects, such as general stimulation of the immune system. Oligonucleotides bind non-specifically to proteins, are degraded in the bloodstream, and enter cells very inefficiently. It will be necessary to develop methods to efficiently deliver oligonucleotides into cells. If antisense therapeutics pass the scrutiny of scientific and medical trials, we may indeed have gained a magic molecular bullet to use in the battle against disease.

Female mosquito on human skin.

KEY TERMS

adaptive enzyme, 523
adenyl cyclase, 529
allosteric repressor, 525
antisense RNA, 538
arabinose operon, 529
attenuation, 532
β-galactosidase enzyme, 524
β-galactoside permease, 524
CAP binding site, 529
catabolite activating protein (CAP), 529
catabolite repression, 529
cis-acting element, 523
constitutive enzyme, 523
constitutive mutation, 524
cooperative binding, 529
co-repressor, 531

cyclic adenosine monophosphate (cAMP), 529
dengue hemorrhagic fever, 539
early genes (in phage), 536
enzyme repression, 529
equilibrium dialysis, 528
F factor, 526
gratuitous inducers, 524
hairpin loop, 532
inducer, 523
inducible enzyme, 523
isopropylthiogalactoside (IPTG), 524
lactose, 523
late genes (in phage), 536
λ (*cI*) repressor protein, 534
leader sequence, 531
lysogenic pathway, 534

lysis, 534
lytic pathway, 534
middle genes (in phage), 536
negative control, 523
operator region, 525
operon model, 525
polycistronic mRNA, 524
positive control, 523
promoter region, 529
repressible system, 523
repressor gene, 525
repressor molecule, 525
SPO1 phage, 536
structural gene, 523
transacetylase, 524
trans-acting elements, 523
tryptophan synthetase, 531

INSIGHTS AND SOLUTIONS

1. A theoretical operon (*theo*) in *E. coli* contains several structural genes encoding enzymes that are involved sequentially in the biosynthesis of an amino acid. Unlike the *lac* operon, where the repressor gene is separate from the operon, the gene encoding the regulator molecule is contained within the *theo* operon. When the end product (the amino acid) is present, it combines with the regulator molecule, and this complex binds to the operator, repressing the operon. In the absence of the amino acid, the regulatory molecule fails to bind to the operator, and transcription proceeds.

Characterize this operon; then consider the following mutations as well as the situation in which the wild-type gene is present along with the mutant gene in partially diploid cells (F′). In each case, will the operon be active or inactive in transcription, assuming that the mutation affects the regulation of the *theo* operon? Compare each response to the equivalent situation of the *lac* operon.

(a) Mutation in the operator gene.
(b) Mutation in the promoter region.
(c) Mutation in the regulator gene.

Solution: The *theo* operon is repressible and under negative control. When there is no amino acid present in the medium (or the environment), the product of the regulatory gene cannot bind to the operator region, and transcription proceeds under the direction of RNA polymerase. The enzymes necessary for the synthesis of the amino acid are produced, as is the regulator molecule. If the amino acid *is* present, or after sufficient synthesis occurs, the amino acid binds to the regulator, forming a complex that interacts with the operator region, causing repression of transcription of the genes within the operon.

The *theo* operon is similar to the tryptophan system with the exception that the regulator gene is within the operon rather than being located separate from it. Therefore, in the *theo* operon, the regulator gene is itself regulated by the presence or absence of the amino acid.

(a) As in the *lac* operon, a mutation in the *theo* operator gene inhibits binding with the repressor complex, and transcription occurs constitutively. The presence of an F′ plasmid bearing the wild-type allele would have no effect since it is not adjacent to the structural genes.

(b) A mutation in the *theo* promoter region would no doubt inhibit binding to RNA polymerase and therefore inhibit transcription. This would also happen in the *lac* operon. A wild-type allele present in an F′ plasmid would have no effect.

(c) A mutation in the *theo* regulator gene, as in the *lac* system, may either inhibit its binding to the repressor or its binding to the operator gene. In both cases, transcription will be constitutive because the *theo* system is repressible. Both cases result in the failure of the regulator to bind to the operator, allowing transcription to proceed. In the *lac* system, failure to bind the co-repressor lactose would permanently repress the system. The addition of a wild-type allele would restore repressibility, provided that this gene was transcribed constitutively.

2. Analysis of the lysis and lysogeny of *E. coli* by phage λ involved many mutations that disrupted the normal capabilities of the virus. Consider the following mutations.

(a) Phage λ bearing a temperature-sensitive mutation in the *cI* gene is unable to lysogenize *E. coli* (it cannot integrate its chromosome into the host genome). If such a mutant phage is allowed to infect a host bacterial cell that is already lysogenized by a wild-type phage, predict the outcome of infection and explain this outcome.

Solution: The mutant phage will be unable to lyse the cell because the wild-type phage (integrated into the bacterial chromosome) is expressing the normal product of the *cI* gene that represses the genetic system leading to lysis. This repressor is effective for repressing all incoming phages as well as its own genome.

(b) Other phage λ mutants have been isolated that, like the one in part (a), cannot independently lysogenize host bacteria. However, they *can* lyse bacteria that are lysogenized by wild-type phages. Can you offer an explanation for this observation?

Solution: The mutation is not in the *cI* gene. Instead, it must be in the "receptor" gene(s) that normally responds to the repressor. The mutation renders it (them) insensitive to the repressor.

PROBLEMS AND DISCUSSION QUESTIONS

1. Contrast the need for the enzymes involved in the metabolism of lactose and tryptophan in bacteria in the presence and absence of lactose and tryptophan, respectively.

2. Contrast positive and negative control systems.

3. Contrast the role of the repressor in an inducible system and in a repressible system.

4. Describe how the *lac* repressor was isolated. What properties demonstrate it to be a protein? Describe the evidence that it indeed serves as a repressor within the operon scheme.

5. Even though the *lac Z, Y,* and *A* structural genes are transcribed as a single polycistronic mRNA, each gene contains the appropriate initiation and termination signals essential for translation. Predict what will happen when a cell growing in the presence of lactose contains a deletion of one nucleotide early in the Z gene. Early in the A gene?

6. For the following *lac* genotypes (right-hand table), predict whether the structural genes (Z) are constitutive, permanently repressed, or inducible in the presence of lactose.

Genotype	Constitutive	Repressed	Inducible
$I^+O^+Z^+$			×
$I^-O^+Z^+$			
$I^+O^CZ^+$			
$I^-O^+Z^+/F'I^+$			
$I^+O^CZ^+/F'O^+$			
$I^SO^+Z^+$			
$I^SO^+Z^+/F'I^+$			

7. For the following genotypes and condition (lactose present or absent), predict whether functional enzymes are made, nonfunctional enzymes are made, or no enzymes are made.

Genotype	Condition	Functional Enzyme Made	Nonfunctional Enzyme Made	No Enzyme Made
$I^+O^+Z^+$	No lactose	—	—	×
$I^+O^CZ^+$	Lactose			
$I^-O^+Z^-$	No lactose			
$I^-O^+Z^-$	Lactose			
$I^-O^+Z^+/F'I^+$	No lactose			
$I^+O^CZ^+/F'O^+$	Lactose			
$I^+O^+Z^-/F'I^+O^+Z^+$	Lactose			
$I^-O^+Z^-/F'I^+O^+Z^+$	No lactose			
$I^SO^+Z^+/F'O^+$	No lactose			
$I^+O^CZ^+/F'O^+Z^+$	Lactose			

8. If the *cI* gene of phage λ contained a mutation with effects similar to the I^- mutation in *E. coli*, what would be the result?

9. Describe the level of genetic activity of the *lac* operon as well as the status of the *lac* repressor and the CAP protein under the cellular conditions listed in the table to the right:

	Lactose	Glucose
(a)	−	−
(b)	+	−
(c)	−	+
(d)	+	+

EXTRA-SPICY PROBLEMS

10. In a theoretical operon, genes *A, B, C,* and *D* represent the repressor gene, the promoter sequence, the operator gene, and the structural gene, *but not necessarily in that order.* This operon is concerned with the metabolism of a theoretic molecule (tm). From the data given below, first decide if the operon is inducible or repressible. Then assign *A, B, C,* and *D* to the four parts of the operon. Explain your rationale. (AE = active enzyme; IE = inactive enzyme; NE = no enzyme)

Genotype	tm Present	tm Absent
$A^+B^+C^+D^+$	AE	NE
$A^-B^+C^+D^+$	AE	AE
$A^+B^-C^+D^+$	NE	NE
$A^+B^+C^-D^+$	IE	NE
$A^+B^+C^+D^-$	AE	AE
$A^-B^+C^+D^+/F'A^+B^+C^+D^+$	AE	AE
$A^+B^-C^+D^+/F'A^+B^+C^+D^+$	AE	NE
$A^+B^+C^-D^+/F'A^+B^+C^+D^+$	AE + IE	NE
$A^+B^+C^+D^-/F'A^+B^+C^+D^+$	AE	NE

11. A bacterial operon is responsible for the production of the biosynthetic enzymes needed to make a theoretical amino acid tisophane (tis). The operon is regulated by a separate gene, *R*. Deletion of the *R* gene causes the loss of synthesis of the enzymes. In the wild-type condition, when tis is present, no enzymes are made. In the absence of tis, the enzymes are made. Mutations in the operator gene (*O⁻*) result in repression regardless of the presence of tis.

    Is the operon under positive or negative control? Propose a model for (1) repression of the genes in the presence of tis in wild-type cells; and (2) the *O⁻* mutations.

12. A marine bacterium is isolated and shown to contain an inducible operon whose genetic products metabolize oil when it is encountered in the environment. Investigation demonstrates that the operon is under positive control, and that there is a *reg* gene whose product interacts with an operator region (*o*) to regulate the structural genes designated *sg*.

In your attempt to understand how the operon functions, a constitutive mutant strain was isolated as well as several partial diploid strains and these were tested with the results shown below. Draw all possible conclusions about the mutation as well as the nature of regulation of the operon. Is the constitutive mutation in the *trans*-acting *reg* element or in the *cis*-acting *o* (operator) element?

Host Chromosome	F′ Factor	Phenotype
wild type	none	inducible
wild type	*reg* gene from mutant strain	inducible
wild type	operon from mutant strain	constitutive
mutant strain	*reg* gene from wild type	constitutive

13. The SOS repair genes in *E. coli* (see Chapter 14) are negatively regulated by the product of the *lexA* gene: the LexA repressor. When a cell sustains extensive damage to its DNA, the LexA repressor is inactivated by the *recA* gene product (RecA) and transcription of the SOS genes is increased dramatically.

    One of the SOS genes is the *uvrA* gene. You are studying the function of the *uvrA* gene product in DNA excision repair. You isolate a mutant strain that shows constitutive expression of the UvrA protein. You name this mutant strain *uvrAᶜᵒⁿ*. Shown below is a simple diagram of the *lexA* and *uvrA* operons.

(a) Describe two different mutations that would result in a uvrA constitutive phenotype. Indicate the actual genotypes involved.

(b) Outline a series of genetic experiments using partial diploid strains that would allow you to determine which of these two possible mutations you have isolated.

(c) A fellow student considers this problem and argues that there is a more straightforward non-genetic experiment that could be done to differentiate between these two types of mutations. While this experiment requires no fancy genetics, you must be able to easily assay the products of the other SOS genes. What is the experiment?

P^lexA   O^lexA   lex A   P^uvrA   O^uvrA   uvrA

14. In Figure 18.9, numerous base pairing regions of mRNA are depicted (1–4) that play important roles during the process of attenuation in the trp operon. Shown on the right, beginning with the 5′-end, is the sequence of 80 ribonucleotides that make up part of the mRNA whose transcription is affected by attenuation. Within this molecule are the sequences that serve as the basis of the formation of the secondary structures that are critical in the attenuation process. Obtain a large piece of paper (such as manila wrapping paper) and along with several colleagues from your genetics class, work through this base sequence, attempting to identify which parts of the mRNA molecule represent the labeled regions (1–4) that are shown to base pair in Figure 18.9. Draw the configuration of the molecule under high tryptophan levels, as well as under low tryptophan levels.

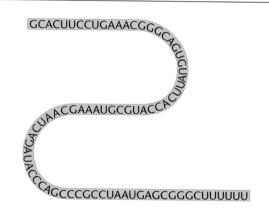

SELECTED READINGS

ANDERSON, J. E., PTASHNE, M., and HARRISON, S. C. 1985. A phage repressor-operator complex at 7Å resolution. *Nature* 316:596–601.

BECKWITH, J., and ROSSOW, P. 1974. Analysis of genetic regulatory mechanisms. *Annu. Rev. Genet.* 8:1–13.

BECKWITH, J. R., and ZIPSER, D., eds. 1970. *The lactose operon.* Cold Spring Harbor, NY: Cold Spring Harbor Laboratory Press.

BERTRAND, K., et al. 1975. New features of the regulation of the tryptophan operon. *Science* 189:22–26.

COHEN, J. S., and HOGAN, M. E. 1994. New genetic medicines. *Sci. Am.* (Dec.) 271:76–82.

ENGLESBERG, E., and WILCOX, G. 1974. Regulation: Positive control. *Annu. Rev. Genet.* 8:219–42.

GILBERT, W., and MÜLLER-HILL, B. 1966. Isolation of the *lac* repressor. *Proc. Natl. Acad. Sci. USA* 56:1891–98.

———. 1967. The *lac* operator is DNA. *Proc. Natl. Acad. Sci. USA* 58:2415–21.

GILBERT, W., and PTASHNE, M. 1970. Genetic repressors. *Sci. Am.* (June) 222:36–44.

HERSHEY, A. D., ed. 1971. *The bacteriophage lambda.* Cold Spring Harbor, NY: Cold Spring Harbor Laboratory Press.

JACOB, F., and MONOD, J. 1961. Genetic regulatory mechanisms in the synthesis of proteins. *J. Mol. Biol.* 3:318–56.

JOHNSON, A. D., et al. 1981. λ repressor and *cro*—Components of an efficient molecular switch. *Nature* 294:217–23.

KELLER, E. B., and CALVO, J. M. 1979. Alternative secondary structures of leader RNAs and the regulation of the *trp, phe, his, thr,* and *leu* operons. *Proc. Natl. Acad. Sci. USA* 76:6186–90.

LEWIN, B. 1974. Interaction of regulator proteins with recognition sequences of DNA. *Cell* 2:1–7.

LEWIS, M., et al. 1996. Crystal structure of the lactose operon repressor and its complexes with DNA and inducer. *Science* 271:1247–54.

MANIATIS, T., and PTASHNE, M. 1976. A DNA operator-repressor system. *Sci. Am.* (Jan.) 234:64–76.

MILLER, J. H., and REZNIKOFF, W. S. 1978. *The operon.* Cold Spring Harbor, NY: Cold Spring Harbor Laboratory Press.

OLSON, K. E., et al. 1996. Genetically engineered resistance to dengue-2 virus transmission in mosquitoes. *Science* 272:884–86.

PETRUSEK, R. H., DUFFY, J., and GEIDUSCHEK, E. 1976. Control of gene action in phage SPO1 development: Phage-specific modifications of RNA polymerase and a mechanism of positive control. In *RNA polymerase*, eds. R. Losick and M. Chamberlin, pp. 567–85. Cold Spring Harbor, NY: Cold Spring Harbor Laboratory Press.

PLATT, T. 1981. Termination of transcription and its regulation in the tryptophan operon of *E. coli. Cell* 24:10–23.

PTASHNE, M. 1986. *A genetic switch.* Palo Alto, CA: Blackwell Scientific.

PTASHNE, M., et al. 1980. How the λ repressor and *cro* work. *Cell* 19:1–11.

PTASHNE, M., and GANN, A. 1990. Activators and targets. *Nature* 346:329–31.

PTASHNE, M., JOHNSON, A. D., and PABO, C. O. 1982. A genetic switch in a bacterial virus. *Sci. Am.* (Nov.) 247:128–40.

STROYNOWSKI, I., and YANOFSKY, C. 1982. Transcript secondary structures regulate transcription termination at the attenuator of *S. marcescens* tryptophan operon. *Nature* 298:34–38.

STUDIER, F. W. 1972. Bacteriophage T7. *Science* 176:367–76.

UMBARGER, H. E. 1978. Amino acid biosynthesis and its regulation. *Annu. Rev. Biochem.* 47:533–606.

WAGNER, R. W. 1994. Gene inhibition using antisense oligodeoxynucleotides. *Nature* 372:333–35.

YANOFSKY, C. 1981. Attenuation in the control of expression of bacterial operons. *Nature* 289:751–58.

YANOFSKY, C., and KOLTER, R. 1982. Attenuation in amino acid biosynthetic operons. *Annu. Rev. Genet.* 16:113–134.

19

Regulation of Gene Expression in Eukaryotes

CHAPTER CONCEPTS

Genetic regulation in eukaryotes can occur at several levels, but transcriptional control is the primary mechanism controlling gene expression. Transcription is modulated by the interaction of regulatory molecules with short DNA sequences most often located upstream from affected genes. Posttranscriptional mechanisms involve the selection of alternative products from a single transcript and the control of mRNA stability.

The discovery of the operon in *Escherichia coli* by Jacob and Monod in 1960 was the first step in unraveling the mechanisms that regulate gene expression. While some of the principles of regulation found in the *lac* operon are also present in eukaryotes, it is clear that eukaryotes have evolved a more complex system of gene regulation.

In multicellular eukaryotes, cells of the pancreas do not make retinal pigment, nor do retinal cells make insulin. In these organisms, differential regulation of gene expression is at the heart of cellular differentiation and function. The question is, how does an organism express a subset of genes in one cell type, and a different subset of genes in an-

other cell type? At the cellular level, this regulation is not accomplished by eliminating unused genetic information; instead, mechanisms have evolved to activate specific portions of the genome, and to repress the expression of other genes. Activation and repression of selected loci represents a delicate balancing act for an organism; expression of a gene at the wrong time, in the wrong cell type, or in abnormal amounts can lead to a deleterious phenotype, even when the gene itself is normal.

In this chapter we will review the general features of eukaryotic regulation, outline the components needed to control

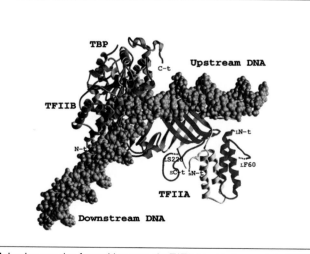

Molecular complex formed between the TATA box-binding protein (TBP), transcription factors TFIIA and TFIIB, and TATA box DNA during the initiation of transcription.

transcription, and discuss how these components interact. We will also consider the role of posttranscriptional mechanisms in regulating gene expression in eukaryotes.

Eukaryotic Gene Regulation: An Overview

Eukaryotic cells contain a much greater amount of genetic information than do prokaryotic cells, and this DNA is complexed with histones and other proteins to form chromatin. Genetic information in eukaryotes is carried on many chromosomes (rather than one), and these chromosomes are enclosed within a nuclear membrane. In eukaryotes, transcription is spatially and temporally separated from translation. The transcripts of eukaryotic genes are processed, cleaved, and ligated before transport to the cytoplasm. The highly differentiated cells of eukaryotes often synthesize large amounts of a restricted number of gene products even though they contain a complete set of genetic information.

Regulation of eukaryotic gene expression can potentially occur at many levels (Figure 19.1). These include (1) transcriptional control, (2) processing of the pre-mRNA, (3) transport to the cytoplasm, (4) stability of the mRNA, (5) selecting which mRNAs are translated, and (6) posttranslational modification of the protein product. Most eukaryotic genes are regulated in part, at the transcriptional level. In the following sections, emphasis is placed on transcriptional control, although other levels of control will also be discussed. There are two main components of transcriptional control of gene expression: short DNA sequences that serve as recognition sites, and regulatory proteins that bind to these sites.

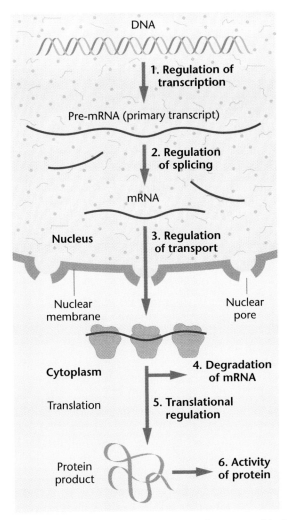

FIGURE 19.1 Various levels of regulation that are possible during the expression of the genetic material.

Regulatory Elements and Eukaryotic Genes

The internal structure of eukaryotic genes was discussed in Chapters 12 and 17. In addition, there are regulatory sequences adjacent to genes that control transcription.

These sequences are of two types: promoters and enhancers (Figure 19.2). These regulatory elements can be on either side of a gene, or at some distance from the gene, and are called *cis* regulators because they are found adjacent to the structural genes they regulate, as opposed to *trans* regulators (e.g., binding proteins), which are not adjacent to the genes they regulate.

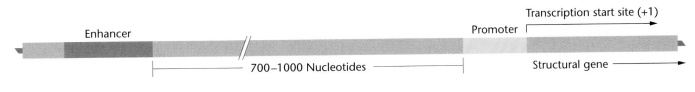

FIGURE 19.2 Organization of typical eukaryotic gene and transcriptional control regions. The promoter region consists of modular DNA sequences, usually located within 100 to 110 bp of the site of transcription initiation. Promoter elements usually consist of modular sequences.

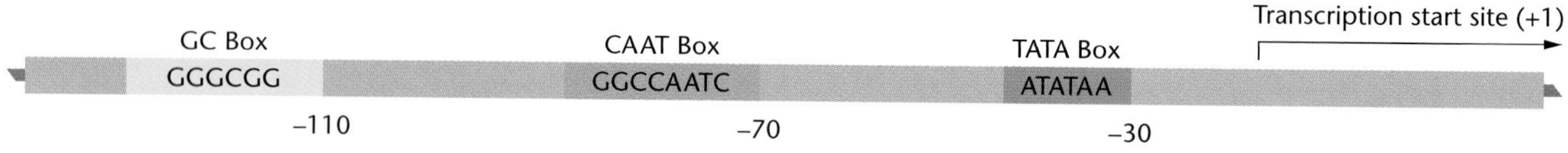

GC Box
GGGCGG

CAAT Box
GGCCAATC

TATA Box
ATATAA

Transcription start site (+1)

−110   −70   −30

FIGURE 19.3 The promoter in eukaryotic genes consists of several modular elements including the TATA box (−25 to −30 bp), the CAAT box (−70 to −80), and the GC box (at about −110).

Promoters

Promoters consist of nucleotide sequences that serve as the recognition point for RNA polymerase binding. Such regions are called **promoters**, as they are in prokaryotes. They represent the region necessary to *initiate* transcription and are located a fixed distance from the site where transcription is initiated. Promoters are located immediately adjacent to the genes they regulate, and are considered to be a part of the gene. Promoter regions are usually several hundred nucleotides in length.

Eukaryotic promoters require the binding of a number of protein factors to initiate transcription. Promoters that are recognized by RNA polymerase II consist of short modular DNA sequences usually located within 100 bp upstream (in the 5′ direction) of the gene. The promoter region of most genes contains several elements. The first is a sequence called the **TATA box** (Figure 19.3). Located about 25 to 30 bases upstream from the initial point of transcription (designated as −25 to −30), it consists of an 8-base-pair consensus sequence (a sequence conserved in most or all genes studied) composed only of T = A pairs, often flanked on either side by G ≡ C-rich regions. Mutations in the TATA box severely reduce transcription, and deletions often alter the initiation point of transcription (Figure 19.4).

Many promoters contain other components; one of these is called the **CAAT box**. Its consensus sequence is CAAT or CCAAT, it frequently appears in the region −70 to −80 bp from the start site, and it can function in a 5′ → 3′ or a 3′ → 5′ orientation. Mutational analysis

suggests that the CAAT box is critical to the promoter's ability to facilitate transcription (Figure 19.4). Mutations on either side of this element have no effect on transcription, whereas mutations within the CAAT sequence dramatically lower the rate of transcription. A third modular element of some promoters is called the **GC box**, which has the consensus sequence GGGCGG and is often found at about position −110. This module is found in either orientation and often occurs in multiple copies.

There is variation in the location and organization of sequences that comprise the upstream regulatory elements in eukaryotic genes (Figure 19.5). Note that in different genes there are significant differences in the number and orientation of promoter elements and the distance between them. In addition, eukaryotic genes use more than one form of RNA polymerase for transcription. These are classified as type I (ribosomal RNAs), type II (mRNAs and snRNAs), and type III (tRNAs, 5S rRNA, and several other small cellular RNAs). Each type of polymerase uses different sequences to recognize transcription factors that bind to the type I, type II, and type III promoters.

Enhancers

In addition to the promoter region, transcription of most, if not all, eukaryotic genes is regulated by additional DNA sequences called **enhancers**. These regions interact with regulatory proteins and can increase the efficiency of transcription initiation or activate the promoter.

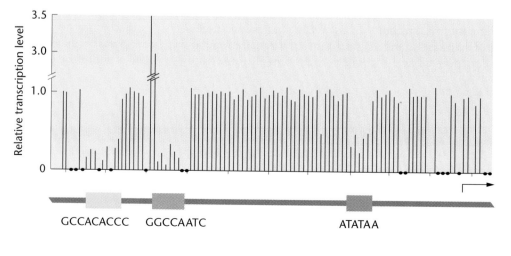

FIGURE 19.4 Summary of the effects of point mutations on transcription of the β-globin gene. Each line represents the level of transcription produced by a single nucleotide mutation (relative to wild type) in the promoter region in a separate experiment. Dots represent nucleotides where no mutation was obtained. Note that mutations within promoter elements have the greatest effect on the level of transcription.

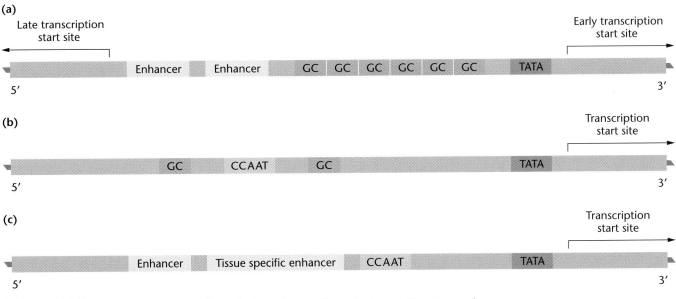

FIGURE 19.5 Upstream organization of several eukaryotic genes, illustrating the variable nature, number, and arrangement of controlling elements. (a) SV40 control region, with early genes to right, late genes to left. (b) Thymidine kinase. (c) Insulin gene.

Thus, there is some analogy between enhancers and operator regions in prokaryotes. However, enhancers appear to be much more complex in both structure and function. Enhancers greatly stimulate the transcriptional activity of a promoter and can be distinguished from promoters by several characteristics:

1. The position of the enhancer need not be fixed; it can be placed upstream, downstream, or within the gene it regulates.

2. Its orientation can be inverted without significant effect on its action.

3. If an enhancer is moved to another location in the genome, or if an unrelated gene is placed near an enhancer, transcription of the adjacent gene is enhanced.

Most eukaryotic genes are under the control of enhancers. In the immunoglobulin heavy chain genes, an enhancer is located in an intron between two coding regions. In this case, the enhancer is located *within* the gene it regulates. It is also active only in the cells in which the immunoglobulin genes are being expressed, indicating that tissue-specific gene expression can be modulated through enhancers.

Internal enhancers have been discovered in other eukaryotic genes, including the immunoglobulin light-chain gene. Downstream enhancers are found in the human beta globin gene and the chicken thymidine kinase gene. In chickens, an enhancer is located between the beta globin and the epsilon globin genes and apparently works in one direction to control the epsilon gene

during embryonic development and in the opposite direction to regulate the beta gene during adult life. In yeast, regulatory sequences similar to enhancers, called **upstream activator sequences (UAS)**, can function upstream at variable distances and in either orientation. They differ from enhancers in that they cannot function downstream of the transcription start point.

Like promoters, enhancers contain modules composed of short DNA sequences. The enhancer of the SV40 virus has a complex structure. It consists of two adjacent 72-bp sequences located some 200 bp upstream from a transcriptional start point (Figure 19.6). Each of the two 72-bp regions contains five sequences that contribute to maximum rates of transcription. If one or the other of these regions is deleted, there is no effect on transcription, but if both are deleted, *in vivo* transcription is greatly reduced. This and other enhancers contain a copy of a sequence (ATGCAAATNA) that is found in both promoters and enhancers, but so far it has proven difficult to determine the basic unit of enhancer structure.

Promoter sequences are essential for transcription, while enhancers are not. Enhancers are able to stimulate levels of transcription at a distance, whereas this is not true of promoters. In addition, enhancers are responsible for tissue-specific gene expression.

As a group, promoters contain modules (the TATA box) that must be located at a fixed site upstream of the transcription start point; enhancers do not contain these modules. Although enhancers and promoters share similar sequences, the sequences in enhancers are most often contiguous; in promoters, they are spaced apart. On the

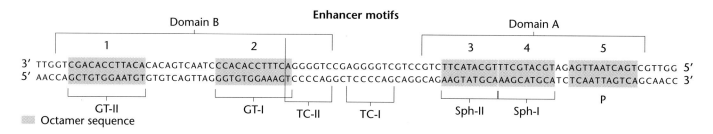

FIGURE 19.6 DNA sequence for the SV40 enhancer. Screened sequences are those required for maximum enhancer effect. The brackets below the sequence are the various sequence motifs within this region. The two domains of the enhancer (A and B) are indicated.

other hand, both enhancers and promoters have some striking similarities: They both have modular organization, with some modular sequences in common, and they both bind transcription factors.

The most intriguing question about enhancers is how they are able to exert control over transcription at a great distance from either promoters or the transcriptional start site. As we will discuss in a later section, transcription factors bind to enhancers and alter the configuration of chromatin by bending or looping the DNA to bring distant enhancers and promoters into direct contact in order to form complexes with transcription factors and polymerases. In the new configuration, transcription is stimulated to a higher level, increasing the overall rate of RNA synthesis.

Transcription Factors Bind to Promoters and Enhancers

As in prokaryotes, transcriptional control in eukaryotes involves interaction between DNA sequences adjacent to genes and DNA-binding proteins. As outlined in Chapter 17, the regulatory sequences adjacent to a wide range of genes have been identified, mapped, and sequenced. Development of DNA protection and gel retardation assays have allowed the detection of DNA-binding proteins in cell extracts, and new affinity chromatography techniques have been used to isolate transcriptional factors present in very low concentrations. The result has been a virtual explosion of information about eukaryotic transcription factors. This section will review some of this new information, including the upstream organization of eukaryotic genes, the characteristics of transcription factors, their interaction with DNA sequences, other regulatory factors, and the initiation of transcription by RNA polymerase.

Proteins that are not part of the RNA polymerase molecule itself, but are needed for the initiation of transcription, are called **transcription factors**. These pro-

teins control where, when, and how genes are expressed. Transcription factors are modular structures with at least two functional domains: one that binds to DNA sequences present in promoters and enhancers (**DNA-binding domain**), and another that activates transcription via protein–protein interaction (***trans*-activating domain**). Protein domains are clusters of amino acids that carry out a specific function. This region binds to RNA polymerase or to other transcription factors. The existence of separate domains in eukaryotic transcription factors was first demonstrated by genetic analysis of mutants in yeast.

Genetic Analysis of Transcription Factors

In yeast, genetic analysis has identified loci that encode the enzymes required for galactose metabolism. Expression of the genes encoding these enzymes is regulated by the presence or absence of galactose. In the absence of galactose, these genes are not transcribed. If galactose is added to the growth medium, immediate transcription of the three genes commences, and the mRNA concentration of these transcripts increases by a thousandfold. A mutation in the *GAL4* gene prevents the activation of these genes in the presence of galactose, indicating that transcription is genetically regulated.

Transcription of the structural genes for galactose metabolism is controlled by a **UAS$_G$** (upstream activating sequence). Recall that UASs are functionally similar to promoters and enhancers in higher eukaryotes. Using recombinant DNA techniques, the Gal4 protein was found to bind to the UAS$_G$ region, and to activate transcription of the genes for galactose metabolism, establishing that the *GAL4* gene encodes a transcription factor.

The *GAL4* gene product is a protein of 881 amino acids, that includes a DNA-binding domain that recognizes and binds to sequences in the UAS$_G$, and a *trans*-activating domain that activates transcription. These functional domains were identified by cloning and expressing truncated *GAL4* genes and assaying the gene products for their ability to bind DNA and to activate transcription

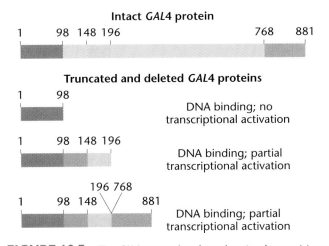

FIGURE 19.7 The *GAL4* protein has three domains that participate in the activation of the *GAL1* and *GAL10* loci. Amino acids 1–98 or 1–147 bind to DNA recognition sites in UAS$_G$. Amino acids 148–196 and 768–881 are required for transcriptional activation. *GAL80* binding activity resides in the amino acid region 851–881.

(Figure 19.7). Because the protein contains two functional domains, it is possible to delete one of these regions, while the other remains intact and functional.

A region at the N-terminus of the protein was found to be the DNA-binding domain (Figure 19.7). Proteins with deletions extending from amino acid 98 to the C-terminus (amino acid 881) retained the ability to bind to the UAS$_G$ region. The DNA-binding domain is responsible for recognition and binding to the nucleotide sequence in UAS$_G$. Similar experiments have identified two regions of the protein involved in activation of transcription: region I, consisting of amino acids 148–196, and region II, encompassing amino acids 768–881 (Figure 19.7). Constructs that contain the DNA-binding domain and region I (amino acids 1–96 and 149–196) or the DNA-binding domain and region II alone (amino acids 1–96 and 768–881) have reduced transcriptional activity.

The nature of the activating regions was explored by starting with a protein that contains the DNA-binding region and activating region I (*GAL4* 1–238) and isolating point mutations in region I that result in increased activation. Most single mutations that increase activation produce amino acid substitutions that increase the negative charge of region I, and most cause a threefold increase of transcription. A multiple mutant that contains four more negative charges than the starting protein causes a ninefold increase in transcriptional activation, and has almost 80 percent of the activity of the complete *GAL4* protein. However, negative charge alone is not responsible for the activating function; some mutants with decreased activity do not have fewer negative charges than the parental *GAL4* 1–238 molecule.

These results strongly suggest that activation involves direct contact between the activating domain of the transcription factor and other proteins. How does this protein–protein contact activate transcription? There are several possibilities, including stabilization of the binding between the promoter DNA and RNA polymerase, increasing the rate at which the double stranded DNA within the transcribed region is unwound, or attracting and stabilizing other factors which bind to the promoter or to the RNA polymerase.

An attractive candidate for an activator target is TFIID, a complex of several proteins that binds to the TATA sequence and is required for transcription. TFIID consists of the TATA binding protein (TBP), and about 10 other proteins called TBP-associated factors, or TAFs (some of which are described in later sections). The idea that TFIID is a target for activation is supported by the finding that some *GAL4* constructs stimulate transcription in mammalian cell extracts and change the conformation of DNA-bound TFIID. In addition, other studies show that TBP alone can bind to the promoter, TAFS must be present for the complex to respond to activators.

Structural Motifs of Transcription Factors

The domains of eukaryotic transcription factors take on several forms. The **DNA-binding domains** have distinctive three-dimensional structural patterns or **motifs**. There are three major types of these structural motifs: **helix-turn-helix (HTH)**, **zinc finger**, and **leucine zippers**. This classification is not exhaustive, and other new groups will undoubtedly be established as new factors are characterized.

The first DNA-binding domain to be discovered was the helix-turn-helix (HTH) motif. In prokaryotes, HTH motifs have been identified in the *cro* repressor (in the DNA-binding domain), the *lac* repressor, *trp* repressor, and other proteins (Figure 19.8). Studies indicate

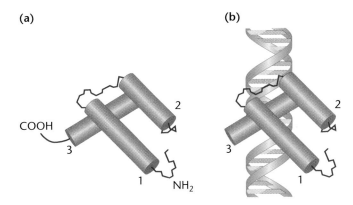

FIGURE 19.8 A *helix-turn-helix* or *homeodomain* where three different planes of the α helix of the protein are established. These domains bind in the grooves of the DNA molecule.

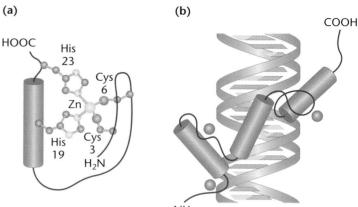

FIGURE 19.9 A *zinc finger*, where cysteine and histidine residues bind to a Zn^{++} atom, looping the amino acid chain out into a fingerlike configuration. Zinc finger domains are common in many DNA regulatory proteins.

that the HTH motif is present in many prokaryotic DNA-binding proteins. This motif is characterized by its geometric conformation rather than a distinctive amino acid sequence. Two adjacent α helices separated by a "turn" of several amino acids enables the protein to bind to DNA (and after which the motif is named). In the *cro* repressor, one of the helices binds within a major groove containing six bases, while the other helix stabilizes the interaction by binding to the DNA backbone. Unlike several of the other DNA-binding motifs, the HTH pattern cannot fold or function alone, but is always part of a larger DNA-binding domain. Amino acid residues outside the HTH motif are important in regulating DNA recognition and binding.

The potential for forming helix-turn-helix geometry has been recognized in distinct regions of a large number of eukaryotic genes known to regulate developmental processes. Present almost universally in eukaryotic organisms and called the **homeobox**, a stretch of 180 bp specifies a 60-amino-acid **homeodomain** sequence that can form a helix-turn-helix structure. Of the 60 amino acids, many are basic (arginine and lysine), and a conserved sequence is found among these many genes. We shall discuss homeobox-containing genes in Chapter 20 because of their significance to developmental processes.

Zinc fingers are one of the major structural families of eukaryotic transcription factors, and they are involved in many aspects of gene regulation. Zinc fingers were originally discovered in the *Xenopus* transcription factor TFIIIA. This structural motif has now been identified in proto-oncogenes (Chapter 21), genes that regulate development in *Drosophila* (the Krüppel gene, see Chapter 20), in proteins whose synthesis is induced by growth factors and differentiation signals, and in transcription factors. There are several types of zinc finger proteins, each with a distinctive structural pattern.

One such zinc finger protein contains clusters of two cysteine and two histidine residues at repeating intervals (Figure 19.9). The consensus amino acid repeat is

Cys–N$_{2-4}$–Cys–N$_{12-14}$–His–N$_3$–His. The interspersed cysteine and histidine residues covalently bind zinc atoms, folding the amino acids into loops known as zinc fingers. Each finger consists of approximately 23 amino acids, with a loop of 12 to 14 amino acids between the Cys and His residues, and a linker between loops consisting of 7 or 8 amino acids. The amino acids in the loop interact with and bind to specific DNA sequences. Studies have shown that zinc fingers bind in the major groove of the DNA helix and wrap at least part way around the DNA. Within the major groove, the zinc finger makes contact with a set of DNA bases and may form hydrogen bonds with the bases, especially in G-rich strands. The number of fingers in a zinc finger transcription factor varies from 2 to 13, as does the length of the DNA-binding sequence.

A third type of domain is represented by the leucine zipper. First seen as a stretch of 35 amino acids in a nuclear protein in rat liver, four leucine residues are spaced seven amino acids apart, and flanked by basic amino acids. The leucine-rich regions form a helix with leucine residues protruding at every other turn. When two such molecules dimerize (Figure 19.10), the leucine residues "zip" together. The dimer contains two alpha helical regions adjacent to the zipper that bind to phosphate residues and specific bases in DNA, making the dimer look like a pair of scissors (Figure 19.10).

In addition to domains that bind DNA, recall that transcription factors contain domains that activate transcription. These regions can occupy from 30 to 100 amino acids and are distinct from the DNA-binding domains. These stretches of amino acids interact with other transcription factors (such as those that bind to the TATA sequence) or directly with the RNA polymerase.

Overall, the picture of transcriptional regulation in eukaryotes is somewhat complex, but a number of generalizations can be drawn. The structural organization of chromatin and alterations in chromatin structure to allow binding of transcription factors are considered to be the primary levels of regulation. The regulation of transcription by protein factors is largely positive, although tran-

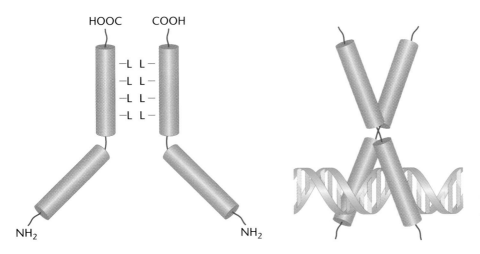

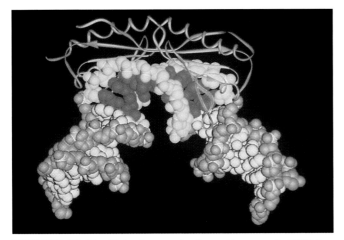

FIGURE 19.10 A *leucine zipper*, where dimers result from leucine residues at every other turn of the α helix in facing stretches of two polypeptide chains. When the α-helical regions form a leucine zipper, the regions beyond the zipper form a Y-shaped region that grips the DNA in a scissors-like configuration.

scription repressors are now being recognized as important components of gene regulation. The binding of one or more factors at promoter regions is a prerequisite to transcriptional activation of a locus. Promoter and enhancer sequences are recognized and bound by transcription factors.

Because of the modular nature of recognition sequences, different transcription factors may compete for binding to a given DNA sequence, or the same site may bind different factors in a cell-specific fashion. For example, sequences in the TATA box and CAAT box can bind to transcription factors that recognize these modules. Finally, because of the variability in the location and modular nature of promoters and enhancers, and the variety of transcription factors, an initiation complex (consisting of transcription factors and RNA polymerase II) can be constructed in a number of ways, all involving protein–protein interaction. Despite this variability, it is possible to provide a general outline that describes how eukaryotic transcription factors operate, and in the next section we will consider a generalized model for the assembly of a transcription complex.

FIGURE 19.11 Molecular complex formed between the TATA-box binding protein (green) and the 8 base-pair TATA box of DNA (red), itself part of DNA (yellow).

Assembling the Transcription Complex

Some insights into the transcriptional apparatus of type II genes (those transcribed by RNA polymerase II) are now available and involve a series of transcriptional factors that are assembled at the promoter in a specific order. To initiate formation of the apparatus, the TFIID complex binds to the TATA promoter via its TBP (Figure 19.11). About 20 base pairs of DNA are involved in binding the TBP, and the other subunits bind to this protein. This has been described as the *commitment* stage (Figure 19.12). TFIID responds to contact with activator proteins by conformational changes that expedite the binding of TFIIA, TFIIB, and polymerase. Unlike the situation in prokaryotes where polymerase binds directly to the DNA promoter region, eukaryotic polymerases bind to transcription factor proteins that are in turn bound to the DNA promoter region. At this point, transcription of the DNA downstream may ensue at a minimal *basal* level. The final stage involves the achievement of the *induced* state, where transcription is stimulated above the basal level. This state is not yet well defined, but involves other areas of the promoter region, enhancers, and numerous transcription factors, which control the assembly of the transcription complex and the rate at which RNA polymerase initiates transcription. Factors bound to enhancers at a distance from the site are thought to interact with the transcription complex, looping out the DNA that separates the enhancer from the complex (Figure 19.13).

How Are Transcription Factors Controlled?

Evidence that gene expression in eukaryotes can be regulated by effector molecules originating outside the cell was elegantly provided by Ulrich Clever and Peter Karlson with their discovery that the steroid hormone **ecdysone** induces specific changes in the puffing pattern in polytene chromosomes (recall from Chapter 2 that puffs represent the transcription of specific genes). In eu-

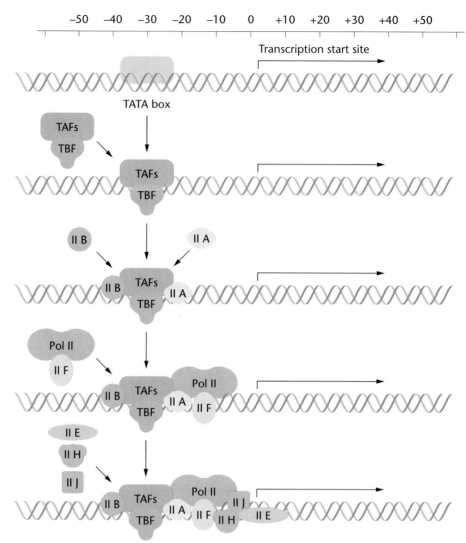

FIGURE 19.12 The assembly of transcription factors required for the initiation of transcription by RNA polymerase II. The first factor binds to the TATA box, and then TFIIA binds, forming a complex with the TATA factor and the DNA. The next factors, TFIIB and TFIIA, bind before RNA polymerase II. The polymerase binds to the transcription factors through protein–protein interactions. Before transcription begins, factors E, H and J bind, and transcription then begins from the initiation site at +1.

karyotes, steroid hormones are used to regulate growth and development and to maintain homeostasis. The major sex hormones are all steroids, as is vitamin D. The homeostatic adrenal hormones (more than 30 different types) that regulate glucose metabolism and mineral utilization are also steroids (Figure 19.14).

Steroid hormones enter the cell by passing through the plasma membrane and binding to a specific cytoplasmic receptor protein. The receptor–hormone complex is translocated to the nucleus and activates transcription of one or more specific genes. These protein receptors are transcription factors. Hormone receptors and the DNA

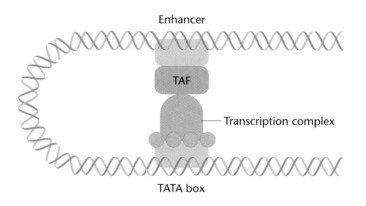

FIGURE 19.13 DNA looping allows factors that bind to enhancers at a distance from the promoter to interact with regulatory proteins in the transcription complex and to maximize transcription.

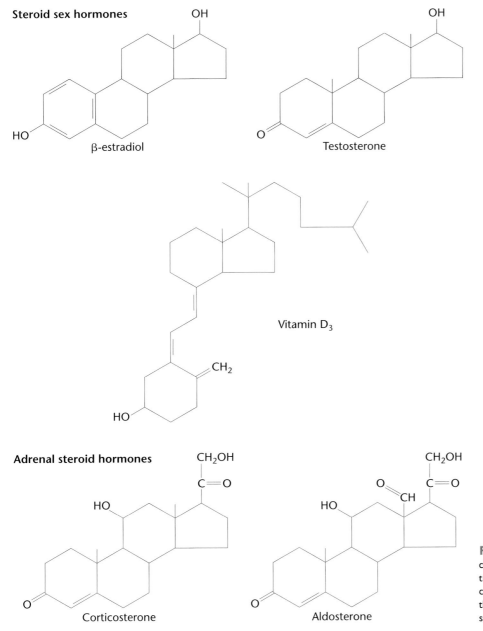

FIGURE 19.14 Structure of several classes of steroid hormones that activate transcription by binding to cytoplasmic receptor proteins that are then transported to the nucleus where they bind to specific DNA sequences.

sequences to which they bind have been intensively studied over the last decade. The structural genes encoding most every known hormone receptor protein have been cloned, and enough information is now available to provide a general picture of how these receptors work. All receptors analyzed to date have three functional domains: a variable N-terminus region that is unique to each receptor; a short, highly conserved central region that binds to the DNA; and a fairly well-conserved C-terminus that binds to the hormone (Figure 19.15). The DNA-binding central region contains two zinc-finger motifs.

In addition to activating the regulated gene, the C-terminal domain of steroid receptors interacts with and

binds to the steroid hormone. (It has been suggested that in the absence of bound hormone, the receptor is unable to recognize the hormone receptor element of DNA). The C-terminal region may also be partially responsible for controlling the translocation of the hormone–receptor complex from the cytoplasm to the nucleus.

The zinc fingers in hormone receptors bind to specific DNA sequences known as **hormone-responsive elements** or **HREs**. In steroid receptors, the finger closer to the N-terminus binds in a base-specific way to the HRE and determines the specificity of binding to a particular target gene. The C-terminus finger contacts the sugar-phosphate residues adjacent to the HRE and may

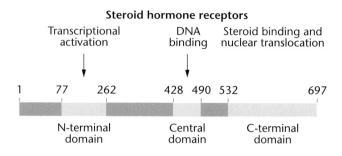

Steroid hormone receptors

FIGURE 19.15 Steroid hormone receptors have three functional domains, as exemplified by the glucocorticoid receptor: the N-terminal domain (amino acids 77–262 in the glucocorticoid receptor); the central domain (amino acids 428–490); and the C-terminal domain (amino acids 532–697). The central domain binds DNA sequences known as the glucocorticoid response element (GRE) by means of two zinc fingers. The C-terminal domain binds to the steroid hormone and may be responsible for allowing the hormone–receptor complex to be translocated to the nucleus. This region is also responsible for regulating the activity of the receptor.

have other functions. In general, HREs share some characteristics with promoters and enhancers. They are composed of short consensus sequences that are related but not always identical (Table 19.1). HREs are often located several hundred bases upstream from the transcription start site, and they may be present in multiple copies. Often, HREs are contained within promoter or enhancer sequences.

Binding of the receptor to the HRE is necessary for activation of a specific gene (Figure 19.16), but it may not be sufficient. Binding of the receptor to the HRE serves to place regions of both the C-terminus and N-terminus in position to activate a specific promoter, but activation may require the participation of other transcription factors, and, as described below, this involves the alteration of chromatin structure in the HRE and adjacent regions.

It is important to note that gene activation by steroid hormones requires alteration of chromatin structure near the regulated gene. Genes in chromatin that are actively transcribed or are able to be transcribed are hypersensitive to digestion with the enzyme DNase I, while inactive or repressed genes are relatively resistant to DNase I. *In vitro* studies of steroid activation of genes in chromatin indicate that following hormone administration, DNase I sensitivity greatly increases in the regulated gene and its

flanking regions, including the hormone receptor binding site. Interpretation of these results is consistent with the idea that the binding of one or more receptor–hormone complexes (receptors may need to act as dimers or tetramers to activate genes) to the HRE causes a change or shift in the nucleosome structure of chromatin at the binding site, making other binding sites, including the promoter, available to transcription factors and RNA polymerase II, resulting in the initiation of transcription.

In some ways, then, the biology of steroid hormone regulation of gene transcription in eukaryotes is similar to the *lac* operon system in prokaryotes in that an external effector binds to a cytoplasmic receptor, changes the configuration of the receptor, and moves the receptor to the DNA, where it acts as a transcription factor. In other ways, however, the regulation of gene expression is quite different, in that the DNA-binding sequence is often at a great distance from the regulated gene, and that alterations in chromatin structure mediate the action of the receptor.

Genomic Alterations and Gene Expression

Alteration of chromatin conformation is one of several ways in which gene expression can be regulated by physical changes in DNA. In this section we will consider two types of changes in chromatin that play a role in gene regulation. One is a chemical modification of DNA that involves adding or removing methyl groups to the bases in DNA, and the other is a change in the number of copies of a DNA sequence brought about by amplifying a specific gene or set of genes in order to increase the level of expression.

DNA Methylation

The DNA of most eukaryotic organisms is modified after replication by the enzyme-mediated addition of methyl groups to bases and sugars. Base methylation most often involves cytosine, and approximately 5 percent of the cytosine residues are methylated in the genome of any given eukaryotic species, although the extent of methylation can be tissue-specific and can vary from less than 2 percent to over 7 percent.

The ability of base methylation to alter gene expression is known from studies on the *lac* operon in *E. coli.* Methylation of DNA in the operator region, even at a single cytosine residue, can cause a marked change in the affinity of the repressor for the operator. Methylation of cytosine occurs at the 5' position, causing the methyl

TABLE 19.1 DNA Receptor Sequences for Hormone Binding Proteins

Hormone	Consensus Sequence	Element Name
Glucocorticoid	GGTACANNNTGTTCT	GRE
Estrogen	AGGTCANNNTGACCT	SRE
Thyroid	AGGTCA . . . TGACCT	TRE

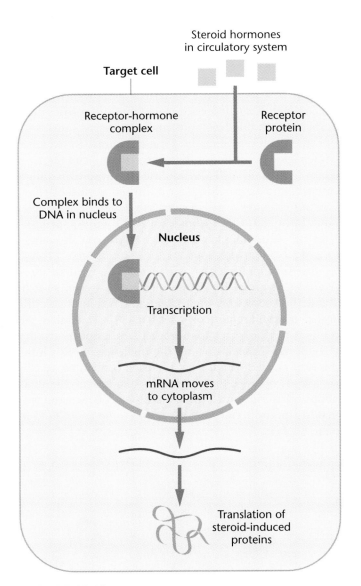

FIGURE 19.16 Stages of steroid hormone effects on gene expression. Steroids in the circulating system pass through the plasma membranes of target cells and bind to cytoplasmic receptor proteins. The steroid–receptor complex moves to the nucleus and binds to DNA at hormone receptor elements, stimulating transcription and translation of steroid-induced genes.

group to protrude into the major groove of the DNA helix where it can alter the binding of proteins to the DNA.

Methylation occurs most often in the cytosine of CG doublets in DNA, usually in both strands:

$$5'\text{-}^m\text{CpG}\text{-}3'$$
$$3'\text{-}\text{GpC}^m\text{-}5'$$

The state of DNA methylation can be determined by restriction enzyme analysis. The enzyme *Hpa*II cleaves at the recognition sequence CCGG; however, if the second cytosine is methylated, the enzyme will not cut the DNA.

The enzyme *Msp*I cuts at the same CCGG site whether or not the second cytosine is methylated. If a segment of DNA is unmethylated, both enzymes produce the same restriction pattern of bands. As shown in Figure 19.17, if one site is methylated, digestion with *Hpa*II produces an altered pattern of fragments. Using this method to analyze the methylation of a given gene in different tissues shows that, in general, if a gene is expressed, it is not methylated, or has a low level of methylation.

Evidence for the role of methylation as a factor in the regulation of eukaryotic gene expression is somewhat indirect and is based on a number of observations. First, an inverse relationship exists between the degree of methylation and the degree of expression. That is, low amounts of methylation are associated with high levels of gene expression, and high levels of methylation are associated with low levels of gene expression. In mammalian females the inactivated X chromosome, which is almost totally inactive in gene expression, has a higher level of methylation than does the active X chromosome. Within the inactive X, those regions that escape inactivation have

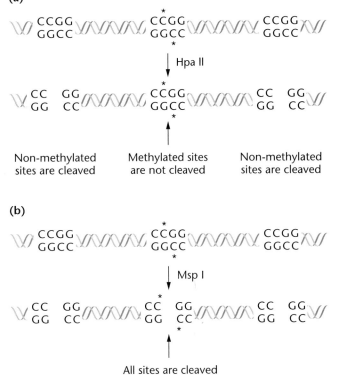

FIGURE 19.17 The restriction enzymes *Hpa*II and *Msp*I recognize and cut at CCGG sequences. (a) If the second cytosine is methylated (indicated by an asterisk), *Hsp*II will not cut. (b) The enzyme *Msp*I cuts at all CCGG sites, whether or not the second cytosine is methylated. Thus the state of methylation of a given gene in a given tissue can be determined by cutting DNA extracted from that tissue with *Hpa*I and by *Msp*II.

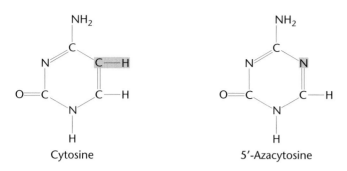

FIGURE 19.18   The base 5′-azacytosine has a nitrogen at the 5′ position. The base 5′-azacytidine can be incorporated into DNA in place of cytidine. The 5′-azacytidine cannot be methylated, causing undermethylation of the CpG dinucleotide wherever it has been incorporated.

much lower levels of methylation than those seen in adjacent, inactive regions.

Second, methylation patterns are tissue-specific and, once established, are heritable for all cells of that tissue. Perhaps the strongest evidence for the role of methylation in gene expression comes from studies using base analogs. The nucleotide 5′-azacytidine is incorporated into DNA in place of cytidine and cannot be methylated (Figure 19.18), causing undermethylation of sites where it is incorporated. Incorporation of 5′-azacytidine causes changes in the pattern of gene expression and can stimulate expression of alleles on inactivated X chromosomes.

Recently, 5′-azacytidine has been used in clinical trials for the treatment of sickle-cell anemia. In this autosomal recessive condition, a mutant hemoglobin protein with a defect in the beta subunit causes changes in shape of red blood cells that, in turn, generate a cascade of clinical symptoms. During embryogenesis, the epsilon (ε) and gamma (γ) globin genes are expressed, but normally become transcriptionally inactive after birth when beta globin synthesis begins. Treatment of affected individuals with 5′-azacytidine causes a reduction in the amount of methylation in the epsilon (ε) and gamma (γ) genes, and initiates reexpression of these embryonic and fetal genes. The epsilon and gamma proteins replace beta-globin in hemoglobin molecules, bringing about a reduction in the amount of sickling in the red blood cells.

As stated above, the available evidence indicates that the absence of methyl groups in DNA is related to increases in gene expression. Methylation cannot, however, be regarded as a general mechanism for gene regulation because methylation is not a general phenomenon in eukaryotes. In *Drosophila*, for example, there is no methylation of DNA. Thus, methylation may represent only one of a number of ways in which gene expression can be regulated by genomic changes, and it seems likely that similar mechanisms remain to be discovered.

Gene Amplification

One way to regulate the amount of a given gene product synthesized over a limited time period is to increase the number of copies of the structural gene that is being transcribed. This process, termed **gene amplification**, plays a role in the developmentally regulated expression of some gene sets, including the ribosomal RNA genes of *Xenopus* (described in Chapter 10) and the chorion genes in *Drosophila*. This amplification results in a high rate of gene expression over a limited period of time. During oogenesis in *Xenopus*, the developing oocyte, containing amplified ribosomal RNA genes, accumulates 10^{12} ribosomes in about 60 to 80 days, at an average rate of 300,000 per second.

In the somatic cells of *Xenopus*, the tandemly repeated ribosomal RNA genes (rDNA) are present in about 1000 copies per diploid genome. Without amplification, it would require over 1000 days to accumulate the number of ribosomes found in a mature oocyte. In the developing *Xenopus* oocyte, the nucleus contains hundreds of nucleoli (Figure 19.19), each containing several thousand copies of the ribosomal gene cluster. Simultaneous transcription of these amplified, extrachromosomal loci permits ribosome synthesis to be completed in 60 to 80 days.

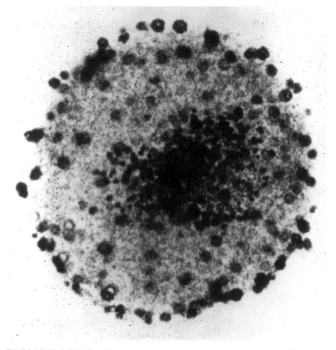

FIGURE 19.19   Photomicrograph of a mature oocyte of *Xenopus laevis*. The dark circles in the interior and at the periphery are amplified nucleoli, each containing thousands of copies of the ribosomal genes. These amplified genes make it possible to synthesize billions of ribosomes over a relatively short time period.

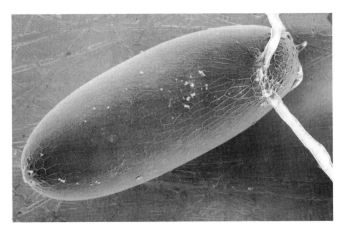

FIGURE 19.20   The chorion of the *Drosophila* egg is secreted by the follicle cells surrounding the developing oocyte. In the follicle cells, the genes for chorion proteins are selectively amplified as a first step in expression.

In *Drosophila*, the amplification of specific protein-encoding genes is a feature of oocyte development. One set of genes encoding eggshell proteins is clustered on chromosome 3. At the time of expression, these genes are amplified up to 60 times in the cells that synthesize the proteinaceous shell of the *Drosophila* egg (Figure 19.20). This amplification extends as a gradient centered on the chorion genes and covers a distance of some 50 kb on either side of this gene set. Sequential amplification of several protein-encoding loci occurs during development in other insects, including the fungal gnat *Sciara*, where salivary gland chromosome puffs contain amplified DNA sequences (Figure 19.21), increasing the copy number up to 16-fold.

In mammals, gene amplification does not appear to play a developmental role, but is often associated with the

TABLE 19.2   Molecular Mechanisms for Gene Amplification

Replication-Based Models	Segregation and Recombination-Based Models
Onionskin model	Deletion plus episome model
Extrachromosomal double rolling circle model	Sister chromatid exchange model
Chromosomal spiral model	

appearance of drug resistance in malignant tumors during chemotherapy, and in generating the multiple copies of tumor-associated genes known as oncogenes (see Chapter 21) that are associated with the progressive changes occurring during tumor growth. Amplified DNA sequences are often visible in such mammalian cells as chromosome structures known as **extended chromosome regions (ECRs)** and **homogeneously staining regions (HSRs)**, or as extrachromosomal elements known as **double minute (DM) chromosomes**.

Several mechanisms have been proposed to explain gene amplification, and they have been grouped into two types: mechanisms associated with DNA replication (discussed above) and those associated with recombination and segregation (Table 19.2).

Perhaps the most straightforward model is that of recombination- and segregation-driven amplification: the sister chromatid exchange model (Figure 19.22). If two sister chromatids engage in an unequal exchange, segments of DNA will be duplicated on one chromatid and

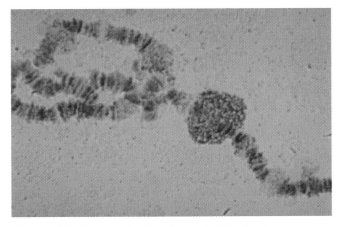

FIGURE 19.21   DNA puffs in *Sciara coprophila*. A polytene chromosome from the larval salivary gland of *Sciara* shows two prominent DNA puffs. In these puffs, DNA segments are amplified thousands of times, thus permitting an increase in the amount of transcription of selected genes over a limited time span.

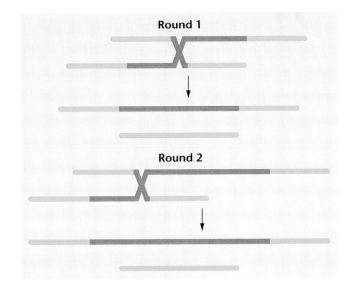

FIGURE 19.22   Sister chromatid recombination model of gene amplification. Misalignment and recombination between sister chromatids result in one chromatid with a duplicated segment (blue portion) and a sister chromatid with a deleted segment. Further rounds of misalignment and recombination involving the chromosome carrying the duplication will result in a linear amplification of the shaded sequence.

will be lost from the other. If this process is repeated for several cell divisions, the outcome will be a cell that contains one chromosome containing an ESR or HSR with many amplified copies of a gene, and another homolog with a single copy of the gene. Cytogenetic analysis of cultured cells that have undergone amplification in response to drug selection confirm this expectation. When drug selection is relaxed, there is often a reduction in the copy number of the amplified gene, and a loss or reduction of the modified chromosomal region. Thus, the sister chromatid model provides for a reversible means of altering the copy number of structural genes. It is hoped that future work on amplification in tumors and cultured cells will provide information about the organization of chromosomes and the chromosomal loci that serve as origins of DNA replication.

Posttranscriptional Regulation of Gene Expression

As outlined above, many opportunities exist along the pathway from DNA to protein where regulation of genetic expression can occur. For example, although transcriptional control is perhaps the most obvious and widely used mode of regulation in eukaryotes, posttranscriptional modes of regulation are also employed in many organisms. Eukaryotic nuclear RNA transcripts are modified prior to translation, noncoding introns are removed, the remaining exons are precisely spliced together, and the mRNA is modified by the addition of a poly-A tail at the 3'-end. Each of these processing steps offers several possibilities for regulation.

Alternative Processing Pathways for mRNA

Alternative splicing can generate different forms of a protein, giving rise to a situation where expression of one gene can give rise to a family of related proteins. Figure 19.23 illustrates an example where the polypeptide products derived from a single type of pre-mRNA are distinct from one another. The initial bovine pre-mRNA transcript is processed into one or two **preprotachykinin mRNAs (PPT mRNAs)**. The precursor mRNA molecule potentially includes the genetic information specifying two neuropeptides called **P** and **K**. These two peptides are members of the family of sensory neurotransmitters referred to as **tachykinins**, and they are believed to play different physiological roles. While the P neuropeptide is largely restricted to tissues of the nervous system, the K neuropeptide is found more predominantly in the intestine and thyroid.

The RNA sequences for both neuropeptides are derived from the same gene. However, processing of the initial RNA transcript can occur in two different ways. In one case, exclusion of the K-exon during processing results in the α-PPT mRNA, which upon translation yields neuropeptide P but not K. Conversely, processing that includes both the P and K exons yields β-PPT mRNA, which upon translation results in the synthesis of both the P and K neuropeptides. Analysis of the relative levels of the two types of RNA has demonstrated striking differences between tissues. In nervous-system tissues, α-PPT mRNA predominates by as much as a threefold factor, while β-PPT mRNA is the predominant type in the thyroid and intestine.

Given the existence of alternative splicing, how many different polypeptides can be derived from the same pre-mRNA? Work on alpha-tropomyosin has provided a partial answer to this question. In rats, the alpha-tropomyosin gene contains a total of 14 exons, 6 of which make up three pairs that are alternatively spliced. Only one member of each pair ends up in the finished mRNA, never both. Alternative splicing of this pre-mRNA results in 10 different forms of alpha-tropomyosin, many of which are expressed in a tissue-specific manner. Another gene, for the muscle form of troponin T, a protein that regulates calcium needed for contraction, produces 64 known forms of the protein from a single pre-mRNA by alternative splicing.

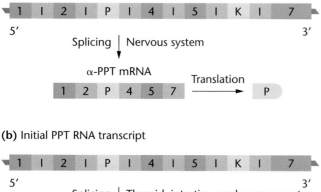

(a) Initial PPT RNA transcript

(b) Initial PPT RNA transcript

FIGURE 19.23 Diagrammatic representation of the alternative splicing of the initial RNA transcript of the preprotachykinin gene (PPT). Introns are unshaded and are labeled I. Exons are either numbered or designated by the letters P or K. Inclusion of the P and K exons leads to β-PPT mRNA, which upon translation yields both the P and K tachykinin neuropeptides. When the K exon is excluded, α-PPT mRNA is produced, and only the P neuropeptide is synthesized.

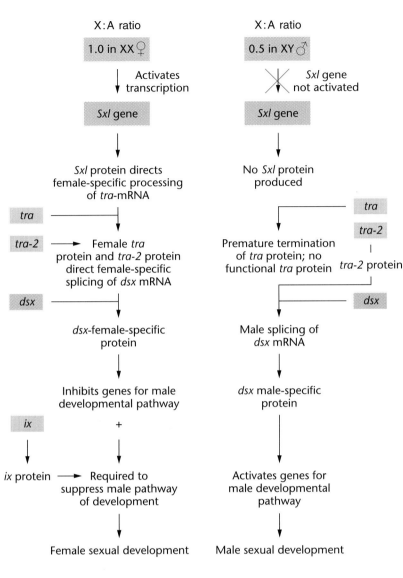

FIGURE 19.24 Hierarchy of gene regulation for sex determination in *Drosophila*. In females, the X:A ratio activates transcription of the *Sxl* gene. The product of this gene binds to pre-mRNA of the *tra* gene and directs its splicing in a female-specific fashion. The female-specific *tra* protein, in combination with the *tra*-2 protein, directs female-specific splicing of *dsx* pre-mRNA, resulting in a female-specific *dsx* protein. This protein, in combination with the *ix* protein, suppresses the pathway of male sexual development and activates the female pathway. In males, the X:A ratio does not activate *Sxl*. The result is a male-specific processing of *tra* pre-mRNA, resulting in no functional *tra* protein. This in turn leads to male-specific processing of *dsx* transcripts, resulting in male-specific *dsx* protein that activates the pathway for male sexual development.

Alternative splicing not only has the potential to produce many different forms of a protein from a single gene, but can affect the outcome of developmental processes. Perhaps the most striking example of this involves sex determination in *Drosophila*, where splicing involving a single exon eventually determines whether the embryo will develop as a male or a female.

As outlined in Chapter 9, sex in *Drosophila* is determined by the ratio of X chromosomes to autosomes (X:A). When the ratio is 0.5 (1X:2A), males are produced, even when no Y chromosome is present; when the ratio is 1.0 (2X:2A), females are produced. At intermediate ratios (2X:3A), intersexes are produced. The chromosomal ratios are interpreted by a small number of genes that initiate a cascade of developmental events resulting in the production of male or female somatic cells and the corresponding male or female phenotypes. The genes in this pathway are *Sex-lethal* (*Sxl*), *transformer* (*tra*), *transformer*-2 (*tra*-2), *doublesex* (*dsx*), and *intersex* (*ix*).

In females, the chromosomal ratio activates transcription of the *Sxl* gene and the production of *Sxl* protein (the mechanism is unknown). This activation involves expression of several intermediate genes, including *sisterless* and *daughterless*, and perhaps a small number of other genes. In males, the chromosomal ratio does not activate production of *Sxl* protein (Figure 19.24). When the *Sxl* protein is present, it binds to and controls the splicing of the pre-mRNA for the next gene in the pathway, *transformer* (*tra*). The pre-mRNA of *tra* includes a termination codon in exon 2. The female-specific *Sxl* protein binds to the *tra* pre-mRNA and directs the splicing to remove exon 2 from the mature mRNA.

In males, where no *Sxl* protein is present, splicing proceeds along the ground state or default pathway, and

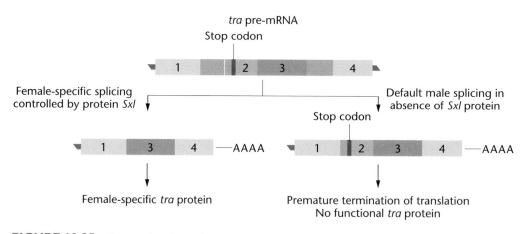

FIGURE 19.25 Sex-specific splicing of *tra* gene transcripts. In females, the *Sxl* protein binds to the pre-mRNA and directs splicing to eliminate exon 2, which contains a stop codon. This allows translation of a female-specific protein that directs female-specific splicing of *dsx* transcripts. In males, no *Sxl* protein is present, and the *tra* pre-mRNA is spliced to include the stop codon in exon 2. The result is a nonfunctional *tra* protein fragment that is unable to direct processing of *dsx* transcripts.

the stop codon is incorporated into the mature mRNA (Figure 19.25). In males, translation of *tra* mRNA is prematurely terminated by the stop codon, resulting in an inactive gene product, whereas in females, translation produces a functional gene product. The result is a female-specific *tra* protein that acts in conjunction with the *tra-2* gene to produce the female phenotype (Figure 19.24). The *tra-2* gene is active in both males and females and contains an RNA-binding domain that binds to the pre-mRNA of the *dsx* gene, but only in the presence of the *tra* protein.

The *dsx* gene is a critical control point in the development of sexual phenotype, for it produces a functional mRNA and protein in both females and males. However, the pre-mRNA is processed in a sex-specific manner to yield different transcripts. In females, the *tra* protein and the *tra-2* protein bind to the *dsx* pre-mRNA and direct its processing in an alternative, female-specific manner. In males, where only the *tra-2* protein is present, default splicing of the *dsx* pre-mRNA results in a male-specific mRNA and protein product. The female *dsx* protein causes female sexual differentiation in somatic tissues by repressing male sexual development through coordinated action with the *ix* gene product (Figure 19.24). The male *dsx* protein brings about male sexual development by suppressing the female pathway. In the absence of *dsx* function, flies develop into phenotypic intersexes.

The *Sxl* gene acts as a switch gene in selecting the pathway of sexual development by eventually controlling the processing of the *dsx* transcript in a female-specific fashion. The *Sxl* gene exerts this control by encoding a splicing factor that causes transcripts from the *tra* gene to be processed in a female-specific fashion. The *tra* protein interacts with the *tra-2* protein to control the splicing of

the *dsx* transcript in a female-specific fashion. The *Sxl* gene is activated only in embryos with an X:A ratio of 1. Failure to control the splicing of the *dsx* transcript in a female mode (failure to initially activate the *Sxl* gene) results in default splicing of the transcript in a male mode, leading to production of a male phenotype.

Controlling mRNA Stability

After mRNA precursors are processed and transported, they enter the population of cytoplasmic mRNA molecules from which messages are recruited for translation. All mRNA molecules have a characteristic life span (called a half-life); they are degraded, usually in the cytoplasm, sometime after they are synthesized and transported to the cytoplasm. The lifetimes of different mRNA molecules varies widely. Some are degraded within minutes after synthesis, whereas others last hours, or even months and years (in the case of mRNAs stored in oocytes).

Although regulation of gene expression at the level of translation may seem inefficient, a growing body of evidence indicates that altering the stability of mRNA engaged in translation may be a significant point for controlling eukaryotic gene expression.

Why would such an inefficient mechanism be present in eukaryotic cells? Clearly, factors that influence the half-life of particular mRNAs have an important effect on the number of protein molecules that can be produced from a single mRNA molecule. Longer lived mRNAs can produce more protein than a short-lived one. Altering the half-life of an mRNA molecule already engaged in translation is one way a cell can respond to rapidly changing conditions inside or outside the cell.

One of the best-studied examples of regulation via mRNA stability is the synthesis of alpha and beta tubulins, the subunit components of eukaryotic microtubules. Treatment of a cell with the drug colchicine leads to a rapid disassembly of microtubules, and an increase in the concentration of the alpha and beta subunits. Under these conditions, synthesis of alpha and beta tubulins drops dramatically. However, when cells are treated with vinblastine, a drug that also causes microtubule disassembly, the synthesis of tubulins is increased. The difference between the two drugs is that in addition to causing microtubule disassembly, vinblastine precipitates the sub-

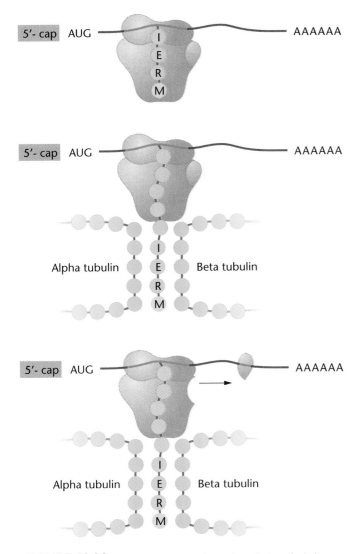

units, lowering the concentration of free alpha and beta subunits. At low concentrations, synthesis of tubulins is stimulated, whereas at higher concentrations, synthesis is inhibited. This type of translational regulation is known as **autoregulation**.

Work by Don Cleveland and his colleagues has described the conditions under which the stability of tubulin mRNA is regulated (Figure 19.26). Fusion of tubulin gene fragments to a cloned thymidine kinase gene shows that the first 13 nucleotides at the 5′ end of the messenger RNA following the transcription start site (encoding the amino terminal region of the protein) cause the hybrid tubulin-thymidine kinase mRNA to be regulated as efficiently as intact tubulin mRNA. These first 13 nucleotides encode the amino acids Met-Arg-Glu-Ile (abbreviated as MREI). Deletions, translocations, or point mutations in this 13-nucleotide segment abolish regulation. Examination of the cytoplasmic distribution of mRNAs demonstrates that only tubulin mRNAs in polysomes are depleted by drug treatment. Copies of tubulin mRNAs not bound to polysomes were protected from degradation. Also, translation must proceed to at least codon 41 of the mRNA in order for regulation to occur.

A model, shown in Figure 19.26, has been proposed to explain these observations. In this model, regulation occurs after the process of translation has begun. The first four amino acids (Met-Arg-Glu-Ile [MREI]) of the tubulin gene product (in this case, beta tubulin) constitute a recognition element to which the regulatory factors bind. The concentration of alpha and beta tubulin subunits in the cytoplasm may serve this regulatory function. This protein–protein interaction invokes the action of an RNase that may be a ribosomal component or a nonspecific cytoplasmic RNase. Action by the RNase degrades the tubulin mRNA in the act of translation, shutting down tubulin biosynthesis. As an alternative, binding of the regulatory element may cause the ribosome to stall in its translocation along the mRNA, leaving it exposed to the action of RNase. Translation must proceed to at least codon 41, because the first 30 to 40 amino acids translated occupy a tunnel within the large ribosomal subunit, and only the translation of this number of amino acids will make the MREI recognition sequence accessible for binding.

Translation-coupled mRNA turnover has been proposed as a regulatory mechanism for other genes, including histones, some transcription factors, lymphokines, and cytokines. This suggests the existence of a regulatory mechanism that acts on mRNAs in the act of translation, probably by protein–protein interaction between a regulatory element and the nascent polypeptide chain, resulting in the activation of RNases.

FIGURE 19.26 Model of posttranslational regulation of tubulin synthesis. Tubulin subunits (or another effector) bind to the amino terminal tubulin tetrapeptide as it emerges from the large ribosomal subunit. Binding activates an RNase (in this case, a ribosomal RNase) that breaks down the polysomal tubulin mRNA.

GENETICS, TECHNOLOGY, AND SOCIETY

Entrapping Genes That Regulate Development

We humans have always sought answers to questions of our origins. We yearn to know how and why we become physically and behaviorally unique organisms. Genetics can tell us that we arise from the simple elements of the universe, through the processes of development and differentiation. However, unravelling the genetic complexities of these events is a daunting prospect. We know that multicellular organisms grow, develop and differentiate from single fertilized egg cells. We also know that the complex choreography of cell functions in a multicellular organism arises from differential gene expression. But how many of the approximately 100,000 genes in a mouse or human are involved in early development? How does the developing organism control the time and place of each gene's expression? And, given that only a small number of those 100,000 genes are cloned, how do we find and clone developmental and tissue-specific genes, when we don't know their location in the genome or their functions?

Until recently, geneticists relied on standard genetic techniques, such as generating mutants and observing developmental defects, to identify developmentally regulated genes. These approaches have contributed significant information; however, many developmentally important genes are essential for viability. Mutations in these genes cause embryonic lethality, thereby limiting the analysis of phenotype.

In order to overcome some of these problems, molecular biologists have developed a set of powerful new genetic tools—the "entrapment vectors." These vectors contain a reporter gene (see the essay in Chapter 15), which either lacks a promoter or has a very weak promoter. The vectors are introduced into an organism and they integrate into the genome at random locations. The reporter gene will be transcribed only if it inserts into a region of the genome that contains an active enhancer or promoter. By monitoring the location and timing of reporter gene activity during embryogenesis, researchers can identify developmentally-active promoters and enhancers.

Two categories of entrapment vectors exist: the "**enhancer traps**" and the "**gene traps**." The design of enhancer traps is based on the knowledge that enhancer elements activate transcription of promoters that are located either nearby or at a distance. Enhancers are also promiscuous, and stimulate transcription of a wide range of promoters. The enhancer trap vector contains a reporter gene (usually *lacZ*, which encodes β-galactosidase) under the control of a weak transcriptional promoter. The vector is introduced into cells of the host—usually an egg cell or very early embryo—and the enhancer trap integrates at random sites in the genome. The embryos or organisms that develop are then stained for β-galactosidase activity, and the cells that express the reporter gene turn dark blue. If the enhancer trap vector integrates into a region of the genome that falls under the influence of a developmental gene enhancer, cells of the embryo stain blue in a pattern that reflects the activity of that developmental gene. Enhancer traps have been extensively used in *Drosophila* genetics, where the *lacZ* reporter gene is incorporated into P element vectors (see Chapter 14). Geneticists have isolated hundreds of

CHAPTER SUMMARY

1. Mechanisms controlling gene regulation in eukaryotes are governed by the properties of multicellularity, expanded genome size, and the spatial and temporal separation between transcription and translation.

2. The roles played by DNA organization and structure, especially chromatin alterations, transcriptional control, and signal transduction, as well as the expression of specific gene sets such as hormone receptors and oncogenes, are studied as models for gene regulation in higher organisms.

3. Transcription in eukaryotes is controlled by regulatory DNA sequences known as promoters and enhancers. Regulatory sequences, including the TATA box and the CCAAT box, are elements of promoters found near the transcriptional starting point. Enhancer elements, which appear to control the degree of transcription, can be located before, after, or within the gene expressed.

4. Transcription factors are proteins that bind to DNA recognition sequences within the promoters and enhancers and activate transcription through protein-protein interaction.

5. Gene expression in eukaryotes can be regulated by effector molecules originating outside the cell. Such signals are transduced by proteins that shuttle from the cytoplasm to the nucleus and regulate patterns of transcription.

6. Gene amplification, a variant of DNA organization, maximizes the accumulation of gene product per unit of time by increasing the number of gene copies necessary during times of cell growth and differentiation.

7. Several types of posttranscriptional control of gene expression are possible in eukaryotes. One such mechanism is alternative processing of a single class of pre-mRNAs to generate different mRNA species. Another mechanism provides a feedback system to mRNA regulation in response to cytoplasmic concentration of the gene product, controlling mRNA stability.

Drosophila strains that express unique temporal and spatial *lacZ* staining patterns resulting from enhancer trap insertions. Enhancer traps have helped researchers isolate genes that are involved in the development of the nervous system, eye structures, and fat cells. Cloning is facilitated by the presence of the reporter gene near the gene of interest. A major advantage of the enhancer trap strategy is that the trap vector need not disrupt a gene for the gene's activity to be detected. In fact, only about 15 percent of enhancer trap insertions lead to embryonic lethal effects, probably due to insertion of the enhancer trap within the coding region of an essential gene.

The second type of entrapment vector is the "gene trap" (or "exon trap") vector. Gene trap vectors are designed to detect fusions of an interesting developmental gene mRNA with the reporter gene mRNA. Therefore, the gene trap must integrate into another gene's transcription unit in order to be expressed. The gene trap vector is similar in structure to the enhancer trap vector, except that the reporter gene (such as *lacZ*) has no promoter and no translation initiation signal. At the 5'-end of

the *lacZ* gene, where the promoter should be located, is a splice acceptor site. The gene trap vector is introduced into cells of the host, and integrations occur randomly through-

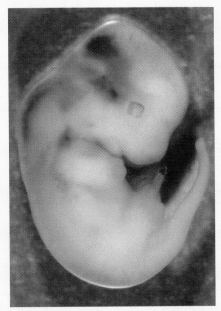

β-galactosidase staining in the central nervous system of a mouse embryo that contains an enhancer trap vector.

out the genome. Occasionally the gene trap will integrate into an intron of a gene, in the correct orientation, so that transcripts originating at the target gene promoter become spliced from the target gene splice donor site to the *lacZ* gene splice acceptor site. This generates a fusion mRNA, which is part target gene and part *lacZ*. If the translation reading frame of the target gene and the *lacZ* gene are maintained after the splicing event, functional β-galactosidase will be expressed in those cells that express the target gene. The β-galactosidase activity is monitored by staining embryos, and developmentally interesting genes are easily cloned because their mRNAs are tagged by fusion with *lacZ*. Unlike enhancer traps, gene traps usually disrupt the normal activity of their target gene and create mutations. Gene traps have been used in genetic studies of mouse and plant development, and have led to the isolation of several novel developmental genes.

Entrapment technologies promise to be powerful additions to the molecular genetics toolbox, helping us to answer fundamental questions of why and how we become unique and complex multicellular organisms.

KEY TERMS

autoregulation, 561
CAAT box, 546
DNA-binding domain, 548
double minute (DM) chromosome, 557
ecdysone, 551
enhancer, 546
enhancer traps, 562
extended chromosome region (ECR), 557
GC box, 546
gene amplification, 556

gene traps, 562
helix-turn-helix motifs, 549
homeobox, 550
homeodomain, 550
homogeneously staining region (HSR), 557
hormone-responsive element (HRE), 553
K neuropeptide, 558
leucine zippers, 549
P neuropeptide, 558

preprotachykinin mRNA (PPT mRNA), 558
promoter, 546
structural motif, 549
tachykinin, 558
TATA box, 546
trans-activating domain, 548
transcription factor, 548
UAS$_G$, 548
upstream activator sequences (UAS), 547
zinc finger, 549

INSIGHTS AND SOLUTIONS

1. Regulatory sites for eukaryotic genes are usually located within a few hundred nucleotides of the transcriptional starting site, but they can be located up to several kilobases away. DNA sequence-specific binding assays have been used to detect and isolate protein factors present at low concentrations in nuclear extracts. In these experiments, DNA-binding sequences are bound to material on a column, and nuclear extracts are passed over the column. The idea is that if proteins that specifically bind to the sequence represented on the column are present, they will bind to the DNA, and they can be recovered from the column after all other nonbinding material has been washed away. Once a DNA-binding protein has been identified and isolated, the problem is to devise a general method for screening cloned libraries for genes encoding these and other DNA-binding factors. Determining the amino acid sequence of the protein and constructing synthetic oligonucleotide probes is time-consuming and useful only for one factor at a time. Knowing of the strong affinity for binding between the protein and its DNA recognition sequence, how would you screen for genes encoding binding factors?

Solution: Several general strategies have been developed, and one of the most promising has been devised by Steve McKnight's laboratory at the Fred Hutchinson

Cancer Center. The cDNA isolated from cells expressing the binding factor is cloned into the lambda vector, gt11. Plaques of this library containing proteins derived from expression of cDNA inserts are adsorbed onto nitrocellulose filters and probed with double-stranded, radioactive DNA corresponding to the binding site. If a fusion protein corresponding to the binding factor is present, it will bind to the DNA probe. After washing off unbound probe, the filter is subjected to autoradiography, and plaques corresponding to the DNA-binding signals can be identified. An added advantage of this strategy is recycling, by washing the bound DNA from the filters for reuse. This ingenious procedure is similar to the colony hybridization and plaque hybridization procedures described for library screening in Chapter 12, and it provides a general method for isolating genes encoding DNA-binding factors.

PROBLEMS AND DISCUSSION QUESTIONS

1. Why is gene regulation assumed to be more complex in a multicellular eukaryote than in a prokaryote? Why is the study of this phenomenon in eukaryotes more difficult?

2. List and define the levels of gene regulation discussed in this chapter.

3. What is the evidence from heterokaryons that suggests that cytoplasmic factors play an important role in regulating genetic activity or inactivity of the nucleus of a cell?

4. Distinguish between the regulatory elements referred to as promoters and enhancers.

5. Is the binding of a transcription factor to its DNA recognition sequence necessary and sufficient for an initiation of transcription at a regulated gene? What else plays a role in this process?

6. You have 10 haploid yeast mutant strains that each fail to grow in the absence of tryptophan (strains A–J). You also have 2 haploid yeast strains (K and L) that do not require tryptophan for growth. The table below lists whether diploid strains made by mating the strains listed in the vertical axis with those on the horizontal axis can grow on media lacking tryptophan. Use these data to answer the following questions:
   (a) Excluding mutant C, how many complementation groups are there, and which mutants are in which groups?
   (b) Mutant C is obviously different from all the other mutations. Assuming that none of the complementation groups in (a) are genetically linked, what is the most likely explanation for the results with strain C? What can you say about the complementation group or groups that mutant C belongs to?

7. Write an essay comparing the control of gene regulation in eukaryotes and prokaryotes at the level of initiation of transcription. How do the regulatory mechanisms work? What are the similarities and differences in these two types of organisms between the specific components of the regulatory mechanisms and address how the difference or similarities relate to the biological context of the control of gene expression.

EXTRA-SPICY PROBLEMS

8. In the autoregulation of tubulin synthesis, two models for the mechanism were proposed: first, that the tubulin subunits bind to the mRNA, or that the subunits interact with the nascent tubulin polypeptides. To distinguish between these two models, Cleveland and colleagues introduced mutations into the 13-base regulatory element of the beta-tubulin gene. Some of the mutations resulted in amino acid substitution, while others did not. In addition, they shifted the reading frame of the intact 13-base sequence. Below are results from a mutagenesis study of the mRNA. Which of the two models is supported by the results? What experiments would you do to confirm this?

	A	B	C	D	E	L
F	−	+	−	+	−	+
G	+	−	−	−	+	+
H	−	+	−	+	−	+
I	−	+	−	+	−	+
J	+	−	−	−	+	+
K	+	+	−	+	+	+

Wild Type	met AUG	arg AGG	glu GAA	lys ATC	Autoregulation +
Second codon mutations		UGG			−
		GGG			−
		CGG			+
		AGA			+
		AGC			−
Third codon mutations			GAC		+
			AAC		−
			UAU		−
			UAC		−

9. Assuming your answer to Problem 8 is correct, predict the results of the following experiment. Site-directed mutagenesis is used to insert four bases just in front of the AUG-initiating codon of the 13-base sequence, keeping the sequence intact, but shifting its position in the mRNA by four bases, and shifting the reading frame to an alternative sequence different from the normal Met-Arg-Glu-Ile sequence. Will tubulin subunits regulate translation of this mRNA? Why or why not?

10. You are interested in studying transcription factors and have developed an in vitro transcription system using a defined segment of DNA that is transcribed under the control of a eukaryotic promoter. Transcription of this DNA occurs when you add purified RNA polymerase II, TFIID (the TATA binding factor), the TFIIB and TFIIE (which bind to RNA polymerase). You perform a series of experiments that compare the efficiency of transcription in this "defined system" with the efficiency of transcription in a crude nuclear extract. You test the two systems with your template DNA and with various deletion templates that you have generated. The results of your study are shown below.

(a) Why is there no transcription from the –11 deletion template?

(b) How do the results for the nuclear extract and the defined system differ with the *undeleted* template? How would you interpret these results?

(c) For the various deleted templates, compare the results with the nuclear extract and the purified system. How would you interpret the results with the deleted templates? Be very specific about what you can conclude from these data.

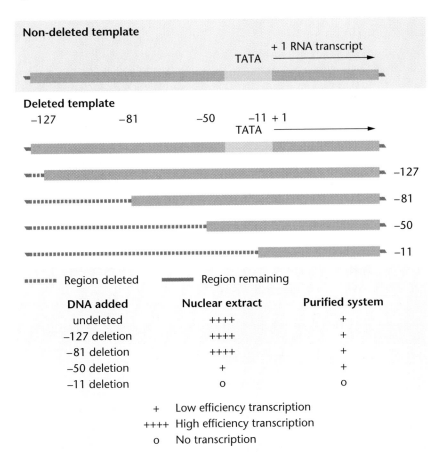

DNA added	Nuclear extract	Purified system
undeleted	++++	+
–127 deletion	++++	+
–81 deletion	++++	+
–50 deletion	+	+
–11 deletion	o	o

+ Low efficiency transcription
++++ High efficiency transcription
o No transcription

SELECTED READINGS

Aso, T., Shilatifard, A., Conaway, J. W., and Conaway, R. C. 1996. Transcription syndromes and the role of RNA polymerase II general transcription factors in human disease. *J. Clin. Investig.* 97:1561–1569.

Ayoubi, T. A., and Van de Ven, W. J. 1996. Regulation of transcription by alternate promoters. *FASEB J.* 10:453–460.

Beato, M., Candau, R., Chavez, S., Mows, C., and Trurs, M. 1996. Interaction of steroid hormone receptors with transcription factors involves chromatin remodelling. *J. Steroid Biochem. Mol. Biol.* 56:47–59.

Brennan, R. 1993. The winged-helix DNA-binding motif: Another helix-turn-helix takeoff. *Cell* 74:773–76.

Buratowski, S. 1995. Mechanisms of gene activation. *Science* 270:1773–74.

BUSCH, S. J., AND SASSONE-CORSI, P. 1990. Dimers, leucine zippers and DNA-binding domains. *Trends Genet.* 6:36–40.

CLEVELAND, D. W. 1988. Autoregulated instability of tubulin mRNAs: A novel eukaryotic regulatory mechanism. *Trends Biochem. Sci.* 13:339–43.

CONAWAY, R., and CONAWAY, J. 1993. General initiation factors for RNA polymerase II. *Annu. Rev. Biochem.* 62:161–90.

DRAPKIN, R., MERINO, A., and REINBERG, D. 1993. Regulation of RNA polymerase II transcription. *Curr. Opin. Cell Biol.* 5:469–76.

DE LA BROUSSE, F. C., and McKNIGHT, S. L. 1993. Glimpses of allostery in the control of eukaryotic gene expression. *Trends Genet.* 9:151–54.

DYNAN, W. S. 1988. Modularity in promoters and enhancers. *Cell* 58:1–4.

GOODRICH, J. A., CUTLER, G., and TJIAN, R. 1996. Contacts in context: promoter specificity and macromolecular interactions in transcription. *Cell* 84:825–30.

JACOBSEN, A., and PELTZ, S. 1996. Interrelationships of the pathway of mRNA decay and translation in eukaryotic cells. *Annu. Rev. Biochem.* 65:693–739.

JACOBSON, R. H., and TIJIAN, R. 1996. Transcription factor IIA: a structure with multiple functions. *Science* 272:830–36.

KAKIDANI, H., and PTASHNE, M. 1988. GAL4 activates gene expression in mammalian cells. *Cell* 52:161–67.

KARIN, M., CASTRILLO, J-L., and THEILL, L. E. 1990. Growth hormone regulation: A paradigm for cell type-specific gene activation. *Trends Genet.* 6:92–96.

LEHMANN, R., and EPHRUSSI, A. 1994. Germ plasm formation and germ cell determination in *Drosophila. Ciba Found. Symp.* 182:282–291.

LEIDEN, J. 1993. Transcriptional regulation of T cell receptor development. *Annu. Rev. Immunol.* 11:539–70.

LEVINE, M., and MANLEY, J. L. 1989. Transcriptional repression of eukaryotic promoters. *Cell* 59:405–8.

LINDAHL, T. 1993. Instability and decay of the primary structure of DNA. *Nature* 362:709–15.

LITTLEWOOD, T. D., and EVANS, G. I. 1994. Transcription factors 2: Helix-loop-helix. *Protein Profile* 1:639–709.

LOPEZ, H. J. 1995. Developmental role of transcription factor isoforms generated by alternate splicing. *Dev. Bio.* 172:396–411.

MacDOUGALL, C., HARBISON, D., and BOWNES, M. 1995. The developmental consequences of alternate splicing in sex determination and differentiation in *Drosophila. Dev. Biol.* 172:353–376.

McKNIGHT, S. L. 1995. Transcription revisited: a commentary on the 1995 Cold Spring Harbor Meeting, Mechanisms of Eukaryotic Transcription. *Genes Dev.* 10:367–381.

MANIATIS, T., GOODBOURN, S., and FISCHER, J. A. 1987. Regulation of inducible and tissue-specific expression. *Science* 236:1237–45.

MEEHAN, R., LEWIS, J., CROSS, S., NAN, X., JEPPESEN, P., and BIRD, A. 1992. Transcriptional repression by methylation of CpG. *J. Cell Sci. Suppl.* 16:9–14.

MITCHELL, P. J., and TJIAN, R. 1989. Transcriptional regulation in mammalian cells by sequence-specific DNA binding proteins. *Science* 245:371–78.

MIZEJEWSKI, G. 1993. An apparent dimerization motif in the third domain of alpha-fetoprotein: Molecular mimicry of the steroid/thyroid nuclear receptor superfamily. *BioEssays* 15:427–32.

O'HALLORAN, T. 1993. Transition metals in control of gene expression. *Science* 261:715–25.

PIELER, T., and THEUNISSEN, O. 1993. TFIIIA: Nine fingers—Three hands? *Trends Biochem. Sci.* 18:226–30.

PTASHNE, M., and GANN, A. A. F. 1990. Activators and targets. *Nature* 346:329–31.

SCHIMKE, R. T. 1989. The discovery of gene amplification in mammalian cells: To be in the right place at the right time. *BioEssays* 11:69–73.

SHARP, Z. D., and MORGAN, W. W. 1996. Brain POU-er. *BioEssays* 18: 347–350.

SIBLEY, E., KASTELIC, T., KELLY, T. J., and LANE, M. D. 1989. Characterization of the mouse insulin receptor gene promoter. *Proc. Natl. Acad. Sci. USA* 86:9732–36.

STARK, G. R., DEBATISSE, M., GIULOTTO, E., and WAHL, G. 1989. Recent progress in understanding mechanisms of gene amplification. *Cell* 57:901–8.

STRINGER, K. F., INGLES, C. J., and GREENBLATT, J. 1990. Direct and selective binding of an acidic transcriptional activation domain to the TATA-box factor TFIID. *Nature* 345:783–86.

STRUHL, K. 1993. Yeast transcription factors. *Curr. Opin. Cell Biol.* 5:513–20.

TATE, P., and BIRD, A. 1993. Effects of DNA methylation on DNA-binding proteins and gene expression. *Curr. Opin. Genet. Dev.* 3:226–31.

TJIAN, R. 1995. Molecular machines that control genes. *Sci. Am.* 272:54-61.

WERNER, J. H., GRONENBORN, A. M., and CLORE, G. M. 1996. Intercalation, DNA kinking and the control of transcription. *Science* 271:778–784.

YEN, T. J., GAY, D. A., PACHTER, J. S., and CLEVELAND, D. W. 1988. Autoregulated changes in stability of polyribosome-bound beta-tubulin mRNAs are specified by the first 13 translated nucleotides. *Mol. Cell Biol.* 8:1224–35.

20

Developmental Genetics

CHAPTER CONCEPTS

The genetic basis of development is differential gene action. Gene actions vary over time in the same tissue, and between different tissues present at the same developmental stage. Gene action is required to bring about and maintain adult structures. Processes such as regeneration involve a re-iteration of the embryonic program of gene action.

In multicellular plants and animals, a fertilized egg, without further stimulus, begins a cycle of developmental events that ultimately give rise to an adult member of the species from which the egg and sperm were derived. Thousands, millions, or even billions of cells are organized into a cohesive and coordinated unit that we perceive as a living organism. The series of events whereby organisms attain their final adult form is studied by developmental biologists. This area of study is perhaps the most intriguing in biology because comprehension of developmental processes requires knowledge of many biological disciplines.

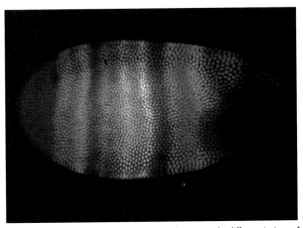

Regional expression of various genes during early differentiation of the *Drosophila* embryo.

In the past hundred years, investigations in embryology, genetics, biochemistry, molecular biology, cell physiology, and biophysics have contributed to the study of development. Largely, the findings have pointed out the tremendous complexity of developmental processes. Unfortunately, a description of what actually happens does not answer the "why" and "how" of development. Over the last two decades, genetic analysis and molecular biology have identified genes that regulate developmental processes. We are now beginning to understand how the action and interaction of these genes control basic developmental processes in a wide range of prokaryotic and eukaryotic organisms.

In this chapter the primary emphasis will be on the role of gene action in regulating development. Because genetic information directs cellular function and determines the cellular phenotype, the genetics of development is being actively studied using a wide range of experimental organisms and techniques.

Developmental Concepts

The current thrust in developmental genetics is to provide a molecular explanation of developmental processes including temporal changes in gene expression and cellular differentiation in order to establish a causal relationship between the presence or absence of molecules, receptors, transcriptional events, cell and tissue interactions, and the observable morphological events that accompany the process of development. To explain the sequence of developmental events taking place in any organism, geneticists rely on certain heuristic concepts. These include the theory of variable gene activity, and the principles of differential transcription and the stability of the differentiated state. These concepts have been derived from the study of a wide range of organisms used as model systems, and these ideas form the theoretical framework that is used to explain how the developmental potential present in a single cell becomes transformed into a recognizable yet individual form that we identify as an adult organism.

The Variable Gene Activity Theory

From a genetic perspective, development may be described as the attainment of a differentiated state. For example, an erythrocyte active in hemoglobin synthesis is differentiated, but a cell in a blastula-stage embryo is undifferentiated. In order to accomplish this specialization, certain genes are actively transcribed but many other genes are not. Because most eukaryotic organisms are composed of a large number of cell types, differential transcription patterns characterize functionally diverse cells.

The concept of differential transcription has led to the **variable gene activity hypothesis** of differentiation. This theory, first entertained by Thomas H. Morgan in 1934 and later proposed by Edgar and Ellen Stedman as well as Alfred Mirsky in the 1950s, is articulated in modern form by Eric Davidson in his book *Gene Activity in Early Development*. The theory holds that the differentiated state assumed by any specific cell type is qualitatively determined by those genes that are actively transcribed in

that cell. Its underlying assumption is that each cell contains an entire genome and that differential transcription of selected genes controls the development and differentiation of that cell.

The variable gene activity theory is a very useful model system for experimental design. The remainder of this chapter examines the validity of this theory and its premises and provides examples that offer experimental support from both prokaryote and eukaryote systems.

Differential Transcription in Development: Prokaryotes

Certain gram-positive bacilli such as *Bacillus subtilis* form spores as a mechanism of survival under adverse conditions. Spores are highly resistant to heat, desiccation, and toxic conditions and are induced to form in nutritionally depleted environments. Development of the spore takes place over a period of several hours in a structure known as a **sporangium**, which consists of two chambers, the mother cell and the **forespore** (Figure 20.1). The two compartments contain identical copies of the chromosome but undertake different programs of transcription. Recent work has demonstrated that there are also interactions between the two compartments during development that coordinate the program of gene expression in both the forespore and the mother cell.

As a model system for the study of development, sporulation in gram-positive bacteria such as *B. subtilis* offers several advantages. Only two cell types are involved. The developmental events occur over a restricted time period (8–10 hours), and these events can be studied by genetic analysis and the techniques of molecular biology. Even a relatively simple developmental event such as sporulation turns out to be a genetically complex event, involving more than 80 genes and multiple levels of control. The primary level of regulation is transcriptional, involving cell-specific transcription factors that control gene expression. The small set of regulatory genes encoding transcription factors are themselves subject to multiple levels of control.

The genes controlling this developmental process have been identified by mutagenesis followed by genetic screens to recover mutants of sporulation. The largest class (over 50 genes), required specifically for spore formation, are called *spo* genes. Most of these genes have been cloned and sequenced, and information about the functions of the gene products and their temporal expression and cellular localization is available. Not surprisingly, one subclass of these *spo* genes are themselves transcription factors (called **σ-factors**) that bind to RNA

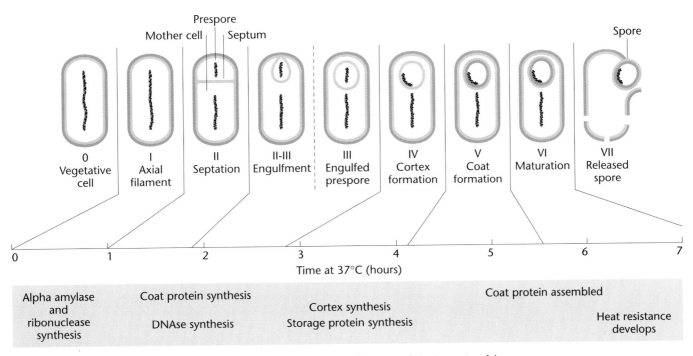

0	I	II	II-III	III	IV	V	VI	VII
Vegetative cell	Axial filament	Septation	Engulfment	Engulfed prespore	Cortex formation	Coat formation	Maturation	Released spore

Time at 37°C (hours)

| Alpha amylase and ribonuclease synthesis | Coat protein synthesis | | Coat protein assembled | |
| DNAse synthesis | Cortex synthesis | Storage protein synthesis | Heat resistance develops |

FIGURE 20.1 Morphological stages and some of the biochemical events during sporulation in species of the bacterium *Bacillus*.

polymerase and confer specificity for a set of promoters. The σ-factors are DNA-binding proteins that recognize clusters of DNA sequences upstream from promoters.

Genes encoding two σ-factors, σ^G and σ^K, are major regulatory loci that determine cell type (forespore or mother cell). The σ^G and σ^K factors control the transcription of other genes, including compartment-specific σ-factors which activate other gene sets during later stages of development. The σ^G is encoded by the gene *spo*IIIG, and in early stages, before septum formation, transcription of *spo*IIIG is activated at a low level, perhaps by another sigma factor. The σ^K factor is encoded by two truncated genes, *spo*IVCB and *spo*IIIC. During sporulation, site-specific recombination in the mother cell involving short, repeated DNA sequences located within each of these genes generates an intact σ^K gene (Figure 20.2). The recombinase enzyme that catalyzes this rearrangement is encoded by the *spo*IVCA gene located in the excised sequence. The rearrangement and transcription of the *spo*IVCB/*spo*IIIC composite gene in mother cells accounts for both the compartmentalized expression of σ^K and the activation of the mother-cell genes.

Many of the aspects of gene regulation found in sporulation are also found in developmental processes in multicellular eukaryotes. Of particular interest are the mechanisms by which cells with identical genomes begin patterns of transcription that lead to different outcomes in cell structure and function. The following sections will explore how these events are regulated in eukaryotic organisms.

Differential Transcription in Development: Eukaryotes

More complex eukaryotic organisms develop from a **zygote**, a cell formed by the fusion of sperm and oocyte. The oocyte is a single cell with a cytoplasm that is heterogeneous and nonuniform in distribution. Following fertilization and early cell division, the nuclei of progeny cells find themselves in different environments as the maternal cytoplasm is distributed into the new cells. Evidence suggests that the cytoplasm exerts influences on the genetic material of different cells, causing differential transcription at specific points during development. Although such cells show no immediate evidence of structural or functional specialization, the combination of their localized cytoplasmic components and their position in the developing embryo seems to determine the ultimate form they will assume. It is as if their fate has been programmed prior to the actual events leading to specialization.

Early gene products synthesized by the zygote further alter the cytoplasm of each cell, producing a still different cellular environment that may in turn lead to the activation of other genes, and so on. In addition, as the number of cells increases, they influence one another in a process of **cell–cell interaction.** The total environment acting on the genetic material, therefore, now includes the cell's individual cytoplasm as well as

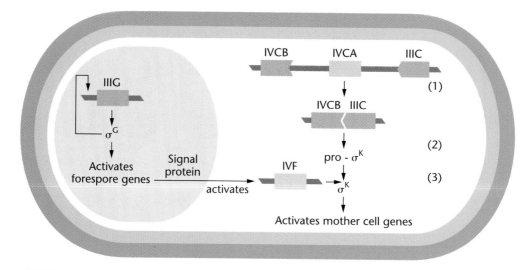

FIGURE 20.2 Levels of gene regulation during sporulation in *B. subtilis*. Differential gene expression in the two compartments is governed by compartment-specific transcription factors, σ^G and σ^K. In the forespore, activation of σ^G transcription allows the σ^G protein to autoregulate expression of the σ^G (*spoIIIG*) gene. In the mother cell, recombination (1) produces an intact σ^K gene that is transcribed to produce an inactive precursor protein (2), called pro-σ^K. Conversion of the precursor to an active transcription factor occurs in response to events in the forespore, where a signal is generated that crosses the septum and activates the *spoIVF* gene in the mother cell. The gene product of the *spoIVF* gene converts inactive pro-σ^K protein into the active σ^K-factor (3), which then initiates transcription of mother-cell-specific genes.

the influence of other cells. As the external and internal environments change, cells embark on pathways of differentiation as development proceeds. **Differentiation** is therefore the final result of a series of stages, the most important of which is **determination**. This is the stage at which a specific developmental fate for a cell becomes fixed prior to the actual events of differentiation. As the developing organism becomes more complex, so does the total environment. In this way, different forms of determination and differentiation occur during development.

In the following sections we will examine how the processes of determination, differentiation, and cell–cell interaction relate to patterns of gene expression in development. We will begin by examining the evidence that cells of multicellular organisms contain an intact genome and that development in eukaryotes involves transcription, translation, morphogenesis, and supracellular assembly in different embryonic cells. We will next examine how cytoplasm influences the program of gene expression in the early stages of embryonic development in *Drosophila*, how activation of key switch genes is accomplished by transcription factors, and how processing of extracellular signals begins the process of determination in early embryos. Later, we will discuss how development is accompanied by progressive restriction of the developmental options available to cells, and we will consider the role of cell–cell interactions in the development of multicellular organisms.

Genome Equivalence

The variable gene activity hypothesis is based on the premise that the somatic cells of multicellular organisms each contain a complete set of genetic information. While biochemical and cytophotometric analysis has shown the quantity of DNA in each cell within an organism to be equivalent and equal to the diploid content in cells of most species studied, other investigations have approached the question differently. Is it possible to show that differentiated cells are genetically equivalent? In other words, is a complete set of genes present in each cell?

One approach is to show that a differentiated nucleus is **totipotent** and thus capable of giving rise to a complete organism under the proper conditions. In the early part of the twentieth century, Hans Spemann demonstrated that a nucleus from a cell of the 16-cell stage of a newt embryo was capable of supporting development of a complete organism. He first constricted a newly fertilized egg prior to the first cell division, forcing the zygote nucleus into one half of the cell. The half with the nucleus divided and produced a cluster of 16 cells. He then loosened the constriction, and one of the 16 nuclei was allowed to pass back into the nonnucleated cytoplasm on the other side. Both halves of the embryo, one with 15 nuclei and one with a single nucleus, subsequently produced complete but separate organisms. Therefore, at the 16-cell stage in this organism, genomic equivalence was demonstrated.

Beginning in the 1950s, more sophisticated experiments were performed by Robert Briggs and Thomas King using the grass frog *Rana pipiens*, and by John Gurdon in the 1960s using the African frog *Xenopus laevis*. After inactivating or surgically removing the nucleus from an egg, these investigators were able to test the developmental capacity of somatic nuclei by transplanting nuclei isolated from cells at various stages of differentiation. Nuclei derived from the blastula stage of development were capable of supporting the development of complete and normal adults when transplanted into enucleated oocytes. In *Rana* (Figure 20.3), transplanted nuclei derived from later stages such as the gastrula and neurula usually allowed only partial development. In *Xenopus*, however, Gurdon's experiments showed that nuclei from epithelial gut cells of tadpoles were able to support the development of an adult frog when transplanted into enucleated oocytes. In both cases, it is clear that differentiated cells do carry a complete copy of the genome, and that, under normal developmental conditions, progressive restrictions that prevent the expression of totipotency are placed on differentiating nuclei.

To test whether the nucleus of a highly differentiated adult cell is irreversibly specialized or can support the development of a normal embryo, Gurdon used nuclei from adult frog skin cells in serial transplant experiments (Figure 20.4). In these experiments, a donor nucleus was transplanted into an enucleated egg, and the recipient

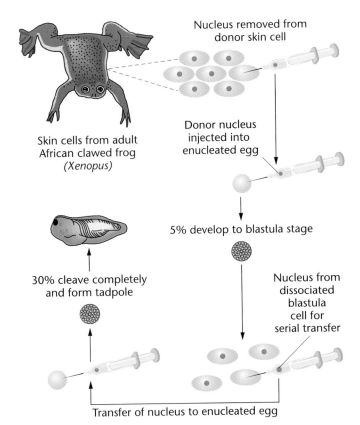

FIGURE 20.4 Serial transplantation experiments using adult frog skin cell nuclei. Serial transplantation dramatically increases the percentage (from 5% to 30%) of recipient eggs that undergo complete cleavage.

was allowed to develop for a short time, say to the blastula stage. The blastula cells were then dissociated and a nucleus removed from one of them. This nucleus was transplanted into still another enucleated egg and development was allowed to occur. Such serial transfers were repeated a number of times. Subsequently, the blastula was not dissociated, but instead was allowed to continue development. Gurdon found that over 30 percent of the serial nuclear transplants resulted in the formation of tadpoles (Table 20.1). Because nuclei from fully differentiated adult epidermal cells can eventually direct the synthesis of gene products such as myosin, hemoglobin, and crystallin and promote the organization of cells and tissues into a tadpole, we infer that such cells contain a complete copy of the genome. These experiments show that under the proper circumstances, genes not expressed in more specialized cells can become reactivated.

In plants, the totipotency of terminally differentiated, nonmeristematic cells has been demonstrated using the carrot. Frederick Steward observed that when individual phloem cells are explanted into a liquid culture medium, each cell divides and eventually forms a mass called a *callus*. Under appropriate conditions, this cell mass will differentiate into a mature plant.

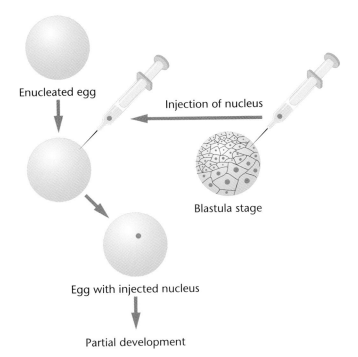

FIGURE 20.3 The process of nuclear transplantation in the frog. The unfertilized egg is activated by needle puncture, and the nucleus is removed with a micropipette. A nucleus from a blastula embryo is injected.

TABLE 20.1 First Transfer Versus Serial Transplantation in *Xenopus laevis*

Donor Cells	Percentage Reaching Tadpole Stage with Differentiated Cell Types	
	First Transfer	Serial Transfer
Intestinal epithelial cells from tadpoles	1.5	7
Cells from adult skin	0.037	8
Blastula or gastrula	36	57

Source: From Gurdon, 1974, p. 24.

These studies all convincingly argue that differentiated adult cells have not lost any of the genetic information present in the zygote. Instead, the majority of genes in any given cell type fail to be activated or are repressed, but they can be reprogrammed to direct normal development. We are only beginning to understand the nature of the molecular processes controlling nuclear differentiation during cellular specialization, but the nuclear transplant experiments indicate that the cytoplasm plays a role in controlling gene expression.

Binary Switch Genes

As part of the progressive restriction of transcriptional capacity that accompanies development, certain genes act as switch points, decreasing the number of alternative developmental pathways. Each decision point is usually binary, that is, there are two alternatives present, and the action of a switch gene programs the cell to follow one of these pathways.

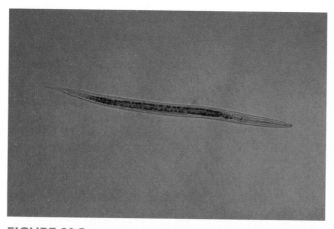

FIGURE 20.5 An adult *Caenorhabditis elegans*, an organism used in research on development.

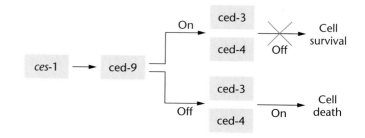

FIGURE 20.6 In the genetic pathway controlling cell death, the gene *ced-9* acts as a binary switch. If *ced-9* is active, it represses the expression of *ced-3* and *ced-4*, and the cell lives. If *ced-9* is inactive, *ced-3* and *ced-4* are expressed, and the cell dies.

We will briefly describe two such switch genes: (1) genes that control cell death in the nematode *Caenorhabditis elegans*, which depends on the action of a single gene, and (2) *myo*D-related proteins, which control muscle cell determination and differentiation in both invertebrates and vertebrates.

The nematode *Caenorhabditis elegans* (Figure 20.5) is widely used to study the genetic control of development. This organism has two advantages for such studies: the genetics of the organism are well known, and adults have a small number of cells that follow a developmental program that is unchanged from individual to individual. The life history of this worm will be discussed in a later section (see below). Development in *C. elegans* involves programmed cell death for a number of cells. The number of cells that die is always the same: 131 of 1090 in hermaphrodites, and 147 of 1178 in males; the time during development at which a given cell dies is always the same, and the identity of the cells that die is always the same.

Genetic analysis of mutants indicates that although cells with different developmental origins undergo programmed cell death, all the cells that die use the same genetic pathway. The expression of two genes, *ced-3* and *ced-4*, are necessary for cell death; mutations that inactivate either of these genes result in the survival of cells that normally die. Expression of *ced-3* and *ced-4* is controlled by the gene *ced-9*. Gain-of-function mutations that lead to constitutive expression or overexpression of *ced-9* prevent cell deaths during development. Conversely, loss-of-function mutants that inactivate *ced-9* cause embryonic lethality. The *ced-9* gene works by preventing expression of *ced-3* and *ced-4* in cells that survive (Figure 20.6). This means that *ced-9* acts as a binary switch to regulate programmed cell death: Cells that express *ced-9* survive, and those that do not, die.

The *ced-9* gene of *C. elegans* has a human homolog, the *bcl-2* proto-oncogene, which plays a role in programmed cell death in mammals. Overexpression of *bcl-2* prevents cell death in cells that would normally die. In hu-

TABLE 20.2 The *myo*D Gene Family

Xenopus	Rodent	Human
*XMyo*D	*myo*D	*myf*3
XMyogenin	*myogenin*	*myf*4
*XMyf*5	*myf*5	*myf*5
*XMRF*4	*MRF*4	*myf*6

mans, this overexpression is associated with follicular lymphoma, a form of cancer. Transfer of a cloned *bcl-2* gene into *ced-9* mutant *C. elegans* embryos prevents programmed cell death, indicating that nematodes and mammals share a common pathway for programmed cell death.

One of the best-characterized classes of switch genes includes genes that control muscle cell formation in both invertebrates and vertebrates. These genes are the **myoblast-determining** or ***myo*D** genes. This family of four genes (Table 20.2) is highly conserved in the vertebrates but is more distantly related to the *myo*D-related genes of invertebrates. The *myo*D genes are major regulatory genes that normally initiate events leading to skeletal muscle cell differentiation in embryonic cells known as *myoblasts*. Experimentally induced expression of *myo*D genes in nonmuscle-forming cell types, including fibroblasts (connective tissue cells of the skin), adipocytes (fat cells), neurons (nerve cells), and melanocytes (pigment cells), results in the activation of muscle-specific genes and, in some cases, causes the formation of completely differentiated muscle cells that contain contractile fibers. The ability to switch the phenotype of differentiated cells by the activation of a single gene provides strong evidence for the central role of *myo*D genes in the regulation of muscle cell formation.

Members of the *myo*D gene family encode DNA-binding proteins with a helix-loop-helix (HLH) motif (see Chapter 19 for a review of eukaryotic transcription factors). Members of the *myo*D class of proteins contain an 80-amino-acid basic region (Figure 20.7) that mediates DNA binding and confers specificity for transcriptional activation. These proteins form heterodimers with proteins known as E-box proteins and recognize the sequence CANNTG (where N is any nucleotide) present in the enhancers and promoters of most skeletal muscle genes.

Results from cell culture studies on myoblasts indicate that *myo*D proteins themselves are under the control of cell growth factors. Myoblasts cultured in the presence of high concentrations of growth factors undergo growth and division, and expressions of *myo*D and *myf*5 are repressed, or are expressed only at very low levels. When growth factors are withdrawn from the culture medium, the myoblasts exit the cell cycle, enter a G0 state, and *myo*D and/or *myf*5 are derepressed. Action of these transcription factors activates expression of **myogenin**, which in turn triggers expression of a cascade of contractile protein genes, causing the myoblast to begin differentiation as a muscle cell. Differences in the timing of expression for the *myo*D-related proteins suggest that they each have separate and distinct functions. According to this model: (1) *myf*5 is the earliest factor to be expressed and is the switch gene involved in determination; (2) *myo*D works with *myf*5 to maintain the myoblast identity; (3) all four factors coordinately initiate the process of terminal differentiation into muscle; and (4) *myf*4 works to maintain the differentiated adult phenotype.

Genetics of Embryonic Development in *Drosophila*

In the preceding discussion, we examined factors responsible for making cells different from one another, (*i.e.*, differential transcription). However, why a certain cell is fated to turn on or off specific genes during development is a broader and more critical issue, and is a central question in developmental biology. This question is distinct from that discussed in Chapter 19, where we asked what actually turns any given gene on or off.

As alluded to earlier, there is no simple answer to this question. However, information derived from the study of many different organisms provides a starting point. We shall examine one of these examples, embryonic development in *Drosophila*, which reflects the direction and scope of our knowledge in this area. We will begin with an overview of development, and then describe how the analysis of mutations has provided insight into the number and action of genes that regulate the processes of determination and differentiation.

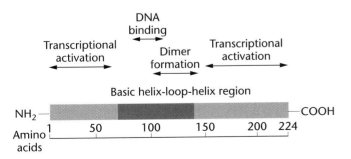

FIGURE 20.7 Structure of myogenin, a 224-amino-acid protein belonging to the *myo*D family. This protein has a characteristic basic region and helix-loop-helix domain near the center. An adjacent region controls dimer formation necessary for activity. Regions near either end control transcriptional activation.

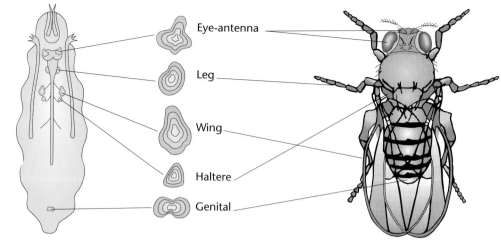

FIGURE 20.8 Imaginal disks of the *Drosophila* larva and the adult structures that are derived from them.

Eye-antenna

Leg

Wing

Haltere

Genital

Overview of *Drosophila* Development

Beginning with the fertilized egg, *Drosophila* passes through a pre-adult period of development with several distinct phases: the embryo, three larval stages and the pupal stage. The adult fly emerges from the pupal case 10–14 days after fertilization.

The adult body of *Drosophila* is composed of head, thoracic, and abdominal segments. Each segment is formed from the descendants of cells set aside during embryogenesis and incorporated into discrete structures called **imaginal disks** (Figure 20.8). These disks, formed from hollow sacs of cells, are determined to form specific parts of the adult body. There are 12 bilaterally paired disks and one genital disk. For example, there are eye-antennal disks, leg disks, wing disks, and so forth. Other cells in the embryo form the internal and external structures of the larva. During metamorphosis, most of the larval body parts histolyze, or break down, and the imaginal disks undergo mitosis and differentiate to form the head, thorax, and abdomen of the adult fly.

Externally, the *Drosophila* egg has a number of structures that delineate the anterior, posterior, dorsal and ventral regions (Figure 20.9). The anterior end of the egg contains the micropyle, a specialized conical structure for the entrance of sperm into the egg, while the posterior end is rounded and marked by a series of aeropyles (openings that allow gas exchange during development). The dorsal side of the egg is flattened and contains the chorionic appendages, while the ventral side is curved.

Internally, the egg cytoplasm is organized into a series of maternally derived molecular gradients. These gradients play a key role in establishing the developmental fates of nuclei that migrate into specific regions of the embryo.

Immediately after fertilization, the zygote nucleus undergoes a series of divisions. After nine rounds of division, most of the approximately 512 nuclei migrate to the egg's outer surface or cortex, where further divisions take place. The nuclei then become enclosed in membranes, forming a single layer of cells over the embryo surface (Figure 20.10). This is the **blastoderm** stage of embryonic development. As nuclei migrate into different regions of the egg's cytoplasm, maternally derived transcripts and gene products localized in these regions initiate a program of gene expression in the nuclei. These transcriptional programs result in the formation of the anterior-posterior axis and the dorsal-ventral axis of the embryo. The formation of these axes requires the action of maternal gene products and zygotic genes, whose discovery is discussed below. Table 20.3 lists some of the maternal genes that are required for formation of the anterior-posterior and dorsal-ventral embryonic axes.

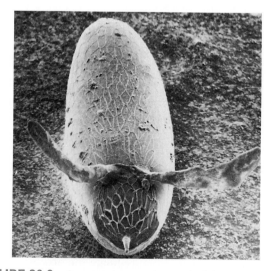

FIGURE 20.9 Scanning electron micrograph of the anterior pole of a newly oviposited *Drosophila* egg.

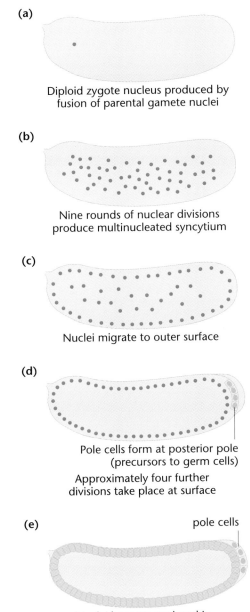

(a)

Diploid zygote nucleus produced by fusion of parental gamete nuclei

(b)

Nine rounds of nuclear divisions produce multinucleated syncytium

(c)

Nuclei migrate to outer surface

(d)

Pole cells form at posterior pole (precursors to germ cells)

Approximately four further divisions take place at surface

(e)   pole cells

Nuclei become enclosed in membranes, forming a single layer of cells over embryo surface

FIGURE 20.10 Early stages of embryonic development in *Drosophila*. (a) Fertilized egg with zygotic nucleus, about 30 minutes after fertilization. (b) Nuclear divisions occur about every 10 minutes, producing a multinucleate cell. (c) After approximately nine divisions (512 nuclei), the nuclei migrate to the outer surface or cortex of the egg. (d) At the surface, four additional rounds of nuclear division occur. A small cluster of cells, the pole cells, form at the posterior pole about $2^{1}/_{2}$ hours after fertilization. These cells will form the germ cells of the adult. (e) About 3 hours after fertilization, the nuclei become enclosed in membranes, forming a single layer of cells over the embryo surface. This stage is known as the blastoderm.

TABLE 20.3 Maternally Transcribed and Zygotically Transcribed Genes That Control the Anterior-Posterior Axis of the *Drosophila* Embryo

	Anterior	Posterior	Terminal
Maternal	*bicoid* *exuperantia* *swallow*	*staufen* *oskar* *vasa* *valois* *tudor* *mago nasbi* *nanos* *pumilio*	*trunk* *fs*(1)*Nasrat* *fs*(1)*polehole* *torso* *torso-like* *l*(1)*polehole*
Zygotic	*hunchback*	*knirps* *giant*	*tailless* *huckebein*

The body axis of the larva, pupa and adult forms along the anterior-posterior axis. As soon as the blastoderm forms, cells that will form adult structures are set aside. Two-dimensional maps of the adult structures that will be formed by regions of the embryo are known as **fate maps** (Figure 20.11). The existence of a fate map implies that the developmental outcome for a given nucleus is governed by the location to which it migrates during embryonic development, and that positional information (where a given nucleus ends up along the anterior-posterior axis and dorsal-ventral axis) directs the nucleus to initiate a specific developmental program.

At a stage after blastoderm formation, the embryo becomes organized into a series of segments (Figure 20.12). Within each segment, cells first become determined to form either an anterior or a posterior portion (called a **compartment**) of the segment. In the following stages of development, further restrictions of developmental options occur so that cells in each compartment are eventually restricted to forming a single structure in the adult body.

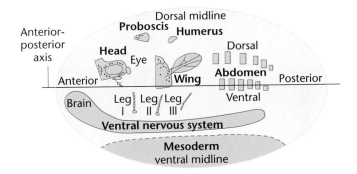

FIGURE 20.11 Diagram showing the origin of some *Drosophila* structures from the embryonic blastoderm.

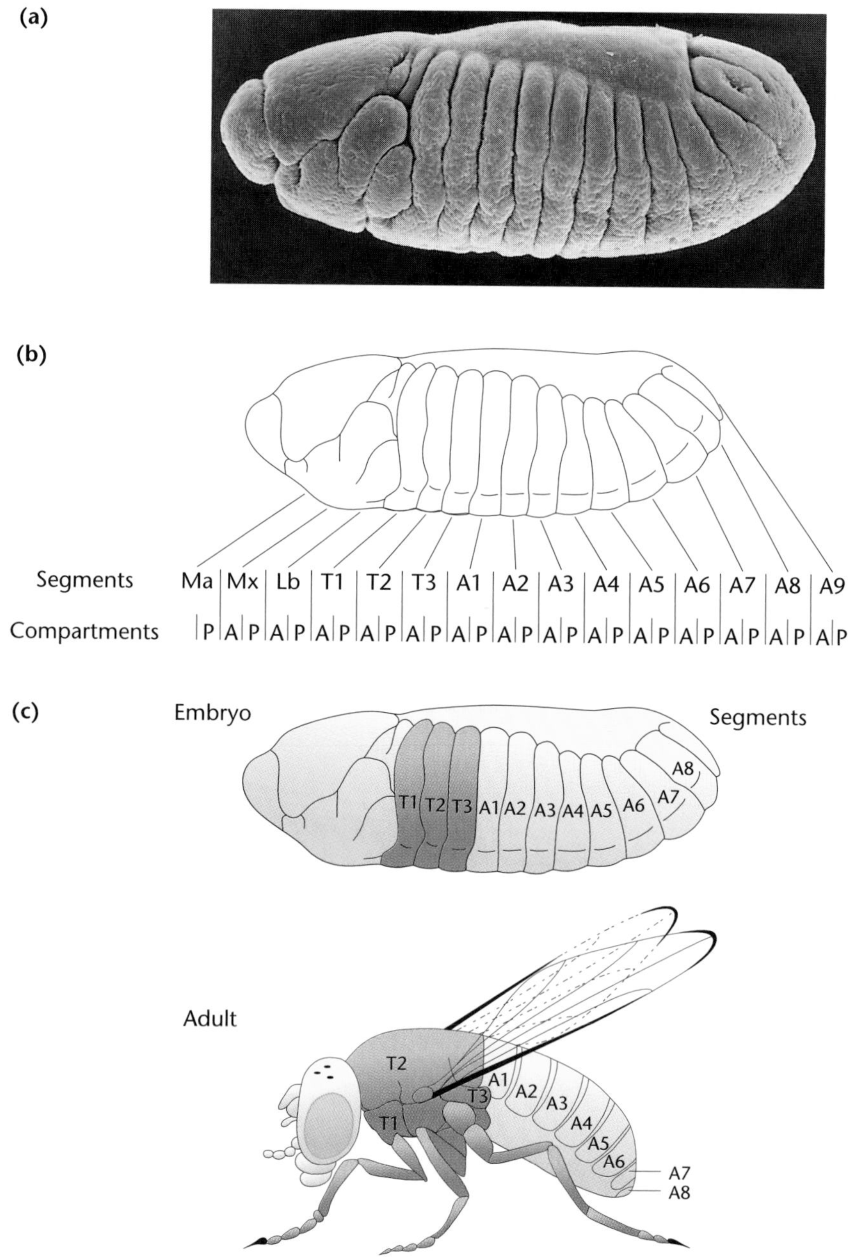
(a)
(b)
Segments
Ma Mx Lb T1 T2 T3 A1 A2 A3 A4 A5 A6 A7 A8 A9
Compartments
P A P A P A P A P A P A P A P A P A P A P A P A P A P A P
(c)
Embryo
Segments
T1 T2 T3 A1 A2 A3 A4 A5 A6 A7 A8
Adult
T2 T1 T3 A1 A2 A3 A4 A5 A6 A7 A8

FIGURE 20.12 Segmentation in *Drosophila*. (a) Scanning electron micrograph of *Drosophila* embryo at about 10 hours after fertilization. By this stage, the segmentation pattern of the body is clearly established. (b) The segments, compartments, and parasegments of the *Drosophila* embryo. Ma, Mx, and Lb represent the segments that will form head structures. T1–T3 are thoracic segments and A1–A9 are abdominal segments. Each segment is divided into anterior (A) and posterior (P) compartments. Parasegments represent an early pattern specification that is later refined into the segmental plan of the body. Note that all the parasegments are shifted forward by one compartment to form the segments. (c) The segmented embryo and the adult structures that will form each segment.

Genetic Analysis of Embryogenesis

Although development in *Drosophila* has been examined anatomically (as described above) and biochemically, genetic analysis has provided the most information about the events in embryogenesis.

Genes that control embryonic development fall into two classes: **maternal-effect genes** and **zygotic genes.** Maternal-effect genes are those whose gene products (mRNA and proteins) are deposited in the developing egg. These products may be distributed in a gradient, or concentrated in specific regions of the egg. The phenotype of flies with mutations in maternal-effect genes is expected to be female sterility, since none of the embryos of females homozygous for a recessive mutation would receive wild-type gene products from their mother, and would not develop normally. In *Drosophila*, these maternal-effect genes encode transcription factors, receptors and proteins that regulate translation. During embryonic development, these gene products work to activate or repress the expression of zygotic genes in a temporal and spatial sequence.

Zygotic genes are those that encode the proteins produced by the developing embryo itself, and are therefore, transcribed after fertilization. The phenotype of mutations in this class show embryonic lethality: In a cross between two flies heterozygous for a recessive zygotic mutation, one-fourth of the embryos (the homozygotes) will fail to develop. In *Drosophila*, many of the zygotic genes are transcribed in patterns that depend on the distribution of maternal-effect proteins.

To identify as many zygotic genes as possible, Christiane Nusslein-Volhard and Eric Weischaus systematically screened for mutations that affect embryonic development. The F₂ offspring of mutagenized flies were examined for recessive embryonic lethal mutations with

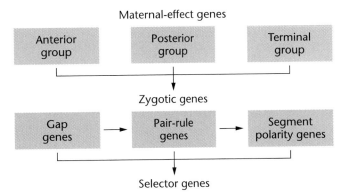

FIGURE 20.14 The hierarchy of genes involved in embryogenesis in *Drosophila*. Gene products from the maternal genes regulate the expression of the first zygotic (segmentation) genes, which in turn control expression of the selector genes.

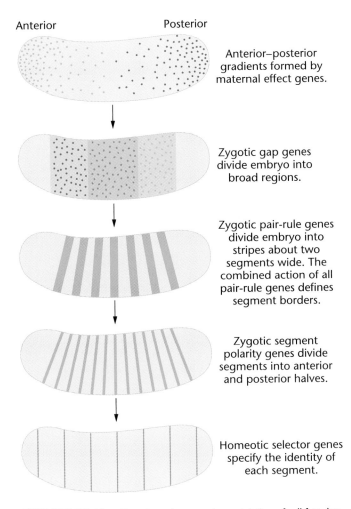

FIGURE 20.13 Overview of progressive restriction of cell fate during development in *Drosophila*. The process begins with the anterior-posterior axis laid down in oogenesis and activated immediately after fertilization. This axis restricts cells to form either anterior or posterior structures. Gene products from this gradient activate transcription of the gap genes that divide the embryo into a limited number of broad regions. Gap proteins are transcription factors that activate pair-rule genes whose products divide the embryo into regions about two segments wide. The combined action of all gap genes defines segment borders. The pair-rule genes in turn activate the segment polarity genes that divide the segments into anterior and posterior compartments. The collective action of the maternal genes that form the anterior-posterior axis and the segmentation genes define fields of action for the homeotic selector genes that specify the identity of each segment.

The labels in Figure 20.13 read:

Anterior — Posterior

Anterior–posterior gradients formed by maternal effect genes.

Zygotic gap genes divide embryo into broad regions.

Zygotic pair-rule genes divide embryo into stripes about two segments wide. The combined action of all pair-rule genes defines segment borders.

Zygotic segment polarity genes divide segments into anterior and posterior halves.

Homeotic selector genes specify the identity of each segment.

defects in external structures. Thousands of dead embryos were screened, and the mutants recovered fell into three classes, which were named *gap*, *pair-rule*, and *segment polarity*. In a paper published in 1980, they proposed a model in which embryonic development is initiated by gradients of maternal-effect gene products (Figure 20.13).

The positional information laid down by molecular gradients along the anterior-posterior axis of the embryo is interpreted by two gene sets: (1) zygotic **segmentation genes**, which divide the embryo into a series of stripes or segments and define the number, size, and polarity of each segment; and (2) **selector genes**, which specify the identity or fate of each segment but do not affect the number or size of the segments (Figure 20.14).

In a second screening, Weischaus and Trudi Schupbach screened thousands of flies for maternal-effect mutations that affected the external structures of the embryo. Based on these exhaustive screens for mutants, it is estimated that there are about 40 maternal effect genes and 50–60 zygotic genes. This means that normal embryonic development is controlled by a relatively small number of genes (about 100), a surprisingly low number, given the complexity of the structures formed from the fertilized egg. For their work on genetic control of development in *Drosophila*, Nusslein-Volhard and Weischaus, along with Ed Lewis, were awarded the 1995 Nobel Prize for Physiology or Medicine.

Maternal-Effect Genes and the Basic Body Plan in *Drosophila*

The maternal genes are those transcribed by the maternal genotype during oogenesis and transported as mRNA or as translated protein into the oocyte for storage. These transcripts and gene products are utilized during early stages of development, and regulate expression of zygotic genes.

Formation of the Anterior-Posterior Axis

The maternal-effect genes that control the anterior-posterior axis are grouped into three classes: anterior, posterior, and terminal. The following discussion will emphasize the formation of the anterior-posterior axis as an example of the interaction of maternal gene products and the zygotic genes.

Of the genes required to form the anterior portion of the embryo, the most important is *bicoid* (Table 20.3). Embryos of *bicoid* mutants lack head and thoracic structures and also have abnormalities of the first four segments of the abdomen. During oogenesis and the first hours of embryonic development, *bicoid* mRNA is localized to a small region at the anterior pole of the egg [Figure 20.15(a)]. The actions of two genes, *swallow* and *exuperantia*, are involved in localizing the *bicoid* mRNA by docking it to anterior cytoskeletal components. In mutants of these genes, the *bicoid* mRNA is distributed to more posterior areas of the egg, and the mutant embryos form enlarged anterior structures.

During early cleavages, *bicoid* mRNA is translated, and the *bicoid* protein becomes distributed in a gradient along the anterior-posterior axis, with very low levels in the posterior region of the egg [Figure 20.15(b)]. This gradient plays a critical role in determining the developmental pattern of the egg. The *bicoid* protein is a transcription factor that contains a homeodomain (see Chapter 19 for a review of these domains). The *bicoid* protein binds to the upstream regulatory region of *hunchback*, one

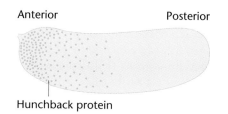
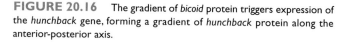

FIGURE 20.16 The gradient of *bicoid* protein triggers expression of the *hunchback* gene, forming a gradient of *hunchback* protein along the anterior-posterior axis.

of the first zygotic genes to be activated early in development, and which is expressed in the anterior half of the embryo. The upstream promoter region of *hunchback* contains five sites, each with the consensus sequence 5′-TCTAATCCC-3′, which bind to the *bicoid* protein. The presence of the consensus binding sequence causes the gene to respond to binding of the *bicoid* transcription factor, and the number of sites occupied by the *bicoid* protein determines the degree of response. In regions of the gradient, where there is a higher concentration of *bicoid* protein, there will be a higher response by the *hunchback* gene, leading to the formation of a gradient of *hunchback* protein along the anterior-posterior axis (Figure 20.16).

This interaction between the *bicoid* protein and the promoter region of *hunchback* was demonstrated by fusing a reporter gene encoding the enzyme chloramphenicol acetyltransferase (CAT) to promoter sites of the *hunchback* gene (Figure 20.17). When these constructs were injected into *bicoid* mutant embryos, no CAT was produced. When injected into wild-type embryos, the CAT constructs were expressed. Full expression of the CAT gene required the presence of three of the five promoter sites. These experiments demonstrated that the bicoid protein determines the spatial pattern of expression of the *hunchback* gene.

The gradient of *bicoid* protein causes transcription of *hunchback* in the anterior half of the embryo, but not in the posterior half. The *hunchback* protein then stimulates transcription of genes forming the head and thorax, and represses activity of genes that form the abdomen.

The Posterior and Terminal Gene Sets

The posterior of the embryo is formed through the action of eight genes (Table 20.3). Mutants of these genes all have a similar phenotype: The posterior portion of the body is dwarfish and abdominal segments are lacking. The names of these mutants reflect this phenotype: *oskar* (the dwarf in the Günter Grass novel, *The Tin Drum*) and *nanos*, derived from the Spanish word for dwarf. The *nanos* gene is transcribed by the maternal genome during oogenesis, and its mRNA is stored at the posterior pole of the egg (Figure 20.18). The *nanos* protein is produced soon after fertiliza-

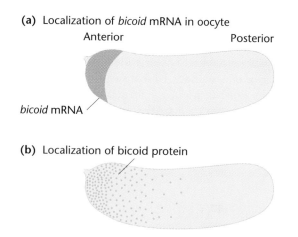

(a) Localization of *bicoid* mRNA in oocyte

Anterior   Posterior

bicoid mRNA

(b) Localization of bicoid protein

FIGURE 20.15 Formation of the anterior-posterior axis of the *Drosophila* egg depends on the action of three independent systems. The anterior portion of the axis is formed by a gradient of *bicoid* mRNA (a) that is translated into a transcription factor which distributes itself in a gradient at the anterior end of the embryo (b).

(a)

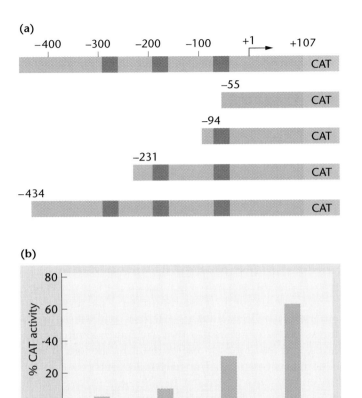

(b)

FIGURE 20.17 (a) Fusion genes between a reporter gene for chloramphenicol acetyltransferase (CAT) with various promoter regions upstream from the *hunchback* gene. Each construct was tested by injection into bicoid embryos and wild-type embryos. When injected into bicoid embryos, the fusion gene was not transcribed, and no CAT protein was produced. When injected into wild-type embryos (b), full expression of the CAT protein required the presence of three of the five *hunchback* promoter sites.

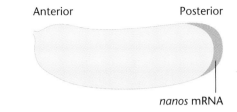

FIGURE 20.18 The posterior end of the anterior-posterior axis is formed by action of the *nanos* gene. The mRNA from *nanos* transcribed during oogenesis is stored at the posterior pole of the oocyte. After fertilization, the mRNA is distributed throughout the posterior region of the embryo by action of the *pumpilo* gene. The *nanos* protein acts to suppress the action of *hunchback* protein, ensuring the formation of abdominal structures in the posterior region.

tion, and it controls formation of posterior structures in the developing embryo, but it does so in a way different from the *bicoid* protein. In contrast to the action of anterior genes, the posterior genes work by a negative effect on a maternal gene transcript. The *nanos* protein acts to prevent the translation of any *hunchback* mRNA that might be present in the posterior region of the embryo, ensuring the expression of abdomen-forming genes (Figure 20.19). The *hunchback* gene, therefore, is regulated by maternal-effect gene products from both the anterior and posterior parts of the developing embryo.

The other genes involved in posterior formation ensure that the *nanos* mRNA is packaged and stored at the posterior pole (*oskar, staufen, valois, vasa*) and is distributed properly (*pumilio*).

The terminal genes represent a third set of genes controlling the formation of the anterior-posterior axis (Table 20.3). Mutations in these genes result in the loss of the unsegmented structures that develop from the anterior and posterior poles of the embryo. Cloning of two of the terminal genes, *torso* and *fs*(1)*polehole*, suggests that this group contains genes that encode transmembrane signal receptors associated with the reception and transduction of an extracellular signal.

Together, these three groups of maternal genes establish the anterior-posterior axis of the embryo as a spatial pattern of gene products. The anterior of the embryo

FIGURE 20.19 Coordinated action of *bicoid* and *nanos* proteins produces the anterior-posterior axis of the early embryo. Throughout the anterior end, *bicoid* protein activates the *hunchback* gene. The *hunchback* protein activates transcription of head and thorax genes. In the posterior region, the *nanos* protein inhibits production of *hunchback* protein, allowing the expression of abdominal genes.

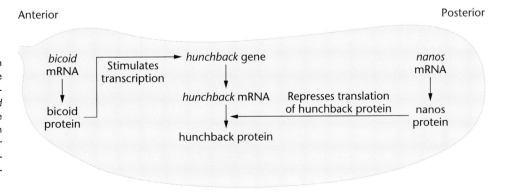

is organized by a gradient of the *bicoid* protein that acts to activate anterior-specific genes and to repress posterior-specific genes. The posterior of the embryo is formed through the action of the *nanos* protein, which after migration from the posterior pole suppresses the formation of the *hunchback* protein, relieving the *hunchback*-mediated inhibition of abdominal-specific gene expression. The critical gene in formation of the terminal region of the embryo is the *torso* gene product, which allows expression of the genes that form the unsegmented terminal structures. Two of these systems function by sequestering a specific mRNA at the poles of the egg (*bicoid* at the anterior, *nanos* at the posterior) and by the formation of a gradient of protein. The third system works via a signal transduction system coupled to the anterior-posterior axis system during oogenesis.

The dorsal-ventral axis of the embryo is specified by a family of maternal genes that depends on the formation of a single protein gradient (dorsal) along the dorsal-ventral axis.

Genes forming the anterior-posterior axis regulate transcriptional activities of cells along the anterior-posterior and dorsal-ventral axes and illustrate the critical role of cytoplasmic gradients in determining the fates of cells in different embryonic regions. The next section describes how components of the anterior-posterior axes control transcription of gene sets that regulate further development of the embryo.

Zygotic Genes and Segment Formation

The zygotic genes are activated or repressed in a positional gradient by the maternal-effect gene products, and they divide the embryo into a series of segments along the anterior-posterior axis. These segmentation genes are transcribed in the developing embryo, and mutations of these genes are seen as embryonic lethal phenotypes. Action of these genes brings about determination of cell fate.

TABLE 20.4 Segmentation Genes in *Drosophila*

Gap Genes	Pair-Rule Genes	Segment Polarity Genes
Krüppel	*hairy*	*engrailed*
knirps	*even-skipped*	*wingless*
hunchback	*runt*	*cubitis interruptus*[D]
giant	*fushi-tarazu*	*hedgehog*
tailless	*odd-paired*	*fused*
huckebein	*odd-skipped*	*armadillo*
	sloppy-paired	*patched*
		gooseberry
		paired
		naked
		disheveled

The segmentation genes are expressed soon after the *bicoid* gradient has been established. Over 20 segmentation loci have been identified (Table 20.4). They are classified on the basis of their mutant phenotypes: (1) the gap genes (*Krüppel*, *knirps*) delete a group of adjacent segments; (2) the pair-rule genes (*even-skipped*, *fushi-tarazu*) affect every other segment and eliminate a specific part of each affected segment; and (3) the segment polarity genes (*hedgehog*, *gooseberry*) cause defects in homologous portions of each segment.

Gap Genes

Transcription of the **gap genes** is activated or inactivated by gene products previously expressed along the anterior-posterior axis from *bicoid* and the other genes of the maternal gradient systems. When mutated, these genes produce large gaps in the segmentation pattern of the embryo. *Hunchback* mutants lose head and thorax structures, *Krüppel* mutants lose thoracic and abdominal structures, and mutations in *knirps* result in the loss of most abdominal structures. Transcription of gap genes divides the embryo into a series of broad regions (roughly, the head, thorax, abdomen) within which different combinations of gene activity will eventually specify both the type of segment that will form and the proper order of segments in the body of the larva, pupa, and adult. To date, gap genes that have been cloned all encode transcription factors with zinc-finger DNA-binding motifs. Expression of the gap genes correlates roughly with their mutant phenotypes (*hunchback* at the anterior, *Krüppel* in the middle, and *knirps* at the posterior). Gap genes control transcription of pair-rule genes.

Pair-Rule Genes

The **pair-rule genes** divide the broad regions established by the gap genes into regions about one segment wide. Mutations in pair-rule genes eliminate segment-size regions at every other segment. The pair-rule genes are expressed in narrow bands or stripes of nuclei that extend circumferentially around the embryo. Expression of this gene set first establishes the boundaries of segments, then establishes the developmental fate of the cells within each segment by controlling the segment polarity genes. At least eight pair-rule genes act to divide the embryo into a series of stripes. However, the boundaries of these stripes overlap, meaning that cells in the stripes express different combinations of pair-rule genes in an overlapping fashion (Figure 20.20).

Many pair-rule genes encode transcription factors containing helix-turn-helix homeodomains. Transcription of the pair-rule genes is mediated by the action of

(a)

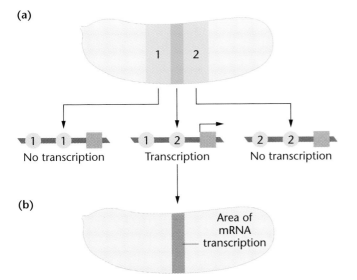

(b)

FIGURE 20.20   New patterns of gene expression can be generated by overlapping regions containing gene products. (a) Transcription factors 1 and 2 are present in an overlapping region of expression. If both transcription factors must bind to a promoter in order to trigger expression of a target gene, the gene will be active only in cells containing both factors (most likely in the zone of overlap). (b) Expression of the target gene in the restricted region of the embryo.

gap gene products, but the pattern is resolved into highly delineated stripes by the interaction among the gene products of the pair-rule genes themselves (Figure 20.21).

Segment-Polarity Genes

Expression of the **segment-polarity** genes is controlled by transcription factors encoded by the pair-rule genes. Each segment polarity gene becomes active in a single

FIGURE 20.21   Stripe pattern of gene expression in *Drosophila* embryo. This embryo is stained to show patterns of expression of the pair-rule genes *even-skipped* and *fushi-tarazu*.

band of cells within each segment that extends around the embryo. This concerted action divides the embryo into 14 segments and controls the cellular identity within each segment. Some of the segment polarity genes, including *engrailed*, encode transcription factors. Rather than activating transcription, however, the *engrailed* protein competitively inhibits activation by other homeodomain proteins, resulting in the establishment of a segment border.

To summarize, the genes that control development in *Drosophila* act in a temporally and spatially ordered cascade, beginning with the genes that establish the anterior-posterior and dorsal-ventral axes of the egg and early embryo. The gradients of mRNAs and proteins along the anterior-posterior axis activate the gap genes, which subdivide the embryo into broad bands. The gap genes in turn activate the pair-rule genes, which divide the embryo into segments. Finally, the segment polarity genes divide each segment into anterior and posterior regions arranged linearly along the anterior-posterior axis. This progressive restriction of the developmental potential of the cells in the *Drosophila* embryo, all of which occurs during the first one-third of embryogenesis, involves transcriptional control of gene sets and is accomplished by regulating the activity of proteins that bind to promoter and enhancer regions of genes and most act as transcription factors.

Selector Genes

As segment boundaries are being established by the action of the segmentation genes, the **selector genes** are activated. Expression of these genes determines the structures to be formed by each segment, including the antennae, mouth parts, legs, wings, thorax, and abdomen. Mutants of these genes are known as **homeotic mutants**. For example, the wild-type allele of the *Antennapedia* (*Antp*) gene is required to specify structures in the second thoracic segment (which carries a leg). Dominant gain of function *Antp* mutations lead to expression of the gene in the head and thorax, and the antennae of the fly are transformed into legs (Figure 20.22).

The *Drosophila* homeotic selector genes are found in two clusters on chromosome 3. The *Antp* complex contains five genes (Table 20.5) required for specification of structures in the head and first two thoracic segments. The *bithorax* (*BX-C*) complex contains three genes required for specification of structures formed by the posterior portion of the second thoracic segment, the entire third thoracic segment, and the abdominal segments (Figures 20.23 and 20.24).

Activation of the selector genes is under the control of the gap genes and the pair-rule genes. Although selec-

FIGURE 20.22 *Antennapedia* (*Antp*) mutation in *Drosophila*. (a) Head from wild-type *Drosophila* showing the structure of the antenna and other head parts. (b) Head from an *Antp* mutant, showing the replacement of normal antenna structures with a leg. This is caused by activation of the *Antp* gene in the head region.

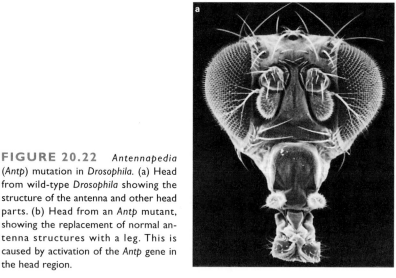

FIGURE 20.23 Genes of the *Antennapedia* complex and the adult structures they specify. The *labial* (*lab*) and *Deformed* (*Dfd*) genes control the formation of head segments. The *Sex comb reduced* (*Scr*) and *Antennapedia* (*Ant*) genes specify the identity of the first two thoracic segments. The remaining gene in the complex, *proboscipedia*, may not act during embryogenesis, but may be required to maintain the differentiated state in adults. In mutants, the labial palps are transformed into legs.

tor gene expression is confined to certain domains in the embryo, the timing and patterns of expression are rather complex and involve interactions between segmentation and selector genes as well as interactions among the various selector genes. For example, the *Antennapedia* gene has two promoters (P1 and P2) and can be transcribed to produce two different pre-mRNAs (Figure 20.25). Transcription from P1 is stimulated by *Krüppel* protein (a gap-gene product) and repressed by *Ultrabithorax* protein (a selector-gene product). Transcription from P2 is activated by *hunchback* and *fushi-tarazu* proteins (maternal-effect

and gap-gene products) and inhibited by *oskar* protein (a maternal-effect gene product). Paradoxically, the protein produced from these two pre-mRNAs is identical.

Regulation of expression of selector genes is also accomplished by alternative splicing of pre-mRNAs to yield different transcripts (Figure 20.26). The *Ultrabithorax* gene transcripts produced during early embryogenesis can be processed at varying 5′ sites in exon 1 and will include two "micro-exons," producing a number of related proteins. Transcripts from the *Ultrabithorax* gene later in development (during formation of the nervous system) include one or neither of the micro-exons.

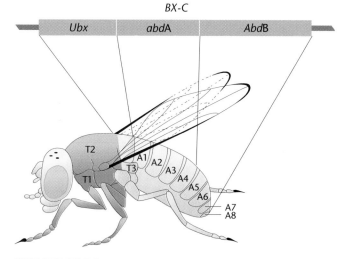

FIGURE 20.24 Genes of the *bithorax* complex and the adult structures they specify. *Ultrabithorax* (*Ubx*) controls structures in the posterior compartment of T2 and structures in T3. The two other genes, *abdominal A* (*abdA*) and *Abdominal B* (*AbdB*), specify the segmental identities of the eight abdominal segments (A1–A8).

FIGURE 20.25 Transcription and processing of mRNA from the *Antp* gene. Transcription from P1 or P2 in the *Antp* gene results in two different pre-mRNAs.

A number of genes that control the expression of selector genes have been identified, including *extra sex combs* (*esc*), *Polycomb* (*Pc*), *super sex combs* (*sxc*), and *trithorax* (*trx*). In the second chromosomal mutant *extra sex combs* (*esc*), some of the head and all of the thoracic and abdominal segments develop as posterior abdominal segments (Figure 20.27), indicating that this gene normally controls the expression of *BX-C* genes in all body segments. The mutation does not affect either the number or the polarity of segments. It does affect their developmental fate, indicating that the *esc*⁺ gene product that is synthesized and stored in the egg by the maternal genome may be required for the correct interpretation of the information gradient in the egg cortex.

E. B. Lewis has proposed that genes in the *bithorax* complex, and perhaps other genes involved in segmentation, arose from a common ancestral gene by tandem duplication and subsequent divergence of structure and function. In fact, each of the homeotic selector genes listed in Table 20.5 encodes a transcription factor that includes a DNA-binding domain encoded by a 180-bp sequence known as a **homeobox**. The homeobox encodes a 60-amino-acid sequence known as a **homeodomain**. Similar sequences have been found in the genomes of other eukaryotes with segmented body plans, including *Xenopus*, chicken, mice, and humans. In mammals, the complexes of selector genes are called ***Hox* gene clusters** (Figure 20.28). Homeodomains from all organisms examined to date are very similar in amino acid sequence and encode a protein associated with the transcriptional regulation of a specific gene set. This suggests that the metameric or segmented body plan may have evolved only once.

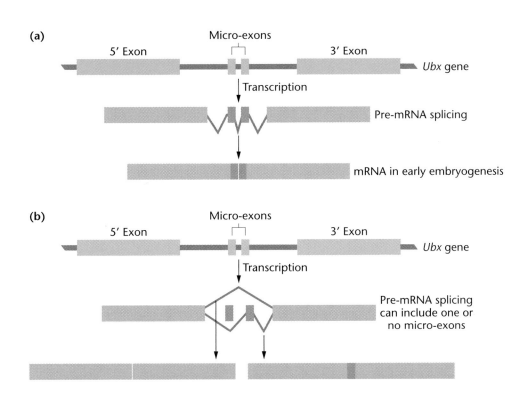

FIGURE 20.26 Alternative processing of *Ubx* pre-mRNA. (a) In early embryogenesis, splicing includes both micro-exons in the mature mRNA. (b) During development of the nervous system, splicing of pre-mRNA includes either no micro-exons or one micro-exon, producing different but related proteins upon translation.

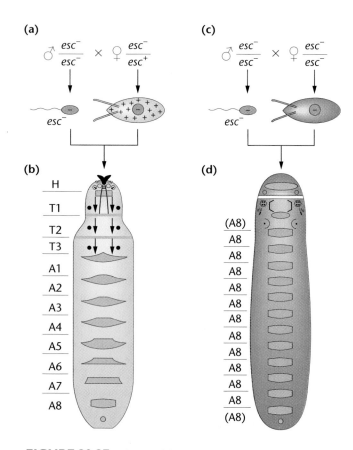

(a)

$\male \dfrac{esc^-}{esc^-} \times \female \dfrac{esc^-}{esc^+}$

esc^-

(b)

H
T1
T2
T3
A1
A2
A3
A4
A5
A6
A7
A8

(c)

$\male \dfrac{esc^-}{esc^-} \times \female \dfrac{esc^-}{esc^-}$

esc^-

(d)

(A8)
A8
A8
A8
A8
A8
A8
A8
A8
A8
A8
A8
A8
(A8)

FIGURE 20.27 Action of the *extra sex combs* mutation in *Drosophila*. (a) Heterozygous females form wild-type *esc* gene product and store it in the oocyte. (b) Fertilization of *esc*⁻ egg formed at meiosis by the heterozygous female by *esc*⁻ sperm still produces wild-type larva with normal segmentation pattern (maternal rescue). (c) Homozygous *esc*⁻ females produce defective eggs, which, when fertilized by *esc*⁻ sperm, (d) produce a larva in which most of the segments of the head, thorax, and abdomen are transformed into the eighth abdominal segment. Maternal rescue demonstrates that the *esc*⁻ gene product is produced by the maternal genome and is stored in the oocyte for use in the embryo. (H, head; T1–3, thoracic segments; A1–8, abdominal segments). Borderlines between the head and thorax and between thorax and abdomen are marked with arrows.

TABLE 20.5 Homeotic Selector Genes of *Drosophila*

Antennapedia Complex	*Bithorax* Complex
labial	*Ultrabithorax*
Antennapedia	*abdominal A*
Sex comb reduced	*Abdominal B*
Deformed	
proboscipedia	

Cell–Cell Interactions in *C. elegans* Development

During development in multicellular organisms, cells influence the transcriptional patterns and developmental fate of neighboring cells; these interactions involve the generation and reception of signal molecules. **Cell–cell interaction** is an important process in the development of most eukaryotic organisms, including *Drosophila* as well as vertebrates such as *Xenopus*, mice, and humans. To study the role of individual genes in developmental processes, the role of higher-level processes such as cell–cell interactions, and the genesis of behavior, Sidney Brenner began working with the soil nematode *Caenorhabditis elegans*.

Overview of *C. elegans* Development

This adult worm is about 1 mm long and matures from a fertilized egg in about two days (Figure 20.29). The life cycle consists of an embryonic stage (about 16 hours), four larval stages (L1 through L4), and the adult stage. The diploid chromosome number is 12 (two X chromosomes

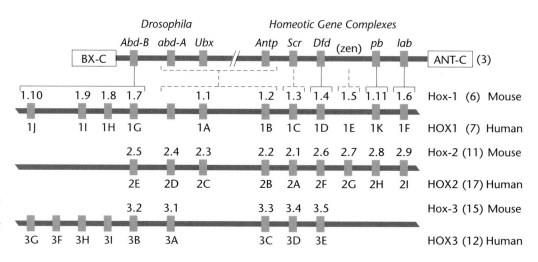

FIGURE 20.28 The *Hox* genes of mammals and their organizational alignment with the *BX-C* and *ANT-C* complexes of *Drosophila*.

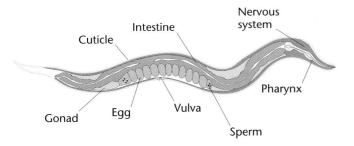

FIGURE 20.29   An adult *Caenorhabditis elegans*. This nematode, about 1 mm in length, consists of 959 cells and has been used to study many aspects of the genetic control of development.

and five pairs of autosomes), and it is estimated that the haploid genome consists of about 15,000 genes. Adults are of two sexes—XX self-fertilizing hermaphrodites that can make both eggs and sperm, and XO males. Self-crossing mutagen-treated hermaphrodites quickly results in homozygous stocks of mutant strains, and hundreds of mutants have been generated, catalogued, and mapped.

The adult hermaphrodite consists of 959 somatic cells (and about 2000 germ cells), whose exact cell lineage from fertilized egg to adult has been mapped (Figure 20.30) and is invariant from individual to individual. With knowledge of the lineage of each cell, it is easy to follow the events that result from mutational alterations in cell fate or the killing of cells using laser microbeams or ultraviolet irradiation.

In *C. elegans* the fate of cells in the development of the female reproductive system is determined by cell–cell interactions and provides some insight into how gene expression and cell–cell interactions are linked in the specification of developmental pathways.

Genetic Analysis of Vulva Formation

Adult hemaphrodites lay eggs through the vulva, an opening located about midbody. The vulva is formed in stages during larval development and involves several rounds of cell–cell interactions.

During development in *C. elegans*, two neighboring cells, Z1.ppp and Z4.aaa, normally have different developmental fates. These cells interact with each other so that one becomes the gonadal anchor cell and the other becomes a precursor to the ventral uterus (Figure 20.31). This decision is made during the second larval stage (L2), and is controlled by the *lin*-12 gene, which encodes a cell surface receptor protein. In recessive *lin*-12(0) mutants (a loss-of-function mutant) both cells become anchor cells. The dominant mutant *lin*-12(d) (a gain-of-function mutation) causes both to become uterine precursors. Thus it appears that the *lin*-12 gene causes the selection of the uterine pathway, since in the absence of the *lin*-12 protein, both cells become anchor cells. However, both cells normally begin to synthesize and secrete a chemical signal (as yet unknown) for uterine differentiation and synthesize the *lin*-12 protein, which is a receptor for the signal. At a critical time in L2, the cell that by chance is secreting more of the signal molecule causes its neighbor to increase transcription of the *lin*-12 gene, thus increasing production of the receptor. The cell with more *lin*-12 receptors becomes the ventral uterine precursor, and the other cell becomes the anchor cell. The critical factor in this first round of cell–cell interaction and determination is the *lin*-12 gene.

A second round of cell–cell interactions in L3 involves the anchor cell, located in the gonad, and six precursors located in the skin (hypodermis) adjacent to the gonad. The precursor cells are named P3.p, P4.p, P5.p, P6.p, P7.p, and P8.p and collectively are called Pn.p cells. The fate of each of the Pn.p cells is specified

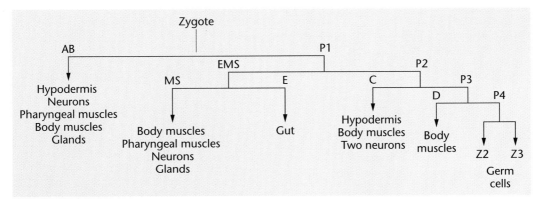

FIGURE 20.30   A truncated cell lineage chart for *C. elegans* showing early divisions and the tissues and organs. The cells shown refer to those present in the first-stage larva L1. During subsequent larval stages, further cell divisions will produce the 959 somatic cells of the adult hermaphrodite worm.

(a)

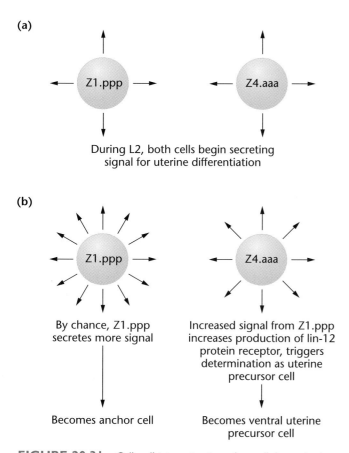

During L2, both cells begin secreting
signal for uterine differentiation

(b)

By chance, Z1.ppp
secretes more signal

Increased signal from Z1.ppp
increases production of lin-12
protein receptor, triggers
determination as uterine
precursor cell

Becomes anchor cell

Becomes ventral uterine
precursor cell

FIGURE 20.31 Cell–cell interaction in anchor cell determination. (a) During L2, two neighboring cells begin secretion of chemical signals for induction of uterine differentiation. (b) By chance, cell Z1.ppp secretes more of these signals, causing cell Z4.aaa to increase production of the receptor for signals. Action of increased signals causes Z4.aaa to become the ventral uterine precursor cell and allows Z1.ppp to become the anchor cell.

by their position relative to the anchor cell (Figure 20.32). Sometime in L3, the *lin-3* gene in the anchor cell is activated and secretes a protein structurally related to the vertebrate **epidermal growth factor (EGF)**. All six of the Pn.p cells express a cell surface receptor encoded by the *let-23* gene. Binding of the *lin-3* protein to the *let-23* receptor triggers an intracellular cascade of events that causes the Pn.p cells to become determined to form either vulval precursor cells or secondary vulval cells. The *let-23* gene is required for establishing the primary and secondary fates of Pn.p cells. Recessive loss-of-function mutations in *let-23* cause all the Pn.p cells to develop as hypodermis (a tertiary fate). In *let-23* mutants, the Pn.p cells act as though they have not received a signal from the anchor cell, and no vulva is formed.

The signal transduction cascade from the anchor cells to the Pn.p cells also involves the *let-60* gene. Reces-

sive mutations in *let-60* cause the Pn.p cells to develop as though they have not received a signal from the anchor cell. Other dominant *let-60* alleles have the opposite phenotype: They cause all the Pn.p cells to respond to form vulval cells, forming multiple vulvas. The data on *let-23* and *let-60* suggest that *let-23* encodes the receptor for the anchor cell signal, and that the *let-60* gene product acts to transfer the signal to other parts of the signal cascade. The *let-60* gene has been cloned, and is the *C. elegans* homolog of the *ras* gene, which in mammals acts downstream of receptor proteins in signal transduction. The *let-60* dominant gain-of-function mutant that causes multiple vulva formation has a Gly-Glu mutation at amino acid 13, the same mutation that converts *c-ras* to an oncogene (see Chapter 21).

The cell (P6.p) closest to the anchor cell receives the strongest signal. In P6.p, expression of the *Vulvaless (Vul)* gene (the gene is named for its mutant phenotype) is activated; the primary fate of this cell is to divide three times to produce vulva cells. The two neighboring cells receive a lower amount of signal, which specifies a secondary fate: assymetric division to form more vulva cells (Figure 20.32). To reinforce this determination, the primary vulval precursor activates the *lin-12* gene in the two neighboring cells. This lateral inhibition signal prevents the neighboring secondary cells from adopting the division pattern of the primary cell. In other words, cells in which both *Vulvaless* and *lin-12* are active cannot become primary vulvar cells (Figure 20.32). The three remaining Pn.p cells receive no signal from the anchor cell. In these cells (P3.p, P4.p, and P8.p) the *Multivulva (Muv)* gene is expressed; *Muv* represses *Vul*, and the three cells develop as hypodermal (skin) cells.

Thus, three levels of cell–cell interaction are required to specify the developmental pathway leading to vulva formation in *C. elegans*. First, two neighboring cells interact during L2 to establish the identity of the anchor cell. Second, in L3 the anchor cell interacts with three Pn.p cells to establish the primary vulvar precursor cell and two secondary cells. Third, the primary vulvar cell interacts with the secondary cells to suppress their selection of the primary vulvar pathway. Each of these interactions is accompanied by the secretion of molecular signals, and by the reception and processing of these signals by neighboring cells.

This example of cell–cell interactions acting in a spatial and temporal cascade to specify the developmental fates of individual cells is a developmental theme repeated over and over in developing organisms, ranging from prokaryotic prespore cells in *Bacillus* to higher vertebrates, including mice and humans.

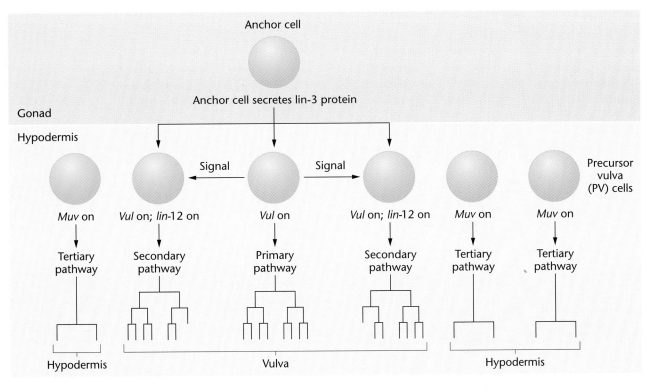

FIGURE 20.32   Cell lineage determination in *C. elegans* vulva formation. A signal from the anchor cell in the form of *lin-3* protein is received by three PV cells. The cells closest to the anchor cell become primary vulval precursor cells, and adjacent cells become secondary precursor cells. Primary cells secrete a signal that activates the *lin-12* gene in secondary cells, preventing them from becoming primary cells. Flanking PV cells, which receive no signal from the anchor cell activity of the *Muv* gene, cause these cells to become hypodermis cells instead of vulval cells.

CHAPTER SUMMARY

1. The role of genetic information during development and differentiation has been studied extensively using a simple model system: bacterial sporulation. In this system, environmental conditions induce specific gene activity, which in turn leads to altered development. By isolating developmental mutations, researchers have identified a number of the regulatory elements involved in controlling this process.

2. The variable gene activity theory, which applies to higher organisms, assumes that all somatic cells in an organism contain equivalent genetic information, but that the information is differentially expressed. Studies conducted on genomic equivalence have demonstrated that somatic cells undergoing differentiation retain the entire gene set, which can be reprogrammed to direct the development of the entire organism.

3. Determination is the regulatory event whereby cells become progressively restricted in developmental capac-

ity during early development. Determination becomes more specific with time and precedes the actual differentiation or specialization of distinctive cell types.

4. During embryogenesis, specific gene activity appears to be affected by the environment of the cell, with early regulation mediated by the maternal cytoplasm, which is then mediated by zygotic gene expression. As development proceeds, the cell's environment becomes further altered by the presence of other cells and early gene products.

5. In *Drosophila*, both genetic and molecular studies have confirmed that the egg contains information that specifies the body plan of the larva and adult, and that interactions of embryonic nuclei with the maternal cytoplasm initiate transcriptional programs characteristic of specific developmental pathways.

KEY TERMS

INSIGHTS AND SOLUTIONS

1. In the slime mold *Dictyostelium*, experimental evidence suggests that cyclic AMP (cAMP) plays a central role in the developmental program leading to spore formation. The genes encoding the cAMP cell-surface receptor have been cloned, and the amino acid sequence of the protein components is known. To form reproductive structures, free-living individual cells aggregate together and then differentiate into one of two cell types: prespore cells or prestalk cells. Aggregating cells secrete waves or oscillations of cAMP to foster aggregation of cells, and then continuously secrete cAMP to activate genes in the aggregated cells at later stages of development. It has been proposed that cAMP plays a central role in cell–cell interaction and gene expression in this developmental program. It is important to test this hypothesis by using several different experimental techniques. What different approaches can you devise to test this hypothesis, and what specific experimental systems would you employ to test them?

Solution: Two of the most powerful forms of analysis in biology involve the use of biochemical analogs or inhibitors to block gene transcription or the action of gene products in a predictable way, and the use of mutations to alter the gene and its products. These two approaches can be used to study the role of cAMP in the developmental program of *Dictyostelium*. First, compounds chemically related to cAMP, such as GTP and GDP, can be used to test whether they have any effect on the processes controlled by cAMP. In fact, both GTP and GDP lower the affinity of cell-surface receptors for cAMP, effectively blocking the action of cAMP. To inhibit the synthesis of the cAMP receptor, it is possible to construct a vector that contains a DNA sequence that transcribes an antisense RNA (a molecule that has a base sequence complementary to the mRNA). Antisense RNA forms a double-stranded structure with the mRNA, preventing it from being transcribed. If normal cells are transformed with a vector that expresses antisense RNA, no cAMP receptors will be produced. It is possible to predict that such cells will fail to respond to a gradient of cAMP and, consequently, will not migrate to an aggregation center. In fact, that is what happens. Such cells remain dispersed and nonmigratory in the presence of cAMP. Similarly, it is possible to determine whether this response to cAMP is necessary to trigger changes in the transcriptional program by assaying for the expression of developmentally regulated genes in cells expressing this antisense RNA.

Mutational analysis can be used to dissect components of the cAMP receptor system. One way to approach this is to use transformation with wild-type genes to restore mutant function. Similarly, because the genes for the receptor proteins have been cloned, it is possible to construct mutants with known alterations in the component proteins and transform them into cells to assess their effects.

PROBLEMS AND DISCUSSION QUESTIONS

1. Carefully distinguish between the terms *differentiation* and *determination*. Which phenomenon occurs initially during development?

2. The *Drosophila* mutant *spineless aristapedia* (*ss^a*) results in the formation of a miniature tarsal structure (normally part of the leg) on the end of the antenna. Such a mutation is referred to as *homeotic*. From your knowledge of imaginal disks, what insight is provided by *ss^a* concerning the role of genes during determination?

3. In the sea urchin, early development up to gastrulation may occur even in the presence of actinomycin D, which inhibits RNA synthesis. However, if actinomycin D is removed at the end of blastula formation, gastrulation does not proceed. In fact, if actinomycin D is present only between the 6th and 11th hours of development, gastrulation (normally occurring at the 15th hour) is arrested. What conclusions can be drawn concerning the role of gene transcription between hours 6 and 15?

4. How can you determine whether a particular gene is being transcribed in different cell types?

5. Observing that a particular gene is being transcribed during development, how can you tell whether expression of this gene is under transcriptional or translational control?

6. Both the *ftz* gene and the *engrailed* gene encode homeobox transcription factors and are capable of eliciting the expression of other genes. Both genes work at about the same time during development and in the same region to specify cell fate in body segments. The question is: Does *ftz* regulate the expression of *engrailed*, or does *engrailed* regulate *ftz*? Or are they both regulated by another gene? To answer these questions, mutant analysis is performed. In *ftz*⁻ embryos (*ftz/ftz*), *engrailed* protein is absent; in *engrailed*⁻ embryos (*eng/eng*), *ftz* expression is normal. What does this tell you about the regulation of these two genes? Does the *engrailed* gene regulate *ftz*? Does the *ftz* gene regulate *engrailed*?

7. In *Bacillus*, expression of mother cell genes depends on recombination of the truncated σ^K gene by action of the *spoIVCA* gene. If a cloned copy of the rearranged σ^K gene is introduced into the chromosome of a *spoIVCA*⁻ cell and restores the capacity of the mutant cell to sporulate, what does this tell you about the essential functions of the *spoIVCA* gene?

8. Nuclei from almost any source may be injected into *Xenopus* oocytes. Studies have shown that these nuclei remain active in transcription and translation. How can such an experimental system be useful in developmental genetic studies?

9. The concept of epigenesis indicates that an organism develops by forming cells that acquire new structures and functions, which become greater in number and complexity as development proceeds. This theory is in contrast to the preformationist doctrine that miniature adult entities are contained in the egg that must merely unfold and grow to give rise to a mature organism. What sorts of isolated evidence presented in this chapter might have led to the preformation doctrine? Why is the epigenetic theory held as correct today?

10. You are given two snails: a female snail that has a *d/d* genotype and a *d* phenotype, and a male snail that has a *D/D* genotype and a *D* phenotype. What genetic crosses are required to generate a snail that has a *d/d* genotype but a *D* phenotype?

EXTRA-SPICY PROBLEMS

11. In studying gene action during development, it is desirable to be able to position genes in a hierarchy or pathway of action, to establish which genes are primary and in what order genes act. There are several ways of doing this. One way is to make double mutants and study the outcome. The gene *fushi-tarazu* (*ftz*) is expressed in early embryos at the seven-stripe stage). All genes involved in forming the anterior-posterior pattern affect the expression of this gene, as do the gap genes. However, expression of segment-polarity genes is affected by *ftz*. What is the location of *ftz* in this hierarchy?

12. (a) The identification and characterization of genes that control sex determination has been another focus of investigators working with *C. elegans*. As with *Drosophila*, sex in this organism is determined by the ratio of X chromosomes to sets of autosomes. A diploid wild-type male has one X chromosome and a diploid wild-type hermaphrodite has two X chromosomes. Many different mutations have been identified that affect sex determination. Loss-of-function mutations (that is, mutations that knock out gene function) in a gene called *her-1* cause an XO animal to develop into a hermaphrodite and have no effect on XX development (that is, XX animals are normal hermaphrodites). In contrast, loss-of-function mutations in a gene called *tra-1* cause an XX animal to develop into a male. From this information, deduce the roles of these genes in wild-type sex determination.

(b) Based on phenotypes of single and double mutant strains, a model for sex determination in *C. elegans* has been generated. This model proposes that the *her-1* gene controls sex determination by determining the level of activity of the *tra-1* gene, which in turn controls the expression of genes involved in generating the various sexually dimorphic tissues. Given the information above, does the *her-1* gene product have a negative or a positive effect on the activity of the *tra-1* gene? What would be the phenotype of a *tra-1; her-1* double mutant?

SELECTED READINGS

BONCINELLI, E., and MALLAMACI, A. 1995. Homeobox genes in vertebrate gastrulation. *Curr. Opin. Genet. Dev.* 5: 619–27.

BRIGGS, R., and KING, T. 1952. Transplantation of living nuclei from blastula cells into enucleated frog eggs. *Proc. Natl. Acad. Sci. USA* 38:455–63.

DESCHAMPS, J., and MEIJLINK, F. 1992. Mammalian homeobox genes in normal development and neoplasia. *Crit. Rev. Oncog.* 3:117–73.

DRIEVER, W., THOMA, G., and NUSSLEIN-VOLHARD, C. 1989. Determination of spatial domains of zygotic gene expression in the *Drosophila* embryo by the affinity of binding sites for the bicoid morphogen. *Nature* 340:363–37.

EDMONDSON, D., and OLSON, E. 1993. Helix-loop-helix proteins as regulators of muscle-specific transcription. *J. Biol. Chem.* 268:755–58.

ERRINGTON, J. *Bacillus subtilis* sporulation: Regulation of gene expression and control of morphogenesis. *Microbiol. Rev.* 57:1–33.

FOSTER, S. J., and JOHNSTONE, K. 1990. Pulling the trigger: The mechanism of bacterial spore germination. *Mol. Microbiol.* 4:137–41.

GAUNT, S. 1991. Expression patterns of mouse *Hox* genes: Clues to an understanding of development and evolutionary strategies. *BioEssays* 13:505–13.

GEHRING, W. 1968. The stability of the differentiated state in cultures of imaginal disks in *Drosophila*. In *The stability of the differentiated state*, ed. H. Unsprung. New York: Springer-Verlag.

GILBERT, S. 1991. *Developmental biology*, 3rd ed. Sunderland, MA: Sinauer Associates.

GROSSMAN, A. 1991. Integration of developmental signals and the initiation of sporulation in *B. subtilis*. *Cell* 65:5–8.

GURDON, J. B. 1968. Transplanted nuclei and cell differentiation. *Sci. Am.* (Dec.) 219:24–35.

GURDON, J. B., LASKEY, R. A., and REEVES, O. R. 1975. The developmental capacity of nuclei transplanted from keratinized skin cells of adult frogs. *J. Embryol. Exp. Morphol.* 34:93–112.

GURDON, J. B., MITCHELL, A., and MAHONY, D. 1995. Direct and continuous assessment by cells of their position in a morphogen gradient. *Nature* 376:520–21.

HENGARTNER, M. 1995. Life and death decisions: *ced-9* and programmed cell death in *Caenorhabditis elegans*. *Science* 270:931.

HENGARTNER, M. O., and HORVITZ, H. R. 1994. The ins and outs of programmed cell death during *C. elegans* development. *Phil. Trans. R. Soc. London* B 345:243–46.

HOLLAND, P. W., and GARCIA-FERNANDEZ, J. 1996. Hox genes and chordate evolution. *Dev. Biol.* 173:382–95.

HUNT, P., and KRUMLAUF, R. 1992. *Hox* codes and positional specification in vertebrate embryonic axes. *Annu. Rev. Cell Biol.* 8:227–56.

JÄCKLE, H., HOCH, M., PANKRATZ, M., GERWIN, N., SAUER, F., and BRÖNNER, G. 1992. Transcriptional control by *Drosophila* gap genes. *J. Cell Sci. Suppl.* 16:39–51.

KAPPEN, C., SCHUGHART, K., and RUDDLE, F. H. 1989. Organization and expression of homeobox genes in mouse and man. *Ann. NY Acad. Sci.* 567:243–52.

KAUFMANN, T., SEEGER, M., and OLSON, G. 1990. Molecular organization of the *Antennapedia* gene cluster of *Drosophila melanogaster*. *Adv. Genet.* 27:309–62.

KENNISON, J., and TAMKUN, J. 1992. Trans-regulation of homeotic genes in *Drosophila*. *New Biol.* 4:91–96.

KING, T. J. 1966. Nuclear transplantation in amphibia. In *Methods in cell physiology*, ed. D. Prescott, Vol. 2. Orlando, FL: Academic Press.

KRAUSE, M. 1995. MyoD and myogenesis in *C. elegans*. *Bio-Essays* 17:219–28.

KRUMLAUFF, R., and GOULD, A. 1992. Homeobox cooperativity. *Trends Genet.* 8:297–300.

KYRAICOU, C. 1992. Sex variations. *Trends Genet.* 8:261–67.

LAWRENCE, P. 1992. *The making of a fly: The genetics of animal design*. Oxford: Blackwell Scientific Publications.

LEVINE, M., RUBIN, G., and TJIAN, R. 1984. Human DNA sequences homologous to a protein coding region conserved between homoeotic genes of *Drosophila*. *Cell* 38:667–73.

LEWIS, E. B. 1976. A gene complex controlling segmentation in *Drosophila*. *Nature* 276:565–70.

———. 1994. Homeosis: the first 100 years. *Trends Genet.* 10:341–43.

MANSEAU, L. J., and SCHÜPBACH, T. 1989. The egg came first, of course! Anterior-posterior pattern formation in *Drosophila* embryogenesis and oogenesis. *Trends Genet.* 5:400–5.

MCGINNIS, W., HART, C. P., GEHRING, W. J., and RUDDLE, F. H. 1984. Molecular cloning and chromosome mapping of a mouse DNA sequence homologous to homoeotic genes of *Drosophila*. *Cell* 38:675–80.

MCGINNIS, W., and KRUMLAUF R. 1992. Homeobox genes and axial patterning. *Cell* 68:283–302.

NUSSLEIN-VOLHARD, C. 1994. Of flies and fishes. *Science* 266:572–74.

NUSSLEIN-VOLHARD, C., and WEISCHAUS, E. 1980. Mutations affecting segment number and polarity in *Drosophila*. *Nature* 287:795–801.

OLSON, E. 1990. *MyoD* family: A paradigm for development? *Genes Dev.* 4:1454–61.

RUDDLE, F. H., HART, C. P., and MCGINNIS, W. 1985. Structural and functional aspects of the mammalian homeobox sequences. *Trends Genet.* 1:46–50.

RUDNICKI, M. A., and JAENISCH, R. 1995. The MyoD family of transcription factors and skeletal myogenesis. *BioEssays* 17:203–9.

SANCHEZ-HERRERO, E., and AKAM, M. 1989. Spatially ordered transcription of regulatory DNA in the *bithorax* complex of *Drosophila*. *Development* 107:321–29.

SATHE, S. S., and HARTE, P. J. 1995. *Drosophila* extra sex combs protien contains WD motifs essential for its function as a repressor of homeotic genes. *Mech. Dev.* 52:77–87.

SMALL, S., and LEVINE, M. 1991. The initiation of pair-rule stripes in the *Drosophila* blastoderm. *Curr. Opin. Genet. Dev.* 1:255–60.

ST. JOHNSTON, D., and NUSSLEIN-VOLHARD, C. 1992. The origin of pattern and polarity in the *Drosophila* embryo. *Cell* 68:201–19.

STRUHL, G. 1981. A gene product required for correct initiation of segmental determination in *Drosophila*. *Nature* 293:36–41.

TAUTZ, D. 1996. Selector genes, polymorphisms, and evolution. *Science* 271:160–61.

WEINTRAUB, J., DAVIS, R., TAPSCOTT, S., THAYER, M., KRAUSE, M., BENEZRA, R., BLACKWELL, T., TURNER, et al. 1991. The *myo*D gene family: Nodal point during specification of the muscle cell lineage. *Science* 251:761–66.

WEISCHAUS, E. 1996. Embryonic transcription and the control of developmental pathways. *Genetics* 142:5–10.

WRIGHT, W. 1992. Muscle basic helix-loop-helix proteins and the regulation of myogenesis. *Curr. Opin. Genet. Dev.* 2:243–48.

21

Genetics and Cancer

CHAPTER CONCEPTS

Cancer is now recognized as a genetic disorder at the cellular level that involves the mutation of a small number of genes. Many of these genes normally act to suppress or stimulate progression through the cell cycle, and loss or inactivation of these genes causes uncontrolled cell division and tumor formation. Environmental factors and viruses also play a role in the genetic alterations that are necessary to transform normal cells into cancerous cells.

Although often viewed as a single disease, cancer is actually a complex group of diseases affecting a wide range of cells and tissues. A genetic link to cancer was first proposed early in the twentieth century, and this idea has served as one of the foundations of cancer research. Mutations that alter the genome or gene expression are now regarded as a common feature of all cancers. In some cases, such mutations are part of the germ line and are inherited. More often, mutations arise in somatic cells and are not passed on to future generations through the germ cells. Sometimes, the inherited mutation must be accompanied by a somatic mutation at the homologous locus, creating ho-

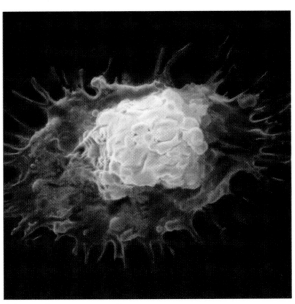

A human breast cancer cell.

mozygosity. Whichever the case, *cancer is now considered a genetic disorder at the cellular level.*

Genomic alterations associated with cancer can involve small-scale changes such as a single nucleotide substitution or large-scale events that include chromosome rearrangement, chromosome gain or loss, or even the integration of viral genomes into chromosomal sites. Large-scale genomic alterations are a common feature of cancer; the majority of human tumors are characterized by visible chromosomal changes. Some of these chromosomal changes, particularly in leukemia, are so characteristic they can be used to diagnose and classify the disorder and to make accurate predictions about the severity and course of the disease.

Familial forms of cancer have been known for over two hundred years. In many of these cases, no well-defined pattern of inheritance can be established. In a small number of cases, however, a Mendelian pattern of dominant or recessive inheritance can be established, indicating the hereditary nature of the cancer.

Because mutation is the underlying cause of cancer, there will always be a baseline rate of cancer, because there is a background rate of spontaneous mutation. Over and above this baseline rate, environmental agents that promote mutation also play a role in the development of cancer. Environmental factors such as ionizing radiation, chemicals, and viruses are cancer-causing agents, and almost all of these act by generating mutations. Given that mutations play a role in cancer, this chapter will explore answers to a series of questions about how mutations convert normal cells into malignant tumors, which mutant genes are most likely to result in cancer, and how many mutations are required to cause cancer.

One approach to answering these questions is to consider what properties of cancer cells distinguish them from normal cells and to ask what genes control these properties. Cancer cells have two properties in common: (1) uncontrolled growth and (2) the ability to spread or metastasize from their original site to other locations in the body. Cell division is the result of cells traversing the cell cycle; in cancer cells, control over the cell cycle is lost, and cells proliferate rapidly. Investigations into the genetic control of the cell cycle are now providing insights into the origins of cancer.

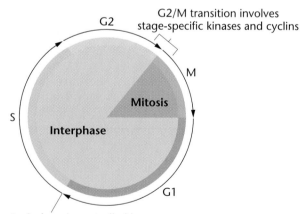

Entry to S phase is controlled by stage-specific kinases and cyclins

FIGURE 21.1   The cell cycle is controlled at two and possibly more checkpoints, one at the G2/M transition, and another in late G1 phase before entry into S phase. These checkpoints involve interaction between transitory proteins, called cyclins, and kinases that add phosphate groups to proteins. Phosphorylation of target proteins triggers a cascade of events allowing progress through the cell cycle.

The metastasis of cancer cells is controlled by gene products that become localized on the cell surface, and the genetics of metastasis is related to an understanding of how cells interact with the extracellular matrix and with other cells through cell surface molecules. Though this field is less well developed than the study of the cell cycle, it is beginning to provide some insights into the secondary events in tumor progression.

This chapter will consider the relationship between genes and cancer, with emphasis on the relationship between the cell cycle and genetic disorders associated with cancer. We will also examine the relationship between mutation and cancer, the identification of genes that when mutated, initiate the transformation of cells, and estimate the number of mutations involved in tumor formation. We will also discuss the relationship between chromosomal changes and cancer, and the role of environmental agents in the genesis of cancer.

The Cell Cycle and Cancer

As described in Chapter 2, the cell cycle represents the sequence of events occurring between mitotic divisions in a eukaryotic cell. In years past, work on control of the cell cycle has been conducted mainly by two groups: (1) geneticists working with yeasts, especially *Saccharomyces cerevisiae* and *Schizosaccharomyces pombe*; and (2) developmental biologists, studying the newly fertilized eggs of organisms such as frogs, sea urchins, and newts. In the last few years, these groups have succeeded in identifying and characterizing genes involved in the cell cycle, and their work is now converging and overlapping with important areas of cancer biology, particularly studies on growth factors and the genes that suppress or promote tumor formation. This convergence has had a synergistic effect, resulting in new insights into the processes that control cell division and how regulation of the cell cycle is coupled to the transcription of selected genes. Because of these recent developments, it is necessary to spend some time describing what is known about the steps of the cell cycle and the genes that regulate progression through the cycle. We will then undertake a discussion of the genetics of cancer.

The Cell Cycle

Stripped to its essentials, the cell cycle progresses from a period of chromosomal DNA replication (S phase) to the segregation of chromosomes into two nuclei during mitosis (M phase). Interspersed between these stages are two gaps, called G1 and G2. Together, G1, S, and G2 make up the interphase portion of the cell cycle (Figure

21.1). The G1 stage begins after mitosis; synthesis of many cytoplasmic elements including ribosomes, enzymes, and membrane-derived organelles occurs at this time. In the S phase, DNA replication takes place, producing a duplicate copy of each chromosome. Then a second period of growth and synthesis known as G2 occurs as a prelude to mitosis.

Because mitosis occurs rapidly, usually lasting less than an hour, cells spend most of the cell cycle in interphase. However, the duration of the cell cycle (the time between mitotic divisions) can vary widely among cells in the life cycle of an organism and among different cell types in the same organism. For example, animal cells exhibit *in vivo* cycles ranging from a few minutes to several months. Most of this variation can be traced to the time spent in G1. The time necessary to complete S and G2 remains relatively constant in most cell types.

While some cells such as meristematic cells in plants and dermal cells in human skin cycle continuously, other cell types, including many nerve cells, withdraw from the G1 phase, and permanently enter a nondividing state known as G0. Still other cell types such as white blood cells can be recruited to return from G0 and reenter the cell cycle.

Taken together, these observations suggest that the cell cycle is tightly regulated and is dependent on the life history and differentiated state of a given cell. A great deal of information about the regulation of the cell cycle has become available in the last few years. A summary of what is currently known about the genetic regulation of the cell cycle will serve as a prelude to an overview of the genetics of cancer.

Checkpoints and Control of the Cell Cycle

As work on the cell cycle in several organisms has converged, a universal model of the cell cycle and its regulation is beginning to emerge. Although details of all molecular events or even the exact number and sequence of steps are not yet known, all eukaryotic cells probably employ a common series of biochemical pathways to regulate events in the cell cycle. If the universal aspect of the current model of the cell cycle is upheld, results gathered from the study of yeasts or clam eggs can be used to understand and predict events in normal human cells and in mutated cells that have become cancerous. As a result, discoveries in cell cycle research will probably continue to be fast-moving and spectacular.

The emerging picture indicates that the cell cycle is regulated at the G2/M transition and at a point within G1. Events at the G2/M transition have been well documented, but the regulatory point within late G1, several hours before the initiation of S phase, is not yet well characterized.

At each of these points, a decision is made to proceed or halt progression through the cell cycle. This decision is controlled through the interaction of two classes of proteins. One is a group of enzymes called **protein kinases** that selectively phosphorylate target proteins. Although a large number of different protein kinases exist in the cell, only a few are involved in regulation of the cell cycle. In the second group, proteins involved in controlling progression through the cell cycle are called **cyclins**. These proteins, first identified in the embryos of developing invertebrates, are synthesized and degraded in a synchronous pattern throughout the cell cycle (Figure 21.2). Several cyclins, including C, D1, D2, and E, accumulate during the G1 phase, with only D1 persisting after the S phase. Cyclin A accumulates during late S and persists until the G2/M transition; cyclin B accumulates during G2 and persists until the end of M. Coupling of a kinase and a cyclin produces a regulatory molecule that controls the movement of the cell from G2 into M and from G1 into S.

The onset of M in most eukaryotic cells is controlled by a kinase called *CDK1* (cyclin-dependent kinase), which was biochemically characterized in maturing amphibian eggs and genetically identified in yeast as the product of the *cdc2* gene. This protein has a highly conserved sequence in all eukaryotes examined; its function is necessary for entry into M phase.

Several events mark the entry from G2 into mitosis (M), including condensation of the chromatin to form chromosomes, breakdown of the nuclear membrane, and reorganization of the cytoskeleton. Major events in this transition are regulated by the formation of an active CDK1/cyclin B complex. When bound to cyclin B, the

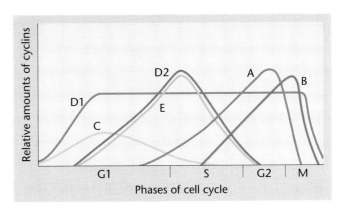

FIGURE 21.2 Relative expression times and amounts of cyclins during the cell cycle. D1 accumulates early in G1 and is expressed at a constant level through most of the cycle. Cyclin C accumulates in G1, reaches a peak, and declines by mid S phase. Cyclins D2 and E begin accumulating in the last half of G1, reach a peak just after the beginning of S, and then decline by early G2. Cyclin A appears in late G1, accumulates through S, reaches a peak in G2, and is degraded rapidly as M phase begins. Cyclin B appears in mid S phase, peaks at the G2/M transition, and is rapidly degraded.

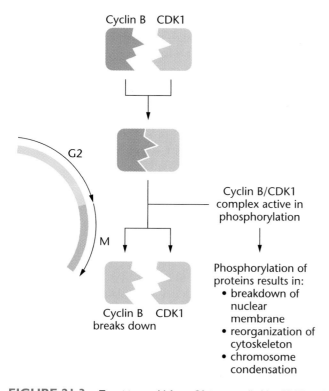

FIGURE 21.3 Transition to M from G2 is controlled by CDK1 and cyclin B. These molecules interact to form a complex that adds phosphate groups to cellular components that break down the nuclear membrane (lamins A, B, and C), reorganize the cytoskeleton (caldesmon), and initiate chromosome condensation (histone H1). Cyclin B may specify cellular localization or target molecules. Other cyclins (especially cyclin A) are thought to be involved at this stage, but these functions are not known.

CDK1 component catalyzes phosphorylation, which brings about nuclear membrane breakdown and re-arrangement of the cytoskeleton. It also phosphorylates histone H1, which may play a role in chromatin condensation (Figure 21.3). The function of cyclin B in this complex is not clear, but it may control the cellular localization or target molecule specificity. Although several lines of experiments suggest that cyclin A is also involved in the progression from G2 to M, its functions are not clearly understood.

While the action of the B cyclin and CDK1 are responsible for passage into M, it appears that several cyclins, including D and E, can move cells from G1 into S. Experiments indicate that the CDK1 kinase is also active in phosphorylation in G1, but that instead of combining with cyclin B (which is not present), CDK1 combines with G1 cyclins, directing the phosphorylation of G1 stage-specific protein substrates.

Altogether, almost a dozen different cyclins have been identified, and a growing number of cyclin-dependent kinases are being described, indicating that multiple control points in the cell cycle exist, or that these kinases and cyclins have multiple functions.

Cell-Cycle Regulation and Cancer

Mutations that disrupt any step in cell cycle regulation are candidates for the study of cancer-causing genes. For example, mutations in genes that encode the kinases and cyclins may be important in generating malignant transformation in cells. Evidence is accumulating that the G1 checkpoint is aberrant in many forms of cancer, and mutant G1 cyclins and kinases are thought to be the best candidates for cancer-promoting genes. Recently, a form of cyclin D called D1 has been shown to be identical to a gene product that is overexpressed in certain forms of leukemia.

The next section summarizes what is known about the genetics of selected cancers and, wherever possible, relates this information to what has been discovered about the cell cycle and its regulation.

Genes and Cancer

Genetic studies of several different cancers have identified a small number of genes that must be mutated in order to trigger the development of cancer or maintain the growth of malignant cells. It is clear that the two main properties of cancer, uncontrolled cell division and the ability to spread or metastasize, are the result of genetic alterations. As mentioned previously, these alterations can involve large-scale genomic instability, resulting in chromosome loss, rearrangement, or the insertion of foreign (often viral) DNA sequences into loci on human chromosomes. Smaller-scale alterations, such as changes in a nucleotide sequence, or more subtle modifications that alter only the amount of a gene product that is present or the time over which the gene product is active may also be involved. Two interrelated questions arise from considering cancer as a genetic disorder at the cellular level: (1) Are there mutant alleles that predispose an organism or specific cell types to cancer, and (2) if so, how many mutational events are necessary to cause cancer?

Genes That Predispose to Cancer

Single genes do indeed predispose cells to becoming malignant, and it is possible to identify families in which certain forms of cancer are inherited. Many studies have documented families with high frequencies of certain types of cancer, such as breast, colon, or kidney cancers. In most cases, however, it is difficult to identify a clear, simple pattern of inheritance. One example of such a predisposition is the inheritance of **retinoblastoma (RB)**, a cancer of the retinal cells of the eye. Retinoblastoma occurs with a frequency that ranges from 1 in 14,000 to 1 in 20,000, and most often appears between the ages of 1 and

TABLE 21.1 Dominantly Inherited Predispositions to Tumors

Tumor Predisposition Gene	Chromosome
Early onset familial breast cancer	17q
Familial adenomatous polyposis	5q
Familial melanoma	9p
Gorlin syndrome	9q
Hereditary nonpolyposis colon cancer	2p
Li-Fraumeni syndrome	17p
Multiple endocrine neoplasia, type 1	11q
Multiple endocrine neoplasia, type 2	22q
Neurofibromatous, type 1	17q
Neurofibromatous, type 2	22q
Retinoblastoma	13q
von Hippel–Lindau syndrome	3p
Wilms tumor	11p

3 years. Two forms of retinoblastoma are known. One is a familial form (about 40 percent of all cases) inherited as an autosomal dominant trait, though the mutation itself is recessive, which will be explained below. Because the trait is dominant, those who inherit the mutant RB allele (50 percent of family members) are predisposed to develop eye tumors, and in fact, 90 percent of these individuals will develop retinal tumors, usually in both eyes. In addition, those family members who inherit the mutant allele are predisposed to developing other forms of cancer, such as osteosarcoma, a bone cancer, even if they do not develop retinoblastoma.

A second form of retinoblastoma is also known, amounting to 60 percent of all cases. This form is not familial, and tumors develop spontaneously. The sporadic form is characterized by the appearance of tumors only in one eye, and onset occurs at a much later age than in the familial form.

Other types of dominantly inherited familial cancer are also known (Table 21.1). These include **Wilms tumor (WT)**, a cancer of the kidney inherited as an autosomal dominant condition, and **Li-Fraumeni syndrome**, a rare autosomal dominant condition that predisposes to a number of different cancers, including breast cancer.

How Many Mutations Are Needed?

The study of genes that predispose an individual to cancer has allowed scientists to estimate the number and sequence of mutational events necessary to trigger different forms of cancer. By studying the two different types of retinoblastoma, Alfred Knudson and his colleagues developed a model that requires the presence of two mutated copies of the RB gene in the same retinal cell for tumor development. (*i.e.*, the mutant allele is recessive.) In the familial form, one mutant RB allele is inherited, and is carried by all cells of the body, including cells of the retina (Figure 21.4). If the normal allele of the RB gene becomes mutated in a retinal cell, formation of retinal tumors will result. Individuals carrying an inherited mutation of the RB gene are predisposed to develop retinoblastoma, as only one additional mutational event is required to cause tumor formation. This does not happen in all cases, however; some 10 percent of those inheriting a mutant RB gene do not develop cancer. This model explains how the mutation itself can act as a recessive trait (both alleles must be inactive to see the mutant phenotype), and how carrying one mutant allele acts as a dominant trait in *predisposing* an individual to cancer (they need only one additional mutation).

In sporadic cases, two independent mutations of the RB gene must occur in the same retinal cell for a tumor to develop. As might be expected, these events are less frequent and occur at a later age, so that sporadic forms of retinoblastoma are more likely to occur in a single eye, and at a later age than familial forms.

Similar studies on the predisposition to other cancers have led to the conclusion that the number of mutations necessary for the development of cancer ranges from 2 to perhaps as many as 20 (Table 21.2).

Tumor Suppressor Genes

In general, mitosis can be regulated in two ways: (1) by genes that normally function to suppress cell division, and (2) by genes that normally function to promote cell division. The first class, called **tumor suppressor genes**, inactivates or represses passage through the cell cycle and the resulting cell division. These genes and/or their gene products must be absent or inactive for cell division to take place. If these genes become permanently inactivated or lost through mutation, control over cell division is lost, and the cell begins to proliferate in an uncontrollable fashion.

Genes of the second class, called **proto-oncogenes**, normally function to promote cell division. These genes can be "off" or "on," and when they are "on," they promote cell division. To halt cell division, these genes and/or their gene products must be inactivated. If proto-oncogenes become permanently switched on, then uncontrolled cell division occurs, leading to tumor formation. The mutant forms of proto-oncogenes are known as **oncogenes**.

We shall consider some examples of how mutations in tumor suppressor genes can lead to a loss of control over cell division and the development of cancer. Following that discussion, we will consider proto-oncogenes and oncogenes, how the normal alleles act to regulate cell division, and how the mutant oncogenes work to promote tumor formation.

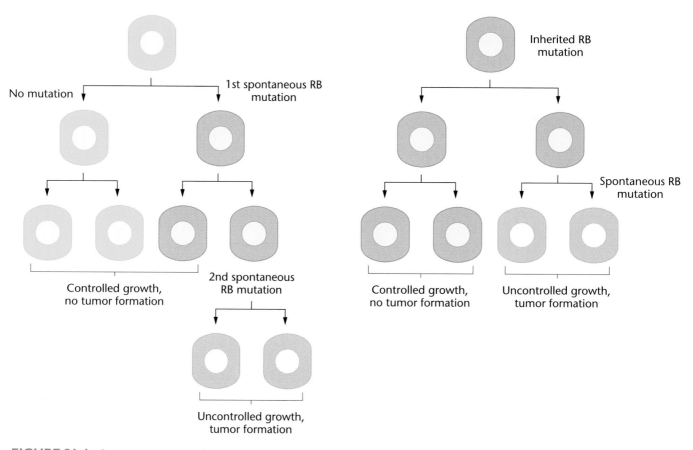

FIGURE 21.4   In spontaneous cases of retinoblastoma (left), two mutations in the retinoblastoma gene are acquired in a single cell, causing uncontrolled cell growth and division, resulting in tumor formation. In familial cases of retinoblastoma (at right), one mutation is inherited and present in all cells. A second mutation at the retinoblastoma locus in any retinal cell will result in uncontrolled cell growth and tumor formation.

Retinoblastoma (RB)

As described previously, retinoblastoma (RB) is a tumor of the retinal layer of the eye. The RB gene, located on chromosome 13, encodes a protein (pRb) 928 amino acids long, which is confined to the nucleus. The pRb is present in all cell and tissue types examined so far, and is found in both resting (G0) cells and actively dividing cells. Moreover, pRb is present at all stages of the cell cycle. Because the gene is expressed at all times, regulation occurs by modifying pRb by adding or removing phosphate groups. The pRb is phosphorylated in the S

and the G2/M stages of the cell cycle, but is not phosphorylated in the G0 and G1 stages of the cycle.

The observation that pRb alternately has phosphate groups added and removed, and that this pattern of modification occurs in synchrony with phase of the cell cycle, suggests that pRb may be part of a control point in G1. In resting cells (G0 or G1), when pRb is not phosphorylated, the protein is active and stops passage through the cell cycle. Just after the cell enters S phase, pRb is modified by the addition of phosphate groups and becomes inactive. When pRb activity is suppressed, cells pass through the S phase and the subsequent events leading to

TABLE 21.2   Number of Mutations Associated with Some Cancers

Cancer	Chromosome Sites	Minimum Number of Mutations Required
Retinoblastoma	13q	2
Wilms tumor	11p	2
Colon cancer	5p, 12p, 17p, 18q	4–5
Small-cell lung cancer	3p, 11p, 13q, 17p	10–15

mitosis. Direct evidence for the role of pRb in regulating the cell cycle comes from several types of experiments.

Cultured osteosarcoma (a form of bone cancer) cells carry RB mutations, and these cells do not produce pRb. When osteosarcoma cells are injected into a cancer-prone strain of mice, tumors are formed. If a normal RB gene is transferred to the cancer cells, pRb is produced and no tumors are formed when these genetically modified cells are injected into mice.

In a separate two-step experiment, a normal RB gene was transferred into osteosarcoma cells grown in culture. These cells produced pRb and stopped cell division. When D or E cyclins were added to these pRb-blocked cells, cell division resumed. Analysis of pRb after the addition of cyclins indicates that the addition of cyclins causes phosphorylation of pRb, presumably by activating CDK1 or another kinase. These results demonstrate that the presence of active pRb stops cell division, that pRb is a target of a G1 cyclin/CDK complex, and that inactivation of pRb allows passage through the cell cycle and subsequent division, confirming the role of pRb in G1 control.

At the molecular level, pRb, which is found only in the nucleus, interacts with a protein called E2F, a well-characterized transcription factor (Figure 21.5). In G1, pRb binds to and inactivates E2F (recall from Chapter 19 that transcription factors bind to specific regions of DNA and regulate transcription). As the cell moves from G1 to S, the CDK/cyclin complexes add phosphate groups to pRb, causing it to release E2F. The E2F is then free to transcribe genes required for progression through the cell cycle. This makes pRb an important link between the cell cycle and gene transcription. If pRb is missing or inacti-

vated, as in retinoblastoma, E2F is not regulated and remains active. This allows permanent expression of genes required for progression through S and cell division, resulting in uncontrolled cell growth.

More recent results suggest, however, that while this explanation for the mechanism of pRb action is consistent with its involvement in tumor formation, it may not explain all the functions of pRb. Knockout mice (see Chapter 16) with no RB genes develop normally for 12 or 13 days (normally mice are born after 21 days of development). Because these embryos pass through many rounds of cell division, control of the cell cycle in G1 may be shared by pRb and other proteins, or other proteins can assume this role in the absence of pRb.

Wilms Tumor

Wilms tumor (WT) is a cancer of the kidney found primarily in children. It occurs with a frequency of about 1 in 10,000 births, and like retinoblastoma, it is found in two forms, a noninherited spontaneous form and a familial form conferring a predisposition to WT. The familial predisposition is inherited as an autosomal dominant trait. According to the model developed by Alfred Knudson and colleagues, familial cases inherit one mutant allele through the germ line and develop a second mutation in the remaining normal allele in a somatic cell. Sporadic cases, on the other hand, require two independent mutations of the WT gene within the same cell, and they more often develop tumors involving only one kidney. Because mutation of the gene and/or the loss of a functional gene product are both associated with the development of tumors, the normal gene is regarded as a tumor suppressor.

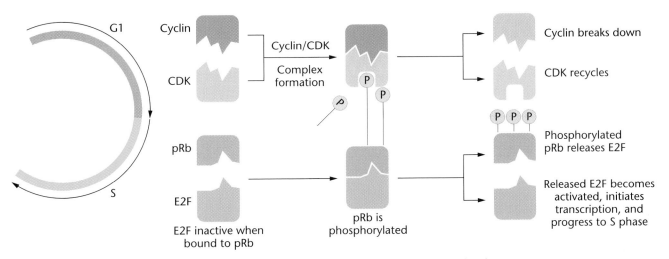

FIGURE 21.5   In the nucleus during G1, pRb, the product of the retinoblastoma gene, interacts with and inactivates transcription factor E2F. As the cell moves from G1 to S, a CDK/cyclin complex forms and adds phosphate groups to pRb. As pRb becomes hyperphosphorylated, E2F is released and becomes transcriptionally active, allowing the cell to pass through S phase. Phosphorylation of pRb is transitory; as cyclin is degraded, phosphorylation declines. The identity of the G1 cyclin is uncertain.

The WT gene has been mapped to the short arm of chromosome 11. The gene product encoded by the WT gene contains four contiguous zinc finger domains, motifs characteristic of DNA binding proteins that regulate transcription. Further, the amino acid sequence upstream from the zinc fingers is similar to other proteins that are known transcription factors.

In contrast to RB, the WT gene has a very restricted pattern of expression; it is active only in a single cell type: the mesenchymal cells of the fetal kidney, and only during the brief time when the nephron (the basic filtration unit of the kidney) is being formed. The only other cell type to express the WT gene is the tumorous nephroblastoma cell. Expression of this gene is barely detectable in cells of the adult kidney and is absent in all other adult cell and tissue types tested.

The structure of the WT gene product, its pattern and time of expression, and its mutant phenotype all suggest that the WT gene encodes a nuclear protein that functions to turn off genes that sustain cell proliferation. Alternatively, the gene product may switch on genes that begin the process of differentiation of the mesenchymal cells into kidney structures. In either case, in Wilms tumor, the mutant gene is switched on at the appropriate time in the appropriate cells, but the altered gene product is unable to regulate its target genes, resulting in continued proliferation of the mesenchymal cells, aberrant differentiation, and tumor formation (Figure 21.6).

Although both the RB and WT genes encode proteins restricted to the nucleus and normally involved in the suppression of tumor formation, conspicuous differences exist in their properties and modes of action. The RB protein does not bind to DNA; instead, it interacts with other nuclear proteins that are transcription factors. The WT protein has all the characteristics of a DNA-binding protein that regulates cell division by acting as a transcription factor. The RB protein is expressed in all dividing cells, whereas expression of WT is restricted to a single cell type during a restricted period of prenatal development. RB is a general regulator of cell division; WT

is a cell- or tissue-specific regulator of gene activity during fetal or neonatal development.

Breast Cancer

The recent discovery of a breast cancer gene (*BRCA1*) and its gene product demonstrates that tumor suppressor genes can act outside the nucleus to regulate cell division. Mutations in *BRCA1*, a gene mapped to chromosome 17q, are associated with about half of all hereditary forms of breast cancer (which account for about 10 percent of all cases). Among women with *BRCA1* mutations, about 80 to 90 percent will develop breast cancer; these women are also at increased risk for ovarian cancer.

BRCA1, discovered in 1994, encodes a protein 1865 amino acids in length. The protein is expressed in epithelial cells of the mammary gland and in several other tissues. The newly synthesized protein passes through the Golgi apparatus, is packaged into vesicles, and is secreted from the cell. The secreted protein binds to receptors on the surface of epithelial cells and generates an intracellular signal cascade that inhibits cell division. The protein also acts to inhibit growth of ovarian cells. Loss of the protein by mutation can contribute to the formation of breast tumors.

A second breast cancer gene, *BRCA2*, which maps to chromosome 13q, encodes a protein similar to the *BRCA1* gene product, indicating that these genes may belong to a new class of tumor suppressors that control the cell cycle at the plasma membrane.

The *p53* Gene and the Cell Cycle

The *p53* gene encodes a nuclear protein that acts as a transcription factor. Normally, *p53* is a tumor-suppressor gene that controls passage of the cell from G1 into S phase of the cell cycle. Mutations of *p53* are found in a wide range of cancers, including breast, lung, bladder, and colon cancers. It is estimated that over half of all cancers are associated with mutations in the *p53* gene, sug-

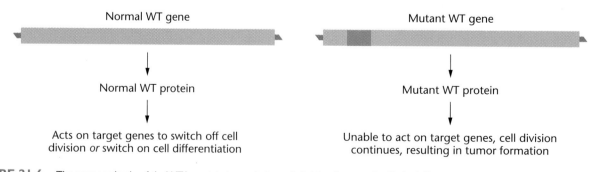

FIGURE 21.6 The proposed role of the WT1 protein in regulating cell division. In normal cells (at left), the WT1 protein is produced and acts on a set of target genes to switch off cell division or switch on cell differentiation. Any mutation that causes the loss or inactivation of the WT1 protein (at right) will result in a failure to regulate the target genes, allowing cell division and tumor formation to occur.

gesting that *p53* controls a key event in cell proliferation and is not involved in a cell- or tissue-specific form of regulation. Inherited mutations in the *p53* gene are associated with the Li-Fraumeni syndrome, an autosomal dominant trait associated with a predisposition to develop cancers in several tissues at a high frequency.

Normal cells contain low levels of *p53* protein, but levels rise dramatically after irradiation of cells by ultraviolet light (UV). Irradiated cells temporarily arrest in G1 to allow repair of DNA damaged by UV light. Cells lacking functional *p53* protein are unable to arrest in G1 following irradiation, and they move immediately from G1 into S. These cells do not repair DNA damage caused by UV; as a result, they have a high rate of mutation. Studies have led to the conclusion that *p53* controls passage through the cell cycle to ensure that DNA damage is repaired before the cell enters S phase. In this role, *p53* is often referred to as the guardian of the genome.

Recently, *p53* has been shown to have a role in cell death following UV irradiation. Following exposure to UV light some cells enter a programmed series of steps leading to cell death or **apoptosis**. This program occurs under the direction of the *p53* gene, which kills the irradiated cell instead of repairing its damaged genome. This genetic program should be familiar to anyone who has suffered a severe sunburn, followed by peeling skin a few days later. The dead cells have undergone apoptosis induced by exposure to the ultraviolet rays of the sun.

In cells lacking a functional *p53* gene, UV irradiation is not followed by apoptosis. Such cells may survive, escape cell-cycle control, and undergo further mutations leading to the formation of a cancerous lesion on the skin.

The central role of the *p53* gene in controlling the cell cycle emphasizes the relationship between cancer and the cell cycle outlined in the chapter introduction. It also highlights the relationship between genes that regulate cell growth and cancer.

Oncogenes

Oncogenes induce or maintain uncontrolled cellular proliferation associated with cancer. The existence of specific genes associated with the transformation of normal cells into cancerous cells was inferred by Francis Peyton Rous in 1910.

Rous Sarcoma Virus and Oncogenes

Using cells from a tumor of chickens known as a **sarcoma**, Rous injected cell-free extracts from these tumors into healthy chickens and induced the formation of sarcomas. He postulated the existence of a "filterable agent" that was responsible for transmitting the disease.

Decades later, his filterable agent was shown by other investigators to be a virus, now known as the **Rous sarcoma virus (RSV)**. Somewhat belatedly, Rous received the Nobel Prize in 1966 for his pioneering work establishing a relationship between viruses and cancer.

The Rous sarcoma virus belongs to a group of oncogenic viruses in which single-stranded RNA molecules serve as the genetic material. These viruses use an enzyme called **reverse transcriptase** to convert the RNA genome into a single-stranded DNA molecule. The enzyme then uses the single-stranded DNA as a template to synthesize the complementary strand, creating a double-stranded DNA molecule. The DNA can be integrated into the genome of the infected cell, forming a **provirus.** At a later time, the DNA can be transcribed into RNA, which can be translated into viral proteins. Packaging of the RNA molecules into these proteins forms new virus particles. Because these viruses "reverse" the flow of genetic information, they are called **retroviruses.**

In RSV, the tumor-forming ability results from a single gene, called the *src* gene, present in the viral genome. This gene, responsible for the induction of tumor formation in chicken cells, is designated as an oncogene, and retroviruses that carry oncogenes are known as **acute transforming viruses**. Other retroviruses can cause tumor formation but do not carry oncogenes. These retroviruses induce the activity of cellular genes that bring about tumor formation. They are designated as **nonacute** or **nondefective viruses**.

Oncogenes carried by acute transforming viruses are acquired from the host's genome during infection, when a portion of the viral genome is exchanged for a cellular proto-oncogene (Figure 21.7). The oncogenes (*onc*) carried by retroviruses are called *v-onc*, or an oncogene, and the cellular version of the gene is called a *c-onc* gene or proto-oncogene. Retroviruses that carry a *v-onc* gene are able to infect and transform a specific type of host cell into a tumor cell. For RSV, the oncogene captured from the chicken genome is called *v-src*, and it confers the ability to transform chicken cells into tumorous growths known as sarcomas. The cellular version of the same gene, found in the chicken genome, is called *c-src*. More than 20 oncogenes have been identified by their presence in retroviral genomes, and over 50 oncogenes have been identified. Some of these are listed in Table 21.3.

Mutations and Oncogenes

Two questions about oncogenes come to mind: (1) Are the *v-onc* and *c-onc* versions of a gene different, and (2) how do oncogenes bring about cellular transformation and tumor formation? In some cases, comparison of *v-onc* DNA sequences with the corresponding *c-onc* sequence (such as *v-ras, v-mos*) shows only minimal differences, probably generated by point mutations. In other cases, segments of

Nonacute retrovirus

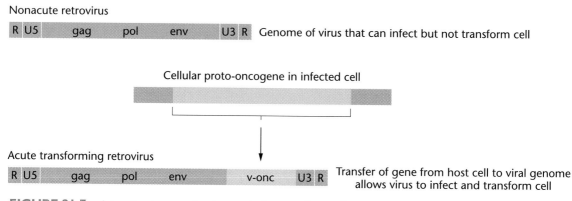

Cellular proto-oncogene in infected cell

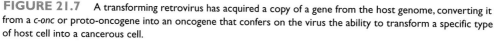

FIGURE 21.7   A transforming retrovirus has acquired a copy of a gene from the host genome, converting it from a *c-onc* or proto-oncogene into an oncogene that confers on the virus the ability to transform a specific type of host cell into a cancerous cell.

the *c-onc* sequence have been replaced or lost. In *v-src*, a string of 19 amino acids of the *c-onc* sequence has been replaced with a sequence of 12 different amino acids. In *v-erb*, the *v-onc* version has lost the N-terminus portion of the gene (see Table 21.4). Mutations, therefore, play a role in differentiating some *v-onc* sequences from their *c-onc* counterparts. However, not all oncogenes are mutant versions of cellular genes carried by retroviruses. We must, therefore, consider a broad range of mechanisms in oncogenetic events.

Oncogenes and Gene Expression

At least three mechanisms can explain how proto-oncogenes are converted into oncogenes. These include: **point mutations**, **translocations**, and **overexpression** (see Table 21.5). Although some of these are mediated by viruses, others are generated solely by intracellular events in the absence of retroviruses.

The *ras* gene family, which encodes a 189 amino acid protein involved in the transduction of signals across the plasma membrane, illustrates how single nucleotide changes can convert a *c-onc* sequence into the oncogenic version of the gene. The normal *ras* protein functions as a molecular switch, alternating between an on-and-off state (Figure 21.8). In some cases, the mutant *ras* protein becomes stuck in the "on" position, stimulating cell growth. In some tumors, this event is the result of a somatic mutation and is not mediated by a retrovirus. Comparison of the amino acid sequence of *ras* proteins from a number of different human carcinomas reveals that *ras* mutations involve single amino acid substitutions at either position 12 or 61 (Figure 21.9).

One of the well-characterized examples of oncogene activation by translocation is that of *c-abl*, an oncogene associated with chronic myelogenous leukemia (CML). In this case, described in detail in a later section, the translocation results in altered gene activity that causes tumor formation.

TABLE 21.3   Oncogenes

c-onc	Origin	Species	Human Chromosome Location
A. Cellular Oncogenes with Viral Equivalents			
src	Rous sarcoma virus	Chicken	20
fos	FBJ osteosarcoma virus	Mouse	2
sis	Simian sarcoma virus	Monkey	22
fes	ST feline virus	Cat	15
abf	Abelson murine leukemia virus	Mouse	9
erbB	Avian erythrobeastosis virus	Chicken	7
Ha-ras-1	Harvey murine sarcoma virus	Mouse	11
myc	Avian MC29 myelocytomatosis virus	Chicken	8
B. Oncogenes Without Viral Equivalents			
N-ras	Neuroblastoma, leukemia	Human	1
N-myc	Neuroblastoma	Human	12
neu	Neuroglioblastoma	Human, rat	17
man	Mammary carcinoma	Human, mouse	2

TABLE 21.4   Structural Alterations in Oncogenes

Gene	Number of Codons	Number of Changed Amino Acids from *c-onc* gene	Region Missing in *v-onc* Protein
H-ras	189	3	None
K-ras	189	7	None
mos	369	11	None
myc	417	2	None
erbA	408	22	Deletion at N terminus
src	533	16	Deletion at C terminus
erbB	1210	49	Deletion at N terminus and C terminus

TABLE 21.5   Conversion of Proto-Oncogenes to Oncogenes

Mechanism	Example
Point mutation	*ras*
Translocation	*abl*
Overexpression of gene product	
New promoter by viral insertion	*mos, myb*
New enhancer by viral insertion	*myc*
Amplification of proto-oncogene	*myc*

FIGURE 21.8   A three-dimensional computer-generated image of *ras* proteins in two different conformations. Normal *ras* proteins act as molecular switches controlling cell growth and differentiation. The switch is "on" when GTP binds to the protein, and "off" when the GTP is hydrolyzed to GDP. Switching the protein between states alters the conformation of the protein in the two regions shown in blue and yellow. Oncogenic mutations of *ras* are stuck in the "on" state, continuously signaling for cell growth.

At least three separate mechanisms of proto-oncogene activation are associated with overexpression. First, the *c-onc* may acquire a new promoter, causing an increase in the level of transcript production, or causing a silent locus to become activated. This is the case in avian leukosis, where strong viral promoters are inserted adjacent to a proto-oncogene, causing an increase in mRNA production and the amount of the gene product. A second mechanism of overexpression involves the acquisition of new upstream regulatory sequences including enhancers. The third mechanism involves the amplification of the proto-oncogene. In human tumors, members of the *myc* family of oncogenes are frequently amplified. The *c-myc* proto-oncogene is found amplified up to several hundred copies in some human tumors.

The protein products of proto-oncogenes are found in the plasma membrane, cytoplasm, and nucleus (Table 21.6). Although their specific functions vary widely, all products characterized to date alter gene expression in a direct or an indirect manner.

Metastasis Is Genetically Controlled

Cancer cells have lost the ability to regulate their own growth and division. As a result, they can develop into malignant tumors that invade neighboring tissue. Sometimes, cancer cells detach from the primary tumor and settle elsewhere in the body, where they grow and divide, producing secondary tumors. This process, called **metastasis**, is often the cause of death of cancer patients.

The Spread of Cancer Cells

A metastatic cancer cell can spread from a primary tumor by entering the blood or lymphatic circulatory system. These cells are carried in the circulation until they become lodged in a capillary bed. Normally, more than 99 percent of the cells die. The surviving cells invade tissue adjacent to the capillary bed and begin dividing to form a secondary tumor. To reach a new site, the tumor cells pass through the layer of epithelial cells lining the interior

TABLE 21.6 Cellular Location of *c-onc* and *v-onc* Proteins

Gene	Location of *c-onc* Protein	Location of *v-onc* Protein
src	Membranes	Membranes
ras	Membranes	Membranes
myc	Nucleus	Nucleus
fps	Cytoplasm	Cytoplasm and membranes
abl	Nucleus	Cytoplasm
erbB	Plasma membrane	Plasma membrane and Golgi

wall of the capillary (or lymph vessel) and penetrate the surrounding extracellular matrices.

The **extracellular matrix** is a meshwork of proteins and carbohydrate molecules separating tissues; it acts as a scaffold for tissue growth and inhibits migration of cells. To establish a secondary tumor, metastatic cells secrete enzymes that digest proteins in the basement membrane, creating holes through which they can move. The cell "tunnels" through the matrix, enters new tissue, and establishes a secondary tumor.

The invasive ability of tumor cells is also a property of some normal cell types. The implantation of the embryo in the uterine wall during pregnancy requires cell migration across the extracellular matrix. Normally, white blood cells reach the site of infection by penetrating capillary walls. The mechanism of invasion is probably the same in normal cells as it is in cancer cells. The difference is that in normal cells, the invasive ability is controlled and tightly regulated, while in tumor cells, regulation has been lost.

Metastasis and Abnormal Gene Regulation

Metastasis may be regulated by genes that encode or regulate matrix-cutting enzymes. Recent studies have shown that protein-cutting enzymes called **metalloproteinases** are necessary for cell invasion. Tumor cell lines with high metastatic ability produce greater amounts of certain metalloproteinases than do tumor cell lines with low invasive properties. Several studies have shown that activated metalloproteinases are inhibited by the presence of a tissue inhibitor of metalloproteinase (TIMP). Normal cells produce TIMP and thereby suppress inappropriate activity of metalloproteinase. In tumor cells, TIMP inhibits metalloproteinase only when the number of TIMP molecules is greater than the number of metallopro-

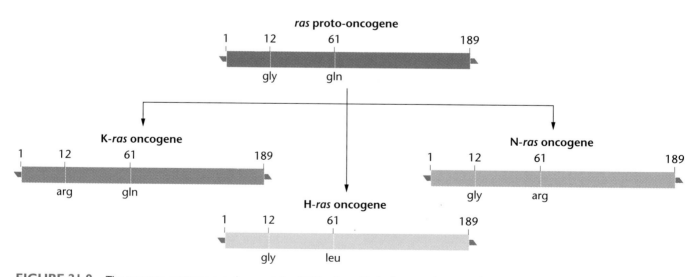

FIGURE 21.9 The *ras* proto-oncogene encodes a protein of 189 amino acids. In the normal protein, glycine is encoded at position 12, and glutamine at position 61. Analysis of *ras* oncogene proteins from several tumors shows a single amino acid substitution at one of these positions. A single amino acid substitution resulting from a single base change can convert a proto-oncogene into a tumor-promoting oncogene.

teinase molecules. On the basis of its action and regulation, TIMP can be considered as a protein that suppresses metastasis.

A Genetic Model for Colon Cancer

Even in the small number of cases where it has been studied in detail, it is clear that cancer is a multistep process resulting from a number of specific genetic alterations. Although the studies of tumors such as retinoblastoma and Wilms tumor have been useful in establishing that a limited number of steps are required to transform a normal cell into a malignant one and in identifying some of the molecules involved, these tumors are of limited value in establishing the order of the genetic events leading to tumor formation and metastasis. For these questions, the study of colon cancer offers several advantages. First, malignant tumors of the colon and rectum develop from preexisting benign tumors, and cells from tumors at all stages of development are available for study. Moreover, two forms of colon cancer are known: one in which a predisposition is inherited in an autosomal dominant fashion (known as **familial adenomatous polyposis** or **FAP**), and one that is completely spontaneous, making it possible to study the interaction of genetic and environmental factors in the genesis of tumors.

Through an analysis of mutations in tumors at various stages, ranging from small benign growths known as **adenomas**, through intermediate stages, to **malignant tumors** and **metastatic tumors**, it has been possible to define the number and nature of the genetic and molecular steps involved in changing normal intestinal epithelial cells into tumor cells and to develop a genetic model for colon cancer. This model is shown in Figure 21.10. The first feature of this model is that multiple mutations are required. Mutations in at least four specific genes are needed to bring about malignant growth. If fewer changes are present, benign growths or intermediate stages of tumor formation result. Second, based on the analysis of many tumors, the order of mutations usually follows a *preferred* sequence (as indicated in the figure). Ultimately, however, it is the accumulation of a critical number of specific mutations that is more important than the order in which they occur.

Colon Cancer Develops in Stages

The first mutation in the sequence occurs in a normal epithelial cell and results in the formation of one or more benign tumors. In cases of FAP, a first mutation is inherited and results in the development of dozens or hundreds of benign adenomas in the colon and rectum. In spontaneous cases, evidence suggests that the initial mutational event takes place in a single cell and that the resulting adenoma consists of a clone of cells, all of which carry the mutation. This first mutation takes place in a gene called **APC** located on the long arm of chromosome 5. The loss of the corresponding allele on the homologous copy of chromosome 5 is *not* necessary for proliferation and adenoma formation. The relative order of subsequent mutations is shown in Figure 21.10. Mutations in the *ras* oncogene may precede or follow the loss of a segment of chromosome 18p. In either case, the accumulation of these two mutations in adenoma cells with a preexisting mutation on chromosome 5 causes the adenoma to grow larger and develop a number of fingerlike villous outgrowths. Finally, a mutation on 17p, involving the loss or inactivation of *p53*, causes the transition to a cancerous cell. As discussed earlier, mutations in the *p53* gene are pivotal to the development of a number of cancers, including lung, brain, and breast cancers. Recall that the normal allele of *p53* is a tumor suppressor gene that reg-

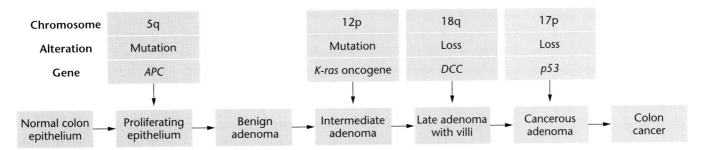

FIGURE 21.10 A model for the step-wise production of colon cancer. The first step is the loss or inactivation of both alleles of the APC gene on chromosome 5. In familial cases, one mutation of the APC gene is inherited. Loss of both alleles leads to the formation of benign adenomas. Subsequent mutations involving genes on chromosomes 12, 17, and 18 in cells of the benign adenomas can lead to a malignant transformation resulting in colon cancer. Although the mutations in chromosomes 12, 17, and 18 usually occur at a later stage than those involving chromosome 5, the sum of the changes is more important than the order in which they occur.

ulates the passage of cells from late G1 to S phase. Metastasis occurs after the formation of colon cancer, and it involves an unknown number of mutational steps.

Genetic and Environmental Factors in Colon Cancer

In sum, the genetic model for colon cancer involves sequential mutations in oncogenes, tumor suppressor genes, and disruption of the cell cycle at a specific transition point, although the nature and function of normal and mutant gene products of the *p53* gene have not yet been identified with certainty. In cases of an inherited predisposition to colon cancer, the first mutation is genetically transmitted; the rest occur by action of environmental agents, pointing up the role of the environment in the development of cancer. This multistep model has applications to other forms of cancer, but many questions remain unanswered, including the normal functions of the genes involved and the molecular mechanisms by which the mutated genes and/or their gene products bring about tumor formation.

Genomic Changes and Cancer

Alterations in chromosome structure and/or number are associated with most forms of cancer. In many cases, the relationship between changes in chromosome number or arrangement and the development of cancer is not clear. For example, individuals with Down syndrome carry an extra copy of chromosome 21. This quantitative alteration in genetic content is associated with a 20-fold increased risk of leukemia as compared to the general population.

TABLE 21.7 Specific Chromosome Aberration and Cancer

Cancer	Chromosome Alteration
Chronic myelogenous leukemia	t(9;22)
Acute promyelocytic leukemia	t(15;17)
Acute lymphocytic leukemia	t(4;11)
Acute myelogenous leukemia	t(8;21)
Prostate cancer	del(10q)
Synovial sarcoma	t(X;18)
Testicular cancer	i(12p)
Retinoblastoma	del(13q)
Wilms tumor	del(11p)

t = translocation, del = deletion, i = inversion

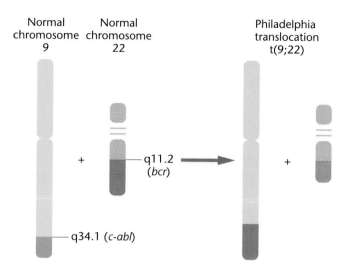

FIGURE 21.11 A reciprocal translocation involving the long arms of chromosomes 9 and 22 results in the production of a characteristic chromosome, the Philadelphia chromosome, which is associated with cases of chronic myelogenous leukemia (CML).

Chromosome Rearrangements and Cancer

In selected cases where specific chromosome rearrangements are associated with cancer, the relationship between the chromosomal aberration and the development and/or maintenance of the cancerous state is known.

The connection between chromosome aberrations and cancer is most clearly seen in leukemias, where the presence of specific chromosome changes is well defined and often diagnostic (Table 21.7). One of the best-studied examples of an association between a chromosome rearrangement and the development of cancer is the translocation between chromosomes 9 and 22 that is associated with **chronic myelocytic leukemia (CML)** (Figure 21.11). Originally, this translocation was described as an abnormal chromosome 21 and called the **Philadelphia chromosome**. Later, it was shown by Janet Rowley that the Philadelphia chromosome results from the exchange of genetic material between chromosomes 9 and 22. This translocation is never seen in a normal cell and is found only in the white blood cells involved in CML. These observations indicate that the translocation is a primary and causal event in the generation of CML, and that this form of cancer may originate from a single cell bearing this translocated chromosome.

Translocations and Hybrid Genes

By examining a large number of cases involving the Philadelphia chromosome, it was possible to establish the exact location of the break points on chromosomes 9 and 22 (Figure 21.12). Genetic mapping studies using recombinant DNA techniques established that a proto-oncogene *c-abl* maps to the breakpoint region on chromosome 9, and

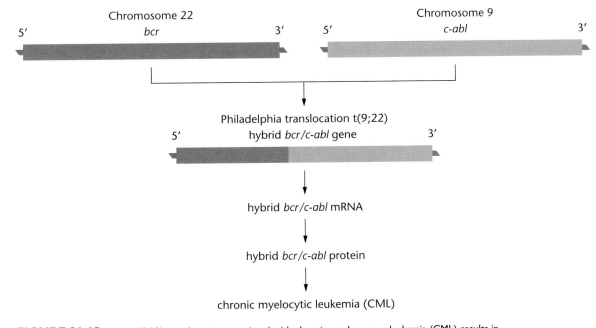

FIGURE 21.12 The t(9;22) translocation associated with chronic myelogenous leukemia (CML) results in the fusion of the *c-abl* oncogene on chromosome 9 with the *bcr* gene on chromosome 22. The normal *c-abl* protein is a kinase and the normal *bcr* protein activates a phosphorylation reaction. The fusion protein is a powerful hybrid that allows cells to escape control of the cell cycle, resulting in leukemia.

that the gene *bcr* maps near the breakpoint on chromosome 22. In the translocation event, all or most of the *c-abl* gene is translocated to a region within the *bcr* gene. This generates a hybrid *bcr/c-abl* gene that is transcriptionally active and produces a hybrid 200-kDa protein product (Figure 21.12) that has been implicated in the generation of CML.

Both cytogenetic and molecular approaches have been applied to the cytogenetic events in translocations involving chromosome 8 [these include t(8;14), t(8;22), and t(2;8)] in a disease related to leukemia known as **lymphoma**. The breakpoint on chromosome 8 in all these tranlocations is the same, and the proto-oncogene *c-myc* has been mapped to this locus. The loci at the breakpoint on the other chromosomes involved in this series of translocations all have immunoglobulin genes at the breakpoints. Movement of the *c-myc* gene to a position near these immunoglobulin genes and activation of the *c-myc* gene lead to the transformation of the lymphoid cells into leukemic cells.

Other genes at translocation breakpoints have also been isolated and characterized. In these cases and in CML, the translocation results in the formation of an abnormal gene product that causes the cell to undergo a malignant transformation even though a second and presumably normal copy of the gene is present and active in synthesizing a normal gene product. If the abnormal gene products at other translocation loci associated with leukemia can be identified, it is hoped that therapeutic strategies can be developed to administer high levels of the normal gene product, or that the abnormal gene product can be inactivated.

Genomic Instability and Cancer

The term *genomic instability* is used to describe events resulting in genomic alterations characteristic of cancer cells. At least three classes of genetic defects can lead to genomic instability: defects in DNA repair and replication, aberrant chromosome segregation, and defects in cell-cycle control. Cultured cells from individuals with Fanconi anemia, ataxia telangiectasia (AT), and Werner syndrome all show an increase in chromosome breaks and translocations, indicating that genomic instability is an heritable trait. The main defect in these syndromes appears to be in DNA repair and/or faulty control of DNA replication.

The model for the development of colon cancer in familial adenomatous polyposis (FAP) illustrates the role of genetic instability in cancer, as evidenced by a number of mutations distributed throughout the genome. A more dramatic relationship between genomic instability and colon cancer has been recently discovered for another form of colon cancer, not associated with polyp formation. This form, hereditary nonpolyposis colon cancer, accounts for up to 15 percent of all cases of colon cancer. The gene responsible for susceptibility to this type of colon cancer, called *FCC*, has been mapped to chromosome 2.

Somewhat surprisingly, malignant cells from affected individuals do not show alterations at a site on chromosome 2, but instead show changes in short, repetitive DNA sequences (called microsatellite DNA or variable nucleotide tandem repeats; see Chapter 17) scattered throughout the genome. These large-scale genomic alterations, representing perhaps thousands of alterations in the genome, indicate that the *FCC* gene may affect the accuracy of DNA replication; when mutant, the *FCC* gene causes widespread genomic instability. The gene has recently been identified and cloned, and it is estimated that the mutant gene may be carried by 1 in 200 individuals in the Western world, making it one of the most common genetic disorders known.

Results from studies on tumor development *in vivo* have established that, in many cases, genomic instability precedes the formation of tumors. Individuals with a chronic regurgitation of stomach acid into the esophagus have a high risk of esophageal cancer. Cytogenetic studies on esophageal cells of those affected by the reflux of stomach acid show that the first detectable change is a progressive alteration in cell cycle control in small clones of cells (descended from a single ancestral cell), followed by the development of aneuploidy, and finally by the appearance of cancer. In addition, preliminary studies on the *FCC* gene indicate that microsatellite instability precedes the formation of colon tumors and may be an early event in cancer development.

Cancer and Environmental Agents

The relationship between environmental agents and the genesis of cancer is often elusive, and in the early stages of investigation this relationship is based on indirect evidence. Such studies often begin with an epidemiological survey, comparing the cancer death rates among different geographic populations. When differences are found in the death rate for a given type of cancer, or a given age group, or a cluster of related occupations such as chemical workers, further work is necessary to identify one or more environmental factors that correlate with these cancer deaths. These correlations are not conclusive, but they serve to identify factors that, upon additional investigation, may turn out to be directly related to the development of cancer. Finally, extensive laboratory investigations may establish the mechanism by which an environmental agent generates cancer. In other cases, such as certain viruses, the relationship between cancer and the environment is more straightforward.

Hepatitis B and Cancer

Epidemiological surveys have shown that individuals who develop a form of liver cancer known as **hepatocellular carcinoma (HCC)** are infected with the **hepatitis B virus (HBV)** (Figure 21.13). In fact, for those carrying HBV, the risk of cancer is increased by a factor of 100. Aside from the risk of cancer, HBV infection is a public health risk affecting some 300 million people worldwide, and is most prevalent in Asia and tropical regions of Africa. Infection produces a wide range of responses, from a chronic, self-limiting infection with few symptoms, to active hepatitis and cirrhosis, to fatal liver disorders and hepatocellular carcinoma. In the last few years, efforts have centered on understanding the mechanism by which HBV replicates and its role in causing liver cancer. The genome of HBV is a mostly double-stranded DNA molecule of 3200 nucleotides. After cellular infection, the DNA moves to the nucleus where it inserts into a chromosome. The viral DNA is transcribed into an RNA molecule and packaged into a viral capsid. The capsid containing the RNA pregenome moves to the cytoplasm where it is reverse-transcribed into a DNA strand, which in turn is made double-stranded. The copied DNA genome is repackaged into a new capsid for release from the cell or returns to the nucleus for another round of replication.

The key to the role of HBV in carcinogenesis apparently resides in its entry into the nucleus. While in the nucleus, the HBV genome can insert itself at many different sites into human chromosomes (Figure 21.14). Insertion often results in cytogenetic alterations of the host genome that include translocations, deletions, or amplification of adjacent regions. As outlined earlier, most forms of cancer are associated with chromosomal rearrangements of these types, and it is possible that such rearrangements trigger the development of a cancerous transformation.

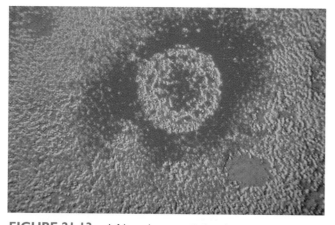

FIGURE 21.13 A false color transmission electron micrograph of a hepatitis B virus. Infection with this virus often results in cancer of the liver. Worldwide, viral infections of all kinds are thought to be responsible for about 15 percent of the total number of cancers.

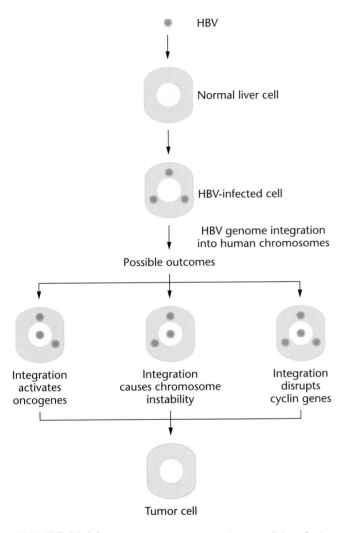

HBV

Normal liver cell

HBV-infected cell

HBV genome integration into human chromosomes

Possible outcomes

Integration activates oncogenes

Integration causes chromosome instability

Integration disrupts cyclin genes

Tumor cell

FIGURE 21.14 The possible outcomes following cellular infection with hepatitis B virus (HBV). Following integration of the viral chromosome into the human genome, oncogenes can be activated, resulting in tumor formation. Integration of the HBV genome can also cause chromosome instability, including the production of translocations and deletions that trigger a malignant cyclin A gene, triggering abnormal regulation of the cell cycle and cellular proliferation, resulting in cancer formation.

Alternatively, the HBV genome has been shown to integrate into the cyclin A gene, and abnormal expression of this gene can disrupt the cell cycle and normal control of proliferation. Similarly, HBV integration has resulted in activation of proto-oncogene loci, and altered expression of such loci has been clearly implicated in the development of malignant cell growth. Thus, a strong link exists between the HBV virus as an environmental agent and the development of cancer in infected individuals.

Environmental Agents

Many surveys of cancer incidence have pointed to the role of the physical environment and personal behavior as factors contributing to the development of cancer. The pre-cise role of the environment in the genesis of cancer can be difficult to ascertain. Obviously, induction of cancer can involve an interaction between the genotype and environmental agents (Table 21.8). The differential expression of a specific genotype in various environments is often neglected in making calculations about the role of the environment in cancer, but it has been estimated that at least 50 percent of all cancers are environmentally induced. Environmental agents responsible for cancer include background levels of radiation, occupational exposure to physical and chemical agents, exposure to sunlight, and personal behavior such as diet and the use of tobacco. To show how environmental factors are identified, let us consider a recent study on the risk of colon cancer.

Epidemiological surveys reveal that cancer of the colon occurs with a much higher frequency in North America and Western Europe than in Asia, Africa, or other parts of the world. When people migrate from low-risk areas to high-risk areas, their rate of colon cancer increases to match that of the native residents. This finding suggests that environmental factors, including diet, may play a role in the incidence of colon cancer. In a long-range health study of more than 88,000 registered nurses across the United States, dietary habits have been monitored by questionnaires since 1980. In that population, 150 cases of colon cancer were observed through 1986. Detailed analysis of dietary intake adjusted for diet composition indicated that the risk of colon cancer was positively associated with the intake of animal fat. Nurses who had daily meals of pork, beef, or lamb had a 2.5-fold higher risk of colon cancer than did those who ate such meals less than once a month. This correlation strongly suggests that animal fat in the diet is an environmental risk factor for colon cancer.

TABLE 21.8 Epidemiology of Cancer in Various Countries

Country	Death Rate per 100,000	
	Male	Female
Australia	212	125
Canada	214	136
Dominican Republic	54	48
Egypt	39	18
England	248	156
Greece	188	103
Israel	174	145
Japan	190	109
Nicaragua	22	35
Portugal	180	108
Singapore	249	130
United States	216	137
Venezuela	135	128

Because a negative correlation existed between eating skinless chicken meat and the incidence of colon cancer, what remains to be sorted out from this study is the role of red meat and/or fat alone as risk factors. Finally, if animal fat is identified as an important risk factor for colon cancer, laboratory studies to identify the mechanisms by which fat brings about a cancerous transformation of the intestinal epithelium will be undertaken. As described in an earlier section, these transformations are associated with mutations involving five or six different genes. However, even in the absence of information about the mechanism of action, it would seem prudent to reduce intake of animal fat in order to reduce the risk of colon cancer.

The accumulated results of research on the role of external factors as a cause of cancer indicate that diet, tobacco, and other agents, including medical and dental X-rays, viruses, drugs, and ultraviolet light, play significant roles in cancer development. It is estimated that 50 percent of all cases of cancer are caused by these agents. It is not the environment in general and pollution in particular that is responsible for a large fraction of cancer cases. Instead, personal choices (e.g., the composition of food in the diet, tobacco use, suntanning) are responsible for a large proportion of all cancers. Education and more judicious choices on the part of individuals could lead to the prevention of a high percentage of all human cancers.

GENETICS, TECHNOLOGY, AND SOCIETY

The Cancer-Yeast Connection: The Seattle Project

Cancer is the second leading cause of death (after heart disease) in North America, with over 500,000 cancer deaths occurring per year in the United States. Since 1971, approximately $25 billion has been earmarked by Congress to combat the disease, and—some critics say—we have little to show for it. In the last 20 years, the cancer death rate has changed very little. Although there have been improvements in some types of therapies and life expectancies following diagnosis have been extended, physicians are limited to three main cancer treatments: surgery, radiation, and chemotherapy.

Cancer chemotherapy, like radiation treatment, is based on the observation that cancer cells are generally more sensitive than normal cells to these toxic agents. The development of new cancer drugs remains an empirical process—testing the toxicity of an interesting compound in normal and cancer cell lines, then in animals, then in clinical trials. The success of any chemotherapeutic drug varies considerably from one cancer type to another. There has been no way to preselect a specific cancer drug to target a specific defect in a specific tumor type. As we shall see below, there is an exciting new prospect on the horizon that may make this possible.

One reason that it has been so difficult to develop anti-cancer drugs is that each cancer contains a unique combination of genetic defects. Anywhere from two to 20 or more mutations are required to transform a cell into a cancerous one. At normal mutation rate (1 in 10^6/cell division/gene), there is a very low probability that an individual should develop cancer. In reality, about 1 in 3 people can expect to get cancer sometime in their lives. Therefore, cells that are destined for cancerous transformation must be more prone to genetic damage than normal cells, accumulating more than the average number of mutations.

Recent research suggests that the initial trigger that propels a cell down the cancerous pathway may be the loss of its ability to repair DNA. Normal cells contain elaborate feedback mechanisms that stop the cell from progressing through the cell cycle until all of its DNA has been replicated and any damage or mutations have been repaired. The checkpoints (see Chapter 2) are dependent on many gene products, such as p53 and pRb. In addition, other gene products detect DNA damage inflicted by radiation, chemicals or spontaneous mutations, repair these defects, and signal the DNA repair status to the checkpoint mechanisms. A mutation in a gene that is necessary for excision repair, mismatch repair, or a cell cycle checkpoint will result in a genetically unstable cell which is prone to accumulate high numbers of mutations in other genes, including those that transform cells to the malignant state.

One of the more surprising outcomes of the cloning of human genes is that many of the repair, checkpoint, and cell cycle genes are very similar in all eukaryotes—from yeast to humans. For example, two human genes have been identified that, when mutated, cause hereditary non-polyposis colon cancer (HNPCC). The products of these genes (*hMLH1, hMLH2*) are components of the DNA mismatch repair pathway. Yeast cells contain two homologous genes that are also involved in

CHAPTER SUMMARY

1. Cancer can be defined as a genetic disorder at the cellular level that can result from the mutation of a given subset of genes, or from alterations in the timing and amount of gene expression. Although some forms of cancer show familial patterns of inheritance, few show clear evidence for Mendelian inheritance.

2. Cancer results from the uncontrolled proliferation of cells and from the ability of such cells to metastasize or migrate to other sites and form secondary tumors. Mutant forms of genes involved in the regulation of the cell cycle are obvious candidates for cancer-causing genes. The cell cycle is apparently regulated at two sites, with two types of gene products primarily involved: cyclins and kinases. The actions of these genes and the genes they control are important in regulating cell division, and links between these genes and the process of tumor formation are being discovered.

3. Although Mendelian patterns of cancer development are not clear-cut, there are genes that predispose to cancer. Study of these genes, such as retinoblastoma, has provided insight into the number and sequence of mutations that result in the development of tumors.

4. Tumor suppressor genes normally act to suppress cell division. When these genes are mutated or have altered expression, control over cell division is lost. Molecular analysis of two such genes, retinoblastoma (RB) and Wilms tumor (WT), indicates that these genes work to control gene expression in different ways, either by affecting the activity of transcription factors or by themselves acting as transcription factors.

mismatch repair. A second example is that of the *ATM* gene. Mutations in the human *ATM* gene lead to ataxia telangiectasia, an autosomal hereditary disease that exhibits a number of defects including cerebellar degeneration, immune deficiencies, radiation sensitivity, chromosome abnormalities, and a 100-fold increase in cancer susceptibility. It is estimated that about one percent of the population carry a mutated *ATM* gene that may increase cancer risk. When the ATM gene was cloned, scientists realized that its sequence resembled that of the yeast *MEC1* and *rad3* genes. These yeast genes are necessary for cell cycle checkpoints. Yeast cells with mutations in *rad3* are radiation sensitive and do not arrest in G2 in response to DNA damage. There is now an ever-lengthening list of human cancer genes that have structural and functional homologs in yeast.

This concordance of structure and function between yeast and human genes has led to a novel approach to cancer drug screening. The proposal—called the "**Seattle Project**"—was devised by Leland Hartwell and Stephen Friend. The project takes advantage of the ease with which genetic manipulation can be accomplished in yeast and proposes to use yeast cells as simple models for the defects found in cancer cells. The idea is to generate strains of yeast that are defective in one (or a few) of the genes involved in tumor development in mammalian cells (e.g., *MEC1*). The yeast cells will exhibit defects in DNA repair or cell cycle checkpoints that may make them susceptible to specific types of chemotherapeutic drug therapy. The researchers hope to determine which drugs work against specific genetic defects, by screening as many as 40,000 potential anti-cancer drugs against about 30 specific mutant yeast strains. Because yeast cells are much easier to grow than cancer cell lines, the whole process can be automated using equipment that can read the amount of yeast cell growth that occurs after a drug is added to the yeast culture. Once initial screening identifies promising specific drugs, the drugs will be further tested in cancer cell lines. The ultimate goal of the yeast-cancer screen is to match specific chemotherapeutic compounds with specific cell cycle defects and thereby increase the specificity of cancer cell killing for each type of cancer. Ultimately, it is anticipated that what is learned using yeast will be applicable to cancer therapy in humans.

The Seattle Project, which is just being launched, is an example of how basic research is being applied to solve clinically important problems. If successful, it could lead to a day when scientists will be able to analyze each tumor individually for its cell cycle and DNA repair defects, and then select chemotherapeutic drugs that target those defects.

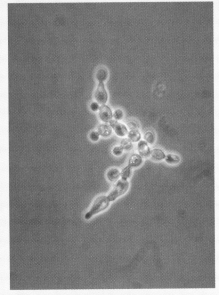

Mutant yeast cells blocked in the transition from G2 to M.

5. Oncogenes are genes that normally function to maintain cell division, and these genes must be mutated or inactivated to halt cell division. If these genes escape control and become permanently switched on, cell division occurs in an uncontrolled fashion.

6. The cells of most tumors have visible chromosomal alterations, and the study of these aberrations has provided some insight into the steps involved in the development of cancer. This relationship between cancer and chromosome alterations has been best studied in leukemias, where the formation of hybrid genes or substitution of regulatory sequences is associated with the transformation of normal cells into malignant tumors.

7. Although cancer is the result of genetic alterations, it appears that the environment is an important factor in cancer induction. Environmental agents include occupational exposure to physical or chemical substances, viruses, diet, and other personal choices such as the use of tobacco and suntanning.

KEY TERMS

acute transforming viruses, 599
adenoma, 603
apoptosis, 599
chronic myelocytic leukemia (CML), 604
cyclins, 593
extracellular matrix, 602
familial adenomatous polyposis (FAP), 603
hepatitis B virus (HBV), 606
hepatocellular carcinoma (HCC), 606

Li-Fraumeni syndrome, 595
lymphoma, 605
malignant tumor, 603
metalloproteinases, 602
metastatic tumor, 603
metastasis, 601
nonacute (nondefective) viruses, 599
oncogenes, 595
overexpression, 600
Philadelphia chromosome, 604
point mutations, 600

protein kinases, 593
proto-oncogene, 595
provirus, 599
retinoblastoma (RB), 594
retrovirus, 599
reverse transcriptase, 599
Rous sarcoma virus, 599
sarcoma, 599
translocations, 600
tumor suppressor genes, 595
Wilms tumor (WT), 595

INSIGHTS AND SOLUTIONS

1. In disorders such as retinoblastoma, a mutation in one allele of the retinoblastoma (RB) gene can be inherited from the germ line, causing an autosomal dominant predisposition to the development of eye tumors. To develop tumors, a somatic mutation in the second copy of the RB gene is necessary, indicating that the mutation itself acts as a recessive trait. In sporadic cases, two independent mutational events, involving both RB alleles, are necessary for tumor formation. Given that the first mutation can be inherited, what are the ways in which a second mutational event can occur?

Solution: In considering how this second mutation can arise, several levels of mutational events need to be considered, including changes in nucleotide sequence and events involving whole chromosomes or chromosome parts. Retinoblastoma results when both copies of the RB locus have been lost or inactivated. With this in mind, perhaps the best way to proceed is to prepare a list of the phenomena that can result in a mutational loss or inactivation of a gene.

One way for the second RB mutation to occur is by a nucleotide alteration that converts the remaining normal RB allele to a mutant form of the gene. This alteration may occur through a base substitution or by a frameshift mutation caused by the insertion or deletion of one or more nucleotides. A second mechanism of mutation can involve loss of the chromosome carrying the normal allele. This event would take place during mitosis, resulting in chromosome 13 monosomy, leaving the mutant copy of the gene as the only RB allele. This mechanism does not necessarily involve loss of the entire chromosome; deletion of the long arm (RB is on 13q) or an interstitial deletion involving the RB locus and some surrounding material would have the same result.

As an alternative, a chromosome aberration involving loss of the normal copy of the RB gene might be followed by a duplication of the chromosome carrying the mutant allele, restoring two copies of chromosome 13 to the cell, but no normal allele of the RB gene would be present. Finally, a recombination event followed by chromosome segregation could produce a homozygous combination of mutant RB alleles.

More can be discovered about the mechanisms involved in RB by analyzing the cells from tumors using a combination of cytogenetic and molecular techniques (such as RFLP analysis and hybridizations to look for deletions) to see which mechanisms are actually found in tumors and to what extent they play a role in generating the second mutation. Although such analysis is still in the preliminary stages, all mechanisms proposed have been found in tumors, indicating that a variety of spontaneous events can bring about the second mutation that triggers retinoblastoma.

PROBLEMS AND DISCUSSION QUESTIONS

1. As a genetic counselor, you are asked to assess the risk for a couple who plans to have children, but where there is a family history of retinoblastoma. In this case, both the husband and wife are phenotypically normal, but the husband has a sister with familial retinoblastoma in both eyes. What is the probability that this couple will have a child with retinoblastoma? Are there any tests that you could recommend to help in this assessment?

2. Review the stages of the cell cycle. What events occur in each of these stages? Which stage is most variable in length?

3. Where are the major regulatory points in the cell cycle?

4. Progression through the cell cycle depends on the interaction of two types of regulatory proteins: kinases and cyclins. List the functions of each and describe how they interact with each other to cause cells to move through the cell cycle.

5. What is the difference between saying that cancer is inherited and saying that predisposition to cancer is inherited?

6. Define tumor suppressor genes. Why are most tumor suppressor genes expected to be recessive?

7. Review the differences among transcriptional activity, tissue distribution, and function of the tumor suppressor genes associated with retinoblastoma and Wilms tumor. What properties do they have in common? Which are most different?

8. Distinguish between oncogenes and proto-oncogenes. In what ways can proto-oncogenes be converted to oncogenes?

9. How do translocations such as the Philadelphia chromosome lead to oncogenesis?

10. Given that 50 percent of all cancers are environmentally induced, and most of these are caused by actions such as smoking, suntanning, and diet choices, how much of the money spent on cancer research do you think should be devoted to research and education on the prevention of cancer rather than on finding a cure?

11. The compound benzo[*a*]pyrene is found in cigarette smoke. This compound chemically modifies guanine bases in DNA. Such abnormal bases are typically removed by an enzyme which hydrolyzes the base leaving an apurinic site. If such a site is left unrepaired, an adenine is preferentially inserted *across* from the apurinic site. In a study of lung cancer patients, tumor cells from 15 out of 25 patients had a G to T transversion in a gene called *p53* which has a known role in cancer formation. You are asked to testify as an expert witness in the following court case: A widow of a man who died of lung cancer is suing R. J. Reynolds for selling tobacco products that killed her husband (who was a lifelong smoker). What do you tell the jury? (Science only; no personal expositions on lawyers or the legal system.) [*Reference: Nature* 350:377–78 (1991).]

EXTRA-SPICY PROBLEMS

About 5 to 10 percent of breast cancer cases are caused by an inherited susceptibility. An *inherited* mutation in a gene called *BRCA1* is thought to account for approximately 80 percent of families with a high incidence of both early-onset breast and ovarian cancer. Table 1 summarizes some of the data that has been collected on *BRCA1* mutations in such families. Table 2 shows neutral polymorphisms found in control families (not showing an increased frequency of breast and ovarian cancer). Study this data and then answer the questions listed below.

12. (a) Note the coding effect of the mutation found in kindred group 2082. This results from a single base pair substitution. Starting with a drawing of the normal double-stranded DNA sequence for this codon (with the 5′ and 3′ ends labeled), show the sequence of events that generated this mutation, assuming that it resulted from an uncorrected mismatch event during DNA replication.

(b) Examine the types of mutations that are listed in Table 1. Is the *BRCA1* gene likely to be a tumor suppressor gene or an oncogene?

TABLE I Predisposing Mutations in *BRCA1*

| | | Mutation | | Frequency in Control Chromosomes |
Kindred	Codon	Nucleotide Change	Coding Effect	
1901	24	−11 bp	Frameshift or splice	0/180
2082	1313		Gln → Stop	0/170
1910	1756	Extra C	Frameshift	0/162
2099	1775	T → G	Met → Arg	0/120
2035	NA*	?	Loss of transcript	NA*

* NA indicates not applicable, as the regulatory mutation is inferred and the position has not been identified.

(c) Although the mutations described in Table 1 are clearly deleterious, causing breast cancer in women at very young ages, each of the kindred groups examined had at least one woman who carried the mutation but lived until age 80 without developing cancer. Name at least two different mechanisms (or variables) that could underlie variation in the expression of a mutant phenotype. Then, *in the context of the current model for cancer formation*, propose an explanation for the incomplete penetrance of this mutation. How do these variables relate to this explanation? [*Reference: Science* 266:66–71 (1994).]

13. (a) Examine Table 2. What is meant by a neutral polymorphism? What is the significance of this table in the context of examining a family or population for *BRCA1* mutations that predispose an individual to cancer?
    (b) Examine Table 2. Is the polymorphism PM2 likely to result in a neutral missense mutation or a silent mutation?
    (c) Answer question (c) for the polymorphism PM3. [*Reference: Science* 266:66–71 (1994).]

TABLE 2 Neutral Polymorphisms in *BRCA1*

Name	Codon Location	Base in Codon[2]	Frequency in Control Chromosomes[1]			
			A	C	G	T
PM1	317	2	152	0	10	0
PM6	878	2	0	55	0	100
PM7	1190	2	109	0	53	0
PM2	1443	3	0	115	0	58
PM3	1619	1	116	0	52	0

[1] The number of chromosomes with a particular base at the indicated polymorphic site (A, C, G, or T) is shown.
[2] That is, position 1, 2, or 3 of the codon.

SELECTED READINGS

AMES, B., MAGRAW, R., and GOLD, L. 1990. Ranking possible cancer hazards. *Science* 236:71–80.

AMES, B., PROFET, M., and GOLD, L. 1990. Dietary pesticides (99.99% all natural). *Proc. Natl. Acad. Sci. USA* 87:7777–81.

BEIJERSBERGEN, R. L., and BERNARDS, R. 1996. Cell cycle regulation by the retinoblastoma family of growth inhibitory proteins. *Biochim. Biophys. Acta* 1287:103–20.

BENEDICT, W., XU, H., HU, S., and TAKAHASHI, R. 1990. Role of the retinoblastoma gene in the initiation and progression of a human cancer. *J. Clin. Invest.* 85:988–93.

BIECHE, I., and LIDEREAU, R. 1995. Genetic alterations in breast cancer. *Genes Chromosomes Cancer* 14:227–51.

COTTER, F. 1993. Molecular pathology of lymphomas. *Cancer Surv.* 16:157–74.

DAMM, K. 1993. ErbA: Tumor suppressor turned oncogene? *FASEB J.* 7:904–909.

EASTON, D., BISHOP, D., FORD, D., CROCKFORD, G., and THE BREAST CANCER LINKAGE CONSORTIUM. 1993. Genetic linkage analysis in familial breast and ovarian cancer: Results from 214 families. *Am. J. Hum. Genet.* 52:678–701.

FEUNTEUN, J., and LENOIR, G. M. 1966. BRCA1, a gene involved in inherited predisposition to breast cancer. *Biochim. Biophys. Acta* 1242:177–180.

FULKA, J., JR., BRADSHAW, J., and MOOR, R. 1994. Meiotic cell cycle checkpoints in mammalian oocytes. *Zygote* 2:351–354.

GALLION, H. H., PIRETTI, M., DEPRIETS, P. D., and VAN NAGELL, J. R. 1995. The molecular basis of ovarian cancer. *Cancer* 15:(Supplement 10) 1992–1997.

GIACCONE, G. 1996. Oncogenes and antioncogenes in lung tumorigenesis. *Chest* 109:(Supplement 5) 130S–35S.

GRIGNANI, F., FAGIOLI, M., FERRUCCI, P., ALCALAY, M., and PELICCI, P. 1993. The molecular genetics of acute promyelocytic leukemia. *Blood Rev.* 7:87–93.

HABER, D., and BUCKLER, A. 1992. WT1: A novel tumor suppressor gene inactivated in Wilms tumor. *New Biol.* 4:97–106.

HASTIE, N. 1993. Wilms' tumour gene and function. *Curr. Opin. Genet. Dev.* 3:408–13.

HATKEYAMA, M., HERRERA, R. A., MAKELA, T., DOWDY, S. F., JACKS, T., and WEINBERG, R. A. 1994. The cancer cell and the cell cycle clock. *Cold Spring Harbor Symp. Quant. Biol.* 59:1–10.

HOROWITZ, J. M. 1993. Regulation of transcription by the retinoblastoma protein. *Genes Chromosomes Cancer* 6:124–131.

HUAN, B., and SIDDIQUE, A. 1993. Regulation of hepatitis B virus gene expression. *J. Hepatol.* 17:Suppl. 3:S20–23.

LASKO, D., CAVANEE, W., and NORDENSKJOLD, M. 1991. Loss of constitutional heterozygosity in human cancer. *Annu. Rev. Genet.* 25:281–314.

LEAKE, R. 1996. The cell cycle and regulation of cancer cell growth. *Ann. NY Acad. Sci.* 784: 252–262.

LOTEM, J., and SACHS, L. 1996. control of apoptosis in hematopoiesis and leukemia by cytokines, tumor suppressor and oncogenes. *Leukemia* 10:925–931.

MURRAY, A., and KIRSCHNER, M. 1991. What controls the cell cycle? *Sci. Am.* (March) 264:56–63.

NORBY, C., and NURSE, P. 1992. Animal cell cycles and their control. *Annu. Rev. Biochem.* 61:441–70.

OLSSON, H., and BORG, A. 1996. Genetic predisposition to breast cancer. *Acta Oncol.* 35:1–8.

PELTOMÄKI, P., AALTONEN, L., SISTONEN, P., PYLKKÄNEN, L., MECKLIN, J. P., JÄRVINEN, H., GREEN, J., JASS, J., WEBER, J., LEACH, F., PETERSEN, G., HAMILTON, S., DE LA CHAPELLE, A., and VOGELSTEIN, B. 1993. Genetic mapping of a locus predisposing to human colorectal cancer. *Science* 260:810–12.

PINES, L. 1996. Cyclin from sea urchin to HeLas: making the human cell cycle. *Biochem. Soc. Trans.* 24:15–33.

RUSTGI, A., and PODOLSKY, D. 1992. The molecular basis of colon cancer. *Annu. Rev. Med.* 43:61–68.

SMITH, M. L., and FORNACE, A. J., JR., 1996. The two faces of tumor suppressor p53. *Am. J. Pathol.* 148:1019–22.

SOLOMON, E., VOSS, R., HALL, V., BODMER, W., JARS, J., JEFFREYS, A., LUCIBELLO, F., PATEL, I., and RIDER, S. 1987. Chromosome 5 allele loss in human colorectal carcinomas. *Nature* 328:616–29.

STAHL, A., LEVY, N., WADZYNSKA, T., SUSSAN, J. M., JOURDAN-FONTA, D., and SARRACCO, J. B. 1994. The genetics of retinoblastoma. *Ann. Genet.* 37:172–178.

STANBRIDGE, E. 1992. Functional evidence for human tumor suppressor genes: Chromosome and molecular genetic studies. *Cancer Summ.* 12:43–57.

TIMAR, J., DICZHAZI, C., LADANYI, A., RASO, E., HORNEBECK, W., ROBERT, J., and LAPIS, K. 1995. Interaction of tumor cells with elastin and the metastatic phenotype. *Ciba Found. Symp.* 192:321–335.

VON LINDERN, M., FORNEROD, M., SOEKARMAN, N., VAN BAAL, S., JAEGLE, M., HAGEMEIJER, A., BOOTSMA, D., and Grosveld, G. 1992. Translocation t(6;9) in acute non-lymphocytic leukaemia results in the formation of a DEK-CAN fusion gene. *Baillieres Clin. Haematol.* 5:857–79.

WEINBERG, R. A. 1995. The molecular basis of oncogenes and tumor suppressor genes. *Ann. NY Acad. Sci.* 758:331–338.

———. 1996. E2F and cell proliferation: A world turned upside down. *Cell* 85:457–459.

22

Genetic Basis of the Immune Response

CHAPTER CONCEPTS

The immune system of vertebrates is a genetically controlled mechanism that defends the body against disease-causing organisms. Extensive genetic recombination in somatic cells reshuffles a limited number of genes into new combinations that produce literally millions of different antibodies and cell-surface receptors. Cell-surface antigens play a role in determining blood types and in the success of transfusions and organ transplants. The proliferation, differentiation, and function of cells in the immune system are all under genetic control, and mutations in these genes result in disorders of the immune system.

As vertebrates evolved, they developed a unique and complex genetic mechanism crucial to their survival. This mechanism, called the **immune system**, coordinates a series of defensive responses to the entry of foreign substances or the invasion of viruses and microorganisms into the body. These responses are specific and involves two phases: a *primary response* to the initial exposure, and a *secondary response* to subsequent exposures to the same agent.

From a genetic standpoint, the immune system has two fundamental characteristics. First, in every individual, it must recognize "self" so that an organism's cells and tissues are not attacked by its own immune system. This

Scanning electron micrograph of a human lymphocyte infected by HIV viruses (red dots).

recognition involves a set of genes that, in humans, is located on chromosome 6. Second, the immune system must be able to produce specific cells (killer cells and others) and gene products (antibodies) that neutralize and subsequently destroy "nonself" agents (antigens).

Antibody production against potential antigens is a remarkable genetic accomplishment. From a set of less than 300 genes, recombination events within developing antibody-producing cells create a vast immune potential. As a result of shuffling this limited number of genes, tens of millions of different antibodies can be produced.

In this chapter we will examine the cells that make up

the immune system and how these cells and their gene products are mobilized to mount an immune response. We will also examine the role of the immune system in determining blood groups and mother–fetus incompatibility, how cell-surface markers are used in transplants, and how these markers can be used in a predictive way to determine risk factors for a wide range of diseases. Finally, we will consider a number of genetic disorders of the immune system, including mutations in single genes that inactivate the entire immune response. We will also describe how acquired immunodeficiency syndrome (AIDS) associated with infection by human immunodeficiency virus (HIV) acts to cripple the immune response of infected individuals. Taken together, work on the immune system constitutes one of the most exciting areas of genetics, one where new information and new breakthroughs occur with increasing frequency.

Components of the Immune System

The immune system is an effective barrier against the successful invasion of potentially harmful foreign substances. Invading viruses, bacteria, fungi, and parasites are recognized as nonself and subsequently sequestered, inactivated, and destroyed by the immune system. This **immune response** usually involves **antibodies** specific to foreign substances. Agents that elicit antibody production are termed **antigens**. Antibodies are proteins produced and secreted by specific cells of the immune system. Many different molecules can act as antigens, including proteins, polysaccharides and, rarely, other molecules such as nucleic acids. Usually, a distinctive structural feature of an antigen, called an **epitope**, stimulates antibody production. Invading antigens may be free molecules, or they may be part of the surface of a cell, microorganism or virus. Whatever the case, organisms with an immune system have the capacity to make antibodies against any antigen they encounter.

The two main branches of the immune system are: (1) **antibody-mediated immunity**, associated with two types of blood cells, the antigen-recognizing **T cells** and the antibody-producing **B cells**; and (2) **cell-mediated immunity** involving a class of T cells known as **cytotoxic** or **killer T cells**.

Cells of the Immune System

One of the key cells in the immune system is the **phagocyte**, a white blood cell that engulfs and destroys foreign molecules and microorganisms (Figure 22.1). One type of phagocyte called a **macrophage** plays a critical role in both antibody-mediated and cell-mediated immunity by signaling other cells of the immune system that foreign antigens are present. In this signaling role, macrophages are known as *antigen-presenting cells*. All phagocytes arise from stem cells, which are mitotically active cells found in bone marrow (Figure 22.2). Mature phagocytes have the ability to leave the circulatory system by squeezing through capillary walls to enter inflamed or damaged tissues to identify, engulf, and destroy antigens.

A second cell type, the T cell, is also produced by mitotic division of stem cells (Figure 22.2). Newly produced, immature T cells migrate to the thymus gland where they develop into several different subtypes, including *helper cells, suppressor cells*, and *killer cells*. During this period of maturation, T cells become programmed to produce a unique type of cell-surface receptor. Each T cell produces one and only one type of receptor, called an **antigen receptor**, which in turn will bind to only one type of antigen. There are literally tens or hundreds of millions of potential antigens, and there are a similar number of differently programmed T cells. The array of T cell receptors is produced from a small number of genes by genetic recombination events.

From the thymus, mature T cells become distributed throughout the circulatory and lymph systems, and they also become sequestered in the lymph nodes and spleen. Mature T cells can divide, and all the offspring of a single T cell form a family group or clone. There are two general classes of T cells: T4 helper/inducer cells and T8 cytotoxic/suppressor cells. The T4 helper cells are the master switch for the immune system; they turn on the immune response. The T8 suppressor cells are the "off"

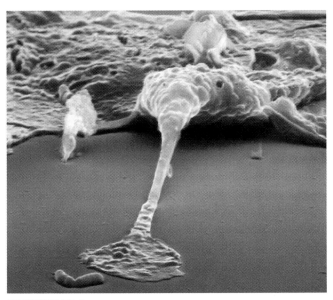

FIGURE 22.1 A phagocyte with a cell process that will engulf a rod-shaped bacterium in the foreground.

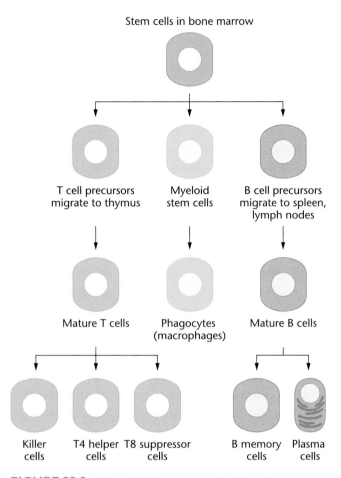

Stem cells in bone marrow

T cell precursors migrate to thymus

Myeloid stem cells

B cell precursors migrate to spleen, lymph nodes

Mature T cells

Phagocytes (macrophages)

Mature B cells

Killer cells

T4 helper cells

T8 suppressor cells

B memory cells

Plasma cells

FIGURE 22.2   Stem cells in bone marrow give rise to the precursors of T cells, macrophages, and B cells of the immune system.

TABLE 22.1   Cells of the Immune System

Cell Type	Function
T4 cells	Also called helper T cells. Role in both cell-mediated and antibody-mediated immune responses. Master switch of the immune system.
T8 suppressor cells	A sub-class of T cells that participate in cell- and antibody-mediated immune response. Slow down and/or turn off the immune response.
B cells	Precursors to plasma cells, B cells recognize antigens, and are activated by T4 cells.
Plasma cells	Synthesize large quantities of a single antibody. Produced mitotically from an activated B cell.
Memory cells	Allow a rapid response to a second encounter with a specific antigen.
Killer cells	A sub-class of T cells that can detect and destroy virus-infected cells.

switch of the immune system; they stop or slow down the immune response (Table 22.1).

A third component of the immune system, the B cells, originate in bone marrow (Figure 22.2) and migrate to the spleen and lymph nodes. As they mature, these cells become genetically programmed to produce antibodies; each B cell produces only one type of antibody. Antibody production is triggered in B cells by interaction with T4 helper cells. These activated B cells proliferate to become **plasma cells** that produce large amounts of the antibody. Memory cells represent a subpopulation of B cells that participate in subsequent encounters with a specific antigen.

The Immune Response

The immune response involves four stages: (1) the detection of foreign antigens, (2) activation of cells of the immune system, (3) inactivation of the antigen, and (4) shutdown of the immune response. Each of these stages is directed by at least one specific cell type. After exposure to an antigen and the generation of an immune response, subsequent challenges to the immune system by the same antigen are directed by a *secondary immune response*, also called *immunological memory*. This secondary immune response is also directed by specific cell types. We will first examine antibody- and cell-mediated immune responses and then consider how immunological memory protects the body against reexposure to an antigen.

Antibody-Mediated Immunity

Macrophages wander through the circulatory system and inflamed tissues and are capable of responding to the presence of foreign molecules, viruses, or microorganisms. When an antigen is encountered, it is engulfed and ingested by the macrophage (Figure 22.3). After ingestion, fragments of the antigen are incorporated into the plasma membrane of the macrophage and displayed on its outer surface in a specific molecular context. As this antigen-presenting macrophage moves through the circulatory system or intercellular spaces, it may encounter a helper T4 cell with a complementary antigen receptor. The receptors on the surface of the T4 cell react with the antigen, activating the helper T cell. In turn, the activated T cell identifies and mobilizes B cells that synthesize an antibody against the antigen presented to the T cell. The stimulated B cells begin to divide and differentiate to form two types of daughter cells. One is the **memory cell**, and the other is the **plasma cell**, which synthesizes and secretes 2000 to 20,000 antibody molecules per *second* into the bloodstream (Figure 22.4).

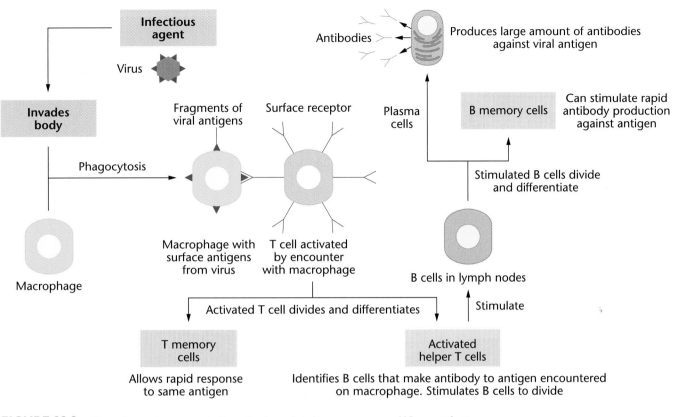

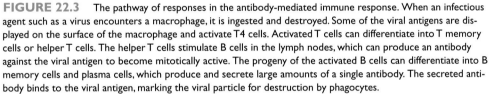

FIGURE 22.3 The pathway of responses in the antibody-mediated immune response. When an infectious agent such as a virus encounters a macrophage, it is ingested and destroyed. Some of the viral antigens are displayed on the surface of the macrophage and activate T4 cells. Activated T cells can differentiate into T memory cells or helper T cells. The helper T cells stimulate B cells in the lymph nodes, which can produce an antibody against the viral antigen to become mitotically active. The progeny of the activated B cells can differentiate into B memory cells and plasma cells, which produce and secrete large amounts of a single antibody. The secreted antibody binds to the viral antigen, marking the viral particle for destruction by phagocytes.

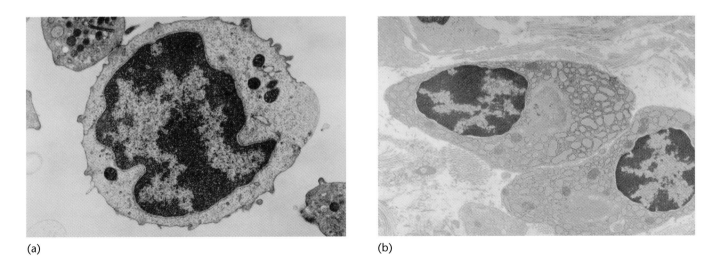

(a)

(b)

FIGURE 22.4 Transmission electron micrograph of (a) an unactivated B cell, and (b) a plasma cell, derived by activation of a B cell. The enlarged plasma cell is filled with endoplasmic reticulum that synthesizes antibodies.

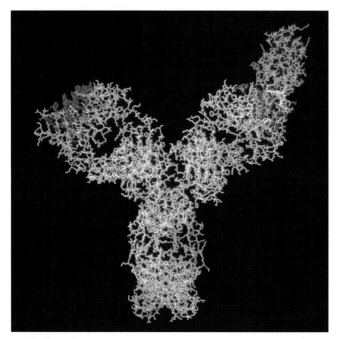

FIGURE 22.5 Computer graphic representation of the structure of an antibody. The backbone of the antibody molecule is represented in green, and the antigen-combining site is shown in blue.

Antibodies (Figure 22.5) are effective against extracellular antigens such as bacterial cells, free virus particles, secreted toxins, and protozoan parasites in the bloodstream, or virus-infected cells that display viral antigens on their surface. Antibodies work in several ways to inactivate antigens. They can interact with the antigen to form antigen–antibody complexes. Clumping of the antigen–antibody complex tags the antigens for destruction, and they are ingested and destroyed by phagocytes. The combination of certain antibodies with antigens activates **complement**, a proteolytic system present in blood serum that lyses invading bacterial cells. Antibodies can also interact with antigens to destroy their ability to function. For example, some viruses have proteins on their surface that help attach the virus to the cell surface as a prelude to viral infection. Antibodies can react with this attachment protein and inactivate it, leaving the virus unable to infect cells and exposing it to ingestion and destruction by phagocytes.

The memory cells produced by activated B cells also produce antibodies, but instead of a life span of several days, as is the case for plasma cells, memory cells have an extended life span of several months to years. These cells play an important role in the secondary immune response (discussed below).

The entire immune response is monitored by T8 suppressor cells, and when detectable antigens have been inactivated or destroyed, the T8 cells stop proliferation of B cells and turn off antibody production. Thus, T8 cells act as the "off" switch for the immune system.

Cell-Mediated Immunity

In contrast to the action of antibodies, which tag antigens for destruction by phagocytes, direct cell–cell interactions called **cell-mediated immunity**, carried out by a class of T cells, can also result in antigen inactivation. The T cells that participate in this form of the immune reaction are known as *killer* T cells or *cytotoxic* T cells (Figure 22.6). Like other T cells, killer cells originate in bone marrow and mature in the thymus gland.

When a cell is infected by a virus, small fragments of the viral coat protein become incorporated into the cell's plasma membrane. These protein fragments are displayed as a new set of antigens on the surface of the infected cell, allowing killer T cells to identify and kill virus-infected cells. Recognition by the killer cells involves both the viral antigen and a set of cell-surface markers known as **histocompatibility antigens**. These antigens are encoded by the human leukocyte antigen (HLA) complex on the short arm of chromosome 6. These markers also play a crucial role in grafts and transplants. Their role in cell-mediated immunity helps ensure that only virally infected cells are attacked by killer cells.

Once identified, killer cells attach to their target cells (Figure 22.7) and secrete a protein called *perforin*, which inserts into the plasma membrane of the virus-infected cell. The cylindrical perforin molecules link together to form pores in the plasma membrane. The cytoplasm of

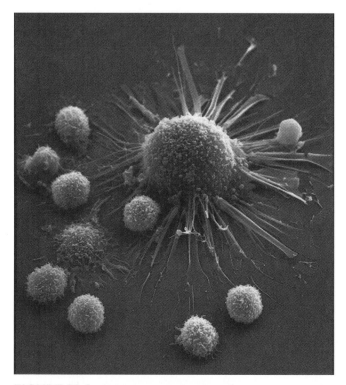

FIGURE 22.6 A large cancer cell, with cellular processes extended, has been detected and surrounded by an array of killer cells.

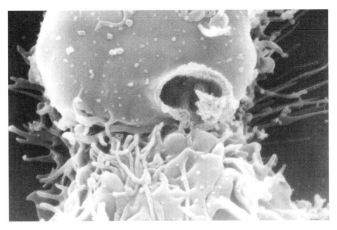

FIGURE 22.7 A killer cell (at bottom) has attacked a cancer cell and punctured its membrane. The contents of the tumor cell will leak out through this hole, and the cell will die. The remaining debris will be scavenged by phagocytes.

the target cell leaks out through these pores, and the virus-infected cell dies. The killer cell detaches from the dying cell, and it can identify and kill other infected cells. In addition to killing virus-infected cells, cytotoxic T cells and a second type of killer cell known as a *natural killer cell* attack and kill cancer cells, fungi, and some types of parasites. Cytotoxic T cells will also attack and kill cells introduced into the body during tissue or organ transplants if the transplanted cells are recognized as foreign.

Immunological Memory and Immunization

Even in ancient times it was known that people exposed to certain diseases cannot contract the disease a second time. For example, people who have had measles (an infectious disease caused by a virus) will not get measles again, even when they are exposed to infected individuals. Resistance to second infections is controlled by the *secondary immune response* and the production of B and T memory cells during the first exposure to the antigen. Because of memory cells, a second exposure to an antigen results in an immediate, large-scale production of antibodies and killer T cells. Because of the presence of the memory cells, the secondary immune response is more rapid, on a larger scale, and lasts longer than the primary immune response.

Existence of the secondary immune response is the basis for vaccination against a number of infectious diseases. A vaccine is a preparation designed to stimulate the production of memory cells against a pathogenic agent. In this procedure, the antigen is administered orally or by injection. Once the antigen is in the body, it provokes a primary immune response, including the production of memory cells. Often, a second or booster dose is administered to elicit a response that boosts the number of memory cells. Vaccines can be prepared from killed pathogens

or weakened strains that are able to stimulate the immune system but unable to produce symptoms of the disease. Recombinant DNA techniques are also used to produce vaccines. For example, fragments of the gene coding for the protein coat of the hepatitis B virus are used to produce a noninfectious antigen that is used as a vaccine.

Genetic Diversity in the Immune System

Among vertebrates, each individual can produce millions of different types of antibodies, each responding to a different antigen. Coincidently, there are millions of different antigen recognition sites on the surface of T cells. The basis for this molecular diversity lies in the amino acid sequences composing the backbones of these proteins, called *T-cell receptors* (TCR). TCRs are encoded by a family of genes, and the rearrangement of these genes in the development of B and T cells is the ultimate basis of their diversity.

The histocompatibility antigens, a third group of proteins involved in the immune response, are also encoded by a family of genes in the HLA complex, but here diversity is achieved by a large number of alleles at each locus in the family, rather than by gene rearrangement. In the following sections we will examine the molecular basis for diversity in antibodies and TCRs and consider diversity in the HLA complex in a later section.

Antibodies

In humans, antibodies are produced by plasma cells, a type of B cell. Five classes of antibodies or **immunoglobulins (Ig)** are recognized: **IgG, IgA, IgM, IgD,** and **IgE** (Table 22.2). The first class, IgG, represents about 80 percent of the antibodies found in the blood and is the most extensively characterized class of antibodies. IgG is associated with immunological memory. The IgA class of immunoglobulins (found in breast milk) can be secreted across plasma membranes and is associated with im-

TABLE 22.2 Categories and Components of Immunoglobins

Ig Class	Light Chain	Heavy Chain	Tetramers	
IgA		α	$\kappa_2\alpha_2$	$\lambda_2\alpha_2$
IgD		δ	$\kappa_2\delta_2$	$\lambda_2\delta_2$
IgE		ε	$\kappa_2\varepsilon_2$	$\lambda_2\varepsilon_2$
IgG	κ or λ	γ	$\kappa_2\gamma_2$	$\lambda_2\gamma_2$
IgM		μ	$\kappa_2\mu_2$	$\lambda_2\mu_2$

munological resistance to infections of the respiratory and digestive tracts. IgM antibodies are usually the first secreted by a plasma cell in response to an antigen, and they are associated with the early stages of the immune response. At this time, little is known about the role of the IgD class of immunoglobulins. They are associated with the surface of B cells, and may regulate their action. IgE antibodies are involved with fighting parasitic infections. IgE antibodies are also associated with allergic responses.

A typical IgG molecule (Figure 22.8) consists of two different polypeptide chains, each present in two copies. The larger or **heavy chain (H)** contains approximately 440 amino acids. The sequence of the first 110 amino acids at the N-terminus differs between different heavy chains, and is known as the **variable region (V_H)**. The remaining C-terminal amino acids are the same in all H chains, and make up the **constant region (C_H)**. In humans, the H chains are encoded by genes on the long arm of chromosome 14.

The **light chain (L)** is composed of 220 amino acids, with the first 110 amino acids making up the variable region (V_L) (Figure 22.8). The rest of the amino acids at the C-terminus make up the constant region (V_H) of the light chain. There are two types of L chains: **kappa chains**, encoded by genes on human chromosome 2, and **lambda chains**, encoded by genes on the long arm of chromosome 22.

Two heavy and two light chains make up a functional IgG molecule, and these are held together by disulfide bonds. The variable regions of the heavy and light chains form the **antibody combining site.** The combining site of each antibody has a unique conformation that allows it to bind to a specific antigen, like a key into a lock.

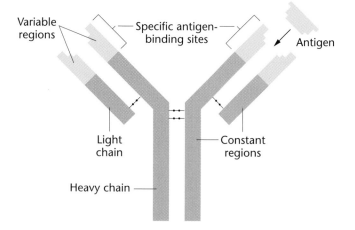

FIGURE 22.8   A typical antibody (immunoglobulin) molecule. The molecule is Y-shaped, and contains four polypeptide chains. The longer arms are heavy chains, and the shorter arms are the light chains. Each chain contains a constant region and a variable region. The variable regions form an antibody combining site that interacts with a specific antigen similar to a lock and key mechanism. The chains are joined together by disulfide bonds indicated as −•−•−.

Theories of Antibody Formation

One of the dominant features of the immune system is the generation of new cells that contain different combinations of antibodies and cell-surface receptors. Because there are billions of such combinations, it is impossible that each combination is directly coded in the genome. There simply is not enough DNA in the human genome to encode tens or hundreds of millions of antibodies. One of the long-standing questions in immunogenetics is how this vast array of molecular variability in antibodies and antigen receptors *is* encoded in the genome. Over the years, several ideas have been proposed to explain how antibodies are encoded. One of these ideas, called the **recombination theory**, has received the greatest amount of experimental support and will be considered in some detail.

The diversity of antibodies is the result of genetic recombination in the three clusters of antibody genes: the H chain genes on chromosome 14, the kappa L genes on chromosome 2, and the lambda L genes on chromosome 22. At each locus, there is a limited number of genes present, encoding various portions of the H and L chains. As antibody-forming B cells mature, DNA recombination rearranges these genes so that each mature B lymphocyte comes to encode, synthesize, and secrete only one specific type of antibody. Each mature B cell can make one type of light chain (kappa or lambda) and one type of heavy chain. When an antigen is present, it stimulates the B cell, which encodes an antibody against that antigen to divide and differentiate. As a result, populations of differentiated plasma cells are produced, all of which synthesize one type of antibody that interacts with the antigen.

Clear evidence for the recombination theory was first provided by Susumu Tonegawa and his colleagues in the mid-1970s. By comparing the size of restriction fragments, these researchers demonstrated that DNA segments coding for parts of the light chain gene are far away from each other in embryonic cells, but are adjacent to one another in antibody-producing cells. In their experiments, they isolated and characterized a cloned DNA fragment specifying a lambda V chain derived from embryonic cells. The organization of this cloned fragment was compared with the equivalent region isolated from an antibody-producing cell. In the embryo DNA, the variable region of an L-chain gene was separated by 4.5 kilobases (kb) of DNA from the region encoding another part of the L chain (Figure 22.9). However, in the DNA isolated from the antibody-producing cell, these regions were joined together to form a single transcription unit encoding a specific L chain. This observation strongly supports the hypothesis that the diverse array of immunoglobulin genes found in antibody-producing cells are formed through a process of somatic recombination.

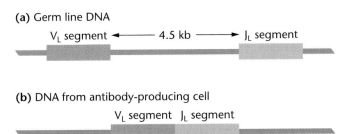

(a) Germ line DNA

V$_L$ segment ←――― 4.5 kb ―――→ J$_L$ segment

(b) DNA from antibody-producing cell

V$_L$ segment   J$_L$ segment

FIGURE 22.9 (a) In DNA isolated and cloned from embryonic DNA, the V region of an L gene was located 4.5 kb away from the DNA encoding the J region. (b) In DNA isolated from an antibody-producing cell, the V and J regions were contiguous, forming a single transcription unit. This evidence provided support for the hypothesis that recombination is involved in forming antibody genes in mature B cells.

Organization of the Immunoglobulin Genes

The work of Tonegawa and his colleagues inspired a large-scale research effort by many laboratories to study the organization and rearrangement of the immunoglobulin genes in mice and humans. From this beginning, a general picture of the origin of antibody diversity has emerged. The coding sequences for kappa and lambda genes have now been isolated from germ line and B-cell DNA in a variety of higher organisms. Comparative studies indicate that the basic organizational plan and mechanism of formation is similar in most mammals. In humans, the kappa L genes (Figure 22.10) consist of several elements: L-V (leader-variable) regions, J (joining) regions, and a C (constant) region. There are 70 to 300 V-L segments in each light-chain gene, each with a dif-

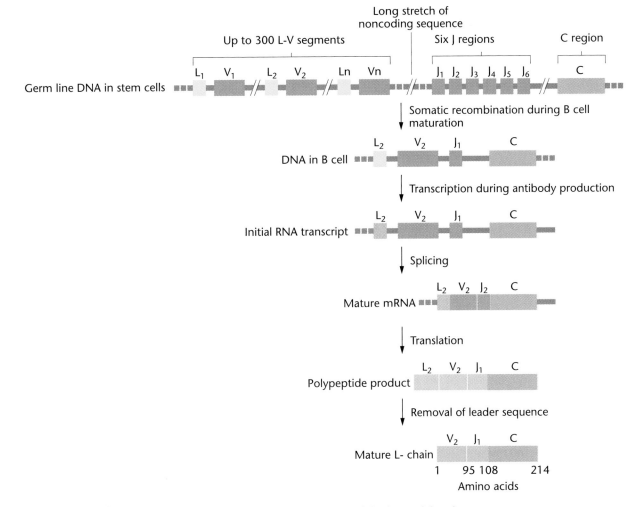

FIGURE 22.10 Formation of the DNA segments encoding a human κ light chain and the subsequent transcription, mRNA splicing, and translation leading to the final polypeptide chain. In germ line DNA, up to 300 different L-V (Leader-Variable) segments are present. These are separated from the J regions by a long noncoding sequence. The J regions are separated from a single C gene by an intervening sequence (intron) that must be spliced out of the initial mRNA transcript. Following translation, the amino acid sequence derived from leader RNA is cleaved off as the mature polypeptide chain passes across the cell membrane.

ferent nucleotide sequence. The J region contains six different segments, and there is one C segment.

During maturation in a B cell, one of the 300 L-V regions is randomly joined by a recombination event to one of the six **J genes** to form a functional light chain gene (Figure 22.10). The remaining gene segments are excised and destroyed. This form of recombination is different from the reciprocal recombination occurring in meiosis. This newly created L-V-J region becomes an exon in a gene that includes the C region. This light-chain gene is transcribed and translated to form kappa L chains that becomes part of an antibody molecule. This rearranged gene is stable, and is passed on to all progeny of the B cell.

Diversity in kappa L chain production is the result of combining any of the 300 **V genes** with any of the six J genes. The joining of one of 300 V genes with any of the six J segments generates about 1800 different kappa genes. The organization of the lambda L chain genes is somewhat different from that of the kappa genes, but recombination events also generate about 1800 different lambda genes.

The heavy-chain genes in humans extend over a large region of DNA and include four types of segments: V (variable), D (diversity), J (joining), and C (constant) regions. There are approximately 300 different V genes, 10 to 50 different D genes, and four different J genes. In addition, there are five different C genes, one for each class of immunoglobulins (Figure 22.11). During B-cell maturation, recombination randomly joins a V region with one of the D sequences and one of the J sequences.

The V-D-J composite assembled by recombination lies adjacent to the C regions of the five classes of heavy chains. A V-D-J segment can be joined to any one of the C segments to form a gene for an IgM, IgD, IgG, IgE, or IgA antibody, respectively. This joining can occur in several ways. One is by transcription of a long pre-mRNA molecule that begins at the 5′-end of the V region and terminates at the 3′-end of the C_A segment. Splicing of this pre-mRNA yields mRNA molecules that can encode the same V-D-J sequence in combination with any of the five different C sequences. An alternative method involves a second round of recombination and excision in the H gene that places one of the C segments adjacent to the joined V-D-J segment and eliminates the other C regions (Figure 22.12).

The random recombination of H-gene components generates a large number of H-chain proteins. If we as-

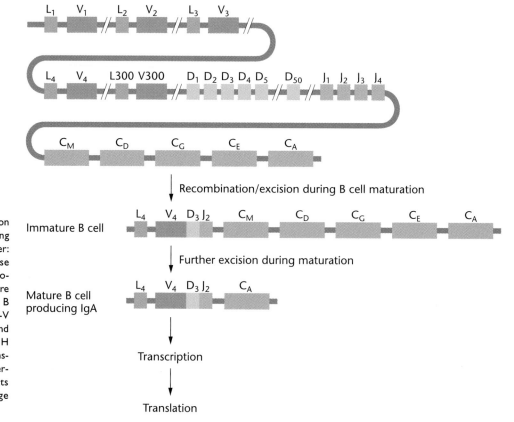

Germ-line DNA (300 L-V segments, 10–50 D segments, 4 J segments, 5 C segments)

FIGURE 22.11 The variable region of the H chain is assembled by joining three different DNA segments together: L-V, D, and J. In embryonic DNA, these segments occur in clusters on chromosomes separated by long intervals and are adjacent to the C region. In a maturing B cell, a random combination of one L-V region with one of the 20 D regions and one of the four J regions produces an H chain gene that is transcribed and translated in the B cell. In other B cells, different combinations of H chain segments are joined together, producing a large number of different H chains.

Somatic DNA in mature B cell

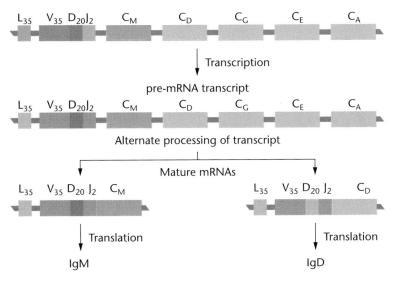

FIGURE 22.12 An alternative method of producing heavy chains with the same variable regions but different constant regions involves alternative splicing of heavy-chain transcripts during B cell maturation. In the somatic cell DNA, the rearranged (L-V)-D-J segment is adjacent to all five heavy chain constant regions. Transcription and alternate processing of the transcripts lead to the production of IgM and IgD containing different constant regions but the same variable region. This phenomenon is called class switching.

sume there are 25 D regions, then combining these with any of the 300 V regions and 4 J regions, 30,000 possible H chains can be generated. The potential for overall antibody diversity can be estimated by calculating the combination of all heavy-chain genes (30,000) with all light-chain genes (3600), resulting in 108 million possible antibody genes (30,000 $\times$ 3600) from a few hundred coding sequences.

Organization of T-Cell Receptors

The T cells play a role in both humoral and cell-mediated immunity. Although something of an oversimplification, T4 cells and T8 cells can be viewed as switches that turn the humoral immune response on and off, respectively. In cell-mediated immunity, killer T cells kill virally infected cells and tumor cells. The ability of T cells to participate in an immune response depends on the action of cell surface proteins called **T cell receptors (TCRs)**. These proteins interact with antigenic fragments on the surface of antigen-presenting cells or virus-infected cells (Figure 22.13). There are two classes of receptors: alpha/beta and gamma/delta. The alpha/beta receptors are found on almost all T cells, and the gamma/delta receptors are found only on about 5 percent of all T cells. The alpha/beta receptors are composed of two polypeptide chains. Each chain has a variable portion that protrudes from the cell surface, and a C-terminal constant region attached to the plasma membrane. The variable regions of the two chains form the antigen-binding site.

Diversity in TCRs is accomplished through somatic recombination events that occur during T-cell maturation, in a manner similar to the generation of antibody diversity in B-cell maturation. The beta gene

(Figure 22.14) is composed of about 25 different V genes (V_β), one D gene (D_β), and seven J genes (J_β) associated with each of the two C genes (C_β). Organization of the alpha, gamma, and beta regions is similar to those of the beta gene.

The diversity of T-cell antigen receptors derives from the recombination of a single V gene with one D and one J gene and the association of the VDJ segments with one of the C regions. These recombination events take place during T-cell maturation, and they involve excision and loss of intervening DNA sequences. There is some imprecision in the cutting and joining of segments, which contributes to receptor diversity, but the overall diversity of TCRs is somewhat lower than that of immunoglobulin genes.

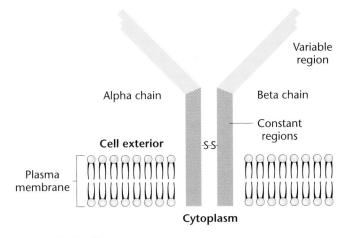

FIGURE 22.13 T cells produce surface receptors that recognize antigens. This T-cell receptor is composed of two chains, an alpha chain and a beta chain. Each chain has a constant region and a variable region, and they are encoded by separate genes containing L-V, D, J, and C segments.

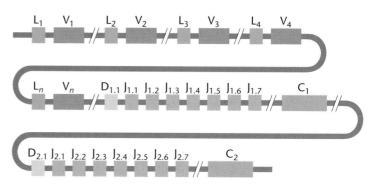

FIGURE 22.14   Organization of the T-cell beta receptor gene. There are about 25 different L-V gene segments, one D gene, and seven J genes associated with each of the two C genes. During T-cell differentiation, recombination and excision combines one L-V segment, one D segment, and one J segment adjacent to each C gene.

Recombination in the Immune System

The process of **V(D)J recombination** in the formation of Ig and TCR genes is complex and involves a number of participating proteins. Of these, two closely linked loci, named **recombination-activating genes RAG-1** and **RAG-2**, have been identified in the human genome. These adjacent loci on chromosome 11 are co-transcribed, contain no introns, and are both required for recombination activity. These genes are normally expressed only in lymphocytes and only in those stages of maturation when V(D)J recombination is taking place. Transfer of these genes to fibroblasts or other cell lines results in the expression of recombination in cells that do not normally exhibit the assembly of V, D, and J segments. RAG-1 or RAG-2 deficient mice have no mature T or B cells, and maturation of these cells is arrested at the stage during which V(D)J recombination takes place. This and other evidence indicates that RAG-1 and RAG-2 directly participate in the somatic recombination events in maturing cells of the immune system during which genomic sequences are lost from somatic cells. It seems likely that these genes work in combination with other factors and regulatory genes to coordinate recombination in the immune system, but their mechanism of action is not yet known.

Blood Groups

Thus far we have concentrated on the genetic basis of antibody production and cell-mediated immunity as components of an immunological defense system. We now turn our attention to the genetic basis of antigens that underlie the immunological concept of self. As we shall see, this is an equally important, closely related topic. Knowledge about the molecules found on the surface of blood cells is needed to carry out blood transfusions and organ transplantations. To date, more than 30 such genes have been identified in humans, and each (along with its alleles) constitutes a blood group or blood type. To be safely transfused, the blood types of the donor and recip-

ient must be matched. If the transfused red blood cells have a foreign antigen on their surface, the body of the recipient will produce antibodies against this antigen, causing the transfused red cells to clump together, blocking circulation in capillaries and other blood vessels, with severe and often fatal consequences. Although there are a large number of blood groups, only two are of major immunological significance for transfusions: (1) the ABO system, and (2) the Rh-blood group.

ABO System

The genetics of the ABO system were described in Chapter 4 as an example of a gene with multiple alleles. To review this system briefly, recall that the gene I (for isoagglutinin) encodes an enzyme that modifies a cell-surface glycolipid and has three alleles: I^A, I^B, and I^O. The I^A and I^B alleles are codominant, each producing a slightly different version of the gene product, while I^O is a recessive null allele. Individuals with type A blood (genotypes AA and AO) carry the A antigen on red blood cells, so they do not have antibodies against this cell surface marker, but have antibodies against the antigen encoded by the B allele (Table 22.3). Those with type B blood carry the B antigen on their red cells and make antibodies against the A antigen. Individuals with AB blood carry both antigens and do not make either antibody, while those with type O blood have neither antigen and carry antibodies against both the A and B antigen. In transfusions, AB individuals do not make serum antibodies against A or B and can receive blood of any type, whereas type O individuals carry neither red cell antigen and thus are universal donors.

Rh Incompatibility

The Rh blood group (named for the rhesus monkey in which it was discovered) consists of at least three genes (C, D, and E), each with two alleles (C, c; D, d; and E, e), making a large number of genotypic combinations possible. However, because the antigens encoded by the C and E genes are weak, they invoke only a minimal immune re-

TABLE 22.3   Antigens, Antibodies of ABO System

Blood Type	Antigens on Red Blood Cells	Antibodies in Blood	Safe Transfusions To	Safe Transfusions From
A	A	B	A, AB	A, O
B	B	A	B, AB	B, O
AB	A, B	—	AB	A, B, AB
O	—	A, B	A, B, AB, O	O

sponse, and for practical purposes, the Rh system can be considered as a simple one-gene, two-allele system involving gene D. Those of genotypes DD or Dd are said to be *Rh positive* (Rh$^+$), and those of genotype dd are said to be *Rh negative* (Rh$^-$).

Although the Rh blood group can play a role in transfusions, it is of major concern in immunological incompatibility between mother and fetus, a condition known as *hemolytic disease of the newborn* (*HDN*). This condition occurs when the mother is Rh$^-$ (genotype dd) and the fetus is Rh$^+$ (genotypes DD or Dd). If the mother is Rh$^-$, and Rh$^+$ blood from the fetus enters the maternal circulation, antibodies against the Rh antigen will be made (Figure 22.15). The most common way fetal blood enters the maternal circulation is during the process of birth, so that usually the first Rh$^+$ child is not affected. However, the maternal circulation now contains antibodies against the antigen, and a subsequent

Rh$^+$ fetus is affected because these antibodies cross the placenta and destroy the red blood cells of the fetus. The resulting anemia can kill the fetus before birth, or the hemoglobin released from the lysed red blood cells can be converted into a degradation product that builds up in the brain causing neonatal death or, in survivors, severe mental retardation.

To circumvent this problem, Rh$^-$ mothers are given an Rh-antibody preparation before the birth of the first Rh$^+$ child, and in all subsequent Rh$^+$ pregnancies. These antibodies move through the maternal circulatory system and destroy any fetal cells before the maternal immune system has a chance to make its own antibodies against the Rh antigen. The antibody preparation is administered to Rh$^-$ mothers before the first birth because she can also be sensitized by a miscarriage, abortion, or blood transfusions with Rh$^+$ blood.

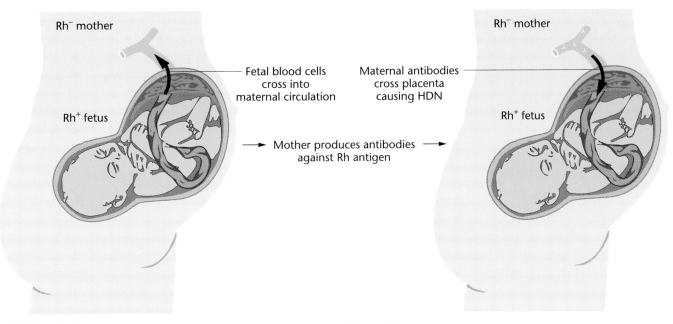

FIGURE 22.15   The development of hemolytic disease of the newborn (HDN). (a) When the mother is Rh$^-$ and the fetus is Rh$^+$, any fetal blood that enters the maternal circulation will cause the production of antibodies against Rh$^+$ red blood cells. Usually this happens during birth, so the first Rh$^+$ child escapes HDN. (b) In a subsequent pregnancy involving an Rh$^+$ child, antibodies against Rh$^+$ cells cross the placenta from the maternal circulation and destroy the red blood cells of the fetus. This causes HDN, which is characterized by anemia, jaundice, and, in extreme cases, can result in mental retardation and/or death.

The HLA System

Organ transplants and skin grafts between unrelated individuals are usually rejected within a few weeks. Second attempts are rejected in a matter of days. The speed of the second rejection indicates that the immune system is involved in accepting or rejecting grafts and transplants. The interaction of genetically encoded cell-surface antigens of the donor with the immune system of the recipient determines whether grafts will be accepted or rejected. In laboratory mice, these antigens, known as **histocompatibility antigens**, are encoded by 20 to 40 genes, many of which have a large number of alleles, producing an enormous number of genotypic combinations.

Inbred strains of mice have been used to study the genetics of histocompatibility. Some strains are so highly inbred that all members of the strain can donate or receive grafts from any other member of the same strain. Crosses with other strains have allowed the identification and isolation of the histocompatibility genes. Results of such experiments indicate that one group of histocompatibility genes, known as the *major histocompatibility complex* (*MHC*) is the primary set of genes responsible for success or failure in transplants. Other minor genes are scattered over several loci in the genome.

In humans, a group of closely linked genes on chromosome 6, known as the HLA (human leukocyte antigen) complex (similar to the mouse MHC), plays a critical role in histocompatible transplants.

HLA Genes

The HLA complex in humans consists of four closely linked major genes: HLA-A, HLA-B, HLA-C, and HLA-D (Figure 22.16). These genes encode antigens that fall into

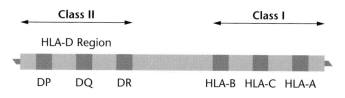

FIGURE 22.16 The HLA region on human chromosome 6, showing the organization of the Class I and Class II regions.

two classes: Class I antigens found on the surface of all cells in the body (HLA-A, HLA-B, and HLA-C), and Class II antigens represented by HLA-D, which encodes antigens found only on cells of the immune system, such as B lymphocytes, activated T lymphocytes, and macrophages. HLA-D is subdivided into the HLA-DR, HLA-DQ, and HLA-DP genes. Class I antigens are glycoproteins; Class II antigens are composed of two polypeptide chains that enable cells of the immune system to identify each other.

A large number of alleles have been identified in each of the HLA genes; it is one of the most highly polymorphic gene systems in the human genome. There are at least 23 alleles in HLA-A, 47 in HLA-B, 8 in HLA-C, 14 alleles in HLA-DR, 3 in HLA-DQ, and 6 in HLA-DP, making literally millions of genotypic combinations possible. Each of these alleles encodes a distinct antigen identified by a letter and number. For example, A6 is allele number 6 at the HLA-A locus, B2 is allele 2 at the HLA-B locus, etc.

The genes of the HLA system are closely linked and inherited in a codominant fashion. This linkage means that recombination in this region is a rare event, and that the allelic combination on a single chromosome tends to be inherited as a unit. The array of HLA alleles on a given copy of chromosome 6 is known as a **haplotype**. Since

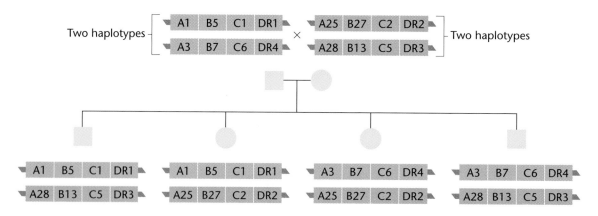

FIGURE 22.17 Transmission of HLA haplotypes. The cluster of HLA alleles on a single chromosome 6 is called a haplotype, each containing four major loci, encoding a different cell-surface antigen. The arrangement of HLA haplotypes and their distribution to the offspring result in new combinations of haplotypes in each generation.

humans have two copies of chromosome 6, we each have two HLA haplotypes (Figure 22.17). The codominant pattern of inheritance means that each haplotype is fully and completely expressed.

The inheritance of HLA haplotypes is shown in Figure 22.17. Because of the large number of alleles that are possible, it is rare that anyone will be homozygous at any of these loci. In the example shown, the unrelated parents have completely different haplotypes, and each child receives one haplotype from each parent, carried by the copy of chromosome 6, which is incorporated into the parental sperm or egg. The result is four new haplotype combinations possible in the offspring. Siblings, therefore, have a one-in-four chance of sharing the same combination of haplotypes.

Organ and Tissue Transplantation

Successful transplantation of organs and tissues depends to a large extent on matching HLA haplotypes between donor and recipient. Because of the number of HLA alleles, the best chance for a match is between siblings or close relatives. Identical twins always have a perfect match. Parents have only one haplotype in common with a child, whereas siblings have a one-in-four chance of a perfect match. Thus, the order of preference for organ and tissue donors among relatives is: identical twin > sibling > parent > unrelated donor. Among unrelated donors and recipients, the chances for a successful match are between 1 in 100,000 and 1 in 200,000. Because HLA allele frequency differs widely between racial and ethnic groups (for example, B27 is found in 8 percent of American whites, but only 4 percent of American blacks), matches between groups is often difficult.

When HLA types are matched, the survival of transplanted organs is improved dramatically. Figure 22.18 shows the 4-year survival rates for matched and unmatched kidney transplants. In HLA-matched transplants, over 90 percent of the transplanted kidneys survived the 4-year period, but in unmatched transplants, less than 50 percent of the kidneys were functional after 4 years. The major causes of rejection are mismatching the HLA and/or ABO alleles. Other causes are more subtle and often difficult to define. For example, if the recipient has had blood transfusions, memory cells against HLA antigens might be present. If these antibodies act against HLA antigens on the grafted tissue, rejection is more likely.

More recently, drugs have been used to improve the survival of transplants even when HLA matching is somewhat imperfect. The most widely used drug is cyclosporine, first isolated from a soil fungus. Its mechanism of action is not known with certainty, but it selectively inactivates the T-cell subpopulation (the killer T

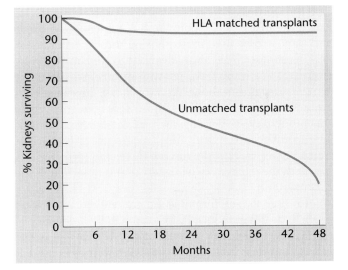

FIGURE 22.18 The survival of kidney transplants when they are HLA matched surpasses that of unmatched transplants.

cells) that is most active in tissue rejection, while not impairing other cells of the immune system. Other drugs to suppress the immune rejection of transplants are now under development, and if successful they may reduce the need for precise HLA matching for transplants, making it possible for more individuals to receive needed organs without prolonged waits for a proper donor.

HLA and Disease

In a number of diseases, certain HLA haplotypes are found at a much higher frequency than would be expected by chance alone. For example, although allele B27 is found in 8 percent of the U.S. white population, more than 90 percent of those afflicted with a chronic inflammatory condition known as *ankylosing spondylitis* carry the B27 allele. Similarly, 70 percent of those with rheumatoid arthritis carry allele DR4, while only 28 percent of those in the general population have this allele. A number of diseases and their HLA associations are shown in Table 22.4.

TABLE 22.4 HLA Alleles and Disease

Disease	HLA Allele	Risk Factor
Ankylosing spondylitis	B27	>100
Systemic lupus erythrematosus	DR3	3
Psoriasis	B17	6
Rheumatoid arthritis	DR4	6
Reiter's syndrome	B27	50
Multiple sclerosis	A3	3
Chronic active hepatitis	B8	6

The nature of the relationship between HLA alleles and these diseases is unknown, but a correlation exists and can often be used in diagnosing the condition. Several hypotheses have been advanced to explain this relationship, but none have proven entirely satisfactory. One idea suggests that a direct relationship exists between specific HLA alleles and disease. According to this model, those carrying such an allele have a genetic susceptibility to a given disease and are therefore at much higher risk for the disease. A second hypothesis is based on the *linkage disequilibrium* that is observed in the HLA system.

Linkage disequilibrium is the nonrandom association of alleles at different loci. If, for example, the frequency of the HLA-A1 allele in a population is 0.17 and that of the HLA-B8 allele is 0.11, then if the alleles were associated at random, the frequency of their being found together in the population would be the product of their independent frequencies: $0.17 \times 0.11 = 0.019$ (1.9%). In fact, the observed frequency of this combination is much higher (about 0.09, or 9%), implying that selection is favoring this combination of alleles. If, however, another nearby locus outside the HLA complex carried an allele-conferring susceptibility to a disease, it would be inherited along with the HLA-A1/HLA-B8 combination, showing up as a disease association, even though the HLA alleles themselves have nothing to do with the disease.

Another hypothesis has suggested that the association between HLA alleles and disease may result from an interaction between certain alleles and environmental factors such as infection.

Disorders of the Immune System

Genetically determined immunodeficiency disorders are caused by mutations that inactivate or destroy some component of the immune system. Most often, these disorders cause deficiencies in the production and functioning of one or more of the cell types in the immune system (Table 22.5). Much has been learned about how the immune system works by studying disorders that remove or disable single components. Unfortunately, not all the ge-

netic disorders of the immune system have been described in enough detail to allow insights into the underlying molecular mechanisms. We will consider three disorders, one that affects B cells, one that affects T cells, and one that affects both B and T cells.

Other disorders of the immune system are due to factors that include infections, cancer chemotherapy, and developmental errors. Two of these, developmental errors and infection, are considered next.

The Genetics of Immunodeficiency

A genetic disorder in which B cells are missing is **X-linked agammaglobulinemia,** or Bruton disease. Affected individuals are almost always male, with an age of onset somewhere between 6 and 12 months. Both B cells and plasma cells are completely absent or, alternatively, immature B cells are present, but they never become functional. The result is a complete lack of circulating antibodies and an inability to make antibodies. Consequently, affected individuals have no antibody-mediated immunity and are highly susceptible to infections from microorganisms such as bacteria and fungi. However, these same individuals have normal levels of T cells and retain an intact cell-mediated immune system that confers immunity to most viral infections. Currently, the most effective method of treatment is a bone marrow transplant to provide a stem-cell population that can give rise to functional B cells. Once these are established, they confer antibody-mediated immunity, which permanently corrects the condition.

A genetically controlled form of **T-cell immunodeficiency** has been described, although the number of reported cases is very small. Affected individuals have a deficiency of the enzyme nucleoside phosphorylase, part of a biochemical pathway that salvages nucleotides. The number of B and T cells in these individuals is normal at birth, but thereafter a gradual decrease occurs in the number of T cells and cell-mediated immunity. The number of B cells is not affected, and antibody-mediated immunity is maintained. Without cell-mediated immunity, there is an increased frequency of viral infections and a high risk of certain forms of cancer. It has been suggested that the decline in T cells is caused by the

TABLE 22.5 Cell Types Affected in Immune Disorders

Immune Disorder	Affected Cell Type
X-linked agammaglobulinemia	B cells missing
Nucleoside phosphorylase deficiency	T cells reduced after birth
Severe combined immunodeficiency	T and/or B cells missing or nonfunctional
DiGeorge syndrome	T cells missing
Acquired immunodeficiency syndrome	T cells decline after infection

FIGURE 22.19 David, a boy born with severe combined immuno-deficiency (SCID), survived by being isolated in a germ-free environment. He died at the age of 12 following a bone marrow transplant undertaken to provide him with a functional immune system.

build-up of purines as a result of the enzyme deficiency and that the resulting toxicity kills T cells but does not affect B cells. The structural gene for nucleoside phos-phorylase maps to the long arm of human chromosome 14, and the condition is inherited as an autosomal domi-nant trait.

In **severe combined immunodeficiency syndrome (SCID)**, T- and B-cell populations are absent or nonfunc-tional, and affected individuals have neither antibody-mediated nor cell-mediated immunity. As a result, they are susceptible to recurring and severe bacterial, viral, and fungal infections. There are autosomal and X-linked forms of SCID, indicating that this disorder is genetically heterogenous. About 50 percent of the cases of SCID in the United States are X linked, 35 percent are autosomal recessive with an unknown cause, and 15 percent are au-tosomal recessive cases associated with a deficiency of the enzyme adenosine deaminase. In the X-linked form (XSCID), there are persistent infections beginning about 6 months after birth, with no T cells present. There are normal or even elevated levels of B cells present, but there is no antibody production by these cells. Experimental re-sults indicate that the XSCID gene product is necessary in both T and B cells and is required for B-cell maturation. The longest-surviving individual with this form of SCID was a boy named David, who was born in Texas and iso-lated from the outside world by being placed in germ-free isolation (Figure 22.19). David died at age 12 of compli-cations following a bone marrow transplant.

In cases of autosomally inherited SCID associated with a deficiency of the enzyme *adenosine deaminase* (*ADA*), there are no T cells present, and B cells (if present) are not functional. Again, the result is a complete lack of cell-mediated and antibody-mediated immunity, and se-vere, recurring infections lead to death, usually by the age

of 7 months. Children affected with ADA-deficient SCID are currently receiving gene therapy to provide them with a normal copy of the gene. In this procedure, samples of their white blood cells are removed from the body and transfected with a retrovirus containing a copy of the nor-mal gene for ADA. The cells are maintained in culture to ensure that the gene is active, and then the genetically modified white blood cells are transferred to the circula-tory system with the hope that ADA expression will stim-ulate the development of functional T and B cells and at least partially restore a functional immune system. This treatment represents the first attempt at gene therapy in humans; a technique that will undoubtedly become widely used as a treatment for certain genetic disorders in the fu-ture. The recombinant DNA techniques used in gene therapy were discussed in Chapter 16.

Acquired Immunodeficiences: DiGeorge Syndrome and AIDS

A nongenetic form of T-cell deficiency called **DiGeorge syndrome** is the result of a mishap early in prenatal de-velopment. In this condition, the thymus and parathyroid glands fail to develop. The result is a complete lack of T cells that depend on the cellular environment of the thy-mus in order to mature. As predicted, no cell-mediated immunity results.

Acquired immunodeficiency syndrome (AIDS) is a collection of disorders that develop as a result of infec-tion with a retrovirus known as the human immunodefi-ciency virus (HIV). The HIV virus consists of a protein coat, enclosing an RNA molecule that serves as the ge-netic material, and an enzyme, reverse transcriptase. The entire viral particle is enclosed in a lipid coat derived from the plasma membrane of a T cell (Figure 22.20).

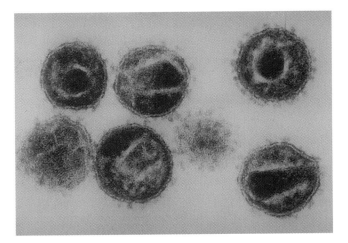

FIGURE 22.20 Transmission electron micrograph of the human im-munodeficiency virus (HIV).

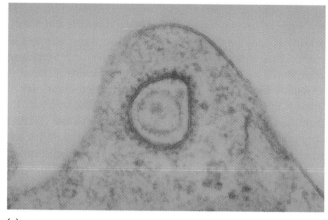
(a)

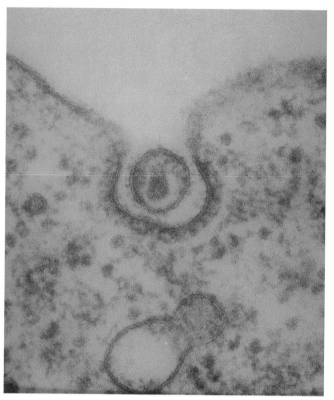
(b)

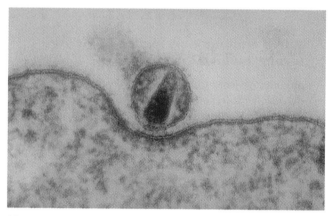
(c)

FIGURE 22.21   Release of new HIV particles from an infected cell. (a) The virus approaches the cell surface. (b) The virus buds off from the cell surface. (c) The virus, free from the cell surface, can now infect another T cell. Each round of replication releases several dozen new virus particles, gradually decreasing the number of T4 cells and reducing the body's ability to mount an immune response.

The virus selectively infects the T4 helper cells of the immune system. Inside the cell, the RNA is transcribed into a DNA molecule by reverse transcriptase, and the viral DNA is inserted into a human chromosome, where it can remain for months or years.

At a later time, when the infected T cell is called upon to participate in an immune response, the cellular and viral DNAs are transcribed. The viral RNA transcript is translated into viral proteins, and new viral particles are formed (Figure 22.21). These bud off the surface of the T cell, rupturing and killing the cell and setting off a new round of T-cell infection. Gradually, over the course of HIV infection, there is a decrease in the number of helper T4 cells. Recall that these cells act as the master "on" switch for the immune system. As the T4 cell population falls, there is a decrease in the ability to mount

an immune response. The results are increased susceptibility to infection and increased risk of certain forms of cancer. Eventually, the outcome is premature death brought about by any of a number of diseases that overwhelm the body and its compromised immune system. The relationship between the loss of T4 cells and the progression of HIV infection is strong and can be used to monitor the status of infected individuals (Figure 22.22).

HIV is transmitted through body fluids from infected individuals to noninfected individuals; these fluids include blood, semen, vaginal secretions, and breast milk. The virus is not viable for more than 1 or 2 hours outside the body and cannot be transmitted by food, water, or casual contact. At this time, treatment options are rather scant and include the use of drugs that limit the reproduction of the virus.

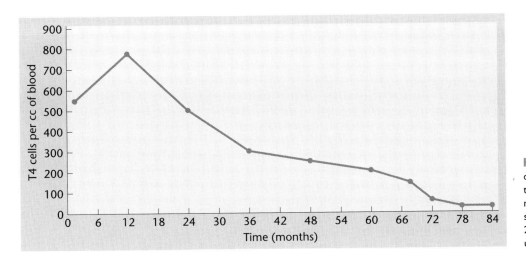

FIGURE 22.22   T4-cell levels during the course of an HIV infection. Following HIV infection, the number of T4 cells rise, then begin a slow decline. When levels fall below 200, the clinical symptoms of AIDS usually appear.

Autoimmunity

As the immune system matures, a state of **immune tolerance** develops between cells of the body and those of the immune system. This distinction between self and nonself prevents the immune system from attacking and destroying body tissues. Occasionally, immune tolerance breaks down, causing an autoimmune response in which the immune system turns against cells, tissues, and/or organs in the body. This breakdown of immune tolerance can take place in two ways: (1) A new antigen, previously unknown to the immune system, can appear on cells, or (2) a new antibody produced against a foreign antigen ends up also attacking a preexisting antigen on cells of the body. For example, normally the eye is an immunologically privileged site, meaning that many proteins in the eye are never exposed to the immune system when immunological tolerance develops, and thus are not recognized by the immune system as self. In a traumatic injury, however, antigens from the eye can be exposed to the immune system, and since these are not recognized as self, antibodies against eye proteins are made. These antibodies may attack and cause blindness even in the uninjured normal eye, a condition known as *sympathetic ophthalmia*.

In other cases, such as infection with certain strains of streptococcus, the antibodies produced against the bacterial antigens are capable of attacking and destroying body cells that carry antigens similar to those of streptococcus. In rheumatic fever, an infection with streptococcus causes antibodies to be produced that kill the invading bacteria, but these antibodies also attack cells in the valves of the heart, causing the heart problems typically associated with rheumatic fever. Some autoimmune disorders are systemic, affecting many organs, whereas others are specific to individual tissues and organs. There appears to be a large number of autoimmune disorders that affect connective tissues, and these may form the underlying basis for many forms of arthritis and other connective tissue diseases.

One autoimmune disease with a specific target is insulin-dependent diabetes (IDDM), or juvenile diabetes. As the name implies, the disease starts in childhood and requires daily injections of insulin as a therapy. Insulin is a hormone produced by clusters of cells, called *islets*, embedded in the pancreas. IDDM develops when the immune system attacks and destroys the islet cells that produce insulin. Without insulin, there is no control over sugar levels in the blood. Initial symptoms include thirst and high levels of blood sugar. Without treatment, the disease progresses to kidney failure, blindness, heart disease, and premature death. More than 50 percent of affected individuals die within 40 years of onset.

The HLA alleles DR-3 and DR-4 are associated with an increased risk for IDDM. Each allele is associated with a 15-fold greater risk, and the combination of DR-3 and DR-4 is associated with a 30-fold greater risk. The mechanism by which the immune system attacks is not yet clear, and other factors are involved because carrying a DR-3 or DR-4 allele alone is not sufficient to cause IDDM. At least one more gene, and possibly two, outside the HLA locus may play a role in stimulating the immune system to destroy the islet cells.

GENETICS, TECHNOLOGY, AND SOCIETY

Why Is There Still No Effective Vaccine against AIDS?

When AIDS (acquired immunodeficiency syndrome) was found to be caused by a virus in 1984, hopes were immediately raised that a vaccine could soon be developed to stem the spread of this disease. Optimism that a vaccine would soon be available seemed justified; after all, vaccines against other viral diseases, including smallpox and polio, have led to their virtual eradication. But more than a decade has now passed and an effective AIDS vaccine is still not available, despite a massive effort to develop one. The reasons for this are many, some having to do with the stealthy way the HIV virus infects its human hosts. However, the main reason a vaccine remains elusive is that the HIV virus mutates so rapidly into variant forms that antibodies directed against it do not remain effective for long.

The ultimate cause of AIDS is infection with two distinct but related viruses called HIV-1 and HIV-2. These are retroviruses belonging to the lentivirus subclass. Each virus particle consists of two single-stranded RNA molecules, both about 9500 nucleotides long, enclosed by a protein core, which is itself surrounded by a glycoprotein envelope. The genome of the HIV viruses contains three genes (*gag*, *pol*, and *env*), typical of all retroviruses, as well as five short regulatory genes (*tat*, *rev*, *nef*, *vpu*, and *vif*) characteristic of the lentivirus subgroup.

The two genes most responsible for the difficulty in developing a vaccine are *pol* and *env*. The *pol* gene encodes the reverse transcriptase (RT) enzyme that transcribes the viral RNA to a double-stranded DNA molecule, which can then integrate into the DNA of the host cell. The *env* gene encodes a precursor protein, called gp160, which is cleaved into a surface glycoprotein, gp120, and a transmembrane glycoprotein, gp41. The gp120 protein, residing on the surface of the viral particle, is the major antigenic determinant of the virus.

Studies in the late 1980s revealed the RT of HIV-1 to be extremely error-prone, introducing one or more mutations into the genome during each round of replication. This error rate is about 1,000,000 times as high as that of eukaryotic DNA polymerases, and is 10 times as high as most other retroviral RTs. The many mutations created by the HIV-1 RT are not randomly distributed throughout the genome. The highest rate of nucleotide substitution occurs in certain parts of the *env* gene, resulting in regions of gp120 that are highly variable in their amino acid sequence as well as their antigenicity.

We may think that such a high rate of mutation would be harmful to the HIV virus, as it would be to most prokaryotic and eukaryotic cells, but instead it appears to confer a selective advantage. Within the first few days after infection, new variants of the virus are created with different amino acid sequences in the variable regions of gp120. The immune system of the host quickly produces antibodies directed against gp120, but the diversity of the viruses that descend from the original virus, each with a variant gp120, allows some of them to escape destruction and go on replicating and infecting more cells. Eventually, mutations in gp120 may contribute to the appearance of more virulent types of the virus capable of infecting and killing the so-called CD4+ T helper cells, triggering the collapse of the host's immunological defenses and the progression to full-blown AIDS.

Most efforts to develop an anti-HIV vaccine have focused on gp120, the primary antigenic determinant of the virus. Some attempts have involved genetically engineering a harmless virus, such as vaccinia, to produce a peptide matching part of the HIV gp120 on its surface. Injecting these

CHAPTER SUMMARY

1. The immune system is composed of two branches; one directs cell-mediated immunity and T cells, and the other involves B cells and antibody-mediated immunity. The cells of the immune system originate from stem cells in bone marrow, and after maturation they move through the circulatory and lymphatic system as components of the immune system.

2. One of the most complex families of genes specifying proteins is the one participating in the immune response. Human antibodies or immunoglobulins are characterized by amino acid sequence diversity. DNA recombination during maturation of B cells produces the genetic variation required to match millions of different antigens with corresponding antibodies.

3. The immune response depends on the functional interrelationships of macrophages, which present foreign antigens to helper T cells, which, in turn, activate appropriate B cells to divide and produce large quantities of antibodies. The progress of the immune response is monitored by T8 suppressor cells that modulate the response and shut it down when it is no longer needed. Both T and B memory cells remain in circulation after the primary immune response, and they serve as the basis for a rapid and massive response if the immune system is challenged by the same antigen again. Formation of the memory cells is the basis for vaccination against pathogenic organisms.

recombinant viruses should then stimulate the production of anti-gp120 antibodies, which, in theory, will protect the host against subsequent infection with authentic HIV. Unfortunately, all the potential vaccines made this way have proved ineffective. The problem seems to be that the presence of antibodies against one fragment of gp120 are simply ineffectual against the diversity of gp120 proteins present soon after HIV infection. Other attempts at immunization using whole virus particles that have been inactivated have not been successful, probably for the same reason.

A great deal has been learned about the biology of HIV from these efforts, knowledge that is being used to devise alternative strategies. One approach is to base vaccines on the conserved regions of gp120, which remain unchanged by mutation. Another tack is to design a vaccine that will stimulate the infected person's T cells rather than to target the virus directly. Finally, infection with the less virulent HIV-2 virus was found to protect the host from subsequent infection with HIV-1, leading to suggestions that vaccines based on an inactivated HIV-2 particle might defend against HIV-1.

While promising anti-HIV treatments have recently been devised using combinations of drugs that inhibit the reproduction of the virus, all of the experimental vaccines tested to date have proved ineffective. Despite these discouraging results, the first truly large-scale trials of two gp120 vaccines are about to begin in Thailand, where 1 in 60 is infected with HIV. Whatever optimism there once was about quickly developing an anti-HIV vaccine has long since faded, however, and some AIDS researchers now fear that an effective, fully tested vaccine may be 25 years in the future. In the meantime, another 10,000 people throughout the world become infected with HIV every day.

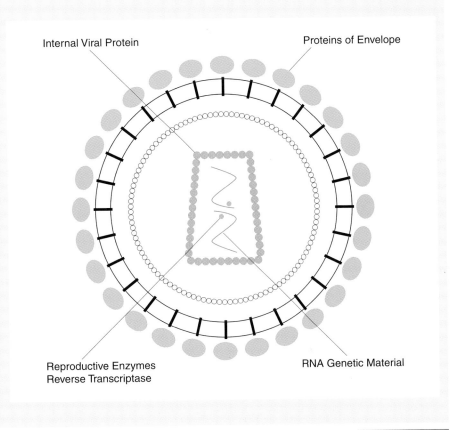

Internal Viral Protein

Proteins of Envelope

Reproductive Enzymes
Reverse Transcriptase

RNA Genetic Material

4. The genetic basis of blood groups is the presence of cell-surface antigens and the ability to make antibodies against foreign antigens. The success or failure of blood transfusions depends on matching blood types. Rh incompatibility between mother and fetus can result from antibodies produced in the maternal circulation against antigens on the blood cells of the fetus.

5. A class of cell-surface antigens is produced by a complex of genes on human chromosome 6 called the HLA complex. With a large number of alleles, the combination carried by an individual as haplotypes constitutes a genetic signature for each individual. Successful tissue and organ transplantation depends on matching the HLA haplotypes of the donor and recipient. Certain HLA alleles are associated with specific diseases and can be used to diagnose and predict risk factors for such diseases.

6. Mutations that disrupt the immune system work by altering the development and/or function of the component cells that participate in the immune response. These mutations can affect one cell type, a subsystem, or the entire immune system. In addition to genetic forms of disruption, the immune system can be inactivated by external factors, most notably by infection with the human immunodeficiency virus (HIV). This virus selectively infects and destroys the T cells that act as the "on" switch for the immune system, producing a gradual loss in the ability of an infected individual to mount an immune response. As a result, infectious diseases that would ordinarily be suppressed by the immune system can pose a serious risk to the life of HIV-infected individuals.

KEY TERMS

INSIGHTS AND SOLUTIONS

1. An antibody molecule contains two identical H chains and two identical L chains. This results in antibody specificity, with the antibody binding to a specific antigen. Recall that there are five classes of H genes and two classes of L genes. Because of the high degree of variability in the genes encoding the H and L chains, it is possible (even likely) that an antibody-producing cell contains two different alleles of the H gene and two different alleles of the L gene. Yet the antibodies produced by the plasma cell contain only a single type of H chain and a single type of L chain. How can you account for this, based on the number of classes of H and L genes and the possibility of heterozygosity?

Solution: Although an antibody-producing cell contains different classes of H and L genes, and although it is entirely possible and even likely that a given antibody-producing cell will contain different alleles for the H gene and for the L gene, a phenomenon known as *allelic exclusion* allows expression of only one H allele and one L allele in any given plasma cell at a given time. That is not to say that the same antibody is produced over the life span of the antibody-producing cell, however. At first, many plasma cells produce antibodies of the IgM class, and at later times they may produce antibodies of a different class (e.g., IgG). Even in switching between different H genes, allelic exclusion is maintained, with only one allele of an H gene (or L gene) being expressed at a given time.

2. In the mouse κ L gene, there are 250 V regions, 4 J regions, and 3 D regions. H chains have 250 V, 10 D, and 4 J regions. How many combinations of these immunoglobulin components are possible?

Solution: First, calculate the number of different L combinations, the number of different H combinations, then the number of H and L combinations. For the L genes: 250 V × 4 J × 3 D = 3000. For the H genes: 250 V × 10 D × 4 J = 10,000. The combination of 3000 L genes with 10,000 H genes gives a total of 3×10^6 combinations of immunoglobulins. This actually is a great underestimate, for it does not take into account several other sources of diversity, including the following: a phenomenon called junctional diversity that produces imprecise joining between VJ or VDJ segments; the insertion of extra nucleotides into the VD or VJ segments of H chains (N regions), lambda L genes, the five classes of H chains, or the high level of point mutations that arise in splice antibody genes. When these are added in, there is probably a 10,000-fold increase in diversity for about 3×10^{11} different combinations of immunoglobulins.

PROBLEMS AND DISCUSSION QUESTIONS

1. What do the following symbols represent in immunoglobulin structure: V_L, C_H, IgG, J, and D? What is the most acceptable theory for the generation of antibody diversity?

2. Figure 22.11 shows nine unique gene sequences formed by recombination. How many other unique sequences can be formed?

3. If germ line DNA contains 10 V, 30 D, 50 J, and 3 C genes, how many unique DNA sequences can be formed by recombination?

4. If there are 5 V, 10 D, and 20 J genes available to form a heavy-chain gene, and 10 V and 100 J genes available to form a light chain, how many unique antibodies can be formed?

5. Distinguish between an antigen and an antibody.
6. What are the functions of helper T cells and suppressor T cells?
7. If a woman has a child with hemolytic disease of the newborn (HDN), what are the possible genotypes of the mother, father, and child? After discovering that her child has HDN, the mother requests treatment with antibodies in the belief that it will prevent HDN in subsequent children. Is she correct? Why?
8. Why is type AB blood regarded as a universal recipient, and type O blood regarded as the universal donor? Which is more important in blood transfusion matching: the antibodies of the donor and recipient, or the antigens of the donor or recipient? Why?
9. How many polypeptide chains are present in an IgG antibody molecule? How many different polypeptide chains are represented in this molecule? How many gene segments from a germ line cell were combined to produce the polypeptide chains in this molecule?
10. What is an HLA haplotype? How many of these haplotypes do you carry? How are these haplotypes inherited?
11. Describe the immunological processes involved in the rejection of an organ transplant.

12. Why is the idea that every cell carries a complete set of genetic information not true for lymphocytes?

EXTRA-SPICY PROBLEMS

13. A couple comes to you with the following story. They have a teenage daughter who suffers from chronic leukemia. Her only hope of a cure is a bone marrow transplant from a matched donor. But rejection of the transplant will occur if tissue from the donor possesses any histocompatibility antigens foreign to the recipient. Her only sibling is a brother who is not genetically compatible. This couple has conceived a third child in the hope that this child will have the same histocompatibility antigens as their daughter and therefore would serve as the perfect donor of bone marrow cells.

They ask you to calculate the probability that their third child will be compatible (that is, share an identical set of histocompatibility antigens) with their daughter. What is the probability? (This is a true story and the third child was compatible. Did they beat the odds?)

SELECTED READINGS

BEATTY, P. G., MORI, M., and MILFORD, E. 1995. Impact of racial genetic polymorphism on the probability of finding an HLA-matched donor. *Transplantation* 60:778–83.

CHEN, J., and ALT, F. 1993. Gene rearrangements and B-cell development. *Curr. Opin. Immunol.* 5:194–200.

COHEN, J. 1996. A shot in the dark. *Discover* (June) 17:66–73.

CORZO, D., SALAZAR, M., GRANJA, C. B., and YUNIS, E. S. 1995. Advances in HLA genetics. *Exp. Clin. Immunogenet.* 12:156–70.

FAUCI, A. S. 1995. AIDS: Newer concepts in the immunopathogenic mechanisms of human immunodeficiency virus disease. *Proc. Assoc. Am. Physicians* 107:1–17.

FRENCH, D., LASKOV, R., and SCHARFF, M. 1989. The role of somatic hypermutation in the generation of antibody diversity. *Science* 244:1152–57.

GOLDE, D. 1991. The stem cell. *Sci. Am.* (Dec.) 265:86–93.

HUBER, B. T. 1992. Mls genes and self-superantigens. *Trends Genet.* 8:399–402.

JUST, J. J. 1995. Genetic predisposition to HIV-1 infection and acquired immunodeficiency virus syndrome. A review of the literature examining associations with HLA. *Hum. Immunol.* 44:156–69.

KING, L., and ASHWELL, J. 1993. Signalling for the death of lymphoid cells. *Curr. Opin. Immunol.* 5:368–73.

LIEBER, M. R. 1991. Site-specific recombination in the immune system. *FASEB J.* 5:2934–44.

———. 1992. The mechanism of V(D)J recombination: A balance of diversity, specificity, and stability. *Cell* 70:873–76.

LIEBER, M. R., CHANG, C.P., GALLO, M., GAUSS, G., GERSTEIN, R., and ISLAS, A. 1994. The mechanism of V(D)J recombination: site specificity, reaction fidelity and immunologic diversity. *Semin. Immunol.* 6:143–53.

LEIDEN, J. 1992. Transcriptional regulation during T-cell development: The alpha TCR gene as a molecular model. *Immunol. Today* 12:22–30.

MAIZELS, N. 1995. Somatic rhypermutation: how many mechanisms diversify V region sequences? *Cell* 83:9–12.

MARCHALONIS, J. T., and SCHLUTER, S. F. 1994. Development of an immune system. *Ann. NY Acad. Sci.* 712:1–12.

MARTIN, S., and DYER, P. A. 1993. The case for matching MHC genes in human organ transplantation. *Nat. Genet.* 5: 210–13.

MASUCCI, M., GAVIOLI, R., DE CAMPOS-LIMA, P., AHZNG, Q., TRIVEDI, P., and DOLCETTI, R. 1993. Transformation-associated Epstein-Barr virus antigens as targets for immune attack. *Ann. NY Acad. Sci.* 690:86–100.

MURRAY, J. E. 1994. The Nobel Lectures in Immunology: The Nobel Prize for Physiology or Medicine, 1990. The first successful organ transplants in man. *Scand. J. Immunol.* 39:1–11.

NOWAK, M. A., and MCMICHAEL, A. J. 1995. How HIV defeats the immune system. *Sci. Am.* (Aug.) 272:58–65.

SCHATZ, D. G., OETTINGER, M. A., and SCHLISSEL, M. S. 1992. V(D)J recombination: Molecular biology and regulation. *Annu. Rev. Immunol.* 10:359–83.

SCHUURMAN, J. J. 1994. Molecular mechanisms of transplant rejection. *Clin. Investig.* 72:715–18.

SCHWARTZ, R. S. 1995. Jumping genes and the immunoglobin V gene system. *N. Engl. J. Med.* 333:42–44.

THOMSON, G. 1988. HLA disease associations: Models for insulin-dependent diabetes mellitus and the study of complex human genetic disorders. *Annu. Rev. Genet.* 22:31–50.

TOMLINSON, I. P., and BODMER, W. F. 1995. The HLA system and the analysis of multifactorial genetic disease. *Trends Genet.* 11:493–98.

VAN BOEHMER, H., and KISIELOW, P. 1991. How the immune system learns about itself. *Sci. Am.* (Oct.) 265:74–86.

YAMAMOTO, F., CLAUSEN, H., WHITE, T., MARKEN, J., and HAKAMORI, S. 1990. Molecular genetic basis of the histo-blood group ABO system. *Nature* 345:229–33.

YOUNG, J., and COHN, Z. 1988. How killer cells kill. *Sci. Am.* (Jan.) 258:38–44.

23

The Genetics of Behavior

CHAPTER CONCEPTS

Behavior in many prokaryotic and eukaryotic organisms is controlled by both single genes and polygenic systems. These genes work by controlling cascades of metabolic reactions such as phosphorylation, methylation, and ion flux and produce effects at the cellular level that are translated into behavioral responses.

Behavior is generally defined as a reaction to stimuli or environment. In broad terms, every action, reaction, and response represents a type of behavior. Animals run, remain still, or counterattack in the presence of a predator; birds build complex and distinctive nests; fruit flies execute intricate courtship rituals; plants bend toward light; and humans reflexively avoid painful stimuli as well as "behave" in a variety of ways as guided by their intellect, emotions, and culture.

Even though clear-cut cases of genetic influence on behavior were known in the early 1900s, the study of behavior was of greater interest to psychologists, who were concerned with learning and conditioning. Although some traits were recognized as innate or instinctive, behavior that could be modified by prior experience received the most attention. Such traits or patterns of behavior were thought to reflect the previous environmental setting to the exclusion of the organism's

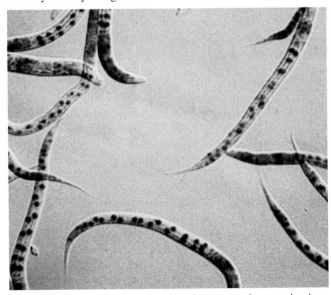

Genetically engineered *Caenorhabditis elegans* roundworms that have turned blue in response to environmental stress, such as toxins or heat.

genotype. This philosophy served as the basis of the **behaviorist school**.

Such thinking provided a somewhat distorted view of the nature of behavioral patterns. It is logical that a genotype can be expressed in a hierarchy, each with a unique environmental setting (e.g., cell, tissue, organ, organism, population, surrounding environment) and that a behavioral pattern must rely on the expression of the individual genotype for its execution. Nevertheless, the so-called **nature–nurture controversy** flourished well into the 1950s. By that time it became clear that while certain behavioral patterns, particularly in less complex animals, seemed to be innate, others were the result of environmental modifications limited by genetic influences. The latter condition is particularly true in organisms with more complicated nervous systems.

Since about 1950, studies of the genetic component of behavioral patterns have intensified, and support for the importance of genetics in understanding behavior has increased. The prevailing view is that all behavior patterns are influenced both genetically and environmentally. The genotype provides the physical basis and/or mental ability essential to execute the behavior and further determines the limitations of environmental influences.

Behavior genetics has blossomed into a distinct specialty within genetics as more and more behaviors have been found to be under genetic control. This chapter provides an overview of the role of genes in behavior. There seems to be no question that this topic will be one of the most exciting areas of genetics in the years to come.

The Methodology of Behavior Genetics

Three different approaches have been used to study the genetics of behavior. The first involves the identification of behavioral differences between genetic strains of the same species or between closely related species. If closely related organisms exist in similar environments and their survival needs are identical, any observed behavioral differences may be due to genotypic differences. In the second approach, a variant behavioral trait is selected from a genetically heterogeneous population. If this trait can be transferred to another strain by genetic crosses, the positive influence of the genotype is established. The first two approaches identify behavioral patterns as being under the control of genes. Genetic crosses can then be used to establish inheritance patterns. In many cases, this approach has shown that the variant behavior does not have

a simple Mendelian pattern of inheritance. Instead, many behavioral traits appear to be polygenic (i.e., many genes controlling one trait).

The third approach, used extensively today, studies the effects of single genes on behavior. Using inbred strains of organisms with relatively constant genotypes, spontaneous or induced mutants with behavioral deviations are collected and analyzed. Although a behavioral trait may be controlled by many genes, mutation in a single gene can alter the behavior. Genetic analysis of the mutant allele can provide insight into the genetic control of this single aspect of a behavioral trait. This approach offers the most objective information concerning the role of genes in behavior. In selected organisms, defining a gene-controlling behavior is a prelude to the isolation, cloning, and molecular characterization of such genes.

The Comparative Approach

Alcohol preference in mice compares behavior in two or more closely related strains, and illustrates the first of the three approaches to the study of behavior genetics discussed above. The second approach, involving selection for a modification of behavior, will be described by studies of learning in rats and geotaxis in *Drosophila*.

Alcohol Preference

Studies of **alcohol preference** in mice have compared preference or aversion to ethanol in different strains. In a typical study, alcohol consumption rates in four strains of inbred mice were measured over a period of 3 weeks. Each strain was presented with seven vessels containing either pure water or alcohol varying in strength from 2.5 to 15.0 percent. Daily consumption was measured. Table 23.1 shows the proportion of alcohol to total liquid consumed on a weekly basis.

Examination of these data shows that the C57BL and C3H/2 strains exhibit a preference for alcohol, while BALB/c and A/3 demonstrate an aversion toward it. As the mice have been raised over many generations in a constant environment, the differences in alcohol preference are attributed to genotypic differences between the strains. Presumably, these strains vary only by the fixation of different alleles at particular loci.

Despite the observed differences in alcohol consumption between mouse strains, the underlying mechanisms remain unclear. It has been suggested that differences in alcohol preference, metabolism, and severity of withdrawal symptoms seen in various mouse strains are related to differences in the major enzymes of alcohol metabolism, particularly **alcohol dehydrogenase (ADH)** and **acetaldehyde dehydrogenase (AHD)**. The major enzymes and their isozymes have been isolated and charac-

TABLE 23.1 Alcohol Consumption in Mice

Strain	Week	Proportion of Absolute Alcohol to Total Liquids	$\bar{x}$
C57BL	1	0.085	
	2	0.093	9.4% alcohol
	3	0.104	
C3H/2	1	0.065	
	2	0.066	6.9% alcohol
	3	0.075	
BALB/c	1	0.024	
	2	0.019	2.0% alcohol
	3	0.018	
A/3	1	0.021	
	2	0.016	1.7% alcohol
	3	0.015	

Source: Modified from Rogers and McClearn, 1962. Reprinted by permission from *Quarterly Journal of Studies on Alcohol*, Vol. 23, pp. 26–33, 1962. Copyrighted by Journal of Studies on Alcohol, Inc. New Brunswick, NJ 08903.

terized with respect to biochemical and kinetic properties, and their distribution and activities in different mouse strains have been catalogued. Genetic variants of these enzymes present in different strains of inbred mice have been used to try to establish an association between a given form of the enzyme and a form of alcohol-related behavior. To date, no correlation between alcohol preference or metabolism and these biochemical markers has been established. Other genetic markers, including the neuropharmacological effects of alcohol, may be needed to establish a link between specific genes and alcohol-related behavior in mice.

Open-Field Behavior

Open-field tests are used to study exploratory and emotional behavior in mice. When placed in a new environment, mice normally explore their surroundings, but they are a bit cautious or "nervous" about the new setting. The latter response is evidenced by their elevated rate of defecation and urination. To study this behavior in the laboratory, an enclosed, brightly illuminated box with the floor marked into squares is used. Exploration is measured by counting movements into different squares, and emotion is measured by counting the number of defecations.

As in the study of alcohol preference, different inbred strains of mice vary significantly in their response to the open-field setting. John C. DeFries and his associates have centered their work on two strains, BALB/cJ and C57BL/6j. The BALB strain is homozygous for a coat color allele, *c*, and is albino, whereas the C57 strain has normal pigmentation (*CC*). BALB demonstrates low exploratory activity and is very emotional, whereas C57 is active in exploration and relatively nonemotional.

To test for genotypic differences, the two strains were crossed and then interbred for several generations. Each generation beyond the F_1 contained albino and nonalbino mice, and these were tested for open field behavior. In all cases, pigmented mice behaved as strain C57, whereas albino mice behaved as BALB. The general conclusion is that the *c* allele behaves pleiotropically, affecting both coat color and behavior.

Heritability (review heritability and its measurement in Chapter 7) analysis has been used to assess the input of the *c* gene to these behavioral patterns. Analysis indicates that that this locus accounts for 12 percent of the variance in open-field activity and 26 percent of the variance in defecation-related emotion. The heritability values indicate that these behaviors are polygenically controlled.

What might be the relation between albinism and behavior? To answer this, albino and nonalbino mice were tested for open-field behavior under white light and red light. Red light provides little visual stimulation to mice. The behavioral differences between mice with the two types of coat pigmentation disappeared under red light, indicating that the open-field responses of albino mice are visually mediated. This is not surprising; albino mice are photophobic, and lack of pigmentation in albinos extends to the iris as well as to the coat.

Artificial Selection

The second major approach in behavior genetics—selection for modified behavior from a genetically heterogenous population and subsequent inbreeding—leads to the production of strains with significant differences in behavior. The end result is similar to the previous approach of initially comparing inbred lines. The study of **maze learning** in rats is an example of this approach.

Maze Learning in Rats

The first experiment of this kind was reported by E. C. Tolman in 1924. He began with 82 white rats of heterozygous ancestry and measured their ability to "learn" to obtain food at the end of a multiple T-maze (Figure 23.1) by recording the number of errors and trials. When first exposed to the maze, a rat explores all alleys and eventually arrives at the end, to be rewarded with food. In succeeding trials, fewer and fewer mistakes are made as the rat learns the correct route. Eventually, a hungry rat may proceed to the food with no errors.

From the initial 82 rats, nine pairs of the "brightest" and "dullest" rats were selected and mated to produce two lines. In each generation, selection was continued. Even in the first generation, Tolman demonstrated that he could select and breed rats whose offspring performed more efficiently in the maze. Subsequently, his approach was pursued by others, notably R. C. Tryon, who in 1942 published results of 18 generations of selection.

FIGURE 23.1 A multiple T-maze used in learning studies with rats.

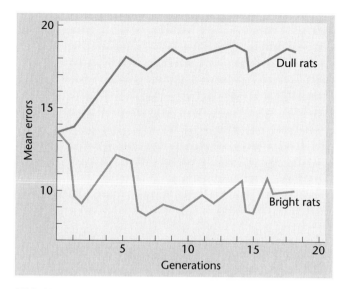

FIGURE 23.2   Selection for the ability and inability of rats to learn to negotiate a maze.

As shown in Figure 23.2, two lines with significant differences in maze-learning ability were established. There is some variation around the mean, but by the eighth generation there was no overlap between lines. That is, the dullest of the bright rats were superior to the brightest of the dull rats by the eighth generation.

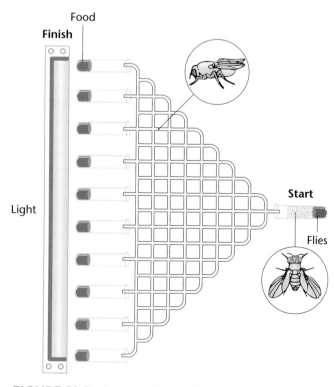

FIGURE 23.3   Schematic drawing of a maze used to study geotaxis in *Drosophila.*

These two lines were also used to study other behavior traits, with varying results. Bright rats were found to be better in solving hunger-motivation problems but inferior in escape-from-water tests. Bright rats were also found to be more emotional in open-field experiments. These results show that selection of genetic strains superior in certain traits is possible, but care must be taken not to generalize such studies to overall intelligence, which is composed of many learning parameters.

Geotaxis in Drosophila

A **taxis** is the movement toward or away from an external stimulus. The response may be positive or negative, and the sources of stimulation may include chemicals (chemotaxis), gravity (geotaxis), light (phototaxis), and so on.

To investigate **geotaxis** in *Drosophila*, Jerry Hirsch and his colleagues designed a mass-screening device that tests about 200 flies per trial, as shown in Figure 23.3. In this test, flies are added to a vertical maze. Flies that turn up at each intersection will arrive at the top of the maze; those that always turn down will arrive at the bottom; and those making both "up" and "down" decisions will end up somewhere in between. Flies could be selected for both positive and negative geotropism, establishing the genetic influence on this behavioral response.

As shown in Figure 23.4, mean scores may vary from +4.0 to −6.0, corresponding to the number of T-junctions the fly encounters. These data show that selection for negative geotropism is stronger than for positive geotropism. The two lines have now undergone selection for almost 30 years, encompassing over 500 generations, and the testing of more than 80,000 flies. Throughout the experiment, clear-cut but fluctuating differences have been observed.

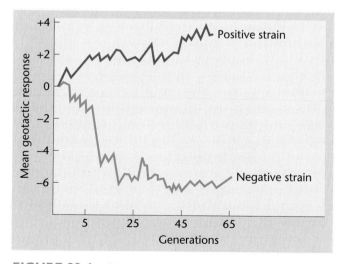

FIGURE 23.4   Selection for positive and negative geotaxis in *Drosophila* over many generations.

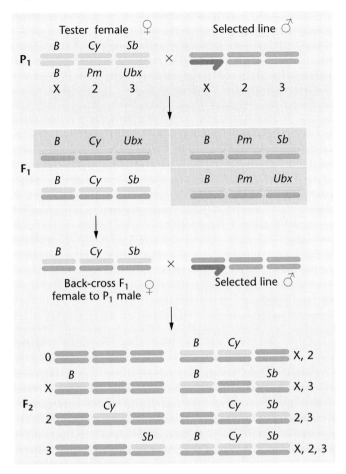

FIGURE 23.5 In this mating scheme in *Drosophila*, the effect of genes located on specific chromosomes that contribute to geotaxis can be assessed. The progeny produced by backcrossing the F₁ female contain all combinations of chromosomes. Examining the phenotypes of these flies makes it possible to determine which chromosomes from the selected strain are present. Subsequent testing for geotaxis is then performed (*B*—*Bar* eyes; *Cy*—*Curly* wing; *Sb*—*Stubble* bristles; *Ubx*—ultrabithorax). The designations alongside each genotype (X, 2, 3, etc.) indicate which chromosomes are heterozygous.

These results indicate that geotropism in *Drosophila* is controlled in a polygenic fashion.

Hirsch and his colleagues have analyzed the relative contribution of loci on different chromosomes to geotaxis. Loci on chromosomes 2 and 3 and the X were identified in an ingenious way. Crosses were performed to produce flies that were either heterozygous or homozygous for a given chromosome. Figure 23.5 shows how this is accomplished. From a selected line, a male is crossed to a "tester" female whose chromosomes are each marked with a dominant mutation. Each marked chromosome carries an inversion to suppress the recovery of any crossover products. One of the F₁ females heterozygous for each chromosome is backcrossed to a male from the original line. The resulting female offspring contain all combinations of chromosomes. The dominant mutations make it possible to recognize which chromosomes from the selected lines are present in homozygous or heterozygous configurations.

Combinations O and X (Figure 23.5) are homozygous and heterozygous, respectively, for the X chromosome. Similar combinations exist for chromosome 2 (0 and 2) and chromosome 3 (0 and 3). By testing these flies in the maze, it is possible to assess the behavioral influence of genes on any given chromosome.

For negatively geotaxic flies (the ones that go up in the maze), genes on the second chromosome make the largest contribution to the phenotype, followed by loci on chromosome 3 and the X (a gradient of chromosomes 2 > 3 > X). For the positive line (flies that go down in the maze), the reverse arrangement (a gradient of chromosomes X > 3 > 2) is true. The overall results indicate that geotaxis is under polygenic control, and that the loci controlling this trait are distributed on all three major chromosomes of *Drosophila*.

Further genetic testing in which each chromosome from a selected line has been isolated in homozygous form in an unselected background has been used to estimate the number of genes controlling the geotaxic response in *Drosophila*. This work indicates that a small number of genes, perhaps two to four loci, are responsible for this behavior.

Analysis of Single Gene Effects

The most definitive information about genetic control of behavior has come from the study of the effects of single alleles on behavior. In this approach, either spontaneous or induced mutations are analyzed in order to infer general principles about how normal behavior is created and regulated.

Nest-Cleaning in Bees

Honeybee nests are frequently infected with *Bacillus larvae*, a bacterium causing foulbrood disease. The disease may be controlled by **hygienic behavior** on the part of worker bees. To combat infection, worker bees open the cells containing infected larvae and remove them from the hive. Hives in which the workers exhibit hygienic behavior are resistant to infection, whereas hives containing workers that do not display removal behavior are susceptible to the disease.

In 1964, Walter Rothenbuhler published results of his cross between a hygienic (Brown) line with a nonhygienic (Van Scoy) line. This work strongly favors the idea that either a gene complex or two independently assorting recessive alleles (*u* and *r*) are responsible for hygienic behavior.

The F₁ offspring of a cross between hygienic and nonhygienic bees were all nonhygienic. However, when F₁ males were backcrossed to hygienic females, four pheno-

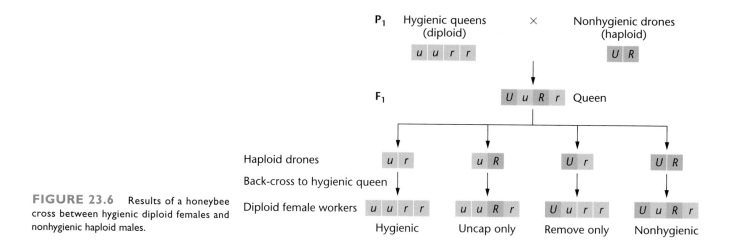

FIGURE 23.6 Results of a honeybee cross between hygienic diploid females and nonhygienic haploid males.

types were produced in roughly equal proportions, as shown in Figure 23.6. While one phenotypic class was hygienic and one was nonhygienic, the other two classes had intermediate phenotypes. One could uncap cells but not remove infected larvae. The other class was able to remove larvae if the cells were artificially uncapped. It appears that one gene pair (u/u) controls uncapping behavior, and a second gene (r/r) determines ability to remove larvae.

Molecular Motors in Bacteria

Behavioral responses also exist in single-celled organisms such as bacteria, which lack an organized nervous system. Bacteria exhibit **chemotaxis** and are attracted to or repelled by a variety of chemical stimuli. These responses have now been analyzed in some detail, and mutants that disrupt normal behavior have been isolated and characterized.

Motile bacteria such as *Escherichia coli* and *Salmonella* respond to gradients of chemicals by moving along the gradient. This movement is controlled by flagellar action (Figure 23.7). As they move in response to a chemical gradient, the cells exhibit two forms of motion. The first

type of movement consists of periods of smooth swimming called *runs* generated by counterclockwise (CCW) rotation of flagella, which act like propellers, driving the bacteria through the medium. During runs, the flagella coil together to form a cohesive bundle. Movement of this bundle drives the cell in a single direction. Also during runs, which have an average duration of 1 second, the bacteria swim about 10 to 20 body lengths. Runs alternate with the second form of movement, called *tumbles*. During tumbles, the flagella switch direction and rotate clockwise (CW). In clockwise rotation, the flagella become dispersed, and each acts independently. As a result, during tumbles, the cell covers little or no distance (Figure 23.8). Tumbles, which last only about 0.1 second, serve to change the direction of the cell. In the presence of a chemical attractant, runs that carry the cell up the gradient are extended, whereas those that move the cell down the gradient are not (Figure 23.9). In the presence of a chemical repellent, the parameters are inverted.

The molecular details of the chemotactic response are becoming known, with the major components of the system identified and the role of several gene products described. The response of bacteria to chemical stimuli

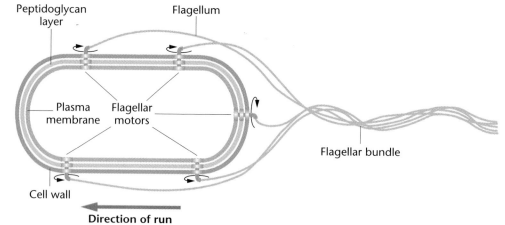

FIGURE 23.7 When the flagellar motors rotate in a counterclockwise (CCW) direction (as shown), the flagella form a bundle that works to push the cell along a mostly linear path called a *run*. When the flagellar motors rotate in a clockwise direction (CW) (not shown), the flagella operate independently, causing the cell to *tumble*.

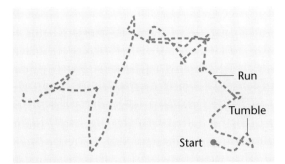

FIGURE 23.8   Runs and tumbles executed by an *E. coli* cell in a medium where no chemical gradient is present. Tracking of the cell began at the large dot at lower right. The cell moves via a series of runs interspersed with shorter periods of tumbling.

serves as a model for the molecular mechanisms by which cells process and transduce sensory signals. In this case, chemical signals are converted into the kinetic action of flagella, moving the cell toward or away from a stimulus. The *E. coli* has at least four types of membrane-spanning receptors that sense chemicals in the environment and initiate intracellular responses (Figure 23.10). These receptor proteins are products of a gene family called **transducers.** Each protein in the family has a unique receptor domain extending into the periplasmic space and a conserved domain that extends into the cytoplasm. The cytoplasmic surface of the chemoreceptor protein forms a complex with two cytoplasmic proteins, *Che*A and *Che*W. *Che*A and the chemoreceptor interact directly, and *Che*W enhances this association. *Che*A integrates the response of the cell to chemical stimuli. The *Che*A mutants are nonchemotactic.

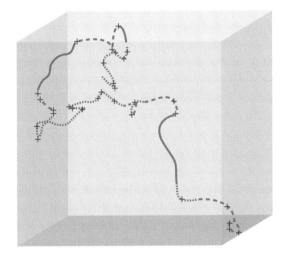

FIGURE 23.9   Runs and tumbles executed by a single *E. coli* cell in an exponential gradient of attractant (highest concentration at top). Movement up the gradient occurred in 35 runs and 34 tumbles. Two extended runs (indicated by the solid lines) represent most of the vertical distance traveled.

A change in the conformation of the chemoreceptor caused by the binding or release of a sensory molecule generates an intracellular signal. This signal is transmitted by changes in the phosphorylation of a group of cytoplasmic proteins. These proteins act as a molecular switch to change the direction (clockwise or counterclockwise) of flagellar rotation. In the presence of a repellent molecule, the rate of *Che*A phosphorylation is increased (Figure 23.10). Phosphorylated *Che*A, in turn, leads to the phosphorylation of *Che*Y. When *Che*Y is phosphorylated, it binds to the base of the flagellar motor and favors clockwise rotation of the flagellum. Recall that clockwise rotation causes the cell to tumble, during which the cell switches direction. Another protein, *Che*Z, turns off the action of *Che*Y by accelerating the dephosphorylation of *Che*Y.

In the presence of an attractant, the cycle of phosphorylation is reversed, decreasing the rate of *Che*A phosphorylation, which in turn slows the phosphorylation of *Che*Y. Dephosphorylated *Che*Y favors counterclockwise rotation of the flagella, moving the cell on a run toward the attractant. The result of this behavior is detection of chemical gradients, and movement toward attractants (food) and away from repellents.

The chemoreceptors are aggregated into clusters, located at the poles of the cell. They are not clustered near the base of the flagellar motors. As expected, *Che*A and *Che*W proteins are also localized in the cytoplasm at the poles, with the *Che*A protein associated with the inner membrane of the cell. In mutants lacking all four receptors, *Che*A and *Che*W proteins are randomly distributed throughout the cytoplasm, indicating that the polar localization of these gene products depends on the presence of receptors. Studies of *Che*W mutants indicate that this protein is also required for aggregation and polar localization of the receptors. In *Che*W deletion mutants, the *Che*A receptor complexes are randomly distributed. These results suggest that signal transduction is not initiated by the formation of the complex, but rather through conformational changes in the complex brought about by sensory molecules.

The chemotactic behavior of *E. coli* involves a number of gene products that receive and process signals and generate a response through a metabolic network that involves protein phosphorylation. At the present time, chemotaxis in bacteria represents the best-known example of the relationship between a behavior and its underlying molecular mechanism.

Behavior in Nematodes

In 1968, Sydney Brenner began an investigation of behavior genetics in the nematode *Caenorhabditis elegans*. In his earlier work, Brenner made important con-

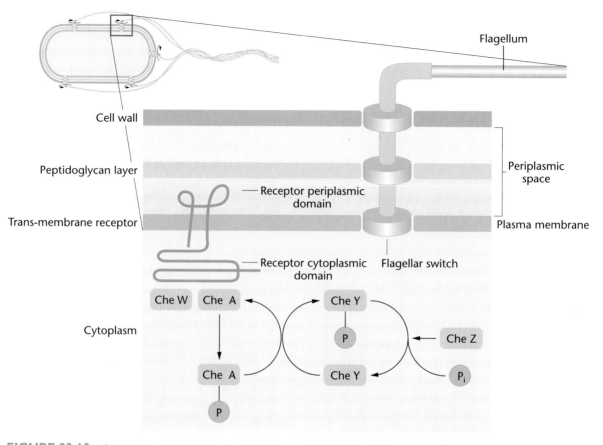

FIGURE 23.10 Sensory transduction during bacterial chemotaxis. Stimulation of the transmembrane receptor leads to the phosphorylation of *Che*A with the aid of *Che*W. Activated *Che*A transfers a phosphate group to *Che*Y. Phosphorylated *Che*Y interacts with the flagellar switch to alter direction of flagellar rotation. *Che*Z inactivates *Che*Y by removing the phosphate group.

tributions to the understanding of DNA replication, F factors, mRNA, and the genetic code. When he turned to the study of behavior, he hoped that it would be possible to dissect genetically the nervous system of *C. elegans* using techniques previously applied successfully to other organisms.

He chose this nematode because it has a relatively simple nervous system. Adult worms are about 1 mm long and contain only 959 somatic cells, about 350 of which are neurons. As a result, it is possible to cut serial sections of an embedded organism and reconstruct the entire organism and its nervous system in the form of a three-dimensional model. Brenner then hoped to induce large numbers of behavioral mutations and to correlate aberrant behavior with structural and biochemical alterations in the nervous system. Some progress has been made toward these goals, particularly in isolating large numbers of mutations. In early experiments, three general types of behavioral mutants were characterized. Worms are positively chemotactic to a variety of stimuli (cyclic AMP and GMP; anions such as Cl^-, Br^-, and I^-; and cations such as Na^+, Li^+, K^+, and Mg^{++}). As shown in Figure 23.11,

positive attraction can be tracked in gradients on agar plates. The study of chemotactic mutants has shown that sensory receptors in the head alone mediate the orientation responses to attractants.

A second class of behavior studied involves *thermotaxis*. Cryophilic mutants move toward cooler temperatures, and thermophilic mutants move toward warmer temperatures. However, this behavior has not yet been correlated with the responsible component of the nervous system.

A third class of behavior involves generalized movement on the surface of an agar plate. Of 300 induced mutations, 77 affected the movement of the animal. Whereas wild-type worms move with a smooth, sinuous pattern, mutants are either **uncoordinated** (*unc*) or **rollers** (*rol*). Those that are uncoordinated vary from the display of partial paralysis to small aberrations of movement, including twitching. Rollers move by rotating along their long axis, creating circular tracks on an agar surface. Many of these mutants have been correlated with defects in the dorsal or ventral nerve cord or in the body musculature.

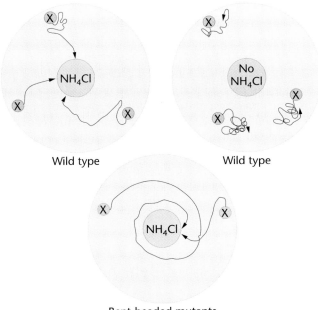

FIGURE 23.11 Chemotactic response to ammonium chloride (NH₄Cl) of wild-type and mutant *Caenorhabditis*.

Linkage mapping has also begun. Mutants are distributed on six linkage groups, corresponding to the haploid number of chromosomes characteristic of this organism. Numerous *unc* mutants are found on each of the six chromosomes, indicating extensive genetic control of nervous system development.

As the genetics of *C. elegans* has developed, analyses of the links among genes, the nervous system, and behavior have become more refined, beginning with a complex behavioral repertoire and isolating genes that control this behavior. As an example, feeding behavior in the worm can be used as a system to dissect genetic components of behavior. In *C. elegans*, the pharynx is a self-contained feeding pump composed of three parts: (1) the corpus, which ingests bacteria, and (2) the isthmus, which conducts bacteria to (3) the terminal bulb, where bacteria are ground up and passed into the intestine (Figure 23.12). The pharynx is surrounded by a basement membrane and contains a total of 80 cells (80 out of 959 in the organism), 20 of which are neurons making up the pharyngeal nervous system.

The pharyngeal nervous system is somewhat self-contained, with only one pair of nerves connecting it to the rest of the nervous system; this connection can be severed without impairing the action of the pharyngeal nerves. In fact, the pharynx can be dissected from the body and still function *in vitro*. Only one of the 20 neurons, M4, which innervates the posterior isthmus, is essential for life. When M4 is missing, the isthmus remains closed and bacteria are not transported for grinding and

digestion, and starvation ensues. Worms with an intact M4 neuron, but missing the 19 others, are viable, although these neurons are necessary for normal patterns of feeding.

In an ongoing study of the genetics of feeding, Leon Avery has set out to isolate and characterize the genes that control the presence or absence, the developmental fate, patterns of innervation, and function of the 20 neurons responsible for feeding behavior. Almost 38,000 progeny of mutagenized worms were screened for feeding-defective mutants. By linkage and mapping studies, 52 mutations were assigned to 35 genes located on all six chromosomes. Based on the number of mutants recovered and the number of genes involved, it is estimated that at least 60 genes are involved in feeding behavior in *C. elegans*, meaning that roughly half the genes remain to be identified.

The 52 mutations recovered to date fall into three broad phenotypic classes. The *eat* mutants affect the motion of the pharyngeal muscles; in addition, some *eat* mutants also affect the function of body-wall muscles. The *pha* mutants have misshapen pharynxes that prevent normal feeding behavior. The third class of mutants, *phm* mutants, all have weak and irregular pharyngeal muscle contractions. The distribution of mutants is shown in Table 23.2. Preliminary work indicates that many of the *eat* mutants may affect function of the nervous system, or muscle functions that control contraction. In addition, *phm* mutants may have defective nervous control of muscle function. The *pha* mutants may represent defects in morphogenesis or specification of cell fate.

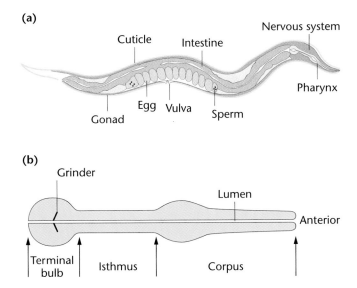

FIGURE 23.12 The pharynx of *C. elegans*. (a) Body plan of intact animal, showing location and arrangement of internal organs, including pharynx. (b) Pharynx, showing three regional divisions and associated structures.

TABLE 23.2 Distribution of Pharynx Mutants in *C. elegans* by Phenotypic Class

Mutations/Gene	*eat*	*pha*	*phm*
1	16	2	5
2	7	0	2
3	1	0	0
4	2	0	0
Total genes:	26	2	7
Total mutations:	41	2	9

Source: Modified from Avery, 1993.

Work is now proceeding on the isolation, cloning, and characterization of genes affecting muscle excitability and neuron function. Results from this study will provide insight into molecular events through which behavior is mediated by both the nervous system and the muscular system.

Locomotory Behavior in Mice

One of the oldest recorded behavior mutants is the *waltzer* mutation in the mouse, recorded in 80 B.C. in China. This mutant was described as a mouse "found dancing with its tail in its mouth." Waltzing mice run in a tight circle and demonstrate both horizontal and vertical head shaking as well as hyperirritability. Some waltzers are also deaf.

Genetic crosses between mutant and normal house mice (*Mus musculus*) reveal a simple recessive inheritance pattern characteristic of Mendel's monohybrid matings. Investigation of the inner ear of mutants has revealed degeneration of both the cochlea and semicircular canals, accounting for the deafness and circling behavior. This is an example of a mutation causing a structural anomaly that, in turn, alters behavior.

Many other single-gene behavior effects are known in the mouse, an organism particularly well characterized genetically. Of over 300 mutants discovered representing

TABLE 23.3 Inherited Neurological Defects in Mice

Group	Name
Waltzer-shaker	*Waltzer, shaker, pirouette, jerker, fidget, twirler, zig-zag*
Convulsive	*Trembler, tottering, spastic*
Incoordination	*Jumping, quaking, reeler, staggerer, leaner, Purkinje cell degeneration, cerebral degeneration*

about 250 loci, over 90 are neurological in nature and alter behavior. These are classified into three groups of syndromes, as shown in Table 23.3. The neurological bases of many of these have now been determined.

The incoordination mutants *quaking* and *jumping* are due to faulty myelination of nervous tissue. *Quaking* is an autosomal recessive mutation, and *jumping* is inherited as an X-linked recessive. The former cannot synthesize adequate myelin, while the latter demonstrates a degenerative process involving myelin. *Cerebral degeneration* is a mutant exhibiting progressive deterioration of behavior. As its name implies, part of the brain deteriorates.

The study of such mutants clearly establishes the genetic basis of normal development.

Genetic Analysis of Behavior in *Drosophila*

We have already discussed geotaxis in *Drosophila* as an example of the selection methodology, but much more information on the behavior genetics of this organism is available. This is not at all surprising because of our extensive knowledge of its genetics and the ease with which *Drosophila* can be manipulated experimentally.

As early as 1915, Alfred Sturtevant observed that the X-linked recessive gene *yellow* affects mating preference in *Drosophila* females, in addition to its effects on body pigmentation. Sturtevant found that both wild-type and *yellow* females, when given the choice of wild-type or *yellow* males, prefer to mate with wild type. Both wild-type and *yellow* males prefer to mate with *yellow* females. These conclusions were based on measurements of mating success in all combinations of *yellow* and wild-type (gray-bodied) males and females.

In 1956, Margaret Bastock extended these observations by investigating which, if any, component of courtship behavior was affected by the *yellow* mutant gene. Courtship in *Drosophila* is a complex ritual. The male first shows **orientation**. The male follows the female, often circling her, and then orients at a right angle to her body and taps her on the abdomen. Once he has her attention, males begin wing display. He raises the wing closest to the female and vibrates this wing rapidly for several seconds. He then moves behind her, and makes contact with her genitalia. If she signals acceptance by remaining in place, he mounts her and copulation occurs.

Bastock compared the courtship ritual in wild-type and *yellow* males. The *yellow* males prolong orientation but spend much less time in the vibrating and genital contact phases. Her observations indicate that the *yellow* mutation alters the pattern of male courtship, making these males less successful in mating. Selection of mates by females is a powerful selective force in evolution, and will be discussed in Chapter 25.

Mosaics

In 1967, Seymour Benzer and his colleagues initiated a study of behavior genetics in *Drosophila*. Benzer's approach uses mutant alleles of behavior to "dissect" a complex biological phenomenon into its simpler components. This approach has several goals:

1. Isolate mutants that disrupt normal behavior.
2. Identify the mutant genes by chromosome localization and mapping.
3. Determine the level and structural component at which the gene expression influences the behavioral response.
4. Establish a causal link between the mutant allele and its behavioral phenotype.

All four steps can be illustrated in a discussion of **phototaxis**, one of the first behaviors studied by Benzer. Normal flies are positively phototactic; that is, they move toward a light source. Mutations were induced by feeding male flies sugar water containing **EMS** (ethylmethanesulfonate, a potent mutagen) and mating them to attached-X virgin females. As shown in Figure 23.13, the F_1 males receive their X chromosome from their fathers. Because they are hemizygous, induced X-linked recessive mutations are phenotypically expressed.

The F_1 males were tested for phototaxic responses, and those with abnormal behavior were isolated. Benzer found *runner* mutants, which move quickly to and from light; *negative phototactic* mutants, which move away from light; and *nonphototactic* mutants, which show no preference for light or darkness. The genetic basis of these behavioral changes was confirmed by mating these F_1 males to attached-X virgin females. Male progeny of this cross also showed the abnormal phototactic responses, indicating that these behavioral alterations are the result of X-linked recessive mutations.

Nonphototactic mutants walk normally in the dark, but show no phototactic response to light. Benzer and Yoshiki Hotta tested electrical activity at the surface of mutant eyes in response to a flash of light. The pattern of electrical activity was recorded as an **electroretinogram**. Various types of abnormal responses were detected in different nonphototactic mutant strains. None of the mutants had a normal pattern of electrical activity. When these mutations were mapped, they were not all allelic; instead, they were shown to occupy several loci on the X

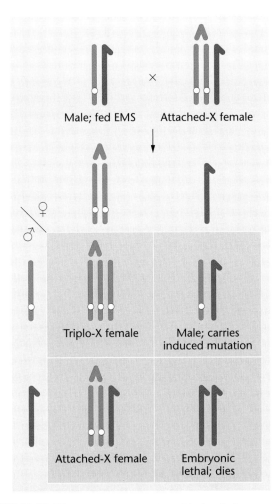

FIGURE 23.13 Genetic cross in *Drosophila* that facilitates recovery of X-linked induced mutations. The female parent contains two X chromosomes that are attached, in addition to a Y chromosome. In a cross between this female and a normal male that has been fed the mutagen ethylmethanesulfonate, all surviving males receive their X chromosomes from their father and express all mutations induced on that chromosome.

chromosome. This analysis indicates that several gene products contribute to the phototactic response.

Where, within the fly, must gene expression occur to produce a normal electroretinogram? In an ingenious approach to answer this question, Benzer turned to the use of mosaics. In mosaic flies, some tissues are mutant and others are wild type. If it can be ascertained which part must be mutant in order to yield the abnormal behavior, the **primary focus** of the genetic alteration can be determined.

To produce mosaic flies, Benzer used a *Drosophila* strain that carries one of its X chromosomes in an unstable ring shape. When the ring-X is present in a zygote undergoing cell division, it is frequently lost by nondisjunction. If the zygote is female and has two X chromosomes (one normal and one ring-X), loss of the ring-X at the first mitotic division will result in two cells—one with

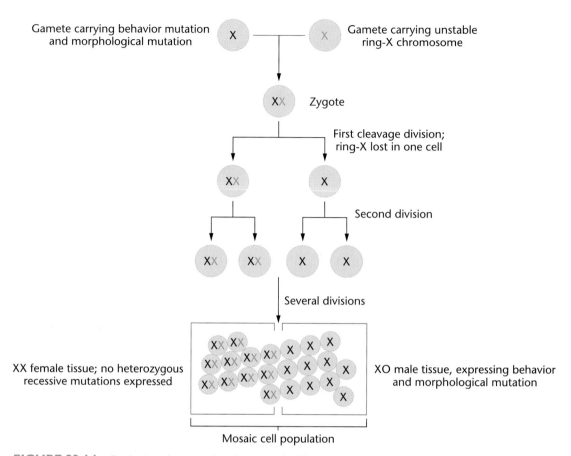

Gamete carrying behavior mutation and morphological mutation

Gamete carrying unstable ring-X chromosome

Zygote

First cleavage division; ring-X lost in one cell

Second division

Several divisions

XX female tissue; no heterozygous recessive mutations expressed

XO male tissue, expressing behavior and morphological mutation

Mosaic cell population

FIGURE 23.14 Production of a mosaic fruit fly as a result of fertilization by a gamete carrying an unstable ring-X chromosome (shown in color). If this chromosome is lost in one of the two cells following the first mitotic division, the body of the fly will consist of one part that is male (XO) and the other part that is female (XX). The male side will express all mutations contained on the X chromosome.

a single X (normal X) and one with two X chromosomes (one normal and one ring-X). The former cell goes on to produce male tissue (XO) and expresses all alleles on the remaining X, whereas the latter produces female tissue and does not express heterozygous recessive X-linked genes. This is illustrated in Figure 23.14. Loss of the ring-X results in mosaic flies with male and female parts that differ in the expression of X-linked recessive genes.

When and where the ring-X is lost in development determines the pattern of mosaicism. The loss usually occurs early in development, before the cells migrate to the embryo surface. As shown in Figure 23.15, different types of mosaics are produced, depending on the orientation of the spindle when loss of the ring-X takes place. If the normal X chromosome carries the behavior mutation and an obvious mutant allele (*yellow*, for example), the pattern of mosaicism will be easy to distinguish. By examining the distribution of body color, flies with a combination of normal and mutant structures can be identified, such as a fly with a mutant head on a wild-type body, or a normal head on a mutant body, or a fly with one normal and one mutant eye on a normal or mutant body, and so on.

When *nonphototactic* mosaics were studied, the focus of the genetic defect was found to be in the eye itself. In mosaics where every part of the fly except the eye was normal, abnormal behavior was still detected. When one eye was mutant, and the other normal, the fly had an unusual behavior. Instead of crawling straight up toward light as the normal fly does, the mosaic fly with one mutant eye crawls upward to light in a spiral pattern. In the dark, this fly moves in a straight line.

Using a combination of genetics, physiology, and biochemistry, Benzer and other workers in the field have identified and analyzed a large number of genes affecting behavior in *Drosophila*. As shown in Table 23.4, mutants that affect locomotion, response to stress, circadian rhythm, sexual behavior, visual behavior, and even learning have been isolated. Some of these mutations have received very descriptive and often humorous names.

Many mutations have been analyzed with the mosaic technique to localize the focus of gene expression. While it was easy to predict that the focus of the *nonphototactic* mutant would be in the eye, other mutants are not so predictable. For example, the focus of mutants affecting cir-

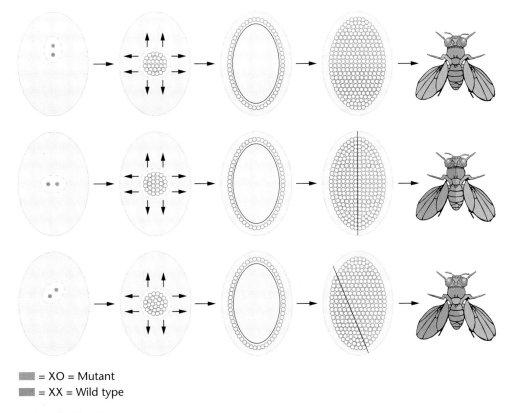

= XO = Mutant

= XX = Wild type

FIGURE 23.15   The effect of spindle orientation on the production of mosaic flies as shown in Figure 23.14.

TABLE 23.4   Some Behavioral Mutants of *Drosophila*

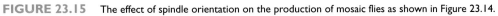

Class	Name	Characteristics
Locomotor	*sluggish*	Moves slowly
	hyperkinetic	High consumption of O_2; shaking of legs; early death
	wings up	Wings perpendicular to body
	flightless	Does not fly well, although wings are well developed
	uncoordinated	Lacks coordinated movements
	nonclimbing	Fails to climb
Response to stress	*easily shocked*	Mechanical shock induces "coma"
	stoned	Stagger induced by mechanical shock
	shaker	Vibrates all legs while etherized
	freaked out	Grotesque, random gyrations under the influence of ether
	paralyzed	Collapses above a critical temperature
	parched	Dies quickly in low humidity conditions
	tko	Epileptic-like response
	comatose	Paralyzed by cool temperature
	out-cold	Similar to comatose
Circadian rhythm	*period^o*	Eclosion at any time; locomotor activity spread randomly over the day
	period^s	19-hour cycle rather than 24-hour cycle
	period^l	28-hour cycle rather than 24-hour cycle
Sexual	*savoir-faire*	Males unsuccessful in courtship
	fruitless	Males pursue each other
	stuck	Male is often unable to withdraw after copulation
	coitus interruptus	Males disengage in about half the normal time
Visual	*nonphototactic*	Blind
	negatively phototactic	Moves away from light
Learning	*dunce*	Fails to learn conditioned response

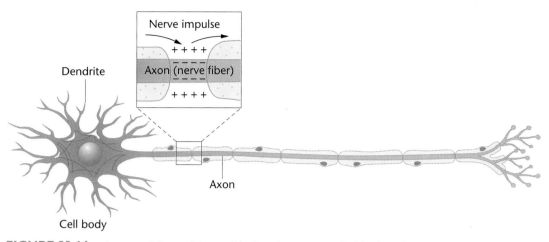

FIGURE 23.16 A nerve cell (neuron) has a cell body and extensions called *dendrites* that carry impulses toward the cell body, and one or more *axons* that carry impulses away from the cell body. Electrodes placed on either side of the plasma membrane can record the electric potential across the membrane. At rest, there is more sodium outside the cell, and more potassium inside the cell. When a nerve impulse is generated, sodium moves into the cell, and potassium moves out. As the impulse moves away, ions are pumped across the membrane to restore the electric potential.

cadian rhythms has been located in the head, presumably in the brain. A mutant with a *wings up* phenotype might have a defect in the wings, muscle attachments in the thorax, abnormal muscle formation, or have a neuromuscular defect. Analysis of mosaics established that the defect is in flight muscles of the thorax. Cytological studies have confirmed this finding, showing a complete lack of myofibrils in these muscles.

The mosaic technique has also been used to determine which regions of the brain are associated with sex-specific aspects of courtship and mating behavior. Jeffrey Hall and his associates have shown that mosaics with male cells in the most dorsal region of the brain, the protocerebrum, exhibit the initial stages of male courtship toward females. Later stages of male sexual behavior including wing vibrations and attempted copulations require male cells in the thoracic ganglion. Similar studies of female-specific sexual behavior have shown that the ability of a mosaic to induce courtship by a male depends on female cells in the posterior thorax or abdominal region. A region of the brain within the protocerebrum must be female for receptivity to copulation. Anatomical studies have confirmed that there are fine structural differences in the brains of male and female *Drosophila*, indicating that some forms of behavior may be dependent on the development and maturation of specific parts of the nervous system.

Neurogenetics

Analysis of behavior mutants in *Drosophila* has led to an understanding of fundamental mechanisms in the animal nervous system. Nerve impulses are generated at one end

of a neuron and move along the cell to the opposite end (Figure 23.16). During this process, sodium and potassium ions move across the plasma membrane. The movement of these electrically charged ions can be monitored by measuring changes in the electrical potential of the neuron's membrane. To screen for genes that control the generation and transmission of nerve impulses, Barry Ganetzky and his colleagues screened behavioral mutants of *Drosophila* to identify those with electrophysiological abnormalities in the generation and propagation of nerve impulses. Two general classes of such mutants have been isolated: those with defects in the movement of sodium, and those with defects in potassium transport (Table 23.5).

One mutant identified in this screening procedure is a temperature-sensitive allele called *paralytic*. Flies homozygous for this mutant allele becomes paralyzed when exposed to temperatures at or above 29°C, but they recover rapidly when the temperature is lowered to 25°C. Mosaic studies revealed that both the brain and thoracic ganglia are the focus of this abnormal behavior. Electrophysiological studies showed that mutant flies have defective sodium transport associated with the conduction of nerve impulses. Subsequently, Ganetzsky and his colleagues mapped, isolated and cloned the *paralytic* gene. This locus encodes a protein, called the sodium channel, that controls the movement of sodium across the membrane of nerve cells.

A second mutant, called *Shaker*, originally isolated over 40 years ago as a behavioral mutant, encodes a potassium channel gene that has also been cloned and characterized. Because the mechanism of nerve impulse conduction has been highly conserved during animal evolution,

TABLE 23.5 Behavioral Mutants of *Drosophila* Affecting Nerve Impulse Transmission

Mutation	Map Location	Ion Channel Affected	Phenotype
nap^{ts}	2–56.2	sodium	Adults, larvae paralyzed at 37.5°C. Reversible at 25°C.
$para^{ts}$	1–53.9	sodium	Adults paralyzed at 29°C, larvae at 37°C. Reversible at 25°C.
tip-E	3–13.5	sodium	Adults, larvae paralyzed at 39–40°C. Reversible at 25°C.
sei^{ts}	2–106	sodium	Adults paralyzed at 38°C, larvae unaffected. Adults recover at 25°C.
Sh	1–57.7	potassium	Aberrant leg shaking in adults exposed to ether.
eag	1–50.0	potassium	Aberrant leg shaking in adults exposed to ether.
Hk	1–30.0	potassium	Ether-induced leg shaking.
sio	3–85.0	potassium	At 22°C, adults are weak fliers; at 38°C, adults are weak, uncoordinated.

the cloned *Drosophila* genes were used as probes to isolate the equivalent human ion channel genes. The cloned human genes are being used to provide new insights into the molecular basis of neuronal activity. One of the human ion channel genes first identified in *Drosophila* is defective in a heritable form of cardiac arrhythmia. Identification and cloning of the human gene now makes it possible to screen for family members at risk for this potentially fatal condition.

Learning in *Drosophila*

To study the genetics of a complex behavior such as learning, it would be advantageous to use an organism like *Drosophila*. However, the first question is: Can *Drosophila* learn? Work from a number of laboratories has shown that *Drosophila* can learn. To demonstrate this, flies are presented with a pair of olfactory cues, one of which is associated with an electrical shock. Flies quickly learn to avoid the odor associated with the shock. For several reasons, this response is thought to be learned: Performance is associated with the pairing of a stimulus/response with a reinforcement, and in addition, the response is reversible. Flies can be trained to select an odor they previously avoided; third, flies exhibit short-term memory for the training they have received.

The demonstration that *Drosophila* can learn opens the way to selecting mutants that are defective in learning and memory. To accomplish this, males from an inbred wild-type strain are mutagenized and mated to females from the same strain. Their progeny are recovered and mated to produce populations of flies, all of which carry a mutagenized X chromosome. Mutants that affect learning are selected by testing for response in the olfactory/shock apparatus. A number of learning-deficient mutants including *dunce, turnip, rutabaga,* and *cabbage* have been

recovered. In addition, a memory-deficient mutant, *amnesiac*, which learns normally but forgets four times faster than normal, has been recovered. Each of these mutations represents a single gene defect that affects a specific form of behavior. Because of the method used to recover them, all the mutants found so far are X-linked genes. Presumably similar genes controlling behavior are also located on autosomes.

Since in many cases, mutation results in the alteration or abolition of a single protein, the nervous system mutants described above can provide a link between behavior and molecular biology. One group of *Drosophila* learning mutants involves defects in a cellular signal transduction system.

In cells of the nervous system, and in many other cell types, cyclic AMP (cAMP) is produced from ATP in the cytoplasm in response to signals received at the cell surface. Cyclic AMP activates protein kinases, which, in turn, phosphorylate proteins and initiate a cascade of metabolic effects that control gene expression. Behavioral mutants of *Drosophila* were among the first to show the link between cyclic AMP and learning. The *rutabaga* locus encodes one form of adenylyl cyclase, the enzyme that synthesizes cyclic AMP from ATP. In the mutant allele of *rutabaga*, a missense mutation destroys the catalytic activity of adenylyl cyclase and is associated with a learning deficiency in homozygous flies. The *dunce* locus encodes the structural gene for the enzyme cyclic AMP phosphodiesterase, which degrades cAMP. The *turnip* mutation occurs in a gene encoding a G protein, a class of molecules that bind GTP and, in turn, activate adenylyl cyclase. The clustering of these independently derived mutations in a signal transduction pathway involving cyclic nucleotides supports the idea that these molecules play a role in learning and provide the basis for further studies on the molecular impetus of learning and memory in flies as well as humans.

Human Behavioral Genetics

Genetic control of **behavior** in humans has proven more difficult to characterize than that of other organisms. Not only are humans unavailable as experimental subjects in genetic investigations, but the types of responses considered to be the most interesting forms of behavior including some aspects of **intelligence**, **language**, **personality**, or **emotion** are difficult to study. Two problems arise in studying such behaviors. First, all are difficult to define objectively and to measure quantitatively. Second, they are affected by environmental factors. In each case, the environment is extremely important in shaping, limiting, or facilitating the final phenotype for each trait.

Historically, the study of human behavior genetics has been hampered by other factors. Many studies of human behavior were performed by psychologists without adequate input from geneticists. Second, traits involving intelligence, personality, and emotion have the greatest social and political significance. As such, these traits are more likely to be the subject of sensationalism when reported to the lay public. Because the study of these traits comes closest to infringing upon individual liberties such as the right to privacy, they are the basis of the most controversial investigations.

In lamenting the gulf between psychology and genetics in explaining human behavior genetics, C. C. Darlington in 1963 wrote, "Human behavior has thus become a happy hunting ground for literary amateurs. And the reason is that psychology and genetics, whose business it is to explain behavior, have failed to face the task together." Since 1963, some progress has been made in bridging this gap, but the genetics of human behavior remains a controversial area.

Single Genes

Many genetic disorders in humans result in a behavioral abnormality. One of the most prominent examples is **Huntington disease (HD)**. Inherited as an autosomal dominant disorder, HD affects the nervous system, including the brain. Symptoms usually appear in the fifth decade of life with a gradual loss of motor function and coordination. Structural degeneration of the nervous system is progressive, and personality changes occur. Most victims die within 10 to 15 years after onset of the disease. Because HD usually appears after a family has been started, all children of an affected person must live with the knowledge that they face a 50 percent probability of developing the disorder (affected individuals are usually heterozygotes). HD is associated with elevated brain levels of quinolinic acid, a naturally occurring neurotoxin. The HD gene has been identified and cloned. It encodes a large protein (348 kd) that is unrelated to any known gene product. The mutant form of the gene is associated with the presence of extra CAG trinucleotide repeats near the 5′-end (see Chapter 14). Normal individuals have 11 to 24 repeats, but those affected by HD carry 42 to 86 CAG repeats. The number of CAG repeats in HD is inversely related to the age of onset in a large number of individuals, but there seems to be no correlation between the number of repeats and the severity of the behavioral and psychiatric symptoms. The expansion of a trinucleotide repeat in HD is similar to expansions of other repeats in several disorders that affect the brain and nervous system, indicating that such "stutter" mutations may be a common form of mutation in neurobehavioral traits.

Monoamine oxidases are enzymes that degrade chemical signals (called neurotransmitters) in the ner-

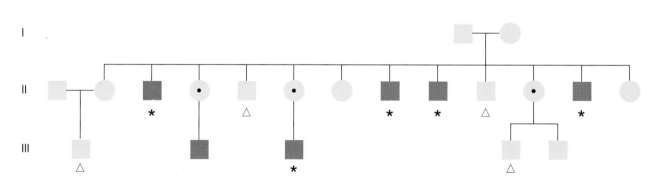

FIGURE 23.17 Partial pedigree of family in which mental retardation and aggressive behavior segregates with a mutation in the monoamine oxidase A gene (MAOA). Affected males are indicated by filled symbols. Those marked with an asterisk are hemizygous for a mutant allele of MAOA. Those marked with a triangle have the normal allele. Female heterozygotes have one normal and one mutant allele.

vous system. Recently, studies have established a link between mutation in monoamine oxidase A (MAOA), an X-linked gene, and a syndrome that involves mild mental retardation and control of aggressive behavior. A pedigree of one affected family is shown in Figure 23.17. Six males are affected, and all showed a characteristic pattern of aggressive behavior and lack of impulse control. Using a variety of RFLP markers, the locus for this behavior was mapped to the short arm of the X chromosome, near the MAOA locus. Analysis of the structural gene for MAOA in five of the six affected males showed that they carry a single nucleotide substitution that introduces a stop codon causing premature termination of translation, resulting in a lack of MAOA activity. This mutation is carried by heterozygous females, but not by normal males. Other studies show that mutations and polymorphisms in the MAOA gene are not common, strengthening the link between a mutation in this gene and an abnormal form of behavior.

Other metabolic disorders are also known to affect mental function. These include **Lesch–Nyhan syndrome**, an X-linked recessive disorder. Onset is within the first year, and the disease is most often fatal early in childhood. The disorder involves a defect in purine biosynthesis. Affected individuals lack an enzyme, hypoxanthine-guanine phosphoribosyltransferase (HGPRT), and accumulate high levels of uric acid. Mental and physical retardation occurs, and these individuals demonstrate uncontrolled self-mutilation. **Tay-Sachs disease**, an autosomal recessive disorder that maps to chromosome 15, involves severe mental retardation. The disease is apparent soon after birth and is fatal. The autosomal recessive disease **phenylketonuria**, unless detected and treated early, results in mental retardation. All of these disorders alter the normal biochemistry of the affected individuals and are inherited in a Mendelian fashion.

Chromosome abnormalities also produce syndromes with behavioral components. **Down syndrome** (trisomy 21) results in varying degrees of mental retardation. Although there is a wide range of variability, the mean IQ of affected individuals is estimated to be 25 to 50. The onset of walking and talking is often delayed until 4 to 5 years of age.

Multifactorial Traits

Other aspects of human behavior, notably **schizophrenia** and **manic–depressive illness**, have been the subject of extensive investigations. Twin studies and studies on adopted and natural siblings have supported the idea that manic depression has a genetic component. Mapping studies originally identified loci on the X chromosome and chromosome 11 as being associated with manic depression, but these studies were later invalidated. Several laboratories are using RFLP markers to screen large pedigrees in which manic depression is segregating in an attempt to find the loci for this disorder, which affects about 1 percent of the U.S. population.

Schizophrenia is a mental disorder characterized by withdrawn, bizarre, and sometimes delusional behavior. Those affected by the disease are unable to lead organized lives and are periodically disabled by the condition. It is clearly a familial disorder, with relatives of schizophrenics having a much higher incidence of this disorder than the general population. Furthermore, the closer the relationship to the index case or proband, the greater is the probability of the disorder occurring.

The concordance of schizophrenia in monozygotic and dizygotic twins has been the subject of many studies. In almost every investigation, concordance has been higher in monozygotic twins than in dizygotic twins reared together. Although these results suggest that a genetic component exists, they do not reveal the precise genetic basis of schizophrenia. Simple monohybrid and dihybrid inheritance as well as multiple gene control have been proposed for schizophrenia. It seems unlikely that only one or two loci are involved, nor is it likely that the control is strictly quantitative, as in polygenic inheritance. As with manic depression, a large-scale collaborative effort uses DNA markers to identify genomic regions that may contain loci controlling this behavioral disorder. Once identified, these regions will be studied in detail to identify, isolate, and clone genes for schizophrenia.

GENETICS, TECHNOLOGY, AND SOCIETY

Coming to Terms with the Heritability of IQ

Of the many areas where genetics interfaces with society, few have been as contentious as the genetics of intelligence. This is as true today as it was during the heyday of the eugenics movement, when eliminating "feeblemindedness" was one of the overriding concerns. More recently, the extent to which intellectual ability is inherited has been considered as one of the numerous factors to be taken into account during the formulation of educational policy. What can genetic analysis tell us about the hereditary basis of intelligence?

To answer this question, the first challenge is to define and assess intelligence. Is it a single entity, and, if so, can it be measured? Obviously, intelligence is not a physical characteristic like height or weight, which are easy to see and easy to measure. However, performance on IQ tests has been taken as a measure of intelligence since Alfred Binet's original test was adapted and introduced in the U.S. in 1916. Despite numerous objections, this practice is unlikely to change. The debate on the nature of intelligence, whether or not it is an innate, mental process that can be reliably assessed, is certain to continue for some time.

What about the genetic basis for IQ score? For a complex human trait such as intelligence, there is no readily discernible chain of cause-and-effect events leading from gene to phenotype, as there is for, say, albinism or hemophilia. Genetic analysis of intelligence therefore requires the assumption that it is a quantitative trait, whereby the final phenotype is influenced by the additive contributions of many genes, as well as a variety of environmental factors. Statistical techniques can then be employed that were first developed for the analysis of quantitative traits in plants and animals. The most frequently used statistic is heritability.

Heritability of a trait is the fraction of the total variance in the phenotypic expression within a specific population that is due to genetic factors rather than varying environmental conditions. The heritability of IQ score would thus seem to be an indication of the extent to which intelligence is genetically based. But it is not as simple as that, and we need to understand why.

Since the breeding techniques used to estimate the heritability of traits in plants and animals obviously cannot be adapted for humans, alternative methods have been devised. These involve comparing closely related individuals to each other and more distantly related individuals to each other. If the more closely related people, who have a higher proportion of genes in common, are more similar to each other than are the more distantly related people, who have a lower proportion of genes in common, it suggests that genes play a significant role in defining the phenotype in question. From these comparisons, the heritability of the trait can be estimated.

One method involves examining sets of identical twins and fraternal twins of the same sex. Identical, or monozygotic, twins develop from a single zygote and thus are genetically identical, whereas fraternal, or dizygotic, twins are no more alike genetically than any other siblings, sharing, on average, 50% of their genes. If the trait in question has a genetic basis, given common environments, identical twins should resemble each other more closely than do fraternal twins. A still better way is to study identical twins who have been separated very early in life and reared apart (a method first suggested by Francis Galton). In theory, this allows the effects of a common genetic inheritance to be distinguished from the effects of different environments.

Reflecting the inherent uncertainties of such approaches, these and similar methods have yielded a range of heritability values for IQ scores over the years, from a low of around 0.30 to a high of around 0.75. A commonly accepted heritability for IQ score is 0.60. So let's consider what a heritability of 0.60 for IQ score really means, and, perhaps more important, what it does *not* mean.

CHAPTER SUMMARY

1. Behavioral genetics has emerged as an important specialty within genetics because both genotype and environment have been found to have an impact in determining an organism's behavioral response.

2. Historically, three approaches have been used in studying the genetic control of behavior: the use of closely related organisms from similar environments whose survival needs appear to be identical; transfer of behavioral traits from one strain to another; and the identification and isolation of single genes that have an effect on behavior patterns.

3. Studies done on alcohol preference and open-field behavior in mice illustrate behaviors strongly influenced by the genotype.

4. Studies of maze learning in rats and geotaxis in *Drosophila* have successfully established both bright and dull lines of rats and either positively or negatively geotaxic *Drosophila*. These results have led researchers to ascribe the relative contributions of genes on a specific chromosome to these behaviors.

Strictly speaking, a heritability of 0.60 for IQ score means that *within the population under study*, 60% of the variance in IQ scores can be attributed to different genotypes within the population.

It should be understood, first of all, that heritability is a statistic pertaining to variation within a population and has no meaning when applied to an individual. It is therefore *not correct* to say that 60% of a person's IQ is determined by genes and the other 40% by environment.

Second, heritability estimates pertain to variation *within* a specific population and not to variation *between* populations. A heritability value of 0.60 for IQ score *does not* mean that 60% of the difference in IQ scores between two populations is due to different genes. Heritability estimates say nothing at all about the reasons that differences in IQ score have been observed in different populations.

Third, the heritability of a trait is not a permanent characteristic of a population. In fact, a heritability value, whether for IQ score or anything else, pertains to that particular population *at a specific time*, under the conditions then prevailing. Heritability values may change as environmental factors change, and may thus vary from one year to the next. "Heritable" does not therefore equal "inevitable." A high heritability value *does not* mean that the trait cannot be modified substantially by environmental factors. The genotype may determine the range of possi-

ble phenotypes, but exactly what phenotype occurs depends on interactions with the environment.

Attempts to elucidate the genetic basis of human intelligence remain controversial and for good reason. The results of such studies, especially when misinterpreted, can be taken as a biological justification of discriminatory social policies. Some people prefer to believe that the problems of the disadvantaged in our society are not the result of social forces beyond their control but of their genes. Recently, heritability estimates of IQ scores have been used to bolster claims that intelligence is largely genetic and not significantly influenced by the environment in which a child finds himself or herself. On the basis of such faulty reasoning, it has been concluded that efforts to boost IQ through better ed-

ucation and social interventions are doomed to failure.

It should be clear, in light of the uncertainties about the very nature of intelligence, the skepticism regarding IQ tests as its measure, and the difficulty in determining the genetic basis of complex human traits, that the heritability of intelligence, even an estimate as high as 0.60, is an inadequate foundation on which to base such claims and upon which to construct social policy. This is not to say that intelligence, however it is defined and measured, is not influenced to a substantial degree by the genes each of us has inherited, or that differences in IQ score will not be found in populations that are studied. However, we cannot allow the use of heritability of IQ score to serve as the rationale for designing policies that discriminate against any individual in our society.

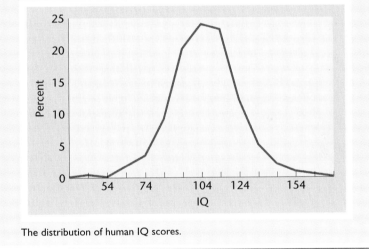

The distribution of human IQ scores.

5. By isolating mutations that cause deviations from normal behavior, the role of a corresponding wild-type allele in the respective response is established. Numerous examples have been examined, including honeybee hygienic behavior; chemotaxis, thermotaxis, and general movement in nematodes; neurological mutations in mice; and a variety of behaviors in *Drosophila*.

6. Any aspect of human behavior is difficult to study because the individual's environment makes an important contribution toward trait development. In humans, studies using twins have shown that, while a family may have a predisposition to schizophrenia, the expression of this disorder may be modified by the environment.

KEY TERMS

INSIGHTS AND SOLUTIONS

1. Manic depression is an affective disorder associated with recurring mood changes. It is estimated that one in four individuals will suffer from some form of affective disorder at least once in his or her lifetime. Genetic studies indicate that manic depression is familial, and that single genes may play a major role in controlling this behavioral disorder. In 1987 two separate studies using RFLP analysis and other genetic markers reported linkage between manic depression and markers on the X chromosome and to the short arm of chromosome 11. At the time, these reports were hailed as landmark discoveries, opening the way to the isolation and characterization of genes that control specific forms of behavior, and to the development of therapeutic strategies based on knowledge of the nature of the gene product and its action. However, further work on the same populations (reported in 1989 and 1990) demonstrated that the original results were invalid, and it was concluded that no linkage to markers on the X chromosome and to markers on chromosome 11 could be established. These findings do not exclude the role of major genes on the X chromosome and autosomes in manic depression, but they do exclude linkage to the markers used in the original reports.

This setback not only caused embarrassment and confusion but it also forced a reexamination of the validity of the methods used in mapping human genes, and the analysis of data from linkage studies involving complex behavioral traits. Several factors have been proposed to explain the flawed conclusions reported in the original studies. What do you suppose some of these factors to be?

Solution: While some of the criticisms were directed at the choice of markers, most of the factors at work in this situation appear to be related to the phenotype of manic depression. At least three confounding elements have been identified. One is age of onset. There is a positive correlation between age and the appearance of manic depression. Therefore, at the time of a pedigree study, younger individuals who will be affected later in life may not show any signs of manic depression. Another confusing factor is that in the populations studied, there may in fact be more than one major X-linked gene and more than one major autosomal gene that trigger manic depression. A third factor relates to the diagnosis of manic depression itself. The phenotype is complex and not as easily quantified as height or weight. In addition, mood swings are a universal part of everyday life, and it is not always easy to distinguish transient mood alterations and the role of environmental factors from affective disorders having a biological and/or genetic basis.

These factors point up the difficulty in researching the genetic basis of complex behavioral traits. Individually or in combinations, the factors described above might skew the results enough so that guidelines for proof that are adequate for other traits are not stringent enough for behavioral traits with complex phenotypes and complex underlying causes.

PROBLEMS AND DISCUSSION QUESTIONS

1. Contrast the methodologies used in studying behavior genetics. What are the advantages of each method with respect to the type of information gained?
2. Contrast the advantages of using *Drosophila* versus *Caenorhabditis* for studying behavior genetics.
3. If a haploid honeybee drone of the genotype *uR* is mated to a queen of genotype *UuRR*, what ratio of behavioral phenotypes will be observed in the hive in the offspring with respect to American foulbrood disease? For *Ur × UURr*? For *uR × Uurr*?

4. Assume that you discovered a fruit fly that walked with a limp and was continually off balance as it moved. Describe how you would determine whether this behavior is due to an injury (induced in the environment) or is an inherited trait. Assuming that it is inherited, what are the various possibilities for the focus of gene expression causing the imbalance? Describe how you would locate the focus experimentally if it were X-linked.

5. In humans, the chemical phenylthiocarbamide (PTC) is either tasted or not. When the offspring of various combinations of taster and nontaster parents are examined, the following data are obtained:

Parents	Offspring
Both tasters	All tasters
Both tasters	1/2 tasters 1/2 nontasters
Both tasters	3/4 tasters 1/4 nontasters
One taster One nontaster	All tasters
One taster One nontaster	1/2 tasters 1/2 nontasters
Both nontasters	All nontasters

Based on these data, how is PTC tasting behavior inherited?

6. Discuss why the study of human behavior genetics has lagged behind that of other organisms.

7. J. P. Scott and J. L. Fuller studied 50 traits in five pure breeds of dogs. Almost all traits varied significantly in the five breeds, but very few bred true in crosses. What can you conclude with respect to the genetic control of these behavioral traits?

EXTRA-SPICY PROBLEMS

8. This chapter introduces *C. elegans* as a model system for studying the genetic specification of the nervous system. The life cycle of this organism makes it an excellent choice for genetic dissection of many biological processes. This species has two natural sexes, hermaphrodite and male. The hermaphrodite is essentially a female that can generate sperm as well as oocytes, so reproduction can occur by hermaphrodite self-fertilization or hermaphrodite–male mating. In the context of studying mutations in the nervous system, what is the advantage of hermaphrodite self-fertilization with respect to identification of recessive mutations and propagation of mutant strains?

9. Theoretical data concerning the genetic effect of *Drosophila* chromosomes on geotaxis are shown below. What conclusions can you draw?

	Chromosome		
	X	2	3
Positive geotaxis	+0.2	+0.1	+3.0
Unselected	+0.1	−0.2	+1.0
Negative geotaxis	−0.1	−2.6	+0.1

SELECTED READINGS

AVERY, L. 1993. The genetics of feeding in *Caenorhabditis elegans*. *Genetics* 133:897–917.

BASTOCK, M. 1956. A gene which changes a behavior pattern. *Evolution* 10:421–39.

BENZER, S. 1973. Genetic dissection of behavior. *Sci. Am.* (Dec.) 229:24–37.

BERG, H. C. 1988. A physicist looks at bacterial chemotaxis. *Cold Spring Harbor Symp. Quant. Biol.* 53:(Part 1) 1–9.

BRENNER, S. 1974. The genetics of *Caenorhabditis elegans*. *Genetics* 77:71–94.

BUCK, K. J., 1995. Strategies for mapping and identifying quantitative trait loci specifying behavioral responses to alcohol. *Alcohol Clin. Exp. Res.* 19:795–801.

BYERS, D., DAVID, R. L., and KIGER, J. A., JR. 1981. Defect in the cyclic AMP phosphodiesterase due to the *dunce* mutation of learning in *Drosophila*. *Nature* 289:79–81.

DEVOR, E. J., and CLONINGER, C. R. 1989. Genetics of alcoholism. *Annu. Rev. Genet.* 23:19–36.

FULLER, J. C., and THOMPSON, W. R. 1978. *Foundations of behavior genetics*. St. Louis: C. V. Mosby.

GAILEY, D. A., HALL, J. C., and SIEGEL, R. W. 1985. Reduced reproductive success for a conditioning mutant in experimental populations of *Drosophila melanogaster*. *Genetics* 111:795–804.

GANETZKY, B., WARMKE, J. W., ROBERTSON, G., and PALLANCK, L. 1995. New potassium channel gene families in flies and mammals: From mutants to molecules. *Soc. Gen. Physiol Ser.* 50:29–39.

GANETZKY, B. 1994. Cysteine strings, calcium channels and synaptic transmission. *BioEssays* 16:461–63.

GERHARD, D. S., LABUDA, M. C., BLAND, S. D., ALLEN, C., EGELAND, J. A., and PAULS, D. L. 1995. Initial report of a genome search for the affective disorder predisposition gene in the old order Amish pedigrees: chromosomes 1 and 11. *Am. J. Med. Genet.* 54:398–404.

HAY, D. A. 1985. *Essentials of behaviour genetics*. Palo Alto, CA: Blackwell Scientific.

HAZELBANER, G. L., BERG, H. C., and MATSUMURA, P. 1993. Bacterial motility and signal transduction. *Cell* 73:15–22.

HOTTA, Y., and BENZER, S. 1972. Mapping behavior of *Drosophila* mosaics. *Nature* 240:527–35.

HUNTINGTON'S DISEASE COLLABORATIVE GROUP. 1993. A novel gene containing a trinucleotide repeat that is expanded and unstable on Huntington's disease chromosomes. *Cell* 72:971–83.

KIDD, K. K., and CAVALLI-SFORZA, L. L. 1973. An analysis of the genetics of schizophrenia. *Social Biol.* 20:254–65.

LIVINGSTONE, K. S. 1985. Genetic dissection of *Drosophila* adenylate cyclase. *Proc. Natl. Acad. Sci. USA* 82:5795–99.

McINNES, L. A., and FREIMER, N. B. 1995. Mapping genes for psychiatric disorders and behavioral traits. *Curr. Opin. Genet. Dev.* 5:376–81.

MADDOCK, J. R., and SHAPIRO, L. 1993. Polar location of the chemoreceptor complex in the *E. coli* cell. *Science* 259:1717–23.

MELO, J. A., SHENDURE, J., POCIASK, K., and SILVER, L. M. 1996. Identification of sex-specific quantitative trait loci controlling alcohol preference in C57BL/6 mice. *Nature Genet.* 13:147–53.

MERRELL, D. J. 1965. Methodology in behavior genetics. *J. Hered.* 56:263–66.

PARKINSON, J. S. 1993. Signal transduction schemes of bacteria. *Cell* 73:857–71.

PARKINSON, J. S., and BLAIR, D. 1993. Does *E. coli* have a nose? *Science* 259:1701–2.

PETRONIS, A., SHERRINGTON, R. P., PATERSON, A. D., and KENNEDY, J. L. 1995. Genetic anticipation in schizophrenia: pro and con. *Clin. Neurosci.* 3:76–80.

PLOMIN, R. 1990. *Nature and nurture: An introduction to human behavioral genetics.* Pacific Grove, CA: Brooks/Cole Publishing Company.

PLOMIN, R., DeFRIES, J. C., and McCLEARN, G. E. 1990. *Behavioral genetics: A primer*, 2nd ed. New York: W. H. Freeman.

QUINN, W. B., and GREENSPAN, R. J. 1984. Learning and courtship in *Drosophila*: Two stories with mutants. *Annu. Rev. Neurosci.* 7:67–93.

RAIZEN, D. M., LEE, R. Y., and AVERY, L. 1995. Interacting genes required for pharyngeal excitation by motor neuron MC in *Caenorhabditis elegans*. *Genetics* 141:1365–82.

RICKER, J. P., and HIRSCH, J. 1988(a). Genetic changes occurring over 500 generations in lines of *Drosophila melanogaster* selected divergently for geotaxis. *Behav. Genet.* 18:13–24.

———. 1988(b). Reversal of genetic homeostasis in laboratory populations of *Drosophila melanogaster* under long-term selection for geotaxis and estimates of gene correlates: Evolution of behavior-genetic systems. *J. Comp. Psychol.* 102:203–14.

RIDDLE, D. L. 1978. The genetics of development and behavior in *Caenorhabditis elegans*. *J. Nematol.* 10:1–15.

ROTHENBUHLER, W. C., KULINCEVIC, J. M., and KERR, W. E. 1968. Bee genetics. *Annu. Rev. Genet.* 2:413–38.

SIEGEL, R. W., HALL, J. C., GAILEY, D. A., and KYRIACOU, C. P. 1984. Genetic elements of courtship in *Drosophila*: Mosaics and learning mutants. *Behav. Genet.* 14:383–410.

STOLTENBERG, S. F., HIRSCH, J., and BERLOCHER, S. H. 1995. Analyzing correlations of three types in selected lines of *Drosophila melanogaster* that have evolved stable extreme geotactic performance. *J. Comp. Psychol.* 105: 85–94.

TERMAN, L. M., and MERRILL, M. A. 1973. *Stanford-Binet intelligence scale: 1972 norms edition.* Boston: Houghton-Mifflin.

TRYON, R. C. 1942. Individual differences. In *Comparative psychology*, 2nd ed., ed. F. A. Moss. Englewood Cliffs, NJ: Prentice-Hall.

24

Population Genetics

CHAPTER CONCEPTS

*Individuals can carry only two different alleles of a given gene. A group of individuals can carry a
larger number of different alleles, giving rise to a reservoir of genetic diversity. The diversity contained in the population can be measured by the Hardy-Weinberg law. Mutation is the ultimate
source of genetic variation, and other factors such as drift, migration, and selection can alter the
amount of genetic variation in populations.*

When Charles Darwin published *The Origin of Species*
in 1859 following work co-authored with Alfred
Russel Wallace on the likely mechanism of natural selection, they established the foundation for the modern interpretation of evolution. Although organisms are capable of
reproducing in an exponential fashion, Wallace and Darwin observed that this growth potential of species is not
realized. Instead, population numbers remain relatively
constant in nature. Both Wallace and Darwin deduced
that some form of competi-
tive struggle for survival must
therefore occur. Darwin observed that there is a natural
variation between individuals
within a species; on this observation, he based his theory
of natural selection: "... that
any being, if it vary however
slightly in any manner prof-
itable to itself ... will have a
better chance of surviving."
Darwin included in his concept of survival "not only the
life of the individual, but success in leaving progeny."

A population of king penguins (along with a fur seal and an elephant seal)
in Antarctica.

While Gregor Mendel was familiar with Darwin's
work, Darwin was unaware of any underlying mechanism
to account for the morphological variation he had observed. However, with the development of the concept of
genes and alleles, the genetic basis of inherited variation
was established.

As others pursued the study of evolution, it became
apparent that the population rather than the individual was
the unit of study in this process. In order to study the role
of genetics in the process of
evolution, therefore, it was
necessary to consider allele
frequencies in populations
rather than offspring from
individual matings. Thus
arose the discipline of **population genetics**.

Early in the twentieth
century a number of work-
ers, including Gudny Yule,
William Castle, Godfrey
Hardy, and Wilhelm Wein-
berg, formulated the basic
principles of this field. In
the early years of population

genetics, the emphasis was on theory and the development of mathematical models to describe the genetic structure of populations. Workers in this field include Sewall Wright, Ronald Fisher, and J.B.S. Haldane. Following their work, experimentalists and field workers have used biochemical and molecular techniques to directly measure variation at the protein and DNA levels in order to test these theories and models. Allele frequencies and the forces that alter these frequencies, such as mutation, migration, selection, and random genetic drift, have been and are being examined. In this chapter we shall consider some general aspects of population genetics and also discuss other areas of genetics relating to evolution.

Populations and Gene Pools

Members of a species are often distributed over a wide geographic range. A **population** is a local group belonging to a single species, within which mating is actually or potentially occurring. The set of genetic information carried by all interbreeding members of a population is called the **gene pool**. For a given locus, this pool includes all the alleles of that gene present in the population. In population genetics the focus is on groups rather than on individuals, and on the measurement of allele and genotype frequencies in succeeding generations rather than on the distribution of genotypes resulting from a single mating. The term **allele frequency** will be used throughout the chapter and will represent the frequency of alleles in contrast to genotype frequencies. Gametes produced by one generation form the zygotes of the next generation. This new generation has a reconstituted gene pool that may differ from that of the preceding generation.

Populations are dynamic; they may grow and expand or diminish and contract through changes in birth or death rates, by migration, or by merging with other populations. This has important consequences and, over time, can lead to changes in the genetic structure of the population.

Calculating Allele Frequencies

One approach used to study a population's genetic structure is to measure the frequency of a given allele. This is possible once the mode of inheritance and the number of different alleles of this gene present in the population have been established. Allele frequencies cannot always be determined directly because in many cases only phenotypes, and not genotypes, can be observed. If, however, alleles expressed in a codominant fashion are considered, there is a direct relationship between phenotypes and genotypes with each phenotype having a unique geno-

TABLE 24.1 MN Blood Groups

Genotype	Blood Type	Reaction with Antibodies	
		Anti-M	Anti-N
$L^M L^M$	M	+	−
$L^M L^N$	MN	+	+
$L^N L^N$	N	−	+

type. Such is the case with the autosomally inherited **MN blood group** in humans.

In this case, the gene L on chromosome 2 has two alleles, L^M and L^N (often referred to as M and N, respectively).* Each allele controls the production of a distinct antigen on the surface of red blood cells. Thus, the genotype of any individual in a population may be type M ($L^M L^M$), N ($L^N L^N$), or MN ($L^M L^N$).

The genotypes, phenotypes, and immunological reactions of the MN blood group are shown in Table 24.1. Because they are codominant (and genotypes may be inferred from phenotypes), the frequency of the M and N alleles in a population can be determined simply by counting the number of individuals with each phenotype. As an example, consider a population of 100 individuals of which 36 are type M, 48 are type MN, and 16 are type N. The 36 type M individuals represent 72 M alleles, and the 48 MN heterozygotes represent an additional 48 M alleles. This means there are 72 + 48 or 120 M alleles in the population, out of a total of 200 alleles at this locus (100 individuals, each with two alleles). Therefore, the frequency of the M allele in this population is 0.6 (120/200 = 0.6 = 60%). The frequency of the N allele can be estimated in a similar fashion (80/200 = 0.4 = 40%) and is 0.4.

Table 24.2 illustrates two methods for computing the frequency of M and N alleles in a hypothetical population of 100 individuals, and Table 24.3 lists the frequencies of M and N alleles actually measured in several human populations.

The Hardy-Weinberg Law

In the case of the MN blood group, M and N are codominant alleles. If, on the other hand, one allele had been recessive, the heterozygotes would have been phenotypically identical to the homozygous dominant individuals, and the frequency of the alleles could not have been directly determined. However, a mathematical model developed independently by the British mathematician Godfrey H. Hardy and the German physician Wilhelm Weinberg can be used to calculate allele fre-

---

*L stands for Karl Landsteiner, the geneticist for whom the locus is named.

TABLE 24.2 Methods of Determining Allele Frequencies for Codominant Alleles

A. Counting Alleles

Genotype/Phenotype	MM	MN	NN	Total
Number of individuals	36	48	16	100
Number of *M* alleles	72	48	0	120
Number of *N* alleles	0	48	32	80
Total number of alleles	72	96	32	200

Frequency of *M* in population: $\dfrac{120}{200} = 0.6 = 60\%$

Frequency of *N* in population: $\dfrac{80}{200} = 0.4 = 40\%$

B. From Genotypes

Genotype/Phenotype	MM	MN	NN	Total
Number of individuals	36	48	16	100
Genotype frequency	36/100 = 0.36	48/100 = 0.48	16/100 = 0.16	1.00

Frequency of *M* in population: 0.36 + (1/2)0.48 = 0.36 + 0.24 = 0.60 = 60%

Frequency of *N* in population: 0.16 + (1/2)0.48 = 0.16 + 0.24 = 0.40 = 40%

TABLE 24.3 Frequencies of M and N Alleles in Various Populations

Population	Genotype Frequency (%)			Allele Frequency	
	MM	MN	NN	M	N
Eskimos (Greenland)	83.48	15.64	0.88	0.913	0.087
U.S. Indians	60.00	35.12	4.88	0.776	0.224
U.S. whites	29.16	49.38	21.26	0.540	0.460
U.S. blacks	28.42	49.64	21.94	0.532	0.468
Ainus (Japan)	17.86	50.20	31.94	0.430	0.570
Aborigines (Australia)	3.00	29.60	67.40	0.178	0.822

quencies in this case. The **Hardy-Weinberg law (HWL)** is one of the fundamental concepts in population genetics. The HWL has three important properties:

1. allele frequencies predict genotype frequencies;

2. at equilibrium, allele and genotype frequencies do not change from generation to generation; and

3. equilibrium is reached in one generation of random mating.

Assumptions for the Hardy-Weinberg Law

In the Hardy-Weinberg law, the following conditions are presumed:

1. The population is infinitely large, which in practical terms means that the population is large enough that sampling errors and random effects are negligible.

2. Mating within the population occurs at random.

3. There is no selective advantage for any genotype; that is, all genotypes produced by random mating are equally viable and fertile.

4. There is an absence of other factors including mutation, migration, and random genetic drift.

In such an ideal population, suppose that a locus has two alleles, *A* and *a*. The frequency of the allele *A* in both eggs and sperm is represented by p, and the frequency of *a* in gametes is represented by q. Because the sum of p and q represents 100 percent of the alleles for that gene in the population, then $p + q = 1$. A diagram can be used to represent the random combination of gametes containing these alleles and the resulting genotypes (Figure 24.1).

In the random combination of gametes in the population, the probability that sperm and egg both contain the *A* allele is $p \times p = p^2$. Similarly, the chance that gametes will carry unlike alleles is $(p \times q) + (p \times q) = 2pq$, and the chance that a homozygous recessive individual will result is $q \times q = q^2$. Note that these terms also de-

Sperm

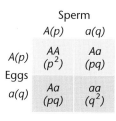

FIGURE 24.1 Gametes represent withdrawals from the gene pool to form the genotypes of the next generation. In this example, males and females have the same frequency (*p*) of the dominant allele *A*, and the same frequency (*q*) for the recessive allele *a*. After mating, the three genotypes *AA*, *Aa*, and *aa* have the frequencies of p^2, $2pq$ and q^2, respectively.

scribe one of the characteristics of the HWL: Allele frequencies determine genotype frequencies. In other words, while the value p^2 is the probability that both gametes in a fertilization event will carry the *A* allele, it is also a measure of the frequency of *AA* homozygotes in the ensuing generation. In a similar way, $2pq$ describes the frequency of *Aa* heterozygotes, and q^2 is a measure of the frequency of homozygous recessive (*aa*) zygotes. Thus, the distribution of homozygous and heterozygous genotypes in the next generation can be expressed as

$$p^2 + 2pq + q^2 = 1 \qquad (24.1)$$

Next, consider a population in which 70 percent of the alleles for a given gene are *A*, and 30 percent are *a*. Thus, $p = 0.7$ and $q = 0.3$, and $p(0.7) + q(0.3) = 1$. The distribution of genotypes produced by random mating is shown in Figure 24.2. In the new generation, 49 percent (p^2) of the individuals will be homozygous dominant, 42 percent ($2pq$) will be heterozygous, and 9 percent (q^2) will be homozygous recessive. The frequency of the *A* allele in the new generation can be calculated as

$$\begin{aligned} p^2 &+ \tfrac{1}{2}(2pq) \qquad (24.2) \\ 0.49 &+ \tfrac{1}{2}(0.42) \\ 0.49 &+ 0.21 = 0.70 \end{aligned}$$

Sperm

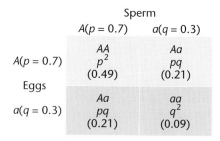

FIGURE 24.2 In this population, the frequency of the *A* allele is *p* = 0.7, and the frequency of the *a* allele is *q* = 0.3. Using the formula $p^2 + 2pq + q^2$, the frequencies of the genotypes in the next generation can be calculated as *AA* = 0.49, *Aa* = 0.42 and *aa* = 0.09. The frequencies of *A* and *a* remain constant from generation to generation.

For *a*, the frequency is

$$\begin{aligned} q^2 &+ \tfrac{1}{2}(2pq) \qquad (24.3) \\ 0.09 &+ \tfrac{1}{2}(0.42) \\ 0.09 &+ 0.21 = 0.30 \end{aligned}$$

Since $p + q = 1$, the value for *a* could have been calculated as

$$\begin{aligned} q &= 1 - p \qquad (24.4) \\ q &= 1 - 0.70 = 0.30 \end{aligned}$$

The frequencies of *A* and *a* in the new generation are the same as in the previous generation.

If Hardy-Weinberg conditions are assumed, allele frequencies and the frequencies of other genotypes can be calculated from knowing only the frequency of one genotype (such as the homozygous recessive). In human genetics, this relationship is used to calculate the frequency of heterozygotes carrying a recessive allele for a genetic disorder (see below).

A population in which the frequency of a given allele remains constant from generation to generation is said to be in a state of **genetic equilibrium**. In this case, the frequencies of *A* and *a* remain constant.

Several important points are relevant to the preceding example. First, while the hypothetical alleles we have considered were at equilibrium, not all alleles in a population are. This is particularly true when the assumptions made in the Hardy-Weinberg law do not hold. Second, the examples illustrate why dominant traits do not tend to increase in frequency as new generations are produced. Finally, the examples demonstrate that genetic equilibrium and **genetic variability** can be maintained in a population. Once established in a population, allelic frequencies remain unchanged during equilibrium—a factor important to the evolutionary process.

Testing for Equilibrium

The Hardy-Weinberg Law can also be used to determine whether genotypes in a given population are in equilibrium. In a natural population, any of the assumptions of the HWL (size, random mating, no selection) may not be valid. To do this, one must be able to phenotypically identify heterozygotes. If this is possible, then it must be ascertained whether the existing population fits a $p^2 + 2pq + q^2 = 1$ relationship. If not, some factor (presumably selection, mutation, migration) is causing allele frequencies to shift with each successive generation.

MN blood group distribution of allele frequencies of the Australian aborigines is a good example of this application of the Hardy-Weinberg law. From the values of allele frequencies in Table 24.3 (0.178 for M and 0.822 for N), the expected frequencies of blood types M, MN, and N can be calculated to determine if the population is in equilibrium:

Expected frequency of type M $= p^2$ (24.5)
$$= (0.178)^2 = 0.032$$
$$= 3.2\%$$
Expected frequency of type MN $= 2pq$
$$= 2(0.178)(0.822) = 0.292$$
$$= 29.2\%$$
Expected frequency of type N $= q^2$
$$= (0.822)^2 = 0.676$$
$$= 67.6\%$$

These **expected frequencies** are nearly identical to the **observed frequencies** shown in Table 24.3, confirming that the population is in equilibrium. If there were a question as to whether the observed frequencies varied significantly from the expected frequencies, a chi-square analysis could be performed (see Chapter 3).

On the other hand, if the Hardy-Weinberg test demonstrates that the population is not in equilibrium, one or more of these necessary conditions are not being met. Using the values listed for M and N frequencies in Table 24.3, we can construct a hypothetical mixed population of 500 Australian aborigines and 500 American Indians. The frequencies for M and N for this hypothetical population are shown in Table 24.4. In such a mixed population, the frequency of M (p) would be

$$0.315 + 1/2(0.324) = 0.477 \qquad (24.6)$$

and that of N (q) would be

$$0.361 + 1/2(0.324) = 0.523 \qquad (24.7)$$

If these allelic frequencies were derived by random mating, then the proportions of phenotypes we should expect are

	♂	
	$M(p = 0.477)$	$N(q = 0.523)$
$M(p = 0.477)$ ♀	MM p^2 0.228	MN pq 0.249
$N(q = 0.523)$	MN pq 0.249	NN q^2 0.274

FIGURE 24.3 If mating in a population with differences in allele frequencies follows the expectations of the Hardy-Weinberg, an equilibrium will be reached in one generation. Compare the genotype frequencies after one generation of mating with those in Table 24.4 for a population in equilibrium.

MM (p^2), 22.8 percent; MN ($2pq$), 49.8 percent; and NN (q^2), 27.4 percent. Because the expected genotype frequencies do not fit those observed (and a chi-square test would confirm this), we would conclude that the population is in a state of nonequilibrium brought about, obviously, by a lack of random mating in this hypothetical population.

What would happen, however, if such a mixed population were to be established on an island with no previous inhabitants and mating did take place at random? How long would it take for equilibrium to be established? Applying the Hardy-Weinberg equation, we can predict that after one generation the observed and expected genotype frequencies would converge, illustrating the third important characteristic of the HWL. Using the p and q values just calculated, and remembering that the allele frequencies represent those of the total gene pool of the first generation mixed population, the frequencies after one generation of random mating are shown in Figure 24.3. We can see that after one generation, the observed genotype frequencies of 22.8 percent for MM (p^2); 49.8 percent for MN ($2pq$); and 27.4 percent for NN (q^2) are those expected for an equilibrium population. The allele frequencies of $M = 0.477$ and $N = 0.523$ are also those expected for an equilibrium population. However, to demonstrate HWL equilibrium rigorously, we would need to calculate the frequency of matings.

Extensions of the Hardy-Weinberg Law

In considering genotype and allele frequencies for autosomal loci using the Hardy-Weinberg equation, we assumed that the frequency of A is the same in both sperm and eggs. What about X-linked genes? In species that have two X chromosomes, such as humans and *Drosophila*, females carry two copies of all genes on the X chromosome, whereas males carry only one copy of all X-linked genes.

TABLE 24.4 Frequencies of M and N Alleles and Genotypes in Natural and Artificial Populations

Population	Genotype Frequencies			Allele Frequencies	
	MM	MN	NN	p	q
Australian aborigines	0.03	0.296	0.674	0.178	0.822
American Indians	0.60	0.351	0.049	0.776	0.224
Mixed population (500 + 500):					
Observed	0.315	0.324	0.361	0.477	0.523
Expected	0.228	0.498	0.274		

Genes on the X chromosome are therefore distributed unequally in the population, and when there are equal numbers of males and females, the females carry two-thirds of all X-linked genes, and males carry one-third of the total number. Can the Hardy-Weinberg law be applied to calculate allelic frequencies and genotypic frequencies in such circumstances?

X-Linked Genes

It is easy to determine frequencies for X-linked genes in males. Because they have only one copy of all genes on this chromosome, the phenotype reveals both dominant and recessive alleles. Therefore, in males, the frequency of an X-linked allele is the same as the phenotypic frequency. In Western Europe, a form of X-linked color blindness occurs with a frequency of 8 percent in males ($q = 0.08$). Because females have two doses of all genes on the X chromosome, color-blind females exhibit the phenotypic frequencies expected for complete dominance, and the genotype and allele frequencies can be calculated by using the standard Hardy-Weinberg equation. For example, as color blindness in males has a frequency of 0.08, at equilibrium, the expected frequency in females is q^2, or 0.0064. This means that 800 out of 10,000 males surveyed would be expected to be color-blind, but only 64 of 10,000 females would show this trait. Expected values of the frequency of X-linked traits in males and females are compared in Table 24.5.

If the frequency of an X-linked allele differs between males and females, then the population is not in equilibrium. In contrast to alleles of autosomal loci, equilibrium in each group will not be reached in a single generation, but will be approached over a series of succeeding generations. Because males inherit their X chromosome maternally, the allelic frequency in females will determine the frequency in males of the next generation. Daughters will inherit both a maternal and a paternal X, and their allelic frequency is the average of that found in the parents. Even though the population's overall allele frequency remains constant, there is an oscillation in frequencies for

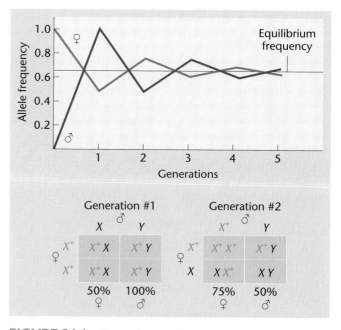

FIGURE 24.4 Approach to equilibrium for an X-linked trait with an initial frequency of 1.0. The fluctuations in frequency in the first two generations are shown. In the parents of generation 1, the X-linked trait (x^+) is present in 100 percent of the females and 0 percent of the males. In the progeny, the trait is present in 50 percent of all female X chromosomes and in 100 percent of all male X chromosomes. Thus, as the graph shows, the frequency in females drops from 1.0 to 0.5 in a generation, while the frequency in males rises from 0.0 to 1.0. In generation 2, the frequency in females is 0.75, and in males 0.5. This oscillation continues until eventually an equilibrium is reached.

the two sexes in each generation, with the differences being halved in each succeeding generation until an equilibrium is reached. This concept is illustrated in Figure 24.4 for the simple case where an allele has an initial frequency of 1.0 in females and 0.0 in males.

Multiple Alleles

In addition to autosomal recessive and sex-linked genes, it is common to find several alleles of a single locus in a population. The ABO blood group in humans is such an example. The locus I (isoagglutinin) has three alleles (I^A, I^B, and I^O), yielding six possible genotypic combinations (I^AI^A, I^BI^B, I^OI^O, I^AI^B, I^AI^O, I^BI^O). Recall that in this case A and B are codominant alleles, and both of these are dominant to O. The result is that homozygous AA and heterozygous AO individuals are phenotypically identical, as are BB and BO individuals, so we can distinguish only four phenotypic combinations.

By adding another variable to the Hardy-Weinberg equation, we can calculate both genotype and allele frequencies for the situation involving three alleles. In an

TABLE 24.5 Expected Relative Frequency of X-Linked Traits in Males and Females

Frequency of Males with Trait	Expected Frequency in Females
90/100	81/100
50/100	25/100
10/100	1/100
1/100	1/10,000
1/1000	1/1,000,000
1/10,000	1/100,000,000

equilibrium population, the frequency of the three alleles can be described by

$$p(A) + q(B) + r(O) = 1 \qquad (24.8)$$

and the distribution of genotypes will be given by

$$(p + q + r)^2 \qquad (24.9)$$

In our hypothetical population, the genotypes AA, AB, AO, BB, BO, and OO will be found in the ratio

$$p^2(AA) + 2\,pq(AB) + 2pr(AO) + q^2(BB) \qquad (24.10)$$
$$+ 2qr(BO) + r^2(OO) = 1$$

If we know the frequencies of blood types for a population, we can then estimate the frequencies for the three alleles of the ABO system. For example, in one population sampled, the following blood types were observed: A = 0.53, B = 0.13, AB = 0.08, and O = 0.26. Because the O allele is recessive, the frequency of type O blood in the population is equal to the frequency of the recessive genotype r^2. Thus,

$$r^2 = 0.26$$
$$r = \sqrt{0.26}$$
$$= 0.51$$

By using the value estimated for r, we can estimate the allele frequencies for the A (p) and B (q) alleles. The A allele is present in two genotypes, AA and AO. The frequency of the AA genotype is represented by p^2, and the AO genotype by $2pr$. Therefore,

$$\text{Frequency of } A + O = p^2 + 2pr + r^2$$

where A and O represent the phenotypic frequencies.

TABLE 24.6   Calculating Genotypic Frequencies for Multiple Alleles Where the Frequency of A (p) Is 0.38, B (q) Is 0.11, and O (r) Is 0.51

Genotype	Genotypic Frequency	Phenotype	Phenotypic Frequency
AA	$p^2 = (0.38)^2 = 0.14$	A	0.53
AO	$2pr = 2(0.38 \times 0.51) = 0.39$		
BB	$q^2 = (0.11)^2 = 0.01$	B	0.12
BO	$2qr = 2(0.11 \times 0.51) = 0.11$		
AB	$2pq = 2(0.38 \times 0.11) = 0.084$	AB	0.08
OO	$r^2 = (0.51)^2 = 0.26$	O	0.26

This can be rearranged to give

$$p = \sqrt{\text{Frequency of } A + O} - r$$
$$= \sqrt{0.53 + 0.26} - 0.51$$
$$= 0.89 - 0.51 = 0.38$$

Having estimated the frequencies for A (p) and O (r), the frequency for the B allele can be estimated from the following relationship:

$$p + q + r = 1$$
$$q = 1 - (p + r)$$
$$= 1 - (0.38 + 0.51)$$
$$= 1 - 0.89$$
$$= 0.11$$

The phenotypic frequencies and genotypic frequencies for this population are summarized in Table 24.6.

Using the Hardy-Weinberg Law: Calculating Heterozygote Frequency

One of the practical applications of the Hardy-Weinberg law is the estimation of heterozygote frequency in a population. The frequency of a recessive phenotype usually can be determined by counting such individuals in a sample of the population. This information and the Hardy-Weinberg equation can be used to calculate the allele and genotype frequencies.

Albinism, an autosomal recessive trait, has an incidence of about 1/10,000 (0.0001) in some populations. Albinos are easily distinguished from the population at large by a lack of pigment in skin, hair, and iris. Because this is a recessive trait, albino individuals must be homozygous. Their frequency in a population is represented by q^2, provided that mating has been at random and all Hardy-Weinberg conditions have been met in the previous generation.

The frequency of the recessive allele therefore is

$$\sqrt{q^2} = \sqrt{0.0001} \qquad (24.11)$$
$$q = 0.01, \text{ or } 1/100$$

Since $p + q = 1$, then the frequency of p is

$$p = 1 - q \qquad (24.12)$$
$$= 1 - 0.01$$
$$= 0.99, \text{ or } 99/100$$

In the Hardy-Weinberg equation, the frequency of heterozygotes is given as $2pq$. Therefore, we have

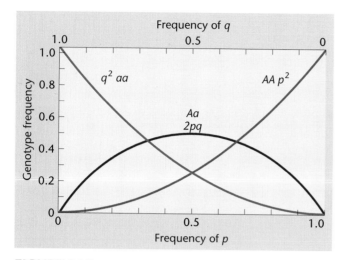

FIGURE 24.5 Relationship between genotype frequency and allele frequencies derived from the Hardy-Weinberg equation.

$$
\begin{aligned}
\text{Heterozygote frequency} &= 2pq && (24.13)\\
&= 2[(0.99)(0.01)]\\
&= 0.02, \text{ or } 2\%, \text{ or } 1/50
\end{aligned}
$$

Thus, heterozygotes for albinism are rather common in the population (2%), even though the incidence of homozygous recessives is only 1/10,000.

In general, the frequencies of all three genotypes can be estimated once the frequency of either allele is known and Hardy-Weinberg conditions are assumed. The relationship between genotype frequency and allele frequency is shown in Figure 24.5. It is important to note how fast heterozygotes increase in a population as the values of p and q move away from zero. This observation confirms our conclusion that when a recessive trait like albinism is rare, the majority of those carrying the allele are heterozygotes. In cases where the frequencies of p and q are between 0.33 and 0.67, heterozygotes actually constitute the major class in the population.

Factors That Alter Allele Frequencies in Populations

The Hardy-Weinberg law establishes a set of ideal conditions that allows us to estimate allele frequencies and genotype frequencies in populations in which initial assumptions about random mating, absence of selection and mutation, and equal viability and fecundity hold true. Obviously, it is difficult to find natural populations in which all these conditions are met. In nature, populations in nature are dynamic, and changes in size and structure are part of their life cycles. This situation al-

lows us to use the Hardy-Weinberg law to investigate exceptions to the initial assumptions. In this section we will discuss factors that prevent populations from reaching equilibrium, or drive populations toward a different equilibrium, and the relative contribution of these factors to evolutionary change.

Mutation

Within a population, the gene pool is reshuffled each generation to produce new combinations in the genotypes of the offspring. Because the number of possible genotypic combinations is so large, the members of a population alive at any given time represent only a fraction of all possible genotypes. Although an enormous genetic reserve is present in the gene pool, new genotypic combinations are produced by Mendelian assortment and recombination, but these processes do not produce any new alleles. **Mutation** alone acts to create new alleles, but in the absence of other forces, mutation has a negligible effect on allele frequencies.

For our purposes, we need only consider that mutational events occur at random—that is, without regard for any possible benefit or disadvantage to the organism. Here we will discuss only the generation of mutant alleles and in a later section will consider the spread and distribution of new alleles through the population.

To know whether mutation is a significant force in changing allele frequencies, we must measure the rate at which mutations are produced. As most mutations are recessive, it is difficult to observe mutation rates directly in diploid organisms. Indirect methods using probability and statistics or large-scale screening programs must be employed. For certain dominant mutations, however, a direct method of measurement can be used. To ensure accuracy, several conditions must be met:

1. The trait must produce a distinctive phenotype that can be distinguished from similar traits produced by recessive alleles.

2. The trait must be fully expressed or completely penetrant so that mutant individuals can be identified.

3. An identical phenotype must never be produced by nongenetic agents such as drugs or chemicals.

Mutation rates can be expressed as the number of new mutant alleles per given number of gametes. Suppose for a given gene that undergoes mutation to a dominant allele, 2 of 100,000 births exhibit a mutant phenotype. In these two cases, the parents are phenotypically normal. Because the zygotes that produced these births each carry two copies of the gene, we have actually surveyed 200,000 copies of the gene (or 200,000 gametes). If we assume that the affected births are each heterozygous, we have uncovered 2 mutant alleles out of 200,000. Thus the mutation rate is 2/200,000 or 1/100,000, which in scientific notation would be written as 1×10^{-5}.

In humans, a dominant form of dwarfism known as **achondroplasia** fulfills the requirements outlined above for the measurement of mutation rates. Individuals with this skeletal disorder have an enlarged skull, short arms and legs, and can be diagnosed by radiologic examination at birth. In a survey of almost 250,000 births, the mutation rate (μ) for achondroplasia has been calculated as

$$1.4 \times 10^{-5} \pm 0.5 \times 10^{-5} \quad (24.14)$$

Knowing the rate of mutation, we can then estimate the change in frequency for each generation. If initially only the normal d allele exists (i.e., all individuals are homozygous dd), the frequency of d (q_0) is 1.0 and the frequency of D (dwarf) is $p_0 = 0$. If the rate of mutation from d to D is μ, then in the next generation,

$$\text{Frequency of } D = p_1 = q_0\mu \quad (24.15)$$

The new frequency for d, which will be lowered by the rate of mutation, can be expressed as

$$\text{Frequency of } d = q_1 = p_0\mu \quad (24.16)$$

Although the rate of mutation is constant from generation to generation, the rate of change in the mutant allele is initially high, but declines to zero at mutational equilibrium.

As a more general example, let us assume that the mutation rate for genes in the human genome is 1.0×10^{-5}. With such a low mutation rate, changes in allele frequency brought about by mutation alone are very small. Figure

24.6 shows that if a population begins with only one allele (A) at a locus ($p = 1$), and a mutation rate of 1.0×10^{-5} for A to a, it would require about 70,000 generations to reduce the frequency of A to 0.5. Even if the rate of mutation were to increase through exposure to higher levels of radioactivity or chemical mutagens, the impact of mutation on allele frequencies would be extremely weak. This example once again emphasizes that mutation is a major force in generating genetic variability, but by itself plays a relatively insignificant role in changing allele frequencies.

Migration

Frequently, a species of plant or animal becomes divided into subpopulations that to some extent are geographically separated. Differences in mutation rate and selective pressures can establish different allele frequencies in the subpopulations. **Migration** occurs when individuals move between these populations. Consider a single pair of alleles, A (p) and a (q). The change in the frequency of A in one generation can be expressed as

$$\Delta p = m(p_m - p) \quad (24.17)$$

where

p = frequency of A in existing population
p_m = frequency of A in immigrants
Δp = change in one generation
m = coefficient of migration (proportion of migrant genes entering the population per generation)

For example, assume that $p = 0.4$ and $p_m = 0.6$, and that 10 percent of the parents giving rise to the next generation are immigrants ($m = 0.1$). Then, the change in the frequency of A in one generation is

$$\begin{aligned}
\Delta p &= m(p_m - p) \\
&= 0.1(0.6 - 0.4) \\
&= 0.1(0.2) \\
&= 0.02
\end{aligned}$$

In the next generation, the frequency of A (p_1) will increase as follows:

$$\begin{aligned}
p_1 &= p + \Delta p \\
&= 0.40 + 0.02 \\
&= 0.42
\end{aligned} \quad (24.18)$$

If either m is large or if p is very different from p_m, then a rather large change in the frequency of A will occur in a single generation. All other factors being equal (including migration), an equilibrium will be attained when $p = p_m$. These calculations reveal that the change in allele fre-

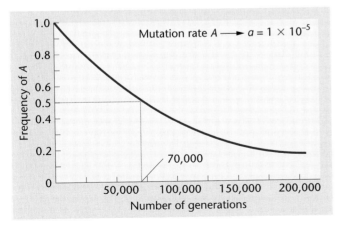

FIGURE 24.6 Rate of replacement of an allele by mutation alone, assuming an average mutation rate of 1.0×10^{-5}.

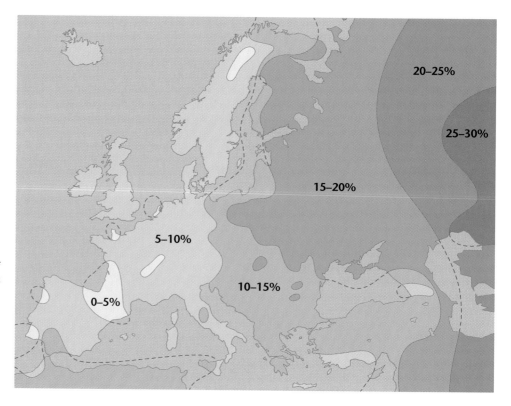

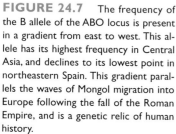

FIGURE 24.7 The frequency of the B allele of the ABO locus is present in a gradient from east to west. This allele has its highest frequency in Central Asia, and declines to its lowest point in northeastern Spain. This gradient parallels the waves of Mongol migration into Europe following the fall of the Roman Empire, and is a genetic relic of human history.

20–25%

25–30%

15–20%

5–10%

10–15%

0–5%

quency attributable to migration is proportional to the differences in allele frequency between the donor and recipient populations and to the rate of migration. As the migration coefficient can have a wide range of values, the effect of migration can substantially alter allele frequencies in populations (Figure 24.7). Although migration can be somewhat difficult to quantify, it can often be estimated.

Migration can also be regarded as the flow of genes between two populations that were once but are no longer geographically isolated. For example, most of the blacks in the United States are descended from ancestors in West Africa. In Africa, the frequency of the Duffy blood group allele Fy^0 is almost 100 percent. In Europe, the source of most U.S. whites, the frequency is close to zero percent, and almost all individuals are Fy^a or Fy^b. By measuring the frequency of Fy^a or Fy^b among U.S. blacks, we can estimate the amount of gene flow into the black population. Figure 24.8 shows the frequency of the Fy^a allele among blacks in three African nations, in several regions of the United States along with the frequency in a white population. Using the average frequency and assuming an equal rate of gene flow in each generation, we can estimate the migration of the Fy^a allele at about 5 percent per generation.

Natural Selection

Mutation and migration introduce new alleles into populations. **Natural selection**, on the other hand, is the principal force that shifts allele frequencies within large

populations and is one of the most important factors in evolutionary change.

Darwin's main contribution to the study of evolution was his recognition that natural selection and reproductive isolation are the mechanisms that lead to the divergence and eventual separation of populations into distinct species. In any population at a given time, there are individuals with different genotypes. Because of these inherent genetic differences, some of the individuals in this population will be better adapted to the environment than will others, leading to the differential survival and reproduction of some genotypes over others. Natural selection is thus the consequence of the differential reproduction of genotypes. Therefore, allelic frequencies will change over time. Natural selection, then, is a major force in changing allelic frequencies and therefore represents a departure from the Hardy-Weinberg assumption that all genotypes have equal viability and fertility.

Because selection is a consequence of the organism's genotypic/phenotypic combination, polygenic or quantitative traits (those that are controlled by a number of genes and that might be susceptible to environmental influences) also respond to selection. Such quantitative traits, including adult body height and weight in humans, often demonstrate a continuously varying distribution resembling a bell-shaped curve. Selection for such traits can be classified as (1) directional, (2) stabilizing, or (3) disruptive.

In **directional selection**, important to plant and animal breeders, desirable traits, often representing pheno-

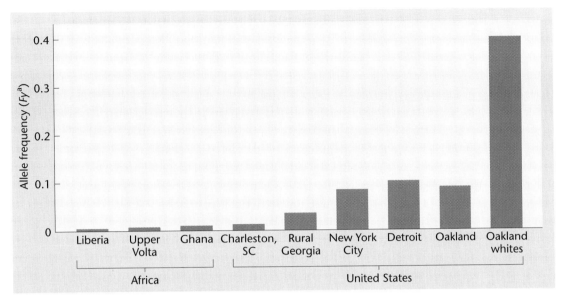

FIGURE 24.8   Frequencies of the Fy^a allele in African and U.S. black populations.

typic extremes, are selected. In fact, if the trait is polygenic, the most extreme phenotypes that the genotype can express will appear in the population only after prolonged selection. An example of directional selection is the long-running experiment at the State Agricultural Laboratory in Illinois selecting for high and low oil content in corn kernels. Beginning in 1896, a population of 163 corn ears was surveyed, and the 24 ears highest in oil content were used as the parents of the next generation. In each succeeding generation, the ears with the highest oil content were used for breeding. In this example of upward selection, the oil content was raised after 50 generations from about 4 percent to just over 16 percent, with no sign that a plateau has been reached (Figure 24.9). This experiment also selected 12 ears with the lowest oil content and downward selection from this parental group has lowered the oil content from 4 percent to less than 1

percent over the period of 50 generations (Figure 24.9). In this case, the high and low lines have diverged to the point where the distribution of oil content in each line does not overlap with that of the other line.

In upward selection, alleles for high oil content increase in frequency, and replace alleles for low oil content. At some point, all individuals in the population will have the genotype for the highest oil content, and all will express the most extreme phenotype for high oil content. In downward selection, the opposite situation will occur, eventually producing a population with a genotype and phenotype for the lowest oil content.

When the lines fail to respond to further selection, there will be no remaining genetic variance for oil content, heritability will be zero, and each line will have a homozygous genotype for oil content.

In nature, directional selection can occur when one of the phenotypic extremes becomes selected for or against, usually as a result of changes in the environment.

Stabilizing selection, on the other hand, tends to favor intermediate types, with both extreme phenotypes being selected against. One of the clearest demonstrations of stabilizing selection is the data of Mary Karn and Sheldon Penrose on human birth weight and survival. Figure 24.10 shows the distribution of birth weight for 13,730 children born over an 11-year period and the percentage of mortality at 4 weeks of age. Infant mortality increases dramatically on either side of the optimal birth weight of 7.5 pounds. At the genetic level, stabilizing selection acts to keep a population well adapted to its environment. In this situation, individuals closer to the average for a given trait will have higher fitness (a concept discussed below).

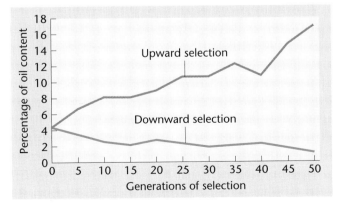

FIGURE 24.9   Long-term selection to alter the oil content of corn kernels.

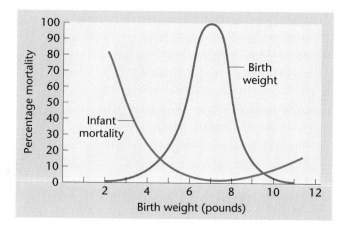

FIGURE 24.10 Relationship between birth weight and mortality in humans.

Disruptive selection is selection against intermediates and for both phenotypic extremes. It may be viewed as the opposite of stabilizing selection because the intermediate types are selected against. In one set of experiments, John Thoday applied disruptive selection to progeny of *Drosophila* strains with high and low bristle number. Matings were carried out using females from one strain and males from the other. Progeny were selected for high and low bristle number, and matings were carried out for a number of generations, and the two lines diverged rapidly (Figure 24.11). In natural populations, such a situation might exist for a population in a heterogeneous environment.

The types and effects of selection are summarized in Figure 24.12.

Fitness and Selection

When a particular genotype/phenotype combination confers an advantage to organisms in competition with others harboring an alternate combination, selection occurs. The relative strength of selection varies with the amount of advantage provided. The probability that a particular phenotype will survive and leave offspring is a measure of its **fitness**. Thus, fitness refers to a total reproductive potential or efficiency. The concept of fitness is usually expressed in relative terms by comparing a particular genotypic/phenotypic combination with one regarded as optimal. Fitness is a relative concept because as environmental conditions change, so does the advantage conferred by a particular genotype.

Mathematically, the difference between the fitness of a given genotype and another, reference genotype is called the **selection coefficient (s)**. For a phenotype conferred by the genotype *aa*, when 99 of every 100 organisms successfully reproduce, $s = 0.01$. If the *aa* geno-

type is a homozygous lethal and *AA* and *Aa* have equal fitness, then $s = 1.0$, and the *a* allele is propagated only in the heterozygous carrier state. Starting with any original frequencies of p_0 and q_0 in a population, when $s = 1.0$, the effect of selection on any successive generation can be calculated using the formula

$$q_n = \frac{q_0}{1 + nq_0} \qquad (24.19)$$

where *n* is the number of generations elapsing since p_0 and q_0.

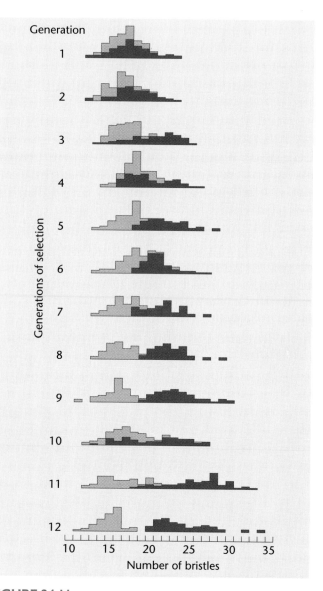

FIGURE 24.11 The effect of disruptive selection on bristle number in *Drosophila*. By selecting lines with the highest and lowest bristle number, the population showed a nonoverlapping divergence in only 12 generations.

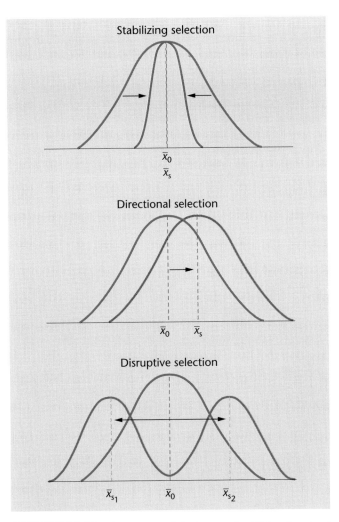

FIGURE 24.12 Comparison of the impact of stabilizing, directional, and disruptive selection. In each case, $\bar{x}_0$ is the mean of an original population, and $\bar{x}_s$ is the mean of the population following selection.

When s is less than 1.0, it is possible to calculate the effects of selection on each successive generation with any s, p_0, and q_0 values using the formula

$$q_1 = \frac{q_0 - sq_0^2}{1 - sq_0^2} \qquad (24.20)$$

Figure 24.14 demonstrates a variety of initial conditions and the change of the frequency of q (a) through a large number of generations. If q_0 is initially 0.5 and $s = 0.5$, five generations must elapse before q is halved. Then, another nine generations must elapse to cut q in half again. Figure 24.14 shows the relative trends as q_0 and s change.

Selection in Natural Populations

There has been much study of selection on both laboratory and natural populations. A classic example of selection in natural populations is that of the peppered moth,

Generation	p	q	p^2	$2pq$	q^2
0	0.50	0.50	0.25	0.50	0.25
1	0.67	0.33	0.44	0.44	0.12
2	0.75	0.25	0.56	0.38	0.06
3	0.80	0.20	0.64	0.32	0.04
4	0.83	0.17	0.69	0.28	0.03
5	0.86	0.14	0.73	0.25	0.02
6	0.88	0.12	0.77	0.21	0.01
10	0.91	0.09	0.84	0.15	0.01
20	0.95	0.05	0.91	0.09	<0.01
40	0.98	0.02	0.95	0.05	<0.01
70	0.99	0.01	0.98	0.02	<0.01
100	0.99	0.01	0.98	0.02	<0.01

Let us consider another example. Figure 24.13 demonstrates the change in allele frequencies when A (p_0) = a (q_0) = 0.5. Initially, because of the high percentage of aa genotypes, the frequency of the a allele is reduced rapidly. The frequency of a is halved (0.25) in only two generations. By the sixth generation, the frequency of a is again reduced twofold (0.12). By now, however, the majority of a alleles are carried by heterozygotes. Because selection operates on phenotypes and not on components of the genotype, this means that heterozygotes are not selected against. Subsequent reductions in frequency occur very slowly in successive generations and depend on the rate at which heterozygotes are removed from the population. For this reason, it is difficult to eliminate a recessive allele from the population by selection as long as heterozygotes continue to mate.

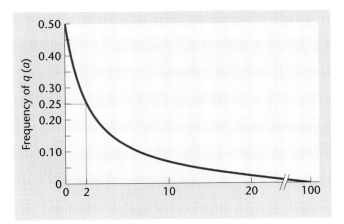

FIGURE 24.13 Changes in allele frequency under selection where $s = 1.0$. The frequency of a is halved in two generations, and halved again by the sixth generation. Subsequent reductions occur slowly because the majority of a alleles are carried by heterozygotes.

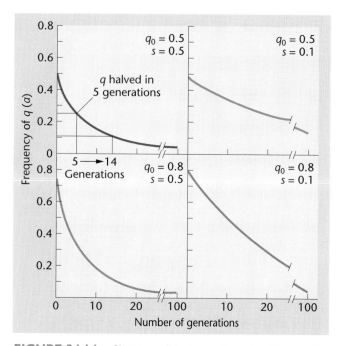

FIGURE 24.14   Changes in allele frequencies under different selection coefficients and different initial gene frequencies. When $q_0 = 0.5$ and $s = 0.5$, the frequency of q is halved in 5 generations. When $q_0 = 0.8$ and $s = 0.5$, the frequency of q is halved in 5 generations and halved again in an additional 5 generations.

Biston betularia, in England. Before 1850, 99 percent of the moth population was light colored, allowing the nocturnal moths to rest undetected during the day on lichen-covered trees. As toxic gases produced by industry killed the lichens growing on trees and buildings, and soot deposits darkened the landscape, the light-colored moths became easy targets for their predators. The rare dark-colored (melanic) moths suddenly gained a great selective advantage because of their natural camouflage. A rapid shift in frequency of this phenotype occurred, probably in 50 generations or less.

Laboratory experiments have demonstrated that the dark form is due to a single dominant allele, C. Figure 24.15 demonstrates the protective advantage from predators provided by the dark phenotype.

Following the adoption of laws in the mid-1960s to restrict environmental pollution, the frequency of non-melanic forms of the moth has started to increase in many areas of England, thus underscoring the close relationship between the environment and selection. In one region south of Liverpool, the frequency of the dark-colored form decreased from over 90 percent to just over 50 percent in the period from 1959 to 1985, showing that genotype frequencies can respond quickly to a shift in the environment.

Genetic Drift

In laboratory crosses, one condition essential to realizing theoretical genetic ratios (e.g., 3:1, 1:2:1, 9:3:3:1) is a fairly large sample size. This sample size is also important to the study of population genetics as allele and genotype frequencies are examined or predicted. For example, if a population consists of 1000 randomly mating heterozygotes (Aa), the next generation will consist of approximately 25 percent AA, 50 percent Aa, and 25 percent aa genotypes. Provided that the initial and subsequent populations are large, there will be at most only minor deviations from this mathematical ratio, and the frequency of A and a will remain about equal (0.50).

However, if a population is formed from only one set of heterozygous parents and they produce only two offspring, allele frequency can change drastically. In this case, genotypes and allele frequencies in the offspring can be predicted, as shown in Table 24.7. As shown, in 10 of 16 times such a cross is made, the new allele frequencies would be altered. In 2 of 16 times, either the A or a allele would be eliminated in a single generation.

FIGURE 24.15   The speckled and melanic forms of the peppered moth, *Biston,* seen on a normal, lichen-covered tree trunk (left) and on a darkened, soot-covered trunk (right).

TABLE 24.7 All Possible Pairs of Offspring Produced by Two Heterozygous Parents (Aa × Aa)

Possible Genotypes of Two Offspring	Probability	Allele Frequency	
		A	a
AA and AA	$(1/4)(1/4) = 1/16$	1.00	0.00
aa and aa	$(1/4)(1/4) = 1/16$	0.00	1.00
AA and aa	$2(1/4)(1/4) = 2/16$	0.50	0.50
Aa and Aa	$(2/4)(2/4) = 4/16$	0.50	0.50
AA and Aa	$2(2/4)(1/4) = 4/16$	0.75	0.25
aa and Aa	$2(1/4)(2/4) = 4/16$	0.25	0.75

Using only one set of heterozygous parents that produce two offspring is an extreme example, but has been chosen to illustrate the point that large interbreeding populations are essential to the Hardy-Weinberg equilibrium. In small populations, significant random fluctuations in allelic frequencies are possible by chance deviation. The degree of fluctuation will increase as the population size decreases. Such changes illustrate the concept of **genetic drift**. In the extreme case, genetic drift may lead to the chance fixation of one allele to the exclusion of another allele.

To study genetic drift in laboratory populations of *Drosophila melanogaster*, Warwick Kerr and Sewall Wright set up over 100 lines, with four males and four females as the parents for each line. Within each line, the frequency of the sex-linked bristle mutant *forked* (*f*) and its wild-type allele (*f*⁺) was 0.5. In each generation, four males and four females were chosen at random to be parents of the next generation. After 16 generations, fixation had occurred in 70 lines—29 in which only the *forked* allele was present and 41 in which the wild-type allele had become fixed. The rest of the lines were still segregating the two alleles or had been lost. If fixation had occurred randomly, then an equal number of lines should have become fixed for each allele. In fact, the experimental results do not differ statistically from the expected ratio of 35:35, demonstrating that alleles can spread through a population and eliminate other alleles by chance alone.

How are small populations created in nature? In one instance, a large group of organisms may be split by some natural event, creating a small, isolated subpopulation. A natural disaster such as an epidemic might also occur, leaving a small number of survivors to constitute the breeding population. Third, a small group might emigrate from the larger population, as founders, to a new environment, such as a volcanically created island.

Allelic frequencies in certain human isolates best support the role of drift as an evolutionary force in natural populations. The Pingelap atoll in the western Pacific Ocean (lat. 6° N, long. 160° E) has in the past been devastated by typhoons and famine, and in about 1780 there were only about 9 surviving males. Today there are fewer than 2000 inhabitants, and their ancestry can be traced to the typhoon survivors. From 4–10 percent of the current population is blind from infancy. These people are affected by an autosomal recessive disorder, **achromatopsia**, which causes ocular disturbances, a form of color blindness and cataract formation. This disorder is extremely rare in the human population as a whole. However, the mutant allele is present at a relatively high frequency in the Pingelap population. From genealogical reconstruction, it was found that one of the original survivors (about 30 individuals) was a chief who was heterozygous for the condition. If we assume that he was the only carrier in the founding population, the initial gene frequency was 1/60, or 0.016. On average, about 7 percent of the current population is affected (a homozygous recessive genotype), the frequency of the allele in the present population is calculated as 0.26.

Another example of genetic drift involves the Dunkers, a small, isolated, religious community who emigrated from the German Rhineland to Pennsylvania. Because their religious beliefs do not permit marriage with outsiders, the population has grown only through marriage within the group. When the frequencies of the ABO and MN blood group alleles are compared, significant differences are found among the Dunkers, the German population, and the U.S. population. The frequency of blood group A in the Dunkers is about 60 percent, in contrast to 45 percent in the U.S. and German populations. The I^B allele is nearly absent in the Dunkers. Type M blood is found in about 45 percent of the Dunkers, compared with about 30 percent in the U.S. and German populations. As there is no evidence for a selective advantage of these alleles, it is apparent that their observed frequencies are the result of chance events in a relatively small isolated population.

Inbreeding

One of the assumptions in the Hardy-Weinberg law is random mating in a large population. In the previous section we considered how allele frequencies in small populations can be altered by drift, leading to a reduction in genetic variability, and a decrease in heterozygosity. The same effect can be produced by inbreeding and nonrandom mating. In a small population, potential mates are more likely to be related to each other than in large populations. In large populations distributed over a wide geographic range, individuals tend to mate with those nearby, rather than those living at a great distance. If the mobility of individuals is restricted, a pattern of nonrandom mating can cause genetic drift, and subdivide the population

P$_1$

Self-fertilization
Aa

	AA	Aa	aa
F$_1$	0.250	0.500	0.250
F$_2$	0.375	0.250	0.375
F$_3$	0.437	0.125	0.437
F$_4$	0.468	0.063	0.468
F$_n$	$\dfrac{1-\frac{1}{2}n}{2}$	$\dfrac{1}{2^n}$	$\dfrac{1-\frac{1}{2}n}{2}$

FIGURE 24.16 Reduction in heterozygote frequency brought about by self-fertilization. After n generations, the frequencies of the genotypes can be calculated according to the formulas in the bottom row.

into a number of smaller, interbreeding subpopulations, differing from each other in the frequency of some alleles, and leading to the chance elimination of alleles.

Nonrandom Mating

Nonrandom mating occurs in many species, including humans. One form of nonrandom mating is called **assortative**. In humans, pair bonds may be established by religious practices, physical characteristics, professional interests, and so forth on a nonrandom basis. In nature, phenotypic similarity plays the same role.

Another form of nonrandom mating is **inbreeding**, where mating occurs between relatives. One of the consequences of inbreeding is an increase in the chance that an individual will be homozygous for a recessive deleterious allele. For a given allele, inbreeding increases the proportion of homozygotes in the population. Over time, with complete inbreeding, only homozygotes will remain in the population. To illustrate this concept, we shall consider the most extreme form of inbreeding, **self-fertilization**.

Figure 24.16 shows the results of four generations of self-fertilization starting with a single individual heterozygous for one pair of alleles. By the fourth generation, only about 6 percent of the individuals are still heterozygous, and 94 percent of the population is homozygous. Note, however, that the frequencies of A and a still remain at 50 percent.

In humans, inbreeding (called **consanguineous marriages**) is related to population size, mobility, and social customs governing marriages among relatives. To determine the probability that two alleles at the same locus in an individual are derived from a common ancestor, Sewall Wright devised the **coefficient of inbreeding**. Expressed as F, the coefficient can be defined as the probability that two alleles of a given gene in an individual are derived from a common allele in an ancestor. For example, an F$_2$ generation produced by brother-sister matings will have an F value of 1/4. If $F = 1$, all genotypes are homozygous and both alleles are derived from the same ancestor. If $F = 0$, no alleles present are derived from a common ancestor. Shown in Figure 24.17 are pedigrees of first- and second-

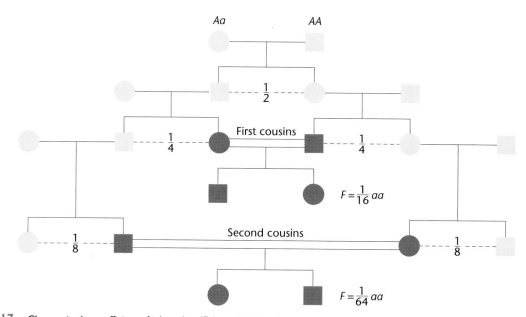

FIGURE 24.17 Changes in the coefficient of inbreeding (F) brought about by consanguineous matings. Numbers indicate the probability of carrying the recessive allele a.

cousin marriages. A lethal allele, *a*, is assumed to be present in one heterozygous parent. For each relative, the probability of the presence of this lethal allele is shown. The pedigrees then culminate in the final probability, *F*, that the lethal allele will become homozygous in the offspring of first or second cousins. Because every population carries as its genetic burden a number of heterozygous recessive alleles that are lethal or deleterious in homozygotes, mortality will rise in matings between relatives.

Genetic Effects of Inbreeding

From an evolutionary standpoint, if a population is split into smaller subpopulations, the effect of inbreeding will become evident. Inbreeding results in the production of individuals homozygous for recessive alleles that were previously concealed in heterozygotes. Because many recessive alleles are deleterious when homozygous, inbred populations will usually have a lowered fitness. **Inbreeding depression** is a measure of the loss of fitness caused by inbreeding. In domesticated plants and animals, inbreeding and selection have been used for thousands of years, and these organisms already have a high degree of homozygosity at many loci. Further inbreeding will usually produce only a small loss of fitness. However, inbreeding among individuals from large, randomly mating populations can produce high levels of inbreeding depression. This effect can be seen by examining the mortality rates in offspring of inbred animals in zoo populations (Table 24.8). To counteract this effect of inbreeding, many zoos are using DNA fingerprinting to estimate the relatedness of animals used in breeding programs, and choosing the most unrelated animals as parents.

As natural populations of endangered species become smaller, concerns about inbreeding are a factor in designing programs to restore these species. For example, the black rhinoceros population in Africa declined from 65,000 in

1970 to less than 4,500 in 1986. These remaining rhinos are distributed in small, isolated populations subject to inbreeding depression. Early in this century, taxonomists used small morphological differences to divide black rhinos into several sub-species. DNA analysis of the mitochondrial genome was used to determine whether the subspecies should be managed separately to maintain genetic diversity, or whether all remaining black rhinos should be regarded as a single population for breeding purposes. The results indicate that there is little divergence in the mitochondrial genomes of these subspecies, and that the rhinos can be grouped into a single population for breeding programs.

In humans, the effects of inbreeding are an increase in the risks for spontaneous abortions, neonatal deaths, and congenital deformities and recessive genetic disorders. Although less common than in the past, inbreeding occurs in many regions of the world, where social custom favors marriage between a man and his first cousin (his father's brother's daughter). Over many generations, inbreeding should eventually reduce the frequency of deleterious recessive alleles (if homozygotes die before reproducing). However, some studies indicate that such families tend to have more children to compensate for those lost to such disorders. On average, two-thirds and these are heterozygotes, carrying the deleterious allele.

Inbreeding is not always harmful, however, and it has been used in breeding programs for domesticated plants and animals. When an inbreeding program is initiated, homozygosity increases, and some breeding stocks become fixed for favorable alleles and others for unfavorable alleles. By selecting the more viable and vigorous plants or animals, the proportion of individuals carrying desirable traits can be increased.

If members of two favorable inbred lines are mated, hybrid offspring are often more vigorous in desirable traits than is either of the parental lines. This phenomenon has been called **hybrid vigor**. Such an approach was used in the breeding programs established for corn, and crop yields increased tremendously. Unfortunately, the hybrid vigor extends only through the first generation. Many hybrid lines are sterile, and those that are fertile show subsequent declines in yield. Consequently, the hybrids must be regenerated each time by crossing the inbred parental lines.

Hybrid vigor has been explained in two ways. The first hypothesis, the **dominance hypothesis**, incorporates the obvious reversal of inbreeding depression, which inevitably must occur in out-crossing. Consider a cross between two strains of corn with the following genotypes:

Strain A   Strain B
aaBBCCddee × *AAbbccDDEE*
↓
AaBbCcDdEe

TABLE 24.8   Mortality in offspring of inbred zoo animals

Species	*N*		Non-inbred	Inbred	Inbreeding Coefficient
Zebra	32	Lived:	20	3	0.250
		Died:	7	2	
Eld's deer	24	Lived:	13	0	0.250
		Died:	4	7	
Giraffe	19	Lived:	11	2	0.250
		Died:	3	3	
Oryx	42	Lived:	35	0	0.250
		Died:	2	5	
Dorcas gazelle	92	Lived:	36	17	0.269
		Died:	14	25	

GENETICS, TECHNOLOGY, AND SOCIETY

The Failure of the Eugenics Movement: What Did We Learn?

The eugenics movement had its origins in the ideas of the English scientist Francis Galton, who became convinced from his study of the appearance of geniuses within families (including his own) that intelligence is inherited. Galton concluded in his 1869 book *Hereditary Genius* that it would be possible to "produce a highly gifted race of men by judicious marriages during several consecutive generations." The term eugenics, coined by Galton in 1883, refers to the improvement of the human species by such selective mating. Once Mendel's principles were rediscovered in 1900, the eugenics movement flourished.

The eugenicists believed that a wide range of human attributes were inherited as Mendelian traits, including many aspects of behavior, intelligence, and moral character. Their overriding concern was that the genetically "feebleminded" and "immoral" in the population were reproducing faster than the genetically superior, and that this differential birthrate would result in the progressive deterioration of the intellectual capacity and moral fiber of the human race. Several remedies were proposed. *Positive eugenics* called for the encouragement of especially "fit" parents to have more children. More central to the goals of the eugenicists, however, was the negative eugenics approach aimed at discouraging the reproduction of the genetically inferior or, better yet, eliminating it altogether.

In the U.S., the eugenics movement was highly organized, and for a time it had a significant impact on public policy. Partially at the urging of prominent eugenicists, 30 states passed laws compelling the sterilization of criminals, epileptics, and inmates in mental institutions; most states enacted laws invalidating marriages between the "feebleminded" and others considered eugenically unfit. The crowning legislative achievement of the eugenics movement, however, was the passage of the Immigration Restriction Act of 1924, which severely limited the entry of immigrants from eastern and southern Europe due to their perceived mental inferiority.

Throughout the first two decades of the century, many geneticists passively accepted the views of eugenicists, but by the 1930s critics recognized that the goals of the eugenics movement were determined more by racism, class prejudice, and anti-immigrant sentiment than by sound genetics. Increasingly, prominent geneticists began to speak out against the eugenics movement, among them William Castle, Thomas Hunt Morgan, and Hermann Muller. When the horrific extremes to which the Nazis took eugenics became known, a strong reaction developed that all but ended the eugenics movement.

Paradoxically, the eugenics movement arose at the same time basic Mendelian principles were being developed, principles that eventually undermined the theoretical foundation of eugenics. Today, any student who has completed an introductory course in genetics should be able to identify several fundamental mistakes the eugenicists made:

1. They assumed that complex human traits such as intelligence and personality are strictly inherited, completely disregarding any environmental contribution to the phenotype. Their reasoning was that because certain traits ran in families, they must be genetically determined.

The F_1 hybrids are heterozygotes at all loci shown. Deleterious recessive alleles present in the homozygous form in the parents will be masked by the more favorable dominant alleles in the hybrids. Such masking is thought to cause hybrid vigor.

The second theory, **overdominance**, holds that in many cases the heterozygote is superior to either homozygote. This may be related to the fact that in the heterozygote, there may be two forms of a gene product present, providing a form of biochemical diversity. Thus, the cumulative effect of heterozygosity at many loci accounts for the hybrid vigor. Most likely, such vigor results from a combination of phenomena explained by both hypotheses.

CHAPTER SUMMARY

1. The idea that the population rather than the individual is the effective unit in the process of evolution has led to the study of allele frequencies and to the emergence of the discipline of population genetics.

2. The Hardy-Weinberg law established the conditions for genetic equilibrium within a population. These conditions include a large, randomly breeding population, and the absence of selection, mutation, migration and drift. If any of these conditions is not met, allele or genotype frequencies will change from generation to generation.

2. They assumed that these complex traits are determined by single genes with dominant and recessive alleles. This belief persisted despite research showing that multiple gene pairs contribute to many phenotypes.

3. They assumed that a single ideal genotype exists in humans. Presumably, such a genotype would be highly homozygous in order to be sustained. This precept runs counter to current evidence suggesting that a high level of homozygosity is often deleterious, supporting the superiority of the heterozygote.

4. They assumed that the frequency of inherited defects in the population could be significantly lowered by preventing those affected from reproducing. In fact, for recessive traits that are relatively rare, most of the recessive alleles in the population are carried by asymptomatic heterozygotes who are spared from such selection. Negative eugenic practices, no matter how harsh, are relatively ineffective at eliminating such traits.

5. They assumed that those deemed genetically unfit in the population were out-reproducing those thought to be genetically fit. This is contrary to the Darwinian concept of evolution that equates reproductive success with fitness. (Galton should have understood this, being Darwin's first cousin!)

After more than seven decades since the eugenics movement was in full bloom, we have a much more sophisticated understanding of genetics, as well as a greater awareness of its potential misuses. But the application of new genetic technologies makes possible a "new eugenics" of a scope and power that Francis Galton could not have imagined. In particular, prenatal genetic screening and *in vitro* fertilization enable the selection of children according to their genotype, a power that will dramatically increase as more and more genes are associated with inherited diseases and perhaps even behaviors.

As we move into this new genetic age, we must not forget the mistakes made by the early eugenicists. We must remember that phenotype is a complex interaction between the genotype and the environment, and not lapse into a new hereditarianism that treats a person as only a collection of genes. We must keep in mind that multiple genes often contribute to a particular phenotype, whether a disease or a behavior, and that the alleles of these genes may interact in unpredictable ways. We must not fall prey to the assumption that there is an ideal genotype when all others are less desirable. And most of all, we must not use genetic information to advance ideological goals. We may find that there is a fine line between the legitimate uses of genetic technologies such as having healthy children, and other eugenic practices. It will be up to us to decide exactly where the line falls.

English scientist Francis Galton.

3. The Hardy-Weinberg formula is also useful in demonstrating whether or not a population is in equilibrium and in calculating the degree of heterozygosity within a population for any particular locus.

4. The process of evolution requires changes in allele frequencies. The mathematical derivation of the Hardy-Weinberg equilibrium allows assessment of the forces of evolution: selection, mutation, migration, genetic drift, and inbreeding.

5. Mutation and migration introduce new alleles into a population, but the retention or loss of these alleles depends on the degree of fitness conferred and the action of selection.

6. Natural selection is the most powerful force altering a population's genetic constitution. In small populations, however, inbreeding and genetic drift can be important factors.

KEY TERMS

INSIGHTS AND SOLUTIONS

1. Color blindness is caused by a recessive allele on the X chromosome. In a survey, 16 women out of 20,000 were found to be color blind. What is the expected frequency of color-blind males in this population?

   Solution: Remember that for X-linked traits, females exhibit the phenotypic frequencies expected for complete dominance. This means that the frequency of color blindness in women ($16/20,000 = 0.0008$) is equal to q^2. If $0.0008 = q^2$, then $q = 0.0282$, the value for color blindness in males. The expected frequency in males would be 2.82 percent or about 1 in 34.

2. Eugenics is the term employed for the selective breeding of humans to bring about improvements in the species. As a eugenic measure, it has been suggested (sometimes by force of law) that individuals suffering from serious genetic disorders should be prevented from reproducing (by sterilization, if necessary) in order to reduce the frequency of the disorder in future generations. Suppose that such a trait were present in the population at a frequency of 1 in 40,000, and that affected individuals did not reproduce. In 100 generations, or about 2500 years, what would be the frequency of the condition? What are your assumptions about this condition? Are the eugenic measures effective in this case?

   Solution: Let us assume that the trait is recessive, and that homozygous recessive individuals are prevented from reproducing. In this case, selection against the homozygous recessive phenotype is equal to 1. Using formula 24.19, we can calculate the frequency after 100 generations as follows:

   $$q_n = \frac{q_0}{1 + nq_0}$$

   $$= \frac{0.000025}{1 + 100(0.000025)}$$

   $$= \frac{0.000025}{1.0025}$$

   where

   $$q_n = 1/41,666$$
   $$s = 1$$
   $$n = 100$$
   $$q_0 = 0.000025$$

   The frequency of the condition has been reduced from 1 in 40,000 to 1 in 41,666 in 100 generations, indicating that this eugenic measure is ineffective.

PROBLEMS AND DISCUSSION QUESTIONS

1. The ability to taste the compound PTC is controlled by a dominant allele T, while individuals homozygous for the recessive allele t are unable to taste this compound. In a genetics class of 125 students, 88 were able to taste PTC, 37 could not. Calculate the frequency of the T and t alleles in this population, and the frequency of the genotypes.

2. Calculate the frequencies of the AA, Aa, and aa genotypes after one generation if the initial population consists of 0.2 AA, 0.6 Aa, and 0.2 aa genotypes. What frequencies will occur after a second generation?

3. Consider rare disorders in a population caused by an autosomal recessive mutation. From the following frequencies of the disorder in the population, calculate the percentage of heterozygous carriers:
   (a) 0.0064
   (b) 0.000081
   (c) 0.09
   (d) 0.01
   (e) 0.10

4. What must be assumed in order to validate the answers in Problem 3?

5. In a population where only the total number of individuals with the dominant phenotype is known, how can you calculate the percentage of carriers and homozygous recessives?

6. For the following two sets of data, determine whether they represent populations that are in equilibrium. Use chi-square analysis if necessary.
   (a) MN blood groups: MM, 60%; MN, 35.1%; NN, 4.9%
   (b) Sickle-cell hemoglobin: AA, 75.6%; AS, 24.2%; SS, 0.2%

7. If 4 percent of the population in equilibrium expresses a recessive trait, what is the probability that the offspring of two individuals who do not express the trait will express it?

8. Consider the allele frequencies $p = 0.7$ and $q = 0.3$ where selection occurs against the homozygous recessive genotype. What will be the allele frequencies after one generation if

(a) $s = 1.0$?

(b) $s = 0.5$?

(c) $s = 0.1$?

(d) $s = 0.01$?

9. If initial allele frequencies are $p = 0.5$ and $q = 0.5$, and $s = 1.0$, what will be the frequencies after 1, 5, 10, 25, 100, and 1000 generations?

10. Determine the frequency of allele A after one generation under the following conditions of migration:

(a) $p = 0.6$; $p_m = 0.1$; $m = 0.2$

(b) $p = 0.2$; $p_m = 0.7$; $m = 0.3$

(c) $p = 0.1$; $p_m = 0.2$; $m = 0.1$

11. Assume that a recessive autosomal disorder occurs in one of 10,000 individuals (0.0001) in the general population. Assume that in this population about 2 percent (0.02) of the individuals are carriers for the disorder. Estimate the probability of this disorder occurring in the offspring of a marriage between first cousins and between second cousins. Compare these probabilities to the population at large.

12. What is the basis of inbreeding depression?

13. Does inbreeding cause an increase in frequency of recessive alleles within a population?

14. Describe how inbreeding may be used in the domestication of plants and animals. Discuss the theories underlying these techniques.

15. Evaluate the following statement: Inbreeding increases the frequency of recessive alleles in a population.

16. In a breeding program to improve crop plants, which of the following mating systems should be employed to produce a homozygous line in the shortest possible time?

(a) self-fertilization

(b) brother–sister matings

(c) first-cousin matings

(d) random matings

Illustrate your choice with pedigree diagrams.

17. If the recessive trait albinism (a) is present in 1/10,000 individuals, calculate the frequency of

(a) The recessive mutant allele.

(b) The normal dominant allele.

(c) Heterozygotes in the population.

(d) Matings between heterozygotes.

18. One of the first Mendelian traits identified in man was a dominant condition known as *brachydactyly*. This gene causes an abnormal shortening of the fingers and/or toes. At the time, it was thought by some that this dominant trait would spread until 75 percent of the population would be affected (since the phenotypic ratio of dominant to recessive is 3:1). Show how this line of reasoning is incorrect.

19. Achondroplasia is a dominant trait that causes a characteristic form of dwarfism. In a survey of 50,000 births, five infants with achondroplasia were identified. Three of the affected infants had affected parents, while the rest had normal parents. Calculate the mutation rate for achondroplasia and express the rate as the number of mutant genes per given number of gametes.

20. A prospective groom, who is normal, has a sister with cystic fibrosis, an autosomal recessive disease state. Both of his parents are normal. He plans to marry a woman who has no history of cystic fibrosis (CF) in her family. What is the probability that they will produce a CF child? They are both Caucasian and the overall frequency of CF in the Caucasian population is 1/2500; that is, one affected child per 2500. (Assume the population meets the Hardy-Weinberg conditions.)

EXTRA-SPICY PROBLEMS

21. A form of dwarfism known as Ellis-van Creveld syndrome was first discovered in the late 1930s, when Richard Ellis and Simon van Creveld shared a train compartment on the way to a pediatrics meeting. In the course of conversation, they discovered they each had a patient with this syndrome. They published a description of the syndrome in 1940. Affected individuals have a short-limbed form of dwarfism and often have defects of the lips and teeth, and polydactyly (extra fingers). The largest pedigree for this condition was reported in an Old Order Amish population in eastern Pennsylvania by Victor McKusick and his colleagues (1964). In this community, about 5 per 1000 births are affected, and in the population of 8000, the observed frequency is 2 per 1000. All affected individuals have unaffected parents, and all affected cases can trace their ancestry to Samuel King and his wife, who arrived in the area in 1774. It is known that neither King nor his wife were affected with this disorder. There are no cases of this disorder in other Amish communities, such as those in Ohio or Indiana.

(a) From the information provided, derive the most likely mode of inheritance of this disorder. Using the Hardy-Weinberg law, calculate the frequency of the mutant allele in the population and the frequency of heterozygotes, assuming Hardy-Weinberg conditions.

(b) What is the most likely explanation for the high frequency of this disorder in the Pennsylvania Amish community, and its absence in other Amish communities?

22. The following graph shows the variation in the frequency of a particular allele (A) that occurred over time in two relatively small, independent populations exposed to very similar environmental conditions. A student has analyzed the graph and concluded that the best explanation for these data is that the selection for allele A is occurring in Population 1. Is the student's conclusion correct? Explain.

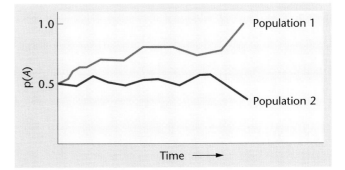

SELECTED READINGS

BALLOU, J., and RALLS, K. 1982. Inbreeding and juvenile mortality in small populations of ungulates: a detailed analysis. *Biol. Conserv.* 24:239–72.

BARTON, N. H., and TURELLI, M. 1989. Evolutionary quantitative genetics: How little do we know? *Annu. Rev. Genet.* 23: 337–70.

BODMER, W. F., and CAVALLI-SFORZA, L. L. 1976. *Genetics, evolution and man.* New York: W. H. Freeman.

COOK, L. M., ASKEW, R. R., and BISHOP, J. A. 1970. Increasing frequency of the typical form of the peppered moth in Manchester. *Nature* 227:1155.

CROW, J. F. 1986. *Basic concepts in population, quantitative and evolutionary genetics.* New York: W. H. Freeman.

CROW, J. F., and KIMURA, M. 1970. *An introduction to population genetic theory.* New York: Harper & Row.

DAVID, J. R., and CAPY, P. 1988. Genetic variation of *Drosophila melanogaster* natural populations. *Trends Genet.* 4:106–11.

DUDLEY, J. W. 1977. 76 generations of selection for oil and protein percentage in maize. Eds. E. Pollack, O. Kempthorne, and T. B. Bailey. In *Proceedings of the International Conference on Quantitative Genetics*, pp. 459–73. Ames: Iowa State University Press.

EDWARDS, J. H. 1989. Familarity, recessivity and germline mosaicism. *Ann. Hum. Genet.* 53:33–47.

FISHER, R. A. 1930. *The genetical theory of natural selection.* Oxford: Clarendon Press. (Reprinted in 1958 by Dover Press.)

FREIRE-MAIA, N. 1990. Five landmarks in inbreeding studies. *Am. J. Med. Genet.* 35:118–20.

GAYLE, J. S. 1990. *Theoretical population genetics.* Boston: Unwin-Hyman.

HARTL, D. L. 1988. *A primer of population genetics*, 2nd ed. Sunderland, MA: Sinauer Associates.

JONES, J. S. 1981. How different are human races? *Nature* 293:188–90.

KARN, M. N., and PENROSE, L. S. 1951. Birth weight and gestation time in relation to maternal age, parity and infant survival. *Ann. Eugen.* 16:147–64.

KERR, W. E., and WRIGHT, S. 1954. Experimental studies of the distribution of gene frequencies in very small populations of *Drosophila melanogaster.* I. *Forked. Evolution* 8: 172–77.

KETTLEWELL, H. B. D. 1961. The phenomenon of industrial melanism in *Lepidoptera. Annu. Rev. Entomol.* 6:245–62.

———. 1973. *The evolution of melanism: The study of a recurring necessity, with special reference to industrial melanism in the Lepidoptera.* London: Oxford University Press.

KEVLES, D. J. 1985. *In the name of eugenics: Genetics and the uses of human heredity.* Berkeley: University of California Press.

KHOURY, M. J., and FLANDERS, W. D. 1989. On the measurement of susceptibility to genetic factors. *Genet. Epidemiol.* 6: 699–701.

LI, C. C. 1977. *First course in population genetics.* Pacific Groves, CA: Boxwood Press.

METTLER, L. E., GREGG, T., and SCHAFFER, H. E. 1988. *Population genetics and evolution*, 2nd ed. Englewood Cliffs, NJ: Prentice-Hall.

REED, T. E. 1969. Caucasian genes in American Negroes. *Science* 165:762–68.

RENFEW, C. 1994. World linguistic diversity. *Sci. Am.* 270: 116–23.

ROBERTS, D. F. 1988. Migration and genetic change. *Hum. Biol.* 60:521–39.

SPIESS, E. B. 1989. *Genes in populations*, 2nd ed. New York: Wiley.

WALLACE, B. 1989. One selectionist's perspective. *Quart. Rev. Biol.* 64:127–45.

WOODWORTH, C. M., LENG, E. R., and JUGENHEIMER, R. W. 1952. Fifty generations of selection for protein and oil in corn. *Agron. J.* 44:60–66.

ZHU, J., NESTOR, K. E., and MORITSU, Y. 1996. Relationship between band sharing levels of DNA fingerprints and inbreeding coefficients and estimation of true inbreeding in turkey lines. *Poult. Sci.* 75:25–28.

25

Genetics and Evolution

CHAPTER CONCEPTS

Evolution is brought about by natural selection acting on the gene pool of a population. Selection leads to changes in genetic diversity and the ability to undergo evolutionary divergence. The process of speciation divides one gene pool into two or more reproductively separate gene pools. This division may be accompanied by changes in morphology, physiology, and adaptation to the environment.

In Chapter 24 we described populations in terms of allele frequencies and genetic equilibria, and outlined the forces that alter the genetic makeup of populations. Mutation, migration, selection, and drift individually and collectively alter allele frequencies and bring about evolutionary divergence and species formation. This process depends not only upon genetic divergence but also upon environmental or ecological diversity. If a population is spread over a geographic range encompassing a number of subenvironments or **niches**, populations occupying these niches will adapt and become genetically differentiated. Over time, this process may lead to a change in allele frequencies and, eventually, the formation of a phenotypic

Lady-bird beetles capping a daisy in the Chiricahua Mountains in Arizona.

variant of the species. Because both the formation and the maintenance of these variants depend ultimately on the interaction of the genotype and environment, these variant populations are dynamic entities that may remain in existence, become extinct, merge with the parental population, or continue to diverge from the parental population until they become reproductively isolated and form new species.

It is difficult to define the exact moment of transition when a new species forms. A **species** is defined as one or more groups of interbreeding or potentially interbreeding organisms that are reproductively isolated in nature from all other organisms. In sexually reproducing organisms, the process of **speciation** divides a single gene pool into two or more reproductively isolated subunits (separate gene pools). Changes in morphology, physiology, and adaptation to an ecological niche may also occur, but are not necessary components of the speciation event. Speciation can take place gradually over a long time period or within a few generations. Throughout this process, the degree of genetic diversity is the key factor. Individuals of a species are members of a common gene pool that distinguishes them from members of other species.

In this chapter we will focus on three topics: first, the role of natural selection in the process of species formation; second, analysis of genetic variation and its role in evolution; and third, an example of evolution in *Drosophila*. In addition, we will consider molecular aspects of human evolution.

Wallace, Darwin and the Origin of Species

At the age of 22, Charles Darwin accepted a position as a naturalist aboard the *Beagle*, a British ship that sailed around the world between 1831 and 1836 on a surveying expedition. During the voyage, Darwin collected plants, animals, and fossils, and he kept a detailed journal about the geology and geographic distribution of the organisms he saw and collected. When he returned home, he began to sort through his journals and collected specimens, with the intention of investigating the origin of species. Some 22 years later, in 1859, he published his book, *The Origin of Species by Means of Natural Selection*, which outlined a mechanism for evolution, and which provided a detailed and meticulous examination of the evidence for how new species arose.

While Darwin labored over the problem of evolution, Alfred Russel Wallace, a self-taught naturalist, spent the years between 1848 and 1852 exploring the Amazon Basin, collecting animals and plants in an attempt to explain how new species arose. He then spent seven years in the islands of Indonesia and Malaysia and formulated a theory on the role of natural selection in species formation, which he sent to Darwin. Their work was read at a joint meeting of the Linnean Society in London in July 1858. The work of Wallace motivated Darwin to write his book, published in the following year.

For both Wallace and Darwin, the idea that natural selection is the driving force of evolution was developed by careful observation of plants and animals in their environment, coupled with readings in geology and an essay by Thomas Malthus on the factors that limit human population growth. Wallace and Darwin extracted from this essay the idea that when resources are limited, the organisms that survive and reproduce, on average, are better adapted to the environment.

The Wallace-Darwin theory of evolution by natural selection as a force in evolution can be summarized in a series of statements:

1. Among individuals of a species, minor variations in phenotype exist. These can include differences in size, agility, coloration, or ability to obtain food. Most of these variations, although small and seemingly insignificant, are heritable, and passed on to offspring.

2. Organisms tend to reproduce in an exponential fashion. More offspring are produced than can survive. This causes members of a species to engage in a struggle for survival, competing with other members of the species for scarce resources.

3. In the struggle for survival, some individuals will be more successful than others, allowing them to survive and reproduce.

4. Those individuals that survive to reproduce because of some favorable phenotypic variation leave behind more offspring than those with less favorable variations, eventually eliminating some variants from the species.

5. Over time, these heritable variations in phenotype can transform a species into a new species, similar to, but distinct from, the species it has replaced or from which it has diverged.

Although Wallace and Darwin proposed that natural selection can explain how evolution occurs, they could not explain the origin of the variations that provided the raw material for evolution, nor how such variation is maintained in a population or a species. In the twentieth century, as the principles of Mendelian genetics were applied to populations, the source of variation (mutation) and the maintenance of variation (allele frequencies) were both explained, and evolution was seen as changes in the genetic structure of populations over time. This union of population genetics with natural selection generated a new view of the evolutionary process, called neo-Darwinism.

Models of Speciation

The process of splitting a genetically homogeneous population into two or more populations that undergo genetic differentiation and eventual reproductive isolation is called **speciation**. According to Ernst Mayr, species originate predominantly in two ways. In the first, often called **phyletic evolution** or **anagenesis**, over a long period of time, species A becomes transformed into species B. In **cladogenesis**, the second way that species originate, one species gives rise to two or more species (Figure 25.1). This second process can occur over a long period of time or, rarely, in a generation or two.

Allopatric Speciation

The most developed model of speciation is **geographic** or **allopatric speciation**, first proposed by Moritz Wagner in 1868. According to Wagner, geographic features such as lakes, rivers, or mountains act as barriers to gene flow between populations, and physical isolation is the first step in the process of evolution. In a second step, the isolated populations undergo independent genetic changes and diverge to produce two distinct species.

Twentieth-century workers such as Mayr and Theodosius Dobzhansky have refined and updated this model but have retained the general features of Wagner's hypothesis. In the refined model, the first step requires that populations become separated; that is, gene flow (interbreeding) must be interrupted. The absence or interruption of gene flow is a prerequisite for the development of genetic differences brought about by adaptation to local conditions. Genetic diversity arising by mutation, or by random genetic drift, will be reflected in the presence of new alleles, changes in allele frequency, or the presence of new chromosomal arrangements. Eventually, a point will be reached when the populations have enough genetic differences that they can be identified as distinct races or semi-species. This process may continue uninterrupted until two or more species are present.

If at any time during the process of genetic divergence, the conditions that prevent gene flow between the populations are removed, two outcomes are possible: (1) the two populations may fuse into a single gene pool, because hybridization does not necessarily reduce fertility or viability; or (2) the gene pools of the populations may have diverged to the point where biological isolating

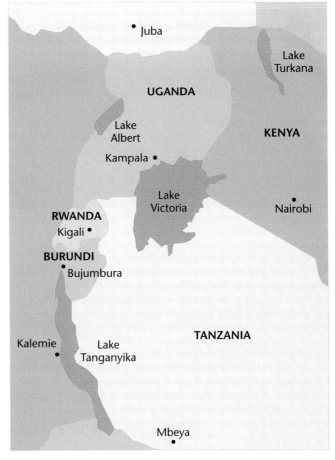

FIGURE 25.2 Lake Victoria in the Rift Valley of East Africa is home to more than 200 species of cichlids.

mechanisms may have arisen, and the two populations cannot interbreed, and two species are present.

Sympatric Speciation

A second model of speciation, called **sympatric speciation**, involves the formation of species from populations that live in the same geographic range and do not become geographically isolated. In this model, the first step is the formation of a population with a variant phenotype (with a degree of genetic divergence), and a shift to a new subenvironment or niche. Whether these populations can diverge to the point of becoming separate species depends on several factors, including how quickly reproductive isolation can be achieved.

There is agreement that the development of variant phenotypes and shifts to new niches occurs frequently within a species range, but there is some disagreement on how often this results in sympatric speciation. The development of over 200 species of cichlid fish in Lake Victoria in the Rift Valley of East Africa is often cited as an example of sympatric speciation (Figure 25.2). The lake is less

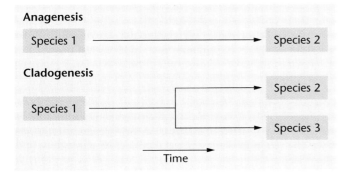

FIGURE 25.1 In anagenesis or phyletic evolution, one species is transformed over time into another species. At all times, only one species exists. In cladogenesis, one species splits into two or more species.

FIGURE 25.3   Cichlids occupy a diverse array of niches, and each species is specialized for a distinct food source.

than a million years old, but is home to over 200 species of cichlids (Figure 25.3). These species have some morphological diversity, but are highly specialized for different niches. Some eat algae floating on the water surface, others are bottom feeders, insect feeders, mollusc eaters, and predators on other fish species. The possibility that these species evolved from a single ancestral species is supported by an examination of protein variation and nucleotide variation (both of which are discussed below) in 14 species from Lake Victoria, and in species from nearby lakes. The lack of variation between species in Lake Victoria, and the high degree of variation between species in Lake Victoria and nearby lakes, supports the idea that the species in Lake Victoria have a single ancestor.

Statispatric Speciation

The third model of speciation that we will discuss, **statispatric speciation**, was originally derived to account for the evolution of flightless grasshoppers in Australia. In this model, a chromosomal aberration such as a translocation arises by chance in a small population. If heterozy-

FIGURE 25.4   Members of a colony of naked mole rats from Kenya.

gotes for the translocation have a slightly reduced fitness, perhaps caused by abnormal meiosis, selection will favor either homokaryotype (two copies of the translocation or two normal chromosomes) by divergent selection. A translocation homokaryotype, with a chromosome number different from the original, may arise, spread, and partially displace the ancestral population. Semi-species would then be reproductively isolated because hybrids would have an unbalanced chromosome complement and a lowered fitness. In this model, reproductive isolation *precedes* the development of genetic variability. In addition to grasshoppers, the statispatric model has been applied to explain the origin of closely related species of naked mole rats, the *Spalax ehrenbergi* complex (Figure 25.4), which differ in chromosome number and do not interbreed.

Isolating Mechanisms

By whatever model speciation occurs, the final step involves the development of mechanisms that prevent interbreeding between individuals of different species. Physiological, behavioral, or mechanical barriers are effective at preventing interbreeding. The biological and behavioral properties of organisms that prevent or reduce interbreeding are called **reproductive isolating mechanisms**. These mechanisms are classified in Table 25.1. For example, genetic divergence may have reached the stage where the viability or fertility of hybrids is reduced. Hybrid zygotes may be formed, but all or most may be inviable. Alternatively, the hybrids may be viable but be sterile or have reduced fertility. In another possibility, the hybrids themselves may be fertile, but their progeny may have lowered viability or fertility. These **postzygotic** mechanisms act at or beyond the level of the zygote and are generated by genetic divergence. Postzygotic isolating mechanisms waste gametes and zygotes and lower the

TABLE 25.1   Reproductive isolating mechanisms

Prezygotic Mechanisms (prevent fertilization and zygote formation

1. Geographic or ecological: The populations live in the same regions but occupy different habitats.
2. Seasonal or temporal: The populations live in the same regions but are sexually mature at different times.
3. Behavioral (only in animals): The populations are isolated by different and incompatible behavior before mating.
4. Mechanical: Cross-fertilization is prevented or restricted by differences in reproductive structures (genitalia in animals, flowers in plants).
5. Physiological: Gametes fail to survive in alien reproductive tracts.

Postzygotic Mechanisms (fertilization takes place and hybrid zygotes are formed, but these are nonviable or give rise to weak or sterile hybrids)

1. Hybrid nonviability or weakness.
2. Developmental hybrid sterility: Hybrids are sterile because gonads develop abnormally or meiosis breaks down before completion.
3. Segregational hybrid sterility: Hybrids are sterile because of abnormal segregation to the gametes of whole chromosomes, chromosome segments, or combinations of genes.
4. F_2 breakdown: F_1 hybrids are normal, vigorous, and fertile, but F_2 contains many weak or sterile individuals.

Source: From G. Ledyard Stebbins, *Processes of Organic Evolution*, 3rd edition, copyright 1977, p. 143. Reprinted by permission of Prentice-Hall, Upper Saddle River, NJ.

reproductive fitness of hybrid survivors. Selection will therefore favor the spread of alleles that reduce the formation of hybrids, leading to the development of **prezygotic isolating mechanisms**. These mechanisms prevent breeding and the formation of hybrid zygotes and offspring. In animal evolution, the most effective of these is behavioral isolation, involving courtship behavior.

The Rate of Speciation

At the heart of the allopatric model of speciation is the concept of **phyletic gradualism**. According to this concept, speciation is a microevolutionary event resulting from the accumulation of many small gene differences over a long period of time under the influence of natural selection. Recently, however, more **stochastic or catastrophic models of speciation** have been proposed. These models emphasize the role of evolutionary events that occur suddenly and intermittently and are therefore referred to as **quantum speciation**. We will briefly discuss one model derived from the fossil record, and an example that depends on chromosome rearrangements as

an isolating mechanism. Finally, we will consider a mechanism that has long been known to exist in angiosperm plants—speciation through polyploidy.

Drawing on evidence from the fossil record, Niles Eldredge and Stephen Jay Gould have proposed a mechanism of species formation known as **punctuated equilibrium**. According to this model, evolutionary changes are not gradual and continuous, but occur intermittently and rapidly as events that punctuate or interrupt long periods of evolutionary equilibrium during which little or no change occurs. In paleontological terms, *rapidly* means within thousands or even hundreds of thousands of years. Results of such evolutionary alterations are thought to cause abrupt changes in the fossil record, as relatively few fossils of a given type will be formed during such changes, compared to the large numbers of fossils formed during long periods (extending into millions of years) of evolutionary equilibrium.

The evolutionary spurts that punctuate periods of stasis are thought to be related to catastrophic events that produce mass extinctions, or events that take place in small populations located at the fringes of the species range. In these locales, environmental stresses disrupt genetic stability, eventually causing the emergence of new species. Proponents of this hypothesis point out that such changes leading to speciation take place through natural selection of individual variations and do not invoke any unusual genetic mechanisms.

Evidence for punctuated evolution caused by the appearance of rare, beneficial mutations has recently been gathered in laboratory studies using the bacterium, *Escherichia coli*. In a study that measured changes in cell size over 3,000 generations of bacteria in a constant environment, researchers found that periods of stasis were interrupted by periods of rapid change. Cultures of bacteria, each founded from a single cell, were grown in a continuous culture, and samples were frozen at regular intervals. Measurements of cell size were done every 100 generations; size remained constant for long periods. However, several major increases in cell size occurred over the course of 3,000 generations. It is thought that these changes in cell size were the result of direct selection (or a side effect) for a rare, beneficial mutation that caused an increase in cell size. This mutation swept through the population, producing a change in cell size in 100 generations or less. The patterns of change observed in these laboratory cultures of asexually reproducing bacteria are similar to those seen in the fossil record of some eukaryotes. Whether these patterns have similar underlying causes, and the role, if any, for punctuated evolution in eukaryotes has yet to be determined.

A second model of quantum speciation, proposed by Hampton Carson, is called the **founder–flush theory**,

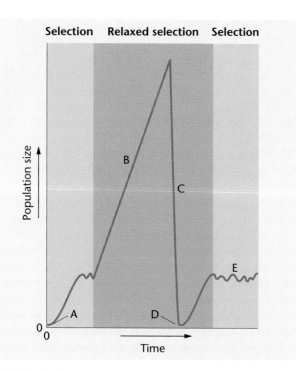

FIGURE 25.5 The flush–crash cycle. Small populations descended from a single founder (A) undergo a population flush (B), followed by a crash (C). A single survivor in the form of a fertilized female (D) builds a new population (E).

and is derived from his studies on Hawaiian *Drosophila*. According to this model, populations and even species can be started by a single individual. The founder principle is not new, as this idea was advanced much earlier by Ernst Mayr. In Carson's proposal, however, gradualism is not a necessary component of speciation; more importantly, reproductive isolation *precedes* adaptation rather than being a consequence of genetic diversification. In other words, Carson has changed the order of steps in the classical model of speciation.

According to the founder–flush theory, a single fertilized female or a mating pair can colonize an isolated territory previously unoccupied by members of this species. If conditions are favorable, the population founded by this individual will undergo a **flush**, or rapid expansion. After several generations, it is likely that the population growth will outstrip the environment's carrying capacity, causing a **population crash**. The crash causes the death or dispersal of almost the entire population. A few survivors may rebuild the population, which eventually undergoes several flush–crash cycles before coming to equilibrium with the environment. This cycle is diagrammed in Figure 25.5. Through the genetic revolution it has undergone, the colony will, in all probability, have acquired adaptations making it unable to interbreed with the parental population.

According to Carson's theory, genetic changes are brought about in two ways. First, the founding of the population by a very few individuals can establish allele frequencies very different from those in the ancestral population. Second and most important, selection is relaxed during a population flush, because resources are in excess. If descendants of the founder can invade a new niche, they expand their numbers in a flush. Carson proposes that normally, some blocks of genes on chromosomes are tightly linked or "closed" to recombination because of some selective advantage conferred by this configuration. New genotypes produced by recombination within these closed areas have reduced fitness. In fact, the advantage conferred by balanced polymorphisms for inversion heterozygotes may derive from the protection of such closed gene complexes (Figure 25.6).

During the flush period, genotypes produced by recombination in closed regions of the genome may survive because selection is relaxed at such times. These new genotypic combinations may be incompatible with the normal closed system of the species, but they survive because there is reduced competition for resources. After a crash, the survivor's reshuffled genome is acted upon by selection to produce a new combination of open and closed gene groups adapted to the environment. These periods of genomic reorganization may affect the timing and/or order of gene expression, bringing about developmental changes that accelerate the pace of evolution.

Several such cycles can produce enough genetic differences so that crosses between the ancestral population and the colony populations produce hybrids with lowered fitness commonly displayed by interspecific hybrids. Carson views cycles of disorganization and reorganization of the genomes as the essence of speciation rather than as the gradual genetic divergence of isolated populations over long periods of time.

The evolution of *Drosophila* species in the Hawaiian Islands appears to have followed such a founder–flush cycle. Geologic evidence indicates that the northwesternmost islands are the oldest, and that the southeastern island of Hawaii (produced by volcanic action) is the youngest at

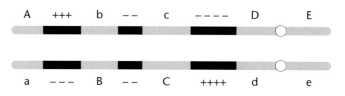

FIGURE 25.6 Model for open and closed regions of chromosomes. Products of crossing over anywhere within the open system (blue) result in offspring with high fitness, whereas crossovers in the closed regions (black) produce zygotes with low fitness. The letters represent genes or polygenes, and the pluses and minuses represent internally balanced gene complexes. Recombination within these blocks produces unfit gene combinations.

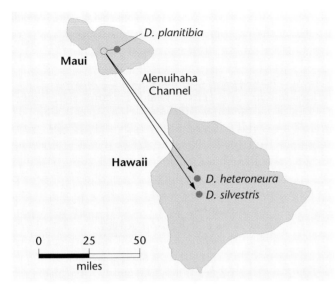

FIGURE 25.7   Proposed pathway of colonization of Hawaii by members of the *D. planitibia* species complex. Open circles represent a population ancestral to the three present-day species.

about 700,000 years old. The relationships among species of *Drosophila* can be traced by mapping the location and frequency of inversions in the banded polytene chromosomes of larval salivary glands. One group, the *planitibia* complex, has three species with the same basic set of chromosome inversions: *D. planitibia*, *D. heteroneura*, and *D. silvestris*. The *D. planitibia* is found on the older island of Maui, and the other two are found on the younger island of Hawaii. It is postulated that a fertilized female belonging to an ancestral stock on Maui (chromosomally related to the present-day *D. planitibia*) crossed the Alenuihaha Channel to Hawaii. Subsequent flush–crash cycles led to the reconstruction of the colony's genotype, giving rise to the two species found on Hawaii—*D. heteroneura* and *D. silvestris* (Figure 25.7). An alternative hypothesis proposes that one of the species on Hawaii could have arisen as the result of a second colonization from Maui. Similar evidence indicates that two other groups on Maui have given rise to a total of five species on Hawaii.

In laboratory tests of this theory, Jeffrey Powell has found that 15 generations after several flush–crash cycles, laboratory strains of *Drosophila* species showed significant behavioral (prezygotic) isolation from other strains. Subsequent testing several months later showed that these differences were stable. These experimental results are important in establishing that the first stages of speciation can occur rapidly under certain circumstances.

The final example of quantum speciation that we will discuss involves polyploidy in plants. Formation of animal species by polyploidy is rare, but polyploidy is an important factor in the evolution of plants. It is estimated that one-half of all flowering plants have evolved by this

mechanism. One form of polyploidy is **allopolyploidy** (see Chapter 9), produced by doubling the chromosome number in an interspecific hybrid.

If two species of related plants have the genetic constitution *SS* and *TT* (where *S* and *T* represent the haploid set of chromosomes in each species), then the F_1 hybrid would have the chromosome constitution *ST*. Normally such a plant would be sterile because there are few or no homologous chromosome pairs, and aberrations would arise during meiosis. However, if the hybrid undergoes a spontaneous doubling of chromosome number, a tetraploid *SSTT* plant would be produced. This chromosomal aberration might occur during mitosis in somatic tissue, giving rise to a partially tetraploid plant that would produce some tetraploid flowers. Alternatively, aberrant meiotic events may produce *ST* gametes, which when fertilized would yield *SSTT* zygotes. The *SSTT* plants would be fertile because they would possess homologous chromosomes producing viable *ST* gametes. This new, true-breeding tetraploid would have a combination of characters derived from the parental species, and would be reproductively isolated from them because F_1 hybrids would be triploids and consequently sterile.

The tobacco plant *Nicotiana tabacum* ($2n = 48$) is the result of the doubling of the chromosome number in the hybrid between *N. otophora* ($2n = 24$) and *N. silvestris* ($2n = 24$) (Figure 25.8).

FIGURE 25.8   The cultivated tobacco plant, *Nicotiana tabacum* is the result of hybridization between two other species.

Measuring Genetic Variation

If a population of organisms constituting a species is well suited to inhabit and reproduce successfully in its environment, how much genetic variation does it possess? We might assume that the members of a well-adapted population are genetically homogeneous because the most favorable allele at each locus has become fixed. Certainly, an examination of most populations of plants and animals reveals them to be phenotypically similar. However, current evidence supports the opinion that a high degree of heterozygosity is maintained within the gene pool of a diploid population. This built-in diversity is concealed, so to speak, because it is not necessarily apparent in the phenotype. Such diversity may better adapt a population to the inevitable changes in the environment and may promote the survival of the species. The detection of this concealed genetic variation is not a simple task. Nevertheless, such investigation has been successful using several techniques, discussed in the following sections.

Inbreeding Depression

A genetic approach to measuring allelic diversity can be used to determine what fraction of the alleles carried by individuals of a population will be deleterious. Deleterious alleles, such as lethals, can be detected by monitoring the effects of increases in homozygosity (death or disease) that result from inbreeding. Because inbreeding involves mating between relatives, mates have genotypic similarities that extend across all segregating loci. As a result, inbreeding causes recessive alleles to become homozygous at a high frequency in the offspring. If these alleles are deleterious, inbreeding depression is evident in successive generations (review inbreeding depression in Chapter 24).

In using inbreeding depression to study genetic diversity, Theodosius Dobzhansky and his colleagues succeeded in making groups of alleles and sometimes entire chromosomes of *Drosophila* species completely homozygous. When this is accomplished, viability is depressed. As shown in Table 25.2, such studies reveal that lethal, semilethal, subvital, and male and female sterility alleles are prominent in natural populations of different species. In *Drosophila pseudoobscura*, almost 100 percent of the copies of chromosomes 2 and 3 taken from a sample of the population cause a depression in viability when made homozygous. In *Drosophila willistoni*, almost 70 percent of chromosomes 2 and 3 contain male sterile alleles.

Although these studies reveal that organisms carry deleterious genes (mostly in the heterozygous condition), we discuss them here as examples of genetic variation maintained in natural populations. If detrimental alleles are maintained in a heterozygous state, certainly alleles that are effectively neutral or potentially advantageous must also be retained in the heterozygous condition. As recessive alleles, they represent concealed genetic variation within the "normal" heterozygous genotypes characterizing the population.

Protein Polymorphisms

Gel electrophoresis (see Appendix A) is a simple technique that can be used to separate protein molecules on the basis of differences in size and electrical charge. If a

TABLE 25.2 Percentage of Homozygous Chromosomes Tested That Reveal Recessive Mutations of Viability and Fertility

Drosophila Species	Chromosome	Lethals and Sublethals	Subvitals	Sterility	Female Sterility	Male Sterility
D. pseudoobscura	2	33.0	62.6	<0.1	10.6	8.3
(Mather, California)	3	25.0	58.7	<0.1	13.6	10.5
	4	22.7	51.8	<0.1	4.3	11.8
D. persimilis	2	25.5	49.8	0.2	18.3	13.2
(Mather, California)	3	22.7	61.7	2.1	14.3	15.7
	4	28.1	70.7	0.3	18.3	8.4
D. Prosultans	2	32.6	33.4	<0.1	9.2	11.0
(Brazil)	3	9.5	14.5	3.0	6.6	4.2
D. willistoni	2	38.8	57.5	<0.1	40.5	64.8
(Venezuela, British Guiana, and Trinidad)	3	34.7	47.1	1.0	40.5	66.7

Note: Lethal alleles cause premature death, often before sexual maturity and reproduction; subvital mutations kill less than 50 percent of carriers before maturation; supervital mutations increase viability above the wild-type level.

Source: From Spiess, 1977, p. 288.

TABLE 25.3   Heterozygosity at the Molecular Level

Species	Number of Populations Studied	Number of Loci Examined	Loci per Population	Heterozygosity per Locus
Homo sapiens (humans)	1	71	28	6.7
Mus musculus (mouse)	4	41	29	9.1
Drosophila pseudoobscura (fruit fly)	10	24	43	12.8
Limulus polyphemus (horseshoe crab)	4	25	25	6.1

Source: From Lewontin, 1974, p. 117.

nucleotide variation in a structural gene results in the substitution of a charged amino acid such as glutamic acid for an uncharged amino acid such as glycine, the net electrical charge on the protein will be altered. This difference in charge can be detected as a change in the rate at which proteins migrate through an electrical field in electrophoresis. In the mid-1960s, John Hubby and Richard Lewontin used gel electrophoresis to measure protein variation in natural populations of *Drosophila*. Since then, genetic variation has been studied using gel electrophoresis in a wide range of organisms (Table 25.3).

The electrophoretic forms of a protein produced by different alleles are designated as **allozymes.** As shown in Table 25.3, a surprisingly large percentage of loci examined from diverse species produce distinct allozymes. Of the populations shown in Table 25.3, approximately 30 loci per species were examined, and about 30 percent of the loci were polymorphic. An approximate average of 10 percent allozyme heterozygosity per diploid genome was revealed.

These values apply only to genetic variation detectable by altered protein migration in an electric field. Electrophoresis probably detects only about 30 percent of the actual variation that is due to amino acid substitutions because many substitutions do not change the net electric charge on the molecule. Richard Lewontin has estimated that about two-thirds of all loci are polymorphic in a population, and that in any individual within

that group, about one-third of the loci exhibit genetic variation in the form of heterozygosity.

There is some controversy over the significance of genetic variation as detected by electrophoresis. Some argue that allozymes are biochemically equivalent and therefore do not play any role in evolution. We shall address this argument later in the chapter.

Variations in Nucleotide Sequence

The most direct way to estimate genetic variation is to compare the nucleotide sequence of genes carried by individuals in a population. With the development of techniques for cloning and sequencing DNA, nucleotide sequence variations have been catalogued for an increasing number of gene systems. Using restriction endonucleases to detect polymorphisms, Alec Jeffrey surveyed 60 unrelated individuals to estimate the total number of DNA sequence variants in humans. His results show that within the genes of the beta-globin cluster, 1 in 100 base pairs shows polymorphic variation. If this region is representative of the genome, this indicates that at least 3×10^7 nucleotide variants per genome are possible.

In another study, Martin Kreitman studied the alcohol dehydrogenase locus (*Adh*) in *Drosophila melanogaster* (Figure 25.9). This locus encodes two allozymic variants: the *Adh-f* and the *Adh-s* alleles. These differ by a single amino acid (*thr* versus *lys* at codon 192). To determine whether the amount of genetic variation detectable at the protein level (one amino acid difference) corresponds to the variation at the nucleotide level, Kreitman cloned and sequenced *Adh* loci from 5 natural populations of *Drosophila*.

The 11 cloned loci contained 43 nucleotide variations from the consensus *Adh* sequence of 2721 base pairs. These variations are distributed throughout the locus: 14 in the exon coding regions, 18 in the introns, and 11 in the nontranslated and flanking regions. Of the 14 in the coding regions, only one leads to an amino acid replacement, accounting for the two observed elec-

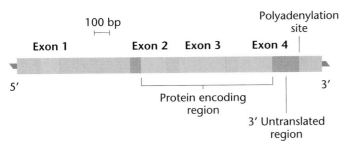

FIGURE 25.9   The *Adh* gene of *Drosophila melanogaster*.

trophoretic variants. The other 13 nucleotide substitutions do not lead to amino acid replacements. Are the differences in the number of, allozyme and nucleotide variants the result of natural selection? If so, of what significance is this? These issues will be discussed below.

Studies of other organisms including the rat and the mouse have produced similar estimates of nucleotide diversity, indicating that there is an enormous reservoir of genetic variability within a population, and that at the level of DNA, most and perhaps all genes exhibit diversity from individual to individual.

Evolution and Genetic Variation: A Dilemma

A population of interbreeding organisms is more or less adapted to its surrounding environment. Phenotypes that enhance adaptation to a given environment are controlled by the genotypes carried by the individuals constituting the group. Ideally, each individual should possess a genotype and phenotype best suited to the immediate environment. However, the preceding discussion on genetic variability in natural populations suggests that there is a wide variation in genotypes within populations.

Evolution depends on the presence of genetic variation within a population of organisms and the action of natural selection on that variation. Although genetic variation can be measured in a variety of ways, there is some controversy about whether it is important to maintain a large amount of variation in a population. This controversy centers particularly on the variation detected at the molecular level. On the one hand, natural selection should favor homozygotes carrying the most favorable alleles at each locus, so that organisms can be best adapted to their environment. Homozygosity, therefore, should be the rule rather than the exception. If this is true, why is heterozygosity so prevalent?

On the other hand, alleles that at one point in time are detrimental to individuals may be of great value to the population in future generations, as shown by the melanic forms of the peppered moth, discussed in Chapter 24. Under changing environmental conditions, previously insignificant or even detrimental alleles may become essential to the maintenance of fitness. Thus, selection should favor the maintenance of a high degree of heterozygosity in populations, so that members of the population have the genetic diversity to respond to changes in environmental conditions. In addition, heterozygotes are often superior in many ways to homozygotes (see the discussion of hybrid vigor in Chapter 24).

The apparent dilemma is that organisms should have low levels of heterozygosity to be well-adapted to their environment, but should have a high level of heterozygosity to respond to changes in the environment. This problem has generated several points of view about the significance of variation. Two of these will be discussed below.

Neutralists and Variation

One point of view argues that the existence of a high degree of genetic variation does not prove that it is important in evolutionary change. The **neutralist hypothesis**, originally proposed by James Crow and Motoo Kimura, argues that mutations leading to amino acid substitutions are rarely favorable. They are sometimes detrimental, but most often neutral or genetically equivalent to the allele that is replaced, and they are, therefore, unimportant from an evolutionary perspective. Polymorphisms that are favorable or detrimental are either preserved or removed from the population, respectively, by natural selection. However, neutral genetic changes will not be affected by selection and will instead randomly accumulate in the population. Their frequency will be determined by the rate of mutation and the principles of random genetic drift.

Selectionists and Variation

Opposed to the neutralist theory are the **selectionists**. These geneticists point out examples where enzyme or protein polymorphism is associated with adaptation to certain environmental conditions. The well-known advantage of sickle-cell anemia heterozygotes to infection by malarial parasites is such an example.

Selectionists also stress that enzyme polymorphisms may often appear to offer no advantage, but exist in such a frequency that it is impossible to explain as a random occurrence. Thus, even though no currently available analytical technique can detect any physiological difference, some slight advantage associated with any given amino acid substitution may exist. There may in fact, be an advantage to having two forms of a given protein, allowing optimum performance under a wider range of cellular conditions.

Even though this controversy seems highly theoretical and somewhat esoteric, it is important that we not lose sight of several factors. First, the neutralists do not discount natural selection as a guiding force in evolution. Rather, they suggest that some features of organisms' genotypes are nonadaptive, fluctuate randomly, and may have been fixed by genetic drift. On the other hand, the selectionists certainly do not deny that genetic drift is an important factor in establishing differences in allele frequencies.

Finally, it should be pointed out that the two theories are not mutually exclusive. It is difficult to argue against the notion that some genetic variation must be neutral. The difference between the two theories is in the degree of neutrality that exists. Although current data are insufficient to resolve the problem, one important point has emerged from the arguments. In considering natural selection, there are clearly two levels that must be examined: the phenotypic level, including all morphological and physiological characteristics imparted by the genotype, and the molecular level, represented by the precise nucleotide and amino acid sequence of DNA and proteins. There is no question about selection acting at the phenotypic level, but at the genotypic level, does it act on individual loci, or the entire genome?

Formation of Species

Although the allopatric model of speciation is well developed, it has been more difficult to observe directly the processes involved. Diversification of isolated populations occurs gradually over thousands or hundreds of thousands of years. In addition, the geographic changes that paralleled this divergence may be complex or unknown. In most cases, therefore, the formation of species is an historical event, and biologists studying this process must rely on the present-day distribution of races, subspecies, and sibling species to reconstruct stages in the evolutionary process.

To study speciation, biologists must find examples in nature where all or most of the stages of race formation and speciation can be documented. The intensive studies carried out on natural populations of *Drosophila* provide a good example of the stages involved in **allopatric speciation**.

To illustrate the first step, namely the formation of populations with substantial genetic differences (race formation), we shall consider studies on *Drosophila pseudoobscura* conducted by Theodosius Dobzhansky and his col-

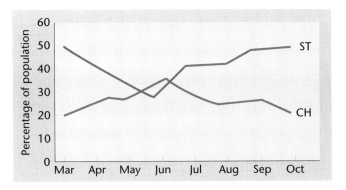

FIGURE 25.11 Changes in the ST and CH arrangements in *D. pseudoobscura* throughout the year.

leagues. This species is found over a wide range of environmental habitats, including the western and southwestern United States. Although the flies throughout this range are morphologically similar, Dobzhansky discovered that populations from different locations vary in the arrangement of genes on chromosome 3. He found several different inversions in this chromosome that can be detected by loop formations in larval polytene chromosomes. Each inversion sequence was named after the locale in which it was first discovered (e.g., AR = Arrowhead, British Columbia; CH = Chiricahua Mountains, etc.). The inversion sequences were compared with one standard sequence, designated ST.

Figure 25.10 shows a comparison of three arrangements found in populations at three different elevations in the Sierra Nevada chain in California. The ST arrangement is most common at low elevations; at 8000 feet, AR is the most common and ST least common. In these populations, the frequency of the CH arrangement gradually increases with elevation. The gradual change in inversion frequencies is probably the result of natural selection and parallels the gradual environmental changes occurring at ascending elevations. As the populations along this gradient show a continuous and gradual change in inversion frequencies, it is difficult to classify a fly as a member of one racial group.

Dobzhansky also found that if populations were collected at a single site throughout the year, inversion frequencies also changed. Through the seasons, cyclic variation in chromosome arrangements occurred, as shown in Figure 25.11. Such variation was consistently observed over a period of several years. The frequency of ST always declined during the spring, with a concomitant increase in CH during the same period.

To test the hypothesis that this cyclic change is a response to natural selection, Dobzhansky devised the following laboratory experiment: He constructed a large population cage from which samples of *D. pseudoobscura* could be periodically removed and studied. He began with a population of a known inversion frequency, 88 per-

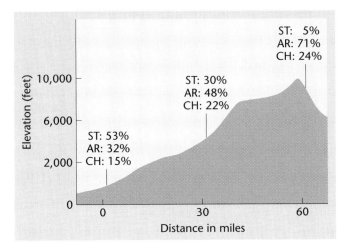

FIGURE 25.10 Inversions in chromosome 3 of *D. pseudoobscura* at different elevations in the Sierra Nevada range near Yosemite National Park.

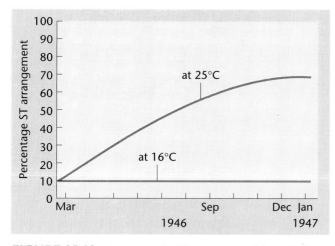

FIGURE 25.12 Increase in the ST arrangement of *D. pseudoobscura* in population cages under laboratory conditions.

cent CH and 12 percent ST. He reared it at 25°C and sampled it over a 1-year period. As shown in Figure 25.12, the frequency of ST increased gradually until it was present at a level of 70 percent. At this point, an equilibrium between ST and CH was reached. When the same experiment was performed at 16°C, no change in inversion frequency occurred. It can be concluded that the equilibrium reached at 25°C was in response to the elevated temperature, the only variable in the experiment.

The results are evidence that a balance in the frequency of the two inversions and their respective gene arrangements in a population is superior to either inversion by itself. The equilibrium attained presumably represents the greatest degree of fitness in the population under controlled laboratory conditions. This interpretation of the experiment suggests that natural selection is the driving force toward equilibrium.

In a more extensive study, Dobzhansky and his colleagues sampled *D. pseudoobscura* populations over a broader geographic range. Twenty-two different chromosome arrangements were found in populations from 12 locations. In Figure 25.13, the frequencies of five of these inversions are shown according to geographic location. The differences are largely quantitative, with most populations differing only in relative frequencies of inversions.

Because these locations represent varied environments and because inversions preserve different gene combinations, we can conclude that numerous races of *D. pseudoobscura* have been formed.

For speciation to occur, the development of races must be followed by a second step, reproductive isolation. We might then ask whether the evolution of *D. pseudoobscura* has gone beyond the formation of races. Dobzhansky investigated this question by examining the chromosome structure of other, closely related species called **sibling species**. Sibling species are reproductively isolated from one another, but remain very similar morphologically.

Drosophila persimilis and *D. pseudoobscura* each have five pairs of chromosomes and carry inversions within chromosome 3. When Dobzhansky examined the chromosome 3 inversions carefully, one arrangement (ST) was found in both species. Using this common inversion as a starting point, a phylogenetic sequence for all arrangements found in both species was constructed. Only one hypothetical arrangement is necessary to complete the continuity of the tree. It appears that an ancestral population with the ST arrangement gave rise to many different inversions. Some were incorporated into members of the *D. pseudoobscura* species, and others gave rise to the *D. persimilis* species (Figure 25.14).

Today, even when the geographic distributions of these sibling species overlap (Figure 25.15), several isolating mechanisms keep them from interbreeding. The species are isolated by prezygotic mechanisms such as habitat selection, with *D. persimilis* preferring high eleva-

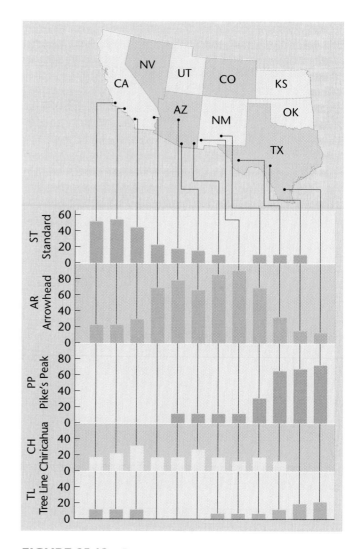

FIGURE 25.13 Frequencies of five chromosomal inversions in *D. pseudoobscura* in different geographic regions.

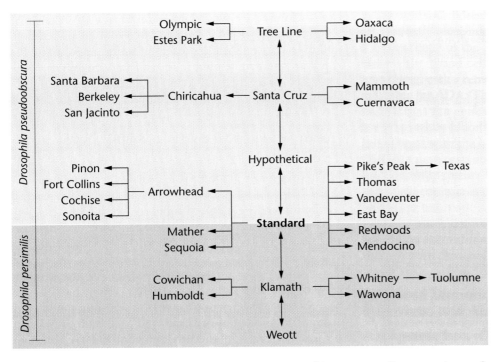

FIGURE 25.14 Inversion phylogeny for arrangements of chromosome 3 in *D. pseudoobscura* and *D. persimilis*. The ST arrangement is shared by both species.

tions and cooler temperatures. Differences in courtship rituals allow females to distinguish between males of the two species and to choose only males of their own species for mating. Even the time of day when courtship and mating occur is different in the species, with *D. persimilis* tending to court in the morning and *D. pseudoobscura* more active in the evening.

Postzygotic mechanisms also maintain reproductive isolation in these species. When cross-fertilized in the laboratory, the species produce sterile F_1 hybrid males, with male sterility being associated with interactions between the X chromosome and chromosome 2. Backcrosses between F_1 females and parental males exhibit hybrid breakdown through lowered viability of the offspring.

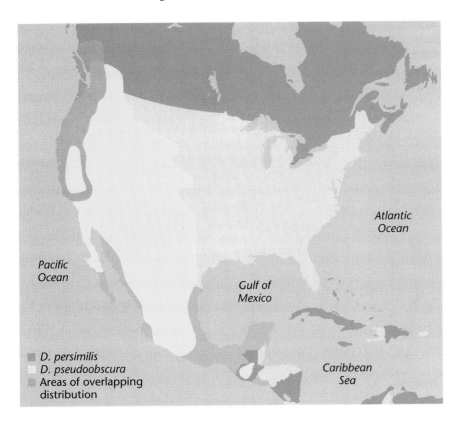

FIGURE 25.15 Although the geographic ranges of *Drosophila persimilis* and *Drosophila pseudoobscura* overlap, there is no interbreeding between the species.

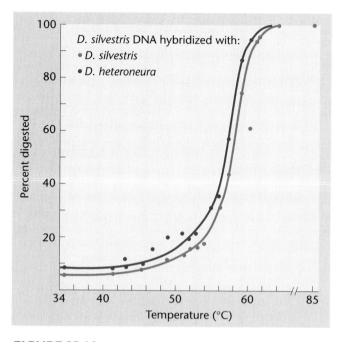

FIGURE 25.16 Nucleotide sequence diversity in the *Drosophila planitibia* species complex. The degree of shift to the left by the heterologous hybrids is an indication of the degree of nucleotide sequence divergence.

Using Molecular Techniques to Study Evolution

The degree of differences between two species can be measured in several ways. Taxonomists often use different ways of evaluating differences, and determining whether two organisms being compared are phenotypic variants of the same species, or members of different species. In some cases, morphological differences are used to define differences and classify organisms; also, cytogenetics, geographical distribution, and behavior are used as measures of differences.

Molecular biology and recombinant DNA technology are now being used to measure evolutionary relationships. These methods offer an increase in resolving power, and can be used to measure the degree of evolutionary divergence and to extrapolate the time scale over which genetic differences developed.

Measuring the Genetic Distance between Species

The evidence from studies on *D. persimilis* and *D. pseudoobscura* indicates that chromosome rearrangements help maintain genetic differences between species, but they cannot provide information about the degree of genetic difference between these species. In other cases, cytogenetic studies of closely related species are uninformative, and genetic analysis has been used to measure the genetic differences between species.

Drosophila heteroneura and *D. silvestris*, found only on the island of Hawaii, are estimated to have diverged from a common ancestral species only about 300,000 years ago, but it is difficult to demonstrate significant differences between these species in chromosomal inversion patterns or protein polymorphisms. When DNA hybridization studies are carried out on these species, the sequence diversity between the two is only about 0.55 percent (Figure 25.16). Thus, nucleotide sequence diversity may precede the development of protein or chromosomal polymorphisms.

The fact that *D. heteroneura* and *D. silvestris* share identical chromosome arrangements, that they have almost no detectable protein differences, are 99 percent homologous in DNA sequence, and yet are classified as separate species may seem paradoxical. However, the two species are clearly separated from each other by different and incompatible courtship and mating behaviors (a prezygotic isolating mechanism), by morphology, and by pigmentation of the body and wings (Figure 25.17). Ge-

(a)

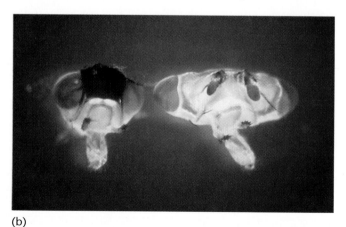

(b)

FIGURE 25.17 (a) Differences in pigmentation patterns in *D. sylvestris* (left) and *D. heteroneura* (right). (b) Head morphology in *D. sylvestris* (left) and *D. heteroneura* (right).

(a)   (b)

FIGURE 25.18   The flowers of two closely related species of monkey flowers. (a) *Mimulus cardinalis* and (b) *Mimulus lewisii*.

netic evidence suggests that these differences are controlled by a relatively small number of genes. Only about 15 to 19 major loci may be responsible for the morphological differences between these species, demonstrating that the process of speciation need only involve a small number of genes.

More recent studies using two closely related species of monkey flowers, a plant that grows in the Rocky Mountains and areas west of the mountains, confirm that species can be separated by only a few genes. One species, *Mimulus cardinalis* is fertilized by hummingbirds and does not interbreed with *Mimulus lewisii*, which is fertilized by bumblebees. Toby Bradshaw and his colleagues studied genetic differences related to reproduction in the two species: flower shape, size, and color, and nectar production. For each trait, a difference in a single gene provided at least 25 percent of the variation observed. *M. cardinalis* makes 80 times more nectar than does *M. lewisii*, and a single gene is responsible for at least half the difference. A single gene also controls a large part of the differences in flower color between the two species (Figure 25.18). In this case, as in the Hawaiian *Drosophila*, species differences can be traced to a small number of genes.

Protein Evolution

The degree of evolutionary relatedness between two species can be measured by comparing the amino acid sequences of proteins found in both species. The first protein to be sequenced was **insulin**, composed of only 51 amino acids. Comparison of the amino acid sequence in a variety of mammals, including cattle, pigs, horses, sperm whales, and sheep, shows that, with the exception of a stretch of three amino acids, the protein is identical in each of these species.

Cytochrome *c* is a respiratory pigment found in the mitochondria of eukaryotes. The amino acid sequence of cytochrome *c* has changed very slowly during evolution. The amino acid sequence in humans and chimpanzees is identical; between humans and rhesus monkeys only one amino acid is different. This is remarkable considering that the fossil record indicates that the lines leading to humans and monkeys diverged from a common ancestral species approximately 20 million years ago.

Table 25.4 shows the number of amino acid differences in cytochrome *c* among a variety of organisms. Distantly related organisms such as humans and yeasts have 38 amino acid differences (out of 104 amino acids), but more closely related species have few, if any, differences.

TABLE 25.4   A Comparison of (a) the Number of Amino Acid Differences and (b) the Minimal Mutational Distance in Cytochrome *c*

Organism	(a) Number of Amino Acid Differences	(b) Minimal Mutational Distance
Human	0	0
Chimpanzee	0	0
Rhesus monkey	1	1
Rabbit	9	12
Pig	10	13
Dog	10	13
Horse	12	17
Penguin	11	18
Moth	24	36
Yeast	38	56

Source: From W. M. Fitch and E. Margoliash, Construction of Phylogenetic Trees, *Science* 155:279–84, 20 January 1967. Copyright 1967 by the American Association for the Advancement of Science.

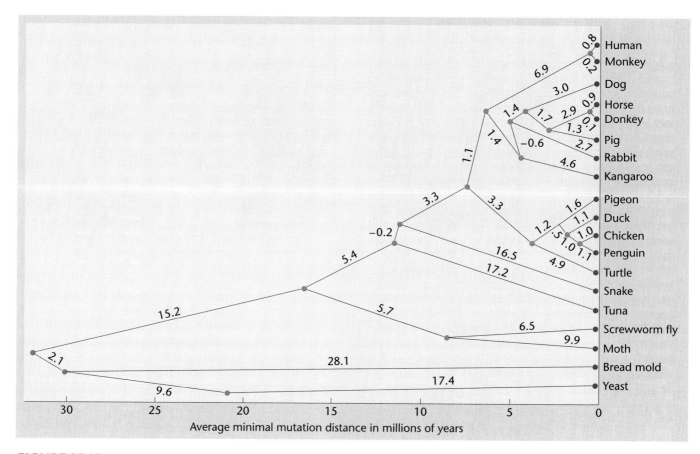

FIGURE 25.19 Phylogenetic sequence constructed by comparison of homologies in cytochrome *c* amino acid sequences.

The Molecular Clock

In addition to comparing amino acid differences, it is possible to assess evolutionary relationships by measuring the minimum number of nucleotide changes that have occurred during the evolution of a protein. This requires sequencing the gene encoding a protein in a number of different species, and then comparing the degree of nucleotide differences. This method is more complex than measuring amino acid differences. More than one nucleotide change may be required to change a given amino acid. When the nucleotide changes necessary for all amino acid differences observed in a protein are totaled, the **minimal mutational distance** between any two species is established. Table 25.4 shows such an analysis of the genes coding for cytochrome *c* in 10 organisms. As expected, these values are larger than the corresponding number of amino acids separating humans from the other nine organisms.

Data on amino acid substitutions and mutational distance can be combined with paleontology to construct a **molecular clock**. Information on amino acid or DNA differences measures the number of mutational events that have accumulated since any two organisms shared a

common ancestor. The fossil record provides information about the time that has elapsed since the two organisms shared a common ancestor. Assuming that the rate of amino acid replacements occurred at a regular rate proportional to absolute time, differences in amino acid content can be used as a molecular clock, measuring the time since the two species diverged.

Phylogenetic Trees

On the basis of information provided by a molecular clock, it is possible to construct **divergence dendrograms** or **phylogenetic trees** based on the analysis of amino acid sequences of a single protein from diverse organisms. The underlying assumption in this analysis is that all present-day sequences from different species represent gene products that diverged from common ancestral sequences at various points in evolutionary time. By determining minimal mutational distances among all species under analysis, and taking into account which amino acids have changed, the most likely relationships among the species can be determined. This analysis can also establish the point in these relationships at which a

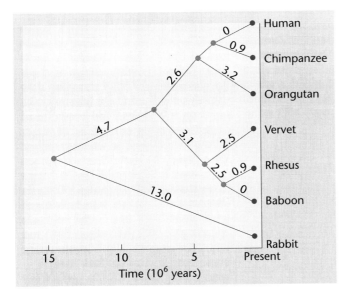

FIGURE 25.20   Phylogenetic sequence of carbonic anhydrase amino acid substitutions.

now-extinct ancestral sequence must have existed in order to lead to evolutionary divergence.

This information is usually summarized in the form of a phylogenetic tree (Figure 25.19). This tree (based on the data of Table 25.4) is plotted so that the ordinate represents proportional amounts of evolutionary distance. If a constant substitution rate is assumed, the ordinate represents a relative estimate of geologic time.

Phylogenetic trees agree remarkably well with trees constructed using more conventional approaches such as morphological and paleontological evidence. Once a number of proteins from a variety of organisms are sequenced and analyzed simultaneously, even more accurate trees can be constructed.

When closely related species are to be examined in this way, proteins that have evolved more rapidly than cytochrome *c* are more useful. The 115-amino-acid protein carbonic anhydrase has been used to analyze more accurately the relationship between humans and several other primates (Figure 25.20).

Molecular Studies on Human Evolution

The hominoid primates include chimpanzees, gorillas, orangutans, gibbons, and humans. Studies using protein differences have failed to resolve the taxonomic relationships among the chimpanzee, gorilla, and humans, because they are so closely related. Using hybridization of DNA sequences from a large number of individuals and calibrations derived from the fossil record, Charles Sibley and Jon Ahlquist have clarified the evolutionary branching pattern in the hominoids and have estimated the times at which divergences occurred (Figure 25.21). According to their model, humans and chimpanzees are more closely related than either is to the gorilla. The measure of relatedness is called the $\Delta T_{50}H$, and is related to the nucleotide matching of hybrid DNA molecules formed when single-stranded DNA from the two species is mixed. Between

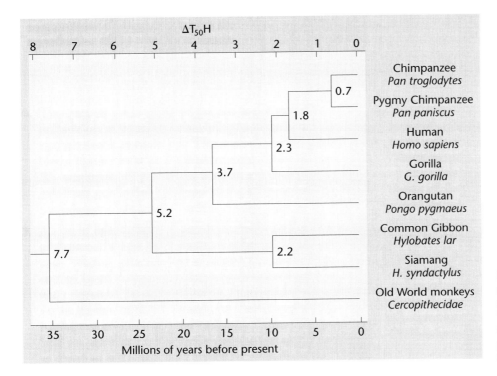

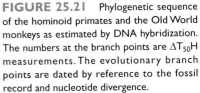

FIGURE 25.21   Phylogenetic sequence of the hominoid primates and the Old World monkeys as estimated by DNA hybridization. The numbers at the branch points are $\Delta T_{50}H$ measurements. The evolutionary branch points are dated by reference to the fossil record and nucleotide divergence.

humans and chimpanzees, the $\Delta T_{50}H$ is 1.8, somewhat less than the value of 2.3 for the distance between the gorilla line and the chimpanzee/human line. Using the fossil record, this places the separation of the line leading to the gorilla at 8 million to 10 million years (MY) ago. The lines leading to the chimpanzees diverged about 6.3 to 7.7 MY ago, and the chimpanzee and pygmy chimpanzee diverged about 2.4 to 3.0 MY ago.

The close evolutionary relationship between the two species of chimpanzees and the human species revealed by DNA hybridization poses an interesting taxonomic problem. In other areas of taxonomy, differences of less than 2 in the $\Delta T_{50}H$ for two species are usually accompanied by classification in the same genus and family. As it stands, current taxonomy places humans not only in a different genus, but also in a different family than gorillas and chimpanzees (humans in Hominidae, gorillas and chimps in Pongidae). The problems and challenges posed by this anthropomorphic inconsistency and the evolutionary relationship between humans and chimpanzees are explored in the book *The Third Chimpanzee* by Jared Diamond.

The question of where the modern human species (*Homo sapiens*) originated has generated several theories. One model holds that there was no single site of origin, and that a gradual transition to *H. sapiens* occurred simultaneously in geographically dispersed but genetically linked populations of a precursor species, *Homo erectus*. Another model suggests that *H. sapiens* was present in southern Africa more than 100,000 years ago, and was found in Asia at least 50,000 years ago, and that *H. sapiens* replaced Neanderthals in Europe about 30,000 to 40,000 years ago. This model suggests a single point of origin for *H. sapiens*, in Africa, with subsequent migrations to other regions, where *H. erectus* was displaced.

There is little dispute over the evidence that earlier ancestors of the human species arose in Africa, and that an ancestral species, *H. erectus*, spread out of Africa to populate Europe and Asia beginning about 1 million years ago. What is in dispute is how and where *H. sapiens* originated.

Over the past two decades, molecular techniques, including restriction fragment-length polymorphism (RFLP) analysis, have been used to provide evidence for

GENETICS, TECHNOLOGY, AND SOCIETY

DNA from Fossils: The Quest for Dinosaur Genes

Evolutionary studies often rely on the analysis of fossils. Detailed structural comparisons between fossilized specimens have provided a basis for establishing evolutionary relationships between extinct species and their living descendants. In recent years a new approach has been taken in unraveling evolutionary relationships that involves comparisons not between fossils but rather between nucleotide sequences of genes. The basic assumption is that as species diverged, so did the sequences of their genes; the more distantly related are two species, the more differences will exist between their gene sequences. Thus, comparing the nucleotide divergence between genes of different species allows the historical relationship among the species to be inferred. In practice, the sequences of specific genes from several related species are entered into a computer, and sophisticated programs are used to compare the sequences and reconstruct gene phylogenies, branching histories

of the genes. Not only does this show the likely relationships between the genes of extant species, but it postulates the sequence of the genes of the extinct species that were their progenitors. What would be particularly valuable would be more direct evidence, sequences of actual genes from extinct species. Using the PCR technique, it may now be possible to acquire such information by isolating genes from extinct species and looking directly at their nucleotide sequences.

When ancient plants and animals are preserved under exactly the right conditions, protected from water, oxygen, radiation, and extremes of heat and pressure, a small number of cells may still be present that contain a few nanograms of DNA, enough to amplify using the PCR reaction. This DNA is likely to be broken into very short fragments, averaging 100–200 base pairs, but a few longer molecules may also exist that will serve as a template and allow full-length gene sequences to be amplified.

Extreme precaution is necessary to prevent contaminating the ancient DNA with contemporary DNA, since

it also would serve as a template in the PCR reaction. This problem is especially acute when dealing with the tiny amounts of DNA recovered from fossils. The contaminating DNA can come from skin cells shed by archaeologists and lab workers, from previous experiments in the same laboratory, or even from dust particles in the air.

Using the PCR technique, gene sequences have been recovered from several extinct species, including a quagga (a zebra-like animal), a marsupial wolf, and a magnolia plant. The Holy Grail in the study of ancient DNA, however, is an intact gene from a dinosaur. Dinosaurs continue to capture our imaginations like no other extinct animals, but they have been gone from the earth so long, about 60 million years, that many wonder whether any dinosaur DNA will ever be recovered.

A great deal of interest was thus aroused in 1994 when Dr. Scott Woodward and his colleagues at Brigham Young University reported that they had recovered DNA sequences from bone fragments found in an underground coal mine in Utah. These bone the origin and spread of the human species across the globe. In Chapter 16 we discussed the use of RFLPs to map genes to specific chromosomes and/or chromosomal regions. The RFLPs can be generated by single nucleotide base changes in a DNA sequence that is recognized by a restriction enzyme. These polymorphisms are inherited in a codominant Mendelian fashion. Also, RFLP analysis has been used to construct both a molecular clock and phyletic trees in order to reconstruct events in human evolution.

Phyletic trees constructed from RFLPs in mitochondrial genomic DNA suggest that the modern human lineage arose in Africa about 200,000 years ago. According to the results of these studies, the maternal lineage of humans (in the form of mitochondrial DNA) has an ultimate common female ancestor, and the modern human species, *H. sapiens*, arose in Africa. From here, modern *H. sapiens* spread throughout the Old World, replacing the several human lineages descended from *H. erectus*.

The opposing hypothesis agrees with the idea that humans originated in Africa and spread throughout Europe and Asia as *H. erectus*. However, this hypothesis holds that the transitions in the fossil record in Asia and the Middle East from *H. erectus* to *H. sapiens* support the idea that modern humans arose in multiple regions of the Old World as part of an interbreeding network of lineages descended from *H. erectus*. This debate is often contentious and bitterly argued. Proponents of the multi-regional model have called into question the accuracy of the molecular clock as measured by nucleotide substitutions in mitochondria as well as the methods used to construct the mitochondrial tree.

Data gathered from the larger and more complex nuclear genome are now being used to resolve the debate. Recent studies have used RFLP sites on the Y chromosome and in the region surrounding the CD4 locus on chromosome 12. Two polymorphisms at the CD4 locus were studied in more than 1600 individuals from 42 geographically diverse populations. Results support the out of Africa model for the origin of *H. sapiens* and place the migration from Africa at about 100,000 to 300,000 years ago, dates that are much more recent than suggested by the multi-regional model. Results from studies on the Y chromosome give similar results. Thus, studies on the nuclear genome support the results from studies on the mitochondrial genome, and both suggest a recent, common origin in Africa for all human populations.

---

fragments were found to be from the Cretaceous Period, roughly 80 million years ago, and Woodward believed them to be from dinosaurs, although they were too small to be positively identified. The DNA sequences that were amplified appeared to be part of a gene for cytochrome *b*, which is found in the mitochondrial genome. Unfortunately, careful analysis of the amplified sequences showed them to be more similar to sequences of present-day mammals than to birds or reptiles, suggesting they are not derived from dinosaur genes. The amplified sequences likely resulted from contamination with human DNA.

Since Woodward's report, further work has cast serious doubt that any intact DNA can be recovered from fossils millions of years old. Even under the most favorable of conditions, the breakdown of DNA, primarily by the loss of purine residues, proceeds at such a rate that any DNA recovered from such specimens is almost certainly to be a recent contaminant. Unless a specimen is preserved in amber, protecting it completely from water, the upper age limit for intact DNA may be about 100,000 years. So although the quest for dinosaur genes goes on, hopes have dimmed that bona fide dinosaur sequences will be isolated any time soon, if ever.

Once genes from extinct species are isolated, what can be learned from them? First, comparisons of these genes with those of presumed descendant species will reveal a great deal about the evolution of genes. If the fossil that yielded the DNA can be dated accurately, it would be possible to eliminate the rate of nucleotide sequence change in one particular gene directly, rather than estimating it from comparisons of genes of present-day species. If an entire dinosaur gene is isolated, its promoters could be compared to that in living species to get some idea about how the regulation of genes changed over millions of years. With an intact gene, it might also be possible to synthesize a dinosaur protein *in vitro* and compare its properties to that of existing species. A dinosaur gene could even be introduced into a living animal to see how it would function. In this way, one small part of an extinct species could be recovered.

This is very different from reanimating the species itself, which would require the recovery of tens of thousands of separate genes with their transcriptional control sequences, the reassembly of these genes into functional chromosomes, and the insertion of these chromosomes into the nucleus of a compatible cell, all far beyond the capabilities of our present technology. Thus, most experts in the study of ancient DNA doubt that it will ever be possible to clone a dinosaur, despite future advances in molecular technology. Even the power of the PCR technique can't overcome extinction. For the foreseeable future, only our imaginations will be capable of that.

Fossilized dinosaur bone.

CHAPTER SUMMARY

1. After extensive observation of natural populations, Alfred Russel Wallace and Charles Darwin formulated a theory on the role of natural selection in the formation of species. The genetic basis of evolution and the role of natural selection in changing allele frequencies were discovered in the twentieth century.

2. Several models of speciation have been proposed, differing in the order and rate at which evolutionary processes take place.

3. The amount of genetic variation present in a population can be studied by genetic analysis of inbreeding depression, protein polymorphisms revealed by gel electrophoresis, or by DNA restriction mapping and sequencing.

4. There are two main views about whether the nucleotide and amino acid changes present in a population are adaptive. Some (the neutralists) believe that the majority of these changes are genetically equivalent, while others (selectionists) believe that all variations have a selective value.

5. Speciation involves the partitioning of a population's gene pool into two or more gene pools that become reproductively isolated. The stages of species formation have been carefully studied in *Drosophila*.

6. Two approaches have been especially fruitful in studying evolution at the level of the genome: (1) comparison of amino acid substitutions in proteins common to a number of organisms, and (2) comparison of complementary nucleotide sequences present in the DNA of related organisms. These studies help measure evolutionary relatedness and allow the amount of genetic variation present in different individuals to be measured.

KEY TERMS

allopatric speciation, 691
allopolyploidy, 687
allozymes, 689
cladogenesis, 683
cytochrome *c*, 695
divergence dendrogram, 696
founder–flush theory, 685
geographic (allopatric) speciation, 683
insulin, 695
minimal mutational distance, 696

molecular clock, 696
neutralist hypothesis, 690
niches, 681
phyletic evolution (anagenesis), 682
phyletic gradualism, 685
phylogenetic trees, 696
population crash, 686
postzygotic isolating mechanisms, 684
prezygotic isolating mechanisms, 685
punctuated equilibrium, 685

quantum speciation, 685
reproductive isolating mechanisms, 684
selectionists, 690
sibling species, 692
speciation, 682
species, 681
statispatric speciation, 684
stochastic (catastrophic) models of speciation, 685
sympatric speciation, 683

INSIGHTS AND SOLUTIONS

1. Sequence analysis of DNA can be accomplished by a number of techniques (see Appendix A for a detailed description). Protein sequencing, on the other hand, is made more complex by the fact that 20 different subunits need to be unambiguously identified and enumerated, rather than the four nucleotides of DNA. Because of their unique properties, the N-terminal and C-terminal amino acids in a protein are easy to identify, but the array in between offer a difficult challenge, since many proteins contain hundreds of amino acids. How is it that protein sequencing is accomplished?

Solution: The strategy for protein sequencing is the same as for DNA sequencing: divide and conquer. To accomplish this, specific enzymes are used that reproducibly cleave proteins between certain amino acids. The use of different enzymes produces overlapping fragments. Each fragment is isolated and its amino acid sequence is determined by chemical means. Sequences from overlapping fragments are then assembled to give sequence for the entire protein.

2. A single plant twice the size of others in the same population suddenly appears. Normally, plants of this species reproduce by self-fertilization and by cross-fertilization. Is this new giant plant simply a variant, or could it be a new species? How would you determine this?

Solution: One of the most widespread mechanisms of speciation in higher plants is polyploidy, the multiplication of entire chromosome sets. The result of polyploidy is usually a larger plant with larger flowers and seeds. There are two ways of testing this new variant to determine whether it is a new species. First, the giant plant should be crossed with a normal-sized plant to see whether it produces viable, fertile offspring. If it does not, the two different types of plants would appear to be reproductively isolated. Second, the giant plant should be cytogenetically screened to examine its chromosome complement. If it has twice the number of its normal-sized neighbors, it is a tetraploid that may have arisen spontaneously. If the chromosome number differs by a factor of two, and the new plant is reproductively isolated from its normal-sized neighbors, it is a new species.

PROBLEMS AND DISCUSSION QUESTIONS

1. Discuss the rationale behind the statement that inversions in chromosome 3 of *Drosophila pseudoobscura* represent genetic variation.
2. Describe how populations with substantial genetic differences can form. What is the role of natural selection?
3. Contrast the classical and balance hypotheses.
4. What types of nucleotide substitutions will not be detected by electrophoretic studies of a gene's protein product?
5. In a sequencing experiment (using the numbers 1 through 6 to represent amino acids), the following two sets of peptide fragments were obtained in independent experiments with different proteolytic enzymes:

Proteins	Fragments			
Enzyme I	624	246	35	135
Enzyme II	136	356	524	24

Determine the sequence of fragments and amino acids in the protein.
6. Shown below are two homologous lengths of the alpha and beta chains of human hemoglobin. Consult the genetic code dictionary (Figure 12.6) and determine how many amino acid substitutions may have occurred as a result of a single nucleotide substitution. For any that cannot occur as the result of a single change, determine the minimal mutational distance.

Alpha:	Ala	Val	Ala	His	Val	Asp	Asp	Met	Pro
Beta:	Gly	Leu	Ala	His	Leu	Asp	Asn	Leu	Lys

7. Determine the minimal mutational distances between the following amino acid sequences of cytochrome c from various organisms. Compare the distance between humans and each organism.

Human:	Lys	Glu	Glu	Arg	Ala	Asp
Horse:	Lys	Thr	Glu	Arg	Glu	Asp
Pig:	Lys	Gly	Glu	Arg	Glu	Asp
Dog:	Thr	Gly	Glu	Arg	Glu	Asp
Chicken:	Lys	Ser	Glu	Arg	Val	Asp
Bullfrog:	Lys	Gly	Glu	Arg	Glu	Asp
Fungus:	Ala	Lys	Asp	Arg	Asn	Asp

8. The genetic difference between two species of *Drosophila*, *D. heteroneura* and *D. sylvestris* as measured by nucleotide diversity, is about 1.8 percent. The difference between chimpanzees (*Pan troglodytes*) and humans (*Homo sapiens*) is about the same, yet the latter species are classified in different genera. In your opinion, is this valid? If so, why; if not, why not?
9. As an extension of the previous question, consider the following: In sorting out the complex taxonomic relationships among birds, species with $\Delta T_{50}H$ values of 4.0 are placed in the same genus, even by traditional taxonomy based on morphology. Using the data in Figure 25.17, construct a phylogeny that obeys this rule, using any of the appropriate genus names (*Pongo, Pan, Homo*), or constructing new ones.
10. The use of nucleotide sequence data to measure genetic variability is complicated by the fact that the genes of higher eukaryotes are complex in organization and contain 5′ and 3′ flanking regions as well as introns. Slightom and colleagues have compared the nucleotide sequence of two cloned alleles of the gamma-globin gene from a single individual and found a variation of 1 percent. Those differences include 13 substitutions of one nucleotide for another, and three short DNA segments that have been inserted in one allele or deleted in the other. None of the changes take place in the exons (coding regions) of the gene. Why do you think this is so, and should it change the concept of genetic variation?
11. Discuss the arguments supporting the neutralist hypothesis. What counterarguments are proposed by the selectionists?
12. Of what value to our understanding of genetic variation and evolution is the debate concerning the neutralist hypothesis?

EXTRA-SPICY PROBLEMS

13. Native tribes of both North and South America are descendants of one or a few groups of mongoloid peoples who immigrated from the Asian landmass sometime between 11,000 and 40,000 years ago.

   The HLA genes in humans code for the histocompatibility antigens found on the surface of most cells. These genes are the most polymorphic genes known in humans. For example, over 40 alleles of the *HLA B* gene have been identified. In contrast to Old World populations, however, native Americans do not show this phenomenal allelic diversity. Instead, a limited set of alleles of all of the HLA genes are found in varying frequency in all tribes, regardless of location of the tribe. What are the various forces that determine allele frequencies? What is the best explanation for these observations?

Another interesting observation about the HLA alleles of native Americans is that South American tribes have alleles of the *HLA B* gene that are not found in the present-day Asian population. Provide two possible explanations for this observation.

14. Nauru is a remote Pacific atoll occupied by 5000 Micronesians. Colonization by Britain, Australia, and New Zealand, as well as income from phosphate mining has changed the lifestyle of the Nauruans. Although these people formerly depended on fishing and subsistence farming and had an active lifestyle, food is now imported, high in energy content and obesity occurs at a high frequency in the population. The multifactorial disease NIDDM (non-insulin-dependent diabetes mellitus) results from a combination of genetic and environmental influences. This disease used to be nonexistent on this island. But after 1950, the prevalence of the disease increased from <1 percent in the 1950s to 21 percent in the mid-1970s and then dropped to 9 percent in 1987. At one point in the 1970s, a severe form striking many young adults reached epidemic proportions in this population. Diabetic women had more stillbirths and less than half as many live births as non-diabetic women. Although the NIDDM epidemic on Nauru has passed its peak, this outcome cannot be attributed to a decline in environmental risk factors since their lifestyle has not changed since the 1970s.

Consider this data carefully and then propose an explanation for the increase and then subsequent decrease in the NIDDM incidence in this population.

SELECTED READINGS

ANDERSON W., et al. 1975. Genetics of natural populations: XLII. Three decades of genetic change in *Drosophila pseudoobscura*. *Evolution* 29:24–36.

ARMOUR, J. A., ANTTINEN, T., MAY, C. A., VEGA, E. E., SAJANTILA, A., KIDD, J. R., KIDD, K. K., BERTRANPETIT, J., PAABO, S., and JEFFREYS, A. J. 1996. Minisatellite diversity supports a recent African origin for modern humans. *Nat. Genet.* 13:154–60.

AVISE, J. C. 1990. Flocks of African fishes. *Nature* 347: 512–13.

AYALA, F. J. 1976. *Molecular evolution.* Sunderland, MA: Sinauer Associates.

———. 1984. Molecular polymorphism: How much is there, and why is there so much? *Dev. Genet.* 4:379–91.

BARTON, N. H., and HEWITT, G. M. 1989. Adaptation, speciation and hybrid zones. *Nature* 341:497–503.

BOWCOCK, A. M., RUIZ-LINARES, A., TOMFOHRDE, J., MINCH, E., KIDD, J. R., and CAVALLI-SFORZA, L. L. 1994. High resolution of human evolutionary trees with polymorphic microsatellites. *Nature* 368:455–57.

BULT, C., WHITE, O., OLSEN, G. J., ZHOU, L., FLEISCHMANN, R. D., SUTTON, G. G., BLAKE, J. A., FITZGERALD, L. M., et al. 1996. Complete genome sequence of the methanogenic Archeon, *Methanococcus jannaschii*. *Science* 273:1058–73.

CARSON, H. 1970. Chromosome tracers of the origin of species. *Science* 168:1414–18.

———. 1975. The genetics of speciation at the diploid level. *Am. Natural.* 109:83–92.

COYNE, J. A. 1992. Genetics and speciation. *Nature* 355: 511–15.

DAYHOFF, M. O. 1969. Computer analysis of protein evolution. *Sci. Am.* (July) 221:86–95.

DIAMOND, J. 1992. *The third chimpanzee: The evolution and future of the human animal.* New York: HarperCollins.

DOBZHANSKY, T. 1947. Adaptive changes induced by natural selection in wild populations of *Drosophila*. *Evolution* 1:1–16.

———. 1948. Genetics of natural populations, XVI. Altitudinal and seasonal changes produced by natural selection in certain populations of *Drosophila pseudoobscura* and *Drosophila persimilis*. *Genetics* 33:158–76.

———. 1955. Genetics of the evolutionary process. New York: Columbia University Press.

DOBZHANSKY, T., et al. 1966. Genetics of natural populations: XXXVIII. Continuity and change in populations of *Drosophila pseudoobscura* in western United States. *Evolution* 20: 418–27.

ELDREDGE, N. 1985. *Time frames: The evolution of punctuated equilibria.* Princeton, NJ: Princeton University Press.

ELENA, S. F., COOPER, V. S., and LENSKI, R. E. 1996. Punctuated evolution caused by selection of rare beneficial mutations. *Science* 272:1802–4.

FITCH, W. M. 1973. Aspects of molecular evolution. *Annu. Rev. Genet.* 7:343–80.

FITCH, W. M., and MARGOLIASH, E. 1967. Construction of phylogenetic trees. *Science* 155:279–84.

———. 1970. The usefulness of amino acid and nucleotide sequences in evolutionary studies. *Evol. Biol.* 4:67–109.

GILLESPIE, J. H. 1992. The causes of molecular evolution. New York: Oxford University Press.

GOULD, S. J. 1982. Darwinism and the expansion of evolutionary theory. *Science* 216:380–87.

HUNT, J., et al. 1981. Evolution distance in Hawaiian Drosophila. *J. Mol. Evol.* 17:361–67.

KIMURA, M. 1979a. Model of effectively neutral mutations in which selective constraint is incorporated. *Proc. Natl. Acad. Sci. USA* 76:3440–44.

———. 1979b. The neutral theory of molecular evolution. *Sci. Am.* (Nov.) 241:98–126.

———. 1989. The neutral theory of molecular evolution and the world view of the neutralists. *Genome* 31:24–31.

KING, M. C., and WILSON, A. C. 1975. Evolution at two levels: Molecular similarities and biological differences between humans and chimpanzees. *Science* 188:107–16.

KREITMAN, M. 1983. Nucleotide polymorphism at the alcohol dehydrogenase locus of *Drosophila melanogaster*. *Nature* 304:412–17.

LANDE, R. 1989. Fisherian and Wrightian theories of speciation. *Genome* 31:221–27.

LEWIN, R. 1993. *Human evolution*, 3rd ed. Cambridge, MA: Blackwell Scientific.

LEWONTIN, R. C., and HUBBY, J. L. 1966. A molecular approach to the study of genic heterozygosity in natural populations: II. Amount of variation and degree of heterozygosity in natural populations of *Drosophila pseudoobscura*. *Genetics* 54:595–609.

MAYR, E. 1963. *Animal species and evolution*. Cambridge, MA: Harvard University Press.

MEYER, A., KOCHER, T. D., BASASIBWAKI, P., and WILSON, A. C. 1990. Monophyletic origin of Lake Victoria cichlid fishes suggested by mitochondrial DNA sequences. *Nature* 347:550–53.

PAABO, S. 1993. Ancient DNA. *Sci. Am.* (Nov.) 269:86–92.

PAGEL, M. D., and HARVEY, P. H. 1989. Comparative methods for examining adaptation depend on evolutionary models. *Folia Primatol.* 53:203–20.

POWELL, J. 1978. The founder–flush speciation theory: An experimental approach. *Evolution* 32:465–74.

RIDLEY, M. 1993. *Evolution*. Cambridge, Blackwell Scientific.

SIBLEY, C., and AHLQUIST, J. 1984. The phylogeny of the hominoid primates, as indicated by DNA-DNA hybridization. *J. Mol. Evol.* 20:2–15.

SIBLEY, C. G., COMSTOCK, J. A., and AHLQUIST, J. E. 1990. DNA evidence of hominoid phylogeny: A re-analysis of the data. *J. Mol. Evol.* 30:202–236.

SMITHIES, O., BLECHL, A. E., SHEN, S., SLIGHTOM, J. L., and VANIN, E. F. 1981. Co-evolution and control of globin genes. In *Levels of genetic control in development*, eds. S. Subtelny and U. Abbot, pp. 185–200. New York: Alan R. Liss.

STEBBINS, G. L. 1977. *Processes of organic evolution*, 3rd ed. Englewood Cliffs, NJ: Prentice-Hall.

STONEKING, M. 1995. Ancient DNA: How Do You Know When You Have It and What Can You Do With It? *Am. J. Hum. Genet.* 57:1259–62.

ASHIAN, R. E., and CARTER, N. D. 1976. Biochemical genetics of carbonic anhydrase. In *Advances in human genetics*, eds. H. Harris and K. Hirschhorn, pp. 1–56. New York: Plenum Press.

TEMPLETON, A. R. 1985. Phylogeny of the hominoid primates: A statistical analysis of the DNA-RNA hybridization data. *Mol. Biol. Evol.* 2:420–33.

———. 1994. The role of molecular genetics in speciation studies. *EXS* 69:455–57.

THORNE, A. G., and WOLPOFF, M. H. 1992. The multiregional evolution of humans. *Sci. Am.* (April) 266:76–83.

TISHKOFF, S. A., DIETZSCH, E., SPEED, W., PAKSTIS, A. J., KIDD, J. R., CHEUNG, K., BONNE-TAMIR, B., SANTACHIARA-BENERECETTI, A. S., et al. 1996. Global patterns of linkage disequilibrium at the CD4 locus and modern human origins. *Science* 271:1380–87.

VAL, F. C. 1977. Genetic analysis of the morphological differences between two interfertile species of Hawaiian Drosophila. *Evolution* 31:611–29.

WHITE, M. J. D. 1977. *Modes of speciation*. New York: W. H. Freeman.

WILSON, A. C., and CANN R. L. 1992. The recent African genesis of humans. *Sci. Am.* (April) 266:68–73.

YUNIS, J. J., and PRAKASH, O. 1982. The origin of man: A chromosomal pictorial legacy. *Science* 215:1525–30.

In addition to the techniques of genetic analysis, physical and chemical techniques for the separation and analysis of macromolecular components of the cell nucleus and cytoplasm have been instrumental in advancing our understanding of genetics at the molecular level. In this appendix we will describe the background and theoretical basis of some techniques that have been important in molecular genetics.

Isotopes

Isotopes are forms of an element that have the same number of protons and electrons but differ in the number of neutrons contained in the atomic nucleus. For example, the most common form of carbon has an atomic number of 6 (the number of protons in the nucleus) and an atomic weight of 12 (the sum of the protons and neutrons in the nucleus). In a very small percentage of carbon atoms, a seventh neutron is present, producing an atom with an atomic weight of 13. This is an example of a so-called **heavy isotope**. Since the number of protons and electrons, and thus the net charge, has not changed, the atom has the same chemical properties as **carbon-12 (^{12}C)** and differs only in mass. **Carbon-13 (^{13}C)** is thus a stable, heavy isotope of carbon.

Although the addition of neutrons does not alter the chemical properties of an atom, it can produce instabilities in the atomic nucleus. If another neutron is added to a carbon-13 atom, the isotope **carbon-14 (^{14}C)** results. However, the presence of eight neutrons and six protons is an unstable condition, and the atom undergoes a nuclear reaction in which radiation is emitted during the transition to a more stable condition. Therefore, carbon-14 is a **radioactive isotope** of carbon.

The type of radiation emitted and the rate at which these nuclear events take place are characteristic of the element. Table A.1 lists types of radioactivity. The rate at which a radioactive isotope emits radiation is expressed as its **half-life**, which is the time required for a given amount of a radioactive substance to lose one-half of its radioactivity. Table A.2 lists some of the isotopes available for use in research.

TABLE A.2 Some Isotopes Used in Research

Element	Isotope	Half-life	Radiation
H	^2H	—	Stable
	^3H	12.3 years	β
C	^{13}C	—	Stable
	^{14}C	5700 years	β
N	^{15}N	—	Stable
O	^{18}O	—	Stable
P	^{32}P	14 days	β
S	^{35}S	87 days	β
K	^{40}K	1.2×10^9 years	β, gamma
Fe	^{59}Fe	45 days	β, gamma
I	^{125}I	60 days	β, gamma
	^{131}I	8 days	β, gamma

Detection of Isotopes

The choice of which isotope to use as a tracer in biological experiments depends on a combination of its physical and chemical properties, which enable the investigator to quantitate the amount of radioactivity or measure the ratio of heavy to light isotopes. For the detection of heavy isotopes, two methods are commonly employed: **mass spectrometry** and **equilibrium density gradient centrifugation** (discussed in the following section). Although the use of heavy isotopes has been more restricted than that of radioisotopes, they have been instrumental in several basic advances in molecular biology [e.g., demonstrating the semiconservative nature of DNA replication and the existence of messenger RNA (mRNA)].

There are a number of methods to detect radioisotopes, the foremost being **liquid scintillation spectrometry**, which provides quantitative information about the amount of radioactive isotope present in a sample, and **autoradiography**, which is used to demonstrate the cytological distribution and localization of radioactively labeled molecules.

In recording radioactivity by liquid scintillation counting, a small sample of the material to be counted is solubilized and immersed in a solution containing a **phosphor**, an organic compound that emits a flash of light after it absorbs energy released by decay of the ra-

TABLE A.1 Properties of Ionizing Radiation

Type	Relative Penetration	Relative Ionization	Range in Biological Tissue
Alpha particle (2 protons + 2 neutrons)	1	10,000	Microns
Beta particle (electron)	100	100	Microns–mm
Gamma ray	>1000	<1	∞

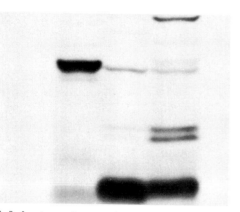

FIGURE A.1   Autoradiogram of radioactive bacterial proteins synthesized in a minicell system.

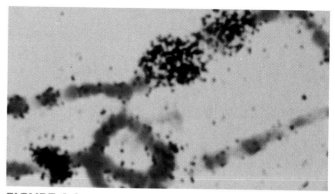

FIGURE A.2   Autoradiogram of RNA synthesis in salivary glands of *Drosophila* larva. Silver grains are deposited over sites of RNA synthesis at chromosome puffs.

dioactive compound. The counting chamber of the liquid scintillation spectrometer is equipped with very sensitive photomultiplier tubes that record the light flashes emitted by the phosphor. The data are recorded as counts of radioactivity per unit time and are displayed on a printout or can be fed into a computer for storage.

In autoradiography, a gel, chromatogram, or plant or animal part is placed against a sheet of photographic film. Radioactive decay from the incorporated isotope behaves just as light energy does and reduces silver grains in the emulsion. After exposure, the sheet or film is developed and fixed, revealing a deposit of silver grains corresponding to the location of the radioactive substance (Figure A.1).

Alternatively, to record the subcellular localization of an incorporated labeled isotope, cells or chromosomal preparations that have been incubated with radioactively labeled compounds are affixed to microscope slides and covered with a thin layer of liquid photographic emulsion. After they are exposed, the slides are processed to develop and fix the reduced silver grains in the emulsion. After staining, microscopic examination reveals the location and extent of labeled isotope incorporation (Figure A.2).

Centrifugation Techniques

The centrifugation of biological macromolecules is widely employed to provide information about their physical characteristics (e.g., size, shape, density, and molecular weight) and to purify and concentrate cells, organelles, and their molecular components.

Differential centrifugation is commonly used to separate materials such as cell homogenates according to size. Initially, the homogenate is distributed uniformly in the centrifuge tube. After a period of centrifugation, the pellet obtained is enriched for the largest and most dense particles, such as nuclei, in the homogenate. After each step, the supernatant can be recentrifuged at higher speeds to pellet the next heavier component. A typical fractionation scheme for cell homogenates is shown in Figure A.3. Further fractionation using density gradient techniques can be used to purify any of the fractions obtained by differential centrifugation.

Rate zonal centrifugation is used to separate particles on the basis of differences in their sedimentation rates. It may be used to separate mixtures of macromolecules such as proteins or nucleic acids and cellular organelles such as mitochondria. In this technique, which employs a medium of increasing density, the rate at which particles sediment depends on size, shape, density, and the frictional resistance of the solvent.

In addition to the preparation and purification of macromolecules and cellular components, rate zonal centrifugation can be used to determine the **sedimentation coefficients** and **molecular weights** of biological macromolecules. If a purified molecule such as a protein is spun in a centrifugal field, the molecule will eventually sediment toward the bottom at a constant velocity. At this point, the molecular weight (M) can be calculated as

$$M = f \times v/\omega^2 r$$

where f is the frictional coefficient of the solvent system (which has been calculated from other measurements) and $v/\omega^2 r$ is the rate of sedimentation per unit applied centrifugal field. The latter value is given the symbol S, or sedimentation coefficient. The S value for most proteins is between 1×10^{-13} sec and 2×10^{-11} sec. A sedimentation coefficient of 1×10^{-13} is defined as one **Svedberg unit (S)**; this unit is named for The Svedberg, a pioneer in the field of centrifugation. Thus, a protein with a sedimentation value of 2×10^{-11} sec would have a value of $200S$. Figure A.4 shows the S values of selected molecules and particles.

Isopycnic centrifugation, or **equilibrium density gradient centrifugation**, is one of the most widely used techniques in genetics and molecular biology. In this technique, the solvent varies in density from one end of the tube to the other. A mixture of molecules layered on top and centrifuged through this gradient will migrate toward the bottom of the tube until each particle reaches its isopycnic point—that is, the place in the gradient where the density of the solvent equals the buoyant density of the particle. When each molecular species in the mixture migrates to its own characteristic isopycnic point, it no longer moves and is at equilibrium no matter how much longer the centrifugal field is applied. After separation by this method, components may be recovered by puncturing the bottom of the tube and collecting fractions. Two methods

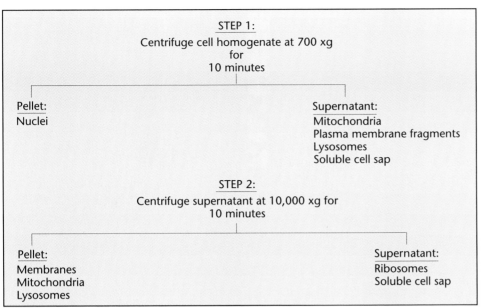

STEP 1:
Centrifuge cell homogenate at 700 xg
for
10 minutes

Pellet:
Nuclei

Supernatant:
Mitochondria
Plasma membrane fragments
Lysosomes
Soluble cell sap

STEP 2:
Centrifuge supernatant at 10,000 xg for
10 minutes

Pellet:
Membranes
Mitochondria
Lysosomes

Supernatant:
Ribosomes
Soluble cell sap

FIGURE A.3   Fractionation scheme for cell homogenates.

of forming density gradients are commonly employed (Figure A.5), one using preformed gradients of sucrose, or soluble salts of heavy metals such as cesium chloride or cesium sulfate. In the second method, the gradient is formed by the action of the centrifugal field on the salt solution.

Renaturation and Hybridization of Nucleic Acids

The ability of separated complementary strands of nucleic acids to unite and form stable, double-stranded molecules has been used to measure the relatedness of nucleic acids

from different parts of the cell, different organs, and even different species. If both strands are DNA, the process is known as **renaturation** or **reassociation**. If one strand is RNA and the other DNA, it is known as **hybridization**. Renaturation and hybridization involve two steps: (1) a rate-limiting step, in which collision or nucleation between two homologous strands initiates base pairing, and (2) a rapid pairing of complementary bases, or "zippering" of the strands, to form a double-stranded molecule. For DNA renaturation, the formation of double-stranded molecules can be assayed at any time during an experiment. A sample is passed over a column of **hydroxyapatite**, which selectively binds double-stranded DNA but allows single-stranded molecules to pass through. The

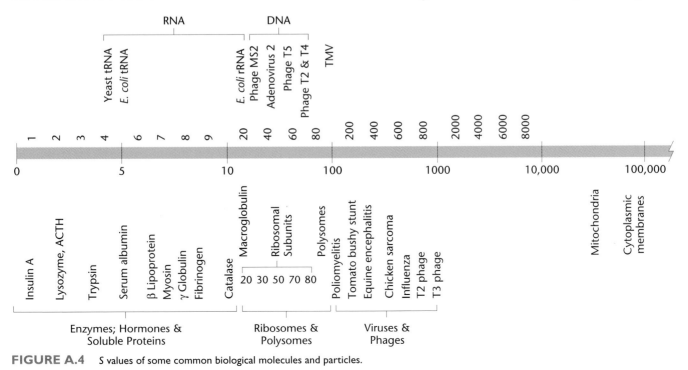

FIGURE A.4   S values of some common biological molecules and particles.

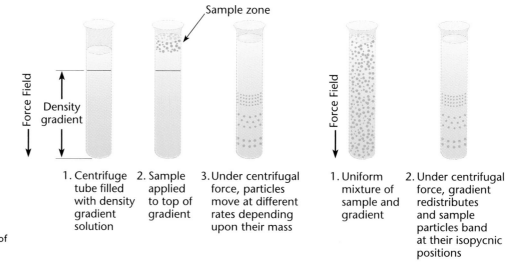

FIGURE A.5 Two methods of forming density gradients.

1. Centrifuge tube filled with density gradient solution
2. Sample applied to top of gradient
3. Under centrifugal force, particles move at different rates depending upon their mass
1. Uniform mixture of sample and gradient
2. Under centrifugal force, gradient redistributes and sample particles band at their isopycnic positions

double-stranded molecules can be released from the column by raising the temperature or salt concentration.

The process of renaturation follows second-order kinetics according to the equation

$$\frac{C}{C_0} = \frac{1}{1 + kC_0t}$$

where C is the single-stranded DNA concentration at time t, C_0 is the total DNA concentration, and k is a second-order rate constant. It is usually convenient to express the data from renaturation experiments as the fraction of single-stranded DNA at any time t versus the product of total DNA concentration and time, as shown in Figure A.6.

In the process of renaturation, the DNA is initially single-stranded and is renatured to double-stranded structures in the final state. The time necessary to reassociate half of the DNA in a sample at a given concentration should be proportional to the number of different pieces of DNA present. Consequently, the half-reassociation time should be proportional to the DNA content of the genome, with smaller genomes having shorter half-renaturation rates. Figure A.7 confirms this expectation and shows increases in C_0t values as the size of the genome increases. This proportional relationship between C_0t and genome size is valid only in cases where repetitive DNA sequences are absent

from the DNA being studied. DNA from calf thymus (and many other eukaryotic sources) exhibits a complex pattern of reassociation, indicating that bovine DNA contains some sequences (in this case, 40% of the total DNA) that reassociate rapidly and others that reassociate more slowly. The rapidly reassociating fraction must therefore contain sequences that are present in many copies. The more slowly renaturing DNA, however, contains sequences present in only one copy per genome. The *E. coli* DNA shown in Figure A.8 renatures with a pattern close to that of an ideal second-order reaction, indicating the absence of significant amounts of repeated DNA sequences.

In the case of hybridization between DNA and RNA, two approaches can be used—either the RNA or the DNA can be in excess. RNA-excess hybridization is usually preferred because most double-stranded molecules that are formed are RNA:DNA hybrids. Since DNA is present in low concentration, and since single-stranded RNA molecules lack complementary RNA strands in the mixture, the number of RNA:RNA and DNA:DNA hybrids is negligible.

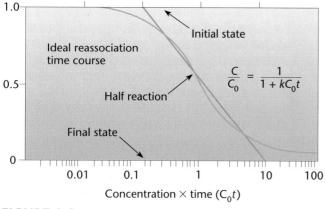

FIGURE A.6 Idealized C_0t curve.

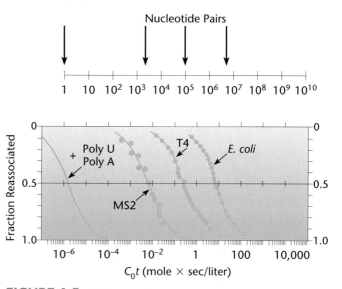

FIGURE A.7 Changes in C_0t value as genome size increases.

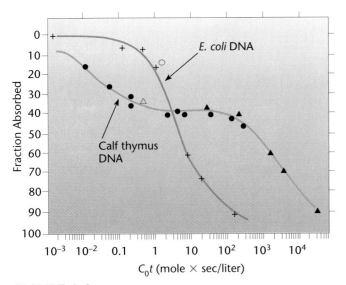

FIGURE A.8 Renaturation curve for *E. coli* DNA, containing no repetitive DNA sequences, and for calf thymus DNA, containing several families of repetitive DNA.

The extent of hybridization can be followed by using hydroxyapatite columns or by using radioactively labeled RNA or DNA. In practice, one of the components of the hybridization reaction is usually immobilized on a substrate such as nitrocellulose paper. For example, DNA may be sheared to a uniform size, denatured to single strands, immobilized on nitrocellulose filters, and hybridized with an excess of labeled RNA. After hybridization, the unbound RNA is removed by washing, and the hybrids assayed by liquid scintillation counting. DNA:RNA hybridization can also be performed using cytological preparations, a technique called *in situ* **hybridization**. DNA, which is a part of an intact chromosome preparation fixed to a slide, can be denatured and hybridized to radioactive RNA. Hybrid formation is detected using autoradiography (Figure A.9).

Electrophoresis

Electrophoresis is a technique that measures the rate of migration of charged molecules in a liquid-containing medium when an electrical field is applied to the liquid. Negatively charged molecules (anions) will migrate toward the positive electrode (anode) and positively charged molecules (cations) will migrate toward the negative electrode (cathode). Several factors affect the rate of migration, including the strength of the electrical field and the molecular sieving action of the medium (paper, starch, or gel) in which migration takes place. Since proteins and nucleic acids are electrically charged, electrophoresis has been used extensively to provide information about the size, conformation, and net charge of these macromolecules. On a larger scale, electrophoresis provides a method of fractionation that can be used to isolate individual components in mixtures of proteins or nucleic acids.

More recently, the analytical separation of proteins has been enhanced with the development of two-dimensional electrophoresis techniques, in which separation in the first dimension is by net charge, and in the second dimension by size and molecular weight.

In practice, electrophoresis employs a buffer system; a medium (paper, cellulose acetate, starch gel, agarose gel, or polyacrylamide gel); and a source of direct current. Samples are applied, and current is passed through the system for an appropriate time. Following migration of the molecules, the gel or paper may be treated with selective stains to reveal the location of the separated components (Figure A.10).

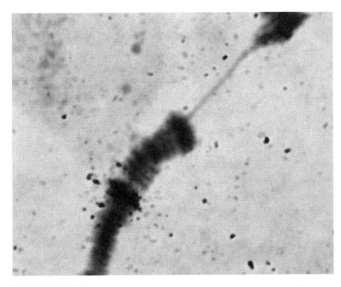

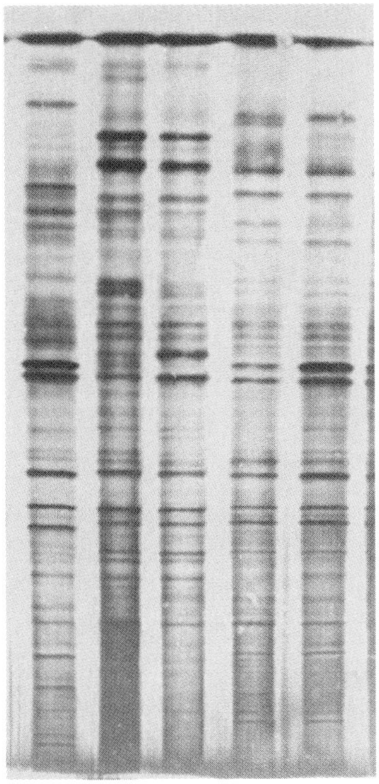

FIGURE A.9 Light micrograph of *in situ* hybridization of radioactive 5S RNA to a single band of polytene chromosome in the salivary gland of a *Drosophila* larva.

FIGURE A.10 Coomassie blue-stained protein slab gel.

DNA Sequencing

The ability to determine directly the sequence of DNA segments has added immensely to our understanding of gene structure and the mechanisms of gene regulation. Although techniques were available in the 1940s to determine the base composition of DNA, it was not until the 1960s that methods providing direct chemical analysis of nucleotide sequence were developed and used. These first methods were based on those used in determining the amino acid sequence of proteins, and were time-consuming and laborious. For example, in 1965 Robert Holley determined the sequence of a tRNA molecule consisting of 74 nucleotides, a task that required almost a year of concentrated effort.

In the 1970s more efficient and direct methods of analysis of nucleotide sequencing were developed. These methods developed in parallel with recombinant DNA techniques that allow the isolation of large quantities of purified DNA segments from any organism. A chemical method, developed by Allan Maxam and Walter Gilbert, cleaves DNA at specific bases. A second method, developed by Fred Sanger and colleagues, synthesizes a stretch of DNA that terminates at a given base. In both methods, the DNA to be sequenced is subjected to four individual reactions (one for each base). The products of the four reactions are a series of DNA fragments that differ in length by only one nucleotide. These reaction products are electrophoretically separated in four adjacent lanes on a gel. Each band on the gel corresponds to a base, and the sequence of the DNA segment can be read from the bands on the gel.

In the Maxam–Gilbert method, the complementary strands of the DNA fragment to be sequenced are separated and recovered. The strand to be sequenced is labeled at its 5′-end with radioactive ^{32}P using the enzyme polynucleotide kinase. This provides a means of identifying specific DNA fragments after gel electrophoresis. In the next step, aliquots of the DNA are subjected to each

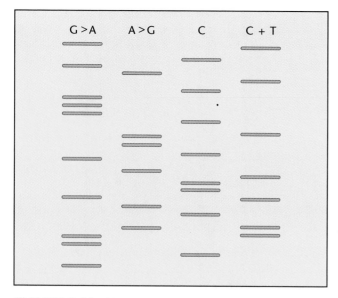

FIGURE A.11   Nucleotide sequence derived by the Maxam–Gilbert method.

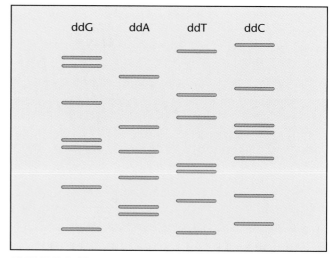

FIGURE A.12   Nucleotide sequence derived by the Sanger method.

of four chemical treatments that cleave the strand at a specific nucleotide. The reaction is carried out for a limited time so that any given molecule is cleaved at only a small number of the target nucleotides. The result is a collection of fragments, all labeled at the 5′-end, but differing in length, depending on the point of cleavage.

Four different reactions cleave DNA at guanine (G > A), adenine (A > G), cytosine alone (C), or cytosine and thymine (C + T). In these reactions, purines are cleaved by using dimethylsulfate. This reagent methylates guanine far more efficiently than adenine, and when heat is applied, the strand is broken at the methylated site, producing a DNA fragment most frequently cleaved at a G residue (G > A). This can be reversed by cleaving the strand in acid, producing fragments most often cleaved at an A residue (A > G). The reaction for pyrimidines uses hydrazine, which cleaves both cytosine and thymine; however, in high salt (2M NaCl), only cytosine reacts. Thus, in the two reactions, one represents only cytosine (C), and the other represents cytosine and thymine (C + T).

For each set of reactions, the fragments produced are then subjected to gel electrophoresis in adjacent lanes on a gel. Each reaction contains a series of fragments in which strand breakage has occurred at one of the target bases, resulting in a series of fragments that differ in length. Under the influence of the electric field during electrophoresis, the different-sized fragments separate from each other, with the smallest fragments migrating the farthest. Since the DNA fragments contain a radioactive 5′-end, the gel is subjected to autoradiography by placing a sheet of X-ray film over the gel for an appropriate exposure time. The DNA sequence is then analyzed by reading the bands on the gel from the bottom up, reading across all four lanes. In the example shown in Figure A.11, the first bases on the gel read GCGG.

In the Sanger method for DNA sequencing, which relies on an initial enzymatic treatment, the DNA strand to be sequenced is used as a template for the synthesis of a new DNA strand catalyzed by DNA polymerase I. This method employs a modified dideoxynucleoside triphosphate to generate a series of DNA fragments. The dideoxynucleoside triphosphates lack a 3'-OH group. This allows them to be added to a DNA strand undergoing synthesis, but since they lack a 3'-OH group, no nucleotide can be added to them, causing termination of strand synthesis, producing a DNA fragment. In use, four separate reaction mixtures are set up, each with a template DNA strand, a primer, all four radioactive nucleoside triphosphates, and a small amount of a single dideoxyribonucleoside triphosphate. Each of the four reactions contains a different dideoxynucleoside triphosphate that acts as a chain terminator. Because only a small amount of the modified nucleoside is used in each reaction, the newly synthesized strands are randomly terminated, producing a collection of fragments.

After synthesis, the radioactive fragments are electrophoresed in four adjacent lanes, one corresponding to each of the reactions. The fragments are visualized by autoradiography, and the gel is analyzed by reading the sequence from the bottom, as in the Maxam–Gilbert reaction. In the example shown in Figure A.12, the first band is in the lane with ddT, so it is a T residue, and the next few bands have the sequence GCAATCG.

DNA sequencing methods have generated a great deal of information about the structure and organization of many genes in a wide range of organisms. In the case of some viruses, the sequence of the entire genome is known, and large portions of the genome of other organisms, including *E. coli*, have been sequenced. To date, only a very small portion of the human genome has been sequenced, but the U.S. Department of Energy and the National Institutes of Health are coordinating efforts to develop the technology to sequence the more than 3 billion bases that constitute the haploid human genome.

The Polymerase Chain Reaction

The polymerase chain reaction (PCR) technique, described in Chapter 15, relies on the fact that the enzyme DNA polymerase requires a double-stranded primer section of DNA in order to initiate DNA synthesis. In the PCR technique, that primer is supplied in the form of a synthetic oligonucleotide, allowing the replication of a specific region of DNA. In other words, DNA polymerase can be directed to replicate only one specific region from an entire genome. Because these same primer sites are copied onto the newly synthesized strands of DNA, after each round of replication they can be used as primers. By repeating the replication process, a region of DNA can be amplified millions or billions of times, producing enough copies of the DNA for experimental purposes without the need for cloning (Figure A.13).

Instead of a single technique, the PCR reaction has become a versatile tool used in combination with other methods and has become the basis of a technical revolution in molecular genetics. The PCR method has been used to detect the presence of mutations, produce *in vitro* mutations, diagnose genetic disorders, prepare DNA for sequencing, identify viruses and bacteria in infectious dis-

eases, amplify DNA from fossils, analyze genetic defects in gametes and single cells from human embryos, and a host of other applications.

Mutation Detection

A PCR-based method called single-strand conformation polymorphism (SSCP) has been used to screen for mutations caused by single base substitutions. Under certain conditions, single-stranded DNA fragments fold into nucleotide-sequence-dependent conformations. These conformations affect the mobility of the fragment during electrophoresis. Single base substitutions change the conformation of the DNA, altering its electrophoretic mobility. The method is fast, simple, and does not require DNA sequencing to detect single base changes.

To detect mutations using SSCP, the DNA of interest is first amplified by PCR. The double-stranded amplified DNA is then converted to a single-stranded form by boiling and is loaded onto a gel for electrophoresis. DNA from a normal gene is similarly amplified, denatured, and loaded in adjacent lanes. After electrophoresis, any variations in base sequence in the DNA being tested can be detected as a shifted band. The method gives best results when the DNA fragments being tested are around 200 base pairs, so screening an entire gene requires the use of several to many fragments, each about 200 base pairs in length.

In the development of new PCR-based techniques, one technique is often piggy-backed onto another. While SSCP is a rapid and simple technique, its accuracy is somewhat limited. DNA sequencing is a more accurate technique for detecting single base changes, but it is relatively slow and laborious. Recently, these two techniques have been combined into a method called dideoxy fingerprinting (ddF). The technique begins by amplifying a DNA segment by PCR. The amplified segment is used as a template for one of the Sanger dideoxy sequencing reactions, and the gel is run under conditions where the position of the band reflects both size (like DNA sequencing) and conformation (like SSCP). This combined technique is three times faster than sequencing alone and although more complex than SSCP, has a degree of accuracy approaching 100 percent.

PCR Cycle Sequencing

Conventional DNA sequencing requires subcloning of fragments into vectors such as a plasmid, followed by growth of a host colony and extraction and purification of the cloned DNA. This DNA is then used as a template in the four reactions for DNA sequencing, followed by gel electrophoresis. The technique of cycle sequencing incorporates parts of the PCR reaction with parts of the standard reaction for the dideoxy method of DNA sequencing. The double-stranded DNA segment to be sequenced is first heated to form single strands that serve as templates. Primers are added, and DNA polymerase then begins the process of DNA replication. Extension of the primer continues until a labeled dideoxynucleotide is incorporated. The labeled strand is separated from the template (by heating) and is then used in another round of replication. The amplified extension products are then analyzed by gel electrophoresis. This adaptation of DNA sequencing is faster than conventional sequencing, requires only a very small amount of template, and can be used to directly sequence plasmid or phage clones. To further simplify the process, commercial kits are available that provide all materials and solutions for this procedure.

Versatility: The Advantage of PCR

The examples above illustrate the versatility of the polymerase chain reaction and some of its applications in molecular genetics. This remarkable technique has so many adaptations and variations that several journals are devoted to reports on PCR. As evidence of its impact on molecular genetics, between 1991 and 1993 there were over 16,000 research reports on PCR.

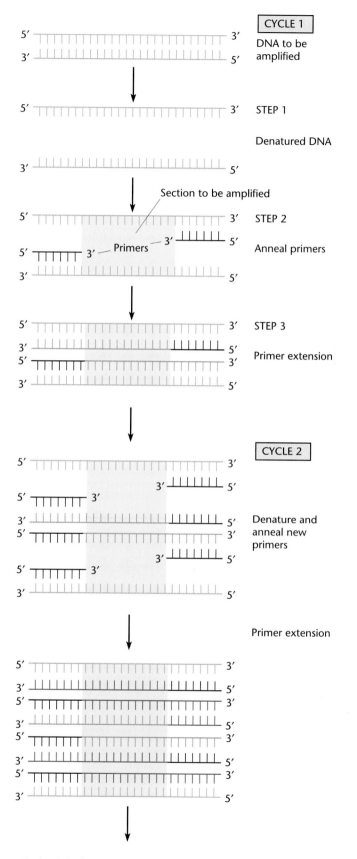

FIGURE A.13   PCR amplification. In the polymerase chain reaction (PCR) method, the DNA to be amplified is first denatured into single strands, and then each strand is annealed to a different primer. The primers are synthetic oligonucleotides complementary to sequences flanking the region to be amplified. DNA polymerase is added along with nucleotides to extend the primers in the 3′ direction, resulting in a double-stranded DNA molecule with the primers incorporated into the newly synthesized strand. In a second PCR cycle, the products of the first cycle are denatured into single strands, primers are added, and they are annealed and extended by DNA polymerase. Repeated cycles can amplify the original DNA sequence by more than a million times.

A-DNA An alternative form of the right-handed double-helical structure of DNA in which the helix is more tightly coiled, with 11 base pairs per full turn of the helix. In this form, the bases in the helix are displaced laterally and tilted in relation to the longitudinal axis. It is not yet clear whether this form has biological significance.

abortive transduction An event in which transducing DNA fails to be incorporated into the recipient chromosome. (See *transduction.*)

acentric chromosome Chromosome or chromosome fragment with no centromere.

acquired immunodeficiency syndrome (AIDS) An infectious disease caused by a retrovirus designated as human immunodeficiency virus (HIV). The disease is characterized by a gradual depletion of T lymphocytes, recurring fever, weight loss, multiple opportunistic infections, and rare forms of pneumonia and cancer associated with collapse of the immune system.

acridine dyes A class of organic compounds that bind to DNA and intercalate into the double-stranded structure, producing local disruptions of base pairing. These disruptions result in additions or deletions in the next round of replication.

acrocentric chromosome Chromosome with the centromere located very close to one end. Human chromosomes 13, 14, 15, 21, and 22 are acrocentric.

active immunity Immunity gained by direct exposure to antigens followed by antibody production.

active site That portion of a protein, usually an enzyme, whose structural integrity is required for function (e.g., the substrate binding site of an enzyme).

adaptation A heritable component of the phenotype that confers an advantage in survival and reproductive success. The process by which organisms adapt to the current environmental conditions.

additive genes See *polygenic inheritance.*

additive variance Genetic variance that is attributed to the substitution of one allele for another at a given locus. This variance can be used to predict the rate of response to phenotypic selection in quantitative traits.

albinism A condition caused by the lack of melanin production in the iris, hair, and skin. In humans, most often inherited as an autosomal recessive.

aleurone layer In seeds, the outer layer of the endosperm.

alkaptonuria An autosomal recessive condition in humans caused by the lack of an enzyme, homogentisic acid oxidase. Urine of homozygous individuals turns dark upon standing due to oxidation of excreted homogentisic acid. The cartilage of homozygous adults blackens from deposition of a pigment derived from homogentisic acid. Such individuals often develop arthritic conditions.

allele One of the possible mutational states of a gene, distinguished from other alleles by phenotypic effects.

allele frequency Measurement of the proportion of individuals in a population carrying a particular allele.

allele-specific nucleotide (ASO) Synthetic nucleotides, usually 15 to 20 bp in length that by hybridization under carefully controlled conditions will hybridize only to a complementary sequence with a perfect match. Under these same conditions, ASOs with a one-nucleotide mismatch will not hybridize.

allelic exclusion In plasma cell heterozygous for an immunoglobulin gene, the selective action of only one allele.

allelism test See *complementation test.*

allopatric speciation Process of speciation associated with geographic isolation.

allopolyploid Polyploid condition formed by the union of two or more distinct chromosome sets with a subsequent doubling of chromosome number.

allosteric effect Conformational change in the active site of a protein brought about by interaction with an effector molecule.

allotetraploid Diploid for two genomes derived from different species.

allozyme An allelic form of a protein that can be distinguished from other forms by electrophoresis.

alpha fetoprotein (AFP) A 70-kd glycoprotein synthesized in embryonic development by the yolk sac. High levels of this protein in the amniotic fluid are associated with neural tube defects such as spina bifida. Lower than normal levels may be associated with Down syndrome.

***Alu* sequence** An interspersed DNA sequence of approximately 300 bp found in the genome of primates that is cleaved by the restriction enzyme *Alu* I. *Alu* sequences are composed of a head-to-tail dimer, with the first monomer approximately 140 bp and the second approximately 170 bp. In humans, they are dispersed throughout the genome and are present in 300,000 to 600,000 copies, constituting some 3 to 6 percent of the genome. See *SINEs.*

amber codon The codon UAG, which does not code for an amino acid but for chain termination.

Ames test An assay developed by Bruce Ames to detect mutagenic and carcinogenic compounds using reversion to histidine independence in the bacterium *Salmonella typhimurium.*

amino acid Any of the subunit building blocks that are covalently linked to form proteins.

aminoacyl tRNA Covalently linked combination of an amino acid and a tRNA molecule.

amniocentesis A procedure used to test for fetal defects in which fluid and fetal cells are withdrawn from the amniotic layer surrounding the fetus.

amphidiploid See *allotetraploid*.

anabolism The metabolic synthesis of complex molecules from less complex precursors.

analogue A chemical compound structurally similar to another, but differing by a single functional group (e.g., 5-bromodeoxyuridine is an analogue of thymidine).

anaphase Stage of cell division in which chromosomes begin moving to opposite poles of the cell.

aneuploidy A condition in which the chromosome number is not an exact multiple of the haploid set.

angstrom Unit of length equal to 10^{-10} meter. Abbreviated Å.

antibody Protein (immunoglobulin) produced in response to an antigenic stimulus with the capacity to bind specifically to the antigen.

anticipation A phenomenon first observed in myotonic dystrophy, where the severity of the symptoms increases from generation to generation and the age of onset decreases from generation to generation. This phenomenon is caused by the expansion of trinucleotide repeats within or near a gene.

anticodon The nucleotide triplet in a tRNA molecule that is complementary to, and binds to, the codon triplet in an mRNA molecule.

antigen A molecule, often a cell surface protein, that is capable of eliciting the formation of antibodies.

antiparallel Describing molecules in parallel alignment, but running in opposite directions. Most commonly used to describe the opposite orientations of the two strands of a DNA molecule.

apoenzyme The protein portion of an enzyme that requires a cofactor or prosthetic group to be functional.

apoptosis A genetically controlled program of cell death, activated as part of normal development, or as a result of cell damage.

ascospore A meiotic spore produced in certain fungi.

ascus In fungi, the sac enclosing the four or eight ascospores.

asexual reproduction Production of offspring in the absence of any sexual process.

assortative mating Nonrandom mating between males and females of a species. Selection of mates with the same genotype is positive; selection of mates with opposite genotypes is negative.

ATP Adenosine triphosphate.

attached-X chromosome Two conjoined X chromosomes that share a single centromere.

attenuator A nucleotide sequence between the promoter and the structural gene of some operons that can act to regulate the transit of RNA polymerase and thus control transcription of the structural gene.

autogamy A process of self-fertilization resulting in homozygosis.

autoimmune disease The production of antibodies that results from an immune response to one's own molecules, cells, or tissues. Such a response results from the inability of the immune system to distinguish self from nonself. Diseases such as arthritis, scleroderma, systemic lupus erythematosus, and perhaps diabetes are considered to be autoimmune diseases.

autopolyploidy Polyploid condition resulting from the replication of one diploid set of chromosomes.

autoradiography Production of a photographic image by radioactive decay. Used to localize radioactively labeled compounds within cells and tissues.

autosomes Chromosomes other than the sex chromosomes. In humans, there are 22 pairs of autosomes.

autotetraploid An autopolyploid condition composed of four similar genomes. In this situation, genes with two alleles (*A* and *a*) can have five genotypic classes: *AAAA* (quadraplex), *AAAa* (triplex), *AAaa* (duplex), *Aaaa* (simplex), and *aaaa* (nulliplex).

auxotroph A mutant microorganism or cell line which requires a substance for growth that can be synthesized by wild-type strains.

B-DNA See *double helix*.

back-cross A cross involving an F_1 heterozygote and one of the P_1 parents (or an organism with a genotype identical to one of the parents).

bacteriophage A virus that infects bacteria (synonym is *phage*).

bacteriophage mu A group of phages whose genetic material behaves as an insertion sequence that can cause inactivation of host genes and rearrangement of host chromosomes.

bacteriostatic A compound that inhibits the growth of bacteria, but does not kill them.

balanced lethals Recessive, nonallelic lethal genes, each carried on different homologous chromosomes. When organisms carrying balanced lethal genes are interbred, only organisms with genotypes identical to the parents (heterozygotes) survive.

balanced polymorphism Genetic polymorphism maintained in a population by natural selection.

Barr body Densely staining nuclear mass seen in the somatic nuclei of mammalian females. Discovered by Murray Barr, this body is thought to represent an inactivated X chromosome.

base analogue See *analogue*.

base substitution A single base change in a DNA molecule that produces a mutation. There are two types of substitutions: *transitions*, in which a purine is substituted for a purine or a pyrimidine for a pyrimidine; and *transversions*, in which a purine is substituted for a pyrimidine, or vice versa.

bidirectional replication A mechanism of DNA replication in which two replication forks move in opposite directions from a common origin of replication.

biometry The application of statistics and statistical methods to biological problems.

biotechnology Commercial and/or industrial processes that utilize biological organisms or products.

bivalents Synapsed homologous chromosomes in the first prophase of meiosis.

Bombay phenotype A rare variant of the ABO system in which affected individuals do not have A or B antigens, and thus appear as blood type O, even though their genotype may carry unexpressed alleles for the A and/or B antigens.

BrdU (5-bromodeoxyuridine) A mutagenically active analogue of thymidine in which the methyl group at the 5′ position in thymine is replaced by bromine.

buoyant density A property of particles (and molecules) that depends upon their actual density, as determined by partial specific volume and degree of hydration. Provides the basis for density gradient separation of molecules or particles.

CAAT box A highly conserved DNA sequence found about 75 base pairs 5′ to the site of transcription in eukaryotic genes.

cAMP Cyclic adenosine monophosphate. An important regulatory molecule in both prokaryotic and eukaryotic organisms.

canonical sequence See *consensus sequence*.

CAP Catabolite activator protein. A protein that binds cAMP and regulates the activation of inducible operons.

carcinogen A physical or chemical agent that causes cancer.

carrier An individual heterozygous for a recessive trait.

cassette model First proposed to explain mating type interconversion in yeast, this model proposes that both genes for mating types, *a* and *alpha* are present as silent or unexpressed genes in transposable DNA segments (cassettes) that are activated (played) by transposition to the mating type locus.

catabolism A metabolic reaction in which complex molecules are broken down into simpler forms, often accompanied by the release of energy.

catabolite activator protein See *CAP*.

catabolite repression The selective inactivation of an operon by a metabolic product of the enzymes encoded by the operon.

cdc **mutation** A class of mutations in yeast that affect the timing and progression through the cell cycle.

cDNA DNA synthesized from an RNA template by the enzyme reverse transcriptase.

cell cycle Sum of the phases of growth of an individual cell type; divided into G1 (gap 1), S (DNA synthesis), G2 (gap 2), and M (mitosis).

cell-free extract A preparation of the soluble fraction of cells, made by lysing cells and removing the particulate matter, such as nuclei, membranes, and organelles. Often used to carry out the synthesis of proteins by the addition of specific, exogenous mRNA molecules.

CEN In yeast, fragments of chromosomal DNA, about 120 bp in length, that when inserted into plasmids confer the ability to segregate during mitosis. These segments contain at least three types of sequence elements associated with centromere function.

centimeter A unit of length equal to 10^{-2} meter. Abbreviated cm.

centimorgan A unit of distance between genes on chromosomes. One centimorgan represents a value of 1 percent crossing over between two genes.

central dogma The concept that information flow progresses from DNA to RNA to proteins. Although exceptions are known, this idea is central to an understanding of gene function.

centric fusion See *Robertsonian translocation*.

centriole A cytoplasmic organelle composed of nine groups of microtubules, generally arranged in triplets. Centrioles function in the generation of cilia and flagella and serve as foci for the spindles in cell division.

centromere Specialized region of a chromosome to which the spindle fibers attach during cell division. Location of the centromere determines the shape of the chromosome during the anaphase portion of cell division. Also known as the primary constriction.

centrosome Region of the cytoplasm containing the centriole.

character An observable phenotypic attribute of an organism.

charon phages A group of genetically modified lambda phages designed to be used as vectors for cloning foreign DNA. Named after the ferryman in Greek mythology who carried the souls of the dead across the River Styx.

chemotaxis Negative or positive response to a chemical gradient.

chiasma (pl., **chiasmata**) The crossed strands of nonsister chromatids seen in diplotene of the first meiotic division. Regarded as the cytological evidence for exchange of chromosomal material, or crossing over.

chiasmatype theory The theory that crossing over between nonsister chromatids is the cause of chiasma formation.

chi-square (χ^2) analysis Statistical test to determine if an observed set of data fits a theoretical expectation.

chloroplast A cytoplasmic self-replicating organelle containing chlorophyll. The site of photosynthesis.

chorionic villus sampling (CVS) A technique of prenatal diagnosis that intravaginally retrieves fetal cells from the chorion and uses them to detect cytogenetic and biochemical defects in the embryo.

chromatid One of the longitudinal subunits of a replicated chromosome, joined to its sister chromatid at the centromere.

chromatin Term used to describe the complex of DNA, RNA, histones, and nonhistone proteins that make up chromosomes.

chromatography Technique for the separation of a mixture of solubilized molecules by their differential migration over a substrate.

chromocenter   An aggregation of centromeres and heterochromatic elements of polytene chromosomes.

chromomere   A coiled, beadlike region of a chromosome most easily visible during cell division. The aligned chromomeres of polytene chromosomes are responsible for their distinctive banding pattern.

chromosomal aberration   Any change resulting in the duplication, deletion, or rearrangement of chromosomal material.

chromosomal mutation   See *chromosomal aberration.*

chromosomal polymorphism   Alternative structures or arrangements of a chromosome that are carried by members of a population.

chromosome   In prokaryotes, an intact DNA molecule containing the genome; in eukaryotes, a DNA molecule complexed with RNA and proteins to form a threadlike structure containing genetic information arranged in a linear sequence.

chromosome banding   Technique for the differential staining of mitotic or meiotic chromosomes to produce a characteristic banding pattern or selective staining of certain chromosomal regions such as centromeres, the nucleolus organizer regions, and GC- or AT-rich regions. Not to be confused with the banding pattern present in unstained polytene chromosomes, which is produced by the alignment of chromomeres.

chromosome map   A diagram showing the location of genes on chromosomes.

chromosome puff   A localized uncoiling and swelling in a polytene chromosome, usually regarded as a sign of active transcription.

chromosome theory of inheritance   The idea put forward by Walter Sutton and Theodore Boveri that chromosomes are the carriers of genes and the basis for the Mendelian mechanisms of segregation and independent assortment.

chromosome walking   A method for analyzing long stretches of DNA, in which the end of a cloned segment of DNA is subcloned and used as a probe to identify other clones that overlap the first clone.

***cis* configuration**   The arrangement of two mutant sites within a gene on the same homolog, such as

$$\frac{a^1 \; a^2}{+ \; +}$$

Contrasts with a *trans* arrangement, where the mutant alleles are located on opposite homologs.

***cis* dominance**   The ability of a gene to affect the expression of other genes adjacent to it on the chromosome.

***cis-trans* test**   A genetic test to determine whether two mutations are located within the same cistron.

cistron   That portion of a DNA molecule that codes for a single polypeptide chain; defined by a genetic test as a region within which two mutations cannot complement each other.

cline   A gradient of genotype or phenotype distributed over a geographic range.

clonal selection   Theory of the immune system that proposes that antibody diversity precedes exposure to the antigen, and that the antigen functions to select the cells containing its specific antibody to undergo proliferation.

clone   Genetically identical cells or organisms all derived from a single ancestor by asexual or parasexual methods. For example, a DNA segment that has been enzymatically inserted into a plasmid or chromosome of a phage or a bacterium and replicated to form many copies.

cloned library   A collection of cloned DNA molecules representing all or part of an individual's genome.

code   See *genetic code.*

codominance   Condition in which the phenotypic effects of a gene's alleles are fully and simultaneously expressed in the heterozygote.

codon   A triplet of bases in a DNA or RNA molecule that specifies or encodes the information for a single amino acid.

coefficient of coincidence   A ratio of the observed number of double-crossovers divided by the expected number of such crossovers.

coefficient of inbreeding   The probability that two alleles present in a zygote are descended from a common ancestor.

coefficient of selection   A measurement of the reproductive disadvantage of a given genotype in a population. If for genotype *aa*, only 99 of 100 individuals reproduce, then the selection coefficient (*s*) is 0.1.

colchicine   An alkaloid compound that inhibits spindle formation during cell division. Used in the preparation of karyotypes to collect a large population of cells inhibited at the metaphase stage of mitosis.

colicin   A bacteriocidal protein produced by certain strains of *E. coli* and other closely related bacterial species.

colinearity   The linear relationship between the nucleotide sequence in a gene (or the RNA transcribed from it) and the order of amino acids in the polypeptide chain specified by the gene.

competence   In bacteria, the transient state or condition during which the cell can bind and internalize exogenous DNA molecules, making transformation possible.

complementarity   Chemical affinity between nitrogenous bases as a result of hydrogen bonding. Responsible for the base pairing between the strands of the DNA double helix.

complementation test   A genetic test to determine whether two mutations occur within the same gene. If two mutations are introduced into a cell simultaneously and produce a wild-type phenotype (i.e., they complement each other), they are often nonallelic. If a mutant phenotype is produced, the mutations are noncomplementing and are often allelic.

complete linkage   A condition in which two genes are located so close to each other that no recombination occurs between them.

complexity   The total number of nucleotides or nucleotide pairs in a population of nucleic acid molecules as determined by reassociation kinetics.

complex locus   A gene within which a set of functionally related pseudoalleles can be identified by recombinational analysis (e.g., the *bithorax* locus in *Drosophila*).

concatemer A chain or linear series of subunits linked together. The process of forming a concatemer is called concatenation (e.g., multiple units of a phage genome produced during replication).

concordance Pairs or groups of individuals identical in their phenotype. In twin studies, a condition in which both twins exhibit or fail to exhibit a trait under investigation.

conditional mutation A mutation that expresses a wild-type phenotype under certain (permissive) conditions and a mutant phenotype under other (restrictive) conditions.

conjugation Temporary fusion of two single-celled organisms for the sexual exchange of genetic material.

consanguine Related by a common ancestor within the previous few generations.

consensus sequence A nucleotide sequence most often found in a defined segment of DNA.

continuous variation Phenotype variation exhibited by quantitative traits distributed from one phenotypic extreme to another in an overlapping or continuous fashion.

cosmid A vector designed to allow cloning of large segments of foreign DNA. Cosmids are hybrids composed of the cos sites of lambda inserted into a plasmid. In cloning, the recombinant DNA molecules are packaged into phage protein coats, and after infection of bacterial cells, the recombinant molecule replicates and can be maintained as a plasmid.

coupling conformation See *cis configuration*.

covalent bond A nonionic chemical bond formed by the sharing of electrons.

cri-du-chat syndrome A clinical syndrome in humans produced by a deletion of a portion of the short arm of chromosome 5. Afflicted infants have a distinctive cry which sounds like that of a cat.

crossing over The exchange of chromosomal material (parts of chromosomal arms) between homologous chromosomes by breakage and reunion. The exchange of material between nonsister chromatids during meiosis is the basis of genetic recombination.

cross-reacting material (CRM) Nonfunctional form of an enzyme, produced by a mutant gene, which is recognized by antibodies made against the normal enzyme.

C-terminal amino acid The terminal amino acid in a peptide chain which carries a free carboxyl group.

C terminus The end of a polypeptide that carries a free carboxyl group of the last amino acid. By convention, the structural formula of polypeptides is written with the C terminus at the right.

C **value** The haploid amount of DNA present in a genome.

C **value paradox** The apparent paradox that there is no relationship between the size of the genome and the evolutionary complexity of species. For example, the *C* value (haploid genome size) of amphibians varies by a factor of 100.

cyclins A class of proteins found in eukaryotic cells that are synthesized and degraded in synchrony with the cell cycle, and regulate passage through stages of the cycle.

cytogenetics A branch of biology in which the techniques of both cytology and genetics are used to study heredity.

cytokinesis The division or separation of the cytoplasm during mitosis or meiosis.

cytological map A diagram showing the location of genes at particular chromosomal sites.

cytoplasmic inheritance Non-Mendelian form of inheritance involving genetic information transmitted by self-replicating cytoplasmic organelles such as mitochondria, chloroplasts, etc.

cytoskeleton An internal array of microtubules, microfilaments, and intermediate filaments that confers shape and the ability to move on a eukaryotic cell.

dalton A unit of mass equal to that of the hydrogen atom, which is 1.67×10^{-24} gram. A unit used in designating molecular weights.

Darwinian fitness See *fitness*.

deficiency (deletion) A chromosomal mutation involving the loss or deletion of chromosomal material.

degenerate code Term used to describe the genetic code, in which a given amino acid may be represented by more than one codon.

deletion See *deficiency*.

deme A local interbreeding population.

denatured DNA DNA molecules that have been separated into single strands.

de novo Newly arising; synthesized from less complex precursors rather than having been produced by modification of an existing molecule.

density gradient centrifugation A method of separating macromolecular mixtures by the use of centrifugal force and solvents of varying density. In sedimentation velocity centrifugation, macromolecules are separated by the velocity of sedimentation through a preformed gradient such as sucrose. In density gradient equilibrium centrifugation, macromolecules in a cesium salt solution are centrifuged until the cesium solution establishes a gradient under the influence of the centrifugal field, and the macromolecules sediment until the density of the solvent equals their own.

deoxyribonuclease A class of enzymes that breaks down DNA into oligonucleotide fragments by introducing single-stranded breaks into the double helix.

deoxyribonucleic acid (DNA) A macromolecule usually consisting of antiparallel polynucleotide chains held together by hydrogen bonds, in which the sugar residues are deoxyribose. The primary carrier of genetic information.

deoxyribose The five-carbon sugar associated with the deoxyribonucleotides found in DNA.

dermatoglyphics The study of the surface ridges of the skin, especially of the hands and feet.

determination A regulatory event that establishes a specific pattern of gene activity and developmental fate for a given cell.

diakinesis The final stage of meiotic prophase I in which the chromosomes become tightly coiled and compacted and move toward the periphery of the nucleus.

dicentric chromosome A chromosome having two centromeres.

differentiation The process of complex changes by which cells and tissues attain their adult structure and functional capacity.

dihybrid cross A genetic cross involving two characters in which the parents possess different forms of each character (e.g., tall, round × short, wrinkled peas).

diploid A condition in which each chromosome exists in pairs; having two of each chromosome.

diplotene A stage of meiotic prophase I immediately after pachytene. In diplotene, one pair of sister chromatids begins separating from the other, and chiasmata become visible. These overlaps move laterally toward the ends of the chromatids (terminalization).

directional selection A selective force that changes the frequency of an allele in a given direction, either toward fixation or toward elimination.

discontinuous replication of DNA The synthesis of DNA in discontinuous fragments on the lagging strand of the replication fork. The fragments, known as Okazaki fragments, are joined by DNA ligase to form a continuous strand.

discontinuous variation Phenotypic data that fall into two or more distinct classes that do not overlap.

discordance In twin studies, a situation where one twin shows a trait but the other does not.

disjunction The separation of chromosomes at the anaphase stage of cell division.

disruptive selection Simultaneous selection for phenotypic extremes in a population, usually resulting in the production of two discontinuous strains.

dizygotic twins Twins produced from separate fertilization events; two ova fertilized independently. Also known as fraternal twins.

DNA See *deoxyribonucleic acid*.

DNA footprinting See *footprinting*.

DNA gyrase One of the DNA topoisomerases that functions during DNA replication to reduce molecular tension caused by supercoiling. DNA gyrase produces, then seals, double-stranded breaks.

DNA ligase An enzyme that forms a covalent bond between the 5′-end of one polynucleotide chain and the 3′-end of another polynucleotide chain. Also called polynucleotide-joining enzyme.

DNA polymerase An enzyme that catalyzes the synthesis of DNA from deoxyribonucleotides and a template DNA molecule.

DNase Deoxyribonucleosidase, an enzyme that degrades or breaks down DNA into fragments or constitutive nucleotides.

dominance The expression of a trait in the heterozygous condition.

dosage compensation A genetic mechanism that regulates the levels of gene products at certain autosomal loci; this results in homozygous dominants and heterozygotes having the same amount of a gene product. In mammals, random inactivation of one X chromosome in females leads to equal levels of X chromosome-coded gene products in males and females.

double-crossover Two separate events of chromosome breakage and exchange occurring within the same tetrad.

double helix The model for DNA structure proposed by James Watson and Francis Crick, involving two antiparallel, hydrogen-bonded polynucleotide chains wound into a right-handed helical configuration, with 10 base pairs per full turn of the double helix. Often called B-DNA.

Duchenne muscular dystrophy An X-linked recessive genetic disorder caused by a mutation in the gene for dystrophin, a protein found in muscle cells.

duplication A chromosomal aberration in which a segment of the chromosome is repeated.

dyad The products of tetrad separation or disjunction at the first meiotic prophase. Consists of two sister chromatids joined at the centromere.

dystrophin See *Duchenne muscular dystrophy*.

effector molecule Small, biologically active molecule that acts to regulate the activity of a protein by binding to a specific receptor site on the protein.

electrophoresis A technique used to separate a mixture of molecules by their differential migration through a stationary phase in an electrical field.

endocytosis The uptake by a cell of fluids, macromolecules, or particles by pinocytosis, phagocytosis, or receptor-mediated endocytosis.

endogenote The segment of the chromosome in a partially diploid bacterial cell (merozygote) that is homologous to the chromosome transmitted by the donor cell.

endomitosis Chromosomal replication that is not accompanied by either nuclear or cytoplasmic division.

endonuclease An enzyme that hydrolyzes internal phosphodiester bonds in a polynucleotide chain or nucleic acid molecule.

endoplasmic reticulum A membraneous organelle system in the cytoplasm of eukaryotic cells. The outer surface of the membranes may be ribosome-studded (rough ER) or smooth ER.

endopolyploidy The increase in chromosome sets that results from endomitotic replication within somatic nuclei.

endosymbiont theory The proposal that self-replicating cellular organelles such as mitochondria and chloroplasts were originally free-living organisms that entered into a symbiotic relationship with nucleated cells.

enhancer Originally, a 72-bp sequence in the genome of the virus, SV40, that increases the transcriptional activity of nearby structural genes. Similar sequences that enhance transcription have been identified in the genomes of eukaryotic cells. Enhancers can act over a distance of thousands of base pairs and can be located 5′ or 3′ to the gene they affect, and thus are different from promoters.

enhanson The DNA sequence that represents the core sequence of an enhancer.

environment The complex of geographic, climatic, and biotic factors within which an organism lives.

enzyme A protein or complex of proteins that catalyzes a specific biochemical reaction.

epigenesis The idea that an organism develops by the appearance and growth of new structures. Opposed to preformationism, which holds that development is the growth of structures already present in the egg.

episome A circular genetic element in bacterial cells that can replicate independently of the bacterial chromosome or integrate and replicate as part of the chromosome.

epistasis Nonreciprocal interaction between genes such that one gene interferes with or prevents the expression of another gene. For example, in *Drosophila*, the recessive gene *eyeless*, when homozygous, prevents the expression of eye color genes.

epitope That portion of a macromolecule or cell that acts to elicit an antibody response; an antigenic determinant. A complex molecule or cell can contain several such sites.

equational division A division of each chromosome into longitudinal halves that are distributed into two daughter nuclei. Chromosome division in mitosis is an example of equational division.

equatorial plate See *metaphase plate.*

euchromatin Chromatin or chromosomal regions that are lightly staining and are relatively uncoiled during the interphase portion of the cell cycle. The region of the chromosomes thought to contain most of the structural genes.

eugenics The improvement of the human species by selective breeding. Positive eugenics refers to the promotion of breeding of those with favorable genes, and negative eugenics refers to the discouragement of breeding among those with undesirable traits.

eukaryotes Those organisms having true nuclei and membranous organelles and whose cells demonstrate mitosis and meiosis.

euphenics Medical or genetic intervention to reduce the impact of defective genotypes.

euploid Polyploid with a chromosome number that is an exact multiple of a basic chromosome set.

evolution The origin of plants and animals from pre-existing types. Descent with modifications.

excision repair Repair of DNA lesions by removal of a polynucleotide segment and its replacement with a newly synthesized, corrected segment.

exogenote In merozygotes, the segment of the bacterial chromosome contributed by the donor cell.

exon (extron) The DNA segment(s) of a gene that are transcribed and translated into protein.

exonuclease An enzyme that breaks down nucleic acid molecules by breaking the phosphodiester bonds at the 3′ or 5′ terminal nucleotides.

expression vector Plasmids or phages carrying promoter regions designed to cause expression of cloned DNA sequences.

expressivity The degree or range in which a phenotype for a given trait is expressed.

extranuclear inheritance Transmission of traits by genetic information contained in cytoplasmic organelles such as mitochondria and chloroplasts.

F⁺ cell A bacterial cell having a fertility (F) factor. Acts as a donor in bacterial conjugation.

F⁻ cell A bacterial cell that does not contain a fertility (F) factor. Acts as a recipient in bacterial conjugation.

F factor An episome in bacterial cells that confers the ability to act as a donor in conjugation.

F′ factor A fertility (F) factor that contains a portion of the bacterial chromosome.

F₁ generation First filial generation; the progeny resulting from the first cross in a series.

F₂ generation Second filial generation; the progeny resulting from a cross of the F₁ generation.

F pilus See *pilus.*

facultative heterochromatin Chromatin that may alternate in form between euchromatic and heterochromatic. The Y chromosome of many species contains facultative heterochromatin.

familial trait A trait transmitted through and expressed by members of a family.

fate map A diagram or "map" of an embryo showing the location of cells whose development fate is known.

fertility (F) factor See *F factor.*

filial generations See *F₁, F₂ generations.*

fingerprint The pattern of ridges and whorls on the tip of a finger. The pattern obtained by enzymatically cleaving a protein or nucleic acid and subjecting the digest to two-dimensional chromatography or electrophoresis.

FISH See *fluorescence in situ hybridization.*

fitness A measure of the relative survival and reproductive success of a given individual or genotype.

fixation In population genetics, a condition in which all members of a population are homozygous for a given allele.

fixity of species The idea that members of a species can give rise only to other members of the species, thus implying that all species are independently created.

fluctuation test A statistical test developed by Salvadore Luria and Max Delbruck to determine whether bacterial mutations arise spontaneously or are produced in response to selective agents.

fluorescence *in situ* hybridization (FISH) A method of *in situ* hybridization that utilizes probes labeled with a fluorescent tag, causing the site of hybridization to fluoresce when viewed under the microscope.

fMet See *formylmethionine.*

folded-fiber model A model of eukaryotic chromosome organization in which each sister chromatid consists of a single fiber, composed of double-stranded DNA and protein, which is wound up like a tightly coiled skein of yarn.

footprinting A technique for identifying a DNA sequence that binds to a particular protein, based on the idea that the phosphodiester bonds in the region covered by the protein are protected from digestion by deoxyribonucleases.

formylmethionine (fMet) A molecule derived from the amino acid methionine by attachment of a formyl group to its terminal amino group. This is the first amino acid inserted in all bacterial polypeptides. Also known as *N*-formyl methionine.

founder effect A form of genetic drift. The establishment of a population by a small number of individuals whose genotypes carry only a fraction of the different kinds of alleles in the parental population.

fragile site A heritable gap or nonstaining region of a chromosome that can be induced to generate chromosome breaks.

fragile X syndrome A genetic disorder caused by the expansion of a CGG trinucleotide repeat and a fragile site at Xq27.3, within the FMR-1 gene.

frameshift mutation A mutational event leading to the insertion of one or more base pairs in a gene, shifting the codon reading frame in all codons following the mutational site.

fraternal twins See *dizygotic twins.*

G1 checkpoint A point in the G1 phase of the cell cycle when a cell becomes committed to initiate DNA synthesis and continue the cycle or withdraw into the G0 resting stage.

G0 The G-zero stage of the cell cycle is a point in G1 where cells withdraw from the cell cycle and enter a resting stage.

gamete A specialized reproductive cell with a haploid number of chromosomes.

gap genes Genes expressed in contiguous domains along the anterior–posterior axis of the *Drosophila* embryo, which regulate the process of segmentation in each domain.

gene The fundamental physical unit of heredity whose existence can be confirmed by allelic variants and which occupies a specific chromosomal locus. A DNA sequence coding for a single polypeptide.

gene amplification The process by which gene sequences are selected for differential replication either extrachromosomally or intrachromosomally.

gene conversion A directional process that changes one allele into another specific allele. In fungi, this process involves meiosis and a recombinational step, but in other organisms, such as trypanosomes, meiosis may not be required.

gene duplication An event in replication leading to the production of a tandem repeat of a gene sequence.

gene flow The gradual exchange of genes between two populations, brought about by the dispersal of gametes or the migration of individuals.

gene frequency The percentage of alleles of a given type in a population.

gene interaction Production of novel phenotypes by the interaction of alleles of different genes.

gene mutation See *point mutation.*

gene pool The total of all genes possessed by reproductive members of a population.

generalized transduction The transduction of any gene in the bacterial genome by a phage.

genetic anticipation The phenomenon of a progressively earlier age of onset for a genetic disorder in successive generations.

genetic background All genes carried in the genome other than the one being studied.

genetic burden Average number of recessive lethal genes carried in the heterozygous condition by an individual in a population. Also called genetic load.

genetic code The nucleotide triplets that code for the 20 amino acids or for chain initiation or termination.

genetic counseling Analysis of risk for genetic defects in a family and the presentation of options available to avoid or ameliorate possible risks.

genetic drift Random variation in gene frequency from generation to generation. Most often observed in small populations.

genetic engineering The technique of altering the genetic constitution of cells or individuals by the selective removal, insertion, or modification of individual genes or gene sets.

genetic equilibrium Maintenance of allele frequencies at the same value in successive generations. A condition in which allele frequencies are neither increasing nor decreasing.

genetic fine structure Intragenic recombinational analysis that provides mapping information at the level of individual nucleotides.

genetic load See *genetic burden.*

genetic polymorphism The stable coexistence of two or more discontinuous genotypes in a population. When the frequencies of two alleles are carried to an equilibrium, the condition is called balanced polymorphism.

genetics The branch of biology that deals with heredity and the expression of inherited traits.

genome The array of genes carried by an individual.

genomic imprinting A condition where the expression of a trait depends on whether the trait has been inherited from a male or a female parent.

genotype The specific allelic or genetic constitution of an organism; often, the allelic composition of one or a limited number of genes under investigation.

germplasm Hereditary material transmitted from generation to generation.

Goldberg–Hogness box A short nucleotide sequence 20 to 30 bp 5′ to the initiation site of eukaryotic genes to which RNA polymerase II binds. The consensus sequence is TATAAAA. Also known as TATA box.

graft versus host disease (GVHD) In transplants, reaction by immunologically competent cells of the donor against the antigens present on the cells of the host. In human bone marrow transplants, often a fatal condition.

gynandromorph An individual composed of cells with both male and female genotypes.

gyrase One of a class of enzymes known as topoisomerases. Gyrase converts closed circular DNA to a negatively supercoiled form prior to replication, transcription, or recombination.

H substance The carbohydrate group present on the surface of red blood cells. When unmodified, it results in blood type O; when modified by the addition of monosaccharides, it results in type A, B, and AB.

haploid A cell or organism having a single set of unpaired chromosomes. The gametic chromosome number.

haplotype The set of alleles from closely linked loci carried by an individual and usually inherited as a unit.

Hardy-Weinberg law The principle that both gene and genotype frequencies will remain in equilibrium in an infinitely large population in the absence of mutation, migration, selection, and nonrandom mating.

heat shock A transient response following exposure of cells or organisms to elevated temperatures. The response involves activation of a small number of loci, inactivation of previously active loci, and selective translation of heat shock mRNA. Appears to be a nearly universal phenomenon observed in organisms ranging from bacteria to humans.

helicase An enzyme that participates in DNA replication by unwinding the double helix near the replication fork.

helix-turn-helix motif The structure of a region of DNA-binding proteins in which a turn of four amino acids holds two alpha helices at right angles to each other.

hemizygous Conditions where a gene is present in a single dose. Usually applied to genes on the X chromosome in heterogametic males.

hemoglobin (Hb) An iron-containing, conjugated respiratory protein occurring chiefly in the red blood cells of vertebrates.

hemophilia An X-linked trait in humans associated with defective blood-clotting mechanisms.

heredity Transmission of traits from one generation to another.

heritability A measure of the degree to which observed phenotypic differences for a trait are genetic.

heterochromatin The heavily staining, late replicating regions of chromosomes that are condensed in interphase. Thought to be devoid of structural genes.

heteroduplex A double-stranded nucleic acid molecule in which each polynucleotide chain has a different origin. These structures may be produced as intermediates in a recombinational event, or by the *in vitro* reannealing of single-stranded, complementary molecules.

heterogametic sex The sex that produces gametes containing unlike sex chromosomes.

heterogeneous nuclear RNA (hnRNA) The collection of RNA transcripts in the nucleus, representing precursors and processing intermediates to rRNA, mRNA, and tRNA. Also represents RNA transcripts that will not be transported to the cytoplasm, such as snRNA (small nuclear RNA).

heterogenote A bacterial merozygote in which the donor (exogenote) chromosome segment carries different alleles than does the chromosome of the recipient (endogenote). A heterozygous merozygote.

heterokaryon A somatic cell containing nuclei from two different sources.

heterosis The superiority of a heterozygote over either homozygote for a given trait.

heterozygote An individual with different alleles at one or more loci. Such individuals will produce unlike gametes and therefore will not breed true.

Hfr A strain of bacteria exhibiting a high frequency of recombination. These strains have a chromosomally integrated F factor that is able to mobilize and transfer all or part of the chromosome to a recipient F⁻ cell.

histocompatibility antigens See *HLA*.

histones Proteins complexed with DNA in the nucleus. They are rich in the basic amino acids arginine and lysine and function in the coiling of DNA to form nucleosomes.

HLA Cell surface proteins, produced by histocompatibility loci, which are involved in the acceptance or rejection of tissue and organ grafts and transplants.

hnRNA See *heterogeneous nuclear RNA*.

holandric A trait transmitted from males to males. In humans, genes on the Y chromosome are holandric.

Holliday structure An intermediate in bidirectional DNA replication seen in the transmission electron microscope as an X-shaped structure showing four single-stranded DNA regions.

homeobox A sequence of about 180 nucleotides that encodes a 60-amino-acid sequence called a homeodomain, which is a DNA-binding protein that acts as a transcription factor.

homeotic mutation A mutation that causes a tissue normally determined to form a specific organ or body part to alter its differentiation and form another structure. Alternatively spelled: homoeotic.

homogametic sex The sex that produces gametes that do not differ with respect to sex chromosome content; in mammals, the female is homogametic.

homogeneously staining regions (hsr) Segments of mammalian chromosomes that stain lightly with Giemsa following exposure of cells to a selective agent. These regions arise in conjunction with gene amplification and are regarded as the structural locus for the amplified gene.

homogenote A bacterial merozygote in which the donor (exogenote) chromosome carries the same alleles as the chromosome of the recipient (endogenote). A homozygous merozygote.

homologous chromosomes Chromosomes that synapse or pair during meiosis. Chromosomes that are identical with respect to their genetic loci and centromere placement.

homozygote An individual with identical alleles at one or more loci. Such individuals will produce identical gametes and will therefore breed true.

homunculus The miniature individual imagined by preformationists to be contained within the sperm or egg.

human immunodeficiency virus (HIV) A human retrovirus associated with the onset and progression of acquired immunodeficiency syndrome (AIDS).

hybrid An individual produced by crossing two parents of different genotypes.

hybridoma A somatic cell hybrid produced by the fusion of an antibody-producing cell and a cancer cell, specifically, a myeloma. The cancer cell contributes the ability to divide indefinitely, and the antibody cell confers the ability to synthesize large amounts of a single antibody.

hybrid vigor See *heterosis*.

hydrogen bond An electrostatic attraction between a hydrogen atom bonded to a strongly electronegative atom such as oxygen or nitrogen and another atom that is electronegative or contains an unshared electron pair.

identical twins See *monozygotic twins*.

Ig See *immunoglobulin*.

imaginal disk Discrete groups of cells set aside during embryogenesis in holometabolous insects which are determined to form the external body parts of the adult.

immunoglobulin The class of serum proteins having the properties of antibodies.

inborn error of metabolism A biochemical disorder that is genetically controlled; usually an enzyme defect that produces a clinical syndrome.

inbreeding Mating between closely related organisms.

inbreeding depression A loss of reduction in fitness that usually accompanies inbreeding of organisms.

incomplete dominance Expression of heterozygous phenotype which is distinct from, and often intermediate to, that of either parent.

incomplete linkage Occasional separation of two genes on the same chromosome by a recombinational event.

independent assortment The independent behavior of each pair of homologous chromosomes during their segregation in meiosis I. The random distribution of genes on different chromosomes into gametes.

inducer An effector molecule that activates transcription.

inducible enzyme system An enzyme system under the control of a regulatory molecule, or inducer, which acts to block a repressor and allow transcription.

initiation codon The triplet of nucleotides (AUG) in an mRNA molecule that codes for the insertion of the amino acid methionine as the first amino acid in a polypeptide chain.

insertion sequence See *IS element*.

***in situ* hybridization** A technique for the cytological localization of DNA sequences complementary to a given nucleic acid or polynucleotide.

intercalary deletion A form of chromosome deletion where material is lost from within the chromosome. Deletions that involve the end of the chromosome are called terminal deletions.

intercalating agent A compound that inserts between bases in a DNA molecule, disrupting the alignments and pairing of bases in the complementary strands (e.g., acridine dyes).

interference A measure of the degree to which one crossover affects the incidence of another crossover in an adjacent region of the same chromatid. Positive interference increases the chances of another crossover; negative interference reduces the probability of a second crossover event.

interferon One of a family of proteins that acts to inhibit viral replication in higher organisms. Some interferons may have anticancer properties.

interphase That portion of the cell cycle between divisions.

intervening sequence See *intron*.

intron A portion of DNA between coding regions in a gene which is transcribed, but which does not appear in the mRNA product.

inversion A chromosomal aberration in which the order of a chromosomal segment has been reversed.

inversion loop The chromosomal configuration resulting from the synapsis of homologous chromosomes, one of which carries an inversion.

in vitro Literally, in glass; outside the living organism; occurring in an artificial environment.

in vivo Literally, in the living; occurring within the living body of an organism.

IS element A mobile DNA segment that is transposable to any of a number of sites in the genome.

isoagglutinogen An antigenic factor or substance present on the surface of cells that is capable of inducing the formation of an antibody.

isochromosome An aberrant chromosome with two identical arms and homologous loci.

isolating mechanism Any barrier to the exchange of genes between different populations of a group of organisms. In general, isolation can be classified as spatial, environmental, or reproductive.

isotopes Forms of a chemical element that have the same number of protons and electrons, but differ in the number of neutrons contained in the atomic nucleus. Unstable isotopes undergo a transition to a more stable form with the release of radioactivity.

isozyme Any of two or more distinct forms of an enzyme that have identical or nearly identical chemical properties, but differ in some property such as net electrical charge, pH optima, number and type of subunits, or substrate concentration.

kappa particles DNA-containing cytoplasmic particles found in certain strains of *Paramecium aurelia*. When these self-reproducing particles are transferred into the growth medium, they release a toxin, paramecin, which kills other sensitive strains. A nuclear gene, K, is responsible for the maintenance of kappa particles in the cytoplasm.

karyokinesis The process of nuclear division.

karyotype The chromosome complement of a cell or an individual. Often used to refer to the arrangement of metaphase chromosomes in a sequence according to length and position of the centromere.

kilobase A unit of length consisting of 1000 nucleotides. Abbreviated kb.

kinetochore A fibrous structure with a size of about 400 nm, located within the centromere. It appears to be the location to which microtubules attach.

Klenow fragment A part of bacterial DNA polymerase that lacks exonuclease activity, but retains polymerase activity. It is produced by enzymatic digestion of the intact enzyme.

Klinefelter syndrome A genetic disease in human males caused by the presence of an extra X chromosome. Klinefelter males are XXY instead of XY. This syndrome is associated with enlarged breasts, small testes, sterility, and, occasionally, mental retardation.

knockout mice in producing knockout mice, a cloned normal gene is inactivated by the insertion of a marker, such as an antibiotic resistance gene. The altered gene is transferred to embryonic stem cells, where the altered gene will replace the normal gene (in some cells). These cells are injected into a blastomere embryo, producing a mouse that is bred to yield mice that are homozygous for the mutated gene.

lagging strand In DNA replication, the strand synthesized in a discontinuous fashion, 5′ to 3′ away from the replication fork. Each short piece of DNA synthesized in this fashion is called an Okazaki fragment.

lampbrush chromosomes Meiotic chromosomes characterized by extended lateral loops, which reach maximum extension during diplotene. Although most intensively studied in amphibians, these structures occur in meiotic cells of organisms ranging from insects through humans.

leader sequence That portion of an mRNA molecule from the 5′-end to the beginning codon; may contain regulatory or ribosome binding sites.

leading strand During DNA replication, the strand synthesized continuously 5′ to 3′ toward the replication fork.

lectins Carbohydrate-binding proteins found in the seeds of leguminous plants such as soybeans. Although their physiological role in plants is unclear, they are useful as probes for cell surface carbohydrates and glycoproteins in animal cells. Proteins with similar properties have also been isolated from organisms such as snails and eels.

leptotene The initial stage of meiotic prophase I, during which the chromosomes become visible and are often arranged in a bouquet configuration, with one or both ends of the chromosomes gathered at one spot on the inner nuclear membrane.

lethal gene A gene whose expression results in death.

leucine zipper A structural motif in a DNA-binding protein that is characterized by a stretch of leucine residues spaced at every seventh amino acid residue, with adjacent regions of positively charged amino acids. Leucine zippers on two polypeptides may interact to form a dimer that binds to DNA.

LINEs Long interspersed elements are repetitive sequences found in the genomes of higher organisms, such as the 6-kb Kpnl sequences found in primate genomes.

linkage Condition in which two or more nonallelic genes tend to be inherited together. Linked genes have their loci along the same chromosome, do not assort independently, but can be separated by crossing over.

linkage group A group of genes that have their loci on the same chromosome.

linking number The number of times that two strands of a closed, circular DNA duplex cross over each other.

locus The site or place on a chromosome where a particular gene is located.

lod score A statistical method used to determine whether two loci are linked or unlinked. A lod (log of the odds) score of +3 by convention indicates linkage.

long period interspersion Pattern of genome organization in which long stretches of single copy DNA are interspersed with long segments of repetitive DNA. This pattern of genome organization is found in *Drosophila* and the honeybee.

long terminal repeat (LTR) Sequence of several hundred base pairs found at the ends of retroviral DNAs.

Lutheran blood group One of a number of blood group systems inherited independently of the ABO, MN, and Rh systems. Alleles of this group determine the presence or absence of antigens on the surface of red blood cells. Gene is on human chromosome 19.

Lyon hypothesis The idea proposed by Mary Lyon that random inactivation of one X chromosome in the somatic cells of mammalian females is responsible for dosage compensation and mosaicism.

lysis The disintegration of a cell brought about by the rupture of its membrane.

lysogenic bacterium A bacterial cell carrying a temperate bacteriophage integrated into its chromosome.

lysogeny The process by which the DNA of an infecting phage becomes repressed and integrated into the chromosome of the bacterial cell it infects.

lytic phase The condition in which a temperate bacteriophage loses its integrated status in the host chromosome (becomes induced), replicates, and lyses the bacterial cell.

major histocompatibility loci See *MHC*.

map unit A measure of the genetic distance between two genes, corresponding to a recombination frequency of 1 percent. See *centimorgan*.

maternal effect Phenotypic effects on the offspring produced by the maternal genome. Factors transmitted through the egg cytoplasm that produce a phenotypic effect in the progeny.

maternal influence See *maternal effect*.

maternal inheritance The transmission of traits via cytoplasmic genetic factors such as mitochondria or chloroplasts.

mean The arithmetic average.

median The value in a group of numbers below and above which there is an equal number of data points or measurements.

meiosis The process in gametogenesis or sporogenesis during which one replication of the chromosomes is followed by two nuclear divisions to produce four haploid cells.

melting profile See T_m.

merozygote A partially diploid bacterial cell containing, in addition to its own chromosome, a chromosome fragment introduced into the cell by transformation, transduction, or conjugation.

messenger RNA See *mRNA*.

metabolism The sum of chemical changes in living organisms by which energy is generated and used.

metacentric chromosome A chromosome with a centrally located centromere, producing chromosome arms of equal lengths.

metafemale In *Drosophila*, a poorly developed female of low viability in which the ratio of X chromosomes to sets of autosomes exceeds 1.0. Previously called a superfemale.

metamale In *Drosophila*, a poorly developed male of low viability in which the ratio of X chromosomes to sets of autosomes is less than 0.5. Previously called a supermale.

metaphase The stage of cell division in which the condensed chromosomes lie in a central plane between the two poles of the cell, and in which the chromosomes become attached to the spindle fibers.

metaphase plate The arrangement of mitotic or meiotic chromosomes at the equator of the cell during metaphase.

MHC Major histocompatibility loci. In humans, the HLA complex; and in mice, the H2 complex.

micrometer A unit of length equal to 1×10^{-6} meter. Previously called a micron. Abbreviated μm.

micron See *micrometer*.

migration coefficient An expression of the proportion of migrant genes entering the population per generation.

millimeter A unit of length equal to 1×10^{-3} meter. Abbreviated mm.

minimal medium A medium containing only those nutrients that will support the growth and reproduction of wild-type strains of an organism.

mismatch repair DNA repair by a mechanism known as cut and patch. In this repair, the defective single-stranded region is excised, followed by the synthesis of a new segment, using the complementary strand as a template.

missense mutation A mutation that alters a codon to that of another amino acid, causing an altered translation product to be made.

mitochondrion Found in the cells of eukaryotes, a cytoplasmic, self-reproducing organelle that is the site of ATP synthesis.

mitogen A substance that stimulates mitosis in nondividing cells (e.g., phytohemagglutinin).

mitosis A form of cell division resulting in the production of two cells, each with the same chromosome and genetic complement as the parent cell.

mode In a set of data, the value occurring in the greatest frequency.

monohybrid cross A genetic cross between two individuals involving only one character (e.g., *AA* × *aa*).

monosomic An aneuploid condition in which one member of a chromosome pair is missing; having a chromosome number of $2n - 1$.

monozygotic twins Twins produced from a single fertilization event; the first division of the zygote produces two cells, each of which develops into an embryo. Also known as identical twins.

mRNA An RNA molecule transcribed from DNA and translated into the amino acid sequence of a polypeptide.

mtDNA Mitochondrial DNA.

multigene family A gene set descended from a common ancestor by duplication and subsequent divergence from a common ancestor. The globin genes represent a multigene family.

multiple alleles Three or more alleles of the same gene.

multiple-factor inheritance See *polygenic inheritance*.

multiple infection Simultaneous infection of a bacterial cell by more than one bacteriophage, often of different genotypes.

mu phage A phage group in which the genetic material behaves like an insertion sequence, capable of insertion, excision, transposition, inactivation of host genes, and induction of chromosomal rearrangements.

mutagen Any agent that causes an increase in the rate of mutation.

mutant A cell or organism carrying an altered or mutant gene.

mutation The process that produces an alteration in DNA or chromosome structure; the source of most alleles.

mutation rate The frequency with which mutations take place at a given locus or in a population.

muton The smallest unit of mutation in a gene, corresponding to a single base change.

nanometer A unit of length equal to 1×10^{-9} meter. Abbreviated nm.

natural selection Differential reproduction of some members of a species resulting from variable fitness conferred by genotypic differences.

nearest-neighbor analysis A molecular technique used to determine the frequency with which nucleotides are adjacent to each other in polynucleotide chains.

neutral mutation A mutation with no immediate adaptive significance or phenotypic effect.

noncrossover gamete A gamete that contains no chromosomes that have undergone genetic recombination.

nondisjunction An accident of cell division in which the homologous chromosomes (in meiosis) or the sister chromatids (in mitosis) fail to separate and migrate to opposite poles; responsible for defects such as monosomy and trisomy.

nonsense codon The nucleotide triplet in an mRNA molecule that signals the termination of translation. Three such codons are known: UGA, UAG, and UAA.

nonsense mutation A mutation that alters a codon to one which encodes no amino acid, that is, UAG (amber codon), UAA (ochre codon), or UGA (opal codon). Leads to premature termination during the translation of mRNA.

NOR See *nucleolar organizer region*.

normal distribution A probability function that approximates the distribution of random variables. The normal curve, also known as a Gaussian or bell-shaped curve, is the graphic display of the normal distribution.

np See *nucleotide pair.*

N-terminal amino acid The terminal amino acid in a peptide chain that carries a free amino group.

N terminus The end of a polypeptide that carries a free amino group of the first amino acid. By convention, the structural formula of polypeptides is written with the N terminus at the left.

nu body See *nucleosome.*

nuclease An enzyme that breaks bonds in nucleic acid molecules.

nucleoid The DNA-containing region within the cytoplasm in prokaryotic cells.

nucleolar organizer region (NOR) A chromosomal region containing the genes for rRNA; most often found in physical association with the nucleolus.

nucleolus A nuclear organelle that is the site of ribosome biosynthesis; usually associated with or formed in association with the NOR.

nucleoside A purine or pyrimidine base covalently linked to a ribose or deoxyribose sugar molecule.

nucleosome A complex of four histone molecules, each present in duplicate, wrapped by two turns of a DNA molecule. One of the basic units of eukaryotic chromosome structure. Also known as a nu body.

nucleotide A nucleoside covalently linked to a phosphate group. Nucleotides are the basic building blocks of nucleic acids. The nucleotides commonly found in DNA are deoxyadenylic acid, deoxycytidylic acid, deoxyguanylic acid, and deoxythymidylic acid. The nucleotides in RNA are adenylic acid, cytidylic acid, guanylic acid, and uridylic acid.

nucleotide pair The pair of nucleotides (A and T, or G and C) in opposite strands of the DNA molecule that are hydrogen-bonded to each other.

nucleus The membrane-bounded cytoplasmic organelle of eukaryotic cells that contains the chromosomes and nucleolus.

nullisomic Describes an individual with a chromosomal aberration in which both members of a chromosome pair are missing.

ochre codon A codon that does not code for the insertion of an amino acid into a polypeptide chain, but signals chain termination. The ochre codon is UAA.

Okazaki fragment The small, discontinuous strands of DNA produced during DNA synthesis.

oligonucleotide A linear sequence of nucleotides (up to 20) connected by 5′ to 3′ phosphodiester bonds.

oncogene A gene whose activity promotes uncontrolled proliferation in eukaryotic cells.

open reading frame (ORF) The interval between the start and stop codon that encodes amino acids for insertion into polypeptide chains.

operator region A region of a DNA molecule that interacts with a specific repressor protein to control the expression of an adjacent gene or gene set.

operon A genetic unit that consists of one or more structural genes (that code for polypeptides) and an adjacent operator gene that controls the transcriptional activity of the structural gene or genes.

overdominance The phenomenon where heterozygotes have a phenotype that is more extreme than either homozygous genotype.

overlapping code A genetic code first proposed by George Gamow in which any given nucleotide is shared by two adjacent codons.

pachytene The stage in meiotic prophase I when the synapsed homologous chromosomes split longitudinally (except at the centromere), producing a group of four chromatids called a tetrad.

pair-rule genes Genes expressed as stripes around the blastoderm embryo during development of the *Drosophila* embryo.

palindrome A word, number, verse, or sentence that reads the same backward or forward (e.g., *able was I ere I saw elba*). In nucleic acids, a sequence in which the base pairs read the same on complementary strands (5′→3′). For example: 5′GAATTC3′, 3′CTTAAG5′. These often occur as sites for restriction endonuclease recognition and cutting.

pangenesis A discarded theory of development that postulated the existence of pangenes, small particles from all parts of the body that concentrated in the gametes, passing traits from generation to generation, blending the traits of the parents in the offspring.

paracentric inversion A chromosomal inversion that does not include the centromere.

parasexual Condition describing recombination of genes from different individuals that does not involve meiosis, gamete formation, or zygote production. The formation of somatic cell hybrids is an example.

parental gamete See *noncrossover gamete.*

parthenogenesis Development of an egg without fertilization.

partial diploids See *merozygote.*

partial dominance See *incomplete dominance.*

passive immunity A form of immunity produced by receipt of antibodies synthesized by another individual.

patroclinous inheritance A form of genetic transmission in which the offspring have the phenotype of the father.

pedigree In human genetics, a diagram showing the ancestral relationships and transmission of genetic traits over several generations in a family.

P element Transposable DNA elements found in *Drosophila* that are responsible for hybrid dysgenesis.

penetrance The frequency (expressed as a percentage) with which individuals of a given genotype manifest at least some degree of a specific mutant phenotype associated with a trait.

peptide bond The covalent bond between the amino group of one amino acid and the carboxyl group of another amino acid.

pericentric inversion A chromosomal inversion that involves both arms of the chromosome and thus involves the centromere.

permissive condition Environmental conditions under which a conditional mutation (such as temperature-sensitive mutant) expresses the wild-type phenotype.

phage See *bacteriophage*.

phenocopy An environmentally induced phenotype (nonheritable) that closely resembles the phenotype produced by a known gene.

phenotype The observable properties of an organism that are genetically controlled.

phenylketonuria (PKU) A hereditary condition in humans associated with the inability to metabolize the amino acid phenylalanine. The most common form is caused by the lack of the liver enzyme phenylalanine hydroxylase.

phosphodiester bond In nucleic acids, the covalent bond between a phosphate group and adjacent nucleotides, extending from the 5′ carbon of one pentose (ribose or deoxyribose) to the 3′ carbon of the pentose in the neighboring nucleotide. Phosphodiester bonds form the backbone of nucleic acid molecules.

photoreactivation enzyme (PRE) An exonuclease that catalyzes the light-activated excision of ultraviolet-induced thymine dimers from DNA.

photoreactivation repair Light-induced repair of damage caused by exposure to ultraviolet light. Associated with an intracellular enzyme system.

phyletic evolution The gradual transformation of one species into another over time; vertical evolution.

pilus A filamentlike projection from the surface of a bacterial cell. Often associated with cells possessing F factors.

plaque A clear area on an otherwise opaque bacterial lawn caused by the growth and reproduction of phages.

plasmid An extrachromosomal, circular DNA molecule (often carrying genetic information) that replicates independently of the host chromosome.

platysome Term originally used in electron and X-ray diffraction studies to describe the flattened appearance of the DNA in the nucleosome core.

pleiotropy Condition in which a single mutation simultaneously affects several characters.

ploidy Term referring to the basic chromosome set or to multiples of that set.

point mutation A mutation that can be mapped to a single locus. At the molecular level, a mutation that results in the substitution of one nucleotide for another.

polar body A cell produced at either the first or second meiotic division in females that contains almost no cytoplasm as a result of an unequal cytokinesis.

polycistronic mRNA A messenger RNA molecule that encodes the amino acid sequence of two or more polypeptide chains in adjacent structural genes.

polygenic inheritance The transmission of a phenotypic trait whose expression depends on the additive effect of a number of genes.

polylinker A segment of DNA that has been engineered to contain multiple sites for restriction enzyme digestion. Polylinkers are usually found in engineered vectors such as plasmids.

polymerase chain reaction (PCR) A method for amplifying DNA segments that uses cycles of denaturation, annealing to primers, and DNA polymerase–directed DNA synthesis.

polymerases The enzymes that catalyze the formation of DNA and RNA from deoxynucleotides and ribonucleotides, respectively.

polymorphism The existence of two or more discontinuous, segregating phenotypes in a population.

polynucleotide A linear sequence of more than 20 nucleotides, joined by 5′-to-3′ phosphodiester bonds. See *oligonucleotide*.

polypeptide A molecule made up of amino acids joined by covalent peptide bonds. This term is used to denote the amino acid chain before it assumes its functional three-dimensional configuration.

polyploid A cell or individual having more than two sets of chromosomes.

polyribosome See *polysome*.

polysome A structure composed of two or more ribosomes associated with mRNA, engaged in translation. Formerly called polyribosome.

polytene chromosome A chromosome that has undergone several rounds of DNA replication without separation of the replicated chromosomes, forming a giant, thick chromosome with aligned chromomeres producing a characteristic banding pattern.

population A local group of individuals belonging to the same species, which are actually or potentially interbreeding.

position effect Change in expression of a gene associated with a change in the gene's location within the genome.

postzygotic isolation mechanism Factors that prevent or reduce inbreeding by acting after fertilization to produce nonviable, sterile hybrids or hybrids of lowered fitness.

preadaptive mutation A mutational event that later becomes of adaptive significance.

preformationism The discredited idea that an organism develops by growth of structures already present in the egg or sperm.

prezygotic isolation mechanism Factors that reduce inbreeding by preventing courtship, mating, or fertilization.

Pribnow box A 6-bp sequence 5′ to the beginning of transcription in prokaryotic genes, to which the sigma subunit of RNA polymerase binds. The consensus sequence for this box is TATAAT.

primary protein structure Refers to the sequence of amino acids in a polypeptide chain.

primary sex ratio Ratio of males to females at fertilization.

primer In nucleic acids, a short length of RNA or single-stranded DNA that is necessary for the functioning of polymerases.

prion An infectious pathogenic agent devoid of nucleic acid and composed mainly of a protein, PrP, with a molecular weight of 27,000 to 30,000 daltons. Prions are known to cause scrapie, a degenerative neurological disease in sheep, and are thought to cause similar diseases in humans, such as Kuru and Creutzfeldt-Jakob disease.

probability Ratio of the frequency of a given event to the frequency of all possible events.

proband See *propositus*.

probe A macromolecule such as DNA or RNA that has been labeled and can be detected by an assay such as autoradiography or fluorescence microscopy. Probes are used to identify target molecules, genes, or gene products.

product law The law that holds that the probability of two independent events occurring simultaneously is the product of their independent probabilities.

proflavin An acridine dye that acts as a mutagen. See *acridine dyes*.

progeny The offspring produced from a mating.

prokaryotes Organisms lacking nuclear membranes, meiosis, and mitosis. Bacteria and blue-green algae are examples of prokaryotic organisms.

promoter Region having a regulatory function and to which RNA polymerase binds prior to the initiation of transcription.

proofreading A molecular mechanism for correcting errors in replication, transcription, or translation. Also known as editing.

prophage A phage genome integrated into a bacterial chromosome. Bacterial cells carrying prophages are said to be lysogenic.

propositus (female, **proposita**) An individual in whom a genetically determined trait of interest is first detected. Also known as a proband.

protein A molecule composed of one or more polypeptides, each composed of amino acids covalently linked together.

proto-oncogene A cellular gene that normally functions to control cellular proliferation. Proto-oncogenes can be converted to oncogenes by alterations in structure or expression.

protoplast A bacterial or plant cell with the cell wall removed. Sometimes called a spheroplast.

prototroph A strain (usually microorganisms) that is capable of growth on a defined, minimal medium. Wild-type strains are usually regarded as prototrophs.

pseudoalleles Genes that behave as alleles to one another by complementation, but that can be separated from one another by recombination.

pseudodominance The expression of a recessive allele on one homolog caused by the deletion of the dominant allele on the other homolog.

pseudogene A nonfunctional gene with sequence homology to a known structural gene present elsewhere in the genome. They differ from their functional relatives by insertions or deletions and by the presence of flanking direct repeat sequences of 10 to 20 nucleotides.

puff See *chromosome puff*.

punctuated equilibrium A pattern in the fossil record of brief periods of species divergence punctuated with long periods of species stability.

quantitative inheritance See *polygenic inheritance*.

quantitative trait loci (QTL) Two or more genes that act on a single polygenic trait.

quantum speciation Formation of a new species within a single or a few generations by a combination of selection and drift.

quaternary protein structure Types and modes of interaction between two or more polypeptide chains within a protein molecule.

R point The point (also known as the restriction point) during the G1 stage of the cell cycle when either a commitment is made to DNA synthesis and another cell cycle, or the cell withdraws from the cycle and becomes quiescent.

race A phenotypically or geographically distinct subgroup within a species.

rad A unit of absorbed dose of radiation with an energy equal to 100 ergs per gram of irradiated tissue.

radioactive isotope One of the forms of an element, differing in atomic weight and possessing an unstable nucleus that emits ionizing radiation.

random mating Mating between individuals without regard to genotype.

reading frame Linear sequence of codons (groups of three nucleotides) in a nucleic acid.

reannealing Formation of double-stranded DNA molecules from dissociated single strands.

recessive Term describing an allele that is not expressed in the heterozygous condition.

reciprocal cross A paired cross in which the genotype of the female in the first cross is present as the genotype of the male in the second cross, and vice versa.

reciprocal translocation A chromosomal aberration in which nonhomologous chromosomes exchange parts.

recombinant DNA A DNA molecule formed by the joining of two heterologous molecules. Usually applied to DNA molecules produced by *in vitro* ligation of DNA from two different organisms.

recombinant gamete A gamete containing a new combination of genes produced by crossing over during meiosis.

recombination The process that leads to the formation of new gene combinations on chromosomes.

recon A term used by Seymour Benzer to denote the smallest genetic units between which recombination can occur.

reductional division The chromosome division that halves the diploid chromosome number. The first division of meiosis is a reductional division.

redundant genes Gene sequences present in more than one copy per haploid genome (e.g., ribosomal genes).

regulatory site A DNA sequence that is involved in the control of expression of other genes, usually involving an interaction with another molecule.

rem Radiation equivalent in man; the dosage of radiation that will cause the same biological effect as one roentgen of X-rays.

renaturation The process by which a denatured protein or nucleic acid returns to its normal three-dimensional structure.

repetitive DNA sequences DNA sequences present in many copies in the haploid genome.

replicating form (RF) Double-stranded nucleic acid molecules present as an intermediate during the reproduction of certain viruses.

replication The process of DNA synthesis.

replication fork The Y-shaped region of a chromosome associated with the site of replication.

replicon A chromosomal region or free genetic element containing the DNA sequences necessary for the initiation of DNA replication.

replisome The term used to describe the complex of proteins, including DNA polymerase, that assembles at the bacterial replication fork to synthesize DNA.

repressible enzyme system An enzyme or group of enzymes whose synthesis is regulated by the intracellular concentration of certain metabolites.

repressor A protein that binds to a regulatory sequence adjacent to a gene and blocks transcription of the gene.

reproductive isolation Absence of interbreeding between populations, subspecies, or species. Reproductive isolation can be brought about by extrinsic factors, such as behavior, and intrinsic barriers, such as hybrid inviability.

resistance transfer factor (RTF) A component of R plasmids that confers the ability for cell-to-cell transfer of the R plasmid by conjugation.

resolution In an optical system, the shortest distance between two points or lines at which they can be perceived to be two points or lines.

restriction endonuclease Nuclease that recognizes specific nucleotide sequences in a DNA molecule, and cleaves or nicks the DNA at that site. Derived from a variety of microorganisms, those enzymes that cleave both strands of the DNA are used in the construction of recombinant DNA molecules.

restriction fragment length polymorphism (RFLP) Variation in the length of DNA fragments generated by a restriction endonuclease. These variations are caused by mutations that create or abolish cutting sites for restriction enzymes. RFLPs are inherited in a codominant fashion and can be used as genetic markers.

restrictive condition Environmental conditions under which a conditional mutation (such as a temperature-sensitive mutant) expresses the mutant phenotype.

restrictive transduction See *specialized transduction*.

retrovirus Viruses with RNA as genetic material that utilize the enzyme reverse transcriptase during their life cycle.

reverse transcriptase A polymerase that uses RNA as a template to transcribe a single-stranded DNA molecule as a product.

reversion A mutation that restores the wild-type phenotype.

R factor (R plasmid) Bacterial plasmids that carry antibiotic resistance genes. Most R plasmids have two components: an r-determinant, which carries the antibiotic resistance genes, and the resistance transfer factor (RTF).

RFLP See *restriction fragment length polymorphism*.

Rh factor An antigenic system first described in the rhesus monkey. Recessive *r/r* individuals produce no Rh antigens and are Rh negative, while *R/R* and *R/r* individuals have Rh antigens on the surface of their red blood cells and are classified as Rh positive.

ribonucleic acid A nucleic acid characterized by the sugar ribose and the pyrimidine uracil, usually a single-stranded polynucleotide. Several forms are recognized, including ribosomal RNA, messenger RNA, transfer RNA, and heterogeneous nuclear RNA.

ribose The five-carbon sugar associated with the ribonucleotides found in RNA.

ribosomal RNA See *rRNA*.

ribosome A ribonucleoprotein organelle consisting of two subunits, each containing RNA and protein. Ribosomes are the site of translation of mRNA codons into the amino acid sequence of a polypeptide chain.

RNA See *ribonucleic acid*.

RNA editing Alteration of the nucleotide sequence of an mRNA molecule after transcription and before translation. There are two main types of editing: substitution editing, which changes individual nucleotides, and insertion/deletion editing, where individual nucleotides are added or deleted.

RNA polymerase An enzyme that catalyzes the formation of an RNA polynucleotide strand using the base sequence of a DNA molecule as a template.

RNase A class of enzymes that hydrolyze RNA molecules.

Robertsonian translocation A form of chromosomal aberration that involves the fusion of long arms of acrocentric chromosomes at the centromere.

roentgen A unit of measure of the amount of radiation corresponding to the generation of 2.083×10^9 ion pairs in one cubic centimeter of air at 0°C at an atmospheric pressure of 760 mm of mercury. Abbreviated R.

rolling circle model A model of DNA replication in which the growing point or replication fork rolls around a circular template strand; in each pass around the circle, the newly synthesized strand displaces the strand from the previous replication, producing a series of contiguous copies of the template strand.

rRNA The RNA molecules that are the structural components of the ribosomal subunits. In prokaryotes, these are the 16*S*, 23*S*, and 5*S* molecules; and in eukaryotes, they are the 18*S*, 28*S*, and 5*S* molecules.

RTF See *resistance transfer factor*.

S₁ nuclease A deoxyribonuclease that cuts and degrades single-stranded molecules of DNA.

satellite DNA DNA that forms a minor band when genomic DNA is centrifuged in a cesium salt gradient. This DNA usually consists of short sequences repeated many times in the genome.

SCE See *sister chromatid exchange*.

secondary protein structure The alpha helical or pleated-sheet form of a protein molecule brought about by the formation of hydrogen bonds between amino acids.

secondary sex ratio The ratio of males to females at birth.

secretor An individual having soluble forms of the blood group antigens A and/or B present in saliva and other body fluids. This condition is caused by a dominant, autosomal gene unlinked to the *ABO* locus (*I* locus).

sedimentation coefficient See *Svedberg coefficient unit*.

segment polarity genes Genes that regulate the spatial pattern of differentiation within each segment of the developing *Drosophila* embryo.

segregation The separation of homologous chromosomes into different gametes during meiosis.

selection The force that brings about changes in the frequency of alleles and genotypes in populations through differential reproduction.

selection coefficient (*s*) A quantitative measure of the relative fitness of one genotype compared with another.

selfing In plant genetics, the fertilization of ovules of a plant by pollen produced by the same plant. Reproduction by self-fertilization.

semiconservative replication A model of DNA replication in which a double-stranded molecule replicates in such a way that the daughter molecules are composed of one parental (old) and one newly synthesized strand.

semisterility A condition in which a proportion of all zygotes are inviable.

sex chromatin body See *Barr body*.

sex chromosome A chromosome, such as the X or Y in humans, which is involved in sex determination.

sexduction Transmission of chromosomal genes from a donor bacterium to a recipient cell by the F factor.

sex-influenced inheritance Phenotypic expression that is conditioned by the sex of the individual. A heterozygote may express one phenotype in one sex and the alternate phenotype in the other sex.

sex-limited inheritance A trait that is expressed in only one sex even though the trait may not be X-linked.

sex ratio See *primary* and *secondary sex ratio*.

sexual reproduction Reproduction through the fusion of gametes, which are the haploid products of meiosis.

Shine–Dalgarno sequence The nucleotides AGGAGG present in the leader sequence of prokaryotic genes that serve as a ribosome binding site. The 16S RNA of the small ribosomal subunit contains a complementary sequence to which the mRNA binds.

short period interspersion Pattern of genome organization in which stretches of single-copy DNA (about 1000 bp) are interspersed with short segments of repetitive DNA (300 bp). This pattern is found in *Xenopus*, humans, and the majority of organisms examined to date.

shotgun experiment The cloning of random fragments of genomic DNA into a vehicle such as a plasmid or phage, usually to produce a bank or library of clones from which clones of specific interest will be selected.

sibling species Species that are morphologically almost identical, but which are reproductively isolated from one another.

sickle-cell anemia A genetic disease in humans caused by an autosomal recessive gene, usually fatal in the homozygous condition. Caused by an alteration in the amino acid sequence of the beta chain of globin.

sickle-cell trait The phenotype exhibited by individuals heterozygous for the sickle-cell gene.

sigma factor A polypeptide subunit of the RNA polymerase that recognizes the binding site for the initiation of transcription.

SINEs Short interspersed elements are repetitive sequences found in the genomes of higher organisms, such as the 300-bp *Alu* sequence.

sister chromatid exchange (SCE) A crossing over event that can occur in meiotic and mitotic cells; involves the reciprocal exchange of chromosomal material between sister chromatids (joined by a common centromere). Such exchanges can be detected cytologically after BrdU incorporation into the replicating chromosomes.

site-directed mutagenesis A process that uses a synthetic oligonucleotide containing a mutant base or sequence as a primer for inducing a mutation at a specific site in a cloned gene.

small nuclear RNA (snRNA) Species of RNA molecules ranging in size from 90 to 400 nucleotides. The abundant snRNAs are present in 1×10^4 to 1×10^6 copies per cell. snRNAs are associated with proteins and form RNP particles known as snRNPs or snurps. Six uridine-rich snRNAs known as U1–U6 are located in the nucleoplasm, and the complete nucleotide sequence of these is known. snRNAs have been implicated in the processing of pre-mRNA and may have a range of cleavage and ligation functions.

snurps See *small nuclear RNA (snRNA)*.

solenoid structure A level of eukaryotic chromosome structure generated by the supercoiling of nucleosomes.

somatic cell genetics The use of cultured somatic cells to investigate genetic phenomena by parasexual techniques.

somatic cells All cells other than the germ cells or gametes in an organism.

somatic mutation A mutational event occurring in a somatic cell. In other words, such mutations are not heritable.

somatic pairing The pairing of homologous chromosomes in somatic cells.

SOS response The induction of enzymes to repair damaged DNA in *E. coli*. The response involves activation of an enzyme that cleaves a repressor, activating a series of genes involved in DNA repair.

spacer DNA DNA sequences found between genes, usually repetitive DNA segments.

special creation An idea that each species originated through a special act of creation by a divine force.

specialized transduction Genetic transfer of only specific host genes by transducing phages.

speciation The process by which new species of plants and animals arise.

species A group of actually or potentially interbreeding individuals that is reproductively isolated from other such groups.

spheroplast See *protoplast*.

spindle fibers Cytoplasmic fibrils formed during cell division that are involved with the separation of chromatids at anaphase and their movement toward opposite poles in the cell.

spliceosome The nuclear macromolecule complex within which splicing reactions occur to remove introns from pre-mRNAs.

spontaneous generation The origin of living systems from nonliving matter.

spontaneous mutation A mutation that is not induced by a mutagenic agent.

spore A unicellular body or cell encased in a protective coat that is produced by some bacteria, plants, and invertebrates; is capable of survival in unfavorable environmental conditions; and can give rise to a new individual upon germination. In plants, spores are the haploid products of meiosis.

SRY A gene (sex-determining region of the Y) found near the pseudoautosomal boundary of the Y chromosome. Accumulated evidence indicates that this gene is the testis-determining factor (TDF).

stabilizing selection Preferential reproduction of those individuals having genotypes close to the mean for the population. A selective elimination of genotypes at both extremes.

standard deviation A quantitative measure of the amount of variation in a sample of measurements from a population.

standard error A quantitative measure of the amount of variation in a sample of measurements from a population.

sterility The condition of being unable to reproduce; free from contaminating microorganisms.

strain A group with common ancestry that has physiological or morphological characteristics of interest for genetic study or domestication.

structural gene A gene that encodes the amino acid sequence of a polypeptide chain.

sublethal gene A mutation causing lowered viability, with death before maturity in less than 50 percent of the individuals carrying the gene.

submetacentric chromosome A chromosome with the centromere placed so that one arm of the chromosome is slightly longer than the other.

subspecies A morphologically or geographically distinct interbreeding population of a species.

sum law The law that holds that the probability of one or the other of two mutually exclusive events occurring is the sum of their individual probabilities.

supercoiled DNA A form of DNA structure in which the helix is coiled upon itself. Such structures can exist in stable forms only when the ends of the DNA are not free, as in a covalently closed circular DNA molecule.

superfemale See *metafemale*.

supermale See *metamale*.

suppressor mutation A mutation that acts to restore (completely or partially) the function lost by a previous mutation at another site.

Svedberg coefficient unit A unit of measure for the rate at which particles (molecules) sediment in a centrifugal field. This unit is a function of several physico-chemical properties, including size and shape. A sedimentation value of 1×10^{-13} sec is defined as one Svedberg coefficient (*S*) unit.

symbiont An organism coexisting in a mutually beneficial relationship with another organism.

sympatric speciation Process of speciation involving populations that inhabit, at least in part, the same geographic range.

synapsis The pairing of homologous chromosomes at meiosis.

synaptonemal complex (SC) An organelle consisting of a tripartite nucleoprotein ribbon that forms between the paired homologous chromosomes in the pachytene stage of the first meiotic division.

syndrome A group of signs or symptoms that occur together and characterize a disease or abnormality.

synkaryon The nucleus of a zygote that results from the fusion of two gametic nuclei. Also used in somatic cell genetics to describe the product of nuclear fusion.

syntenic test In somatic cell genetics, a method for determining whether or not two genes are on the same chromosome.

T_m The temperature at which a population of double-stranded nucleic acid molecules is half-dissociated into single strands. This is taken to be the melting temperature for that species of nucleic acid.

target theory In radiation biology, a theory stating that damage and death from radiation is caused by the inactivation of specific targets within the organism.

TATA box See *Goldberg–Hogness box*.

tautomeric shift A reversible isomerization in a molecule brought about by a shift in the localization of a hydrogen atom. In nucleic acids, tautomeric shifts in the bases of nucleotides can cause changes in other bases at replication and are a source of mutations.

TDF (testis-determining factor) A gene on the Y chromosome that controls the developmental switch point for the development of the indifferent gonad into a testis. See *SRY*.

telocentric chromosome A chromosome in which the centromere is located at the end of the chromosome.

telomerase The enzyme that adds short, tandemly repeated DNA sequences to the ends of eukaryotic chromosomes.

telomere The terminal chromomere of a chromosome.

telophase The stage of cell division in which the daughter chromosomes reach the opposite poles of the cell and reform nuclei. Telophase ends with the completion of cytokinesis.

temperate phage A bacteriophage that can become a prophage and confer lysogeny upon the host bacterial cell.

temperature-sensitive mutation A conditional mutation that produces a mutant phenotype at one temperature range and a wild-type phenotype at another temperature range.

template The single-stranded DNA or RNA molecule that specifies the nucleotide sequence of a strand synthesized by a polymerase molecule.

teratocarcinoma Embryonal tumors that arise in the yolk sac or gonads and are able to undergo differentiation into a wide variety of cell types. These tumors are used to investigate the regulatory mechanisms underlying development.

terminalization The movement of chiasmata toward the ends of chromosomes during the diplotene stage of the first meiotic division.

tertiary protein structure The three-dimensional structure of a polypeptide chain brought about by folding upon itself.

test cross A cross between an individual whose genotype at one or more loci may be unknown and an individual who is homozygous recessive for the genes in question.

tetrad The four chromatids that make up paired homologs in the prophase of the first meiotic division. The four haploid cells produced by a single meiotic division.

tetrad analysis Method for the analysis of gene linkage and recombination using the four haploid cells produced in a single meiotic division.

tetranucleotide hypothesis An early theory of DNA structure proposing that the molecule was composed of repeating units, each consisting of the four nucleotides adenosine, thymidine, cytosine, and guanine.

tetraparental mouse A mouse produced from an embryo that was derived by the fusion of two separate blastulas.

theta structure An intermediate in the bidirectional replication of circular DNA molecules. At about midway through the cycle of replication, the intermediate resembles the Greek letter theta.

thymine dimer A pair of adjacent thymine bases in a single polynucleotide strand between which chemical bonds have formed. This lesion, usually the result of damage caused by exposure to ultraviolet light, inhibits DNA replication unless repaired by the appropriate enzyme system.

topoisomerase A class of enzymes that convert DNA from one topological form to another. During DNA replication, these enzymes facilitate the unwinding of the double-helical structure of DNA.

totipotent The ability of a cell or embryo part to give rise to all adult structures. This capacity is usually progressively restricted during development.

trailer sequence A transcribed but nontranslated region of a gene or its mRNA that follows the termination signal.

trait Any detectable phenotypic variation of a particular inherited character.

***trans* configuration** The arrangement of two mutant sites on opposite homologs, such as

$$\frac{a^1\ +}{+\ a^2}$$

Contrasts with a *cis* arrangement, where they are located on the same homolog.

transcription Transfer of genetic information from DNA by the synthesis of an RNA molecule copied from a DNA template.

transdetermination Change in developmental fate of a cell or group of cells.

transduction Virally mediated genes transfer from one bacterium to another, or the transfer of eukaryotic genes mediated by retrovirus.

transfer RNA See *tRNA*.

transformation Heritable change in a cell or an organism brought about by exogenous DNA.

transgenic organism An organism whose genome has been modified by the introduction of external DNA sequences into the germ line.

transition A mutational event in which one purine is replaced by another, or one pyrimidine is replaced by another.

translation The derivation of the amino acid sequence of a polypeptide from the base sequence of an mRNA molecule in association with a ribosome.

translocation A chromosomal mutation associated with the transfer of a chromosomal segment from one chromosome to another. Also used to denote the movement of mRNA through the ribosome during translation.

transmission genetics The field of genetics concerned with the mechanisms by which genes are transferred from parent to offspring.

transposable element A defined length of DNA that translocates to other sites in the genome, essentially independent of sequence homology. Usually such elements are flanked by short, inverted repeats of 20 to 40 base pairs at each end. Insertion into a structural gene can produce a mutant phenotype. Insertion and excision of transposable elements depends on two enzymes, transposase and resolvase. Such elements have been identified in both prokaryotes and eukaryotes.

transversion A mutational event in which a purine is replaced by a pyrimidine, or a pyrimidine is replaced by a purine.

trinucleotide repeat A tandemly repeated cluster of three nucleotides (such as CTG) in or near a gene, that undergo an expansion in copy number, resulting in a disease phenotype.

triploidy The condition in which a cell or organism possesses three haploid sets of chromosomes.

trisomy The condition in which a cell or organism possesses two copies of each chromosome, except for one, which is present in three copies. The general form for trisomy is therefore $2n + 1$.

tritium (^{3}H) A radioactive isotope of hydrogen, with a half-life of 12.46 years.

trivalent An association between three homologous chromosomes.

tRNA Transfer RNA; a small ribonucleic acid molecule that contains a three-base segment (anticodon) that recognizes a codon in mRNA, a binding site for a specific amino acid, and recognition sites for interaction with the ribosomes and the enzyme that links it to its specific amino acid.

tumor suppressor gene A gene that encodes a gene product that normally functions to suppress cell division. Mutations in tumor suppressor genes result in the activation of cell division and tumor formation.

Turner syndrome A genetic condition in human females caused by a 45,X genotype (XO). Such individuals are phenotypically female but are sterile because of undeveloped ovaries.

unequal crossing over A crossover between two improperly aligned homologs, producing one homolog with three copies of a region and the other with one copy of that region.

unique DNA DNA sequences that are present only once per genome. Single copy DNA.

universal code The assumption that the genetic code is used by all life forms. In general, this is true; some exceptions are found in mitochondria, ciliates, and mycoplasmas.

unwinding proteins Nuclear proteins that act during DNA replication to destabilize and unwind the DNA helix ahead of the replicating fork.

up promoter A promoter sequence, often mutant, that increases the rate of transcription initiation. Also known as strong promoter.

variable number tandem repeats (VNTRs) Short DNA sequences (2–20 nucleotides) present as tandem repeats between two restriction enzyme sites. Variations in the number of repeats creates DNA fragments of differing lengths following restriction enzyme digestion.

variable region Portion of an immunoglobulin molecule that exhibits many amino acid sequence differences between antibodies of differing specificities.

variance A statistical measure of the variation of values from a central value, calculated as the square of the standard deviation.

variegation Patches of differing phenotypes, such as color, in a tissue.

vector In recombinant DNA, an agent such as a phage or plasmid into which a foreign DNA segment will be inserted.

viability The measure of the number of individuals in a given phenotypic class that survive, relative to another class (usually wild type).

virulent phage A bacteriophage that infects and lyses the host bacterial cell.

VNTR See *variable number tandem repeats.*

W, Z chromosomes Sex chromosomes in species where the female is the heterogametic sex (WZ).

western blot A technique in which proteins are separated by gel electrophoresis and transferred by capillary action to a nylon membrane or nitrocellulose sheet. A specific protein can be identified through hybridization to a labeled antibody.

wild type The most commonly observed phenotype or genotype, designated as the norm or standard.

wobble hypothesis An idea proposed by Francis Crick stating that the third base in an anticodon can align in several ways to allow it to recognize more than one base in the codons of mRNA.

writhing number The number of times that the axis of a DNA duplex crosses itself by supercoiling.

X inactivation In mammalian females, the random cessation of transcriptional activity of one X chromosome. This event, which occurs early in development, is a mechanism of dosage compensation. Molecular basis of inactivation is unknown, but involves a region called the X-inactivation center (XIC) on the proximal end of the p arm. Some loci on the tip of the short arm of the X can escape inactivation. See *Barr body, Lyon hypothesis.*

XIST A locus in the X-chromosome inactivation center that may control inactivation of the X chromosome in mammalian females.

X linkage The pattern of inheritance resulting from genes located on the X chromosome.

X-ray crystallography A technique to determine the three-dimensional structure of molecules through diffraction patterns produced by X-ray scattering by crystals of the molecule under study.

YAC A cloning vector in the form of a yeast artificial chromosome, constructed using chromosomal elements including telomeres (from a ciliate), centromeres, origin of replication, and marker genes from yeast. YACs are used to clone long stretches of eukaryotic DNA.

Y chromosome Sex chromosome in species where the male is heterogametic (XY).

Y linkage Mode of inheritance shown by genes located on the Y chromosome.

Z-DNA An alternative structure of DNA in which the two antiparallel polynucleotide chains form a left-handed double helix. Z-DNA has been shown to be present along with B-DNA in chromosomes and may have a role in regulation of gene expression.

zein Principal storage protein of corn endosperm, consisting of two major proteins, with molecular weights of 19,000 and 21,000 daltons.

zinc finger A DNA-binding domain of a protein that has a characteristic pattern of cysteine and histidine residues that complex with zinc ions, throwing intermediate amino acid residues into a series of loops or fingers.

zygote The diploid cell produced by the fusion of haploid gametic nuclei.

zygotene A stage of meiotic prophase I in which the homologous chromosomes synapse and pair along their entire length, forming bivalents. The synaptonemal complex forms at this stage.

A P P E N D I X C

SOLUTIONS to SELECTED EVEN-NUMBERED PROBLEMS and DISCUSSION QUESTIONS

CHAPTER 1 An Introduction to Genetics

2. *Epigenesis* refers to the theory that organisms are derived from the assembly and reorganization of substances in the egg which eventually lead to the development of the adult. *Preformationism* is a seventeenth-century theory that states that the sex cells (eggs or sperm) contain miniature adults.

4. A significant gap in Darwin's theory was a lack of understanding of the sources of genetic variation and the mechanism of inheritance. Darwin proposed that such gemmules could adapt to an individual's environment.

6. *Transmission* genetics is the most classical approach in which the patterns of inheritance are studied through selective matings or the results of natural matings. With the discovery of mitotic and meiotic processes, and the knowledge that genes are located on chromosomes, much interest centers on the *cytological investigation* of chromosomes. *Molecular and biochemical* analysis of the genetic material has recently evolved into one of the most exciting and rapidly growing subdisciplines of genetics. *Recombinant DNA technology* has had a significant impact in this area as well as others. In *population* genetics, the interest is in the behavior of genes in groups of organisms (populations).

8. Norman Borlaug applied Mendelian principles of hybridization and trait selection to the development of superior varieties of wheat.

10. For over 20 years, Lysenko directed Soviet genetics based on the idea of inheritance of acquired characteristics: that plant productivity could be improved in an inherited fashion by changes in the environment.

CHAPTER 2 Cell Division and Chromosomes

2. Chromosomes which are homologous share many properties including: *overall length, position of the centromere, banding patterns, type and location of genes*, and *autoradiographic pattern*. *Diploidy* means that both members of a homologous pair of chromosomes are present. *Haploidy* specifically refers to the fact that each haploid cell contains *one chromosome of each homologous pair of chromosomes.*

6. Refer to Figure 2.4 for an explanation. Pay special attention to the right-hand portion of the figure and notice the different anaphase shapes.

8. Refer to Figure 2.8 for a diagram of mitosis. *Autoradiography* is a technique which can be used to determine that there is DNA synthesis during the interphase. Complexed CDK and cyclin proteins provide the mechanism for passage from one stage of the cell cycle to the next.

10. Compared with mitosis which maintains a chromosomal constancy, meiosis provides for a reduction in chromosome number, and an opportunity for exchange of genetic material from homologous chromosomes (see Figure 2.11).

12. Sister chromatids are genetically identical, except where mutations may have occurred during DNA replication. Nonsister chromatids are genetically similar if on homologous chromosomes or genetically dissimilar if on nonhomologous chromosomes. If crossing over occurs, chromatids attached to the same centromere will no longer be identical.

14. (a) 8 tetrads, (b) 8 dyads, (c) 8 monads, (d) 2^8

16. Through independent assortment of chromosomes at anaphase I of meiosis, daughter cells (secondary spermatocytes and secondary oocytes) may contain different sets of maternally and paternally derived chromosomes. Examine the diagram below. Notice that there are several ways in which the maternally and paternally derived chromosomes may align. Second, crossing over, which happens at a much higher frequency in meiotic cells as compared to mitotic cells, allows maternally and paternally derived chromosomes to exchange segments, thereby increasing the likelihood that daughter cells (secondary spermatocytes and secondary oocytes) are genetically unique.

18. One-half of each tetrad will have a maternal homolog: $(1/2)^{10}$.

22. The transition is at the end of interphase and the beginning of mitosis (prophase) when the chromosomes are in the condensation process.

24. Puffs represent active genes as evidenced by staining and uptake of labeled RNA precursors as assayed by autoradiography. (See Figure 2.18.)

26. Lampbrush chromosomes are diploid. Polytene chromosomes have high levels of ploidy.

28. CDK is the symbol given to a cell division cycle protein which is a kinase that couples with various cyclins to direct cells past certain critical points (checkpoints) in the cell cycle (G1/S and G2/M). p53 is a 53-kilodalton protein which acts as a tumor suppressor by regulating the transition between G1 and S.

CHAPTER 3 Mendelian Genetics

2. (a) Parents are *Aa*. (b) Male = either the *AA* or *Aa*. Female = *aa*. Children = *Aa*. The male *could* be *Aa*. Under that circumstance, the likelihood of having six children, all normal, is 1/64. (c) Male = *Aa*, female = *aa*. (d) The 1:1 ratio of albino to normal in the last generation theoretically results because the mother is *Aa* and the father is *aa*.

6. *P* = checkered; *p* = plain

	F_1 *Progeny*	
P₁ Cross	*Checkered*	*Plain*
(a) *PP × PP*	*PP*	
(b) *PP × pp*	*Pp*	
(c) *pp × pp*		*pp*
(d) *PP × pp*	*Pp*	
(e) *Pp × pp*	*Pp*	*pp*
(f) *Pp × Pp*	*PP, Pp*	*pp*
(g) *PP × Pp*	*PP, Pp*	

8. *WWgg* = 1/16

10. A test cross involves a fully heterozygous organism mated with a fully homozygous recessive organism. In Problem 9, (d) fits this description.

12. 1. Factors occur in pairs. 2. Some genes are dominant to their alleles. 3. Alleles segregate from each other during gamete formation. 4. One gene pair separates independently from other gene pairs.

16. (a) 4: AB, Ab, aB, ab; (b) 2: AB, aB; (c) 8: ABC, ABc, AbC, Abc, aBC, aBc, abC, abc; (d) 2: ABc, aBc; (e) 4: ABc, Abc, aBc, abc; (f) $2^5 = 32$.

18. G = yellow seeds; g = green seeds

	Phenotypes	Genotypes
P_1:	yellow × green	$GG \times gg$
F_1:	all yellow	Gg
F_2:	6022 yellow	1/4 GG; 2/4 Gg
	2001 green	1/4 gg

Of the yellow F_2 offspring, notice that 1/3 of them are GG and 2/3 are Gg. If you selfed the 1/3 GG types, then all the offspring (the 166) would breed true, whereas the others (353 which are Gg) should produce offspring in a 3:1 ratio when selfed.

20. Symbols:

Seed shape	Seed color
W = round	G = yellow
w = wrinkled	g = green

P_1: $WWgg \times wwGG$
F_1: $WwGg$ cross to $wwgg$
The offspring will occur in a typical 1:1:1:1.

22. (a) χ^2 = .064; p between 0.9 and 0.5. (b) $\chi^2 = 0.39$; p between 0.9 and 0.5.

24. $\chi^2 = 33.3$; $p < 0.01$ for 1 degree of freedom.

26. Recessive
I-1 (aa), I-2 (Aa), I-3 (Aa), I-4 (Aa)
II-1 (aa), II-2 (Aa), II-3 (aa), II-4 (Aa), II-5 (Aa), II-6 (aa), II-7 (AA or Aa), II-8 (AA or Aa)
III-1 (AA or Aa), III-2 (AA or Aa), III-3 (AA or Aa), III-4 (aa), III-5 (probably AA), III-6 (aa)
IV-1 through IV-7 are all Aa.

28. (a) There are two possibilities. Either the trait is dominant, in which case I-1 is heterozygous as are II-2 and II-3, or the trait is recessive and I-1 is homozygous and I-2 is heterozygous recessive. Under the condition of recessiveness, II-1 and II-4 would be heterozygous, II-2 and II-3 homozygous. (b) Recessive: Parents Aa, Aa. (c) Recessive: Parents Aa, Aa. (d) Recessive: Parents AA (probably), aa. Second pedigree: Recessive or dominant, not X-linked, if recessive, parents Aa, aa. (e) See initial explanation in this problem. It is identical to the first pedigree.

30. The probability is zero.

32. (a) 1/6, (b) 1/36, (c) 1/18

34. 1/9

38. (b) 1/12, (c) 6/12, (d) 5/12

40. Incomplete dominance or codominance

CHAPTER 4 Modification of Mendelian Ratios

4. S = short; s = long
Cross 1: $Ss \times ss \to$ 1/2 Ss (short), 1/2 ss (long)
Cross 2: $Ss \times Ss \to$ 1/4 SS (lethal), 2/4 Ss (short), 1/4 ss (long)

6. Male Parent: $I^B I^O$; Female Parent: $I^A I^O$
Offspring: 1:1:1:1

8. S = secretor; s = nonsecretor
Cross 1: $SS \times SS \to SS$
Cross 2: $ss \times ss \to ss$
Cross 3: $SS \times ss \to Ss$
Cross 4: $Ss \times Ss \to$ 1/4 SS; 2/4 Ss; 1/4 ss

10. (a) $c^a c^a \times c^{ch} c^a \to$ 1/2 chinchilla; 1/2 albino
(b) $c^a c^a \times C c^a \to$ 1/2 full color; 1/2 albino
(c) $c^h c^a \times c^h c^a \to$ 3/4 Himalayan; 1/4 albino

12. (a) $RRPPDD \times rrppdd \to RrPpDd$ (pink, personate, tall)
(b) 18/64

14. (a) $C^{ch}C^{ch}$ = chestnut; $C^c C^c$ = cremello; $C^{ch}C^c$ = palomino
(b) F_1 = palomino; F_2 = 1:2:1

16. (a) 9:3:4, (b) $AACc$, $AaCC$, $AaCc$

18. (a) gray, (b) gray, (c) 16:9:3:3:1, (d) 9:3:4, (e) 3:1:4

20. (a) Cross A:
P_1: $AABB \times aaBB$
F_1: $AaBB$
F_2: 3/4 A–BB:1/4 $aaBB$

Cross B:
P_1: $AABB \times AAbb$
F_1: $AABb$
F_2: 3/4 AAB–:1/4 $AAbb$

Cross C:
P_1: $aaBB \times AAbb$
F_1: $AaBb$
F_2: 9/16 A–B–:3/16 A–bb:3/16 aaB–:1/16 $aabb$

(b) The genotype of the unknown P_1 individual would be $AAbb$ (brown) while the F_1 would be $AaBb$ (green).

22. (a) The F_1 plants are yellow and the F_2 will occur in a 9:3:3:1 ratio. (b) The F_1 should be red ($aaBb$). The F_2 would be as follows: 3/4 aaB– (red):1/4 $aabb$ (mauve).

24. (a) 40 cm, (b) 1:4:6:4:1, (c) The possibilities for a 25-cm plant are: $A^1 a^2 b^1 b^2$, $a^1 A^2 b^1 b^2$, $a^1 a^2 B^1 b^2$, and $a^1 a^2 b^1 B^2$. The possibilities for a 35-cm plant are: $a^1 A^2 B^1 B^2$, $A^1 a^2 B^1 B^2$, $A^1 A^2 b^1 B^2$, and $A^1 A^2 B^1 b^2$.

26. (a,b) four gene pairs, (c) 3 cm for each increment, (d) An example: $AABBccdd \times aabbCCDD$, (e) many possibilities for an 18-cm plant: $AAbbccdd$, $AaBbccdd$, $aaBbCcdd$, etc. Any plant with seven upper-case letters will be 33 cm tall.

28. (a) It is likely that there are three gene pairs in this cross. The genotypes of the parents would be combinations of alleles which would produce a 6-cm ($aabbcc$) tail and a 30-cm ($AABBCC$) tail while the 18-cm offspring would have a genotype of $AaBbCc$. (b) A mating of an $AaBbCc$ (for example) pig with the 6-cm $aabbcc$ pig would result in a 1:3:3:1 ratio.

30. RG = normal vision; rg = color-blind
(a) 1/4, (b) 1/2, (c) 1/4, (d) zero

32. (a) P_1: $X^{sd}X^{sd} \times X^+/Y$
F_1: 1/2 X^+X^{sd} (female, normal), 1/2 X^{sd}/Y (male, scalloped)
F_2: 1/4 X^+X^{sd} (female, normal)
1/4 $X^{sd}X^{sd}$ (female, scalloped)
1/4 X^+/Y (male, normal)
1/4 X^{sd}/Y (male, scalloped)

34. 8/16 wild-type females; 5/16 wild-type males (4/16 because they have no *vermilion* gene and 1/4 because the X-linked, hemizygous *vermilion* gene is suppressed by *su-v/su-v*); 3/16 vermilion males.

36. (a) 3/16 males wild, 4/16 males white, 1/16 males sepia, 3/16 females wild, 4/16 females white, 1/16 females sepia, (b) 3/16 males wild, 4/16 males white, 1/16 males sepia, 6/16 females wild, 2/16 females sepia.

38. P$_1$: female: *RR* (red) × male: *rr* (mahogany)
F$_1$: 1/2 females (red), 1/2 males (mahogany)
F$_2$: 1/4 *RR*; 2/4 *Rr*; 1/4 *rr*

40. The fact that the father in couple #2 has hemophilia would not predispose his son to hemophilia.

42. (a) *AAB–* × *aaBB* (other configurations possible but each must give all offspring with *A* and *B* dominant alleles), (b) *AaB–* × *aaBB* (other configurations are possible but no *bb* types can be produced), (c) *AABb* × *aaBb*, (d) *AABB* × *aabb*, (e) *AaBb* × *Aabb*, (f) *AaBb* × *aabb*, (g) *aaBb* × *aaBb*, (h) *AaBb* × *AaBb*. Those genotypes which will breed true will be as follows: black = *AABB*, golden = all genotypes which are *bb*, brown = *aaBB*.

44. 27(purple):37(white)

46. (a) Hypothesize that two gene pairs are involved in the inheritance of one trait while one gene pair is involved in the other. (b) Because there is a (9:4:3) ratio regarding eye color, some gene interaction (epistasis) is indicated. (c,d) Symbolism: *R–* = rib-it; *rr* = knee-deep. For the purple class, "a 3/16 group" use the *A–bb* genotypes. The "4/16" class (green) would be the *aaB–* and the *aabb* groups. (e) F$_1$ of *AABbRr* would be blue-eyed and rib-it. The F$_2$ will follow a pattern of a 9:3:3:1 ratio. (f) The following genotypes can define the green phenotype: *aaBB*, *aaBb*, *aabb*, (g) Both parents would be *AabbRr*.

CHAPTER 5 Linkage, Crossing Over, and Chromosome Mapping

2. Proximity through synapsis, chiasmata.

4. Because crossing over occurs at the four-strand stage of the cell cycle (that is, after S phase) notice that each single crossover involves only two of the four chromatids.

6. Physical constraints

8.   *dp—cl————————ap*
      3 mu   39 mu

10. *RY/ry* × *ry/ry* = 10 map units

12. *PZ/pz* × *pz/pz* = 14 map units

14.

	Female A	Female B	Frequency
NCO	3, 4	7, 8	first
SCO	1, 2	3, 4	second
SCO	7, 8	5, 6	third
DCO	6, 5	1, 2	fourth

16. (a)   *y w +/+ + ct* × *y w +/Y*
(b)
   y————w————————ct
   0.0   1.5   20.0
(c) There were .185 × .015 × 1000 = 2.775 double crossovers expected. (d) No.

18. 0.20 wild type; 0.05 ebony; 0.05 pink; 0.20 pink, ebony; 0.20 dumpy; 0.05 dumpy, ebony; 0.05 dumpy, pink; 0.20 dumpy, pink, ebony. For the reciprocal cross: .25 wild type; .25 pink, ebony; .25 dumpy; .25 dumpy, pink, ebony.

20. (a,b) + *b c*/ *a* + +, *a – b* = 7 map units, *b – c* = 2 map units; (c) The progeny phenotypes that are missing are + + *c* and *a b* +, which, of 1000 offspring, 1.4 (.07 × .02 × 1000) would be expected.

24. (a) There would be $2^n = 8$ genotypic and phenotypic classes and they would occur in a 1:1:1:1:1:1:1:1 ratio. (b) There would be two classes and they would occur in a 1:1 ratio. (c) There are 10 map units between the *A* and *B* loci and locus *C* assorts independently from both *A* and *B* loci.

26. Progeny A: *Ro/rO* × *rroo* = 10 map units
Progeny B: *RO/ro* × *rroo* = 10 map units

28. *a – b* = 11 map units; *b – c* = 6 map units

30. (a) The *short* gene is on chromosome 2 with the *black* gene.
(b)   Females: *b sh p*   ×   Males: *b sh p*
          + + +   *b sh p*
The map distance between the two genes must be 15.

CHAPTER 6 Recombination and Mapping in Bacteria and Bacteriophages

2. (a) By placing a filter in a U-tube. (b) By treating cells with streptomycin, an antibiotic, it was shown that recombination would not occur if one of the two bacterial strains was inactivated. (c) An F$^+$ bacterium contains a circular, double-stranded, structurally independent, DNA molecule which can direct recombination.

6. The F$^+$ element can enter the host bacterial chromosome and upon returning to its independent state, it may pick up a piece of a bacterial chromosome. When combined with a bacterium with a complete chromosome, a partial diploid, or merozygote, is formed.

8. No linkage

10. A filter was placed between the two auxotrophic strains which would not allow contact. The treatment with DNase showed that the filterable agent was not naked DNA.

12. In *generalized transduction* virtually any genetic element from a host strain may be included in the phage coat and thereby be transduced. In *specialized transduction* only those genetic elements of the host which are closely linked to the insertion point of the phage can be transduced. Specialized transduction involves the process of lysogeny.

14. Viral recombination occurs when there is a sufficiently high number of infecting viruses so that there is a high likelihood that more than one type of phage will infect a given bacterium.

16. (a) The concentration of phage is greater than 10^4. (b) The concentration of phage is around 1.4×10^6. (c) The concentration of phage is less than 10^6.

18. (a) 2 × 3 = no lysis; 2 × 4 = lysis; 3 × 4 = lysis. (b) Major alteration which influences both regions. A deletion which overlaps both cistrons could cause such a major alteration. (c) 5 × 10^{-4} (d) Because mutant 6 complemented mutations 2 and 3, it is likely to be in the cistron with mutants 1,4, and 5. A lack of recombinants with mutant 4 indicates that mutant 6 is a deletion which overlaps mutation 4. Recombinants with 1 and 5 indicates that the deletion does not overlap these mutations.

20. (a)

Combination	Complementation
1, 2	–
1, 3	+
2, 4	+
4, 5	–

(b) $2(8 \times 10^2/4 \times 10^7) = 4 \times 10^{-5}$

(c) The dilution would be 10^{-3} and the colony number would be 8×10^3. (d) Mutant 7 might well be a deletion spanning parts of both A and B cistrons.

22. (a) Rifampicin eliminates the donor strain which is rif^s

(b) _b a_ _____ _c_←————→_c_ F

(c) The interrupted mating experiment is conducted as usual on an ampicillin-containing medium but the recombinants must be replated on a rifampicin medium to determine which ones are sensitive.

CHAPTER 7 Extensions of Genetic Analysis

2. Height, general body structure, skin color, and perhaps most common behavioral traits including intelligence.

6. (a) Offspring from this cross can range from very tall _RrSSTTUu_ (12 "tall" units) to very short _rrssttuu_ (8 "small" units). (b) If the individual with a minimum height, _rrssttuu_, is married to an individual of intermediate height, _RrSsTtUu_, the offspring can be no taller than the height of the tallest parent.

8. $h^2 = (7.5 - 8.5/6.0 - 8.5) = 0.4$. Selection will have little relative influence on olfactory learning in _Drosophila_.

10. 10 map units

12. For Cross 1: 50 map units, not linked; for Cross 2: 12 map units.

14.

(a)

Tetrad in Problem	Class
1	NP
2	T
3	P
4	NP
5	T
6	P
7	T

(b) The genes are linked.

(c) For the centromere to _c_ distance: 7.2 map units. For the centromere to _d_ distance: 15.9 map units. The map would be, according to these figures:

$$\textbf{C}\text{———}c\text{————}d$$
$$\leftarrow\!7.2\!\rightarrow\!\leftarrow 15.9 \longrightarrow$$

(d) 20 map units

(e)

$$\textbf{C}\text{———}c\text{————}d$$
$$\leftarrow\!7.9\!\rightarrow$$
$$\longleftarrow 16.4 \longrightarrow$$

16. (a) mean = 140 cm, (b) = 374.18, (c) The _standard deviation_ is the square root of the variance or 19.34. (d) The _standard error of the mean_ is about 0.70.

18. (a) The simplest explanation would involve two gene pairs, with each additive gene contributing about 1.2 mm to the phenotype. (b) The fit to this backcross supports the original hypothesis. (c) These data do not support the simple hypothesis provided in part (a). (d) With these data, one can see no distinct phenotypic classes suggesting that the environment may play a role in eye development or that there are more genes involved.

CHAPTER 8 Extranuclear Inheritance

2. All of the offspring must have the phenotype of the mother's genotype, which is dextral.

4. (a) green, (b) white, (c) variegated, (d) green

6. The _petite_ gene is recessive.

8. (a) $Kk \times kk$, (b) any case where there is no kk such as: $KK \times KK$ or $KK \times kk$, (c) $Kk \times Kk$

10. Parents: $Dd \times dd$

Offspring (F_1): 1/2 Dd, 1/2 dd (all dextral because of the maternal genotype)

Progeny (F_2): All those from Dd parents will be dextral, while all those from dd parents will be sinistral.

12. Since there is no evidence for segregation patterns typical of chromosomal genes and Mendelian traits, some form of extranuclear inheritance seems possible. If the _lethargic_ gene is dominant then a maternal effect may be involved. In that case, some of the F_2 progeny would be hyperactive, because maternal effects are only temporary, affecting only the immediate progeny. If the lethargic condition is caused by some infective agent, then perhaps injection experiments could be used. If caused by a mitochondrial defect, then the condition would persist in all offspring of lethargic mothers, through more than one generation.

14. (a) The presence of bcd^-/bcd^- males can be explained by the maternal effect: mothers were bcd^+/bcd^+. (b) The cross female $bcd^+/bcd^- \times$ male bcd^-/bcd^- will produce an F_1 with normal embryogenesis because of the maternal effect. In the F_2, any cross having bcd^+/bcd^- mothers will have phenotypically normal embryos.

16. (a) A locus, Segregation Distortion (_SD_), is present on the wild-type chromosome. (b) One could use this _SD_ chromosome in a variety of crosses and determine that the abnormal segregation is based on a particular chromosomal element. (c) Segregation Distortion describes a condition in which typical Mendelian segregation is distorted from the 50:50.

CHAPTER 9 Chromosome Variation and Sex Determination

2. Fertilization, by a Y-bearing sperm cell, of those female gametes with two X chromosomes would produce the XXY Klinefelter syndrome. Fertilization of the "no-X" female gamete with a normal X-bearing sperm will produce the Turner syndrome.

4. There will be one fewer Barr body than number of X chromosomes.

8. Phenotypic mosaicism is dependent on the heterozygous condition of genes on the two X chromosomes.

16. While several trisomies (for chromosomes 21, 18, 13, the X and Y) are tolerated, monosomy for the autosomes is not tolerated. The delicate genetic balance produced by millions of years of evolution must be maintained in order for any organism (but especially animals) to develop normally. Monosomy leads to the exposure of recessive, deleterious genes, thus producing developmental abnormalities. Dosage compensation of the sex chromosomes and the relative paucity of Y-linked genes probably contribute to the survival of sex-chromosome aneuploidy.

18. Because an allotetraploid has a possibility of producing bivalents at meiosis I, it would be considered the most fertile of the three.

20. If the diploid chromosome number is 18, $2n = 18$, then in the somatic nuclei of haploid individuals $n = 9$, triploid ($3n$) = 27, and tetraploid ($4n$) = 36.

22. The section which has no homolog will "loop out" as in Figure 9.21.

24. Examine Figure 9.26 and notice that in (a) there are two genetically balanced chromatids (normal and inverted) and two, those resulting from a single crossover in the inversion loop, which are genetically unbalanced and abnormal (dicentric and acentric). The dicentric chromatid will often break, thereby producing highly abnormal fragments whereas the acentric fragment is often lost in the meiotic process. In part (b) all the chromatids have centromeres, but the two chromatids involved in the crossover are genetically unbalanced. The balanced chromatids are of normal or inverted sequence.

26. In a work entitled *Evolution by Gene Duplication*, Ohno suggests that gene duplication has been essential in the origin of new genes.

28. If the father had hemophilia it is likely that the Turner syndrome individual inherited the X chromosome from the father and no sex chromosome from the mother. If nondisjunction occurred in the mother, either during meiosis I or meiosis II, an egg with no X chromosome can be the result.

30. Individual organisms with 27 chromosomes ($3n$) are more likely to be sterile because there are trivalents at meiosis I, which causes a relatively high number of unbalanced gametes to be formed.

32. Let b = bent bristles; b^+ = normal bristles

(a)

F_1:

$-/b^+$ = normal bristles
b/b^+ = normal bristles

F_2:

$-/b^+ \times b/b^+$
↓
$-/b^+$ = normal bristles
$-/b$ = bent bristles
b^+/b^+ = normal bristles
b/b^+ = normal bristles

(b)

$-/b^+ \times b/b$
↓

F_1:

$-/b$ = bent bristles
b/b^+ = normal bristles

F_2:

$-/b \times b/b^+$
↓
$-/b^+$ = normal bristles
$-/b$ = bent bristles
b^+/b = normal bristles
b/b = bent bristles

34. 35 W and 1 w

36. (a) In all probability, crossing over in the inversion loop of an inversion (in the heterozygous state) had produced defective, unbalanced chromatids. (b) It is probable that a significant proportion (perhaps 50%) of the children of the man will be similarly influenced by the inversion. (c) Since the karyotypic abnormality is observable, it may be possible to detect some of the abnormal chromosomes of the fetus by amniocentesis or CVS.

38. Because the Klinefelter son is $X^{g1}X^{g2}Y$, he must have obtained the X^{g1} allele and the Y chromosome from the father. Thus nondisjunction must have occurred during meiosis I in the father.

40. The presence of the Y chromosome provides a factor (or factors) which leads to the initial specification of maleness. Subsequent expression of secondary sex characteristics must be dependent on the interaction of the normal X-linked *Tfm* allele with testosterone. Without such interaction, differentiation takes the female path.

42. (a) Reciprocal translocation. (c) Notice that all chromosomal segments are present and there is no apparent loss of chromosomal material. However, if the breakpoints for the translocation occurred within genes then an abnormal phenotype may be the result. In addition, a gene's function is sometimes influenced by its position, its neighbors in other words. If such "position effects" occur then a different phenotype may result. (d) It is likely that the translocation is the cause of the miscarriages.

CHAPTER 10 Structure and Analysis of DNA and RNA

2. Proteins are composed of as many as 20 different subunits (amino acids) thereby providing ample structural and functional variation for the multiple tasks which must be accomplished by the genetic material. The tetranucleotide hypothesis (structure) provided insufficient variability to account for the diverse roles of the genetic material.

4. DNase eliminates DNA and transformation, therefore it must be the transforming principle.

6. The T2 phage, in its mature state, contains very little if any RNA, therefore DNA would be interpreted as being the genetic material in T2 phage.

10. Linkages among the three components require the removal of water (H_2O).

12. Guanine: 2-amino-6-oxypurine, Cytosine: 2-oxy-4-amino-pyrimidine, Thymine: 2,4-dioxy-5-methylpyrimidine, Uracil: 2,4-dioxypyrimidine

16. The lack of pairing of these bases favors a single-stranded structure or some other nonhydrogen-bonded structure. Alternatively, from the data it would appear that A=G and T=C which would require purines to pair with purines and pyrimidines to pair with pyrimidines.

18. (1) Uracil in RNA replaces thymine in DNA, (2) ribose in RNA replaces deoxyribose in DNA, and (3) RNA often occurs as both single- and double-stranded forms, whereas DNA most often occurs in a double-stranded form.

20. The nitrogenous bases of nucleic acids (nucleosides, nucleotides, and single- and double-stranded polynucleotides), absorb UV light maximally at wavelengths 254 to 260 nm. One can often determine the presence and concentration of nucleic acids in a mixture.

22. Because G-C base pairs are more compact, they are more dense than A-T pairs. The percentage of G-C pairs in DNA is thus proportional to the buoyant density of the molecule.

24. For curve A in the problem, there is evidence for a rapidly renaturing species (repetitive) and a slowly renaturing species (unique). The fraction which reassociates faster than the *E. coli* DNA is highly repetitive and the last fraction (with the highest $C_0t_{1/2}$ value) contains primarily unique sequences. Fraction B contains mostly unique, relatively complex DNA.

26. Because G-C base pairs are formed with three hydrogen bonds while A-T base pairs by two such bonds, it takes more energy (higher temperature) to separate G-C pairs.

28. In one sentence of Watson and Crick's paper in *Nature*, they state: "It has not escaped our notice that the specific pairing we have postulated immediately suggests a possible copying mechanism for the genetic material."

30. For MS2, X = 200 base pairs. For *E. coli*, X = 2×10^6 base pairs.

32. Left side (a) = left; right side (b) = right.

34. Under this condition, the hydrolyzed 5-methylcytosine becomes thymine.

36. (a) Heat application would yield a hyperchromic shift if the DNA is double-stranded. One could also get a rough estimation of the GC content from the kinetics of denaturation and the degree of sequence complexity from comparative renaturation studies. (b) estimation of base content by hydrolysis and chromatography, (c) Antibodies for Z-DNA could be used to determine the degree of left-handed structures, if present. (d) Sequencing the DNA from both viruses would indicate sequence homology. In addition, through various electronic searches readily available on the Internet (*blast@ncbi.nlm.nih.gov*, for example) one could determine whether similar sequences exist in other viruses or in other organisms.

CHAPTER 11 DNA Replication and Recombination

2. A comparison of the density of DNA samples at various times in the experiment (initial ^{15}N culture, and subsequent cultures grown in the ^{14}N medium), showed that after one round of replication in the ^{14}N medium, the DNA was half as dense (intermediate) as the DNA from bacteria grown only in the ^{15}N medium. In a sample taken after two rounds of replication in the ^{14}N medium, half of the DNA was of the intermediate density and the other half was as dense as DNA containing only ^{14}N DNA.

4. Those cells which pass through the S phase in the presence of the ^{3}H-thymidine are labeled and that each double helix (per chromatid) is "half-labeled." (a) Under a conservative scheme all of the newly labeled DNA will go to one sister chromatid, while the other sister chromatid will remain unlabeled. In contrast to a semiconservative scheme, the first replicative round would produce one sister chromatid which has label on both strands of the double helix. (b) Under a dispersive scheme all of the newly labeled DNA will be interspersed with unlabeled DNA.

6. The *in vitro* replication requires a DNA template, a divalent cation (Mg^{++}), and all four of the deoxyribonucleoside triphosphates: dATP, dCTP, dTTP, and dGTP. The lower case "d" refers to the deoxyribose sugar.

8. By comparing *nearest neighbor frequencies*, Kornberg determined that there is a very high likelihood that the product was of the same base sequence as the template.

12. *Biologically active* DNA implies that the DNA is capable of supporting typical metabolic activities of the cell or organism and is capable of faithful reproduction.

18. Given a stretch of double-stranded DNA, one could initiate synthesis at a given point and either replicate strands in one direction only (unidirectional) or in both directions (bidirectional), as shown below. Notice that in Figure 11.13, the synthesis of complementary strands occurs in a *continuous* 5'→3' mode on the leading strand in the direction of the replication fork, and in a *discontinuous* 5'→3' mode on the lagging strand.

20. *Okazaki fragments* are relatively short (1000 to 2000 bases in prokaryotes) DNA fragments which are synthesized in a discontinuous fashion on the lagging strand during DNA replication. *DNA ligase* is required to form phosphodiester linkages in gaps which are generated when DNA polymerase I removes RNA primer and meets newly synthesized DNA ahead of it. *Primer*

RNA is formed by RNA primase to serve as an initiation point for the production of DNA strands on a DNA template.

22. Eukaryotic DNA is replicated in a manner which is very similar to that of *E. coli*. Synthesis is bidirectional, continuous on one strand and discontinuous on the other, and the requirements of synthesis (four deoxyribonucleoside triphosphates, divalent cation, template, and primer) are the same. Okazaki fragments of eukaryotes are about one-tenth the size of those in bacteria.

24. (a) In *E. coli*, 100 kb are added to each growing chain per minute. Therefore the chain should be about 4,000,000 bp. (b) 1.36×10^6 nm or 1.3 mm

26. *Gene conversion* is likely to be a consequence of genetic recombination in which nonreciprocal recombination yields products in which it appears that one allele is "converted" to another.

28. About 8,800 replication sites

30. If replication is conservative, the first autoradiographs would have label distributed only on one side (chromatid) of the metaphase chromosome.

CHAPTER 12 Storage and Expression of Genetic Information

2. No.

4.

Codons:		CUA	CUA	CUA	CUA...
Amino Acids:		leu	leu	leu	leu...
		UAC	UAC	UAC	UAC...
		tyr	tyr	tyr	tyr...
		ACU	ACU	ACU	ACU...
		thr	thr	thr	thr...
Codons:	ACG	UAC	GUA	CGU	ACG...
Amino Acids:	thr	tyr	val	arg	thr...
	CGU	ACG	UAC	GUA	CGU...
	arg	thr	tyr	val	arg...
	GUA	CGU	ACG	UAC	GUA..
	val	arg	thr	tyr	val...
	UAC	GUA	CGU	ACG	UAC...
	tyr	val	arg	thr	tyr...

6. Threonine would have the codon ACA.

8. A relatively large initiation complex is formed which contains the ribosome, the tRNA, and the trinucleotide. This complex is trapped in the filter whereas the components by themselves are not trapped. If the amino acid on a charged, trapped tRNA is radioactive, then the filter becomes radioactive.

10. Apply the most conservative pathway of change.

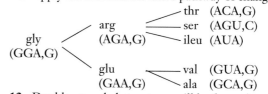

12. Double-stranded structures will be formed.

14. (b) TCCGCGGCTGAGATGA (use complementary bases, substituting T for U), (c) GCU, (d) arg-arg-arg-leu-tyr

16. (a) met-his-thr-tyr-glu-thr-leu-gly *and* met-arg-pro-leu-gly, (b) In the shorter of the two reading sequences (the one using the internal AUG triplet), a UGA triplet was introduced at the second codon.

18. The central dogma of molecular genetics and to some, all of biology, states that DNA produces, through transcription, RNA, which is "decoded" (during translation) to produce proteins.

22. A functional polyribosome will contain the following components: mRNA, charged tRNA, large and small ribosomal subunits, elongation and perhaps initiation factors, peptidyl transferase, GTP, Mg^{++}, nascent proteins, possibly GTP-dependent release factors.

24. With an adaptor molecule, specific hydrogen bonding could occur between nucleic acids, and specific covalent bonding could occur between an amino acid and a nucleic acid tRNA.

26. Approximately 20

28. attachment of the specific amino acid, interaction with the aminoacyl tRNA synthetase, interaction with the ribosome, and interaction with the codon (anticodon)

30. (a) Sequence 1: GAAAAAACGGUA
Sequence 2: UGUAGUUAUUGA
Sequence 3: AUGUUCCCAAGA
(b) Sequence 1: glu-lys-thr-val
Sequence 2: cys-ser-tyr
Sequence 3: met-phe-pro-arg
(c) Sequence 1 = middle. Sequence 2 = terminal portion. Sequence 3 = initial portion. (d) Apply complementary bases: GAAAAAACGGTA.

32. The amino acid is not involved in recognition of the anticodon.

CHAPTER 13 Proteins: The End Product of Genetic Expression

2. Even though individuals with PKU cannot convert phenylalanine to tyrosine, it is obtained from the diet.

4. The blocks are ordered: *b, c, a.*

6. (a) The F_1 is *AaBb* and pigmented (purple). The typical F_2 ratio would be as follows:

9/16	*A–B–*	purple
3/16	*aaB–*	white
3/16	*A–bb*	white
1/16	*aabb*	white

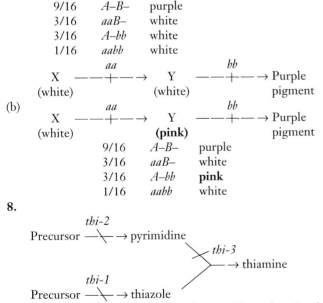

(b)

9/16	*A–B–*	purple
3/16	*aaB–*	white
3/16	*A–bb*	**pink**
1/16	*aabb*	white

8.

14. It is possible for an amino acid to change without changing the electrophoretic mobility of a protein under standard conditions.

16. Mutations occurring early in a gene will produce proteins with defects near the N-terminus. In this problem, the lesions cause chain termination; therefore, the nearer the mutations are to the 5′-end of the mRNA, the shorter will be the polypeptide product.

22. All of the substitutions involve one base change.

24. Because cross (a) is essentially a monohybrid cross, there would be no difference in the results if crossing over occurred (or did not occur) between the *a* and *b* loci.

26. A cross of the following nature would give results to satisfy the data:

$$AABBCC \times aabbcc$$

pink $\xrightarrow[]{c}$ rose $\xrightarrow[]{b}$ orange $\xrightarrow[]{a}$ purple

CHAPTER 14 Gene Mutation, DNA Repair, and Transposable Elements

2. Mutations are the "windows" through which geneticists look at the normal function of genes, cells, and organisms.

4. *All* mutations may not be deleterious. Those few, rare variations which are beneficial will provide a basis for possible differential propagation of the variation.

6. A diploid organism possesses at least two copies of each gene (except for "hemizygous" genes) and in most cases, the amount of product from one gene of each pair is sufficient for production of a normal phenotype, which under one environmental condition leads to premature death of the organism.

10. The rate of induced mutation for each essential X-linked gene is less than 1/500.

12. Tautomeric forms, caused by single proton shifts, could exist for the nitrogenous bases of DNA. Such shifts could result in mutations by allowing hydrogen bonding of normally non-complementary bases.

14. Frameshift mutations are likely to change more than one amino acid in a protein. In addition, there is the possibility that a nonsense triplet could be introduced, thus causing premature chain termination. If a single pyrimidine or purine has been substituted, then only one amino acid is influenced.

16. X-rays penetrate surface layers of cells and thus can affect gamete-forming tissues in multicellular organisms. UV light generates pyrimidine dimers, primarily thymine, which distort the normal conformation of DNA and inhibit normal function.

18. Because X-rays are known to be mutagenic, it has been suggested that frequent mammograms may do harm.

22. Studies with heterokaryons provided evidence for complementation, indicating that there may be as many as seven different genes involved. The photoreactivation repair enzyme appears to be involved.

24. 1.5×10^{-6}

26. It is probable that the IS occupied or interrupted normal function of a controlling region related to the galactose genes.

28.

Group 1	Group 2	Group 3
XP1	*XP4*	*XP5*
XP2		*XP6*
XP3		*XP7*

30. Mutations which radically alter the structure of the protein (frameshift, splicing, nonsense, deletions, duplications, etc.) would probably have more influence on protein function than those which cause relatively minor amino acid substitutions, although this generalization does not always hold true. A protein with multiple functional domains would be expected to react to mutational insult in a variety of ways.

CHAPTER 15 Recombinant DNA Technology

2. The poly-dT segment provides a double-stranded section which serves to prime the production of the complementary strand.

4. 1.38×10^5

6. Cuts will be made such that a four-base single-stranded set of sticky ends will be produced. For the antibiotic resistance to be present, the ligation will reform the plasmid into its original form. However, two of the plasmids can join to form a dimer.

8. Generally for large genomes, it is best to use a vector which will accept relatively large fragments.

10. Assuming a random distribution of all four bases, the four-base sequence would occur (on average) every 256 base pairs (4^4), the six-base sequence every 4096 base pairs (4^6), and the eight-base sequence every 65,536 base pairs (4^8). One might use an eight-base restriction enzyme to produce a relatively few large fragments.

12. A filter is used to bind the DNA from the colonies and a labeled probe is used to detect, through hybridization, the DNA of interest.

18. With repeated sequences in the genome, chromosome walking is complicated because the clone hybridizes to multiple regions. One can "chromosome jump" over repeated sequences.

20. (a) Starting with zero at the top, the various patterns tell us that there is an E site at 1000 bp, an A site at 500 bp, and a B site at 2500 bp. (b) The probe hybridizes consistently to the 2000 bp fragment between the A and B restriction sites.

CHAPTER 16 Applications of Recombinant DNA Technology

4. (a,b) (1) Integration into the host must be cell specific so as not to damage non-target cells. (2) Retroviral integration into host cell genomes only occurs if the host cell is replicating. (3) Insertion of the viral genome might influence non-target but essential genes. (4) Retroviral genomes have a low cloning capacity and cannot carry large inserted sequences as are many human genes. (5) There is a possibility that recombination with host viruses will produce an infectious virus which may do harm.

6. Positional cloning is a technique whereby the linkage group of a genetic disorder is determined by association of certain RFLPs (as markers). A more precise location is obtained by association (linkage studies in kindreds) with additional RFLP markers. It is unlikely that this technique will be successfully applied to genetically complex traits in the near future.

8. Even though you have developed a method for screening seven of the mutations described, it is possible that negative results can occur even though the person carries the gene for CF.

10. In the case of haplo-insufficient mutations, gene therapy holds promise; however in "gain-of-function" mutations, in all probability the mutant gene's activity or product must be compromised.

12. One method is to use the amino acid sequence of the protein to produce the gene synthetically. Alternatively, since the introns are spliced out of the hnRNA in the production of mRNA, if mRNA can be obtained, it can be used to make DNA through the use of reverse transcriptase.

14. (a) Y-linked excluded, X-linked recessive excluded, autosomal recessive possible but unlikely, X-linked dominant possible if heterozygous, autosomal dominant possible. (b) Chromosome 21 with the B1 marker probably contains the mutation. (c) The disease gene is segregating with some certainty with the B1 RFLP marker in the family. However, this prediction is not completely accurate because a crossover in the mother could put the undesirable gene with the B3 marker. (d) A crossover between the restriction sites in the father giving a B1 chromosome, or a mutation eliminating either the B2 or B3 restriction site.

CHAPTER 17 Genomic Organization of DNA

4. Digestion of chromatin with endonucleases, such as micrococcal nuclease, gives DNA fragments of approximately 200 base pairs or multiples of such. X-ray diffraction data indicated a regular spacing of DNA in chromatin. Regularly spaced beadlike structures (nucleosomes) were identified by electron microscopy.

8. Volume of DNA: 1.57×10^8 Å^3

Volume of capsid: 2.67×10^8 Å^3

The DNA will fit into the capsid.

10. *Exon shuffling* is a term used to describe the likely phenomenon whereby the exons of genes, each coding for functional domains of proteins, can be "shuffled" or rearranged to facilitate the evolution of a new protein.

14. A "C value" is the amount of DNA contained in a haploid genome. The *C value paradox* recognizes that with evolutionary divergence there has been a dramatic divergence in the amount of DNA among different taxa.

16. With consideration given to the 5′–3′ pairing restrictions required of the RNA/DNA hybrid, there should be single-stranded "loops" of DNA in places where the introns have been spliced out of the RNA.

CHAPTER 18 Regulation of Gene Expression in Bacteria and Phages

2. Under *negative* control, the regulatory molecule interferes with transcription while in *positive* control, the regulatory molecule stimulates transcription. Catabolite repression and a portion of the *arabinose* regulatory systems are examples of positive control.

4. The IPTG-binding protein was labeled with sulfur-containing amino acids and mixed with DNA from λ phage which contained the *lac* O^+ section of DNA.

6.

$I^+O^+Z^+$	= Inducible
$I^-O^+Z^+$	= Constitutive
$I^+O^CZ^+$	= Constitutive
$I^-O^+Z^+/$ F′I^+	= Inducible
$I^+O^CZ^+/$ F′O^+	= Constitutive
$I^SO^+Z^+$	= Repressed
$I^SO^+Z^+/$F′I^+	= Repressed

8. The *cI* gene is responsible for repressing the genes which control the lytic cycle of λ phage. If the *cI* gene is mutant then the lysogenic cycle could not occur.

10. C codes for the **structural gene**, B must be the **promoter**.

12. Because the operon by itself (when mutant as in strain 3) gives constitutive synthesis of the structural genes, *cis*-acting system is also supported. The *cis*-acting element is most likely part of the operon.

14. You will need to identify the complementary regions, then arrange them as in Figure 18.9. You will find four regions which "fit." To get started, find the CACUUCC sequence. It pairs, with one mismatch with a second region. Hint: The third region is composed of seven bases and starts with an A.

CHAPTER 19 Regulation of Gene Expression in Eukaryotes

4. *Promoters* are conserved DNA sequences which influence transcription from the "upstream" side (5′) of mRNA coding genes. *Enhancers* are *cis*-acting sequences of DNA which stimulate the transcription from most, if not all, promoters.

6. (a) There are two complementation groups:

AFEHI and BDGJ

(b) Strain "C" probably produces a controlling element which in some way exerts "dominant" control. (c) If "C" acts as a dominant (which it probably does), it may be in either or neither complementation group. It would be difficult to tell from the data given.

8. Only the engineered mRNA sequences which caused an amino acid substitution negated the autoregulation, indicating that it is the sequence of the amino acids, not the mRNA, which is critical in the process of autoregulation. One might stabilize the proposed MREI-protein complex with "crosslinkers," treat with RNase to digest mRNA and to break up polysomes, then isolate individual ribosomes. One may use some specific antibody or other method to determine whether tubulin subunits contaminate the ribosome population.

10. (a) There is no place for the TFIID to bind. (b) There is more transcription in the nuclear extracts. (c) There is a region probably in the −81 to −50 area which responds to a component in the nuclear extract to bring about high efficiency transcription.

CHAPTER 20 Developmental Genetics

2. One could consider that one "selector" gene distinguishes aristal from tarsal structures. Notice that a "one-step" change is involved in the interchange of leg and antennal structures.

4. If a labeled probe can be obtained which contains base sequences that are complementary to the transcribed RNA, then such probes will hybridize to that RNA if present in different tissues. This technique is called *in situ* hybridization.

6. The *ftz* gene product regulates, either directly or indirectly, *eng*.

8. Combinations of injected nuclei may reveal nuclear–nuclear interactions which could not normally be studied by other methods.

10.

Female *dd* × Male *DD*

all offspring *Dd* (phenotypically "d")

Female *Dd* × Male *Dd*

1/4 *DD* 1/2 *Dd* 1/4 *dd*

(phenotypically "D")

12. (a) The *her-1⁺* gene produces a product which suppresses hermaphrodite development, while the *tra-1⁺* gene product is needed for hermaphrodite development. (b) The double mutant should be male.

CHAPTER 21 Genetics and Cancer

2. The G1 stage begins after mitosis and is involved in the synthesis of many cytoplasmic elements. In the S phase DNA synthesis occurs. G2 is a period of growth and preparation for mitosis. Most cell cycle time variation is caused by changes in the duration of G1. G0 is the non-dividing state.

4. Kinases regulate other proteins by adding phosphate groups. Cyclins bind to the kinases, switching them on and off. Several cyclins, including D and E, can move cells from G1 to S. At the G2/mitosis border a CDK1 (cyclin dependent kinase) combines with another cyclin (cyclin B). Phosphorylation occurs, bringing about a series of changes in the nuclear membrane, cytoskeleton, and histone 1.

6. A tumor suppressor gene is a gene that normally functions to suppress cell division. If a tumor suppressor gene makes a product that regulates the cell cycle favorably, cellular conditions have evolved in such a way that sufficient quantities of this gene product are made from just one gene (of the two present in each diploid individual) to provide normal function.

8. Oncogenes induce or maintain uncontrolled cellular proliferation associated with cancer. They are mutant forms of proto-oncogenes which normally function to regulate cell division.

12. (a) If the G mutated to an A (transition), then the transcribing DNA strand would be

3′-ATC(T)-5′

which would cause a UAG(A) triplet to be produced and this would cause the stop. (b) Tumor suppression. (c) Some women may carry genes (perhaps mutant) which "spare" for the *BRCA1* gene product. Some women may have immune systems which recognize and destroy precancerous cells or they may have mutations in breast signal transduction genes so that cell division suppression occurs in the absence of *BRCA1*.

CHAPTER 22 Genetic Basis of the Immune Response

2. There would be a possibility of 18 more combinations, 27 total possible.

4. 10^6

6. A *helper T cell* is a subtype of T cell which serves as the master switch for the immune system, serving to "turn on" the immune response. *Suppressor T cells* are the "off" switch of the immune system; they stop or slow down the immune response.

8. Individuals with type AB blood do not produce antibodies against the A or B antigens, therefore they can receive blood from individuals with either or both of these antigens. Since type O blood does not have A or B antigens, it can be given to all other blood types (assuming other antigens or factors are compatible).

10. The *HLA haplotype* is an array of HLA alleles on a given copy of chromosome 6 in humans. We each carry two copies of chromosome 6 and therefore we each have two haplotypes.

12. Antibody production involves somatic recombination which results in a change in DNA positioning.

CHAPTER 23 The Genetics of Behavior

4. Mapping the primary focus of the gene could be accomplished with patience and the use of the unstable ring-X chromosome to generate gynandromorphs. Given that the gene is X-linked, one would use classical recombination methods to place a recessive X-linked marker, such as *singed bristles*, on the X chromosome with the gene causing the limp.

6. Several problems in the study of human behavioral genetics would include the following:

1. With a relatively small number of offspring produced per mating, standard genetic methods of analysis are difficult.
2. Records on family illnesses, especially behavioral illnesses, are difficult to obtain.
3. The long generation time makes longitudinal (transmission genetics) studies difficult.
4. The scientist cannot direct matings which will provide the most informative results.
5. The scientists cannot always subject humans to the same types of experimental treatments as with other organisms.
6. Traits which are of interest to study are often extremely complex and difficult to quantify.

8. Self-fertilization, the ultimate form of inbreeding, greatly enhances the likelihood that recessive genes will become homozygous.

CHAPTER 24 Population Genetics

2. $AA = .25$ or 25%, $Aa = .5$ or 50%, $aa = .25$ or 25%.
The initial population was not in equilibrium; however, after one generation of mating under the Hardy-Weinberg assumptions, the population is in equilibrium.

4. In order for the Hardy-Weinberg equations to apply, the population must be in equilibrium.

6. (a) $MM = .6014$ or 60.14%, $MN = .3482$ or 34.82%, $NN = .0504$ or 5.04%. The population is in equilibrium. (b) $AA = .7691$ or 76.91%, $AS = .2157$ or 21.57%, $SS = .0151$ or 1.51%. $\chi^2 = 1.47$, the frequencies of AA, AS, SS sampled a population which is in equilibrium.

8. (a) $q_1 = .23$, $p_1 = .77$; (b) $q_1 = .267$ $p_1 = .733$; (c) $q_1 = .293$, $p_1 = .707$; (d) $q_1 = .299$, $p_1 = .701$

10. (a) $p_1 = 0.6 + 0.2(0.1 - 0.6) = 0.5$
(b) $p_1 = 0.2 + 0.3(0.7 - 0.2) = 0.35$
(c) $p_1 = 0.1 + 0.1(0.2 - 0.1) = 0.11$

12. *Inbreeding depression* refers to the reduction in fitness observed in populations which are inbred.

14. Inbreeding schemes will often be used to render strains homozygous so that such recessive genes can be expressed.

16. (a) *self-fertilize* or brother-sister matings
(b) With self-fertilization, the percentage of homozygous individuals increases dramatically.

18. The frequency of a gene is determined by a number of factors including the fitness it confers, mutation rate, and input from migration. There is no tendency for a gene to reach any artificial frequency such as 0.5.

20. The overall probability of the couple producing a CF child is $98/2500 \times 2/3 \times 1/4$.

22. Given small populations and very similar environmental conditions, it is more likely that "sampling error" or genetic drift is operating.

CHAPTER 25 Genetics and Evolution

4. Because of degeneracy in the code, there are some nucleotide substitutions, especially in the third base, which do not change amino acids. In addition, if there is no change in the overall charge of the protein, it is likely that electrophoresis will not separate the variants. If a positively charged amino acid is replaced by an amino acid of like charge, then the overall charge on the protein is unchanged. The same may be said for other, negatively charged and neutral amino acid substitutions.

6. All of the amino acid substitutions (Ala → Gly, Val → Leu, Asp → Asn, Met → Leu) require only one nucleotide change. The last change from Pro (CC–) → Lys (AAA,G) requires two changes (the minimal mutational distance).

8. The classification of organisms into different species is based on evidence (morphological, genetic, ecological, etc.) that they are reproductively isolated. Classifications above the species level (genus, family, etc.) are not based on such empirical data. DNA sequence divergence is not always directly proportional to morphological, behavioral, or ecological divergence. As more information is gained on the meaning of DNA sequence differences (ΔT_m) in comparison to morphological factors, many phylogenetic relationships will be reconsidered.

10. There are many sections of DNA in a eukaryotic genome which are not reflected in a protein product. Indeed, there are many sections of DNA which are not even transcribed and/or have no apparent physiological role.

12. It is likely that some genes (like histones) will not tolerate nucleotide substitutions to a significant degree and the neutral mutation theory will not hold. However, there are other genes which produce quite variable products and provide support for the neutral mutation theory. It is controversy which stimulates a desire to seek answers.

14. The increased frequency of diabetics seems to correspond with an increase in sugar intake perhaps compounded genotypes which are very efficient in sugar conversions. In the post-1970 period, the decline may be due to selection against the gene. Diabetic women pass fewer of the diabetic genes to the next generation.

C R E D I T S

INDEX

Bold page number indicates a figure. *Italic* page number indicates a table.

WATSON-CRICK MODEL OF DNA

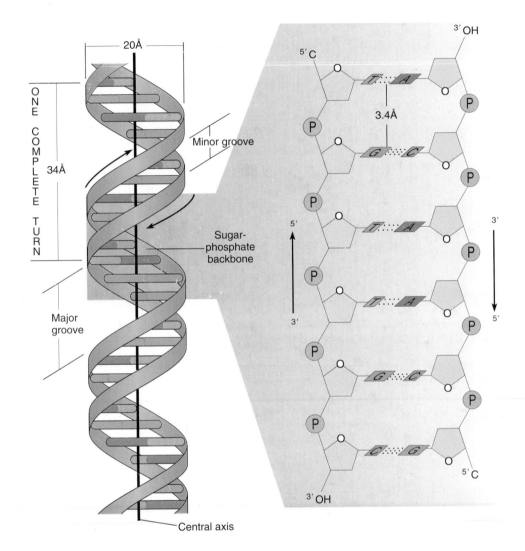